Introduction to Functional Analysis

Gabriel N. Gatica

Introduction to Functional Analysis

Theory and Applications

Springer

Gabriel N. Gatica
Centro de Investigación en Ingeniería Matemática
and Departamento de Ingeniería Matemática
Universidad de Concepción
Concepción, VIII - Concepción, Chile

ISBN 978-3-032-37414-1 ISBN 978-3-032-37415-8 (eBook)
https://doi.org/10.1007/978-3-032-37415-8

Translation from the Spanish language edition: "Introducción al análisis funcional. Teoría y aplicaciones" by Gabriel N. Gatica, © Editorial Reverté 2024. Published by Editorial Reverté.

Mathematics Subject Classification: 46: Functional Analysis, 65: Numerical Analysis, 35: Partial Differential Equatio, 47: Operator Theory

This Springer imprint is published by the registered company Springer Nature Switzerland AG
The registered company address is: Gewerbestrasse 11, 6330 Cham, Switzerland

To my inspiring teachers: *Benjamín Jorquera, Hipólito Monsalves, Henry Rivas, Hermann Alder, María Cristina Bustos, Hubert Mennickent, Patricio Basso* and *George C. Hsiao.*

In memory of my parents, *Armida* and *Temístocles*, and to my dearest loved ones: *Maggie, Gabriel and Gonzalo.*

To my compadre, great friend and exceptional educator, or as my children would lovingly say, *to uncle Freddy*, on the occasion of his 72th birthday

Preface to the present English edition

This book constitutes the English translation of the second edition of its original Spanish version, both published by Reverté Editorial, Barcelona, Spain, in September 2024 and November 2013, respectively. The corresponding prefaces, properly translated as well, are provided in the following pages. However, in all fairness, this was not merely a matter of translating the latest Spanish version. The opportunity was also taken to improve and refine the presentation of several topics, to correct previously identified typos, to unify and simplify notation, and even to add new exercises and subjects of interest that have emerged in recent years.

Accordingly, this English edition fully preserves the main features of its predecessors in Spanish. Among them, I would like to emphasize once again that it constantly seeks to explicitly connect, whenever feasible, the results and proofs of each chapter with those of the others. Consequently, instead of the usual presentation of apparently isolated subjects, the aim is to develop a much more unified and interconnected text. This means, for example, that some relevant topics are not treated exclusively in a single chapter, but are instead discussed, with varying degrees of depth and detail, in different places throughout the book. In a similar spirit, the results concerning Banach spaces and those concerning Hilbert spaces are not kept strictly separate, but are rather suitably interwoven so that, as the reading progresses, the reader may gradually assimilate both the similarities and the differences between them. The only exception to this approach is now provided by Chapters 4 and 5, in which such a separation is indeed adopted.

Certainly, the above does not necessarily lead to the writing of a standard book on Functional Analysis. Moreover, although the aforementioned features may not be fully appreciated by some readers, I must acknowledge that my intention was never really to write yet another typical book on the subject. Along the same lines, I also seek to incorporate as many original exercises as possible, many of them stemming from my own research work, as well as from that of collaborators and other researchers. In this regard, I would like to point out once again, as in the Spanish versions of the text, that the not-so-few exercises motivated by a specific bibliography are identified by including the corresponding references at the beginning of their statements. In this way, by consulting such references, readers may find both explicit and implicit indications that should help them solve the exercises.

To summarize, with a nod to my eldest son Gabriel, who is an architect, and drawing a parallel with the notion of "unfinished architecture", which, as I understand it, characterizes the work of the Chilean architect Smiljan Radi'c Clarke, recently awarded the 2026 Pritzker

Prize, one could say, certainly keeping in mind the proper proportions, that this book does not necessarily seek absolute perfection, but rather aims to coexist with imperfection and to preserve a certain unfinished character. In this way, the possibility of future editions and further refinements always remains open.

I conclude by thanking once again all the people and academic institutions that made the Spanish versions of the book possible, whose names are listed in the respective prefaces. Additionally, and more specifically regarding the present English edition, I would like to acknowledge the AI system ChatGPT Plus for providing a preliminary rough version of the corresponding translation, which was subsequently carefully revised, first by the former undergraduate students from my Functional Analysis course, *Vicente Cartes*, *Cristofer Mamani*, and *Benjamín Mendieta*, then by my former research assistants, *Alonso Bustos* and *Benjamín Venegas*, and finally by myself. My sincere and heartfelt thanks to these young and promising mathematicians.

Last but not least, I would like to express my profound gratitude to the reviewers of this book, whose diverse comments and suggestions contributed significantly to improving its final version.

Concepción, Chile

Gabriel N. Gatica
July 2026

Preface to the second Spanish edition

Having passed a little more than a decade since the initial publication of this work, and thanks to the publisher's confirmation of a positive reception of the same within the Latin American scholar community, the author believes, by virtue of his own academic experiences during this time, that conditions and antecedents have arisen that fully justify the introduction of a new edition. In particular, his research work, both with national and foreign collaborators, as well as with thesis students of the Mathematical Civil Engineering degree and the Doctorate Program in Applied Sciences with a mention in Mathematical Engineering, both from the University of Concepción, has motivated the use of new theoretical tools in the discipline, whether extracted from other sources or originating, precisely, from those investigations, which certainly constitute very suitable material to be included in this new version. Part of this refers to extensions or complements of results already contained in the original text, which has given rise to new sections in some of its chapters, while a broader theme, precisely that which has been present in practically all of the author's scientific productivity in the last 5 years, has led to the inclusion of a complete new chapter. It must be recognized, however, that what has been described has changed the plans committed to in the preface to the first edition, in which the future incorporation of topics other than those that are finally being added here was announced. On the other hand, the growing interest of students in the subjects of functional analysis and related subjects, as well as the significant increase in their number, have generated in the undersigned a permanent self-demand for better and more innovative examples of illustration and application of the respective theoretical results, thus giving rise to a set of additional exercises of greater validity and relevance.

In accordance with the above, and together with clearly establishing *a priori* that the original objectives and characteristics of the book are maintained, we proceed below to detail the main novelties of this second edition. Firstly, and as can be seen from the preceding paragraph, several of the lists of exercises at the end of each chapter have been renewed, which implied modifying or excluding some of those that were already there, and certainly incorporating new proposals. By virtue of the latter, and given that the great majority of them arise from research work by the author and his collaborators, or even from other researchers, there has also been a natural update of the bibliography used in the book. In fact, we take this opportunity to remind you, as mentioned in the preface to the first edition, that **the exercises that have been motivated by a specific bibliography are identified by including the corresponding references at the beginning of their writing**. On the other hand, an important aspect has been, in turn, the correction of all the *fe de errata* that had been detected to date. Without prejudice to this, it is very likely that the present edition is not exempt from similar errors, so that, in addition to apologizing for this in advance, the author once again thanks in advance for the respective corrections that are found. In what follows, only the new specific contents of those chapters that actually contain them are described. To view all the topics in each chapter, simply read the general index or the preface to the first edition.

Chapter 3 (linear operators) presents additional details and results on the consequences of the Open Mapping Theorem, particularly those related to topological complements; interpolation spline functions; the characterization of operators with closed rank, including a case related to the concept of reflexivity; and unbounded operators and their adjoints. Chapter 4 (variational problems in Hilbert spaces) discusses complementary details and results on variational inequalities (Lions-Stampacchia Theorem); and traces and normal traces. In particular, the latter concerns orthogonal decomposition of spaces of the type $\mathbf{H}(\text{div}; \Omega)$; restrictions and extensions of trace spaces and their duals over parts of the boundary; and direct sum decompositions of trace spaces and their duals. In addition, this chapter contains an application of Babuška-Brezzi Theory to Neumann problems and the main theoretical aspects of a nonlinear case of it. In the new Chapter 5 (variational problems in Banach spaces), which extends and complements the work developed in Chapter 4, the continuous and discrete versions of the generalized Lax-Milgram Lemma, both classical and general Babuška-Brezzi Theory, and what are known as perturbed saddle-point problems are addressed, all in the context of Banach spaces. Examples of the application of the last two topics mentioned, as well as a variant of the second one, are also described here. In a sense, Chapters 4 and 5 now constitute the exception to the rule stipulated in the first edition of the book, in that "*the results for Hilbert and Banach spaces are not strictly separated, but rather they are blend together*". In Chapter 6 (ex-5) (COMPACT OPERATORS) the section on convergence and compactness is complemented with additional details and results related to weak convergence*. Likewise, a by-product is provided, which is given by the characterization of the existence of what is called "preadjoint", a concept that is previously introduced here. Finally, Chapter 7 (ex-6) (REFLEXIVITY AND SEPARABILITY) incorporates additional details and results on reflexivity, and its application, in particular, to Banach spaces defined in terms of those of Lebesgue.

As with the first version, we conclude this preface by thanking all the people, organizations, and institutions that have made this new edition possible. First, once again, a very special mention to all the students in Mathematical Civil Engineering and the PhD program who have taken my Functional Analysis courses since 2014. As always, the interaction with them through their questions and comments, both in and out of class, generates a very valuable mutual learning experience. At the same time, I would also like to sincerely acknowledge all my thesis students, both undergraduate and graduate, during the aforementioned period. They, as I have stated on more than one occasion, are the fundamental cause and effect of much of what one does in research. At the risk of forgetting a name, I refer, in alphabetical order, to: *Javier Almonacid, Mario Álvarez, Nicolás Barnafi, Ernesto Cáceres, Sergio Carrasco, Jorge Castillo, Sergio Caucao, Eligio Colmenares, Claudio Correa, Hugo Díaz, Carolina Domínguez, Sebastián Domínguez, Carlos García, Bryan Gómez, Cristián Inzunza, Saulo Medrado, Franco Milanese, Willian Miranda, Sebastián Moraga, Mauricio Munar, Nicolás Nuñez, Juan P. Ortega, Juan C. Rojas, Felipe Sandoval, Filánder Sequeira, and Juan P. Silva.* Many thanks to all of them for having deigned to choose me to direct their respective thesis projects. In a similar vein, but more in a category that could be called "grandchildren" or "adopted children," my gratitude also goes to: *Isaac Bermúdez, Gonzalo Benavides, Esteban Henríquez, Alejandro Hopper*, and *Nestor Sánchez*. On the other hand, the undeniable influence of the author's most frequent collaborators in the writing of this book remains as strong as ever, so that, in addition to those already mentioned in the first preface, I would like to explicitly thank on this occasion those who have honored me with their co-authorships in more recent years, namely: *Julio Careaga, Sergio Caucao, Daniel Hurtado, Ricardo Ruiz-Baier*, and *Yuri D. Sobral.*

My gratitude also goes to my distinguished former students in the undergraduate course of Functional Analysis, and current research assistants, *Alonso Bustos* and *Benjamín Venegas*, for taking the time to thoroughly read some of the new sections to detect possible corrections, and for several LaTeX "*tips*" that improved the presentation of the text. At an institutional level, my thanks to the ongoing support of the University of Concepción, primarily from the Vice-Rector's Office for Research and Development, the CI^{2}MA Center, the Department of Mathematical Engineering, and the Faculty of Physical and Mathematical Sciences. At the same time, the completion of this second edition has been made possible thanks to the author's participation as a researcher in the Basal Project FB210005 (Center for Mathematical Modeling, CMM) and the Anillo Project ACT 210087 (Ring of Computational Mathematics for Desalination Processes), both initiatives funded by ANID, Chile, to which I express my gratitude through their respective directors, *Héctor Ramírez* and *Ricardo Oyarzúa.*

Finally, and reiterating what was stated in the first edition, my deep gratitude to my loved ones and my inspiring teachers, to whom I dedicate this book with total love and affection.

Concepción, Chile

Gabriel N. Gatica
September 2024

Preface to the first Spanish edition

The present book, which may serve both as a textbook and as a reference, deals with the main results of Functional Analysis and its various applications, both in Mathematics itself and in other disciplines (mainly Physics and Engineering). It is therefore addressed to undergraduate and graduate students and professors in mathematics (Mathematical Civil Engineering, Bachelor in Mathematics, and other related fields), as well as master's and doctoral programs in mathematics, physics, and engineering sciences. The specific courses in which it could be used include: *Fundamentals of Applied Mathematics, Theoretical and/or Applied Functional Analysis (at both undergraduate and graduate levels), Optimization, Partial Differential Equations, Numerical Analysis of Partial Differential Equations, Finite Element Methods, Mixed Finite Element Methods, Mathematical Methods in Physics, Mathematical Methods in Engineering, and other related ones.*

The respective chapters address most of the classical topics, namely: *duality, linear operators, variational problems, compact operators, spectral theory, reflexivity and separability, weak topologies, distribution theory, and Sobolev spaces.* In this regard, the most significant feature of the presentation lies in the fact that, together with an extreme care in the details of the various proofs and justifications of each theoretical result, we attempt not to lose sight of the concrete applicability of the vast majority of them. In other words, the main motivation for writing this book is based on the desire to provide two key aspects: maximum clarity in the presentation, for a greater and faster understanding by the readers, and eventual usefulness and applicability, both in mathematics and in other areas, of the results presented. We believe that the first aspect is supported by the author's many years of experience teaching the courses Fundamentals of Applied Mathematics I and II, and Functional Analysis and Applications I and II to students of the Bachelor in Mathematics and Mathematical Civil Engineering programs at the Universidad de Concepción, as well as the Functional Analysis course to students of the PhD Program in Applied Sciences with specialization in Mathematical Engineering, of the same university. In turn, the second aspect has been permanently strengthened by the research carried out by the author with several collaborators from Chile and abroad (mainly the USA and Europe) in the areas of NUMERICAL ANALYSIS OF PARTIAL DIFFERENTIAL EQUATIONS and FUNCTIONAL ANALYSIS itself. In fact, this experience suggested the inclusion in this book of some topics that are usually not fully developed in the classical Functional Analysis texts. As an example, Chapter 4 on Variational Problems, whose subjects are not available in the existing Spanish bibliography with the detail, generality, and variety of examples described here, as well as many of the proposed exercises at the end of each chapter, have been motivated in one way or another by the aforementioned research work. In particular, such problems find their essence and origin in the statement or proof of a lemma or theorem contained in relatively recent articles in these areas, and in particular in some published by the author. Moreover, the vast majority of these exercises have already been included and successfully tested in the respective evaluations applied in the courses mentioned above. These facts have enriched not only the contents of those courses but also of other related ones such as: Mixed Finite Element Method (at the advanced undergraduate level), and Finite Element Theory and Partial Differential Equations (at the graduate level).

Another interesting aspect of this book is the permanent attempt to explicitly connect, whenever appropriate, the results and proofs of a given chapter with those of both the previous and the subsequent ones. In this way, instead of the typical presentation of apparently isolated topics, the elaboration of a much more unified and connected text is favored. Consistently with this, some relevant topics in applications, such as distribution theory and Sobolev spaces, are not treated in a single chapter each, but rather are addressed, with different degrees of depth and detail, in more than one place throughout the text. At the same time, no strict separation is made between results valid for Banach spaces and those for Hilbert spaces, but instead they are suitably interwoven so that the reader gradually assimilates both the similarities and the differences between them.

The remainder of the book is divided into 9 chapters and is organized as follows. In Chapter 1, which is introductory in nature, we present some application examples that illustrate the use (or the need to use) various tools of Functional Analysis, and introduce the basic concepts related to metric spaces, normed spaces, and distributions. Chapter 2 addresses the main concepts and results on functionals and dual spaces, including several classical theorems such as the Best Approximation Theorem, the Orthogonal Decomposition Theorem, the Riesz Representation Theorem, and both the analytic and geometric versions of the Hahn-Banach Theorem. The development of the fundamental aspects of the theory of bounded and unbounded linear operators is the main objective of Chapter 3. This includes a variety of concepts such as adjoint operator, annihilators, orthogonals, null spaces, ranges and graphs, as well as most of the classical results associated with them, in particular the Bounded Inverse Theorem, the Closed Graph Theorem, the Open Mapping Theorem, and the Banach-Steinhaus Theorem (also known as the Uniform Boundedness Theorem). In Chapter 4 we establish and prove various results that allow us to conclude the solvability and approximation of a wide range of variational problems. The respective analysis focuses on formulations posed in Hilbert spaces and includes the renowned Lax-Milgram Lemma and the Babuška-Brezzi Theory for saddle-point problems. On the other hand, in Chapter 5 we study compact operators, which are a generalization of the concept of finite-rank operators. The main results to be discussed here include Fredholm's alternative, its connection with the stability of Galerkin methods, and the various convergence properties of these operators. In Chapter 6 the concepts of reflexivity and separability are introduced, their most important associated results are proved, and their various connections with weak topologies are established. Chapter 7 studies some properties of the spectrum of bounded and unbounded operators, including, in particular, the compact and self-adjoint cases. In turn, Chapter 8 completes the study initiated in Chapter 1 on the main concepts and results of distribution theory. These refer, among others, to regularization of functions, compactly supported distributions, convolution of distributions, tempered distributions, and Fourier transform. Finally, in Chapter 9 certain essential aspects of Sobolev Space Theory are discussed. More precisely, the presentation here is restricted to some results that have had and still have a significant presence in applications. We conclude this content description by emphasizing that at the end of each chapter a selection of exercises on the topics covered therein is provided. In this regard, **it is important to note that the problems that**

have been directly or indirectly motivated by the partial or total content of some specific bibliography are identified by including the respective references at the beginning of the corresponding statement. The resolution of these problems may, in some cases, be facilitated by consulting those references.

On the other hand, it is important to point out that, very likely, this book will not be free of typographical errors, notational inconsistencies or confusions, nor of mathematical mistakes, and at the same time, it is undeniable that many topics of equal or greater importance than those addressed here will be missing. Regarding the former, the author apologizes to the readers in advance and thanks them for letting him know, in due course, the respective corrections that may be detected. As for the latter, and although it also reflects the obvious fact of having to finish the writing at some point, in addition to the author's own taste and motivation, he commits himself to overcome the most important shortcomings in possible future editions of the book. For this purpose, the following complementary topics are already being considered, for example: *analysis of nonlinear operators, evolutionary variational problems, spectral decomposition of positive operators, other aspects of Sobolev space Theory, and new applications.*

We conclude this preface by thanking all the people, instances, and institutions that have made the completion of this book possible. First of all, a special acknowledgment to all the undergraduate and graduate students who have taken my Functional Analysis courses since the 1990s. Many approaches, contents, and exercises of this text have been motivated/generated by the permanent interaction with them. In fact, this experience has been very enriching for the author, who, despite teaching these courses for almost 20 years, has always had the good fortune to learn something more (precisely as a consequence of such interactions) each time he had the opportunity to transmit his knowledge to a new group of students. On the other hand, the writing of this book has also been enormously benefited by the undeniable influence of the author's most frequent collaborators during the last years, namely: Luis F. Gatica, Norbert Heuer, George C. Hsiao, Antonio Márquez, Salim Meddahi, Ricardo Oyarzúa, and Francisco J. Sayas. In turn, the constant concern of my colleague, compadre, and great friend, Freddy Paiva, for the various aspects of Mathematics teaching, and his consequent interest in examining practically all possible websites on the subject, have contributed, thanks to the immense generosity that characterizes him as a person, to significantly improving the presentation of several of the topics contained in this text. To each one of them, and as Cantinflas would probably say: "*some more and some less, but all the same*," my infinite thanks for all I have learned from them.

On another note, my gratitude also to my current student, Filánder Sequeira, for preparing all the figures included in the text, for his various LaTeX "*tips*" and for taking the time to read some of the chapters to detect possible corrections. At the same time, my acknowledgment to Mrs. Angelina Fritz, Secretary of the Center for Research in Mathematical Engineering (CI^{2}MA) at the Universidad de Concepción, for having typed the LaTeX version of a large part of this book. In another equally valuable area, my thanks to the Universidad de Concepción, represented by its Office of Teaching, Office of Graduate Studies, Vice-Rectory and Vice-Rectory of Research and Development, for co-financing the publication of this work together with Editorial Reverté. The completion of this initiative has also been possible thanks to the constant support of Conicyt-Chile, through the Fondap and Basal projects with the CMM of the Universidad de Chile and our Anillo ACT 1118 (ANANUM) project. Likewise, we acknowledge the enthusiastic support of both the CI^{2}MA Center and the Department of Mathematical Engineering and the Faculty of Physical and Mathematical Sciences of my university.

Finally, but not less importantly, my deepest gratitude to my dearest loved ones and to my inspiring professors, to whom I dedicate this book with total affection.

Concepción, Chile

Gabriel N. Gatica
November 2013

Competing Interests The author has no competing interests to declare that are relevant to the content of this manuscript.

Ethics Approval The author declares that this manuscript does not include primary studies involving animals.

Contents

Main Symbols and Spaces

The following list contains the principal symbols and spaces that are used throughout the book.

$\mathbb{K}$	Underlying scalar field ($\mathbb{R}$ or $\mathbb{C}$).
X, Y, V	Normed or Hilbert spaces.
W, H, Q	Normed or Hilbert spaces.
Ω	Open subset of $\mathbb{R}^n$.
Γ	Boundary, or a distinguished portion of the boundary, of Ω.
$C(\Omega)$	Space of continuous functions on Ω.
$\text{supp}(\varphi)$	Support of a function $\varphi \in C(\Omega)$.
$C(\bar{\Omega})$	Space of continuous functions on $\bar{\Omega}$.
∂^α	Derivative of multi-index order α.
$C^m(\Omega)$	Space of functions $u : \Omega \to \mathbb{K}$ such that $\partial^\alpha u \in C(\Omega)\ \forall\, \alpha,\ \lvert\alpha\rvert \leqslant m$.
$C^\infty(\Omega)$	Space of functions $u : \Omega \to \mathbb{K}$ such that $\partial^\alpha u \in C(\Omega)\ \forall\, \alpha$.
$C_0^\infty(\Omega)$	Space of functions in $C^\infty(\Omega)$ with compact support contained in Ω.
X'	Dual space of X.
X''	Dual space of X', or Bidual space of X.
$\mathcal{J}_X$	Canonical injection from X into X''.
$\langle \cdot, \cdot \rangle$	Duality pairing or inner product, according to the context.
$U^\perp$	Orthogonal of a set or subspace U of an inner product space.
P_U	Orthogonal projector onto a closed subspace U.
$\mathcal{R}$	Riesz operator associated with the Riesz Representation Theorem.
$\mathcal{L}(X, Y)$	Space of bounded linear operators from X into Y.
$\mathcal{L}(X)$	Space of bounded linear operators from X into X.
$\mathcal{D}(A)$	Domain of an operator A.
$N(A)$	Null space of an operator A.
$R(A)$	Range of an operator A.
S^o	Annihilator of a subset S of a normed space X.
${}^oT$	Annihilator of a subset T of a dual space X'.
$\mathcal{G}(A)$	Graph of an operator A.
A'	Banach adjoint of an operator A.
A^*	Hilbert adjoint of an operator A.
$a(\cdot, \cdot)$	Bilinear form in a variational formulation.
$b(\cdot, \cdot)$	Constraint bilinear form in a saddle-point formulation.

F	Linear functional in a variational problem.
α	Usual coercivity or ellipticity constant.
β	Usual inf-sup constant.
$\mathrm{L}^p(\Omega)$, $\mathrm{L}^p(\Gamma)$	Lebesgue spaces on Ω and Γ, respectively.
$\mathbf{L}^p(\Omega)$	Vector version of $\mathrm{L}^p(\Omega)$.
$\mathbb{L}^p(\Omega)$	Tensor version of $\mathrm{L}^p(\Omega)$.
$\mathrm{W}^{m,p}(\Omega)$	Sobolev space of order m and exponent p.
$\mathrm{W}_0^{m,p}(\Omega)$	Closure of $C_0^\infty(\Omega)$ in $\mathrm{W}^{m,p}(\Omega)$.
$\mathrm{H}^m(\Omega)$	Sobolev space $\mathrm{W}^{m,2}(\Omega)$.
$\mathrm{H}_0^m(\Omega)$	Closure of $C_0^\infty(\Omega)$ in $\mathrm{H}^m(\Omega)$.
$\mathrm{H}^1(\Omega)$	Sobolev space of order 1.
γ_0	Trace operator acting from $\mathrm{H}^1(\Omega)$ into $\mathrm{L}^2(\Gamma)$.
$\mathrm{H}^{1/2}(\Gamma) := R(\gamma_0)$	Range of the operator γ_0.
$\mathrm{H}_0^1(\Omega) := N(\gamma_0)$	Null space of γ_0.
$\mathrm{H}^{-1/2}(\Gamma)$	Dual space of $\mathrm{H}^{1/2}(\Gamma)$.
$\mathbf{H}(\mathrm{div};\Omega)$	Space of vector fields in $\mathbf{L}^2(\Omega)$ with divergence in $\mathrm{L}^2(\Omega)$.
$\mathbf{H}(\mathrm{div}_\varrho;\Omega)$	Space of vector fields in $\mathbf{L}^2(\Omega)$ with divergence in $\mathrm{L}^\varrho(\Omega)$.
$\gamma_{\boldsymbol{\nu}}$	Normal trace operator acting from $\mathbf{H}(\mathrm{div};\Omega)$ onto $\mathrm{H}^{-1/2}(\Gamma)$.
$\gamma_{\boldsymbol{\nu}}$	Normal trace operator acting from $\mathbf{H}(\mathrm{div}_\varrho;\Omega)$ onto $\mathrm{H}^{-1/2}(\Gamma)$.
$\boldsymbol{\tau}\cdot\boldsymbol{\nu} := \gamma_{\boldsymbol{\nu}}(\boldsymbol{\tau})$	Operator $\gamma_{\boldsymbol{\nu}}$ applied to $\boldsymbol{\tau}\in\mathbf{H}(\mathrm{div};\Omega)$ (or $\boldsymbol{\tau}\in\mathbf{H}(\mathrm{div}_\varrho;\Omega)$).
$\mathbf{H}^t(\mathrm{div}_t;\Omega)$	Space of vector fields in $\mathbf{L}^t(\Omega)$ with divergence in $\mathrm{L}^t(\Omega)$.
$\mathbb{H}(\mathbf{div};\Omega)$	Tensor version of $\mathbf{H}(\mathrm{div};\Omega)$.
$\mathbf{H}^{1/2}(\Gamma)$	Vector version of $\mathrm{H}^{1/2}(\Gamma)$.
$\mathbf{H}^{-1/2}(\Gamma)$	Dual space of $\mathbf{H}^{1/2}(\Gamma)$.
$\boldsymbol{\gamma_\nu}$	Normal trace operator acting from $\mathbb{H}(\mathbf{div};\Omega)$ onto $\mathbf{H}^{-1/2}(\Gamma)$.
$\boldsymbol{\tau}\boldsymbol{\nu} := \boldsymbol{\gamma_\nu}(\boldsymbol{\tau})$	Operator $\boldsymbol{\gamma_\nu}$ applied to $\boldsymbol{\tau}\in\mathbb{H}(\mathbf{div};\Omega)$.
$\mathbf{H}(\mathrm{rot};\Omega)$	Space of vector fields in $\mathbf{L}^2(\Omega)$ with rot in $\mathrm{L}^2(\Omega)$.
$\mathcal{K}(X,Y)$	Space of compact linear operators from X into Y.
$\sigma(X,X')$	Weak topology on X.
$\sigma(X',X)$	Weak∗ topology on X'.
$x_n \overset{\omega}{\to} x$	Weak convergence of x_n to x in a normed space X.
$F_n \overset{\omega^*}{\to} F$	Weak* convergence of F_n to F in a dual space X'.
$\boldsymbol{\sigma}(A)$	Spectrum of an operator A.
$\rho(A)$	Resolvent set of an operator A.
$\sigma_p(A)$	Point spectrum of an operator A.
$\mathfrak{R}_A(\lambda)$	Resolvent operator $(A-\lambda I)^{-1}$ of an operator A.
$r_{\boldsymbol{\sigma}}(A)$	Spectral radius of an operator A.
$\mathcal{D}(\Omega)$	Space of test functions on Ω: $C_0^\infty(\Omega)$ provided with a topology.
$\mathrm{supp}(\varphi)$	Support of a function $\varphi\in\mathcal{D}(\Omega)$.
$\mathcal{D}'(\Omega)$	Space of distributions on Ω.
$\mathcal{E}(\Omega)$	Space of smooth functions on Ω: $C^\infty(\Omega)$ provided with a topology.
$\mathcal{E}'(\Omega)$	Space of compactly supported distributions on Ω.
$\mathcal{S}(\mathbb{R}^n)$	Schwartz space: rapidly decreasing test functions.
$\mathcal{S}'(\mathbb{R}^n)$	Space of tempered distributions.
δ_x	Dirac distribution at the point x.
$\mathrm{supp}(u)$	Support of a distribution u.
$u*\varphi$	Convolution of a distribution u and a test function φ.
$\widehat{\varphi} := \mathcal{F}(\varphi)$	Fourier transform $\mathcal{F}$ of a test function $\varphi\in\mathcal{S}(\mathbb{R}^n)$.
$\widehat{u} := \mathcal{F}(u)$	Fourier transform $\mathcal{F}$ of a tempered distribution $u\in\mathcal{S}'(\mathbb{R}^n)$.

(X_0, X_1)	Compatible pair of normed spaces.
$X_0 + X_1$	Sum space associated with a compatible pair.
$X_0 \cap X_1$	Intersection space associated with a compatible pair.
$K(t, u)$	Real interpolation K-functional.
$(X_0, X_1)_{\delta,q}$	Real interpolation space obtained by the K-method.
$\mathrm{H}^s(\mathbb{R}^n)$	Sobolev space of order $s \in \mathbb{R}$, over $\mathbb{R}^n$.
$\mathrm{H}^s(\mathbb{R}^n)$	Interpolation space $\big(\mathrm{H}^{s_0}(\mathbb{R}^n), \mathrm{H}^{s_1}(\mathbb{R}^n)\big)_{\delta,2}$, where $s = (1-\delta)s_0 + \delta s_1$.
$\mathrm{H}^s(\Omega)$	Distributional restrictions to Ω of functions $u \in \mathrm{H}^s(\mathbb{R}^n)$.
$\mathrm{H}^s(\Omega)$	Interpolation space $\big(\mathrm{H}^{s_0}(\Omega), \mathrm{H}^{s_1}(\Omega)\big)_{\delta,2}$, where $s = (1-\delta)s_0 + \delta s_1$.

1

Introduction

In this chapter, some application examples are presented to illustrate the use (or the need to use) various tools from Functional Analysis, and the most important basic concepts related to metric spaces, normed spaces, Banach spaces, Hilbert spaces, and distributions are introduced. Some of the material developed below is based on the approaches adopted in [54], [100], and [119].

1.1 An initial value problem

Given an open subset A of $\mathbb{R}^2$, $(t_0, y_0) \in A$, $\delta > 0$, and a function $f : A \to \mathbb{R}$, let us consider the initial value problem: Find $\psi \in C([t_0, t_0 + \delta])$, with continuous derivative in $(t_0, t_0 + \delta)$, such that

$$\psi'(t) = f(t, \psi(t)) \qquad \forall\, t \in (t_0, t_0 + \delta)\,, \quad \psi(t_0) = y_0\,. \tag{1.1}$$

According to the Fundamental Theorem of Calculus it follows that any solution ψ of (1.1) satisfies

$$\psi(t) = \psi(t_0) + \int_{t_0}^{t} \psi'(s)\, ds \qquad \forall\, t \in [t_0, t_0 + \delta]\,,$$

or, replacing $\psi'(s)$ by $f(s, \psi(s))$ and $\psi(t_0)$ by y_0,

$$\psi(t) = y_0 + \int_{t_0}^{t} f(s, \psi(s))\, ds \qquad \forall\, t \in [t_0, t_0 + \delta]\,. \tag{1.2}$$

Conversely, it is not difficult to see that if ψ is a solution of (1.2), then, evaluating at $t = t_0$ and differentiating under the integral sign, one deduces that ψ is a solution of the initial value problem (1.1), and thus (1.1) and (1.2) turn out to be equivalent. Motivated by the above, one initially defines the set

$$U := \big\{\varphi \in C([t_0, t_0 + \delta]) : \quad \varphi(t_0) = y_0\big\} \tag{1.3}$$

and the mapping $T : U \to U$, where

$$T(\varphi)(t) := y_0 + \int_{t_0}^{t} f(s, \varphi(s))\, ds \qquad \forall\, t \in [t_0, t_0 + \delta]\,, \quad \forall\, \varphi \in U\,. \tag{1.4}$$

Nevertheless, it is important to note that for the definition of T to make sense it is required that f be continuous in A and that $(s, \varphi(s)) \in A$ for all $s \in [t_0, t_0 + \delta]\,, \quad \forall\, \varphi \in U$. The

G. N. Gatica, *Introduction to Functional Analysis*,
https://doi.org/10.1007/978-3-032-37415-8_1

former is assumed from now on, and the latter requires a modification of the set U, as detailed below. Indeed, from now on it is assumed that there exists $M > 0$ such that

$$|f(t,y)| \leqslant M \quad \forall\,(t,y) \in A \quad \text{and} \quad [t_0, t_0+\delta] \times [y_0 - M\delta, y_0 + M\delta] \subseteq A\,,$$

and one redefines

$$U := \big\{\varphi \in C([t_0, t_0+\delta]) : \quad \varphi(t_0) = y_0\,, \quad |\varphi(t) - y_0| \leqslant M\delta \quad \forall\, t \in [t_0, t_0+\delta]\big\}\,.$$

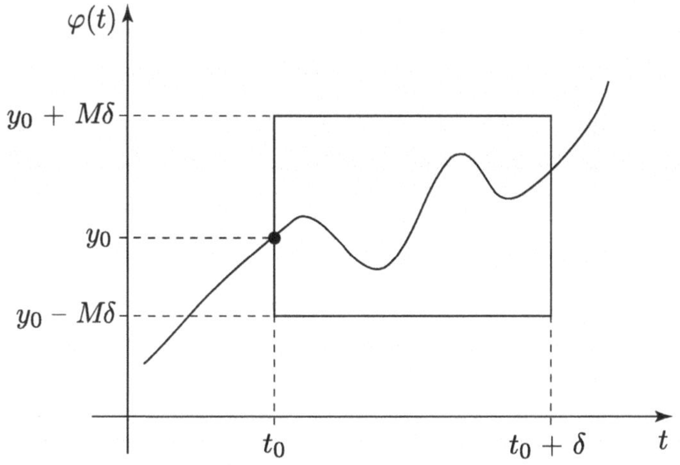

Fig. 1.1. *Example of* $\varphi \in U$

Figure 1.1 illustrates the new restriction on the functions of U. In this way, it is observed that T is indeed well defined, which means, in particular, that $T(\varphi) \in U$ for all $\varphi \in U$, and therefore it is clear that solving (1.2) is equivalent to finding $\psi \in U$ such that

$$\psi(t) = T(\psi)(t) \qquad \forall\, t \in [t_0, t_0+\delta]\,,$$

that is, finding $\psi \in U$ such that

$$\psi = T(\psi)\,. \tag{1.5}$$

By virtue of the fixed point equation for T given by (1.5), our next purpose is to introduce the respective theorem due to Banach, for which several definitions are needed.

Definition 1.1. *Given an arbitrary set U, a mapping $\rho : U \times U \to \mathbb{R}$ is called a* METRIC (*or* DISTANCE) *on U if it satisfies the following properties:*

i) $\rho(u,v) \geqslant 0$ *for all* $u,\, v \in U$.
ii) $\rho(u,v) = 0$ *if and only if* $u = v$.
iii) $\rho(u,v) = \rho(v,u)$ *for all* $u,\, v \in U$.
iv) $\rho(u,v) \leqslant \rho(u,w) + \rho(w,v)$ *for all* $u,\, v,\, w \in U$ (TRIANGLE INEQUALITY).

In this case, the pair (U,ρ) is called a METRIC SPACE.

Before introducing further useful concepts, we point out that property i) in Definition 1.1 actually follows from the other three. In fact, knowing from ii) that $\rho(u,u) = 0$, we deduce from iv) and iii) that

$$0 = \rho(u,u) \leqslant \rho(u,v) + \rho(v,u) = 2\,\rho(u,v)\,,$$

which yields i).

Definition 1.2. *A sequence $\{u_n\}_{n\in\mathbb{N}}$ in a metric space (U,ρ) is said to be a Cauchy sequence if $\lim\limits_{n,m\to\infty} \rho(u_n,u_m) = 0$, that is, if*

$$\forall\,\varepsilon > 0, \quad \exists\, N \in \mathbb{N} \quad \text{such that} \quad \rho(u_n,u_m) < \varepsilon \quad \forall\, n,m \geqslant N\,.$$

Definition 1.3. *A sequence $\{u_n\}_{n\in\mathbb{N}}$ in a metric space (U,ρ) is said to be convergent if there exists $u \in U$ such that $\lim\limits_{n\to\infty} \rho(u_n,u) = 0$, that is, if*

$$\forall\,\varepsilon > 0, \quad \exists\, N \in \mathbb{N} \quad \text{such that} \quad \rho(u_n,u) < \varepsilon \quad \forall\, n \geqslant N\,.$$

Note that the concept of convergence of a sequence implies the existence of a unique limit for it. Moreover, the following result shows that a metric ρ is sequentially continuous in each of its components.

Lemma 1.4. *Let (U,ρ) be a metric space, and let $\{u_n\}_{n\in\mathbb{N}}$ be a sequence in U that converges to $u \in U$. Then $\lim\limits_{n\to\infty} \rho(u_n,v) = \rho(u,v)$ for all $v \in U$.*

Proof. Using the triangle inequality (property iv)) and the symmetry of ρ (property iii)), one easily obtains

$$\rho(u_n,v) - \rho(u,v) \leqslant \rho(u_n,u) \qquad \text{and} \qquad \rho(u,v) - \rho(u_n,v) \leqslant \rho(u_n,u)\,,$$

which is equivalent to

$$|\,\rho(u_n,v) - \rho(u,v)\,| \leqslant \rho(u_n,u)\,,$$

and then, taking the limit as $n \to \infty$, the proof is concluded.

□

The following concepts are required to establish the already announced theorem and to apply it later.

Definition 1.5. *A metric space (U,ρ) is said to be complete if every Cauchy sequence in U is convergent.*

Definition 1.6. *Let (U,ρ) be a metric space, and let $V \subseteq U$. We say that V is a closed subset of U if the limit of every convergent sequence in V belongs to V.*

As a consequence of these definitions, it follows immediately that every closed subset of a complete metric space is also complete.

Definition 1.7. *Let (U,ρ) be a metric space, and let $T : U \to U$. We say that the mapping T is a* CONTRACTION *if there exists $\kappa \in (0,1)$ such that*

$$\rho(T(u),T(v)) \leqslant \kappa\,\rho(u,v) \qquad \forall\, u,\, v \in U\,.$$

Theorem 1.8 (Banach Fixed Point Theorem). *Let (U, ρ) be a complete metric space, and let $T : U \to U$ be a contraction. Then there exists a unique $w \in U$ such that $T(w) = w$.*

Proof. Let w_0 be any element of U, and define the sequence $\{w_n\}_{n\in\mathbb{N}}$, where $w_1 = T(w_0)$, $w_2 = T(w_1)$, and in general, $w_{n+1} = T(w_n)$ for all $n \in \mathbb{N}$. Thus, if $\kappa \in (0,1)$ denotes the contraction constant of T, we observe that

$$\rho(w_n, w_{n+1}) = \rho(T(w_{n-1}), T(w_n)) \leqslant \kappa\, \rho(w_{n-1}, w_n)\,,$$

which, applied recursively, implies that

$$\rho(w_n, w_{n+1}) \leqslant \kappa^n\, \rho(w_0, w_1) \qquad \forall\, n \in \mathbb{N}\,. \tag{1.6}$$

Now, given p, $n \in \mathbb{N}$, we use the triangle inequality for the metric ρ and the estimate (1.6), and obtain

$$\rho(w_n, w_{n+p}) \leqslant \sum_{j=0}^{p-1} \rho(w_{n+j}, w_{n+j+1}) \leqslant \sum_{j=0}^{p-1} \kappa^{n+j}\, \rho(w_0, w_1)$$

$$= \kappa^n \sum_{j=0}^{p-1} \kappa^j\, \rho(w_0, w_1) = \frac{\kappa^n\,(1-\kappa^p)}{1-\kappa}\, \rho(w_0, w_1)\,,$$

from which

$$\rho(w_n, w_{n+p}) \leqslant \frac{\kappa^n}{1-\kappa}\, \rho(w_0, w_1) \qquad \forall\, p,\, n \in \mathbb{N}\,. \tag{1.7}$$

It is easy to see from (1.7) that $\{w_n\}_{n\in\mathbb{N}}$ is a Cauchy sequence in U, and therefore there exists $w \in U$ such that $\lim\limits_{n\to\infty} \rho(w_n, w) = 0$. Moreover, fixing n in (1.7), taking the limit as $p \to \infty$, and applying Lemma 1.4, it follows that

$$\rho(w_n, w) \leqslant \frac{\kappa^n}{1-\kappa}\, \rho(w_0, w_1) \qquad \forall\, n \in \mathbb{N}\,. \tag{1.8}$$

Let us now show that w is the unique fixed point of T. Indeed, using once again the triangle inequality, we have

$$\rho(T(w), w) \leqslant \rho(T(w), T(w_n)) + \rho(T(w_n), w)$$

$$= \rho(T(w), T(w_n)) + \rho(w_{n+1}, w) \leqslant \kappa\, \rho(w, w_n) + \rho(w_{n+1}, w)\,,$$

and taking the limit as $n \to \infty$ yields $\rho(T(w), w) = 0$, which means that $T(w) = w$. For uniqueness, let $\widetilde{w}$ be any other element of U such that $T(\widetilde{w}) = \widetilde{w}$. Then, applying once more that T is a contraction, we can write $\rho(w, \widetilde{w}) = \rho(T(w), T(\widetilde{w})) \leqslant \kappa\, \rho(w, \widetilde{w})$, that is, $(1-\kappa)\, \rho(w, \widetilde{w}) \leqslant 0$, from which necessarily $w = \widetilde{w}$.

□

As a historical remark, let us briefly mention that the Banach Fixed Point Theorem was first published in [9].

It is interesting to observe here that the proof of the Banach Fixed Point Theorem (cf. Theorem 1.8) provides a fixed point iterative scheme, starting from any initial element $w_0 \in U$, to construct a sequence that converges to $w \in U$. Moreover, the eventual proximity or distance of the elements of this sequence to the exact solution w is measured by (1.8). In particular, from this error estimate it follows that the minimum number n of iterations that guarantees a priori that $\rho(w_n, w)$ is less than a given tolerance $\varepsilon > 0$, is given by the first natural number n that satisfies

$$n \geqslant \frac{\log\left(\frac{\varepsilon(1-\kappa)}{\rho(w_0,w_1)}\right)}{\log \kappa}.$$

The next objective is to apply Theorem 1.8 to $T : U \to U$ defined by (1.4). To this end, let us first recall that the set $C([t_0, t_0 + \delta])$ endowed with the uniform metric

$$\rho(\psi, \varphi) := \max_{t\in[t_0,t_0+\delta]} |\psi(t) - \varphi(t)| \qquad \forall\, \psi,\, \varphi \in C([t_0, t_0 + \delta])$$

is a complete metric space , and since U is a closed subset of $C([t_0, t_0 + \delta])$, it follows that (U, ρ) is also complete. Furthermore, we now introduce the additional assumption that f is Lipschitz continuous in its second variable, which means assuming the existence of $K > 0$ such that

$$|\, f(t, y_1) - f(t, y_2)\,| \leqslant K\,|y_1 - y_2| \qquad \forall\,(t, y_1),\,(t, y_2) \in A. \tag{1.9}$$

Then, using (1.9) and the definitions of T and ρ, we observe that

$$|T(\psi)(t) - T(\varphi)(t)| = \left|\int_{t_0}^{t} \{f(s, \psi(s)) - f(s, \varphi(s))\}\, ds\right| \leqslant \int_{t_0}^{t} |f(s, \psi(s)) - f(s, \varphi(s))|\, ds$$

$$\leqslant K \int_{t_0}^{t} |\psi(s) - \varphi(s)|\, ds \leqslant K\,(t - t_0)\,\rho(\psi, \varphi) \qquad \forall\, t \in [t_0, t_0 + \delta], \tag{1.10}$$

which proves that

$$\rho(T(\psi), T(\varphi)) \leqslant K\,\delta\,\rho(\psi, \varphi) \qquad \forall\, \psi,\, \varphi \in U.$$

This inequality shows that a sufficient condition for T to be a contraction is that $K\,\delta < 1$, for which reason we can assert that the above analysis has established the following result.

Theorem 1.9. *Let $f : A \subseteq \mathbb{R}^2 \to \mathbb{R}$ be continuous, and suppose that there exist M, $K > 0$ such that*

i) $|f(t, y)| \leqslant M$ *for all* $(t, y) \in A$.
ii) $|\, f(t, y_1) - f(t, y_2)\,| \leqslant K\,|y_1 - y_2|$ *for all* $(t, y_1),\,(t, y_2) \in A$.

Moreover, let $(t_0, y_0) \in A$ and $\delta > 0$ such that

iii) $\delta < \dfrac{1}{K}$ *and* $[t_0, t_0 + \delta] \times [y_0 - M\delta, y_0 + M\delta] \subseteq A$.

Then the integral equation (1.2) *has a unique solution $\psi \in U$. Equivalently, the initial value problem* (1.1) *has a unique solution $\psi \in C([t_0, t_0 + \delta])$ with continuous derivative in $(t_0, t_0 + \delta)$.*

Certainly, Theorem 1.9 guarantees the unique solvability of (1.1) only in a local sense, since this result occurs on an interval $[t_0, t_0 + \delta]$ whose length is bounded above by $\frac{1}{K}$. Because of this, and with the purpose of obtaining existence and uniqueness of solution of (1.1) in a global sense, we now introduce an improved version of the Banach Fixed Point Theorem.

Theorem 1.10. *Let (U, ρ) be a complete metric space, and let $T : U \to U$ be such that T^m is a contraction for some $m \in \mathbb{N}$. Then T^m and T have the same unique fixed point.*

Proof. First note that the powers of T are defined recursively by

$$T^1(u) = T(u) \quad \text{and} \quad T^n(u) = T(T^{n-1}(u)) \quad \forall\, u \in U\,, \quad \forall\, n \in \mathbb{N},\, n \geqslant 2\,,$$

which implies that $T^m(u) \in U$ for all $u \in U$. Now, according to the hypothesis on T^m and Theorem 1.8, it follows that there exists a unique element $w \in U$ such that $T^m(w) = w$. Let us see that w is also a fixed point of T and that it is unique. Indeed, if $\kappa \in (0,1)$ denotes the contraction constant of T^m, we have

$$\rho(T(w), w) = \rho(T^{m+1}(w), T^m(w)) = \rho(T^m(T(w)), T^m(w)) \leqslant \kappa\, \rho(T(w), w)\,,$$

which yields $\rho(T(w), w) = 0$, and therefore $T(w) = w$. For uniqueness, let $\widetilde{w}$ be any other element of U such that $T(\widetilde{w}) = \widetilde{w}$. Then, applying T to this identity successively ($m-1$ times), we deduce that $T^m(\widetilde{w}) = \widetilde{w}$, and consequently $\widetilde{w} = w$.

□

A slight variant of Theorem 1.10 is provided later in Exercise 1.7.

From the previous theorem it is clear that it is not necessary for T itself to be a contraction, but only that some power of it is. In what follows we show that this is precisely what happens with the mapping defined by (1.4). To this end, we first prove the following result.

Lemma 1.11. *Given ψ, $\varphi \in U$, one has*

$$|\, T^n(\psi)(t) - T^n(\varphi)(t)\,| \leqslant \frac{K^n\,(t-t_0)^n}{n!}\,\rho(\psi,\varphi) \qquad \forall\, t \in [t_0, t_0+\delta], \quad \forall\, n \in \mathbb{N}\,.$$

Proof. We argue by induction on n. For $n = 1$ the result was already established in (1.10). Assume now its validity for n and prove that it is also true for $n+1$. Indeed, according to the definition of T and its powers, we have

$$\begin{aligned} |\, T^{n+1}(\psi)(t) - T^{n+1}(\varphi)(t)\,| &= |\, T(T^n(\psi))(t) - T(T^n(\varphi))(t)\,| \\ &= \left| \int_{t_0}^{t} \{ f(s, T^n(\psi)(s)) - f(s, T^n(\varphi)(s)) \}\, ds \right|, \end{aligned}$$

from which, applying the Lipschitz-continuity of f and the induction hypothesis, it follows that

$$\begin{aligned} |\, T^{n+1}(\psi)(t) - T^{n+1}(\varphi)(t)\,| &\leqslant K \int_{t_0}^{t} |\, T^n(\psi)(s) - T^n(\varphi)(s)\,|\, ds \\ &\leqslant \frac{K^{n+1}}{n!} \int_{t_0}^{t} (s-t_0)^n\, ds\, \rho(\psi,\varphi) = \frac{K^{n+1}\,(t-t_0)^{n+1}}{(n+1)!}\,\rho(\psi,\varphi)\,, \end{aligned}$$

which completes the proof.

□

From Lemma 1.11 and the definition of the metric ρ it follows immediately that

$$\rho(T^n(\psi), T^n(\varphi)) \leqslant \frac{(K\,\delta)^n}{n!}\,\rho(\psi,\varphi) \qquad \forall\, \psi,\, \varphi \in U, \quad \forall\, n \in \mathbb{N}\,,$$

and since

$$\lim_{n\to\infty} \frac{(K\,\delta)^n}{n!} = 0 \qquad \forall\, K,\, \delta \in \mathbb{R}\,,$$

it follows that there exists $m \in \mathbb{N}$ such that

$$\frac{(K\,\delta)^n}{n!} < 1 \qquad \forall\, n \geqslant m\,,$$

which proves, in particular, that T^m is a contraction. Consequently, we have thus proved the following improved version of Theorem 1.9.

Theorem 1.12. *Let $f : A \subseteq \mathbb{R}^2 \to \mathbb{R}$ be continuous, and suppose that there exist M, $K > 0$ such that*

i) $|f(t,y)| \leqslant M$ *for all* $(t,y) \in A$.
ii) $|\,f(t,y_1) - f(t,y_2)\,| \leqslant K\,|y_1 - y_2|$ *for all* (t,y_1), $(t,y_2) \in A$.

Moreover, let $(t_0,y_0) \in A$ and $\delta > 0$ be such that

iii) $[t_0, t_0+\delta] \times [y_0 - M\,\delta, y_0 + M\,\delta] \subseteq A$.

Then the integral equation (1.2) *has a unique solution $\psi \in U$. Equivalently, the initial value problem* (1.1) *has a unique solution $\psi \in C([t_0,t_0+\delta])$ with continuous derivative in $(t_0,t_0+\delta)$.*

Naturally, Theorem 1.12 does not guarantee global existence of a solution to problem (1.1), but at least reduces the restrictions on δ only to those stipulated in iii), namely $[t_0,t_0+\delta] \times [y_0 - M\delta, y_0 + M\delta] \subseteq A$. However, when $A = \mathbb{R}^2$ this condition is trivially satisfied for any $\delta > 0$, which allows us to establish the following global existence result for problem (1.1).

Theorem 1.13. *Let $f : \mathbb{R}^2 \to \mathbb{R}$ be continuous, let $(t_0,y_0) \in \mathbb{R}^2$, and suppose that there exist M, $K > 0$ such that*

i) $|f(t,y)| \leqslant M$ *for all* $(t,y) \in \mathbb{R}^2$.
ii) $|\,f(t,y_1) - f(t,y_2)\,| \leqslant K\,|y_1 - y_2|$ *for all* (t,y_1), $(t,y_2) \in \mathbb{R}^2$.

Then, for any $\delta > 0$ the integral equation (1.2) *has a unique solution $\psi \in U$. Equivalently, the initial value problem* (1.1) *has a unique solution $\psi \in C([t_0,t_0+\delta])$ with continuous derivative in $(t_0,t_0+\delta)$.*

Even in the case when $A = \mathbb{R}^2$ the previous result can be further improved by applying once again Theorem 1.10 to the integral operator T on the set U originally defined by (1.3). In this way, a second version for the global existence result of the initial value problem (1.1) is given as follows.

Theorem 1.14. *Let $f : \mathbb{R}^2 \to \mathbb{R}$ be continuous, $(t_0,y_0) \in \mathbb{R}^2$, and suppose that there exists $K > 0$ such that $|f(t,y_1) - f(t,y_2)| \leqslant K\,|y_1 - y_2|$ for all (t,y_1), $(t,y_2) \in \mathbb{R}^2$. Then, for any $\delta > 0$ the integral equation* (1.2) *has a unique solution $\psi \in U$. Equivalently, the initial value problem* (1.1) *has a unique solution $\psi \in C([t_0,t_0+\delta])$ with continuous derivative in $(t_0,t_0+\delta)$.*

Other applications of Banach's Fixed Point Theorem and related results are found later in Example 4.14 and in the exercises of Chapter 4 identified as 4.35, 4.37, 4.41, and 4.39.

1.2 Some concepts and preliminary results

Throughout this section we assume that the concept of vector space is known. Thus, given a vector space X over the field $\mathbb{K}$, which may be $\mathbb{C}$ or $\mathbb{R}$, with zero element $0 \in X$, and with operations $+ : X \times X \to X$ (vector addition) and $\cdot : \mathbb{K} \times X \to X$ (scalar multiplication), some basic definitions are introduced below.

Definition 1.15. *A mapping $\|\cdot\| : X \to \mathbb{R}$ is called a* NORM *on X if it satisfies the following properties:*

i) $\|x\| \geqslant 0$ *for all* $x \in X$.
ii) $\|x\| = 0$ *if and only if* $x = 0$.
iii) $\|\alpha x\| = |\alpha| \|x\|$ *for all* $\alpha \in \mathbb{K}$, *for all* $x \in X$.
iv) $\|x + y\| \leqslant \|x\| + \|y\|$ *for all* $x, y \in X$ (TRIANGLE INEQUALITY).

X endowed with the norm $\|\cdot\|$ is called a NORMED VECTOR SPACE *and is denoted $(X, \|\cdot\|)$.*

Regarding the above definition, we say that $(X, \|\cdot\|)$ is real (resp. complex) if the field $\mathbb{K}$ is $\mathbb{R}$ (resp. $\mathbb{C}$), which naturally leads to certain differences between the two concepts. In particular, we will see later in Definition 1.22 that, when defining an inner product on a complex normed space, it is conjugate symmetric instead of symmetric, as in the real case, and its second component becomes sesquilinear instead of linear, as in the real case. Similar differences arise, for example, when discussing some properties of the Riesz and Hilbert adjoint operators later on in Chapters 2 (cf. (2.15), (2.17)) and 3 (cf. Lemma 3.17, ii)), respectively. For further comparisons between real and complex Banach spaces, we refer to [108].

On the other hand, similarly to Definition 1.1 (cf. Section 1.1), we point out that property i) in Definition 1.15 actually follows by taking $y = -x$ in iv), and then using ii) and iii). Furthermore, $(X, \|\cdot\|)$ also constitutes a metric space (X, ρ), where $\rho(x, y) := \|x - y\|$ for all $x, y \in X$. Hence, all the definitions, remarks, and results given in Section 1.1 for such spaces hold in the present context as well. Nevertheless, for the sake of clarity in the presentation, some of those definitions are rewritten below in terms of the norm $\|\cdot\|$.

Definition 1.16. *A sequence $\{x_n\}_{n\in\mathbb{N}}$ in the normed vector space $(X, \|\cdot\|)$ is called a Cauchy sequence if $\lim\limits_{n,m\to\infty} \|x_n - x_m\| = 0$, that is, if*

$$\forall \varepsilon > 0, \quad \exists N \in \mathbb{N} \quad \textit{such that} \quad \|x_n - x_m\| < \varepsilon \quad \forall n, m \geqslant N .$$

Definition 1.17. *A sequence $\{x_n\}_{n\in\mathbb{N}}$ in the normed vector space $(X, \|\cdot\|)$ is called convergent if there exists $x \in X$ such that $\lim\limits_{n\to\infty} \|x_n - x\| = 0$, that is, if*

$$\forall \varepsilon > 0, \quad \exists N \in \mathbb{N} \quad \textit{such that} \quad \|x_n - x\| < \varepsilon \quad \forall n \geqslant N .$$

It is easy to see that every convergent sequence is a Cauchy sequence, but the converse is not always true, which gives rise to the following definition.

Definition 1.18. *The normed vector space $(X, \|\cdot\|)$ is said to be* COMPLETE *(or a* BANACH SPACE*) if every Cauchy sequence in X is convergent.*

In this regard, the following lemma establishes a necessary and sufficient condition for the convergence of a Cauchy sequence. Clearly, sufficiency is of much greater interest and applicability than the necessity of this condition.

Lemma 1.19. *Let* $\{x_n\}_{n\in\mathbb{N}}$ *be a Cauchy sequence in* $(X, \|\cdot\|)$. *Then* $\{x_n\}_{n\in\mathbb{N}}$ *is convergent if and only if it has a convergent subsequence.*

Proof. It is clear that if $\{x_n\}_{n\in\mathbb{N}}$ is convergent, then all its subsequences are also convergent. Conversely, let $\{x_n^{(1)}\}_{n\in\mathbb{N}}$ be a subsequence of $\{x_n\}_{n\in\mathbb{N}}$ converging to $x \in X$. Then, given $\varepsilon > 0$, there exists $N_1 \in \mathbb{N}$ such that $\|x_n^{(1)} - x\| < \varepsilon$ for all $n \geqslant N_1$. On the other hand, since $\{x_n\}_{n\in\mathbb{N}}$ is Cauchy, there exists $N_2 \in \mathbb{N}$ such that $\|x_n - x_m\| < \varepsilon$ for all $n, m \geqslant N_2$. Now, for each $n \in \mathbb{N}$ there exists $m \geqslant n$ such that $x_n^{(1)}$ coincides with x_m. Then, defining $N := \max\{N_1, N_2\}$, it follows that

$$\|x_n - x\| \leqslant \|x_n - x_n^{(1)}\| + \|x_n^{(1)} - x\| < 2\varepsilon \qquad \forall\, n \geqslant N,$$

which shows that $\{x_n\}_{n\in\mathbb{N}}$ also converges to x.

□

As a matter of notation, we now introduce

$$x_n \overset{n\to\infty}{\longrightarrow} x$$

to denote, from now on, the convergence of a sequence $\{x_n\}_{n\in\mathbb{N}}$ in a normed space X to an element $x \in X$ with respect to the corresponding norm, which is also called STRONG CONVERGENCE. Throughout the rest of the book we will similarly use the symbols $\overset{w}{\to}$ and $\overset{w^*}{\to}$ to denote weak and weak* convergence, respectively, concepts that are introduced later on in Definitions 6.27 and 6.28; the latter is repeated in Definition 7.10.

Example 1.20 The following are examples of Banach spaces.

1. $\mathbb{C}^n := \{\mathbf{x} := (x_1, x_2, \ldots, x_n) : \; x_i \in \mathbb{C} \;\; \forall\, i \in \{1, \ldots, n\}\}$, with $\mathbb{K} = \mathbb{C}$, endowed with the norm
$$\|\mathbf{x}\| := \sum_{j=1}^{n} |x_j| \qquad \forall\, \mathbf{x} \in \mathbb{C}^n .$$
2. $C([-1,1]) := \{f : [-1,1] \to \mathbb{R} : \; f \text{ continuous}\}$, with $\mathbb{K} = \mathbb{R}$, endowed with the norm
$$\|f\| := \max_{x\in[-1,1]} |f(x)| \qquad \forall\, f \in C([-1,1]) .$$
3. $\ell_\infty(\mathbb{C}) := \{\mathbf{u} := \{u_n\}_{n\in\mathbb{N}} : \; u_n \in \mathbb{C}, \; \sup_{n\in\mathbb{N}} |u_n| < \infty\}$, with $\mathbb{K} = \mathbb{C}$, endowed with the norm
$$\|\mathbf{u}\| := \sup_{n\in\mathbb{N}} |u_n| \qquad \forall\, \mathbf{u} \in \ell_\infty(\mathbb{C}) .$$
4. $\mathrm{L}^2((a,b))$, with $\mathbb{K} = \mathbb{R}$, endowed with the norm
$$\|f\| = \left(\int_a^b |f|^2\right)^{1/2} \qquad \forall\, f \in \mathrm{L}^2((a,b)) ,$$
where, given $a, b \in \mathbb{R}$ with $a < b$, $\mathrm{L}^2((a,b))$ denotes the space of all (equivalence classes of) measurable functions $f : (a,b) \to \mathbb{R}$ such that
$$\int_a^b |f|^2 < \infty .$$
These equivalence classes are formed by functions that coincide almost everywhere in (a,b).

Example 1.21 Here we give an example of a normed vector space that is not Banach. Indeed, consider again the real space $C([-1,1])$, but now endowed with the $L^2(-1,1)$ norm, that is

$$\|f\| := \left\{ \int_{-1}^{1} |f|^2 \right\}^{1/2} \qquad \forall\, f \in C([-1,1]),$$

which is well defined since $C([-1,1]) \subseteq L^2(-1,1)$. Now, let $\{f_n\}_{n\in\mathbb{N}}$ be the sequence in $C([-1,1])$ given by

$$f_n(t) := \begin{cases} 0\,, & -1 \leqslant t < 0 \\ nt\,, & 0 \leqslant t < \frac{1}{n} \\ 1\,, & \frac{1}{n} \leqslant t \leqslant 1 \end{cases}.$$

Then, given $n,\, m \in \mathbb{N}$, $n > m$, we have $\frac{1}{n} < \frac{1}{m}$ and hence

$$\|f_n - f_m\|^2 = \int_0^{1/n} (n-m)^2\, t^2\, dt + \int_{1/n}^{1/m} (1-mt)^2\, dt = -\frac{2}{3n} + \frac{m}{3n^2} + \frac{1}{3m} < \frac{1}{3m} - \frac{1}{3n},$$

which proves that $\{f_n\}_{n\in\mathbb{N}}$ is Cauchy in $(C([-1,1]), \|\cdot\|)$. On the other hand, let $\widetilde{f} : [-1,1] \to \mathbb{R}$ be the function in $L^2(-1,1)$ defined by

$$\widetilde{f}(t) := \begin{cases} 0\,, & -1 \leqslant t \leqslant 0 \\ 1\,, & 0 < t \leqslant 1 \end{cases},$$

and note that $\{f_n\}_{n\in\mathbb{N}}$ converges to $\widetilde{f}$ in the $L^2(-1,1)$ norm since

$$\|f_n - \widetilde{f}\| = \int_0^{1/n} (nt-1)^2\, dt = \frac{1}{3n} \qquad \forall\, n \in \mathbb{N}.$$

Then, if we assume that there exists $f \in C([-1,1])$ such that $\{f_n\}_{n\in\mathbb{N}}$ converges to f in the $L^2(-1,1)$ norm, we must necessarily have $f = \widetilde{f}$, which is a contradiction since $\widetilde{f}$ is not continuous. This proves that $\{f_n\}_{n\in\mathbb{N}}$ is not convergent and therefore $(C([-1,1]), \|\cdot\|)$ is not a Banach space.

Next, we introduce the concepts of inner product spaces and Hilbert spaces, for which we first require the definition, precisely, of an inner product. In what follows, $\bar{z}$, $\mathrm{Re}(z)$, and $\mathrm{Im}(z)$ denote the conjugate, the real part, and the imaginary part, respectively, of $z \in \mathbb{C}$. Moreover, $\bar{z} = \mathrm{Re}(z) = z$, and $\mathrm{Im}(z) = 0$, whenever $z \in \mathbb{R}$.

Definition 1.22. *Let X be a vector space over $\mathbb{K}$ ($\mathbb{R}$ or $\mathbb{C}$). A mapping $\langle\cdot,\cdot\rangle : X \times X \to \mathbb{K}$ is called an* INNER PRODUCT *on X, if it satisfies the following properties:*

i) $\langle x, x\rangle \geqslant 0$ *for all* $x \in X$ *and* $\langle x, x\rangle = 0$ *if and only if* $x = 0$ (POSITIVITY).
ii) $\langle x, y\rangle = \overline{\langle y, x\rangle}$ *for all* $x,\, y \in X$ (CONJUGATE SYMMETRY).
iii) $\langle \alpha x + \beta y, z\rangle = \alpha\langle x, z\rangle + \beta\langle y, z\rangle$ *for all* $\alpha,\, \beta \in \mathbb{K}$, *for all* $x,\, y,\, z \in X$ (LINEARITY IN THE FIRST COMPONENT).

The vector space X endowed with an inner product $\langle\cdot,\cdot\rangle$ is called an INNER PRODUCT SPACE *and is denoted by $(X, \langle\cdot,\cdot\rangle)$.*

It is important to note here that, in the real case, property ii) becomes

ii) $\langle x, y\rangle = \langle y, x\rangle$ for all $x,\, y \in X$ (SYMMETRY),

which, together with iii), guarantees linearity of the inner product in the second component as well. However, in the complex case this does not occur, since ii) and iii) only imply that

$$\langle x, \alpha y + \beta z\rangle = \bar{\alpha}\langle x, y\rangle + \bar{\beta}\langle x, z\rangle \quad \forall\, \alpha, \beta \in \mathbb{C}, \quad \forall\, x,\, y,\, z \in X,$$

which is known as SESQUILINEARITY. A direct consequence of linearity or sesquilinearity is the fact that the inner product of the zero vector with any vector in X is zero, that is

$$\langle x, 0\rangle = \langle 0, x\rangle = 0 \qquad \forall\, x \in X\,. \tag{1.11}$$

One of the most important results on these spaces is given by the following classical inequality.

Theorem 1.23 (Cauchy–Schwarz's Inequality). *Let $(X, \langle\cdot,\cdot\rangle)$ be an inner product space over the field $\mathbb{K}$. Then we have*

$$|\langle x, y\rangle| \leqslant \langle x, x\rangle^{1/2}\,\langle y, y\rangle^{1/2} \qquad \forall\, x,\ y \in X\,.$$

Proof. If $x = 0$ or $y = 0$, then the inequality is trivially satisfied thanks to (1.11). Now, given $x,\ y \in X - \{0\}$, and $\alpha \in \mathbb{K}$, the properties of the inner product imply that

$$\begin{aligned} 0 \leqslant \langle x + \alpha\, y, x + \alpha\, y\rangle &= \langle x, x\rangle + \alpha\,\langle y, x\rangle + \bar{\alpha}\,\langle x, y\rangle + \alpha\,\bar{\alpha}\,\langle y, y\rangle \\ &= \langle x, x\rangle + 2\,\mathrm{Re}(\bar{\alpha}\,\langle x, y\rangle) + |\alpha|^2\,\langle y, y\rangle\,. \end{aligned}$$

In particular, for $\alpha := -\langle x, y\rangle/\langle y, y\rangle$ we obtain

$$0 \leqslant \langle x, x\rangle - 2\,\frac{|\langle x, y\rangle|^2}{\langle y, y\rangle} + \frac{|\langle x, y\rangle|^2}{\langle y, y\rangle} = \langle x, x\rangle - \frac{|\langle x, y\rangle|^2}{\langle y, y\rangle}\,,$$

from which, multiplying by $\langle y, y\rangle$ and adding $|\langle x, y\rangle|^2$ to both sides of the previous inequality, the proof is completed.

□

In addition to the previous theorem, it is interesting to note that every inner product induces a norm on the respective vector space X. Indeed, we have the following result.

Theorem 1.24. *Let $(X, \langle\cdot,\cdot\rangle)$ be an inner product space over the field $\mathbb{K}$, and consider the mapping $\|\cdot\| : X \to \mathbb{R}$ defined by $\|x\| := \langle x, x\rangle^{1/2}$ for all $x \in X$. Then $\|\cdot\|$ is a norm on X, which is called the* INDUCED NORM *by $\langle\cdot,\cdot\rangle$.*

Proof. The positivity of $\|\cdot\|$ follows from the corresponding property of the inner product. On the other hand, using the linearity and sesquilinearity of $\langle\cdot,\cdot\rangle$, we obtain that $\|\alpha\, x\|^2 = \langle \alpha\, x, \alpha\, x\rangle = \alpha\,\bar{\alpha}\,\langle x, x\rangle = |\alpha|^2\,\|x\|^2$, and therefore

$$\|\alpha\, x\| = |\alpha|\,\|x\| \qquad \forall\, \alpha \in \mathbb{K}, \quad \forall\, x \in X\,.$$

Finally, applying the Cauchy-Schwarz inequality we deduce that

$$\begin{aligned} \|x + y\|^2 &= \langle x + y, x + y\rangle = \|x\|^2 + \|y\|^2 + 2\,\mathrm{Re}(\langle x, y\rangle) \\ &\leqslant \|x\|^2 + \|y\|^2 + 2\,\|x\|\,\|y\| = \{\,\|x\| + \|y\|\,\}^2 \qquad \forall\, x,\ y \in X\,, \end{aligned}$$

which proves the triangle inequality for $\|\cdot\|$.

□

Definition 1.25. *An inner product space $(X, \langle\cdot,\cdot\rangle)$ is called a* HILBERT SPACE *if it is complete with the norm induced by its inner product.*

Example 1.26 The following are examples of Hilbert spaces.

1. $\mathbb{R}^n := \{\mathbf{x} := (x_1, x_2, ..., x_n) : \quad x_i \in \mathbb{R} \quad \forall\, i \in \{1, ..., n\}\}$, with $\mathbb{K} = \mathbb{R}$, endowed with the inner product
$$\langle \mathbf{x}, \mathbf{y} \rangle := \sum_{i=1}^{n} x_i\, y_i \qquad \forall\, \mathbf{x},\, \mathbf{y} \in \mathbb{R}^n .$$
2. $\mathbb{C}^n := \{\mathbf{x} := (x_1, x_2, ..., x_n) : \quad x_i \in \mathbb{C} \quad \forall\, i \in \{1, ..., n\}\}$, with $\mathbb{K} = \mathbb{C}$, endowed with the inner product
$$\langle \mathbf{x}, \mathbf{y} \rangle := \sum_{i=1}^{n} x_i\, \bar{y}_i \qquad \forall\, \mathbf{x},\, \mathbf{y} \in \mathbb{C}^n .$$
3. $\ell_2(\mathbb{C}) := \Big\{\mathbf{u} := \{u_n\}_{n\in\mathbb{N}} : \quad u_n \in \mathbb{C}, \quad \sum_{n=1}^{\infty} |u_n|^2 < \infty\Big\}$, with $\mathbb{K} = \mathbb{R}$ or $\mathbb{C}$, endowed with the inner product
$$\langle \mathbf{u}, \mathbf{v} \rangle := \sum_{n=1}^{\infty} u_n\, \bar{v}_n \qquad \forall\, \mathbf{u},\, \mathbf{v} \in \ell_2(\mathbb{C}) .$$
4. $\mathrm{L}^2(\Omega)$, with $\mathbb{K} = \mathbb{C}$, endowed with the inner product
$$\langle f, g \rangle_{0,\Omega} := \int_{\Omega} f\, \bar{g} \qquad \forall\, f,\, g \in \mathrm{L}^2(\Omega), \tag{1.12}$$
and induced norm
$$\|f\|_{0,\Omega} := \left\{ \int_{\Omega} |f|^2 \right\}^{1/2} \qquad \forall\, f \in \mathrm{L}^2(\Omega), \tag{1.13}$$
where, given an open subset Ω of $\mathbb{R}^n$, $\mathrm{L}^2(\Omega)$ denotes the space of all (equivalence classes of) measurable functions $f : \Omega \to \mathbb{C}$ such that
$$\int_{\Omega} |f|^2 < \infty .$$
These equivalence classes are formed by functions that coincide almost everywhere in Ω.
5. $\mathcal{M}_{n\times n}(\mathbb{R}) := \big\{A = (a_{ij})_{i,j=1,\dots,n} : \quad a_{ij} \in \mathbb{R}\big\}$, with $\mathbb{K} = \mathbb{R}$, endowed with the inner product
$$\langle A, B \rangle := \operatorname{tr}(A^t\, B) \qquad \forall\, A,\, B \in \mathcal{M}_{n\times n}(\mathbb{R}),$$
where $\operatorname{tr}(C)$ and C^t denote the trace and the transpose, respectively, of a given $C \in \mathcal{M}_{n\times n}(\mathbb{R})$. It is interesting to note here that the explicit computation of this inner product reduces to
$$\langle A, B \rangle = \sum_{i,j=1}^{n} a_{ij}\, b_{ij} \qquad \forall\, A := (a_{ij})_{n\times n},\, \forall\, B := (b_{ij})_{n\times n} \in \mathcal{M}_{n\times n}(\mathbb{R}),$$
which amounts to arranging the coefficients of these matrices into two vectors in $\mathbb{R}^{n^2}$ and then taking their usual inner product.

In connection with Example 4 above, we stress that the definition of $\mathrm{L}^2(\Omega)$ is readily extended later, in Exercise 1.10, to the space $\mathrm{L}^p(\Omega)$, with $p \in [1, +\infty)$. In turn, regarding Examples 1, 2, and 5 above, we can now establish the following more general result.

Lemma 1.27. *Let X be a finite-dimensional vector space over the field $\mathbb{K}$. Then there exists an inner product on X with respect to which it becomes a Hilbert space.*

Proof. Let $N \in \mathbb{N}$ be the dimension of X, let $\{u_1, \ldots, u_N\}$ be a basis of X, and define the mapping $T : X \to \mathbb{K}^N$ that assigns to each $x \in X$ its coordinates with respect to this basis. Since T is linear and bijective, the mapping $\langle \cdot, \cdot \rangle : X \times X \to \mathbb{K}$ given by

$$\langle x, y \rangle := \langle T(x), T(y) \rangle_N \qquad \forall\, x,\, y \in X\,,$$

where $\langle \cdot, \cdot \rangle_N$ denotes the usual inner product on $\mathbb{K}^N$ (see items 1 and 2 in Example 1.26), constitutes an inner product on X. Since $(\mathbb{K}^N, \langle \cdot, \cdot \rangle_N)$ is a Hilbert space, it follows readily that $(X, \langle \cdot, \cdot \rangle)$ is also a Hilbert space.

□

On the other hand, it is clear that every Hilbert space is a Banach space, but the converse is not necessarily true. In other words, there exist complete normed vector spaces whose norms do not arise from any inner product. A classical result that is usually employed to confirm this fact is the following.

Lemma 1.28 (Parallelogram law). *Let $(X, \langle \cdot, \cdot \rangle)$ be an inner product space over the field $\mathbb{K}$, and let $\| \cdot \|$ be the induced norm by the inner product. Then*

$$\|x + y\|^2 + \|x - y\|^2 = 2 \left\{ \|x\|^2 + \|y\|^2 \right\} \quad \forall\, x,\, y \in X\,.$$

Proof. It is enough to observe that

$$\|x \pm y\|^2 = \langle x \pm y, x \pm y \rangle = \|x\|^2 + \|y\|^2 \pm 2\,\mathrm{Re}(\langle x, y \rangle)$$

and then add the expressions with both signs.

□

In virtue of the previous lemma, if a norm does not satisfy the parallelogram law, then it is not induced by any inner product. An example of this fact is the following.

Example 1.29 Consider again the complex Banach space $(\ell_\infty(\mathbb{C}), \| \cdot \|)$ introduced in Example 1.20, that is

$$\ell_\infty(\mathbb{C}) := \left\{ \mathbf{u} := \{u_n\}_{n\in\mathbb{N}} : \ u_n \in \mathbb{C},\ \sup_{n\in\mathbb{N}} |u_n| < \infty \right\} \quad \text{and} \quad \|\mathbf{u}\| := \sup_{n\in\mathbb{N}} |u_n| \quad \forall\, \mathbf{u} \in \ell_\infty(\mathbb{C})\,.$$

To see that the norm of $\ell_\infty(\mathbb{C})$ is not generated by an inner product, it suffices to show a concrete example of elements of $\ell_\infty(\mathbb{C})$ that do not satisfy the parallelogram law. In particular, take the sequences $\mathbf{u} := \{u_n\}_{n\in\mathbb{N}}$ and $\mathbf{v} := \{v_n\}_{n\in\mathbb{N}}$, where

$$u_n := \begin{cases} 1\,, & \text{if} \quad n = 1 \\ \frac{1}{2}\,, & \text{if} \quad n = 2 \\ 0\,, & \text{if} \quad n \geqslant 3 \end{cases} \quad \text{and} \quad v_n := \begin{cases} 0\,, & \text{if} \quad n = 1 \\ \frac{1}{2}\,, & \text{if} \quad n = 2 \\ 1\,, & \text{if} \quad n \geqslant 3 \end{cases}.$$

It follows that

$$\|\mathbf{u}\| = \|\mathbf{v}\| = \|\mathbf{u} + \mathbf{v}\| = \|\mathbf{u} - \mathbf{v}\| = 1\,,$$

and therefore

$$2 \left\{ \|\mathbf{u}\|^2 + \|\mathbf{v}\|^2 \right\} - \|\mathbf{u} + \mathbf{v}\|^2 - \|\mathbf{u} - \mathbf{v}\|^2 = 2 \neq 0\,,$$

which proves that $\| \cdot \|$ does not satisfy the mentioned law. In fact, it is not difficult to see that there exists an infinity of counterexamples (besides the one presented here).

1.3 A boundary value problem in $\mathbb{R}$

Given $\Omega := (0,1)$ and $f \in C(\Omega)$, consider the problem: Find $u \in C^2(\Omega) \cap C(\bar{\Omega})$ such that

$$- u'' = f \quad \text{in} \quad \Omega, \quad u(0) = u(1) = 0. \tag{1.14}$$

Hereafter, $C(\bar{\Omega})$ stands for the vector space of all continuous functions $v : \bar{\Omega} \to \mathbb{R}$. Our goal is to present a method of solution for (1.14), different from the classical scheme, and in which, instead of requiring the differential equation to be satisfied at each point $x \in \Omega$, it is required only to verify some appropriate integral identity. For this purpose, we need some preliminary concepts. Given an open set Ω of $\mathbb{R}$, $C^\infty(\Omega)$ denotes the space of real-valued functions infinitely differentiable defined on Ω. In turn, for each $\varphi \in C(\Omega)$ its support, denoted $\text{supp}(\varphi)$, is defined as the closure of $\Omega - \varphi^{-1}(\{0\})$, that is

$$\text{supp}(\varphi) := \overline{\{x \in \Omega : \quad \varphi(x) \neq 0\}}, \tag{1.15}$$

with which one introduces the space of functions

$$C_0^\infty(\Omega) := \{\varphi \in C^\infty(\Omega) : \quad \text{supp}(\varphi) \quad \text{is compact} \quad \text{and} \quad \text{supp}(\varphi) \subseteq \Omega\}.$$

Note that the requirement of compactness for the support guarantees that the functions of $C_0^\infty(\Omega)$ vanish in some neighborhood of the boundary of Ω. Then, multiplying the differential equation of (1.14) by a function $\varphi \in C_0^\infty(\Omega)$ and integrating by parts in Ω, we obtain

$$\int_0^1 f \, \varphi \, dx = - \int_0^1 u'' \, \varphi \, dx = \int_0^1 u' \, \varphi' \, dx - u'(1) \, \varphi(1) + u'(0) \, \varphi(0),$$

from which, using that $\varphi(0) = \varphi(1) = 0$, we get

$$\int_0^1 u' \, \varphi' \, dx = \int_0^1 f \, \varphi \, dx \qquad \forall \, \varphi \in C_0^\infty(\Omega). \tag{1.16}$$

To continue the deduction of a new scheme of solution for (1.14), we need the following concept linked to the theory of distributions (see also Section 1.5).

Definition 1.30. *Given $v \in L^2(\Omega)$, we say that $v' \in L^2(\Omega)$, in the distributional sense, if there exists $z \in L^2(\Omega)$ such that*

$$- \int_\Omega v \, \varphi' \, dx = \int_\Omega z \, \varphi \, dx \qquad \forall \, \varphi \in C_0^\infty(\Omega),$$

and in this case we write $v' := z$.

Example 1.31 Consider $\Omega = (0,2)$ and the function $v(x) = \begin{cases} x, & 0 \leqslant x \leqslant 1 \\ 1, & 1 \leqslant x \leqslant 2 \end{cases}$. Then, given $\varphi \in C_0^\infty(\Omega)$ we have $\varphi(0) = \varphi(2) = 0$ and thus

$$- \int_\Omega v \, \varphi' \, dx = - \int_0^1 x \, \varphi' \, dx - \int_1^2 \varphi' \, dx$$

$$= \int_0^1 \varphi \, dx - \varphi(1) - (\varphi(2) - \varphi(1)) = \int_0^1 \varphi \, dx = \int_\Omega z \, \varphi \, dx,$$

where

$$z(x) = \begin{cases} 1, & 0 \leqslant x < 1 \\ 0, & 1 < x \leqslant 2 \end{cases},$$

which proves that $v' = z \in L^2(\Omega)$.

Motivated by the previous definition, we introduce the SOBOLEV SPACE

$$\mathrm{H}^1(\Omega) := \{ v \in \mathrm{L}^2(\Omega) : \quad v' \in \mathrm{L}^2(\Omega) \}, \tag{1.17}$$

which is a Hilbert space endowed with the inner product

$$\langle v, w \rangle_{1,\Omega} := \int_\Omega (v' w' + v w)\, dx \qquad \forall\, v, w \in \mathrm{H}^1(\Omega),$$

that is (cf. (1.12))

$$\langle v, w \rangle_{1,\Omega} := \langle v', w' \rangle_{0,\Omega} + \langle v, w \rangle_{0,\Omega} \quad \forall\, v, w \in \mathrm{H}^1(\Omega).$$

It follows that the induced norm $\| \cdot \|_{1,\Omega} : \mathrm{H}^1(\Omega) \to \mathbb{R}$ is given by

$$\|v\|_{1,\Omega} := \left\{ \int_\Omega \left((v')^2 + v^2 \right) dx \right\}^{1/2} \qquad \forall\, v \in \mathrm{H}^1(\Omega),$$

or, equivalently (cf. (1.13))

$$\|v\|_{1,\Omega} := \left\{ |v|_{1,\Omega}^2 + \|v\|_{0,\Omega}^2 \right\}^{1/2} \qquad \forall\, v \in \mathrm{H}^1(\Omega),$$

where $| \cdot |_{1,\Omega}$, defined as

$$|v|_{1,\Omega} := \|v'\|_{0,\Omega} \qquad \forall\, v \in \mathrm{H}^1(\Omega), \tag{1.18}$$

denotes from now on the seminorm of $\mathrm{H}^1(\Omega)$. The fact that $\mathrm{H}^1(\Omega)$ is indeed a Hilbert space is proposed as a particular case of Exercise 1.10. A direct proof of the more general result for $\mathrm{H}^m(\Omega)$, with $m \in \mathbb{N}$, is given later in Theorem 10.1.

On the other hand, we denote by $\mathrm{H}_0^1(\Omega)$ the closure of $C_0^\infty(\Omega)$ in $\mathrm{H}^1(\Omega)$, which can also be characterized as the subspace of elements of $\mathrm{H}^1(\Omega)$ that vanish on the boundary of Ω, that is

$$\mathrm{H}_0^1(\Omega) := \{ v \in \mathrm{H}^1(\Omega) : \quad v(0) = v(1) = 0 \}.$$

With these notations, and in virtue of identity (1.16), we can state the following solution scheme for our boundary value problem (1.14): Find $u \in \mathrm{H}_0^1(\Omega)$ such that

$$\int_\Omega u' \varphi' \, dx = \int_\Omega f \varphi \, dx \qquad \forall\, \varphi \in C_0^\infty(\Omega). \tag{1.19}$$

Moreover, in the following lemma we prove, using the density of $C_0^\infty(\Omega)$ in $\mathrm{H}_0^1(\Omega)$, that (1.19) can be reformulated as: Find $u \in \mathrm{H}_0^1(\Omega)$ such that

$$\int_\Omega u' v' \, dx = \int_\Omega f v \, dx \qquad \forall\, v \in \mathrm{H}_0^1(\Omega). \tag{1.20}$$

Lemma 1.32. *Formulations* (1.19) *and* (1.20) *are equivalent.*

Proof. Let $u \in \mathrm{H}_0^1(\Omega)$ be a solution of (1.20). Since $C_0^\infty(\Omega) \subseteq \mathrm{H}_0^1(\Omega)$, it is clear that u is also a solution of (1.19). Conversely, let $u \in \mathrm{H}_0^1(\Omega)$ be a solution of (1.19) and consider $v \in \mathrm{H}_0^1(\Omega)$. Since $C_0^\infty(\Omega)$ is dense in $\mathrm{H}_0^1(\Omega)$, it follows that there exists a sequence $\{\varphi_n\}_{n\in\mathbb{N}} \subseteq$

$C_0^\infty(\Omega)$ such that $\|v-\varphi_n\|_{\mathrm{H}^1(\Omega)} \rightarrow 0$ as $n \rightarrow +\infty$, and in this case, according to (1.19), we have

$$\int_\Omega u'\,\varphi_n'\,dx \;=\; \int_\Omega f\,\varphi_n\,dx \qquad \forall\, n \in \mathbb{N}\,. \tag{1.21}$$

Now, applying the Cauchy–Schwarz inequality in $\mathrm{L}^2(\Omega)$, we obtain

$$\left|\int_\Omega u'\,v'\,dx \;-\; \int_\Omega u'\,\varphi_n'\,dx\right| \;=\; \left|\int_\Omega u'(v'-\varphi_n')\,dx\right|$$

$$\leqslant\; |u|_{1,\Omega}\,|v-\varphi_n|_{1,\Omega} \;\leqslant\; \|u\|_{1,\Omega}\,\|v-\varphi_n\|_{1,\Omega}\,,$$

which implies that $\lim\limits_{n\to\infty}\int_\Omega u'\,\varphi_n'\,dx \;=\; \int_\Omega u'\,v'\,dx$. Similarly, one shows that

$$\lim_{n\to\infty}\int_\Omega f\,\varphi_n\,dx \;=\; \int_\Omega f\,v\,dx\,,$$

and thus, taking the limit as $n \rightarrow \infty$ in (1.21), we conclude that

$$\int_\Omega u'\,v'\,dx \;=\; \int_\Omega f\,v\,dx\,,$$

which proves that u is a solution of (1.20).

□

On the other hand, defining $H := \mathrm{H}_0^1(\Omega)$, and the mappings $A : H \times H \rightarrow \mathbb{R}$ and $F : H \rightarrow \mathbb{R}$ given by

$$A(w,v) := \int_\Omega w'\,v'\,dx \quad\text{and}\quad F(v) := \int_\Omega f\,v\,dx \qquad \forall\, w,v \in H\,,$$

we observe that (1.20) reduces to: Find $u \in H$ such that

$$A(u,v) \;=\; F(v) \qquad \forall\, v \in H\,. \tag{1.22}$$

Consequently, regardless of the boundary value problem (1.14) that gave rise to (1.20), we may think of studying the solvability of abstract formulations of the type (1.22). Precisely, in one of the following chapters of this text we introduce the LAX-MILGRAM LEMMA, which provides sufficient conditions on H, A and F for (1.22) to have a unique solution. Naturally, in order to establish and prove such a theorem, we will need several preliminary results from functional analysis.

1.4 A boundary value problem in $\mathbb{R}^n$

Let $n \in \mathbb{N}$ and Ω be a bounded open subset of $\mathbb{R}^n$ with smooth boundary $\Gamma := \partial\Omega$ of class C^1. The precise definition of the degree of smoothness of the boundary, including certainly C^1 and other classes of functions, can be found in [22], [39], or [112]. Alternatively, we can refer to Definition 10.9. Then, given a function $f \in C(\Omega)$, we are interested in the following boundary value problem: Find $u \in C^2(\Omega) \cap C(\bar{\Omega})$ such that

$$-\Delta u \;=\; f \quad\text{in}\quad \Omega\,, \qquad u \;=\; 0 \quad\text{in}\quad \Gamma\,. \tag{1.23}$$

In order to derive a solution scheme for (1.23), analogous to that of the previous section, we now deduce the GREEN'S IDENTITIES. To this end, we first recall the Gauss Divergence Theorem, which states that for a vector field $G \in [C^1(\Omega) \cap C(\bar{\Omega})]^n$, one has

$$\int_\Omega \operatorname{div}(G)\, d\mathbf{x} = \int_\Gamma G \cdot \boldsymbol{\nu}\, ds\,,$$

where $\operatorname{div}(G) := \sum_{i=1}^{n} \frac{\partial G_i}{\partial x_i}$ and $\boldsymbol{\nu} := (\nu_1, ..., \nu_n)^{\mathrm{t}}$ is the outward unit normal vector to Ω. In particular, if G has all components equal to zero except the i-th one, which is equal to uv, with $u, v \in C^1(\Omega) \cap C(\bar{\Omega})$, then we obtain

$$\int_\Omega \frac{\partial}{\partial x_i}(uv)\, d\mathbf{x} = \int_\Gamma uv\, \nu_i\, ds\,,$$

or equivalently

$$\int_\Omega v\, \frac{\partial u}{\partial x_i}\, d\mathbf{x} = -\int_\Omega u\, \frac{\partial v}{\partial x_i}\, d\mathbf{x} + \int_\Gamma uv\, \nu_i\, ds \quad \forall\, i \in \{1, ..., n\}\,,$$

which is known as the INTEGRATION BY PARTS FORMULA. Replacing u by $\frac{\partial w}{\partial x_i}$ in the previous equation, we obtain

$$\int_\Omega v\, \frac{\partial^2 w}{\partial x_i^2}\, d\mathbf{x} = -\int_\Omega \frac{\partial w}{\partial x_i}\, \frac{\partial v}{\partial x_i}\, d\mathbf{x} + \int_\Gamma \frac{\partial w}{\partial x_i} v\, \nu_i\, ds \quad \forall\, i \in \{1, ..., n\}\,,$$

and then, summing over all indices i, we conclude the FIRST GREEN'S IDENTITY

$$\int_\Omega v\, \Delta w\, d\mathbf{x} = -\int_\Omega \nabla w \cdot \nabla v\, d\mathbf{x} + \int_\Gamma \frac{\partial w}{\partial \boldsymbol{\nu}}\, v\, ds\,, \tag{1.24}$$

which holds for all $v \in C^1(\Omega) \cap C(\bar{\Omega})$ and for all $w \in C^2(\Omega) \cap C^1(\bar{\Omega})$. Here, $\frac{\partial w}{\partial \boldsymbol{\nu}} := \sum_{i=1}^{N} \frac{\partial w}{\partial x_i}\, \nu_i$ denotes the normal derivative on Γ. Interchanging the roles of v and w in (1.24) and then subtracting the resulting equation from (1.24), we obtain the SECOND GREEN'S IDENTITY:

$$\int_\Omega (\, v\, \Delta w - w\, \Delta v\,)\, d\mathbf{x} = \int_\Gamma \left(v\, \frac{\partial w}{\partial \boldsymbol{\nu}} - w\, \frac{\partial v}{\partial \boldsymbol{\nu}} \right) ds\,,$$

for all $v, w \in C^2(\Omega) \cap C^1(\bar{\Omega})$. Now, we consider again the space $C_0^\infty(\Omega)$, whose definition in the n-dimensional case is analogous to that given in Section 1.3 for an open set Ω of $\mathbb{R}$. Multiplying the partial differential equation from (1.23) by a function $\varphi \in C_0^\infty(\Omega)$, applying the FIRST GREEN'S IDENTITY (1.24), and using the fact that $\varphi = 0$ on Γ, we obtain

$$\int_\Omega \nabla u \cdot \nabla \varphi\, d\mathbf{x} = \int_\Omega f \varphi\, d\mathbf{x} \qquad \forall\, \varphi \in C_0^\infty(\Omega)\,. \tag{1.25}$$

Definition 1.33. *Given* $v \in \mathrm{L}^2(\Omega)$, *we say that* $\frac{\partial v}{\partial x_i} \in \mathrm{L}^2(\Omega)$ *in the distributional sense, if there exists* $z_i \in \mathrm{L}^2(\Omega)$ *such that*

$$-\int_\Omega v \frac{\partial \varphi}{\partial x_i}\,d\mathbf{x} = \int_\Omega z_i\,\varphi\,d\mathbf{x} \qquad \forall\,\varphi \in C_0^\infty(\Omega),$$

and in that case we write $\dfrac{\partial v}{\partial x_i} = z_i$.

According to the previous definition, we can introduce the SOBOLEV SPACE analogous to that given in (1.17),

$$\mathrm{H}^1(\Omega) := \left\{ v \in \mathrm{L}^2(\Omega) : \quad \frac{\partial v}{\partial x_i} \in \mathrm{L}^2(\Omega) \quad \forall\, i \in \{1,\dots,n\} \right\}, \tag{1.26}$$

which is also a Hilbert space endowed with the following inner product

$$\langle v, w\rangle_{1,\Omega} := \int_\Omega (\nabla v \cdot \nabla w + v\,w)\,d\mathbf{x} \qquad \forall\, v,\, w \in \mathrm{H}^1(\Omega). \tag{1.27}$$

Then, the norm $\|\cdot\|_{1,\Omega} : \mathrm{H}^1(\Omega) \to \mathbb{R}$ is given by

$$\|v\|_{1,\Omega} := \left\{ |v|_{1,\Omega}^2 + \|v\|_{0,\Omega}^2 \right\}^{1/2} \qquad \forall\, v \in \mathrm{H}^1(\Omega), \tag{1.28}$$

where $|\cdot|_{1,\Omega}$ is the associated seminorm defined as

$$|v|_{1,\Omega} := \left\{ \int_\Omega \|\nabla v\|^2\,d\mathbf{x} \right\}^{1/2} = \left\{ \sum_{i=1}^n \left\| \frac{\partial v}{\partial x_i} \right\|_{0,\Omega}^2 \right\}^{1/2} \qquad \forall\, v \in \mathrm{H}^1(\Omega),$$

and $\|\cdot\|_{0,\Omega}$ is already described by (1.13). Similarly to the one-dimensional case considered in Section 1.3, we refer to Exercise 1.10 or Theorem 10.1 for a proof of the completeness of $\mathrm{H}^1(\Omega)$. In turn, following also the approach of Section 1.3, the space $\mathrm{H}_0^1(\Omega)$ is defined as the closure of $C_0^\infty(\Omega)$ in $\mathrm{H}^1(\Omega)$. It can be shown (see Chapter 1 in [112]) that $\mathrm{H}_0^1(\Omega)$ coincides with the subspace of elements of $\mathrm{H}^1(\Omega)$ that vanish on the boundary in the sense of traces, that is

$$\mathrm{H}_0^1(\Omega) := \left\{ v \in \mathrm{H}^1(\Omega) : v = 0 \quad \text{in} \quad \partial\Omega \right\}.$$

We will refer in more details to the concept of traces in Chapters 4 (cf. Section 4.5) and 10 (cf. Section 10.2). In view of identity (1.25) and of the density of $C_0^\infty(\Omega)$ in $\mathrm{H}_0^1(\Omega)$, which allows one to prove a result analogous to Lemma 1.32, the alternative solution scheme for the boundary value problem (1.23) reduces to: Find $u \in \mathrm{H}_0^1(\Omega)$ such that

$$\int_\Omega \nabla u \cdot \nabla v\,d\mathbf{x} = \int_\Omega f v\,d\mathbf{x} \qquad \forall\, v \in \mathrm{H}_0^1(\Omega). \tag{1.29}$$

Then, defining $H := \mathrm{H}_0^1(\Omega)$, and the mappings $A : H \times H \to \mathbb{R}$ and $F : H \to \mathbb{R}$ given by

$$A(w,v) := \int_\Omega \nabla w \cdot \nabla v\,d\mathbf{x} \quad \text{and} \quad F(v) := \int_\Omega f v\,d\mathbf{x} \quad \forall\, w,\, v \in H,$$

we observe that (1.29) reduces to: Find $u \in H$ such that

$$A(u,v) = F(v) \qquad \forall\, v \in H. \tag{1.30}$$

Certainly, the abstract structure of (1.30) coincides with that of problem (1.22), and therefore the same comments given at the end of Section 1.3 apply in this case. In particular, it is reiterated once again that the LAX-MILGRAM LEMMA will provide us with sufficient conditions on H, A, and F for formulations of this type to be well-posed.

1.5 Basic notions of distributions

In this section we present some basic results on distributions, which formalize several of the arguments given in the two previous sections and also provide some necessary foundations for the following chapters. To this end, let us consider $n \in \mathbb{N}$ and an open set Ω of $\mathbb{R}^n$.

Definition 1.34. *A multi-index is an n-tuple $\alpha := (\alpha_1, ..., \alpha_n)$ where each α_i is a non-negative integer. The length (or order) of the n-tuple α is defined as $|\alpha| := \sum_{i=1}^{n} \alpha_i$. Given two n-tuples α, β, their sum is defined componentwise, that is, if $\beta := (\beta_1, ..., \beta_n)$, then $\alpha + \beta := (\alpha_1 + \beta_1, ..., \alpha_n + \beta_n)$. Also, we say that $\alpha \leqslant \beta$ if $\alpha_i \leqslant \beta_i$ for all $i \in \{1, \ldots, n\}$. In turn, for a function $u \in C^{|\alpha|}(\Omega)$, we write*

$$\partial^\alpha u := \frac{\partial^{|\alpha|} u}{\partial x_1^{\alpha_1} \partial x_2^{\alpha_2} \cdots \partial x_n^{\alpha_n}}.$$

A specific example of functions in $C_0^\infty(\Omega)$ is given below.

Example 1.35 Let us consider the function $\psi : \mathbb{R} \to \mathbb{R}$ defined by (see Figure 1.2)

$$\psi(t) := \begin{cases} 0 & \text{if} \quad t \leqslant 0, \\ e^{-1/t} & \text{if} \quad t > 0. \end{cases}$$

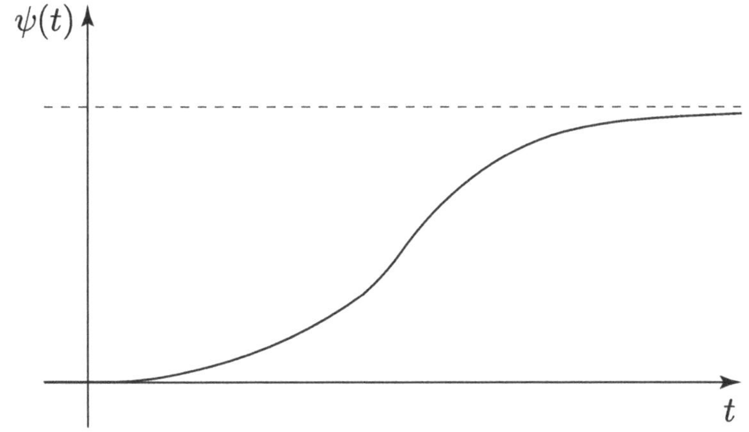

Fig. 1.2. *Illustration of ψ*

It is not difficult to prove that ψ belongs to $C^\infty(\mathbb{R})$, the space of infinitely differentiable functions defined on $\mathbb{R}$. This fact suggests then defining the function $\varphi : \mathbb{R}^n \to \mathbb{R}$ given by $\varphi(x) := \psi(1 - \|x\|^2)$ for all $x \in \mathbb{R}^n$, which, being the composition of ψ with a polynomial function, belongs to $C^\infty(\mathbb{R}^n)$. Moreover, it follows that

$$\varphi(x) := \begin{cases} 0 & \text{if} \quad \|x\| \geqslant 1, \\ e^{-1/(1-\|x\|^2)} & \text{if} \quad \|x\| < 1, \end{cases}$$

which indicates that $\text{supp}(\varphi)$ is the unit ball $\bar{B}(0,1) := \{x \in \mathbb{R}^n : \|x\| \leqslant 1\}$, and therefore $\varphi \in C_0^\infty(\Omega)$ for any open set Ω containing $\bar{B}(0,1)$. In general, given $\varepsilon > 0$ and $x_0 \in \mathbb{R}^n$, the function $\varphi_{\varepsilon,x_0} : \mathbb{R}^n \to \mathbb{R}$ defined by

$$\varphi_{\varepsilon,x_0}(x) := \psi(1-\|x-x_0\|^2/\varepsilon^2) := \begin{cases} 0 & \text{if} \quad \|x-x_0\| \geqslant \varepsilon\,, \\ e^{-\varepsilon^2/(\varepsilon^2-\|x-x_0\|^2)} & \text{if} \quad \|x-x_0\| < \varepsilon\,, \end{cases}$$

also belongs to $C^\infty(\mathbb{R}^n)$ with $\mathrm{supp}(\varphi_{\varepsilon,x_0}) = \bar{B}(x_0,\varepsilon) := \{x \in \mathbb{R}^n : \quad \|x-x_0\| \leqslant \varepsilon\}$. We thus conclude that $\varphi_{\varepsilon,x_0} \in C_0^\infty(\Omega)$ for any open set Ω containing $\bar{B}(x_0,\varepsilon)$.

The previous example shows that one can construct functions in $C_0^\infty(\Omega)$ on any open subset Ω of $\mathbb{R}^n$, with supports as small as desired, or as large as Ω allows.

Definition 1.36. *Given a compact subset K of Ω, we let $\mathcal{D}_K(\Omega)$ denote the topological vector space of functions $\varphi \in C_0^\infty(\Omega)$ satisfying $\mathrm{supp}(\varphi) \subseteq K$, which is endowed with the topology $\mathfrak{T}_K$ induced by the countable family of seminorms $\{p_{K,m}\}_{m\in\mathbb{N}\cup\{0\}}$, where*

$$p_{K,m}(\varphi) := \sum_{|\alpha|\leqslant m} \sup_{x\in K} |\partial^\alpha \varphi(x)| \qquad \forall\, \varphi \in C_0^\infty(\Omega)\,. \tag{1.31}$$

Recall that a topological vector space is a vector space endowed with a topology with respect to which vector addition and scalar multiplication are continuous mappings.

We now introduce

$$\mathcal{B}_0 := \Big\{W : \quad W \subseteq C_0^\infty(\Omega)\,, \quad W \text{ convex and balanced}\,,$$
$$W \cap \mathcal{D}_K(\Omega) \in \mathfrak{T}_K \quad \forall \text{compact subset } K \text{ of } \Omega\Big\}\,,$$

where W is said to be balanced if $\kappa W \subseteq W$ for all $\kappa \in \mathbb{C}$, $|\kappa| \leqslant 1$, and define

$$\mathfrak{T} := \Big\{V : \quad V \subseteq C_0^\infty(\Omega)\,, \quad \text{there exists an arbitrary set of indexes } I$$
$$\text{such that} \quad V = \bigcup_{i\in I} (\varphi_i + W_i)\,, \quad \varphi_i \in C_0^\infty(\Omega)\,, \quad W_i \in \mathcal{B}_0\Big\}\,.$$

Then, it can be proved (see, e.g. [114, Theorem 6.4]) that $\mathfrak{T}$ is a topology on $C_0^\infty(\Omega)$, and that

$$\mathcal{B} := \Big\{V : \quad V = \varphi + W\,, \quad \varphi \in C_0^\infty(\Omega)\,, \quad W \in \mathcal{B}_0\Big\}$$

is a basis for $\mathfrak{T}$. Moreover, since $\mathcal{B}$ consists of convex sets, we can say that $\big(C_0^\infty(\Omega), \mathfrak{T}\big)$ is a locally convex topological vector space. It is also Hausdorff, the latter meaning that for all φ, $\psi \in C_0^\infty(\Omega)$, $\varphi \neq \psi$, there exist V_φ, $V_\psi \in \mathfrak{T}$ such that $\varphi \in V_\varphi$, $\psi \in V_\psi$, and $V_\varphi \cap V_\psi = \varnothing$.

For further details on these topological concepts, including the definition of a topology induced by a family of seminorms, as well as proofs of the results mentioned above, we refer to [114, Chapter 1, Section 1.46, and Chapter 6]. In this regard, we remark that, besides $\mathfrak{T}_K$, two additional examples of topologies induced by countable families of seminorms will arise in Sections 9.7 and 9.9, where the corresponding compatible metrics are introduced.

Definition 1.37. *The vector space $C_0^\infty(\Omega)$, endowed with the above-mentioned topology $\mathfrak{T}$, is denoted $\mathcal{D}(\Omega)$ and is called the* SPACE OF TEST FUNCTIONS.

We now introduce the following general definitions.

Definition 1.38. *Given topological spaces X and Y, endowed with topologies $\mathcal{T}_X$ and $\mathcal{T}_Y$, respectively, we say that a mapping $u : (X, \mathcal{T}_X) \to (Y, \mathcal{T}_Y)$ is* CONTINUOUS *if $u^{-1}(V) \in \mathcal{T}_X$ for all $V \in \mathcal{T}_Y$.*

Definition 1.39. *Given topological vector spaces X and Y over the same field $\mathbb{K}$, endowed with topologies $\mathcal{T}_X$ and $\mathcal{T}_Y$, respectively, we say that a mapping $u : (X, \mathcal{T}_X) \to (Y, \mathcal{T}_Y)$ is* LINEAR *if*

$$u(\alpha\varphi + \beta\psi) = \alpha\, u(\varphi) + \beta\, u(\psi) \qquad \forall\, \alpha,\, \beta \in \mathbb{K}, \quad \forall\, \varphi,\, \psi \in X\,.$$

In the particular case when $Y = \mathbb{K}$, we say that $u : (X, \mathcal{T}_X) \to \mathbb{K}$ is a LINEAR FORM.

In connection with these definitions, and as an important consequence of the topology of $\mathcal{D}(\Omega)$, we stress that a **linear** mapping from $\mathcal{D}(\Omega)$ into any topological vector space X is continuous (cf. Definition 1.38) if and only if it is sequentially continuous. The proof of this result is certainly not trivial, since the topology $\mathfrak{T}$ of $\mathcal{D}(\Omega)$ is not metrizable. For the particular case of a locally convex target space X, we refer to [114, Theorem 6.6]. In addition, we emphasize the necessity of linearity by observing that there may exist nonlinear mappings from $\mathcal{D}(\Omega)$ into X that are sequentially continuous but not continuous. Accordingly, we now introduce the following definitions.

Definition 1.40. *Let $\{\varphi_k\}_{k\in\mathbb{N}}$ be a sequence in $\mathcal{D}(\Omega)$. We say that $\{\varphi_k\}_{k\in\mathbb{N}}$ converges to the zero function in $\mathcal{D}(\Omega)$ if there exists a compact set $K_0 \subset \Omega$ such that $\operatorname{supp}(\varphi_k) \subseteq K_0$ for all $k \in \mathbb{N}$, and for every n-tuple α we have that $\{\partial^\alpha \varphi_k\}_{k\in\mathbb{N}}$ converges uniformly to zero on K_0, that is,*

$$\lim_{k\to\infty} \sup_{x\in K_0} |\partial^\alpha \varphi_k(x)| = 0\,.$$

Definition 1.41. *A linear form $u : \mathcal{D}(\Omega) \to \mathbb{C}$ is called a* DISTRIBUTION ON Ω *if it is continuous (equivalently, sequentially continuous). In other words, $u : \mathcal{D}(\Omega) \to \mathbb{C}$ is said to be a distribution if for every sequence $\{\varphi_k\}_{k\in\mathbb{N}} \subseteq \mathcal{D}(\Omega)$ that converges to the zero function, we have that $u(\varphi_k)$ converges to zero as $k \to \infty$. The space of all distributions on Ω is denoted by $\mathcal{D}'(\Omega)$. Moreover, given $u \in \mathcal{D}'(\Omega)$ and $\varphi \in \mathcal{D}(\Omega)$, we adopt the notation $u(\varphi) = \langle u, \varphi\rangle$.*

A characterization result for distributions, easier to verify than Definition 1.41, is given by the following theorem.

Theorem 1.42. *A linear form $u : \mathcal{D}(\Omega) \to \mathbb{C}$ is a distribution on Ω if and only if for each compact set $K \subseteq \Omega$, there exist a nonnegative integer N and a positive constant C, such that*

$$|\langle u, \varphi\rangle| \leqslant C \sum_{|\alpha|\leqslant N} \sup_{x\in K} |\partial^\alpha \varphi(x)| \qquad \forall\, \varphi \in \mathcal{D}_K(\Omega)\,. \tag{1.32}$$

Proof. For the first implication, suppose by contradiction that there exists a compact set $K \subseteq \Omega$ such that for every nonnegative integer N, and for every $C > 0$, there exists $\widetilde{\varphi} \in \mathcal{D}_K(\Omega)$ such that

$$|\langle u, \widetilde{\varphi}\rangle| > C \sum_{|\alpha|\leqslant N} \sup_{x\in K} |\partial^\alpha \widetilde{\varphi}(x)|\,.$$

In particular, for $C = N$, there exists $\widetilde{\varphi}_N \in \mathcal{D}_K(\Omega)$ such that

$$|\langle u, \widetilde{\varphi}_N\rangle| > N \sum_{|\alpha|\leqslant N} \sup_{x\in K} |\partial^\alpha \widetilde{\varphi}_N(x)|\,.$$

For each N, let us define the function

$$\varphi_N := \frac{\widetilde{\varphi}_N}{|\langle u, \widetilde{\varphi}_N\rangle|}\,.$$

It follows that

$$|\langle u, \varphi_N\rangle| = 1 \qquad \text{and} \qquad \frac{1}{N} > \sum_{|\alpha|\leqslant N} \sup_{x\in K} |\partial^\alpha \varphi_N(x)|\,. \tag{1.33}$$

Notice now that $\operatorname{supp}(\varphi_N) = \operatorname{supp}(\widetilde{\varphi}_N) \subseteq K$. Moreover, given an n-tuple β, it follows from (1.33) that for every $N \geqslant |\beta|$ we have

$$\sup_{x\in K} |\partial^\beta \varphi_N(x)| \leqslant \frac{1}{N}\,,$$

and therefore

$$\lim_{N\to\infty} \sup_{x\in K} |\partial^\beta \varphi_N(x)| = 0 \quad \forall\, \beta\,.$$

This proves that the sequence $\{\varphi_N\}_{N\in\mathbb{N}}$ converges to the zero function in $\mathcal{D}(\Omega)$. However, due to (1.33), $\{\langle u, \varphi_N\rangle\}_{N\in\mathbb{N}}$ does not converge to zero in $\mathbb{C}$, which contradicts the fact that u is continuous. Conversely, suppose that for every compact set $K \subseteq \Omega$ there exist $N \in \mathbb{N} \cup \{0\}$ and $C > 0$, such that (1.32) holds, and let $\{\varphi_k\}_{k\in\mathbb{N}} \subseteq \mathcal{D}(\Omega)$ be a sequence that converges to zero in $\mathcal{D}(\Omega)$. This means that there exists a compact set $K_0 \subseteq \Omega$ such that $\operatorname{supp}(\varphi_k) \subseteq K_0$ for all $k \in \mathbb{N}$, and for every n-tuple α we have

$$\lim_{k\to\infty} \sup_{x\in K} |\partial^\alpha \varphi_k(x)| = 0\,.$$

Then, applying (1.32) with $K = K_0$, we conclude that $\{\langle u, \varphi_k\rangle\}_{k\in\mathbb{N}}$ converges to zero in $\mathbb{C}$, thus proving the continuity of u.

□

Definition 1.43. *The space of locally integrable functions on Ω, denoted by $L^1_{loc}(\Omega)$, is defined as the set of all (equivalence classes of) measurable functions $f : \Omega \to \mathbb{C}$ such that*

$$\int_K |f| < \infty \qquad \forall \text{ compact } K \subseteq \Omega\,.$$

These equivalence classes are formed by functions that coincide almost everywhere in Ω.

Example 1.44 Locally integrable functions provide the simplest examples of distributions. Indeed, given $f \in L^1_{loc}(\Omega)$, the induced distribution, which we denote again by f, is defined as

$$\langle f, \varphi\rangle := \int_\Omega f\varphi\, dx \quad \forall \varphi \in \mathcal{D}(\Omega)\,.$$

Let us check that $f \in \mathcal{D}'(\Omega)$. Linearity is a direct consequence of the same property of the integral. For continuity, let us consider a compact set $K \subseteq \Omega$ and take $\varphi \in \mathcal{D}_K(\Omega)$. It follows that

$$\begin{aligned} |\langle f, \varphi\rangle| &= \left|\int_\Omega f\varphi\, dx\right| = \left|\int_K f\varphi\, dx\right| \\ &\leqslant \int_K |f|\,|\varphi|\, dx \leqslant \int_K |f|\, dx \sup_{x\in K} |\varphi(x)|\,. \end{aligned}$$

Therefore, by virtue of Theorem 1.42, it suffices to choose $C := \displaystyle\int_K |f|\, dx$ and $N = 0$, yielding

$$|\langle f, \varphi\rangle| \leqslant C \sum_{|\alpha|\leqslant N} \sup_{x\in K} |\partial^\alpha \varphi(x)|\,.$$

The previous example shows that the space $\mathrm{L}^1_{loc}(\Omega)$ can be identified with a subspace of $\mathcal{D}'(\Omega)$, which leads to the following definition.

Definition 1.45. *A distribution $u \in \mathcal{D}'(\Omega)$ is said to be* REGULAR *if it is generated by a locally integrable function. Otherwise, u is called* SINGULAR.

The classical example of a singular distribution is given by $\delta \in \mathcal{D}'(\mathbb{R}^n)$, called the DIRAC DELTA, and defined as

$$\langle \delta, \varphi \rangle := \varphi(0) \qquad \forall\, \varphi \in \mathcal{D}(\mathbb{R}^n).$$

More generally, if $x_0 \in \Omega$, one also defines $\delta_{x_0} \in \mathcal{D}'(\Omega)$ as

$$\langle \delta_{x_0}, \varphi \rangle := \varphi(x_0) \qquad \forall\, \varphi \in \mathcal{D}(\Omega).$$

Lemma 1.46. *The distribution $\delta \in \mathcal{D}'(\mathbb{R}^n)$ is singular.*

Proof. For simplicity, let us consider the case $n = 1$. The proof for $n > 1$ is analogous. Given the test function $\varphi_{\varepsilon,0} \in \mathcal{D}(\mathbb{R})$ (defined in Example 1.35), with $\varepsilon \in (0,1)$, we first observe that $\langle \delta, \varphi_{\varepsilon,0} \rangle = \varphi_{\varepsilon,0}(0) = e^{-1}$, $|\varphi_{\varepsilon,0}(x)| \leqslant e^{-1}$ for all $x \in \mathbb{R}$, and $\sup \varphi_{\varepsilon,0} = [-\varepsilon, \varepsilon]$. Suppose now, by contradiction, that there exists $f \in \mathrm{L}^1_{loc}(\mathbb{R})$ such that $\langle \delta, \varphi \rangle = \int_{\mathbb{R}} f\,\varphi\,dx$ for all $\varphi \in \mathcal{D}(\mathbb{R})$. It then follows that

$$e^{-1} = |\langle \delta, \varphi_{\varepsilon,0} \rangle| = \left| \int_{-\varepsilon}^{\varepsilon} f\,\varphi_{\varepsilon,0}\,dx \right| \leqslant e^{-1} \int_{-\varepsilon}^{\varepsilon} |f|\,dx = e^{-1} \int_{-1}^{1} \mathbf{1}_{[-\varepsilon,\varepsilon]}\,|f|\,dx,$$

where $\mathbf{1}_{[-\varepsilon,\varepsilon]}$ denotes the characteristic function of the interval $[-\varepsilon, \varepsilon]$. Then, taking the limit as $\varepsilon \to 0$ and applying Lebesgue's Dominated Convergence Theorem (cf. [113, Theorem 1.34], [120, Theorem 5.36]), we obtain $e^{-1} \leqslant 0$, which is absurd. This proves that δ cannot be regular.

□

On the other hand, one of the most important notions about distributions concerns their differentiation. To understand the origin of this concept, let us now consider an open set Ω in $\mathbb{R}^n$ and a function $u \in C^1(\Omega)$. Since for each $i \in \{1, \ldots, n\}$, the partial derivative $\dfrac{\partial u}{\partial x_i}$ is a continuous function, it follows that it induces the distribution given by

$$\left\langle \frac{\partial u}{\partial x_i}, \varphi \right\rangle = \int_{\Omega} \frac{\partial u}{\partial x_i}\,\varphi\,dx \qquad \forall\, \varphi \in \mathcal{D}(\Omega).$$

However, using the integration by parts formula we obtain

$$\left\langle \frac{\partial u}{\partial x_i}, \varphi \right\rangle = -\int_{\Omega} u\,\frac{\partial \varphi}{\partial x_i} = -\left\langle u, \frac{\partial \varphi}{\partial x_i} \right\rangle \qquad \forall\, \varphi \in \mathcal{D}(\Omega),$$

and we observe thus that the differentiability of u is not a necessary condition for this last term to be well defined. Indeed, it suffices that u defines a distribution for this expression to make sense. Similarly, one deduces that if β is a multi-index and $u \in C^{|\beta|}(\Omega)$, then applying integration by parts $|\beta|$ times, we obtain

$$\langle \partial^\beta u, \varphi \rangle = (-1)^{|\beta|} \langle u, \partial^\beta \varphi \rangle \qquad \forall\, \varphi \in \mathcal{D}(\Omega).$$

The previous comments lead to the following definition.

Definition 1.47. *Let $u \in \mathcal{D}'(\Omega)$, and let β be a multi-index. The β-th derivative of u, denoted $\partial^\beta u$, is defined as the linear form on $\mathcal{D}(\Omega)$ given by*

$$(\partial^\beta u)(\varphi) := (-1)^{|\beta|} \langle u, \partial^\beta \varphi \rangle \qquad \forall \varphi \in \mathcal{D}(\Omega).$$

Let us now see that for any multi-index β, $\partial^\beta u$ is also a distribution on Ω. Indeed, by virtue of Theorem 1.42 and the fact that $u \in \mathcal{D}'(\Omega)$, we know that given a compact set $K \subseteq \Omega$, there exist a non-negative integer N and a positive constant C such that

$$|\langle u, \varphi \rangle| \leqslant C \sum_{|\alpha| \leqslant N} \sup_{x \in K} |\partial^\alpha \varphi(x)| \qquad \forall \varphi \in \mathcal{D}_K(\Omega).$$

It follows that for each $\varphi \in \mathcal{D}_K(\Omega)$

$$|\langle \partial^\beta u, \varphi \rangle| = |\langle u, \partial^\beta \varphi \rangle| \leqslant C \sum_{|\alpha| \leqslant N} \sup_{x \in K} |\partial^{\alpha+\beta} \varphi(x)| \leqslant C \sum_{|\alpha| \leqslant N + |\beta|} \sup_{x \in K} |\partial^\alpha \varphi(x)|,$$

which, according again to Theorem 1.42, implies that $\partial^\beta u \in \mathcal{D}'(\Omega)$.

A detailed discussion of several other fundamental aspects of the theory of distributions is presented later on in Chapter 9.

1.6 Exercises

▸ **1.1** *Let X be a normed vector space, and let $S := \{x \in X : \|x\| = 1\}$. Prove that S is complete if and only if X is Banach.*

▸ **1.2** *Let $I := (t_0, t_0 + \tau)$, $f_j : \bar{I} \times \mathbb{R}^n \to \mathbb{R}$, $j \in \{1, \ldots, n\}$, continuous functions, and consider the system of first-order ordinary differential equations: Find $u := (u_1, \cdots, u_n)^{\mathrm{t}} \in [C^1(I) \times C(\bar{I})]^n$ such that*

$$\begin{cases} \dfrac{du_j}{dt} = f_j(t, u_1(t), \cdots, u_n(t)) \quad \forall t \in I \quad \text{and} \\[2ex] u_j(t_0) = \eta_j \quad \forall j \in \{1, \ldots, n\}, \end{cases} \tag{1.34}$$

where t_0, τ, η_j, $j = 1, \cdots, n$ are given real constants and τ is positive. Suppose in addition that there exists $M > 0$ such that

$$|f_j(t, z) - f_j(t, w)| \leqslant M \|z - w\|_\infty \quad \forall z, w \in \mathbb{R}^n, \quad \forall t \in \bar{I}.$$

Prove that (1.34) has a unique solution $u(t)$ for all $t \in \bar{I}$.

▸ **1.3** *Prove Theorem 1.14.*

▸ **1.4** *Let $(H, \langle \cdot, \cdot \rangle)$ be a Hilbert space, and consider a sequence $\{x_n\}_{n \in \mathbb{N}}$ in H such that $\langle x_n, x_m \rangle = \begin{cases} 1 & \text{if } n = m \\ 0 & \text{if } n \neq m \end{cases}$. Prove that $\sum_{n=1}^{\infty} |\langle x, x_n \rangle|^2 \leqslant \|x\|^2$ for all $x \in H$. Moreover, given a sequence of scalars $\{\alpha_n\}_{n \in \mathbb{N}}$, prove that $\sum_{n=1}^{\infty} \alpha_n x_n$ converges in H if and only if $\sum_{n=1}^{\infty} |\alpha_n|^2 < +\infty$.*

▸ **1.5** *A Banach space X is said to be* UNIFORMLY CONVEX *if for every $\varepsilon > 0$ there exists $\delta > 0$ such that*

$$\Big(x, y \in X\,,\ \|x\| \leqslant 1\,,\ \|y\| \leqslant 1 \quad \text{and} \quad \|x - y\| > \varepsilon\Big) \ \Rightarrow\ \left\|\frac{x+y}{2}\right\| < 1 - \delta\,.$$

Prove that every Hilbert space is uniformly convex.

▸ **1.6** ([90], Exercise 4.6) *The Brouwer Fixed Point Theorem states that if S is a compact, convex, nonempty subset of a finite-dimensional vector space, and T is a continuous mapping from S into S, then T has at least one fixed point. Use this Brouwer result to prove that if X is a finite-dimensional Hilbert space with inner product $\langle \cdot, \cdot \rangle$ and norm $\|\cdot\|$, and if F is a continuous mapping from X into X such that, for some $\mu > 0$, $\langle F(u), u\rangle \geqslant 0$ for all $u \in X$ with $\|u\| = \mu$, then there exists $u_0 \in X$, $\|u_0\| \leqslant \mu$, such that $F(u_0) = 0$.*

▸ **1.7** *Prove that if (U, ρ) is a metric space and $T : U \to U$ is a mapping such that T^m has a unique fixed point u for some $m \in \mathbb{N}$, then u is also the unique fixed point of T.*

▸ **1.8** ([90]) *Let Ω be a smooth domain of $\mathbb{R}^2$ with boundary Γ, and let $\mathbf{f} \in \mathbf{L}^2(\Omega) := [\mathrm{L}^2(\Omega)]^2$. The* NAVIER-STOKES PROBLEM, *a problem of fundamental importance in fluid mechanics, consists in finding the velocity vector $\mathbf{u} := (u_1, u_2)^{\mathrm{t}}$ and the pressure p of a fluid, such that*

$$\begin{gathered} -\Delta \mathbf{u} + \sum_{j=1}^{2} u_j \frac{\partial \mathbf{u}}{\partial x_j} + \nabla p = \mathbf{f} \quad in \quad \Omega\,, \quad \operatorname{div}(\mathbf{u}) = 0 \quad in \quad \Omega\,, \\ \mathbf{u} = 0 \quad on \quad \Gamma\,, \qquad \int_\Omega p\,dx = 0\,. \end{gathered} \tag{1.35}$$

Define the spaces $H = \mathbf{H}_0^1(\Omega) := [\mathrm{H}_0^1(\Omega)]^2$, $Q := \mathrm{L}_0^2(\Omega) := \Big\{ q \in \mathrm{L}^2(\Omega) : \displaystyle\int_\Omega q\,dx = 0 \Big\}$, and prove that the weak formulation of (1.35) reduces to: Find $(\mathbf{u}, p) \in H \times Q$ such that

$$\begin{aligned} a(\mathbf{u}; \mathbf{u}, \mathbf{v}) + b(\mathbf{v}, p) &= f(\mathbf{v}) \qquad \forall\, \mathbf{v} \in H \\ b(\mathbf{u}, q) &= 0 \qquad \forall\, q \in Q\,, \end{aligned}$$

where $a : H \times H \times H \to \mathbb{R}$, $b : H \times Q \to \mathbb{R}$, and $f : H \to \mathbb{R}$, are defined by

$$a(\mathbf{w}; \mathbf{u}, \mathbf{v}) := \sum_{i=1}^{2} \int_\Omega \nabla u_i \cdot \nabla v_i\,dx + \sum_{i,j=1}^{2} \int_\Omega w_j \frac{\partial u_i}{\partial x_j} v_i\,dx\,,$$

$$b(\mathbf{v}, p) := -\int_\Omega p \operatorname{div}(\mathbf{v})\,dx \quad , \quad f(\mathbf{v}) = \int_\Omega \mathbf{f} \cdot \mathbf{v}\,dx\,.$$

▸ **1.9** ([90]) *Let Ω be a smooth domain of $\mathbb{R}^2$ with boundary Γ, and let $\mathbf{f} \in \mathbf{L}^2(\Omega)$. The* STOKES PROBLEM *corresponds to a linearized version of the Navier-Stokes problem and consists in finding the velocity vector $\mathbf{u} := (u_1, u_2)^{\mathrm{t}}$ and the pressure p of a fluid, such that*

$$\begin{gathered} -\Delta \mathbf{u} + \nabla p = \mathbf{f} \quad in \quad \Omega\,, \quad \nabla \cdot \mathbf{u} = 0 \quad in \quad \Omega\,, \\ \mathbf{u} = 0 \quad on \quad \Gamma\,, \qquad \int_\Omega p\,dx = 0\,. \end{gathered} \tag{1.36}$$

Use the previous exercise to deduce the weak formulation of (1.36).

▸ **1.10** *Given a domain Ω of $\mathbb{R}^n$, $p \in [1, +\infty)$, and $\mathbb{K} = \mathbb{R}$ or $\mathbb{C}$, we proceed analogously to Example 1.26, part 4, and let $\mathrm{L}^p(\Omega)$ be the space of all (equivalence classes of) measurable functions $f : \Omega \to \mathbb{K}$ such that*

$$\int_{\Omega} |f|^p < \infty .$$

As before, these equivalence classes are formed by functions that coincide almost everywhere in Ω. Then, it can be proved that $\mathrm{L}^p(\Omega)$, endowed with the norm $\|f\|_{0,p;\Omega} := \left\{ \int_{\Omega} |f|^p \, dx \right\}^{1/p}$ is a Banach space. Moreover, given $f \in \mathrm{L}^p(\Omega)$ and $g \in \mathrm{L}^q(\Omega)$, with $\frac{1}{p} + \frac{1}{q} = 1$, we have Hölder's inequality

$$\int_{\Omega} |f\, g| \, dx \leqslant \|f\|_{0,p;\Omega} \, \|g\|_{0,q;\Omega} .$$

Then, given $m \in \mathbb{N}$, the SOBOLEV SPACE *of order (m,p) is defined as*

$$\mathrm{W}^{m,p}(\Omega) := \Big\{ u \in \mathrm{L}^p(\Omega) : \quad \partial^{\alpha} u \in \mathrm{L}^p(\Omega) \quad \forall\, \alpha, \, |\alpha| \leqslant m \Big\},$$

endowed with the norm $\|u\|_{m,p;\Omega} := \left\{ \sum_{|\alpha| \leqslant m} \|\partial^{\alpha} u\|_{0,p;\Omega}^p \right\}^{1/p}$. Prove that $\mathrm{W}^{m,p}(\Omega)$ is a Banach space, and conclude, taking in particular $p = 2$, that

$$\mathrm{H}^m(\Omega) \;:=\; \mathrm{W}^{m,2}(\Omega) \;=\; \Big\{ u \in \mathrm{L}^2(\Omega) : \quad \partial^{\alpha} u \in \mathrm{L}^2(\Omega) \quad \forall\, \alpha, \, |\alpha| \leqslant m \Big\}, \qquad (1.37)$$

is a Hilbert space, identifying the respective inner product. A direct proof for this result is provided later on in Theorem 10.1.

▸ **1.11** *Let Ω be a bounded open subset of $\mathbb{R}^2$, and let $\mathcal{T} := \big\{ T_1, \ldots, T_N \big\}$ be a triangulation of Ω, that is*

i) *$\bar{T}_j$ is a triangle with nonempty interior for all $j \in \{1, ..., N\}$,*
ii) *$T_i \cap T_j = \emptyset$ for all $i \neq j$, and*
iii) *$\bar{\Omega} = \bigcup \{ \bar{T}_j : \quad j \in \{1, ..., N\} \}$.*

Define the subspace of $\mathbf{L}^2(\Omega)$

$$H := \{ \boldsymbol{\tau} \in \mathbf{L}^2(\Omega) : \quad \mathrm{div}(\boldsymbol{\tau}) \in \mathrm{L}^2(T_j) \quad \forall\, j \in \{1, ..., N\} \},$$

where the local membership $\mathrm{div}(\boldsymbol{\tau}) \in \mathrm{L}^2(T_j)$ is given in the distributional sense, that is, in $\mathcal{D}'(T_j)$, which means that there exists $z_j \in \mathrm{L}^2(T_j)$ such that

$$- \int_{T_j} \nabla \varphi \cdot \boldsymbol{\tau} \, dx = \int_{T_j} z_j \, \varphi \, dx \qquad \forall\, \varphi \in C_0^{\infty}(T_j) ,$$

and in this case we write $z_j = \mathrm{div}(\boldsymbol{\tau})$ in T_j. Prove that H, endowed with the norm

$$\|\boldsymbol{\tau}\| \;:=\; \left\{ \|\boldsymbol{\tau}\|_{0,\Omega}^2 + \sum_{j=1}^{N} \|\mathrm{div}(\boldsymbol{\tau})\|_{0,T_j}^2 \right\}^{1/2} \qquad \forall\, \boldsymbol{\tau} \in H ,$$

is a real Hilbert space.

2

Duality

In this chapter we mainly follow the approaches of [23], [115] and [123] to introduce the main concepts and results on functionals and dual spaces, including the classical theorems of best approximation, orthogonal decomposition, Riesz representation, and both the analytic and geometric versions of the Hahn-Banach Theorem.

2.1 Preliminaries

Let $(V, \|\cdot\|)$ be a normed vector space over a field $\mathbb{K}$ ($\mathbb{R}$ or $\mathbb{C}$), which, unless explicitly stated otherwise, is assumed henceforth to be nontrivial. Every mapping $F : V \to \mathbb{K}$ is called a FUNCTIONAL. Note, for example, that any norm on V is a functional. Next, we define the concepts of linearity and boundedness associated with a functional.

Definition 2.1. *A functional $F : V \to \mathbb{K}$ is said to be linear if*

$$F(\alpha\, v + \beta\, w) = \alpha\, F(v) + \beta\, F(w) \qquad \forall\, \alpha\,,\, \beta \in \mathbb{K}\,, \quad \forall\, v,\, w \in V\,.$$

Definition 2.2. *A linear functional $F : V \to \mathbb{K}$ is said to be bounded with respect to a given norm $\|\cdot\|$ on V if there exists a constant $M > 0$ such that*

$$|F(v)| \leqslant M\, \|v\| \qquad \forall\, v \in V\,.$$

Example 2.3 The following are simple examples of functionals.

1. Given a normed vector space $(V, \|\cdot\|)$, the norm $\|\cdot\| : V \to \mathbb{R}$ is a nonlinear functional. In particular, observe that for $\alpha < 0$ and $x \in V$, $x \neq 0$, one has $\|\alpha\, x\| = |\alpha|\, \|x\| \neq \alpha\, \|x\|$.
2. Given a Hilbert space $(V, \langle\cdot,\cdot\rangle)$ and $x \in V$, the mapping that assigns to each $v \in V$ the value $\langle v, x\rangle \in \mathbb{K}$ is a linear functional. Moreover, by using the Cauchy-Schwarz inequality it is easy to prove that this functional is bounded. In this regard, it is important to highlight here that the Riesz Representation Theorem, a result that will be proved later in Section 2.2, is precisely the converse of this fact. In other words, that theorem states that every bounded linear functional F defined on V admits a unique *representative* $x \in V$ through the relation: $F(v) = \langle v, x\rangle$ for all $v \in V$.

The set of all bounded linear functionals defined on V is called the **dual** of V and is denoted by V'. On V' we define the following operations:

$$\begin{aligned} + : V' \times V' &\to V' \\ (F, G) &\to F + G\,, \qquad (F+G)(v) := F(v) + G(v) \quad \forall\, v \in V\,, \end{aligned} \tag{2.1}$$

G. N. Gatica, *Introduction to Functional Analysis*,
https://doi.org/10.1007/978-3-032-37415-8_2

and

$$\begin{aligned} \cdot : \mathbb{K} \times V' &\to V' \\ (\lambda, F) &\to \lambda F, \qquad (\lambda F)(v) := \lambda F(v) \quad \forall\, v \in V. \end{aligned} \tag{2.2}$$

It then follows easily that $(V', +, \cdot)$ is a vector space over $\mathbb{K}$, whose neutral element for addition is the zero functional, which assigns the zero vector to each $v \in V$.

Definition 2.4. *Given $F \in V'$ we define $\|F\|_{V'}$ as the infimum of all constants $M > 0$ that satisfy the boundedness condition of F as indicated in Definition 2.2, that is*

$$\|F\|_{V'} := \inf \mathcal{M}(F), \quad \textit{where} \quad \mathcal{M}(F) := \Big\{ M \geqslant 0 : \quad |F(v)| \leqslant M\, \|v\|_V \quad \forall\, v \in V \Big\}.$$

Moreover, we have the identity provided by the following lemma, which is considered from now on as the standard definition of $\|F\|_{V'}$.

Lemma 2.5.

$$\|F\|_{V'} = \sup_{0 \neq v \in V} \frac{|F(v)|}{\|v\|_V} \qquad \forall\, F \in V'. \tag{2.3}$$

Proof. Given $F \in V'$, its boundedness guarantees that the supremum in (2.3) does exist, so that we denote it as $S(F)$. It follows that $|F(v)| \leqslant S(F)\,\|v\|$ for all $v \in V$, $v \neq 0$, which yields $S(F) \in \mathcal{M}(F)$, and thus $\|F\|_{V'} \leqslant S(F)$. In turn, given $\varepsilon > 0$, there exists $M_\varepsilon \in \mathcal{M}(F)$ such that $\|F\|_{V'} \leqslant M_\varepsilon < \|F\|_{V'} + \varepsilon$, and hence $|F(v)| \leqslant M_\varepsilon\,\|v\| < \big(\|F\|_{V'} + \varepsilon\big)\,\|v\|$ for all $v \in V$, $v \neq 0$, which implies $S(F) < \|F\|_{V'} + \varepsilon$. In this way, since ε is arbitrary, we conclude that $S(F) \leqslant \|F\|_{V'}$, which ends the proof.

□

Likewise, $\|F\|_{V'}$ can also be defined as

$$\|F\|_{V'} := \sup_{\substack{v \in V \\ \|v\| \leqslant 1}} |F(v)|,$$

or equivalently

$$\|F\|_{V'} := \sup_{\substack{v \in V \\ \|v\| = 1}} |F(v)|.$$

In any case, it is easy to prove that indeed $\|\cdot\|_{V'}$ is a norm on V', so that $(V', +, \cdot, \|\cdot\|_{V'})$ is a normed vector space. Moreover, one can show, using the completeness of the field $\mathbb{K}$, that V' is a Banach space. A more general result, of which the completeness of dual spaces is a particular case, will be proved later in Chapter 3 (see Theorem 3.9). In what follows, and whenever there is no risk of confusion, we will simply write $\|v\|$ and $\|F\|$ instead of $\|v\|_V$ and $\|F\|_{V'}$, for each $v \in V$ and $F \in V'$. At the end of this section we may ask whether it is possible to obtain a dual identity to the one established in (2.3), that is, whether

$$\|v\| = \sup_{0 \neq F \in V'} \frac{|F(v)|}{\|F\|} \qquad \forall\, v \in V. \tag{2.4}$$

The answer to this simple question is positive and will be given later as one of the many consequences of the Hahn-Banach Theorem, one of the most important classical results of Functional Analysis.

2.2 Riesz Representation Theorem

The purpose of this section is to prove the Riesz Representation Theorem, which was already announced in Example 2.3. For this purpose, we need some preliminary results.

Theorem 2.6 (First Best Approximation Theorem). *Let $U \neq \emptyset$ be a closed subspace of a Hilbert space $(V, \langle\cdot,\cdot\rangle)$. Then, for every $x \in V$ there exists a unique $z \in U$ such that*

$$\|x - z\| = \inf_{u \in U} \|x - u\|, \tag{2.5}$$

which is called the best approximation of x by elements of U. Moreover, $z \in U$ is characterized by the orthogonality relation

$$\langle x - z, u\rangle = 0 \qquad \forall\, u \in U. \tag{2.6}$$

Proof. Let $d := \inf_{u \in U} \|x - u\|$. Then, for every $\varepsilon > 0$ there exists $u \in U$ such that

$$d \leqslant \|x - u\| < d + \varepsilon .$$

In particular, taking $\varepsilon = \frac{1}{n}$ with $n \in \mathbb{N}$, it follows that there exists a sequence $\{z_n\}_{n \in \mathbb{N}} \subseteq U$ such that $d \leqslant \|x - z_n\| < d + \frac{1}{n}$ and therefore $\|x - z_n\| \to d$ as $n \to +\infty$ (see Figure 2.1 for a geometric illustration of this convergence). Now, applying the parallelogram law (cf. Lemma 1.28) to $(x - z_n)$ and $(x - z_m)$, with $m, n \in \mathbb{N}$, we obtain

$$\|(x - z_n) + (x - z_m)\|^2 + \|(x - z_n) - (x - z_m)\|^2 = 2\left\{\|x - z_n\|^2 + \|x - z_m\|^2\right\},$$

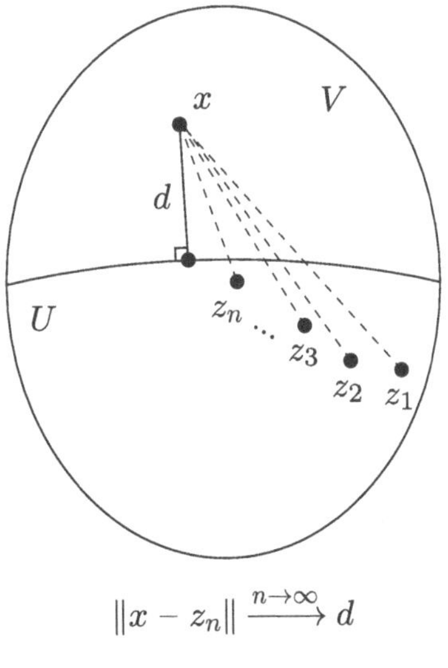

Fig. 2.1. *Graphical illustration of* $\|x - z_n\| \xrightarrow{n\to\infty} d$

or equivalently

$$4\left\|x - \frac{(z_n + z_m)}{2}\right\|^2 + \|z_n - z_m\|^2 = 2\|x - z_n\|^2 + 2\|x - z_m\|^2,$$

from which

$$\|z_n - z_m\|^2 = 2\|x - z_n\|^2 + 2\|x - z_m\|^2 - 4\left\|x - \frac{(z_n + z_m)}{2}\right\|^2 .$$

Since $\frac{z_n+z_m}{2} \in U$, it follows that $\|x - \frac{(z_n+z_m)}{2}\| \geqslant d$, and then

$$\|z_n - z_m\|^2 \leqslant 2\|x - z_n\|^2 + 2\|x - z_m\|^2 - 4d^2 .$$

Thus, taking the limit as $m, n \to +\infty$, we obtain $\|z_n - z_m\| \to 0$, which means that $\{z_n\}_{n\in\mathbb{N}}$ is Cauchy in V. Since U is closed, it follows that there exists $z \in U$ such that $z_n \to z$ and therefore

$$d = \lim_{n\to+\infty} \|x - z_n\| = \|x - z\| .$$

For uniqueness, let $z_1, z_2 \in U$ be such that $\|x - z_1\| = \|x - z_2\| = d$. Applying again the same previous computations, we obtain

$$0 \leqslant \|z_1 - z_2\|^2 = 2\|x - z_1\|^2 + 2\|x - z_2\|^2 - 4\left\|x - \frac{(z_1 + z_2)}{2}\right\|^2$$

$$= 4d^2 - 4\left\|x - \frac{(z_1 + z_2)}{2}\right\|^2 \leqslant 0 ,$$

and thus $\|z_1 - z_2\| = 0$. Now we prove the equivalence between (2.5) and (2.6). Let $z \in U$ such that

$$d := \|x - z\| = \inf_{u\in U} \|x - u\| .$$

It follows that $d^2 \leqslant \|x - u\|^2$ for every $u \in U$, and in particular, since U is a subspace,

$$d^2 \leqslant \|x - (z - \alpha u)\|^2 = \|(x - z) + \alpha u\|^2 \qquad \forall \alpha \in \mathbb{R}, \quad \forall u \in U . \tag{2.7}$$

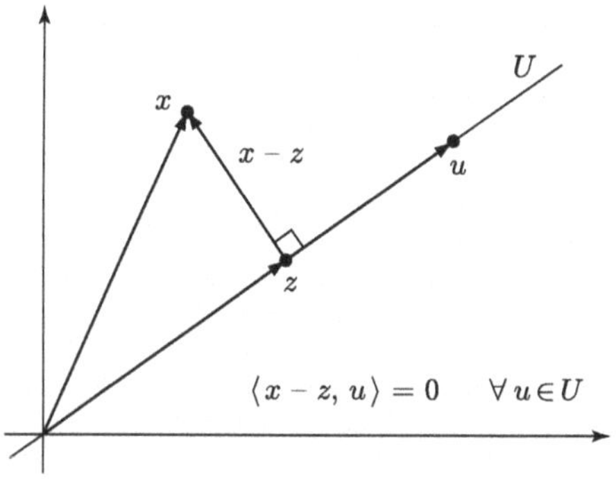

Fig. 2.2. *The orthogonality condition in* $\mathbb{R}^2$

If V is real, the expansion of (2.7) yields

$$d^2 \leqslant \|x - z\|^2 + 2\alpha\langle x - z, u\rangle + \alpha^2\|u\|^2 = d^2 + 2\alpha\langle x - z, u\rangle + \alpha^2\|u\|^2 ,$$

from which it follows that

$$0 \leqslant 2\alpha\langle x-z,u\rangle + \alpha^2\|u\|^2 \qquad \forall\,\alpha\in\mathbb{R}, \quad \forall\,u\in U. \tag{2.8}$$

Then, dividing by $\alpha>0$ (resp. $\alpha<0$) and taking the limit as $\alpha\to 0^+$ (resp. $\alpha\to 0^-$), it follows that $\langle x-z,u\rangle = 0$ for all $u\in U$. In turn, if V is complex, we get instead of (2.8)

$$0 \leqslant 2\alpha\,\mathrm{Re}\{\langle x-z,u\rangle\} + \alpha^2\|u\|^2 \qquad \forall\,\alpha\in\mathbb{R}, \quad \forall\,u\in U,$$

so that, proceeding exactly as before, we deduce that

$$\mathrm{Re}\{\langle x-z,u\rangle\} = 0 \qquad \forall\,u\in U. \tag{2.9}$$

Next, taking in particular $i\,u\in U$ in (2.9), and noting that there actually holds $\mathrm{Re}\{\langle x-z,i\,u\rangle\} = \mathrm{Im}\{\langle x-z,u\rangle\}$, we deduce that the latter expression vanishes as well, whence $\langle x-z,u\rangle = 0$ for all $u\in U$. Needless to say, the above shows that in the complex case this last identity is equivalent to (2.9). Conversely, let $z\in U$ be such that $\langle x-z,u\rangle = 0$ for all $u\in U$. It then follows easily, for both the real and complex case, that

$$\|x-u\|^2 = \|(x-z)-(u-z)\|^2 = \|x-z\|^2 + \|u-z\|^2,$$

from which $\|x-z\| \leqslant \|x-u\|$ for all $u\in U$ and $\|x-z\| = \|x-u\|$ if and only if $u=z$. We finish the proof by noting that Figure 2.2 illustrates precisely the orthogonality condition in $\mathbb{R}^2$.

□

Furthermore, it is important to note here that the same proof used to show the existence and uniqueness of the best approximation $z\in U$ remains valid in the case when U is merely a closed convex subset of the Hilbert space V. In fact, the only argument that relies on the fact that U is a subspace of V is the one that allows us to conclude that the average of two elements of U still belongs to U, which naturally also holds if U is convex. However, convexity of U is not sufficient to prove the equivalence between (2.5) and (2.6). In fact, the corresponding characterization theorem, which will be detailed later in the chapter on variational problems in Hilbert spaces (cf. Chapter 4), involves an inequality instead of the orthogonality relation (2.6).

In accordance with the above, we can now state the following result.

Theorem 2.7 (Second Best Approximation Theorem). *Let $U \neq \varnothing$ be a closed convex subset of a Hilbert space $(V,\langle\cdot,\cdot\rangle)$. Then, for every $x\in V$ there exists a unique $z\in U$ such that*

$$\|x-z\| = \inf_{u\in U}\|x-u\|,$$

which is called the best approximation of x by elements of U.

Example 2.8 This example illustrates the finite-dimensional case of Theorem 2.6.

1. If U is a finite-dimensional subspace of V, then the computation of the best approximation reduces to solving a system of linear equations. Indeed, if $\{u_1,\ldots,u_N\}$ is a basis of U, then (2.6) is equivalent to

$$\langle x-z,u_j\rangle = 0 \qquad \forall\,j\in\{1,2,\ldots,N\}.$$

Thus, if $z = \sum_{i=1}^{N}\alpha_i\,u_i$, then we can write

$$\langle x, u_j \rangle = \langle z, u_j \rangle = \sum_{i=1}^{N} \alpha_i \langle u_i, u_j \rangle,$$

or, equivalently, in matrix form we obtain the Gramm system

$$\boldsymbol{A}\boldsymbol{\alpha} = \boldsymbol{b}, \tag{2.10}$$

where $\boldsymbol{A} := (a_{ij})_{N\times N}$, with $a_{ij} := \langle u_i, u_j\rangle$, $\boldsymbol{\alpha} := (\alpha_1, \alpha_2, ..., \alpha_N)^{\mathrm{t}}$ and $\boldsymbol{b} := (b_1, b_2, ..., b_N)^{\mathrm{t}}$, with $b_j := \langle x, u_j\rangle$.

2. If the basis $\{u_1, \ldots, u_N\}$ of U is orthogonal, i.e. if $\langle u_i, u_j\rangle = 0$ for $i \neq j$, then solving (2.10) is straightforward, yielding in this case

$$\alpha_i = \frac{b_i}{\|u_i\|^2} = \frac{\langle x, u_i\rangle}{\|u_i\|^2} \qquad \forall\, i \in \{1, \ldots, N\}, \tag{2.11}$$

so that the best approximation is given by

$$z = \sum_{i=1}^{N} \frac{\langle x, u_i\rangle}{\|u_i\|^2}\, u_i\,.$$

3. A particular example of 2. is given by the finite Fourier series. Indeed, in this case we consider $V := \mathrm{L}^2(\Omega)$ with $\Omega := (-\pi, \pi)$, which is endowed with its usual inner product (cf. (1.12)) $\langle f, g\rangle_{0,\Omega} := \int_{-\pi}^{\pi} f(x)\, g(x)\, dx$, for all $f, g \in \mathrm{L}^2(\Omega)$, and we choose U as the subspace generated by the trigonometric functions $\{\varphi_0, \varphi_1, \varphi_2, ..., \varphi_N, \psi_1, \psi_2, ..., \psi_N\}$, where $\varphi_j(x) := \cos(j\, x)$ for all $j \in \{0, 1, 2, ..., N\}$, and $\psi_j(x) := \sin(j\, x)$ for all $j \in \{1, 2, ..., N\}$, for all $x \in \Omega$. Thus, given $f \in \mathrm{L}^2(\Omega)$, its best approximation by elements of U is given by

$$f_N(x) = \sum_{i=0}^{N} \alpha_i \cos(i\, x) + \sum_{i=1}^{N} \beta_i \sin(i\, x) \qquad \forall\, x \in (-\pi, \pi),$$

where the scalars α_i and β_i, called Fourier coefficients, are defined according to (2.11), namely

$$\alpha_i := \frac{\int_{-\pi}^{\pi} f(x) \cos(i\, x)\, dx}{\int_{-\pi}^{\pi} \cos^2(i\, x)\, dx} \quad \text{and} \quad \beta_i := \frac{\int_{-\pi}^{\pi} f(x) \sin(i\, x)\, dx}{\int_{-\pi}^{\pi} \sin^2(i\, x)\, dx}.$$

Definition 2.9. *Given a subset S of a Hilbert space $(V, \langle\cdot,\cdot\rangle)$, the orthogonal of S, denoted by $S^\perp$, is defined as*

$$S^\perp := \big\{x \in V : \quad \langle x, v\rangle = 0 \quad \forall\, v \in S\big\},$$

which is a closed subspace of V. Note that $S \cap S^\perp = \{0\}$ for every $S \subseteq V$ such that $0 \in S$.

One of the most important consequences of Theorem 2.6 is given by the following result, which establishes that every Hilbert space can be written as the direct sum of any closed subspace with its orthogonal.

Theorem 2.10 (Orthogonal Decomposition Theorem). *Let $U \neq \emptyset$ be a closed subspace of a Hilbert space $(V, \langle\cdot,\cdot\rangle)$. Then*

$$V = U \oplus U^\perp. \tag{2.12}$$

Proof. Given $x \in V$, we can write $x = z + (x - z)$, where $z \in U$ is the best approximation of x by elements of U and, thanks to the orthogonality relation (2.6), $(x - z) \in U^{\perp}$. Then, since the zero vector is the only vector orthogonal to itself, we deduce that $U \cap U^{\perp} = \{0\}$, which proves that the above decomposition is unique. □

The following lemma constitutes another interesting corollary of Theorem 2.6. A second proof of it, employing identities between orthogonal and annihilator sets, will be given later on in Chapter 3 (cf. Lemma 3.25).

Lemma 2.11. *Let $U \neq \emptyset$ be a closed subspace of a Hilbert space $(V, \langle\cdot,\cdot\rangle)$. Then there holds*

$$\left(U^{\perp}\right)^{\perp} = U .$$

Proof. Given $x \in U$, it is clear that $\langle x, v\rangle = 0$ for all $v \in U^{\perp}$, which implies that $x \in \left(U^{\perp}\right)^{\perp}$. Let us now assume that $x \notin U$ and prove that $x \notin \left(U^{\perp}\right)^{\perp}$. In fact, since U is closed, we know from Theorem 2.6 that there exists a unique $z \in U$ such that z is the best approximation of x by elements of U, which is characterized as $x - z \in U^{\perp}$. Then, bearing in mind that $\operatorname{dist}(x, U) = \|x - z\| > 0$ and that $\langle x - z, z\rangle = 0$, we deduce that

$$0 < \|x - z\|^2 = \langle x - z, x - z\rangle = \langle x - z, x\rangle ,$$

which shows that $\langle x - z, x\rangle \neq 0$, and hence $x \notin \left(U^{\perp}\right)^{\perp}$, thus ending the proof. □

Since $U^{\perp}$ is also a closed subspace of the Hilbert space V, we observe that $z \in U$ is the best approximation of x (by elements of U) if and only if $(x - z) \in U^{\perp}$ is the best approximation of x (by elements of $U^{\perp}$). Equivalently, if P_S is the mapping (later referred to as an *operator*) that assigns to each $x \in V$ its best approximation by elements of a closed subspace S of V, then (2.12) can be rewritten as

$$I = P_U + P_{U^{\perp}} ,$$

where I is the identity mapping from V to V. In this way, and due to the orthogonality relation (2.6), the mappings $P_U : V \to U$ and $P_{U^{\perp}} : V \to U^{\perp}$ are called ORTHOGONAL PROJECTORS onto U and $U^{\perp}$, respectively (see Figure 2.3 for the case $V = \mathbb{R}^2$). According to this, Theorem 2.6 is also known as the Orthogonal Projection Theorem.

Example 2.12 The following examples illustrate the Orthogonal Decomposition Theorem (cf. Theorem 2.10).

1. Let Ω be a bounded open subset of $\mathbb{R}^n$, and consider the Lebesgue space $V := \mathrm{L}^2(\Omega)$ over the field $\mathbb{R}$ with the usual inner product (cf. (1.12)) $\langle f, g\rangle_{0,\Omega} := \int_\Omega f\, g$ for all $f,\, g \in V$. In turn, define the subspace

$$U := \{\varphi : \Omega \to \mathbb{R} : \quad \varphi \text{ is constant}\} = \operatorname{span}\{\varphi_0\} ,$$

where $\varphi_0 \in V$ is such that $\varphi_0(x) = 1$ for all $x \in \Omega$. It follows that

$$U^{\perp} = \{f \in V : \quad \langle f, \varphi_0\rangle_{0,\Omega} = 0\} = \left\{ f \in V : \quad \int_\Omega f = 0\right\} .$$

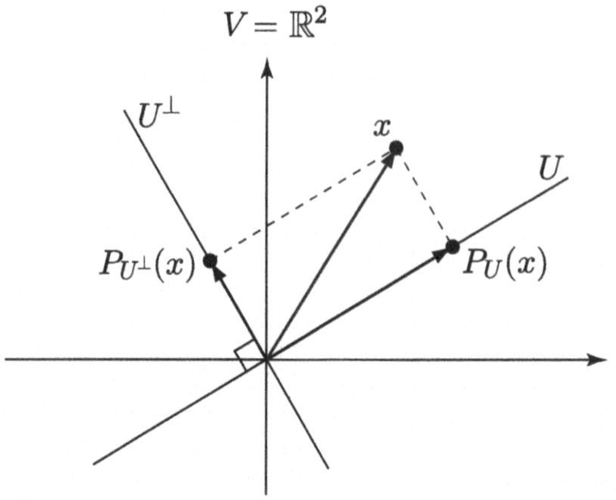

Fig. 2.3. *Orthogonal projections onto U and $U^\perp$ in $\mathbb{R}^2$*

Then, given $f \in V$, its best approximation by elements of U reduces to the constant function

$$f_U(x) := \frac{\langle f, \varphi_0 \rangle_{0,\Omega}}{\langle \varphi_0, \varphi_0 \rangle_{0,\Omega}} \varphi_0(x) = \frac{1}{|\Omega|} \int_\Omega f \qquad \forall\, x \in \Omega .$$

In this way, each $f \in V$ can be uniquely decomposed as

$$f = f_U + f_{U^\perp}, \quad \text{with} \quad f_U \in U \quad \text{and} \quad f_{U^\perp} := \left(f - \frac{1}{|\Omega|} \int_\Omega f \right) \in U^\perp .$$

The usefulness of this decomposition can be illustrated, for example, by the unknown p arising in the Navier–Stokes and Stokes problems considered in Exercises 1.8 and 1.9, respectively. Indeed, as can be seen from (1.35) and (1.36), imposing the condition $\int_\Omega p\, dx = 0$, which ensures the uniqueness of the pressure, is equivalent to saying that only the $U^\perp$-component of p can be determined.

2. Let $V := \mathbb{R}^{n\times n}$ be the space of real square matrices of order n, equipped with the inner product

$$\langle A, B \rangle := \mathrm{tr}(B^{\mathrm{t}} A) \qquad \forall\, A, B \in V.$$

We then define the subspaces

$$U_1 := \{ \kappa \mathbb{I} : \quad \kappa \in \mathbb{R} \} = \mathrm{span}\{\mathbb{I}\} \quad \text{and} \quad U_2 := \{ A \in V : \quad A = A^{\mathrm{t}} \},$$

where $\mathbb{I}$ denotes the identity matrix of V. It follows that

$$U_1^\perp = \{ A \in V : \quad \mathrm{tr}(A) = 0 \} \quad \text{and} \quad U_2^\perp = \{ A \in V : \quad A^{\mathrm{t}} = -A \}.$$

Then, given $A \in V$, it can be shown that its best approximations by elements of U_1 and U_2 reduce, respectively, to

$$A_1 := \frac{1}{n} \mathrm{tr}(A) \mathbb{I} \quad \text{and} \quad A_2 := \frac{1}{2} \left(A + A^{\mathrm{t}} \right).$$

Hence, each $A \in V$ can be uniquely decomposed as

$$A = A_1 + A_1^\perp, \quad \text{with} \quad A_1 \in U_1 \quad \text{and} \quad A_1^\perp := \left(A - \frac{1}{n} \mathrm{tr}(A) \mathbb{I} \right) \in U_1^\perp ,$$

and also as

$$A = A_2 + A_2^{\perp}\,, \quad \text{with} \quad A_2 \in U_2 \quad \text{and} \quad A_2^{\perp} := \frac{1}{2}\big(A - A^{\mathrm{t}}\big) \in U_2^{\perp}\,.$$

A particular case of this second decomposition, of great interest in continuum mechanics, concerns the gradient of the displacement vector $\mathbf{u}$ of a solid subject to certain volume and surface forces. In this case, it is usually written as

$$\nabla\mathbf{u} := \mathbf{e}(\mathbf{u}) + \gamma(\mathbf{u})\,,$$

where

$$\mathbf{e}(\mathbf{u}) := \frac{1}{2}\,\big(\nabla\mathbf{u} + (\nabla\mathbf{u})^{\mathrm{t}}\big) \quad \text{and} \quad \gamma(\mathbf{u}) := \frac{1}{2}\,\big(\nabla\mathbf{u} - (\nabla\mathbf{u})^{\mathrm{t}}\big)$$

are the strain tensor of small deformations and the rotation tensor, respectively.

3. Let Ω be a bounded open subset of $\mathbb{R}^n$, and consider the real Hilbert space given by $V = \mathbb{L}^2(\Omega) := [\mathrm{L}^2(\Omega)]^{n\times n}$ with the inner product $\langle \boldsymbol{\sigma}, \boldsymbol{\tau}\rangle_{0,\Omega} := \int_\Omega \boldsymbol{\sigma} : \boldsymbol{\tau}$, where $\boldsymbol{\sigma} : \boldsymbol{\tau} := \sum_{i,j=1}^{n} \sigma_{ij}\,\tau_{ij}$ for all $\boldsymbol{\sigma} := (\sigma_{ij})_{n\times n}$, $\boldsymbol{\tau} := (\tau_{ij})_{n\times n} \in V$. Note that we use the same notation $\langle\cdot,\cdot\rangle_{0,\Omega}$ for the inner products of both $\mathrm{L}^2(\Omega)$ (cf. (1.12)) and its tensor version $\mathbb{L}^2(\Omega)$, so that we adopt the same for the corresponding vector version $\mathbf{L}^2(\Omega) := [\mathrm{L}^2(\Omega)]^n$. Equivalently, it is easy to see that $\boldsymbol{\sigma} : \boldsymbol{\tau} = \langle \boldsymbol{\sigma}, \boldsymbol{\tau}\rangle$, where $\langle\cdot,\cdot\rangle$ is the inner product used in 2. In turn, we define the subspace

$$U := \{\kappa\,\mathbb{I} : \quad \kappa \in \mathbb{R}\} = \mathrm{span}\{\mathbb{I}\}\,,$$

where $\mathbb{I}$ is the identity tensor of V. It follows that

$$U^{\perp} = \left\{\boldsymbol{\sigma} \in V : \quad \int_\Omega \mathrm{tr}(\boldsymbol{\sigma}) = 0\right\}.$$

Then, given $\boldsymbol{\sigma} \in V$, its best approximations by elements of U and $U^{\perp}$ reduce, respectively, to

$$\boldsymbol{\sigma}_U := \frac{1}{n\,|\Omega|}\int_\Omega \mathrm{tr}(\boldsymbol{\sigma})\,\mathbb{I} \quad \text{and} \quad \boldsymbol{\sigma}_{U^{\perp}} := \boldsymbol{\sigma} - \frac{1}{n\,|\Omega|}\int_\Omega \mathrm{tr}(\boldsymbol{\sigma})\,\mathbb{I}\,,$$

so that each $\boldsymbol{\sigma} \in V$ can be uniquely decomposed as the sum of a tensor $\boldsymbol{\sigma}_U$, a multiple of the identity, and a tensor $\boldsymbol{\sigma}_{U^{\perp}}$, whose matrix trace has zero mean in Ω. This result, analogous to the first decomposition in 2, is also of great usefulness in continuum mechanics, especially for the analysis of mixed variational formulations of some problems in elasticity.

4. Let Ω be a bounded open subset of $\mathbb{R}^n$, and consider the real Sobolev space $V := \mathrm{H}^1(\Omega)$ equipped with its usual inner product (cf. (1.27)) $\langle w, v\rangle_{1,\Omega} := \int_\Omega \big\{\nabla w \cdot \nabla v + w\,v\big\}$ for all $w, v \in V$. Moreover, let $U := \mathrm{H}_0^1(\Omega)$ be the closure of $C_0^\infty(\Omega)$ with respect to the norm $\|\cdot\|_{1,\Omega}$ induced by $\langle\cdot,\cdot\rangle_{1,\Omega}$ (cf. (1.28)). Then, it is easy to see that

$$U^{\perp} = \big\{v \in V : \quad -\Delta v + v = 0 \quad \text{in} \quad \mathcal{D}'(\Omega)\big\}.$$

Consequently, given $w \in V$, its best approximation by elements of U is the unique element w_U of this subspace such that

$$-\,\Delta\, w_U + w_U = f \quad \text{in} \quad \mathcal{D}'(\Omega)\,,$$

where $f := -\,\Delta\, w + w$.

The following lemma is a direct corollary of Theorem 2.10.

Lemma 2.13. *Let $U \neq \emptyset$ be a proper closed subspace of a Hilbert space V. Then, there exists $\widetilde{x} \in V$, $\widetilde{x} \neq 0$, such that $\widetilde{x} \in U^{\perp}$.*

We are now in a position to prove the main theorem of this section.

Theorem 2.14 (Riesz Representation Theorem). *Let $(V, \langle\cdot,\cdot\rangle)$ be a Hilbert space, and let $F \in V'$. Then, there exists a unique $x \in V$ such that*

$$F(v) = \langle v, x\rangle \qquad \forall\, v \in V\,, \tag{2.13}$$

and

$$\|F\| = \|x\|\,. \tag{2.14}$$

Proof. Given $F \in V'$, define $U := \{v \in V : F(v) = 0\}$. If $U = V$ then F is the zero functional, and in such case it suffices to take $x = 0$. Suppose now that $U \neq V$. Since U is a closed subspace of V, it follows, according to Lemma 2.13, that there exists $\widetilde{x} \in V$, $\widetilde{x} \neq 0$ such that $\widetilde{x} \in U^\perp$. It is then clear that $\widetilde{x} \notin U$, and therefore $F(\widetilde{x}) \neq 0$. Moreover, for each $v \in V$ we observe that $F(\widetilde{x})\,v - F(v)\,\widetilde{x} \in U$, and since $\widetilde{x} \in U^\perp$ it follows that

$$0 = \langle F(\widetilde{x})\,v - F(v)\,\widetilde{x}, \widetilde{x}\rangle \qquad \forall\, v \in V\,,$$

or equivalently

$$\langle F(\widetilde{x})\,v, \widetilde{x}\rangle = \langle F(v)\,\widetilde{x}, \widetilde{x}\rangle = F(v)\,\|\widetilde{x}\|^2\,,$$

from which

$$F(v) = \langle v, x\rangle \qquad \forall\, v \in V\,,$$

with $\quad x := \dfrac{\overline{F(\widetilde{x})}\,\widetilde{x}}{\|\widetilde{x}\|^2}$. For the uniqueness of x, suppose that there exists $z \in V$ such that

$$F(v) = \langle v, x\rangle = \langle v, z\rangle \qquad \forall\, v \in V\,.$$

It follows that $x - z \in V^\perp = \{0\}$, and therefore $x = z$. On the other hand, using the Cauchy-Schwarz inequality we obtain

$$\|F\| := \sup_{0 \neq v \in V} \frac{|F(v)|}{\|v\|} = \sup_{0 \neq v \in V} \frac{|\langle v, x\rangle|}{\|v\|} \leqslant \|x\|\,,$$

and moreover, by taking $v = x$, it follows that

$$\|F\| := \sup_{0 \neq v \in V} \frac{|F(v)|}{\|v\|} \geqslant \frac{|F(x)|}{\|x\|} = \|x\|\,,$$

from which we conclude that $\|F\| = \|x\|$.

□

We stress here that the Riesz Representation Theorem (cf. Theorem 2.14),, induces the definition of the RIESZ MAPPING, also called the RIESZ OPERATOR,

$$\begin{aligned} \mathcal{R} : V' &\to V \\ F &\mapsto \mathcal{R}(F) \end{aligned} \tag{2.15}$$

where $\mathcal{R}(F)$ is the unique element of V, called the REPRESENTATIVE OF F, such that

$$F(v) = \langle v, \mathcal{R}(F)\rangle \qquad \forall\, v \in V\,. \tag{2.16}$$

Note that $\mathcal{R}$ is an isometric bijection. Moreover, given scalars α, $\beta \in \mathbb{K}$ and functionals F, $G \in V'$, it follows from (2.16) that

$$(\alpha F + \beta G)(v) = \alpha F(v) + \beta G(v) = \alpha \langle v, \mathcal{R}(F) \rangle + \beta \langle v, \mathcal{R}(G) \rangle$$
$$= \langle v, \bar{\alpha} \mathcal{R}(F) + \bar{\beta} \mathcal{R}(G) \rangle \qquad \forall v \in V,$$

thereby showing that

$$\mathcal{R}(\alpha F + \beta G) = \bar{\alpha} \mathcal{R}(F) + \bar{\beta} \mathcal{R}(G). \tag{2.17}$$

This identity establishes the linearity of $\mathcal{R}$ in the real case and its conjugate-linearity in the complex case.

2.3 Hahn-Banach Theorem

In this section we prove one of the most classical theorems of Functional Analysis. For this purpose, in what follows we consider $(V, \|\cdot\|)$ a normed vector space over $\mathbb{R}$, and we ask the essential question that motivates the development of the Hahn-Banach Theorem:

Does there exist any nonzero element of the dual space V'?

To partially answer this question, we take an arbitrary element $x_0 \in V$, $x_0 \neq 0$, and define the subspace of V

$$V_0 := \{\alpha x_0 : \quad \alpha \in \mathbb{R}\} = \text{span}\{x_0\},$$

which is clearly of dimension 1. Then, we define the linear functional

$$\begin{aligned} f : \ V_0 &\to \mathbb{R} \\ \alpha x_0 &\to f(\alpha x_0) := \alpha . \end{aligned}$$

It is easy to see that f is linear. Moreover, given $x := \alpha x_0 \in V_0$, we have

$$|f(x)| = |\alpha| = \frac{|\alpha| \, \|x_0\|}{\|x_0\|} = \frac{1}{\|x_0\|} \|x\|,$$

which proves that f is bounded, and $\|f\|_{V_0'} = \dfrac{1}{\|x_0\|}$. Therefore, the answer to the essential question will be affirmative if f, which is a nonzero functional of V_0', can be extended to a functional $F \in V'$. Precisely, the existence of this extension is what the Hahn-Banach Theorem guarantees.

2.3.1 Preliminaries

In this section we provide some definitions and preliminary results needed for the proof of the aforementioned theorem.

Definition 2.15. *A set $M \neq \emptyset$ is said to be partially ordered if there exists an order relation $(\leqslant)$ defined on a subset of $M \times M$, such that*

- *$x \leqslant x$ for all $x \in M$* (REFLEXIVITY).
- *$x \leqslant y$ and $y \leqslant x \Rightarrow x = y$* (ANTISYMMETRY).
- *$x \leqslant y$ and $y \leqslant z \Rightarrow x \leqslant z$* (TRANSITIVITY).

The term *partially* emphasizes the fact that M may contain elements x and y for which neither $x \leqslant y$ nor $y \leqslant x$ hold. In such a case, x and y are said to be *incomparable*. Otherwise, that is, if $x \leqslant y$ or if $y \leqslant x$, then x and y are said to be *comparable*.

Example 2.16 *Given a set A, consider $\mathcal{P}(A) := \{X : \quad X \subseteq A\}$ and define the order relation $X \leqslant Y$ if and only if $X \subseteq Y$, with $X, Y \in \mathcal{P}(A)$. In general, $\mathcal{P}(A)$ is partially ordered.*

Definition 2.17. *Let $(M, \leqslant)$ be a nonempty partially ordered set.*

- *A subset S of M is said to be* totally ordered *(or a* chain*) if it has no incomparable elements.*
- *An element $z \in M$ is said to be an* upper bound *of a nonempty subset S of M if $x \leqslant z$ for all $x \in S$.*
- *An element $z \in M$ is said to be* maximal *if whenever $x \in M$ satisfies $z \leqslant x$, then necessarily $x = z$.*

Notice that M may or may not possess maximal elements. Moreover, a maximal element is not necessarily an upper bound. The converse, however, is true.

Lemma 2.18 (Zorn's Lemma). *Let $(M, \leqslant)$ be a nonempty partially ordered set, and suppose that every chain in M has an upper bound. Then M possesses at least one maximal element.*

Zorn's Lemma, which is employed in the next section to prove Theorem 2.20, is actually an axiom, due to its equivalence with the Axiom of Choice, stated below.

Axiom of Choice. *Given any family $\{A_i\}_{i\in I}$ of nonempty subsets A_i of a set X, there exists a function $f : I \to X$, called a choice function, such that $f(i) \in A_i$ for all $i \in I$.*

2.3.2 Analytic form of the Hahn-Banach Theorem

This version of the theorem requires the concept of a sublinear functional.

Definition 2.19. *Let V be a vector space over $\mathbb{R}$. A functional $p : V \to \mathbb{R}$ is said to be sublinear if*

- *$p(x+y) \leqslant p(x) + p(y)$ for all $x, y \in V$.*
- *$p(\alpha x) = \alpha p(x)$ for all $x \in V$, for all $\alpha > 0$.*

Theorem 2.20 (Hahn-Banach Theorem). *Let V be a vector space over $\mathbb{R}$, and let $p : V \to \mathbb{R}$ be a sublinear functional. Let U be a subspace of V, and let $f : U \to \mathbb{R}$ be a linear functional such that*

$$f(x) \leqslant p(x) \qquad \forall x \in U.$$

Then, there exists a linear functional $F : V \to \mathbb{R}$ such that

$$F(x) = f(x) \qquad \forall x \in U \qquad \text{and} \qquad F(x) \leqslant p(x) \qquad \forall x \in V.$$

Proof. The proof will make use of Zorn's Lemma. For this purpose, we first define the family $\mathcal{F}$ of all linear extensions of f bounded by p, that is

$$\mathcal{F} := \Big\{ g : \mathcal{D}(g) \to \mathbb{R} \text{ linear} : \quad \mathcal{D}(g) \text{ subspace of } V \text{ such that } U \subseteq \mathcal{D}(g), \\ g(x) = f(x) \quad \forall x \in U, \quad g(x) \leqslant p(x) \qquad \forall x \in \mathcal{D}(g) \Big\}.$$

Note that $\mathcal{F} \neq \varnothing$ since clearly $f \in \mathcal{F}$. Then, on $\mathcal{F}$ we define an order relation as follows: given g, $h \in \mathcal{F}$ we say that $g \leqslant h$ if and only if h is a linear extension of g, that is,

$\mathcal{D}(g) \subseteq \mathcal{D}(h)$ and $h(x) = g(x)$ for all $x \in \mathcal{D}(g)$. Now, given a chain $\mathcal{C} \subseteq \mathcal{F}$, define the set

$$\widehat{\mathcal{D}} := \cup \{\mathcal{D}(g) : \quad g \in \mathcal{C}\},$$

which turns out to be a vector subspace of V. Indeed, first note that $\widehat{\mathcal{D}} \neq \emptyset$ since $0 \in \widehat{\mathcal{D}}$. Moreover, given $\lambda \in \mathbb{R}$ and $x \in \widehat{\mathcal{D}}$, one has $x \in \mathcal{D}(g_0)$ for some $g_0 \in \mathcal{C}$, and therefore $\lambda x \in \mathcal{D}(g_0)$, so that $\lambda x \in \widehat{\mathcal{D}}$. Now, given $x, y \in \widehat{\mathcal{D}}$, there exist $g_1, g_2 \in \mathcal{C}$ such that $x \in \mathcal{D}(g_1)$ and $y \in \mathcal{D}(g_2)$. Since $\mathcal{C}$ is a chain, we have that $g_1 \leqslant g_2$ or $g_2 \leqslant g_1$, and in either case it follows that $x + y \in \widehat{\mathcal{D}}$. We then define

$$\begin{aligned} \widehat{g} : \widehat{\mathcal{D}} &\to \mathbb{R} \\ x &\to \widehat{g}(x) := g(x), \quad \text{if}\, x \in \mathcal{D}(g)\,, \text{with}\, g \in \mathcal{C}\,, \end{aligned}$$

and we show that $\widehat{g}$ is well-defined. Indeed, if $x \in \mathcal{D}(g_1) \cap \mathcal{D}(g_2)$, with $g_1, g_2 \in \mathcal{C}$, we have that $g_1 \leqslant g_2$ or $g_2 \leqslant g_1$, and in both cases it follows that $g_1(x) = g_2(x)$. Since $g(x) \leqslant p(x)$ for all $x \in \mathcal{D}(g)$ and for all $g \in \mathcal{C}$, it follows that $\widehat{g}(x) \leqslant p(x)$ for all $x \in \mathcal{D}(\widehat{g})$. Moreover, it is clear that $\widehat{g}$ is a linear extension of every $g \in \mathcal{C}$, which shows that $g \leqslant \widehat{g}$ for all $g \in \mathcal{C}$, and therefore $\widehat{g}$ constitutes an upper bound of the chain $\mathcal{C}$. Consequently, by virtue of Zorn's Lemma, it follows that the family $\mathcal{F}$ has a maximal element F. According to the definition of $\mathcal{F}$, this means that F is a linear extension of f, whose domain $\mathcal{D}(F)$ is a subspace of V, such that

$$U \subseteq \mathcal{D}(F)\,, \quad F(x) \leqslant p(x) \quad \forall\, x \in \mathcal{D}(F)\,, \quad \text{and} \quad F(x) = f(x) \quad \forall\, x \in U\,.$$

Also, if $g \in \mathcal{F}$ is such that $F \leqslant g$, then necessarily $F = g$. It only remains to prove that $\mathcal{D}(F) = V$ in order to complete the proof. Suppose, by contradiction, that there exists $y_1 \in V$ such that $y_1 \notin \mathcal{D}(F)$, and consider the subspace $Y_1 := \mathcal{D}(F) + \text{span}\{y_1\}$, that is,

$$Y_1 := \big\{x := y + \alpha\, y_1 : \quad y \in \mathcal{D}(F)\,, \quad \alpha \in \mathbb{R}\big\}\,.$$

Note that $y_1 \neq 0$. Moreover, the representation of the elements of Y_1 is unique, that is, $Y_1 := \mathcal{D}(F) \oplus \text{span}\{y_1\}$ since $\mathcal{D}(F) \cap \text{span}\{y_1\} = \{0\}$. Thus, we can define the linear functional

$$\begin{aligned} g_1 : \quad & Y_1 \to \mathbb{R} \\ & x := y + \alpha\, y_1 \to g_1(x) := F(y) + \gamma\, \alpha\,, \end{aligned}$$

where $\gamma \in \mathbb{R}$ is a parameter to be determined. It follows that $g_1(x) = F(x)$ for all $x \in \mathcal{D}(F)$, which shows that g_1 is a proper extension of F (since $\mathcal{D}(F)$ is a proper subspace of Y_1). The next goal is to prove that γ can be chosen suitably so that $g_1(x) \leqslant p(x)$ for all $x \in Y_1$. This would show that $g_1 \in \mathcal{F}$ and that $F \leqslant g_1$, with $g_1 \neq F$, which contradicts the fact that F is maximal, and hence necessarily $\mathcal{D}(F) = V$. Given $y, z \in \mathcal{D}(F)$, we can write

$$F(y) - F(z) = F(y - z) \leqslant p(y - z) = p((y + y_1) + (-y_1 - z))\,,$$

and using the sublinearity of p, we obtain

$$F(y) - F(z) \leqslant p(y + y_1) + p(-y_1 - z)\,,$$

from which it follows that

$$-\,p(-y_1 - z) - F(z) \leqslant p(y + y_1) - F(y) \qquad \forall\, y\,, z \in \mathcal{D}(F)\,.$$

The previous inequality suggests defining the constants

$$m^- := \sup_{z\in\mathcal{D}(F)} \Big\{ -\, p(-y_1 - z) \,-\, F(z)\Big\},$$

and

$$m^+ := \inf_{y\in\mathcal{D}(F)} \Big\{ p(y + y_1) \,-\, F(y)\Big\},$$

for which it is clear that $m^- \leqslant m^+$. Then, given $\gamma \in [m^-, m^+]$ we have

$$-\,p(-y_1 - z) \,-\, F(z) \leqslant \gamma \qquad \forall\, z \in \mathcal{D}(F)\,, \tag{2.18}$$

and

$$\gamma \leqslant p(y + y_1) \,-\, F(y) \qquad \forall\, y \in \mathcal{D}(F)\,. \tag{2.19}$$

Now, let $x := y + \alpha\, y_1 \in Y_1$, with $y \in \mathcal{D}(F)$ and $\alpha \in \mathbb{R}$. To prove that $g_1(x) \leqslant p(x)$ we consider three possible cases. If $\alpha = 0$ then $x \in \mathcal{D}(F)$ and therefore the inequality is immediate. If $\alpha < 0$ we take $z = \alpha^{-1} y$ in (2.18) and obtain

$$-\,p(-y_1 - \alpha^{-1} y) \,-\, F(\alpha^{-1} y) \leqslant \gamma,$$

from which, multiplying by $(-\alpha) > 0$, it follows that

$$\alpha\, p(-y_1 - \alpha^{-1} y) \,+\, F(y) \leqslant -\,\alpha\,\gamma,$$

or equivalently

$$g_1(x) := F(y) \,+\, \alpha\,\gamma \leqslant (-\alpha)\, p(-y_1 - \alpha^{-1} y) \,=\, p(y + \alpha\, y_1) \,=\, p(x)\,.$$

Similarly, if $\alpha > 0$ we take $y = \alpha^{-1} y$ in (2.19) and then multiply by α. This completes the proof.

□

The following theorem considers the particular case in which V is a normed vector space and constitutes one of the most well-known versions of the Hahn-Banach Theorem.

Theorem 2.21 (Hahn-Banach Theorem in normed spaces). *Let $(V, \|\cdot\|)$ be a normed vector space over $\mathbb{R}$, and let $f \in U'$, where U is a subspace of V. Then, there exists $F \in V'$ such that*

$$F(x) \,=\, f(x) \quad \forall\, x \in U \quad \text{and} \quad \|F\|_{V'} \,=\, \|f\|_{U'}\,.$$

Proof. Applying Theorem 2.20 to $f \in U'$ and to the sublinear functional $p : V \to \mathbb{R}$ defined by $p(x) = \|f\|_{U'}\, \|x\|$ for all $x \in V$, we find that there exists a linear map $F : V \to \mathbb{R}$ such that

$$F(x) \,=\, f(x) \quad \forall\, x \in U \quad \text{and} \quad F(x) \leqslant \|f\|_{U'}\, \|x\| \quad \forall\, x \in V\,.$$

It follows that

$$F(-x) \leqslant \|f\|_{U'}\, \|-x\| \,=\, \|f\|_{U'}\, \|x\|\,,$$

so that, using the linearity of F,

$$F(x) \geqslant -\,\|f\|_{U'}\, \|x\|\,,$$

and hence

$$|\,F(x)\,| \leqslant \|f\|_{U'}\, \|x\| \quad \forall\, x \in V\,.$$

This inequality shows that $F \in V'$ and $\|F\|_{V'} \leqslant \|f\|_{U'}$. On the other hand, it is clear that

$$\|F\|_{V'} = \sup_{\substack{v\in V\\ v\neq 0}} \frac{|F(v)|}{\|v\|} \geqslant \sup_{\substack{v\in U\\ v\neq 0}} \frac{|F(v)|}{\|v\|} = \sup_{\substack{v\in U\\ v\neq 0}} \frac{|f(v)|}{\|v\|} = \|f\|_{U'},$$

which completes the proof. □

We end this section with the complex version of Theorem 2.20.

Theorem 2.22 (Hahn-Banach Theorem (complex version)). *Let V be a vector space over $\mathbb{C}$, and let $p : V \to \mathbb{R}$ be a sublinear functional. Let U be a subspace of V, and let $f : U \to \mathbb{C}$ be a linear functional such that*

$$|f(x)| \leqslant p(x) \qquad \forall\, x \in U\,.$$

Then, there exists a linear functional $F : V \to \mathbb{C}$ such that

$$F(x) = f(x) \qquad \forall\, x \in U \qquad \textit{and} \qquad |F(x)| \leqslant p(x) \qquad \forall\, x \in V\,.$$

Proof. Let $f = f_0 + i\, g_0$, where $f_0,\, g_0 : U \to \mathbb{R}$ denote the respective real and imaginary parts of f. Then, viewing U as a vector space over $\mathbb{R}$, the mapping

$$f_0 : U \to \mathbb{R}$$

becomes a linear functional. Moreover, it follows immediately from the assumptions on f that

$$f_0(x) \leqslant |f(x)| \leqslant p(x) \qquad \forall\, x \in U\,,$$

and hence, applying Theorem 2.20 to V regarded as a real vector space, we deduce that there exists a linear functional $F_0 : V \to \mathbb{R}$ such that

$$F_0(x) = f_0(x) \quad \forall\, x \in U \quad \text{and} \quad F_0(x) \leqslant p(x) \quad \forall\, x \in V\,. \tag{2.20}$$

Next, using the definition and linearity of f, we observe that for every $x \in U$ there hold

$$i\, f(x) = f(i\, x) = f_0(i\, x) + i\, g_0(i\, x)$$

and

$$i\, f(x) = -g_0(x) + i\, f_0(x)\,,$$

from which it follows that

$$g_0(x) = -f_0(i\, x) \qquad \forall\, x \in U\,. \tag{2.21}$$

This identity suggests defining the sought functional $F : V \to \mathbb{C}$ by

$$F(x) := F_0(x) - i\, F_0(i\, x) \qquad \forall\, x \in V\,. \tag{2.22}$$

In fact, we begin by observing, according to (2.22) and the linearity of F_0, that for every $x \in V$ there holds

$$F(i\, x) = F_0(i\, x) - i\, F_0(i^2 x) = F_0(i\, x) - i\, F_0(-x) = F_0(i\, x) + i\, F_0(x)\,,$$

whereas multiplying (2.22) by i, we obtain

$$i\, F(x) = i\, F_0(x) + F_0(i\, x)\,,$$

so that, comparing the previous identities, we conclude that

$$F(i\,x) \,=\, i\,F(x) \qquad \forall\, x \in V\,. \tag{2.23}$$

We emphasize here that the fact that F_0 is linear only over the field $\mathbb{R}$ prevents us from incorrectly claiming that $F_0(i\,x)$ coincides with $i\,F_0(x)$. In turn, it follows readily from (2.22) and the linearity of F_0 that for every $\alpha \in \mathbb{R}$ and for all $x,\, z \in V$ there hold

$$F(\alpha\,x) \,=\, \alpha\,F(x) \quad \text{and} \quad F(x+z) \,=\, F(x) \,+\, F(z)\,. \tag{2.24}$$

Thus, a combined use of (2.23) and (2.24) yields, for every $\alpha,\, \beta \in \mathbb{R}$ and each $x \in V$,

$$F\big((\alpha + i\beta)\,x\big) = F(\alpha\,x) + F(i\,\beta\,x) = \alpha\,F(x) + i\,F(\beta\,x) = (\alpha + i\beta)\,F(x)\,,$$

thus completing the proof of linearity of F over the field $\mathbb{C}$. On the other hand, for each $x \in U$ we obviously have $i\,x \in U$, and hence, bearing in mind (2.22), the identity in (2.20), and (2.21), we find that

$$F(x) \,=\, F_0(x) - i\,F_0(i\,x) \,=\, f_0(x) - i\,f_0(i\,x) \,=\, f_0(x) + i\,g_0(x) \,=\, f(x)\,,$$

which proves that F and f coincide on U. Finally, given $x \in V$, we write $F(x)$ in polar form as

$$F(x) \,=\, |F(x)|\,\exp(i\,\vartheta(x))\,,$$

which, together with (2.22), the fact that $|F(x)|$ is obviously real, and the inequality in (2.20), implies that

$$|F(x)| \,=\, F\big(\exp(-i\,\vartheta(x))\,x\big) \,=\, F_0\big(\exp(-i\,\vartheta(x))\,x\big) \,\leqslant\, p\big(\exp(-i\,\vartheta(x))\,x\big) \,=\, p(x)\,,$$

thus finishing the proof.

□

2.3.3 Other consequences of the Hahn-Banach Theorem

We now present some results that follow from the Hahn-Banach Theorem (cf. Theorem 2.21) and that are applied quite frequently in various contexts.

Theorem 2.23. *Let $(V, \|\cdot\|)$ be a normed vector space over $\mathbb{R}$, and let $x_0 \in V$ such that $x_0 \neq 0$. Then, there exists $F \in V'$ such that $\|F\| \,=\, 1$ and $F(x_0) \,=\, \|x_0\|$.*

Proof. Let $S := \operatorname{span}\{x_0\} \,:=\, \big\{\alpha\,x_0 : \quad \alpha \in \mathbb{R}\big\}$ and define the functional $f : S \to \mathbb{R}$ by

$$f(x) \,:=\, \alpha\,\|x_0\| \qquad \forall\, x = \alpha\,x_0 \in S\,.$$

It is clear that f is linear. Moreover, $|f(\alpha\,x_0)| \,=\, |\alpha|\,\|x_0\| \,=\, \|\alpha\,x_0\|$, which shows that f is bounded and $\|f\|_{S'} \,=\, 1$. Thus, applying Theorem 2.21 it follows that there exists $F \in V'$ such that $F(x) \,=\, f(x)$ for all $x \in S$ and $\|F\|_{V'} \,=\, \|f\|_{S'} \,=\, 1$. In particular, for $x_0 \in S$ we have

$$F(x_0) \,=\, f(x_0) \,=\, \|x_0\|\,,$$

which completes the proof.

□

Observe here that if V' is strictly convex then the functional $F \in V'$ from the previous theorem is unique. Indeed, recall first that the strict convexity of the dual means that for all $F_1, F_2 \in V'$ such that $F_1 \neq F_2$ and $\|F_1\| = \|F_2\| = 1$, there holds

$$\|t\,F_1 + (1-t)\,F_2\| < 1 \qquad \forall\, t \in (0,1)\,.$$

Then, if we suppose that there exist distinct $F,\, G \in V'$ such that $\|F\| = \|G\| = 1$ and $F(x_0) = G(x_0) = \|x_0\|$, we obtain

$$\begin{aligned} 1 > \|tF + (1-t)G\|_{V'} &= \sup_{\substack{x \in V \\ x \neq 0}} \frac{tF(x) + (1-t)G(x)}{\|x\|} \\ &\geqslant \frac{tF(x_0) + (1-t)G(x_0)}{\|x_0\|} = 1\,, \end{aligned}$$

which is a contradiction.

Lemma 2.24. *Let $(V, \|\cdot\|)$ be a normed vector space over $\mathbb{R}$, and let $v \in V$ such that $F(v) = 0$ for all $F \in V'$. Then $v = 0$.*

Proof. This follows directly from Theorem 2.23. □

Another important consequence of Theorem 2.23 is given by the following result, which was already announced at the end of Section 2.1 (see (2.4)).

Lemma 2.25. *Let $(V, \|\cdot\|)$ be a normed vector space. Then*

$$\|v\| = \sup_{0 \neq F \in V'} \frac{|F(v)|}{\|F\|} = \max_{0 \neq F \in V'} \frac{|F(v)|}{\|F\|} \qquad \forall\, v \in V\,.$$

Proof. If $v = 0$ the result is trivial. If $v \neq 0$, it follows from Theorem 2.23 that there exists $\bar{F} \in V'$ such that $\|\bar{F}\| = 1$ and $\bar{F}(v) = \|v\|$, whereby

$$\sup_{0 \neq F \in V'} \frac{|F(v)|}{\|F\|} \geqslant \frac{|\bar{F}(v)|}{\|\bar{F}\|} = \|v\|\,.$$

On the other hand, it is clear that

$$\sup_{0 \neq F \in V'} \frac{|F(v)|}{\|F\|} \leqslant \|v\|\,,$$

which completes the proof. □

We conclude the present section with a result that will also be used with some frequency in the following chapters.

Theorem 2.26. *Let S be a subspace of a normed vector space $(V, \|\cdot\|)$, and let $x_0 \in V$ such that $d := \operatorname{dist}(x_0, S) := \inf_{x \in S} \|x_0 - x\| > 0$. Then there exists $F \in V'$ such that $\|F\| = 1$, $F(x_0) = d$ and $F(x) = 0$ for all $x \in S$.*

Proof. Let $S_1 := S + \text{span}\{x_0\}$, that is

$$S_1 := \{z = x + \alpha\, x_0 : \quad x \in S, \quad \alpha \in \mathbb{R}\}.$$

Note that the representation of the members of S_1 is unique, that is $S_1 := S \oplus \text{span}\{x_0\}$. Indeed, if $x \in S \cap \text{span}\{x_0\}$ then $x \in S$ and $x = \alpha\, x_0$, with $\alpha \in \mathbb{R}$. It follows necessarily that $\alpha = 0$ and thus $x = 0$, since otherwise we would have $x_0 = \frac{1}{\alpha} x \in S$, which contradicts the fact that $\text{dist}(x_0, S) > 0$. According to the above we can define the following functional

$$\begin{aligned} f : \quad & S_1 \quad \to \mathbb{R} \\ & z = x + \alpha\, x_0 \to f(z) := \alpha\, d, \end{aligned}$$

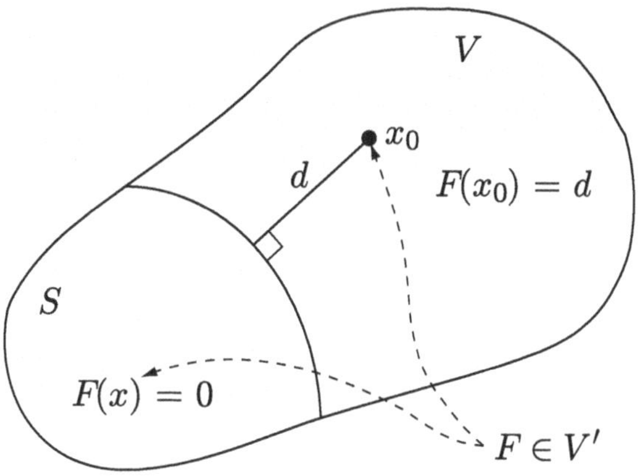

Fig. 2.4. *Graphical illustration of the properties of* $F \in V'$

which is easily shown to be linear and satisfies $f(z) = 0$ for all $z \in S$. Moreover, given an arbitrary $z = x + \alpha\, x_0 \in S_1$, with $\alpha \neq 0$, we have that $\frac{1}{\alpha} x \in S$ and then

$$|f(z)| = |\alpha\, d| = |\alpha|\, d \leqslant |\alpha| \left\| x_0 + \frac{1}{\alpha} x \right\| = \|z\|,$$

which shows that f is bounded and $\|f\|_{S_1'} \leqslant 1$. On the other hand, given $\varepsilon > 0$ there exists $x_1 \in S$ such that $\| -x_1 + x_0\| < d + \varepsilon$. It follows that $f(-x_1 + x_0) = d$ and hence

$$\frac{|f(-x_1 + x_0)|}{\| -x_1 + x_0\|} > \frac{d}{d + \varepsilon},$$

so that

$$\|f\|_{S_1'} := \sup_{\substack{z \in S_1 \\ z \neq 0}} \frac{|f(z)|}{\|z\|} \geqslant \frac{|f(-x_1 + x_0)|}{\| -x_1 + x_0\|} > \frac{d}{d + \varepsilon} \qquad \forall\, \varepsilon > 0,$$

and therefore, taking the limit as $\varepsilon \to 0$, we obtain $\|f\|_{S_1'} \geqslant 1$. Thus, $\|f\|_{S_1'} = 1$ and applying Theorem 2.21 it follows that there exists $F \in V'$ such that $\|F\|_{V'} = \|f\|_{S_1'} = 1$ and $F(z) = f(z)$ for all $z \in S_1$. In particular, $F(x_0) = d$ and $F(x) = 0$ for all $x \in S$. Finally, note that Figure 2.4 graphically illustrates the properties of F.

□

We end this section with some comments on the historical origin of the analytic version of the Hahn–Banach Theorem. Indeed, this result was first published for the case of a real normed space by the Polish mathematician Stefan Banach in [10]. Independently of this work, but in the less general setting of a Banach space, it had been published earlier by the Austrian mathematician Hans Hahn in [92], which, in turn, was a generalization of a previous related result by the same author published in 1922. This certainly explains why the theorem bears the names of both authors. There are actually several versions of this result (see, e.g. [114] for details). In particular, the complex version was proved later, also independently, in [118] and [21]. For further information on the history of this and other fundamental theorems of functional analysis, we refer to [41] and the references therein.

2.3.4 Geometric form of the Hahn-Banach Theorem

This version of the theorem is concerned with the separation of convex sets. In what follows, E denotes a normed vector space over $\mathbb{R}$. Moreover, given a subset S of E, we write S^c for its complement in E, that is, $S^c := E \backslash S$.

Definition 2.27. *Given $\alpha \in \mathbb{R}$ and a linear functional $f : E \to \mathbb{R}$, nonzero and not necessarily bounded, we define an affine hyperplane as*

$$H := \big\{ x \in E : \quad f(x) = \alpha \big\},$$

and we say that H is the hyperplane with equation $[f = \alpha]$.

Lemma 2.28. *The hyperplane $[f = \alpha]$ is closed if and only if f is bounded.*

Proof. If f is continuous then clearly $H = f^{-1}(\{\alpha\})$ is closed because $\{\alpha\}$ is closed in $\mathbb{R}$. Conversely, suppose that H is closed. It follows that H^c is open and nonempty since f is nonzero. Let then $x_0 \in H^c$ and suppose, without loss of generality, that $f(x_0) < \alpha$. Choose $r > 0$ such that $B(x_0, r) \subset H^c$. We claim that $f(x) < \alpha$ for all $x \in B(x_0, r)$. Indeed, suppose by contradiction that there exists $x_1 \in B(x_0, r)$ such that $f(x_1) > \alpha$. Note that $f(x_1) \neq \alpha$ since $x_1 \notin H$. It is clear that

$$\big\{ (1-t)x_1 + t x_0 : \quad t \in [0,1] \big\} \subseteq B(x_0, r),$$

and therefore

$$f((1-t)x_1 + t x_0) \neq \alpha \quad \forall t \in [0,1].$$

However, since $f(x_0) < \alpha < f(x_1)$ it follows that

$$\tilde{t} := \frac{f(x_1) - \alpha}{f(x_1) - f(x_0)} \in (0,1)$$

and

$$f((1-\tilde{t})\, x_1 + \tilde{t}\, x_0) = (1-\tilde{t})\, f(x_1) + \tilde{t}\, f(x_0) = \alpha,$$

which is a contradiction. Thus, necessarily $f(x) < \alpha$ for all $x \in B(x_0, r)$, or equivalently $f(x_0 + r z) < \alpha$ for all $z \in B(0,1)$, from which

$$f\left(x_0 \pm r\,\varepsilon \frac{x}{\|x\|} \right) < \alpha \qquad \forall\, x \in E,\, x \neq 0,\, \forall\, \varepsilon \in (0,1),$$

that is

$$|f(x)| < \frac{1}{r\varepsilon}(\alpha - f(x_0))\,\|x\| \qquad \forall x \in E, \quad \forall \varepsilon \in (0,1).$$

Then, taking the limit as $\varepsilon \to 1^-$ we obtain

$$|f(x)| \leqslant \frac{1}{r}(\alpha - f(x_0))\,\|x\| \quad \forall x \in E,$$

which proves that f is bounded and $\|f\|_{E'} \leqslant \dfrac{1}{r}(\alpha - f(x_0))$.

□

Note that one can give an alternative proof of the previous lemma by using Exercise 2.1 of the present chapter, which establishes that, given a linear functional $f : E \to \mathbb{R}$, its null space $N(f) := \{x \in E : \quad f(x) = 0\}$ is closed if and only if $f \in E'$. Accordingly, since for any element x_0 in the affine hyperplane $[f = \alpha]$ we have $[f = \alpha] = \{x_0\} + N(f)$, the proof follows from the fact that $[f = \alpha]$ is closed if and only if $N(f)$ is.

Definition 2.29. *Let A, $B \subseteq E$. We say that the affine hyperplane $[f = \alpha]$ separates A and B in the* BROAD *sense if*

$$f(x) \leqslant \alpha \quad \forall x \in A \qquad \textit{and} \qquad f(x) \geqslant \alpha \quad \forall x \in B.$$

In turn, we say that this hyperplane separates A and B in the STRICT *sense if there exists $\varepsilon > 0$ such that*

$$f(x) \leqslant \alpha - \varepsilon \quad \forall x \in A \qquad \textit{and} \qquad f(x) \geqslant \alpha + \varepsilon \quad \forall x \in B.$$

Figure 2.5 illustrates this second situation.

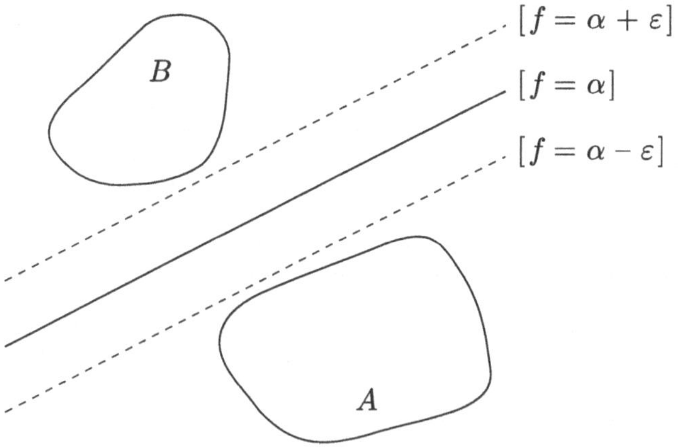

Fig. 2.5. *Strict separation in $\mathbb{R}^2$*

Lemma 2.30 (Gauge of a convex set). *Let $S \subseteq E$ be convex, open, such that $0 \in S$, and define the Minkowski functional $p : E \to \mathbb{R}$ as*

$$p(x) = \inf\{\alpha > 0 : \quad \alpha^{-1} x \in S\} \qquad \forall x \in E .$$

Then p is sublinear and there exists $M \geqslant 0$ such that $0 \leqslant p(x) \leqslant M\|x\|$ for all $x \in E$. Moreover,

$$S = \{x \in E : \quad p(x) < 1\} .$$

Proof. Since the zero vector belongs to the open set S, it follows that there exists $r > 0$ such that $\bar{B}(0,r) \subseteq S$. Then, for each $x \in E$, $x \neq 0$, we have that $\frac{r}{\|x\|} x \in S$, which shows that the scalar $\frac{\|x\|}{r}$ belongs to the set $\{\alpha > 0 : \quad \alpha^{-1} x \in S\}$, and therefore $p(x)$ is well defined and $p(x) \leqslant \frac{\|x\|}{r}$ for all $x \in E$, $x \neq 0$. This inequality and the fact that $p(0) = 0$ imply the required estimate for p with $M = \frac{1}{r}$. Moreover, from the definition of p one easily sees that

$$p(\lambda x) = \lambda\, p(x) \quad \forall\, \lambda > 0, \forall x \in E .$$

Next we prove that $S = \{x \in E : \quad p(x) < 1\}$. Indeed, let $x \in S$. Since S is open there exists $\varepsilon > 0$, sufficiently small, such that $(x + \varepsilon x) = (1+\varepsilon)\, x$ belongs to S, and then $1 \geqslant p((1+\varepsilon)x) = (1+\varepsilon)\, p(x)$, from which

$$p(x) \leqslant \frac{1}{1+\varepsilon} < 1 .$$

Conversely, let $x \in E$ such that $p(x) < 1$. It follows that there exists $\alpha \in (0,1)$ such that $\alpha^{-1} x \in S$ and therefore, since S is convex, $\alpha\,(\alpha^{-1} x) + (1-\alpha)\, 0 = x \in S$. It remains to prove the triangle-type inequality that characterizes the sublinearity of the functional p (cf. Definition 2.19). Indeed, given x, $y \in E$ and $\varepsilon > 0$, we first observe that

$$p\left(\frac{x}{p(x) + \varepsilon}\right) = \frac{p(x)}{p(x)+\varepsilon} < 1 \quad \text{and} \quad p\left(\frac{y}{p(y) + \varepsilon}\right) = \frac{p(y)}{p(y)+\varepsilon} < 1 ,$$

which means that $\dfrac{x}{p(x)+\varepsilon}, \dfrac{y}{p(y)+\varepsilon} \in S$, and hence, by the convexity of S, we obtain

$$\frac{t\, x}{p(x)+\varepsilon} + \frac{(1-t)\, y}{p(y)+\varepsilon} \in S \quad \forall\, t \in [0,1] .$$

In particular, for $t := \dfrac{p(x) + \varepsilon}{p(x)+p(y)+2\varepsilon} \in (0,1)$, we have $\dfrac{x+y}{p(x)+p(y)+2\varepsilon} \in S$, and thus

$$1 > p\left(\frac{x+y}{p(x)+p(y)+2\varepsilon}\right) = \frac{p(x+y)}{p(x)+p(y)+2\varepsilon} \qquad \forall \varepsilon > 0 ,$$

which yields $p(x+y) \leqslant p(x)+p(y)$.

□

Lemma 2.31. *Let $S \subseteq E$ be convex, open, non-empty, and let $x_0 \in E$ such that $x_0 \notin S$. Then, there exists $f \in E'$, $f \neq 0$, such that $f(x) < f(x_0)$ for all $x \in S$, which means that the hyperplane of equation $[f = f(x_0)]$ separates $\{x_0\}$ and S in the* BROAD *sense.*

Proof. Assume, without loss of generality, that $0 \in S$, and define the Minkowski functional $p : E \to \mathbb{R}$ by

$$p(x) := \inf\{\alpha > 0 : \quad \alpha^{-1} x \in S\} \quad \forall x \in E .$$

Let G be the space generated by x_0, that is $G := \{t\, x_0 : \quad t \in \mathbb{R}\}$, and let $g : G \to \mathbb{R}$ be the linear functional defined by $g(t\, x_0) := t$ for all $t \in \mathbb{R}$. Since $x_0 \notin S$, it follows from Lemma 2.30 that $p(x_0) \geqslant 1$. Then, for $t > 0$ we have

$$g(t\,x_0) \;=\; t \leqslant t\,p(x_0) \;=\; p(t\,x_0)\,.$$

On the other hand, if $t \leqslant 0$ we obtain

$$g(t\,x_0) \;=\; t \leqslant 0 \leqslant p(t\,x_0)\,,$$

and consequently

$$g(x) \leqslant p(x) \quad \forall\, x \in G\,.$$

Using the analytic version of the Hahn-Banach Theorem we deduce that there exists a linear functional $f : E \to \mathbb{R}$ such that $f(x) = g(x)$ for all $x \in G$ and $f(x) \leqslant p(x)$ for all $x \in E$. In particular, we have $f(x_0) = g(x_0) = 1$, which confirms that $f \neq 0$. Moreover, according to Lemma 2.30 we know that $0 \leqslant p(x) \leqslant M\,\|x\|$ for all $x \in E$, and then $f(x) \leqslant M\,\|x\|$ for all $x \in E$. On the other hand, $-f(x) = f(-x) \leqslant M\,\|x\|$, so that $f(x) \geqslant -M\,\|x\|$, and therefore

$$|f(x)| \leqslant M\,\|x\| \quad \forall\, x \in E\,,$$

thus proving that $f \in E'$. Finally, using the characterization of the convex set S given in Lemma 2.30, we conclude that $f(x) \leqslant p(x) < 1 = f(x_0)$ for all $x \in S$, which completes the proof in the first case. In the case when the zero vector does not belong to S, consider an arbitrary element $\widetilde{x} \in S$ and define the translated set $\widetilde{S} := \big\{x - \widetilde{x} : \quad x \in S\big\}$, which, besides being convex and open, does contain 0. Then, applying the previous analysis to $\widetilde{S}$ and to the vector $\widetilde{x}_0 := x_0 - \widetilde{x}$ which does not belong to $\widetilde{S}$, we deduce that there exists $f \in E'$ such that $f(z) < f(\widetilde{x}_0)$ for all $z \in \widetilde{S}$, that is $f(x) < f(x_0)$ for all $x \in S$.

□

It is important to note here that the first part of the above proof could also have been carried out by directly applying the Hahn-Banach Theorem in normed spaces (cf. Theorem 2.21) instead of the original analytic version of this result, since, being p bounded by a multiple of the norm (cf. Lemma 2.30), it follows directly that g belongs to the dual of G.

We are now in a position to state and prove both geometric versions of the Hahn-Banach Theorem, which establish broad and strict separations, respectively, of convex sets.

Theorem 2.32 (First geometric version/Hahn-Banach Theorem). *Let $A, B \subseteq E$ be convex, non-empty and disjoint sets, such that A is open. Then there exists a closed hyperplane $[f = \alpha]$, with $f \neq 0$, that separates A and B in the broad sense.*

Proof. Let $S = A - B := \big\{x - y : \quad x \in A, y \in B\big\}$. It is clear that S is convex since A and B are. Moreover, S is open since $S = \cup\big\{A - \{y\} : \quad y \in B\big\}$ and the set $A - \{y\} := \big\{x - y : \quad x \in A\big\}$ is open because A is. Also $0 \notin S$ because $A \cap B = \emptyset$. Therefore, applying Lemma 2.31 we deduce that there exists $f \in E'$, $f \neq 0$, such that $f(z) < f(0) = 0$ for all $z \in S$, that is

$$f(x) < f(y) \quad \forall\, x \in A, \quad \forall\, y \in B\,.$$

Now define

$$m := \sup_{x\in A} f(x) \quad \text{and} \quad M := \inf_{y\in B} f(y)\,.$$

It follows that for any $\alpha \in [m, M]$, the hyperplane H of equation $[f = \alpha]$ separates A and B in the broad sense.

□

Theorem 2.33 (Second geometric version/Hahn-Banach Theorem). *Let A, $B \subseteq E$ be convex, nonempty and disjoint sets, with A closed and B compact. Then there exists a closed hyperplane $[f = \alpha]$, with $f \neq 0$, that separates A and B in the strict sense.*

Proof. Given $\varepsilon > 0$, define the nonempty sets

$$A_\varepsilon := A + B(0,\varepsilon) \quad \text{and} \quad B_\varepsilon := B + B(0,\varepsilon),$$

where $B(0,\varepsilon)$ is the open ball of radius ε centered at 0. It is easy to see, from the convexity of A, B, and $B(0,\varepsilon)$, that A_ε and B_ε are convex. Moreover, A_ε and B_ε are open since the sets $\{x\} + B(0,\varepsilon)$ and $\{y\} + B(0,\varepsilon)$ are open for all $x \in A$, $y \in B$, and we clearly have

$$A_\varepsilon = \cup\big\{\{x\} + B(0,\varepsilon) : \quad x \in A\big\} \quad \text{and} \quad B_\varepsilon = \cup\big\{\{y\} + B(0,\varepsilon) : \quad y \in B\big\}.$$

Let us now prove that there exists $\varepsilon > 0$ such that A_ε and B_ε are disjoint. Indeed, suppose by contradiction that this is not the case. Then, for every $n \in \mathbb{N}$, the set $A_{1/n} \cap B_{1/n}$ is nonempty, which implies that there exist $x_n \in A$, $y_n \in B$, $u_n, v_n \in B(0,1/n)$, such that $x_n + u_n = y_n + v_n$. It follows that

$$\|x_n - y_n\| = \|u_n - v_n\| < \frac{2}{n}.$$

Now, since B is compact, there exists a subsequence $\{y_n^{(1)}\}_{n\in\mathbb{N}} \subseteq \{y_n\}_{n\in\mathbb{N}}$ and a point $y \in B$ such that $y_n^{(1)} \overset{n\to\infty}{\to} y$, and by the inequality above, it follows that $\{x_n^{(1)}\}_{n\in\mathbb{N}}$ also converges to y. Then, since A is closed, necessarily $y \in A$, which contradicts the fact that A and B are disjoint. Therefore, applying the first geometric version of the Hahn-Banach Theorem (cf. Theorem 2.32) to the disjoint sets A_ε and B_ε, it follows that there exists a closed hyperplane with equation $[f = \alpha]$, with $f \neq 0$, that separates them in the broad sense. This means that

$$f(x+u) \leqslant \alpha \leqslant f(y+v) \qquad \forall\, x \in A, \quad \forall\, y \in B, \quad \forall\, u, v \in B(0,\varepsilon),$$

and in particular

$$f(x+\varepsilon z) \leqslant \alpha \leqslant f(y-\varepsilon z) \qquad \forall\, x \in A, \quad \forall\, y \in B, \quad \forall\, z \in B(0,1),$$

or equivalently

$$f(x) + \varepsilon f(z) \leqslant \alpha \leqslant f(y) - \varepsilon f(z) \qquad \forall\, x \in A, \quad \forall\, y \in B, \quad \forall\, z \in B(0,1).$$

Finally, choosing $\bar{z} \in B(0,1)$ such that $f(\bar{z}) > 0$, it follows that

$$f(x) \leqslant \alpha - \varepsilon f(\bar{z}) \qquad \text{and} \qquad f(y) \geqslant \alpha + \varepsilon f(\bar{z}) \qquad \forall\, x \in A, \quad \forall\, y \in B,$$

which proves that the hyperplane $[f = \alpha]$ separates A and B in the strict sense. □

Note that the proof of Theorem 2.33 (essentially taken from [23]) can be simplified by working only with the sets A_ε and B (instead of B_ε). In that case, by following exactly the same line of reasoning, one concludes that the sets A and B are separated in the strict sense by the hyperplane $[f = \alpha - \frac{\varepsilon}{2} f(\bar{z})]$. This remark motivated Exercise 2.31.

In turn, it is interesting to observe that the compactness assumption on the set B cannot be dropped. Indeed, the following example in $E = \mathbb{R}^2$, with

$$A := \left\{x := (x_1, x_2) \in \mathbb{R}^2 : \quad x_1 \geqslant 1 , \quad -1 \leqslant x_2 \leqslant -\frac{1}{x_1} \right\},$$

and

$$B := \left\{x := (x_1, x_2) \in \mathbb{R}^2 : \quad x_1 \geqslant 1 , \quad \frac{1}{x_1} \leqslant x_2 \leqslant 1 \right\}$$

provides two closed sets A and B, neither of which is compact, for which there exists no hyperplane that strictly separates them (see Figure 2.6).

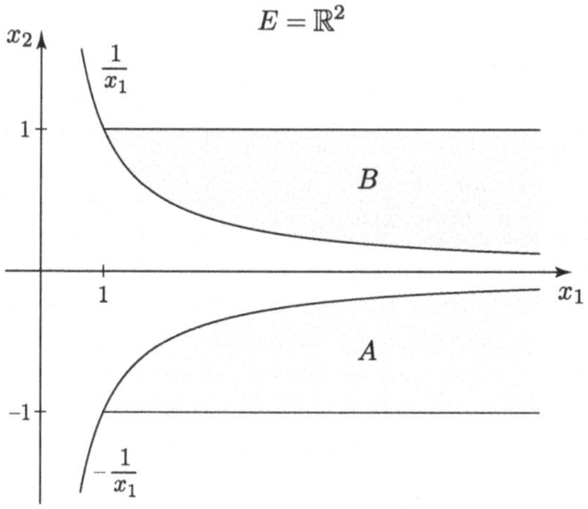

Fig. 2.6. *Separation of the sets A and B by $[f = 0]$ only in the broad sense*

2.4 Hahn-Banach Theorem in Hilbert spaces

In this section we assume a Hilbertian framework to reconsider the analytical version of the Hahn–Banach Theorem in normed spaces and its consequences, which gives rise, in particular, to more specific versions of Theorems 2.21 and 2.26. The main novelty of them is the uniqueness of the resulting functionals.

We begin with the following extension result, which, while stated in Hilbert spaces, is actually valid in normed spaces as well.

Lemma 2.34. *Let $\big(V, \langle\cdot,\cdot\rangle\big)$ be a Hilbert space, S a subspace of V, and $g \in S'$. Then g can be uniquely extended to a functional $\bar{g} \in \bar{S}'$ such that $\|\bar{g}\|_{\bar{S}'} = \|g\|_{S'}$ and $\bar{g}|_S = g$.*

Proof. Let $x \in \bar{S}$, and let $\{x_n\}_{n\in\mathbb{N}} \subseteq S$ be such that $x_n \overset{n\to+\infty}{\longrightarrow} x$. Since $g \in S'$ is bounded, we have $|g(x_n) - g(x_m)| \leqslant \|g\|_{S'} \|x_n - x_m\|$, which implies that $\{g(x_n)\}_{n\in\mathbb{N}}$ is a Cauchy sequence in $\mathbb{R}$, and therefore there exists $\ell \in \mathbb{R}$ such that $\ell = \lim\limits_{n\to+\infty} g(x_n)$. We now show that this limit ℓ is independent of the sequence $\{x_n\}_{n\in\mathbb{N}}$. Indeed, let $\{\widetilde{x}_n\}_{n\in\mathbb{N}}$ be another sequence in S converging to x. Then, using again that $g \in S'$, we have

$$|g(x_n) - g(\widetilde{x}_n)| \leqslant \|g\|_{S'} \|x_n - \widetilde{x}_n\| \overset{n\to+\infty}{\longrightarrow} 0,$$

from which it follows that $\lim_{n\to+\infty} g(x_n) = \lim_{n\to+\infty} g(\tilde{x}_n)$. Therefore, we can now define the functional $\bar{g} : \bar{S} \to \mathbb{R}$ by

$$\bar{g}(x) := \lim_{n\to+\infty} g(x_n),$$

where $\{x_n\}_{n\in\mathbb{N}}$ is any sequence in S converging to the given x. Using this definition, it is easy to see that $\bar{g}$, as defined, is linear and that $\bar{g}|_S = g$. Furthermore, taking the limit as $n \to +\infty$ in the inequality $|g(x_n)| \leqslant \|g\|_{S'} \|x_n\|$, we obtain

$$|\bar{g}(x)| \leqslant \|g\|_{S'} \|x\| \qquad \forall\, x \in \bar{S},$$

which shows that $\bar{g}$ is bounded, i.e. $\bar{g} \in \bar{S}'$, and that $\|\bar{g}\|_{\bar{S}'} \leqslant \|g\|_{S'}$. On the other hand,

$$\|\bar{g}\|_{\bar{S}'} := \sup_{\substack{x\in\bar{S}\\ x\neq 0}} \frac{|\bar{g}(x)|}{\|x\|} \geqslant \sup_{\substack{x\in S\\ x\neq 0}} \frac{|\bar{g}(x)|}{\|x\|} = \sup_{\substack{x\in S\\ x\neq 0}} \frac{|g(x)|}{\|x\|} = \|g\|_{S'},$$

which implies $\|\bar{g}\|_{\bar{S}'} = \|g\|_{S'}$. We now show that this extension $\bar{g}$ is unique. Indeed, let $\tilde{g} \in \bar{S}'$ such that $\tilde{g}|_S = g$ and $\|\tilde{g}\|_{\bar{S}'} = \|g\|_{S'}$. Then, for any $x \in \bar{S}$ and any sequence $\{x_n\}_{n\in\mathbb{N}}$ in S such that $x_n \overset{n\to+\infty}{\to} x$, we have, by continuity of $\tilde{g}$,

$$\tilde{g}(x) = \lim_{n\to+\infty} \tilde{g}(x_n) = \lim_{n\to+\infty} g(x_n) = \bar{g}(x), \tag{2.25}$$

which shows that $\tilde{g} = \bar{g}$.

□

Alternatively, the first part of the above proof can also be obtained by applying the Hahn–Banach Theorem in its normed-space version. However, the idea of defining $\bar{g}$ through the density argument explained here is more intuitive and does not rely on that theorem.

We prove next the Hilbertian versions of Theorems 2.21 and 2.26.

Theorem 2.35 (Hahn-Banach Theorem in Hilbert spaces). *Let $(V, \langle\cdot,\cdot\rangle)$ be a real Hilbert space, and let $f \in U'$, where U is a subspace of V. Then, there exists a unique $F \in V'$ such that*

$$F(x) = f(x) \quad \forall\, x \in U \quad \text{and} \quad \|F\|_{V'} = \|f\|_{U'}.$$

Proof. According to Lemma 2.34, we consider the unique extension $\bar{f} \in \bar{U}'$ such that $\bar{f}|_U = f$ and $\|\bar{f}\|_{\bar{U}'} = \|f\|_{U'}$. Since $\bar{U}$ is obviously closed, it follows that $(\bar{U}, \langle\cdot,\cdot\rangle)$ is a Hilbert space. Therefore, by applying the Riesz Representation Theorem (cf. Theorem 2.14) to $\bar{f} \in \bar{U}'$, we deduce that there exists a unique $\bar{x} \in \bar{U}$ such that $\bar{f}(z) = \langle \bar{x}, z\rangle$ for all $z \in \bar{U}$, and $\|\bar{f}\|_{\bar{U}'} = \|\bar{x}\|$. Thus, defining $\bar{F} \in V'$ by $\bar{F}(z) := \langle \bar{x}, z\rangle$ for all $z \in V$, we have

$$\|\bar{F}\|_{V'} = \|\bar{x}\| = \|\bar{f}\|_{\bar{U}'} = \|f\|_{U'} \quad \text{and} \quad \bar{F}|_{\bar{U}} = \bar{f},$$

from which $\bar{F}|_U = f$, thereby proving existence. For uniqueness, let $\widetilde{F} \in V'$ with Riesz representative $\tilde{x} \in V$ such that

$$\|\tilde{x}\| = \|\widetilde{F}\|_{V'} = \|f\|_{U'} \quad \text{and} \quad \widetilde{F}|_U = f.$$

Analogously to what was done in (2.25), it follows from the continuity of $\widetilde{F}$ that there holds $\widetilde{F}(x) = \bar{f}(x)$ for all $x \in \bar{U}$, and hence

$$\begin{aligned}\|\widetilde{x}-\bar{x}\|^2 &= \|\widetilde{x}\|^2+\|\bar{x}\|^2-2\langle\widetilde{x},\bar{x}\rangle = \|\widetilde{x}\|^2+\|\bar{x}\|^2-2\widetilde{F}(\bar{x}) \\ &= \|\widetilde{x}\|^2+\|\bar{x}\|^2-2\bar{f}(\bar{x}) = \|\widetilde{x}\|^2+\|\bar{x}\|^2-2\|\bar{x}\|^2 = \|\widetilde{x}\|^2-\|\bar{x}\|^2 = 0\,,\end{aligned}$$

which shows that $\widetilde{x}=\bar{x}$, thus yielding $\widetilde{F}=\bar{F}$.

□

Theorem 2.36. *Let S be a subspace of a real Hilbert space $(V,\langle\cdot,\cdot\rangle)$, and let $x_0\in V$ such that $d := \operatorname{dist}(x_0,S) := \inf\limits_{x\in S}\|x_0-x\| > 0$. Then there exists a unique $F\in V'$ such that $\|F\| = 1$, $F(x_0) = d$ and $F(x) = 0$ for all $x\in S$.*

Proof. Since $\bar{S}$ is certainly a closed subspace of V, we let $z\in\bar{S}$ be the best approximation of x_0 by elements of $\bar{S}$; that is, z is the unique element in $\bar{S}$ such that $\langle x_0-z,u\rangle = 0$ for all $u\in\bar{S}$. Then, defining $F: V\to\mathbb{R}$ by $F(x) := \langle\frac{x_0-z}{\|x_0-z\|},x\rangle$ for all $x\in V$, we have $\|F\|_{V'} = \left\|\frac{x_0-z}{\|x_0-z\|}\right\| = 1$, $F(u) = \langle\frac{x_0-z}{\|x_0-z\|},u\rangle = 0$ for all $u\in S$, and

$$F(x_0) = \left\langle\frac{x_0-z}{\|x_0-z\|},x_0\right\rangle = \left\langle\frac{x_0-z}{\|x_0-z\|},x_0-z\right\rangle = \|x_0-z\| = \operatorname{dist}(x_0,\bar{S})\,.$$

Moreover, it is easy to see that $d := \operatorname{dist}(x_0,S) = \operatorname{dist}(x_0,\bar{S})$. Indeed, since clearly $S\subset\bar{S}$, we have $\operatorname{dist}(x_0,\bar{S}) := \min\limits_{\bar{u}\in\bar{S}}\|x_0-\bar{u}\| \leqslant \|x_0-u\|$ for all $u\in S$, and taking the infimum over $u\in S$ yields $\operatorname{dist}(x_0,\bar{S}) \leqslant \operatorname{dist}(x_0,S)$. Conversely, for any $\bar{u}\in\bar{S}$, we can take a sequence $\{u_n\}_{n\in\mathbb{N}}$ in S such that $u_n \overset{n\to+\infty}{\to} \bar{u}$, and observe that $\operatorname{dist}(x_0,S) \leqslant \|x_0-u_n\|$ for all $n\in\mathbb{N}$. Taking the limit as $n\to+\infty$ we obtain $\operatorname{dist}(x_0,S) \leqslant \|x_0-\bar{u}\|$ for all $\bar{u}\in\bar{S}$, and hence, taking the infimum over $\bar{u}\in\bar{S}$, we deduce that $\operatorname{dist}(x_0,S) \leqslant \operatorname{dist}(x_0,\bar{S})$, thus proving the equality. Now, suppose there exists another functional $\widetilde{F}\in V'$, with representative $\widetilde{x}\in V$, that is, $\widetilde{F}(x) = \langle\widetilde{x},x\rangle$ for all $x\in V$, such that $\|\widetilde{x}\| = \|\widetilde{F}\|_{V'} = 1$, $\widetilde{F}(x_0) = \langle\widetilde{x},x_0\rangle = d$, and $\widetilde{F}(u) = 0$ for all $u\in S$. By the continuity of $\widetilde{F}$, it follows that $\widetilde{F}(z) = 0$, and therefore

$$\begin{aligned}\left\|\widetilde{x}-\tfrac{(x_0-z)}{\|x_0-z\|}\right\|^2 &= 2-2\,\tfrac{\langle\widetilde{x},x_0\rangle}{\|x_0-z\|}+2\,\tfrac{\langle\widetilde{x},z\rangle}{\|x_0-z\|} \\ &= 2-2\,\tfrac{\widetilde{F}(x_0)}{\|x_0-z\|}+2\,\tfrac{\widetilde{F}(z)}{\|x_0-z\|} = 2-2\,\tfrac{d}{\|x_0-z\|} = 0\,,\end{aligned}$$

that is, $\widetilde{x} = \frac{(x_0-z)}{\|x_0-z\|}$, thus establishing uniqueness.

□

2.5 A simple exercise with three solutions

In this section we consider a very simple exercise, apparently harmless, which, however, gave rise to three quite different solutions, two of them using results from the present chapter. The respective authors are the author of these notes and former students of the course FUNCTIONAL ANALYSIS AND APPLICATIONS I at the Department of Mathematical Engineering of the University of Concepción, Mr. Miguel Silva and Mr. Sebastián Niklitschek. The exercise is described as follows. Given $m,n\in\mathbb{N}$, a matrix $A\in\mathbb{R}^{m\times n}$ and a vector $b\in\mathbb{R}^m$, $b\neq 0$, we define the solution set

$$S(A,b) := \{x\in\mathbb{R}^n: \quad Ax = b\}\,,$$

and assume that $S(A,b) \neq \varnothing$. The objective is to prove that there exists $z \in \mathbb{R}^n$ such that

$$\inf_{x\in S(A,b)} \langle z, x \rangle > 0,$$

where $\langle \cdot, \cdot \rangle$ denotes the usual inner product in $\mathbb{R}^n$.

2.5.1 Solution 1

Define the set $\mathcal{A} := S(A,b)$, which is clearly convex, closed, and nonempty (as stated in the hypothesis). In turn, let $\mathcal{B}$ be the convex, compact set disjoint from $\mathcal{A}$ given by $\mathcal{B} := \{0\}$. Then, by applying the second geometric version of the Hahn-Banach Theorem (cf. Theorem 2.33), it follows that there exist $f \in (\mathbb{R}^n)'$ and $\alpha \in \mathbb{R}$ such that the hyperplane $[f = \alpha]$ separates $\mathcal{A}$ and $\mathcal{B}$ in the strict sense. This means that there exist $\widetilde{z} \in \mathbb{R}^n$ and $\varepsilon > 0$ such that

$$f(x) = \langle \widetilde{z}, x \rangle \quad \forall\, x \in \mathbb{R}^n,$$

and moreover

$$f(x) = \langle \widetilde{z}, x \rangle \leqslant \alpha - \varepsilon < \alpha + \varepsilon \leqslant f(0) = 0 \quad \forall\, x \in \mathcal{A}.$$

Then, defining $z := -\widetilde{z}$, we obtain:

$$\langle z, x \rangle \geqslant -(\alpha - \varepsilon) > -(\alpha + \varepsilon) \geqslant 0 \quad \forall\, x \in \mathcal{A},$$

from which it follows that

$$\inf_{x\in\mathcal{A}} \langle z, x \rangle > 0.$$

2.5.2 Solution 2

Let $N(A)$ be the null space associated with the matrix A, that is

$$N(A) := \{x \in \mathbb{R}^n : \quad A\,x = 0\}.$$

Since $S(A,b) \neq \varnothing$, there exists $x_0 \in \mathbb{R}^n$ such that $A\,x_0 = b$, and then it is easy to see that

$$S(A,b) = \{x_0\} + N(A) \tag{2.26}$$

Now, let $P : \mathbb{R}^n \to N(A)$ and $(I - P) : \mathbb{R}^n \to N(A)^\perp$ be the respective orthogonal projectors, and define $z := (I - P)(x_0) \in N(A)^\perp$. It is clear that $z \neq 0$ since, otherwise, we would have $x_0 = P(x_0) \in N(A)$, which contradicts the fact that $b = A(x_0) \neq 0$. It follows that

$$\inf_{x\in S(A,b)} \langle z, x \rangle = \inf_{y\in N(A)} \langle (I-P)(x_0), x_0 + y \rangle = \inf_{y\in N(A)} \langle (I-P)(x_0), x_0 \rangle$$

$$= \langle (I-P)(x_0), x_0 \rangle = \|(I-P)(x_0)\|^2 = \|z\|^2 > 0.$$

2.5.3 Solution 3

We use again that $S(A,b)$ is convex and closed. In particular, note that the latter follows from the identity (2.26) and from the fact that $N(A)$ is obviously closed. Moreover, it is clear that $0 \notin S(A,b)$ since $b \neq 0$. Thus, applying the best approximation result over a closed convex set (cf. Theorem 2.7) , it follows that there exists a unique $z \in S(A,b)$ such that

$$\mathrm{dist}(0, S(A,b)) := \inf_{x \in S(A,b)} \|x\| = \|z\| > 0.$$

Moreover, the characterization result of this best approximation, which will be established later in Theorem 4.15, states that

$$0 \geqslant \langle 0 - z, x - z \rangle = \|z\|^2 - \langle z, x \rangle \quad \forall x \in S(A,b),$$

that is,

$$\langle z, x \rangle \geqslant \|z\|^2 \quad \forall x \in S(A,b),$$

from which it is concluded that

$$\inf_{x \in S(A,b)} \langle z, x \rangle = \|z\|^2 > 0.$$

2.6 Exercises

▸ **2.1** *Let X be a normed vector space, and let $f : X \to \mathbb{R}$ be a linear functional. Prove that $f \in X'$ if and only if $N(f)$, the null space of f, is a closed subspace of X.*

▸ **2.2** *Let X be a Hilbert space over $\mathbb{K}$ ($\mathbb{R}$ or $\mathbb{C}$), and let $F : X \to \mathbb{K}$ be a linear functional such that its null space $N(F) := \{x \in X : \; F(x) = 0\}$ is not dense in X. Prove that $F \in X'$ and conclude that $N(F)$ is a proper closed subspace of X.*

▸ **2.3** *Let H be a real Hilbert space, and let $F : H \longrightarrow \mathbb{R}$ be a linear functional with null space*

$$N(F) := \Big\{x \in H : \quad F(x) = 0\Big\}$$

different from H. Prove that F is not bounded if and only if $N(F)$ is dense in H.

▸ **2.4** *Let M be a proper closed subspace of a Banach space X, and let $x_0 \in X - M$. Apply the second geometric form of the Hahn-Banach Theorem to prove that there exist $F_1, F_2 \in M^o$ such that $\|F_1\| = 1$ and $F_2(x_0) = \mathrm{dist}(x_0, M)$.*

▸ **2.5** *Let U be a finite-dimensional subspace of a normed vector space $(X, \|\cdot\|)$, and let $\{u_1, u_2, \ldots, u_n\}$ be a basis of U. Prove that there exists a set*

$$\mathcal{F} := \{f_1, f_2, \ldots, f_n\} \subseteq U'$$

such that $f_i(u_j) = \delta_{ij}$ for all $i, j \in \{1, 2, \ldots, n\}$. Then, prove that

$$u = \sum_{i=1}^{n} f_i(u)\, u_i \qquad \forall u \in U,$$

and deduce, using this identity, that $\mathcal{F}$ is a basis of U', thus concluding that the dimension of U' is also n.

▸ **2.6** *Let X be a real normed vector space, and let $\{p_1, \ldots, p_n\}$ be a basis of a subspace of X. Prove that there exists a set of functionals $\{F_1, \ldots, F_n\} \subseteq X'$, such that*

$$F_1(p_1) = F_2(p_2) = \cdots = F_n(p_n) \quad \text{and} \quad F_j|_{U_j} = 0 \quad \forall\, j \in \{1, \ldots, n\},$$

where

$$U_j := \langle \{p_i\}_{i \neq j} \rangle \qquad \forall\, j \in \{1, \ldots, n\}.$$

What additional consequence could be added to the above if X is a Hilbert space?

▸ **2.7** *Let H be a real normed vector space. Prove that if the norm of H satisfies*

$$\|x + y\|^2 + \|x - y\|^2 = 2\,\{\, \|x\|^2 + \|y\|^2 \,\} \quad \forall\, x, y \in H,$$

then it comes from an inner product.

▸ **2.8** *Let $X := C([0,1])$ endowed with the uniform norm*

$$\|u\| := \max_{t \in [0,1]} |u(t)| \qquad \forall\, u \in X, \tag{2.27}$$

and given $f \in X$, fixed, define the linear functional $F : X \to \mathbb{R}$ as

$$F(u) := \int_0^1 u(t)\, f(t)\, dt \qquad \forall\, u \in X.$$

Prove that $F \in X'$ and $\|F\| = \displaystyle\int_0^1 |f(t)|\, dt$.

HINT: *For each $n \in \mathbb{N}$ consider the function $x_n \in X$ given by $x_n(t) := u_n(f(t))$ for all $t \in [0,1]$, where $u_n : \mathbb{R} \to \mathbb{R}$ is the continuous function*

$$u_n(t) := \begin{cases} 1 \text{ if } t \geqslant 1/n, \\ -1 \text{ if } t \leqslant -1/n, \\ nt \text{ if } -1/n \leqslant t \leqslant 1/n, \end{cases}$$

and then use x_n to prove that $\quad \|F\| \geqslant \displaystyle\int_0^1 |f(t)|\, dt - \frac{1}{n}$.

▸ **2.9** ([24], [112]) *Let Ω be a bounded open set of $\mathbb{R}^n$ with sufficiently smooth boundary Γ, recall that $\mathbf{L}^2(\Omega) := [\mathrm{L}^2(\Omega)]^n$, and define the space*

$$\mathbf{H}(\mathrm{div};\Omega) := \left\{ \boldsymbol{\tau} := (\tau_1, \ldots, \tau_n) \in \mathbf{L}^2(\Omega) : \quad \mathrm{div}(\boldsymbol{\tau}) := \sum_{i=1}^n \frac{\partial \tau_i}{\partial x_i} \in \mathrm{L}^2(\Omega) \right\}$$

endowed with the inner product

$$\langle \boldsymbol{\tau}, \boldsymbol{\zeta} \rangle_{\mathrm{div};\Omega} := \int_\Omega \boldsymbol{\tau} \cdot \boldsymbol{\zeta} + \int_\Omega \mathrm{div}(\boldsymbol{\tau})\, \mathrm{div}(\boldsymbol{\zeta}) \quad \forall\, \boldsymbol{\tau}, \boldsymbol{\zeta} \in \mathbf{H}(\mathrm{div};\Omega).$$

a) *Prove that $\big(\mathbf{H}(\mathrm{div};\Omega); \langle \cdot, \cdot \rangle_{\mathrm{div};\Omega}\big)$ is a Hilbert space.*
b) *Show that for every $\mathbf{g} \in \mathbf{L}^2(\Omega)$ there exists a unique $\boldsymbol{\tau}_{\mathbf{g}} \in \mathbf{H}(\mathrm{div};\Omega)$ such that*

$$\int_\Omega \boldsymbol{\tau}_{\mathbf{g}} \cdot \boldsymbol{\zeta} + \int_\Omega \mathrm{div}(\boldsymbol{\tau}_{\mathbf{g}})\, \mathrm{div}(\boldsymbol{\zeta}) = \int_\Omega \mathbf{g} \cdot \boldsymbol{\zeta} \quad \forall\, \boldsymbol{\zeta} \in \mathbf{H}(\mathrm{div};\Omega).$$

HINT: *Given $\boldsymbol{\tau} \in \mathbf{L}^2(\Omega)$, we say that $\mathrm{div}(\boldsymbol{\tau}) \in \mathrm{L}^2(\Omega)$ if there exists $z \in \mathrm{L}^2(\Omega)$ such that*

$$-\int_\Omega \boldsymbol{\tau} \cdot \nabla \varphi = \int_\Omega z\, \varphi \quad \forall\, \varphi \in C_0^\infty(\Omega).$$

▸ **2.10** ([24], [112]) *Let Ω be a bounded open set of $\mathbb{R}^2$ with sufficiently smooth boundary Γ, and define the space*

$$\mathbf{H}(\mathrm{rot};\Omega) := \left\{ \boldsymbol{\tau} := (\tau_1,\tau_2) \in \mathbf{L}^2(\Omega) : \quad \mathrm{rot}(\boldsymbol{\tau}) := \frac{\partial \tau_2}{\partial x_1} - \frac{\partial \tau_1}{\partial x_2} \in \mathrm{L}^2(\Omega) \right\},$$

endowed with the inner product

$$\langle \boldsymbol{\tau}, \boldsymbol{\zeta} \rangle_{\mathrm{rot};\Omega} := \int_\Omega \boldsymbol{\tau}\cdot\boldsymbol{\zeta} + \int_\Omega \mathrm{rot}(\boldsymbol{\tau})\,\mathrm{rot}(\boldsymbol{\zeta}) \quad \forall\, \boldsymbol{\tau},\, \boldsymbol{\zeta} \in \mathbf{H}(\mathrm{rot};\Omega).$$

a) *Prove that $\big(\mathbf{H}(\mathrm{rot};\Omega); \langle\cdot,\cdot\rangle_{\mathrm{rot};\Omega}\big)$ is a Hilbert space.*
b) *Show that for every $\mathbf{g} \in \mathbf{H}(\mathrm{rot};\Omega)$ there exists a unique $\boldsymbol{\tau}_{\mathbf{g}} \in \mathbf{H}(\mathrm{rot};\Omega)$ such that*

$$\int_\Omega \boldsymbol{\tau}_{\mathbf{g}}\cdot\boldsymbol{\zeta} + \int_\Omega \mathrm{rot}(\boldsymbol{\tau}_{\mathbf{g}})\,\mathrm{rot}(\boldsymbol{\zeta}) = \int_\Omega \mathrm{rot}(\mathbf{g})\,\mathrm{rot}(\boldsymbol{\zeta}) \quad \forall\, \boldsymbol{\zeta} \in \mathbf{H}(\mathrm{rot};\Omega).$$

HINT: *Note that $\boldsymbol{\tau} \in \mathbf{H}(\mathrm{rot};\Omega)$ if and only if $(\tau_2, -\tau_1) \in \mathbf{H}(\mathrm{div};\Omega)$. Also, given $\boldsymbol{\tau} \in \mathbf{L}^2(\Omega)$, we say that $\mathrm{rot}(\boldsymbol{\tau}) \in \mathrm{L}^2(\Omega)$ if there exists $z \in \mathrm{L}^2(\Omega)$ such that*

$$-\int_\Omega \tau_2 \frac{\partial \varphi}{\partial x_1} + \int_\Omega \tau_1 \frac{\partial \varphi}{\partial x_2} = \int_\Omega z\,\varphi \qquad \forall\, \varphi \in C_0^\infty(\Omega).$$

▸ **2.11** *Prove that every Hilbert space is strictly convex.*

▸ **2.12** *Prove that for every $f \in (\mathrm{H}^m(\Omega))'$ there exist functions $f_\alpha \in \mathrm{L}^2(\Omega)$, $|\alpha| \leqslant m$, such that*

$$f(v) = \sum_{|\alpha|\leqslant m} \int_\Omega f_\alpha\, \partial^\alpha v\, dx \quad \forall\, v \in \mathrm{H}^m(\Omega).$$

▸ **2.13** *Let X be a normed vector space.*

a) *Let Y be a non-dense subspace of X. Prove that there exists a nonzero functional $F \in X'$ such that $F(x) = 0$ for all $x \in Y$.*
b) *Let $x_0, x_1 \in X$ such that $x_0 \neq x_1$. Prove that there exists a sequence $\{F_n\}_{n\in\mathbb{N}} \subseteq X'$ such that $\|F_n\| = \|x_0 - x_1\|^n$ and $F_n(x_1) = F_n(x_0) + \|x_1 - x_0\|^{n+1}$ for all $n \in \mathbb{N}$.*

▸ **2.14** *Let $(V, \langle\cdot,\cdot\rangle)$ be a Hilbert space, and let $J : V \to \mathbb{R}$ be a nonlinear functional. We say that J admits a directional derivative at $v \in V$, in the direction $\varphi \in V$, if the expression $\frac{J(v+\varepsilon\varphi)-J(v)}{\varepsilon}$ admits a limit when $\varepsilon \to 0$. The value of this limit is denoted $DJ(v,\varphi)$. Then, J is said to be differentiable in the sense of Gâteaux (or G-differentiable) at $v \in V$, if $DJ(v,\varphi)$ exists for all $\varphi \in V$. Now, if J is G-differentiable at $v \in V$ and if $DJ(v,\cdot) \in V'$, it follows, by the Riesz Representation Theorem (cf. Theorem 2.14), that there exists a unique element $z \in V$ such that $DJ(v,\varphi) = \langle z, \varphi\rangle$ for all $\varphi \in V$. In this case, we denote $z := J'(v)$ and call it the* GRADIENT *of J at v. Moreover, we say that J has a second derivative in the sense of Gâteaux at $v \in V$, in the directions φ and $\psi \in V$, if the expression $\frac{DJ(v+\varepsilon\psi,\varphi) - DJ(v,\varphi)}{\varepsilon}$ admits a limit when $\varepsilon \to 0$. The value of this limit is denoted $D^2J(v,\varphi,\psi)$.*

a) *Let Ω be a bounded open set of $\mathbb{R}^n$, and consider $V := \mathrm{L}^2(\Omega)$. Define $J_0 : V \to \mathbb{R}$ by*

$$J_0(v) := \int_\Omega (v^2 + v) \quad \forall\, v \in V$$

and compute $J_0'(v)$ for all $v \in V$. What can be said about $J_0'(v)$ if J_0 is restricted to the Sobolev space $\mathrm{H}^1(\Omega)$ (cf. (1.26)) ?
b) *Suppose that J is G-differentiable at $(v + \alpha\varphi)$, in the direction φ, for all $\alpha \in [0,1]$. Prove that there exists $\beta \in (0,1)$ such that*

$$J(v+\varphi) = J(v) + DJ(v + \beta\,\varphi, \varphi).$$

c) *Suppose that J is α-convex, that is, there exists $\alpha > 0$ such that for all $u, v \in V$ and for all $\beta \in [0,1]$:*

$$J((1-\beta)\,u + \beta\,v) \leqslant (1-\beta)\,J(u) + \beta\,J(v) - \frac{\alpha}{2}\,\beta\,(1-\beta)\,\|u-v\|^2 .$$

Moreover, assume that J is G-differentiable at every $v \in V$. Prove that for all $u, v \in V$ one has

$$DJ(u, u-v) - DJ(v, u-v) \geqslant \alpha\,\|u-v\|^2$$

and

$$J(v) \geqslant J(u) + DJ(u, v-u) + \frac{\alpha}{2}\,\|v-u\|^2 .$$

d) *Suppose that J is G-differentiable at $v \in V$ and that, given a direction $\varphi \in V$, $D^2 J(v+\alpha\varphi, \varphi, \varphi)$ exists for all $\alpha \in [0,1]$. Prove that there exists a constant $\beta \in (0,1)$ such that*

$$J(v+\varphi) = J(v) + DJ(v,\varphi) + \frac{1}{2}\,D^2 J(v+\beta\varphi, \varphi, \varphi) .$$

▸ **2.15** *Let $(H, \langle\cdot,\cdot\rangle)$ be a Hilbert space, and let V be a closed subspace of H. The annihilator of V is denoted by V^o and is defined as*

$$V^o := \big\{F \in H' : \quad F(x) = 0 \quad \forall\, x \in V\big\} .$$

Prove that

$$H = V \oplus \mathcal{R}(V^o) ,$$

where $\mathcal{R} : H' \to H$ denotes the Riesz operator.

▸ **2.16** *Let X be a normed vector space, and let $x_0 \in X$ such that $|\,F(x_0)\,| \leqslant C_0$ for all $F \in X'$ with $\|F\|_{X'} = 1$. Prove that $\|x_0\| \leqslant C_0$.*

▸ **2.17** *Let V be a subspace of a Hilbert space $(Y, \langle\cdot,\cdot\rangle)$, and define*

$$V^\perp := \big\{y \in Y : \quad \langle y, z\rangle = 0 \quad \forall\, z \in V\big\} .$$

If $\bar{V}$ denotes the closure of V, prove that $\bar{V}^\perp = V^\perp$. Conclude that V is dense in Y if and only if $V^\perp = \{0\}$.

▸ **2.18**

a) *Let S be a subset of a Hilbert space H, and let M be the closed subspace generated by S. Prove that $S^\perp$ is a closed subspace of H, $M^\perp = S^\perp$, and $M = (S^\perp)^\perp$.*

b) *Let V be a subspace of a Hilbert space H. Prove that $H = \bar{V} \oplus V^\perp$.*

▸ **2.19** *Let Ω be a bounded open set of $\mathbb{R}^n$ with smooth boundary Γ, and consider the Sobolev space $\mathrm{H}^1(\Omega)$. Find and characterize a subspace V of $\mathrm{H}^1(\Omega)$ such that $\mathrm{H}^1(\Omega) = \mathrm{H}^1_0(\Omega) \oplus V$.*

▸ **2.20** *Let Ω be a bounded open set of $\mathbb{R}^n$, and let $\kappa : \Omega \to \mathbb{R}$ be a continuous function for which there exist constants M, $\beta > 0$, such that $\beta \leqslant \kappa(x) \leqslant M$ for all $x \in \Omega$. Define the set*

$$S := \Big\{ v \in \mathrm{H}^1_0(\Omega) : \quad \|v\|_{\mathrm{H}^1(\Omega)} \leqslant 1 \Big\} ,$$

and prove that for all $u \in \mathrm{H}^1_0(\Omega)$ there exists a unique $g \in S$ such that

$$\int_\Omega \kappa(x)\,\|\nabla u(x) - \nabla g(x)\|^2_{\mathbb{R}^n}\,dx = \min_{v\in S} \int_\Omega \kappa(x)\,\|\nabla u(x) - \nabla v(x)\|^2_{\mathbb{R}^n}\,dx .$$

▸ **2.21** *Let $(H, \langle\cdot,\cdot\rangle)$ be a* COMPLEX *Hilbert space, and let u, $v \in H$, $u \neq v$, such that $\|u\| = \|v\|$. Define $w := u - v$ and consider the orthogonal projection $\mathbf{P} : H \to S^{\perp}$, where S is the subspace generated by w. Prove that*

$$\mathbf{P}(u) = \mathbf{P}(v) = \frac{u+v}{2} + \left\{ \frac{\mathrm{i}\,\mathrm{Im}(\langle u, v\rangle)}{\|w\|^2} \right\} w\,.$$

What happens when H is a REAL *Hilbert space? Give a graphical interpretation.*

▸ **2.22** ([24], [112]) *Given $\Omega := (0,1) \times (0,1)$, consider again the space $(\mathbf{H}(\mathrm{div};\Omega), \langle\cdot,\cdot\rangle_{\mathrm{div};\Omega})$ (cf. Exercise 2.9), where*

$$\mathbf{H}(\mathrm{div};\Omega) := \left\{ \boldsymbol{\tau} \in \mathbf{L}^2(\Omega) : \quad \mathrm{div}(\boldsymbol{\tau}) \in \mathrm{L}^2(\Omega) \right\}$$

and

$$\langle \boldsymbol{\zeta}, \boldsymbol{\tau}\rangle_{\mathrm{div};\Omega} := \int_\Omega \boldsymbol{\zeta} \cdot \boldsymbol{\tau} + \int_\Omega \mathrm{div}(\boldsymbol{\zeta})\,\mathrm{div}(\boldsymbol{\tau}) \qquad \forall\, \boldsymbol{\zeta}, \boldsymbol{\tau} \in \mathbf{H}(\mathrm{div};\Omega)\,.$$

Moreover, let S be the subspace of $\mathbf{H}(\mathrm{div};\Omega)$ defined by

$$S := \left\{ \boldsymbol{\tau} : \ \boldsymbol{\tau}(x) = (\alpha + \gamma x_1, \beta + \gamma x_2) \ \forall\, x := (x_1, x_2) \in \Omega; \ \text{with}\, \alpha,\, \beta,\, \gamma \in \mathbb{R} \right\},$$

and let $\boldsymbol{\sigma} \in \mathbf{H}(\mathrm{div};\Omega)$ such that $\boldsymbol{\sigma}(x) = (x_1 x_2, x_1 + x_2)$ for all $x := (x_1, x_2) \in \Omega$. Apply the respective characterization theorem and find the best approximation of $\boldsymbol{\sigma}$ by elements of S, with respect to the norm induced by $\langle\cdot,\cdot\rangle$.

▸ **2.23** *Let $n \in \mathbb{N}$ and let $X := \mathcal{M}_{n\times n}(\mathbb{R})$ denote the vector space of real square matrices of order n, endowed with the inner product $\langle A, B\rangle := \mathrm{tr}(A^t B)$ for all A, $B \in X$.*

a) *Prove that $X = X_{\mathtt{sym}} \oplus X_{\mathtt{skew}}$, where*

$$X_{\mathtt{sym}} = \left\{ A \in X : \quad A^t = A \right\} \quad \text{and} \quad X_{\mathtt{skew}} = \left\{ A \in X : \quad A^t = -A \right\}.$$

b) *Let $C := (c_{ij})_{n\times n} \in X$ such that $c_{ij} = 1$ for all $i \geqslant j$ and $c_{ij} = 0$ for all $i < j$. Find the best approximations of C by matrices of $X_{\mathtt{sym}}$ and $X_{\mathtt{skew}}$, with respect to the norm induced by $\langle\cdot,\cdot\rangle$.*

▸ **2.24** *Let $(X, \|\cdot\|)$ be a normed vector space, and let $\{x_1, x_2, ..., x_n\}$ be a basis of a subspace U of X.*

a) *Prove that there exist $F_1, F_2, ..., F_n \in X'$ such that $F_j(x_i) = \delta_{ij}$ for all $i, j \in \{1, ..., n\}$.*
b) *Prove that $X = U \oplus V$, where $V := \left\{ x \in X : \quad F_j(x) = 0 \quad \forall\, j \in \{1, ..., n\} \right\}$.*

▸ **2.25**

a) *Given $(X, \langle\cdot,\cdot\rangle)$ a Hilbert space over $\mathbb{R}$ with Riesz map $\mathcal{R} : X' \to X$, define $[\cdot,\cdot] : X' \times X' \to \mathbb{R}$ by*

$$[F, G] := \langle \mathcal{R}(F), \mathcal{R}(G)\rangle \qquad \forall\, F, G \in X',$$

and prove that $(X', [\cdot,\cdot])$ is a Hilbert space.

b) *Let X be a normed vector space over $\mathbb{R}$ and denote by X'' the dual of the dual X', that is,*

$$X'' := \left\{ \mathcal{F} : X' \to \mathbb{R} : \quad \mathcal{F} \text{ is linear and bounded} \right\}.$$

Prove that for each $x \in X$ the functional $J(x) : X' \to \mathbb{R}$ defined by $J(x)(F) := F(x)$ for all $F \in X'$ is an element of X'', and that the resulting mapping $J : X \to X''$ is injective and isometric. Then, use a) to prove that if X is a Hilbert space, then J is bijective.

▸ **2.26** *Let Ω be a bounded open subset of $\mathbb{R}^n$, and let $\kappa : \Omega \to \mathbb{R}$ be a continuous function for which there exist constants M, $\beta > 0$ such that $\beta \leqslant \kappa(x) \leqslant M$ for all $x \in \Omega$. Prove, using some duality result, that for every $f \in \mathrm{L}^2(\Omega)$ there exists a unique $u \in \mathrm{H}^1(\Omega)$ such that*

$$\int_\Omega \left\{ \kappa \nabla u \cdot \nabla v + \frac{1}{\kappa} u\, v \right\} = \int_\Omega f\, v \qquad \forall\, v \in \mathrm{H}^1(\Omega),$$

and

$$\|u\|_{1,\Omega} \leqslant \left\{ \frac{M}{\min\{\beta, \frac{1}{M}\}} \right\}^{1/2} \|f\|_{0,\Omega}.$$

▸ **2.27** *Let $(X, \|\cdot\|)$ be a normed vector space, and let $\bar{B}(0,1) := \Big\{ x \in X : \quad \|x\| \leqslant 1 \Big\}$ be its closed unit ball. Moreover, given $N \in \mathbb{N}$, $A \in \mathcal{L}(X, \mathbb{R}^N)$, and $\boldsymbol{\beta} \in \mathbb{R}^N$, define the functional $(\boldsymbol{\beta}\cdot A) : X \to \mathbb{R}$ by $(\boldsymbol{\beta}\cdot A)(x) := \langle \boldsymbol{\beta}, A(x)\rangle_N$ for all $x \in X$, where $\langle\cdot,\cdot\rangle_N$ is the usual inner product of $\mathbb{R}^N$. Prove that a vector $\boldsymbol{\alpha}$ in $\mathbb{R}^N$ belongs to $\overline{A(\bar{B}(0,1))}$ if and only if $|\langle \boldsymbol{\beta}, \boldsymbol{\alpha}\rangle_N| \leqslant \|\boldsymbol{\beta}\cdot A\|_{X'}$ for all $\boldsymbol{\beta} \in \mathbb{R}^N$.*
[HINT: *For the reciprocal implication argue by contradiction*].

▸ **2.28** *Given $n \in \mathbb{N}$, consider a partition $0 = t_0 < t_1 \ldots < t_n = 1$ of $\Omega := (0,1)$ and define the subspaces of $\mathrm{L}^2(\Omega)$ given by*

$$\mathrm{H}^1_n(\Omega) := \Big\{ u \in \mathrm{L}^2(\Omega) : \quad u|_{]t_{j-1},t_j[} \in \mathrm{H}^1(]t_{j-1}, t_j[) \quad \forall\, j \in \{1, 2, \ldots, n\} \Big\},$$

and

$$S_n := \Big\{ u \in \mathrm{L}^2(\Omega) : \quad u|_{]t_{j-1},t_j[} \text{ is constant} \,\forall\, j \in \{1, 2, \ldots, n\} \Big\}.$$

Then, given $u \in \mathrm{H}^1_n(\Omega)$, find its best approximation by elements of $S_n^\perp$ with respect to the inner product

$$\langle v, w\rangle := \sum_{j=1}^n \int_{t_{j-1}}^{t_j} p_j\, v\, w \qquad \forall\, v,\, w \in \mathrm{H}^1_n(\Omega),$$

where p_j is the polynomial defined by $p_j(t) := (t - t_{j-1})$ for all $t \in]t_{j-1}, t_j[$. What happens with such best approximation if $\langle\cdot,\cdot\rangle$ is replaced by

$$\langle\!\langle v, w\rangle\!\rangle := \sum_{j=1}^n \int_{t_{j-1}}^{t_j} \Big\{ p_j\, v\, w + v'\, w' \Big\} \qquad \forall\, v,\, w \in \mathrm{H}^1_n(\Omega)?$$

▸ **2.29** *Let Ω be a bounded open subset of $\mathbb{R}^2$, and let $\mathcal{T}$ be a finite set of triangles T, whose interiors are pairwise disjoint, such that $\bar{\Omega} = \cup\{T : \quad T \in \mathcal{T}\}$. Define*

$$\mathrm{H}^1_{\mathcal{T}}(\Omega) := \Big\{ u \in \mathrm{L}^2(\Omega) : \quad u|_T \in \mathrm{H}^1(T) \quad \forall\, T \in \mathcal{T} \Big\}, \quad \text{and}$$

$$S_{\mathcal{T}} := \Big\{ u \in \mathrm{L}^2(\Omega) : \quad u|_T \quad \text{is constant} \quad \forall\, T \in \mathcal{T} \Big\}.$$

Then, given $u \in \mathrm{H}^1_{\mathcal{T}}(\Omega)$, find its best approximation by elements of $S_{\mathcal{T}}^\perp$ with respect to the inner product

$$\langle v, w\rangle := \sum_{T\in\mathcal{T}} \int_T \Big\{ v\, w + \nabla v \cdot \nabla w \Big\} \qquad \forall\, v,\, w \in \mathrm{H}^1_{\mathcal{T}}(\Omega).$$

What happens with such best approximation if $\langle\cdot,\cdot\rangle$ is replaced by

$$\langle\!\langle v, w\rangle\!\rangle := \sum_{T\in\mathcal{T}} \int_T v\, w \qquad \forall\, v,\, w \in \mathrm{H}^1_{\mathcal{T}}(\Omega)?$$

▸ **2.30** *Given $k \in \mathbb{N}$ and $a, b \in \mathbb{R}$ such that $a < b$, denote by $P_k([a,b])$ the space of polynomials of degree $\leqslant k$ defined on $[a,b]$. Equivalently, $p \in P_k([a,b])$ if and only if there exist $\alpha_0, \alpha_1, \ldots, \alpha_k \in \mathbb{R}$ such that $p(x) = \sum_{j=0}^{k} \alpha_j x^j$ for all $x \in [a,b]$. Prove that for each $n \in \mathbb{N}$ there exist $p_n \in P_n([a,b])$ and $\delta_n > 0$ such that*

$$\inf_{q \in P_{n-1}([a,b])} \int_a^b p_n(x)\Big\{x^n + q(x)\Big\}\, dx \geqslant \delta_n .$$

▸ **2.31** *Provide an alternative proof of Theorem 2.33 by following the idea suggested immediately after its proof in the text, namely, by defining the sets A_ε and B instead of B_ε.*

▸ **2.32** *Let U and V be subspaces of a normed vector space $(X, \|\cdot\|)$ over $\mathbb{R}$, such that the subspace $U \cap V$ is nontrivial, and the sets $U \backslash \bar{V}$ and $V \backslash \bar{U}$ are nonempty. Prove that there exists $F \in X'$ such that*

$$F|_{U \cap V} = 0, \quad F|_{U\backslash \bar{V}} \neq 0, \quad F|_{V \backslash \bar{U}} \neq 0, \quad \text{and} \quad \|F\| \leqslant 2 .$$

▸ **2.33** *Given $k \in \mathbb{N}$, define the finite-dimensional subspaces of $\mathrm{L}^2((-\pi,\pi))$:*

$$E_k := \langle\big\{\varphi_0, \varphi_1, \ldots, \varphi_k\big\}\rangle \qquad \text{and} \qquad F_k := \langle\big\{\psi_1, \ldots, \psi_k\big\}\rangle,$$

where $\varphi_j(x) := \cos(jx)$ and $\psi_j(x) := \sin(jx)$, for all $j \in \mathbb{N} \cup \{0\}$, for all $x \in (-\pi,\pi)$. Prove that for each $n \in \mathbb{N}$ there exist $(r_n, s_n) \in E_n \times F_n$ and $\varepsilon_n > 0$, such that

$$\inf_{(\varphi,\psi) \in E_{n-1} \times F_{n-1}} \int_{-\pi}^{\pi} \Big\{r_n + s_n\Big\}\Big\{\varphi_n - \varphi + \psi_n - \psi\Big\} \geqslant \varepsilon_n .$$

▸ **2.34** *Given normed vector spaces X and Y over $\mathbb{R}$ and $N \in \mathbb{N}$, consider subsets $X_N := \big\{x_1, x_2, \ldots, x_N\big\}$ and $Y_N := \big\{y_1, y_2, \ldots, y_N\big\}$ of X and Y, respectively, such that X_N is linearly independent. Prove that there exist $A \in \mathcal{L}(X,Y)$ and a constant $C_N > 0$, which depends on X_N, such that*

$$A(x_j) = y_j \quad \forall\, j \in \{1, 2, \ldots, N\} \quad \text{and} \quad \|A\| \leqslant C_N \sum_{j=1}^{N} \|y_j\| .$$

▸ **2.35** *Given a normed vector space $(X, \|\cdot\|)$ over $\mathbb{R}$ and $N \in \mathbb{N}$, consider N finite-dimensional subspaces $U_1, U_2, \ldots, U_N$ of X such that for each $I \subseteq \{1, 2, ..., N\}$ and for each $j \notin I$ one has $U_j \cap \sum_{i \in I} U_i = \{0\}$. Moreover, let $\big\{f_1, f_2, \ldots, f_N\big\}$ be a set of functionals such that $f_j \in U_j'$ for all $j \in \{1, 2, \cdots, N\}$. Prove that there exist $F \in X'$ and a constant $C > 0$ such that*

$$F|_{U_j} = f_j \qquad \forall\, j \in \{1, 2, ..., N\} \qquad \text{and} \qquad \|F\|_{X'} \leqslant C \max_{j \in \{1,2,\ldots,N\}} \|f_j\|_{U_j'} .$$

[A VARIANT OF THE ABOVE IS GIVEN LATER IN EXERCISE 3.65].

▸ **2.36** *Let $(X, \|\cdot\|_X)$ be a normed vector space, $n \in \mathbb{N}$, and consider $\big\{x_1, x_2, \ldots, x_n\big\} \subseteq X$ and $\boldsymbol{\beta} := (\beta_1, \beta_2, \ldots, \beta_n) \in \mathbb{R}^n$ such that*

$$|F(x_j)| \leqslant \|\beta_j\, F\|_{X'} \quad \forall\, F \in X', \quad \forall\, j \in \big\{1, 2, \ldots, n\big\} .$$

Prove that for each $x \in V := \langle\big\{x_1, x_2, \ldots, x_n\big\}\rangle$ there exists $\boldsymbol{\alpha} \in \mathbb{R}^n$ such that $\|x\|_X \leqslant \|\boldsymbol{\alpha}\|\,\|\boldsymbol{\beta}\|$, where $\|\cdot\|$ denotes the usual Euclidean norm in $\mathbb{R}^n$.

▸ **2.37** *Let $(\mathrm{H}, \langle\cdot,\cdot\rangle_{\mathrm{H}})$ and $(\mathrm{Q}, \langle\cdot,\cdot\rangle_{\mathrm{Q}})$ be real Hilbert spaces, with respective Riesz operators given by $\mathcal{R}_{\mathrm{H}}$ and $\mathcal{R}_{\mathrm{Q}}$, and define the product spaces*

$$X = \mathrm{H}^n := \left\{ \boldsymbol{\tau} = (\tau_i)_{i=\overline{1,n}} : \quad \tau_i \in \mathrm{H} \quad \forall\, i = \overline{1,n} \right\}, \quad \textit{and}$$
$$Y = \mathrm{Q}^{n\times n} := \left\{ \mathbf{u} = (u_{ij})_{n\times n} : \quad u_{ij} \in \mathrm{Q} \quad \forall\, i,j = \overline{1,n} \right\},$$

which are endowed with the inner products

$$\langle \boldsymbol{\zeta}, \boldsymbol{\tau} \rangle_X := \sum_{i=1}^{n} \langle \zeta_i, \tau_i \rangle_{\mathrm{H}} \qquad \forall\, \boldsymbol{\zeta} = (\zeta_i)_{i=\overline{1,n}},\ \boldsymbol{\tau} = (\tau_i)_{i=\overline{1,n}} \in X, \quad \textit{and}$$
$$\langle \mathbf{u}, \mathbf{v} \rangle_Y := \sum_{i,j=1}^{n} \langle u_{ij}, v_{ij} \rangle_{\mathrm{Q}} \qquad \forall\, \mathbf{u} = (u_{ij})_{n\times n},\ \mathbf{v} = (v_{ij})_{n\times n} \in Y.$$

Express the Riesz operators of X (denoted $\mathcal{R}_X$) and Y (denoted $\mathcal{R}_Y$) in terms of $\mathcal{R}_{\mathrm{H}}$ and $\mathcal{R}_{\mathrm{Q}}$, respectively.

▸ **2.38** *Given Ω a bounded open subset of $\mathbb{R}^2$, denote $\mathbf{L}^2(\Omega) := [\mathrm{L}^2(\Omega)]^2$ and define the space*

$$\mathbf{H}_{\mathrm{r,d}}(\Omega) := \Big\{ \boldsymbol{\tau} := (\tau_1, \tau_2) \in \mathbf{L}^2(\Omega) : \quad \mathrm{rot}(\boldsymbol{\tau}) := \frac{\partial \tau_2}{\partial x_1} - \frac{\partial \tau_1}{\partial x_2} \in \mathrm{L}^2(\Omega)$$
$$\textit{and} \qquad \mathrm{div}(\boldsymbol{\tau}) := \frac{\partial \tau_1}{\partial x_1} + \frac{\partial \tau_2}{\partial x_2} \in \mathrm{L}^2(\Omega) \Big\}$$

with rot *and* div *in the distributional sense, endowed with the inner product*

$$\langle \boldsymbol{\sigma}, \boldsymbol{\tau} \rangle_{\mathrm{r,d}} := \int_\Omega \Big\{ \boldsymbol{\sigma} \cdot \boldsymbol{\tau} + \mathrm{rot}(\boldsymbol{\sigma})\, \mathrm{rot}(\boldsymbol{\tau}) + \mathrm{div}(\boldsymbol{\sigma})\, \mathrm{div}(\boldsymbol{\tau}) \Big\} \qquad \forall\, \boldsymbol{\sigma},\, \boldsymbol{\tau} \in \mathbf{H}_{\mathrm{r,d}}(\Omega),$$

and induced norm $\|\cdot\|_{\mathrm{r,d}} := \langle \cdot, \cdot \rangle_{\mathrm{r,d}}^{1/2}$.

a) *Prove that for every $\boldsymbol{\tau} \in [\mathcal{D}'(\Omega)]^2$, its distributional curl and divergence reduce to $\langle \mathrm{rot}(\boldsymbol{\tau}), \varphi \rangle = \langle \boldsymbol{\tau}, \mathbf{curl}(\varphi) \rangle$ and $\langle \mathrm{div}(\boldsymbol{\tau}), \varphi \rangle = -\langle \boldsymbol{\tau}, \nabla \varphi \rangle$ for all $\varphi \in \mathcal{D}(\Omega)$, where $\mathbf{curl}(\varphi) := (\frac{\partial \varphi}{\partial x_2}, -\frac{\partial \varphi}{\partial x_1})$. In particular, prove that if $\boldsymbol{\tau} \in \mathbf{L}^2(\Omega)$ then*

$$\langle \mathrm{rot}(\boldsymbol{\tau}), \varphi \rangle = \int_\Omega \boldsymbol{\tau} \cdot \mathbf{curl}(\varphi) \quad \textit{and} \quad \langle \mathrm{div}(\boldsymbol{\tau}), \varphi \rangle = -\int_\Omega \boldsymbol{\tau} \cdot \nabla \varphi \quad \forall\, \varphi \in \mathcal{D}(\Omega).$$

b) *Use a) together with the fact that $\mathrm{L}^2(\Omega)$ and $\mathbf{L}^2(\Omega)$ are Hilbert spaces to prove that $\big(\mathbf{H}_{\mathrm{r,d}}(\Omega), \langle \cdot, \cdot \rangle_{\mathrm{r,d}}\big)$ is also a Hilbert space.*

c) *Let $a \in C(\bar{\Omega})$ and suppose there exist constants α and β such that for all $x \in \bar{\Omega}$ one has $0 < \alpha \leqslant a(x) \leqslant \beta$. Prove that for each $\mathbf{v} \in \mathrm{L}^2(\Omega)$ there exist functions $\boldsymbol{\zeta} \in \mathbf{L}^2(\Omega)$, $f \in \mathrm{L}^2(\Omega)$, and $g \in \mathrm{L}^2(\Omega)$ such that*

$$\int_\Omega \Big\{ a\, \boldsymbol{\zeta} \cdot \boldsymbol{\tau} - f\, \mathrm{rot}(\boldsymbol{\tau}) + g\, \mathrm{div}(\boldsymbol{\tau}) \Big\} = \int_\Omega \mathbf{v} \cdot \boldsymbol{\tau} \qquad \forall\, \boldsymbol{\tau} \in \mathbf{H}_{\mathrm{r,d}}(\Omega),$$

and deduce from this identity that there exists a constant $C > 0$, depending only on α, such that $\|\boldsymbol{\zeta}\|_{0,\Omega} + \|f\|_{0,\Omega} + \|g\|_{0,\Omega} \leqslant C\, \|\mathbf{v}\|_{0,\Omega}$.

▸ **2.39** *Let $\big(X_1, \|\cdot\|_1\big)$ and $\big(X_2, \|\cdot\|_2\big)$ be normed vector spaces over $\mathbb{R}$, and consider the products $X_1 \times X_2$ and $X_1' \times X_2'$ endowed with the corresponding componentwise vector sums and scalar multiplications, together with the norms*

$$\|(x_1, x_2)\| := \|x_1\|_1 + \|x_2\|_2 \qquad \forall\, (x_1, x_2) \in X_1 \times X_2 \quad \textit{and}$$
$$\|(F_1, F_2)\| := \|F_1\|_{X_1'} + \|F_2\|_{X_2'} \qquad \forall\, (F_1, F_2) \in X_1' \times X_2'.$$

a) *Prove that there exists $A \in \mathcal{L}\big((X_1 \times X_2)', X_1' \times X_2'\big)$ bijective.*

b) *Given U_1 and U_2 nondense subspaces of X_1 and X_2, respectively, prove in two different (although similar) ways that there exist nonzero functionals $F \in \big(X_1 \times X_2\big)'$, $F_1 \in X_1'$ and $F_2 \in X_2'$, such that*

$$F(u_1, 0) = F_1(u_1) = 0 \qquad \forall\, u_1 \in U_1 \quad \textit{and}$$
$$F(0, u_2) = F_2(u_2) = 0 \qquad \forall\, u_2 \in U_2.$$

c) *Extend a) and b) to the case of N normed vector spaces $X_1, \dots, X_N$ over $\mathbb{R}$ and respective nondense subspaces $U_1, \dots, U_N$.*

▸ **2.40** *Given $T > 0$, $p \in (1, +\infty)$, and a separable*[1] *Banach space V endowed with the norm $\|\cdot\|_V$, we let $L^p(0,T;V)$ be the space of functions $f : (0,T) \to V$ such that $F \circ f : (0,T) \to \mathbb{R}$ is measurable for all $F \in V'$, and*

$$\|f\|_{L^p(0,T;V)} := \left\{ \int_0^T \|f(t)\|_V^p \, dt \right\}^{1/p} < \infty .$$

In addition, given $f \in L^p(0,T;V)$, we say that $f' \in L^p(0,T;V)$ in a distributional sense, if there exists $g \in L^p(0,T;V)$ such that

$$\int_0^T F(f(t))\, \varphi'(t)\, dt = - \int_0^T F(g(t))\, \varphi(t)\, dt \qquad \forall\, (F, \varphi) \in V' \times C_0^\infty(0,T),$$

in which case we simply write $f' = g$. Assume now that f is differentiable in the classical sense in $(0,T)$, that is, there exists a continuous function $\eta : (0,T) \to V$ such that

$$\lim_{h \to 0} \frac{f(t+h) - f(t)}{h} = \eta(t) \qquad \forall\, t \in (0,T),$$

and use integration by parts, along with the continuity of each $F \in V'$, to prove that $f' = \eta$.

▸ **2.41** [THE PRESENT IS AN EXTENSION OF EXERCISE 2.32]. *Let $U_1, U_2, \dots, U_n$ be subspaces of a real normed vector space $(X, \|\cdot\|)$ such that $U := U_1 \cap U_2 \cap \dots \cap U_n$ is nontrivial, and suppose that for each $i \in \{1, 2, \dots, n\}$ one has $U_i \backslash \bar{S}_i \neq \phi$, where $S_i := \sum_{j \neq i} U_j$. Prove that there exists $F \in X'$ such that $F|_U = 0$, $F|_{U_i \backslash \bar{S}_i} \neq 0$ for each i, and $\|F\| \leq n$.*

▸ **2.42** *Given Ω a bounded open subset of $\mathbb{R}^2$, denote $\mathbf{L}^2(\Omega) := [\mathrm{L}^2(\Omega)]^2$ and define the space*

$$\mathrm{H}_{\mathfrak{L},\Delta}(\Omega) := \Big\{ v \in \mathrm{L}^2(\Omega) : \quad \mathfrak{L}(v) := \left(\frac{\partial v}{\partial x_2}, \frac{\partial^2 v}{\partial x_1 \partial x_2} \right) \in \mathbf{L}^2(\Omega) \quad \text{and} \quad \Delta v := \frac{\partial^2 v}{\partial x_1^2} + \frac{\partial^2 v}{\partial x_2^2} \in \mathrm{L}^2(\Omega) \Big\}$$

with $\mathfrak{L}$ and Δ in the distributional sense, endowed with the inner product

$$\langle v, w \rangle_{\mathfrak{L},\Delta} := \int_\Omega \Big\{ v\, w + \mathfrak{L}(v) \cdot \mathfrak{L}(w) + \Delta v\, \Delta w \Big\} \quad \forall\, v,\, w \in \mathrm{H}_{\mathfrak{L},\Delta}(\Omega),$$

and induced norm $\|\cdot\|_{\mathfrak{L},\Delta} := \langle \cdot, \cdot \rangle_{\mathfrak{L},\Delta}^{1/2}$.

a) *Prove that for every $v \in \mathcal{D}'(\Omega)$, its distributional $\mathfrak{L}$ and Δ reduce to*

$$\langle \mathfrak{L}(v), \boldsymbol{\varphi} \rangle = \langle v, \widetilde{\mathfrak{L}}(\boldsymbol{\varphi}) \rangle \qquad \forall\, \boldsymbol{\varphi} := (\varphi_1, \varphi_2) \in [\mathcal{D}(\Omega)]^2,$$

where $\widetilde{\mathfrak{L}}(\boldsymbol{\varphi}) := \frac{\partial^2 \varphi_2}{\partial x_1 \partial x_2} - \frac{\partial \varphi_1}{\partial x_2}$, and

$$\langle \Delta v, \varphi \rangle = \langle v, \Delta \varphi \rangle \qquad \forall\, \varphi \in \mathcal{D}(\Omega).$$

In particular, prove that if $v \in \mathrm{L}^2(\Omega)$ one has

$$\langle \mathfrak{L}(v), \boldsymbol{\varphi} \rangle = \int_\Omega v\, \widetilde{\mathfrak{L}}(\boldsymbol{\varphi}) \qquad \forall\, \boldsymbol{\varphi} := (\varphi_1, \varphi_2) \in [\mathcal{D}(\Omega)]^2, \quad \text{and}$$

$$\langle \Delta v, \varphi \rangle = \int_\Omega v\, \Delta \varphi \qquad \forall\, \varphi \in \mathcal{D}(\Omega).$$

---

[1] while it is not explicitly needed for the exercise, we refer to Definition 7.3 in Chapter 7 for the separability concept

b) *Use a) together with the fact that* $\mathrm{L}^2(\Omega)$ *and* $\mathbf{L}^2(\Omega)$ *are Hilbert spaces to prove that* $\big(\mathrm{H}_{\mathfrak{L},\Delta}(\Omega), \langle\cdot,\cdot\rangle_{\mathfrak{L},\Delta}\big)$ *is also a Hilbert space.*

c) *Consider in particular the square* $\Omega = (0,1)\times(0,1)$ *and the polynomial* p *defined by* $p(\mathbf{x}) := x_1 + x_2$ *for all* $\mathbf{x} := (x_1, x_2) \in \Omega$*, and define the subspace*

$$U_p := \left\{ v \in \mathrm{H}_{\mathfrak{L},\Delta}(\Omega) : \quad \int_\Omega \left(p\, v + \frac{\partial v}{\partial x_2} \right) = 0 \right\}.$$

Then, given $u \in \mathrm{H}_{\mathfrak{L},\Delta}(\Omega) \backslash U_p$*, find its best approximation by elements of* U_p *with respect to the norm* $\|\cdot\|_{\mathfrak{L},\Delta}$.

▸ **2.43** *Let* $\big(H, \langle\cdot,\cdot\rangle_H\big)$ *be a real Hilbert space, with induced norm* $\|\cdot\|_H$*, and let* $\{u_1, u_2, \ldots, u_n\}$ *be a set of linearly independent vectors in* H.

a) *Derive an algorithm that explicitly computes the best approximation – with respect to* $\|\cdot\|_H$ *– of any* $u \in H$ *by elements of the subspace*

$$V_n := \left\{ v \in H : \quad \langle v, u_1\rangle_H = \langle v, u_2\rangle_H = \cdots = \langle v, u_n\rangle_H \right\}.$$

b) *Consider the particular case* $H = L^2(\Omega)$*, with* $\Omega = (0,1)$ *and inner product* $\langle v, w\rangle_H := \int_0^1 v\, w$ *for all* $v, w \in H$*, and then apply the result from a) to compute the best approximation of the function* $u \in H$ *defined by* $u(x) := x^2$ *for all* $x \in \Omega$*, by elements of* V_2*, where*

$$u_1(x) := 1 \quad \text{and} \quad u_2(x) := x \quad \forall\, x \in \Omega\,.$$

3

Linear Operators

The objective of this chapter, one of the most extensive in this book, is to study the main aspects of the theory of bounded and unbounded linear operators. This includes a variety of concepts such as adjoint operator, annihilators, orthogonals, null spaces, ranges, and graphs, as well as most of the classical results associated with them, in particular the theorems of the bounded inverse, the closed graph, the open mapping, and the uniform boundedness principle. The material presented below is largely based on the respective developments in [6], [23], and [115].

3.1 Preliminaries

Let $(X, \|\cdot\|_X)$ and $(Y, \|\cdot\|_Y)$ be normed vector spaces over the same field $\mathbb{K}$, whose zero vectors, although certainly distinct, are both denoted by 0. A mapping $A : \mathcal{D}(A) \subseteq X \to Y$ that assigns to each $x \in \mathcal{D}(A)$ a unique vector $y \in Y$ is called an operator (or transformation) from X into Y. In this case, we write $y = A(x)$, and refer to $A(x)$ as the image of x under A. The set $\mathcal{D}(A)$ on which A acts is called the domain of the operator.

Definition 3.1. *An operator $A : \mathcal{D}(A) \subseteq X \to Y$ is said to be* LINEAR *if*

- *$\mathcal{D}(A)$ is a subspace of X.*
- *$A(\alpha\, x + \beta\, z) \,=\, \alpha\, A(x) + \beta\, A(z)$ for all $\alpha,\, \beta \in \mathbb{K}$, for all $x,\, z \in \mathcal{D}(A)$.*

Note that if $A : \mathcal{D}(A) \subseteq X \to Y$ is linear, then clearly $A(0) = 0$.

Definition 3.2. *A linear operator $A : \mathcal{D}(A) \subseteq X \to Y$ is said to be* BOUNDED *if there exists a constant $M > 0$ such that*

$$\|A(x)\|_Y \,\leqslant\, M\,\|x\|_X \qquad \forall\, x \in \mathcal{D}(A)\,.$$

Example 3.3

1. Given a real Hilbert space $(H, \langle\cdot,\cdot\rangle)$, the Riesz operator (cf. (2.15))

$$\begin{aligned} \mathcal{R} : H' &\to H \\ F &\mapsto \mathcal{R}(F) \end{aligned}$$

is linear and bounded. In fact, linearity was already established in (2.17), whereas boundedness follows from the identity (2.14) (cf. Theorem 2.14), which can be rewritten as

$$\|\mathcal{R}(F)\| \,=\, \|F\| \qquad \forall\, F \in H'\,.$$

G. N. Gatica, *Introduction to Functional Analysis*,
https://doi.org/10.1007/978-3-032-37415-8_3

2. Given a fixed constant $\kappa > 0$, for each $f \in \mathrm{L}^2(\Omega)$ we let $u \in \mathrm{H}^1(\Omega)$ be the unique element of this space such that

$$\int_\Omega \Big\{ \kappa\, u\, v + \nabla u \cdot \nabla v \Big\} = \int_\Omega f\, v \qquad \forall\, v \in \mathrm{H}^1(\Omega)\,. \tag{3.1}$$

   Then one can prove that the operator A defined by

$$\begin{aligned} A : \mathrm{L}^2(\Omega) &\to \mathrm{H}^1(\Omega) \\ f \quad &\mapsto A(f) := u \end{aligned}$$

   is linear and bounded. Indeed, it suffices to observe that the left-hand side of (3.1) defines an inner product on $\mathrm{H}^1(\Omega)$ whose induced norm is equivalent to the usual norm $\|\cdot\|_{1,\Omega}$ (cf. (1.28)), and then apply the Riesz Representation Theorem (cf. Theorem 2.14) to conclude.
3. Given a closed subspace U of a Hilbert space V, the orthogonal projectors

$$P_U : V \to U \qquad \text{and} \qquad P_{U^\perp} : V \to U^\perp$$

   defined in Section 2.2, which assign to each $x \in V$ its best approximations $P_U(x)$ and $P_{U^\perp}(x)$ by elements of U and $U^\perp$, respectively, are easily seen to be linear and bounded.

In what follows, unless explicitly stated otherwise, we assume that $\mathcal{D}(A) = X$.

Definition 3.4. *The set of all bounded linear operators from X to Y is denoted by $\mathcal{L}(X,Y)$ (or $\mathcal{B}(X,Y)$). In particular, when $X = Y$ we simply write $\mathcal{L}(X)$ instead of $\mathcal{L}(X,X)$. Moreover, note that $X' = \mathcal{L}(X,\mathbb{K})$.*

Now, on the set $\mathcal{L}(X,Y)$ we define the following operations:

$$\begin{aligned} + : \mathcal{L}(X,Y) \times \mathcal{L}(X,Y) &\to \mathcal{L}(X,Y) \\ (A,B) \qquad &\to A + B\,, \quad (A+B)(x) := A(x) + B(x) \quad \forall\, x \in X\,, \end{aligned} \tag{3.2}$$

and

$$\begin{aligned} \cdot : \mathbb{K} \times \mathcal{L}(X,Y) &\to \mathcal{L}(X,Y) \\ (\lambda, A) \quad &\to \lambda\, A\,, \qquad (\lambda\, A)(x) := \lambda\, A(x) \quad \forall\, x \in X\,. \end{aligned} \tag{3.3}$$

Then, it is easily shown that $(\mathcal{L}(X,Y), +, \cdot)$ is a vector space over $\mathbb{K}$, whose additive identity is the zero operator that assigns to each $x \in X$ the zero vector of Y.

Definition 3.5. *Given $A \in \mathcal{L}(X,Y)$ we define $\|A\|_{\mathcal{L}(X,Y)}$ as the infimum of all constants $M > 0$ satisfying the boundedness condition of A as indicated in Definition* 3.2, *that is*

$$\|A\|_{\mathcal{L}(X,Y)} := \inf \Big\{ M > 0 : \quad \|A(x)\|_Y \leqslant M\, \|x\|_X \quad \forall\, x \in X \Big\}.$$

It is easy to see that $\|\cdot\|_{\mathcal{L}(X,Y)}$ is a norm on $\mathcal{L}(X,Y)$, and hence $(\mathcal{L}(X,Y), +, \cdot, \|\cdot\|_{\mathcal{L}(X,Y)})$ is a normed vector space. Moreover, it can be shown that $\|\cdot\|_{\mathcal{L}(X,Y)}$ is also given by

$$\|A\|_{\mathcal{L}(X,Y)} = \sup_{0 \neq x \in X} \frac{\|A(x)\|_Y}{\|x\|_X} \qquad \forall\, A \in \mathcal{L}(X,Y)\,, \tag{3.4}$$

or equivalently

$$\|A\|_{\mathcal{L}(X,Y)} := \sup_{\substack{x \in X \\ \|x\|_X \leqslant 1}} \|A(x)\|_Y = \sup_{\substack{x \in X \\ \|x\|_X = 1}} \|A(x)\|_Y\,.$$

In particular, if $(X, \langle \cdot, \cdot \rangle)$ is a Hilbert space and $A \in \mathcal{L}(X)$, one can prove that (cf. Exercise 3.1)

$$\|A\|_{\mathcal{L}(X)} = \sup_{\substack{x,\, z \in X \\ \|x\|,\, \|z\| \leqslant 1}} |\langle A(x), z \rangle| .$$

In general, given $A \in \mathcal{L}(X, Y)$, we clearly have:

$$\|A(x)\|_Y \leqslant \|A\|_{\mathcal{L}(X,Y)} \|x\|_X \qquad \forall\, x \in X ,$$

and if $M > 0$ is such that $\|A(x)\|_Y \leqslant M \|x\|_X$ for all $x \in X$, then $\|A\|_{\mathcal{L}(X,Y)} \leqslant M$. It is worth noting here that the analogue of Lemma 2.25, namely the identity

$$\|x\|_X = \max_{\substack{A \in \mathcal{L}(X,Y) \\ A \neq 0}} \frac{\|A(x)\|_Y}{\|A\|_{\mathcal{L}(X,Y)}} \qquad \forall\, x \in X , \tag{3.5}$$

also holds in this case. The corresponding proof is given at the end of Section 3.2. In what follows, and when there is no risk of confusion, the respective subscripts of each norm will be omitted.

Definition 3.6. *An operator $A : X \to Y$ is said to be continuous at $x_0 \in X$ if for every sequence $\{x_n\}_{n \in \mathbb{N}}$ in X that converges to x_0, it holds that $\{A(x_n)\}_{n \in \mathbb{N}}$ converges to $A(x_0)$ in Y, that is $\|A(x_n) - A(x_0)\| \to 0$ as $n \to \infty$. Equivalently, A is continuous at x_0 if for all $\varepsilon > 0$ there exists $\delta > 0$ such that*

$$\|A(x) - A(x_0)\| < \varepsilon \quad \forall\, x \in B(x_0, \delta) := \big\{ x \in X : \quad \|x - x_0\| < \delta \big\} .$$

It is important to observe that if $A : X \to Y$ is a bounded linear operator, that is $A \in \mathcal{L}(X, Y)$, then A is continuous at every $x_0 \in X$. Indeed, given $\{x_n\}_{n \in \mathbb{N}}$ in X converging to $x_0 \in X$, we have

$$\|A(x_n) - A(x_0)\| = \|A(x_n - x_0)\| \leqslant \|A\| \, \|x_n - x_0\| ,$$

which proves that $\{A(x_n)\}_{n \in \mathbb{N}}$ converges to $A(x_0)$ in Y.

Conversely, we have the following result, somewhat surprising, which establishes that continuity at a single vector of X guarantees the boundedness of a linear operator and therefore its continuity on the whole space.

Lemma 3.7. *Let $A : X \to Y$ be a linear operator, and let $x_0 \in X$ be such that A is continuous at x_0. Then A is bounded and therefore continuous on the whole of X.*

Proof. Suppose, by contradiction, that A is not bounded. Then, for every $M > 0$ there exists $x_M \in X$ such that $\|A(x_M)\| > M \|x_M\|$. In particular, there exists a sequence $\{x_n\}_{n \in \mathbb{N}} \subseteq X$ such that $\|A(x_n)\| > n \|x_n\|$ for all $n \in \mathbb{N}$. Note that $x_n \neq 0$ for all $n \in \mathbb{N}$, since otherwise the above inequality would lead to a contradiction. Then, defining

$$z_n := \frac{x_n}{n \|x_n\|} + x_0 \qquad \forall\, n \in \mathbb{N} ,$$

we get that $\|z_n - x_0\| = \dfrac{1}{n}$, which shows that $\{z_n\}_{n \in \mathbb{N}}$ converges to x_0. However, using the linearity of A we obtain

$$A(z_n) = \frac{A(x_n)}{n \|x_n\|} + A(x_0) ,$$

and thus

$$\|A(z_n) - A(x_0)\| = \frac{\|A(x_n)\|}{n\,\|x_n\|} > 1,$$

which contradicts the continuity of A at x_0.

□

3.2 Characterization of $\mathcal{L}(X, Y)$

The following purpose is to characterize the possible completeness of the space $\mathcal{L}(X, Y)$. More precisely, we will prove that $\mathcal{L}(X, Y)$ is Banach if and only if Y is Banach. In particular, when Y is the field $\mathbb{K}$, the above will prove what was stated at the beginning of Chapter 2 regarding the fact that all duals are complete. We need the following preliminary result.

Lemma 3.8. *Let X and Y be normed vector spaces over $\mathbb{K}$, and let $x_0 \in X$ and $y_0 \in Y$ be such that $x_0 \neq 0$. Then there exists $A \in \mathcal{L}(X, Y)$ such that*

$$A(x_0) = y_0 \qquad \textit{and} \qquad \|A\| = \frac{\|y_0\|}{\|x_0\|}.$$

Proof. According to Theorem 2.23 (cf. Section 2.3.3 on other consequences of the Hahn-Banach Theorem), we know that there exists a functional $F \in X'$ such that $\|F\| = 1$ and $F(x_0) = \|x_0\|$. Then, we can define the operator $A : X \to Y$ given by

$$A(x) := \frac{F(x)}{\|x_0\|}\, y_0 \qquad \forall\, x \in X.$$

It is clear that A is linear and bounded, because F is, and moreover $A(x_0) = y_0$. In turn, we have

$$\|A\| = \sup_{\substack{x \in X\\ x \neq 0}} \frac{\|A(x)\|}{\|x\|} = \sup_{\substack{x \in X\\ x \neq 0}} \frac{|F(x)|\,\|y_0\|}{\|x\|\,\|x_0\|} = \frac{\|y_0\|}{\|x_0\|} \sup_{\substack{x \in X\\ x \neq 0}} \frac{|F(x)|}{\|x\|} = \frac{\|y_0\|}{\|x_0\|}\,\|F\| = \frac{\|y_0\|}{\|x_0\|},$$

which completes the proof.

□

We are now in a position to prove the announced result.

Theorem 3.9. *Let X and Y be normed vector spaces over the field $\mathbb{K}$. Then $\mathcal{L}(X, Y)$ is Banach if and only if Y is Banach.*

Proof. Suppose that $\mathcal{L}(X, Y)$ is Banach, and let $\{y_n\}_{n \in \mathbb{N}}$ be a Cauchy sequence in Y. Also, let $x_0 \in X$ be such that $x_0 \neq 0$. Then, applying Lemma 3.8 we deduce the existence of a sequence $\{A_n\}_{n \in \mathbb{N}} \subseteq \mathcal{L}(X, Y)$ such that

$$A_n(x_0) = y_n \quad \text{and} \quad \|A_n\| = \frac{\|y_n\|}{\|x_0\|} \qquad \forall\, n \in \mathbb{N}.$$

More precisely, according to the proof of the aforementioned lemma, we know that

$$A_n(x) := \frac{F(x)}{\|x_0\|}\, y_n \qquad \forall\, x \in X,$$

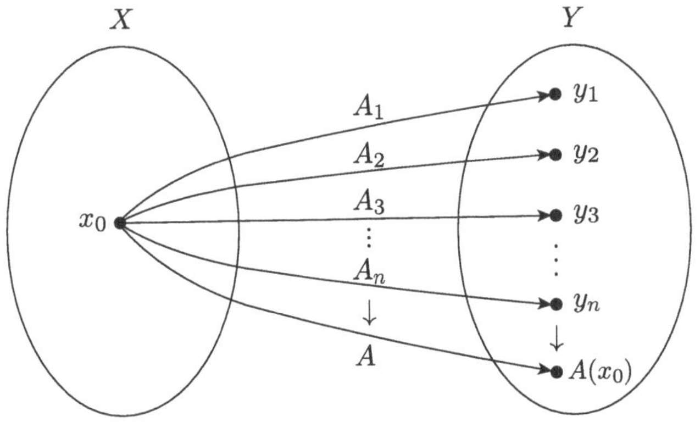

Fig. 3.1. *Illustration of the implication: $\mathcal{L}(X,Y)$ Banach $\Longrightarrow$ Y Banach*

where $F \in X'$ is such that $\|F\| = 1$ and $F(x_0) = \|x_0\|$. It follows that

$$(A_n - A_m)(x) = \frac{F(x)}{\|x_0\|}(y_n - y_m) \quad \text{and} \quad \|A_n - A_m\| = \frac{\|y_n - y_m\|}{\|x_0\|} \quad \forall\, x \in X\,,\ \forall\, n\,,\, m \in \mathbb{N}\,,$$

so that $\{A_n\}_{n\in\mathbb{N}}$ is Cauchy in $\mathcal{L}(X,Y)$. Consequently, if $A \in \mathcal{L}(X,Y)$ is the limit of this sequence, we obtain that

$$\|y_n - A(x_0)\| = \|A_n(x_0) - A(x_0)\| = \|(A_n - A)(x_0)\| \leqslant \|A_n - A\|\,\|x_0\| \overset{n\to\infty}{\longrightarrow} 0\,,$$

and therefore $\{y_n\}_{n\in\mathbb{N}}$ converges to $A(x_0)$ in Y. Figure 3.1 graphically illustrates the above reasoning. Conversely, suppose that Y is Banach and let $\{A_n\}_{n\in\mathbb{N}}$ be a Cauchy sequence in $\mathcal{L}(X,Y)$. This means that given $\varepsilon > 0$, there exists $N \in \mathbb{N}$ such that

$$\|A_n - A_m\| < \varepsilon \qquad \forall\, n\,,\, m \geqslant N\,,$$

which, by Definition 3.5 and equation (3.4), is equivalent to writing:

$$\|A_n(x) - A_m(x)\| \leqslant \varepsilon\,\|x\| \qquad \forall\, x \in X\,, \qquad \forall\, n\,,\ m \geqslant N\,. \tag{3.6}$$

It follows from (3.6) that for each $x \in X$, $\{A_n(x)\}_{n\in\mathbb{N}}$ is a Cauchy sequence and therefore convergent in Y. Thus, we can define the operator $A : X \to Y$ given by

$$A(x) := \lim_{m\to\infty} A_m(x) \qquad \forall\, x \in X\,,$$

which is clearly linear. Moreover, fixing $n \geqslant N$ and applying the triangle inequality and (3.6), we obtain

$$\|A_n(x) - A(x)\| \leqslant \varepsilon\,\|x\| + \|A_m(x) - A(x)\| \qquad \forall\, m \geqslant N\,,$$

from which, taking the limit as $m \to \infty$, it follows:

$$\|A_n(x) - A(x)\| \leqslant \varepsilon\,\|x\| \qquad \forall\, x \in X\,, \qquad \forall\, n \geqslant N\,. \tag{3.7}$$

In particular, for $\varepsilon = 1$ there exists $M \in \mathbb{N}$ such that

$$\|A_n(x) - A(x)\| \leqslant \|x\| \qquad \forall\, x \in X\,, \qquad \forall\, n \geqslant M\,,$$

and consequently

$$\|A(x)\| \leqslant \|A(x) - A_M(x)\| + \|A_M(x)\| \leqslant (1 + \|A_M\|)\,\|x\| \qquad \forall\, x \in X\,,$$

which proves that A is bounded and thus $A \in \mathcal{L}(X,Y)$. Finally, from (3.7) it follows that

$$\|A_n - A\| \leqslant \varepsilon \qquad \forall\, n \geqslant N\,,$$

and therefore $\{A_n\}_{n\in\mathbb{N}}$ converges to A in $\mathcal{L}(X,Y)$.

□

We conclude this section with the proof of identity (3.5).

Lemma 3.10. *Let X and Y be normed vector spaces over the field $\mathbb{K}$, such that Y is nontrivial. Then*

$$\|x\| = \sup_{\substack{A\in\mathcal{L}(X,Y)\\ A\neq 0}} \frac{\|A(x)\|}{\|A\|} = \max_{\substack{A\in\mathcal{L}(X,Y)\\ A\neq 0}} \frac{\|A(x)\|}{\|A\|} \qquad \forall\, x \in X\,.$$

Proof. If $x = 0$ the identity is immediate since $A(0) = 0$ for all $A \in \mathcal{L}(X,Y)$. If $x \neq 0$, we choose a vector $y_0 \in Y$, $y_0 \neq 0$, and apply Lemma 3.8. Thus, it follows that there exists $A_0 \in \mathcal{L}(X,Y)$ such that $A_0(x) = y_0$ and $\|A_0\| = \frac{\|y_0\|}{\|x\|}$, which implies that

$$\sup_{\substack{A\in\mathcal{L}(X,Y)\\ A\neq 0}} \frac{\|A(x)\|}{\|A\|} \geqslant \frac{\|A_0(x)\|}{\|A_0\|} = \|x\| \qquad \forall\, x \in X\,.$$

Finally, it is clear that

$$\sup_{\substack{A\in\mathcal{L}(X,Y)\\ A\neq 0}} \frac{\|A(x)\|}{\|A\|} \leqslant \|x\| \qquad \forall\, x \in X\,,$$

which completes the proof.

□

3.3 The adjoint operator in normed spaces

Let X, Y be normed vector spaces over the field $\mathbb{K}$, and let $A \in \mathcal{L}(X,Y)$. Then, given $G \in Y'$, we define the functional $G \circ A : X \longrightarrow \mathbb{K}$ by

$$(G \circ A)(x) := G(A(x)) \qquad \forall\, x \in X\,.$$

It is clear that $G \circ A$ is linear and bounded because both G and A are, and therefore $G \circ A \in X'$. This induces the definition of the operator

$$\begin{aligned} A' : Y' &\longrightarrow X' \\ G &\longrightarrow A'(G) := G \circ A\,, \end{aligned}$$

which is called the ADJOINT OPERATOR of A. Since most of the results on adjoint operators discussed in the rest of the chapter hold for Banach spaces, we may also refer to A' as the BANACH ADJOINT of A.

Lemma 3.11. *Given* $A \in \mathcal{L}(X,Y)$ *we have* $A' \in \mathcal{L}(Y',X')$ *and* $\|A\| = \|A'\|$.

Proof. The linearity of A' follows from the definitions of addition and scalar multiplication in the dual spaces X' and Y' (see (2.1) and (2.2) for the general case). For the boundedness of A' we first observe that

$$\|A'(G)\| := \|G \circ A\| = \sup_{\substack{x \in X \\ x \neq 0}} \frac{|G(A(x))|}{\|x\|} \leqslant \|A\| \, \|G\| \qquad \forall \, G \in Y',$$

which proves that A' is indeed bounded and $\|A'\| \leqslant \|A\|$. On the other hand, applying the consequence of the Hahn-Banach Theorem given by Lemma 2.25, we obtain

$$\begin{aligned} \|A(x)\| &= \sup_{\substack{G \in Y' \\ G \neq 0}} \frac{|G(A(x))|}{\|G\|} = \sup_{\substack{G \in Y' \\ G \neq 0}} \frac{|A'(G)(x)|}{\|G\|} \\ &\leqslant \sup_{\substack{G \in Y' \\ G \neq 0}} \frac{\|A'(G)\| \, \|x\|}{\|G\|} \leqslant \|A'\| \, \|x\| \qquad \forall \, x \in X, \end{aligned}$$

from which $\|A\| \leqslant \|A'\|$, which completes the proof.

□

The following lemma, whose proof follows directly from the definitions of the adjoint operator and of addition and scalar multiplication in $\mathcal{L}(X,Y)$ (see (3.2) and (3.3)), is left to the reader.

Lemma 3.12. *Let* X, Y, *and* Z *be normed vector spaces. Then*

i) $(A+B)' = A' + B'$ *for all* A, $B \in \mathcal{L}(X,Y)$.
ii) $(\alpha\, A)' = \alpha\, A'$ *for all* $\alpha \in \mathbb{K}$, *for all* $A \in \mathcal{L}(X,Y)$.
iii) $(AB)' = B'\, A'$ *for all* $A \in \mathcal{L}(X,Y)$, *for all* $B \in \mathcal{L}(Z,X)$.

The following example illustrates the explicit computation of an adjoint operator.

Example 3.13 Let X be the real vector space of continuous functions on $[0,1]$, endowed with the uniform norm (cf. (2.27)), and let $\{p_0, p_1, ..., p_n\}$ be the $n+1$ monomial polynomials in X, that is, $p_j(t) := t^j$ for all $t \in [0,1]$, for all $j \in \{0,1,...,n\}$. Then, we define the operator $A : X \longrightarrow X$ as

$$A(u) := \sum_{j=0}^{n} \left\{ \int_0^1 u(t)\, p_j(t)\, dt \right\} p_j \qquad \forall \, u \in X.$$

The goal is to prove that $A \in \mathcal{L}(X)$ and to find its adjoint A'. To this end, given $j \in \{0,1,...,n\}$, we define the functional $F_j : X \longrightarrow \mathbb{R}$ by

$$F_j(u) := \int_0^1 u(t)\, p_j(t)\, dt \qquad \forall \, u \in X.$$

It is clear that F_j is linear and bounded. In particular, for the boundedness we easily have

$$|F_j(u)| \leqslant \int_0^1 |u(t)|\, |p_j(t)|\, dt \leqslant \int_0^1 |p_j(t)|\, dt\, \|u\| \leqslant \|u\|,$$

which proves that $F_j \in X'$ and $\|F_j\| \leqslant 1$. Using these functionals, we can rewrite A as

$$A(u) := \sum_{j=0}^{n} F_j(u)\, p_j \qquad \forall \, u \in X,$$

which is clearly linear and bounded since each F_j is. Thus, given $G \in X'$ and $u \in X$, we obtain

$$\begin{aligned} A'(G)(u) &= (G \circ A)(u) = G(A(u)) = G\left(\sum_{j=0}^{n} F_j(u)\, p_j\right) \\ &= \sum_{j=0}^{n} F_j(u)\, G(p_j) = \sum_{j=0}^{n} G(p_j)\, F_j(u) = \left(\sum_{j=0}^{n} G(p_j)\, F_j\right)(u), \end{aligned}$$

and hence

$$A'(G) = \sum_{j=0}^{n} G(p_j)\, F_j \qquad \forall\, G \in X'.$$

3.4 The adjoint operator in Hilbert spaces

Let $(X, \langle\cdot,\cdot\rangle_X)$ and $(Y, \langle\cdot,\cdot\rangle_Y)$ be Hilbert spaces over $\mathbb{K}$, and let $A \in \mathcal{L}(X,Y)$. Given $y \in Y$, define the functional

$$\begin{aligned} X &\to \mathbb{K} \\ x &\to \langle A(x), y\rangle_Y, \end{aligned}$$

which is clearly linear and bounded. Indeed, linearity follows from the fact that A and the first component of the inner product $\langle\cdot,\cdot\rangle_Y$ are linear. Moreover, applying the Cauchy-Schwarz inequality we obtain

$$|\langle A(x), y\rangle_Y| \leqslant \|A(x)\|\,\|y\| \leqslant \|A\|\,\|x\|\,\|y\|,$$

which shows that this functional is bounded and that its norm is less than or equal to $\|A\|\,\|y\|$. Then, applying the Riesz Representation Theorem (cf. Theorem 2.14), it follows that there exists a unique $z \in X$ such that

$$\langle A(x), y\rangle_Y = \langle x, z\rangle_X \qquad \forall\, x \in X.$$

This analysis induces the definition of the operator

$$\begin{aligned} A^* : Y &\to X \\ y &\to A^*(y) := z, \end{aligned}$$

which is called the HILBERT ADJOINT of A and is characterized by the relation

$$\langle A(x), y\rangle_Y = \langle x, A^*(y)\rangle_X \qquad \forall\, x \in X, \quad \forall\, y \in Y. \tag{3.8}$$

The analogue of Lemma 3.11 is proved below.

Lemma 3.14. *Given Hilbert spaces X and Y and $A \in \mathcal{L}(X,Y)$, one has $A^* \in \mathcal{L}(Y,X)$ and $\|A^*\| = \|A\|$.*

Proof. The linearity of A^* follows from the characterization (3.8) and from the properties of inner products. Indeed, given y_1, $y_2 \in Y$ and α_1, $\alpha_2 \in \mathbb{K}$, we have

$$\begin{aligned} \langle x, A^*(\alpha_1 y_1 + \alpha_2 y_2)\rangle_X &= \langle A(x), \alpha_1 y_1 + \alpha_2 y_2\rangle_Y = \bar{\alpha}_1 \langle A(x), y_1\rangle_Y + \bar{\alpha}_2 \langle A(x), y_2\rangle_Y \\ &= \bar{\alpha}_1 \langle x, A^*(y_1)\rangle_X + \bar{\alpha}_2 \langle x, A^*(y_2)\rangle_X = \langle x, \alpha_1 A^*(y_1) + \alpha_2 A^*(y_2)\rangle_X \qquad \forall\, x \in X, \end{aligned}$$

which proves that $A^*(\alpha_1 y_1 + \alpha_2 y_2) = \alpha_1 A^*(y_1) + \alpha_2 A^*(y_2)$. Furthermore, using the Riesz Representation Theorem (cf. Theorem 2.14) in X, the identity (3.8), and the Cauchy-Schwarz inequality, we obtain

$$\|A^*(y)\| = \sup_{\substack{x \in X \\ x \neq 0}} \frac{\langle x, A^*(y)\rangle_X}{\|x\|} = \sup_{\substack{x \in X \\ x \neq 0}} \frac{\langle A(x), y\rangle_Y}{\|x\|} \leqslant \|A\| \, \|y\| \qquad \forall\, y \in Y ,$$

which shows that A^* is bounded and $\|A^*\| \leqslant \|A\|$. Proceeding in an analogous way, now using Theorem 2.14 in Y, together with identity (3.8) and the Cauchy-Schwarz inequality, one shows that $\|A\| \leqslant \|A^*\|$, which completes the proof.

□

On the other hand, note that the operators A^* and A' are related according to the following commutative diagram:

$$\begin{array}{ccc} X & \overset{A}{\underset{A^*}{\rightleftarrows}} & Y \\ \mathcal{R}_X \uparrow & & \uparrow \mathcal{R}_Y \\ X' & \overset{A'}{\longleftarrow} & Y' , \end{array}$$

that is,

$$A^* = \mathcal{R}_X \circ A' \circ \mathcal{R}_Y^{-1} ,$$

or equivalently,

$$A' = \mathcal{R}_X^{-1} \circ A^* \circ \mathcal{R}_Y .$$

Definition 3.15. *Let X be a Hilbert space, and let $A \in \mathcal{L}(X, X)$. We say that the operator A is* SELF-ADJOINT *if $A = A^*$.*

Example 3.16 Let $(X, \langle\cdot,\cdot\rangle_X)$ and $(Y, \langle\cdot,\cdot\rangle_Y)$ be Hilbert spaces, and, given $m \in \mathbb{N}$, consider sets $\{F_1, F_2, ..., F_m\} \subseteq X'$, $\{x_1, x_2, ..., x_m\} \subseteq X$, and $\{y_1, y_2, ..., y_m\} \subseteq Y$. Then we define the operators $A : X \to Y$ and $B : X \to X$ by

$$A(x) := \sum_{j=1}^{m} F_j(x)\, y_j \qquad \forall\, x \in X ,$$

and

$$B(x) := \sum_{j=1}^{m} \langle x, x_j\rangle_X \, x_j \quad \forall\, x \in X .$$

The goal of this example is to compute A^* and then to show that B is self-adjoint. Indeed, according to the Riesz Representation Theorem, for each $j \in \{1, 2, ..., m\}$ there exists a unique $z_j \in X$ such that $F_j(x) := \langle x, z_j\rangle_X$ for all $x \in X$. It follows that

$$A(x) := \sum_{j=1}^{m} \langle x, z_j\rangle_X \, y_j \qquad \forall\, x \in X , \tag{3.9}$$

and then, given $y \in Y$, we have

$$\langle A(x), y\rangle_Y = \left\langle \sum_{j=1}^{m} \langle x, z_j\rangle_X \, y_j, y \right\rangle_Y = \sum_{j=1}^{m} \langle x, z_j\rangle_X \, \langle y_j, y\rangle_Y = \left\langle x, \sum_{j=1}^{m} \langle y, y_j\rangle_Y \, z_j \right\rangle_X ,$$

from which we deduce that

$$A^*(y) := \sum_{j=1}^{m} \langle y, y_j \rangle_Y \, z_j \,. \tag{3.10}$$

On the other hand, since B is a particular case of the operator A (written in the form (3.9) with $Y = X$ and $z_j = y_j = x_j$), we obtain directly from (3.10) that

$$B^*(x) := \sum_{j=1}^{m} \langle x, x_j \rangle_X \, x_j = B(x) \,,$$

which proves that B is self-adjoint.

The Hilbert space version of Lemma 3.12, whose proof is also left to the reader, reads as follows.

Lemma 3.17. *Let X, Y, and Z be Hilbert spaces. Then*

i) $(A + B)^* = A^* + B^*$ *for all* $A,\ B \in \mathcal{L}(X, Y)$.
ii) $(\alpha A)^* = \bar{\alpha}\, A^*$ *for all* $\alpha \in \mathbb{K}$, *for all* $A \in \mathcal{L}(X, Y)$.
iii) $(AB)^* = B^* A^*$ *for all* $A \in \mathcal{L}(X, Y)$, *for all* $B \in \mathcal{L}(Z, X)$.

We emphasize that the only difference between the above properties of A^* and the corresponding ones of A' lies in ii) of Lemmas 3.12 and 3.17 in the complex case. In the latter lemma, this property is referred to as the conjugate-linearity of the Hilbert adjoint. An analogous property for the Riesz operator was established in (2.17).

3.5 The fundamental equation

3.5.1 Preliminaries

Let X and Y be normed vector spaces over the field $\mathbb{K}$, and let $A \in \mathcal{L}(X, Y)$. Given $y \in Y$, the FUNDAMENTAL EQUATION is:

$$\text{Find } x \in X \text{ such that } \quad A(x) = y \,, \tag{3.11}$$

that is, to find a preimage of y through the operator A. In order to analyze this problem, we introduce the following sets.

Definition 3.18. *The null space or kernel of A is defined as*

$$N(A) := \big\{ x \in X : \quad A(x) = 0 \big\} \,.$$

Definition 3.19. *The range of A is defined as*

$$R(A) := \big\{ y \in Y : \quad \text{there exists } x \in X \text{ such that } A(x) = y \big\} \,.$$

It is easy to see that $N(A)$ and $R(A)$ are subspaces of X and Y, respectively. Moreover, since $N(A)$ is the inverse image of $\{0\}$ under A, it follows that $N(A)$ is closed. On the other hand, it is clear that the linear operator A is INJECTIVE if and only if $N(A) = \{0\}$, which means that, whenever a solution to (3.11) exists, it is unique. In turn, the operator A is SURJECTIVE if and only if $R(A) = Y$, which means that (3.11) always has at least one solution.

The next goal is to characterize $R(A)$. To this end, consider an element $y \in R(A)$. It follows that there exists $x \in X$ such that $A(x) = y$, and then, given $G \in N(A')$, we have

$$G(y) = G(A(x)) = A'(G)(x) = 0 \,. \tag{3.12}$$

This motivates the following section.

3.5.2 Annihilators and orthogonals

We begin with the classical definitions of annihilators.

Definition 3.20. *Let X be a normed vector space, and let $S \subseteq X$. A functional $F \in X'$ is said to be an* ANNIHILATOR *of S if $F(x) = 0$ for all $x \in S$. In this case, the annihilator set of S is introduced as*

$$S^\circ := \big\{ F \in X' : \quad F(x) = 0 \quad \forall\, x \in S \big\}.$$

Definition 3.21. *Let X be a normed vector space, and let $T \subseteq X'$. An element $x \in X$ is said to be an* ANNIHILATOR *of T if $F(x) = 0$ for all $F \in T$. In this case, the annihilator set of T is introduced as*

$${}^\circ T := \big\{ x \in X : \quad F(x) = 0 \quad \forall\, F \in T \big\}.$$

According to the above, equation (3.12) is equivalent to saying that $y \in {}^\circ N(A')$, and therefore it has been proved that

$$R(A) \subseteq {}^\circ N(A'). \tag{3.13}$$

In the case where $\big(X, \langle \cdot, \cdot \rangle_X\big)$ and $\big(Y, \langle \cdot, \cdot \rangle_Y\big)$ are Hilbert spaces and $A \in \mathcal{L}\big(X, Y\big)$, the above reduces to (cf. Definition 2.9)

$$R(A) \subseteq N(A^*)^\perp. \tag{3.14}$$

Indeed, given $y \in R(A)$ there exists $x \in X$ such that $y = A(x)$. Then, if $z \in N(A^*)$ we obtain

$$\langle y, z \rangle_Y = \langle A(x), z \rangle_Y = \langle x, A^*(z) \rangle_X = \langle x, 0 \rangle_X = 0,$$

which shows that $y \in N(A^*)^\perp$. In order to deduce possible reciprocals of the inclusions (3.13) and (3.14), some additional results on annihilators and orthogonals are needed.

Lemma 3.22. *Let $\big(X, \|\cdot\|\big)$ be a normed vector space, and let $S \subseteq X$ and $T \subseteq X'$. Then S° and ${}^\circ T$ are closed subspaces of X' and X, respectively.*

Proof. We prove it only for ${}^\circ T$. The case of S° is analogous. Since $F(0) = 0$ for all $F \in X'$, it is easily seen that $0 \in {}^\circ T$. In turn, given $\alpha, \beta \in \mathbb{R}$, $x, \widetilde{x} \in {}^\circ T$ and $F \in T$, we obtain

$$F(\alpha\, x + \beta\, \widetilde{x}) = \alpha\, F(x) + \beta\, F(\widetilde{x}) = \alpha \cdot 0 + \beta \cdot 0 = 0,$$

which implies $\alpha\, x + \beta\, \widetilde{x} \in {}^\circ T$, thus proving that ${}^\circ T$ is a subspace of X. To see that ${}^\circ T$ is closed, consider a sequence $\{x_n\}_{n \in \mathbb{N}} \subseteq {}^\circ T$ that converges to $x \in X$. Then, given $F \in T$, we have:

$$|F(x)| = |F(x) - F(x_n)| = |F(x - x_n)| \leqslant \|F\| \, \|x - x_n\| \overset{n \to \infty}{\longrightarrow} 0,$$

and therefore $x \in {}^\circ T$.

□

Lemma 3.23. *Let $\big(X, \langle \cdot, \cdot \rangle_X\big)$ be a Hilbert space with Riesz operator $\mathcal{R} : X' \to X$, and let $S \subseteq X$ and $T \subseteq X'$. Then*

$$S^\perp = \mathcal{R}(S^\circ) \quad \text{and} \quad {}^\circ T = \mathcal{R}(T)^\perp.$$

Proof. Clearly,

$$\begin{aligned} x \in S^{\perp} \Leftrightarrow \quad & \langle s\,, x\rangle = 0 \qquad \forall\, s \in S \\ \Leftrightarrow \quad & \mathcal{R}^{-1}(x)\,(s) = 0 \qquad \forall\, s \in S \\ \Leftrightarrow \quad & \mathcal{R}^{-1}(x) \in S^{\circ} \\ \Leftrightarrow \quad & x \in \mathcal{R}(S^{\circ}). \end{aligned}$$

On the other hand,

$$\begin{aligned} x \in {}^{\circ}T \Leftrightarrow \quad & F(x) = 0 \qquad \forall\, F \in T \\ \Leftrightarrow \quad & \langle x, \mathcal{R}(F)\rangle = 0 \qquad \forall\, F \in T \\ \Leftrightarrow \quad & x \perp \mathcal{R}(F) \qquad \forall\, F \in T \\ \Leftrightarrow \quad & x \in \mathcal{R}(T)^{\perp}. \end{aligned}$$

□

Note that the identity $S^{\perp} = \mathcal{R}(S^{\circ})$, along with Lemma 3.22 and the fact that $\mathcal{R}$ is an isometric bijection, guarantees that $S^{\perp}$ is also a closed subspace of X.

The following result is usually known as the Bipolar Theorem.

Lemma 3.24. *Let M be a closed subspace of a normed vector space $\big(X \| \cdot \|\big)$. Then*

$$^{\circ}(M^{\circ}) \;=\; M\,.$$

Proof. We proceed by double inclusion. Given $x \in M$, it is clear that $F(x) = 0$ for all $F \in M^{\circ}$, which indicates that $x \in {}^{\circ}(M^{\circ})$, and therefore $M \subseteq {}^{\circ}(M^{\circ})$. For the other inclusion we use the counterreciprocal. Indeed, let $\widetilde{x} \in X$ be such that $\widetilde{x} \notin M$. Since M is closed, we have that $\mathrm{dist}(\widetilde{x}, M) > 0$. Then, by applying the consequence of the Hahn-Banach Theorem given by Theorem 2.26, we deduce that there exists $\widetilde{F} \in X'$ such that $\|\widetilde{F}\| = 1$, $\widetilde{F}(\widetilde{x}) = \mathrm{dist}(\widetilde{x}, M)$ and $\widetilde{F}(z) = 0$ for all $z \in M$. This implies that $\widetilde{F} \in M^{\circ}$, and since $\widetilde{F}(\widetilde{x}) \neq 0$, we obtain that $\widetilde{x} \notin {}^{\circ}(M^{\circ})$. Hence we conclude that ${}^{\circ}(M^{\circ}) \subseteq M$, which completes the proof.

□

As a corollary of Lemmas 3.23 and 3.24, we now restate Lemma 2.11 and provide an alternative proof of it.

Lemma 3.25. *Let M be a closed subspace of a Hilbert space $\big(X, \langle \cdot\,, \cdot \rangle\big)$. Then*

$$\big(M^{\perp}\big)^{\perp} \;=\; M\,.$$

Proof. Using the result of Lemma 3.24 and the identities of Lemma 3.23, we obtain

$$M \;=\; {}^{\circ}(M^{\circ}) \;=\; {}^{\circ}(\mathcal{R}^{-1}(M^{\perp})) \;=\; \big\{\mathcal{R}\big(\mathcal{R}^{-1}(M^{\perp})\big)\big\}^{\perp} \;=\; (M^{\perp})^{\perp}\,.$$

□

Next we establish a more general result than Lemma 3.24, in which we do not consider a closed subspace of the normed vector space, but only a subset of the normed vector space.

Lemma 3.26. *Let $(X, \|\cdot\|)$ be a normed vector space, and let $W \subseteq X$ and S be the closed subspace generated by W. Then $W^\circ = S^\circ$ and $S = {}^\circ(W^\circ)$.*

Proof. It suffices to show that $W^\circ = S^\circ$ since then, according to Lemma 3.24, we obtain ${}^\circ(W^\circ) = {}^\circ(S^\circ) = S$. Indeed, since $W \subseteq S$, it is clear that $S^\circ \subseteq W^\circ$. It remains to prove that $W^\circ \subseteq S^\circ$. For this, let $F \in W^\circ$, that is, $F \in X'$ and $F(x) = 0$ for all $x \in W$. Now, given $x \in \langle W \rangle$, the subspace generated by W, there exist $N \in \mathbb{N}$, scalars $\alpha_1, \alpha_2, \cdots, \alpha_N \in \mathbb{C}$, and vectors $x_1, x_2, \cdots, x_N \in W$, such that $x = \sum_{j=1}^{N} \alpha_j x_j$. It follows that $F(x) = \sum_{j=1}^{N} \alpha_j F(x_j) = \sum_{j=1}^{N} \alpha_j \cdot 0 = 0$, which shows that $F \in \langle W \rangle^\circ$. Finally, for $x \in S = \overline{\langle W \rangle}$ there exists a sequence $\{x_n\}_{n \in \mathbb{N}} \subseteq \langle W \rangle$ such that $x_n \overset{n\to\infty}{\longrightarrow} x$, and then $|F(x)| = |F(x) - F(x_n)| = |F(x - x_n)| \leqslant \|F\| \, \|x - x_n\| \overset{n\to\infty}{\longrightarrow} 0$, whence $F(x) = 0$. This shows that $F \in S^\circ$, and therefore $W^\circ \subseteq S^\circ$.

□

The analogue of the previous result for the Hilbert case is established as follows.

Lemma 3.27. *Let $(X, \langle \cdot, \cdot \rangle)$ be a Hilbert space, let $W \subseteq X$, and let S be the closed subspace generated by W. Then $W^\perp = S^\perp$ and $S = (W^\perp)^\perp$.*

Proof. Applying the identities given by Lemmas 3.23, 3.26 and 3.25, we easily deduce that

$$W^\perp = \mathcal{R}(W^\circ) = \mathcal{R}(S^\circ) = S^\perp \quad \text{and} \quad S = (S^\perp)^\perp = (W^\perp)^\perp .$$

□

We conclude this section with a couple of key results on the range of a bounded linear operator.

Theorem 3.28. *Let $(X, \|\cdot\|_X)$ and $(Y, \|\cdot\|_Y)$ be normed vector spaces, and let $A \in \mathcal{L}(X, Y)$. Then:*

i) $R(A)^\circ = N(A')$
ii) $\overline{R(A)} = {}^\circ N(A')$
iii) $R(A)$ *is closed if and only if* $R(A) = {}^\circ N(A')$.

Proof. For i) it is easily observed that

$$\begin{aligned} G \in R(A)^\circ \Leftrightarrow \quad & G \in Y' \quad \text{and} \quad G(z) = 0 \quad \forall z \in R(A) \\ \Leftrightarrow \quad & G \in Y' \quad \text{and} \quad G(A(x)) = 0 \quad \forall x \in X \\ \Leftrightarrow \quad & G \in Y' \quad \text{and} \quad A'(G)(x) = 0 \quad \forall x \in X \\ \Leftrightarrow \quad & G \in Y' \quad \text{and} \quad A'(G) = 0 \in X' \\ \Leftrightarrow \quad & G \in N(A'). \end{aligned}$$

Then, applying the second identity of Lemma 3.26, and noting that $\overline{R(A)}$ is the closed subspace generated by $R(A)$, it follows that

$${}^\circ N(A') = {}^\circ(R(A)^\circ) = \overline{R(A)},$$

which proves ii). Finally, iii) follows from ii) and from the fact that, according to Lemma 3.22, ${}^\circ T$ is a closed subspace of Y for all $T \subseteq Y'$.

□

The Hilbert space version of the previous theorem is given as follows.

Theorem 3.29. *Let* $\big(X, \langle\cdot,\cdot\rangle_X\big)$ *and* $\big(Y, \langle\cdot,\cdot\rangle_Y\big)$ *be Hilbert spaces, and let* $A \in \mathcal{L}(X,Y)$. *Then:*

i) $R(A)^\perp = N(A^*)$
ii) $\overline{R(A)} = N(A^*)^\perp$
iii) $R(A)$ *is closed if and only if* $R(A) = N(A^*)^\perp$.

Proof. Part i) can be proved directly, in an analogous way to the proof of i) in Theorem 3.28. Alternatively, if $\mathcal{R}_X : X' \to X$ and $\mathcal{R}_Y : Y' \to Y$ are the corresponding Riesz maps, let us first recall from Section 3.4 that $A' = \mathcal{R}_X^{-1} \circ A^* \circ \mathcal{R}_Y$. It easily follows that $N(A') = \mathcal{R}_Y^{-1}(N(A^*))$, and then, applying the first identity of Lemma 3.23 and Theorem 3.28, part i), we obtain

$$R(A)^\perp = \mathcal{R}_Y\left(R(A)^\circ\right) = \mathcal{R}_Y(N(A')) = \mathcal{R}_Y\left(\mathcal{R}_Y^{-1}(N(A^*))\right) = N(A^*),$$

which also proves i). Then, using now the second identity of Lemma 3.27, it follows that $N(A^*)^\perp = (R(A)^\perp)^\perp = \overline{R(A)}$, which proves ii). Finally, iii) is a consequence of ii) and of the fact that $N(A^*)^\perp$ is a closed subspace of Y.

□

In the case when Y is finite-dimensional, $R(A)$ is certainly closed, and therefore, since $N(A^*)^\perp$ is also closed, the previous theorem reduces to the identity $R(A) = N(A^*)^\perp$.

Example 3.30 This is a classical example in finite dimension, which concerns the solution of a linear system of equations. Indeed, let $\mathbb{R}^{n\times m}$ be the space of $n \times m$ matrices with real entries, and let $A \in \mathbb{R}^{n\times m}$ and $b \in \mathbb{R}^n$. Then, consider the linear system: find $x \in \mathbb{R}^m$ such that $Ax = b$. Equivalently, if we denote $X = \mathbb{R}^m$, $Y = \mathbb{R}^n$ and $\mathcal{A} : X \to Y$ the bounded linear operator induced by A, that is, $\mathcal{A}(x) := A\,x$ for all $x \in X$, then solving the system reduces to checking whether b belongs or not to $R(\mathcal{A})$. For this purpose, recall that $\mathcal{A}^* : Y \to X$ is defined by $\mathcal{A}^*(y) := A^t y$ for all $y \in Y$. Then, by virtue of Theorem 3.29, which in finite dimension reduces simply to the identity $R(\mathcal{A}) = N(\mathcal{A}^*)^\perp$, it follows that $Ax = b$ has a solution

$$\begin{aligned}
&\Leftrightarrow \quad b \in \quad R(\mathcal{A})\\
&\Leftrightarrow \quad b \in \quad N(\mathcal{A}^*)^\perp\\
&\Leftrightarrow \quad \langle b, z\rangle_{\mathbb{R}^n} = 0 \quad \forall\, z \in N(\mathcal{A}^*)\\
&\Leftrightarrow \quad \langle b, z\rangle_{\mathbb{R}^n} = 0 \quad \forall\, z \in \mathbb{R}^n \quad \text{such that} \quad A^t z = 0.
\end{aligned}$$

That is, the linear system is solvable if and only if the right-hand side b is orthogonal to all the solutions of the homogeneous transposed system. Now, in the particular case when $n = m$, that is $X = Y = \mathbb{R}^n$, we have $|A| = |A^t|$, and therefore

$$\begin{aligned}
R(\mathcal{A}) = Y \Leftrightarrow N(\mathcal{A}^*)^\perp = Y \Leftrightarrow N(\mathcal{A}^*) = \{0\}\\
\Leftrightarrow |A^t| \neq 0 \Leftrightarrow |A| \neq 0 \Leftrightarrow N(\mathcal{A}) = \{0\},
\end{aligned}$$

which proves that $\mathcal{A}$ is SURJECTIVE if and only if $\mathcal{A}$ is INJECTIVE. This result, known as the FREDHOLM ALTERNATIVE in finite dimension, can also be deduced from the Dimension Theorem in linear algebra (also called the Rank–Nullity Theorem), which states that

$$n = \dim N(\mathcal{A}) + \dim R(\mathcal{A}).$$

3.6 The inverse operator

Let X and Y be normed vector spaces, and let $A \in \mathcal{L}(X,Y)$. It is clear that the fundamental equation $A(x) = y$ has a solution for every $y \in R(A)$, and this solution is unique if $N(A) = \{0\}$. These facts lead to the following definition.

Definition 3.31. *Let $A \in \mathcal{L}(X,Y)$ such that $R(A) = Y$ and $N(A) = \{0\}$. Then, we denote by $A^{-1} : Y \to X$, and call* INVERSE *of A, the operator that assigns to each $y \in Y$ the unique $x \in X$ such that $A(x) = y$. In this case, one writes $x = A^{-1}(y)$ for all $y \in Y$.*

The main goal of this section is to discuss the Bounded Inverse Theorem, which provides sufficient conditions for the boundedness of an inverse operator. To this end, we first prove that this result is equivalent to the Closed Graph Theorem. Both theorems and the aforementioned equivalence are stated and proved later on. Before that, it is necessary to introduce the concept of a closed operator.

In what follows, $A : \mathcal{D}(A) \subseteq X \to Y$ is a linear operator, not necessarily bounded, whose domain $\mathcal{D}(A)$ is a subspace of X.

Definition 3.32. *Let $A : \mathcal{D}(A) \subseteq X \to Y$ be linear. We say that A is* CLOSED *if for every sequence $\{x_n\}_{n\in\mathbb{N}} \subseteq \mathcal{D}(A)$ such that $x_n \xrightarrow{n\to\infty} x \in X$ and $A(x_n) \xrightarrow{n\to\infty} y \in Y$, it follows necessarily that $x \in \mathcal{D}(A)$ and $A(x) = y$.*

Next, we give an equivalent definition of a closed operator. Indeed, let us first recall that the product space $X \times Y := \{(x,y) : \quad x \in X,\, y \in Y\}$, endowed with the operations:

$$\begin{aligned} + : (X \times Y) \times (X \times Y) &\longrightarrow X \times Y \\ ((x_1\,,\, y_1), (x_2\,,\, y_2)) &\longrightarrow (x_1\,,\, y_1) + (x_2\,,\, y_2) := (x_1 + x_2\,,\, y_1 + y_2) \end{aligned}$$

$$\begin{aligned} \cdot : \mathbb{R} \times (X \times Y) &\longrightarrow X \times Y \\ (\alpha\,,\, (x\,,\, y)) &\longrightarrow \alpha \cdot (x\,,\, y) := (\alpha\, x\,,\, \alpha\, y) \end{aligned}$$

is a vector space. Moreover, endowed with the product norm (or any equivalent one)

$$\|(x,y)\| := \|x\| + \|y\| \quad \forall\, (x,y) \in X \times Y\,,$$

$X \times Y$ is a normed vector space. Furthermore, if X and Y are Banach spaces, then the product $X \times Y$ is also a Banach space.

On the other hand, the graph of the linear operator $A : \mathcal{D}(A) \subseteq X \to Y$, denoted by $\mathcal{G}(A)$, is defined as the subspace of $X \times Y$ given by

$$\mathcal{G}(A) := \{(x, A(x)) : \quad x \in \mathcal{D}(A)\}\,.$$

Note that the zero element of $X \times Y$ is $(0,0)$, and since A is linear we have $0 \in \mathcal{D}(A)$ and $A(0) = 0$, which shows that $(0,0) \in \mathcal{G}(A)$.

By virtue of the above and Definition 3.32, it is easy to see that the linear operator $A : \mathcal{D}(A) \subseteq X \to Y$ is closed if and only if $\mathcal{G}(A)$ is a closed subspace of $X \times Y$. In particular, note that if A is bounded, that is, $A \in \mathcal{L}(X,Y)$, then clearly A is closed. The following example, which exhibits a closed operator that is not bounded, confirms that the converse is not necessarily true.

Example 3.33 Let $X = C([0,1])$ endowed with the uniform norm (cf. (2.27)), with which we know that X is a Banach space. Define the operator given by the classical derivative

$$\begin{aligned} A : \mathcal{D}(A) \subseteq X &\to X \\ u &\to A(u) = u'. \end{aligned}$$

Note that $\mathcal{D}(A) = C^1([0,1])$. Next, we prove that A is not bounded and that, nevertheless, A is closed. Indeed, given $n \in \mathbb{N}$, let $p_n \in X$ be the polynomial defined by $p_n(t) := t^n$ for all $t \in [0,1]$. It follows that $A(p_n) = q_n$, where $q_n(t) := n\, t^{n-1}$ for all $t \in [0,1]$, and then

$$\sup_{\substack{x \in \mathcal{D}(A) \\ x \neq 0}} \frac{\| A(x) \|}{\| x \|} \geqslant \frac{\| A(p_n) \|}{\| p_n \|} = \frac{\| q_n \|}{\| p_n \|} = n \qquad \forall\, n \in \mathbb{N},$$

from which it follows that A is not bounded. Now, to see that A is indeed closed, consider a sequence $\{x_n\}_{n \in \mathbb{N}} \subseteq \mathcal{D}(A)$ such that $x_n \overset{n\to\infty}{\longrightarrow} x \in X$ and $A(x_n) \overset{n\to\infty}{\longrightarrow} y \in X$, that is $\| x_n - x\| \overset{n\to\infty}{\longrightarrow} 0$ and $\| x_n' - y\| \overset{n\to\infty}{\longrightarrow} 0$. Then, for each $t \in [0,1]$ we have

$$\begin{aligned} \left| \int_0^t x_n'(s)\, ds - \int_0^t y(s)\, ds \right| &= \left| \int_0^t \{x_n'(s) - y(s)\}\, ds \right| \\ &\leqslant \int_0^t |x_n'(s) - y(s)|\, ds \leqslant \| x_n' - y\| \overset{n\to\infty}{\longrightarrow} 0, \end{aligned}$$

which proves that

$$\int_0^t y(s)\, ds = \lim_{n\to\infty} \int_0^t x_n'(s)\, ds.$$

Thus, using the pointwise convergence of x_n, we obtain

$$\int_0^t y(s)\, ds = \lim_{n\to\infty} \{x_n(t) - x_n(0)\} = x(t) - x(0),$$

from which

$$x(t) = x(0) + \int_0^t y(s)\, ds \qquad \forall\, t \in [0,1].$$

This identity shows that $x \in \mathcal{D}(A)$ and $A(x) := x' = y$.

We now present the Bounded Inverse Theorem and the Closed Graph Theorem, and prove that they are equivalent.

Theorem 3.34 (Bounded Inverse Theorem). *Let X and Y be Banach spaces, and let $A \in \mathcal{L}(X,Y)$ be such that $R(A) = Y$ and $N(A) = \{0\}$. Then $A^{-1} \in \mathcal{L}(Y,X)$.*

Theorem 3.35 (Closed Graph Theorem). *Let X and Y be Banach spaces, and let $A : \mathcal{D}(A) \subseteq X \to Y$ be a closed linear operator such that $\mathcal{D}(A) = X$. Then $A \in \mathcal{L}(X,Y)$.*

Theorem 3.36. *Bounded Inverse Theorem $\Leftrightarrow$ Closed Graph Theorem*

Proof.
($\Leftarrow$) Assume the Closed Graph Theorem holds, and let X and Y be Banach spaces, and $A \in \mathcal{L}(X,Y)$ such that $R(A) = Y$ and $N(A) = \{0\}$. Then, the inverse operator $A^{-1} : Y \to X$ exists, which is linear and clearly satisfies $\mathcal{D}(A^{-1}) = Y$. In order to apply the Closed Graph Theorem and conclude that $A^{-1} \in \mathcal{L}(Y,X)$, it only remains to show that A^{-1} is closed. To this end, let $\{y_n\}_{n \in \mathbb{N}} \subseteq Y$ be such that

$$y_n \overset{n\to\infty}{\longrightarrow} y \in Y \quad \text{and} \quad A^{-1}(y_n) \overset{n\to\infty}{\longrightarrow} x \in X.$$

Since A is bounded, it follows that $y_n = A(A^{-1}(y_n)) \overset{n\to\infty}{\longrightarrow} A(x)$, and by the uniqueness of limits we deduce that $y = A(x)$, or equivalently $x = A^{-1}(y)$, which proves that A^{-1} is closed.

($\Rightarrow$) Assume the Bounded Inverse Theorem holds, and let X and Y be Banach spaces. Also, let $A : \mathcal{D}(A) \subseteq X \to Y$ be linear and closed, such that $\mathcal{D}(A) = X$, and define the operator $E : \mathcal{G}(A) \to X$ by

$$E(x, A(x)) = x \quad \forall\, x \in \mathcal{D}(A) = X.$$

Since A is closed, $\mathcal{G}(A)$ is a closed subspace of the Banach space $X \times Y$, and therefore $\mathcal{G}(A)$ is also Banach. Moreover, it is clear that $R(E) = X$ and

$$N(E) = \big\{(x, A(x)) : \quad x = 0\big\} = \big\{(0,0)\big\}.$$

Furthermore,

$$\|E(x, A(x))\| = \|x\| \leqslant \|x\| + \|A(x)\| = \|(x, A(x))\| \quad \forall\,(x, A(x)) \in \mathcal{G}(A),$$

which shows that E is bounded and $\| E \| \leqslant 1$. Summarizing, $E \in \mathcal{L}(\mathcal{G}(A), X)$ and E is bijective, so the Bounded Inverse Theorem implies that $E^{-1} \in \mathcal{L}(X, \mathcal{G}(A))$. This means that there exists $C > 0$ such that

$$\| E^{-1}(x) \| \leqslant C \| x \| \qquad \forall\, x \in X,$$

that is

$$\|(x, A(x))\| = \|x\| + \|A(x)\| \leqslant C\,\|x\| \qquad \forall\, x \in X,$$

from which $\|A(x)\| \leqslant C\,\|x\|$ for all $x \in X$, thus proving that $A \in \mathcal{L}(X,Y)$.

□

Before proving either one of the two theorems (Bounded Inverse Theorem or Closed Graph Theorem), we now establish a more general version of the Bounded Inverse Theorem.

Theorem 3.37 (Improved Bounded Inverse Theorem). *Let X and Y be Banach spaces, and let $A : \mathcal{D}(A) \subseteq X \to Y$ be a closed linear operator such that $N(A) = \{0\}$ and $R(A) = Y$. Then $A^{-1} \in \mathcal{L}(Y, X)$.*

Proof. First, let us define $\| \cdot \|_A : \mathcal{D}(A) \to \mathbb{R}$ by

$$\|x\|_A := \|x\| + \|A(x)\| \quad \forall\, x \in \mathcal{D}(A),$$

which is easily seen to constitute a norm on $\mathcal{D}(A)$. Let us prove that $(\mathcal{D}(A), \| \cdot \|_A)$ is complete. Indeed, let $\{x_n\}_{n\in\mathbb{N}}$ be a Cauchy sequence in $\mathcal{D}(A)$ with respect to $\| \cdot \|_A$. It follows that $\{x_n\}_{n\in\mathbb{N}}$ and $\{A(x_n)\}_{n\in\mathbb{N}}$ are Cauchy sequences with respect to the norms of X and Y, respectively. Since X and Y are Banach spaces, we deduce that there exist $x \in X$ and $y \in Y$ such that $x_n \overset{n\to\infty}{\longrightarrow} x$ and $A(x_n) \overset{n\to\infty}{\longrightarrow} y$, which, thanks to the fact that A is closed, implies that $x \in \mathcal{D}(A)$ and $A(x) = y$, whence

$$\|x_n - x\|_A = \|x_n - x\| + \|A(x_n) - A(x)\| = \|x_n - x\| + \|A(x_n) - y\| \overset{n\to\infty}{\longrightarrow} 0.$$

Therefore, since $(\mathcal{D}(A), \| \cdot \|_A)$ is now known to be a Banach space, we define the operator $\widetilde{A} : (\mathcal{D}(A), \| \cdot \|_A) \to Y$ by $\widetilde{A}(x) = A(x)$ for all $x \in \mathcal{D}(A)$. It is clear that $\widetilde{A}$ is linear, and moreover

$$\|\widetilde{A}(x)\| = \|A(x)\| \leqslant \|x\| + \|A(x)\| = \|x\|_A \qquad \forall\, x \in \mathcal{D}(A),$$

which shows that $\widetilde{A}$ is bounded and $\|\widetilde{A}\| \leqslant 1$. Notice also that $N(\widetilde{A}) = N(A) = \{0\}$ and $R(\widetilde{A}) = R(A) = Y$. Therefore, applying the Bounded Inverse Theorem (cf. Theorem 3.34) we deduce that $\widetilde{A}^{-1} \in \mathcal{L}(Y, (\mathcal{D}(A), \|\cdot\|_A))$, which means that there exists $C > 0$ such that

$$\|\widetilde{A}^{-1}(y)\|_A \leqslant C\,\|y\| \quad \forall\, y \in Y,$$

that is,

$$\|A^{-1}(y)\| + \|y\| \leqslant C\,\|y\| \quad \forall\, y \in Y,$$

which yields $\|A^{-1}(y)\| \leqslant C\,\|y\|$ for all $y \in Y$, thus proving that $A^{-1} \in \mathcal{L}(Y, X)$.

□

It is important to note, according to the above proof, that the Improved Bounded Inverse Theorem is a consequence of the original Bounded Inverse Theorem (see Theorem 3.34). Since clearly the former implies the latter, we deduce that both theorems are equivalent to each other, and therefore also equivalent to the Closed Graph Theorem. In fact, it is easy to see that in the first implication of Theorem 3.36 it suffices to assume that $A : \mathcal{D}(A) \subseteq X \to Y$ is linear and closed, such that $R(A) = Y$ and $N(A) = \{0\}$. Indeed, given the sequence $\{y_n\}_{n\in\mathbb{N}} \subseteq Y$ such that

$$y_n \overset{n\to\infty}{\longrightarrow} y \in Y \quad \text{and} \quad A^{-1}(y_n) \overset{n\to\infty}{\longrightarrow} x \in X,$$

it follows in this case that $x \in \mathcal{D}(A)$ and $A(x) = y$, whence $x = A^{-1}(y)$, thus proving that A^{-1} is closed. In other words, a linear closed bijective operator A implies that A^{-1} is also closed. In this way, the first implication of Theorem 3.36, modified as indicated here, constitutes an alternative proof of the Improved Bounded Inverse Theorem, now starting from the Closed Graph Theorem (cf. Theorem 3.35).

The above theorems and observations are summarized in the following result.

Theorem 3.38. *Improved Bounded Inverse Theorem* $\Leftrightarrow$ *Bounded Inverse Theorem* $\Leftrightarrow$ *Closed Graph Theorem*

3.7 Two other classical theorems

In this section we prove two other classical results, namely: the Open Mapping Theorem, which, among other consequences, implies the Bounded Inverse Theorem, and the Banach-Steinhaus Theorem, also known as the Uniform Boundedness Theorem. The respective statement of the first one is as follows.

Theorem 3.39 (Open Mapping Theorem). *Let X and Y be Banach spaces, and let $A \in \mathcal{L}(X, Y)$ be such that $R(A) = Y$. Then, there exists $r > 0$ such that*

$$B_Y(0, r) \subseteq A(B_X(0, 1)). \tag{3.15}$$

We begin the analysis by proving that the inclusion (3.15) implies that A maps open sets of X onto open sets of Y, which justifies the name of the previous theorem. More precisely, we have the following result.

Lemma 3.40. *Let X and Y be Banach spaces, and let $A \in \mathcal{L}(X, Y)$. Suppose, in addition, that there exists $r > 0$ such that $B_Y(0, r) \subseteq A(B_X(0, 1))$. Then, for every open set U of X, we have that $A(U)$ is open in Y.*

Proof. Let U be an open set of X, and let $y_0 = A(x_0) \in A(U)$ with $x_0 \in U$. It follows that there exists $\delta > 0$ such that $B_X(x_0, \delta) \subseteq U$, or equivalently $x_0 + B_X(0, \delta) \subseteq U$, and hence, applying the operator A, we obtain

$$y_0 + A(B_X(0, \delta)) \subseteq A(U).$$

Now, from the hypothesis we have that $B_Y(0, \delta r) \subseteq A(B_X(0, \delta))$. Indeed,

$$\| y \| < \delta r \Rightarrow \frac{y}{\delta} \in B_Y(0, r) \Rightarrow \frac{y}{\delta} \in A(B_X(0, 1)) \Rightarrow y \in A(B_X(0, \delta)).$$

Therefore, we deduce that $y \in B_Y(y_0, \delta r) \Rightarrow y - y_0 \in B_Y(0, \delta r) \Rightarrow y - y_0 \in A(B_X(0, \delta)) \Rightarrow y \in y_0 + A(B_X(0, \delta)) \Rightarrow y \in A(U)$, which shows that $B_Y(y_0, \delta r) \subseteq A(U)$, and consequently $A(U)$ is open in Y.

□

Next we prove the Bounded Inverse Theorem using precisely the Open Mapping Theorem (cf. Theorem 3.39).

Theorem 3.41 (Bounded Inverse Theorem with proof). *Let X and Y be Banach spaces, and let $A \in \mathcal{L}(X, Y)$ be such that $R(A) = Y$ and $N(A) = \{0\}$. Then $A^{-1} \in \mathcal{L}(X, Y)$.*

Proof. Since A is surjective we may apply Theorem 3.39, thus deducing that there exists $r > 0$ such that $B_Y(0, r) \subseteq A(B_X(0, 1))$. Then, given $x \in X$ such that $\| A(x) \| < r$, we have $A(x) \in A(B_X(0, 1))$, which means that there exists $z \in X$, $\| z \| < 1$, such that $A(x) = A(z)$. Since A is injective it follows that $x = z$ and therefore $\| x \| < 1$. Now, given $x \in X$ with $x \neq 0$, we clearly have

$$\left\| A\left(\frac{\varepsilon r}{\| A(x) \|} x \right) \right\| = \varepsilon r < r \quad \forall \varepsilon \in (0, 1),$$

and hence, by the previous analysis, we obtain

$$\left\| \frac{\varepsilon r}{\|A(x)\|} x \right\| < 1 \quad \forall \varepsilon \in (0, 1),$$

or equivalently

$$\|x\| < \frac{1}{\varepsilon r} \|A(x)\| \quad \forall \varepsilon \in (0, 1).$$

Thus, taking the limit as $\varepsilon \to 1^-$, we conclude that

$$\|x\| \leqslant \frac{1}{r} \|A(x)\| \quad \forall x \in X,$$

which proves that A^{-1} is bounded and $\| A^{-1} \| \leqslant \frac{1}{r}$.

□

The following example constitutes an interesting application of the Bounded Inverse Theorem.

Example 3.42 Let X be a vector space, and let $\|\cdot\|_1$, $\|\cdot\|_2$ be norms on X such that $(X, \|\cdot\|_1)$ and $(X, \|\cdot\|_2)$ are Banach spaces. Suppose, moreover, that there exists $C > 0$ such that $\|x\|_1 \leqslant C\,\|x\|_2$ for all $x \in X$. Then $\|\cdot\|_1$ and $\|\cdot\|_2$ are equivalent, that is, there exists another constant $c > 0$ such that

$$\|x\|_2 \leqslant c\,\|x\|_1 \quad \forall\, x \in X.$$

To see this, consider the identity operator $I : (X, \|\cdot\|_2) \to (X, \|\cdot\|_1)$, which is clearly linear, bijective and bounded, since

$$\|\, I(x)\,\|_1 = \|x\|_1 \leqslant C\,\|x\|_2 \quad \forall\, x \in X.$$

Applying Theorem 3.34 we deduce that I^{-1} is also bounded, which means that there exists $c > 0$ such that

$$\|\, I^{-1}(x)\,\|_2 = \|x\|_2 \leqslant c\,\|x\|_1 \quad \forall\, x \in X.$$

The next goal is the proof of the Open Mapping Theorem, which is split into Lemmas 3.44 and 3.45 proved below. The classical Baire Category Theorem, stated below, is used in the proof of the first one.

Theorem 3.43 (Baire Category Theorem). *Let X be a complete metric space, and let $\{X_n\}_{n\in\mathbb{N}}$ be a sequence of closed subsets of X such that $\overset{\circ}{X_n} = \varnothing$ for all $n \in \mathbb{N}$. Then,*

$$\overset{\circ}{\overbrace{\bigcup_{n\in\mathbb{N}} X_n}} = \varnothing\,.$$

It is important to note here that this result is usually applied in its contrapositive form. In particular, if X is a complete metric space and $\{X_n\}_{n\in\mathbb{N}}$ is a sequence of closed subsets of X such that $\bigcup_{n\in\mathbb{N}} X_n = X$, then necessarily there exists $k \in \mathbb{N}$ such that $\overset{\circ}{X_k} \neq \varnothing$.

We now proceed to state and prove the lemmas that imply Theorem 3.39. When there is no risk of confusion we omit the subscripts of the open balls.

Lemma 3.44. *Let X and Y be Banach spaces, and let $A : X \to Y$ be linear such that $R(A) = Y$. Then, there exists $r > 0$ such that $B(0, 2r) \subseteq \overline{A(B(0,1))}$.*

Proof. For each $n \in \mathbb{N}$ define $Y_n = n\,\overline{A(B(0,1))}$. Since A is surjective, given $y \in Y$ there exists $x \in X$ such that $A(x) = y$. If $N \in \mathbb{N}$ is such that $\|x\| < N$, then

$$y = N\,A\Big(\frac{x}{N}\Big) = N\,A(z) \quad \text{with} \quad \|z\| < 1\,,$$

which shows that $y \in Y_N$ and therefore $Y = \bigcup_{n\in\mathbb{N}} Y_n$. Thus, applying Theorem 3.43 we deduce that there exists $k \in \mathbb{N}$ such that $\overset{\circ}{Y_k} \neq \varnothing$, or equivalently, $\overset{\circ}{\overline{A(B(0,1))}} \neq \varnothing$. It follows that there exist $y_0 \in Y$ and $r > 0$ such that

$$B(y_0, 4r) \subseteq \overline{A(B(0,1))}. \tag{3.16}$$

In particular, since clearly $y_0 \in \overline{A(B(0,1))}$, there exists a sequence $\{z_n\}_{n\in\mathbb{N}} \subseteq B(0,1)$ such that $y_0 = \lim_{n\to\infty} A(z_n)$, and hence $(-y_0) = \lim_{n\to\infty} A(-z_n)$, which shows that $(-y_0)$ also belongs to $\overline{A(B(0,1))}$. From this fact and the inclusion (3.16) we obtain

$$B(0, 4r) \subseteq \overline{A(B(0,1))} + \overline{A(B(0,1))} = 2\,\overline{A(B(0,1))}. \tag{3.17}$$

Indeed, $z \in B(0,4r) \Rightarrow z = z + y_0 + (-y_0)$ with $(z + y_0) \in B(y_0, 4r) \subseteq \overline{A(B(0,1))}$ and $(-y_0) \in \overline{A(B(0,1))}$. Moreover, it is easy to see that the equality in (3.17) follows from the convexity of $\overline{A(B(0,1))}$. Finally, from (3.17) we deduce that $B(0,2r) \subseteq \overline{A(B(0,1))}$, which completes the proof.

□

Lemma 3.45. *Let X and Y be Banach spaces, and let $A \in \mathcal{L}(X,Y)$. Suppose there exists $r > 0$ such that $B(0,2r) \subseteq \overline{A(B(0,1))}$. Then,* $\quad B(0,r) \subseteq A(B(0,1))$.

Proof. Let $y \in Y$ such that $\|y\| < r$. We must prove that there exists $x \in X$ such that $\|x\| < 1$ and $A(x) = y$. Let $\delta > 0$ be such that $y_\delta := (1+\delta)y \in B(0,r)$. Since $2y_\delta \in B(0,2r) \subseteq \overline{A(B(0,1))}$, it follows that there exists $\tilde{z} \in X$, $\|\tilde{z}\| < 1$, such that $\|2y_\delta - A(\tilde{z})\| < r$, or, defining $z_1 = \frac{1}{2}\tilde{z}$, we have $\|z_1\| < 1/2$ and $\|y_\delta - A(z_1\|) < r/2$. Similarly, since $4(y_\delta - A(z_1)) \in B(0,2r)$, there exists another $\tilde{z} \in X$, $\|\tilde{z}\| < 1$, such that

$$\|4(y_\delta - A(z_1)) - A(\tilde{z})\| < r,$$

whence, defining $z_2 = \frac{1}{4}\tilde{z}$, we obtain $\|z_2\| < 1/4$ and $\|y_\delta - A(z_1) - A(z_2)\| < r/4$. Proceeding in this way we construct by induction a sequence $\{z_n\}_{n\in\mathbb{N}} \subseteq X$ such that $\|z_n\| < \dfrac{1}{2^n}$ and $\|y_\delta - A(x_n)\| < \dfrac{r}{2^n}$ for all $n \in \mathbb{N}$, where $x_n = \sum\limits_{j=1}^{n} z_j$. Since the series $\sum\limits_{n=1}^{\infty} \dfrac{1}{2^n}$ is convergent, it follows that $\{x_n\}_{n\in\mathbb{N}}$ is Cauchy in X, and therefore there exists $x_\delta \in X$ such that $x_n \overset{n\to\infty}{\longrightarrow} x_\delta$. Moreover,

$$\|x_\delta\| = \lim_{n\to\infty} \|x_n\| = \lim_{n\to\infty} \left\| \sum_{j=1}^{n} z_j \right\| \leqslant \lim_{n\to\infty} \sum_{j=1}^{n} \|z_j\| \leqslant \lim_{n\to\infty} \sum_{j=1}^{n} \frac{1}{2^j} = 1,$$

that is, $\|x_\delta\| \leqslant 1$, and from the relation $\|y_\delta - A(x_n)\| < r/2^n$ for all $n \in \mathbb{N}$ we conclude that $y_\delta = A(x_\delta)$, that is $y = A\left(\dfrac{x_\delta}{1+\delta}\right)$, or equivalently $y = A(x)$ with $x = \dfrac{x_\delta}{1+\delta}$ such that

$$\|x\| = \frac{1}{1+\delta}\|x_\delta\| \leqslant \frac{1}{1+\delta} < 1.$$

□

We finish this section with the Banach-Steinhaus Theorem, which is also a consequence of Theorem 3.43.

Theorem 3.46 (Banach-Steinhaus Theorem). *Let X be a Banach space, let Y be a normed space, and let $\{A_i\}_{i\in I}$ be an arbitrary family (not necessarily countable) in $\mathcal{L}(X,Y)$. Suppose that*

$$\sup_{i\in I} \|A_i(x)\| < \infty \qquad \forall x \in X.$$

Then, there exists $M > 0$ such that $\|A_i\| \leqslant M$ for every $i \in I$.

Proof. For each $n \in \mathbb{N}$ define the set

$$X_n = \{x \in X : \|A_i(x)\| \leqslant n \;\forall i \in I\}.$$

It is easy to see that X_n is closed. Indeed, let $\{x_k\}_{k\in\mathbb{N}} \subseteq X_n$ be such that $x_n \overset{n\to\infty}{\to} x \in X$. Then, for every $i \in I$ we have:

$$\|A_i(x)\| = \lim_{k\to\infty} \|A_i(x_k)\| \leqslant \lim_{k\to\infty} n = n,$$

which proves that $x \in X_n$. Alternatively, the fact that X_n is closed follows from the identity $X_n = \bigcap_{i\in I} A_i^{-1}(\bar{B}(0,n))$. Now, since $\sup_{i\in I} \|A_i(x)\| < \infty$ for all $x \in X$, it follows that each $x \in X$ belongs to some set X_n, and therefore $X = \bigcup_{n\in\mathbb{N}} X_n$. By Theorem 3.43, there exists $N \in \mathbb{N}$ such that $\overset{\circ}{X}_N \neq \emptyset$, which means that there exist $x_0 \in X$ and $r > 0$ such that $B(x_0, r) \subseteq X_N$. Thus, given $x \in X, x \neq 0$, define $z = x_0 + \dfrac{r}{2\|x\|}x$, which belongs to the ball $B(x_0, r)$, and therefore to X_N, since $\|z - x_0\| = r/2 < r$. According to the above and to the fact that, obviously, $x_0 \in X_N$, we have $\|A_i(z)\| \leqslant N$ and $\|A_i(x_0)\| \leqslant N$ for all $i \in I$, which implies that

$$\begin{aligned}\|A_i(x)\| &= \left\|A_i\left(\frac{2\|x\|}{r}(z - x_0)\right)\right\| = \frac{2\|x\|}{r}\,\|A_i(z - x_0)\| \\ &\leqslant \frac{2\|x\|}{r}\Big\{\|A_i(z)\| + \|A_i(x_0)\|\Big\} \leqslant M\,\|x\| \quad \forall x \in X,\ \forall i \in I,\end{aligned}$$

with the constant $M = \dfrac{4N}{r}$.

□

Two interesting corollaries of the Banach-Steinhaus Theorem (cf. Theorem 3.46) are given by the following lemmas.

Lemma 3.47. *Let X and Y be normed spaces such that X is Banach, and let $\{A_n\}_{n\in\mathbb{N}}$ be a sequence in $\mathcal{L}(X,Y)$ such that $\{A_n(x)\}_{n\in\mathbb{N}}$ is convergent in Y for each $x \in X$. Then*

a) $\sup_{n\in\mathbb{N}} \|A_n\| < +\infty$.
b) *The operator $A : X \to Y$ defined by $A(x) := \lim_{n\to\infty} A_n(x)$ for all $x \in X$, belongs to $\mathcal{L}(X,Y)$.*
c) $\|A\| \leqslant \liminf_{n\to\infty} \|A_n\|$.

Proof. Since for each $x \in X$ the sequence $\{A_n(x)\}_{n\in\mathbb{N}}$ is convergent, it is bounded in Y, that is

$$\sup_{n\in\mathbb{N}} \|A_n(x)\| < +\infty \quad \forall x \in X,$$

and therefore, a direct application of Theorem 3.46 implies a). This means that there exists $M > 0$ such that

$$\|A_n(x)\| \leqslant M\|x\| \quad \forall x \in X,\ \forall n \in \mathbb{N},$$

whence, taking the limit as $n \to \infty$, we obtain $\|A(x)\| \leqslant M\,\|x\|$ for all $x \in X$. Since A is clearly linear (by the linearity of each A_n), the above proves that $A \in \mathcal{L}(X,Y)$, that is b). Finally, from the relation $\|A_n(x)\| \leqslant \|A_n\|\,\|x\|$ for all $x \in X$, for all $n \in \mathbb{N}$, we obtain

$$\begin{aligned}\|A(x)\| &= \lim_{n\to\infty} \|A_n(x)\| = \liminf_{n\to\infty} \|A_n(x)\| \\ &\leqslant \liminf_{n\to\infty} \|A_n\|\,\|x\| = \Big\{\liminf_{n\to\infty} \|A_n\|\Big\}\,\|x\| \qquad \forall x \in X,\end{aligned}$$

which proves c).

□

Lemma 3.48. *Let S be a subset of a normed vector space X over $\mathbb{R}$, and suppose that for each $F \in X'$ the set $\{F(x) : x \in S\}$ is bounded in $\mathbb{R}$. Then S is bounded, that is, there exists $C > 0$ such that $\|x\| \leqslant C$ for all $x \in S$.*

Proof. We apply Theorem 3.46 to the following context of spaces

$$X \leftarrow X', \quad Y \leftarrow \mathbb{R}, \quad I \leftarrow S,$$

and for each $x \in S$ we define the operator $A_x : X' \to \mathbb{R}$ by $A_x(F) := F(x)$ for all $F \in X'$. According to the hypothesis we have

$$\sup_{x \in S} |A_x(F)| = \sup_{x \in S} |F(x)| < +\infty,$$

and then Theorem 3.46 implies that there exists $\widetilde{C} > 0$ such that

$$|A_x(F)| = |F(x)| \leqslant \widetilde{C}\,\|F\| \qquad \forall\, F \in X',\ \forall\, x \in S.$$

Thus, applying a consequence of the Hahn-Banach Theorem (see Lemma 2.25), it follows that for each $x \in S$

$$\|x\| = \sup_{\substack{F \in X' \\ F \neq 0}} \frac{|F(x)|}{\|F\|} \leqslant \widetilde{C},$$

which shows the boundedness of S.

□

Some brief historical remarks regarding the fundamental theorems already discussed in this chapter, all of which appeared during the early development of Banach space theory, are now in order. We begin by recalling that a preliminary version of the Banach-Steinhaus Theorem was originally published in [13], a joint work by Banach and the Polish mathematician Hugo Steinhaus. Later on, in the monograph [12] authored by Banach, a generalization of this result, based on the Baire Category Theorem (cf. Theorem 3.43), was established. Needless to say, this corresponds essentially to our present Theorem 3.46. In turn, the Open Mapping Theorem (cf. Theorem 3.39), which also follows from the Baire Category Theorem, as shown in the proof of Lemma 3.44, the Bounded Inverse Theorem (cf. Theorem 3.34), which constitutes a direct consequence of the former, as shown in the proof of Theorem 3.41, and the Closed Graph Theorem (cf. Theorem 3.35) were first systematically presented in the monograph [12] as well.

3.8 Uniform boundedness without Baire Category Theorem

In this section we provide the details of an alternative proof of the Banach-Steinhaus Theorem (cf. Theorem 3.46) which does not make use of Theorem 3.43 and, therefore, is self-contained. This result was recently published in [117]. We start with the following preliminary lemma.

Lemma 3.49. *Let X and Y be normed vector spaces, and let $A \in \mathcal{L}(X,Y)$. Then, for every $x \in X$ and for every $r > 0$ we have*

$$\sup_{y \in B(x,r)} \|A(y)\| \geqslant \|A\|\, r.$$

Proof. Let $x \in X$ and $r > 0$. Then, for each $z \in B(0, r)$ we have

$$\begin{aligned} \|A(z)\| &= \tfrac{1}{2}\|A(z) + A(z)\| = \tfrac{1}{2}\|A(x+z) - A(x-z)\| \\ &\leqslant \tfrac{1}{2}\left\{\|A(x+z)\| + \|A(x-z)\|\right\} \leqslant \sup_{y \in B(x,r)} \|A(y)\|, \end{aligned}$$

and taking the supremum over $z \in B(0, r)$, we get

$$\sup_{z \in B(0,r)} \|A(z)\| \leqslant \sup_{y \in B(x,r)} \|A(y)\|$$

Thus, it suffices to see that

$$\sup_{z \in B(0,r)} \|A(z)\| = r \sup_{z \in B(0,r)} \frac{1}{r}\|A(z)\| = r \sup_{z \in B(0,1)} \|A(z)\| = r\|A\|.$$

□

Next, we state again and prove, using only Lemma 3.49, the Uniform Boundedness Theorem (or Banach-Steinhaus Theorem) .

Theorem 3.50. *Let X and Y be normed vector spaces, with X Banach, and let $\{A_i\}_{i \in I}$ be a family, not necessarily countable, in $\mathcal{L}(X, Y)$. Suppose that*

$$\sup_{i \in I} \|A_i(x)\| < \infty \quad \forall x \in X.$$

Then, there exists $M > 0$ such that

$$\|A_i(x)\| \leqslant M\|x\| \qquad \forall x \in X, \quad \forall i \in I.$$

Proof. Suppose, by contradiction, that $\sup_{i \in I} \|A_i\| = +\infty$. Then there exists a sequence $\{A_n\}_{n \in \mathbb{N}} \subseteq \{A_i\}_{i \in I}$, such that $\|A_n\| \geqslant 4^n$ for all $n \in \mathbb{N}$. Applying the previous lemma to $A = A_1$, $x = 0$ and $r = 3^{-1}$, we have

$$\sup_{y \in B(0,3^{-1})} \|A_1(y)\| \geqslant 3^{-1}\|A_1\|,$$

and then, by definition of the supremum, given $\varepsilon > 0$, there exists $x_1 \in B(0, 3^{-1})$ such that

$$\|A_1(x_1)\| + \varepsilon \geqslant \sup_{y \in B(0,3^{-1})} \|A_1(y)\| \geqslant 3^{-1}\|A_1\|.$$

In particular, for $\varepsilon = 3^{-2}\|A_1\|$, we get

$$\|A_1(x_1)\| \geqslant \frac{2}{3} 3^{-1}\|A_1\| \quad \text{and} \quad \|x_1 - x_0\| < 3^{-1},$$

with $x_0 = 0$. In turn, applying again the lemma to $A = A_2$, $x = x_1$ and $r = 3^{-2}$, we obtain

$$\sup_{y \in B(x_1,3^{-2})} \|A_2(y)\| \geqslant 3^{-2}\|A_2\|,$$

and thus, given $\varepsilon = 3^{-3}\|A_2\|$, there exists $x_2 \in B(x_1, 3^{-2})$ such that

$$\|A_2(x_2)\| + 3^{-3}\|A_2\| \geqslant 3^{-2}\|A_2\|,$$

that is,

$$\|A_2(x_2)\| \geqslant \frac{2}{3}3^{-2}\|A_2\| \quad \text{and} \quad \|x_2 - x_1\| < 3^{-2}.$$

Following this inductive reasoning we construct a sequence $\{x_n\}_{n\in\mathbb{N}} \subseteq X$ such that

$$\|A_n(x_n)\| \geqslant \frac{2}{3}3^{-n}\|A_n\| \quad \text{and} \quad \|x_n - x_{n-1}\| < 3^{-n} \quad \forall\, n \in \mathbb{N}.$$

Let us now show that $\{x_n\}_{n\in\mathbb{N}}$ is a Cauchy sequence. Indeed, given $n, k \in \mathbb{N}$ with $n > k$, we have

$$\begin{aligned}
\|x_n - x_k\| &\leqslant \|x_{k+1} - x_k\| + \|x_{k+2} - x_{k+1}\| + \cdots + \|x_n - x_{n-1}\| \\
&\leqslant 3^{-(k+1)} + 3^{-(k+2)} + \cdots + 3^{-(k+n-k)} \\
&= 3^{-k}\left(3^{-1} + 3^{-2} + \cdots + 3^{-(n-k)}\right) \\
&= 3^{-k}\frac{1}{2}\left\{1 - \left(\frac{1}{3}\right)^{n-k}\right\} \leqslant \frac{1}{2}3^{-k} \overset{k\to\infty}{\longrightarrow} 0.
\end{aligned}$$

Then, since X is Banach, there exists $\bar{x} \in X$ such that $\|x_n - \bar{x}\| \overset{n\to\infty}{\longrightarrow} 0$. Moreover, from the same estimate above it follows that for all $n, k \in \mathbb{N}$

$$\|x_{n+k} - x_n\| \leqslant \frac{1}{2}3^{-n},$$

and hence, keeping $n \in \mathbb{N}$ fixed and letting $k \to \infty$, we get

$$\|\bar{x} - x_n\| \leqslant \frac{1}{2}3^{-n} \quad \forall\, n \in \mathbb{N}.$$

Finally, we conclude that

$$\begin{aligned}
\|A_n(\bar{x})\| &= \|A_n(x_n) - A_n(x_n - \bar{x})\| \geqslant \|A_n(x_n)\| - \|A_n\|\,\|x_n - \bar{x}\| \\
&\geqslant \frac{2}{3}3^{-n}\|A_n\| - \|A_n\|\frac{1}{2}3^{-n} = \frac{1}{6}3^{-n}\|A_n\| \geqslant \frac{1}{6}\left(\frac{4}{3}\right)^n \overset{n\to\infty}{\longrightarrow} \infty,
\end{aligned}$$

which contradicts the fact that $\sup_{i\in I}\|A_i(\bar{x})\| < +\infty$.

□

3.9 Other consequences of the Open Mapping Theorem

In this section, we show some additional results that follow from the Open Mapping Theorem (cf. Theorem 3.39). They relate to certain properties of boundedness of the sum of closed subspaces and the connection between right or left inverses and the existence of topological complements . We start with the following lemma.

Lemma 3.51. *Let X be a Banach space, and let U and V be closed subspaces of X such that $U + V$ is closed. Then there exists $C > 0$ such that for each $z \in U + V$, there exist $x \in U$ and $y \in V$ satisfying*

$$z = x + y, \quad \|x\| \leqslant C\|z\| \quad \text{and} \quad \|y\| \leqslant C\|z\|.$$

Proof. Consider the spaces $U \times V$ and $U + V$ equipped with the product norm

$$|||(x, y)||| := \|x\| + \|y\| \qquad \forall\, (x, y) \in U \times V,$$

and the usual norm $\|\cdot\|$ on X, respectively. Define $T : U \times V \to U + V$ by

$$T(x, y) := x + y \quad \forall (x, y) \in U \times V.$$

It is clear that T is linear and surjective. Moreover, since

$$\|T(x, y)\| = \|x + y\| \leqslant \|x\| + \|y\| =: |||(x, y)|||,$$

T is obviously bounded. Since $U \times V$ and $U + V$ are Banach spaces, a direct application of the Open Mapping Theorem (cf. Theorem 3.39) implies that there exists $r > 0$ such that

$$B_{U+V}(0, r) \subseteq T(B_{U\times V}(0, 1)).$$

This means that for all $\bar{z} \in U + V$ with $\|\bar{z}\| < r$, there exists $(\bar{x}, \bar{y}) \in U \times V$ such that

$$|||(\bar{x}, \bar{y})||| := \|\bar{x}\| + \|\bar{y}\| < 1 \quad \text{and} \quad \bar{z} = T(\bar{x}, \bar{y}) := \bar{x} + \bar{y}.$$

In particular, given $z \in U + V$, $z \neq 0$, and $\delta \in (0, 1)$, we have

$$\left\| \delta r \frac{z}{\|z\|} \right\| = \delta r < r,$$

so that there exists $(\bar{x}, \bar{y}) \in U \times V$ such that

$$\|\bar{x}\| + \|\bar{y}\| < 1, \qquad \delta r \frac{z}{\|z\|} = \bar{x} + \bar{y}.$$

It then follows that $z = x + y$, where

$$x := \frac{\|z\|}{\delta r}\bar{x} \in U, \quad y := \frac{\|z\|}{\delta r}\bar{y} \in V,$$

and

$$\|x\| < \frac{1}{\delta r}\|z\|, \quad \|y\| < \frac{1}{\delta r}\|z\|,$$

which completes the proof with $C := 1/\delta r$.

□

Under the same hypotheses of the previous lemma, it is further shown that the distance to the intersection of two closed subspaces of a Banach space is equivalent to the sum of the distances to these subspaces.

Lemma 3.52. *Let X be a Banach space, and let U and V be closed subspaces of X such that $U + V$ is closed. Then there exist C_1, $C_2 > 0$ such that for every $x \in X$, we have*

$$C_1 \operatorname{dist}(x, U \cap V) \leqslant \operatorname{dist}(x, U) + \operatorname{dist}(x, V) \leqslant C_2 \operatorname{dist}(x, U \cap V).$$

Proof. First, note that since $U \cap V \subseteq U$ and $U \cap V \subseteq V$, it follows that

$$\operatorname{dist}(x, U) \leqslant \operatorname{dist}(x, U \cap V) \quad \text{and} \quad \operatorname{dist}(x, V) \leqslant \operatorname{dist}(x, U \cap V) \quad \forall\, x \in X,$$

which implies the second inequality with $C_2 = 2$. Next, given $x \in X$ and $\varepsilon > 0$, there exist $u \in U$ and $v \in V$ such that

$$\|x - u\| \leqslant \operatorname{dist}(x, U) + \varepsilon, \quad \text{and} \quad \|x - v\| \leqslant \operatorname{dist}(x, V) + \varepsilon. \tag{3.18}$$

Applying the previous lemma to $z := u - v \in U + V$, there exist $\widetilde{u} \in U$ and $\widetilde{v} \in V$ such that

$$u - v = \widetilde{u} + \widetilde{v}, \quad \|\widetilde{u}\| \leqslant C\,\|u - v\|, \quad \text{and} \quad \|\widetilde{v}\| \leqslant C\,\|u - v\|.$$

In particular, $u - \widetilde{u} = v + \widetilde{v} \in U \cap V$, and thus

$$\begin{aligned}\operatorname{dist}(x, U \cap V) &\leqslant \|x - (u - \widetilde{u})\| \leqslant \|x - u\| + \|\widetilde{u}\| \\ &\leqslant \|x - u\| + C\,\|u - v\| \leqslant (1 + C)\,\|x - u\| + C\,\|x - v\|,\end{aligned}$$

and using (3.18), it follows that

$$\operatorname{dist}(x, U \cap V) \leqslant (1 + C)\,\big\{\operatorname{dist}(x, U) + \operatorname{dist}(x, V)\big\} + (1 + 2C)\varepsilon.$$

Letting $\varepsilon \to 0^+$, we conclude the first inequality with $C_1 = 1/(1 + C)$.

□

The converse of Lemma 3.52 is also true. More precisely, the following result holds.

Lemma 3.53. *Let U and V be closed subspaces of a Banach space X such that*

$$\operatorname{dist}(x, U \cap V) \leqslant C\,\big\{\operatorname{dist}(x, U) + \operatorname{dist}(x, V)\big\} \quad \forall\, x \in X\,.$$

Then $U + V$ is closed.

Proof. Let us consider the operator $T : X \longrightarrow X/U$ defined by $T(x) = [x]$ for all $x \in X$, which is clearly linear, bounded, and surjective (concepts related to quotient spaces are reviewed later in Section 3.10). Moreover, define $S := T|_V : V \longrightarrow X/U$. It follows that

$$N(S) := \big\{v \in V : [v] = [0]\big\} = \big\{v \in V : v \in U\big\} = U \cap V\,,$$

where $[0]$ denotes the equivalence class of the zero element. For each $v \in V$, we have

$$\begin{aligned}\operatorname{dist}(v, N(S)) &= \operatorname{dist}(v, U \cap V) \leqslant C\,\big\{\operatorname{dist}(v, U) + \operatorname{dist}(v, V)\big\} \\ &= C\operatorname{dist}(v, U) = C\,\|[v]\| = C\,\|S(v)\|\,,\end{aligned}$$

which implies

$$\operatorname{dist}(v, N(S)) \leqslant C\,\|S(v)\| \quad \forall v \in V,$$

and, according to Theorem 3.61, this proves that $S(V)$ is closed in X/U. Since T is bounded, it follows that $T^{-1}(S(V))$ is also closed. Finally, we have

$$\begin{aligned}T^{-1}(S(V)) &= \big\{x \in X : [x] \in S(V)\big\} = \big\{x \in X : [x] = [v],\, v \in V\big\} \\ &= \big\{x \in X : x - v \in U,\, v \in V\big\} = \big\{x \in X : x \in U + V\big\} = U + V\,,\end{aligned}$$

which completes the proof.

□

For the subsequent results, the following concept is required.

Definition 3.54. *Let U be a closed subspace of a normed vector space X. A subspace V of X is called a* **topological complement** *(or* **topological supplement***) of U if*

a) V is closed, and
b) $X = U \oplus V$.

In this case, we say that U is a COMPLEMENTED *subspace of X. Otherwise, U is said to be* UNCOMPLEMENTED.

In the case described by the above definition, it is obvious that every $x \in X$ can be uniquely written as $x = u + v$, with $u \in U$ and $v \in V$. It then follows from Lemma 3.51 that there exists a constant $C > 0$ such that

$$\|u\| \leqslant C\|x\| \quad \text{and} \quad \|v\| \leqslant C\|x\|,$$

which means the operators (called projectors in this case):

$$\begin{aligned} P : X &\longrightarrow U \\ x &\longrightarrow P(x) := u, \end{aligned}$$

and

$$\begin{aligned} Q : X &\longrightarrow V \\ x &\longrightarrow Q(x) := v, \end{aligned}$$

are linear and bounded.

Furthermore, when X is a Hilbert space, the Orthogonal Decomposition Theorem (cf. Theorem 2.10) precisely states that every closed subspace U admits its orthogonal $U^\perp$ as a topological complement, so that the corresponding operators P and Q are now orthogonal projections.

On the other hand, the following lemma demonstrates that every finite-dimensional subspace admits a topological complement.

Lemma 3.55. *Let U be a finite-dimensional subspace of a normed vector space X. Then there exists a closed subspace V of X such that*

$$X = U \oplus V.$$

Proof. Let $\{e_1, \ldots, e_n\}$ be a basis of U. Clearly, for each $x \in U$, there exist unique scalars $\alpha_1, \alpha_2, \ldots, \alpha_n \in \mathbb{K}$ such that

$$x = \sum_{j=1}^{n} \alpha_j e_j,$$

which induces the definition of the linear bounded functionals

$$\begin{aligned} \varphi_i : U &\longrightarrow \mathbb{K} \\ x &\longrightarrow \varphi_i(x) := \alpha_i \end{aligned}$$

for all $i \in \{1, 2, \ldots, n\}$. Then, applying the analytic version of the Hahn-Banach Theorem (cf. Theorem 2.21), it follows that there exist $\psi_i \in X'$, for all i, such that

$$\|\psi_i\|_{X'} = \|\varphi_i\|_{U'} \quad \text{and} \quad \psi_i|_U = \varphi_i.$$

Now, define the subspace

$$V := \bigcap_{i=1}^{n} \psi_i^{-1}(\{0\}) = \bigcap_{i=1}^{n} N(\psi_i),$$

and we will prove that V is a topological complement of U. Indeed, because it is the intersection of n closed subspaces of X, V is also closed. Also, for any $x \in U \cap V$, we can write $x = \sum_{j=1}^{n} \alpha_j x_j$, and it follows that

$$0 = \psi_i(x) = \varphi_i(x) = \alpha_i \quad \forall i \in \{1, \ldots, n\},$$

which implies that $x = 0$, and so $U \cap V = \{0\}$. Furthermore, for any $x \in X$, it is clear that

$$x_U := \sum_{j=1}^{n} \psi_j(x) e_j \in U,$$

and

$$\begin{aligned}\psi_i(x - x_U) &= \psi_i(x) - \sum_{j=1}^{n} \psi_j(x)\psi_i(e_j) = \psi_i(x) - \sum_{j=1}^{n} \psi_j(x)\varphi_i(e_j)\\ &= \psi_i(x) - \sum_{j=1}^{n} \psi_j(x)\delta_{ij} = \psi_i(x) - \psi_i(x) = 0\,,\end{aligned}$$

for all $i \in \{1, \ldots, n\}$, thus showing that $x - x_U \in V$. Therefore, $x = x_U + (x - x_U) \in U + V$, and consequently $X = U \oplus V$.

□

Now, let $A \in \mathcal{L}(X, Y)$ be such that A is surjective. From the Open Mapping Theorem we know that there exists $r > 0$ such that $B(0, r) \subseteq A(B(0, 1))$. Then, given $y \in Y$, it is clear that $\frac{r}{2}\frac{y}{\|y\|} \in B(0, r)$ and therefore there exists $\bar{x} \in B(0, 1)$ such that $\frac{r}{2}\frac{y}{\|y\|} = A(\bar{x})$. Then, defining $x = \frac{2\,\|y\|}{r}\,\bar{x}$, we have that

$$A(x) = y \qquad \text{and} \qquad \|x\| \leqslant \frac{2}{r}\,\|y\|, \tag{3.19}$$

which makes it natural to ask whether it is possible to construct an operator $B \in \mathcal{L}(Y, X)$ such that $A \circ B = I : Y \longrightarrow Y$. In this case, B is said to be a right-inverse of A. The following theorem shows that the existence of such B is equivalent to the fact that $N(A)$ has a topological complement in X.

Theorem 3.56. *Let X and Y be Banach spaces, and let $A \in \mathcal{L}(X, Y)$ be surjective, that is, $R(A) = Y$. Then the following properties are equivalent:*

i) *A admits a right-inverse.*
ii) *$N(A)$ has a topological complement in X.*

Proof. Let $B \in \mathcal{L}(Y, X)$ be a right-inverse of A, and let us prove that $R(B)$ is a topological complement of $N(A)$. In fact, let $\{y_n\}_{n\in\mathbb{N}} \subseteq Y$ be such that $B(y_n) \overset{n\to\infty}{\longrightarrow} x \in X$. Applying the operator A, and using that $A \circ B = I : Y \longrightarrow Y$, it follows that $y_n \overset{n\to\infty}{\longrightarrow} A(x)$, and thus $B(y_n) \overset{n\to\infty}{\longrightarrow} B(A(x))$. By the uniqueness of the limit, we have $x = B(A(x)) \in R(B)$, and consequently, $R(B)$ is closed. Now, given $x \in N(A) \cap R(B)$, it follows that $A(x) = 0$, and there exists $y \in Y$ such that $x = B(y)$. It follows that $0 = A(x) = A(B(y)) = y$, and therefore, $x = B(y) = B(0) = 0$, which proves that $N(A) \cap R(B) = \{0\}$. Lastly, given $x \in X$, we have

$$A(x - B(A(x))) = A(x) - A \circ B(A(x)) = 0,$$

which shows that $x - B(A(x)) \in N(A)$, and therefore

$$x = x - B(A(x)) + B(A(x)) \in N(A) + R(B),$$

which implies $X = N(A) \oplus R(B)$. Conversely, let V be a topological complement of $N(A)$, that is, V is a closed subspace of X and $X = N(A) \oplus V$. Define the projection $P : X \longrightarrow V$, which projects onto V, meaning that for each $x \in X$, $P(x)$ is the unique element in V such that $x - P(x) \in N(A)$. It then follows that

$$A(x) = A(P(x)) \quad \forall\, x \in X.$$

Now, given $y \in Y$, consider $x \in X$ such that $A(x) = y$, and define $B(y) := P(x)$. Note, thanks to the derivation of (3.19), that x can be chosen so that $\|x\| \leqslant \frac{2}{r}\|y\|$. Let's verify that $B : Y \longrightarrow X$ is well-defined. In fact, let $\bar{x} \in X$ such that $A(\bar{x}) = y = A(x)$. It then follows that $x - \bar{x} \in N(A)$, and thus $P(x - \bar{x}) = 0$, which implies $P(x) = P(\bar{x})$. Furthermore, it is clear that B is linear and bounded, since, according to previous estimates, we have

$$\|B(y)\| = \|P(x)\| \leqslant C\|x\| \leqslant \widetilde{C}\|y\| \qquad \forall y \in Y.$$

Moreover, $(A \circ B)(y) = A(B(y)) = A(P(x)) = A(x) = y$ for all $y \in Y$, which shows that B is a right-inverse of A.

□

Note that the proof of the existence of a right-inverse, starting from the decomposition $X = N(A) \oplus V$, can also be obtained using the Bounded Inverse Theorem. Indeed, consider $\widetilde{A} := A|_V : V \to Y$, the restriction of A to V. It is clear that $\widetilde{A}$ is bijective and V is a Banach space, which implies, thanks to the mentioned theorem, that $\widetilde{A}^{-1} \in \mathcal{L}(Y, V)$. In this way, $B := \widetilde{A}^{-1}$ serves as the right-inverse of A. In the proof of Theorem 3.56, it is specified that $\widetilde{A}^{-1}(y) = P(x)$ for all $y \in Y$, where $x \in X$ is any preimage of y under A.

A similar result to the previous theorem is given by the following statement, whose proof is left as an exercise (see Exercise 3.67).

Theorem 3.57. *Let X and Y be Banach spaces, and let $A \in \mathcal{L}(X, Y)$ satisfy $N(A) = \{0\}$. Then the following properties are equivalent:*

i) *A admits a left-inverse; that is, there exists $B \in \mathcal{L}(Y, X)$ such that $B \circ A = I : X \to X$.*
ii) *$R(A)$ is closed and possesses a topological complement in Y.*

We end this section by pointing out that the question of whether an arbitrary non-trivial and proper closed subspace U of a Banach space X is complemented or not constitutes what is known as the COMPLEMENTED SUBSPACE PROBLEM. Specific examples for which the answer is positive were given in Lemma 3.55, and Theorems 3.56 and 3.57. In turn, the existence of both complemented and uncomplemented subspaces of given Banach spaces has also been established in the literature. In particular, it has been proved that every infinite-dimensional Banach space that is not isomorphic to a Hilbert space contains a closed uncomplemented subspace, whereas it is also known that there exist uncountably many complemented subspaces of the Lebesgue spaces L^p. For further details on the aforementioned problem and its open questions, as well as on the above results, we refer to [51] and the references therein.

3.10 Operators with closed range

The goal of this section is to characterize linear operators with closed range. In this regard, let us first recall that if X and Y are normed vector spaces and $A \in \mathcal{L}(X,Y)$, then Theorem 3.28 states that $R(A)$ is closed if and only if $R(A) = {}^{\circ}N(A')$. Similarly, in the case where X and Y are Hilbert spaces, it is known that $R(A)$ is closed if and only if $R(A) = N(A^*)^{\perp}$ (see Theorem 3.29). Next, we derive other necessary and sufficient conditions, easier to verify, for $R(A)$ to be a closed subspace of Y. To this end, we first separate the discussion into the injective and non-injective cases, and then provide further related results involving the concept of reflexive spaces.

3.10.1 The injective case

Let $A : \mathcal{D}(A) \subseteq X \to Y$ be a closed and injective linear operator, meaning that $N(A) = \{0\}$. If Y is Banach and $R(A)$ is closed, then $(R(A), \|\cdot\|_Y)$ is clearly a Banach space. Suppose further that X is Banach. Then, applying the Improved Bounded Inverse Theorem (see Theorem 3.37), we deduce that $A^{-1} \in \mathcal{L}(R(A), X)$, which means that there exists $C > 0$ such that

$$\|A^{-1}(y)\| \leqslant C\,\|y\| \quad \forall\, y \in R(A),$$

or equivalently,

$$\|x\| \leqslant C\,\|A(x)\| \quad \forall\, x \in \mathcal{D}(A). \tag{3.20}$$

Thus, we have shown that (3.20) is a necessary condition for $R(A)$ to be closed. Conversely, suppose that $A : \mathcal{D}(A) \subseteq X \to Y$ is a linear, closed and injective operator, and that there exists $C > 0$ such that (3.20) holds. Furthermore, let $\{A(x_n)\}_{n\in\mathbb{N}}$ be a sequence in $R(A)$, with $\{x_n\}_{n\in\mathbb{N}} \subseteq \mathcal{D}(A)$, such that $A(x_n) \overset{n\to\infty}{\longrightarrow} y \in Y$. Applying the inequality (3.20) and the linearity of A, it follows that

$$\|x_n - x_m\| \leqslant C\|A(x_n) - A(x_m)\| \to 0 \quad \text{as } n, m \to \infty,$$

which proves that $\{x_n\}_{n\in\mathbb{N}}$ is a Cauchy sequence in X, and therefore there exists $x \in X$ such that $x_n \overset{n\to\infty}{\longrightarrow} x$. Finally, since A is closed, it follows that $x \in \mathcal{D}(A)$ and $A(x) = y \in R(A)$, which implies that $R(A)$ is closed.

In summary, we have thus proved the following characterization result.

Theorem 3.58. *Let X and Y be Banach spaces, and let $A : \mathcal{D}(A) \subseteq X \to Y$ be a closed linear operator such that $N(A) = \{0\}$. Then $R(A)$ is a closed subspace of Y if and only if there exists $C > 0$ such that*

$$\|x\| \leqslant C\,\|A(x)\| \qquad \forall\, x \in \mathcal{D}(A).$$

3.10.2 The non-injective case

We now extend the previous characterization result to the non-injective case. Hence, in what follows, we consider X and Y Banach spaces, and $A : \mathcal{D}(A) \subseteq X \to Y$ a closed linear operator such that $N(A) \neq \{0\}$. The corresponding analysis requires some preliminary results and concepts.

Lemma 3.59. *Let X and Y be normed vector spaces, and let $A : \mathcal{D}(A) \subseteq X \to Y$ be a closed linear operator. Then, $N(A)$ is a closed subspace of X.*

Proof. Let $\{x_n\}_{n\in\mathbb{N}} \subseteq N(A)$ be such that $x_n \overset{n\to\infty}{\longrightarrow} x \in X$. It is clear that

$$A(x_n) = 0 \overset{n\to\infty}{\longrightarrow} 0 \in Y.$$

Since A is closed, it follows that $x \in \mathcal{D}(A)$ and $A(x) = 0$, which proves that $x \in N(A)$. □

The concept of quotient space is key in what follows. More precisely, let X be a normed vector space, and let M be a closed subspace of X. Then we define the equivalence relation

$$\sim: \quad u, v \in X, \quad u \sim v \quad \text{if and only if} \quad u - v \in M.$$

Thus, given $u \in X$, its equivalence class is given by

$$[u] := \{v \in X : \quad u - v \in M\},$$

and we say that u is a (non-unique) representative of $[u]$. Note that $[0] = M$. Moreover, it is clear that $[u] = [v]$ if and only if $u \sim v$. Also, $[u] \cap [v] = \varnothing$ if and only if $u \nsim v$. The set of all equivalence classes is called the QUOTIENT SPACE and is denoted by X/M, that is:

$$X/M := \{[u] : \quad u \in X\}.$$

In turn, X/M is endowed with a vector space structure with the following operations:

$$\begin{aligned} + : X/M \times X/M &\to X/M \\ ([u],[v]) &\to [u] + [v] := [u+v] \end{aligned}$$

$$\begin{aligned} \cdot : \mathbb{R} \times X/M &\longrightarrow X/M \\ (\lambda,[u]) &\longrightarrow \lambda \cdot [u] := [\lambda\, u] \end{aligned}$$

Let us check that $+$ and $\cdot$ are well defined in the sense that they do not depend on the representatives of each equivalence class. Indeed, let u, $\tilde{u}$, v, $\tilde{v} \in X$ be such that $[u] = [\tilde{u}]$ and $[v] = [\tilde{v}]$. It follows that $u - \tilde{u} \in M$ and $v - \tilde{v} \in M$, and hence $(u - \tilde{u}) + (v - \tilde{v}) = (u + v) - (\tilde{u} + \tilde{v}) \in M$, from which $(u + v) \sim (\tilde{u} + \tilde{v})$, or equivalently $[u + v] = [\tilde{u} + \tilde{v}]$. Similarly, if $\lambda \in \mathbb{R}$ and $u, \tilde{u} \in X$ are such that $[u] = [\tilde{u}]$, then $u - \tilde{u} \in M$, and therefore $\lambda\,(u - \tilde{u}) = \lambda\, u - \lambda\, \tilde{u} \in M$, which proves that $[\lambda\, u] = [\lambda\, \tilde{u}]$.

According to the above, it is easy to prove that $(X/M, +, \cdot)$ is a real vector space with zero element $[0] = M$. Moreover, with the mapping $\|\cdot\| : X/M \to \mathbb{R}$ defined by

$$\|[u]\| := \operatorname{dist}(u, M) = \inf_{v\in M} \|u - v\| \quad \forall\, [u] \in X/M, \tag{3.21}$$

X/M is a normed vector space. Indeed, let us verify that (3.21) is well defined and constitutes a norm on X/M. Let u, $\tilde{u} \in X$ be such that $[u] = [\tilde{u}]$. It follows that $u - \tilde{u} \in M$, and hence

$$\operatorname{dist}(u, M) = \inf_{v\in M} \|u - v\| = \inf_{v\in M} \|\,\tilde{u} - (v + \tilde{u} - u)\,\| = \inf_{w\in M} \|\,\tilde{u} - w\,\| = \operatorname{dist}(\tilde{u}, M),$$

which proves that there is no ambiguity, with respect to the chosen representative, in the definition of $\|[u]\|$. Let us now verify that the properties of a norm are satisfied. First, it is clear that $\|[u]\| \geqslant 0$ for all $[u] \in X/M$ since the set $\{\|u - v\| : v \in M\}$ is bounded below by 0. Moreover,

$$\|[u]\| = 0 \Leftrightarrow \text{dist}(u, M) = 0 \Leftrightarrow u \in M \Leftrightarrow [u] = [0].$$

Now, given $[u], [z] \in X/M$, we have

$$\begin{aligned} \|[u] + [z]\| = \|[u + z]\| &= \inf_{v \in M} \|u + z - v\| \\ &\leq \|u + z - (v + w)\| \quad \forall\, v,\, w \in M \\ &\leq \|u - v\| + \|z - w\| \quad \forall\, v,\, w \in M, \end{aligned}$$

from which, taking the infimum over $v \in M$ and then over $w \in M$, it follows that

$$\|[u] + [z]\| \leq \inf_{v \in M} \|u - v\| + \inf_{w \in M} \|z - w\| = \|[u]\| + \|[z]\|,$$

which proves the triangle inequality. Finally, let $\lambda \in \mathbb{K}$ and $[u] \in X/M$. If $\lambda = 0$, it is clear that

$$\|\lambda[u]\| = \|[\lambda u]\| = \|[0]\| = 0 = \lambda \cdot \|[u]\|,$$

and if $\lambda \neq 0$ we have

$$\begin{aligned} \|\lambda[u]\| = \|[\lambda u]\| &= \inf_{v \in M} \|\lambda u - v\| = \inf_{v \in M} \|\lambda(u - v/\lambda)\| \\ &= \inf_{v \in M} |\lambda|\, \|u - v/\lambda\| = |\lambda| \inf_{w \in M} \|u - w\| = |\lambda|\, \|[u]\|. \end{aligned}$$

Thus, we have proved that $(X/M, \|[\cdot]\|)$ is a normed vector space. Moreover, we have the following result, whose proof will be required later in one implication of Exercise 3.8. Alternatively, for an explicit proof of Theorem 3.60 below we refer to [115, Chapter III, Theorem 5.2].

Theorem 3.60. *Let M be a closed subspace of a Banach space X. Then, the quotient space $(X/M, \|[\cdot]\|)$ is also a Banach space.*

Having established the above, we are now in a position to return to our original interest of characterizing operators, not necessarily injective, with closed range. To this end, recall that X and Y are Banach spaces, and $A : \mathcal{D}(A) \subseteq X \to Y$ is a closed linear operator. Moreover, since $N(A)$ is a closed subspace of X (see Lemma 3.59), by Theorem 3.60 it follows that $X/N(A)$ is a Banach space. We then define the operator

$$\widehat{A} : \mathcal{D}(\widehat{A}) \subseteq X/N(A) \to Y$$

by

$$\widehat{A}([u]) = A(u) \quad \forall\, [u] \in \mathcal{D}(\widehat{A}),$$

where $\mathcal{D}(\widehat{A}) := \{[u] : \ u \in \mathcal{D}(A)\}$. Note that if $[u] = [z]$ then $A(u) = A(z)$, since $u - z \in N(A)$, which shows that $\widehat{A}$ is well defined. We also observe that $\mathcal{D}(\widehat{A})$ is a subspace of $X/N(A)$ (since $\mathcal{D}(A)$ is a subspace of X) and therefore $\widehat{A}$ is linear (since A is linear). In addition, it is clear that $R(\widehat{A}) = R(A)$.

On the other hand, let us show that $\widehat{A}$ is injective and closed. Indeed, let $[u] \in \mathcal{D}(\widehat{A})$ be such that $\widehat{A}([u]) = 0$. It follows that $0 = A(u)$, which implies $u \in N(A)$, and hence $[u] = [0]$. Therefore, $N(\widehat{A}) = \{[0]\}$. To see that $\widehat{A}$ is closed, consider a sequence $\{[u_n]\}_{n \in \mathbb{N}} \subseteq \mathcal{D}(\widehat{A})$ such that $[u_n] \overset{n \to \infty}{\longrightarrow} [u] \in X/M$ and $\widehat{A}([u_n]) = A(u_n) \overset{n \to \infty}{\longrightarrow} y \in Y$, that is,

$$\|[u_n] - [u]\| := \text{dist}\big(u - u_n, N(A)\big) \overset{n \to \infty}{\longrightarrow} 0 \quad \text{and} \quad \|A(u_n) - y\| \overset{n \to \infty}{\longrightarrow} 0. \tag{3.22}$$

From the definition of $\operatorname{dist}\big(u - u_n, N(A)\big)$ it follows that for each $n \in \mathbb{N}$ there exists $z_n \in N(A)$ such that

$$\operatorname{dist}\big(u - u_n, N(A)\big) \leqslant \|u_n - u - z_n\| < \operatorname{dist}\big(u - u_n, N(A)\big) + \frac{1}{n},$$

which, together with the first convergence in (3.22), yields $(u_n - z_n) \overset{n\to\infty}{\longrightarrow} u$. Moreover, noting that $A(z_n) = 0$ for all $n \in \mathbb{N}$, and using the second convergence in (3.22), it follows that $A(u_n - z_n) = A(u_n) \overset{n\to\infty}{\longrightarrow} y \in Y$, so that, since A is closed, we conclude that $u \in \mathcal{D}(A)$ and $A(u) = y$, that is, $[u] \in \mathcal{D}(\widehat{A})$ and $\widehat{A}([u]) = y$.

Summarizing, we have that $X/N(A)$ and Y are Banach spaces, and $\widehat{A} : X/N(A) \to Y$ is a closed linear operator with $N(\widehat{A}) = \{[0]\}$ and $R(\widehat{A}) = R(A)$. Consequently, applying Theorem 3.58 to the present situation, it follows that $R(A)$ is a closed subspace of Y if and only if there exists $C > 0$ such that

$$\|[u]\| \leqslant C\,\|\widehat{A}([u])\| \quad \forall\, [u] \in \mathcal{D}(\widehat{A}),$$

that is,

$$\operatorname{dist}(u, N(A)) \leqslant C\,\|A(u)\| \quad \forall\, u \in \mathcal{D}(A).$$

We have thus proved the following result.

Theorem 3.61. *Let X and Y be Banach spaces, and let $A : \mathcal{D}(A) \subseteq X \to Y$ be a closed linear operator. Then $R(A)$ is closed in Y if and only if there exists $C > 0$ such that*

$$\operatorname{dist}(u, N(A)) \leqslant C\,\|A(u)\| \quad \forall\, u \in \mathcal{D}(A)\,. \tag{3.23}$$

Note that if $N(A) = \{0\}$ then $\operatorname{dist}(u, N(A)) = \|u\|$, and in such case (3.23) becomes the particular characterization given by Theorem 3.58.

Within the context of closed-range operators, it is also important to establish that this property is inherited by the adjoint operator. More precisely, we first have the following result for bounded operators. Later on, we address the case of closed operators (cf. Theorem 3.85).

Theorem 3.62. *Let X and Y be Banach spaces, and let $A \in \mathcal{L}(X, Y)$ be such that $R(A)$ is closed in Y. Then $R(A') = N(A)^\circ$ and, therefore, $R(A')$ is a closed subspace of X'.*

Proof. We prove this by double inclusion. First, given $F \in R(A')$, there exists $G \in Y'$ such that $F = A'(G)$. Then, for each $x \in N(A)$ we have

$$F(x) = A'(G)(x) = G(A(x)) = G(0) = 0,$$

which proves that $F \in N(A)^\circ$. Conversely, let $F \in N(A)^\circ$, that is, $F \in X'$ and $F(x) = 0$ for all $x \in N(A)$. We want to prove that there exists $G \in Y'$ such that $F = A'(G)$. To this end, we first define the functional $g : R(A) \to \mathbb{R}$ by

$$g(y) = F(x) \quad \forall\, y \in R(A),$$

where $x \in X$ is such that $A(x) = y$. Given another preimage of y under A, that is, $\widetilde{x} \in X$ such that $A(\widetilde{x}) = y$, it follows that $x - \widetilde{x} \in N(A)$ and therefore $F(x) = F(\widetilde{x})$, which shows that g is well defined. Let us now see that g is linear. Indeed, given $\alpha, \widetilde{\alpha} \in \mathbb{R}$, $y, \widetilde{y} \in R(A)$ and $x, \widetilde{x} \in X$ such that $y = A(x)$ and $\widetilde{y} = A(\widetilde{x})$, the linearity of A yields $\alpha\, y + \widetilde{\alpha}\, \widetilde{y} = A(\alpha\, x + \widetilde{\alpha}\, \widetilde{x})$. Hence,

$$g(\alpha\, y + \widetilde{\alpha}\, \widetilde{y}) = F(\alpha\, x + \widetilde{\alpha}\, \widetilde{x}) = \alpha\, F(x) + \widetilde{\alpha}\, F(\widetilde{x}) = \alpha\, g(y) + \widetilde{\alpha}\, g(\widetilde{y}).$$

On the other hand, given $y = A(x) \in R(A)$ and $z \in N(A)$, we obtain

$$|g(y)| = |F(x)| = |F(x - z)| \leqslant \|F\|\, \|x - z\|\,,$$

whence

$$|g(y)| \leqslant \|F\| \inf_{z \in N(A)} \|x - z\| = \|F\|\, \mathrm{dist}(x, N(A))\,,$$

which, using that $R(A)$ is closed and Theorem 3.61, implies that

$$|g(y)| \leqslant C\, \|F\|\, \|A(x)\| = C \|F\|\, \|y\| \quad \forall\, y \in R(A).$$

Thus, $g \in R(A)'$, and therefore, applying the Hahn-Banach Theorem (analytic version), we deduce that there exists $G \in Y'$ such that $\|G\|_{Y'} = \|g\|_{R(A)'}$ and $G|_{R(A)} = g$. This means that $G(y) = g(y)$ for all $y \in R(A)$, or equivalently, for each $x \in X$

$$F(x) = g(A(x)) = G(A(x)) = A'(G)(x)\,,$$

which proves that $F = A'(G)$.

□

The Hilbert space version of the previous theorem is given as follows.

Theorem 3.63. *Let X and Y be Hilbert spaces, and let $A \in \mathcal{L}(X, Y)$ be such that $R(A)$ is closed in Y. Then $R(A^*) = N(A)^{\perp}$ and therefore $R(A^*)$ is a closed subspace of X.*

Proof. We employ Theorem 3.12, the identity $A^* = R_X \circ A' \circ R_Y^{-1}$, where R_X and R_Y are the respective Riesz maps, and Lemma 3.23, which states that $R_X(S^\circ) = S^{\perp}$ for all $S \subseteq X$. Indeed, since $R(A)$ is closed we have that $R(A') = N(A)^\circ$. Then $x \in R(A^*)$

$$\begin{aligned}
&\Leftrightarrow \quad \exists\, y \in Y \quad \text{such that} \quad x = R_X(A'(R_Y^{-1}(y)))\\
&\Leftrightarrow \quad \exists\, y \in Y \quad \text{such that} \quad R_X^{-1}(x) = A'(R_Y^{-1}(y))\\
&\Leftrightarrow \quad \exists\, G := R_Y^{-1}(y) \in Y' \quad \text{such that} \quad R_X^{-1}(x) = A'(G)\\
&\Leftrightarrow \quad R_X^{-1}(x) \in R(A') = N(A)^\circ\\
&\Leftrightarrow \quad x \in R_X(N(A)^\circ) = N(A)^{\perp}.
\end{aligned}$$

□

3.10.3 Further results involving reflexive spaces

Let us begin by observing that another characterization result of operators with closed range, which follows from Theorem 3.61 and the Open Mapping Theorem (cf. Theorem 3.39), is given by the following lemma.

Lemma 3.64. *Let X and Y be Banach spaces, and let $A \in \mathcal{L}(X, Y)$. Then the following statements are equivalent:*

i) *$R(A)$ is closed.*
ii) *There exists $\alpha > 0$ such that, for every $y \in R(A)$, there exists $z \in X$ satisfying*

$$y = A(z) \qquad \text{and} \qquad \alpha\, \|z\| \leqslant \|y\|\,.$$

Proof. Suppose that i) holds. Since Y is Banach, $R(A)$ is also Banach, and therefore we can apply the Open Mapping Theorem (cf. Theorem 3.39) to the operator $A \in \mathcal{L}(X, R(A))$. It follows that there exists $r > 0$ such that $B_{R(A)}(0, r) \subseteq A(B_X(0,1))$. Then, given $y \in R(A)$, $y \neq 0$, it is clear that $\frac{r}{2}\frac{y}{\|y\|} \in B_{R(A)}(0, r)$, so that using the previous inclusion we deduce that there exists $x \in B_X(0,1)$ such that $A(x) = \frac{r}{2}\frac{y}{\|y\|}$. Thus, denoting $z = \frac{2\|y\|}{r}\,x$, we obtain $A(z) = y$ and $\|z\| \leqslant \frac{2}{r}\|y\|$, which proves ii) with $\alpha = \frac{r}{2}$. Conversely, suppose ii) holds. Then, given $x \in X$ such that $A(x) \neq 0$, we have by hypothesis that there exists $z \in X$ such that $A(x) = A(z)$ and $\alpha\|z\| \leqslant \|A(x)\|$. Then, since $x - z \in N(A)$, it follows that

$$\|A(x)\| \geqslant \alpha\|z\| = \alpha\|x - (x - z)\| \geqslant \alpha \operatorname{dist}(x, N(A)),$$

which, thanks to Theorem 3.61, implies that $R(A)$ is closed.

□

The result to be established next is related to the previous lemma and makes use of the concept of a reflexive Banach space. While this definition is formally introduced in Definition 7.1 (cf. Section 7.1) and studied in detail throughout the remainder of Chapter 7, we anticipate for the moment that, given an arbitrary normed vector space Z, the operator $\mathcal{J}_Z : Z \to Z''$ defined by

$$\mathcal{J}_Z(z)(F) := F(z) \qquad \forall\, z \in Z, \quad \forall\, F \in Z', \tag{3.24}$$

is always an isometric injection. Hence, Z is said to be reflexive if, in addition, $\mathcal{J}_Z$ is surjective. Alternatively, one may simply say that Z is reflexive if and only if $\mathcal{J}_Z$ is a bijection, a characterization used, for example, in Section 3.12.1.

We also note in advance that, in proving the converse of the lemma presented below, we will make use of a general result that will be established later for closed linear operators (cf. Theorem 3.85), whose particular version for bounded operators in reflexive spaces is proved at the end of this section (cf. Theorem 3.66). A slightly different argument for this converse is provided in [52, Lemma A.42], which is, in turn, based on the original proof given in [7].

Lemma 3.65. *Let X and Y be Banach spaces, and let $A \in \mathcal{L}(X,Y)$ be surjective. Then there exists $\alpha > 0$ such that*

i) *for each $y \in R(A) = Y$ there exists $z \in X$ such that $A(z) = y$, and $\alpha\|z\| \leqslant \|y\|$, which implies that*
ii) *$\|A'(G)\|_{X'} \geqslant \alpha\|G\|_{Y'}$ for all $G \in Y'$.*

Conversely, if X is reflexive, i) *follows from* ii) *with the same constant α.*

Proof. The surjectivity of A obviously guarantees that $R(A)$ is closed, so that by applying Lemma 3.64 we deduce the existence of a constant $\alpha > 0$ for which i) holds. Then, for each $y \in Y$, $y \neq 0$, there exists $z \in X$ such that $A(z) = y$ and $\alpha\|z\| \leqslant \|y\|$, which implies, given $G \in Y'$, that

$$\frac{|G(y)|}{\|y\|} = \frac{|G(A(z))|}{\|y\|} = \frac{|A'(G)(z)|}{\|y\|} \leqslant \frac{1}{\alpha}\|A'(G)\|,$$

and therefore it easily follows that

$$\|G\| = \sup_{\substack{y \in Y \\ y \neq 0}} \frac{|G(y)|}{\|y\|} \leqslant \frac{1}{\alpha}\|A'(G)\|,$$

which proves ii). Conversely, suppose that X is reflexive and that ii) is valid with a constant $\alpha > 0$. Equivalently, this means, by Theorem 3.58, that $A' \in \mathcal{L}(Y', X')$ is injective and of closed range, and consequently (cf. Theorem 3.85) $R(A') = N(A)^\circ$. It follows from this identity and from ii) that $A' : Y' \longrightarrow N(A)^\circ$ is bijective and that $\|(A')^{-1}\| \leqslant \frac{1}{\alpha}$. Moreover, $A'' : \big(N(A)^\circ\big)' \longrightarrow Y''$ is also bijective and

$$\|(A'')^{-1}\| = \|((A')^{-1})'\| = \|(A')^{-1}\| \leqslant \frac{1}{\alpha}.$$

Now, given $y \in Y$, we consider $\mathcal{J}_Y(y) \in Y''$, for which there exists $\mathcal{F}_0 \in \big(N(A)^\circ\big)'$ such that $A''(\mathcal{F}_0) = \mathcal{J}_Y(y)$. Since $N(A)^\circ$ is a subspace of X', the Hahn-Banach Theorem for normed spaces (cf. Theorem 2.21) guarantees the existence of $\mathcal{F} \in X''$ such that $\mathcal{F}\big|_{N(A)^\circ} = \mathcal{F}_0$ and $\|\mathcal{F}\|_{X''} = \|\mathcal{F}_0\|_{(N(A)^\circ)'}$. Then, denoting $z = J_X^{-1}(\mathcal{F}) \in X$, we have for each $G \in Y'$

$$\begin{aligned} G(y) &= \mathcal{J}_Y(y)(G) = A''(\mathcal{F}_0)(G) = (A')'(\mathcal{F}_0)(G) = \mathcal{F}_0(A'(G)) \\ &= \mathcal{F}(A'(G)) = \mathcal{J}_X(z)(A'(G)) = A'(G)(z) = G(A(z)), \end{aligned}$$

from which it follows, thanks to the consequence of the Hahn-Banach Theorem given by Lemma 2.24, that $y = A(z)$. Finally, we have

$$\begin{aligned} \|z\| &= \|\mathcal{J}_X(z)\| = \|\mathcal{F}\|_{X''} = \|\mathcal{F}_0\|_{(N(A)^\circ)'} \\ &= \|(A'')^{-1}(\mathcal{J}_Y(y))\|_{(N(A)^\circ)'} \leqslant \|(A'')^{-1}\| \, \|y\| \leqslant \frac{1}{\alpha} \|y\|, \end{aligned}$$

which completes i) and ends the proof.

□

We conclude this section with a result that includes, in particular, the converse of Theorem 3.62, and which constitutes the version for reflexive Banach spaces of the general equivalence to be established later in Theorem 3.85. More precisely, we have the following theorem.

Theorem 3.66. *Let X and Y be reflexive Banach spaces, and let $A \in \mathcal{L}(X, Y)$. Then the following statements are equivalent:*

a) *$R(A)$ is closed.*
b) *$R(A') = N(A)^\circ$.*
c) *$R(A')$ is closed.*
d) *$R(A) = {}^\circ N(A')$.*

Proof. We first note that the implications given by a) $\Rightarrow$ b), b) $\Rightarrow$ c), and d) $\Rightarrow$ a) follow directly from Theorem 3.62, and from the fact (cf. Lemma 3.22) that both right and left annihilators are closed subspaces. It remains to establish c) $\Rightarrow$ d), for which we first prove that the diagram

$$\begin{array}{ccc} X & \xrightarrow{A} & Y \\ \mathcal{J}_X \big\downarrow & & \big\downarrow \mathcal{J}_Y \\ X'' & \xrightarrow{A''} & Y'' \end{array} \tag{3.25}$$

is commutative, that is

$$A'' \circ \mathcal{J}_X \;=\; \mathcal{J}_Y \circ A\,, \tag{3.26}$$

where $\mathcal{J}_X : X \to X''$ and $\mathcal{J}_Y : Y \to Y''$ are the isomorphisms characterizing the reflexivity of X and Y, respectively. Indeed, using the concept of adjoint operator and the definitions of $\mathcal{J}_X$ and $\mathcal{J}_Y$ (cf. (3.24)), we have for each $x \in X$, and for each $G \in Y'$

$$\begin{aligned}\Big((A'' \circ \mathcal{J}_X)(x)\Big)(G) \;&=\; (A')'(\mathcal{J}_X(x))(G) \;=\; \mathcal{J}_X(x)\big(A'(G)\big) \;=\; \big(A'(G)\big)(x)\\ &=\; G\big(A(x)\big) \;=\; \mathcal{J}_Y\big(A(x)\big)(G) \;=\; \Big((\mathcal{J}_Y \circ A)(x)\Big)(G)\,,\end{aligned}$$

from which (3.26) follows. Thus, by the bijectivity of $\mathcal{J}_X$ and $\mathcal{J}_Y$, it follows that

$$A \;=\; \mathcal{J}_Y^{-1} \circ A'' \circ \mathcal{J}_X\,, \tag{3.27}$$

so that, now assuming statement c), that is, $R(A')$ closed, we deduce, thanks to Theorem 3.62 and the identity (3.27), that $R(A'')$, and hence $R(A) \;=\; \mathcal{J}_Y^{-1}\big(R(A'')\big)$, are also closed. Finally, a direct application of part iii) of Theorem 3.28 completes the proof.

□

3.11 A problem in approximation theory

In this section we apply some of the previous results of this chapter to the problem in approximation theory given by the characterization of INTERPOLATION SPLINE FUNCTIONS. More precisely, let $\Omega := (a,b), n \in \mathbb{N}$, and let $\{t_1, \ldots, t_n\} \subseteq \Omega$ be such that $a < t_1 < \ldots < t_n < b$. Then, given $m \in \mathbb{N}$, we consider the real spaces:

$$X = \mathrm{H}^m(\Omega), \quad Y = \mathrm{L}^2(\Omega), \quad Z = \mathbb{R}^n,$$

and define the operators $T : X \to Y$ and $A : X \to Z$, respectively, by

$$T(u) \;=\; u^{(m)} \quad \forall\, u \in X, \tag{3.28}$$

and

$$A(u) \;=\; \begin{bmatrix} u(t_1) \\ u(t_2) \\ \vdots \\ u(t_n) \end{bmatrix} \qquad \forall\, u \in X. \tag{3.29}$$

It is easy to see that T and A are linear and bounded . In particular, for the boundedness of T we have:

$$\|T(u)\|_{0,\Omega} = \|u^{(m)}\|_{0,\Omega} \leqslant \left\{ \sum_{j=0}^{m} \|u^{(j)}\|_{0,\Omega}^2 \right\}^{1/2} =: \|u\|_{m,\Omega} \qquad \forall\, u \in X\,,$$

where $\|\cdot\|_{m,\Omega}$ denotes the norm induced by the usual inner product $\langle\cdot,\cdot\rangle_{m,\Omega}$ of $\mathrm{H}^m(\Omega)$, the Sobolev space of order m (cf. Exercise 1.10, eq. (1.37)), which, in this one-dimensional case, reduces to

$$\mathrm{H}^m(\Omega) := \Big\{ v \in \mathrm{L}^2(\Omega) : \quad v^{(j)} \in \mathrm{L}^2(\Omega) \quad \forall\, j \in \{1, \ldots, m\} \Big\}. \tag{3.30}$$

Moreover, applying the Trace Theorem in bounded open subsets of $\mathbb{R}$ (see Theorem 4.26 for the general case of $\mathbb{R}^n$), we deduce that

$$|u(t_j)| \leqslant c\,\|u\|_{1,]a,t_j[} \leqslant c\,\|u\|_{m,\Omega} \quad \forall\, u \in X,\ \forall\, j \in \{1,\cdots,n\},$$

which shows that each of the component operators of A, and therefore A, are bounded. On the other hand, one can also prove that T and A are surjective, i.e. $R(T) = Y$ and $R(A) = Z$. For the surjectivity of A, we first observe, according to the Riesz Representation Theorem (cf. Theorem 2.14), that there exists a set $\{u_1, u_2, \cdots, u_n\} \subseteq X = \mathrm{H}^m(\Omega)$ such that

$$u(t_j) = \langle u, u_j\rangle_{m,\Omega} \quad \forall\, u \in X, \quad \forall\, j \in \{1,\cdots,n\}, \tag{3.31}$$

that is

$$A(u) = \sum_{j=1}^{n} \langle u, u_j\rangle_{m,\Omega}\, e_j \quad \forall\, u \in X,$$

where $\{e_1, e_2 \cdots, e_n\}$ is the canonical basis of $Z = \mathbb{R}^n$. Next, we note that $\{u_1, u_2 \cdots, u_n\}$ is linearly independent. Indeed, let $\alpha_1, \ldots, \alpha_n \in \mathbb{R}$ such that $\sum_{i=1}^{n} \alpha_i\, u_i = 0 \in \mathrm{H}^m(\Omega)$. For each $j \in \{1, 2, \cdots, n\}$ we can construct a function $\varphi_j \in C_0^\infty(\Omega) \subseteq \mathrm{H}^m(\Omega)$ such that $\varphi_j(t_j) = 1$ and $\mathrm{supp}\varphi_j = \bar{B}(t_j, r_j) \subseteq]t_{j-1}, t_{j+1}[$ (cf. Example 1.35), where we denote $t_0 = a$ and $t_{n+1} = b$. In this way we obtain for each $j \in \{1, 2, \cdots, n\}$

$$0 = \langle \varphi_j, \sum_{i=1}^{n} \alpha_i, u_i\rangle_{m,\Omega} = \sum_{i=1}^{n} \alpha_i \,\langle \varphi_j, u_i\rangle_{m,\Omega} = \sum_{i=1}^{n} \alpha_i\, \varphi_j(t_i) = \alpha_j\,,$$

which confirms this linear independence. Consequently, given that it is easily shown that

$$A^*(z) = \sum_{j=1}^{n} \langle e_j, z\rangle_{\mathbb{R}^n}\, u_j = \sum_{j=1}^{n} z_j\, u_j \quad \forall\, z \in Z = \mathbb{R}^n, \tag{3.32}$$

where $\langle\cdot,\cdot\rangle_{\mathbb{R}^n}$ stands for the usual inner product of $\mathbb{R}^n$, it follows that A^* is injective and therefore $R(A) = N(A^*)^\perp = \mathbb{R}^n$. On the other hand, for the surjectivity of T we first assume that m is even. Then, given $v \in Y := \mathrm{L}^2(\Omega)$, we apply Example 4.6 (cf. Chapter 4) with $\Omega := (a, b)$ instead of $\Omega := (0, 1)$, to deduce that there exists a unique $u_1 \in \mathrm{H}_0^1(\Omega)$ such that $u_1^{(2)} := u_1'' = v$, whence $u_1 \in \mathrm{H}^2(\Omega)$. In the same way, we obtain a unique $u_2 \in \mathrm{H}_0^1(\Omega)$ such that $u_2^{(2)} = u_1$, so that $u_2 \in \mathrm{H}^4(\Omega)$ and $u_2^{(4)} = v$. Following this recursive argument, we conclude, after $m/2$ steps, that there exists a unique $u := u_{m/2} \in \mathrm{H}^m(\Omega)$ such that $T(u) := u^{(m)} = v$. Now, if m is odd, we again consider $v \in Y := \mathrm{L}^2(\Omega)$, define $z(x) := \int_0^x v(t)\, dt$ for all $x \in \Omega$, and note that clearly $z \in \mathrm{L}^2(\Omega)$ and $z' = v$, so that $z \in \mathrm{H}^1(\Omega)$. Then, applying the previous case to the pair $m - 1$ and z, we find a unique $u \in \mathrm{H}^{m-1}(\Omega)$ such that $u^{(m-1)} = z$, from which $u \in \mathrm{H}^m(\Omega)$ and $T(u) := u^{(m)} = z' = v$. Thus, for all $m \in \mathbb{N}$ we have that $T : \mathrm{H}^m(\Omega) \to \mathrm{L}^2(\Omega)$ is surjective.

Having established the above, we now define the INTERPOLATION SPLINE FUNCTION WITH RESPECT TO T AND A as the solution, if it exists, of the following problem. Given $z \in \mathbb{R}^n$, find $u \in X$ such that

$$\|T(u)\| = \min\Big\{ \|T(v)\| : \quad v \in X, \quad A(v) = z \Big\}, \tag{3.33}$$

or, equivalently, find $u \in X$ such that

$$\int_a^b \{u^{(m)}(t)\}^2\,dt$$
$$= \min\left\{\int_a^b \{v^{(m)}(t)\}^2\,dt : \quad v \in \mathrm{H}^m(\Omega),\ v(t_i) = z_i \quad \forall i \in \{1,2,\cdots,n\}\right\}. \tag{3.34}$$

In other words, among all functions $v \in X$ that interpolate the set of data $\Big\{(t_i, z_i) : \quad i \in \{1, 2, ..., n\}\Big\}$ (see Figure 3.2 for examples of such functions), we seek the one that minimizes the integral over $[a, b]$ of the square of its m-th derivative.

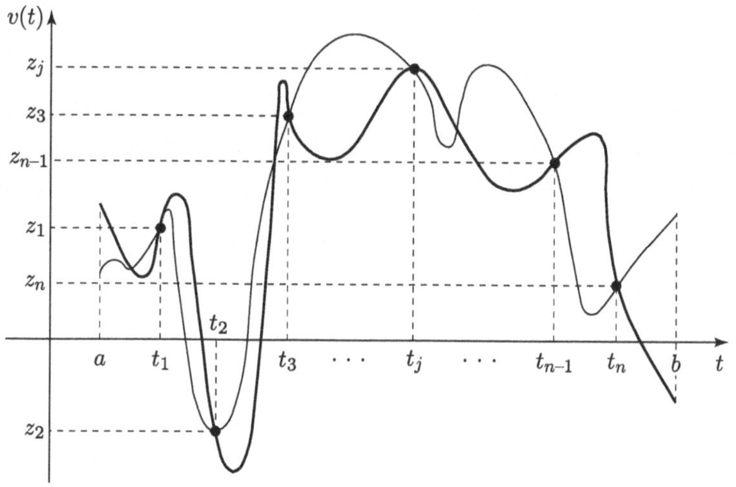

Fig. 3.2. *Example of feasible functions v for problem* (3.34)

In what follows we analyze in detail problem (3.33). In particular, we derive necessary and sufficient conditions for the existence and uniqueness of such $u \in X$, and we also obtain a characterization of this solution, thanks to which it can be computed explicitly.

Let us begin our analysis by observing, thanks to the surjectivity of A, that there exists $v_0 \in X$ such that $A(v_0) = z$, and therefore

$$A^{-1}(\{z\}) := \Big\{v \in X : A(v) = z\Big\} = \{v_0\} + N(A).$$

It follows that $A^{-1}(\{z\})$ is a closed subspace of X (since $N(A)$ is), and moreover it is easy to see that $T(A^{-1}(\{z\}))$ is convex. In this way, problem (3.33) can be reformulated, equivalently, as

$$\min_{v\in A^{-1}(\{z\})} \|T(v)\| = \min_{y\in T(A^{-1}(\{z\}))} \|y\|, \tag{3.35}$$

where:

$$T(A^{-1}(\{z\})) = \{T(v_0)\} + T(N(A)).$$

In view of the above, we can state the following preliminary result.

Lemma 3.67. *Problem* (3.33) (*see also* (3.35)) *has a solution for every* $z \in Z$ *if and only if* $T(N(A))$ *is closed in* Y.

Proof.

($\Leftarrow$) Suppose that $T(N(A))$ is closed. It follows that $T(A^{-1}(\{z\}))$ is also closed for all $z \in Z$, and since we already observed that this set is convex, the respective best approximation result (see Theorem 2.7), implies that there exists a unique $\widetilde{y} \in T(A^{-1}(\{z\}))$ such that

$$\|\widetilde{y}\| = \min_{y \in T(A^{-1}(\{z\}))} \|y\|.$$

Then, any element $v \in A^{-1}(\{z\})$ such that $T(v) = \widetilde{y}$ constitutes a solution of (3.33).

($\Rightarrow$) Now suppose that problem (3.33) has a solution for each $z \in Z$, and assume, by contradiction, that $T(N(A))$ is not closed. Then, there exists $y_0 \in \overline{T(N(A))}$ such that $y_0 \notin T(N(A))$, which implies that the problem

$$\min_{y \in T(N(A))} \|y - y_0\| = \min_{y \in T(N(A))} \|y + y_0\| \tag{3.36}$$

has no solution. Now, since T is surjective, there exists $u_0 \in X$ such that $T(u_0) = y_0$. It follows that $y_0 \in T(A^{-1}\{z_0\})$, where $z_0 = A(u_0) \in Z$. Then, by virtue of the hypothesis of the present implication, there exists $\widetilde{u}_0 \in A^{-1}(\{z_0\})$ such that

$$\|T(\widetilde{u}_0)\| = \min_{v \in A^{-1}(\{z_0\})} \|T(v)\|,$$

or, denoting $\widetilde{y}_0 := T(\widetilde{u}_0)$, we have

$$\|\widetilde{y}_0\| = \min_{y \in T(A^{-1}(\{z_0\}))} \|y\| = \min_{y \in T(N(A))} \|y + y_0\|, \tag{3.37}$$

where we have used that

$$T(A^{-1}(\{z_0\})) = \{T(u_0)\} + T(N(A)) = \{y_0\} + T(N(A)).$$

Since $\widetilde{y}_0 - y_0 = T(\widetilde{u}_0 - u_0) \in T(N(A))$, because $A(\widetilde{u}_0) = A(u_0) = z_0$, it follows that the minimum in (3.37) is attained at $y = \widetilde{y}_0 - y_0$, which contradicts the nonexistence of a solution to problem (3.36).

□

On the other hand, in order to establish another necessary and sufficient condition for $T(N(A))$ to be closed, we need to introduce the bounded linear operator $\widetilde{T} := T|_{N(T)^\perp} : N(T)^\perp \to Y$. It is well known, thanks to the surjectivity of T, that $\widetilde{T}$ is bijective, and therefore the Bounded Inverse Theorem ensures that $\widetilde{T}^{-1} \in \mathcal{L}(Y, N(T)^\perp)$. Moreover, we have the following technical result.

Lemma 3.68.

$$T(N(A)) = \widetilde{T}((N(A) + N(T)) \cap N(T)^\perp).$$

Proof. Let $y = T(v)$ with $v \in N(A)$, and let P and $P^\perp$ be the orthogonal projectors of X onto $N(T)$ and $N(T)^\perp$, respectively. It follows that $v = P(v) + P^\perp(v)$, and then $y = T(v) = T(P^\perp(v)) = \widetilde{T}(P^\perp(v))$, where $P^\perp(v) = v - P(v) \in (N(A) + N(T)) \cap N(T)^\perp$. Conversely, let $y = \widetilde{T}(v)$ with $v \in (N(A) + N(T)) \cap N(T)^\perp$, and let $w \in N(A)$ and $z \in N(T)$ be such that $v = w + z$. Then we obtain

$$y = \widetilde{T}(v) = T(w + z) = T(w) \in T(N(A)).$$

□

We can now prove the following characterization result.

Lemma 3.69. *The space $T(N(A))$ is closed if and only if $N(A)+N(T)$ is closed.*

Proof.

($\Rightarrow$) Suppose that $T(N(A))$ is closed, and let $\{w_n\}_{n\in\mathbb{N}} \subseteq N(A)$ and $\{z_n\}_{n\in\mathbb{N}} \subseteq N(T)$ be such that $v_n := w_n + z_n \overset{n\to\infty}{\longrightarrow} v \in X$. By the continuity of T we have that $T(v_n) = T(w_n) \to T(v) \in Y$, and since $\{T(w_n)\}_{n\in\mathbb{N}} \subseteq T(N(A))$ it follows that $T(v) \in T(N(A))$. Then, there exists $w \in N(A)$ such that $T(v) = T(w)$, from which it results that $v = w + (v-w) \in N(A) + N(T)$.

($\Leftarrow$) Now suppose that $N(A)+N(T)$ is closed. It clearly follows that the space $(N(A) + N(T)) \cap N(T)^{\perp}$ is also closed, and therefore the same property holds for

$$T(N(A)) = (\widetilde{T}^{-1})^{-1}\left((N(A)+N(T)) \cap N(T)^{\perp}\right).$$

□

The following theorem is an immediate consequence of the previous lemmas.

Theorem 3.70. *A necessary and sufficient condition for the existence of at least one interpolation spline function with respect to the operators T and A, for each $z \in Z$, is that $N(A)+N(T)$ be closed in X.*

Moreover, the following theorem characterizes the uniqueness of such an interpolation spline function.

Theorem 3.71. *The interpolation spline function with respect to T and A, whose existence for each $z \in Z$ is guaranteed if $N(A)+N(T)$ is closed in X, is unique if and only if $N(A) \cap N(T) = \{0\}$.*

Proof.

($\Rightarrow$) Suppose the uniqueness of the interpolation spline function for each $z \in Z$, and let $\widetilde{v} \in N(A) \cap N(T)$, that is, $\widetilde{v} \in X$, $A(\widetilde{v}) = 0 \in Z$, and $T(\widetilde{v}) = 0 \in Y$. It follows that

$$\min_{v\in A^{-1}(\{0\})} \|T(v)\| = \|T(\widetilde{v})\| = 0,$$

and by the uniqueness of the solution of this problem, necessarily $\widetilde{v} = 0$.

($\Leftarrow$) Suppose that $N(A) \cap N(T) = \{0\}$ and, given $z \in Z$, let $v_1, v_2 \in A^{-1}(\{z\})$ be such that $T(v_1) = T(v_2) = \widetilde{y}$, where

$$\|\widetilde{y}\| = \min_{y\in T(A^{-1}(\{z\}))} \|y\|.$$

Then, it is clear that $v_1 - v_2 \in N(A) \cap N(T)$, and therefore $v_1 = v_2$.

□

In order to confirm that the operators T and A defined in (3.28) and (3.29) satisfy the hypotheses of Theorems 3.70 and 3.71, we need to establish two preliminary results. First, we recall that the sum of two closed subspaces of a Banach space is closed if one of the subspaces is finite-dimensional. Then, we have the following lemma.

Lemma 3.72. *If either of the spaces $N(A)$ or $N(T)$ has finite dimension or finite codimension, then $N(A) + N(T)$ is closed in X.*

Proof. It suffices to reason with one of these spaces. If $N(T)$ has finite dimension then clearly $N(A) + N(T)$ is closed. Now, suppose that $X/N(T)$ has finite dimension. Since $R(T) = Y$, the linear operator $\widehat{T} : X/N(T) \to Y$ defined by $\widehat{T}([x]) = T(x)$ for all $[x] \in X/N(T)$ is an isomorphism, and therefore Y is also finite-dimensional. It follows that $T(N(A))$ is closed, and by applying Lemma 3.69 we deduce that $N(A) + N(T)$ is closed. □

In the particular case of our operators T and A (cf. (3.28), (3.29)) we have that $R(T) = Y = \mathrm{L}^2(\Omega), R(A) = Z = \mathbb{R}^n$, and $N(T) = \mathbb{P}_{m-1}(\Omega)$, the space of polynomials of degree $\leqslant m-1$ defined on Ω. Note also that $X/N(A)$ is isomorphic to $Z = \mathbb{R}^n$. Thus, either by using that $\dim N(T) = m$ or that $\dim X/N(A) = n$, it follows from Lemma 3.72 that $N(A) + N(T)$ is closed in X. On the other hand, let $v \in N(A) \cap N(T)$. It follows that

$$v \in \mathbb{P}_{m-1}(\Omega) \quad \text{and} \quad v(t_i) = 0 \quad \forall\, i \in \{1, 2, \cdots n\}.$$

Then, by virtue of the Fundamental Theorem of Algebra, it follows that a necessary and sufficient condition for v to vanish in Ω, and therefore conclude that $N(A) \cap N(T) = \{0\}$, is that $n \geqslant m$.

From now on we assume that $n \geqslant m$, which, according to Theorems 3.70 and 3.71, guarantees that for each $z \in Z$ there exists a unique interpolation spline function u with respect to T and A. The next goal is to characterize and explicitly compute u.

Given $z \in Z = \mathbb{R}^n$, let $u \in A^{-1}(\{z\})$ and $\widetilde{y} = T(u) \in T(A^{-1}(\{z\}))$ such that

$$\|T(u)\| = \min_{v \in A^{-1}(\{z\})} \|T(v)\| = \min_{y \in T\,(A^{-1}(\{z\}))} \|y\| = \|\widetilde{y}\|.$$

Since $T\,(A^{-1}(\{z\}))$ is closed and convex, it can be shown (cf. (4.60) in Theorem 4.15) that the projection of 0 onto this set, that is, $\widetilde{y}$, is characterized by the inequality:

$$\langle 0 - \widetilde{y},\, y - \widetilde{y} \rangle_Y \leqslant 0 \qquad \forall\, y \in T(A^{-1}(\{z\})),$$

where $\langle \cdot, \cdot \rangle_Y$ denotes the inner product of $Y = \mathrm{L}^2(\Omega)$. Moreover, it is easy to see that the above relation can be rewritten as

$$\langle T(u), T(v - u) \rangle_Y \geqslant 0 \qquad \forall\, v \in A^{-1}(\{z\}),$$

or equivalently, noting that $A(v) = A(u) = z$,

$$\langle T(u), T(w) \rangle_Y \geqslant 0 \qquad \forall\, w \in N(A),$$

which, using the fact that $N(A)$ is a subspace, reduces, equivalently, to

$$\langle T(u), T(w) \rangle_Y = 0 \qquad \forall\, w \in N(A). \tag{3.38}$$

We have thus shown that, given $z \in Z = \mathbb{R}^n$, $u \in A^{-1}(\{z\})$ is the interpolation spline function with respect to the operators T and A if and only if (3.38) holds, that is:

$$\langle T^*(T(u)), w \rangle_X = 0 \qquad \forall\, w \in N(A),$$

which is equivalent to requiring that

$$T^*(T(u)) \in N(A)^\perp = R(A^*).$$

In this way, using the expression for A^* given in (3.32), it follows that there exists $\lambda := (\lambda_1, \lambda_2 \cdots, \lambda_n)^t \in Z = \mathbb{R}^n$ such that

$$T^*(T(u)) = A^*(\lambda) = \sum_{j=1}^{n} \lambda_j \, u_j .$$

Now, since $R(T) = Y = \mathrm{L}^2(\Omega)$ is obviously closed, we have that $R(T^*)$ is also closed, whence $R(T^*) = N(T)^\perp$, and therefore

$$T^*(T(u)) = \sum_{j=1}^{n} \lambda_j \, u_j \in N(T)^\perp . \tag{3.39}$$

Thus, since $N(T) = \mathbb{P}_{m-1}(\Omega)$, we have that

$$\langle T^*(T(u)), p \rangle_X = 0 \quad \forall\, p \in \mathbb{P}_{m-1}(\Omega). \tag{3.40}$$

On the other hand, for each $v \in X = \mathrm{H}^m(\Omega)$ we can use the Taylor series expansion with integral remainder and write:

$$v(t) = \sum_{j=0}^{m-1} \frac{v^{(j)}(a)(t-a)^j}{j!} + \int_a^b \frac{(t-s)_+^{m-1} \, v^{(m)}(s)}{(m-1)!} \, ds \quad \forall\, t \in [a,b],$$

where s_+ denotes the nonnegative part of $s \in \mathbb{R}$. Note that the expression given by the summation in the previous equation is an element of $\mathbb{P}_{m-1}(\Omega)$, and define:

$$v_+(t) := \int_a^b \frac{(t-s)_+^{m-1} \, v^{(m)}(s)}{(m-1)!} \, ds \qquad \forall\, t \in [a,b].$$

Then, using (3.39), (3.40), and (3.31), we deduce that

$$\begin{aligned}
&\langle T(u), T(v) \rangle_Y = \langle T^*(T(u)), v \rangle_X = \langle T^*(T(u)), v_+ \rangle_X \\
&= \sum_{j=1}^{n} \lambda_j \langle u_j, v_+ \rangle_X = \sum_{j=1}^{n} \lambda_j \langle v_+, u_j \rangle_X = \sum_{j=1}^{n} \lambda_j \, v_+(t_j) \\
&= \sum_{j=1}^{n} \lambda_j \int_a^b \frac{(t_j-s)_+^{m-1} \, v^{(m)}(s)}{(m-1)!} \, ds = \int_a^b \left\{ \sum_{j=1}^{n} \lambda_j \frac{(t_j-s)_+^{m-1}}{(m-1)!} \right\} v^{(m)}(s) \, ds \\
&= \left\langle \sum_{j=1}^{n} \frac{\lambda_j (t_j - \cdot)_+^{m-1}}{(m-1)!}, T(v) \right\rangle_Y \qquad \forall\, v \in X,
\end{aligned}$$

which implies, since $R(T) = Y$, that

$$T(u)(s) := u^{(m)}(s) = \sum_{j=1}^{n} \frac{\lambda_j \, (t_j - s)_+^{m-1}}{(m-1)!} \qquad \forall\, s \in [a,b].$$

The previous identity allows us to prove the following characterization result for the interpolation spline function u.

Theorem 3.73. *The following properties hold:*

i) $u|_{[a\,,\,t_1]} \in \mathbb{P}_{m-1}([a,\,t_1])$ *and* $u|_{[t_n\,,\,b]} \in \mathbb{P}_{m-1}([t_n\,,\,b])$.
ii) $u|_{[t_{i-1}\,,\,t_i]} \in \mathbb{P}_{2m-1}([t_{i-1}\,,\,t_i])$ *for all* $i \in \{2,3,\cdots,n\}$.
iii) $u \in C^{2m-2}(\Omega)$.

Proof. First note, according to (3.39) and (3.40), that

$$0 = \langle T^*(T(u)), p\rangle_X = \sum_{j=1}^{n} \lambda_j\, p(t_j) \qquad \forall\, p \in \mathbb{P}_{m-1}(\Omega). \tag{3.41}$$

Now, given $s \in [a, t_1]$, we have:

$$u^{(m)}(s) = \sum_{j=1}^{n} \frac{\lambda_j (t_j - s)_+^{m-1}}{(m-1)!} = \sum_{j=1}^{n} \frac{\lambda_j (t_j - s)^{m-1}}{(m-1)!} = \sum_{j=1}^{n} \lambda_j\, \widetilde{p}(t_j),$$

with

$$\widetilde{p}(t) := \frac{(t-s)^{m-1}}{(m-1)!} \in \mathbb{P}_{m-1}(\Omega),$$

and then, according to (3.41), $u^{(m)}(s) = 0$, which shows that $u|_{[a,t_1]} \in \mathbb{P}_{m-1}([a,t_1])$. In turn, given $s \in [t_n, b]$, we have

$$u^{(m)}(s) = \sum_{j=1}^{n} \frac{\lambda_j (t_j - s)_+^{m-1}}{(m-1)!} = 0,$$

since $t_j \leqslant s$ for all $j \in \{1,2,\cdots,n\}$, and then $u|_{[t_n,b]} \in \mathbb{P}_{m-1}([t_n,b])$. On the other hand, for $s \in [t_{i-1}, t_i]$, with $i \in \{2,3,\cdots,n\}$, we have

$$u^{(m)}(s) = \sum_{j=1}^{n} \frac{\lambda_j (t_j - s)_+^{m-1}}{(m-1)!} = \sum_{j=i}^{n} \frac{\lambda_j (t_j - s)^{m-1}}{(m-1)!} \in \mathbb{P}_{m-1}(\Omega),$$

and hence $u|_{[t_{i-1},t_i]} \in \mathbb{P}_{2m-1}([t_{i-1},t_i])$. From now on we assume that $m \geqslant 2$. Then, differentiating $u^{(m)}$, $m-2$ times, we obtain

$$\begin{aligned}
&u^{(2m-2)}(s) = 0 && \forall\, s \in [a,t_1],\\
&u^{(2m-2)}(s) = (-1)^{m-2} \sum_{j=i}^{n} \lambda_j (t_j - s) && \forall\, s \in [t_{i-1},t_i], \quad \forall\, i \in \{2,3,\cdots,n\},\\
&u^{(2m-2)}(s) = (-1)^{m-2} \sum_{j=i+1}^{n} \lambda_j (t_j - s) && \forall\, s \in [t_i,t_{i+1}], \quad \forall\, i \in \{1,2,\cdots,n-1\},\\
&u^{(2m-2)}(s) = 0 && \forall\, s \in [t_n,b].
\end{aligned}$$

According to the above identities, and using in the second one of them the orthogonality given by (3.41), we deduce the following limits:

$$
\begin{aligned}
\lim_{s\in t_1^-} u^{(2m-2)}(s) &= 0, \\
\lim_{s\to t_1^+} u^{(2m-2)}(s) &= (-1)^{m-2}\sum_{j=2}^{n}\lambda_j(t_j-t_1) = (-1)^{m-2}\sum_{j=1}^{n}\lambda_j(t_j-t_1) = 0, \\
\lim_{s\to t_i^-} u^{(2m-2)}(s) &= (-1)^{m-2}\sum_{j=i}^{n}\lambda_j(t_j-t_i) = (-1)^{m-2}\sum_{j=i+1}^{n}\lambda_j(t_j-t_i) \\
&\quad \forall\, i\in\{2,3,\cdots,n-1\}, \\
\lim_{s\to t_i^+} u^{(2m-2)}(s) &= (-1)^{m-2}\sum_{j=i+1}^{n}\lambda_j(t_j-t_i) \\
&\quad \forall\, i\in\{2,3,\cdots,n-1\}, \\
\lim_{s\to t_n^-} u^{(2m-2)}(s) &= (-1)^{m-2}\sum_{j=n}^{n}\lambda_j(t_j-t_n) = 0\,, \quad \text{and} \\
\lim_{s\to t_n^+} u^{(2m-2)}(s) &= 0,
\end{aligned}
$$

which, all together, imply that $u^{(2m-2)}$ is continuous at t_i for all $i \in \{1,2,\cdots,n\}$, and therefore $u \in C^{2m-2}(\Omega)$. However, differentiating once more we obtain

$$
u^{(2m-1)}(s) = (-1)^{m-1}\sum_{j=i}^{n}\lambda_j \quad \forall\, s\in[t_{i-1},t_i], \quad \forall\, i\in\{2,3,\cdots,n\},
$$

and

$$
u^{(2m-1)}(s) = (-1)^{m-1}\sum_{j=i+1}^{n}\lambda_j \quad \forall\, s\in[t_i,t_{i+1}], \quad \forall\, i\in\{1,2,\cdots,n-1\},
$$

from which it follows that

$$
\lim_{s\to t_i^-} u^{(2m-1)}(s) - \lim_{s\to t_i^+} u^{(2m-1)}(s) = (-1)^m\lambda_i \quad \forall\, i\in\{2,3,\cdots,n-1\},
$$

thus proving that $u^{(2m-1)}$ is not continuous.

□

Finally, for $m = 1$ we obtain $2m - 2 = 0$, and in this case the continuity of u in Ω follows from the fact that $u \in X = \mathrm{H}^1(\Omega)$ and from the Sobolev Embedding Theorem (cf. Theorem 10.23), which states, precisely for domains $\Omega \subseteq \mathbb{R}$, that $\mathrm{H}^1(\Omega) \subseteq C(\Omega)$.

3.12 Further results on unbounded operators

In this section we provide additional results on closed operators, as well as on some unbounded operators in general. These include a more general definition of adjoint operator and the corresponding connections between ranges and null spaces. We begin with a complementary subsection on annihilators.

3.12.1 Further results on annihilators

Let X be a Banach space, let $S \subseteq X$, and let $T \subseteq X'$. Recall that the respective annihilators are defined as (cf. Definitions 3.20 and 3.21)

$$S^{\circ} := \Big\{F \in X' : \quad F(x) = 0 \quad \forall\, x \in S\Big\}$$

and

$$^{\circ}T := \Big\{x \in X : \quad F(x) = 0 \quad \forall\, F \in T\Big\},$$

which, as shown in Lemma 3.22, are closed subspaces of X' and X, respectively. Moreover, Lemma 3.26 established that

$$^{\circ}(S^{\circ}) = \overline{\langle S\rangle},$$

where $\overline{\langle S\rangle}$ denotes the closed subspace generated by S. In particular, if S is a subspace, then

$$^{\circ}(S^{\circ}) = \bar{S},$$

and if S is also closed, then

$$^{\circ}(S^{\circ}) = S.$$

Similarly, one may ask what relationship exists, if any, between $\overline{\langle T\rangle}$ and $(^{\circ}T)^{\circ}$. In this regard, it is easy to see that $T \subseteq (^{\circ}T)^{\circ}$ and therefore $\langle T\rangle \subseteq (^{\circ}T)^{\circ}$, from which it follows, since $(^{\circ}T)^{\circ}$ is clearly closed, that $\overline{\langle T\rangle} \subseteq (^{\circ}T)^{\circ}$. Next, we attempt to prove the reverse inclusion using the same reasoning employed in the proof of Lemma 3.24. Indeed, suppose by contradiction that there exists $\widetilde{F} \in (^{\circ}T)^{\circ}$ such that $\widetilde{F} \notin \overline{\langle T\rangle}$. By applying the corollary of the Hahn–Banach Theorem given in Theorem 2.26, it follows that there exists $\widetilde{\mathcal{F}} \in X'' := (X')'$ (the dual of the dual, or simply the bidual) such that $\|\widetilde{\mathcal{F}}\| = 1$, $\widetilde{\mathcal{F}}(\widetilde{F}) = \operatorname{dist}(\widetilde{F}, \overline{\langle T\rangle}) > 0$ and $\widetilde{\mathcal{F}}(G) = 0$ for all $G \in \overline{\langle T\rangle}$. The latter indicates that $\widetilde{\mathcal{F}} \in (\overline{\langle T\rangle})^{\circ}$, which is a closed subspace of X'', while the fact that $\widetilde{\mathcal{F}}(\widetilde{F}) > 0$ implies that $\widetilde{F} \notin {}^{\circ}((\overline{\langle T\rangle})^{\circ}) = \overline{\langle T\rangle}$, which produces no contradiction. However, the situation changes if we further assume that X is a REFLEXIVE Banach space, a concept we already discussed through the operator $\mathcal{J}_X : X \to X''$ defined in (3.24), which in the present analysis means that for each functional $\mathcal{H} \in X''$ there exists a unique $x \in X$ such that

$$\mathcal{H}(F) = F(x) \qquad \forall\, F \in X'.$$

In particular, for $\widetilde{\mathcal{F}}$ there exists $\widetilde{x} \in X$ such that $\widetilde{\mathcal{F}}(F) = F(\widetilde{x})$ for all $F \in X'$. It follows that for each $G \in \overline{\langle T\rangle}$ we have

$$0 = \widetilde{\mathcal{F}}(G) = G(\widetilde{x}),$$

which means that $\widetilde{x} \in {}^{\circ}\overline{\langle T\rangle} = {}^{\circ}T$, and since $\widetilde{F} \in (^{\circ}T)^{\circ}$, necessarily $\widetilde{F}(\widetilde{x}) = 0$. However, we also have $\widetilde{F}(\widetilde{x}) = \widetilde{\mathcal{F}}(\widetilde{F}) > 0$, which constitutes a contradiction.

Thus we have proven the following lemma.

Lemma 3.74. *Let X be a reflexive Banach space, and let T be a subset of X'. Then, the following identity holds:* $(^{\circ}T)^{\circ} = \overline{\langle T\rangle}$.

On the other hand, the following results connect the sum and intersection of closed subspaces with their respective annihilators.

Lemma 3.75. *Let U and V be closed subspaces of a Banach space X. Then*

a) $U \cap V = {}^\circ(U^\circ + V^\circ)$.
b) $U^\circ \cap V^\circ = (U + V)^\circ$.

Proof. For a) we proceed by double inclusion. Let $x \in U \cap V$ and $F \in U^\circ + V^\circ$. It follows that $F = F_U + F_V$ with $F_U \in U^\circ$ and $F_V \in V^\circ$, and since $x \in U$ and $x \in V$, we obtain $F(x) = F_U(x) + F_V(x) = 0$, so that $x \in {}^\circ(U^\circ + V^\circ)$. Conversely, it is clear that $U^\circ \subseteq U^\circ + V^\circ$, and therefore ${}^\circ(U^\circ + V^\circ) \subseteq {}^\circ(U^\circ) = U$. Similarly, ${}^\circ(U^\circ + V^\circ) \subseteq V$, and hence ${}^\circ(U^\circ + V^\circ) \subseteq U \cap V$. On the other hand, for b) we also argue by double inclusion. Let $F \in U^\circ \cap V^\circ$ and $x \in U + V$. It follows that $x = u + v$, with $u \in U$ and $v \in V$, and thus $F(x) = F(u) + F(v) = 0$ because $F \in U^\circ$ and $F \in V^\circ$, which proves that $F \in (U + V)^\circ$. Conversely, since obviously $U \subseteq U + V$, we obtain $(U + V)^\circ \subseteq U^\circ$, and similarly $(U + V)^\circ \subseteq V^\circ$, so that $(U + V)^\circ \subseteq U^\circ \cap V^\circ$.

□

Corollary 3.76. *Let U and V be closed subspaces of a Banach space X. Then*

a) $\overline{(U^\circ + V^\circ)} \subseteq (U \cap V)^\circ$.
b) ${}^\circ(U^\circ \cap V^\circ) = \overline{U + V}$.

Proof. Applying part a) of Lemma 3.75 we have

$$(U \cap V)^\circ = ({}^\circ(U^\circ + V^\circ))^\circ,$$

and using that $\bar{T} \subseteq ({}^\circ T)^\circ$ for every subspace T of X', we deduce with $T = U^\circ + V^\circ$ that

$$\overline{(U^\circ + V^\circ)} \subseteq ({}^\circ(U^\circ + V^\circ))^\circ = (U \cap V)^\circ,$$

which proves a). In turn, applying part b) of Lemma 3.75 we obtain

$${}^\circ(U^\circ \cap V^\circ) = {}^\circ((U + V)^\circ) = \overline{U + V},$$

which proves b).

□

We now aim to characterize the closure of the sum of two closed subspaces, for which we need the auxiliary lemmas to be established next.

Lemma 3.77. *Let U be a closed subspace of a Banach space X. Then*

$$\operatorname{dist}(F, U^\circ) = \sup_{\substack{x \in U \\ \|x\| \leqslant 1}} |F(x)| \qquad \forall\, F \in X'. \tag{3.42}$$

Proof. Given $F \in X'$, $x \in U$ with $\|x\| \leqslant 1$, and $G \in U^\circ$, we have

$$|F(x)| = |F(x) - G(x)| \leqslant \sup_{\substack{z \in X \\ \|z\| \leqslant 1}} |F(z) - G(z)| = \|F - G\|,$$

from which it follows that

$$|F(x)| \leqslant \operatorname{dist}(F, U^\circ) \qquad \forall\, x \in U,\ \|x\| \leqslant 1,$$

and therefore

$$\sup_{\substack{x \in U \\ \|x\| \leqslant 1}} |F(x)| \leqslant \operatorname{dist}(F, U^\circ).$$

On the other hand, given $F \in X'$, we denote $F_0 := F|_U \in U'$. By the analytic version of the Hahn-Banach Theorem there exists $\widetilde{F} \in X'$ such that $\widetilde{F}|_U = F_0$ and $\|\widetilde{F}\|_{X'} = \|F_0\|_{U'}$. It follows that $(F - \widetilde{F}) \in U^\circ$ and therefore

$$\operatorname{dist}(F, U^\circ) \leqslant \|F - (F - \widetilde{F})\| = \|\widetilde{F}\| = \|F_0\|_{U'} = \sup_{\substack{x \in U \\ \|x\| \leqslant 1}} |F_0(x)| = \sup_{\substack{x \in U \\ \|x\| \leqslant 1}} |F(x)|,$$

that is,

$$\operatorname{dist}(F, U^\circ) \leqslant \sup_{\substack{x \in U \\ \|x\| \leqslant 1}} |F(x)|,$$

which completes the proof.

□

Note that (3.42) can also be written as

$$\operatorname{dist}(F, U^\circ) = \sup_{\substack{x \in U \\ \|x\| = 1}} |F(x)| = \sup_{\substack{x \in U \\ x \neq 0}} \frac{|F(x)|}{\|x\|} \qquad \forall\, F \in X'.$$

Lemma 3.78. *Let X be a reflexive Banach space, and let T be a closed subspace of X'. Then*

$$\operatorname{dist}(x, {}^\circ T) = \sup_{\substack{F \in T \\ \|F\| \leqslant 1}} |F(x)| \qquad \forall\, x \in X. \tag{3.43}$$

Proof. Given $x \in X$ and $F \in T$ with $\|F\| \leqslant 1$, we have for each $z \in {}^\circ T$

$$|F(x)| = |F(x - z)| \leqslant \|x - z\|,$$

whence

$$|F(x)| \leqslant \operatorname{dist}(x, {}^\circ T)$$

and therefore

$$\sup_{\substack{F \in T \\ \|F\| \leqslant 1}} |F(x)| \leqslant \operatorname{dist}(x, {}^\circ T).$$

On the other hand, if $x \in {}^\circ T$ it is clear that

$$\operatorname{dist}(x, {}^\circ T) = 0 \leqslant \sup_{\substack{F \in T \\ \|F\| \leqslant 1}} |F(x)|.$$

Moreover, if $x \notin {}^\circ T$, then $\operatorname{dist}(x, {}^\circ T) > 0$, and applying a consequence of the Hahn–Banach Theorem (cf. Theorem 2.26) it follows that there exists $\widetilde{F} \in X'$ such that

$$\|\widetilde{F}\| = 1, \quad \widetilde{F}(x) = \operatorname{dist}(x, {}^\circ T), \quad \text{and} \quad \widetilde{F}(z) = 0 \quad \forall\, z \in {}^\circ T.$$

This means that $\widetilde{F} \in ({}^\circ T)^\circ = T$, which follows from the fact that X is reflexive (cf. Lemma 3.74), and consequently

$$\operatorname{dist}(x, {}^\circ T) = \widetilde{F}(x) \leqslant \sup_{\substack{F \in T \\ \|F\| \leqslant 1}} |F(x)|,$$

which completes the proof.

□

Note that (3.43) can also be written as

$$\operatorname{dist}(x, {}^\circ T) = \sup_{\substack{F \in T \\ \|F\|=1}} |F(x)| = \sup_{\substack{F \in T \\ F \neq 0}} \frac{|F(x)|}{\|F\|} \qquad \forall\, x \in X . \tag{3.44}$$

We are now in position to provide the announced characterization result.

Theorem 3.79. *Let U and V be closed subspaces of a Banach space X. The following statements are equivalent:*

a) *$U + V$ is closed in X.*
b) *$U^\circ + V^\circ = (U \cap V)^\circ$.*
c) *$U^\circ + V^\circ$ is closed in X'.*
d) *$U + V = {}^\circ(U^\circ \cap V^\circ)$.*

Proof.

a) $\Rightarrow$ b) Suppose that $U + V$ is closed in X. By Corollary 3.76 we know that

$$U^\circ + V^\circ \subseteq \overline{(U^\circ + V^\circ)} \subseteq (U \cap V)^\circ .$$

It remains to prove that $(U \cap V)^\circ \subseteq U^\circ + V^\circ$. Indeed, let $F \in (U \cap V)^\circ$, that is, $F \in X'$ and $F(x) = 0$ for all $x \in U \cap V$. Define $\varphi : U + V \to \mathbb{R}$ by

$$\varphi(x) = F(u) \qquad \forall\, x := u + v \in U + V,$$

and let us check that there is no ambiguity in this definition. Indeed, if $x = u + v = \widetilde{u} + \widetilde{v}$ with $u, \widetilde{u} \in U$ and $v, \widetilde{v} \in V$, then $u - \widetilde{u} = \widetilde{v} - v \in U \cap V$ and hence $F(u - \widetilde{u}) = 0$, that is $F(u) = F(\widetilde{u})$. It is easy to see that φ is linear. Moreover, by a consequence of the Open Mapping Theorem (cf. Lemma 3.51) we can choose a decomposition $x = u + v$, $u \in U$, $v \in V$, such that $\|u\| \leqslant C\,\|x\|$ and $\|v\| \leqslant C\,\|x\|$ for all $x \in U + V$, so that

$$|\varphi(x)| = |F(u)| \leqslant \|F\|\,\|u\| \leqslant \|F\|\,C\,\|x\| \qquad \forall\, x \in U + V,$$

showing that $\varphi \in (U + V)'$. Thus, by the Hahn–Banach Theorem there exists $\widetilde{\varphi} \in X'$ such that $\|\widetilde{\varphi}\|_{X'} = \|\varphi\|_{(U+V)'}$ and $\widetilde{\varphi}|_{U+V} = \varphi$. It follows that

$$F = (F - \widetilde{\varphi}) + \widetilde{\varphi},$$

where $(F - \widetilde{\varphi}) \in U^\circ$ and $\widetilde{\varphi} \in V^\circ$. Indeed, given $x \in U$ we have $x = x + 0$, $x \in U$, $0 \in V$, and then $\widetilde{\varphi}(x) = \varphi(x) = F(x)$. Likewise, given $x \in V$ we have $x = 0 + x$, $0 \in U$, $x \in V$, so that $\widetilde{\varphi}(x) = \varphi(x) = F(0) = 0$. This proves that $F \in U^\circ + V^\circ$ and hence $(U \cap V)^\circ \subseteq U^\circ + V^\circ$.

b) $\Rightarrow$ c) Suppose that $U^\circ + V^\circ = (U \cap V)^\circ$. Then it is immediate that $U^\circ + V^\circ$ is closed in X'.

c) $\Rightarrow$ a) Suppose that $U^\circ + V^\circ$ is closed in X. We want to prove that $U + V$ is also closed. We know from another consequence of the Open Mapping Theorem (cf. Lemma 3.52) that there exists $C > 0$ such that

$$\operatorname{dist}(F, U^\circ \cap V^\circ) \leqslant C \left\{ \operatorname{dist}(F, U^\circ) + \operatorname{dist}(F, V^\circ) \right\} \quad \forall\, F \in X'. \tag{3.45}$$

Then, using Lemmas 3.75 and 3.77 we have

$$U^\circ \cap V^\circ = (U+V)^\circ = \big(\overline{U+V}\big)^\circ$$

and

$$\operatorname{dist}(F, U^\circ \cap V^\circ) = \operatorname{dist}\big(F, \big(\overline{U+V}\big)^\circ\big) = \sup_{\substack{z\in \overline{U+V}\\ \|z\|\leqslant 1}} |F(z)|.$$

Thus, applying again Lemma 3.77, the inequality (3.45) can be rewritten as

$$\sup_{\substack{z\in \overline{U+V}\\ \|z\|\leqslant 1}} |F(z)| \leqslant C \left\{ \sup_{\substack{x\in U\\ \|x\|\leqslant 1}} |F(x)| + \sup_{\substack{y\in V\\ \|y\|\leqslant 1}} |F(y)| \right\} \qquad \forall\, F \in X'. \tag{3.46}$$

Next we show that from (3.46) it follows that

$$B_{\overline{U+V}}(0,\, 1/C) \subseteq \overline{B_U(0,1) + B_V(0,1)}. \tag{3.47}$$

Indeed, suppose by contradiction that there exists $z_0 \in \overline{U+V}$ such that $\|z_0\| < 1/C$ and $z_0 \notin \overline{B_U(0,1) + B_V(0,1)}$. Then, applying the geometric form of the Hahn–Banach Theorem (cf. Theorem 2.33) to the sets

$$A := \overline{B_U(0,1) + B_V(0,1)} \quad \text{and} \quad B := \{z_0\},$$

it follows that there exist $F_0 \in X'$, $F_0 \neq 0$, and $\alpha \in \mathbb{R}$ such that

$$F_0(z) < \alpha < F_0(z_0) \qquad \forall\, z \in \overline{B_U(0,1) + B_V(0,1)},$$

and hence, in particular,

$$|F_0(x)| + |F_0(y)| < \alpha < F_0(z_0) \qquad \forall\, x \in B_U(0,1), \quad \forall\, y \in B_V(0,1),$$

which implies

$$\sup_{\substack{x\in U\\ \|x\|\leqslant 1}} |F_0(x)| + \sup_{\substack{y\in V\\ \|y\|\leqslant 1}} |F_0(y)| \leqslant \alpha < F_0(z_0) \leqslant \sup_{\substack{z\in \overline{U+V}\\ \|z\|\leqslant 1/C}} |F_0(z)| = \frac{1}{C} \sup_{\substack{z\in \overline{U+V}\\ \|z\|\leqslant 1}} |F_0(z)|,$$

contradicting (3.46). Therefore, the inclusion (3.47) must hold. Now let $Y := U \times V$ and $Z := \overline{U+V}$ with the product norm and the norm of X, respectively, and define the map $T : Y \to Z$ by $T(u,v) := u + v$ for all $(u,v) \in Y$. It is clear that T is linear, bounded, and according to (3.47)

$$B_Z(0, 1/C) \subseteq \overline{T(B_Y(0,1))}$$

since certainly $B_U(0,1) + B_V(0,1) \subseteq T(B_Y(0,1))$. Then, applying the second part of the proof of the Open Mapping Theorem (cf. Lemma 3.45) we deduce that

$$B_Z(0,\, 1/2C) \subseteq T(B_Y(0,1)),$$

whence it easily follows that T is surjective, that is $Z = T(Y)$, which yields

$$\overline{U+V} = U + V,$$

and therefore $U + V$ is closed.

<u>d) ⇔ a)</u> The proof is completed with this equivalence. Indeed, the implication d) ⇒ a) is obvious, whereas, assuming a), that is, $U + V$ closed, we have, by Lemma 3.75, that

$${}^\circ(U^\circ \cap V^\circ) = {}^\circ((U+V)^\circ) = U + V,$$

which proves d).

□

3.12.2 The adjoint of unbounded operators

Let X and Y be Banach spaces, and let $A : \mathcal{D}(A) \subseteq X \longrightarrow Y$, with $\mathcal{D}(A)$ a vector subspace of X, be a (not necessarily bounded) linear operator. Recall in this case that the graph, the range, and the null space of A are defined, respectively, as

$$\begin{aligned} \mathcal{G}(A) &:= \big\{(x, A(x)) \in X \times Y : \quad x \in \mathcal{D}(A)\big\}, \\ R(A) &:= \big\{A(x) \in Y : \quad x \in \mathcal{D}(A)\big\}, \\ N(A) &:= \big\{x \in \mathcal{D}(A) : \quad A(x) = 0 \in Y\big\}. \end{aligned}$$

Moreover, from Definition 3.32 (cf. Section 3.6) we know that A is said to be CLOSED if $\mathcal{G}(A)$ is a closed subspace of $X \times Y$, and that in this case $N(A)$ is a closed subspace of X (cf. Lemma 3.59).

At this point we remark that, in applications, most operators that are unbounded are closed with dense domain in X.

Definition 3.80. *Let $A : \mathcal{D}(A) \subseteq X \longrightarrow Y$ be a linear operator such that $\mathcal{D}(A)$ is dense in X. We now introduce an adjoint operator $A' : \mathcal{D}(A') \subseteq Y' \to X'$, which we define as follows. Set*

$$\mathcal{D}(A') := \Big\{G \in Y' : \quad G \circ A \in \mathcal{D}(A)'\Big\}.$$

It is immediate that $\mathcal{D}(A')$ is a vector subspace of Y'. Now, given $G \in \mathcal{D}(A')$, consider the functional $g := G \circ A \in \mathcal{D}(A)'$, that is, assuming without loss of generality that the spaces are real,

$$\begin{aligned} g : \mathcal{D}(A) &\longrightarrow \mathbb{R} \\ x &\longmapsto g(x) := G(A(x)). \end{aligned}$$

Clearly, $|g(x)| \leqslant c\,\|x\|$ for all $x \in \mathcal{D}(A)$, with $c = \|G \circ A\|_{\mathcal{D}(A)'}$. In addition, given $x \in X$, the density of $\mathcal{D}(A)$ implies the existence of $\{x_n\}_{n\in\mathbb{N}} \subseteq \mathcal{D}(A)$ such that $x_n \overset{n\to\infty}{\longrightarrow} x$. It follows that $\{g(x_n)\}_{n\in\mathbb{N}}$ is a Cauchy sequence in $\mathbb{R}$, and hence we may define

$$F(x) := \lim_{n\to\infty} g(x_n).$$

Moreover, this limit is independent of the sequence converging to x. Indeed, given another sequence $\{\widetilde{x}_n\}_{n\in\mathbb{N}} \subseteq \mathcal{D}(A)$ such that $\widetilde{x}_n \overset{n\to\infty}{\longrightarrow} x$, we have

$$|g(x_n) - g(\widetilde{x}_n)| \leqslant c\,\|x_n - \widetilde{x}_n\| \overset{n\to\infty}{\longrightarrow} 0.$$

Thus, $F : X \longrightarrow \mathbb{R}$ is well defined, clearly linear, and bounded since

$$|F(x)| = \lim_{n\to\infty} |g(x_n)| \leqslant c\,\|x\| \qquad \forall\, x \in X,$$

whence $F \in X'$, and $F = g$ on $\mathcal{D}(A)$. Therefore, the operator $A' : \mathcal{D}(A') \to X'$, defined by $A'(G) = F$, is called the adjoint of A. It is easily verified that A' is linear and satisfies

$$A'(G)(x) = G(A(x)) \qquad \forall\, x \in \mathcal{D}(A), \quad \forall\, G \in \mathcal{D}(A').$$

It is important to observe here that $\mathcal{D}(A')$ is not necessarily dense in Y', even if A is closed. In fact, it can be proved that if A is closed, then $\mathcal{D}(A')$ is dense in Y' only with respect to another topology, namely the weak* topology, a concept that is not required for the present discussion and that will be explained in detail later in Section 7.4.3. In the same Chapter 7 we will prove that if Y is reflexive, then $\mathcal{D}(A')$ is dense in Y' with respect to the strong topology (cf. Theorem 7.59).

Theorem 3.81. *Let $A : \mathcal{D}(A) \subseteq X \longrightarrow Y$ be an unbounded linear operator such that $\mathcal{D}(A)$ is dense in X. Then $A' : \mathcal{D}(A') \subseteq Y' \longrightarrow X'$ is closed, that is, $\mathcal{G}(A')$ is a closed subspace of $Y' \times X'$.*

Proof. Let $\{G_n\}_{n\in\mathbb{N}} \subseteq \mathcal{D}(A')$ be such that

$$G_n \overset{n\to\infty}{\longrightarrow} G \in Y' \quad \text{and} \quad A'(G_n) \overset{n\to\infty}{\longrightarrow} F \in X' .$$

Then, for each $x \in \mathcal{D}(A)$,

$$(G \circ A)(x) = G(A(x)) = \lim_{n\to\infty} G_n(A(x)) = \lim_{n\to\infty} A'(G_n)(x) = F(x),$$

and hence $G \circ A = F|_{\mathcal{D}(A)} \in \mathcal{D}(A)'$. Therefore, $G \in \mathcal{D}(A')$ and $A'(G) = F$, which proves that A' is closed.

□

3.12.3 Relations between graphs, ranges, and null spaces

Let X and Y be Banach spaces, and let $A : \mathcal{D}(A) \subseteq X \longrightarrow Y$ be an unbounded linear operator with $\mathcal{D}(A)$ dense in X. The goal of this section is to establish the relations between the graphs $\mathcal{G}(A)$ and $\mathcal{G}(A')$, as well as the connections between the ranges and null spaces of A and A', respectively. In what follows, given $F \in X'$ and $G \in Y'$, we define $(F, G) \in (X \times Y)' \equiv X' \times Y'$ as

$$(F, G)(x, y) := F(x) + G(y) \quad \forall\, (x, y) \in X \times Y.$$

Lemma 3.82. *Let $\boldsymbol{\mathcal{S}} : Y' \times X' \longrightarrow X' \times Y'$ be defined by $\boldsymbol{\mathcal{S}}(G, F) := (-F, G)$ for every $(G, F) \in Y' \times X'$. Then $\boldsymbol{\mathcal{S}}(\mathcal{G}(A')) = \mathcal{G}(A)^\circ$.*

Proof. Let $(G, A'(G)) \in \mathcal{G}(A')$ with $G \in \mathcal{D}(A')$, and consider $\boldsymbol{\mathcal{S}}(G, A'(G)) = (-A'(G), G)$. Then, given $(x, A(x)) \in \mathcal{G}(A)$ with $x \in \mathcal{D}(A)$, we have

$$\boldsymbol{\mathcal{S}}(G, A'(G))(x, A(x)) = -A'(G)(x) + G(A(x)) = 0,$$

which proves that $\boldsymbol{\mathcal{S}}(G, A'(G)) \in \mathcal{G}(A)^\circ$, and thus $\boldsymbol{\mathcal{S}}(\mathcal{G}(A')) \subseteq \mathcal{G}(A)^\circ$. Conversely, let $(F, G) \in X' \times Y'$ such that $(F, G) \in \mathcal{G}(A)^\circ$. It follows that for every $x \in \mathcal{D}(A)$

$$0 = (F, G)(x, A(x)) := F(x) + G(A(x)),$$

that is, $A'(G) = -F$, from which

$$(F, G) = (-A'(G), G) = \boldsymbol{\mathcal{S}}(G, A'(G)) \in \boldsymbol{\mathcal{S}}(\mathcal{G}(A')),$$

and therefore $\mathcal{G}(A)^\circ \subseteq \boldsymbol{\mathcal{S}}(\mathcal{G}(A'))$.

□

Notice that from the previous lemma one obtains an alternative proof for the fact that A' is closed (cf. Theorem 3.81). Indeed, since $\boldsymbol{\mathcal{S}}$ is a linear isomorphism and $\mathcal{G}(A)^\circ$ is certainly closed, it follows that $\mathcal{G}(A') = \boldsymbol{\mathcal{S}}^{-1}(\mathcal{G}(A)^\circ)$ is a closed subspace of $Y' \times X'$.

Lemma 3.83. *Let U and V be the subspaces of $X \times Y$ given by*

$$U := \mathcal{G}(A) := \Big\{(x, A(x)) : \quad x \in \mathcal{D}(A)\Big\},$$

and

$$V := X \times \{0\} := \Big\{(x, 0) : \quad x \in X\Big\}.$$

Then the following identities hold:

a) $N(A) \times \{0\} = U \cap V$.
b) $X \times R(A) = U + V$.
c) $\{0\} \times N(A') = U^\circ \cap V^\circ$.
d) $R(A') \times Y' = U^\circ + V^\circ$.

Proof.
a) Given $(x, y) \in X \times Y$ we clearly have

$$\begin{aligned} (x, y) \in U \cap V &\Leftrightarrow (x, y) \in U \text{ and } (x, y) \in V \\ &\Leftrightarrow x \in \mathcal{D}(A),\, y = A(x) \quad \text{and} \quad y = 0 \\ &\Leftrightarrow (x, y) \in N(A) \times \{0\}, \end{aligned}$$

which proves the required equality.

b) Given $(x, y) \in U + V$, there exist $z \in \mathcal{D}(A)$ and $\omega \in X$ such that $(x, y) = (z, A(z)) + (\omega, 0)$, that is,

$$(x, y) = (z + \omega, A(z)) \in X \times R(A).$$

Conversely, if $(x, y) \in X \times R(A)$, then $x \in X$ and there exists $z \in \mathcal{D}(A)$ such that $y = A(z)$, from which it follows that

$$(x, y) = (x, A(z)) = (z, A(z)) + (x - z, 0) \in \mathcal{G}(A) + \big(X \times \{0\}\big) = U + V.$$

c) Given $(F, G) \in \{0\} \times N(A')$, we have $F \equiv 0$ and $A'(G) = 0$. Then, for each $x \in \mathcal{D}(A)$,

$$(F, G)(x, A(x)) = F(x) + G(A(x)) = A'(G)(x) = 0,$$

so that $(F, G) \in U^\circ$. Moreover, it is straightforward to see that $(F, G) \in V^\circ$, and consequently $(F, G) \in U^\circ \cap V^\circ$. Conversely, let $(F, G) \in U^\circ \cap V^\circ$. This means that

$$(F, G)(x, A(x)) = 0 \quad \forall\, x \in \mathcal{D}(A) \quad \text{and} \quad (F, G)(x, 0) = 0 \quad \forall\, x \in X,$$

that is, $F(x) + G(A(x)) = 0$ for all $x \in \mathcal{D}(A)$ and $F(x) = 0$ for all $x \in X$. It follows that $F = 0$ and $A'(G) = 0$, which shows that $(F, G) \in \{0\} \times N(A')$.

d) Given $(F, G) \in R(A') \times Y'$, there exists $\widetilde{G} \in Y'$ such that $F = A'(\widetilde{G})$, and then

$$(F, G) = (A'(\widetilde{G}), G) = (A'(\widetilde{G}), -\widetilde{G}) + (0,\, G + \widetilde{G}) \in U^\circ + V^\circ.$$

Indeed, by applying Lemma 3.82 we have

$$(A'(\widetilde{G}), -\widetilde{G}) = -\boldsymbol{\mathcal{S}}(\widetilde{G}, A'(\widetilde{G})) \in \mathcal{G}(A)^\circ = U^\circ,$$

and at the same time it is easy to see that $(0, G+\widetilde{G}) \in V^\circ$. Conversely, let $(F,G) \in X' \times Y'$ such that $(F,G) \in U^\circ + V^\circ$, that is,

$$(F,G) = (F_1,G_1) + (F_2,G_2)\,,$$

where $(F_1,G_1) \in U^\circ$ and $(F_2,G_2) \in V^\circ$. It follows that

$$0 = (F_1,G_1)(x,A(x)) = F_1(x) + G_1(A(x)) = F_1(x) + A'(G_1)(x) \quad \forall\, x \in \mathcal{D}(A)$$

and

$$0 = (F_2,G_2)(x,0) = F_2(x) \quad \forall\, x \in X\,,$$

which implies $F_1 = -A'(G_1)$ and $F_2 = 0$, and therefore

$$(F,G) = (A'(-G_1),G) \in R(A') \times Y'\,.$$

□

Having established the previous lemmas, we now prove the following.

Theorem 3.84. *Let X and Y be Banach spaces, and let $A : \mathcal{D}(A) \subseteq X \longrightarrow Y$ be a closed linear operator with $\overline{\mathcal{D}(A)} = X$. Then*

a) $N(A) = {}^\circ R(A')$.
b) $N(A') = R(A)^\circ$.
c) $N(A)^\circ \supseteq \overline{R(A')}$.
d) ${}^\circ N(A') = \overline{R(A)}$.

Proof. First note that $U := \mathcal{G}(A)$ and $V := X \times \{0\}$ are clearly closed subspaces of $X \times Y$, since A is closed and X is Banach. In what follows, we use the fact that the annihilator of a subset of a product space transfers to each of its components.

a) Applying parts d) and a) of Lemma 3.83, and part a) of Lemma 3.75, we deduce that

$$\begin{aligned} {}^\circ R(A') \times \{0\} &= {}^\circ R(A') \times {}^\circ(Y') \\ &= {}^\circ(R(A') \times Y') = {}^\circ(U^\circ + V^\circ) \\ &= U \cap V = N(A) \times \{0\}\,, \end{aligned}$$

from which ${}^\circ R(A') = N(A)$, proving a).

b) Applying now parts b) and c) of Lemma 3.83 and part b) of Lemma 3.75, we obtain

$$\begin{aligned} \{0\} \times R(A)^\circ &= X^\circ \times R(A)^\circ \\ &= (X \times R(A))^\circ = (U+V)^\circ \\ &= U^\circ \cap V^\circ = \{0\} \times N(A')\,, \end{aligned}$$

whence $R(A)^\circ = N(A')$, proving b).

c) Since $\overline{\langle T \rangle} \subseteq ({}^\circ T)^\circ$ for all T subset of a dual space, it clearly follows that

$$\overline{R(A')} \subseteq ({}^\circ R(A'))^\circ = N(A)^\circ\,,$$

where the last equality uses part a) of this theorem.

d) Recalling that ${}^{\circ}(W^{\circ}) = \overline{\langle W \rangle}$ for all W subset of a normed space (cf. Lemma 3.26) and applying part b) of the present theorem, we obtain

$$\overline{R(A)} = {}^{\circ}(R(A)^{\circ}) = {}^{\circ}N(A'),$$

which proves d).

□

We observe here, by virtue of Lemma 3.74, that if X is reflexive, then c) becomes an equality, that is, $\overline{R(A')} = ({}^{\circ}R(A'))^{\circ} = N(A)^{\circ}$.

The next goal is to complement the analysis of Section 3.10 with some additional results on closed-range operators. In particular, the following theorem generalizes the statement in Theorem 3.66 to the case of a closed operator, without assuming reflexivity of any of the spaces involved.

Theorem 3.85. *Let X and Y be Banach spaces, and let $A : \mathcal{D}(A) \subseteq X \to Y$ be a closed linear operator with $\mathcal{D}(A)$ dense in X. The following statements are equivalent:*

a) *$R(A)$ is closed.*
b) *$R(A') = N(A)^{\circ}$.*
c) *$R(A')$ is closed.*
d) *$R(A) = {}^{\circ}N(A')$.*

Proof. First note that $R(A)$ is closed in Y if and only if $X \times R(A)$ is closed in $X \times Y$, which, according to part b) of Lemma 3.83, is equivalent to $U + V$ being closed. On the other hand, the identity $R(A') = N(A)^{\circ}$ holds if and only if

$$R(A') \times Y' = N(A)^{\circ} \times Y' = N(A)^{\circ} \times \{0\}^{\circ} = (N(A) \times \{0\})^{\circ},$$

which, according to parts a) and d) of the same Lemma 3.83, is equivalent to establishing that $U^{\circ} + V^{\circ} = (U \cap V)^{\circ}$. Similarly, it is clear that $R(A')$ is closed in X' if and only if $R(A') \times Y'$ is closed in $X' \times Y'$, which, according to part d) of Lemma 3.83, is equivalent to $U^{\circ} + V^{\circ}$ being closed in $X' \times Y'$. Finally, the possible equality $R(A) = {}^{\circ}N(A')$ holds if and only if

$$X \times R(A) = X \times {}^{\circ}N(A') = {}^{\circ}\{0\} \times {}^{\circ}N(A') = {}^{\circ}(\{0\} \times N(A')),$$

which, by b) and c) of Lemma 3.83, is equivalent to $U + V = {}^{\circ}(U^{\circ} \cap V^{\circ})$. Consequently, from the above analysis it is clear that the equivalence among the statements of this theorem follows directly from Theorem 3.79.

□

The following two theorems, whose proofs make use of the characterization given in Theorem 3.61, establish the respective equivalences between surjectivity and injectivity and closed range for the case of closed operators and their adjoints.

Theorem 3.86. *Let X and Y be Banach spaces, and let $A : \mathcal{D}(A) \subseteq X \longrightarrow Y$ be a closed linear operator with $\mathcal{D}(A)$ dense in X. The following statements are equivalent:*

a) *A is surjective, that is, $R(A) = Y$.*

b) *There exists* $c \geqslant 0$ *such that*

$$\|G\| \leqslant c\,\|A'(G)\| \qquad \forall\, G \in \mathcal{D}(A').$$

c) $N(A') = \{0\}$ *and* $R(A')$ *is closed.*

Proof.

a) $\Rightarrow$ b). Suppose that $R(A) = Y$. It follows from parts c) and d) of Theorem 3.85 that $R(A')$ is closed and that $Y = \overline{R(A)} = {}^{\circ}N(A')$. Taking the right annihilator we obtain $N(A') = \overline{N(A')} \subseteq \left({}^{\circ}N(A')\right)^{\circ} = Y^{\circ} = \{0\}$, whence $N(A') = \{0\}$. Finally, it suffices to apply to the operator A' the characterization result given by Theorem 3.61.

b) $\Rightarrow$ c). This implication also follows directly from the characterization result given by Theorem 3.61, or in particular, since clearly $N(A') = \{0\}$, it is a consequence of Theorem 3.58.

c) $\Rightarrow$ a). Suppose that $N(A') = \{0\}$ and $R(A')$ is closed in X'. Then, applying part d) of Theorem 3.85 we clearly have

$$R(A) = {}^{\circ}N(A') = {}^{\circ}\{0\} = Y.$$

□

We observe here that the Banach-Steinhaus Theorem (cf. Theorem 3.46) provides an alternative proof of the first implication of the previous theorem. Indeed, suppose that $R(A) = Y$, and consider the set

$$S := \left\{G \in \mathcal{D}(A') : \quad \|A'(G)\| \leqslant 1\right\}.$$

Given $y \in Y$, there exists $x \in \mathcal{D}(A)$ such that $A(x) = y$, and then, for each $G \in S$ we have

$$|G(y)| = |G(A(x))| = |A'(G)(x)| \leqslant \|A'(G)\|\,\|x\| \leqslant \|x\|,$$

whence

$$\sup_{G \in S} |G(y)| \leqslant \|x\| < +\infty \qquad \forall\, y \in Y.$$

Thus, applying Theorem 3.46 to the setting

$$X \leftarrow Y, \quad Y \leftarrow \mathbb{R}, \quad I \leftarrow S, \quad A_i \leftarrow G,$$

we deduce that there exists $M > 0$ such that $\|G\| \leqslant M$ for all $G \in S$, that is, S is a bounded subset of Y'. Now, let $G \in \mathcal{D}(A')$ be such that $A'(G) \neq 0$. It is clear that $\dfrac{G}{\|A'(G)\|} \in S$, and hence $\dfrac{\|G\|}{\|A'(G)\|} \leqslant M$, i.e., $\|G\| \leqslant M\,\|A'(G)\|$. On the other hand, if $G \in \mathcal{D}(A')$ is such that $A'(G) = 0$, then, using part b) of Theorem 3.84, we have $G \in N(A') = R(A)^{\circ} = Y^{\circ} = \{0\}$, which means that $G = 0$, and therefore the inequality is trivially satisfied.

On the other hand, we note in advance that the Hilbert space version of Theorem 3.86 is detailed later, at the beginning of Exercise 3.36, for the case of a bounded operator. In turn, this last result together with the Hilbert space version of Theorem 3.87 (to be stated below) are later summarized in Lemma 4.7 through specific proofs of them.

Theorem 3.87. *Let X and Y be Banach spaces, and let $A : \mathcal{D}(A) \subseteq X \longrightarrow Y$ be a closed linear operator with $\mathcal{D}(A)$ dense in X. The following statements are equivalent:*

a) *A' is surjective, that is, $R(A') = X'$.*
b) *There exists $c \geqslant 0$ such that*

$$\|x\| \leqslant c\|A(x)\| \qquad \forall x \in \mathcal{D}(A).$$

c) *$N(A) = \{0\}$ and $R(A)$ is closed.*

Proof. It is analogous to the proof of the previous theorem. The details are left to be filled by the readers (see Exercise 3.55).

□

It is important to note here, as deduced from the equivalence between a) and c) of Theorem 3.86, or simply from part b) of Theorem 3.84, that the surjectivity of A implies the injectivity of A'. In turn, from the equivalence between a) and c) of Theorem 3.87, or from part a) of Theorem 3.84, it follows that the surjectivity of A' implies the injectivity of A. Nevertheless, it is also clear from the same Theorems 3.86 and 3.87, that in order to obtain the reciprocal implications, and therefore the equivalence between the injectivity of one and the surjectivity of the other, it is necessary that the ranges $R(A)$ and $R(A')$ be both closed. Accordingly, and in virtue of Theorem 3.85, it is observed that a particular case in which this occurs is when X or Y has finite dimension. This is summarized in the following result.

Lemma 3.88. *Let X and Y be Banach spaces, one of them of finite dimension, and let $A : \mathcal{D}(A) \subseteq X \longrightarrow Y$ be a closed linear operator with $\mathcal{D}(A)$ dense in X. Then:*

a) *A is surjective if and only if A' is injective.*
b) *A' is surjective if and only if A is injective.*

We end this section with the equivalence between the eventual boundedness of A and its adjoint A', as well as between the bijectivity of one and the other. More precisely, we have the following theorems.

Theorem 3.89. *Let X and Y be Banach spaces, and let $A : \mathcal{D}(A) \subseteq X \longrightarrow Y$ be a closed linear operator with $\mathcal{D}(A)$ dense in X. The following statements are equivalent:*

a) *$\mathcal{D}(A) = X$.*
b) *A is bounded.*
c) *$\mathcal{D}(A') = Y'$.*
d) *A' is bounded.*

In this case, we necessarily have $\|A\| = \|A'\|$.

Proof.

<u>a) ⇒ b)</u>. This is a direct consequence of the Closed Graph Theorem (cf. Theorem 3.35).

<u>b) ⇒ c)</u>. This follows easily from Definition 3.80 since, being A bounded, all functionals $G \in Y'$ satisfy the boundedness condition required by that definition.

<u>c) ⇒ d)</u>. As in the first implication, this also follows from the Closed Graph Theorem, recalling that A' is a closed operator (cf. Theorem 3.81).

d) $\Rightarrow$ a). Suppose that $A' : \mathcal{D}(A') \subseteq Y' \longrightarrow X'$ is bounded and let us first see that $\mathcal{D}(A')$ is closed. Indeed, let $\{G_n\}_{n\in\mathbb{N}} \subseteq \mathcal{D}(A')$ be such that $G_n \overset{n\to\infty}{\longrightarrow} G \in Y'$. Thanks to the boundedness of A' we have that $\{A'(G_n)\}_{n\in\mathbb{N}}$ is Cauchy in X', and therefore there exists $F \in X'$ such that $A'(G_n) \overset{n\to\infty}{\longrightarrow} F$. Then, since A' is closed it follows that $G \in \mathcal{D}(A')$ and $A'(G) = F$, which shows that $\mathcal{D}(A')$ is indeed a closed subspace of Y'. On the other hand, given the closed subspaces of $X \times Y$

$$U := \mathcal{G}(A) \qquad \text{and} \qquad V := \{0\} \times Y,$$

it is easy to see that

$$U + V = \mathcal{D}(A) \times Y \qquad \text{and} \qquad U^\circ + V^\circ = X' \times \mathcal{D}(A').$$

Then, according to the previous analysis we have that $U^\circ + V^\circ$ is closed, and applying one of the equivalences given by Theorem 3.79, we conclude that $U + V$ is also closed. Thus, it follows that $\mathcal{D}(A)$ is closed, and hence $X = \overline{\mathcal{D}(A)} = \mathcal{D}(A)$. Finally, it remains to see that, under any of the statements of this theorem, we have $\|A\| = \|A'\|$. Indeed, in such case we have $A \in \mathcal{L}(X,Y)$ and $A' \in \mathcal{L}(Y',X')$, and the required identity was already established in Lemma 3.11.

□

Theorem 3.90. *Let X and Y be Banach spaces, and let $A : \mathcal{D}(A) \subseteq X \longrightarrow Y$ be a closed linear operator with $\mathcal{D}(A)$ dense in X. Then A is bijective if and only if A' is also bijective.*

Proof. This follows directly from the equivalences between parts a) and c) of Theorems 3.86 and 3.87, together with the identities b) and d) of Theorem 3.85.

□

3.13 Exercises

▸ **3.1** *Let $(X, \langle\cdot,\cdot\rangle)$ be a Hilbert space, and let $A \in \mathcal{L}(X)$. Prove that*

$$\|A\|_{\mathcal{L}(X)} = \sup_{\substack{x,\, z\in X \\ \|x\|,\, \|z\| \leqslant 1}} |\langle A(x), z\rangle| .$$

▸ **3.2** *Let X, Y, Z be Banach spaces, and let $A \in \mathcal{L}(X,Y)$, $B \in \mathcal{L}(Z,X)$. Prove that there holds $(AB)' = B'\,A'$.*

▸ **3.3** *Let X be the space of continuous functions on $[0,1]$ endowed with the uniform norm (cf. (2.27)). Given the polynomials $p_j \in X$, $j \in \{0,1,\dots,n\}$, with $p_j(t) = t^j$ for all $t \in [0,1]$, define the operator $A : X \to X$ by*

$$A(u) := \sum_{j=0}^{n} \left\{ \int_0^1 u(t)\, p_j(t)\, dt \right\} p_j \qquad \forall\, u \in X .$$

Prove that $A \in \mathcal{L}(X,X)$ and explicitly find the adjoint operator $A' : X' \to X'$.

Hint: *For every $j \in \{0,1,\dots,n\}$ define $F_j : X \to \mathbb{R}$ by $F_j(u) := \displaystyle\int_0^1 u(t)\, p_j(t)\, dt$ for all $u \in X$.*

▸ **3.4** *Consider two normed vector spaces X and Y.*

a) *Let* $A_0 \in \mathcal{L}(X,Y)$ *be such that* $A_0^{-1} \in \mathcal{L}(Y,X)$. *Prove that there exists an operator* $\mathcal{T}_0 \in \mathcal{L}(\mathcal{L}(X,Y),\mathcal{L}(Y,X))$ *such that* $\mathcal{T}_0 A_0 = A_0^{-1}$ *and* $\|\mathcal{T}_0\| = \|A_0^{-1}\|/\|A_0\|$.

b) *Prove that if* Y *is Banach and* $\mathcal{T} \in \mathcal{L}(\mathcal{L}(X,Y),\mathcal{L}(Y',X'))$ *is a bijective operator, then* $\mathcal{T}^{-1} \in \mathcal{L}(\mathcal{L}(Y',X'),\mathcal{L}(X,Y))$.

▸ **3.5** Let X,Y be Banach spaces, and consider $A \in \mathcal{L}(X,Y)$.

a) Prove that $R(A)^0 = N(A')$ and that

$$N(A') = \{0\} \quad \text{if and only if} \quad \overline{R(A)} = Y.$$

b) Prove that if $N(A) = \{0\}$ and $R(A) = Y$, then $(A')^{-1} = (A^{-1})'$.

▸ **3.6** Let X,Y be Hilbert spaces, and consider $A \in \mathcal{L}(X,Y)$ such that $N(A) = \{0\}$ and $R(A) = Y$. Prove that $(A^*)^{-1} = (A^{-1})^*$.

▸ **3.7** Let M be a closed subspace of a normed vector space X. Prove that the annihilator of M is isomorphic to $\left(\frac{X}{M}\right)'$.

▸ **3.8** ([115, Chapter III, Section 5]) Let M be a subspace of a Banach space X. Prove that the quotient space X/M is Banach if and only if M is closed.

▸ **3.9** Let X,Y be normed vector spaces, and let $T : X' \to Y'$ be a closed and bijective linear operator. Prove that $T^{-1} \in \mathcal{L}(Y',X')$.

▸ **3.10** Let X,Y be Banach spaces, and let $A : X \to Y$ be a linear operator. Prove that if $GA \in X'$ for all $G \in Y'$, then $A \in \mathcal{L}(X,Y)$.
Hint: Use the Banach-Steinhaus Theorem.

▸ **3.11** Prove what is indicated in Example 3.42 using the Closed Graph Theorem.

▸ **3.12** Let X be a Banach space, and let $A \in \mathcal{L}(X,X)$ be such that $\|A\| < 1$. Prove that the operator $(I+A) : X \to X$ is bijective, and hence invertible, and that

$$(I+A)^{-1} = \sum_{n=0}^{\infty} (-1)^n A^n,$$

where the series converges absolutely in $\mathcal{L}(X,X)$. Also show that

$$\|(I+A)^{-1}\| \leqslant \frac{1}{1-\|A\|}.$$

▸ **3.13** Let X,Y be Banach spaces, and let $T : \mathcal{D}(T) \subseteq X \to Y$ be a linear operator. The graph of T is denoted by $\mathcal{G}_T$ and is defined as

$$\mathcal{G}_T := \big\{(x,y) \in X \times Y : \quad x \in D(T),\ y = T(x)\big\}.$$

Assume that $\mathcal{D}(T)$ and $\mathcal{G}_T$ are closed subspaces of X and $X \times Y$, respectively. Prove that T is bounded on $\mathcal{D}(T)$.

▸ **3.14** Let X,Y be normed vector spaces, and let $T : D(T) \subseteq X \to Y$ be a linear operator. A linear operator $\widetilde{T} : D(\widetilde{T}) \subseteq X \to Y$ is an *extension of* T if $D(T) \subseteq D(\widetilde{T})$ and $T(x) = \widetilde{T}(x)$ for all $x \in D(T)$.

Definition. *We say that* $\overline{T} : D(\overline{T}) \subseteq X \to Y$ *is the* closure *of* T, *if:*

i) $\overline{T}$ *is a closed linear operator.*
ii) $\overline{T}$ *is an extension of* T.
iii) *If* $\widetilde{T} : D(\widetilde{T}) \subseteq X \to Y$ *is any operator with properties* **i)** *and* **ii)**, *then* $\widetilde{T}$ *is an extension of* $\overline{T}$.

Let X, Y be Banach spaces, and let $T : D(T) \subseteq X \to Y$ be a linear operator. Prove that T has a closure $\bar{T}$ if and only if the following condition holds:

$$\{x_n\}_{n\in\mathbf{N}} \subseteq D(T), \quad x_n \to 0, \quad T(x_n) \to y \qquad \Rightarrow \qquad y = 0.$$

▸ **3.15** *Let X be the space of continuous functions on $[0, \pi]$ endowed with the uniform norm. Given the trigonometric functions $p_j, q_j \in X$, $j \in \{0, 1, ..., n\}$, with $p_j(t) = \sin(jt)$ and $q_j(t) = \cos(jt)$, for all $t \in [0, \pi]$, define the operator $A : X \to X$ by*

$$A(u) := \sum_{j=0}^{n} \left\{ \int_0^{\pi} u(t)\, p_j(t)\, dt \right\} p_j + \sum_{j=0}^{n} \left\{ \int_0^{\pi} u(t)\, q_j(t)\, dt \right\} q_j \qquad \forall\, u \in X\,.$$

Prove that $A \in \mathcal{L}(X, X)$ and explicitly find the adjoint operator $A' : X' \to X'$.

▸ **3.16** *Let X, Y be Hilbert spaces, and consider $A \in \mathcal{L}(X, Y)$. Recall that the Hilbert adjoint operator of A is denoted by $A^* : Y \to X$, where for each $y \in Y$, $A^*y \in X$ is the unique element (given by the Riesz Representation Theorem) such that*

$$\langle A(x), y\rangle_Y = \langle x, A^*(y)\rangle_X \qquad \forall\, x \in X\,.$$

a) *Prove again that $A^* \in \mathcal{L}(Y, X)$ and $\|A^*\| = \|A\|$, and also deduce that $(A^*)^* = A$.*
b) *Prove that A^* is injective if and only if $R(A)$ is dense in Y.*
c) *Suppose that there exists $\beta > 0$ such that*

$$\inf_{z\,\in\, N(A)} \|x - z\| \leqslant \beta\, \|A(x)\| \qquad \forall\, x \in X,$$

and prove that $R(A) = N(A^*)^{\perp}$.
d) *Prove that $A^* = R_X\, A'\, R_Y^{-1}$, where $R_X : X' \to X$ and $R_Y : Y' \to Y$ denote the Riesz mappings. Conclude moreover that ${}^{\circ}N(A') = N(A^*)^{\perp}$.*

▸ **3.17** *Let U be a normed vector space and V a Banach space. Moreover, let $P : U' \to \mathcal{L}(U, V)$ be a closed linear operator such that $N(P)$ is the zero functional on U and $R(P) = \mathcal{L}(U, V)$. Prove that there exists a constant $\alpha > 0$ such that*

$$\alpha\, \|F\|_{U'} \leqslant \|P(F)\|_{\mathcal{L}(U,V)} \quad \forall\, F \in \mathcal{D}(P)\,.$$

▸ **3.18** *Let X, Y be Banach spaces, and let $A : X \to Y$ be a linear operator. Prove that if $GA \in X'$ for all $G \in Y'$, then $A \in \mathcal{L}(X, Y)$.*

HINT: *Provide an alternative proof to the one suggested in Exercise 3.10, this time using the Closed Graph Theorem instead of the Banach-Steinhaus Theorem.*

▸ **3.19**

a) *Let $(H, \langle\cdot,\cdot\rangle)$ be a* HILBERT *space. Given $n \in \mathbb{N}$, $\{p_1, p_2, \ldots, p_n\} \subseteq H$ and a set of functionals $\{F_1, F_2, \ldots, F_n\} \subseteq H'$, define the operator $A : H \to H$ by*

$$A(u) := \sum_{j=1}^{n} F_j(u)\, p_j \qquad \forall\, u \in H\,.$$

Prove that $A \in \mathcal{L}(H, H)$ and explicitly find the operator $A^* : H \to H$.
b) *Let $(H, \langle\cdot,\cdot\rangle)$ be the* HILBERT *space $\mathrm{L}^2(0,1)$ equipped with the inner product*

$$\langle u, v\rangle := \int_0^1 u(t)\, v(t)\, dt \qquad \forall\, u, v \in H\,.$$

Given the polynomials $p_j \in H$, $j \in \{1, ..., n\}$, with $p_j(t) = t^j$ for all $t \in (0, 1)$, define the operator $A : H \to H$ by

$$A(u) := \sum_{j=1}^{n} \left\{ \int_0^1 u(t)\, p_j(t)\, dt \right\} p_j \qquad \forall\, u \in H\,.$$

Prove that A is linear, bounded, and SELF-ADJOINT, *that is, $A \in \mathcal{L}(H, H)$ and $A = A^*$.*

▸ **3.20** *Let M be a closed subspace of a Hilbert space H. By the Orthogonal Decomposition Theorem (cf. Theorem 2.10), each $x \in H$ can be uniquely written in the form $x = y + z$, with $y \in M$ and $z \in M^{\perp}$. The point $y \in M$ is called the* PROJECTION *of x onto M, and the operator $P := X \to M$, $Px = y$, is called the projection onto M. It is also denoted $P := P_M$ and said to be a projection.*

i) *Prove that if P is a projection, then P is self-adjoint, $P^2 = P$, and $\|P\| = 1$ if $P \neq 0$.*
ii) *Prove that if $P \in \mathcal{L}(H, H)$ is self-adjoint and $P^2 = P$, then P is a projection onto some closed subspace of H.*

▸ **3.21** ([24], [112]) *Let $(H, \langle \cdot, \cdot \rangle_H)$ and $(Q, \langle \cdot, \cdot \rangle_Q)$ be Hilbert spaces, and let $\mathbf{B} \in \mathcal{L}(H, Q)$ with null space $V := N(\mathbf{B})$.*

a) *Prove that* $$\sup_{\substack{v \in H \\ v \neq 0}} \frac{\langle \mathbf{B}(v), q \rangle_Q}{\|v\|_H} = \sup_{\substack{v \in V^{\perp} \\ v \neq 0}} \frac{\langle \mathbf{B}(v), q \rangle_Q}{\|v\|_H} \quad \forall\, q \in Q.$$

b) *Suppose there exists $\beta > 0$ such that* $\displaystyle\sup_{\substack{v \in V^{\perp} \\ v \neq 0}} \frac{\langle \mathbf{B}(v), q \rangle_Q}{\|v\|_H} \geqslant \beta \|q\|_Q \quad \forall\, q \in Q$, *and prove that*
$H = R(\mathbf{B}^*) \oplus V$.

▸ **3.22** [MOORE-PENROSE PSEUDOINVERSE]. *Let X, Y be Hilbert spaces, $A \in \mathcal{L}(X, Y)$ such that $\mathcal{R}(A) = Y$, and let $V = N(A)$. Given the orthogonal projection operator $P : X \to V$, consider $B : Y \to X$ such that $B(y) = x - P(x)$ for every $y \in Y$, where $x \in X$ is such that $A(x) = y$.*

a) *Prove that B is well defined and that B is a bounded linear bijection from Y to $V^{\perp}$. Also prove that B is a right inverse of A, that is $AB(y) = y$ for all $y \in Y$.*
b) *Define $A_0 : V^{\perp} \to Y$ as $A_0(x) = A(x)$ for all $x \in V^{\perp}$, i.e. $A_0 = A|_{V^{\perp}}$, and prove that $A_0^{-1} = B$.*
c) *Extend the previous results to the case where $\mathcal{R}(A)$ is a proper* CLOSED *subspace of Y.*

▸ **3.23** ([25], [73]) *Let $(H, \langle \cdot, \cdot \rangle)$ be a Hilbert space, and let $P \in \mathcal{L}(H, H)$, nontrivial. We say that P is a* PROJECTOR *if it satisfies $P^2 = P$. In such case, P is called an* ORTHOGONAL PROJECTOR *if it also satisfies $\langle u, v \rangle = 0$ for all $u \in R(P)$, for all $v \in N(P)$.*

a) *Prove that if $P \in \mathcal{L}(H, H)$ is a projector then $H = N(P) \oplus R(P)$ and $\|P\| \geqslant 1$.*
b) *Prove that the following statements are equivalent:*
  i) *P is an orthogonal projector.*
  ii) *$R(P) = N(P)^{\perp}$.*
  iii) *P is self-adjoint.*
c) *Prove that if $P \in \mathcal{L}(H, H)$ is an orthogonal projector then $\|P\| = 1$.*

▸ **3.24** [CLOSURE OF OPERATORS]. *Let X and Y be Banach spaces. We say that a linear operator $A : \mathcal{D}(A) \subseteq X \to Y$ admits a closure if there exists a linear operator $B : \mathcal{D}(B) \subseteq X \to Y$ such that B is an extension of A and $G(B) = \overline{G(A)}$. Prove that A admits a closure if and only if for every sequence $\{x_n\}_{n \in \mathbb{N}} \subseteq \mathcal{D}(A)$ such that $(x_n, A(x_n)) \overset{n \to \infty}{\to} (0, y)$, with $y \in Y$, it necessarily follows that $y = 0$*

[NOTE THAT THIS STATEMENT IS EQUIVALENT TO THAT OF EXERCISE 3.14].

▸ **3.25** *Let $(H, \langle \cdot, \cdot \rangle)$ be a Hilbert space, and let $A : \mathcal{D}(A) \subseteq H \to H$ be a linear operator such that $\mathcal{D}(A)$ is dense in H.*

a) *Let $y \in H$ and suppose that there exists $\widetilde{y} \in H$ such that*
$$\langle A(x), y \rangle = \langle x, \widetilde{y} \rangle \qquad \forall\, x \in \mathcal{D}(A).$$
Prove that such $\widetilde{y}$ is unique.

b) *Consider the set*

$$\mathcal{D}(\widetilde{A}) := \{y \in H : \text{ there exists } \widetilde{y} \in H \text{ such that } \langle A(x), y\rangle = \langle x, \widetilde{y}\rangle \quad \forall\, x \in \mathcal{D}(A)\},$$

and define the operator $\widetilde{A} : \mathcal{D}(\widetilde{A}) \to H$ *given by* $\widetilde{A}(y) := \widetilde{y}$ *for all* $y \in \mathcal{D}(\widetilde{A})$. *Prove that* $\widetilde{A}$ *is linear and closed.*

▸ **3.26** ([122]) *Let* $(H, \langle\cdot,\cdot\rangle)$ *be a Hilbert space, and let* $\mathbf{P} \in \mathcal{L}(H, H)$, *nontrivial, different from the identity operator* $\mathbf{I} : H \to H$, *such that* $\mathbf{P}^2 = \mathbf{P}$. *The goal of this exercise is to prove that* $\|\mathbf{P}\|_{\mathcal{L}(H,H)} = \|\mathbb{I} - \mathbf{P}\|_{\mathcal{L}(H,H)}$, *by proceeding as follows:*

a) *Let* S *be a 2-dimensional subspace of* H, *and let* $\mathbf{Q} \in \mathcal{L}(S, S)$ *be nontrivial, and different from the identity operator* $\mathbf{I} : S \to S$, *such that* $\mathbf{Q}^2 = \mathbf{Q}$. *Prove that* $S = R(\mathbf{Q}) \oplus R(\mathbb{I} - \mathbf{Q})$ *and conclude that there exist nonzero vectors* $p,\, q,\, r,\, s \in S$ *satisfying* $\langle p, q\rangle = \langle r, s\rangle = 1$, *such that*

$$\mathbf{Q}(v) = \langle q, v\rangle\, p \qquad \text{and} \qquad (\mathbf{I} - \mathbf{Q})(v) = \langle s, v\rangle\, r \qquad \forall\, v \in S\,.$$

b) *From the identity* $v = \mathbf{Q}(v) + (\mathbf{I} - \mathbf{Q})(v)$ *for all* $v \in S$, *deduce that* $\|p\|^2\, \|q\|^2 = \|r\|^2\, \|s\|^2 = 1 - \langle p, r\rangle\langle q, s\rangle$ *and conclude that*

$$\|\mathbf{Q}\|_{\mathcal{L}(S,S)} = \|\mathbf{I} - \mathbf{Q}\|_{\mathcal{L}(S,S)}.$$

c) *Given* $x \in H$ *such that* $\|x\| = 1$, *consider the subspace* S *generated by the vectors* x *and* $\mathbf{P}(x)$ *and define* $\mathbf{Q} := \mathbf{P}|_S$. *Prove that* $\mathbf{Q} \in \mathcal{L}(S, S)$, *note that the dimension of* S *is* $\leqslant 2$, *and then prove, using b), that* $\|(\mathbf{I} - \mathbf{P})(x)\| \leqslant \|\mathbf{P}\|_{\mathcal{L}(H,H)}$.

d) *Conclude, from c), that* $\|\mathbf{P}\|_{\mathcal{L}(H,H)} = \|\mathbf{I} - \mathbf{P}\|_{\mathcal{L}(H,H)}$.

▸ **3.27** *Let* $(H, \langle\cdot,\cdot\rangle_H)$ *be a complex Hilbert space, and consider* $H \times H$ *equipped with the inner product*

$$\langle (u, v), (z, w)\rangle_{H\times H} := \langle u, z\rangle_H + \langle v, w\rangle_H \qquad \forall\, (u, v), (z, w) \in H \times H\,.$$

Moreover, given $A \in \mathcal{L}(H, H)$, *define the operator* $B : H \times H \to H \times H$ *by*

$$B((u, v)) := (\imath\, A(v), -\imath\, A^*(u)) \qquad \forall\, (u, v) \in H \times H\,.$$

Prove that $\|B\| = \|A\|$ *and that* B *is self-adjoint.*

▸ **3.28** *Let* E *and* F *be Banach spaces, and let* $T : E \to F$ *be a closed linear operator with domain* $D(T)$ *and range* $R(T)$. *Prove that the following are equivalent:*

a) *The operator* T *is injective, and* T^{-1} *is bounded on* $R(T)$.
b) *There exists a positive constant* C *such that* $\|Tx\| \geqslant C\|x\|$ *for all* $x \in D(T)$.
c) $R(T)$ *is closed in* F, *and* T *is injective.*

▸ **3.29** *Let* Ω *be a convex and bounded domain of* $\mathbb{R}^2$ *with polygonal boundary* Γ, *and let* $\langle\cdot,\cdot\rangle_{0,\Omega}$ *and* $\langle\cdot,\cdot\rangle_{1,\Omega}$ *be the inner products of* $\mathrm{L}^2(\Omega)$ *and* $\mathrm{H}^1(\Omega)$ *(cf. (1.12), (1.27)), respectively.*

a) *Prove that for every* $r \in \mathrm{L}^2(\Omega)$ *there exists a unique* $z \in \mathrm{H}^1(\Omega)$ *such that*

$$\langle z, w\rangle_{1,\Omega} = \langle r, w\rangle_{0,\Omega} \qquad \forall\, w \in \mathrm{H}^1(\Omega)\,.$$

b) *Deduce that* z *is the unique weak solution of the boundary value problem:*

$$-\Delta z + z = r \quad \text{in} \quad \Omega\,, \qquad \nabla z \cdot \boldsymbol{\nu} = 0 \quad \text{on} \quad \Gamma\,,$$

where $\boldsymbol{\nu}$ *is the outward normal vector on* Γ, *and* **observe** *(do not prove) that the convexity of* Ω *guarantees that* $z \in \mathrm{H}^2(\Omega)$.

c) *Define an appropriate linear operator and prove, using the Closed Graph Theorem, that there exists* $C > 0$ *such that*

$$\|z\|_{2,\Omega} \leqslant C\, \|r\|_{0,\Omega} \qquad \forall\, r \in \mathrm{L}^2(\Omega)\,.$$

▸ **3.30** *Let $(H,\langle\cdot,\cdot\rangle)$ be a real Hilbert space, and let $A\in\mathcal{L}(H,H)$ be the operator induced by a bounded bilinear form $a: H\times H\to\mathbb{R}$. Moreover, let Π be the orthogonal projection of H onto a closed subspace S, and suppose that there exists $\alpha>0$ such that*

$$a(v,v)\geqslant\alpha\langle v,v\rangle\qquad\forall\, v\in S\,.$$

Prove that $\Pi A: S\to S$ is a linear bijection.

▸ **3.31** ([86]) *Let $(H,\langle\cdot,\cdot\rangle)$ be a Hilbert space, and let $V_1,\ldots,V_N$ be mutually orthogonal closed subspaces of H, that is $v_i\perp v_j$ for all $v_i\in V_i$, $v_j\in V_j$, $i\neq j$. Prove that $\mathbb{I}-\mathbb{P}=\mathbb{P}_1+\mathbb{P}_2+\cdots+\mathbb{P}_N$, where $\mathbb{P}: H\longrightarrow V_1^\perp\cap V_2^\perp\cap\cdots\cap V_N^\perp$ and $\mathbb{P}_j: H\longrightarrow V_j$ are the respective orthogonal projections.*

▸ **3.32** *Let Ω be a bounded open subset of $\mathbb{R}^n$, and consider once again the real Hilbert space $V=\mathbf{L}^2(\Omega):=[\mathrm{L}^2(\Omega)]^{n\times n}$ with the inner product*

$$\langle\boldsymbol{\sigma},\boldsymbol{\tau}\rangle_{0,\Omega}\,:=\,\int_\Omega\boldsymbol{\sigma}:\boldsymbol{\tau},$$

where

$$\boldsymbol{\sigma}:\boldsymbol{\tau}:=\sum_{i,j=1}^n\sigma_{ij}\,\tau_{ij}\quad\forall\,\boldsymbol{\sigma}:=(\sigma_{ij})_{n\times n},\quad\boldsymbol{\tau}:=(\tau_{ij})_{n\times n}\in V\,.$$

Define the subspace

$$U:=\left\{\alpha\,\mathbb{A}+\beta\,\mathbb{B}:\quad\alpha,\,\beta\in\mathbb{R}\right\}\;=\;\langle\{\mathbb{A},\mathbb{B}\}\rangle,$$

where $\mathbb{A}:=(a_{ij})_{n\times n}$ and $\mathbb{B}:=(b_{ij})_{n\times n}$ are defined by

$$a_{ij}:=\begin{cases}1 \text{ if } j=i-1\\ 0 \text{ otherwise}\end{cases}\quad\text{and}\quad b_{ij}:=\begin{cases}1 \text{ if } j=i+1\\ 0 \text{ otherwise}\end{cases}.$$

Find $U^\perp$ and explicitly define the orthogonal projections onto U and $U^\perp$. What happens with these projections if U is replaced by $\langle\{\mathbb{A},\mathbb{B},\mathbb{I}\}\rangle$, where $\mathbb{I}$ is the identity tensor of V?

▸ **3.33** *Let $(H_1,\langle\cdot,\cdot\rangle_1)$ and $(H_2,\langle\cdot,\cdot\rangle_2)$ be Hilbert spaces over $\mathbb{C}$ with induced norms $\|\cdot\|_1$ and $\|\cdot\|_2$, respectively, and let $T\in\mathcal{L}(H_1,H_2)$ be such that $\|T(x)\|_2=\|x\|_1$ for all $x\in H_1$. Prove that*

$$\langle T(x),T(y)\rangle_2=\langle x,y\rangle_1\qquad\forall\,x,y\in H_1\,.$$

HINT: *Consider the null expressions $\|T(x+y)\|^2-\|x+y\|^2$ and $\|T(x+iy)\|^2-\|x+iy\|^2$.*

▸ **3.34** *Given Ω a bounded open subset of $\mathbb{R}^n$, define the operator $A:\mathrm{H}^1(\Omega)\longrightarrow\mathrm{L}^2(\Omega)$ by*

$$A(u)\,:=\,\sum_{j=1}^N\left\{\int_\Omega\nabla u\cdot\nabla u_j\right\}v_j\quad\forall\,u\,\in\,\mathrm{H}^1(\Omega)\,,$$

where $\{u_1,u_2,\ldots,u_N\}\subseteq\mathrm{H}^1(\Omega)$ and $\{v_1,v_2,\ldots,v_N\}\subseteq\mathrm{L}^2(\Omega)$. Prove that A is linear and bounded, and find the Hilbert adjoint operator A^*.

▸ **3.35** *Let $X:=C[0,1]$ endowed with the uniform norm (cf. (2.27)), and consider sequences $\{p_j\}_{j\in\mathbb{N}},\ \{q_j\}_{j\in\mathbb{N}}\subseteq X$ such that the series $\sum\limits_{j=1}^\infty p_j(s)\,q_j(t)$ converges uniformly to a continuous function $K:[0,1]\times[0,1]\longrightarrow\mathbb{R}$, that is,*

$$\lim_{N\to+\infty}\left\{\max_{(s,t)\in[0,1]\times[0,1]}\left|K(s,t)\,-\,\sum_{j=1}^N p_j(s)\,q_j(t)\right|\right\}\,=\,0\,.$$

Moreover, let $F_j\in X'$ be defined by $F_j(u):=\displaystyle\int_0^1 q_j(t)\,u(t)\,dt$ for all $u\in X$. Prove that for every $G\in X'$, the series $\sum\limits_{j=1}^\infty G(p_j)\,F_j$ converges in X'.

HINT: *Identify the value of the respective limit in terms of the adjoint of a suitable operator.*

▸ **3.36** ([24], [112]) [THE CONTINUOUS INF-SUP CONDITION]. *Let* $(X, \langle\cdot,\cdot\rangle_X)$ *and* $(Y, \langle\cdot,\cdot\rangle_Y)$ *be Hilbert spaces, and consider* $A \in \mathcal{L}(X,Y)$. *Prove that the following statements are equivalent:*

i) A *is surjective.*
ii) $A^* : Y \to X$ *is injective and has closed range.*
iii) *There exists* $\alpha > 0$ *such that*

$$\|A^*(y)\|_X := \sup_{\substack{x \in X \\ x \neq 0}} \frac{|\langle A(x), y\rangle_Y|}{\|x\|_X} \geqslant \alpha \|y\|_Y \quad \forall\, y \in Y .$$

iv) *There exists an operator* $B \in \mathcal{L}(Y,X)$ *such that* $A\,B = I$ *in* Y *and* $B\,A = I - P$ *in* X*, where* $P : X \to N(A)$ *is the orthogonal projector.*

▸ **3.37** *Given* Ω *a bounded open subset of* $\mathbb{R}^n$*, define the operator* $A : \mathrm{H}^2(\Omega) \longrightarrow \mathrm{H}^1(\Omega)$ *by*

$$A(u) := \sum_{j=1}^{N} \left\{ \int_\Omega \Delta u\, \Delta u_j \right\} v_j \quad \forall\, u \in \mathrm{H}^2(\Omega),$$

where $\{u_1, u_2, ..., u_N\} \subseteq \mathrm{H}^2(\Omega)$ *and* $\{v_1, v_2, ..., v_N\} \subseteq \mathrm{H}^1(\Omega)$. *Prove that* A *is linear and bounded, and find the Hilbert adjoint operator* A^*.

▸ **3.38** *Let* $(H, \langle\cdot,\cdot\rangle)$ *be a complex Hilbert space, and let* $A : H \to H$ *be a linear operator such that* $\langle A(x), y\rangle = \langle x, A(y)\rangle$ *for all* $x,\ y \in H$. *Prove that* A *is bounded.*

▸ **3.39** *Let* V *and* W *be closed subspaces of a Hilbert space* $(H, \langle\cdot,\cdot\rangle)$*, and let* $P : H \to V$ *and* $Q : H \to W$ *be the respective orthogonal projectors. Prove that*

$$\langle (Q - P)(x), x\rangle \geqslant 0 \quad \forall\, x \in H \quad \text{if and only if} \quad V \subseteq W .$$

▸ **3.40** *Given* Ω *a bounded open subset of* $\mathbb{R}^n$*, define the operator* $A : \mathrm{H}^1(\Omega) \longrightarrow \mathrm{H}^2(\Omega)$ *by*

$$A(u) := \sum_{j=1}^{N} \left\{ \int_\Omega \nabla u \cdot \nabla u_j \right\} v_j \quad \forall\, u \in \mathrm{H}^1(\Omega)$$

where $\{u_1, u_2, ..., u_N\} \subseteq \mathrm{H}^1(\Omega)$ *and* $\{v_1, v_2, ..., v_N\} \subseteq \mathrm{H}^2(\Omega)$. *Prove that* A *is linear and bounded, and find the Hilbert adjoint operator* A^*.

▸ **3.41** *This exercise aims to apply the theory developed for the analysis of the interpolating spline function to three other different but similar situations.*

a) *Let* $\Omega := (a,b)$, $n \in \mathbb{N}$*, and consider the partition* $a < t_1 < t_2 < \cdots < t_n < b$. *Moreover, let* $m \in \mathbb{N}$ *and* $\alpha,\ \beta > 0$. *Then, given* $\mathbf{z} := (z_1, z_2, \ldots, z_n)^{\mathrm{t}} \in \mathbb{R}^n$*, analyze the solvability of the following problem: Find* $\mathbf{u} := (u_1, u_2) \in \mathrm{H}^m(\Omega) \times \mathrm{H}^m(\Omega)$ *such that* $u_1(t_i) - u_2(t_i) = z_i$ *for all* $i \in \{1, 2, \ldots, n\}$*, and such that*

$$\begin{aligned} &\alpha \int_\Omega \left(u_1^{(m)}(t)\right)^2 dt + \beta \int_\Omega \left(u_2^{(m)}(t)\right)^2 dt \\ = &\min_{\substack{\mathbf{v} := (v_1, v_2) \in \mathrm{H}^m(\Omega) \times \mathrm{H}^m(\Omega) \\ v_1(t_i) - v_2(t_i) = z_i\ \forall\, i}} \left\{ \alpha \int_\Omega \left(v_1^{(m)}(t)\right)^2 dt + \beta \int_\Omega \left(v_2^{(m)}(t)\right)^2 dt \right\} \end{aligned}$$

Does any change occur in the analysis if the interpolation condition is replaced by $v_1(t_i) = v_2(t_i) = z_i$ *for all* i*? Explain and justify.*

b) *Let S be a finite-dimensional subspace of dimension N of a Hilbert space $(H, \langle\cdot,\cdot\rangle_H)$, and consider a basis $\{p_1, \ldots, p_N\}$ of S. Moreover, let $\Pi : H \to S^\perp$ be the orthogonal projector. Prove that for each $\mathbf{z} := (z_1, z_2, \ldots, z_N)^t \in \mathbb{R}^N$, there exists a unique $u \in H$ such that $\langle u, p_i\rangle_H = z_i$ for all $i \in \{1, 2, \ldots, N\}$, and such that*

$$\|\Pi(u)\|_H = \min_{\substack{v \in H \\ \langle v, p_i\rangle_H = z_i \ \forall i}} \|\Pi(v)\|_H$$

Deduce an algorithm to compute u.

c) *Let $\Omega := (-\pi, \pi)$, and let $(\mathrm{H}^1(\Omega), \langle\cdot,\cdot\rangle_{1,\Omega})$ be the Sobolev space of order 1. Moreover, let $n \in \mathbb{N}$ and consider the sets of functions $\{p_0, p_1, \ldots, p_n\}$ and $\{q_0, q_1, \ldots, q_n\}$ defined by $p_k(t) := \sin(kt)$ and $q_k(t) := \cos(kt)$ for all $t \in [-\pi, \pi]$, for all $k \in \{0, 1, \ldots, n\}$. Then, given $\mathbf{z} := (z_0, z_1, \ldots, z_n)^t \in \mathbb{R}^{n+1}$, analyze the solvability of the following problem: Find $\mathbf{u} := (u_1, u_2) \in \mathrm{H}^1(\Omega) \times \mathrm{H}^1(\Omega)$ such that $\langle u_1, p_k\rangle_{1,\Omega} + \langle u_2, q_k\rangle_{1,\Omega} = z_k$ for all $k \in \{0, 1, \ldots, n\}$, and such that*

$$\int_\Omega (u_1'(t))^2\, dt + \int_\Omega (u_2'(t))^2\, dt$$
$$= \min_{\substack{\mathbf{v} := (v_1, v_2) \in \mathrm{H}^1(\Omega) \times \mathrm{H}^1(\Omega) \\ \langle v_1, p_k\rangle_{1,\Omega} + \langle v_2, q_k\rangle_{1,\Omega} = z_k \ \forall k}} \left\{ \int_\Omega (v_1'(t))^2\, dt + \int_\Omega (v_2'(t))^2\, dt \right\}$$

▸ **3.42** *Let G be a subspace of a normed vector space X, and let N be a subspace of the dual of a reflexive Banach space Y. Prove that*

a) $\mathrm{dist}(f, G^o) = \sup_{\substack{x \in \bar{G} \\ \|x\| \leqslant 1}} |f(x)| \quad \forall f \in X'$.

b) $\bar{N} = ({}^o N)^o$.

c) $\mathrm{dist}(y, {}^o N) = \sup_{\substack{g \in \bar{N} \\ \|g\| \leqslant 1}} |g(y)| \quad \forall y \in Y$.

d) $\mathrm{dist}(g, \bar{N}) = \sup_{\substack{y \in {}^o N \\ \|y\| \leqslant 1}} |g(y)| \quad \forall g \in Y'$.

▸ **3.43** *Let Ω be a bounded open subset of $\mathbb{R}$, and consider the real Hilbert spaces given by $X := \mathrm{L}^2(\Omega)$ and $Y := \mathbb{R}^{2\times 2}$ endowed with the inner products*

$$\langle u, v\rangle_X := \int_\Omega u\, v \quad \forall\, u,\, v \in X\,, \qquad \langle A, B\rangle_Y := \mathrm{tr}(A^t\, B) \quad \forall\, A,\, B \in Y\,,$$

and define the operator $\mathcal{A} : X \to Y$ by

$$\mathcal{A}(u) := \begin{pmatrix} \int_\Omega u & \int_\Omega x\, u \\ \int_\Omega x\, u & \int_\Omega x^2\, u \end{pmatrix} \qquad \forall u \in X\,.$$

i) *Prove that $\mathcal{A}$ is linear and bounded, and compute explicitly the Hilbert adjoint operator $\mathcal{A}^*$ and the spaces $N(\mathcal{A})$, $R(\mathcal{A})$, $N(\mathcal{A}^*)$, and $R(\mathcal{A}^*)$.*

ii) *Find a closed subspace S of X such that $\mathcal{A}|_S : S \to R(\mathcal{A})$ is bijective.*

iii) *Define $\widetilde{Y} := \{B \in Y : \ B^t = -B\}$ and prove that there exists $C > 0$ such that*

$$\|\mathcal{A}^*(A)\|_X \geqslant C \inf_{B \in \widetilde{Y}} \|A - B\|_Y \quad \forall A \in Y\,.$$

iv) *Replace $X := \mathrm{L}^2(\Omega)$ by $X := \mathrm{H}^1(\Omega)$ with its usual inner product, and recalculate the Hilbert adjoint operator $\mathcal{A}^*$.*

▸ **3.44** *Let X, Y_1, and Y_2 be Banach spaces over $\mathbb{K}$, and let $A_1 : \mathcal{D}(A_1) \subseteq X \to Y_1$ and $A_2 : \mathcal{D}(A_2) \subseteq X \to Y_2$ be closed operators. Consider the product space $Y := Y_1 \times Y_2$ and prove that the linear operator $A : \mathcal{D}(A) \subseteq X \to Y$, defined by $A(x) := (A_1(x), A_2(x))$ for all $x \in \mathcal{D}(A) := \mathcal{D}(A_1) \cap \mathcal{D}(A_2)$, is also closed. Conversely, suppose that $A : \mathcal{D}(A) \subseteq X \to Y$ is a closed linear operator, and that for every sequence $\{x_n\}_{n\in\mathbb{N}} \subseteq \mathcal{D}(A)$ such that $\{A_i(x_n)\}_{n\in\mathbb{N}}$ is convergent in Y_i, $i \in \{1,2\}$, there exists a subsequence $\{x_n^{(1)}\}_{n\in\mathbb{N}} \subseteq \{x_n\}_{n\in\mathbb{N}}$ for which $\{A_j(x_n^{(1)})\}_{n\in\mathbb{N}}$ is convergent in Y_j, for all $j \in \{1,2\}$. Prove in this case that for each $i \in \{1,2\}$, $A_i|_{\mathcal{D}(A)}$ is closed.*

▸ **3.45** *Determine,* **with justification**, *whether the following statements are true or false:*

i) *Given Hilbert spaces H and Q, one has that $A \in \mathcal{L}(H,Q)$ is bijective if and only if $A^* \in \mathcal{L}(Q,H)$ is bijective.*
ii) *Given Banach spaces X and Y, the inverse of a linear, closed, and bijective operator $A : \mathcal{D}(A) \subseteq X \to Y$ is also closed.*
iii) *Given Banach spaces X and Y, the inverse of a linear, closed, and bijective operator $A : \mathcal{D}(A) \subseteq X \to Y$ is bounded.*
iv) *Given Banach spaces X and Y, with Y finite-dimensional, and $A \in \mathcal{L}(X,Y)$, the range of the adjoint operator $A' \in \mathcal{L}(Y',X')$ is closed.*

▸ **3.46** *Let $\Omega := (0,1)$, and consider the real Hilbert spaces given by $X = \mathrm{L}^2(\Omega)$ and $Y = \ell_2(\mathbb{R}) := \Big\{ A := \{a_k\}_{k\in\mathbb{N}} : \quad a_k \in \mathbb{R}, \quad \sum_{k=1}^{\infty} |a_k|^2 < \infty \Big\}$, endowed with the usual inner products*

$$\langle u, v\rangle_X := \int_0^1 u\, v \qquad \forall\, u,\, v \in X\,,$$

$$\langle A, B\rangle_Y := \sum_{k=1}^{\infty} a_k\, b_k \qquad \forall\, A := \{a_k\}_{k\in\mathbb{N}}\,,\ B := \{b_k\}_{k\in\mathbb{N}} \in Y,$$

and define the operator $\mathcal{A} : X \to Y$ by

$$\mathcal{A}(u) := \left\{ \frac{1}{k} \int_0^1 x^k\, u \right\}_{k\in\mathbb{N}} \qquad \forall\, u \in X\,.$$

i) *Prove that $\mathcal{A}$ is well-defined and linear and bounded, and then compute the Hilbert adjoint operator $\mathcal{A}^*$ and the spaces $N(\mathcal{A})$ and $\overline{R(\mathcal{A}^*)}$.*
ii) *Find a closed subspace S of X such that $\mathcal{A}|_S : S \to R(\mathcal{A})$ is bijective.*

▸ **3.47** ([86]) *The present exercise aims to characterize the surjectivity of an operator whose range is contained in a product space, in terms of the surjectivity of its respective components.*

a) *Recall first that, given $m, n \in \mathbb{N}$ and a linear operator $\mathcal{B} : \mathbb{R}^n \longrightarrow \mathbb{R}^m$, the Dimension Theorem states that $n = \dim N(\mathcal{B}) + \dim R(\mathcal{B})$. Then, suppose that $m = m_1 + m_2$ and that $\mathcal{B}(x) = (\mathcal{B}_1(x), \mathcal{B}_2(x))$ for all $x \in \mathbb{R}^n$, where $\mathcal{B}_i : \mathbb{R}^n \longrightarrow \mathbb{R}^{m_i}$ is linear for each $i \in \{1,2\}$, and prove that $\mathcal{B}$ is surjective if and only if*

   i) $\mathcal{B}_1$ and $\mathcal{B}_2$ are surjective. ii) $\mathbb{R}^n = N(\mathcal{B}_1) + N(\mathcal{B}_2)$.
b) *Let U and V be closed subspaces of a Hilbert space $(X, \langle\cdot,\cdot\rangle)$. Prove that the following statements are equivalent:*

   i) $U^\perp \subseteq U + V$. ii) $X = U + V$. iii) $U^\perp + V^\perp \subseteq U + V$.
c) *Let X, Y_1, and Y_2 be arbitrary Hilbert spaces, and define $Y := Y_1 \times Y_2$. Then, let $\mathcal{B} \in \mathcal{L}(X,Y)$ be such that $\mathcal{B}(x) = (\mathcal{B}_1(x), \mathcal{B}_2(x))$ for all $x \in X$, where $\mathcal{B}_i \in \mathcal{L}(X,Y_i)$ for all $i \in \{1,2\}$, and prove that $\mathcal{B}$ is surjective if and only if*

   i) $\mathcal{B}_1$ and $\mathcal{B}_2$ are surjective. ii) $X = N(\mathcal{B}_1) + N(\mathcal{B}_2)$.

▸ **3.48** *Let X and Y be Banach spaces for which there exists a bijection $T \in \mathcal{L}(X,Y)$, and let Z be a subspace of X. Prove that every $g \in Z'$ can be "extended" to some $G \in Y'$ such that $\|G\|_{Y'} \cong \|g\|_{Z'}$.*

▸ **3.49** *Let X be a real Banach space, and let $\mathbf{a}$ be a bilinear form on X, that is, $\mathbf{a} : X \times X \to \mathbb{R}$ is a mapping that satisfies:*

i) $\mathbf{a}(\alpha\, x + \beta\, z, y) = \alpha\, \mathbf{a}(x,y) + \beta\, \mathbf{a}(z,y)$ *for all* $\alpha,\ \beta \in \mathbb{R}$, *for all* $x,\ y,\ z \in X$,

ii) $\mathbf{a}(x, \alpha\, y + \beta\, z) = \alpha\, \mathbf{a}(x,y) + \beta\, \mathbf{a}(x,z)$ *for all* $\alpha,\ \beta \in \mathbb{R}$, *for all* $x,\ y,\ z \in X$.

In this case, define $\mathbf{A} : X \to X'$ by $\mathbf{A}(x)(y) := \mathbf{a}(x,y)$ for all $x,\ y \in X$, which is called the LINEAR OPERATOR INDUCED BY $\mathbf{a}$. *Assume now that there exist constants $M,\ m > 0$ such that*

iii) $|\mathbf{a}(x,y)| \leqslant M\,\|x\|\,\|y\|$ *for all* $x,\ y \in X$, *and*

iv) $\mathbf{a}(x,x) \geqslant m\,\|x\|^2$ *for all* $x \in X$,

and prove that $\mathbf{A}$ is bounded, injective, and has closed range $R(\mathbf{A})$. Prove moreover that ${}^\circ R(\mathbf{A}) = \{0\}$, define the adjoint operator $\mathbf{A}' : (X')' \to X'$, and discuss under what additional assumption on the space X one can conclude, from hypotheses i)–iv), that $\mathbf{A}$ is bijective.

▸ **3.50** *Let X and Y be real Banach spaces, and let $\mathcal{A} : X \times Y \to \mathbb{R}$ be a bilinear form, that is, $\mathcal{A}$ satisfies:*

i) $\mathcal{A}(\alpha\, x + \beta\, z, y) = \alpha\, \mathcal{A}(x,y) + \beta\, \mathcal{A}(z,y)$ *for all* $\alpha,\ \beta \in \mathbb{R}$, *for all* $x,\ z \in X$, *for all* $y \in Y$,

ii) $\mathcal{A}(x, \alpha\, y + \beta\, z) = \alpha\, \mathcal{A}(x,y) + \beta\, \mathcal{A}(x,z)$ *for all* $\alpha,\ \beta \in \mathbb{R}$, *for all* $x \in X$, *for all* $y,\ z \in Y$.

Assume additionally that $\mathcal{A}$ is bounded and weakly coercive, that is:

iii) *there exists $M > 0$ such that* $|\mathcal{A}(x,y)| \leqslant M\,\|x\|\,\|y\|$ *for all* $x \in X$, *for all* $y \in Y$,

iv) *there exists $\alpha > 0$ such that* $\displaystyle\sup_{\substack{x\in X\\ x\neq 0}} \frac{\mathcal{A}(x,y)}{\|x\|} \geqslant \alpha\,\|y\| \quad \forall\, y \in Y,$

v) $\displaystyle\sup_{y\in Y} \mathcal{A}(x,y) > 0 \qquad \forall\, x \in X,\ x \neq 0,$

and define the linear operators $\mathbf{A} : X \to Y'$ and $\mathbf{B} : Y \to X'$ as

$$\mathbf{A}(x)(y) := \mathcal{A}(x,y) \quad \text{and} \quad \mathbf{B}(y)(x) := \mathcal{A}(x,y) \quad \forall\, x \in X,\ \forall\, y \in Y\,.$$

a) *Prove that $\mathbf{A}$ and $\mathbf{B}$ are well defined and bounded.*
b) *Show that $\mathbf{A}$ and $\mathbf{B}$ are injective and that $R(\mathbf{B})$ is closed in X'.*
c) *Explicitly define the operators $\mathbf{A}' : Y'' \to X'$ and $\mathbf{B}' : X'' \to Y'$, and deduce that $\mathbf{B}'$ is surjective.*
d) *Suppose in particular that X and Y are Hilbert spaces and prove in this case that $\mathcal{R}_X \circ \mathbf{B} = (\mathcal{R}_Y \circ \mathbf{A})^*$, where $\mathcal{R}_X$ and $\mathcal{R}_Y$ are the respective Riesz maps. Conclude from this identity and iv) that $\mathbf{A}$ is bijective.*

▸ **3.51** *Given $n \in \mathbb{N}$, consider a partition $0 = t_0 < t_1 \ldots < t_n = 1$ of $\Omega := (0,1)$, and define the linear operator $A : \mathrm{H}^1(\Omega) \to \mathbb{R}^n$ by*

$$A(u) := \left(\int_{t_{j-1}}^{t_j} \big\{ t\,u(t) + u'(t) \big\}\, dt \right)_{j=\overline{1,n}} \qquad \forall\, u \in \mathrm{H}^1(\Omega)\,,$$

where the spaces $\mathrm{H}^1(\Omega)$ and $\mathbb{R}^n$ are endowed with their usual inner products. Show that $A \in \mathcal{L}(\mathrm{H}^1(\Omega), \mathbb{R}^n)$, define explicitly the Hilbert adjoint operator $A^* : \mathbb{R}^n \to \mathrm{H}^1(\Omega)$, and prove that $\|A^*\| \leqslant 2/\sqrt{3}$.

▸ **3.52** *Let Ω be a bounded open set of $\mathbb{R}^2$, and let $\mathcal{T}$ be a finite set of triangles T, whose interiors are mutually disjoint, such that $\bar{\Omega} = \cup\{T : \quad T \in \mathcal{T}\}$. Define $A : \mathrm{H}^1(\Omega) \to \mathbb{R}^N$ by*

$$A(u) := \left(\int_T \left\{ u(x) + x \cdot \nabla u(x) \right\} dx \right)_{T \in \mathcal{T}} \qquad \forall\, u \in \mathrm{H}^1(\Omega),$$

where N is the cardinality of $\mathcal{T}$, and the spaces $\mathrm{H}^1(\Omega)$ and $\mathbb{R}^N$ are endowed with their usual inner products. Show that $A \in \mathcal{L}(\mathrm{H}^1(\Omega), \mathbb{R}^N)$ and define explicitly the Hilbert adjoint operator

$$A^* : \mathbb{R}^N \to \mathrm{H}^1(\Omega).$$

▸ **3.53** *Let H and Q be Banach spaces, and let $A : \mathcal{D}(A) \subseteq H \to Q$ be a linear, closed, and injective operator. Prove in two different ways (for each of the respective implications) that*

$$A^{-1} : R(A) \to H \quad \text{is bounded if and only if} \quad R(A) \ \text{is closed in}\ Q.$$

▸ **3.54** *Determine,* **with justification**, *whether the following statements are true, false, or not necessarily true (unless an additional hypothesis is assumed):*

i) *The sum of two closed operators is a closed operator.*
ii) *If X is a Hilbert space and A, $B \in \mathcal{L}(X)$ are self-adjoint and injective, then $R(A) = R(B)$.*
iii) *If X is a normed vector space and $A : X \to X'$ is a linear, closed, bijective operator, then A^{-1} is bounded.*
iv) *If X and Y are Banach spaces and $A : \mathcal{D}(A) \subseteq X \to Y$ is a linear closed unbounded operator, then $\mathcal{D}(A)$ is a proper closed subspace of X.*
v) *The adjoint of every linear, bounded and surjective operator from a Hilbert space X onto a Hilbert space Y is injective.*
vi) *The sum of a bounded operator with a closed one is a closed operator.*
vii) *If X and Y are normed vector spaces and $A : \mathcal{L}(X,Y) \to Y'$ is a linear, closed and bijective operator, then A^{-1} is bounded.*
viii) *If X and Y are Banach spaces and $A : \mathcal{D}(A) \subseteq X \longrightarrow Y$ is a linear closed and injective operator such that $A^{-1} : \mathcal{D}(A^{-1}) \subseteq Y \to X$ is not bounded, then $R(A)$ is a proper closed subspace of Y.*
ix) *The Closed Graph Theorem cannot be proven using the Open Mapping Theorem.*

▸ **3.55** *Prove Theorem 3.87.*

▸ **3.56** *Let $\ell^\infty(\mathrm{R}) := \left\{ (x_n)_{n\in\mathrm{N}} : \quad \sup_{n\in\mathrm{N}} |x_n| < \infty \right\}$, which, endowed with the norm*

$$\|(x_n)_{n\in\mathrm{N}}\| := \sup_{n\in\mathrm{N}} |x_n|,$$

is a Banach space, and consider the subspace

$$X := \left\{ (x_n)_{n\in\mathrm{N}} \in \ell^\infty(\mathrm{R}) : \quad \exists\, m \in \mathrm{N} \ \text{such that}\ x_n = 0 \ \forall\, n > m \right\}.$$

Now, let $A : X \to X$ be the operator defined by $A\big((x_n)_{n\in\mathrm{N}}\big) := \left(\frac{x_n}{n}\right)_{n\in\mathrm{N}}$ for every $(x_n)_{n\in\mathrm{N}} \in X$, and prove that A is linear, bounded and bijective. Then, show that A^{-1} is not bounded, and deduce from this fact, by applying the Bounded Inverse Theorem, that X is not closed.

▸ **3.57** *Given a bounded simply connected open set Ω in $\mathbb{R}^n$ with smooth boundary Γ, consider the Hilbert spaces $\mathrm{L}^2(\Omega)$ (cf. Example 1.26) and*

$$\mathbf{H}(\mathrm{div};\Omega) := \left\{ \boldsymbol{\tau} \in \mathbf{L}^2(\Omega) : \quad \mathrm{div}(\boldsymbol{\tau}) \in \mathrm{L}^2(\Omega) \right\},$$

where $\mathbf{L}^2(\Omega) := [\mathrm{L}^2(\Omega)]^n$, which are endowed, respectively, with the inner products

$$\langle v, w\rangle_{0,\Omega} := \int_\Omega v\,w \quad \text{and} \quad \langle \boldsymbol{\sigma}, \boldsymbol{\tau}\rangle_{\operatorname{div};\Omega} := \int_\Omega \Big\{ \boldsymbol{\sigma}\cdot\boldsymbol{\tau} + \operatorname{div}\boldsymbol{\sigma}\operatorname{div}\boldsymbol{\tau}\Big\},$$

with induced norms $\|\cdot\|_{0,\Omega}$ *and* $\|\cdot\|_{\operatorname{div};\Omega}$. *Prove that for each* $N \in \mathbb{N}$ *and for every set* $\Big\{\boldsymbol{\tau}_1, \boldsymbol{\tau}_2, \ldots, \boldsymbol{\tau}_N\Big\} \subseteq \mathbf{H}(\operatorname{div};\Omega)$ *one has*

$$A_N := \bigcap_{n=1}^{N} \Big\{ \operatorname{div}\boldsymbol{\tau} : \quad \boldsymbol{\tau} \in \mathbf{H}(\operatorname{div};\Omega), \quad \|\boldsymbol{\tau} - \boldsymbol{\tau}_n\|_{\operatorname{div};\Omega} < n \Big\} \quad \text{is open in } \mathrm{L}^2(\Omega).$$

▸ **3.58** *Let* $(H, \langle\cdot,\cdot\rangle_H)$ *and* $(Q, \langle\cdot,\cdot\rangle_Q)$ *be Hilbert spaces over* $\mathbb{R}$ *with induced norms* $\|\cdot\|_H$ *and* $\|\cdot\|_Q$, *respectively, and let* $A : \mathcal{D}(A) = H \to Q$ *be a linear operator. Suppose that for every* $q \in Q$ *there exists* $c_q > 0$ *such that*

$$|\langle q, A(\boldsymbol{\tau})\rangle_Q| \leqslant c_q \|\boldsymbol{\tau}\|_H \qquad \forall\, \boldsymbol{\tau} \in H.$$

Prove in two different ways that A *is bounded.*

▸ **3.59** *Let* S_1 *and* S_2 *be closed subspaces of a Banach space* X. *First, suppose that there exists a sequence* $\{x_n\}_{n\in\mathbb{N}} \subseteq X$ *such that*

$$\operatorname{dist}(x_n, S_1 \cap S_2) = 1 \quad \text{and} \quad \lim_{n\to\infty} \Big\{ \operatorname{dist}(x_n, S_1) + \operatorname{dist}(x_n, S_2)\Big\} = 0,$$

and conclude that $S_1 + S_2$ *is a proper subspace of* $\overline{S_1 + S_2}$. *Next, suppose that there exist a subspace* S *dense in* X *and a constant* $c > 0$ *such that*

$$\operatorname{dist}(x, S_1 \cap S_2) \leqslant c \Big\{ \operatorname{dist}(x, S_1) + \operatorname{dist}(x, S_2)\Big\} \qquad \forall\, x \in S,$$

and prove in this case that $S_1 + S_2$ *is closed.*

▸ **3.60** *Let* A *and* B *be closed subspaces of a Hilbert space* $(H, \langle\cdot,\cdot\rangle)$ *over* $\mathbb{R}$.

6.1) *Prove that* $\operatorname{dist}(x, A^\perp) = \sup\limits_{\substack{a\in A\\ \|a\|\leqslant 1}} |\langle a, x\rangle| \qquad \forall\, x \in H.$

6.2) *Show that the following statements are equivalent:*

a) $A + B$ *is closed in* H.
b) $A^\perp + B^\perp = (A \cap B)^\perp$.
c) $A^\perp + B^\perp$ *is closed in* H.
d) $A + B = (A^\perp \cap B^\perp)^\perp$.

▸ **3.61** *Let* $(H, \langle\cdot,\cdot\rangle_H)$ *and* $(Q, \langle\cdot,\cdot\rangle_Q)$ *be Hilbert spaces over* $\mathbb{R}$, *and let* $A : \mathcal{D}(A) \subseteq H \to Q$ *be a linear operator such that* $\mathcal{D}(A)$ *is dense in* H. *Then, the* ADJOINT OPERATOR *of* A *is introduced as* $A^* : \mathcal{D}(A^*) \subseteq Q \to H$, *where*

$$\mathcal{D}(A^*) := \Big\{ q \in Q : \quad \exists\, \boldsymbol{\tau}^* \in H \text{ such that } \langle A(\boldsymbol{\tau}), q\rangle_Q = \langle \boldsymbol{\tau}, \boldsymbol{\tau}^*\rangle_H \quad \forall\, \boldsymbol{\tau} \in \mathcal{D}(A)\Big\},$$

and in this case one defines $A^*(q) = \boldsymbol{\tau}^*$ *for all* $q \in \mathcal{D}(A^*)$.

a) *Prove that* A^* *is well-defined and is a closed operator, and then deduce the concept of self-adjointness when* $H = Q$.
b) *Suppose in particular that* $Q = Q_1 \times Q_2$, *where* $(Q_1, \langle\cdot,\cdot\rangle_1)$ *and* $(Q_2, \langle\cdot,\cdot\rangle_2)$ *are also real Hilbert spaces, and that*

$$A(\boldsymbol{\tau}) = (A_1(\boldsymbol{\tau}), A_2(\boldsymbol{\tau})) \qquad \forall\, \boldsymbol{\tau} \in \mathcal{D}(A),$$

where $A_1 : \mathcal{D}(A_1) \subseteq H \to Q_1$ *and* $A_2 : \mathcal{D}(A_2) \subseteq H \to Q_2$ *are linear operators with* $\mathcal{D}(A_1) = \mathcal{D}(A_2) = \mathcal{D}(A)$ *dense in* H. *Describe the domains* $\mathcal{D}(A_1^*)$, $\mathcal{D}(A_2^*)$, *and* $\mathcal{D}(A^*)$ *only in terms of* A_1 *and* A_2, *and prove that*

$$A^*(q) = A_1^*(q_1) + A_2^*(q_2) \qquad \forall\, q := (q_1, q_2) \in \mathcal{D}(A_1^*) \times \mathcal{D}(A_2^*) .$$

What relation (if any) exists between $\mathcal{D}(A^*)$ and $\mathcal{D}(A_1^*) \times \mathcal{D}(A_2^*)$? Finally, deduce that if A is closed and surjective then there exists $c > 0$ such that

$$\|A_1^*(q_1) + A_2^*(q_2)\|_H \geqslant c\,\|q\|_Q \qquad \forall\, q := (q_1, q_2) \in \mathcal{D}(A_1^*) \times \mathcal{D}(A_2^*) .$$

▸ **3.62** *Let X and Y be Banach spaces, and let $\mathcal{J} : Y' \times X' \to X' \times Y'$ be the mapping defined by $\mathcal{J}(G, F) = \underline{(-F, G)}$ for all $(G, F) \in Y' \times X'$. In turn, let $A : \mathcal{D}(A) \subseteq X \to Y$ be a closed linear operator with $\overline{\mathcal{D}(A)} = X$, such that $\mathcal{J}(\mathcal{G}(A')) + N(A)^o \times N(A')$ is a closed subspace of $X' \times Y'$. Prove that A is bounded.*

▸ **3.63** *Given a bounded, simply connected open set $\Omega \subset \mathbb{R}^2$ with smooth boundary Γ, consider the Hilbert spaces $\mathrm{L}^2(\Omega)$ and*

$$\mathbf{H}(\mathrm{rot}; \Omega) := \left\{ \boldsymbol{\tau} := (\tau_1, \tau_2) \in \mathbf{L}^2(\Omega) : \quad \mathrm{rot}\, \boldsymbol{\tau} := \frac{\partial \tau_2}{\partial x_1} - \frac{\partial \tau_1}{\partial x_2} \in \mathrm{L}^2(\Omega) \right\},$$

endowed, respectively, with the inner products

$$\langle v, w\rangle_{0,\Omega} := \int_\Omega v\, w \quad \text{and} \quad \langle \boldsymbol{\sigma}, \boldsymbol{\tau}\rangle_{\mathrm{rot};\Omega} := \int_\Omega \left\{ \boldsymbol{\sigma} \cdot \boldsymbol{\tau} + \mathrm{rot}\, \boldsymbol{\sigma}\, \mathrm{rot}\, \boldsymbol{\tau} \right\},$$

with induced norms $\|\cdot\|_{0,\Omega}$ and $\|\cdot\|_{\mathrm{rot};\Omega}$. Show that there exists $\delta \in (0,1)$ such that for each $v \in \mathrm{L}^2(\Omega)$, there exists $\boldsymbol{\zeta}_v \in \mathbf{H}(\mathrm{rot}; \Omega)$ with $\|\boldsymbol{\zeta}_v\|_{\mathrm{rot};\Omega} < \delta^{-1}$ such that $v = \|v\|_{0,\Omega}\, \mathrm{rot}\, \boldsymbol{\zeta}_v$.

▸ **3.64** *Let X and Y be Banach spaces, and let $\{A_n\}_{n\in\mathbb{N}} \subseteq \mathcal{L}(X, Y)$ be a sequence of bijective operators. Assume that $\{A_n^{-1}\}_{n\in\mathbb{N}} \subseteq \mathcal{L}(Y, X)$ is uniformly bounded and that there exists an operator $A \in \mathcal{L}(X, Y)$ such that $A_n \overset{n\to+\infty}{\to} A$ in $\mathcal{L}(X, Y)$, that is $\|A_n - A\|_{\mathcal{L}(X,Y)} \overset{n\to+\infty}{\to} 0$. Prove that A is bijective and that $A_n^{-1} \overset{n\to+\infty}{\to} A^{-1}$ in $\mathcal{L}(Y, X)$.*
HINT: *Note that for all S, $T \in \mathcal{L}(X, Y)$ that are bijective, there holds $S^{-1} - T^{-1} = S^{-1}(T - S)T^{-1}$.*

▸ **3.65** [THE PRESENT IS A VARIANT OF EXERCISE 2.35]. *Given a Banach space $(X, \|\cdot\|)$ over $\mathbb{R}$ and $N \in \mathbb{N}$, consider N closed subspaces $U_1, U_2, \ldots, U_N$ of X such that $U := \sum_{j=1}^{N} U_j$ is also closed, and such that for each $I \subseteq \{1, 2, ..., N\}$ and each $j \notin I$ one has $U_j \cap \sum_{i\in I} U_i = \{0\}$. Moreover, let $\{f_1, f_2, \ldots, f_N\}$ be a set of functionals such that $f_j \in U_j'$ for all $j \in \{1, 2, \cdots, N\}$. Prove that there exist $F \in X'$ and a constant $C > 0$ such that*

$$F|_{U_j} = f_j \qquad \forall\, j \in \{1, 2, ..., N\} \qquad \text{and} \qquad \|F\|_{X'} \leqslant C \max_{j\in\{1,2,\ldots,N\}} \|f_j\|_{U_j'} .$$

HINT: *consider the possibility of eventually applying the Bounded Inverse Theorem.*

▸ **3.66** [THE PRESENT IS A SIMPLIFIED VERSION OF EXERCISE 3.65]. *Let $(X, \|\cdot\|)$ be a Banach space over $\mathbb{R}$, and let U, V, and W be closed subspaces of X such that $Z := U \oplus V \oplus W$ is also closed. Consider the norm $|\!|\!|\cdot|\!|\!|$ on Z defined as*

$$|\!|\!| z |\!|\!| := \|u\| + \|v\| + \|w\| \qquad \forall\, z := u + v + w \in Z ,$$

where (u, v, w) is the unique vector in $U \times V \times W$ such that $z = u + v + w$, and prove that $(Z, |\!|\!|\cdot|\!|\!|)$ is Banach. Then, apply the Bounded Inverse Theorem to conclude that $\|\cdot\|$ and $|\!|\!|\cdot|\!|\!|$ are equivalent on Z. Finally, show that, given $f \in U'$, $g \in V'$, and $h \in W'$, there exist $F \in X'$ and a constant $C > 0$ such that

$$F|_U = f , \quad F|_V = g , \quad F|_W = h$$

and

$$\|F\|_{X'} \leqslant C \left\{ \|f\|_{U'} + \|g\|_{V'} + \|h\|_{W'} \right\} .$$

▸ **3.67** [LEFT INVERSE]. *Let H, Q be Banach spaces, and let $T \in \mathcal{L}(H,Q)$ be such that $N(T) = \{0\}$. Prove that the following statements are equivalent:*

a) *There exists $S \in \mathcal{L}(Q,H)$ such that $ST = I : H \to H$.*
b) *$R(T)$ is closed and has a topological complement in Q.*

Confirm or refute whether, in the case where both spaces are Hilbert, such a topological complement is given by $R(T)^\perp$.

▸ **3.68** *Let U and V be closed subspaces of a real Hilbert space $(H, \langle\cdot,\cdot\rangle)$ such that $U+V$ is closed, and let $P_U : H \longrightarrow U$, $P_V : H \longrightarrow V$, $P_+ : H \longrightarrow U+V$, and $P_\cap : H \to U \cap V$ denote the corresponding orthogonal projections.*

a) *Prove that there exist bounded linear operators $P : H \longrightarrow U^\perp$ and $Q : H \longrightarrow V^\perp$ such that $P_+ = \frac{1}{2}\big(P_U + P_V\big) + \frac{1}{2}\big(P + Q\big)$.*
b) *Show that $P_+ = P_U + P_V$ if and only if $U \subseteq V^\perp$ (equivalently, $V \subseteq U^\perp$).*
c) *Prove that there exist bounded linear operators S, $T : H \to (U \cap V)^\perp$ such that $P_\cap = \frac{1}{2}\big(P_U + P_V\big) + \frac{1}{2}\big(S + T\big)$.*
d) *Show that $P_\cap = \frac{1}{2}\big(P_U + P_V\big)$ if and only if $P_U(x) + P_V(x) \in U \cap V$ for all $x \in H$.*

▸ **3.69** [THE PRESENT IS A PARTICULAR CASE OF THE 2ND PART OF EXERCISE 3.59]. *Let X_1 and X_2 be closed subspaces of a Banach space X, and assume that there exists $C > 0$ such that*

$$\operatorname{dist}(x, X_1 \cap X_2) \leqslant C\Big\{\operatorname{dist}(x, X_1) + \operatorname{dist}(x, X_2)\Big\} \qquad \forall\, x \in X\,.$$

The purpose of this exercise is to prove that $X_1 + X_2$ is also closed, for which it is suggested to proceed as follows:

i) *Define the operator $T : X \longrightarrow X/X_1$ by $T(x) := [x]$ (equivalence class of x) for all $x \in X$, and prove that T is linear, bounded, and surjective.*
ii) *Consider the operator $S := T|_{X_2}$, prove that $N(S) = X_1 \cap X_2$, and then use the hypothesis together with a characterization result to show that $R(S)$ is closed in X/X_1. Finally, show that $X_1 + X_2 = T^{-1}(R(S))$ and conclude.*

▸ **3.70** [THE PRESENT IS THE HILBERT SPACE VERSION OF EXERCISE 3.69]. *Let U and V be closed subspaces of a Hilbert space H, and assume that there exists $C > 0$ such that*

$$\operatorname{dist}(x, U \cap V) \leqslant C\Big\{\operatorname{dist}(x, U) + \operatorname{dist}(x, V)\Big\} \qquad \forall\, x \in H\,. \tag{3.48}$$

The purpose of this exercise is to prove that $U + V$ is also closed, for which it is suggested to consider the orthogonal projection $A : H \longrightarrow U^\perp$ and define its restriction $B := A|_V : V \longrightarrow U^\perp$. Then, prove that $N(B) = U \cap V$, and use (3.48) to deduce that $R(B)$ is closed in $U^\perp$. Finally, show that $U + V = A^{-1}(B(V))$ and conclude.

▸ **3.71** *Given $\Omega := (0,1)$ and $n \in \mathbb{N}$, consider the matrix-valued space over $\mathbb{R}$*

$$X = [\mathrm{H}^1(\Omega)]^{n\times n} := \Big\{\mathbf{u} = (u_{ij})_{n\times n} : \quad u_{ij} \in \mathrm{H}^1(\Omega) \quad \forall\, i,\, j = \overline{1,n}\Big\}\,,$$

which is endowed with the inner product

$$\langle \mathbf{u}, \mathbf{v}\rangle := \sum_{i,j=1}^{n} \langle u_{ij}, v_{ij}\rangle_{1,\Omega} \qquad \forall\, \mathbf{u} = (u_{ij})_{n\times n},\ \mathbf{v} = (v_{ij})_{n\times n} \in X\,,$$

where $\langle\cdot,\cdot\rangle_{1,\Omega}$ denotes the usual inner product of $\mathrm{H}^1(\Omega)$ (cf. (1.27)) [alternatively, we could have simply denoted $(\mathbb{H}^1(\Omega), \langle\cdot,\cdot\rangle_{1,\Omega})$ instead of $(X, \langle\cdot,\cdot\rangle)$].

a) *Prove that* $(X, \langle\cdot,\cdot\rangle)$ *is a Hilbert space, and express its associated Riesz operator* $\mathcal{J}_X$ *in terms of that of* $\mathrm{H}^1(\Omega)$ *(denoted by* $\mathcal{J}$*).*

b) *Define the operator* $A : X \longrightarrow \mathrm{H}^1(\Omega)$ *by* $A(\mathbf{u}) := \operatorname{tr}(\mathbf{u}) = \sum_{i=1}^{n} u_{ii}$ *for all* $\mathbf{u} \in X$*, and compute explicitly its adjoints* A^* *and* A'*.*

c) *Define the operator* $B : X \longrightarrow \mathrm{L}^2(\Omega)$ *by* $B(\mathbf{u}) := \operatorname{tr}(\mathbf{u}) = \sum_{i=1}^{n} u_{ii}$ *for all* $\mathbf{u} \in X$*, and compute explicitly its adjoints* B^* *and* B'*.*

d) *Let* p_j*,* $j = \overline{1,n}$*, be the monic polynomials, that is* $p_j(t) := t^j$ *for all* $t \in (0,1)$*. Denote by* U_j *the subspace of* $\mathrm{H}^1(\Omega)$ *generated by* p_j*, and introduce the subspace of* X *given by*

$$S := \Big\{ \mathbf{u} = (u_{ij})_{n\times n} \in X : \quad u_{ij} = 0 \quad \forall\, i \neq j, \quad \text{and} \quad u_{ii} \in U_i \quad \forall\, i = \overline{1,n} \Big\}.$$

Compute $S^\perp$ *and define as explicitly as possible the orthogonal projector* $\Pi : X \longrightarrow S^\perp$*.*

e) *Prove that the result of Exercises 2.35 and 3.65 can be applied to* $\mathrm{H}^1(\Omega)$ *and its subspaces* U_1*,* U_2*, ... ,* U_n*. Moreover, conclude that in this particular application one has* $\Pi_j\big(\mathcal{J}(F)\big) = \mathcal{J}_j(f_j)$ *for all* $j = \overline{1,n}$*, where* $\Pi_j : \mathrm{H}^1(\Omega) \longrightarrow U_j$ *is the orthogonal projector and* $\mathcal{J}_j : U_j' \longrightarrow U_j$ *is the Riesz operator. Furthermore, show that* $F \in \big(\mathrm{H}^1(\Omega)\big)'$ *can be chosen in such a way that* $\mathcal{J}(F) \in \sum_{j=1}^{n} U_j$*.*

▸ **3.72** [THE PRESENT IS A VARIANT OF EXERCISE 3.71]. *Let* Ω *be a bounded open subset of* $\mathbb{R}^n$*,* $n \in \mathbb{N}$*, and consider the real vector space*

$$X = [\mathbf{H}(\mathrm{div};\Omega)]^n := \Big\{ \boldsymbol{\tau} = (\tau_i)_{i=\overline{1,n}} : \quad \tau_i \in \mathbf{H}(\mathrm{div};\Omega) \quad \forall\, i = \overline{1,n} \Big\},$$

which is endowed with the inner product

$$\langle \boldsymbol{\zeta}, \boldsymbol{\tau} \rangle := \sum_{i=1}^{n} \langle \zeta_i, \tau_i \rangle_{\mathrm{div};\Omega} \qquad \forall\, \boldsymbol{\zeta} = (\zeta_i)_{i=\overline{1,n}}, \ \boldsymbol{\tau} = (\tau_i)_{i=\overline{1,n}} \in X,$$

where $\langle\cdot,\cdot\rangle_{\mathrm{div};\Omega}$ *denotes the usual inner product of* $\mathbf{H}(\mathrm{div};\Omega)$*.*

a) *Prove that* $(X, \langle\cdot,\cdot\rangle)$ *is a Hilbert space, and express its associated Riesz operator* $\mathcal{J}_X$ *in terms of that of* $\mathbf{H}(\mathrm{div};\Omega)$ *(denoted by* $\mathcal{J}$*).*

b) *Define the bounded linear operator* $A : X \longrightarrow \mathbf{H}(\mathrm{div};\Omega)$ *by* $A(\boldsymbol{\tau}) := \sum_{i=1}^{n} \tau_i$ *for all* $\boldsymbol{\tau} = (\tau_i)_{i=\overline{1,n}} \in X$*, and compute explicitly its Hilbert adjoint* A^**.*

c) *Define the bounded linear operator* $B : X \longrightarrow \mathrm{L}^2(\Omega)$ *by* $B(\boldsymbol{\tau}) := \sum_{i=1}^{n} \operatorname{div} \tau_i$ *for all* $\boldsymbol{\tau} = (\tau_i)_{i=\overline{1,n}} \in X$*, and compute explicitly its Hilbert adjoint* B^**.*

d) *Let* p_i*,* $i = \overline{1,n}$*, be the polynomials* $p_i(\mathbf{x}) := x_i^i$ *for all* $\mathbf{x} := (x_1, \ldots, x_n) \in \Omega$*, denote by* U_i *the subspace of* $\mathbf{H}(\mathrm{div};\Omega)$ *generated by* $p_i\, e_i$*, where* $\{e_1, e_2, \ldots, e_n\}$ *is the canonical basis of* $\mathbb{R}^n$*, and introduce the subspace of* X *given by*

$$S := \Big\{ \boldsymbol{\tau} = (\tau_i)_{i=\overline{1,n}} \in X : \quad \tau_i \in U_i \quad \forall\, i = \overline{1,n} \Big\}.$$

Compute $S^\perp$ *and define explicitly the orthogonal projector* $\Pi : X \longrightarrow S^\perp$*.*

▸ **3.73** *Let* $(X, \langle\cdot,\cdot\rangle_X)$ *be a Hilbert space over* $\mathbb{R}$*, and define the operator* $T : X \to X'$ *by* $T(z)(x) := \langle x, z \rangle$ *for all* z*,* $x \in X$*. The main goal of this exercise is to provide a slightly different proof of the Riesz Representation Theorem, compared to the one in Chapter 2, by showing that* T *is bijective. For this, proceed as follows:*

a) *Show that* $T \in \mathcal{L}(X, X')$ *and that* $\|T(z)\|_{X'} \geqslant \|z\|_X$ *for all* $z \in X$, *and conclude from this last fact that* T *is injective.*
b) *For the surjectivity of* T, *take* $F \in X'$, $F \neq 0$, *and choose a vector* $x_0 \in X$ *such that* $F(x_0) = 1$. *Let* V *be the null space of* F, *denote by* $Q : X \to V^\perp$ *the orthogonal projection, and show that* $F(Q(x_0)) = 1$. *Then, decompose each* $x \in X$ *as* $x = \big(x - F(x)Q(x_0)\big) + F(x)Q(x_0)$, *and take the inner product with* $Q(x_0)$ *to express* $F(x)$. *Conclude from this that* T *is surjective.*
c) *Denote* $X'' := (X')'$ *(the dual of the dual), define explicitly the adjoint operator* $T' : X'' \to X'$, *and prove that* $N(T')$ *is the null subspace of* X''.

▸ **3.74** *Let* H *and* Q *be arbitrary Hilbert spaces with Riesz operators* $\mathcal{R}_{\mathrm{H}} : \mathrm{H}' \to \mathrm{H}$ *and* $\mathcal{R}_{\mathrm{Q}} : \mathrm{Q}' \to \mathrm{Q}$, *respectively, and let* $\mathbf{B} \in \mathcal{L}(\mathrm{H}, \mathrm{Q}')$.

a) *Define the operators* $\widetilde{\mathbf{B}} := \mathcal{R}_{\mathrm{Q}}\,\mathbf{B} : \mathrm{H} \to \mathrm{Q}$ *and* $\underline{\mathbf{B}} : \mathrm{Q} \to \mathrm{H}'$ *by*

$$\underline{\mathbf{B}}(v)(\tau) := \mathbf{B}(\tau)(v) \quad \forall\, \tau \in \mathrm{H}, \quad \forall\, v \in \mathrm{Q},$$

and prove that $\widetilde{\mathbf{B}}^* = \mathcal{R}_{\mathrm{H}}\,\underline{\mathbf{B}} : \mathrm{Q} \to \mathrm{H}$.
b) *Suppose that there exists a constant* $\beta > 0$ *such that* $\|\underline{\mathbf{B}}(v)\|_{\mathrm{H}'} \geqslant \beta\,\|v\|_{\mathrm{Q}}$ *for all* $v \in \mathrm{Q}$, *and deduce from this that* $\widetilde{\mathbf{B}}^*$ *is injective and its range* $R(\widetilde{\mathbf{B}}^*)$ *is closed.*
c) *Conclude from b) and from the definition of* $\widetilde{\mathbf{B}}$ *that* $R(\widetilde{\mathbf{B}}^*) = N(\mathbf{B})^\perp$.

▸ **3.75** *Given* Ω *a bounded open subset of* $\mathbb{R}^n$, *consider the spaces*

$$\mathbf{H}(\mathrm{div};\Omega) := \Big\{\boldsymbol{\tau} \in \mathbf{L}^2(\Omega) : \quad \mathrm{div}(\boldsymbol{\tau}) \in \mathrm{L}^2(\Omega)\Big\} \quad \text{and}$$

$$\mathrm{H}^1_\Delta(\Omega) := \Big\{v \in \mathrm{H}^1(\Omega) : \quad \Delta v \in \mathrm{L}^2(\Omega)\Big\},$$

with div *and* Δ *understood in the distributional sense, endowed with the inner products*

$$\langle \boldsymbol{\zeta}, \boldsymbol{\tau}\rangle_{\mathrm{div};\Omega} := \int_\Omega \Big\{\boldsymbol{\zeta}\cdot\boldsymbol{\tau} + \mathrm{div}(\boldsymbol{\zeta})\,\mathrm{div}(\boldsymbol{\tau})\Big\} \quad \forall\, \boldsymbol{\zeta}, \boldsymbol{\tau} \in \mathbf{H}(\mathrm{div};\Omega), \quad \text{and}$$

$$\langle w, v\rangle_\Delta := \int_\Omega \Big\{w\,v + \nabla w\cdot\nabla v + \Delta w\,\Delta v\Big\} \quad \forall\, w,\, v \in \mathrm{H}^1_\Delta(\Omega),$$

whose induced norms are denoted by $\|\cdot\|_{\mathrm{div};\Omega}$ *and* $\|\cdot\|_\Delta$, *respectively.*

a) *Prove that for every* $u \in \mathcal{D}'(\Omega)$, *its distributional Laplacian reduces to*

$$\langle \Delta u, \varphi\rangle = \langle u, \Delta\varphi\rangle \qquad \forall\, \varphi \in \mathcal{D}(\Omega),$$

and that, in particular, for $u \in \mathrm{L}^2(\Omega)$ *we have*

$$\langle \Delta u, \varphi\rangle = \int_\Omega u\,\Delta\varphi \qquad \forall\, \varphi \in \mathcal{D}(\Omega).$$

b) *Use a) together with the fact that* $\mathrm{H}^1(\Omega)$ *and* $\mathrm{L}^2(\Omega)$ *are Hilbert spaces to prove that* $\big(\mathrm{H}^1_\Delta(\Omega), \langle\cdot,\cdot\rangle_\Delta\big)$ *is also a Hilbert space.*
c) *Define the operator* $A : \mathrm{H}^1_\Delta(\Omega) \to \mathbf{H}(\mathrm{div};\Omega)$ *by* $A(v) := \nabla v$ *for all* $v \in \mathrm{H}^1_\Delta(\Omega)$, *and prove that* A *is linear and bounded with* $\|A\| \leqslant 1$.
d) *Prove that the adjoint* $A^* : \mathbf{H}(\mathrm{div};\Omega) \to \mathrm{H}^1_\Delta(\Omega)$ *can be defined as* $A^*(\boldsymbol{\tau}) := \mathcal{R}_\Delta(F_{\boldsymbol{\tau}})$ *for every* $\boldsymbol{\tau} \in \mathbf{H}(\mathrm{div};\Omega)$, *where* $\mathcal{R}_\Delta : \mathrm{H}^1_\Delta(\Omega)' \to \mathrm{H}^1_\Delta(\Omega)$ *is the Riesz operator and* $F_{\boldsymbol{\tau}} \in \mathrm{H}^1_\Delta(\Omega)'$ *is the functional given by*

$$F_{\boldsymbol{\tau}}(v) := \int_\Omega \Big\{\nabla v\cdot\boldsymbol{\tau} + \Delta v\,\mathrm{div}(\boldsymbol{\tau})\Big\} \quad \forall\, v \in \mathrm{H}^1_\Delta(\Omega).$$

▸ **3.76** *Let* X *and* Y *be reflexive Banach spaces, let* $A \in \mathcal{L}(X, Y')$, *and define the operator* $\widetilde{A} \in \mathcal{L}(Y, X')$ *by* $\widetilde{A}(y)(x) := A(x)(y)$ *for all* $y \in Y$, *for all* $x \in X$.

a) *Prove that* $\widetilde{A} = A' \circ \mathcal{J}_Y$ *and* $A = \widetilde{A}' \circ \mathcal{J}_X$, *where* $\mathcal{J}_Y : Y \to Y''$ *and* $\mathcal{J}_X : X \to X''$ *are the bijective linear operators that guarantee the reflexivity of* X *and* Y, *respectively.*
b) *Prove that the following statements are equivalent:*
   i) $R(A)$ *is closed.*
   ii) $R(A') = N(A)^\circ$.
   iii) $R(A')$ *is closed.*
   iv) $R(A) = {}^\circ N(A')$.

   HINT.: *it is suggested to follow the same logical order as the sequence, namely* i) $\Rightarrow$ ii), ii) $\Rightarrow$ iii), iii) $\Rightarrow$ iv), *and* iv) $\Rightarrow$ i). *Moreover, for* iii) $\Rightarrow$ iv) *it is recommended to use the identities from* a) *and the implication* i) $\Rightarrow$ ii).

▸ **3.77** [GENERALIZATION OF EXERCISE 3.29]. *Let* $(H_1, \langle\cdot,\cdot\rangle_1)$ *and* $(H_2, \langle\cdot,\cdot\rangle_2)$ *be Hilbert spaces over* $\mathbb{R}$ *with induced norms denoted by* $\|\cdot\|_1$ *and* $\|\cdot\|_2$, *respectively, such that* $H_2 \subset H_1$, *and suppose there exists a constant* $\widetilde{C} > 0$ *such that* $\|w\|_1 \leqslant \widetilde{C}\,\|w\|_2$ *for all* $w \in H_2$. *Moreover, let* $A : H_1 \times H_1 \to \mathbb{R}$ *be a bounded bilinear form, and assume that for every* $F \in H_1'$ *there exists a unique* $u \in H_2$ *such that* $A(u,v) = F(v)$ *for all* $v \in H_1$. *Use the Closed Graph Theorem to prove that there exists a constant* $C > 0$ *such that, for each* $F \in H_1'$, *the unique solution* u *above satisfies*

$$\|u\|_2 \leqslant C\,\|F\|_{H_1'}\,.$$

▸ **3.78** *Let* $(X, \langle\cdot,\cdot\rangle_X)$ *and* $(Y, \langle\cdot,\cdot\rangle_Y)$ *be Hilbert spaces over* $\mathbb{R}$, *and let* $A \in \mathcal{L}(X,Y)$ *be such that* $R(A)$ *is a closed subspace of* Y. *The goal of this exercise is to prove that* $R(A^*) = N(A)^\perp$, *which implies that* $R(A^*)$ *is also closed. To this end, adapt to the Hilbert space context the proof given in the Banach space setting (cf. Theorem 3.62). Accordingly, follow the logical sequence below:*

a) *Prove by direct inclusion that* $R(A^*) \subseteq N(A)^\perp$.
b) *Given* $z \in N(A)^\perp$, *define the functional* $g : R(A) \to \mathbb{R}$ *by* $g(y) := \langle x, z\rangle_X$ *for all* $y \in R(A)$, *where* $x \in X$ *is such that* $A(x) = y$, *and show that* g *is well defined and linear.*
c) *Use the characterization of operators with closed range to show that* g *is bounded.*
d) *Apply the Riesz Representation Theorem to deduce the existence of a vector* $y_g \in Y$ *such that* $z = A^*(y_g) \in R(A^*)$, *and conclude the reciprocal inclusion to that of a).*

▸ **3.79** [THE PRESENT IS ANALOGOUS TO EXERCISE 3.75] *Let* Ω *be a bounded open subset of* $\mathbb{R}^2$, *and consider the spaces*

$$\mathbf{H}(\mathrm{rot};\Omega) := \left\{\boldsymbol{\tau} := (\tau_1,\tau_2) \in \mathbf{L}^2(\Omega) : \quad \mathrm{rot}(\boldsymbol{\tau}) := \frac{\partial \tau_2}{\partial x_1} - \frac{\partial \tau_1}{\partial x_2} \in \mathrm{L}^2(\Omega)\right\},$$

$$\mathrm{H}^1_{\mathfrak{L}}(\Omega) := \left\{v \in \mathrm{H}^1(\Omega) : \quad \mathfrak{L}(v) := \frac{\partial^2 v}{\partial x_1^2} - \frac{\partial^2 v}{\partial x_2^2} \in \mathrm{L}^2(\Omega)\right\},$$

with rot *and* $\mathfrak{L}$ *in the distributional sense, endowed with the inner products*

$$\langle \boldsymbol{\zeta}, \boldsymbol{\tau}\rangle_{\mathrm{rot};\Omega} := \int_\Omega \Big\{\boldsymbol{\zeta}\cdot\boldsymbol{\tau} + \mathrm{rot}(\boldsymbol{\zeta})\,\mathrm{rot}(\boldsymbol{\tau})\Big\} \quad \forall\, \boldsymbol{\zeta}, \boldsymbol{\tau} \in \mathbf{H}(\mathrm{rot};\Omega)\,, \quad \text{and}$$

$$\langle w, v\rangle_{\mathfrak{L}} := \int_\Omega \Big\{w\,v + \nabla w\cdot\nabla v + \mathfrak{L}(w)\,\mathfrak{L}(v)\Big\} \quad \forall\, w,\, v \in \mathrm{H}^1_{\mathfrak{L}}(\Omega)\,,$$

with induced norms denoted by $\|\cdot\|_{\mathrm{rot};\Omega}$ *and* $\|\cdot\|_{\mathfrak{L}}$, *respectively.*

a) *Prove that for every* $u \in \mathcal{D}'(\Omega)$, $\mathfrak{L}(u)$ *reduces distributionally to* $\langle \mathfrak{L}(u), \varphi\rangle = \langle u, \mathfrak{L}(\varphi)\rangle$ *for all* $\varphi \in \mathcal{D}(\Omega)$, *and that, in particular, for* $u \in \mathrm{L}^2(\Omega)$ *we have* $\langle \mathfrak{L}(u), \varphi\rangle = \int_\Omega u\,\mathfrak{L}(\varphi)$ *for all* $\varphi \in \mathcal{D}(\Omega)$.
b) *Use a) together with the fact that* $\mathrm{H}^1(\Omega)$ *and* $\mathrm{L}^2(\Omega)$ *are Hilbert spaces to prove that* $(\mathrm{H}^1_{\mathfrak{L}}(\Omega), \langle\cdot,\cdot\rangle_{\mathfrak{L}})$ *is also a Hilbert space.*
c) *Define the operator* $T : \mathrm{H}^1_{\mathfrak{L}}(\Omega) \to \mathbf{H}(\mathrm{rot};\Omega)$ *by* $T(v) := \left(\frac{\partial v}{\partial x_2}, \frac{\partial v}{\partial x_1}\right)^{\mathrm{t}}$ *for every* $v \in \mathrm{H}^1_{\mathfrak{L}}(\Omega)$, *and prove that* T *is linear and bounded with* $\|T\| \leqslant 1$.

d) *Show that the adjoint* $T^* : \mathbf{H}(\mathrm{rot};\Omega) \to \mathrm{H}^1_{\mathfrak{L}}(\Omega)$ *can be defined as* $T^*(\tau) := \mathcal{R}_{\mathfrak{L}}(F_\tau)$ *for every* $\tau \in \mathbf{H}(\mathrm{rot};\Omega)$, *where* $\mathcal{R}_{\mathfrak{L}} : \mathrm{H}^1_{\mathfrak{L}}(\Omega)' \to \mathrm{H}^1_{\mathfrak{L}}(\Omega)$ *is the Riesz operator and* $F_\tau \in \mathrm{H}^1_{\mathfrak{L}}(\Omega)'$ *is the functional given by*

$$F_\tau(v) := \int_\Omega \Big\{ T(v)\cdot\tau + \mathfrak{L}(v)\,\mathrm{rot}(\tau) \Big\} \quad \forall\, v \in \mathrm{H}^1_{\mathfrak{L}}(\Omega)\,.$$

▸ **3.80** *Let* Ω *be a bounded open subset of* $\mathbb{R}^n$, *and let* U *be a finite-dimensional subspace of* $\mathrm{H}^1(\Omega)$ *of dimension* N. *Given a basis* $\{u_1, u_2, \ldots, u_N\}$ *of* U, *denote* $\mathbf{w}_j := \nabla u_j$ *for each* $j \in \{1,2,\ldots,N\}$, *observe that* $\{\mathbf{w}_1, \mathbf{w}_2, \ldots, \mathbf{w}_N\} \subseteq \mathbf{L}^2(\Omega)$, *and define the operator* $A : \mathbf{L}^2(\Omega) \longrightarrow \mathrm{H}^1(\Omega)$ *by*

$$A(\mathbf{z}) := \sum_{j=1}^N \left\{ \int_\Omega \mathbf{z}\cdot\mathbf{w}_j \right\} u_j \qquad \forall\, \mathbf{z} \in \mathbf{L}^2(\Omega)\,.$$

a) *Find explicitly the adjoint operator* $A' : \mathrm{H}^1(\Omega)' \longrightarrow \mathbf{L}^2(\Omega)'$.
b) *Prove that the adjoint operator* $A^* : \mathrm{H}^1(\Omega) \longrightarrow \mathbf{L}^2(\Omega)$ *is given by*

$$A^*(v) := \sum_{j=1}^N \left\{ \int_\Omega \Big(u_j\, v + \mathbf{w}_j \cdot \nabla v \Big) \right\} \mathbf{w}_j \qquad \forall\, v \in \mathrm{H}^1(\Omega)\,,$$

and conclude, moreover, that $\|A^*\| \leqslant \sum_{j=1}^N \|u_j\|^2_{1,\Omega}$.

c) *Recall that the inner product of* $\mathbf{L}^2(\Omega)$ *is defined as*

$$\langle \mathbf{z}, \mathbf{w} \rangle_{0,\Omega} := \int_\Omega \mathbf{z}\cdot\mathbf{w} \qquad \forall\, \mathbf{z},\, \mathbf{w} \in \mathbf{L}^2(\Omega)\,,$$

and prove that $N(A)^\perp = \langle \{\mathbf{w}_1, \mathbf{w}_2, \ldots, \mathbf{w}_N\} \rangle$.
d) *Consider the orthogonal projector* $\mathrm{P} : \mathbf{L}^2(\Omega) \longrightarrow N(A)$, *and explain how to compute, for each* $\mathbf{z} \in \mathbf{L}^2(\Omega)\backslash N(A)$, *its best approximation* $\mathrm{P}(\mathbf{z})$. *What happens with this computation if* $\{\mathbf{w}_1, \mathbf{w}_2, \ldots, \mathbf{w}_N\}$ *is linearly independent in* $\mathbf{L}^2(\Omega)$? *What happens in the case when* $\{\mathbf{w}_1, \mathbf{w}_2, \ldots, \mathbf{w}_N\}$ *is an orthogonal set in* $\mathbf{L}^2(\Omega)$?

▸ **3.81** *Let* X *and* Y *be Banach spaces, and consider the product spaces* $X \times Y$ *and* $X' \times Y'$ *with induced norms given by* $\|(x,y)\|_{X\times Y} := \|x\|_X + \|y\|_Y$ *for all* $(x,y) \in X \times Y$, *and* $\|(\mathbf{f},\boldsymbol{g})\|_{X'\times Y'} := \|\mathbf{f}\|_{X'} + \|\boldsymbol{g}\|_{Y'}$ *for all* $(\mathbf{f},\boldsymbol{g}) \in X' \times Y'$.

a) *Prove that every functional* $F \in (X\times Y)'$ *can be uniquely identified with a pair* $(\mathbf{f},\boldsymbol{g}) \in X' \times Y'$ *such that*

$$F\big((x,y)\big) = \mathbf{f}(x) + \boldsymbol{g}(y) \qquad \forall\, (x,y) \in X \times Y\,,$$

thus establishing an equivalence between $(X\times Y)'$ *and* $X' \times Y'$.
b) *Let* $\{x_1,\ldots,x_n\} \subseteq X$, $\{y_1,\ldots,y_n\} \subseteq Y$, $\{f_1,\ldots,f_n\} \subseteq X'$, $\{g_1,\ldots,g_n\} \subseteq Y'$, *and define the operator* $T : X\times Y \longrightarrow X \times Y$ *by*

$$T\big((x,y)\big) := \Big(\sum_{j=1}^n g_j(y)\, x_j,\, \sum_{j=1}^n f_j(x)\, y_j \Big) \qquad \forall\, (x,y) \in X \times Y\,.$$

Use the equivalence described in a) to prove that $T' : (X\times Y)' \to (X\times Y)'$ *is given by*

$$T'(F) := \Big(\sum_{j=1}^n \boldsymbol{g}(y_j)\, f_j,\, \sum_{j=1}^n \mathbf{f}(x_j)\, g_j \Big) \quad \forall\, F := (\mathbf{f},\boldsymbol{g}) \in (X\times Y)' \equiv X' \times Y'\,.$$

▸ **3.82** *Let* X *and* Y *be Banach spaces, let* $\Big\{F_1,\ldots,F_N\Big\} \subseteq X'$, *define the closed subspace of* X *given by* $V := {}^\circ\Big\{F_1,\ldots,F_N\Big\}$, *and consider a linearly independent set* $\Big\{y_1,\ldots,y_N\Big\} \subseteq Y$.

a) *Prove that there exists a positive constant* C_N *such that*

$$\Big\| \sum_{j=1}^{N} F_j(x)\, y_j \Big\|_Y \geqslant C_N \inf_{v \in V} \|x - v\|_X \qquad \forall\, x \in X .$$

b) *Now assume that* $V = \{0\}$ *and prove that* X' *is finite-dimensional with* $\dim X' \leqslant N$.

▸ **3.83** [VARIANT OF EXERCISE 3.29]. *Let* Ω *be a bounded domain of* $\mathbb{R}^n$, *and consider the Hilbert space*

$$\mathbf{H}(\mathrm{div};\Omega) := \Big\{ \boldsymbol{\tau} \in \mathbf{L}^2(\Omega) : \quad \mathrm{div}(\boldsymbol{\tau}) \in \mathrm{L}^2(\Omega) \Big\} .$$

a) *Prove that for each* $v \in \mathrm{H}^1(\Omega)$ *there exists a unique* $\boldsymbol{\sigma} \in \mathbf{H}(\mathrm{div};\Omega)$ *such that*

$$\int_\Omega \Big\{ \boldsymbol{\sigma} \cdot \boldsymbol{\tau} + \mathrm{div}(\boldsymbol{\sigma})\, \mathrm{div}(\boldsymbol{\tau}) \Big\} = \int_\Omega \nabla v \cdot \boldsymbol{\tau} \qquad \forall\, \boldsymbol{\tau} \in \mathbf{H}(\mathrm{div};\Omega) ,$$

and deduce that $\|\boldsymbol{\sigma}\|_{\mathrm{div};\Omega} \leqslant \|v\|_{1,\Omega}$.

b) *Assuming that for each* $v \in \mathrm{H}^1(\Omega)$, $\boldsymbol{\sigma}$ *belongs to* $\mathbf{H}^1(\Omega) := [\mathrm{H}^1(\Omega)]^n$, *define an appropriate linear operator and prove, using the Closed Graph Theorem, that there exists a constant* $C > 0$ *such that*

$$\|\boldsymbol{\sigma}\|_{1,\Omega} \leqslant C\, \|v\|_{1,\Omega} \qquad \forall\, v \in \mathrm{H}^1(\Omega) .$$

▸ **3.84** [THE PRESENT IS A MORE GENERAL VERSION OF EXERCISE 3.82] *Let* X *and* Y *be Banach spaces, and consider the closed subspaces given by the annihilators* $U := {}^\circ\{F_1, \ldots, F_N\}$ *and* $V := {}^\circ\{G_1, \ldots, G_N\}$, *where* $\{F_1, \ldots, F_N\}$ *and* $\{G_1, \ldots, G_N\}$ *are subsets of* X' *and* Y', *respectively. Moreover, let* $\{x_1, \ldots, x_N\} \subseteq X$ *and* $\{y_1, \ldots, y_N\} \subseteq Y$ *be linearly independent sets.*

a) *Prove that there exists a positive constant* C_N *such that*

$$\Big\| \sum_{j=1}^{N} G_j(y)\, x_j \Big\|_X + \Big\| \sum_{j=1}^{N} F_j(x)\, y_j \Big\|_Y \geqslant C_N \Big\{ \inf_{u \in U} \|x - u\|_X + \inf_{v \in V} \|y - v\|_Y \Big\}$$

for all $(x, y) \in X \times Y$.

b) *Now assume that* $U = \{0\}$ *and* $V = \{0\}$, *and prove from this that*

$$\dim (X \times Y)' \leqslant 2N .$$

▸ **3.85** *Let* X *and* Y *be Banach spaces, let* $A \in \mathcal{L}(X, Y)$, *and assume that there exists a constant* $\alpha > 0$ *such that for each* $y \in R(A)$ *there exists* $z \in X$ *such that* $A(z) = y$ *and* $\alpha \|z\| \leqslant \|y\|$. *Use the respective characterization result to prove that* $R(A)$ *is a closed subspace of* Y.

▸ **3.86** *Let* V *and* W *be closed subspaces of a Banach space* $(X, \|\cdot\|)$ *such that* $X = V \oplus W$, *and define the operators* $P : X \longrightarrow X$ *and* $Q : X \longrightarrow X$ *which assign to each* $x \in X$ *the unique elements* $P(x) := x_V \in V$ *and* $Q(x) := x_W \in W$ *such that* $x = x_V + x_W$. *Use a consequence of the Open Mapping Theorem to show that* $P, Q \in \mathcal{L}(X, X)$.

▸ **3.87** [THE PRESENT IS AN ALTERNATIVE VERSION OF EXERCISE 3.72]. *Given* $\Omega \subset \mathbb{R}^n$, $n \in \mathbb{N}$, *recall that* $(\mathbf{H}(\mathrm{div};\Omega), \langle \cdot, \cdot \rangle_{\mathrm{div};\Omega})$ *is a Hilbert space, and define the real vector space*

$$\mathbb{X} := \Big\{ \vec{\boldsymbol{\tau}} = (\tau_1, \ldots, \tau_n) : \quad \tau_i \in \mathbf{H}(\mathrm{div};\Omega) \quad \forall\, i \in \{1, 2, \ldots, n\} \Big\} ,$$

which is also Hilbert with the inner product

$$\langle \vec{\boldsymbol{\zeta}}, \vec{\boldsymbol{\tau}} \rangle := \sum_{i=1}^{n} \langle \zeta_i, \tau_i \rangle_{\mathrm{div};\Omega} \quad \forall\, \vec{\boldsymbol{\zeta}} = (\zeta_1, \ldots, \zeta_n) ,\ \vec{\boldsymbol{\tau}} = (\tau_1, \ldots, \tau_n) \in \mathbb{X} .$$

a) *Define* $A \in \mathcal{L}(\mathbb{X}, \mathbf{H}(\mathrm{div}; \Omega))$ *by* $A(\vec{\tau}) := \sum_{i=1}^{n} \tau_i$ *for all* $\vec{\tau} \in \mathbb{X}$*, and compute explicitly* A^* *and* A'. *In particular, prove that for each* $G \in \mathbf{H}(\mathrm{div}; \Omega)'$ *it holds that* $A'(G) = \mathcal{R}_{\mathbb{X}}^{-1}(\vec{\mathcal{R}}(G))$*, with* $\vec{\mathcal{R}}(G) := (\mathcal{R}(G), \dots, \mathcal{R}(G))$*, where* $\mathcal{R}_{\mathbb{X}} : \mathbb{X}' \to \mathbb{X}$ *and* $\mathcal{R} : \mathbf{H}(\mathrm{div}; \Omega)' \to \mathbf{H}(\mathrm{div}; \Omega)$ *are the respective Riesz operators.*

b) *Define* $B \in \mathcal{L}(\mathbb{X}, \mathrm{L}^2(\Omega))$ *by* $B(\vec{\tau}) := \sum_{i=1}^{n} \mathrm{div}(\tau_i)$ *for all* $\vec{\tau} \in \mathbb{X}$*, and prove that*

$$B^*(v) = (\mathcal{R}(F_v), \dots, \mathcal{R}(F_v)) \qquad \forall\, v \in \mathrm{L}^2(\Omega),$$

where $F_v \in \mathbf{H}(\mathrm{div}; \Omega)'$ *is defined by* $F_v(\tau) := \langle \mathrm{div}(\tau), v \rangle_{0,\Omega}$ *for all* $\tau \in \mathbf{H}(\mathrm{div}; \Omega)$.

4

Variational Problems in Hilbert Spaces

In this chapter we address the continuous and discrete solvability properties of variational problems arising mainly from diverse models involving differential equations. Our analysis, partially based on approaches developed in [8], [23], [24], [39], [110], [90], and [112], will focus on formulations posed in Hilbert spaces. Later, in Chapter 5, we will discuss both theoretical and applied aspects of some formulations defined in terms of Banach spaces.

In what follows, and throughout the rest of the book, we continue to use the notation for inner products, norms, and semi-norms of Sobolev spaces that were introduced in the previous chapters, some of which are later on reintroduced in Chapter 10 as well. In particular, given a domain Ω of $\mathbb{R}^n$, $n \in \{1,2,3\}$, with boundary Γ, we denote by $\langle\cdot,\cdot\rangle_{0,\Omega}$, $\langle\cdot,\cdot\rangle_{1,\Omega}$, $\|\cdot\|_{0,\Omega}$, $\|\cdot\|_{1,\Omega}$, and $|\cdot|_{1,\Omega}$ the inner products and norms of $\mathrm{L}^2(\Omega)$ and $\mathrm{H}^1(\Omega)$, and the semi-norm of the latter, respectively. In turn, given $s \in \{-1/2, 0, 1/2\}$, $\|\cdot\|_{s,\Gamma}$ stands for the norm of $\mathrm{H}^s(\Gamma)$, spaces that are introduced and used extensively in this chapter. Note that, in the particular case $s = 0$, $\langle\cdot,\cdot\rangle_{0,\Gamma}$ denotes the usual inner product of $\mathrm{L}^2(\Gamma)$. In addition, given any generic scalar space S, we denote by $\mathbf{S} := \mathrm{S}^n$ and $\mathbb{S} := \mathrm{S}^{n\times n}$ its corresponding vector and tensor versions.

4.1 The Lax-Milgram Lemma

In order to state and prove this result, the most classical one in the scope of the present chapter, we need some preliminary concepts.

Definition 4.1. *Let $(H_1, \langle\cdot,\cdot\rangle_1)$ and $(H_2, \langle\cdot,\cdot\rangle_2)$ be real Hilbert spaces. We say that a form $B : H_1 \times H_2 \to \mathbb{R}$ is* BILINEAR *if it is linear in each of its arguments, that is*

i) $B(\alpha\, x+\beta\, y, z) = \alpha\, B(x,z)+\beta\, B(y,z)$ *for all* $x,\, y \in H_1$, *for all* $z \in H_2$, *for all* $\alpha,\, \beta \in \mathbb{R}$.
ii) $B(x, \alpha\, y+\beta\, z) = \alpha\, B(x,y)+\beta\, B(x,z)$ *for all* $x \in H_1$, *for all* $y,\, z \in H_2$, *for all* $\alpha,\, \beta \in \mathbb{R}$.

Definition 4.2. *Let $(H_1, \langle\cdot,\cdot\rangle_1)$ and $(H_2, \langle\cdot,\cdot\rangle_2)$ be real Hilbert spaces. A bilinear form $B : H_1 \times H_2 \to \mathbb{R}$ is said to be* BOUNDED *if there exists a constant $M > 0$ such that*

$$|B(x,y)| \leqslant M\, \|x\|_1\, \|y\|_2 \quad \forall (x,y) \in H_1 \times H_2 .$$

Definition 4.3. *Let $(H, \langle\cdot,\cdot\rangle)$ be a real Hilbert space, and let $B : H \times H \to \mathbb{R}$ be a bilinear form. We say that B is* STRONGLY COERCIVE *(or H-*ELLIPTIC*) if there exists a constant $\alpha > 0$ such that*

$$B(x,x) \geqslant \alpha\, \|x\|^2 \qquad \forall\, x \in H .$$

G. N. Gatica, *Introduction to Functional Analysis*,
https://doi.org/10.1007/978-3-032-37415-8_4

Now, given $(H_1, \langle\cdot,\cdot\rangle_1)$ and $(H_2, \langle\cdot,\cdot\rangle_2)$ real Hilbert spaces and $B : H_1 \times H_2 \to \mathbb{R}$ a bounded bilinear form, we are interested in defining the operator $\mathbf{B} : H_1 \to H_2$ induced by B, and conversely. To this end, consider an element $v \in H_1$ and define the functional $F_v : H_2 \to \mathbb{R}$ by

$$F_v(w) := B(v,w) \quad \forall\, w \in H_2.$$

Since B is linear in its second argument (because B is bilinear), we clearly have that F_v is linear. Moreover, since B is bounded (with constant M) it follows that

$$|F_v(w)| \leqslant M\|v\|_1\, \|w\|_2 \quad \forall\, w \in H_2,$$

which implies $F_v \in H_2'$ and

$$\|F_v\| \leqslant M\|v\|_1 \quad \forall\, v \in H_1. \tag{4.1}$$

The above induces the definition of the operator $\mathcal{B} : H_1 \to H_2'$ as

$$\mathcal{B}(v) := F_v \quad \forall\, v \in H_1,$$

which, due to the linearity of B in its first argument and inequality (4.1), is a linear and bounded operator with

$$\|\mathcal{B}\|_{\mathcal{L}(H_1,H_2')} \leqslant M.$$

Finally, if $\mathcal{R}_2 : H_2' \to H_2$ denotes the corresponding Riesz map, the operator $\mathbf{B} : H_1 \to H_2$ induced by B is defined as

$$\mathbf{B} := \mathcal{R}_2 \circ \mathcal{B}, \tag{4.2}$$

or graphically

$$\begin{array}{ccc} H_1 & \xrightarrow{\mathcal{B}} & H_2' \\ & \mathbf{B} \searrow & \downarrow \mathcal{R}_2 \\ & & H_2 \end{array}$$

Note that $\mathbf{B}$ is linear and bounded (since both $\mathcal{R}_2$ and $\mathcal{B}$ are), and moreover:

$$\langle \mathbf{B}(v), w\rangle_2 = \langle \mathcal{R}_2(\mathcal{B}(v)), w\rangle_2 = \mathcal{B}(v)(w) = B(v,w) \quad \forall\, (v,w) \in H_1 \times H_2\,. \tag{4.3}$$

Conversely, if $\mathbf{B} \in \mathcal{L}(H_1, H_2)$ then the bilinear form $B : H_1 \times H_2 \to \mathbb{R}$ induced by $\mathbf{B}$ is defined as

$$B(v,w) := \langle \mathbf{B}(v), w\rangle_2 \quad \forall\, (v,w) \in H_1 \times H_2\,. \tag{4.4}$$

Lemma 4.4 (Lax–Milgram Lemma). *Let $(H, \langle\cdot,\cdot\rangle)$ be a real Hilbert space, and let $B \times H \to \mathbb{R}$ be a bilinear form that is bounded and H-elliptic with constants M and α, respectively. Then, for each $F \in H'$, there exists a unique $u \in H$ such that*

$$B(u,v) = F(v) \qquad \forall\, v \in H\,, \tag{4.5}$$

and

$$\|u\| \leqslant \frac{1}{\alpha}\, \|F\|\,. \tag{4.6}$$

Proof. Let $\mathbf{B} : H \to H$ be the linear and bounded operator induced by B, that is:

$$\langle \mathbf{B}(v), w\rangle = B(v,w) \quad \forall\, (v,w) \in H \times H,$$

and let $\mathcal{R} : H' \to H$ be the corresponding Riesz map. Then, finding a unique $u \in H$ such that (4.5) holds is equivalent to requiring that

$$\langle \mathbf{B}(u), v\rangle = \langle \mathcal{R}(F), v\rangle \quad \forall\, v \in H,$$

or equivalently, such that

$$\mathbf{B}(u) = \mathcal{R}(F). \tag{4.7}$$

Moreover, since this is required for each $F \in H'$, the present proof reduces to showing that $\mathbf{B} : H \to H$ is bijective. To this end, note first, from the H-ellipticity of B, that for each $v \in H$:

$$\alpha \, \|v\|^2 \leqslant B(v,v) = \langle \mathbf{B}(v), v\rangle \leqslant \|\mathbf{B}(v)\| \, \|v\|,$$

whence

$$\alpha \, \|v\| \leqslant \|\mathbf{B}(v)\| \quad \forall\, v \in H. \tag{4.8}$$

It follows, from the characterization result for operators with closed range (see Theorem 3.58), that $\mathbf{B}$ is injective and $R(\mathbf{B})$ is a closed subspace of H. Then, according to the Orthogonal Decomposition Theorem (see Theorem 2.10), we have $H = R(\mathbf{B}) \oplus R(\mathbf{B})^\perp$, and therefore, it only remains to show that $R(\mathbf{B})^\perp = \{0\}$ in order to conclude that $\mathbf{B}$ is surjective. Indeed, given $w \in R(\mathbf{B})^\perp$ we have $\langle\, z, w \,\rangle = 0$ for all $z \in R(\mathbf{B})$, or equivalently, $\langle \mathbf{B}(v), w\rangle = 0$ for all $v \in H$. In particular, for $v = w$ we obtain, using the H-ellipticity of B,

$$0 = \langle \mathbf{B}(w), w\rangle = B(w,w) \geqslant \alpha \, \|w\|^2,$$

whence $w = 0$, thus completing the proof of the bijectivity of $\mathbf{B}$. Consequently, given $F \in H'$ there exists a unique $u \in H$ such that $\mathbf{B}(u) = \mathcal{R}(F)$, that is:

$$B(u,v) = F(v) \quad \forall\, v \in H.$$

Then, taking $v = u$ in (4.8), we obtain

$$\alpha \, \|u\| \leqslant \|\mathbf{B}(u)\| = \|\mathcal{R}(F)\| = \|F\|,$$

that is,

$$\|u\| \leqslant \frac{1}{\alpha} \, \|F\|.$$

□

The Lax-Milgram Lemma and its proof motivate several remarks. First, note that inequality (4.6) represents a result of continuous dependence for problem (4.5). Indeed, given $F_1,\, F_2 \in H'$, let $u_1,\, u_2 \in H$ be the unique elements, guaranteed by the Lax-Milgram Lemma, such that

$$B(u_1, v) = F_1(v) \qquad \forall\, v \in H$$

and

$$B(u_2, v) = F_2(v) \qquad \forall\, v \in H.$$

It follows that $u := u_1 - u_2 \in H$ is the unique solution of the problem

$$B(u,v) = (F_1 - F_2)(v) \qquad \forall\, v \in H,$$

and therefore, inequality (4.6) implies that

$$\|u_1 - u_2\| \leqslant \frac{1}{\alpha} \, \|F_1 - F_2\| \, .$$

The above indicates that the stability of the solution of problem (4.5) is ensured with large values of the ellipticity constant α.

On the other hand, recall from (4.7) that proving the Lax-Milgram Lemma reduces, given $F \in H'$, to showing the existence of a unique $u \in H$ such that $\mathbf{B}(u) = \mathcal{R}(F)$, which, by considering a parameter $\rho > 0$, is equivalent to finding $u \in H$ such that

$$-\rho\left\{\mathbf{B}(u) - \mathcal{R}(F)\right\} = 0,$$

or equivalently, finding $u \in H$ such that

$$T(u) = u,$$

where $T : H \to H$ is the nonlinear operator defined by

$$T(v) := v - \rho\left\{\mathbf{B}(v) - \mathcal{R}(F)\right\} \qquad \forall\, v \in H.$$

Thus, an alternative proof of the Lax-Milgram Lemma consists in proving that T possesses a unique fixed point, which can be achieved, by virtue of the Banach Fixed Point Theorem (cf. Theorem 1.8), by showing that T is a contraction for some $\rho > 0$. Indeed, using the ellipticity and boundedness of $\mathbf{B}$, we obtain

$$\begin{aligned}
&\|T(v) - T(w)\|^2 = \langle T(v) - T(w), T(v) - T(w)\rangle \\
&= \langle (v-w) - \rho\,\mathbf{B}(v-w), (v-w) - \rho\,\mathbf{B}(v-w)\rangle \\
&= \|v-w\|^2 - 2\rho\langle \mathbf{B}(v-w), v-w\rangle + \rho^2\,\|\mathbf{B}(v-w)\|^2 \\
&\leqslant (1 - 2\rho\,\alpha + \rho^2 M^2)\|v-w\|^2
\end{aligned}$$

for all $v, w \in H$, from which it follows that a sufficient condition for the contractivity of T is that $1 - 2\,\rho\,\alpha + \rho^2 M^2 < 1$, that is

$$\rho \in \left(0, \frac{2\alpha}{M^2}\right).$$

Another interesting aspect of problem (4.5) is the resulting analysis for the case in which the bilinear form B is SYMMETRIC. Indeed, under this additional assumption, B becomes another inner product on H, whose induced norm, denoted by $\|\cdot\|_B$, is given by

$$\|v\|_B := B(v,v)^{1/2} \qquad \forall\, v \in H.$$

Moreover, thanks to the H-ellipticity and boundedness of B, we obtain that

$$\alpha\,\|v\|^2 \leqslant B(v,v) = \|v\|_B^2 \leqslant M\,\|v\|^2 \qquad \forall\, v \in H,$$

which proves that $\|\cdot\|$ and $\|\cdot\|_B$ are equivalent norms on H, and therefore $(H, B(\cdot,\cdot))$ is also a Hilbert space. Consequently, given $F \in H'$ (with respect to either of the two norms), the Riesz Representation Theorem (cf. Theorem 2.14) applied to $(H, B(\cdot,\cdot))$ implies the existence of a unique $u \in H$ such that

$$F(v) = B(u,v) \qquad \forall\, v \in H.$$

Thus, the proof of the Lax-Milgram Lemma in the case of a symmetric bilinear form B reduces to a simple application of Theorem 2.14. Another way to view this is that the Lax-Milgram Lemma is nothing more than an extension of Theorem 2.14 to the case of a bounded and H-elliptic bilinear form, not necessarily symmetric.

We now illustrate the applicability of the Lax-Milgram Lemma with a one-dimensional example. For this, we first need to prove the following result.

Lemma 4.5 (Friedrichs-Poincaré inequality). *Let* $\Omega := (a,b) \subseteq \mathbb{R}$ *and*

$$\mathrm{H}_0^1(\Omega) := \left\{ v \in \mathrm{H}^1(\Omega) : \quad v(a) = v(b) = 0 \right\}.$$

Then, we have

$$\|v\|_{0,\Omega}^2 \leqslant \frac{(b-a)^2}{2} \, |v|_{1,\Omega}^2 \qquad \forall\, v \in \mathrm{H}_0^1(\Omega)\,, \tag{4.9}$$

and therefore

$$\|v\|_{1,\Omega}^2 \leqslant \left\{ 1 + \frac{(b-a)^2}{2} \right\} |v|_{1,\Omega}^2 \qquad \forall\, v \in \mathrm{H}_0^1(\Omega). \tag{4.10}$$

Proof. We first prove inequality (4.9) in the space $C_0^\infty(\Omega)$ and then use that $C_0^\infty(\Omega)$ is dense in $\mathrm{H}_0^1(\Omega)$ with respect to the usual norm $\|\cdot\|_{1,\Omega}$ of $\mathrm{H}^1(\Omega)$. Indeed, let $\varphi \in C_0^\infty(\Omega)$. Then, for every $x \in \Omega$ we have

$$\varphi(x) = \int_a^x \varphi'(t)\, dt,$$

which, by applying the Cauchy-Schwarz inequality, implies

$$\begin{aligned} |\varphi(x)|^2 &\leqslant \int_a^x 1^2\, dt \int_a^x (\varphi'(t))^2\, dt = (x-a) \int_a^x (\varphi'(t))^2\, dt \\ &\leqslant (x-a) \int_a^b (\varphi'(t))^2\, dt = (x-a)\, |\varphi|_{1,\Omega}^2. \end{aligned}$$

Then, integrating with respect to $x \in \Omega$, we obtain from the previous estimate

$$\|\varphi\|_{0,\Omega}^2 \leqslant |\varphi|_{1,\Omega}^2 \int_a^b (x-a)\, dx = \frac{(b-a)^2}{2}\, |\varphi|_{1,\Omega}^2\,. \tag{4.11}$$

Now, given $v \in \mathrm{H}_0^1(\Omega)$ we consider a sequence $\{\varphi_n\}_{n\in\mathbb{N}} \subseteq C_0^\infty(\Omega)$ such that

$$\|v - \varphi_n\|_{1,\Omega} \overset{n\to\infty}{\rightarrow} 0. \tag{4.12}$$

It follows from (4.11) that

$$\|\varphi_n\|_{0,\Omega}^2 \leqslant \frac{(b-a)^2}{2}\, |\varphi_n|_{1,\Omega}^2 \quad \forall\, n \in \mathbb{N}\,,$$

and therefore, taking the limit as $n \to \infty$, and using (4.12), we obtain (4.9). Finally, it is clear that (4.10) is a direct consequence of (4.9).

□

Example 4.6 Given $\Omega = (0,1)$ and $f \in \mathrm{L}^2(\Omega)$, consider the boundary value problem:

$$-u'' = f \quad \text{in} \quad \Omega, \quad u(0) = u(1) = 0.$$

We know (see Chapter 1) that the corresponding variational formulation is given by: Find $u \in H := \mathrm{H}_0^1(\Omega)$ such that

$$B(u,v) := \int_0^1 u'v' = F(v) := \int_0^1 fv \quad \forall\, v \in H. \tag{4.13}$$

It is clear that F is linear and also bounded since, by applying Cauchy-Schwarz, we obtain

$$|F(v)| = \left| \int_0^1 fv \right| \leqslant \|f\|_{0,\Omega} \|v\|_{0,\Omega} \leqslant \|f\|_{0,\Omega} \|v\|_{1,\Omega} \quad \forall\, v \in H,$$

which shows that $\|F\| \leqslant \|f\|_{0,\Omega}$. In turn, $B : H \times H \to \mathbb{R}$ is bilinear and also bounded since, using Cauchy-Schwarz again, we obtain

$$|B(w,v)| = \left| \int_0^1 w'v' \right| \leqslant |w|_{1,\Omega} |v|_{1,\Omega} \leqslant \|w\|_{1,\Omega} \|v\|_{1,\Omega} \quad \forall\, w, v \in H.$$

Finally, according to Lemma 4.5, with $a = 0$ and $b = 1$, we can write

$$B(v,v) = \int_0^1 (v')^2 = |v|_{1,\Omega}^2 \geqslant \frac{2}{3} \|v\|_{1,\Omega}^2 \quad \forall\, v \in H,$$

which shows that B is H-elliptic with constant $\alpha = 2/3$. Thus, a direct application of the Lax-Milgram Lemma allows us to conclude that (4.13) has a unique solution $u \in \mathrm{H}_0^1(\Omega)$, which satisfies

$$\|u\|_{1,\Omega} \leqslant \frac{3}{2} \|F\| \leqslant \frac{3}{2} \|f\|_{0,\Omega}.$$

The next goal is to deduce a more general version of the Lax–Milgram Lemma (see Lemma 4.4). To this end, we now consider real Hilbert spaces $(H_1, \langle\cdot,\cdot\rangle_1)$ and $(H_2, \langle\cdot,\cdot\rangle_2)$, a functional $F \in H_2'$, a bounded bilinear form $B : H_1 \times H_2 \to \mathbb{R}$, and we look for $u \in H_1$ such that

$$B(u,v) = F(v) \qquad \forall\, v \in H_2. \tag{4.14}$$

In other words, if $\mathbf{B} : H_1 \to H_2$ is the bounded linear operator induced by B, and $\mathcal{R}_2 : H_2' \to H_2$ is the corresponding Riesz map, we are interested in finding $u \in H_1$ such that

$$\mathbf{B}(u) = \mathcal{R}_2(F).$$

Thus, a necessary and sufficient condition for (4.14) to have a unique solution for each $F \in H_2'$ is that $\mathbf{B}$ be bijective. In turn, the bijectivity of $\mathbf{B}$ can be reformulated according to the equivalences provided by the following lemma, which recovers part of Exercise 3.36 and provides specific proofs of the Hilbert-space versions of Theorems 3.86 and 3.87 in the case of a bounded operator.

Lemma 4.7. *Let $(H_1, \langle\cdot,\cdot\rangle_1)$ and $(H_2, \langle\cdot,\cdot\rangle_2)$ be Hilbert spaces, and let $\mathbf{B} \in \mathcal{L}(H_1, H_2)$. Then:*

a) $\mathbf{B}$ *is surjective if and only if* $\mathbf{B}^*$ *is injective and has closed range, that is, if there exists $\alpha > 0$ such that*

$$\|\mathbf{B}^*(v)\|_1 \geqslant \alpha \|v\|_2 \qquad \forall\, v \in H_2. \tag{4.15}$$

b) $\mathbf{B}$ *is injective if and only if*

$$\sup_{v \in H_2} \langle \mathbf{B}(u), v \rangle_2 > 0 \qquad \forall\, u \in H_1,\ u \neq 0.$$

c) $\mathbf{B}^*$ *is surjective if and only if* $\mathbf{B}$ *is injective and has closed range, that is, if there exists $\alpha > 0$ such that*

$$\|\mathbf{B}(u)\|_2 \geqslant \alpha \|u\|_1 \qquad \forall\, u \in H_1. \tag{4.16}$$

d) $\mathbf{B}^*$ *is injective if and only if*

$$\sup_{u \in H_1} \langle \mathbf{B}(u), v \rangle_2 > 0 \qquad \forall\, v \in H_2,\ v \neq 0.$$

e) $\mathbf{B}$ *is bijective if and only if* $\mathbf{B}^*$ *is bijective.*

Proof.

a) Suppose that $R(\mathbf{B}) = H_2$. Then $R(\mathbf{B}^*)$ is also closed (since $R(\mathbf{B})$ obviously is), and moreover $N(\mathbf{B}^*) = R(\mathbf{B})^\perp = H_2^\perp = \{0\}$. Conversely, if $\mathbf{B}^*$ is injective and has closed range, then the range of $\mathbf{B}$ is also closed, and hence

$$R(\mathbf{B}) = N(\mathbf{B}^*)^\perp = \{0\}^\perp = H_2.$$

The equivalence with (4.15) corresponds exactly to the characterization result given by Theorem 3.58.

b) It suffices to note that $\mathbf{B}$ is injective if and only if $\mathbf{B}(u) \neq 0$ for all $u \in H_1$, $u \neq 0$.

c) and d) These equivalences follow directly from a) and b) by applying them to $\mathbf{B}^*$.

e) Suppose that $\mathbf{B}$ is bijective. From a) it follows that $\mathbf{B}^*$ is injective and has closed range, which implies that

$$R(\mathbf{B}^*) = N(\mathbf{B})^\perp = \{0\}^\perp = H_1,$$

and thus $\mathbf{B}^*$ is bijective. For the converse, it suffices to apply the previous implication to $\mathbf{B}^*$ instead of $\mathbf{B}$.

□

It is important to observe here, according to e), that the pairs of conditions given by a) - b) and c) - d) are equivalent. In this regard, note also that (4.15) and (4.16) can be rewritten as

$$\|\mathbf{B}^*(v)\|_1 := \sup_{\substack{u\in H_1\\ u\neq 0}} \frac{\langle \mathbf{B}(u), v\rangle_2}{\|u\|_1} = \sup_{\substack{u\in H_1\\ u\neq 0}} \frac{B(u,v)}{\|u\|_1} \geqslant \alpha\,\|v\|_2 \quad \forall\, v \in H_2, \tag{4.17}$$

and

$$\|\mathbf{B}(u)\|_2 := \sup_{\substack{v\in H_2\\ v\neq 0}} \frac{\langle \mathbf{B}(u), v\rangle_2}{\|v\|_2} = \sup_{\substack{v\in H_2\\ v\neq 0}} \frac{B(u,v)}{\|v\|_2} \geqslant \alpha\,\|u\|_1 \quad \forall\, u \in H_1, \tag{4.18}$$

or, respectively,

$$\inf_{\substack{v\in H_2\\ v\neq 0}} \sup_{\substack{u\in H_1\\ u\neq 0}} \frac{B(u,v)}{\|u\|_1\,\|v\|_2} \geqslant \alpha, \tag{4.19}$$

and

$$\inf_{\substack{u\in H_1\\ u\neq 0}} \sup_{\substack{v\in H_2\\ v\neq 0}} \frac{B(u,v)}{\|u\|_1\,\|v\|_2} \geqslant \alpha, \tag{4.20}$$

which is why they are referred to as INF-SUP CONDITIONS.

According to the above analysis, we can now establish a more general version of the Lax–Milgram Lemma.

Lemma 4.8 (Generalized Lax-Milgram Lemma). *Let H_1 and H_2 be real Hilbert spaces, and let $B : H_1 \times H_2 \to \mathbb{R}$ be a bounded bilinear form. Assume that*

i) *there exists $\alpha > 0$ such that*

$$\sup_{\substack{v\in H_2\\ v\neq 0}} \frac{B(u,v)}{\|v\|_2} \geqslant \alpha\,\|u\|_1 \qquad \forall\, u \in H_1,$$

ii)
$$\sup_{u\in H_1} B(u,v) > 0 \qquad \forall\, v \in H_2,\ v \neq 0.$$

Then, for each $F \in H_2'$ *there exists a unique* $u \in H_1$ *such that*
$$B(u,v) = F(v) \qquad \forall\, v \in H_2$$

and
$$\|u\|_1 \leqslant \frac{1}{\alpha}\,\|F\|_{H_2'}. \tag{4.21}$$

Moreover, conditions i) *and* ii) *are also necessary.*

Proof. It suffices to observe that i) and ii) correspond to conditions c) and d) of Lemma 4.7, which (together) are equivalent to the bijectivity of $\mathbf{B}^*$ and hence to that of $\mathbf{B}$. The estimate (4.21) follows from (4.16) by noting that $\mathbf{B}(u) = R_2(F)$.

□

Certainly, the previous theorem can be stated equivalently with conditions a) and b) instead of c) and d) of Lemma 4.7. In such a case, the same constant α can be assumed in (4.15) and (4.16), since $\|(\mathbf{B}^*)^{-1}\|$ (bounded by $1/\alpha$ in (4.15)) equals $\|\mathbf{B}^{-1}\|$ (also bounded by $1/\alpha$ in (4.16)), and therefore the estimate (4.21) is also obtained from a) and b), by making use of this identity.

Now, in the particular case where $H_1 = H_2 = H$, the Lax–Milgram Lemma (Lemma 4.4) follows directly from Lemma 4.8. Indeed, if $B : H \times H \to \mathbb{R}$ is a bilinear form, bounded and H-elliptic with constants M and α, respectively, then clearly
$$\sup_{\substack{v\in H\\ v\neq 0}} \frac{B(u,v)}{\|v\|} \geqslant \frac{B(u,u)}{\|u\|} \geqslant \alpha\,\|u\| \qquad \forall\, u \in H,\ u \neq 0,$$

and
$$\sup_{u\in H} B(u,v) \geqslant B(v,v) \geqslant \alpha\,\|v\|^2 > 0 \qquad \forall\, v \in H,\ v \neq 0,$$

which proves, respectively, hypotheses i) and ii) of Lemma 4.8. In turn, if instead of H-ellipticity we assume that $B : H \times H \to \mathbb{R}$ is symmetric, then the operator $\mathbf{B} \in \mathcal{L}(H)$ induced by B turns out to be self-adjoint, and in this case hypothesis ii) of Lemma 4.8 is redundant. This suggests the following symmetric version of the generalized Lax–Milgram Lemma.

Lemma 4.9. *Let* H *be a real Hilbert space, and let* $B : H \times H \to \mathbb{R}$ *be a bounded bilinear form. Assume that*

i) $B(w,v) = B(v,w) \quad \forall\, w, v \in H.$
ii) *there exists* $\alpha > 0$ *such that*
$$\sup_{\substack{v\in H\\ v\neq 0}} \frac{B(u,v)}{\|v\|} \geqslant \alpha\,\|u\| \qquad \forall\, u \in H.$$

Then, for each $F \in H'$ *there exists a unique* $u \in H$ *such that*
$$B(u,v) = F(v) \qquad \forall\, v \in H,$$

and
$$\|u\| \leqslant \frac{1}{\alpha}\,\|F\|_{H'}.$$

Proof. It is a direct corollary of Lemma 4.8.

□

4.2 The Galerkin method

4.2.1 The H-elliptic case

In what follows H is an arbitrary Hilbert space, $B : H \times H \rightarrow \mathbb{R}$ is a bounded and H-elliptic bilinear form, and F is a linear bounded functional in H′. Then, according to the simpler version of the Lax-Milgram Lemma (cf. Lemma 4.4), it follows that there exists a unique $u \in H$ such that

$$B(u,v) = F(v) \qquad \forall\, v \in H\,. \tag{4.22}$$

The next goal is to present a procedure that allows us to approximate the solution of (4.22). To this end, let $\{H_n\}_{n\in\mathbb{N}}$ be a sequence of finite-dimensional subspaces of H, and consider the following problems: *Find* $u_n \in H_n$ *such that*

$$B(u_n, v_n) = F(v_n) \qquad \forall\, v_n \in H_n\,. \tag{4.23}$$

Without loss of generality, we may assume that the dimension of H_n is n. This simple idea of reducing the original problem posed in the infinite-dimensional space H, to a sequence of analogous problems in the *finite-dimensional* subspaces H_n, is what is known as the *Galerkin Method*. The objective will be, then, to choose the H_n in such a way that at least the following property holds

$$\lim_{n\to\infty} \|u - u_n\|_H = 0\,. \tag{4.24}$$

Next we show that (4.23) reduces to a linear system of equations. Indeed, let $\{e_1, \dots, e_n\}$ be a basis of H_n. Then, there exist scalars $\alpha_1, \dots, \alpha_n$ such that

$$u_n = \sum_{j=1}^{n} \alpha_j\, e_j\,.$$

Then, (4.23) becomes: *Find* $\alpha_1, ..., \alpha_n \in \mathbb{R}$ *such that*

$$\sum_{j=1}^{n} \alpha_j\, B(e_j, v_n) = F(v_n) \qquad \forall\, v_n \in H_n\,,$$

which is equivalent to: *Find* $\alpha_1, ..., \alpha_n \in \mathbb{R}$ *such that*

$$\sum_{j=1}^{n} \alpha_j\, B(e_j, e_i) = F(e_i) \qquad \forall\, i \in \{1, ..., n\}\,. \tag{4.25}$$

Thus, defining the matrix $\mathbf{B} := (b_{ij})_{n\times n}$ and the vectors $\boldsymbol{\alpha} := (\alpha_j)_{n\times 1}$ and $\mathbf{F} :=, (f_j)_{n\times 1}$, with

$$b_{ij} := B(e_j, e_i) \qquad \text{and} \qquad f_j := F(e_j)\,,$$

the formulation (4.25) is rewritten as follows: *Find* $\boldsymbol{\alpha} \in \mathbb{R}^n$ *such that*

$$\mathbf{B}\boldsymbol{\alpha} = \mathbf{F}\,. \tag{4.26}$$

Note here that the H-ellipticity of B also guarantees that (4.26) always has a unique solution. In the literature, $\mathbf{B}$ is usually called the *stiffness matrix*, and $\mathbf{F}$ the corresponding *load vector*. These names come from the terminology initially used by engineers, who created a particular case of the Galerkin method to solve the linear elasticity problem, known as the *finite element method*.

Now, it should be noted that the design of efficient algorithms to solve the linear system (4.26) becomes of tremendous importance when the dimension of the subspace H_n tends to ∞. In this respect, we must observe that for large matrices, a *sparse* or *banded* structure is very desirable. Also, if the matrix $\mathbf{B}$ is symmetric and positive definite, then the solution of (4.26) is further simplified. This is precisely the case when the bilinear form B, in addition to being bounded and H-elliptic, is also *symmetric*. Indeed, given $\boldsymbol{\beta} := (\beta_1, \beta_2, ..., \beta_n)^{\mathrm{t}} \in \mathbb{R}^n$, we first define $v := \sum_{j=1}^{n} \beta_j e_j \in H_n \subseteq H$. Then, by virtue of the H-ellipticity of B it follows that

$$\alpha \|v\|_H^2 \leqslant B(v,v) = \sum_{i=1}^{n} \sum_{j=1}^{n} b_{ij} \beta_i \beta_j = \boldsymbol{\beta}^{\mathrm{t}} \mathbf{B} \boldsymbol{\beta},$$

whence $\boldsymbol{\beta}^{\mathrm{t}} \mathbf{B} \boldsymbol{\beta} \geqslant 0$ for all $\boldsymbol{\beta} \in \mathbb{R}^n$, and $\boldsymbol{\beta}^{\mathrm{t}} \mathbf{B} \boldsymbol{\beta} = 0$ if and only if $\boldsymbol{\beta}$ is the zero vector, thus proving the aforementioned property for $\mathbf{B}$.

Example 4.10 Here we present a specific example of a Galerkin scheme associated with the boundary value problem of Example 4.6. For this purpose, we consider a partition $0 = x_0 < x_1 < x_2 < \cdots < x_n < x_{n+1} = 1$ of the domain $\Omega := (0,1)$, and introduce the space

$$H_n := \left\{ v \in C(\bar{\Omega}) \,:\, v(0) = v(1) = 0 \text{ and } v|_{[x_{j-1},x_j]} \in P_1([x_{j-1},x_j]) \ \forall\, j = \overline{1, n+1} \right\},$$

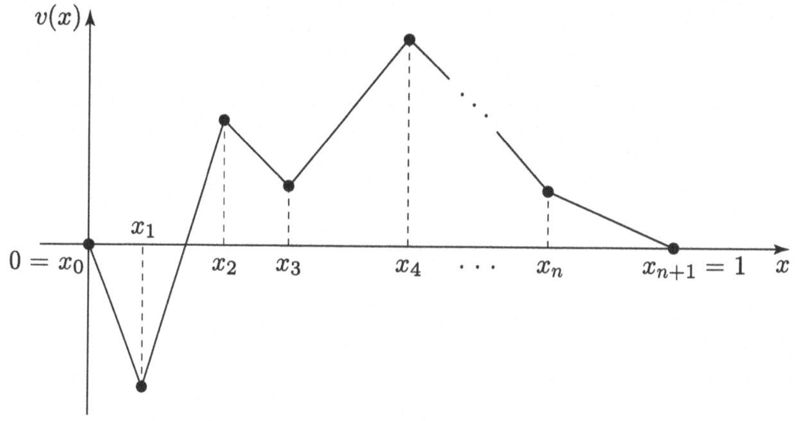

Fig. 4.1. *Graphical example of* $v \in H_n$

where $P_1(S)$ denotes the space of polynomials of degree $\leqslant 1$ defined on any subset S of Ω. It is easy to prove that the functions of H_n are uniquely determined by their values at the nodes x_j of the partition of Ω (see Figure 4.1 for an example of $v \in H_n$). In particular, $\{e_1, e_2, ..., e_n\}$ constitutes a basis of H_n, where $e_i \in H_n$ is such that

$$e_i(x_j) = \begin{cases} 1 & \text{if } j = i\,, \\ 0 & \text{if } j \neq i\,. \end{cases}$$

Such basis functions of H_n are called *hat functions*. Regarding the computation of the stiffness matrix $\mathbf{B}$, one obtains

$$b_{ij} := B(e_i, e_j) = \int_0^1 e_i' \, e_j' = \begin{cases} 0 & \text{if } |i-j| \geqslant 2\,, \\[2mm] -\frac{1}{h_{j+1}} & \text{if } j = i-1\,, \\[2mm] -\frac{1}{h_j} & \text{if } j = i+1\,, \\[2mm] \frac{1}{h_j} + \frac{1}{h_{j+1}} & \text{if } j = i\,, \end{cases}$$

where $h_j := x_j - x_{j-1}$ for all $j = \overline{1, n+1}$, and therefore $\mathbf{B}$ turns out to be tridiagonal. In the case of a uniform partition with $h_j = h := \frac{1}{n+1}$, the linear system (4.26) becomes

$$\frac{1}{h} \begin{bmatrix} 2 & -1 & 0 & \cdot & \cdot & \cdot & 0 \\ -1 & 2 & -1 & \cdot & \cdot & \cdot & 0 \\ 0 & & & & & & \\ \cdot & & \cdot & & & & \cdot \\ \cdot & & & \cdot & & & \cdot \\ \cdot & & & & \cdot & & \cdot \\ & & & & & & 0 \\ 0 & \cdot & \cdot & \cdot & -1 & 2 & -1 \\ 0 & \cdot & \cdot & \cdot & 0 & -1 & 2 \end{bmatrix} \begin{bmatrix} \alpha_1 \\ \alpha_2 \\ \\ \cdot \\ \cdot \\ \cdot \\ \\ \alpha_{n-1} \\ \alpha_n \end{bmatrix} = \begin{bmatrix} f_1 \\ f_2 \\ \\ \cdot \\ \cdot \\ \cdot \\ \\ f_{n-1} \\ f_n \end{bmatrix}, \tag{4.27}$$

where $f_j := F(e_j) = \int_{x_{j-1}}^{x_{j+1}} f\, e_j$. Finally, note that, due to the definition of the subspace H_n, the solution $\alpha_1, \alpha_2, \dots, \alpha_n$ provided by the linear system (4.27) coincides with the values taken by the Galerkin solution u_n at each of the partition nodes. In other words, we have $\alpha_i = u_n(x_i)$ for all $i \in \{1, \dots, n\}$.

The above constitutes one of the simplest possible situations that show the type of linear systems obtained when applying the Galerkin method to problem (4.22). However, the procedure illustrated through Example 4.10 represents the same idea that motivated engineers in the 1960s to create the *Finite Element Method* (FEM) to solve the weak formulation of the elasticity problem. More precisely, the FEM is a particular case of the Galerkin method, in which finite-dimensional subspaces H_n are used, whose elements are piecewise polynomial functions.

Our next purpose is to obtain an upper bound for the *error* that arises from the Galerkin method. This is defined as the norm difference between the exact solution $u \in H$ of (4.22) and the approximate solution $u_n \in H_n$ of (4.23). To this end, we first note that from (4.22) and (4.23) it follows that

$$B(u - u_n, v_n) = 0 \qquad \forall\, v_n \in H_n\,, \tag{4.28}$$

which is known as the *orthogonality relation.* The reason for this name is due to what happens when B is symmetric, in which case B becomes an inner product whose associated norm, defined by

$$\|v\|_E := B(v, v)^{1/2} \qquad \forall\, v \in H\,,$$

and called the *energy norm,* is equivalent to the original norm of the space H. Thus, equation (4.28) is the typical orthogonality condition that characterizes the best approximation of $u \in H$ by elements of the subspace H_n, with respect to $\|\cdot\|_E$ (cf. (2.6) in Theorem 2.6). In other words, (4.28) shows that, in the symmetric case, the Galerkin solution u_n is the best approximation of the exact solution u with respect to the energy norm, or equivalently,

$$\|u - u_n\|_E = \inf_{v_n \in H_n} \|u - v_n\|_E\,. \tag{4.29}$$

Nevertheless, it is important to stress that the *orthogonality relation* (4.28) also holds for non-symmetric B.

Lemma 4.11 (Céa's Lemma). *Suppose that B is a bounded and H-elliptic bilinear form with constants M and α, respectively. Let $u \in H$ and $u_n \in H_n$ be the solutions of (4.22) and (4.23), respectively. Then*

$$\|u - u_n\|_H \leqslant \frac{M}{\alpha} \inf_{v_n \in H_n} \|u - v_n\|_H . \tag{4.30}$$

Moreover, if B is symmetric the above estimate reduces to

$$\|u - u_n\|_H \leqslant \sqrt{\frac{M}{\alpha}} \inf_{v_n \in H_n} \|u - v_n\|_H . \tag{4.31}$$

Proof. Using the H-ellipticity of B, adding and subtracting an arbitrary element $v_n \in H_n$ in the second argument of B, and then applying the orthogonality relation (4.28), we obtain

$$\begin{aligned} \alpha \|u - u_n\|_{\mathrm{H}}^2 &\leqslant B(u - u_n, u - u_n) \\ &= B(u - u_n, u - v_n + v_n - u_n) = B(u - u_n, u - v_n) . \end{aligned}$$

Since B is bounded, the previous inequality implies that

$$\alpha \|u - u_n\|_H^2 \leqslant M \|u - u_n\|_H \|u - v_n\|_H ,$$

from which, assuming $\|u - u_n\|_H \neq 0$, it follows that

$$\|u - u_n\|_H \leqslant \frac{M}{\alpha} \|u - v_n\|_H \qquad \forall\, v_n \in H_n . \tag{4.32}$$

Naturally, (4.32) is also valid for $\|u - u_n\|_H = 0$. Thus, taking the infimum in (4.32) with respect to $v_n \in H_n$ one obtains (4.30). In turn, if we assume that B is symmetric, then the H-ellipticity of B, the characterization given by (4.29), and the boundedness of B imply that

$$\alpha \|u - u_n\|_{\mathrm{H}}^2 \leqslant \|u - u_n\|_E^2 \leqslant \|u - v_n\|_E^2 \leqslant M \|u - v_n\|_{\mathrm{H}}^2 \quad \forall\, v_h \in H_n ,$$

from which, taking square roots, one easily concludes (4.31).

□

The error bounds given by (4.30) and (4.31) are known as *Céa's estimates* and they provide the starting point for finally deriving the convergence properties of finite element methods. As mentioned before, the minimum requirement in each case is condition (4.24), which guarantees that the error goes to zero as the dimension of H_n tends to ∞.

4.2.2 The general case

In what follows, $(H_1, \langle\cdot,\cdot\rangle_1)$ and $(H_2, \langle\cdot,\cdot\rangle_2)$ are real Hilbert spaces, F is a functional in H_2', and $B : H_1 \times H_2 \to \mathbb{R}$ is a bounded bilinear form that satisfies the hypotheses of the Generalized Lax-Milgram Lemma (cf. Lemma 4.8). Then, there exists a unique $u \in H_1$ such that

$$B(u, v) = F(v) \qquad \forall\, v \in H_2. \tag{4.33}$$

The next goal is to define the Galerkin scheme for (4.33), analyze its solvability, and determine the corresponding a priori error bound. To this end, we now consider sequences $\{H_{n,1}\}_{n\in\mathbb{N}}$ and $\{H_{n,2}\}_{n\in\mathbb{N}}$ of finite-dimensional subspaces of H_1 and H_2, respectively, and we state the discrete problem: Find $u_n \in H_{n,1}$ such that

$$B(u_n, v_n) = F(v_n) \qquad \forall\, v_n \in H_{n,2}. \tag{4.34}$$

Equivalently, if $\mathbf{B}_n : H_{n,1} \to H_{n,2}$ is the linear and bounded operator induced by $B|_{H_{n,1}\times H_{n,2}}$, and $\mathcal{R}_{n,2} : H'_{n,2} \to H_{n,2}$ is the corresponding Riesz map, the problem (4.34) is reformulated as: Find $u_n \in H_{n,1}$ such that

$$\mathbf{B}_n(u_n) = \mathcal{R}_{n,2}\Big(F|_{H_{n,2}}\Big).$$

It follows that a necessary and sufficient condition for (4.34) to have a unique solution for each $F \in H_2'$ is that $\mathbf{B}_n$ be bijective. Accordingly, we may use the same equivalences given by Lemma 4.7, which led to Lemma 4.8, or else apply this latter result directly in the present finite-dimensional context. Before doing so, let us first note that for the surjectivity of $\mathbf{B}_n$ it is obviously required that $\dim H_{n,1} \geqslant \dim H_{n,2}$. In turn, if $\mathbf{B}_n$ is injective, by the Dimension Theorem we have

$$\dim H_{n,1} \;=\; \dim N(\mathbf{B}_n) + \dim R(\mathbf{B}_n) \;=\; \dim R(\mathbf{B}_n) \;\leqslant\; \dim H_{n,2},$$

from which it follows that a necessary condition for the bijectivity of $\mathbf{B}_n$ is that $\dim H_{n,1} = \dim H_{n,2}$. Moreover, by virtue of the Fredholm alternative in finite dimension, we know that in this case the surjectivity of $\mathbf{B}_n : H_{n,1} \to H_{n,2}$ is equivalent to its injectivity, and therefore it suffices to assume only one of the conditions i) or ii) of Lemma 4.8. In this way, we deduce the following discrete version of the Generalized Lax-Milgram Lemma.

Lemma 4.12. *In addition to the hypotheses already specified on the bilinear form B and the sequences $\{H_{n,1}\}_{n\in\mathbb{N}}$ and $\{H_{n,2}\}_{n\in\mathbb{N}}$, suppose that*

i) $\dim H_{n,1} = \dim H_{n,2}$ *for all* $n \in \mathbb{N}$.
ii) *there exists* $\alpha_n > 0$ *such that*

$$\sup_{\substack{v_n\in H_{n,2}\\ v_n\neq 0}} \frac{B(w_n, v_n)}{\|v_n\|_2} \geqslant \alpha_n\,\|w_n\|_1 \qquad \forall\, w_n \in H_{n,1},$$

or equivalently,

$$\sup_{\substack{w_n\in H_{n,1}\\ w_n\neq 0}} \frac{B(w_n, v_n)}{\|w_n\|_1} \geqslant \alpha_n\,\|v_n\|_2 \qquad \forall\, v_n \in H_{n,2}.$$

Then, for each $F \in H_2'$ there exists a unique $u_n \in H_{n,1}$ such that

$$B(u_n, v_n) = F(v_n) \qquad \forall\, v_n \in H_{n,2}$$

and

$$\|u_n\|_1 \leqslant \frac{1}{\alpha_n}\,\|F|_{H_{n,2}}\|_{H'_{n,2}}. \tag{4.35}$$

Moreover, conditions i) *and* ii) *are necessary.*

Analogously to what was discussed in Section 4.1 (cf. (4.19), (4.20)), the inequalities in ii) of Lemma 4.12 are called DISCRETE INF-SUP CONDITIONS for B. In turn, and with the purpose of ensuring the stability of the Galerkin scheme (4.34), in both cases it is convenient to assume in addition that the constant α_n can be chosen independently of n, that is, there exists $\tilde{\alpha} > 0$ such that $\alpha_n \geqslant \tilde{\alpha}$ for all $n \in \mathbb{N}$.

Next, we establish an upper bound for the error $\|u - u_n\|_1$, where u and u_n are the unique solutions of (4.33) and (4.34), respectively, which, as in the H-elliptic case (cf. Lemma 4.11), is also called Céa's estimate.

Lemma 4.13 (General Céa's Lemma). *Let $B : H_1 \times H_2 \to \mathbb{R}$ be a bounded bilinear form satisfying the hypotheses of Lemma* 4.8, *and let $\mathbf{B} : H_1 \to H_2$ be the linear and bounded operator induced by B. Moreover, let $\{H_{n,1}\}_{n\in\mathbb{N}}$ and $\{H_{n,2}\}_{n\in\mathbb{N}}$ be sequences of finite-dimensional subspaces of H_1 and H_2, respectively, which satisfy the hypotheses of Lemma* 4.12 *with a constant $\alpha_n > 0$ for all $n \in \mathbb{N}$. Then we have*

$$\|u - u_n\|_1 \leqslant \left\{1 + \frac{\|\mathbf{B}\|}{\alpha_n}\right\} \inf_{z_n \in H_{n,1}} \|u - z_n\|_1. \tag{4.36}$$

Proof. Let us first observe, from (4.33) and (4.34), that in this case the orthogonality relation analogous to (4.28) reduces to

$$B(u - u_n, v_n) = 0 \qquad \forall\, v_n \in H_{n,2}. \tag{4.37}$$

In turn, the triangular inequality obviously implies

$$\|u - u_n\|_1 \leqslant \|u - z_n\|_1 + \|u_n - z_n\|_1 \qquad \forall\, z_n \in H_{n,1}. \tag{4.38}$$

Then, applying the discrete inf-sup condition for B, the equation (4.37), and the fact that $\|\mathbf{B}\|$ is the boundedness constant of B, we obtain

$$\begin{aligned}
\|u_n - z_n\|_1 &\leqslant \frac{1}{\alpha_n} \sup_{\substack{v_n \in H_{n,2} \\ v_n \neq 0}} \frac{B(u_n - z_n, v_n)}{\|v_n\|_2} \\
&= \frac{1}{\alpha_n} \sup_{\substack{v_n \in H_{n,2} \\ v_n \neq 0}} \frac{B(u - z_n, v_n)}{\|v_n\|_2} \\
&\leqslant \frac{\|\mathbf{B}\|}{\alpha_n} \|u - z_n\|_1 \qquad \forall\, z_n \in H_{n,1}.
\end{aligned}$$

Thus, replacing this estimate in (4.38) we obtain

$$\|u - u_n\|_1 \leqslant \left\{1 + \frac{\|\mathbf{B}\|}{\alpha_n}\right\} \|u - z_n\|_1 \qquad \forall\, z_n \in H_{n,1},$$

which implies (4.36) and concludes the proof.

□

Again, it is worth commenting here on the convenience of having $\tilde{\alpha} > 0$ such that $\alpha_n \geqslant \tilde{\alpha}$ for all $n \in \mathbb{N}$, in which case, Céa's estimate (4.36) becomes

$$\|u - u_n\|_1 \leqslant \left\{1 + \frac{\|\mathbf{B}\|}{\tilde{\alpha}}\right\} \inf_{z_n \in H_{n,1}} \|u - z_n\|_1. \tag{4.39}$$

It is also important to observe that, using some properties of projectors, the constant "1" in (4.36) and (4.39) can be omitted. To see this, we now denote by $G_n : H_1 \to H_{n,1}$ the Galerkin projector associated with (4.34), that is, given $z \in H_1$, and under the hypotheses of Lemma 4.13, $G_n(z)$ is the unique element in $H_{n,1}$ such that

$$B(G_n(z), v_n) = B(z, v_n) \qquad \forall\, v_n \in H_{n,2}. \tag{4.40}$$

In other words, $G_n(z) \in H_{n,1}$ corresponds to the solution of the Galerkin scheme (4.34) with the functional $F := B(z, \cdot) \in H_2'$. Accordingly, the stability bound given by (4.35) implies that

$$\|G_n(z)\|_1 \leqslant \frac{\|\mathbf{B}\|}{\alpha_n} \|z\|_1 \qquad \forall\, z \in H_1. \tag{4.41}$$

In turn, it is easy to see from (4.40) that $G_n(z_n) = z_n$ for all $z_n \in H_{n,1}$, whence it is clear that $G_n^2 = G_n$. Consequently, given an arbitrary functional $F \in H_2'$, and denoting again by u and u_n the unique solutions of (4.33) and (4.34), it follows from the previous analysis that $u_n = G_n(u)$, and then, with $I : H_1 \to H_1$ the identity operator, we have

$$\begin{aligned} \|u - u_n\|_1 &= \|(I - G_n)u\|_1 = \|(I - G_n)(u - z_n)\|_1 \\ &\leqslant \|I - G_n\| \, \|u - z_n\|_1 \qquad \forall z_n \in H_{n,1}. \end{aligned} \tag{4.42}$$

Then, applying the identity given in Exercise 3.26, which, inspired by the analysis in [122], establishes that $\|I - G_n\| = \|G_n\|$, we conclude from (4.42) and (4.41) that

$$\|u - u_n\|_1 \leqslant \|G_n\| \, \|u - z_n\|_1 \leqslant \frac{\|\mathbf{B}\|}{\alpha_n} \|u - z_n\|_1 \qquad \forall\, z_n \in H_{n,1},$$

from which

$$\|u - u_n\|_1 \leqslant \frac{\|\mathbf{B}\|}{\alpha_n} \inf_{z_n \in H_{n,1}} \|u - z_n\|_1 \,.$$

4.2.3 Strang's estimate

In this section we briefly refer to a theoretical framework even more general than that given by (4.33) and (4.34), in which, instead of the latter, the following modified Galerkin scheme is considered: Find $u_n \in H_{n,1}$ such that

$$B_n(u_n, v_n) = F_n(v_n) \qquad \forall\, v_n \in H_{n,2}\,, \tag{4.43}$$

where $B_n : H_{n,1} \times H_{n,2} \to \mathbb{R}$ and $F_n : H_{n,2} \to \mathbb{R}$ denote a bounded bilinear form and a bounded linear functional, respectively, that approximate $B|_{H_{n,1} \times H_{n,2}}$ and $F|_{H_{n,2}}$. Typical examples of (4.43) occur when numerical integration formulas or polygonal approximations of the domain are applied in order to compute the corresponding stiffness matrix and load vector. Although the analysis of systems of the type given by (4.33) and (4.43) will be treated in greater depth and generality in Chapter 5, we state here in advance one of its results, mainly to complement Céa's estimate (4.36), and at the same time to show an interesting connection with a fixed point strategy applied to a nonlinear model. In fact, the continuous and discrete formulations of problems of this type may also lead to a similar situation, as illustrated later in Example 4.14.

In what follows, we assume that B_n satisfies the hypotheses of Lemma 4.12, so that (4.43) is uniquely solvable. In this case, and as will be proved later in Chapter 5 for formulations in Banach spaces (cf. Lemma 5.5), the error bound is called STRANG'S ESTIMATE and is given by

$$\|u - u_n\|_1 \leqslant \inf_{z_n \in H_{n,1}} \left\{ \left(1 + \frac{\|\mathbf{B}\|}{\alpha_n}\right) \|u - z_n\|_1 + \frac{1}{\alpha_n} \|(B - B_n)(z_n, \cdot)\|_{H'_{n,2}} \right\} + \frac{1}{\alpha_n} \|F - F_n\|_{H'_{n,2}}, \tag{4.44}$$

where $\|(B - B_n)(z_n, \cdot)\|_{H'_{n,2}}$ and $\|F - F_n\|_{H'_{n,2}}$, called consistency terms, are given by

$$\|(B - B_n)(z_n, \cdot)\|_{H'_{n,2}} := \sup_{\substack{v_n \in H_{n,2} \\ v_n \neq 0}} \frac{(B - B_n)(z_n, v_n)}{\|v_n\|_2} \quad \forall\, z_n \in H_{n,1},$$

and

$$\|F - F_n\|_{H'_{n,2}} := \sup_{\substack{v_n \in H_{n,2} \\ v_n \neq 0}} \frac{(F - F_n)(v_n)}{\|v_n\|_2},$$

which measure how well scheme (4.34) is approximated by (4.43). In turn, in the particular case when $B_n(u, v_n)$ is computable for all $v_n \in H_{n,2}$, one obtains, instead of (4.44), the modified Strang estimate (cf. Lemma 5.6)

$$\|u - u_n\|_1 \leqslant \left(1 + \frac{\|\mathbf{B}_n\|}{\alpha_n}\right) \operatorname{dist}(u, H_{n,1}) + \frac{1}{\alpha_n} \|F - F_n\|_{H'_{n,2}} + \frac{1}{\alpha_n} \|(B - B_n)(u, \cdot)\|_{H'_{n,2}}, \tag{4.45}$$

where $\mathbf{B}_n : H_{n,1} \to H'_{n,2}$ is the bounded linear operator induced by B_n, and the expression $\|(B - B_n)(u, \cdot)\|_{H'_{n,2}}$ is defined analogously to $\|(B - B_n)(z_n, \cdot)\|_{H'_{n,2}}$. Note that in the particular case when $B|_{H_{n,1} \times H_{n,2}} = B_n$ and $F|_{H_{n,2}} = F_n$, (4.44) reduces to (4.36). The same occurs from (4.45) when $(B - B_n)(u, \cdot) = 0$ in $H_{n,2}$ and $F|_{H_{n,2}} = F_n$.

In the following example we describe the use of Banach's Fixed Point Theorem (cf. Theorem 1.8) to analyze the continuous and discrete solvability of a nonlinear boundary value problem, and then apply (4.45) to deduce the corresponding a priori error bound.

Example 4.14 Given Ω a bounded open subset of $\mathbb{R}^n$ with polyhedral boundary Γ, we consider the nonlinear Poisson problem:

$$-\Delta u = f(u)\kappa + g \quad \text{in} \quad \Omega, \quad u = 0 \quad \text{on} \quad \Gamma, \tag{4.46}$$

where $\kappa \in \mathrm{L}^\infty(\Omega)$, $g \in \mathrm{L}^2(\Omega)$, and $f : \mathbb{R} \to \mathbb{R}$ is a Lipschitz-continuous function, which means that there exists a positive constant L_f such that

$$|f(s) - f(t)| \leqslant L_f\, |s - t| \qquad \forall\, s, t \in \mathbb{R}. \tag{4.47}$$

Moreover, it is assumed that $f(v) \in \mathrm{L}^2(\Omega)$ for all $v \in \mathrm{H}^1(\Omega)$. As an example, the function f defined by $f(s) := s^2$ for all $s \in \mathbb{R}$ satisfies this requirement since the embedding of $\mathrm{H}^1(\Omega)$ into $\mathrm{L}^4(\Omega)$ is continuous (cf. [111, Theorem 1.3.4]), Exercise 5.2). Applying the integration by parts formula given later in (4.134) (cf. Section 4.5.4), one deduces that the primal variational formulation of (4.46) is: Find $u \in \mathrm{H}_0^1(\Omega)$ such that

$$\int_\Omega \nabla u \cdot \nabla v = \int_\Omega f(u)\,\kappa\, v + \int_\Omega g\, v \qquad \forall\, v \in \mathrm{H}_0^1(\Omega). \tag{4.48}$$

This suggests defining the operator $T : \mathrm{H}_0^1(\Omega) \to \mathrm{H}_0^1(\Omega)$ by $T(w) := \widetilde{w}$ for all $w \in \mathrm{H}_0^1(\Omega)$, where $\widetilde{w}$ is the unique element in $\mathrm{H}_0^1(\Omega)$ (to be proved below) such that

$$B(\widetilde{w}, v) := \int_\Omega \nabla \widetilde{w} \cdot \nabla v = \int_\Omega f(w)\, \kappa\, v + \int_\Omega g\, v =: F_w(v) \qquad \forall\, v \in \mathrm{H}_0^1(\Omega)\,. \tag{4.49}$$

Thus, it is easy to see that solving (4.48) is equivalent to finding a fixed point of T, i.e.: Find $u \in \mathrm{H}_0^1(\Omega)$ such that

$$T(u) = u\,. \tag{4.50}$$

Let us show that T is well defined. Indeed, the Friedrichs-Poincaré inequality in $\mathbb{R}^n$ (cf. (4.167)) directly implies that the bilinear form B is $\mathrm{H}_0^1(\Omega)$-elliptic with constant $\alpha = C_P$. Moreover, it is clear that the functional F_w is linear and bounded with

$$\|F_w\| \leqslant \|\kappa\|_{0,\infty;\Omega}\, \|f(w)\|_{0,\Omega} + \|g\|_{0,\Omega}\,,$$

which follows from its definition in (4.49) by bounding κ by $\|\kappa\|_{0,\infty;\Omega}$, and applying the triangle and Cauchy–Schwarz inequalities. Thus, the Lax–Milgram Lemma (cf. Lemma 4.4) allows us to conclude that (4.49) has a unique solution $\widetilde{w} \in \mathrm{H}_0^1(\Omega)$, which satisfies

$$\|\widetilde{w}\|_{1,\Omega} \leqslant \frac{1}{\alpha} \Big\{ \|\kappa\|_{0,\infty;\Omega}\, \|f(w)\|_{0,\Omega} + \|g\|_{0,\Omega} \Big\}\,,$$

thus confirming the above statement about T. The next objective is to establish the Lipschitz continuity of this operator. For this purpose, given $w,\, z \in \mathrm{H}_0^1(\Omega)$, we denote $\widetilde{w} := T(w)$ and $\widetilde{z} := T(z)$, so that evaluating at $v := \widetilde{w} - \widetilde{z}$ the respective equations arising from (4.49), and then subtracting them, we arrive at

$$\alpha\, \|\widetilde{w} - \widetilde{z}\|_{1,\Omega}^2 \leqslant |\widetilde{w} - \widetilde{z}|_{1,\Omega}^2 = \int_\Omega \big(f(w) - f(z)\big)\, \kappa\, (\widetilde{w} - \widetilde{z})\,. \tag{4.51}$$

Now, using (4.47) and again the Cauchy–Schwarz inequality, we obtain

$$\begin{aligned} \left| \int_\Omega \big(f(w) - f(z)\big)\, \kappa\, (\widetilde{w} - \widetilde{z}) \right| &\leqslant L_f\, \|\kappa\|_{0,\infty;\Omega}\, \|w - z\|_{0,\Omega}\, \|\widetilde{w} - \widetilde{z}\|_{0,\Omega} \\ &\leqslant L_f\, \|\kappa\|_{0,\infty;\Omega}\, \|w - z\|_{0,\Omega}\, \|\widetilde{w} - \widetilde{z}\|_{1,\Omega}\,, \end{aligned} \tag{4.52}$$

which, replaced into (4.51), gives

$$\begin{aligned} \|T(w) - T(z)\|_{1,\Omega} &\leqslant \frac{L_f}{\alpha}\, \|\kappa\|_{0,\infty;\Omega}\, \|w - z\|_{0,\Omega} \\ &\leqslant \frac{L_f}{\alpha}\, \|\kappa\|_{0,\infty;\Omega}\, \|w - z\|_{1,\Omega} \quad \forall\, w,\, z \in \mathrm{H}_0^1(\Omega)\,. \end{aligned} \tag{4.53}$$

Consequently, assuming now that $\|\kappa\|_{0,\infty;\Omega}$ is sufficiently small such that, for a given $\delta \in (0,1)$, one has

$$\frac{L_f}{\alpha}\, \|\kappa\|_{0,\infty;\Omega} \leqslant \delta\,, \tag{4.54}$$

it follows from (4.53) that T is a contraction, and therefore, by invoking Banach's Theorem (cf. Theorem 1.8), we conclude that T has a unique fixed point $u \in \mathrm{H}_0^1(\Omega)$. Equivalently, (4.48) has a unique solution $u \in \mathrm{H}_0^1(\Omega)$. Moreover, using (4.53) with $w = u = T(u)$ and $z = 0$, and keeping in mind the assumption (4.54), one obtains the a priori bound

$$\|u\|_{1,\Omega} \leqslant (1 - \delta)^{-1}\, \|T(0)\|_{1,\Omega}\,.$$

Although we do not intend to go deeper into the solvability of (4.50), we take this opportunity to observe here that the compact embedding of $\mathrm{H}^1(\Omega)$ into $\mathrm{L}^2(\Omega)$ (cf. [111, Theorem 1.3.5], Theorem 10.18 in Chapter 10) and the first inequality in (4.53), allow us to apply another abstract result, known as Schauder's Theorem (cf. [40, Theorem 9.12-1, part b)]), to conclude the existence, only, of a fixed point of T, but without the need to assume (4.54). On the other hand, given a finite-dimensional subspace H_n of $\mathrm{H}_0^1(\Omega)$, the Galerkin scheme associated with (4.48) consists of: Find $u_n \in H_n$ such that

$$\int_\Omega \nabla u_n \cdot \nabla v_n = \int_\Omega f(u_n)\,\kappa\, v_n + \int_\Omega g\, v_n \qquad \forall\, v_n \in H_n\,. \tag{4.55}$$

Following a fixed-point strategy analogous to the one used for the continuous formulation, that is, defining the operator $T_n : H_n \to H_n$ by $T_n(w_n) := \widetilde{w}_n$, where $\widetilde{w}_n$ is the unique element in H_n such that

$$\int_\Omega \nabla \widetilde{w}_n \cdot \nabla v_n = \int_\Omega f(w_n)\,\kappa\, v_n + \int_\Omega g\, v_n \qquad \forall\, v_n \in H_n\,, \tag{4.56}$$

one can prove, assuming again (4.54), that (4.55) has a unique solution u_n. Note, in particular, that B is H_n-elliptic with the same constant α as before, so that the Lax-Milgram Lemma (cf. Lemma 4.4) also guarantees the unique solvability of (4.56). Then, denoting by F_u and F_{u_n} the linear functionals on the right-hand sides of (4.48) and (4.55), respectively, and applying Strang's estimate (4.45) to the context generated by these formulations, we obtain

$$\|u - u_n\|_1 \leqslant \left(1 + \frac{\|\mathbf{B}\|}{\alpha}\right) \operatorname{dist}(u, H_n) + \frac{1}{\alpha} \|F_u - F_{u_n}\|_{H_n'}\,, \tag{4.57}$$

where

$$\|F_u - F_{u_n}\|_{H_n'} := \sup_{\substack{v_n \in H_n \\ v_n \neq 0}} \frac{(F_u - F_{u_n})(v_n)}{\|v_n\|} = \sup_{\substack{v_n \in H_n \\ v_n \neq 0}} \frac{\displaystyle\int_\Omega \big(f(u) - f(u_n)\big)\,\kappa\, v_n}{\|v_n\|}\,.$$

Then, proceeding as in the derivation of (4.52), one deduces that

$$\|F_u - F_{u_n}\|_{H_n'} \leqslant L_f\, \|\kappa\|_{0,\infty;\Omega}\, \|u - u_n\|_{1,\Omega}\,,$$

so that, replacing this estimate in (4.57), and making use of the assumption (4.54), we finally obtain

$$\|u - u_n\|_1 \leqslant (1 - \delta)^{-1} \left(1 + \frac{\|\mathbf{B}\|}{\alpha}\right) \operatorname{dist}(u, H_n)\,.$$

4.3 The Lions-Stampacchia Theorem

In this section we study the analogue of the simplest version of the Lax-Milgram Lemma (cf. Lemma 4.4) for variational inequalities. As motivation, we consider the obstacle problem, which consists in determining the displacement u of a membrane that is fixed on the boundary Γ of a bounded open set Ω in $\mathbb{R}^2$, and that is laid down over a solid located on Ω, whose contact surface is defined by a function $g \in \mathrm{H}^1(\Omega)$. In addition to this obstacle, it is assumed that a force of magnitude f acts on the membrane. Then, it can be proved that the model in question reduces to: Find $u \in S$ such that

$$B(u, v - u) \geqslant F(v - u) \qquad \forall\, v \in S, \tag{4.58}$$

where

$$S := \left\{ v \in \mathrm{H}_0^1(\Omega) : \quad v \geqslant g \quad \text{in } \Omega \right\},$$

$B : \mathrm{H}_0^1(\Omega) \times \mathrm{H}_0^1(\Omega) \to \mathbb{R}$ is the bilinear form given by

$$B(w, v) := \int_\Omega \nabla w \cdot \nabla v \qquad \forall\, w, v \in \mathrm{H}_0^1(\Omega),$$

and $F : \mathrm{H}_0^1(\Omega) \to \mathbb{R}$ is the linear functional

$$F(v) := \int_\Omega f v \qquad \forall\, v \in \mathrm{H}_0^1(\Omega).$$

With the purpose of clarifying the meaning of the inequality that defines the set S, it is necessary to introduce beforehand the set

$$\mathcal{D}_+(\Omega) := \Big\{\varphi \in \mathcal{D}(\Omega) : \quad \varphi \geqslant 0 \quad \text{in} \quad \Omega\Big\}.$$

Then, we say that $z \in \mathcal{D}'(\Omega)$ satisfies $z \geqslant 0$ in Ω if $\langle z, \varphi\rangle \geqslant 0$ for all $\varphi \in \mathcal{D}_+(\Omega)$, and therefore $v \geqslant g$ in Ω certainly means that $v - g \geqslant 0$ in Ω.

A key aspect of problem (4.58) is the fact that S is a convex and closed subset of $\mathrm{H}_0^1(\Omega)$. Indeed, since convexity is rather simple to visualize, we focus only on proving that S is closed. To this end, let us consider a sequence $\{v_n\}_{n\in\mathbb{N}}$ in S such that it converges to $v \in \mathrm{H}_0^1(\Omega)$. It follows from the definition of S and from the above clarification that

$$\langle v_n - g, \varphi\rangle = \int_\Omega (v_n - g)\,\varphi \geqslant 0 \qquad \forall\, n \in \mathbb{N}, \quad \forall\, \varphi \in \mathcal{D}_+(\Omega),$$

from which, taking the limit as $n \to +\infty$, and using that

$$\|v_n - v\|_{0,\Omega} \leqslant \|v_n - v\|_{1,\Omega} \overset{n\to+\infty}{\longrightarrow} 0,$$

we obtain

$$\int_\Omega (v - g)\,\varphi \geqslant 0 \quad \forall\, \varphi \in \mathcal{D}_+(\Omega),$$

which means that $v \geqslant g$ in Ω, and therefore $v \in S$. According to the above, the starting point of the present analysis is to prove the following result, which first recalls and then complements the Second Best Approximation Theorem given by Theorem 2.7.

Theorem 4.15. *Let S be a nonempty, convex, and closed subset of a Hilbert space $(H, \langle\cdot,\cdot\rangle)$. Then, for every $x \in H$ there exists a unique $z \in S$ such that*

$$\|x - z\| = \min_{y\in S} \|x - y\|, \tag{4.59}$$

which is called the best approximation of x by elements of S. Moreover, z is characterized by the following property

$$z \in S \qquad \text{and} \qquad \mathrm{Re}\big\{\langle x - z, y - z\rangle\big\} \leqslant 0 \qquad \forall\, y \in S. \tag{4.60}$$

Proof. Given $x \in H$, the existence of its best approximation by elements of S corresponds exactly to what is stated by Theorem 2.7. It remains to prove the equivalence between (4.59) and (4.60). Indeed, let $z \in S$ be the unique solution of (4.59). Then, using that S is convex, we have

$$(1-t)z + ty \in S \quad \forall\, y \in S, \quad \forall\, t \in [0,1],$$

and thus

$$\|x - z\| \leqslant \|x - \{(1-t)z + ty\}\| = \|(x - z) - t(y - z)\|,$$

or equivalently

$$\|x - z\|^2 \leqslant \|x - z\|^2 - 2t\,\mathrm{Re}\big\{\langle x - z, y - z\rangle\big\} + t^2\,\|y - z\|^2. \tag{4.61}$$

It follows that

$$2t\,\mathrm{Re}\big\{\langle x - z, y - z\rangle\big\} \leqslant t^2\,\|y - z\|^2 \qquad \forall\, t \in [0,1], \quad \forall\, y \in S,$$

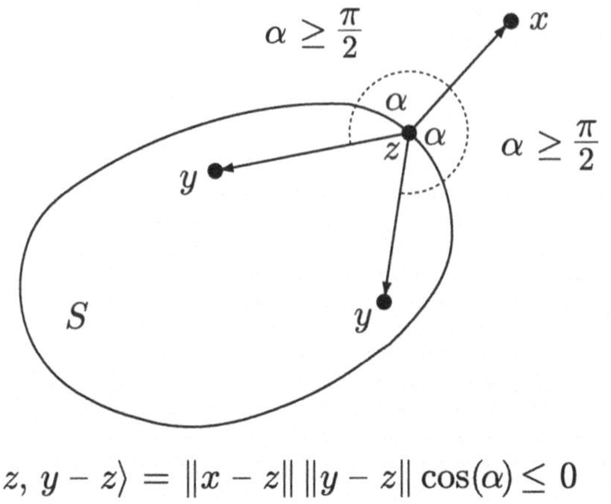

Fig. 4.2. *Illustration in* $\mathbb{R}^2$ *of the characterization* (4.60)

whence

$$\mathrm{Re}\{\langle x - z, y - z\rangle\} \leqslant \frac{t}{2} \|y - z\|^2 \qquad \forall\, t \in (0, 1], \quad \forall\, y \in S, \tag{4.62}$$

and taking the limit as $t \to 0^+$, we obtain (4.60). Conversely, let $z \in S$ be such that (4.60) holds. Then, given $y \in S$ we have

$$\|x - y\|^2 = \|(x - z) - (y - z)\|^2 = \|x - z\|^2 - 2\,\mathrm{Re}\{\langle x - z, y - z\rangle\} + \|y - z\|^2,$$

that is,

$$\|x - z\|^2 - \|x - y\|^2 = 2\,\mathrm{Re}\{\langle x - z, y - z\rangle\} - \|y - z\|^2 < 0 \qquad \forall\, y \neq z,$$

and hence

$$\|x - z\| < \|x - y\| \qquad \forall\, y \in S, \quad y \neq z,$$

which implies that z is the unique solution of (4.59).

□

Needless to say, when H is real the characterization (4.60) becomes

$$z \in S \quad \text{and} \quad \langle x - z, y - z\rangle \leqslant 0 \qquad \forall\, y \in S, \tag{4.63}$$

which is illustrated in Figure 4.2 for the case $H = \mathbb{R}^2$.

The previous theorem induces the definition of the orthogonal projector $P_S : H \to S$ which satisfies the "*pseudo-orthogonal*" condition:

$$\mathrm{Re}\{\langle x - P_S(x), y - P_S(x)\rangle\} \leqslant 0 \quad \forall\, x \in H, \quad \forall\, y \in S. \tag{4.64}$$

Moreover, the following lemma shows that P_S is Lipschitz-continuous with constant 1.

Lemma 4.16.

$$\|P_S(x) - P_S(\widetilde{x})\| \leqslant \|x - \widetilde{x}\| \qquad \forall\, x, \widetilde{x} \in H.$$

Proof. Given x and $\widetilde{x}$ in H, we have, according to (4.64),

$$\mathrm{Re}\big\{\langle x - P_S(x), y - P_S(x)\rangle\big\} \leqslant 0 \qquad \forall\, y \in S \tag{4.65}$$

and

$$\mathrm{Re}\big\{\langle \widetilde{x} - P_S(\widetilde{x}), y - P_S(\widetilde{x})\rangle\big\} \leqslant 0 \qquad \forall\, y \in S. \tag{4.66}$$

In particular, for $y = P_S(\widetilde{x}) \in S$ in (4.65), and $y = P_S(x) \in S$ in (4.66), we obtain respectively

$$\mathrm{Re}\big\{\langle x - P_S(x), P_S(\widetilde{x}) - P_S(x)\rangle\big\} \leqslant 0$$

and

$$\mathrm{Re}\big\{\langle \widetilde{x} - P_S(\widetilde{x}), P_S(x) - P_S(\widetilde{x})\rangle\big\} \leqslant 0.$$

Then, rewriting the first inequality as

$$\mathrm{Re}\big\{\langle P_S(x) - x, P_S(x) - P_S(\widetilde{x})\rangle\big\} \leqslant 0,$$

and adding it to the second, we obtain

$$\mathrm{Re}\big\{\langle \widetilde{x} - x + \{P_S(x) - P_S(\widetilde{x})\}, P_S(x) - P_S(\widetilde{x})\rangle\big\} \leqslant 0,$$

that is,

$$\|P_S(x) - P_S(\widetilde{x})\|^2 \leqslant \mathrm{Re}\big\{\langle x - \widetilde{x}, P_S(x) - P_S(\widetilde{x})\rangle\big\}.$$

Finally, the obvious fact that $|\mathrm{Re}(z)| \leqslant |z|$ for all $z \in \mathbb{C}$, and a simple application of the Cauchy-Schwarz inequality, complete the proof.

□

We are now in a position to state and prove the main result of this section.

Theorem 4.17 (Lions-Stampacchia Theorem). *Let $(H, \langle\cdot,\cdot\rangle)$ be a real Hilbert space, and let $B : H \times H \to \mathbb{R}$ be a bilinear form that is bounded and H-elliptic with constants M and α, respectively. Let $S \neq \varnothing$ be a convex and closed subset of H. Then, for each $F \in H'$ there exists a unique $u \in S$ such that*

$$B(u, v - u) \geqslant F(v - u) \qquad \forall\, v \in S. \tag{4.67}$$

Moreover, if B is symmetric, then u is characterized by the property

$$u \in S \quad \text{and} \quad Q(u) = \min_{v \in S} Q(v),$$

where Q is the quadratic form

$$Q(v) := B(v, v) - 2F(v) \qquad \forall\, v \in S.$$

Proof. The key aspects of this proof are, on the one hand, the transformation of the variational inequality (4.67) into an **equivalent variational equality**, and then the application of the Banach Fixed Point Theorem (see Theorem 1.8 in Chapter 1) to such equality. To this end, we first observe that (4.67) can be rewritten as

$$\langle \mathbf{B}(u), v - u\rangle \geqslant \langle \mathcal{R}(F), v - u\rangle \qquad \forall\, v \in S,$$

where $\mathbf{B} \in \mathcal{L}(H)$ is the operator induced by B, and $\mathcal{R} : H' \to H$ is the corresponding Riesz map. Then, given a parameter $\rho > 0$, and proceeding in an analogous way to the alternative proof of the Lax-Milgram Lemma (see Section 4.1), our variational inequality becomes

$$\langle \rho \mathbf{B}(u) - \rho \mathcal{R}(F), v - u \rangle \geqslant 0 \qquad \forall v \in S,$$

or, multiplying by (-1) and adding and subtracting u,

$$\langle \rho \mathcal{R}(F) - \rho \mathbf{B}(u) + u - u, v - u \rangle \leqslant 0 \qquad \forall v \in S.$$

It follows, by virtue of the characterization of the best approximation over a closed convex set (see Theorem 4.15 or (4.64)), that the above is equivalent to the equation

$$u = P_S(\rho \mathcal{R}(F) - \rho \mathbf{B}(u) + u) =: \quad T(u),$$

where $T : S \to S$ is defined by

$$T(v) := P_S(\rho \mathcal{R}(F) - \rho \mathbf{B}(v) + v) \qquad \forall v \in S.$$

Thus, the present proof now reduces to showing that, for a suitably chosen $\rho > 0$, the map T has a unique fixed point in S. For this purpose, and as announced at the beginning, we now apply the Banach Fixed Point Theorem. Indeed, by Lemma 4.16, we first have that

$$\|T(v) - T(w)\| \leqslant \|(v - w) - \rho (\mathbf{B}(v) - \mathbf{B}(w))\|,$$

from which, using the linearity of the operator $\mathbf{B}$, it follows that

$$\|T(v) - T(w)\|^2 \leqslant \|v - w\|^2 - 2\rho \langle \mathbf{B}(v - w), v - w \rangle + \rho^2 \|\mathbf{B}(v - w)\|^2 \qquad \forall v, w \in S.$$

Now, from the boundedness and H-ellipticity of B, we have, respectively,

$$\|\mathbf{B}(v - w)\| \leqslant M \|v - w\| \quad \text{and} \quad \langle \mathbf{B}(v - w), v - w \rangle \geqslant \alpha \|v - w\|^2,$$

and therefore

$$\|T(v) - T(w)\|^2 \leqslant (1 - 2\rho\alpha + \rho^2 M^2) \|v - w\|^2 \qquad \forall v, w \in S,$$

so that, choosing $\rho > 0$ such that $1 - 2\rho\alpha + \rho^2 M^2 < 1$, that is $\rho \in \left(0, \frac{2\alpha}{M^2}\right)$, we deduce that T is a contraction and therefore there exists a unique $u \in S$ such that $u = T(u)$. This unique fixed point is, in turn, the unique solution of the original variational inequality (4.67).

On the other hand, if we further assume that B is symmetric, then this bilinear form also defines an inner product on H, whose associated norm $\|\cdot\|_B := B(\cdot, \cdot)^{1/2}$ is equivalent to the usual norm. It follows that $(H, B(\cdot, \cdot))$ is also a Hilbert space, and therefore, applying the Riesz Representation Theorem, we deduce the existence of $g \in H$ such that

$$F(v) = B(g, v) \qquad \forall v \in H. \tag{4.68}$$

Consequently, the original variational inequality can be equivalently rewritten as: Find $u \in S$ such that

$$B(g - u, v - u) \leqslant 0 \qquad \forall v \in S,$$

which means, according to condition (4.64) (see also Theorem 4.15), that u is the best approximation of g by elements of S with respect to the norm $\|\cdot\|_B$. Symbolically, this means that $u \in S$ and

$$B(g-u,g-u) = \min_{v\in S} B(g-v,g-v),$$

or, eliminating $B(g,g)$ on both sides,

$$B(u,u) - 2B(g,u) = \min_{v\in S}\Big\{B(v,v) - 2B(g,v)\Big\},$$

which, using (4.68), reduces to:

$$Q(u) = \min_{v\in S} Q(v),$$

thus completing the proof.

□

4.4 Babuška-Brezzi Theory

In this section we present the main results that constitute the Babuška-Brezzi Theory in Hilbert spaces, which allows the analysis of a wide range of variational formulations of mixed type and their respective Galerkin approximations. More specifically, we describe below the type of formulations in which we are interested here.

4.4.1 The operator equation

Let $(H,\langle\cdot,\cdot\rangle_H)$ and $(Q,\langle\cdot,\cdot\rangle_Q)$ be Hilbert spaces over $\mathbb{R}$, and let $a : H\times H \to \mathbb{R}$ and $b : H\times Q \to \mathbb{R}$ be bounded bilinear forms. Then, given $F \in H'$ and $G \in Q'$, we are interested in the following problem: Find $(\sigma,u) \in H\times Q$ such that

$$\begin{aligned} a(\sigma,\tau) + b(\tau,u) &= F(\tau) \qquad \forall\,\tau \in H\,, \\ b(\sigma,v) \qquad\qquad &= G(v) \qquad \forall\,v \in Q\,. \end{aligned} \tag{4.69}$$

Now, let $\mathbf{A} : H \to H$ and $\mathbf{B} : H \to Q$ be the linear and bounded operators induced by a and b, respectively. Equivalently, according to the analysis in Section 4.1 (cf. (4.2), (4.3), and (4.4)), we have

$$\mathbf{A} := \mathcal{R}_H \circ \mathcal{A} \quad \text{and} \quad \mathbf{B} := \mathcal{R}_Q \circ \mathcal{B}\,,$$

where $\mathcal{R}_H : H' \to H$ and $\mathcal{R}_Q : Q' \to Q$ are the respective Riesz maps , and the operators $\mathcal{A} : H \to H'$ and $\mathcal{B} : H \to Q'$ are defined by

$$\mathcal{A}(\sigma)(\tau) := a(\sigma,\tau) \qquad \forall\,\sigma \in H\,, \quad \forall\,\tau \in H\,,$$

and

$$\mathcal{B}(\tau)(v) := b(\tau,v) \qquad \forall\,\tau \in H\,, \quad \forall\,v \in Q\,.$$

It follows that

$$a(\sigma,\tau) = \langle \mathbf{A}(\sigma),\tau\rangle_H \qquad \forall\,(\sigma,\tau) \in H\times H \tag{4.70}$$

and

$$b(\tau,v) = \langle \mathbf{B}(\tau),v\rangle_Q = \langle \mathbf{B}^*(v),\tau\rangle_H \qquad \forall\,(\tau,v) \in H\times Q\,, \tag{4.71}$$

where $\mathbf{B}^* : Q \to H$ is the adjoint operator of $\mathbf{B}$.

Thus, (4.69) can be equivalently rewritten as: Find $(\sigma,u) \in H\times Q$ such that

$$\begin{aligned} \langle \mathbf{A}(\sigma), \tau \rangle_H + \langle \mathbf{B}^*(u), \tau \rangle_H &= \langle \mathcal{R}_H(F), \tau \rangle_H \qquad \forall\, \tau \in H\,, \\ \langle \mathbf{B}(\sigma), v \rangle_Q \qquad\qquad &= \langle \mathcal{R}_Q(G), v \rangle_Q \qquad \forall\, v \in Q\,, \end{aligned}$$

or equivalently: Find $(\sigma, u) \in H \times Q$ such that

$$\begin{aligned} \mathbf{A}(\sigma) + \mathbf{B}^*(u) &= \mathcal{R}_H(F)\,, \\ \mathbf{B}(\sigma) \qquad &= \mathcal{R}_Q(G)\,, \end{aligned} \tag{4.72}$$

which, denoting the zero operator by 0, reduces to the following matrix operator equation: Find $(\sigma, u) \in H \times Q$ such that

$$\begin{pmatrix} \mathbf{A} & \mathbf{B}^* \\ \mathbf{B} & 0 \end{pmatrix} \begin{pmatrix} \sigma \\ u \end{pmatrix} = \begin{pmatrix} \mathcal{R}_H(F) \\ \mathcal{R}_Q(G) \end{pmatrix} . \tag{4.73}$$

The next goal is to provide necessary and sufficient conditions for (4.69) (equivalently (4.72) or (4.73)) to be well-posed.

4.4.2 The inf-sup condition

Let us first recall that this concept was already introduced in Section 4.1 (cf. (4.17), (4.18), (4.19), and (4.20)). Indeed, we say that the bounded bilinear form $b : H \times Q \to \mathbb{R}$ satisfies the continuous inf-sup condition if there exists a constant $\beta > 0$ such that

$$\sup_{\substack{\tau \in H \\ \tau \neq 0}} \frac{b(\tau, v)}{\|\tau\|_H} \geqslant \beta \, \|v\|_Q \qquad \forall\, v \in Q\,. \tag{4.74}$$

Note, just as was established for the pairs (4.17) - (4.19) and (4.18) - (4.20), that (4.74) is equivalent to

$$\inf_{\substack{v \in Q \\ v \neq 0}} \sup_{\substack{\tau \in H \\ \tau \neq 0}} \frac{b(\tau, v)}{\|\tau\|_H \, \|v\|_Q} \geqslant \beta\,,$$

which again explains the term INF-SUP. This hypothesis is also referred to as the LADYZHENSKAYA-BABUŠKA-BREZZI CONDITION, or simply the BABUŠKA-BREZZI CONDITION. Moreover, using the adjoint operator $\mathbf{B}^*$, we have that

$$\sup_{\substack{\tau \in H \\ \tau \neq 0}} \frac{b(\tau, v)}{\|\tau\|_H} = \sup_{\substack{\tau \in H \\ \tau \neq 0}} \frac{\langle \mathbf{B}^*(v), \tau \rangle_H}{\|\tau\|_H} = \|\mathbf{B}^*(v)\|_H\,,$$

and therefore, the condition (4.74) can also be written as

$$\|\mathbf{B}^*(v)\|_H \geqslant \beta \, \|v\|_Q \qquad \forall\, v \in Q\,. \tag{4.75}$$

Furthermore, the following lemma establishes equivalent conditions for the inequality (4.74) (or (4.75)).

Lemma 4.18. *The following statements are equivalent*

i) *There exists* $\beta > 0$ *such that*

$$\sup_{\substack{\tau \in H \\ \tau \neq 0}} \frac{b(\tau, v)}{\|\tau\|_H} \geqslant \beta \, \|v\|_Q \qquad \forall\, v \in Q\,.$$

ii) $\mathbf{B}^*$ *is an isomorphism (linear bijection) from* Q *onto* $N(\mathbf{B})^\perp$ *and*

$$\|\mathbf{B}^*(v)\|_H \geqslant \beta \|v\|_Q \qquad \forall v \in Q .$$

iii) $\mathbf{B}$ *is an isomorphism (linear bijection) from* $N(\mathbf{B})^\perp$ *onto* Q *and*

$$\|\mathbf{B}(\tau)\|_Q \geqslant \beta \|\tau\|_H \qquad \forall \tau \in N(\mathbf{B})^\perp . \tag{4.76}$$

iv) $\mathbf{B} : H \to Q$ *is surjective.*

Proof.
i) $\Rightarrow$ ii): Suppose that there exists $\beta > 0$ such that (4.74) (equivalently, (4.75)) holds. It follows precisely from (4.75) that $N(\mathbf{B}^*) = \{0\}$ and $R(\mathbf{B}^*)$ is closed, which implies that $\mathbf{B}^*$ is injective and $R(\mathbf{B}^*) = N((\mathbf{B}^*)^*)^\perp = N(\mathbf{B})^\perp$. Thus, $\mathbf{B}^*$ is a linear bijection from Q onto $N(\mathbf{B})^\perp$.

ii) $\Rightarrow$ iii): Suppose that $\mathbf{B}^*$ is a linear bijection from Q onto $N(\mathbf{B})^\perp$ and that (4.75) holds. It follows again from (4.75) that $N(\mathbf{B}^*) = \{0\}$ and $R(\mathbf{B}^*)$ is closed. The latter implies, by a result of functional analysis, that $R(\mathbf{B})$ is also closed, and therefore $R(\mathbf{B}) = N(\mathbf{B}^*)^\perp = \{0\}^\perp = Q$. In this way, $\mathbf{B}$ is a linear bijection from $N(\mathbf{B})^\perp$ onto Q. Moreover, it is clear from (4.75) that $\|(\mathbf{B}^*)^{-1}\| \leqslant \frac{1}{\beta}$, and then

$$\|\mathbf{B}^{-1}\| = \|(\mathbf{B}^{-1})^*\| = \|(\mathbf{B}^*)^{-1}\| \leqslant \frac{1}{\beta},$$

which translates into (4.76).

iii) $\Rightarrow$ iv): This follows directly from the fact that $\mathbf{B} : N(\mathbf{B})^\perp \to Q$ is bijective (in particular, surjective), also recalling that $H = N(\mathbf{B}) \oplus N(\mathbf{B})^\perp$.

iv) $\Rightarrow$ i): Suppose that $\mathbf{B} : H \to Q$ is surjective. Since $R(\mathbf{B}) = Q$ is obviously closed, we have that $R(\mathbf{B}^*)$ is also closed (cf. Theorem 3.63). Moreover, applying orthogonal to the identity $Q = R(\mathbf{B}) = N(\mathbf{B}^*)^\perp$ yields $N(\mathbf{B}^*) = \{0\}$, which says that $\mathbf{B}^*$ is injective. In this way, the characterization result given by Theorem 3.58 implies the inequality (4.75), which coincides with i).

□

4.4.3 The main result

The characterization of the inf-sup condition given by Lemma 4.18 is fundamental in the proof of the following theorem, which establishes sufficient conditions for (4.69) to be well-posed. In its statement and in what follows of this section the same notations and definitions from the previous sections are assumed.

Theorem 4.19 (Babuška-Brezzi Theorem). *Let* $V := N(\mathbf{B})$, *and let* $\Pi : H \to V$ *be the orthogonal projection operator. Assume that*

i) $\Pi\,\mathbf{A} : V \to V$ *is a bijection.*
ii) *The bilinear form* b *satisfies the inf-sup condition* (4.74) (*equivalently* (4.75)).

Then, for each pair $(F, G) \in H' \times Q'$ *there exists a unique solution* $(\sigma, u) \in H \times Q$ *of* (4.69) (*equivalently* (4.72) *or* (4.73)). *Moreover, there hold the a priori estimates*

$$\begin{aligned} \|\sigma\|_H &\leqslant \|(\Pi\,\mathbf{A})^{-1}\|\,\|F\|_{H'} + \frac{1}{\beta}\left(1 + \|\mathbf{A}\|\,\|(\Pi\,\mathbf{A})^{-1}\|\right)\|G\|_{Q'}\,, \\ \|u\|_Q &\leqslant \frac{1}{\beta}\left(1 + \|\mathbf{A}\|\,\|(\Pi\,\mathbf{A})^{-1}\|\right)\|F\|_{H'} + \frac{\|\mathbf{A}\|}{\beta^2}\left(1 + \|\mathbf{A}\|\,\|(\Pi\,\mathbf{A})^{-1}\|\right)\|G\|_{Q'}\,. \end{aligned} \tag{4.77}$$

Proof. Since $\mathbf{B}$ is a bijection from $V^{\perp}$ onto Q (as a consequence of (4.74) and Lemma 4.18, part iii)), it follows that there exists a unique $\sigma_g \in V^{\perp}$ such that

$$\mathbf{B}(\sigma_g) \;=\; \mathcal{R}_Q(G)\,, \tag{4.78}$$

and according to (4.76), we have

$$\|\sigma_g\|_H \;\leqslant\; \frac{1}{\beta}\,\|\mathbf{B}(\sigma_g)\|_Q \;=\; \frac{1}{\beta}\,\|G\|_{Q'}\,. \tag{4.79}$$

On the other hand, since $\Pi\,\mathbf{A} : V \to V$ is a bijection and $\Pi(\mathcal{R}_H(F) - \mathbf{A}(\sigma_g))$ belongs to V, there exists a unique $\sigma_0 \in V$ such that $\Pi\,\mathbf{A}(\sigma_0) = \Pi(\mathcal{R}_H(F) - \mathbf{A}(\sigma_g))$. Moreover, the Bounded Inverse Theorem (cf. Theorem 3.34) guarantees the existence of $\widetilde{C} := \|(\Pi\,\mathbf{A})^{-1}\|$ such that

$$\|\sigma_0\|_H \;\leqslant\; \widetilde{C}\,\|\Pi(\mathcal{R}_H(F) - \mathbf{A}(\sigma_g))\|_H \;\leqslant\; \widetilde{C}\,\|\mathcal{R}_H(F) - \mathbf{A}(\sigma_g)\|_H\,,$$

whence, using the bound for $\|\sigma_g\|_H$, it follows that

$$\|\sigma_0\|_H \;\leqslant\; \widetilde{C}\left\{\|F\|_{H'} + \frac{1}{\beta}\,\|\mathbf{A}\|\,\|G\|_{Q'}\right\}. \tag{4.80}$$

Now, according to the orthogonality condition of the projector Π, it is easy to see that the identity $\Pi\,\mathbf{A}(\sigma_0) = \Pi(\mathcal{R}_H(F) - \mathbf{A}(\sigma_g))$ is equivalent to saying that $\mathbf{A}(\sigma_0 + \sigma_g) - \mathcal{R}_H(F)$ belongs to $V^{\perp}$. It follows from Lemma 4.18, part ii), that there exists a unique $u \in Q$ such that

$$\mathbf{B}^*(u) \;=\; \mathcal{R}_H(F) \;-\; \mathbf{A}(\sigma_0 + \sigma_g) \tag{4.81}$$

and

$$\|u\|_Q \;\leqslant\; \frac{1}{\beta}\,\|\mathbf{B}^*(u)\|_H \;=\; \frac{1}{\beta}\,\|\mathcal{R}_H(F) \;-\; \mathbf{A}(\sigma_0 + \sigma_g)\|_H\,,$$

whence

$$\|u\|_Q \;\leqslant\; \frac{1}{\beta}\left\{\|F\|_{H'} + \|\mathbf{A}\|\left(\|\sigma_0\|_H + \|\sigma_g\|_H\right)\right\}. \tag{4.82}$$

Thus, defining $\sigma := \sigma_0 + \sigma_g \in H$, and noting that $\mathbf{B}(\sigma_0) = 0$, it follows from (4.78), (4.81), and from the estimates (4.79), (4.80), and (4.82), that (σ, u) solves (4.72) and satisfies (4.77). For uniqueness, let $(\sigma, u) \in H \times Q$ be a solution of the homogeneous problem

$$\begin{aligned} \mathbf{A}(\sigma) + \mathbf{B}^*(u) &= 0\,, \\ \mathbf{B}(\sigma) \qquad\quad &= 0\,. \end{aligned}$$

From the second equation we have $\sigma \in V$, and by applying the projector Π in the first one, recalling that $\mathbf{B}^*(u) \in V^{\perp}$, we obtain $\Pi\,\mathbf{A}(\sigma) = 0$. Since $\Pi\,\mathbf{A} : V \to V$ is a bijection, it follows that $\sigma = 0$, and then from the first equation we get $\mathbf{B}^*(u) = 0$. Finally, since $\mathbf{B}^* : Q \to V^{\perp}$ is also a bijection, we conclude that $u = 0$.

□

We now show that conditions i) and ii) of Theorem 4.19 are also **necessary**. Indeed, we have the following result.

Theorem 4.20. *Let* $V := N(\mathbf{B})$, *and let* $\Pi : H \to V$ *be the orthogonal projection operator. Assume that for each pair* $(F, G) \in H' \times Q'$ *there exists a unique solution* $(\sigma, u) \in H \times Q$ *of* (4.69) *(equivalently* (4.72) *or* (4.73)*), which moreover satisfies*

$$\|(\sigma, u)\|_{H\times Q} \leqslant C\Big\{ \|F\|_{H'} + \|G\|_{Q'} \Big\},$$

with a constant $C > 0$ *independent of* F *and* G. *Then*

i) $\Pi\,\mathbf{A} : V \to V$ *is a bijection.*
ii) *The bilinear form* b *satisfies the inf-sup condition* (4.74).

Proof. We first prove ii). To this end, by virtue of Lemma 4.18, it suffices to show that $\mathbf{B}$ is surjective. Indeed, given $g \in Q$, we know by hypothesis that there exists a unique pair $(\sigma_g, u_g) \in H \times Q$ such that

$$\begin{aligned} \mathbf{A}(\sigma_g) + \mathbf{B}^*(u_g) &= 0\,, \\ \mathbf{B}(\sigma_g) \qquad\quad &= g\,, \end{aligned}$$

and it is clear that the second equation of this system confirms the surjectivity of $\mathbf{B}$. Thus, knowing that b satisfies the continuous inf-sup condition, we may now use the equivalences given by Lemma 4.18 to prove that $\Pi\,\mathbf{A} : V \to V$ is a bijection. Given $f \in V$, we also know by hypothesis that there exists a unique $(\sigma_f, u_f) \in H \times Q$ such that

$$\begin{aligned} \mathbf{A}(\sigma_f) + \mathbf{B}^*(u_f) &= f\,, \\ \mathbf{B}(\sigma_f) \qquad\quad &= 0\,, \end{aligned}$$

which shows, according to the second equation, that $\sigma_f \in V$. Then, applying the orthogonal projector Π to the first equation, and using, according to part ii) of Lemma 4.18, that $\mathbf{B}^*(u_f) \in V^\perp$, we obtain $\Pi\,\mathbf{A}(\sigma_f) = \Pi(f) = f$, thus proving that $\Pi\,\mathbf{A} : V \to V$ is surjective. On the other hand, let $\sigma_0 \in V$ be such that $\Pi\,\mathbf{A}(\sigma_0) = 0$. It follows that $\mathbf{A}(\sigma_0) \in V^\perp$, and since, according to part ii) of Lemma 4.18, $\mathbf{B}^* : Q \to V^\perp$ is a bijection, it follows that there exists a unique $u_0 \in Q$ such that $\mathbf{B}^*(u_0) = -\mathbf{A}(\sigma_0)$. In this way we obtain

$$\begin{aligned} \mathbf{A}(\sigma_0) + \mathbf{B}^*(u_0) &= 0\,, \\ \mathbf{B}(\sigma_0) \qquad\quad &= 0\,, \end{aligned}$$

and thanks again to the hypothesis it follows that $(\sigma_0, u_0) = (0, 0)$, from which the injectivity of $\Pi\,\mathbf{A} : V \to V$ is concluded.

□

On the other hand, it is easy to see, according to Lemma 4.7 and inequalities (4.17) and (4.18), and by virtue of the orthogonality relation that characterizes the projector $\Pi : H \to V$ (cf. Theorem 2.6, (2.6)), that condition i) in Theorems 4.19 and 4.20 is equivalent to each one of the following pairs of conditions:

i-1) there exists $\alpha > 0$ such that

$$\|(\Pi\,\mathbf{A})(\sigma)\|_H = \sup_{\substack{\tau\in V\\ \tau\neq 0}} \frac{a(\sigma,\tau)}{\|\tau\|_H} \geqslant \alpha\,\|\sigma\|_H \qquad \forall\,\sigma \in V\,, \tag{4.83}$$

i-2) for all $\tau \in V$, $\tau \neq 0$, $\sup_{\sigma \in V} a(\sigma,\tau) > 0$.

and

i-1)' there exists $\alpha > 0$ such that

$$\|(\Pi\,\mathbf{A})^*(\sigma)\|_H = \sup_{\substack{\tau\in V\\ \tau\neq 0}} \frac{a(\tau,\sigma)}{\|\tau\|_H} \geqslant \alpha\,\|\sigma\|_H \qquad \forall\,\sigma \in V\,, \tag{4.84}$$

i-2)' for all $\tau \in V$, $\tau \neq 0$, $\sup_{\sigma \in V} a(\tau,\sigma) > 0$.

More precisely, hypothesis i-1) (resp. i-1)') is an inf-sup condition for the bilinear form a, which is equivalent to requiring that the operator $\Pi\,\mathbf{A}$ (resp. $(\Pi\,\mathbf{A})^*$) be injective and have closed range, which, moreover, is equivalent to the operator $(\Pi\,\mathbf{A})^*$ (resp. $\Pi\,\mathbf{A}$) being surjective. In turn, i-2) (resp. i-2)') is equivalent to the injectivity of $(\Pi\,\mathbf{A})^*$ (resp. $\Pi\,\mathbf{A}$). In this way, being both $\Pi\,\mathbf{A} : V \to V$ and $(\Pi\,\mathbf{A})^* : V \to V$ bijective, and observing that $\|(\Pi\,\mathbf{A})^{-1}\| = \|\big((\Pi\,\mathbf{A})^*\big)^{-1}\|$, which justifies taking the same constant α in (4.83) and (4.84), we deduce that (cf. Lemma 3.14 and Exercise 3.6)

$$\|(\Pi\,\mathbf{A})^{-1}\| = \|\big((\Pi\,\mathbf{A})^*\big)^{-1}\| \leqslant \frac{1}{\alpha}\,,$$

whence the a priori estimates in (4.77) become

$$\begin{aligned} \|\sigma\|_H &\leqslant \frac{1}{\alpha}\,\|F\|_{H'} + \frac{1}{\beta}\Big(1 + \frac{\|\mathbf{A}\|}{\alpha}\Big)\,\|G\|_{Q'}\,,\\ \|u\|_Q &\leqslant \frac{1}{\beta}\Big(1 + \frac{\|\mathbf{A}\|}{\alpha}\Big)\,\|F\|_{H'} + \frac{\|\mathbf{A}\|}{\beta^2}\Big(1 + \frac{\|\mathbf{A}\|}{\alpha}\Big)\,\|G\|_{Q'}\,. \end{aligned} \tag{4.85}$$

We stress here that the supreme on the right-hand sides of (4.83) and (4.84) can be taken in V because both $(\Pi\,\mathbf{A})(\sigma)$ and $(\Pi\,\mathbf{A})^*(\sigma)$ belong to this subspace. Furthermore, when a is a symmetric bilinear form on $V \times V$, the operator $\Pi\,\mathbf{A}$ turns out to be self-adjoint, and in such case i-2) and i-2)' are redundant and therefore unnecessary.

On the other hand, it is important to emphasize that a sufficient (but not necessary) condition for i), which appears very frequently in applications, is the V-ellipticity of the bilinear form a, which, according to Definition 4.3, means that there exists $\alpha > 0$ such that

$$a(\tau,\tau) \geqslant \alpha\,\|\tau\|_H^2 \qquad \forall\,\tau \in V\,. \tag{4.86}$$

In view of the above, the theorem that is usually found in the literature, even more frequently than Theorem 4.19, is the one stated below.

Theorem 4.21. *Let $V := N(\mathbf{B})$, and suppose that*

i) *The bilinear form a is V-elliptic (cf.* (4.86)*).*
ii) *The bilinear form b satisfies the inf-sup condition* (4.74) *(equivalently* (4.75)*).*

Then, for each pair $(F,G) \in H' \times Q'$ there exists a unique solution $(\sigma,u) \in H \times Q$ of (4.69) *(equivalently* (4.72) *or* (4.73)*), which satisfies* (4.85).

Proof. It suffices to note, for instance by virtue of the Lax-Milgram Lemma, that the V-ellipticity of a implies hypothesis i) of Theorem 4.19 with the same constant α from (4.86).

□

4.4.4 The Galerkin scheme

Let $\{H_h\}_{h>0}$ and $\{Q_h\}_{h>0}$ be sequences of finite-dimensional subspaces of H and Q, respectively. Then, given $F \in H'$ and $G \in Q'$, the Galerkin scheme nassociated with (4.69) consists of: Find $(\sigma_h, u_h) \in H_h \times Q_h$ such that

$$\begin{aligned} a(\sigma_h, \tau_h) + b(\tau_h, u_h) &= F(\tau_h) \qquad \forall\, \tau_h \in H_h\,, \\ b(\sigma_h, v_h) \qquad\qquad &= G(v_h) \qquad \forall\, v_h \in Q_h\,. \end{aligned} \tag{4.87}$$

For the solvability analysis of (4.87) we follow the same scheme as in Section 4.4. In particular, let $\mathbf{A}_h : H_h \to H_h$ and $\mathbf{B}_h : H_h \to Q_h$ be the linear and bounded operators induced by a and b on $H_h \times H_h$ and $H_h \times Q_h$, respectively, that is:

$$\mathbf{A}_h := \mathcal{R}_{H_h} \circ \mathcal{A}_h \quad \text{and} \quad \mathbf{B}_h := \mathcal{R}_{Q_h} \circ \mathcal{B}_h\,,$$

where $\mathcal{R}_{H_h} : H_h' \to H_h$ and $\mathcal{R}_{Q_h} : Q_h' \to Q_h$ are the corresponding Riesz maps, and the operators $\mathcal{A}_h : H_h \to H_h'$ and $\mathcal{B}_h : Q_h \to Q_h'$ are defined by

$$\mathcal{A}_h(\sigma_h)(\tau_h) := a(\sigma_h, \tau_h) \qquad \forall\, \sigma_h \in H_h\,, \quad \forall\, \tau_h \in H_h\,,$$

and

$$\mathcal{B}_h(\tau_h)(v_h) := b(\tau_h, v_h) \qquad \forall\, \tau_h \in H_h\,, \quad \forall\, v_h \in Q_h\,.$$

The following theorem provides sufficient conditions for (4.87) to be well-posed.

Theorem 4.22. *Let* $V_h := N(\mathbf{B}_h) = \big\{\tau_h \in H_h : \quad b(\tau_h, v_h) = 0 \quad \forall\, v_h \in Q_h\big\}$, *and let* $\Pi_h : H_h \to V_h$ *be the orthogonal projector. Suppose that*

i) $\Pi_h\, \mathbf{A}_h : V_h \to V_h$ *is injective.*
ii) *there exists* $\beta_h > 0$ *such that*

$$\sup_{\substack{\tau_h \in H_h \\ \tau_h \neq 0}} \frac{b(\tau_h, v_h)}{\|\tau_h\|_H} \geqslant \beta_h\, \|v_h\|_Q \qquad \forall\, v_h \in Q_h\,. \tag{4.88}$$

Then, for each pair $(F, G) \in H' \times Q'$ *there exists a unique* $(\sigma_h, u_h) \in H_h \times Q_h$ *solution of* (4.87). *Moreover, there hold the a priori estimates*

$$\begin{aligned} \|\sigma_h\|_H &\leqslant \|(\Pi_h\, \mathbf{A}_h)^{-1}\|\, \|F_h\|_{H_h'} + \frac{1}{\beta_h}\big(1 + \|\mathbf{A}_h\|\, \|(\Pi_h\, \mathbf{A}_h)^{-1}\|\big)\, \|G_h\|_{Q_h'}\,, \\ \|u_h\|_Q &\leqslant \frac{1}{\beta_h}\big(1 + \|\mathbf{A}_h\|\, \|(\Pi_h\, \mathbf{A}_h)^{-1}\|\big)\, \|F_h\|_{H_h'} + \frac{\|\mathbf{A}_h\|}{\beta_h^2}\big(1 + \|\mathbf{A}_h\|\, \|(\Pi_h\, \mathbf{A}_h)^{-1}\|\big)\, \|G_h\|_{Q_h'}\,, \end{aligned}$$

where $F_h := F|_{H_h}$ *and* $G_h := G|_{Q_h}$.

Proof. Note first that, since V_h is finite-dimensional, the injectivity of $\Pi_h\, \mathbf{A}_h : V_h \to V_h$ is equivalent to its surjectivity, and therefore condition i) is equivalent to requiring that this operator be bijective. Consequently, to conclude the proof, it suffices to apply Theorem 4.19 to the present discrete context.

□

Analogously to what was shown in Theorem 4.20, it can be proved here that conditions i) and ii) of Theorem 4.22 are also **necessary**. Moreover, it is not difficult to see that i) is equivalent to each one of the following inf-sup conditions for a:

i-1) there exists $\alpha_h > 0$ such that

$$\|(\Pi_h \mathbf{A}_h)(\sigma_h)\|_H = \sup_{\substack{\tau_h \in V_h \\ \tau_h \neq 0}} \frac{a(\sigma_h, \tau_h)}{\|\tau_h\|_H} \geqslant \alpha_h \|\sigma_h\|_H \qquad \forall \sigma_h \in V_h, \tag{4.89}$$

i-1)' there exists $\alpha_h > 0$ such that

$$\|(\Pi_h \mathbf{A}_h)^*(\sigma_h)\|_H = \sup_{\substack{\tau_h \in V_h \\ \tau_h \neq 0}} \frac{a(\tau_h, \sigma_h)}{\|\tau_h\|_H} \geqslant \alpha_h \|\sigma_h\|_H \qquad \forall \sigma_h \in V_h, \tag{4.90}$$

from which it follows now that

$$\|(\Pi_h \mathbf{A}_h)^{-1}\| = \|((\Pi_h \mathbf{A}_h)^*)^{-1}\| \leqslant \frac{1}{\alpha_h},$$

whence the a priori bounds from Theorem 4.22 become

$$\begin{aligned} \|\sigma_h\|_H &\leqslant \frac{1}{\alpha_h} \|F_h\|_{H_h'} + \frac{1}{\beta_h}\left(1 + \frac{\|\mathbf{A}_h\|}{\alpha_h}\right) \|G_h\|_{Q_h'}, \\ \|u_h\|_Q &\leqslant \frac{1}{\beta_h}\left(1 + \frac{\|\mathbf{A}_h\|}{\alpha_h}\right) \|F_h\|_{H_h'} + \frac{\|\mathbf{A}_h\|}{\beta_h^2}\left(1 + \frac{\|\mathbf{A}_h\|}{\alpha_h}\right) \|G_h\|_{Q_h'}. \end{aligned} \tag{4.91}$$

Similarly as for the continuous analysis, we notice here that the supreme on the right-hand sides of (4.89) and (4.90) can be taken in V_h because both $(\Pi_h \mathbf{A}_h)(\sigma_h)$ and $(\Pi_h \mathbf{A}_h)^*(\sigma_h)$ belong to this subspace. In addition, since (4.89), (4.90) and (4.88) must be satisfied in finite-dimensional spaces, they are usually called the DISCRETE INF-SUP CONDITIONS for a and b, respectively.

In turn, just as was noted about the V-ellipticity of a, it can be shown that a sufficient (but not necessary) condition for i) is the V_h-ellipticity of the bilinear form a, which means that there exists $\alpha_h > 0$ such that

$$a(\tau_h, \tau_h) \geqslant \alpha_h \|\tau_h\|_H^2 \qquad \forall \tau_h \in V_h.$$

On the other hand, it is important to point out that one usually refers to (4.91) as the STABILITY OF THE GALERKIN SCHEME (4.87). Moreover, it is quite straightforward to see that this inequality implies the boundedness of the GALERKIN PROJECTOR

$$G_h : H \times Q \to H_h \times Q_h,$$

which, given $(\zeta, w) \in H \times Q$, is defined as $G_h(\zeta, w) := (\zeta_h, w_h)$, where $(\zeta_h, w_h) \in H_h \times Q_h$ is the unique solution of the problem:

$$\begin{aligned} a(\zeta_h, \tau_h) + b(\tau_h, w_h) &= a(\zeta, \tau_h) + b(\tau_h, w) \qquad \forall \tau_h \in H_h, \\ b(\zeta_h, v_h) &= b(\zeta, v_h) \qquad \forall v_h \in Q_h. \end{aligned} \tag{4.92}$$

Indeed, it follows from (4.91) and (4.92) that $\|G_h\|$ depends on $\|\mathbf{A}_h\|$, α_h, β_h, $\|\mathbf{A}\|$, and $\|\mathbf{B}\|$. Also, note that

$$G_h(\zeta_h, w_h) = (\zeta_h, w_h) \quad \forall (\zeta_h, w_h) \in H_h \times Q_h, \tag{4.93}$$

and

$$G_h(\sigma, u) = (\sigma_h, u_h), \tag{4.94}$$

where $(\sigma, u) \in H \times Q$ and $(\sigma_h, u_h) \in H_h \times Q_h$ are the unique solutions of (4.69) and (4.87), respectively.

The boundedness of G_h allows us to establish in a very simple way the a priori bound for the error $\|(\sigma, u) - (\sigma_h, u_h)\|_{H\times Q}$, which is known as Céa's estimate.

Theorem 4.23. *Under the same assumptions of Theorems* 4.19 *and* 4.22, *one has*

$$\|(\sigma, u) - (\sigma_h, u_h)\|_{H\times Q} \leqslant \big(1 + \|G_h\|\big) \inf_{(\zeta_h, w_h)\in H_h\times Q_h} \|(\sigma, u) - (\zeta_h, w_h)\|_{H\times Q}.$$

Proof. It suffices to observe, using (4.93) and (4.94), that

$$(\sigma, u) - (\sigma_h, u_h) = (I - G_h)\Big((\sigma, u) - (\zeta_h, w_h)\Big) \qquad \forall\, (\zeta_h, w_h) \in H_h \times Q_h,$$

where I is the identity operator. □

Naturally, in order to conclude the convergence of the Galerkin scheme, that is

$$\lim_{h\to 0} \|(\sigma, u) - (\sigma_h, u_h)\|_{H\times Q} = 0,$$

it will be required that $\|G_h\|$ be independent of h, which translates into requiring that all the constants involved, including the operator norms and the constants of the discrete inf-sup conditions, be independent of the subspace $H_h \times Q_h$. In fact, this need for independence of h is observed much more clearly when, instead of deducing Céa's estimate using the Galerkin projector G_h, it is obtained by analyzing individually each of the errors $\|\sigma - \sigma_h\|_H$ and $\|u - u_h\|_Q$. More precisely, one has the following theorem.

Theorem 4.24. *Under the same assumptions and notations of Theorems* 4.19 *and* 4.22, *there hold*

$$\begin{aligned}\|\sigma - \sigma_h\|_H \leqslant{}& \left(1 + \frac{\|\mathbf{A}\|}{\alpha_h}\right)\left(1 + \frac{\|\mathbf{B}\|}{\beta_h}\right) \inf_{\zeta_h\in H_h} \|\sigma - \zeta_h\|_H \\ &+ \frac{\|\mathbf{B}\|}{\alpha_h} \inf_{w_h\in Q_h} \|u - w_h\|_Q,\end{aligned}$$

and

$$\begin{aligned}\|u - u_h\|_Q \leqslant{}& \frac{\|\mathbf{A}\|}{\beta_h}\left(1 + \frac{\|\mathbf{A}\|}{\alpha_h}\right)\left(1 + \frac{\|\mathbf{B}\|}{\beta_h}\right) \inf_{\zeta_h\in H_h} \|\sigma - \zeta_h\|_H \\ &+ \left(1 + \frac{\|\mathbf{B}\|}{\beta_h} + \frac{\|\mathbf{A}\|\,\|\mathbf{B}\|}{\alpha_h\,\beta_h}\right) \inf_{w_h\in Q_h} \|u - w_h\|_Q.\end{aligned}$$

Proof. We begin by defining the set

$$V_h^g := \{\tau_h \in H_h : \quad b(\tau_h, v_h) = G(v_h) \quad \forall\, v_h \in Q_h\}$$

and observing, according to the second equation of (4.86), that $\sigma_h \in V_h^g$ and that, clearly, $(\sigma_h - \tau_h^g) \in V_h$ for all $\tau_h^g \in V_h^g$. The next goal is to bound $\|\sigma - \sigma_h\|_H$ in terms of $\mathrm{dist}(u, Q_h)$ and $\mathrm{dist}(\sigma, V_h^g)$. To this end, we first apply the triangular inequality and obtain

$$\|\sigma - \sigma_h\|_H \leqslant \|\sigma - \tau_h^g\|_H + \|\sigma_h - \tau_h^g\|_H \qquad \forall\, \tau_h^g \in V_h. \tag{4.95}$$

In turn, by virtue of the discrete inf-sup condition (4.89), which is equivalent to condition i) of Theorem 4.22, we have

$$\alpha_h \, \|\sigma_h - \tau_h^g\|_H \leqslant \sup_{\substack{\tau_h \in V_h \\ \tau_h \neq 0}} \frac{a(\sigma_h - \tau_h^g, \tau_h)}{\|\tau_h\|_H}. \tag{4.96}$$

Now, adding and subtracting σ, and using from the first equations of (4.69) and (4.87) that

$$a(\sigma - \sigma_h, \tau_h) + b(\tau_h, u - u_h) = 0 \qquad \forall\, \tau_h \in H_h, \tag{4.97}$$

and then adding and subtracting $w_h \in Q_h$, we deduce for each $\tau_h \in V_h$ that

$$\begin{aligned} a(\sigma_h - \tau_h^g, \tau_h) &= a(\sigma_h - \sigma, \tau_h) + a(\sigma - \tau_h^g, \tau_h) \\ &= b(\tau_h, u - u_h) + a(\sigma - \tau_h^g, \tau_h) \\ &= b(\tau_h, u - w_h) + b(\tau_h, w_h - u_h) + a(\sigma - \tau_h^g, \tau_h) \\ &= b(\tau_h, u - w_h) + a(\sigma - \tau_h^g, \tau_h) \qquad \forall\, w_h \in Q_h\,. \end{aligned} \tag{4.98}$$

Notice here that the term $b(\tau_h, w_h - u_h)$ vanishes since $\tau_h \in V_h$. Then, replacing the previous expression into (4.96) and observing from (4.70) and (4.71) that $\|\mathbf{A}\|$ and $\|\mathbf{B}\|$ are bounding constants for a and b, respectively, it follows that

$$\begin{aligned} \alpha_h \, \|\sigma_h - \tau_h^g\|_H &\leqslant \sup_{\substack{\tau_h \in V_h \\ \tau_h \neq 0}} \frac{b(\tau_h, u - w_h) + a(\sigma - \tau_h^g, \tau_h)}{\|\tau_h\|_H} \\ &\leqslant \|\mathbf{B}\| \, \|u - w_h\|_Q + \|\mathbf{A}\| \, \|\sigma - \tau_h^g\|_H \qquad \forall\, w_h \in Q_h, \end{aligned}$$

from which

$$\|\sigma_h - \tau_h^g\|_H \leqslant \frac{\|\mathbf{B}\|}{\alpha_h} \operatorname{dist}(u, Q_h) + \frac{\|\mathbf{A}\|}{\alpha_h} \|\sigma - \tau_h^g\|_H \qquad \forall\, \tau_h^g \in V_h^g.$$

Thus, (4.95) and the previous estimate directly imply

$$\|\sigma - \sigma_h\|_H \leqslant \left(1 + \frac{\|\mathbf{A}\|}{\alpha_h}\right) \operatorname{dist}(\sigma, V_h^g) + \frac{\|\mathbf{B}\|}{\alpha_h} \operatorname{dist}(\mathrm{u}, \mathrm{Q}_h). \tag{4.99}$$

Having established (4.99), we now need to bound $\operatorname{dist}(\sigma, V_h^g)$ in order to obtain the a priori estimate for the error $\|\sigma - \sigma_h\|_H$. To this end, we note that the application of Lemma 4.18 to the present finite-dimensional context implies that the discrete inf-sup condition for b given by hypothesis ii) of Theorem 4.22 (cf. (4.88)) is equivalent, in particular, to the fact that $\mathbf{B}_h$ is a bijection from $V_h^{\perp}$ onto Q_h and

$$\|\mathbf{B}_h(\tau_h)\|_Q \geqslant \beta_h \, \|\tau_h\|_H \qquad \forall\, \tau_h \in V_h^{\perp}. \tag{4.100}$$

We remark that the orthogonal complement of V_h is with respect to the space H_h, that is $H_h = V_h \oplus V_h^{\perp}$. Then, given $\zeta_h \in H_h$ there exists a unique $\bar{\zeta}_h \in V_h^{\perp}$ such that

$$\mathbf{B}_h(\bar{\zeta}_h) = \mathbf{B}_h(\sigma_h - \zeta_h),$$

that is

$$b(\bar{\zeta}_h, w_h) = b(\sigma_h - \zeta_h, w_h) \qquad \forall\, w_h \in Q_h\,,$$

and moreover, according to (4.100),

$$\|\bar{\zeta}_h\|_H \leqslant \frac{1}{\beta_h}\, \|\mathbf{B}_h(\sigma_h - \zeta_h)\|_Q.$$

Since $b(\sigma_h - \zeta_h, w_h) = b(\sigma - \zeta_h, w_h)$ for all $w_h \in Q_h$, which follows from the second equations of (4.69) and (4.87), we obtain

$$\begin{aligned}\|\mathbf{B}_h(\sigma_h - \zeta_h)\|_Q &= \sup_{\substack{w_h \in Q_h \\ w_h \neq 0}} \frac{b(\sigma_h - \zeta_h, w_h)}{\|w_h\|_Q} \\ &= \sup_{\substack{w_h \in Q_h \\ w_h \neq 0}} \frac{b(\sigma - \zeta_h, w_h)}{\|w_h\|_Q} \leqslant \|\mathbf{B}\|\, \|\sigma - \zeta_h\|_H\,,\end{aligned}$$

which yields

$$\|\bar{\zeta}_h\|_H \leqslant \frac{\|\mathbf{B}\|}{\beta_h}\, \|\sigma - \zeta_h\|_H\,. \tag{4.101}$$

In turn, when introducing $\bar{\zeta}_h \in V_h^{\perp}$ we observe that $\zeta_h + \bar{\zeta}_h \in V_h^g$, and therefore it follows easily, also using (4.101), that

$$\begin{aligned}\operatorname{dist}(\sigma, V_h^g) &\leqslant \|\sigma - (\zeta_h + \bar{\zeta}_h)\|_H \leqslant \|\sigma - \zeta_h\|_H + \|\bar{\zeta}_h\|_H \\ &\leqslant \left(1 + \frac{\|\mathbf{B}\|}{\beta_h}\right) \|\sigma - \zeta_h\|_H \qquad \forall\, \zeta_h \in H_h\,,\end{aligned}$$

that is

$$\operatorname{dist}(\sigma, V_h^g) \leqslant \left(1 + \frac{\|\mathbf{B}\|}{\beta_h}\right) \operatorname{dist}(\sigma, H_h). \tag{4.102}$$

In this way, (4.99) and (4.102) give us Céa's estimate for $\|\sigma - \sigma_h\|_H$. On the other hand, applying again the triangular inequality we have

$$\|u - u_h\|_Q \leqslant \|u - w_h\|_Q + \|u_h - w_h\|_Q \quad \forall\, w_h \in Q_h. \tag{4.103}$$

Then, using the discrete inf-sup condition for b, subtracting and adding u, and employing the identity (4.97), we obtain

$$\begin{aligned}\beta_h \|u_h - w_h\|_Q &\leqslant \sup_{\substack{\tau_h \in H_h \\ \tau_h \neq 0}} \frac{b(\tau_h, u_h - w_h)}{\|\tau_h\|_H} = \sup_{\substack{\tau_h \in H_h \\ \tau_h \neq 0}} \frac{b(\tau_h, u_h - u) + b(\tau_h, u - w_h)}{\|\tau_h\|_H} \\ &= \sup_{\substack{\tau_h \in H_h \\ \tau_h \neq 0}} \frac{a(\sigma - \sigma_h, \tau_h) + b(\tau_h, u - w_h)}{\|\tau_h\|_H} \leqslant \|\mathbf{A}\|\, \|\sigma - \sigma_h\|_H + \|\mathbf{B}\|\, \|u - w_h\|_Q\end{aligned}$$

for all $w_h \in Q_h$, so that replacing this estimate into (4.103), we conclude directly that

$$\|u - u_h\|_Q \leqslant \left(1 + \frac{\|\mathbf{B}\|}{\beta_h}\right) \operatorname{dist}(u, Q_h) + \frac{\|\mathbf{A}\|}{\beta_h}\, \|\sigma - \sigma_h\|_H. \tag{4.104}$$

Finally, (4.104) and the a priori bound for the error $\|\sigma - \sigma_h\|_H$ imply Céa's estimate for $\|u - u_h\|_Q$ and complete the proof.

□

It is interesting to observe here that if $V_h \subseteq V$, then the expression $b(\tau_h, u - u_h)$ vanishes for all $\tau_h \in V_h$, and therefore it is not necessary to add and subtract $w_h \in Q_h$ in (4.98). In fact, in such case (4.99) reduces simply to

$$\|\sigma - \sigma_h\|_H \leqslant \left(1 + \frac{\|\mathbf{A}\|}{\alpha_h}\right) \operatorname{dist}(\sigma, V_h^g),$$

which, together with (4.102), yields

$$\|\sigma - \sigma_h\|_H \leqslant \left(1 + \frac{\|\mathbf{A}\|}{\alpha_h}\right) \left(1 + \frac{\|\mathbf{B}\|}{\beta_h}\right) \operatorname{dist}(\sigma, H_h).$$

In other words, when the discrete kernel V_h of b is contained in its continuous kernel V, the Céa estimate for $\|\sigma - \sigma_h\|_H$ depends only on $\operatorname{dist}(\sigma, H_h)$ and not on $\operatorname{dist}(u, Q_h)$.

On the other hand, let us also note that the discrete inf-sup condition for b is crucial for the uniqueness of u_h. Indeed, in analogy with the continuous case, we know that this hypothesis can be rewritten as

$$\|\mathbf{B}_h^*(v_h)\|_H \geqslant \beta_h \, \|v_h\|_Q \quad \forall\, v_h \in Q_h,$$

which is equivalent to the injectivity of $\mathbf{B}_h^*$ (since in finite dimension all operator ranges are certainly closed). Therefore, if this condition does not hold, there necessarily exists $w \in Q_h, w \neq 0$, such that $\mathbf{B}_h^*(w) = 0$, and consequently, given any solution $(\sigma_h, u_h) \in H_h \times Q_h$ of the Galerkin scheme (4.87), one has that $(\sigma_h, u_h + c\,w)$ is also a solution for all $c \in \mathbb{R}$.

In concluding this section, it should be emphasized that the subspaces H_h and Q_h that allow us to define the Galerkin scheme (4.87) cannot be chosen arbitrarily, since they must obviously satisfy the various conditions stated in Theorem 4.22. In fact, the most restrictive one is the discrete inf-sup condition for b (cf. (4.88)). In this regard, and in addition to the classical Fortin Lemma (cf. [24] and Exercise 4.15), it is important to note that the occurrence of the continuous and discrete inf-sup conditions has been recently characterized on product spaces in [86].

4.5 Traces and Green's identities

In this section we present some results on traces, integration by parts formulas and Green's identities for certain Sobolev spaces, and in particular for $\mathbf{H}(\operatorname{div}; \Omega)$. For this purpose, given a bounded open set $\Omega \subset \mathbb{R}^n$ with Lipschitz-continuous boundary Γ, we first recall from (1.26) that the Sobolev space of order 1 is defined as

$$\mathrm{H}^1(\Omega) := \left\{ v \in \mathrm{L}^2(\Omega) : \quad \frac{\partial v}{\partial x_i} \in \mathrm{L}^2(\Omega) \quad \forall\, i \in \{1, \ldots, n\} \right\},$$

where the condition $\dfrac{\partial v}{\partial x_i}$ to $\mathrm{L}^2(\Omega)$ must be understood in the distributional sense, which means that there exists $z \in \mathrm{L}^2(\Omega)$ such that

$$-\int_\Omega v \, \frac{\partial \varphi}{\partial x_i} = \int_\Omega z\, \varphi \qquad \forall\, \varphi \in C_0^\infty(\Omega).$$

Moreover, $\mathcal{D}(\bar{\Omega})$ (or $C^\infty(\bar{\Omega})$) denotes the space of restrictions to Ω of functions belonging to $C_0^\infty(\mathbb{R}^n)$.

4.5.1 Traces of $\mathrm{H}^1(\Omega)$

The classical trace theorem in $\mathrm{H}^1(\Omega)$ requires the preliminary result to be stated next. A less general version of it, which is valid for a 1-regular open subset of $\mathbb{R}^n$, is provided later on by Theorem 10.15.

Theorem 4.25 (Trace inequality). *Let Ω be a bounded open set in $\mathbb{R}^n$ with Lipschitz-continuous boundary Γ, and let $\gamma_0 : \mathcal{D}(\bar{\Omega}) \to \mathrm{L}^2(\Gamma)$ be the map defined by*

$$\gamma_0(\varphi) := \varphi|_\Gamma \qquad \forall\, \varphi \in \mathcal{D}(\bar{\Omega}).$$

Then, there exists $C > 0$ such that

$$\|\gamma_0(\varphi)\|_{0,\Gamma} \leqslant C\,\|\varphi\|_{1,\Omega} \qquad \forall\, \varphi \in \mathcal{D}(\bar{\Omega}). \tag{4.105}$$

Theorem 4.26 (Trace Theorem in $\mathrm{H}^1(\Omega)$). *Let Ω be a bounded open set in $\mathbb{R}^n$ with Lipschitz-continuous boundary Γ. Then, the mapping $\gamma_0 : \mathcal{D}(\bar{\Omega}) \to \mathrm{L}^2(\Gamma)$ extends by continuity and density to a linear and bounded mapping $\gamma_0 : \mathrm{H}^1(\Omega) \to \mathrm{L}^2(\Gamma)$, such that $\gamma_0(\varphi) := \varphi|_\Gamma$ for all $\varphi \in \mathcal{D}(\bar{\Omega})$.*

Proof. We use the trace inequality (cf. Theorem 4.25) and the fact that $\mathcal{D}(\bar{\Omega})$ is dense in $\mathrm{H}^1(\Omega)$ with respect to the norm $\|\cdot\|_{1,\Omega}$. Indeed, given $v \in \mathrm{H}^1(\Omega)$, consider a sequence $\{\varphi_j\}_{j\in\mathbb{N}} \subseteq \mathcal{D}(\bar{\Omega})$ such that

$$\|\varphi_j - v\|_{1,\Omega} \to 0 \quad \text{as} \quad j \to +\infty.$$

Then, using the linearity of γ_0 and applying the trace inequality (4.105) to $(\varphi_j - \varphi_k)$, $j, k \in \mathbb{N}$, we obtain

$$\|\gamma_0(\varphi_j) - \gamma_0(\varphi_k)\|_{0,\Gamma} = \|\gamma_0(\varphi_j - \varphi_k)\|_{0,\Gamma} \leqslant C\,\|\varphi_j - \varphi_k\|_{1,\Omega},$$

from which it follows that $\{\gamma_0(\varphi_j)\}_{j\in\mathbb{N}}$ is a Cauchy sequence in $\mathrm{L}^2(\Gamma)$. Since $\mathrm{L}^2(\Gamma)$ is complete, there exists $\varphi \in \mathrm{L}^2(\Gamma)$ such that

$$\|\gamma_0(\varphi_j) - \varphi\|_{0,\Gamma} \to 0 \quad \text{as} \quad j \to +\infty,$$

which induces the definition $\gamma_0(v) := \varphi$. Nevertheless, we must ensure that there is no ambiguity in the function φ, in the sense that it is independent of the chosen sequence. Indeed, consider now a second sequence $\{\widetilde{\varphi}_j\}_{j\in\mathbb{N}} \subseteq \mathcal{D}(\bar{\Omega})$ such that

$$\|\widetilde{\varphi}_j - v\|_{1,\Omega} \to 0 \quad \text{as} \quad j \to +\infty.$$

It follows that

$$\|\gamma_0(\widetilde{\varphi}_j) - \gamma_0(\varphi_j)\|_{0,\Gamma} = \|\gamma_0(\widetilde{\varphi}_j - \varphi_j)\|_{0,\Gamma} \leqslant C\,\|\widetilde{\varphi}_j - \varphi_j\|_{1,\Omega} \to 0 \text{ as } j \to +\infty,$$

which shows that $\{\gamma_0(\widetilde{\varphi}_j)\}_{j\in\mathbb{N}}$ and $\{\gamma_0(\varphi_j)\}_{j\in\mathbb{N}}$ have the same limit φ in $\mathrm{L}^2(\Gamma)$. This indicates that γ_0 is well defined in $\mathrm{H}^1(\Omega)$. Moreover, the same argument on the independence of the sequence easily shows that γ_0 is linear. On the other hand, taking the limit as $j \to +\infty$ in the inequality

$$\|\gamma_0(\varphi_j)\|_{0,\Gamma} \leqslant C\,\|\varphi_j\|_{1,\Omega},$$

yields

$$\|\varphi\|_{0,\Gamma} \leqslant C\,\|v\|_{1,\Omega},$$

which proves that γ_0 is bounded.

□

4.5.2 The space $\mathrm{H}^{1/2}(\Gamma)$

We now denote by $\mathrm{H}^{1/2}(\Gamma)$, and call it the trace space on the boundary, the range of γ_0, that is

$$\mathrm{H}^{1/2}(\Gamma) \;:=\; \gamma_0(\mathrm{H}^1(\Omega)),$$

which is endowed with the norm

$$\|\varphi\|_{1/2,\Gamma} := \inf\left\{\|v\|_{1,\Omega}: \quad v \in \mathrm{H}^1(\Omega) \text{ such that } \gamma_0(v) = \varphi\right\} \qquad \forall\, \varphi \in \mathrm{H}^{1/2}(\Gamma). \quad (4.106)$$

It is immediate from the above definition that

$$\|\gamma_0(w)\|_{1/2,\Gamma} \leqslant \|w\|_{1,\Omega} \qquad \forall\, w \in \mathrm{H}^1(\Omega), \qquad (4.107)$$

which shows that the operator $\gamma_0 : \mathrm{H}^1(\Omega) \to \mathrm{H}^{1/2}(\Gamma)$, besides being linear and surjective, is bounded. Moreover, it can be proved that $N(\gamma_0)$, the null space of γ_0, is given by the closure of $C_0^\infty(\Omega)$ in $\mathrm{H}^1(\Omega)$ with respect to the $\|\cdot\|_{1,\Omega}$ norm, that is

$$N(\gamma_0) \;=\; \mathrm{H}_0^1(\Omega) \;:=\; \overline{C_0^\infty(\Omega)}^{\|\cdot\|_{1,\Omega}}.$$

Equivalently, $v \in N(\gamma_0) = \mathrm{H}_0^1(\Omega)$ if and only if $v \in \mathrm{H}^1(\Omega)$ and there exists a sequence $\{\varphi_j\}_{j\in\mathrm{N}} \subseteq C_0^\infty(\Omega)$ such that $\|\varphi_j - v\|_{1,\Omega} \to 0$ as $j \to +\infty$. Then, according to the Orthogonal Decomposition Theorem (cf. Theorem 2.10), there holds

$$\mathrm{H}^1(\Omega) \;=\; \mathrm{H}_0^1(\Omega) \oplus \mathrm{H}_0^1(\Omega)^\perp, \qquad (4.108)$$

from which we conclude that $\widetilde{\gamma}_0 := \gamma_0|_{\mathrm{H}_0^1(\Omega)^\perp} : \mathrm{H}_0^1(\Omega)^\perp \to \mathrm{H}^{1/2}(\Gamma)$ is a linear and bounded bijection. It follows that $\widetilde{\gamma}_0^{-1} : \mathrm{H}^{1/2}(\Gamma) \to \mathrm{H}_0^1(\Omega)^\perp$ is also a linear bijection and that $\widetilde{\gamma}_0^{-1}$ is a right inverse of γ_0, that is (see Figure 4.3)

$$\gamma_0\, \widetilde{\gamma}_0^{-1} \;=\; I : \mathrm{H}^{1/2}(\Gamma) \to \mathrm{H}^{1/2}(\Gamma).$$

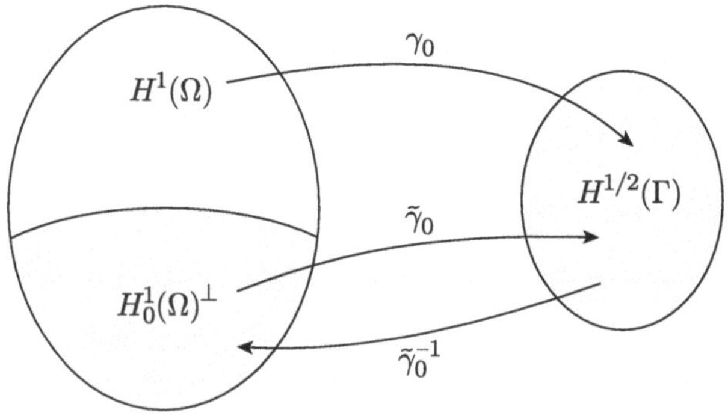

Fig. 4.3. *Illustrative diagram of γ_0 and $\widetilde{\gamma}_0$*

Furthermore, the following lemma implies the boundedness of $\widetilde{\gamma}_0^{-1}$, which, in turn, allows us to conclude the completeness of $(\mathrm{H}^{1/2}(\Gamma), \|\cdot\|_{1/2,\Gamma})$.

Lemma 4.27. *According to the definition given in* (4.106), *there holds*

$$\|\varphi\|_{1/2,\Gamma} = \|\tilde{\gamma}_0^{-1}(\varphi)\|_{1,\Omega} \qquad \forall\, \varphi \in \mathrm{H}^{1/2}(\Gamma)\,. \tag{4.109}$$

Proof. Given $v \in \mathrm{H}^1(\Omega)$, consider its unique decomposition $v = v_0 + v^\perp$, with $v_0 \in \mathrm{H}_0^1(\Omega)$ and $v^\perp \in \mathrm{H}_0^1(\Omega)^\perp$ (cf. (4.108)). If we also assume that $\gamma_0(v) = \varphi$, then it follows that $\varphi = \gamma_0(v^\perp)$, which means that $v^\perp = \tilde{\gamma}_0^{-1}(\varphi)$. Accordingly, we have

$$\inf\left\{\|v\|_{1,\Omega}: \quad v \in \mathrm{H}^1(\Omega) \text{ such that } \gamma_0(v) = \varphi\right\} = \inf_{v_0 \in \mathrm{H}_0^1(\Omega)} \|v_0 + \tilde{\gamma}_0^{-1}(\varphi)\|_{1,\Omega}\,,$$

and since $\tilde{\gamma}_0^{-1}(\varphi) \in \mathrm{H}_0^1(\Omega)^\perp$, this infimum is attained at $v_0 = 0$, which implies (4.109). □

Note that (4.109) suggests defining the inner product on $\mathrm{H}^{1/2}(\Gamma)$ by

$$\langle\varphi,\psi\rangle_{1/2,\Gamma} := \langle\tilde{\gamma}_0^{-1}(\varphi),\tilde{\gamma}_0^{-1}(\psi)\rangle_{1,\Omega} \qquad \forall\, \varphi,\, \psi \in \mathrm{H}^{1/2}(\Gamma)\,, \tag{4.110}$$

whose induced norm is precisely $\|\cdot\|_{1/2,\Gamma}$. Therefore, we obtain the following result.

Corollary 4.28. $(\mathrm{H}^{1/2}(\Gamma),\langle\cdot,\cdot\rangle_{1/2,\Gamma})$ *is a Hilbert space.*

Proof. This follows from identity (4.109) and the fact that $(\mathrm{H}_0^1(\Omega)^\perp, \|\cdot\|_{1,\Omega})$, being a closed subspace of the Hilbert space $\mathrm{H}^1(\Omega)$, is Hilbert as well. □

4.5.3 Integration by parts and Green's identity

The density of $\mathcal{D}(\bar{\Omega})$ in $\mathrm{H}^1(\Omega)$ and the Trace Theorem allow us to prove the following result.

Theorem 4.29 (Integration by parts formula). *Let Ω be a bounded open set of $\mathbb{R}^n$ with Lipschitz-continuous boundary Γ. Then, for all $v,\, w \in \mathrm{H}^1(\Omega)$ we have*

$$\int_\Omega v\,\frac{\partial w}{\partial x_i} = -\int_\Omega w\,\frac{\partial v}{\partial x_i} + \int_\Gamma \gamma_0(v)\,\gamma_0(w)\,\nu_i \qquad \forall\, i \in \{1,2,...,n\}\,, \tag{4.111}$$

where ν_i is the i-th component of the normal vector $\boldsymbol{\nu}$.

Proof. Let $v,\, w \in \mathrm{H}^1(\Omega)$, and consider sequences $\{\varphi_j\}_{j\in\mathbb{N}}, \{\psi_j\}_{j\in\mathbb{N}} \subseteq \mathcal{D}(\bar{\Omega})$ such that

$$\|\varphi_j - v\|_{1,\Omega} \to 0 \quad \text{and} \quad \|\psi_j - w\|_{1,\Omega} \to 0 \quad \text{as} \quad j \to +\infty\,. \tag{4.112}$$

According to the Trace Theorem we know that

$$\|\gamma_0(\varphi_j) - \gamma_0(v)\|_{0,\Gamma} \to 0 \quad \text{and} \quad \|\gamma_0(\psi_j) - \gamma_0(w)\|_{0,\Gamma} \to 0 \quad \text{as} \quad j \to +\infty\,. \tag{4.113}$$

Now, applying the classical integration by parts formula for smooth functions (which follows from Gauss' Divergence Theorem), we obtain

$$\int_\Omega \varphi_j\,\frac{\partial \psi_j}{\partial x_i} = -\int_\Omega \psi_j\,\frac{\partial \varphi_j}{\partial x_i} + \int_\Gamma \gamma_0(\varphi_j)\,\gamma_0(\psi_j)\,\nu_i \qquad \forall\, i \in \{1,2,...,n\}\,. \tag{4.114}$$

Then, taking the limit in (4.114) as $j \to +\infty$, leads to (4.111). Indeed, by adding and subtracting a convenient term, applying the Cauchy–Schwarz inequality, and noting that the sequence $\left\{\dfrac{\partial \psi_j}{\partial x_i}\right\}_{j\in\mathbb{N}}$ is bounded in $\mathrm{L}^2(\Omega)$ (because $\{\psi_j\}_{j\in\mathbb{N}}$ is convergent in $\mathrm{H}^1(\Omega)$), we deduce that

$$\begin{aligned}\left|\int_\Omega \varphi_j \frac{\partial \psi_j}{\partial x_i} - \int_\Omega v \frac{\partial w}{\partial x_i}\right| &\leqslant \left|\int_\Omega (\varphi_j - v)\frac{\partial \psi_j}{\partial x_i}\right| + \left|\int_\Omega v\left(\frac{\partial \psi_j}{\partial x_i} - \frac{\partial w}{\partial x_i}\right)\right| \\ &\leqslant \|\varphi_j - v\|_{0,\Omega}\left\|\frac{\partial \psi_j}{\partial x_i}\right\|_{0,\Omega} + \|v\|_{0,\Omega}\left\|\frac{\partial \psi_j}{\partial x_i} - \frac{\partial w}{\partial x_i}\right\|_{0,\Omega} \\ &\leqslant C\,\|\varphi_j - v\|_{0,\Omega} + \|v\|_{0,\Omega}\,|\psi_j - w|_{1,\Omega},\end{aligned}$$

from which, thanks to (4.112), it follows that

$$\lim_{j\to+\infty}\int_\Omega \varphi_j \frac{\partial \psi_j}{\partial x_i} = \int_\Omega v\frac{\partial w}{\partial x_i}.$$

Similarly, one proves that

$$\lim_{j\to+\infty}\int_\Omega \psi_j \frac{\partial \varphi_j}{\partial x_i} = \int_\Omega w\frac{\partial v}{\partial x_i}.$$

Finally, arguments similar to the above, this time using (4.113), imply that

$$\lim_{j\to+\infty}\int_\Gamma \gamma_0(\varphi_j)\,\gamma_0(\psi_j)\,\nu_i = \int_\Gamma \gamma_0(v)\,\gamma_0(w)\,\nu_i,$$

which completes the proof.

□

A consequence of the previous theorem, which requires first defining the space

$$\mathrm{H}^2(\Omega) := \left\{v\in \mathrm{H}^1(\Omega): \quad \frac{\partial^2 v}{\partial x_i\, x_j}\in \mathrm{L}^2(\Omega) \quad \forall\, i,\, j\in\{1,2,...,n\}\right\},$$

is given as follows.

Corollary 4.30 (Green's identity). *Let Ω be a bounded open set of $\mathbb{R}^n$ with Lipschitz-continuous boundary Γ. Then there holds*

$$\int_\Omega v\,\Delta u = -\int_\Omega \nabla u\cdot\nabla v + \int_\Gamma \gamma_0(v)\,\gamma_0(\nabla u)\cdot\boldsymbol{\nu} \quad \forall\, v\in \mathrm{H}^1(\Omega),\ \forall\, u\in \mathrm{H}^2(\Omega), \tag{4.115}$$

where $\gamma_0(\nabla u)$ is the componentwise application of γ_0 to ∇u.

Proof. Let $v\in \mathrm{H}^1(\Omega)$ and $u\in \mathrm{H}^2(\Omega)$. Then, applying (4.111) with v and $w := \dfrac{\partial u}{\partial x_i}$, we obtain

$$\int_\Omega v\frac{\partial^2 u}{\partial x_i^2} = -\int_\Omega \frac{\partial u}{\partial x_i}\frac{\partial v}{\partial x_i} + \int_\Gamma \gamma_0(v)\,\gamma_0\left(\frac{\partial u}{\partial x_i}\right)\nu_i \qquad \forall\, i\in\{1,2,...,n\},$$

from which, summing over i, we arrive at (4.115).

□

Note that the expression $\gamma_0(\nabla u)\cdot\boldsymbol{\nu}$ is usually written as $\dfrac{\partial u}{\partial \boldsymbol{\nu}}$, and is called the normal derivative of u.

4.5.4 Normal traces of $\mathbf{H}(\mathrm{div};\Omega)$

We begin this section by recalling (see, for instance, Exercise 2.9) that, given a bounded open set $\Omega \subset \mathbb{R}^n$ with sufficiently smooth boundary Γ, the space

$$\mathbf{H}(\mathrm{div};\Omega) := \left\{ \boldsymbol{\tau} := (\tau_1, \ldots, \tau_n) \in \mathbf{L}^2(\Omega) : \quad \mathrm{div}(\boldsymbol{\tau}) := \sum_{i=1}^{n} \frac{\partial \tau_i}{\partial x_i} \in \mathrm{L}^2(\Omega) \right\}, \tag{4.116}$$

is Hilbert with the inner product

$$\langle \boldsymbol{\tau}, \boldsymbol{\zeta} \rangle_{\mathrm{div};\Omega} := \int_\Omega \boldsymbol{\tau} \cdot \boldsymbol{\zeta} + \int_\Omega \mathrm{div}(\boldsymbol{\tau})\, \mathrm{div}(\boldsymbol{\zeta}) \qquad \forall\, \boldsymbol{\tau},\, \boldsymbol{\zeta} \in \mathbf{H}(\mathrm{div};\Omega),$$

whose induced norm reduces to

$$\|\boldsymbol{\tau}\|_{\mathrm{div};\Omega} := \left\{ \|\boldsymbol{\tau}\|_{0,\Omega}^2 + \|\mathrm{div}(\boldsymbol{\tau})\|_{0,\Omega}^2 \right\}^{1/2} \qquad \forall\, \boldsymbol{\tau} \in \mathbf{H}(\mathrm{div};\Omega). \tag{4.117}$$

In what follows, we prove that functions in $\mathbf{H}(\mathrm{div};\Omega)$ possess normal traces on Γ. For this purpose, from now on we denote the dual of $\mathrm{H}^{1/2}(\Gamma)$ by $\mathrm{H}^{-1/2}(\Gamma)$, and its respective norm by $\|\cdot\|_{-1/2,\Gamma}$. Moreover, given a functional $\psi \in \mathrm{H}^{-1/2}(\Gamma)$, its evaluation on an element φ of $\mathrm{H}^{1/2}(\Gamma)$ is usually written as $\langle \psi, \varphi \rangle$, that is

$$\langle \psi, \varphi \rangle := \psi(\varphi) \qquad \forall\, \varphi \in \mathrm{H}^{1/2}(\Gamma),$$

which is why $\langle \cdot, \cdot \rangle$ is here called the dual pairing of $\mathrm{H}^{-1/2}(\Gamma)$ and $\mathrm{H}^{1/2}(\Gamma)$. In turn, bear in mind that $\langle \cdot, \cdot \rangle_{0,\Gamma}$ denotes the inner product of $\mathrm{L}^2(\Gamma)$.

Theorem 4.31 (Normal trace of $\mathbf{H}(\mathrm{div};\Omega)$). *Let Ω be a bounded open set of $\mathbb{R}^n$ with Lipschitz-continuous boundary Γ. Then, there exists a surjective operator $\gamma_{\boldsymbol{\nu}} \in \mathcal{L}\big(\mathbf{H}(\mathrm{div};\Omega), \mathrm{H}^{-1/2}(\Gamma)\big)$ such that for each $\boldsymbol{\tau} \in \mathbf{H}^1(\Omega)$, $\gamma_{\boldsymbol{\nu}}(\boldsymbol{\tau})$ is identified, through $\langle \cdot, \cdot \rangle_{0,\Gamma}$, with $\gamma_0(\boldsymbol{\tau}) \cdot \boldsymbol{\nu}$, where $\gamma_0(\boldsymbol{\tau})$ is the componentwise application of γ_0 to $\boldsymbol{\tau}$.*

Proof. Let $\widetilde{\gamma}_0^{-1} : \mathrm{H}^{1/2}(\Gamma) \to \mathrm{H}_0^1(\Omega)^\perp$ be the right inverse of γ_0. Then, given $\boldsymbol{\tau} \in \mathbf{H}(\mathrm{div};\Omega)$, we define the linear functional $\gamma_{\boldsymbol{\nu}}(\boldsymbol{\tau}) : \mathrm{H}^{1/2}(\Gamma) \to \mathbb{R}$ as

$$\gamma_{\boldsymbol{\nu}}(\boldsymbol{\tau})(\varphi) := \int_\Omega \left\{ \boldsymbol{\tau} \cdot \nabla \widetilde{\gamma}_0^{-1}(\varphi) + \widetilde{\gamma}_0^{-1}(\varphi)\, \mathrm{div}(\boldsymbol{\tau}) \right\} \qquad \forall\, \varphi \in \mathrm{H}^{1/2}(\Gamma). \tag{4.118}$$

Note that the linearity of $\gamma_{\boldsymbol{\nu}}(\boldsymbol{\tau})$ follows from that of $\widetilde{\gamma}_0^{-1}$. Moreover, applying the Cauchy–Schwarz inequality, we obtain

$$\begin{aligned} |\, \gamma_{\boldsymbol{\nu}}(\boldsymbol{\tau})(\varphi)\, | &\leqslant \|\boldsymbol{\tau}\|_{0,\Omega} \, \|\nabla \widetilde{\gamma}_0^{-1}(\varphi)\|_{0,\Omega} + \|\widetilde{\gamma}_0^{-1}(\varphi)\|_{0,\Omega} \, \|\mathrm{div}(\boldsymbol{\tau})\|_{0,\Omega} \\ &\leqslant \|\boldsymbol{\tau}\|_{\mathrm{div};\Omega} \, \|\widetilde{\gamma}_0^{-1}(\varphi)\|_{1,\Omega} = \|\boldsymbol{\tau}\|_{\mathrm{div};\Omega} \, \|\varphi\|_{1/2,\Gamma} \quad \forall\, \varphi \in \mathrm{H}^{1/2}(\Gamma), \end{aligned}$$

which shows that $\gamma_{\boldsymbol{\nu}}(\boldsymbol{\tau})$ is bounded, and therefore belongs to $\mathrm{H}^{-1/2}(\Gamma)$, with

$$\|\gamma_{\boldsymbol{\nu}}(\boldsymbol{\tau})\|_{-1/2,\Gamma} \leqslant \|\boldsymbol{\tau}\|_{\mathrm{div};\Omega}. \tag{4.119}$$

The above analysis justifies the definition of the linear and bounded operator

$$\begin{aligned} \gamma_{\boldsymbol{\nu}} : \mathbf{H}(\mathrm{div};\Omega) &\to \mathrm{H}^{-1/2}(\Gamma) \\ \boldsymbol{\tau} &\to \gamma_{\boldsymbol{\nu}}(\boldsymbol{\tau}), \end{aligned}$$

which, by virtue of (4.119), satisfies $\|\gamma_{\boldsymbol{\nu}}\| \leqslant 1$. Now, if $\boldsymbol{\tau} := (\tau_1, \tau_2, ..., \tau_n)^{\mathrm{t}} \in \mathbf{H}^1(\Omega)$, the integration by parts formula (4.111) allows us to deduce that

$$\begin{aligned}\int_\Omega \widetilde{\gamma}_0^{-1}(\varphi)\,\mathrm{div}(\boldsymbol{\tau}) &= \sum_{i=1}^n \int_\Omega \widetilde{\gamma}_0^{-1}(\varphi)\,\frac{\partial \tau_i}{\partial x_i} = \sum_{i=1}^n \left\{ -\int_\Omega \tau_i \frac{\partial}{\partial x_i}\widetilde{\gamma}_0^{-1}(\varphi) + \int_\Gamma \varphi\,\gamma_0(\tau_i)\cdot\nu_i \right\} \\ &= -\int_\Omega \boldsymbol{\tau}\cdot\nabla\widetilde{\gamma}_0^{-1}(\varphi) + \int_\Gamma \varphi\,\gamma_0(\boldsymbol{\tau})\cdot\boldsymbol{\nu},\end{aligned}$$

so that in this particular case, with $\boldsymbol{\tau} \in \mathbf{H}^1(\Omega)$, (4.118) reduces to

$$\langle \gamma_{\boldsymbol{\nu}}(\boldsymbol{\tau}), \varphi\rangle = \int_\Gamma \gamma_0(\boldsymbol{\tau})\cdot\boldsymbol{\nu}\,\varphi = \langle \gamma_0(\boldsymbol{\tau})\cdot\boldsymbol{\nu}, \varphi\rangle_{0,\Gamma} \qquad \forall\,\varphi \in \mathrm{H}^{1/2}(\Gamma)\,, \tag{4.120}$$

where, as mentioned at the beginning of this chapter, $\langle\cdot,\cdot\rangle_{0,\Gamma}$ denotes the usual $\mathrm{L}^2(\Gamma)$ inner product. It remains to prove the surjectivity of the operator $\gamma_{\boldsymbol{\nu}}$. In other words, given $\psi \in \mathrm{H}^{-1/2}(\Gamma) := (\mathrm{H}^{1/2}(\Gamma))'$, we must show that there exists $\boldsymbol{\tau} \in \mathbf{H}(\mathrm{div};\Omega)$ such that $\psi = \gamma_{\boldsymbol{\nu}}(\boldsymbol{\tau})$. Indeed, let us define the subspace

$$\widetilde{\mathrm{H}}^1(\Omega) := \left\{ v \in \mathrm{H}^1(\Omega) : \quad \int_\Omega v = 0 \right\}, \tag{4.121}$$

and consider the following variational problem: Find $z \in \widetilde{\mathrm{H}}^1(\Omega)$ such that

$$\int_\Omega \nabla z\cdot\nabla w = F(w) \qquad \forall\, w \in \mathrm{H}^1(\Omega)\,, \tag{4.122}$$

where

$$F(w) := -\frac{\langle\psi, \gamma_0(1)\rangle}{|\Omega|}\int_\Omega w + \langle\psi, \gamma_0(w)\rangle \quad \forall\, w \in \mathrm{H}^1(\Omega)\,. \tag{4.123}$$

Note that F is linear (since the integral, ψ, and γ_0 are), and bounded because

$$|F(w)| \leqslant \frac{|\langle\psi, \gamma_0(1)\rangle|}{|\Omega|^{1/2}}\,\|w\|_{0,\Omega} + \|\psi\|_{-1/2,\Gamma}\,\|w\|_{1,\Omega} \leqslant C_\psi\,\|w\|_{1,\Omega} \quad \forall\, w \in \mathrm{H}^1(\Omega)\,,$$

where $C_\psi > 0$ is a constant depending on $\|\psi\|_{-1/2,\Gamma}$, $\|\gamma_0(1)\|_{1/2,\Gamma}$, and $|\Omega|$. Moreover, using the decomposition $\mathrm{H}^1(\Omega) = \widetilde{\mathrm{H}}^1(\Omega) \oplus \mathbb{R}$ and the fact that $F(1) = 0$, it is easily proved that (4.122) is equivalent to: Find $z \in \widetilde{\mathrm{H}}^1(\Omega)$ such that

$$\int_\Omega \nabla z\cdot\nabla w = F(w) \qquad \forall\, w \in \widetilde{\mathrm{H}}^1(\Omega)\,. \tag{4.124}$$

Then, since the norm $\|\cdot\|_{1,\Omega}$ and the seminorm $|\cdot|_{1,\Omega}$ of $\mathrm{H}^1(\Omega)$ are equivalent in $\widetilde{\mathrm{H}}^1(\Omega)$, namely, there exists a constant $\underline{C} > 0$ such that

$$\underline{C}\,\|w\|_{1,\Omega}^2 \leqslant |w|_{1,\Omega}^2 \qquad \forall\, w \in \widetilde{\mathrm{H}}^1(\Omega)\,, \tag{4.125}$$

which follows from the generalized Poincaré inequality (cf. Theorem 10.21 and Example 10.22, part 1), we may apply the Lax–Milgram Lemma (cf. Lemma 4.4) to conclude that (4.124) (and therefore (4.122)) has a unique solution $z \in \widetilde{\mathrm{H}}^1(\Omega)$. We then define $\boldsymbol{\tau} := \nabla z$ in Ω, which, at the first instance, belongs to $\mathbf{L}^2(\Omega)$. However, it follows from (4.122) and (4.123) that

$$\int_\Omega \boldsymbol{\tau} \cdot \nabla w \,=\, -\frac{\langle \psi, \gamma_0(1)\rangle}{|\Omega|} \int_\Omega w \qquad \forall\, w \in C_0^\infty(\Omega)\,,$$

which indicates that

$$\operatorname{div}(\boldsymbol{\tau}) \,=\, \frac{\langle \psi, \gamma_0(1)\rangle}{|\Omega|} \quad \text{in} \quad \mathcal{D}'(\Omega)\,, \tag{4.126}$$

and therefore $\boldsymbol{\tau} \in \mathbf{H}(\operatorname{div};\Omega)$. Thus, formulation (4.122) can be rewritten as

$$\int_\Omega \boldsymbol{\tau} \cdot \nabla w + \int_\Omega w \operatorname{div}(\boldsymbol{\tau}) \,=\, \langle \psi, \gamma_0(w)\rangle \qquad \forall\, w \in \mathrm{H}^1(\Omega)\,,$$

and taking in particular $w = \tilde{\gamma}_0^{-1}(\varphi)$, with $\varphi \in \mathrm{H}^{1/2}(\Gamma)$, we obtain

$$\langle \psi, \varphi\rangle \,=\, \int_\Omega \boldsymbol{\tau} \cdot \nabla \tilde{\gamma}_0^{-1}(\varphi) + \int_\Omega \tilde{\gamma}_0^{-1}(\varphi) \operatorname{div}(\boldsymbol{\tau}) \,=\, \langle \gamma_{\boldsymbol{\nu}}(\boldsymbol{\tau}), \varphi\rangle\,, \tag{4.127}$$

for all $\varphi \in \mathrm{H}^{1/2}(\Gamma)$, which proves that $\psi = \gamma_{\boldsymbol{\nu}}(\boldsymbol{\tau})$.

□

It is interesting to observe here, in virtue of (4.126) and (4.127), that the surjectivity of $\gamma_{\boldsymbol{\nu}}$ actually occurs, more precisely, from $\mathbf{H}(\operatorname{div}_{\mathbb{R}};\Omega)$ onto $\mathrm{H}^{-1/2}(\Gamma)$, where $\mathbf{H}(\operatorname{div}_{\mathbb{R}};\Omega)$ is the closed subspace of $\mathbf{H}(\operatorname{div};\Omega)$ given by those vectors with constant divergence, that is

$$\mathbf{H}(\operatorname{div}_{\mathbb{R}};\Omega) := \Big\{ \boldsymbol{\tau} \in \mathbf{H}(\operatorname{div};\Omega) : \quad \operatorname{div}(\boldsymbol{\tau}) \,=\, c \,\in\, \mathbb{R} \quad \text{in} \quad \mathcal{D}'(\Omega) \Big\}\,. \tag{4.128}$$

Thus, by defining $\hat{\gamma}_{\boldsymbol{\nu}} := \gamma_{\boldsymbol{\nu}}|_{\mathbf{H}(\operatorname{div}_{\mathbb{R}};\Omega)}$, we have that $\hat{\gamma}_{\boldsymbol{\nu}} : \mathbf{H}(\operatorname{div}_{\mathbb{R}};\Omega) \to \mathrm{H}^{-1/2}(\Gamma)$ is surjective, and therefore, introducing

$$\mathbf{H}_0(\operatorname{div}_{\mathbb{R}};\Omega) \,=\, N(\hat{\gamma}_{\boldsymbol{\nu}}) := \Big\{ \boldsymbol{\tau} \in \mathbf{H}(\operatorname{div}_{\mathbb{R}};\Omega) : \quad \hat{\gamma}_{\boldsymbol{\nu}}(\boldsymbol{\tau}) = 0 \Big\}\,,$$

it follows from the Orthogonal Decomposition Theorem (cf. Theorem 2.10) that

$$\mathbf{H}(\operatorname{div}_{\mathbb{R}};\Omega) \,=\, \mathbf{H}_0(\operatorname{div}_{\mathbb{R}};\Omega) \oplus \mathbf{H}_0^{\perp}(\operatorname{div}_{\mathbb{R}};\Omega)\,,$$

where

$$\mathbf{H}_0^{\perp}(\operatorname{div}_{\mathbb{R}};\Omega) := \Big\{ \boldsymbol{\zeta} \in \mathbf{H}(\operatorname{div}_{\mathbb{R}};\Omega) : \quad \langle \boldsymbol{\zeta}, \boldsymbol{\tau}\rangle_{\operatorname{div};\Omega} \,=\, 0 \quad \forall\, \boldsymbol{\tau} \in \mathbf{H}_0(\operatorname{div}_{\mathbb{R}};\Omega) \Big\}\,,$$

and therefore, denoting $\tilde{\gamma}_{\boldsymbol{\nu}} := \hat{\gamma}_{\boldsymbol{\nu}}|_{\mathbf{H}_0^{\perp}(\operatorname{div}_{\mathbb{R}};\Omega)}$, we conclude that

$$\tilde{\gamma}_{\boldsymbol{\nu}} : \mathbf{H}_0^{\perp}(\operatorname{div}_{\mathbb{R}};\Omega) \to \mathrm{H}^{-1/2}(\Gamma) \quad \text{is a bijection.} \tag{4.129}$$

Moreover, $\tilde{\gamma}_{\boldsymbol{\nu}}^{-1}$ is a right inverse of $\hat{\gamma}_{\boldsymbol{\nu}}$, that is, $\hat{\gamma}_{\boldsymbol{\nu}} \circ \tilde{\gamma}_{\boldsymbol{\nu}}^{-1} = I : \mathrm{H}^{-1/2}(\Gamma) \to \mathrm{H}^{-1/2}(\Gamma)$.

A less informative, though interesting, proof of the surjectivity of $\gamma_{\boldsymbol{\nu}}$, is provided later on in Exercise 4.47.

We now show that the functional $\gamma_{\boldsymbol{\nu}}(\boldsymbol{\tau})$ can be defined not only as indicated in (4.118), but also by using any $w \in \mathrm{H}^1(\Omega)$ such that $\gamma_0(w) = \varphi \in \mathrm{H}^{1/2}(\Gamma)$. Indeed, we have the following result

Lemma 4.32 (Green's identity in $\mathbf{H}(\mathrm{div};\Omega)$). *Let Ω be a bounded open subset of $\mathbb{R}^n$ with Lipschitz-continuous boundary Γ. Then, there holds*

$$\langle \gamma_{\boldsymbol{\nu}}(\boldsymbol{\tau}), \gamma_0(w)\rangle = \int_\Omega \Big\{\boldsymbol{\tau}\cdot\nabla w + w\,\mathrm{div}(\boldsymbol{\tau})\Big\} \tag{4.130}$$

for all $w \in \mathrm{H}^1(\Omega)$ and for all $\boldsymbol{\tau} \in \mathbf{H}(\mathrm{div};\Omega)$.

Proof. Given $w \in \mathrm{H}^1(\Omega)$ and $\boldsymbol{\tau} \in \mathbf{H}(\mathrm{div};\Omega)$, we first observe from (4.108) that there exist unique $w_0 \in \mathrm{H}_0^1(\Omega)$ and $w^\perp \in \mathrm{H}_0^1(\Omega)^\perp$, such that $w = w_0 + w^\perp$. Then, letting $\varphi := \gamma_0(w) \in \mathrm{H}^{1/2}(\Gamma)$, it follows that $\varphi = \gamma_0(w^\perp) = \tilde{\gamma}_0(w^\perp)$, which yields $\tilde{\gamma}_0^{-1}(\varphi) = w^\perp$, and hence (4.118) becomes

$$\langle \gamma_{\boldsymbol{\nu}}(\boldsymbol{\tau}), \gamma_0(w)\rangle := \int_\Omega \Big\{\boldsymbol{\tau}\cdot\nabla w^\perp + w^\perp\,\mathrm{div}(\boldsymbol{\tau})\Big\} \qquad \forall\, w \in \mathrm{H}^1(\Omega)\,. \tag{4.131}$$

In turn, knowing that $C_0^\infty(\Omega)$ is dense in $\mathrm{H}_0^1(\Omega)$, there exists a sequence $\{\varphi_k\}_{k\in\mathbb{N}} \subseteq C_0^\infty(\Omega)$ such that $\|w_0 - \varphi_k\|_{1,\Omega} \overset{k\to\infty}{\to} 0$. Thus, the distributional divergence of $\boldsymbol{\tau}$ evaluated at each φ_k reads

$$\langle \mathrm{div}(\boldsymbol{\tau}), \varphi_k\rangle_{\mathcal{D}'(\Omega)\times\mathcal{D}(\Omega)} := -\int_\Omega \boldsymbol{\tau}\cdot\nabla\varphi_k \qquad \forall\, k \in \mathbb{N}\,,$$

which, using that $\mathrm{div}(\boldsymbol{\tau}) \in \mathrm{L}^2(\Omega)$, can be rewritten as

$$0 = \int_\Omega \Big\{\boldsymbol{\tau}\cdot\nabla\varphi_k + \varphi_k\,\mathrm{div}(\boldsymbol{\tau})\Big\} \qquad \forall\, k \in \mathbb{N}\,. \tag{4.132}$$

Next, employing Cauchy-Schwarz's inequality we get

$$\left|\int_\Omega \boldsymbol{\tau}\cdot\nabla w_0 - \int_\Omega \boldsymbol{\tau}\cdot\nabla\varphi_k\right| \leqslant \|\boldsymbol{\tau}\|_{0,\Omega}\,|w_0 - \varphi_k|_{1,\Omega} \overset{k\to+\infty}{\to} 0\,,$$

and

$$\left|\int_\Omega w_0\,\mathrm{div}(\boldsymbol{\tau}) - \int_\Omega \varphi_k\,\mathrm{div}(\boldsymbol{\tau})\right| \leqslant \|w_0 - \varphi_k\|_{0,\Omega}\,\|\mathrm{div}(\boldsymbol{\tau})\|_{0,\Omega} \overset{k\to+\infty}{\to} 0\,,$$

so that, taking the limit as $k \to +\infty$ in (4.132), we obtain

$$0 = \int_\Omega \Big\{\boldsymbol{\tau}\cdot\nabla w_0 + w_0\,\mathrm{div}(\boldsymbol{\tau})\Big\}\,. \tag{4.133}$$

In this way, adding (4.131) and (4.133), we arrive at the required identity (4.130). □

It is important to note here that when $\boldsymbol{\tau} \in \mathbf{H}(\mathrm{div};\Omega)$, the evaluation $\langle \gamma_{\boldsymbol{\nu}}(\boldsymbol{\tau}), \gamma_0(w)\rangle$ cannot be replaced by the expression $\langle \gamma_0(\boldsymbol{\tau})\cdot\boldsymbol{\nu}, \gamma_0(w)\rangle_{0,\Gamma}$ since, as established in (4.120), the latter only makes sense if $\boldsymbol{\tau} \in \mathbf{H}^1(\Omega)$. Nevertheless, knowing that $[C^\infty(\bar{\Omega})]^n$ is dense in $\mathbf{H}(\mathrm{div};\Omega)$, and applying the continuity of $\gamma_{\boldsymbol{\nu}}$ along with (4.120), we find that for each sequence $\{\mathbf{z}_k\}_{k\in\mathbb{N}} \subseteq [C^\infty(\bar{\Omega})]^n$ converging to $\boldsymbol{\tau} \in \mathbf{H}(\mathrm{div};\Omega)$ there holds

$$\langle \gamma_{\boldsymbol{\nu}}(\boldsymbol{\tau}), \gamma_0(w)\rangle = \lim_{k\to+\infty} \langle \gamma_{\boldsymbol{\nu}}(\mathbf{z}_k), \gamma_0(w)\rangle = \lim_{k\to+\infty} \langle \gamma_0(\mathbf{z}_k)\cdot\boldsymbol{\nu}, \gamma_0(w)\rangle_{0,\Gamma} \quad \forall\, w \in \mathrm{H}^1(\Omega)\,,$$

which is the reason why it is usually said that $\langle\cdot,\cdot\rangle$ denotes the duality pairing of $\mathrm{H}^{-1/2}(\Gamma)$ and $\mathrm{H}^{1/2}(\Gamma)$ with respect to the inner product $\langle\cdot,\cdot\rangle_{0,\Gamma}$ of $\mathrm{L}^2(\Gamma)$.

Finally, the following theorem establishes an interesting consequence of the previous results.

Theorem 4.33. *Let Ω be a bounded open subset of $\mathbb{R}^n$ with Lipschitz-continuous boundary Γ, and define the space*

$$\mathrm{H}^1_\Delta(\Omega) := \Big\{ v \in \mathrm{H}^1(\Omega) : \quad \Delta v \in \mathrm{L}^2(\Omega) \Big\} .$$

Then, there exists an operator $\gamma_1 \in \mathcal{L}\big(\mathrm{H}^1_\Delta(\Omega), \mathrm{H}^{-1/2}(\Gamma)\big)$ such that for each $u \in \mathrm{H}^2(\Omega)$, $\gamma_1(u)$ is identified, via the inner product of $\mathrm{L}^2(\Gamma)$, with $\gamma_0(\nabla u)\cdot \boldsymbol{\nu}$, that is

$$\langle \gamma_1(u), \varphi \rangle = \langle \gamma_0(\nabla u)\cdot \boldsymbol{\nu}, \varphi\rangle_{0,\Gamma} \quad \forall\, \varphi \in \mathrm{H}^{1/2}(\Gamma), \quad \forall\, u \in \mathrm{H}^2(\Omega) .$$

Moreover, one has

$$\langle \gamma_1(u), \gamma_0(w) \rangle = \int_\Omega \nabla u \cdot \nabla w + \int_\Omega w\, \Delta u \qquad \forall\, w \in \mathrm{H}^1(\Omega), \quad \forall\, u \in \mathrm{H}^1_\Delta(\Omega) . \tag{4.134}$$

Proof. Since $\nabla u \in \mathbf{H}(\mathrm{div};\Omega)$ for all $u \in \mathrm{H}^1_\Delta(\Omega)$, it suffices to define $\gamma_1(u) := \gamma_{\boldsymbol{\nu}}(\nabla u)$ and then apply Theorem 4.31 and Lemma 4.32.

□

Note that this theorem significantly improves Green's identity (4.115) given in Corollary 4.30, since γ_1 allows us to extend the notion of normal derivative to the entire space $\mathrm{H}^1_\Delta(\Omega)$, which strictly contains $\mathrm{H}^2(\Omega)$.

Example 4.34 Here we use (4.130) to analyze the solvability of the convection–diffusion equation given by

$$-\kappa\,\Delta\varphi + \mathbf{u}\cdot\nabla\varphi = f \quad \text{in} \quad \Omega, \quad \varphi = 0 \quad \text{on} \quad \Gamma , \tag{4.135}$$

where $\mathbf{u}$ is the velocity of a fluid in a porous medium occupying a Lipschitz-continuous and simply connected domain Ω of $\mathbb{R}^n$, $n \in \{2,3\}$, Γ is the boundary of Ω, whose outward normal vector is denoted $\boldsymbol{\nu}$, φ is the temperature of the fluid, κ is the thermal conductivity coefficient, and f is an external scalar heat source. The aforedescribed equation along with Darcy's law for $\mathbf{u}$ and the pressure p of the fluid, constitute a coupled model that will be utilized later on in Section 5.5.5.1 to illustrate the application of the Babuška-Brezzi Theory in Banach spaces to the case in which a mixed approach is adopted for (4.135). In what follows, we assume that $\mathbf{u}$ is given in a suitable space (to be indicated next), and aim to derive the usual primal variational formulation of (4.135). In doing so, we make clear that it is not the equation from (4.135), but a conservative form of it to be displayed below, the expression that we use to perform the corresponding integration by parts procedure. Indeed, we begin by observing, thanks to Cauchy-Schwarz's inequality in $\mathrm{L}^2(\Omega)$ and the continuous injection $\mathrm{i}_4 : \mathrm{H}^1(\Omega) \to \mathrm{L}^4(\Omega)$ (cf. [111, Theorem 1.3.4]), that

$$\left(\int_\Omega |\varphi\,\mathbf{u}|^2\right)^{1/2} \leqslant \|\varphi\|_{0,4;\Omega}\,\|\mathbf{u}\|_{0,4;\Omega} \leqslant \|\mathrm{i}_4\|\,\|\varphi\|_{1,\Omega}\,\|\mathbf{u}\|_{0,4;\Omega} ,$$

where $\|\cdot\|_{0,4;\Omega}$ stands for the norms of both the scalar and vector versions of $\mathrm{L}^4(\Omega)$, from which we deduce that, whenever $\mathbf{u} \in \mathbf{L}^4(\Omega)$ and $\varphi \in \mathrm{H}^1(\Omega)$, there holds $\varphi\,\mathbf{u} \in \mathbf{L}^2(\Omega)$. In turn, bearing in mind that $C^\infty(\bar{\Omega})$ is dense in $\mathrm{H}^1(\Omega)$ (cf. Theorem 10.14), and using the incompressibility of the fluid, which means $\mathrm{div}(\mathbf{u}) = 0$ in $\mathcal{D}'(\Omega)$, we can show that (cf. Example 9.51 in Chapter 9)

$$\mathrm{div}(\phi\,\mathbf{u}) = \mathbf{u}\cdot\nabla\phi \quad \text{in} \quad \mathcal{D}'(\Omega) \qquad \forall\, \phi \in \mathrm{H}^1(\Omega) , \tag{4.136}$$

and thus (4.135) can be rewritten as

$$-\,\mathrm{div}\big(\kappa\,\nabla\varphi - \varphi\,\mathbf{u}\big) = f \quad \text{in} \quad \Omega, \quad \varphi = 0 \quad \text{on} \quad \Gamma . \tag{4.137}$$

In this way, noting that the Dirichlet boundary condition suggests to look for $\varphi \in \mathrm{H}^1_0(\Omega)$, and assuming that f is taken in $\mathrm{L}^2(\Omega)$, it makes sense to consider that $\kappa\,\nabla\varphi - \varphi\,\mathbf{u}$ belongs to $\mathbf{H}(\mathrm{div};\Omega)$.

Then, testing (4.137) against an arbitrary $\psi \in \mathrm{H}_0^1(\Omega)$, and applying (4.130) with $w := \psi$ and $\boldsymbol{\tau} := \kappa \nabla\varphi - \varphi\,\mathbf{u}$, we find that

$$\int_\Omega f\,\psi = -\int_\Omega \psi\,\mathrm{div}\big(\kappa\nabla\varphi - \varphi\,\mathbf{u}\big) = \int_\Omega \nabla\psi\cdot\big(\kappa\nabla\varphi - \varphi\,\mathbf{u}\big),$$

so that, at first instance, given $\mathbf{u} \in \mathbf{L}^4(\Omega)$, the weak formulation of (4.135) reduces to: Find $\varphi \in \mathrm{H}_0^1(\Omega)$ such that

$$A(\varphi,\psi) := \kappa\int_\Omega \nabla\varphi\cdot\nabla\psi - \int_\Omega \varphi\,\mathbf{u}\cdot\nabla\psi = F(\psi) := \int_\Omega f\,\psi \qquad \forall\,\psi \in \mathrm{H}_0^1(\Omega). \tag{4.138}$$

Alternatively, if we decided to test (4.135), we would need to deal, for each $\psi \in \mathrm{H}_0^1(\Omega)$, with the expressions

$$\int_\Omega \psi\,\Delta\varphi \qquad \text{and} \qquad \int_\Omega \mathbf{u}\cdot\nabla\varphi\,\psi = \int_\Omega \psi\,\mathbf{u}\cdot\nabla\varphi.$$

While the latter is easily shown to be well-defined, whence it could be kept as it is, the eventual application of (4.134) to the former would require having $\Delta\varphi \in \mathrm{L}^2(\Omega)$, which does not necessarily hold. This is the reason why we proceeded as indicated, that is, by testing (4.137) instead of (4.135). On the other hand, it is also proved in Example 9.51 that

$$\int_\Omega \varphi\,\mathbf{u}\cdot\nabla\psi = -\int_\Omega \psi\,\mathbf{u}\cdot\nabla\varphi \qquad \forall\,\varphi,\,\psi \in \mathrm{H}_0^1(\Omega), \tag{4.139}$$

which implies

$$\int_\Omega \varphi\,\mathbf{u}\cdot\nabla\psi = \frac{1}{2}\Big(\int_\Omega \varphi\,\mathbf{u}\cdot\nabla\psi - \int_\Omega \psi\,\mathbf{u}\cdot\nabla\varphi\Big) \qquad \forall\,\varphi,\,\psi \in \mathrm{H}_0^1(\Omega),$$

and hence (4.138) can be reformulated, equivalently, as: Find $\varphi \in \mathrm{H}_0^1(\Omega)$ such that

$$B(\varphi,\psi) := \kappa\int_\Omega \nabla\varphi\cdot\nabla\psi + \frac{1}{2}\Big(\int_\Omega \psi\,\mathbf{u}\cdot\nabla\varphi - \int_\Omega \varphi\,\mathbf{u}\cdot\nabla\psi\Big) = F(\psi) \quad \forall\,\psi \in \mathrm{H}_0^1(\Omega). \tag{4.140}$$

Because of the skew-symmetry of the second term defining the bilinear form B, it readily follows that $B(\psi,\psi) = |\psi|_{1,\Omega}^2$ for all $\psi \in \mathrm{H}_0^1(\Omega)$, which, along with the boundedness of F, and the equivalence between $|\cdot|_{1,\Omega}$ and $\|\cdot\|_{1,\Omega}$ in $\mathrm{H}_0^1(\Omega)$ (cf. (4.167)), allows to conclude, thanks to the classical Lax-Milgram Lemma (cf. Lemma 4.4), the well-posedness of (4.140). Needless to say, the above could also have been deduced from (4.138) since, according to (4.139), there clearly holds

$$\int_\Omega \psi\,\mathbf{u}\cdot\nabla\psi = 0 \qquad \forall\,\psi \in \mathrm{H}_0^1(\Omega), \tag{4.141}$$

whence A becomes $\mathrm{H}_0^1(\Omega)$-elliptic independently of $\mathbf{u}$ as well. Nevertheless, differently from B, this property of A does not transfer to an associated Galerkin scheme, mainly due to the fact that the identity (4.141) does not necessarily hold at the discrete level, and therefore, for practical purposes, one would usually stay with (4.140).

4.5.5 Additional results on $\mathbf{H}^{1/2}(\boldsymbol{\Gamma})$ and $\mathbf{H}^{-1/2}(\boldsymbol{\Gamma})$

In this section we partially rely on the analysis given in [55] to present complementary results on the trace space $\mathrm{H}^{1/2}(\Gamma)$ and its dual $\mathrm{H}^{-1/2}(\Gamma)$, which concern restrictions and extensions to and from parts of Γ, respectively, as well as useful decompositions of these spaces.

4.5.5.1 Restrictions of $\mathbf{H}^{1/2}(\Gamma)$ and $\mathbf{H}^{-1/2}(\Gamma)$ on parts of Γ

Let Ω be a bounded domain of $\mathbb{R}^n, n \in \{2, 3\}$, with Lipschitz-continuous boundary Γ, and let Γ_D and Γ_N be disjoint parts of Γ such that $\Gamma = \bar{\Gamma}_D \cup \bar{\Gamma}_N$. Then, recalling that $\gamma_0 : \mathrm{H}^1(\Omega) \rightarrow \mathrm{H}^{1/2}(\Gamma)$ denotes the usual trace operator, we define for each $* \in \{D, N\}$ the space:

$$\mathrm{H}^{1/2}(\Gamma_*) := \left\{\gamma_0(\omega)|_{\Gamma_*} : \quad \omega \in \mathrm{H}^1(\Omega)\right\}$$

endowed with the norm

$$\|\varphi\|_{1/2,\Gamma_*} := \inf\left\{\|\omega\|_{1,\Omega} : \quad \omega \in \mathrm{H}^1(\Omega), \quad \gamma_0(\omega)\Big|_{\Gamma_*} = \varphi\right\} \quad \forall\, \varphi \in \mathrm{H}^{1/2}(\Gamma_*).$$

It is not difficult to prove that $\|\cdot\|_{1/2,\Gamma_*}$ is indeed a norm on $\mathrm{H}^{1/2}(\Gamma_*)$. Moreover, we can equivalently write

$$\|\varphi\|_{1/2,\Gamma_*} = \operatorname{dist}(0, S_\varphi) := \inf_{\omega \in S_\varphi} \|\omega\|_{1,\Omega} \qquad \forall\, \varphi \in \mathrm{H}^{1/2}(\Gamma_*), \tag{4.142}$$

where S_φ is the subset of $\mathrm{H}^1(\Omega)$ given by

$$S_\varphi := \left\{\omega \in \mathrm{H}^1(\Omega) : \quad \gamma_0(\omega)\big|_{\Gamma_*} = \varphi\right\},$$

which is clearly convex and closed. Therefore, applying the best approximation theorem on a convex and closed subset of a Hilbert space (cf. Theorem 2.7), it follows that there exists a unique $\omega_\varphi \in S_\varphi$ such that

$$\|\varphi\|_{1/2,\Gamma_*} = \|\omega_\varphi\|_{1,\Omega} \qquad \forall\, \varphi \in \mathrm{H}^{1/2}(\Gamma_*). \tag{4.143}$$

Note also that the restriction operator $i_* : \mathrm{H}^{1/2}(\Gamma) \rightarrow \mathrm{H}^{1/2}(\Gamma_*)$ defined by $i_*(\psi) := \psi|_{\Gamma_*}$ for all $\psi \in \mathrm{H}^{1/2}(\Gamma)$ is clearly bounded. Indeed, according to the definitions given in (4.106) and (4.142), it is easy to prove that

$$\|\psi|_{\Gamma_*}\|_{1/2,\Gamma_*} \leqslant \|\psi\|_{1/2,\Gamma} \qquad \forall\, \psi \in \mathrm{H}^{1/2}(\Gamma). \tag{4.144}$$

On the other hand, we denote by $E_{*,0} : \mathrm{H}^{1/2}(\Gamma_*) \rightarrow \mathrm{L}^2(\Gamma)$ the extension by zero in $\Gamma \backslash \bar{\Gamma}_*$, and introduce the space:

$$\mathrm{H}_{00}^{1/2}(\Gamma_*) := \left\{\varphi \in \mathrm{H}^{1/2}(\Gamma_*) : \quad E_{*,0}(\varphi) \in \mathrm{H}^{1/2}(\Gamma)\right\} \tag{4.145}$$

endowed with the norm

$$\|\varphi\|_{1/2,00,\Gamma_*} := \|E_{*,0}(\varphi)\|_{1/2,\Gamma} \quad \forall\, \varphi \in \mathrm{H}_{00}^{1/2}(\Gamma). \tag{4.146}$$

Then, using that $E_{*,0}$ is linear, one can prove that $\|\cdot\|_{1/2,00,\Gamma_*}$ is indeed a norm on $\mathrm{H}_{00}^{1/2}(\Gamma_*)$. Now, for each $\lambda \in \mathrm{H}^{-1/2}(\Gamma)$, we define its "restriction" to Γ_* as the functional given by $\lambda|_{\Gamma_*} : \mathrm{H}_{00}^{1/2}(\Gamma_*) \rightarrow \mathbb{R}$, where

$$\lambda|_{\Gamma_*}(\varphi) := \langle \lambda, E_{*,0}(\varphi) \rangle_\Gamma \quad \forall\, \varphi \in \mathrm{H}_{00}^{1/2}(\Gamma_*). \tag{4.147}$$

Note that the linearities of $E_{*,0}$ and λ guarantee that $\lambda|_{\Gamma_*}$ is also linear. Moreover, for each $\varphi \in \mathrm{H}_{00}^{1/2}(\Gamma_*)$ we have

$$\left|\lambda|_{\Gamma_*}(\varphi)\right| \leqslant \|\lambda\|_{-1/2,\Gamma} \|E_{*,0}(\varphi)\|_{1/2,\Gamma} = \|\lambda\|_{-1/2,\Gamma} \|\varphi\|_{1/2,00,\Gamma_*},$$

which shows that $\lambda|_{\Gamma_*}$ is bounded, that is, $\lambda|_{\Gamma_*} \in \mathrm{H}_{00}^{-1/2}(\Gamma_*) := \mathrm{H}_{00}^{1/2}(\Gamma_*)'$, and

$$\|\lambda|_{\Gamma_*}\|_{-1/2,00,\Gamma_*} \leqslant \|\lambda\|_{-1/2,\Gamma}.$$

From now on we denote the evaluation $\lambda|_{\Gamma_*}(\varphi)$ by $\langle\lambda|_{\Gamma_*}, \varphi\rangle_{\Gamma_*}$, which means that $\langle\cdot,\cdot\rangle_{\Gamma_*}$ represents the duality pairing between $\mathrm{H}_{00}^{-1/2}(\Gamma_*)$ and $\mathrm{H}_{00}^{1/2}(\Gamma_*)$.

4.5.5.2 Stable extensions of $\mathbf{H}^{\pm 1/2}$ over parts of $\boldsymbol{\Gamma}$ to $\mathbf{H}^{\pm 1/2}(\boldsymbol{\Gamma})$

The first goal is to define a linear and bounded operator $E_* : \mathrm{H}^{1/2}(\Gamma_*) \to \mathrm{H}^{1/2}(\Gamma)$ such that $E_*(\varphi)|_{\Gamma_*} = \varphi$ for all $\varphi \in \mathrm{H}^{1/2}(\Gamma_*)$. In this regard, and as a first attempt, we could employ the result given by (4.143) to define

$$E_*(\varphi) := \gamma_0(\omega_\varphi) \quad \forall\, \varphi \in \mathrm{H}^{1/2}(\Gamma_*).$$

However, the mapping $\mathrm{H}^{1/2}(\Gamma^*) \ni \varphi \to \omega_\varphi \in \mathrm{H}^1(\Omega)$ is not necessarily linear, and therefore the above definition of E_* does not lead to a linear operator either. In what follows we avoid this difficulty by employing an auxiliary boundary value problem with mixed boundary conditions. More precisely, without loss of generality we now take $* = D$, and given $\varphi \in \mathrm{H}^{1/2}(\Gamma_D)$, we consider the problem:

$$\Delta z_\varphi = 0 \text{ in } \Omega, \quad \gamma_0(z_\varphi) = \varphi \text{ on } \Gamma_D, \quad \gamma_1(z_\varphi) = 0 \text{ on } \Gamma_N, \tag{4.148}$$

where $\gamma_1 := \gamma_{\boldsymbol{\nu}} \circ \nabla : \mathrm{H}^1_\Delta(\Omega) \to \mathrm{H}^{-1/2}(\Gamma)$ is the operator given by Theorem 4.33. We also need the spaces

$$\mathrm{H}^1_{\Gamma_*}(\Omega) := \left\{\omega \in \mathrm{H}^1(\Omega) : \quad \gamma_0(\omega)|_{\Gamma_*} = 0\right\} \qquad \forall * \in \{D, N\}.$$

Then, given $v \in \mathrm{H}^1_{\Gamma_D}(\Omega)$, we first obtain from (4.148) and integrating by parts (cf. (4.134)) that

$$0 = \int_\Omega v\, \Delta z_\varphi = -\int_\Omega \nabla z_\varphi \cdot \nabla v + \langle\gamma_1(z_\varphi), \gamma_0(v)\rangle_\Gamma.$$

In turn, we recall from (4.147) that the Neumann condition $\gamma_1(z_\varphi) = 0$ on Γ_N means that

$$\langle\gamma_1(z_\varphi), E_{N,0}(\varphi)\rangle_\Gamma = 0 \quad \forall\, \varphi \in \mathrm{H}_{00}^{1/2}(\Gamma_N).$$

Then, since clearly $\gamma_0(v)|_{\Gamma_N} \in \mathrm{H}_{00}^{1/2}(\Gamma_N)$ because $E_{N,0}(\gamma_0(v)|_{\Gamma_N}) = \gamma_0(v)$, we deduce that $\langle\gamma_1(z_\varphi), \gamma_0(v)\rangle_\Gamma = 0$, and therefore the primal variational formulation of (4.148) is initially: Find $z_\varphi \in \mathrm{H}^1(\Omega)$ such that $\gamma_0(z_\varphi) = \varphi$ on Γ_D and

$$\int_\Omega \nabla z_\varphi \cdot \nabla v = 0 \quad \forall\, v \in \mathrm{H}^1_{\Gamma_D}(\Omega). \tag{4.149}$$

Moreover, defining the auxiliary unknown $\tilde{z}_\varphi := z_\varphi - \omega_\varphi$, with ω_φ given according to (4.143), we find that (4.149) can be equivalently reformulated as: Find $\tilde{z}_\varphi \in \mathrm{H}^1_{\Gamma_D}(\Omega)$ such that

$$\int_\Omega \nabla\tilde{z}_\varphi \cdot \nabla v = -\int_\Omega \nabla\omega_\varphi \cdot \nabla v \qquad \forall\, v \in \mathrm{H}^1_{\Gamma_D}(\Omega). \tag{4.150}$$

In this way, the fact that $\|\cdot\|_{1,\Omega}$ and $|\cdot|_{1,\Omega}$ are equivalent in $H^1_{\Gamma_D}(\Omega)$, namely, there exists a constant $C_D > 0$ such that

$$C_D \, \|w\|_{1,\Omega}^2 \leqslant |w|_{1,\Omega}^2 \qquad \forall\, w \in H^1_{\Gamma_D}(\Omega)\,, \tag{4.151}$$

which follows from the generalized Poincaré inequality (cf. Theorem 10.21 and Example 10.22, part 2), guarantees, together with the classical Lax-Milgram Lemma (cf. Lemma 4.4), the unique solvability of (4.150) with $\|\widetilde{z}_\varphi\|_{1,\Omega} \leqslant c\,\|\omega_\varphi\|_{1,\Omega} = c\,\|\varphi\|_{1/2,\Gamma_D}$. It follows that (4.149) also has a unique solution $z_\varphi := \widetilde{z}_\varphi + \omega_\varphi$, with $\|z_\varphi\|_{1,\Omega} \leqslant C\,\|\varphi\|_{1/2,\Gamma_D}$. Consequently, the above analysis suggests defining the linear operator $E_D : H^{1/2}(\Gamma_D) \to H^{1/2}(\Gamma)$ as

$$E_D(\varphi) := \gamma_0(z_\varphi) \quad \forall\, \varphi \in H^{1/2}(\Gamma_D).$$

Note here that the linearity of E_D follows directly from the fact that γ_0 and the operator $H^{1/2}(\Gamma_D) \ni \varphi \to z_\varphi \in H^1(\Omega)$ are linear. Moreover, using the trace inequality (4.107) and the upper bound for $\|z_\varphi\|_{1,\Omega}$ already indicated, we obtain

$$\|E_D(\varphi)\|_{1/2,\Gamma} = \|\gamma_0(z_\varphi)\|_{1/2,\Gamma} \leqslant \|z_\varphi\|_{1,\Omega} \leqslant C\|\varphi\|_{1/2,\Gamma_D}\,, \tag{4.152}$$

which proves that E_D is indeed bounded. In what follows we define a stable extension operator $E^* : H^{-1/2}(\Gamma_*) \to H^{-1/2}(\Gamma)$, for which we assume again, without loss of generality, that $* = D$. Then, given $\lambda \in H^{-1/2}(\Gamma_D)$, we introduce the extension of λ by zero on Γ_N as $E^D(\lambda) := \widetilde{\lambda} \in H^{-1/2}(\Gamma)$, that is

$$\langle \widetilde{\lambda}, \widetilde{\varphi} \rangle_\Gamma := \langle \lambda, \widetilde{\varphi}|_{\Gamma_D} \rangle_{\Gamma_D} \quad \forall \widetilde{\varphi} \in H^{1/2}(\Gamma), \tag{4.153}$$

where $\langle \cdot, \cdot \rangle_{\Gamma_D}$ denotes in this case the duality pairing between $H^{-1/2}(\Gamma_D)$ and $H^{1/2}(\Gamma_D)$. Note that for each $\psi \in H^{1/2}_{00}(\Gamma_N)$ one has

$$\langle \widetilde{\lambda}|_{\Gamma_N}, \psi \rangle_{\Gamma_N} := \langle \widetilde{\lambda}, E_{N,0}(\psi) \rangle_\Gamma = \langle \lambda, 0 \rangle_{\Gamma_D} = 0\,,$$

which confirms that, indeed, $\widetilde{\lambda}|_{\Gamma_N} \equiv 0$. Moreover, for each $\psi \in H^{1/2}_{00}(\Gamma_D)$ one has

$$\langle \widetilde{\lambda}|_{\Gamma_D}, \psi \rangle_{\Gamma_D} := \langle \widetilde{\lambda}, E_{D,0}(\psi) \rangle_\Gamma = \langle \lambda, E_{D,0}(\psi)|_{\Gamma_D} \rangle_{\Gamma_D} = \langle \lambda, \psi \rangle_{\Gamma_D}\,,$$

which says, as expected, that $\widetilde{\lambda}|_{\Gamma_D} \equiv \lambda$. On the other hand, since (cf. (4.144))

$$\|\widetilde{\varphi}|_{\Gamma_D}\|_{1/2,\Gamma_D} \leqslant \|\widetilde{\varphi}\|_{1/2,\Gamma} \qquad \forall \widetilde{\varphi} \in H^{1/2}(\Gamma)\,,$$

it follows from (4.153) that

$$\|\widetilde{\lambda}\|_{-1/2,\Gamma} \leqslant \|\lambda\|_{-1/2,\Gamma_D}\,. \tag{4.154}$$

In turn, employing the boundedness of the extension operator E_D (cf. (4.152)), we deduce that for each $\varphi \in H^{1/2}(\Gamma_D)$ one has

$$\begin{aligned} |\langle \lambda, \varphi \rangle_{\Gamma_D}| &= |\langle \lambda, E_D(\varphi)|_{\Gamma_D} \rangle_{\Gamma_D}| = |\langle \widetilde{\lambda}, E_D(\varphi) \rangle_\Gamma| \\ &\leqslant \|\widetilde{\lambda}\|_{-1/2,\Gamma}\, \|E_D(\varphi)\|_{1/2,\Gamma} \leqslant \|E_D\|\, \|\widetilde{\lambda}\|_{-1/2,\Gamma}\, \|\varphi\|_{1/2,\Gamma_D}\,, \end{aligned}$$

which shows that

$$\|\lambda\|_{-1/2,\Gamma_D} \leqslant \|E_D\|\, \|\widetilde{\lambda}\|_{-1/2,\Gamma}\,,$$

thus proving, together with (4.154), that $\|\lambda\|_{-1/2,\Gamma_D}$ and $\|\widetilde{\lambda}\|_{-1/2,\Gamma}$ are equivalent.

4.5.5.3 Decompositions of $\mathrm{H}^{1/2}(\Gamma)$ and $\mathrm{H}^{-1/2}(\Gamma)$

The purpose of this section is to derive convenient decompositions of $\mathrm{H}^{1/2}(\Gamma)$ and its dual $\mathrm{H}^{-1/2}(\Gamma)$. To this end, we first consider $\psi \in \mathrm{H}^{1/2}(\Gamma)$ and define

$$\psi_D := \psi|_{\Gamma_D} \in \mathrm{H}^{1/2}(\Gamma_D) \quad \text{and} \quad \psi_N := (\psi - E_D(\psi_D))|_{\Gamma_N} \in \mathrm{H}^{1/2}(\Gamma_N).$$

Since $\psi|_{\Gamma_D} = \psi_D = E_D(\psi_D)|_{\Gamma_D}$, it follows that $E_{N,0}(\psi_N) = \psi - E_D(\psi_D) \in \mathrm{H}^{1/2}(\Gamma)$, which implies that $\psi_N \in \mathrm{H}_{00}^{1/2}(\Gamma_N)$, and therefore the identity

$$\psi = E_D(\psi_D) + E_{N,0}(\psi_N)$$

proves that

$$\mathrm{H}^{1/2}(\Gamma) = E_D(\mathrm{H}^{1/2}(\Gamma_D)) + E_{N,0}(\mathrm{H}_{00}^{1/2}(\Gamma_N)). \tag{4.155}$$

Moreover, given $\widetilde{\psi} \in E_D(\mathrm{H}^{1/2}(\Gamma_D)) \cap E_{N,0}(\mathrm{H}_{00}^{1/2}(\Gamma_N))$, there exist $\widetilde{\psi}_D \in \mathrm{H}^{1/2}(\Gamma_D)$ and $\widetilde{\psi}_N \in \mathrm{H}_{00}^{1/2}(\Gamma_N)$ such that $\widetilde{\psi} = E_D(\widetilde{\psi}_D) = E_{N,0}(\widetilde{\psi}_N)$. The second equality implies that $\widetilde{\psi} = 0$ on Γ_D, and the fact that $E_D(\widetilde{\psi}_D)|_{\Gamma_D} = \widetilde{\psi}_D$ yields $\widetilde{\psi}_D = 0$, which leads to $\widetilde{\psi} \equiv 0$, thereby showing that the decomposition (4.155) is unique, that is

$$\mathrm{H}^{1/2}(\Gamma) = E_D(\mathrm{H}^{1/2}(\Gamma_D)) \oplus E_{N,0}(\mathrm{H}_{00}^{1/2}(\Gamma_N)). \tag{4.156}$$

As a first consequence of (4.156), we next prove that $\mathrm{H}_{00}^{1/2}(\Gamma_*)$ is a Hilbert space with the inner product

$$\langle \chi, \eta \rangle_{1/2,00,\Gamma_*} := \langle E_{*,0}(\chi), E_{*,0}(\eta) \rangle_{1/2,\Gamma} \qquad \forall\, \chi, \eta \in \mathrm{H}_{00}^{1/2}(\Gamma_*),$$

where $\langle \cdot, \cdot \rangle_{1/2,\Gamma}$ is the inner product of $\mathrm{H}^{1/2}(\Gamma)$ (cf. (4.110)). Note that the norm induced by $\langle \cdot, \cdot \rangle_{1/2,00,\Gamma_*}$ is precisely the one defined by (4.146). Without loss of generality, and for the sake of simplicity of the respective notations, the announced result is established in the following lemma for $* = N$.

Lemma 4.35. $\big(\mathrm{H}_{00}^{1/2}(\Gamma_N), \langle \cdot, \cdot \rangle_{1/2,00,\Gamma_N}\big)$ *is a Hilbert space.*

Proof. Let $\{\varphi_k\}_{k \in \mathbb{N}}$ be a Cauchy sequence in $\mathrm{H}_{00}^{1/2}(\Gamma_N)$, that is,

$$\|\varphi_k - \varphi_\ell\|_{1/2,00,\Gamma_N} = \|E_{N,0}(\varphi_k) - E_{N,0}(\varphi_\ell)\|_{1/2,\Gamma} \overset{k,\ell \to \infty}{\longrightarrow} 0 .$$

Equivalently, $\{E_{N,0}(\varphi_k)\}_{k \in \mathbb{N}}$ is Cauchy in $\mathrm{H}^{1/2}(\Gamma)$, and thus, by Corollary 4.28, there exists $\psi \in \mathrm{H}^{1/2}(\Gamma)$ such that

$$\|E_{N,0}(\varphi_k) - \psi\|_{1/2,\Gamma} \overset{k \to \infty}{\longrightarrow} 0 . \tag{4.157}$$

Then, using (4.156) we have that $\psi = E_D(\psi_D) + E_{N,0}(\psi_N)$, with $\psi_D = \psi|_{\Gamma_D} \in \mathrm{H}^{1/2}(\Gamma_D)$ and $\psi_N \in \mathrm{H}_{00}^{1/2}(\Gamma_N)$. Moreover, by (4.144) and (4.157), it follows that

$$\begin{aligned} \|\psi_D\|_{1/2,\Gamma_D} &= \|E_{N,0}(\varphi_k)|_{\Gamma_D} - \psi|_{\Gamma_D}\|_{1/2,\Gamma_D} = \|\big(E_{N,0}(\varphi_k) - \psi\big)|_{\Gamma_D}\|_{1/2,\Gamma_D} \\ &\leqslant \|E_{N,0}(\varphi_k) - \psi\|_{1/2,\Gamma} \overset{k \to \infty}{\longrightarrow} 0 , \end{aligned}$$

from which we conclude that $\psi_D = 0$. It follows that $\psi = E_{N,0}(\psi_N)$, and therefore (4.157) rewrites as

$$\|\varphi_k - \psi_N\|_{1/2,00,\Gamma_N} = \|E_{N,0}(\varphi_k) - E_{N,0}(\psi_N)\|_{1/2,\Gamma} \overset{k \to \infty}{\longrightarrow} 0 ,$$

thus proving that $\{\varphi_k\}_{k \in \mathbb{N}}$ converges to ψ_N in $\mathrm{H}_{00}^{1/2}(\Gamma_N)$.

□

To describe a second result that arises from (4.156), let us now consider $\lambda \in \mathrm{H}^{-1/2}(\Gamma)$, and let us introduce the linear forms $\lambda_D : \mathrm{H}^{1/2}(\Gamma_D) \to \mathbb{R}$ and $\lambda_N : \mathrm{H}^{1/2}_{00}(\Gamma_N) \to \mathbb{R}$ defined by

$$\lambda_D(\varphi_D) := \langle \lambda, E_D(\varphi_D)\rangle_\Gamma \quad \forall\, \varphi_D \in \mathrm{H}^{1/2}(\Gamma_D) \quad \text{and}$$

$$\lambda_N(\varphi_N) := \langle \lambda, E_{N,0}(\varphi_N)\rangle_\Gamma \quad \forall\, \varphi_N \in \mathrm{H}^{1/2}_{00}(\Gamma_N).$$

We observe that $\lambda_N = \lambda|_{\Gamma_N} \in \mathrm{H}^{-1/2}_{00}(\Gamma_N)$, which was previously defined in (4.147). On the other hand, thanks to the boundedness of E_D (cf. (4.152)), it follows that $\lambda_D \in \mathrm{H}^{-1/2}(\Gamma_D) := \mathrm{H}^{1/2}(\Gamma_D)'$. Consequently, using now the decomposition (4.156), we find that for each $\psi \in \mathrm{H}^{1/2}(\Gamma)$ there exist unique $\psi_D := \psi|_{\Gamma_D} \in \mathrm{H}^{1/2}(\Gamma_D)$ and $\psi_N \in \mathrm{H}^{1/2}_{00}(\Gamma_N)$, such that

$$\psi = E_D(\psi_D) + E_{N,0}(\psi_N),$$

which allows us to write

$$\langle \lambda, \psi\rangle_\Gamma = \langle \lambda, E_D(\psi_D)\rangle_\Gamma + \langle \lambda, E_{N,0}(\psi_N)\rangle_\Gamma,$$

that is,

$$\langle \lambda, \psi\rangle_\Gamma = \lambda_D(\psi|_{\Gamma_D}) + \langle \lambda|_{\Gamma_N}, \psi_N\rangle_{\Gamma_N} \quad \forall\, \psi \in \mathrm{H}^{1/2}(\Gamma). \tag{4.158}$$

In particular, when $\lambda|_{\Gamma_N} = 0$, the above equality reduces to

$$\langle \lambda, \psi\rangle_\Gamma = \lambda_D(\psi|_{\Gamma_D}) \quad \forall\, \psi \in \mathrm{H}^{1/2}(\Gamma),$$

which, denoting by $\langle \cdot, \cdot\rangle_{\Gamma_D}$ the duality pairing between $\mathrm{H}^{-1/2}(\Gamma_D)$ and $\mathrm{H}^{1/2}(\Gamma_D)$, can be rewritten as

$$\langle \lambda, \psi\rangle_\Gamma = \langle \lambda_D, \psi|_{\Gamma_D}\rangle_{\Gamma_D} \quad \forall\, \psi \in \mathrm{H}^{1/2}(\Gamma), \tag{4.159}$$

suggesting in this case that $\lambda|_{\Gamma_D}$ may be identified, via λ_D, with a functional in $\mathrm{H}^{-1/2}(\Gamma_D)$ instead of $\mathrm{H}^{-1/2}_{00}(\Gamma_D)$. Moreover, we next show that the evaluation on the right-hand side of (4.159) can be carried out with any $\varphi \in \mathrm{H}^{1/2}(\Gamma)$ such that $\varphi|_{\Gamma_D} = \psi_D := \psi|_{\Gamma_D}$. Indeed, we then have that $\varphi - E_D(\psi_D) = 0$ on Γ_D, which means that $\big(\varphi - E_D(\psi_D)\big)\big|_{\Gamma_N} \in \mathrm{H}^{1/2}_{00}(\Gamma_N)$ and $E_{N,0}\big((\varphi - E_D(\psi_D))|_{\Gamma_N}\big) = \varphi - E_D(\psi_D)$. Therefore, since $\lambda|_{\Gamma_N} = 0$, we conclude that

$$0 = \langle \lambda|_{\Gamma_N}, (\varphi - E_D(\psi_D))|_{\Gamma_N}\rangle_{\Gamma_N} = \langle \lambda, \varphi - E_D(\psi_D)\rangle_\Gamma,$$

which leads to

$$\langle \lambda, E_D(\psi_D)\rangle_\Gamma = \langle \lambda, \varphi\rangle_\Gamma,$$

that is, using (4.159),

$$\langle \lambda_D, \psi|_{\Gamma_D}\rangle_{\Gamma_D} = \langle \lambda_D, \varphi|_{\Gamma_D}\rangle_{\Gamma_D}.$$

We have thus proved the following result.

Lemma 4.36. *For each $\lambda \in \mathrm{H}^{-1/2}(\Gamma)$ such that $\lambda|_{\Gamma_N} = 0$, it holds that $\lambda|_{\Gamma_D}$ can be identified with a functional $\lambda_D \in \mathrm{H}^{-1/2}(\Gamma_D)$ through the identity* (4.159)*. Moreover, for each $\psi \in \mathrm{H}^{1/2}(\Gamma)$ we have that*

$$\langle \lambda, \psi\rangle_\Gamma = \langle \lambda_D, \varphi|_{\Gamma_D}\rangle_{\Gamma_D}$$

where φ is any function in $\mathrm{H}^{1/2}(\Gamma)$ such that $\varphi|_{\Gamma_D} = \psi|_{\Gamma_D}$.

4.6 Examples of application of the Babuška-Brezzi Theory

In this section, we illustrate the application of the Babuška-Brezzi Theory with the classical examples given by the Poisson problem and the elasticity problem.

4.6.1 The Poisson problem

Let Ω be a bounded open subset of $\mathbb{R}^n$, $n \geqslant 2$, with Lipschitz-continuous boundary Γ. Then, given $f \in \mathrm{L}^2(\Omega)$ and $g \in \mathrm{H}^{1/2}(\Gamma)$, we consider the problem:

$$-\Delta u = f \quad \text{in} \quad \Omega, \quad u = g \quad \text{on} \quad \Gamma. \tag{4.160}$$

Next, we define the additional unknown $\boldsymbol{\sigma} := \nabla u \quad \text{in} \quad \Omega$, so that the problem (4.160) can be rewritten as the first-order system

$$\boldsymbol{\sigma} = \nabla u \quad \text{in} \quad \Omega, \quad \operatorname{div}(\boldsymbol{\sigma}) = -f \quad \text{in} \quad \Omega, \quad u = g \quad \text{on} \quad \Gamma.$$

Then, multiplying the equation $\boldsymbol{\sigma} = \nabla u \quad \text{in} \quad \Omega$ by an element $\boldsymbol{\tau} \in \mathbf{H}(\operatorname{div};\Omega)$, and applying the Green identity given by (4.130) (cf. Lemma 4.32), we obtain

$$\int_\Omega \boldsymbol{\sigma}\cdot\boldsymbol{\tau} = \int_\Omega \nabla u\cdot\boldsymbol{\tau} = -\int_\Omega u\,\operatorname{div}(\boldsymbol{\tau}) + \langle \gamma_{\boldsymbol{\nu}}(\boldsymbol{\tau}), \gamma_0(u)\rangle,$$

from which, using the Dirichlet boundary condition $\gamma_0(u) = g$ on Γ, it follows that

$$\int_\Omega \boldsymbol{\sigma}\cdot\boldsymbol{\tau} + \int_\Omega u\,\operatorname{div}(\boldsymbol{\tau}) = \langle \gamma_{\boldsymbol{\nu}}(\boldsymbol{\tau}), g\rangle \quad \forall\,\boldsymbol{\tau} \in \mathbf{H}(\operatorname{div};\Omega). \tag{4.161}$$

We recall here that $\gamma_0 : \mathrm{H}^1(\Omega) \to \mathrm{H}^{1/2}(\Gamma)$ and $\gamma_{\boldsymbol{\nu}} : \mathbf{H}(\operatorname{div};\Omega) \to \mathrm{H}^{-1/2}(\Gamma)$ are the trace operators studied in Sections 4.5.1 and 4.5.4, and $\langle\cdot,\cdot\rangle$ denotes the duality pairing between $\mathrm{H}^{-1/2}(\Gamma)$ and $\mathrm{H}^{1/2}(\Gamma)$ with respect to the $\mathrm{L}^2(\Gamma)$ inner product. On the other hand, the equilibrium equation $\operatorname{div}(\boldsymbol{\sigma}) = -f \quad \text{in} \quad \Omega$, can be written as

$$\int_\Omega v\,\operatorname{div}(\boldsymbol{\sigma}) = -\int_\Omega f\,v \quad \forall\, v \in \mathrm{L}^2(\Omega). \tag{4.162}$$

Consequently, combining (4.161) and (4.162) we arrive at the mixed variational formulation of (4.160), which reads: Find $(\boldsymbol{\sigma}, u) \in H \times Q$ such that

$$\begin{aligned} a(\boldsymbol{\sigma},\boldsymbol{\tau}) + b(\boldsymbol{\tau}, u) &= F(\boldsymbol{\tau}) \quad \forall\,\boldsymbol{\tau} \in H\,, \\ b(\boldsymbol{\sigma}, v) \qquad\qquad &= G(v) \quad \forall\, v \in Q\,, \end{aligned} \tag{4.163}$$

where

$$H := \mathbf{H}(\operatorname{div};\Omega)\,, \qquad Q := \mathrm{L}^2(\Omega)\,,$$

a and b are the bilinear forms defined by

$$\begin{aligned} a(\boldsymbol{\sigma},\boldsymbol{\tau}) &:= \int_\Omega \boldsymbol{\sigma}\cdot\boldsymbol{\tau} \qquad \forall\,(\boldsymbol{\sigma},\boldsymbol{\tau}) \in H \times H\,, \\ b(\boldsymbol{\tau}, v) &:= \int_\Omega v\,\operatorname{div}(\boldsymbol{\tau}) \quad \forall\,(\boldsymbol{\tau}, v) \in H \times Q\,, \end{aligned}$$

and the linear and bounded functionals $F \in H'$ and $G \in Q'$ are given by

$$F(\boldsymbol{\tau}) := \langle \gamma_{\boldsymbol{\nu}}(\boldsymbol{\tau}), g \rangle \quad \forall\, \boldsymbol{\tau} \in H\,, \qquad G(v) := -\int_\Omega f\, v \quad \forall\, v \in Q\,. \tag{4.164}$$

In what follows, we apply the particular case of the Babuška-Brezzi Theory given by Theorem 4.21. Indeed, let us first note that a and b are bounded with

$$\|\mathbf{A}\| \leqslant 1 \quad \text{and} \quad \|\mathbf{B}\| \leqslant 1\,,$$

where $\mathbf{A} : H \to H$ and $\mathbf{B} : H \to Q$ are the operators induced by a and b, respectively. Moreover, it is clear that $\mathbf{B}(\boldsymbol{\tau}) := \operatorname{div}(\boldsymbol{\tau})$ for all $\boldsymbol{\tau} \in H$, and then

$$V := N(\mathbf{B}) = \Big\{ \boldsymbol{\tau} \in H : \quad \mathbf{B}(\boldsymbol{\tau}) = 0 \Big\} = \Big\{ \boldsymbol{\tau} \in \mathbf{H}(\operatorname{div};\Omega) : \quad \operatorname{div}(\boldsymbol{\tau}) = 0 \quad \text{in} \quad \Omega \Big\}.$$

It follows that

$$a(\boldsymbol{\tau},\boldsymbol{\tau}) = \|\boldsymbol{\tau}\|_{0,\Omega}^2 = \|\boldsymbol{\tau}\|_{\operatorname{div};\Omega}^2 \qquad \forall\, \boldsymbol{\tau} \in V\,, \tag{4.165}$$

which proves that a is V-elliptic with ellipticity constant $\alpha = 1$. In turn, we know from Lemma 4.18 that proving the continuous inf-sup condition for b is equivalent to showing that $\mathbf{B}$ is surjective. For this purpose, given $v \in Q$, we consider the boundary value problem: $-\Delta z = v$ in Ω, $z = 0$ on Γ, whose primal variational formulation reduces to: Find $z \in \mathrm{H}_0^1(\Omega)$ such that

$$\int_\Omega \nabla z \cdot \nabla w = \int_\Omega v\, w \quad \forall\, w \in \mathrm{H}_0^1(\Omega). \tag{4.166}$$

We recall here that the n-dimensional version of the Friedrichs-Poincaré inequality (cf. Lemma 4.5), which follows from Theorem 10.21 (cf. Example 10.22, part 2, with $\Gamma_D = \Gamma$), states that there exists a constant $C_P > 0$ such that

$$C_P\, \|w\|_{1,\Omega}^2 \leqslant |w|_{1,\Omega}^2 \qquad \forall\, w \in \mathrm{H}_0^1(\Omega)\,. \tag{4.167}$$

It follows from the Lax-Milgram Lemma (cf. Lemma 4.4) that (4.166) has a unique solution $z \in \mathrm{H}_0^1(\Omega)$, which satisfies

$$|z|_{1,\Omega} \leqslant \widetilde{C}\, \|v\|_{0,\Omega}, \tag{4.168}$$

where $\widetilde{C} > 0$ is a constant depending on C_P. Then, defining $\widetilde{\boldsymbol{\tau}} := -\nabla z \in \mathbf{L}^2(\Omega)$, we have that $\operatorname{div}(\widetilde{\boldsymbol{\tau}}) = v \in \mathrm{L}^2(\Omega)$, and thus $\widetilde{\boldsymbol{\tau}} \in \mathbf{H}(\operatorname{div};\Omega)$, which proves that $\mathbf{B} := \operatorname{div}$ is surjective and therefore b satisfies the continuous inf-sup condition. Furthermore, thanks to the boundedness of the operator $\gamma_{\boldsymbol{\nu}}$ (cf. Theorem 4.31), the duality pairing $\langle \cdot, \cdot \rangle$, and Cauchy-Schwarz's inequality in $\mathrm{L}^2(\Omega)$, it follows from (4.164) that F and G are bounded with

$$\|F\|_{H'} \leqslant \|g\|_{1/2,\Gamma} \quad \text{and} \quad \|G\|_{Q'} \leqslant \|f\|_{0,\Omega}.$$

Consequently, Theorem 4.21 implies that there exists a unique pair $(\boldsymbol{\sigma}, u) \in H \times Q$ solution of the mixed variational formulation (4.163), which satisfies

$$\|(\boldsymbol{\sigma}, u)\|_{H \times Q} \leqslant C \Big\{ \|g\|_{1/2,\Gamma} + \|f\|_{0,\Omega} \Big\},$$

where, according to the a priori estimates provided by (4.85), C is a positive constant depending on α $(= 1)$, the constant β of the continuous inf-sup condition for b, and $\|\mathbf{A}\|$ $(\leqslant 1)$. In this regard, and in order to have a more precise idea of the value of β, we first observe from the definition of $\widetilde{\boldsymbol{\tau}}$, and using inequality (4.168), that

$$\|\widetilde{\boldsymbol{\tau}}\|_{\operatorname{div};\Omega}^2 = \|\widetilde{\boldsymbol{\tau}}\|_{0,\Omega}^2 + \|\operatorname{div}(\widetilde{\boldsymbol{\tau}})\|_{0,\Omega}^2 = |z|_{1,\Omega}^2 + \|v\|_{0,\Omega}^2 \leqslant (1 + \widetilde{C}^2)\, \|v\|_{0,\Omega}^2\,,$$

so that we obtain

$$\sup_{\substack{\boldsymbol{\tau}\in H\\ \boldsymbol{\tau}\neq 0}} \frac{b(\boldsymbol{\tau},\,v)}{\|\boldsymbol{\tau}\|_H} \geqslant \frac{b(\widetilde{\boldsymbol{\tau}},\,v)}{\|\widetilde{\boldsymbol{\tau}}\|_H} = \frac{\int_\Omega v\,\mathrm{div}(\widetilde{\boldsymbol{\tau}})}{\|\widetilde{\boldsymbol{\tau}}\|_{\mathrm{div};\Omega}} = \frac{\|v\|_{0,\Omega}^2}{\|\widetilde{\boldsymbol{\tau}}\|_{\mathrm{div};\Omega}} \geqslant \frac{1}{(1+\widetilde{C}^2)^{1/2}}\,\|v\|_{0,\Omega}\,,$$

from which it follows that we can take $\beta := \left(1+\widetilde{C}^2\right)^{-1/2}$.

4.6.2 The Poisson problem with Neumann boundary conditions

Let Ω be a bounded open subset of $\mathbb{R}^n$, $n \geqslant 2$, with Lipschitz-continuous boundary Γ. Then, given $f \in \mathrm{L}^2(\Omega)$ and $g \in \mathrm{H}^{-1/2}(\Gamma)$, we consider the Neumann problem

$$-\Delta u = f \quad \text{in} \quad \Omega, \quad \gamma_1(u) = g \quad \text{on} \quad \Gamma, \quad \int_\Omega u = 0, \tag{4.169}$$

where $\gamma_1 := \gamma_{\boldsymbol{\nu}} \circ \nabla : \mathrm{H}^1_\Delta(\Omega) \to \mathrm{H}^{-1/2}(\Gamma)$ is the operator given by Theorem 4.33. Thus, analogously to the analysis in Section 4.6.1, we define the additional unknown $\boldsymbol{\sigma} := \nabla u$ in Ω, so that, using that $\gamma_1(u) = \gamma_{\boldsymbol{\nu}}(\boldsymbol{\sigma})$, the problem (4.169) can be rewritten as the first-order system

$$\begin{aligned} \boldsymbol{\sigma} = \nabla u \quad \text{in} \quad \Omega, \quad \mathrm{div}(\boldsymbol{\sigma}) = -f \quad \text{in} \quad \Omega\,, \\ \gamma_{\boldsymbol{\nu}}(\boldsymbol{\sigma}) = g \quad \text{on} \quad \Gamma, \quad \int_\Omega u = 0\,. \end{aligned} \tag{4.170}$$

Moreover, applying the integration by parts formula given by (4.134) to $u \in \mathrm{H}^1_\Delta(\Omega)$ and $w \equiv 1 \in \mathrm{H}^1(\Omega)$, we obtain

$$\langle \gamma_{\boldsymbol{\nu}}(\boldsymbol{\sigma}), \gamma_0(1)\rangle = \int_\Omega \Delta u = -\int_\Omega f\,,$$

which shows, according to the boundary condition in (4.170), that f and g must satisfy the compatibility hypothesis

$$\int_\Omega f + \langle g, 1\rangle = 0\,, \tag{4.171}$$

which is assumed from now on. Then, multiplying the first equation of (4.170) by an element $\boldsymbol{\tau} \in \mathbf{H}(\mathrm{div};\Omega)$, and applying the Green identity given by (4.130) (cf. Lemma 4.32), we get

$$\int_\Omega \boldsymbol{\sigma}\cdot\boldsymbol{\tau} = \int_\Omega \nabla u\cdot\boldsymbol{\tau} = -\int_\Omega u\,\mathrm{div}(\boldsymbol{\tau}) + \langle \gamma_{\boldsymbol{\nu}}(\boldsymbol{\tau}), \gamma_0(u)\rangle,$$

from which, introducing the auxiliary unknown

$$\varphi := -\gamma_0(u) \in \mathrm{H}^{1/2}(\Gamma)\,,$$

we arrive at

$$\int_\Omega \boldsymbol{\sigma}\cdot\boldsymbol{\tau} + \int_\Omega u\,\mathrm{div}(\boldsymbol{\tau}) + \langle \gamma_{\boldsymbol{\nu}}(\boldsymbol{\tau}), \varphi\rangle = 0 \quad \forall\,\boldsymbol{\tau} \in \mathbf{H}(\mathrm{div};\Omega). \tag{4.172}$$

In turn, the equilibrium equation (cf. the second identity in the first row of (4.170)) is first tested with $v \in \mathrm{L}^2(\Omega)$, that is,

$$\int_\Omega v\,\mathrm{div}(\boldsymbol{\sigma}) = -\int_\Omega f\,v \qquad \forall\, v \in \mathrm{L}^2(\Omega)\,. \tag{4.173}$$

However, invoking the decomposition $\mathrm{L}^2(\Omega) = \mathrm{L}^2_0(\Omega) \oplus \mathbb{R}$, and observing, thanks to (4.171), formula (4.130), and the Neumann condition in (4.170), that (4.173) is always satisfied for $v \equiv c \in \mathbb{R}$, we conclude that requiring this identity to hold for all $v \in \mathrm{L}^2(\Omega)$ is equivalent to requiring it for all $v \in \mathrm{L}^2_0(\Omega)$. Moreover, adding to the resulting equation the weak imposition of the Neumann boundary condition, we obtain

$$\int_\Omega v\,\mathrm{div}(\boldsymbol{\sigma}) + \langle \gamma_{\boldsymbol{\nu}}(\boldsymbol{\sigma}), \psi\rangle = -\int_\Omega f\,v + \langle g, \psi\rangle \tag{4.174}$$

for all $(v,\psi) \in \mathrm{L}^2_0(\Omega) \times \mathrm{H}^{1/2}(\Gamma)$. In this way, combining (4.172) and (4.174), we arrive at the following mixed variational formulation of (4.170): Find $\big(\boldsymbol{\sigma},(u,\varphi)\big) \in H \times Q$ such that

$$\begin{aligned} a(\boldsymbol{\sigma},\boldsymbol{\tau}) + b\big(\boldsymbol{\tau},(u,\varphi)\big) &= F(\boldsymbol{\tau}) && \forall\, \boldsymbol{\tau} \in H\,, \\ b\big(\boldsymbol{\sigma},(v,\psi)\big) &= G(v,\psi) && \forall\, (v,\psi) \in Q\,, \end{aligned} \tag{4.175}$$

where

$$H := \mathbf{H}(\mathrm{div};\Omega)\,, \qquad Q := \mathrm{L}^2_0(\Omega) \times \mathrm{H}^{1/2}(\Gamma)\,,$$

a and b are the bilinear forms defined by

$$\begin{aligned} a(\boldsymbol{\sigma},\boldsymbol{\tau}) &:= \int_\Omega \boldsymbol{\sigma}\cdot\boldsymbol{\tau} && \forall\, (\boldsymbol{\sigma},\boldsymbol{\tau}) \in H \times H\,, \\ b\big(\boldsymbol{\tau},(v,\psi)\big) &:= \int_\Omega v\,\mathrm{div}(\boldsymbol{\tau}) + \langle \gamma_{\boldsymbol{\nu}}(\boldsymbol{\tau}),\psi\rangle && \forall\, \big(\boldsymbol{\tau},(v,\psi)\big) \in H \times Q\,, \end{aligned} \tag{4.176}$$

and the linear and bounded functionals $F \in H'$ and $G \in Q'$ are given by

$$F(\boldsymbol{\tau}) := 0 \quad \forall\, \boldsymbol{\tau} \in H\,, \quad G(v,\psi) := -\int_\Omega f\,v + \langle g,\psi\rangle \quad \forall\,(v,\psi) \in Q\,. \tag{4.177}$$

Using the Cauchy-Schwarz inequality and the boundedness of $\gamma_{\boldsymbol{\nu}}$ (cf. (4.119)), it is easily deduced that

$$\|\mathbf{A}\| \leqslant 1 \quad \text{and} \quad \|\mathbf{B}\| \leqslant 1\,,$$

where $\mathbf{A} : H \to H$ and $\mathbf{B} : H \to Q$ are the bounded linear operators induced by a and b, respectively. Moreover, denoting by $\mathcal{R}_{1/2} : \mathrm{H}^{-1/2}(\Gamma) \to \mathrm{H}^{1/2}(\Gamma)$ the Riesz operator of the Hilbert space $\mathrm{H}^{1/2}(\Gamma)$ (cf. Corollary 4.28), and keeping in mind the definition of the bilinear form b, we have, on the one hand, that (cf. (4.110))

$$\langle \gamma_{\boldsymbol{\nu}}(\boldsymbol{\tau}),\psi\rangle = \langle \mathcal{R}_{1/2}\big(\gamma_{\boldsymbol{\nu}}(\boldsymbol{\tau})\big),\psi\rangle_{1/2,\Gamma} \qquad \forall\,(\boldsymbol{\tau},\psi) \in H \times \mathrm{H}^{1/2}(\Gamma)\,,$$

while, on the other hand,

$$\int_\Omega v\,\mathrm{div}(\boldsymbol{\tau}) = \int_\Omega v\left\{\mathrm{div}(\boldsymbol{\tau}) - \frac{1}{|\Omega|}\int_\Omega \mathrm{div}(\boldsymbol{\tau})\right\} \qquad \forall\,(\boldsymbol{\tau},v) \in H \times \mathrm{L}^2_0(\Omega)\,,$$

so that, recalling that $\mathbf{B}(\boldsymbol{\tau})$ is the unique element in Q such that

$$\langle \mathbf{B}(\boldsymbol{\tau}),(v,\psi)\rangle_Q = b\big(\boldsymbol{\tau},(v,\psi)\big) \qquad \forall\,(v,\psi) \in Q\,,$$

we deduce that

$$\mathbf{B}(\boldsymbol{\tau}) \,=\, \left(\operatorname{div}(\boldsymbol{\tau}) - \frac{1}{|\Omega|}\int_\Omega \operatorname{div}(\boldsymbol{\tau}), \big(\mathcal{R}_{1/2}\circ\gamma_{\boldsymbol{\nu}}\big)(\boldsymbol{\tau})\right) \qquad \forall\, \boldsymbol{\tau}\in H\,.$$

It follows, at first, that

$$V := N(\mathbf{B}) = \Big\{\boldsymbol{\tau}\in H: \quad \operatorname{div}(\boldsymbol{\tau}) = \frac{1}{|\Omega|}\int_\Omega \operatorname{div}(\boldsymbol{\tau}) \in \mathbb{R} \quad \text{in} \quad \Omega\,, \quad \text{and} \quad \gamma_{\boldsymbol{\nu}}(\boldsymbol{\tau}) \,=\, 0 \quad \text{on} \quad \Gamma\Big\}.$$

However, applying (4.130) to $\boldsymbol{\tau}\in V$ and $w\equiv 1$, we obtain

$$0 \,=\, \langle \gamma_{\boldsymbol{\nu}}(\boldsymbol{\tau}), 1\rangle \,=\, \int_\Omega \operatorname{div}(\boldsymbol{\tau}) \,=\, \operatorname{div}(\boldsymbol{\tau})\,|\Omega|\,,$$

and consequently

$$V := N(\mathbf{B}) \,=\, \Big\{\boldsymbol{\tau}\in H: \quad \operatorname{div}(\boldsymbol{\tau}) = 0 \quad \text{in} \quad \Omega, \quad \text{and} \quad \gamma_{\boldsymbol{\nu}}(\boldsymbol{\tau}) \,=\, 0 \quad \text{on} \quad \Gamma\Big\}.$$

Thus, as in Section 4.6.1 (cf. (4.165)), it follows that a is V-elliptic with ellipticity constant $\alpha = 1$. We now address the proof of the continuous inf-sup condition for the bilinear form b. Indeed, given $(v,\psi)\in Q := \mathrm{L}^2_0(\Omega)\times \mathrm{H}^{1/2}(\Gamma)$, we first consider the auxiliary Neumann problem

$$-\Delta z = v \quad \text{in} \quad \Omega, \quad \gamma_1(z) = 0 \quad \text{on} \quad \Gamma\,, \quad \int_\Omega z \,=\, 0\,, \tag{4.178}$$

whose primal formulation, obtained by using (4.134), reduces to (cf. (4.121)): Find $z\in \widetilde{\mathrm{H}}^1(\Omega)$ such that

$$\int_\Omega \nabla z\cdot\nabla w \,=\, \int_\Omega v\,w \qquad \forall\, w\in\widetilde{\mathrm{H}}^1(\Omega)\,. \tag{4.179}$$

Note that the fact that v belongs to $\mathrm{L}^2_0(\Omega)$ implies that the compatibility condition (4.171) is satisfied for the data of (4.178). Moreover, since the norm and the seminorm of $\mathrm{H}^1(\Omega)$ are equivalent in $\widetilde{\mathrm{H}}^1(\Omega)$ (cf. (4.125), Section 4.5.4), it follows, thanks to the Lax-Milgram Lemma (cf. Lemma 4.4), that (4.179) has a unique solution $z\in\widetilde{\mathrm{H}}^1(\Omega)$. In turn, the corresponding a priori estimate establishes that

$$\|z\|_{1,\Omega} \,\leqslant\, \widetilde{C}\,\|v\|_{0,\Omega}\,,$$

where $\widetilde{C}$ is a positive constant that depends on $\underline{C}$ (cf. (4.125)). Then, defining $\widetilde{\boldsymbol{\tau}} := -\nabla z \in \mathbf{L}^2(\Omega)$, we have that $\operatorname{div}(\widetilde{\boldsymbol{\tau}}) = v\in \mathrm{L}^2(\Omega)$, with which $\widetilde{\boldsymbol{\tau}}\in\mathbf{H}(\operatorname{div};\Omega)$ and (cf. (4.117))

$$\|\widetilde{\boldsymbol{\tau}}\|_{\operatorname{div};\Omega} \,=\, \Big\{|z|^2_{1,\Omega} + \|v\|^2_{0,\Omega}\Big\}^{1/2} \,\leqslant\, \Big(1+\widetilde{C}^2\Big)^{1/2}\,\|v\|_{0,\Omega}\,,$$

while, according to the Neumann condition in (4.178), we have that $\gamma_{\boldsymbol{\nu}}(\widetilde{\boldsymbol{\tau}}) = 0$ on Γ. It follows that

$$\sup_{\substack{\boldsymbol{\tau}\in H\\ \boldsymbol{\tau}\neq 0}} \frac{b\big(\boldsymbol{\tau},(v,\psi)\big)}{\|\boldsymbol{\tau}\|_{\operatorname{div};\Omega}} \,\geqslant\, \frac{b\big(\widetilde{\boldsymbol{\tau}},(v,\psi)\big)}{\|\widetilde{\boldsymbol{\tau}}\|_{\operatorname{div};\Omega}} \,=\, \frac{\|v\|^2_{0,\Omega}}{\|\widetilde{\boldsymbol{\tau}}\|_{\operatorname{div};\Omega}} \,\geqslant\, \beta_1\,\|v\|_{0,\Omega}\,, \tag{4.180}$$

with $\beta_1 := \Big(1+\widetilde{C}^2\Big)^{-1/2}$. On the other hand, noting from the definition of b (cf. (4.176)) and from the inclusion $\mathbf{H}^\perp_0(\operatorname{div}_{\mathbb{R}};\Omega)\subseteq \mathbf{H}(\operatorname{div}_{\mathbb{R}};\Omega)$ (cf. (4.128)) that

$$b\big(\boldsymbol{\tau},(v,\psi)\big) = \langle\gamma_{\boldsymbol{\nu}}(\boldsymbol{\tau}),\psi\rangle \qquad \forall\,\boldsymbol{\tau}\in \mathbf{H}_0^{\perp}(\mathrm{div}_{\mathbb{R}};\Omega)\,,$$

and using the bijection $\widetilde{\gamma}_{\boldsymbol{\nu}}:\mathbf{H}_0^{\perp}(\mathrm{div}_{\mathbb{R}};\Omega)\to \mathrm{H}^{-1/2}(\Gamma)$ (cf. (4.129)), we find that

$$\begin{aligned}
\sup_{\substack{\boldsymbol{\tau}\in H\\ \boldsymbol{\tau}\neq 0}} \frac{b\big(\boldsymbol{\tau},(v,\psi)\big)}{\|\boldsymbol{\tau}\|_{\mathrm{div};\Omega}} &\geqslant \sup_{\substack{\boldsymbol{\tau}\in \mathbf{H}_0^{\perp}(\mathrm{div}_{\mathbb{R}};\Omega)\\ \boldsymbol{\tau}\neq 0}} \frac{b\big(\boldsymbol{\tau},(v,\psi)\big)}{\|\boldsymbol{\tau}\|_{\mathrm{div};\Omega}} = \sup_{\substack{\boldsymbol{\tau}\in \mathbf{H}_0^{\perp}(\mathrm{div}_{\mathbb{R}};\Omega)\\ \boldsymbol{\tau}\neq 0}} \frac{\langle\gamma_{\boldsymbol{\nu}}(\boldsymbol{\tau}),\psi\rangle}{\|\boldsymbol{\tau}\|_{\mathrm{div};\Omega}}\\
&= \sup_{\substack{\lambda\in \mathrm{H}^{-1/2}(\Gamma)\\ \lambda\neq 0}} \frac{\langle\gamma_{\boldsymbol{\nu}}(\widetilde{\gamma}_{\boldsymbol{\nu}}^{-1}(\lambda)),\psi\rangle}{\|\widetilde{\gamma}_{\boldsymbol{\nu}}^{-1}(\lambda)\|_{\mathrm{div};\Omega}} = \sup_{\substack{\lambda\in \mathrm{H}^{-1/2}(\Gamma)\\ \lambda\neq 0}} \frac{\langle\lambda,\psi\rangle}{\|\widetilde{\gamma}_{\boldsymbol{\nu}}^{-1}(\lambda)\|_{\mathrm{div};\Omega}} \qquad (4.181)\\
&\geqslant \beta_2 \sup_{\substack{\lambda\in \mathrm{H}^{-1/2}(\Gamma)\\ \lambda\neq 0}} \frac{\langle\lambda,\psi\rangle}{\|\lambda\|_{-1/2,\Gamma}} = \beta_2\,\|\psi\|_{1/2,\Gamma}\,,
\end{aligned}$$

with $\beta_2 := \|\widetilde{\gamma}_{\boldsymbol{\nu}}^{-1}\|^{-1}$. We remark here that the last equality is a consequence of Lemma 2.25. Finally, it follows from (4.180) and (4.181) that b satisfies the continuous inf-sup condition with constant $\beta := \frac{1}{2}\min\{\beta_1,\beta_2\}$. In virtue of the previous analysis, a direct application of Theorem 4.21 allows us to conclude that there exists a unique solution $\big(\boldsymbol{\sigma},(u,\varphi)\big)\in H\times Q$ of the variational formulation (4.175), and that there exists a positive constant C, depending on $\|\mathbf{A}\|\leqslant 1$, $\alpha = 1$, and β, such that

$$\|\boldsymbol{\sigma}\|_H + \|(u,\varphi)\|_Q \leqslant C\Big\{\|f\|_{0,\Omega} + \|g\|_{-1/2,\Gamma}\Big\}\,,$$

where this last inequality makes use of the fact, as deduced from (4.177), that

$$\|F\|_{H'} = 0 \quad\text{and}\quad \|G\|_{Q'} \leqslant \|f\|_{0,\Omega} + \|g\|_{-1/2,\Gamma}\,.$$

4.6.3 The Poisson problem with mixed boundary conditions

Let Ω be a bounded open subset of $\mathbb{R}^n$, $n\geqslant 2$, with Lipschitz-continuous boundary Γ, and let Γ_D and Γ_N be disjoint parts of Γ such that $|\Gamma_D|\neq 0$ and $\Gamma = \bar{\Gamma}_D\cup\bar{\Gamma}_N$. Then, given $f\in \mathrm{L}^2(\Omega)$ and $g\in \mathrm{H}_{00}^{-1/2}(\Gamma_N)$, we are interested in the boundary value problem

$$-\Delta u = f \quad\text{in}\quad \Omega, \qquad u = 0 \quad\text{on}\quad \Gamma_D, \qquad \nabla u\cdot\boldsymbol{\nu} = g \quad\text{on}\quad \Gamma_N\,, \tag{4.182}$$

where $\boldsymbol{\nu}$ is the outward normal vector to Γ. Here we recall from Section 4.5.5.1 that, if $E_{N,0}:\mathrm{H}^{1/2}(\Gamma_N)\to \mathrm{L}^2(\Gamma)$ is the extension operator

$$E_{N,0}(\eta) := \begin{cases}\eta \text{ on } \Gamma_N\\ 0 \text{ on } \Gamma_D\end{cases} \quad \forall\,\eta\in \mathrm{H}^{1/2}(\Gamma_N)\,,$$

then (cf. (4.145))

$$\mathrm{H}_{00}^{1/2}(\Gamma_N) := \Big\{\eta\in \mathrm{H}^{1/2}(\Gamma_N):\quad E_{N,0}(\eta)\in \mathrm{H}^{1/2}(\Gamma)\Big\},$$

which is endowed with the norm (cf. (4.146))

$$\|\eta\|_{1/2,00,\Gamma_N} := \|E_{N,0}(\eta)\|_{1/2,\Gamma} \qquad \forall\,\eta\in \mathrm{H}_{00}^{1/2}(\Gamma_N). \tag{4.183}$$

In turn, $\mathrm{H}_{00}^{-1/2}(\Gamma_N)$ is the dual of $\mathrm{H}_{00}^{1/2}(\Gamma_N)$, and the corresponding duality pairing is denoted by $\langle\cdot,\cdot\rangle_{\Gamma_N}$. Moreover, given $\psi \in \mathrm{H}^{-1/2}(\Gamma)$, its restriction to Γ_N, denoted $\psi|_{\Gamma_N}$ and defined by (cf. (4.147))

$$\langle\psi|_{\Gamma_N},\eta\rangle_{\Gamma_N} := \langle\psi, E_{N,0}(\eta)\rangle \quad \forall\,\eta \in \mathrm{H}_{00}^{1/2}(\Gamma_N)\,, \tag{4.184}$$

where $\langle\cdot,\cdot\rangle$ denotes the duality pairing between $\mathrm{H}^{-1/2}(\Gamma)$ and $\mathrm{H}^{1/2}(\Gamma)$, is clearly an element of $\mathrm{H}_{00}^{-1/2}(\Gamma_N)$. Now, for the mixed variational formulation of (4.182) we proceed analogously to the previous example, and define the additional unknown $\boldsymbol{\sigma} := \nabla u$ in Ω, so that (4.182) is rewritten as

$$\boldsymbol{\sigma} = \nabla u \quad \text{in} \quad \Omega\,, \quad \operatorname{div}(\boldsymbol{\sigma}) = -f \quad \text{in} \quad \Omega\,, \quad u = 0 \quad \text{on} \quad \Gamma_D\,, \quad \boldsymbol{\sigma}\cdot\boldsymbol{\nu} = g \quad \text{on} \quad \Gamma_N\,.$$

Then, applying again Green's identity given by (4.130) (cf. Lemma 4.32), we obtain

$$\int_\Omega \boldsymbol{\sigma}\cdot\boldsymbol{\tau} = \int_\Omega \nabla u\cdot\boldsymbol{\tau} = -\int_\Omega u\operatorname{div}(\boldsymbol{\tau}) + \langle\gamma_{\boldsymbol{\nu}}(\boldsymbol{\tau}), \gamma_0(u)\rangle,$$

from which, introducing the auxiliary unknown $\xi := -\gamma_0(u) \in \mathrm{H}_{00}^{1/2}(\Gamma_N)$, which is justified by the Dirichlet boundary condition, we deduce that

$$\int_\Omega \boldsymbol{\sigma}\cdot\boldsymbol{\tau} + \int_\Omega u\operatorname{div}(\boldsymbol{\tau}) + \langle\gamma_{\boldsymbol{\nu}}(\boldsymbol{\tau})|_{\Gamma_N},\xi\rangle_{\Gamma_N} = 0 \qquad \forall\,\boldsymbol{\tau} \in \mathbf{H}(\operatorname{div};\Omega). \tag{4.185}$$

On the other hand, as in Section 4.6.1, the equation $\operatorname{div}(\boldsymbol{\sigma}) = -f$ in Ω, reduces to

$$\int_\Omega v\operatorname{div}(\boldsymbol{\sigma}) = -\int_\Omega f\,v \qquad \forall \in \mathrm{L}^2(\Omega)\,. \tag{4.186}$$

Finally, since $\gamma_{\boldsymbol{\nu}}(\boldsymbol{\sigma})|_{\Gamma_N} \in \mathrm{H}_{00}^{-1/2}(\Gamma_N)$, the Neumann boundary condition $\boldsymbol{\sigma}\cdot\boldsymbol{\nu} = g$ on Γ_N, is equivalently reformulated as

$$\langle\gamma_{\boldsymbol{\nu}}(\boldsymbol{\sigma})|_{\Gamma_N},\eta\rangle_{\Gamma_N} = \langle g,\eta\rangle_{\Gamma_N} \qquad \forall\,\eta \in \mathrm{H}_{00}^{1/2}(\Gamma_N)\,. \tag{4.187}$$

Consequently, grouping (4.186) and (4.187) into one equation, and combining it with (4.185), we arrive at the mixed variational formulation of (4.182): Find $(\boldsymbol{\sigma},(u,\xi)) \in H\times Q$ such that

$$\begin{aligned} a(\boldsymbol{\sigma},\boldsymbol{\tau}) + b(\boldsymbol{\tau},(u,\xi)) &= F(\boldsymbol{\tau}) \qquad \forall\,\boldsymbol{\tau}\in H,\\ b(\boldsymbol{\sigma},\,(v,\eta)) &= G(v,\eta) \quad \forall\,(v,\eta)\in Q\,, \end{aligned} \tag{4.188}$$

where

$$H := \mathbf{H}(\operatorname{div};\Omega), \quad Q = \mathrm{L}^2(\Omega)\times \mathrm{H}_{00}^{1/2}(\Gamma_N),$$

a is the same bilinear form as in the previous example (cf. (4.176)), $b : H\times Q \to \mathbb{R}$ is defined by

$$b(\boldsymbol{\tau},(v,\eta)) = \int_\Omega v\operatorname{div}(\boldsymbol{\tau}) + \langle\gamma_{\boldsymbol{\nu}}(\boldsymbol{\tau})|_{\Gamma_N},\eta\rangle_{\Gamma_N} \quad \forall\,(\boldsymbol{\tau},(v,\eta)) \in H\times Q\,,$$

and the functionals $F\in H'$ and $G\in Q'$ are given by

$$F(\boldsymbol{\tau}) := 0 \quad \forall\,\boldsymbol{\tau}\in H, \quad G(v,\eta) := -\int_\Omega f\,v + \langle g,\eta\rangle_{\Gamma_N} \quad \forall\,(v,\eta)\in Q.$$

Next, we apply again Theorem 4.21. First, we observe that $\mathbf{B} : H \to Q$, the operator induced by b, is given by

$$\mathbf{B}(\boldsymbol{\tau}) := \big(\operatorname{div}(\boldsymbol{\tau}),\, \mathcal{R}_{00}\, \gamma_{\boldsymbol{\nu}}(\boldsymbol{\tau})|_{\Gamma_N}\big) \quad \forall\, \boldsymbol{\tau} \in H\,, \tag{4.189}$$

where $\mathcal{R}_{00} : \mathrm{H}_{00}^{-1/2}(\Gamma_N) \to \mathrm{H}_{00}^{1/2}(\Gamma_N)$ is the corresponding Riesz mapping (cf. Lemma 4.35). Since $\mathcal{R}_{00}$ is an isometry and $\gamma_{\boldsymbol{\nu}}|_{\Gamma_N}$ (cf. (4.184)) is bounded because $\gamma_{\boldsymbol{\nu}}$ is bounded with $\|\gamma_{\boldsymbol{\nu}}\| \leqslant 1$ (cf. (4.119) in Theorem 4.31), it follows that $\mathbf{B}$ is also bounded with $\|\mathbf{B}\| \leqslant 2$. In turn, it is clear that

$$V := N(\mathbf{B}) = \Big\{\boldsymbol{\tau} \in H : \ \operatorname{div}(\boldsymbol{\tau}) = 0 \quad \text{in} \quad \Omega, \quad \gamma_{\boldsymbol{\nu}}(\boldsymbol{\tau}) = 0 \quad \text{on} \quad \Gamma_N\Big\},$$

and then, for each $\boldsymbol{\tau} \in V$ we have

$$a(\boldsymbol{\tau}, \boldsymbol{\tau}) = \|\boldsymbol{\tau}\|_{0,\Omega}^2 = \|\boldsymbol{\tau}\|_{\mathrm{div};\Omega}^2\,,$$

which proves that a is V-elliptic with ellipticity constant $\alpha = 1$. On the other hand, given $(v,\eta) \in Q$, we consider the boundary value problem

$$\Delta z = v \quad \text{in} \quad \Omega, \qquad z = 0 \quad \text{on} \quad \Gamma_D, \qquad \nabla z \cdot \boldsymbol{\nu} = \mathcal{R}_{00}^{-1}(\eta) \quad \text{on} \quad \Gamma_N\,,$$

whose primal variational formulation reduces to: Find $z \in \mathrm{H}^1_{\Gamma_D}(\Omega)$ such that

$$\int_\Omega \nabla z \cdot \nabla w = -\int_\Omega v\, w + \langle \mathcal{R}_{00}^{-1}(\eta),\, \gamma_0\,(w)\rangle_{\Gamma_N} \quad \forall\, w \in \mathrm{H}^1_{\Gamma_D}(\Omega), \tag{4.190}$$

where

$$\mathrm{H}^1_{\Gamma_D}(\Omega) := \Big\{w \in \mathrm{H}^1(\Omega) : \quad \gamma_0(w) = 0 \text{ on } \Gamma_D\Big\}.$$

The Lax-Milgram Lemma (cf. Lemma 4.4) implies that (4.190) has a unique solution $z \in \mathrm{H}^1_{\Gamma_D}(\Omega)$, which, according to the corresponding result on continuous dependence and the dual pairing between $\mathrm{H}_{00}^{-1/2}(\Gamma_N)$ and $\mathrm{H}_{00}^{1/2}(\Gamma_N)$, satisfies

$$|z|_{1,\Omega} \leqslant \widehat{C}\,\Big\{\|v\|_{0,\Omega} + \|\eta\|_{1/2,00,\Gamma_N}\Big\}\,, \tag{4.191}$$

where $\widehat{C} > 0$ is a constant depending on C_D (cf. (4.151)), an inequality that establishes the equivalence between $\|\cdot\|_{1,\Omega}$ and $|\cdot|_{1,\Omega}$ in $\mathrm{H}^1_{\Gamma_D}(\Omega)$, and on the trace operator. Thus, defining $\widehat{\boldsymbol{\tau}} := \nabla z \in \mathbf{L}^2(\Omega)$, we have that $\operatorname{div}(\widehat{\boldsymbol{\tau}}) = v \in \mathrm{L}^2(\Omega)$, whence $\widehat{\boldsymbol{\tau}} \in \mathbf{H}(\mathrm{div};\Omega)$, and moreover $\gamma_{\boldsymbol{\nu}}(\widehat{\boldsymbol{\tau}})|_{\Gamma_N} = \mathcal{R}_{00}^{-1}(\eta)$ on Γ_N, from which it is clear that $\mathcal{R}_{00}\,\gamma_{\boldsymbol{\nu}}(\widehat{\boldsymbol{\tau}})|_{\Gamma_N} = \eta$ on Γ_N. This shows, according to (4.189), that $\mathbf{B}(\widehat{\boldsymbol{\tau}}) = (v,\eta)$ and therefore $\mathbf{B}$ is surjective. Finally, it is easy to see that G is bounded with

$$\|G\|_{Q'} \leqslant \Big\{\|f\|_{0,\Omega} + \|g\|_{-1/2,00,\Gamma_N}\Big\}\,.$$

Consequently, a direct application of Theorem 4.21 implies that there exists a unique solution $(\boldsymbol{\sigma},(u,\xi)) \in H \times Q$ of problem (4.188), which satisfies

$$\|(\boldsymbol{\sigma},\,(u,\xi))\|_{H\times Q} \leqslant C\Big\{\|f\|_{0,\Omega} + \|g\|_{-1/2,00,\Gamma_N}\Big\}.$$

Similarly to the previous example, we note here that C is a positive constant depending on the constant β of the continuous inf-sup condition of b, and on $\|\mathbf{A}\| = 1$, where $\mathbf{A}$ is the operator induced by a. Now, in order to get a more precise idea of the value of β, we first note from the definition of $\widehat{\boldsymbol{\tau}}$, using also (4.191), that

$$\|\widehat{\boldsymbol{\tau}}\|_{\mathrm{div};\Omega}^2 = |z|_{1,\Omega}^2 + \|v\|_{0,\Omega}^2 \leqslant \left(1 + 2\,\widehat{C}^2\right) \|(v,\eta)\|_Q^2 .$$

Therefore, employing the foregoing inequality and letting $\langle\cdot,\cdot\rangle_Q$ be the inner product of Q, we find that

$$\sup_{\substack{\boldsymbol{\tau}\in H\\ \boldsymbol{\tau}\neq 0}} \frac{b(\boldsymbol{\tau},(v,\eta))}{\|\boldsymbol{\tau}\|_H} \geqslant \frac{b(\widehat{\boldsymbol{\tau}},(v,\eta))}{\|\widehat{\boldsymbol{\tau}}\|_H} = \frac{\displaystyle\int_\Omega v\,\mathrm{div}(\widehat{\boldsymbol{\tau}}) + \langle \gamma_{\boldsymbol{\nu}}(\widehat{\boldsymbol{\tau}}),\eta\rangle_{\Gamma_N}}{\|\widehat{\boldsymbol{\tau}}\|_{\mathrm{div};\Omega}}$$
$$= \frac{\langle \mathbf{B}(\widehat{\boldsymbol{\tau}}),(v,\eta)\rangle_Q}{\|\widehat{\boldsymbol{\tau}}\|_{\mathrm{div};\Omega}} = \frac{\|(v,\eta)\|_Q^2}{\|\widehat{\boldsymbol{\tau}}\|_{\mathrm{div};\Omega}} \geqslant \beta\,\|(v,\eta)\|_Q ,$$

with $\beta := \left(1 + 2\,\widehat{C}^2\right)^{-1/2}$.

4.6.4 The linear elasticity problem

Before defining the problem of interest, and as stated also at the beginning of the chapter, we recall that, given a domain $\mathcal{O}$ and a Lipschitz-continuous curve/surface Σ, both in $\mathbb{R}^n$, $n \in \{2,3\}$, and given $r \in \mathbb{R}$, we write

$$\mathbf{H}^r(\mathcal{O}) := [\mathrm{H}^r(\mathcal{O})]^n , \quad \mathbb{H}^r(\mathcal{O}) := [\mathrm{H}^r(\mathcal{O})]^{n\times n} , \text{ and } \mathbf{H}^r(\Sigma) := [\mathrm{H}^r(\Sigma)]^n ,$$

with corresponding norms $\|\cdot\|_{r,\mathcal{O}}$ (for $\mathrm{H}^r(\mathcal{O})$, $\mathbf{H}^r(\mathcal{O})$ and $\mathbb{H}^r(\mathcal{O})$) and $\|\cdot\|_{r,\Sigma}$ (for $\mathrm{H}^r(\Sigma)$ and $\mathbf{H}^r(\Sigma)$). In turn, in the particular case $r = 0$ one usually writes $\mathbf{L}^2(\mathcal{O})$, $\mathbb{L}^2(\mathcal{O})$ and $\mathbf{L}^2(\Sigma)$ instead of $\mathbf{H}^0(\mathcal{O})$, $\mathbb{H}^0(\mathcal{O})$ and $\mathbf{H}^0(\Sigma)$, respectively.

Now, let Ω be a bounded and simply connected domain of $\mathbb{R}^n$, $n \in \{2,3\}$, with Lipschitz-continuous boundary Γ, and let Γ_D and Γ_N be disjoint parts of Γ such that $|\Gamma_D| \neq 0$ and $\Gamma = \bar{\Gamma}_D \cup \bar{\Gamma}_N$. Then, the linear elasticity problem consists in determining the displacement $\mathbf{u}$ and the stress tensor $\boldsymbol{\sigma}$ of an elastic material occupying the region Ω. More precisely, given $\mathbf{f} \in \mathbf{L}^2(\Omega)$ and $\mathbf{g} \in \mathbf{H}_{00}^{-1/2}(\Gamma_N)$, we seek to find a symmetric tensor $\boldsymbol{\sigma}$ and a vector field $\mathbf{u}$ such that

$$\begin{aligned} \boldsymbol{\sigma} &= \mathcal{C}\,\mathbf{e}(\mathbf{u}) \quad \text{in} \quad \Omega, & \mathbf{div}(\boldsymbol{\sigma}) &= -\mathbf{f} \quad \text{in} \quad \Omega, \\ \mathbf{u} &= 0 \quad \text{on} \quad \Gamma_D, & \boldsymbol{\sigma}\boldsymbol{\nu} &= \mathbf{g} \quad \text{on} \quad \Gamma_N, \end{aligned} \tag{4.192}$$

where $\mathbf{e}(\mathbf{u}) := \frac{1}{2}(\nabla\mathbf{u} + (\nabla\mathbf{u})^{\mathrm{t}})$ is the strain tensor (or the symmetric part of $\nabla\mathbf{u}$), $\mathbf{div}$ is the usual divergence operator div applied to the rows of each tensor, $\boldsymbol{\nu}$ is the unit normal vector to Γ, and $\mathcal{C}$ is the elasticity operator determined by Hooke's law

$$\mathcal{C}\,\zeta := \lambda\,\mathrm{tr}(\zeta)\,\mathbb{I} + 2\mu\,\zeta \quad \forall\,\zeta \in \mathbb{L}^2(\Omega) . \tag{4.193}$$

Here, λ, $\mu > 0$ are the corresponding Lamé constants, $\mathbb{I}$ is the identity matrix of $\mathbb{R}^{n\times n}$, and tr denotes the matrix trace. Applying tr to (4.193), this law can be inverted, which gives

$$\mathcal{C}^{-1}\zeta = \frac{1}{2\mu}\,\zeta - \frac{\lambda}{2\mu\,(n\lambda + 2\mu)}\,\mathrm{tr}(\zeta)\,\mathbb{I} \qquad \forall\,\zeta \in \mathbb{L}^2(\Omega) .$$

4.6.4.1 Dirichlet boundary conditions

In this part we consider the particular case of (4.192) given by $\Gamma_D = \Gamma$, that is,

$$\boldsymbol{\sigma} = \mathcal{C}\,\mathbf{e}(\mathbf{u}) \quad \text{in} \quad \Omega, \quad \mathbf{div}(\boldsymbol{\sigma}) = -\mathbf{f} \quad \text{in} \quad \Omega, \quad \mathbf{u} = 0 \quad \text{on} \quad \Gamma. \tag{4.194}$$

To derive a mixed variational formulation of (4.194), we first observe that

$$\mathcal{C}^{-1}\boldsymbol{\sigma} = \mathbf{e}(\mathbf{u}) = \nabla\mathbf{u} - \boldsymbol{\rho}, \quad \text{where} \tag{4.195}$$

$$\boldsymbol{\rho} := \frac{1}{2}\left(\nabla\mathbf{u} - (\nabla\mathbf{u})^{\mathrm{t}}\right) \tag{4.196}$$

denotes the auxiliary unknown called the solid rotation. In addition, we let $\mathbb{H}(\mathbf{div};\Omega)$ be the tensor version of $\mathbf{H}(\mathrm{div};\Omega)$, that is

$$\mathbb{H}(\mathbf{div};\Omega) := \left\{\boldsymbol{\tau} \in \mathbb{L}^2(\Omega): \quad \mathbf{div}(\boldsymbol{\tau}) \in \mathbf{L}^2(\Omega)\right\},$$

whose inner product and induced norm are defined, respectively, as

$$\langle\boldsymbol{\zeta},\boldsymbol{\tau}\rangle_{\mathbf{div};\Omega} := \int_\Omega \left\{\boldsymbol{\zeta}:\boldsymbol{\tau} + \mathbf{div}(\boldsymbol{\zeta})\cdot\mathbf{div}(\boldsymbol{\tau})\right\} \qquad \forall\,\boldsymbol{\zeta},\,\boldsymbol{\tau} \in \mathbb{H}(\mathbf{div};\Omega),$$

and

$$\|\boldsymbol{\tau}\|_{\mathbf{div};\Omega} := \left\{\|\boldsymbol{\tau}\|_{0,\Omega}^2 + \|\mathbf{div}(\boldsymbol{\tau})\|_{0,\Omega}^2\right\}^{1/2} \qquad \forall\,\boldsymbol{\tau} \in \mathbb{H}(\mathbf{div};\Omega).$$

Then, multiplying tensorially (:) by $\boldsymbol{\tau} \in \mathbb{H}(\mathbf{div};\Omega)$, and applying Green's identity (4.130) through the rows of each tensor, we obtain:

$$\begin{aligned}\int_\Omega \mathcal{C}^{-1}\boldsymbol{\sigma}:\boldsymbol{\tau} &= \int_\Omega \nabla\mathbf{u}:\boldsymbol{\tau} - \int_\Omega \boldsymbol{\rho}:\boldsymbol{\tau}\\ &= -\int_\Omega \mathbf{u}\cdot\mathbf{div}(\boldsymbol{\tau}) + \langle\gamma_{\boldsymbol{\nu}}(\boldsymbol{\tau}),\gamma_0(\mathbf{u})\rangle - \int_\Omega \boldsymbol{\rho}:\boldsymbol{\tau},\end{aligned} \tag{4.197}$$

where $\gamma_{\boldsymbol{\nu}}: \mathbb{H}(\mathbf{div};\Omega) \to \mathbf{H}^{-1/2}(\Gamma)$ and $\gamma_0: \mathbf{H}^1(\Omega) \to \mathbf{H}^{1/2}(\Gamma)$ are the tensor and vector extensions of the respective trace operators defined in Section 4.5. In particular, denoting by $\boldsymbol{\tau}_i$ and u_i the i-th row of $\boldsymbol{\tau}$ and the i-th component of $\mathbf{u}$, respectively, there holds

$$\langle\gamma_{\boldsymbol{\nu}}(\boldsymbol{\tau}),\gamma_0(\mathbf{u})\rangle := \sum_{i=1}^n \langle\gamma_{\boldsymbol{\nu}}(\boldsymbol{\tau}_i),\gamma_0(u_i)\rangle,$$

where $\langle\cdot,\cdot\rangle$ stands for the duality pairing between $\mathbf{H}^{-1/2}(\Gamma)$ and $\mathbf{H}^{1/2}(\Gamma)$, as well as between $\mathrm{H}^{-1/2}(\Gamma)$ and $\mathrm{H}^{1/2}(\Gamma)$, in the left-hand and right-hand sides of the foregoing equation, respectively. Then, using that $\gamma_0(\mathbf{u}) = 0$ on Γ, we obtain

$$\int_\Omega \mathcal{C}^{-1}\boldsymbol{\sigma}:\boldsymbol{\tau} + \int_\Omega \mathbf{u}\cdot\mathbf{div}(\boldsymbol{\tau}) + \int_\Omega \boldsymbol{\rho}:\boldsymbol{\tau} = 0 \qquad \forall\,\boldsymbol{\tau} \in \mathbb{H}(\mathbf{div};\Omega). \tag{4.198}$$

Let us note here that $\boldsymbol{\rho} \in \mathbb{L}^2_{\text{skew}}(\Omega)$, where

$$\mathbb{L}^2_{\text{skew}}(\Omega) := \left\{\boldsymbol{\eta} \in \mathbb{L}^2(\Omega): \quad \boldsymbol{\eta} + \boldsymbol{\eta}^t = 0\right\}.$$

Thus, the symmetry of $\boldsymbol{\sigma}$ is imposed weakly through the equation

$$\int_\Omega \boldsymbol{\sigma} : \boldsymbol{\eta} = 0 \qquad \forall\, \boldsymbol{\eta} \in \mathbb{L}^2_{\text{skew}}(\Omega)\,. \tag{4.199}$$

Finally, the equilibrium equation can be rewritten as

$$\int_\Omega \mathbf{v} \cdot \mathbf{div}(\boldsymbol{\sigma}) = -\int_\Omega \mathbf{f} \cdot \mathbf{v} \qquad \forall\, \mathbf{v} \in \mathbf{L}^2(\Omega). \tag{4.200}$$

In this way, by adding (4.199) and (4.200), and combining the resulting equation with (4.198), we obtain the following mixed variational formulation of (4.194): Find $(\boldsymbol{\sigma}, (\mathbf{u}, \boldsymbol{\rho})) \in H \times Q$ such that

$$\begin{aligned} a(\boldsymbol{\sigma}, \boldsymbol{\tau}) + b(\boldsymbol{\tau}, (\mathbf{u}, \boldsymbol{\rho})) &= F(\boldsymbol{\tau}) && \forall\, \boldsymbol{\tau} \in H, \\ b(\boldsymbol{\sigma}, (\mathbf{v}, \boldsymbol{\eta})) &= G(\mathbf{v}, \boldsymbol{\eta}) && \forall\, (\mathbf{v}, \boldsymbol{\eta}) \in Q, \end{aligned} \tag{4.201}$$

where

$$H := \mathbb{H}(\mathbf{div}; \Omega), \quad Q := \mathbf{L}^2(\Omega) \times \mathbb{L}^2_{\text{skew}}(\Omega),$$

$a : H \times H \to \mathbb{R}$ and $b : H \times Q \to \mathbb{R}$ are the bilinear forms defined by

$$a(\boldsymbol{\zeta}, \boldsymbol{\tau}) := \int_\Omega \mathcal{C}^{-1} \boldsymbol{\zeta} : \boldsymbol{\tau} = \frac{1}{2\mu} \int_\Omega \boldsymbol{\zeta} : \boldsymbol{\tau} - \frac{\lambda}{2\mu(n\lambda + 2\mu)} \int_\Omega \operatorname{tr}(\boldsymbol{\zeta}) \operatorname{tr}(\boldsymbol{\tau}) \tag{4.202}$$

and

$$b(\boldsymbol{\tau}, (\mathbf{v}, \boldsymbol{\eta})) := \int_\Omega \mathbf{v} \cdot \mathbf{div}(\boldsymbol{\tau}) + \int_\Omega \boldsymbol{\eta} : \boldsymbol{\tau},$$

whereas the linear and bounded functionals $F \in H'$ and $G \in Q'$ are given by

$$F(\boldsymbol{\tau}) := 0 \quad \forall\, \boldsymbol{\tau} \in H\,, \qquad G(\mathbf{v}, \boldsymbol{\eta}) := -\int_\Omega \mathbf{f} \cdot \mathbf{v} \quad \forall\, (\mathbf{v}, \boldsymbol{\eta}) \in Q\,. \tag{4.203}$$

In what follows, we analyze the solvability of (4.201) by applying once again Theorem 4.21. However, this theorem will not be applied exactly to (4.201) but to an equivalent formulation. For this, it is first necessary to note some properties of the bilinear forms a and b. Indeed, simple algebraic calculations show that

$$a(\mathbb{I}, \boldsymbol{\tau}) = \frac{1}{(n\,\lambda + 2\mu)} \int_\Omega \operatorname{tr}(\boldsymbol{\tau}) \qquad \forall\, \boldsymbol{\tau} \in H, \tag{4.204}$$

and

$$a(\boldsymbol{\zeta}, \boldsymbol{\tau}) = \frac{1}{2\mu} \int_\Omega \boldsymbol{\zeta}^{\mathrm{d}} : \boldsymbol{\tau}^{\mathrm{d}} + \frac{1}{n\,(n\,\lambda + 2\mu)} \int_\Omega \operatorname{tr}(\boldsymbol{\zeta}) \operatorname{tr}(\boldsymbol{\tau}) \qquad \forall\, \boldsymbol{\zeta}, \boldsymbol{\tau} \in H, \tag{4.205}$$

where, given $\boldsymbol{\tau} \in \mathbb{L}^2(\Omega)$, $\boldsymbol{\tau}^{\mathrm{d}} := \boldsymbol{\tau} - \dfrac{1}{n} \operatorname{tr}(\boldsymbol{\tau})\, \mathbb{I}$ is the corresponding deviatoric tensor. The convenience of writing a in the form (4.205) will be seen later when it is shown that a is elliptic in the kernel of the operator induced by b. In turn, it is easy to see that

$$b(\mathbb{I}, (\mathbf{v}, \boldsymbol{\eta})) = 0 \qquad \forall\, (\mathbf{v}, \boldsymbol{\eta}) \in Q. \tag{4.206}$$

Let us now observe that H can be decomposed as

$$H := H_0 \oplus \mathbb{R}\,\mathbb{I},$$

where

$$H_0 := \left\{ \boldsymbol{\tau} \in \mathbb{H}(\mathbf{div};\Omega) : \quad \int_\Omega \operatorname{tr}(\boldsymbol{\tau}) = 0 \right\}.$$

More specifically, for each $\boldsymbol{\tau} \in H$ there exist unique $\boldsymbol{\tau}_0 \in H_0$ and $d := \dfrac{1}{n\,|\Omega|} \displaystyle\int_\Omega \operatorname{tr}(\boldsymbol{\tau})$ $\in \mathbb{R}$ such that $\boldsymbol{\tau} = \boldsymbol{\tau}_0 + d\,\mathbb{I}$. Then, we consider the problem that arises from (4.201) by replacing H by H_0, that is: Find $(\boldsymbol{\sigma},(\mathbf{u},\boldsymbol{\rho})) \in H_0 \times Q$ such that

$$\begin{aligned} a(\boldsymbol{\sigma},\boldsymbol{\tau}) + b(\boldsymbol{\tau},(\mathbf{u},\boldsymbol{\rho})) &= F(\boldsymbol{\tau}) \qquad \forall\, \boldsymbol{\tau} \in H_0\,, \\ b(\boldsymbol{\sigma},(\mathbf{v},\boldsymbol{\eta})) &= G(\mathbf{v},\boldsymbol{\eta}) \quad \forall\,(\mathbf{v},\boldsymbol{\eta}) \in Q\,. \end{aligned} \tag{4.207}$$

Regarding this, we have the following result that relates (4.201) and (4.207).

Lemma 4.37. *Problems* (4.201) *and* (4.207) *are equivalent, that is,* $(\boldsymbol{\sigma},(\mathbf{u},\boldsymbol{\rho})) \in H \times Q$ *is a solution of* (4.201) *if and only if* $\boldsymbol{\sigma} \in H_0$ *and* $(\boldsymbol{\sigma},(\mathbf{u},\boldsymbol{\rho}))$ *is a solution of* (4.207).

Proof. Let $(\boldsymbol{\sigma},(\mathbf{u},\boldsymbol{\rho})) \in H \times Q$ be a solution of (4.201). Taking $\boldsymbol{\tau} = \mathbb{I}$ in the first equation of (4.201), and using (4.204) and (4.206), we obtain

$$0 = a(\boldsymbol{\sigma},\mathbb{I}) + b(\mathbb{I},(\mathbf{u},\boldsymbol{\rho})) = \frac{1}{(n\lambda + 2\mu)} \int_\Omega \operatorname{tr}(\boldsymbol{\sigma})\,,$$

from which it follows that $\boldsymbol{\sigma} \in H_0$, and then $(\boldsymbol{\sigma},(\mathbf{u},\boldsymbol{\rho})) \in H_0 \times Q$ is a solution of (4.207). Reciprocally, let $(\boldsymbol{\sigma},(\mathbf{u},\boldsymbol{\rho})) \in H_0 \times Q$ be a solution of (4.207). It is clear that $(\boldsymbol{\sigma},(\mathbf{u},\boldsymbol{\rho}))$ satisfies the second equation of (4.201). Now, given $\boldsymbol{\tau} = \boldsymbol{\tau}_0 + d\,\mathbb{I} \in H$, with $\boldsymbol{\tau}_0 \in H_0$ and $d \in \mathbb{R}$, we have, using (4.206) and the first equation of (4.207), that

$$\begin{aligned} a(\boldsymbol{\sigma},\boldsymbol{\tau}) + b(\boldsymbol{\tau},(\mathbf{u},\boldsymbol{\rho})) &= a(\boldsymbol{\sigma},\boldsymbol{\tau}_0) + b(\boldsymbol{\tau}_0,(\mathbf{u},\boldsymbol{\rho})) + d\left\{ a(\boldsymbol{\sigma},\mathbb{I}) + b(\mathbb{I},(\mathbf{u},\boldsymbol{\rho})) \right\} \\ &= a(\boldsymbol{\sigma},\boldsymbol{\tau}_0) + b(\boldsymbol{\tau}_0,(\mathbf{u},\boldsymbol{\rho})) = 0 = F(\boldsymbol{\tau})\,, \end{aligned}$$

which proves that $(\boldsymbol{\sigma},(\mathbf{u},\boldsymbol{\rho}))$ is also a solution of (4.201).

□

According to the equivalence established by the previous lemma, in what follows we focus our analysis on problem (4.207). To this end, let us first note that $\mathbf{B} : H_0 \to Q$, the operator induced by $b : H_0 \times Q \to \mathbb{R}$, is given by

$$\mathbf{B}(\boldsymbol{\tau}) := \left(\mathbf{div}(\boldsymbol{\tau}), \frac{1}{2}(\boldsymbol{\tau} - \boldsymbol{\tau}^{\mathrm{t}}) \right) \qquad \forall\, \boldsymbol{\tau} \in H_0\,, \tag{4.208}$$

from which it is clear that $\mathbf{B}$ is bounded with $\|\mathbf{B}\| \leqslant 1$. In turn, from the definition of the bilinear form a (cf. (4.202)), applying the Cauchy-Schwarz inequality, using that $\|\operatorname{tr}(\boldsymbol{\tau})\|_{0,\Omega} \leqslant \sqrt{n}\,\|\boldsymbol{\tau}\|_{0,\Omega}$ for all $\boldsymbol{\tau} \in \mathbb{L}^2(\Omega)$, and noting that

$$\frac{n\,\lambda}{(n\,\lambda + 2\,\mu)} \leqslant 1 \qquad \forall\, n \in \mathbb{N}\,,$$

it follows that

$$\begin{aligned} |a(\boldsymbol{\zeta},\boldsymbol{\tau})| &= \left| \frac{1}{2\mu} \int_\Omega \boldsymbol{\zeta} : \boldsymbol{\tau} - \frac{\lambda}{2\mu\,(n\,\lambda + 2\mu)} \int_\Omega \operatorname{tr}(\boldsymbol{\zeta})\operatorname{tr}(\boldsymbol{\tau}) \right| \\ &\leqslant \frac{1}{2\mu}\,\|\boldsymbol{\zeta}\|_{0,\Omega}\,\|\boldsymbol{\tau}\|_{0,\Omega} + \frac{1}{2\,\mu}\,\frac{n\lambda}{(n\lambda + 2\mu)}\,\|\boldsymbol{\zeta}\|_{0,\Omega}\,\|\boldsymbol{\tau}\|_{0,\Omega} \\ &\leqslant \frac{1}{\mu}\,\|\boldsymbol{\zeta}\|_{0,\Omega}\,\|\boldsymbol{\tau}\|_{0,\Omega} \leqslant \frac{1}{\mu}\,\|\boldsymbol{\zeta}\|_{\mathbf{div};\Omega}\,\|\boldsymbol{\tau}\|_{\mathbf{div};\Omega} \qquad \forall\, \boldsymbol{\zeta},\, \boldsymbol{\tau} \in H_0\,, \end{aligned}$$

which shows that $\mathbf{A} : H_0 \to H_0$, the operator induced by a, is bounded with $\|\mathbf{A}\| \leqslant \dfrac{1}{\mu}$. On the other hand, from (4.208) we also have that

$$V := N(\mathbf{B}) = \Big\{\boldsymbol{\tau} \in H_0 : \ \mathbf{div}(\boldsymbol{\tau}) = 0 \quad \text{in} \quad \Omega, \quad \boldsymbol{\tau} = \boldsymbol{\tau}^{\mathrm{t}} \quad \text{in} \quad \Omega\Big\},$$

and therefore, from (4.205) we obtain the inequality:

$$a(\boldsymbol{\tau},\boldsymbol{\tau}) = \frac{1}{2\mu}\|\boldsymbol{\tau}^{\mathrm{d}}\|_{0,\Omega}^2 + \frac{1}{n(n\lambda+2\mu)}\|\mathrm{tr}(\boldsymbol{\tau})\|_{0,\Omega}^2 \geqslant \frac{1}{2\mu}\|\boldsymbol{\tau}^{\mathrm{d}}\|_{0,\Omega}^2 \qquad \forall\, \boldsymbol{\tau} \in V, \tag{4.209}$$

which, however, is still insufficient to conclude the V-ellipticity of a. To be able to achieve this, the following result is needed.

Lemma 4.38. *There exists $c_1 > 0$, depending only on Ω, such that*

$$c_1\|\boldsymbol{\tau}\|_{0,\Omega}^2 \leqslant \|\boldsymbol{\tau}^{\mathrm{d}}\|_{0,\Omega}^2 + \|\mathbf{div}(\boldsymbol{\tau})\|_{0,\Omega}^2 \quad \forall\, \boldsymbol{\tau} \in H_0 .$$

Proof. We begin by recalling from [90, Corollary 2.4, Chapter I] (see also Exercise 10.8, which, in turn, makes use of Exercise 10.7) that the operator div is an isomorphism from $W^{\perp}$ onto $\mathrm{L}_0^2(\Omega)$, where

$$W := \Big\{\mathbf{z} \in \mathbf{H}_0^1(\Omega) : \quad \mathrm{div}(\mathbf{z}) = 0 \text{ in } \Omega\Big\}$$

and

$$\mathrm{L}_0^2(\Omega) := \Big\{v \in \mathrm{L}^2(\Omega) : \quad \int_\Omega v = 0\Big\}.$$

Then, given $\boldsymbol{\tau} \in H_0$, we have $\mathrm{tr}(\boldsymbol{\tau}) \in \mathrm{L}_0^2(\Omega)$, and hence there exists a unique $\mathbf{z} \in W^{\perp}$ such that $\mathrm{div}(\mathbf{z}) = \mathrm{tr}(\boldsymbol{\tau})$. Moreover, thanks to the Bounded Inverse Theorem (cf. Theorem 3.34), there holds

$$\|\mathbf{z}\|_{1,\Omega} \leqslant \|(\mathrm{div})^{-1}\|\,\|\mathrm{tr}(\boldsymbol{\tau})\|_{0,\Omega} . \tag{4.210}$$

It follows that

$$\begin{aligned}
\|\mathrm{tr}(\boldsymbol{\tau})\|_{0,\Omega}^2 &= \int_\Omega \mathrm{tr}(\boldsymbol{\tau})\,\mathrm{div}(\mathbf{z}) = \int_\Omega \boldsymbol{\tau} : \mathbb{I}\,\mathrm{tr}(\nabla\mathbf{z}) \\
&= \int_\Omega \boldsymbol{\tau} : \mathrm{tr}(\nabla\mathbf{z})\,\mathbb{I} = n\int_\Omega \boldsymbol{\tau} : \big(\nabla\mathbf{z} - (\nabla\mathbf{z})^{\mathrm{d}}\big) \\
&= n\int_\Omega \boldsymbol{\tau} : \nabla\mathbf{z} - n\int_\Omega \boldsymbol{\tau}^{\mathrm{d}} : \nabla\mathbf{z},
\end{aligned}$$

where the last expression uses that $\boldsymbol{\tau} : \boldsymbol{\zeta}^{\mathrm{d}} = \boldsymbol{\tau}^{\mathrm{d}} : \boldsymbol{\zeta}$ for all $\boldsymbol{\tau}, \boldsymbol{\zeta} \in \mathbb{L}^2(\Omega)$. Then, applying Green's identity (4.130) through the rows of each tensor in the first term, as we did for the derivation of (4.197), noting that $\gamma_0(\mathbf{z}) = 0$ on Γ, and then employing the Cauchy-Schwarz inequality, we obtain

$$\begin{aligned}
\|\mathrm{tr}(\boldsymbol{\tau})\|_{0,\Omega}^2 &= -n\int_\Omega \mathbf{z}\cdot\mathbf{div}(\boldsymbol{\tau}) - n\int_\Omega \boldsymbol{\tau}^{\mathrm{d}} : \nabla\mathbf{z} \\
&\leqslant n\Big\{\|\mathbf{z}\|_{0,\Omega}\|\mathbf{div}(\boldsymbol{\tau})\|_{0,\Omega} + \|\boldsymbol{\tau}^{\mathrm{d}}\|_{0,\Omega}|\mathbf{z}|_{1,\Omega}\Big\} \\
&\leqslant n\|\mathbf{z}\|_{1,\Omega}\Big\{\|\mathbf{div}(\boldsymbol{\tau})\|_{0,\Omega}^2 + \|\boldsymbol{\tau}^{\mathrm{d}}\|_{0,\Omega}^2\Big\}^{1/2},
\end{aligned}$$

which implies, together with (4.210), that

$$\|\mathrm{tr}(\boldsymbol{\tau})\|_{0,\Omega} \leqslant n\,\|(\mathrm{div})^{-1}\| \left\{ \|\mathbf{div}(\boldsymbol{\tau})\|_{0,\Omega}^2 + \|\boldsymbol{\tau}^{\mathrm{d}}\|_{0,\Omega}^2 \right\}^{1/2}.$$

Finally, the proof is completed by noting that

$$\|\boldsymbol{\tau}\|_{0,\Omega}^2 = \|\boldsymbol{\tau}^{\mathrm{d}} + \frac{1}{n}\,\mathrm{tr}(\boldsymbol{\tau})\,\mathbb{I}\|_{0,\Omega}^2 = \|\boldsymbol{\tau}^{\mathrm{d}}\|_{0,\Omega}^2 + \frac{1}{n}\,\|\mathrm{tr}(\boldsymbol{\tau})\|_{0,\Omega}^2\,.$$

□

In this way, applying Lemma 4.38 to $\boldsymbol{\tau} \in V$, it follows from (4.209) that

$$a(\boldsymbol{\tau}, \boldsymbol{\tau}) \geqslant \frac{1}{2\mu}\,\|\boldsymbol{\tau}^{\mathrm{d}}\|_{0,\Omega}^2 \geqslant \frac{c_1}{2\mu}\,\|\boldsymbol{\tau}\|_{0,\Omega}^2 = \frac{c_1}{2\mu}\,\|\boldsymbol{\tau}\|_{\mathbf{div};\Omega}^2\,,$$

which proves that a is V-elliptic with constant $\alpha = c_1/2\mu$. On the other hand, given $(\mathbf{v}, \boldsymbol{\eta}) \in Q$, we consider the boundary value problem:

$$\mathbf{div}\big(\mathbf{e}(\mathbf{z}) + \boldsymbol{\eta}\big) = \mathbf{v} \quad \text{in} \quad \Omega\,, \qquad \mathbf{z} = 0 \text{ on } \Gamma,$$

whose primal variational formulation reduces to: Find $\mathbf{z} \in \mathbf{H}_0^1(\Omega)$ such that

$$\int_\Omega \mathbf{e}(\mathbf{z}) : \mathbf{e}(\mathbf{w}) = -\int_\Omega \mathbf{v}\cdot\mathbf{w} - \int_\Omega \boldsymbol{\eta} : \nabla\mathbf{w} \qquad \forall\,\mathbf{w} \in \mathbf{H}_0^1(\Omega). \tag{4.211}$$

Here we recall that the first Korn inequality states that (cf. Exercise 10.6)

$$\|\mathbf{e}(\mathbf{w})\|_{0\,\Omega}^2 \geqslant \frac{1}{2}\,|\mathbf{w}|_{1,\Omega}^2 \qquad \forall\,\mathbf{w} \in \mathbf{H}_0^1(\Omega)\,,$$

and then, a direct application of the Lax-Milgram Lemma (cf. Lemma 4.4) implies that (4.211) has a unique solution $\mathbf{z} \in \mathbf{H}_0^1(\Omega)$, which, according to the corresponding continuous dependence result, satisfies

$$|\mathbf{z}|_{1,\Omega} \leqslant \widetilde{C}\left\{\|\mathbf{v}\|_{0,\Omega} + \|\boldsymbol{\eta}\|_{0,\Omega}\right\}, \tag{4.212}$$

where $\widetilde{C} > 0$ is a constant that depends on C_P (cf. Friedrichs-Poincaré inequality (4.167)). Then, defining $\hat{\boldsymbol{\tau}} := \mathbf{e}(\mathbf{z}) + \boldsymbol{\eta}$ in Ω, we have $\hat{\boldsymbol{\tau}} \in \mathbb{L}^2(\Omega)$ and $\mathbf{div}(\hat{\boldsymbol{\tau}}) = \mathbf{v}$ in Ω, which means that $\hat{\boldsymbol{\tau}} \in \mathbb{H}(\mathbf{div};\Omega)$. Thus, denoting by $\tilde{\boldsymbol{\tau}}$ the H_0 component of $\hat{\boldsymbol{\tau}}$, it follows that

$$\mathbf{B}(\tilde{\boldsymbol{\tau}}) := \left(\mathbf{div}(\tilde{\boldsymbol{\tau}}),\, \frac{1}{2}(\tilde{\boldsymbol{\tau}} - \tilde{\boldsymbol{\tau}}^{\mathrm{t}})\right) = (\mathbf{v}, \boldsymbol{\eta})\,,$$

which proves that $\mathbf{B}$ is surjective. We note here, using the identity $\mathbf{e}(\mathbf{z}) : \boldsymbol{\eta} = 0$ and inequality (4.212), that

$$\begin{aligned} \|\tilde{\boldsymbol{\tau}}\|_{\mathbf{div};\Omega}^2 \leqslant \|\hat{\boldsymbol{\tau}}\|_{\mathbf{div};\Omega}^2 &= \|\mathbf{e}(\mathbf{z})\|_{0,\Omega}^2 + \|\boldsymbol{\eta}\|_{0,\Omega}^2 + \|\mathbf{v}\|_{0,\Omega}^2 \\ &\leqslant \left(1 + 2\,\widetilde{C}^2\right)\|(\mathbf{v}, \boldsymbol{\eta})\|_Q^2\,. \end{aligned} \tag{4.213}$$

Finally, it follows directly from (4.203) that G is bounded with

$$\|G\| \leqslant \|\mathbf{f}\|_{0,\Omega}.$$

Therefore, Theorem 4.21 implies that problem (4.207) has a unique solution $(\boldsymbol{\sigma}, (\mathbf{u}, \boldsymbol{\rho})) \in H_0 \times Q$, which satisfies

$$\|(\boldsymbol{\sigma}, (\mathbf{u}, \boldsymbol{\rho}))\|_{H\times Q} \leqslant C\,\|\mathbf{f}\|_{0,\Omega},$$

where $C > 0$ depends on the constant β of the continuous inf-sup condition for b, on $\|\mathbf{A}\| \leqslant 1/\mu$, and on the ellipticity constant $\alpha = c_1/2\mu$. Estimate (4.213) allows us, as shown below, to obtain an approximation of β. Indeed, given $(\mathbf{v}, \boldsymbol{\eta}) \in Q$, we have

$$\sup_{\substack{\boldsymbol{\tau}\in H_0\\ \boldsymbol{\tau}\neq 0}} \frac{b(\boldsymbol{\tau}, (\mathbf{v}, \boldsymbol{\eta}))}{\|\boldsymbol{\tau}\|_H} \geqslant \frac{b(\widetilde{\boldsymbol{\tau}}, (\mathbf{v}, \boldsymbol{\eta}))}{\|\widetilde{\boldsymbol{\tau}}\|_H} = \frac{\|(\mathbf{v}, \boldsymbol{\eta})\|_Q^2}{\|\widetilde{\boldsymbol{\tau}}\|_H} \geqslant \frac{1}{(1 + 2\,\widetilde{C}^2)^{1/2}}\,\|(\mathbf{v}, \boldsymbol{\eta})\|_Q\,,$$

from which it follows that we can take $\beta := \left(1 + 2\,\widetilde{C}^2\right)^{-1/2}$.

Before analyzing the problem with mixed boundary conditions, we find it interesting to comment on the situation that arises with non-homogeneous Dirichlet boundary conditions. In this case, given $\mathbf{g} \in \mathbf{H}^{1/2}(\Gamma)$, the functional F in (4.201) and (4.207) is given by

$$F(\boldsymbol{\tau}) := \langle \gamma_{\boldsymbol{\nu}}(\boldsymbol{\tau}), \mathbf{g}\rangle \qquad \forall\, \boldsymbol{\tau} \in H\,, \tag{4.214}$$

and the equivalence given by Lemma 4.37 becomes the following.

Lemma 4.39. *Problems* (4.201) *and* (4.207) *with F given by* (4.214) *are equivalent as explained below.*

i) *Let $(\boldsymbol{\sigma}, (\mathbf{u}, \boldsymbol{\rho})) \in H \times Q$ be a solution of* (4.201), *and let $\boldsymbol{\sigma} := \boldsymbol{\sigma}_0 + c\,\mathbb{I}$, with $\boldsymbol{\sigma}_0 \in H_0$ and $c \in \mathbb{R}$. Then $(\boldsymbol{\sigma}_0, (\mathbf{u}, \boldsymbol{\rho})) \in H_0 \times Q$ is a solution of* (4.207).
ii) *Let $(\boldsymbol{\sigma}_0, (\mathbf{u}, \boldsymbol{\rho})) \in H_0 \times Q$ be a solution of* (4.207), *and define $\boldsymbol{\sigma} := \boldsymbol{\sigma}_0 + c\,\mathbb{I}$, with*

$$c := \frac{(n\,\lambda + 2\mu)}{n\,|\Omega|} \int_\Gamma \mathbf{g}\cdot\boldsymbol{\nu}\,.$$

Then $(\boldsymbol{\sigma}, (\mathbf{u}, \boldsymbol{\rho})) \in H \times Q$ is a solution of (4.201).

Proof. It is similar to that of Lemma 4.37. The details are omitted and it is left as an exercise for the reader (cf. Exercise 4.28).

□

4.6.4.2 Mixed boundary conditions

We now return to the most general case given by (4.192), that is

$$\begin{aligned} \boldsymbol{\sigma} &= \mathcal{C}\,\mathbf{e}(\mathbf{u}) \ \text{ in } \ \Omega\,, & \mathbf{div}(\boldsymbol{\sigma}) &= -\mathbf{f} \ \text{ in } \ \Omega\,,\\ \mathbf{u} &= 0 \ \text{ on } \ \Gamma_D\,, & \boldsymbol{\sigma}\boldsymbol{\nu} &= \mathbf{g} \ \text{ on } \ \Gamma_N\,, \end{aligned} \tag{4.215}$$

where $\mathbf{f} \in \mathbf{L}^2(\Omega)$ and $\mathbf{g} \in \mathbf{H}_{00}^{-1/2}(\Gamma_N)$. Inverting again Hooke's law as in (4.195), defining the rotation $\boldsymbol{\rho}$ (cf. (4.196)) as an additional unknown, and applying Green's identity (4.130), it follows that

$$\int_\Omega \mathcal{C}^{-1}\boldsymbol{\sigma} : \boldsymbol{\tau} = -\int_\Omega \mathbf{u}\cdot\mathbf{div}(\boldsymbol{\tau}) + \langle \gamma_{\boldsymbol{\nu}}(\boldsymbol{\tau}), \gamma_0(\mathbf{u})\rangle - \int_\Omega \boldsymbol{\rho} : \boldsymbol{\tau}\,,$$

from which, defining the auxiliary unknown $\boldsymbol{\xi} := -\gamma_0(\mathbf{u}) \in \mathbf{H}_{00}^{1/2}(\Gamma_N)$, which is justified by the homogeneous condition $\gamma_0(\mathbf{u}) = 0$ on Γ_D, it follows that

$$\int_\Omega \mathcal{C}^{-1}\boldsymbol{\sigma} : \boldsymbol{\tau} + \int_\Omega \mathbf{u}\cdot \mathbf{div}(\boldsymbol{\tau}) + \int_\Omega \boldsymbol{\rho} : \boldsymbol{\tau} + \langle \gamma_{\boldsymbol{\nu}}(\boldsymbol{\tau})|_{\Gamma_N}, \boldsymbol{\xi}\rangle_{\Gamma_N} = 0 \quad \forall\, \boldsymbol{\tau} \in \mathbb{H}(\mathbf{div};\Omega).$$

For the rest of the equations we reason as in Section 4.6.3 (cf. (4.186) and (4.187)), and thus we obtain the following mixed variational formulation of (4.215): Find $(\boldsymbol{\sigma}, (\mathbf{u}, \boldsymbol{\rho}, \boldsymbol{\xi})) \in H \times Q$ such that

$$\begin{aligned} a(\boldsymbol{\sigma},\boldsymbol{\tau}) + b(\boldsymbol{\tau}, (\mathbf{u},\boldsymbol{\rho},\boldsymbol{\xi})) &= F(\boldsymbol{\tau}) && \forall\, \boldsymbol{\tau} \in H, \\ b(\boldsymbol{\sigma}, (\mathbf{v},\boldsymbol{\eta},\boldsymbol{\chi})) &= G((\mathbf{v},\boldsymbol{\eta},\boldsymbol{\chi})) && \forall\, (\mathbf{v},\boldsymbol{\eta},\boldsymbol{\chi}) \in Q, \end{aligned} \tag{4.216}$$

where

$$H := \mathbb{H}(\mathbf{div};\Omega), \quad Q := \mathbf{L}^2(\Omega) \times \mathbb{L}^2_{\mathrm{asim}}(\Omega) \times \mathbf{H}^{1/2}_{00}(\Gamma_N),$$

$a : H \times H \to \mathbb{R}$ is the bilinear form given in (4.202), $b : H \times Q \to \mathbb{R}$ is defined by

$$b(\boldsymbol{\sigma}, (\mathbf{v},\boldsymbol{\eta},\boldsymbol{\chi})) := \int_\Omega \mathbf{v}\cdot \mathbf{div}(\boldsymbol{\tau}) + \int_\Omega \boldsymbol{\eta} : \boldsymbol{\tau} + \langle \gamma_{\boldsymbol{\nu}}(\boldsymbol{\tau})|_{\Gamma_N}, \boldsymbol{\chi}\rangle_{\Gamma_N}, \tag{4.217}$$

$F \in H'$ is the zero functional (cf. (4.203)), and $G \in Q'$ is given by

$$G((\mathbf{v},\boldsymbol{\eta},\boldsymbol{\chi})) := -\int_\Omega \mathbf{f}\cdot \mathbf{v} + \langle \mathbf{g}, \boldsymbol{\chi}\rangle_{\Gamma_N} \qquad \forall\, (\mathbf{v},\boldsymbol{\eta},\boldsymbol{\chi}) \in Q. \tag{4.218}$$

In this case we observe, unlike what is stated in Lemma 4.37, that it is not possible to reformulate (4.216), in an equivalent way, with H_0 instead of H. Indeed, from Hooke's law (4.193) and from the first equation of (4.215) we obtain

$$\mathrm{tr}(\boldsymbol{\sigma}) = (n\lambda + 2\mu)\,\mathrm{tr}(\mathbf{e}(\mathbf{u})) = (n\lambda + 2\mu)\,\mathbf{div}(\mathbf{u}),$$

and therefore, using that $\mathbf{u} = 0$ on Γ_D, it follows that

$$\int_\Omega \mathrm{tr}(\boldsymbol{\sigma}) = (n\lambda + 2\mu)\int_\Omega \mathrm{div}(\mathbf{u}) = (n\lambda + 2\mu)\int_{\Gamma_D} \mathbf{u}\cdot\boldsymbol{\nu}.$$

Equivalently, taking $\boldsymbol{\tau} = \mathbb{I}$ in the first equation of (4.216), and using (4.204), we obtain

$$\frac{1}{(n\lambda + 2\mu)}\int_\Omega \mathrm{tr}(\boldsymbol{\sigma}) + \langle \boldsymbol{\nu}, \boldsymbol{\xi}\rangle_{\Gamma_N} = 0,$$

that is, recalling that $\boldsymbol{\xi} = -\gamma_0(\mathbf{u})$ on Γ_N,

$$\int_\Omega \mathrm{tr}(\boldsymbol{\sigma}) = (n\lambda + 2\mu)\langle \boldsymbol{\nu}, \gamma_0(\mathbf{u})\rangle_{\Gamma_N} = (n\lambda + 2\mu)\int_{\Gamma_N} \mathbf{u}\cdot\boldsymbol{\nu}.$$

In other words, the presence of the Neumann condition on Γ_N does not allow us to guarantee that $\boldsymbol{\sigma} \in H_0$, so the inequality given by Lemma 4.38 cannot be applied in this case. In order to overcome this difficulty, the following result is proved.

Lemma 4.40. *There exists $c_2 > 0$, depending on Γ_N and Ω, such that*

$$c_2\,\|\boldsymbol{\tau}\|^2_{\mathbf{div};\Omega} \leqslant \|\boldsymbol{\tau}_0\|^2_{\mathbf{div};\Omega} \qquad \forall\, \boldsymbol{\tau} := \boldsymbol{\tau}_0 + d\,\mathbb{I} \in \mathbb{H}_{\Gamma_N}(\mathbf{div};\Omega),$$

with $\boldsymbol{\tau}_0 \in H_0$ and $d \in \mathbb{R}$, where

$$\mathbb{H}_{\Gamma_N}(\mathbf{div};\Omega) := \left\{\boldsymbol{\tau} \in \mathbb{H}(\mathbf{div};\Omega) : \quad \gamma_{\boldsymbol{\nu}}(\boldsymbol{\tau})|_{\Gamma_N} = 0\right\}.$$

Proof. Let $\boldsymbol{\tau} := \boldsymbol{\tau}_0 + d\mathbb{I} \in \mathbb{H}_{\Gamma_N}(\mathbf{div};\Omega)$ with $\boldsymbol{\tau}_0 \in H_0$ and $d \in \mathbb{R}$. It follows that

$$0 = \gamma_{\boldsymbol{\nu}}(\boldsymbol{\tau})|_{\Gamma_N} = \gamma_{\boldsymbol{\nu}}(\boldsymbol{\tau}_0)|_{\Gamma_N} + d\,\gamma_{\boldsymbol{\nu}}(\mathbb{I})|_{\Gamma_N} = \gamma_{\boldsymbol{\nu}}(\boldsymbol{\tau}_0)|_{\Gamma_N} + d\,\boldsymbol{\nu},$$

from which, using according to (4.183) and (4.184) that

$$\|\boldsymbol{\psi}|_{\Gamma_N}\|_{-1/2,\,00,\Gamma_N} \leqslant \|\boldsymbol{\psi}\|_{-1/2,\,\Gamma} \qquad \forall\,\boldsymbol{\psi} \in \mathbf{H}^{-1/2}(\Gamma),$$

it follows that

$$|d|\,\|\boldsymbol{\nu}\|_{-1/2,00\,\Gamma_N} = \|\gamma_{\boldsymbol{\nu}}(\boldsymbol{\tau}_0)|_{\Gamma_N}\|_{-1/2,00\,\Gamma_N} \leqslant \|\gamma_{\boldsymbol{\nu}}(\boldsymbol{\tau}_0)\|_{-1/2,\Gamma} \leqslant \|\boldsymbol{\tau}_0\|_{\mathrm{div};\Omega},$$

and therefore

$$|d| \leqslant \frac{1}{\|\boldsymbol{\nu}\|_{-1/2,\,00,\Gamma_N}}\,\|\boldsymbol{\tau}_0\|_{\mathbf{div};\Omega}.$$

Thanks to this estimate and to the fact that

$$\|\boldsymbol{\tau}\|^2_{\mathbf{div};\Omega} = \|\boldsymbol{\tau}_0\|^2_{\mathbf{div};\Omega} + n\,d^2|\Omega|,$$

we conclude that

$$\|\boldsymbol{\tau}\|^2_{\mathbf{div};\Omega} \leqslant \left\{1 + \frac{n|\Omega|}{\|\boldsymbol{\nu}\|^2_{-1/2,00,\Gamma_N}}\right\}\|\boldsymbol{\tau}_0\|^2_{\mathbf{div};\Omega},$$

which completes the proof.

□

Next, we note that $\mathbf{B} : H \to Q$, the operator induced by the present bilinear form $b : H \times Q \to \mathbb{R}$ (cf. (4.217)), is given by

$$\mathbf{B}(\boldsymbol{\tau}) := \left(\mathbf{div}(\boldsymbol{\tau}),\,\frac{1}{2}(\boldsymbol{\tau} - \boldsymbol{\tau}^{\mathrm{t}}),\,\mathcal{R}_{00}\,\gamma_{\boldsymbol{\nu}}(\boldsymbol{\tau})|_{\Gamma_N}\right) \quad \forall\,\boldsymbol{\tau} \in H,$$

where $\mathcal{R}_{00} : \mathbf{H}^{-1/2}_{00}(\Gamma_N) \to \mathbf{H}^{1/2}_{00}(\Gamma_N)$ is the corresponding Riesz operator. Similarly to the analysis in Section 4.6.3, it is shown here that $\mathbf{B}$ is bounded with $\|\mathbf{B}\| \leqslant 2$. Moreover, it is easy to see that

$$V := N(\mathbf{B}) = \left\{\boldsymbol{\tau} \in H : \ \mathbf{div}(\boldsymbol{\tau}) = 0 \text{ in } \Omega, \quad \boldsymbol{\tau} = \boldsymbol{\tau}^{\mathrm{t}} \text{ in } \Omega, \quad \gamma_{\boldsymbol{\nu}}(\boldsymbol{\tau})|_{\Gamma_N} = 0\right\},$$

and then, given $\boldsymbol{\tau} \in V$, we apply (4.209) and Lemmas 4.38 and 4.40, to deduce that

$$\begin{aligned} a(\boldsymbol{\tau},\boldsymbol{\tau}) &\geqslant \frac{1}{2\mu}\,\|\boldsymbol{\tau}^{\mathrm{d}}\|^2_{0,\Omega} = \frac{1}{2\mu}\,\|\boldsymbol{\tau}_0^{\mathrm{d}}\|^2_{0,\Omega} \geqslant \frac{c_1}{2\mu}\,\|\boldsymbol{\tau}_0\|^2_{0,\Omega} \\ &= \frac{c_1}{2\mu}\,\|\boldsymbol{\tau}_0\|^2_{\mathbf{div};\Omega} \geqslant \frac{c_1\,c_2}{2\mu}\,\|\boldsymbol{\tau}\|^2_{\mathbf{div};\Omega}, \end{aligned}$$

thus proving that a is V-elliptic with constant $\alpha = c_1\,c_2/2\mu$. On the other hand, given $(\mathbf{v},\boldsymbol{\eta},\boldsymbol{\chi}) \in Q$, we consider the boundary value problem

$$\begin{aligned} \mathbf{div}(\mathbf{e}(\mathbf{z})) &= \mathbf{v} - \mathbf{div}(\boldsymbol{\eta}) \quad \text{in} \quad \Omega, \quad \mathbf{z} = \mathbf{0} \quad \text{on} \quad \Gamma_D, \\ (\mathbf{e}(\mathbf{z}) + \boldsymbol{\eta})\,\boldsymbol{\nu} &= \mathcal{R}_{00}^{-1}(\boldsymbol{\chi}) \quad \text{on} \quad \Gamma_N, \end{aligned}$$

whose primal formulation reduces to: Find $\mathbf{z} \in \mathbf{H}^1_{\Gamma_D}(\Omega)$ such that

$$\int_\Omega \mathbf{e}(\mathbf{z}) : \mathbf{e}(\mathbf{w}) = -\int_\Omega \mathbf{v} \cdot \mathbf{w} - \int_\Omega \boldsymbol{\eta} : \nabla \mathbf{w} + \langle \mathcal{R}_{00}^{-1}(\boldsymbol{\chi}), \gamma_0(\mathbf{w}) \rangle_{\Gamma_N} \tag{4.219}$$

for all $\mathbf{w} \in \mathbf{H}^1_{\Gamma_D}(\Omega)$, where $\mathbf{H}^1_{\Gamma_D}(\Omega) := \{\mathbf{w} \in \mathbf{H}^1(\Omega) : \quad \gamma_0(\mathbf{w}) = 0 \quad \text{on} \quad \Gamma_D\}$. In this case, Korn's inequality states the existence of $c > 0$ such that

$$\|\mathbf{e}(\mathbf{w})\|^2_{0,\Omega} \geqslant c\,|\mathbf{w}|^2_{1,\Omega} \qquad \forall\, \mathbf{w} \in \mathbf{H}^1_{\Gamma_D}(\Omega),$$

which constitutes a generalization of the first Korn inequality (cf. Exercise 10.6), and then, the Lax-Milgram Lemma (cf. Lemma 4.4) implies that (4.219) has a unique solution $\mathbf{z} \in \mathbf{H}^1_{\Gamma_D}(\Omega)$ which satisfies

$$|\mathbf{z}|_{1,\Omega} \leqslant \hat{C}\left\{\|\mathbf{v}\|_{0,\Omega} + \|\boldsymbol{\eta}\|_{0,\Omega} + \|\boldsymbol{\chi}\|_{1/2,00,\Gamma_N}\right\}, \tag{4.220}$$

where $\hat{C} > 0$ depends on c and C_D (cf. (4.151)). In this way, defining $\hat{\boldsymbol{\tau}} := \mathbf{e}(\mathbf{z}) + \boldsymbol{\eta}$ in Ω, we clearly have $\hat{\boldsymbol{\tau}} \in \mathbb{L}^2(\Omega)$ and $\mathbf{div}(\hat{\boldsymbol{\tau}}) = \mathbf{v}$ in Ω, whence $\hat{\boldsymbol{\tau}} \in \mathbb{H}(\mathbf{div};\Omega)$, and $\gamma_{\boldsymbol{\nu}}(\hat{\boldsymbol{\tau}})|_{\Gamma_N} = \mathcal{R}_{00}^{-1}(\boldsymbol{\chi})$. It follows that $\mathbf{B}(\hat{\boldsymbol{\tau}}) = (\mathbf{v}, \boldsymbol{\eta}, \boldsymbol{\chi})$, thus proving that $\mathbf{B}$ is surjective. Moreover, using the fact that $\mathbf{e}(\mathbf{z}) : \boldsymbol{\eta} = 0$ and inequality (4.220), we obtain

$$\|\hat{\boldsymbol{\tau}}\|^2_{\mathbf{div};\Omega} = \|\mathbf{e}(\mathbf{z})\|^2_{0,\Omega} + \|\boldsymbol{\eta}\|^2_{0,\Omega} + \|\mathbf{v}\|^2_{0,\Omega} \leqslant \left(1 + 3\hat{C}^2\right)\|(\mathbf{v}, \boldsymbol{\eta}, \boldsymbol{\chi})\|^2_Q. \tag{4.221}$$

Then, denoting by $\langle\cdot,\cdot\rangle_Q$ the inner product of Q, we deduce, thanks to (4.221), that

$$\sup_{\substack{\boldsymbol{\tau} \in H \\ \boldsymbol{\tau} \neq 0}} \frac{b(\boldsymbol{\tau}, (\mathbf{v}, \boldsymbol{\eta}, \boldsymbol{\chi}))}{\|\boldsymbol{\tau}\|_H} \geqslant \frac{b(\hat{\boldsymbol{\tau}}, (\mathbf{v}, \boldsymbol{\eta}, \boldsymbol{\chi}))}{\|\hat{\boldsymbol{\tau}}\|_{\mathbf{div};\Omega}} = \frac{\langle \mathbf{B}(\hat{\boldsymbol{\tau}}), (\mathbf{v}, \boldsymbol{\eta}, \boldsymbol{\chi})\rangle_Q}{\|\hat{\boldsymbol{\tau}}\|_{\mathbf{div};\Omega}}$$

$$= \frac{\|(\mathbf{v}, \boldsymbol{\eta}, \boldsymbol{\chi})\|^2_Q}{\|\hat{\boldsymbol{\tau}}\|_{\mathbf{div};\Omega}} \geqslant \frac{1}{(1 + 3\hat{C}^2)^{1/2}} \|(\mathbf{v}, \boldsymbol{\eta}, \boldsymbol{\chi})\|_Q,$$

which yields $\beta := \left(1 + 3\hat{C}^2\right)^{-1/2}$ as the inf-sup constant for b. Consequently, having shown that the hypotheses of Theorem 4.21 regarding the bilinear forms a and b are accomplished, and observing from (4.218) that G is bounded with

$$\|G\| \leqslant \|\mathbf{f}\|_{0,\Omega} + \|\mathbf{g}\|_{-1/2,00,\Gamma_N},$$

a direct application of that abstract result implies that there exists a unique solution $(\boldsymbol{\sigma}, (\mathbf{u}, \boldsymbol{\rho}, \boldsymbol{\xi})) \in H \times Q$ of (4.216). Moreover, there exists a positive constant C, depending on α $(= c_1 c_2/2\mu)$, β, and $\|\mathbf{A}\|$ $(\leqslant 1/\mu)$, such that

$$\|(\boldsymbol{\sigma}, (\mathbf{u}, \boldsymbol{\rho}, \boldsymbol{\xi}))\|_{H \times Q} \leqslant C\left\{\|\mathbf{f}\|_{0,\Omega} + \|\mathbf{g}\|_{-1/2,00,\Gamma_N}\right\}.$$

4.6.5 The Poisson problem in primal-mixed formulation

We conclude these application examples with the Poisson problem analyzed in Section 2.5.1, but now using what we have called the primal-mixed formulation. Let us first recall that the geometry is given by a bounded open set Ω of $\mathbb{R}^n$, $n \geqslant 2$, with a Lipschitz-continuous boundary Γ. Then, given $f \in \mathrm{L}^2(\Omega)$ and $g \in \mathrm{H}^{1/2}(\Gamma)$, we are interested in the boundary value problem

$$-\Delta u = f \quad \text{in} \quad \Omega, \qquad u = g \quad \text{on} \quad \Gamma. \tag{4.222}$$

Now, multiplying the partial differential equation by $v \in \mathrm{H}^1(\Omega)$ and applying the improved Green's identity (4.134) (cf. Theorem 4.33), we obtain

$$\int_\Omega f\,v = -\int_\Omega \Delta u = \int_\Omega \nabla u \cdot \nabla v - \langle \gamma_1(u), \gamma_0(v) \rangle,$$

where $\gamma_1 : \mathrm{H}^1_\Delta(\Omega) \to \mathrm{H}^{-1/2}(\Gamma)$ is the bounded linear operator given by $\gamma_{\boldsymbol{\nu}} \circ \nabla$, and $\langle \cdot, \cdot \rangle$ denotes the duality pairing between $\mathrm{H}^{-1/2}(\Gamma)$ and $\mathrm{H}^{1/2}(\Gamma)$. Then, introducing the auxiliary unknown $\boldsymbol{\xi} := -\gamma_1(u) \in \mathrm{H}^{-1/2}(\Gamma)$, we obtain the equation

$$\int_\Omega \nabla u \cdot \nabla v + \langle \boldsymbol{\xi}, \gamma_0(v) \rangle = \int_\Omega f\,v \qquad \forall\, v \in \mathrm{H}^1(\Omega)\,. \tag{4.223}$$

In turn, the non-homogeneous Dirichlet boundary condition is rewritten in the form $\gamma_0(u) = g$, so that it can be reformulated as

$$\langle \lambda, \gamma_0(u) \rangle = \langle \lambda, g \rangle \qquad \forall\, \lambda \in \mathrm{H}^{-1/2}(\Gamma)\,. \tag{4.224}$$

In this way, combining (4.223) and (4.224) we obtain that the primal-mixed formulation of (4.222) consists of: Find $(u, \boldsymbol{\xi}) \in H \times Q$ such that

$$\begin{aligned} a(u,v) + b(v,\boldsymbol{\xi}) &= F(v) \quad \forall\, v \in H\,, \\ b(u,\lambda) \qquad\quad &= G(\lambda) \quad \forall\, \lambda \in Q\,, \end{aligned} \tag{4.225}$$

where $H := \mathrm{H}^1(\Omega)$, $Q := \mathrm{H}^{-1/2}(\Gamma)$, a and b are the bilinear forms given by

$$\begin{aligned} a(u,v) &:= \int_\Omega \nabla u \cdot \nabla v \qquad \forall\, (u,v) \in H \times H\,, \quad \text{and} \\ b(v,\lambda) &:= \langle \lambda, \gamma_0(v) \rangle \qquad \forall\, (v,\lambda) \in H \times Q\,, \end{aligned}$$

and the functionals $F \in H'$ and $G \in Q'$ are defined as

$$F(v) := \int_\Omega f\,v \quad \forall\, v \in H, \quad \text{and} \quad G(\lambda) := \langle \lambda, g \rangle \quad \forall\, \lambda \in Q\,.$$

It is easy to see that a and b are bounded since, by applying the Cauchy–Schwarz inequality, we get

$$|a(u,v)| \leqslant |u|_{1,\Omega}\, |v|_{1,\Omega} \leqslant \|u\|_{1,\Omega}\, \|v\|_{1,\Omega}\,,$$

and, using the trace inequality, we obtain

$$|b(v,\lambda)| \leqslant \|\lambda\|_{-1/2,\Gamma}\, \|\gamma_0(v)\|_{1/2,\Gamma} \leqslant \|v\|_{1,\Omega}\, \|\lambda\|_{-1/2,\Gamma}\,,$$

which proves that $\|\mathbf{A}\| \leqslant 1$ and $\|\mathbf{B}\| \leqslant 1$, where $\mathbf{A} : H \to H$ and $\mathbf{B} : H \to Q$ are the operators induced by a and b, respectively. In particular, letting $\mathcal{R} : \mathrm{H}^{-1/2}(\Gamma) \to \mathrm{H}^{1/2}(\Gamma)$ be the corresponding Riesz operator, we have

$$b(v,\lambda) = \langle \lambda, \gamma_0(v) \rangle = \langle \mathcal{R}(\lambda), \gamma_0(v) \rangle_{1/2,\Gamma} = \langle \mathcal{R}^* \gamma_0(v), \lambda \rangle_{-1/2,\Gamma}\,,$$

where $\langle \cdot, \cdot \rangle_{r,\Gamma}$ denotes the inner product in $\mathrm{H}^r(\Gamma)$, $r \in \{-1/2,\, 1/2\}$, which shows that the operator $\mathbf{B}$ is given by

$$\mathbf{B}(v) = \mathcal{R}^* \gamma_0(v) \quad \forall\, v \in H\,.$$

In this way, since the adjoint Riesz operator $\mathcal{R}^* : \mathrm{H}^{1/2}(\Gamma) \to \mathrm{H}^{-1/2}(\Gamma)$ is bijective as well, it follows that

$$V := N(\mathbf{B}) = \Big\{ v \in H : \quad \mathbf{B}(v) = 0 \Big\} = \Big\{ v \in \mathrm{H}^1(\Omega) : \quad \gamma_0\,(v) = 0 \Big\} = \mathrm{H}^1_0(\Omega)\,,$$

and then, thanks to the Friedrichs–Poincaré inequality (cf. (4.167)), we find that

$$a(v,v) \,=\, |v|^2_{1,\Omega} \geqslant \alpha\,\|v\|^2_{1,\Omega} \quad \forall\, v \in V\,,$$

with $\alpha = C_P$, thus establishing the V-ellipticity of a. On the other hand, for the surjectivity of $\mathbf{B}$ it suffices to note that this operator results from the composition of the surjective operators $\mathcal{R}^*$ and γ_0. For example, given $\lambda \in \mathrm{H}^{-1/2}(\Gamma)$, the function $z := \widetilde{\gamma}_0^{-1}(\mathcal{R}^*)^{-1}(\lambda)$ belongs to $\mathrm{H}^1_0(\Omega)^\perp$ and satisfies $\mathbf{B}(z) = \lambda$, confirming the previous statement. Moreover, bounding by below with $v = \widetilde{\gamma}_0^{-1}\big(\mathcal{R}(\lambda)\big)$, and using (4.109) (cf. Lemma 4.27), we obtain

$$\begin{aligned}\sup_{\substack{v \in H\\ v\neq 0}} \frac{b(v,\lambda)}{\|v\|_H} &= \sup_{\substack{v \in \mathrm{H}^1(\Omega)\\ v\neq 0}} \frac{\langle \lambda, \gamma_0(v)\rangle}{\|v\|_{1,\Omega}} \geqslant \frac{\langle \lambda, \gamma_0\,\widetilde{\gamma}_0^{-1}\big(\mathcal{R}(\lambda)\big)\rangle}{\|\widetilde{\gamma}_0^{-1}\big(\mathcal{R}(\lambda)\big)\|_{1,\Omega}}\\ &= \frac{\langle \lambda, \mathcal{R}(\lambda)\rangle}{\|\widetilde{\gamma}_0^{-1}\big(\mathcal{R}(\lambda)\big)\|_{1,\Omega}} = \frac{\|\mathcal{R}(\lambda)\|^2_{1/2,\Gamma}}{\|\mathcal{R}(\lambda)\|_{1/2,\Gamma}} = \|\mathcal{R}(\lambda)\|_{1/2,\Gamma} = \|\lambda\|_{-1/2,\Gamma}\,,\end{aligned}$$

which allows us to take $\beta = 1$ as the inf-sup constant for b. Furthermore, employing the Cauchy–Schwarz inequality and the duality pairing between $\mathrm{H}^{-1/2}(\Gamma)$ and $\mathrm{H}^{1/2}(\Gamma)$, it follows easily that F and G are indeed bounded with

$$\|F\| \leqslant \|f\|_{0,\Omega} \quad \text{and} \quad \|G\| \leqslant \|g\|_{1/2,\Gamma}\,.$$

Consequently, by applying once again Theorem 4.21, we deduce that there exists a unique solution $(u,\boldsymbol{\xi}) \in H \times Q$ of the primal-mixed formulation (4.225). In addition, there exists a positive constant C, depending only on α, β, and $\|\mathbf{A}\|$ ($\leqslant 1$), such that

$$\|(u,\boldsymbol{\xi})\|_{H\times Q} \leqslant C\Big\{\|f\|_{0,\Omega} + \|g\|_{1/2,\Gamma}\Big\}\,.$$

4.7 A nonlinear case of the Babuška-Brezzi Theory

In this section we consider a nonlinear case of the particular version of the Babuška–Brezzi Theory given by Theorem 4.21, and we analyze the corresponding continuous and discrete schemes. Moreover, following a reasoning similar to that of the proof of Theorem 4.24, we establish the respective Céa estimate. Although the results detailed below can be obtained as applications of more general abstract theories (cf. [116, Proposition 2.3], [36, Theorem 3.1]), to date we are not aware of specific proofs for them available in the literature, which is the reason why we proceed to present them here. In what follows, given an arbitrary normed vector space X, $[\cdot,\cdot]$ denotes the corresponding duality pairing between X' and X, that is

$$[\mathrm{F}, x] := \mathrm{F}(x) \qquad \forall\,(F,x) \in X' \times X\,.$$

Theorem 4.41. *Let* $\mathbf{H}$ *and* $\mathbf{Q}$ *be real Hilbert spaces, and let* $\mathbf{A}: \mathbf{H} \to \mathbf{H}'$ *be a nonlinear operator, and* $\mathbf{B} \in \mathcal{L}\big(\mathbf{H}, \mathbf{Q}'\big)$. *Moreover, let* $\mathbf{V}$ *be the null space of* $\mathbf{B}$, *and suppose that*

a) *the family of operators* $\mathbf{A}(\cdot + \tilde{\tau}) : \mathbf{V} \to \mathbf{V}'$, *with* $\tilde{\tau} \in \mathbf{H}$, *is uniformly strongly monotone, that is, there exists* $\alpha > 0$ *such that*

$$[\mathbf{A}(\tau + \tilde{\tau}) - \mathbf{A}(\zeta + \tilde{\tau}), \tau - \zeta] \geqslant \alpha \, \|\tau - \zeta\|_{\mathbf{H}}^2 \qquad \forall\, \tilde{\tau} \in \mathbf{H}, \quad \forall\, \tau, \zeta \in \mathbf{V}, \tag{4.226}$$

b) $\mathbf{A} : \mathbf{H} \to \mathbf{H}'$ *is Lipschitz continuous, that is, there exists* $L_{\mathbf{A}} > 0$ *such that*

$$\|\mathbf{A}(\tau) - \mathbf{A}(\zeta)\|_{\mathbf{H}'} \leqslant L_{\mathbf{A}} \, \|\tau - \zeta\|_{\mathbf{H}} \qquad \forall\, \tau, \zeta \in \mathbf{H}, \tag{4.227}$$

c) *and there exists* $\beta > 0$ *such that*

$$\sup_{\substack{\tau \in \mathbf{H} \\ \tau \neq 0}} \frac{[\mathbf{B}(\tau), v]}{\|\tau\|_{\mathbf{H}}} \geqslant \beta \, \|v\|_{\mathbf{Q}} \qquad \forall\, v \in \mathbf{Q}. \tag{4.228}$$

Then, for each pair $(\mathbf{F}, \mathbf{G}) \in \mathbf{H}' \times \mathbf{Q}'$ *there exists a unique* $(\sigma, u) \in \mathbf{H} \times \mathbf{Q}$ *such that*

$$\begin{aligned} [\mathbf{A}(\sigma), \tau] \, + \, [\mathbf{B}(\tau), u] &= \mathbf{F}(\tau) \qquad \forall\, \tau \in \mathbf{H}, \\ [\mathbf{B}(\sigma), v] \qquad\qquad\quad &= \mathbf{G}(v) \qquad \forall\, v \in \mathbf{Q}, \end{aligned} \tag{4.229}$$

and the following a priori estimates hold:

$$\|\sigma\|_{\mathbf{H}} \leqslant \frac{1}{\alpha} \|\mathbf{F}\|_{\mathbf{H}'} + \frac{1}{\beta} \left(1 + \frac{L_{\mathbf{A}}}{\alpha}\right) \|\mathbf{G}\|_{\mathbf{Q}'} + \frac{1}{\alpha} \, \|\mathbf{A}(0)\|_{\mathbf{H}'}, \tag{4.230}$$

and

$$\begin{aligned} \|u\|_{\mathbf{Q}} \leqslant \; & \frac{1}{\beta} \left(1 + \frac{L_{\mathbf{A}}}{\alpha}\right) \|\mathbf{F}\|_{\mathbf{H}'} + \frac{L_{\mathbf{A}}}{\beta^2} \left(1 + \frac{L_{\mathbf{A}}}{\alpha}\right) \|\mathbf{G}\|_{\mathbf{Q}'} \\ & + \; \frac{1}{\beta} \left(1 + \frac{L_{\mathbf{A}}}{\alpha}\right) \|\mathbf{A}(0)\|_{\mathbf{H}'} . \end{aligned} \tag{4.231}$$

Proof. We first introduce the operator $\underline{\mathbf{B}} \in \mathcal{L}\big(\mathbf{Q}, \mathbf{H}'\big)$ defined for each $v \in \mathbf{Q}$ by

$$[\underline{\mathbf{B}}(v), \tau] \, := \, [\mathbf{B}(\tau), v] \qquad \forall, \tau \in \mathbf{H},$$

so that (4.229) can be rewritten as the operator equation system

$$\begin{aligned} \mathbf{A}(\sigma) \, + \, \underline{\mathbf{B}}(u) &= \mathbf{F}, \\ \mathbf{B}(\sigma) \qquad\quad &= \mathbf{G}. \end{aligned} \tag{4.232}$$

On the other hand, denoting by $\mathcal{R}_{\mathbf{H}} : \mathbf{H}' \to \mathbf{H}$ and $\mathcal{R}_{\mathbf{Q}} : \mathbf{Q}' \to \mathbf{Q}$ the respective Riesz operators, and defining $\widetilde{\mathbf{B}} := \mathcal{R}_{\mathbf{Q}} \mathbf{B} : \mathbf{H} \to \mathbf{Q}$, it follows that the Hilbert adjoint of the latter is given by $\widetilde{\mathbf{B}}^* = \mathcal{R}_{\mathbf{H}} \underline{\mathbf{B}} : \mathbf{Q} \to \mathbf{H}$. In turn, it is easy to see that the inf-sup condition (4.228) is equivalent to $\|\underline{\mathbf{B}}(v)\|_{\mathbf{H}'} \geqslant \beta \, \|v\|_{\mathbf{Q}}$ for all $v \in \mathbf{Q}$, which, according to the identity defining $\widetilde{\mathbf{B}}^*$, reduces to $\|\widetilde{\mathbf{B}}^*(v)\|_{\mathbf{H}} \geqslant \beta \, \|v\|_{\mathbf{Q}}$ for all $v \in \mathbf{Q}$. Moreover, noting that the null spaces of $\mathbf{B}$ and $\widetilde{\mathbf{B}}$ coincide, the previous inequality also establishes that $\widetilde{\mathbf{B}}^* : \mathbf{Q} \to \mathbf{V}^\perp$ is an isomorphism, which, thanks to Lemma 4.18, is equivalent to saying that $\widetilde{\mathbf{B}} : \mathbf{V}^\perp \to \mathbf{Q}$ is also an isomorphism, and that $\|\widetilde{\mathbf{B}}(\tau)\|_{\mathbf{Q}} \geqslant \beta \, \|\tau\|_{\mathbf{H}}$ for all $\tau \in \mathbf{V}^\perp$. Therefore, returning to $\mathbf{B}$ and $\underline{\mathbf{B}}$, it follows from this discussion that $\mathbf{B} : \mathbf{V}^\perp \to \mathbf{Q}'$ and $\underline{\mathbf{B}} : \mathbf{Q} \to \mathcal{R}_{\mathbf{H}}^{-1}(\mathbf{V}^\perp)$ are isomorphisms satisfying

$$\begin{aligned} \|\mathbf{B}(\tau)\|_{\mathbf{Q}'} &\geqslant \beta\,\|\tau\|_{\mathbf{H}} \quad \forall\,\tau \in \mathbf{V}^{\perp} \quad \text{and} \\ \|\underline{\mathbf{B}}(v)\|_{\mathbf{H}'} &\geqslant \beta\,\|v\|_{\mathbf{Q}} \quad \forall\,v \in \mathbf{Q}\,. \end{aligned} \tag{4.233}$$

Now, given $(\mathbf{F},\mathbf{G}) \in \mathbf{H}' \times \mathbf{Q}'$, we proceed to construct a solution $(\sigma,u) \in \mathbf{H} \times \mathbf{Q}$ of (4.232). Indeed, we begin by denoting by σ_g the unique element in $\mathbf{V}^{\perp}$ such that $\mathbf{B}(\sigma_g) = \mathbf{G}$ and $\|\sigma_g\|_{\mathbf{H}} \leqslant \dfrac{1}{\beta}\,\|\mathbf{G}\|_{\mathbf{Q}'}$ (cf. first inequality in (4.233)). Then, thanks to assumptions (4.226) and (4.227), a direct application of [109, Theorem 3.3.23] implies that $\mathbf{A}(\cdot + \sigma_g) : \mathbf{V} \to \mathbf{V}'$ is bijective, and therefore there exists a unique $\sigma_0 \in \mathbf{V}$ such that $\mathbf{A}(\sigma_0 + \sigma_g) = \mathbf{F}|_{\mathbf{V}} \in \mathbf{V}'$, that is, $[\mathbf{F} - \mathbf{A}(\sigma_0+\sigma_g), \tau] = 0$ for all $\tau \in \mathbf{V}$, or equivalently, $\mathbf{F} - \mathbf{A}(\sigma_0+\sigma_g) \in \mathbf{V}^{\circ} = \mathcal{R}_{\mathbf{H}}^{-1}(\mathbf{V}^{\perp})$ (cf. first identity in Lemma 3.23). According to the bijectivity of $\underline{\mathbf{B}} : \mathbf{Q} \to \mathcal{R}_{\mathbf{H}}^{-1}(\mathbf{V}^{\perp})$, it follows that there exists a unique $u \in \mathbf{Q}$ such that $\underline{\mathbf{B}}(u) = \mathbf{F} - \mathbf{A}(\sigma_0 + \sigma_g)$ and (cf. second inequality in (4.233))

$$\|u\|_{\mathbf{Q}} \leqslant \frac{1}{\beta}\,\|\mathbf{F} - \mathbf{A}(\sigma_0 + \sigma_g)\|_{\mathbf{H}'}\,. \tag{4.234}$$

Thus, defining $\boldsymbol{\sigma} := \boldsymbol{\sigma}_0 + \boldsymbol{\sigma}_g \in \mathbf{H}$, we have that $\mathbf{A}(\sigma) + \underline{\mathbf{B}}(u) = \mathbf{F}$ and $\mathbf{B}(\sigma) = \mathbf{B}(\sigma_g) = \mathbf{G}$, which proves that (σ,u) is a solution of (4.232). Regarding the boundedness of (σ,u), we first apply (4.226) with $\tilde{\tau} = \sigma_g$, $\tau = \sigma_0$, and $\zeta = 0$, which gives

$$\alpha\,\|\sigma_0\|_{\mathbf{H}}^2 \leqslant [\mathbf{A}(\sigma_0 + \sigma_g) - \mathbf{A}(\sigma_g), \sigma_0] = [\mathbf{F} - \mathbf{A}(\sigma_g), \sigma_0]\,.$$

Then, adding and subtracting $\mathbf{A}(0)$ in the previous inequality, and using (4.227), we obtain

$$\|\sigma_0\|_{\mathbf{H}} \leqslant \frac{1}{\alpha}\Big\{\|\mathbf{F}\|_{\mathbf{H}'} + \|\mathbf{A}(0)\|_{\mathbf{H}'} + L_{\mathbf{A}}\,\|\sigma_g\|_{\mathbf{H}}\Big\}\,. \tag{4.235}$$

On the other hand, proceeding in a similar way for the bound of $\|u\|_{\mathbf{Q}}$, we find from (4.234) that

$$\|u\|_{\mathbf{Q}} \leqslant \frac{1}{\beta}\Big\{\|\mathbf{F}\|_{\mathbf{H}'} + \|\mathbf{A}(0)\|_{\mathbf{H}'} + L_{\mathbf{A}}\,\|\sigma_0\|_{\mathbf{H}} + L_{\mathbf{A}}\,\|\sigma_g\|_{\mathbf{H}}\Big\}\,. \tag{4.236}$$

Consequently, recalling that $\sigma = \sigma_0 + \sigma_g$, the required estimates (4.230) and (4.231) follow directly from (4.235), (4.236), and the bound already established for $\|\sigma_g\|_{\mathbf{H}}$. It remains to show that (σ,u) is the unique solution of (4.229) (equivalently (4.232)). For this purpose, let $(\tilde{\sigma},\tilde{u}) \in \mathbf{H} \times \mathbf{Q}$ be another solution of (4.232). It follows that $\sigma - \tilde{\sigma} \in \mathbf{V}$ and $\mathbf{A}(\sigma) - \mathbf{A}(\tilde{\sigma}) = \underline{\mathbf{B}}(\tilde{u} - u)$, and from this identity we easily deduce that

$$[\mathbf{A}(\sigma) - \mathbf{A}(\tilde{\sigma}), \tau] = 0 \qquad \forall\,\tau \in \mathbf{V}\,.$$

Then, using (4.226) with $\tilde{\tau} = \tilde{\sigma}$, $\tau = \sigma - \tilde{\sigma}$, and $\zeta = 0$, we obtain

$$\alpha\,\|\sigma - \tilde{\sigma}\|_{\mathbf{H}}^2 \leqslant [\mathbf{A}(\sigma) - \mathbf{A}(\tilde{\sigma}), \sigma - \tilde{\sigma}] = 0\,,$$

whence $\sigma = \tilde{\sigma}$, and thus $\underline{\mathbf{B}}(\tilde{u} - u) = 0$. Finally, using again the bijectivity of the operator $\underline{\mathbf{B}} : \mathbf{Q} \to \mathcal{R}_{\mathbf{H}}^{-1}(\mathbf{V}^{\perp})$, we deduce that $u = \tilde{u}$, which completes the uniqueness and ends the proof.

□

The discrete version of Theorem 4.41, including the corresponding Céa estimates, is established as follows.

Theorem 4.42. *Let* $\{\mathbf{H}_h\}_{h>0}$ *and* $\{\mathbf{Q}_h\}_{h>0}$ *be families of finite-dimensional subspaces of* $\mathbf{H}$ *and* $\mathbf{Q}$*, respectively, and let* $\mathbf{V}_h$ *be the null space of* $\mathbf{B}|_{\mathbf{H}_h} : \mathbf{H}_h \to \mathbf{Q}_h'$*, that is,*

$$\mathbf{V}_h := \Big\{ \tau_h \in \mathbf{H}_h : \quad [\mathbf{B}(\tau_h), v_h] = 0 \quad \forall\, v_h \in \mathbf{Q}_h \Big\} .$$

In addition to the hypotheses of Theorem 4.41*, suppose that*

a) *the family of operators* $\mathbf{A}(\cdot + \tilde{\tau}_h) : \mathbf{V}_h \to \mathbf{V}_h'$*, with* $\tilde{\tau}_h \in \mathbf{H}_h$*, is uniformly strongly monotone, that is, there exists* $\alpha_{\mathrm{d}} > 0$*, independent of* h*, such that*

$$[\mathbf{A}(\tau_h + \tilde{\tau}_h) - \mathbf{A}(\zeta_h + \tilde{\tau}_h), \tau_h - \zeta_h] \geqslant \alpha_{\mathrm{d}} \|\tau_h - \zeta_h\|_{\mathbf{H}}^2 \tag{4.237}$$

for all $\tilde{\tau}_h \in \mathbf{H}_h$*, and for all* $\tau_h, \zeta_h \in \mathbf{V}_h$*,*

b) *and there exists* $\beta_{\mathrm{d}} > 0$ *such that*

$$\sup_{\substack{\tau_h \in \mathbf{H}_h \\ \tau_h \neq 0}} \frac{[\mathbf{B}(\tau_h), v_h]}{\|\tau_h\|_{\mathbf{H}}} \geqslant \beta_{\mathrm{d}} \|v_h\|_{\mathbf{Q}} \qquad \forall\, v_h \in \mathbf{Q}_h . \tag{4.238}$$

Then, for each $(\mathbf{F}, \mathbf{G}) \in \mathbf{H}' \times \mathbf{Q}'$ *there exists a unique* $(\sigma_h, u_h) \in \mathbf{H}_h \times \mathbf{Q}_h$ *such that*

$$\begin{aligned} [\mathbf{A}(\sigma_h), \tau_h] + [\mathbf{B}(\tau_h), u_h] &= \mathbf{F}(\tau_h) \qquad \forall\, \tau_h \in \mathbf{H}_h , \\ [\mathbf{B}(\sigma_h), v_h] &= \mathbf{G}(v_h) \qquad \forall\, v_h \in \mathbf{Q}_h , \end{aligned} \tag{4.239}$$

and the following a priori estimates hold:

$$\|\sigma_h\|_{\mathbf{H}} \leqslant \frac{1}{\alpha_{\mathrm{d}}} \|\mathbf{F}\|_{\mathbf{H}'} + \frac{1}{\beta_{\mathrm{d}}} \left(1 + \frac{L_{\mathbf{A}}}{\alpha_{\mathrm{d}}}\right) \|\mathbf{G}\|_{\mathbf{Q}'} + \frac{1}{\alpha_{\mathrm{d}}} \|\mathbf{A}(0)\|_{\mathbf{H}'} , \quad \text{and} \tag{4.240}$$

$$\begin{aligned} \|u_h\|_{\mathbf{Q}} \leqslant\ & \frac{1}{\beta_{\mathrm{d}}} \left(1 + \frac{L_{\mathbf{A}}}{\alpha_{\mathrm{d}}}\right) \|\mathbf{F}\|_{\mathbf{H}'} + \frac{L_{\mathbf{A}}}{\beta_{\mathrm{d}}^2} \left(1 + \frac{L_{\mathbf{A}}}{\alpha_{\mathrm{d}}}\right) \|\mathbf{G}\|_{\mathbf{Q}'} \\ & + \frac{1}{\beta_{\mathrm{d}}} \left(1 + \frac{L_{\mathbf{A}}}{\alpha_{\mathrm{d}}}\right) \|\mathbf{A}(0)\|_{\mathbf{H}'} . \end{aligned} \tag{4.241}$$

Moreover, keeping in mind that $(\sigma, u) \in \mathbf{H} \times \mathbf{Q}$ *is the unique solution of* (4.229)*, the corresponding Céa estimates reduce to*

$$\|\sigma - \sigma_h\|_{\mathbf{H}} \leqslant \left(1 + \frac{L_{\mathbf{A}}}{\alpha_{\mathrm{d}}}\right) \left(1 + \frac{\|\mathbf{B}\|}{\beta_{\mathrm{d}}}\right) \operatorname{dist}(\sigma, \mathbf{H}_h) + \frac{\|\mathbf{B}\|}{\alpha_{\mathrm{d}}} \operatorname{dist}(u, \mathbf{Q}_h) \tag{4.242}$$

and

$$\begin{aligned} \|u - u_h\|_{\mathbf{Q}} \leqslant\ & \frac{L_{\mathbf{A}}}{\beta_{\mathrm{d}}} \left(1 + \frac{L_{\mathbf{A}}}{\alpha_{\mathrm{d}}}\right) \left(1 + \frac{\|\mathbf{B}\|}{\beta_{\mathrm{d}}}\right) \operatorname{dist}(\sigma, \mathbf{H}_h) \\ & + \left(1 + \frac{\|\mathbf{B}\|}{\beta_{\mathrm{d}}} + \frac{L_{\mathbf{A}} \|\mathbf{B}\|}{\alpha_{\mathrm{d}} \beta_{\mathrm{d}}}\right) \operatorname{dist}(u, \mathbf{Q}_h) . \end{aligned} \tag{4.243}$$

Proof. Note first that the Lipschitz continuity of $\mathbf{A}$, which is guaranteed by (4.227) (hypothesis b) of Theorem 4.41), also holds for $\mathbf{A}|_{\mathbf{H}_h} : \mathbf{H}_h \to \mathbf{H}_h'$ with the same constant $L_{\mathbf{A}}$. Then, by virtue of this fact and the current hypotheses a) and b), the existence of a unique solution $(\sigma_h, u_h) \in \mathbf{H}_h \times \mathbf{Q}_h$ of (4.239), and the occurrence of the respective a priori estimates (4.240) and (4.241), follow from a direct application of Theorem 4.41 to the present discrete context. It only remains to derive the Céa estimates (4.242) and (4.243), for which we proceed in a similar way to the proof of the general linear case of the Babuška-Brezzi Theory (cf. Theorem 4.24), even using some of the estimates provided there. Indeed, we first define the affine shifted space of $\mathbf{V}_h$ given by

$$\mathbf{V}_h^g := \Big\{ \tau_h \in \mathbf{H}_h : \quad [\mathbf{B}(\tau_h), v_h] = \mathbf{G}(v_h) \quad \forall\, v_h \in \mathbf{Q}_h \Big\},$$

and observe that σ_h belongs to $\mathbf{V}_h^g$, from which it follows easily that $\sigma_h - \tau_h^g \in \mathbf{V}_h$ for all $\tau_h^g \in \mathbf{V}_h^g$. Then, applying the strong monotonicity assumption (4.237) to $\widetilde{\tau}_h = \tau_h^g \in \mathbf{H}_h$, $\tau_h = \sigma_h - \tau_h^g \in \mathbf{V}_h$, and $\zeta_h = 0 \in \mathbf{V}_h$, we obtain

$$\alpha_{\mathtt{d}} \, \|\sigma_h - \tau_h^g\|_{\mathbf{H}}^2 \leqslant [\mathbf{A}(\sigma_h) - \mathbf{A}(\tau_h^g), \sigma_h - \tau_h^g],$$

from which, adding and subtracting $[\mathbf{A}(\sigma), \sigma_h - \tau_h^g]$ on the right-hand side, and using the first equations of (4.229) and (4.239), as well as the fact that $[\mathbf{B}(\sigma_h - \tau_h^g), v_h] = 0$ for all $v_h \in \mathbf{Q}_h$, we arrive at

$$\alpha_{\mathtt{d}} \, \|\sigma_h - \tau_h^g\|_{\mathbf{H}}^2 \leqslant [\mathbf{A}(\sigma) - \mathbf{A}(\tau_h^g), \sigma_h - \tau_h^g] \,+\, [\mathbf{B}(\sigma_h - \tau_h^g), u - v_h] \quad \forall\, v_h \in \mathbf{Q}_h . \qquad (4.244)$$

This is a key step in the argument, in the sense that, if $\mathbf{V}_h$ were contained in $\mathbf{V}$, then the first term on the right-hand side of (4.244) would vanish, and, contrary to what we show below, the a priori estimate for $\|\sigma - \sigma_h\|_{\mathbf{H}}$ would not depend on $\mathrm{dist}(u, \mathbf{Q}_h)$ but only on $\mathrm{dist}(\sigma, \mathbf{H}_h)$. After finishing the present proof, we shall return to this point. Now, without assuming any a priori relationship between $\mathbf{V}_h$ and $\mathbf{V}$, we simply use the boundedness of $\mathbf{B}$ and the Lipschitz continuity of $\mathbf{A}$ (cf. (4.227)), to deduce from (4.244) that

$$\|\sigma_h - \tau_h^g\|_{\mathbf{H}} \leqslant \frac{L_{\mathbf{A}}}{\alpha_{\mathtt{d}}} \, \|\sigma - \tau_h^g\|_{\mathbf{H}} + \frac{\|\mathbf{B}\|}{\alpha_{\mathtt{d}}} \, \|u - v_h\|_{\mathbf{Q}} \qquad \forall\, v_h \in \mathbf{Q}_h , \quad \forall\, \tau_h^g \in \mathbf{V}_h^g ,$$

so that, by additionally applying the triangle inequality to estimate $\|\sigma - \sigma_h\|_{\mathbf{H}}$, we conclude that

$$\|\sigma - \sigma_h\|_{\mathbf{H}} \leqslant \left(1 + \frac{L_{\mathbf{A}}}{\alpha_{\mathtt{d}}}\right) \mathrm{dist}\big(\sigma, \mathbf{V}_h^g\big) \,+\, \frac{\|\mathbf{B}\|}{\alpha_{\mathtt{d}}} \, \mathrm{dist}\big(u, \mathbf{Q}_h\big) . \qquad (4.245)$$

Here we recall from (4.102) (using $\beta_{\mathtt{d}}$ instead of β_h) that

$$\mathrm{dist}\big(\sigma, \mathbf{V}_h^g\big) \leqslant \left(1 + \frac{\|\mathbf{B}\|}{\beta_{\mathtt{d}}}\right) \mathrm{dist}\big(\sigma, \mathbf{H}_h\big),$$

which, when replaced in (4.245), leads to (4.242). In turn, the preliminary estimate for $\|u - u_h\|_{\mathbf{Q}}$ is obtained in an almost identical way to that used in the proof of Theorem 4.24, except that, in addition to applying the triangle inequality and the discrete inf-sup condition (4.238), the Lipschitz continuity of $\mathbf{A}$ is also required. In this way, we obtain the inequality

$$\|u - u_h\|_{\mathbf{Q}} \leqslant \frac{L_{\mathbf{A}}}{\beta_{\mathtt{d}}} \, \|\sigma - \sigma_h\|_{\mathbf{H}} + \left(1 + \frac{\|\mathbf{B}\|}{\beta_{\mathtt{d}}}\right) \mathrm{dist}\big(u, \mathbf{Q}_h\big),$$

which, together with (4.242), imply (4.243) and complete the proof.

□

We conclude this section by emphasizing, as already pointed out in the previous proof, that in the case where $\mathbf{V}_h$ is contained in $\mathbf{V}$, the a priori bound for the error $\|\sigma - \sigma_h\|_{\mathbf{H}}$

does not depend on $\operatorname{dist}(u, \mathbf{Q}_h)$, and therefore the Céa estimates (4.242) and (4.243) reduce, respectively, to

$$\|\sigma - \sigma_h\|_{\mathbf{H}} \leqslant \left(1 + \frac{L_{\mathbf{A}}}{\alpha_{\mathrm{d}}}\right)\left(1 + \frac{\|\mathbf{B}\|}{\beta_{\mathrm{d}}}\right) \operatorname{dist}(\sigma, \mathbf{H}_h)$$

and

$$\|u - u_h\|_{\mathbf{Q}} \leqslant \frac{L_{\mathbf{A}}}{\beta_{\mathrm{d}}}\left(1 + \frac{L_{\mathbf{A}}}{\alpha_{\mathrm{d}}}\right)\left(1 + \frac{\|\mathbf{B}\|}{\beta_{\mathrm{d}}}\right) \operatorname{dist}(\sigma, \mathbf{H}_h) + \left(1 + \frac{\|\mathbf{B}\|}{\beta_{\mathrm{d}}}\right) \operatorname{dist}(u, \mathbf{Q}_h)\,.$$

At the same time, it is also important to mention that the proofs of Theorems 4.41 and 4.42 were first presented in [70, Appendix B], a work in which these results were applied to analyze the continuous and discrete solvability of nonlinear transmission problems through the combined use of the boundary integral equation and finite element methods.

4.8 Exercises.

▸ **4.1** ([56]) *Let Ω be a bounded open subset of $\mathbb{R}^n$ with smooth boundary Γ, and define the space $V := \mathrm{H}^2(\Omega) \cap \mathrm{H}^1_0(\Omega)$, that is*

$$V = \left\{ v \in \mathrm{H}^2(\Omega) : \quad v = 0 \quad \text{on} \quad \Gamma \right\}.$$

Prove that there exists a constant $C > 0$ such that

$$\|v\|_{1,\Omega} \leqslant C\,\|\Delta v\|_{0,\Omega} \quad \forall\, v \in V\,.$$

Hint: *Given $v \in V$, define $f = -\Delta v \in \mathrm{L}^2(\Omega)$ and apply the Lax-Milgram Lemma to the boundary value problem: $-\Delta u = f$ in Ω, $u = 0$ on Γ.*

▸ **4.2** ([99]) *Let $\Omega := (0,1)$ and consider the boundary value problem*

$$\begin{aligned} -(au')' + bu &= f \quad \text{in} \quad \Omega \\ u'(0) &= u'(1) = 0\,, \end{aligned} \tag{4.246}$$

where $a(x) = x^2 + 2$, $b(x) = 2 + \sin(x)$, and $f(x) = 1$ for all $x \in \Omega$. It can be shown that the weak formulation of (4.246) consists in finding $u \in \mathrm{H}^1(\Omega)$ such that

$$\int_0^1 (a\,u'\,v' + buv) = \int_0^1 f\,v \quad \forall\, v \in \mathrm{H}^1(\Omega)\,. \tag{4.247}$$

Prove that there exists a unique solution $u \in \mathrm{H}^1(\Omega)$ of (4.247).

▸ **4.3** ([99]) *Let H be a Hilbert space, and let $\{H_n\}_{n\in\mathbb{N}}$ be a sequence of finite-dimensional subspaces of H such that $H_{n-1} \subset H_n$ for all $n \in \mathbb{N}$, and $\cup_{n\in\mathbb{N}} H_n$ is dense in H. In addition, let $B : H \times H \to \mathbb{R}$ be a bounded bilinear form, and suppose that for all $f \in \mathrm{H}'$ there exists a unique sequence $\{u_n(f)\}_{n\in\mathbb{N}} \subseteq H$ such that*

$$u_n(f) \in H_n\,, \quad B\big(u_n(f), v_n\big) = f(v_n) \quad \forall\, v_n \in H_n\,,$$

and $\|u_n(f)\| \leqslant C_0\,\|f\|$ for all $n \in \mathbb{N}$, where C_0 is a positive constant independent of n and f. Suppose also that for all $f \in \mathrm{H}'$ there exists a unique $u(f) \in H$ such that

$$B(u(f), v) = f(v) \quad \forall\, v \in H\,.$$

Prove that

$$\lim_{n\to+\infty} \|u(f) - u_n(f)\| = 0 \quad \forall\, f \in \mathrm{H}'\,.$$

Hint: *Define the Galerkin projection $P_n : H \to H_n$, where for all $v \in H$, $P_n(v)$ denotes the unique solution of $B\big(P_n(v), w_n\big) = B(v, w_n)$ for all $w_n \in H_n$, and observe that $P_n\big(u(f)\big) = u_n(f)$.*

▸ **4.4** ([97]) *Let Ω be a bounded domain of $\mathbb{R}^2$ with smooth boundary Γ. Consider the boundary value problem*

$$-\Delta u = f \quad in \quad \Omega, \qquad \frac{\partial u}{\partial \nu} = g \quad on \quad \Gamma, \qquad \int_\Omega u = a_1, \tag{4.248}$$

where $a_1 \in \mathbb{R}$, $f \in \mathrm{L}^2(\Omega)$, and $g \in \mathrm{L}^2(\Gamma)$ satisfy the compatibility condition

$$\int_\Omega f + \int_\Gamma g = 0.$$

Prove that a weak formulation of (4.248) consists in: Find $(u, \lambda) \in \mathrm{H}^1(\Omega) \times \mathbb{R}$ such that

$$\begin{aligned} \int_\Omega \nabla u \cdot \nabla v + \lambda \int_\Omega v &= \int_\Omega f\, v + \int_\Gamma g\, v \qquad \forall\, v \in \mathrm{H}^1(\Omega), \\ \mu \int_\Omega u &= \mu\, a_1 \qquad \forall\, \mu \in \mathbb{R}. \end{aligned}$$

▸ **4.5** ([90]) *Let V be a Hilbert space, and let $a : V \times V \times V \to \mathbb{R}$ be a trilinear form such that for all $w \in V$, the bilinear form $a(w;\cdot,\cdot) : V \times V \to \mathbb{R}$ is bounded. Given $F \in V'$, consider the problem: Find $u \in V$ such that*

$$a(u;u,v) = F(v) \qquad \forall\, v \in V. \tag{4.249}$$

Suppose there exists $\alpha > 0$ such that

$$a(w;v,v) \geqslant \alpha \|v\|_V^2 \qquad \forall\, v, w \in V.$$

Assume also that there exists a constant $L_a > 0$ such that for all w_1, w_2, u, $v \in V$,

$$|a(w_1;u,v) - a(w_2;u,v)| \leqslant L_a \|w_1 + w_2\|_V \|u\|_V \|v\|_V \|w_1 - w_2\|_V.$$

Find a positive constant C such that for all $F \in V'$ with $\|F\|_{V'} \leqslant C$, problem (4.249) admits a unique solution.

▸ **4.6** ([90], Exercise 1.6)

a) *The Brouwer Fixed Point Theorem states that: Given a compact, convex, and non-empty subset S of a finite-dimensional vector space, and a continuous mapping T from S into S, then T has at least one fixed point. Use this Brouwer result to prove that if X is a finite-dimensional Hilbert space with inner product $\langle \cdot,\cdot \rangle$ and norm $\|\cdot\|$, and if F is a continuous mapping from X into X such that, for some $\mu > 0$, $\langle F(u), u\rangle \geqslant 0$ for all $u \in X$ with $\|u\| = \mu$, then there exists $u_0 \in X$, $\|u_0\| \leqslant \mu$, such that $F(u_0) = 0$.*

[NOTE THAT THIS IS THE SAME STATEMENT AS EXERCISE 1.6]

b) *Let Ω be a bounded open subset of $\mathbb{R}^2$ with smooth boundary Γ, and let $\mathbf{f} \in \mathbf{L}^2(\Omega)$. THE NAVIER-STOKES PROBLEM, a problem of great importance in fluid mechanics, consists in finding the velocity vector $\mathbf{u} := (u_1, u_2)$ and the pressure p of a fluid, such that*

$$-\Delta \mathbf{u} + \sum_{j=1}^{2} u_j \frac{\partial \mathbf{u}}{\partial x_j} + \nabla p = \mathbf{f} \quad in \quad \Omega, \qquad \nabla \cdot \mathbf{u} = 0 \quad in \quad \Omega, \tag{4.250}$$

$$\mathbf{u} = 0 \quad on \quad \Gamma, \qquad \int_\Omega p = 0.$$

Define the spaces $H := \mathbf{H}_0^1(\Omega)$, and

$$Q := \mathrm{L}_0^2(\Omega) := \left\{ q \in \mathrm{L}^2(\Omega) : \int_\Omega q = 0 \right\}.$$

Prove that the weak formulation of (4.250) reduces to: Find $(\mathbf{u}, p) \in H \times Q$ *such that*

$$\begin{aligned} a(\mathbf{u}; \mathbf{u}, \mathbf{v}) + b(\mathbf{v}, p) &= f(\mathbf{v}) \quad \forall\, \mathbf{v} \in H \\ b(\mathbf{u}, q) &= 0 \qquad\; \forall\, q \in Q\,, \end{aligned} \tag{4.251}$$

where $a : H \times H \times H \to \mathbb{R}$, $b : H \times Q \to \mathbb{R}$, *and* $f \in \mathrm{H}'$, *are given by*

$$a(\mathbf{w}; \mathbf{u}, \mathbf{v}) := \sum_{i=1}^{2} \int_{\Omega} \nabla u_i \cdot \nabla v_i + \sum_{i,j=1}^{2} \int_{\Omega} w_j \frac{\partial u_i}{\partial x_j} v_i\,,$$

$$b(\mathbf{v}, p) := -\int_{\Omega} p \nabla \cdot \mathbf{v} \quad , \quad f(\mathbf{v}) = \int_{\Omega} \mathbf{f} \cdot \mathbf{v}\,.$$

c) *In what follows, consider a problem of the type (4.251), not necessarily derived from (4.250), and assume that the following hypotheses hold:* b *is a bounded bilinear form,* b *satisfies the continuous Babuška-Brezzi condition, and for each* $\mathbf{w} \in H$ *the mapping* $(\mathbf{u}, \mathbf{v}) \to a(\mathbf{w}; \mathbf{u}, \mathbf{v})$ *is also a bounded bilinear form. In turn, let*

$$V := \Big\{ \mathbf{v} \in H : \quad b(\mathbf{v}, q) = 0 \quad \forall\, q \in Q \Big\}.$$

Prove that (4.251) can be equivalently reduced to the following nonlinear problem: Find $\mathbf{u} \in V$ *such that*

$$a(\mathbf{u}; \mathbf{u}, \mathbf{v}) = f(\mathbf{v}) \quad \forall\, \mathbf{v} \in V\,. \tag{4.252}$$

d) *In addition to what was stated in* **c)**, *suppose now that there exists a constant* $\alpha > 0$ *such that* $a(\mathbf{v}; \mathbf{v}, \mathbf{v}) \geqslant \alpha \|\mathbf{v}\|^2$ *for all* $\mathbf{v} \in V$; V *is separable (see Definition 7.3 in Chapter 7); and for each* $\mathbf{v} \in V$, *the mapping* $\mathbf{u} \to a(\mathbf{u}; \mathbf{u}, \mathbf{v})$ *is sequentially weakly continuous, that is,*

$$\mathbf{u}_n \overset{w}{\to} \mathbf{u} \in V \quad \Rightarrow \quad \lim_{n \to +\infty} a(\mathbf{u}_n; \mathbf{u}_n, \mathbf{v}) = a(\mathbf{u}; \mathbf{u}, \mathbf{v})\,.$$

Prove that the problem (4.252) has at least one solution $\mathbf{u} \in V$.

HINT: *First show that there exists a sequence* $\{\mathbf{v}_n\}_{n \in \mathrm{N}}$ *in* V *such that for every* $m \geqslant 1$ *the set* $\{\mathbf{v}_1, \mathbf{v}_2, \ldots, \mathbf{v}_m\}$ *is linearly independent, and the finite linear combinations of the* $\mathbf{v}_n$ *form a dense set in* V. *Then, denote by* V_m *the subspace of* V *generated by* $\{\mathbf{v}_1, \ldots, \mathbf{v}_m\}$, *and apply the Galerkin method together with what was proven in part* **a)** *to construct a sequence of approximate solutions.*

▸ **4.7** ([24]) *Let* X *and* M *be Hilbert spaces, and let* $a : X \times X \to \mathbb{R}$, $b : X \times M \to \mathbb{R}$ *be two bounded bilinear forms. Suppose further that* a *is symmetric and positive semi-definite on* X. *Given* $F \in X'$, $G \in M'$, *define the operator* $\mathcal{J} : X \times M \to \mathbb{R}$,

$$\mathcal{J}(v, \mu) := \frac{1}{2} a(v, v) + b(v, \mu) - F(v) - G(\mu)\,.$$

Then consider the following variational problems:

Find $(u, \lambda) \in X \times M$ *such that*

$$\begin{aligned} a(u, v) + b(v, \lambda) &= F(v) \quad \forall\, v \in X\,, \\ b(u, \mu) &= G(\mu) \quad \forall\, \mu \in M\,. \end{aligned} \tag{4.253}$$

Find $(u, \lambda) \in X \times M$ *such that*

$$\mathcal{J}(u, \mu) \leqslant \mathcal{J}(u, \lambda) \leqslant \mathcal{J}(v, \lambda) \qquad \forall\, (v, \mu) \in X \times M\,. \tag{4.254}$$

Prove that (u, λ) *is a solution of (4.253) if and only if* (u, λ) *is a solution of (4.254).*

▸ **4.8** ([87], [88]) *Let* $(X, \langle\cdot,\cdot\rangle_X)$ *and* $(Y, \langle\cdot,\cdot\rangle_Y)$ *be Hilbert spaces, and consider operators* $\mathbf{P} \in \mathcal{L}(X,X)$, $\mathbf{Q} \in \mathcal{L}(X,Y)$ *and* $\mathbf{S} \in \mathcal{L}(Y,Y)$. *Suppose that* $\mathbf{S}$ *is positive semi-definite, that is,*

$$\langle \mathbf{S}(y), y\rangle_Y \geqslant 0 \quad \forall\, y \in Y\,,$$

and that there exist constants α, $\beta > 0$ *such that*

$$\langle \mathbf{P}x, x\rangle_X \geqslant \alpha\, \|x\|_X^2 \qquad \forall\, x \in X$$

and

$$\sup_{x\in X\setminus\{0\}} \frac{\langle \mathbf{Q}(x), y\rangle_Y}{\|x\|_X} \geqslant \beta\, \|y\|_Y \quad \forall\, y \in Y\,.$$

Given $f \in X$ *and* $g \in Y$, *prove that there exists a unique pair* $(\widetilde{x}, \widetilde{y}) \in X \times Y$ *such that*

$$\begin{bmatrix} \mathbf{P} & \mathbf{Q}^* \\ \mathbf{Q} & -\mathbf{S} \end{bmatrix} \begin{bmatrix} \widetilde{x} \\ \widetilde{y} \end{bmatrix} = \begin{bmatrix} f \\ g \end{bmatrix}. \tag{4.255}$$

Also, prove that there exists a constant $C > 0$, *depending on* $\|\mathbf{P}\|$, α, β *and* $\|\mathbf{Q}\|$, *such that*

$$\|\widetilde{x}\| + \|\widetilde{y}\| \leqslant C\,\big\{\|f\| + \|g\|\big\}\,.$$

HINT: *Transform (4.255) into an equivalent equation in* Y *and then apply the Lax-Milgram Lemma (cf. Lemma 4.4).*

▸ **4.9** ([87], [88]) *Let* $(X, \langle\cdot,\cdot\rangle_X)$ *and* $(Y, \langle\cdot,\cdot\rangle_Y)$ *be Hilbert spaces, and consider a nonlinear operator* $\mathbf{P} : X \to X$ *and linear operators* $\mathbf{Q} \in \mathcal{L}(X,Y)$ *and* $\mathbf{S} \in \mathcal{L}(Y,Y)$. *Suppose that* $\mathbf{S}$ *is positive semi-definite, that is,*

$$\langle \mathbf{S}(y), y\rangle_Y \geqslant 0 \quad \forall\, y \in Y\,,$$

and that there exist constants M, α, $\beta > 0$ *such that*

$$\|\mathbf{P}x - \mathbf{P}\bar{x}\|_X \leqslant M\,\|x - \bar{x}\|_X \quad , \quad \langle \mathbf{P}x - \mathbf{P}\bar{x}, x - \bar{x}\rangle_X \geqslant \alpha\,\|x - \bar{x}\|_X^2 \quad \forall\, x,\, \bar{x} \in X\,,$$

and

$$\sup_{x\in X\setminus\{0\}} \frac{\langle \mathbf{Q}(x), y\rangle_Y}{\|x\|_X} \geqslant \beta\, \|y\|_Y \quad \forall\, y \in Y\,.$$

Then, given $f \in X$ *and* $g \in Y$, *show that there exists a unique pair* $(\widetilde{x}, \widetilde{y}) \in X \times Y$ *such that*

$$\begin{bmatrix} \mathbf{P} & \mathbf{Q}^* \\ \mathbf{Q} & -\mathbf{S} \end{bmatrix} \begin{bmatrix} \widetilde{x} \\ \widetilde{y} \end{bmatrix} = \begin{bmatrix} f \\ g \end{bmatrix}.$$

Also, prove that there exists a constant $C > 0$, *which depends on* M, α, β, *and* $\|\mathbf{Q}\|$, *such that*

$$\|\widetilde{x}\| + \|\widetilde{y}\| \leqslant C\,\big\{\|f\| + \|g\| + \|\mathbf{P}(0)\|\big\}\,.$$

▸ **4.10** *Let* $\Omega := (0,3)$ *and consider the boundary value problem:* $-u'' + (x+1)\,u = x$ *in* Ω, $u(0) = u(3) = 0$. *Derive the corresponding weak formulation and prove that it has a unique solution* $u \in \mathrm{H}_0^1(\Omega)$. *Consider the uniform partition* $0 = x_0 < x_1 = 1 < x_2 = 2 < x_3 = 3$ *and establish the associated Galerkin system using the subspace*

$$H_2 := \Big\{ v \in C(\bar{\Omega}) : \quad v(0) = v(3) = 0 \quad \text{and} \quad v|_{[x_j, x_{j-1}]} \text{ is a} \\ \text{polynomial of degree} \leqslant 1\,, \forall\, j \in \{1,2,3\} \Big\}\,.$$

HINT: *Note that* $H_2 = \langle\{e_1, e_2\}\rangle$, *where* $e_j \in H_2$ *is such that* $e_j(x_i) = \delta_{ij}$ *for all* i, $j \in \{1,2\}$.

▸ **4.11** *Let Ω be a bounded open subset of $\mathbb{R}^2$ with smooth boundary Γ, and let $\mathbf{f} \in \mathbf{L}^2(\Omega)$.* The Stokes Problem *consists in finding the velocity vector $\mathbf{u} := (u_1, u_2)$ and the pressure p of a fluid, such that*

$$\begin{gathered} -\Delta \mathbf{u} + \nabla p = \mathbf{f} \quad \text{in} \quad \Omega\,, \quad \operatorname{div}(\mathbf{u}) = 0 \quad \text{in} \quad \Omega\,, \\ \mathbf{u} = 0 \quad \text{on} \quad \Gamma\,, \qquad \int_\Omega p = 0\,. \end{gathered} \tag{4.256}$$

Define the spaces $H := \mathbf{H}_0^1(\Omega)$ and $Q := \left\{ q \in \mathrm{L}^2(\Omega) : \int_\Omega q = 0 \right\}$. Prove that the weak formulation of (4.256) reduces to: Find $(\mathbf{u}, p) \in H \times Q$ such that

$$\begin{aligned} A(\mathbf{u}, \mathbf{v}) + B(\mathbf{v}, p) &= F(\mathbf{v}) \qquad \forall\, \mathbf{v} \in H\,, \\ B(\mathbf{u}, q) &= 0 \qquad\quad\;\; \forall\, q \in Q\,, \end{aligned}$$

where $A : H \times H \to \mathbb{R}$, $B : H \times Q \to \mathbb{R}$ and $F \in \mathrm{H}'$, are defined by

$$A(\mathbf{u}, \mathbf{v}) := \sum_{i=1}^{2} \int_\Omega \nabla u_i \cdot \nabla v_i\,,$$

$$B(\mathbf{v}, p) := -\int_\Omega p \operatorname{div}(\mathbf{v}) \quad , \quad F(\mathbf{v}) = \int_\Omega \mathbf{f} \cdot \mathbf{v}\,.$$

▸ **4.12** ([39]) *Let H be a Hilbert space, let $A : H \times H \to \mathbb{R}$ be a bilinear form that is bounded and H-elliptic, and let $F \in H'$. Moreover, let $\{H_n\}_{n \in \mathbb{N}}$ be a sequence of finite-dimensional subspaces of H, and for each $n \in \mathbb{N}$ consider a bounded bilinear form $A_n : H_n \times H_n \to \mathbb{R}$ such that the sequence $\{A_n\}_{n \in \mathbb{N}}$ is uniformly elliptic. This means that there exists $\tilde{\alpha} > 0$, independent of n, such that $A_n(v_n, v_n) \geqslant \tilde{\alpha}\, \|v_n\|_H^2$ for all $v_n \in H_n$, for all $n \in \mathbb{N}$.*

a) *Prove that there exist unique $u \in H$ and $u_n \in H_n$ such that*

$$A(u, v) = F(v) \quad \forall\, v \in H \quad \text{and} \quad A_n(u_n, v_n) = F(v_n) \quad \forall\, v_n \in H_n\,.$$

b) *Show that there exists $C > 0$, independent of $n \in \mathbb{N}$, such that*

$$\|u - u_n\|_H \leqslant C \inf_{v_n \in H_n} \left\{ \|u - v_n\|_H + \sup_{w_n \in H_n \setminus \{0\}} \frac{|A(v_n, w_n) - A_n(v_n, w_n)|}{\|w_n\|_H} \right\}.$$

▸ **4.13** *Let Ω be a bounded open subset of $\mathbb{R}^n$, and let $\kappa : \Omega \to \mathbb{R}$ be a continuous function for which there exist constants $M, \beta > 0$, such that $\beta \leqslant \kappa(x) \leqslant M$ for all $x \in \Omega$. Define the set*

$$S := \left\{ v \in \mathrm{H}_0^1(\Omega) : \quad \|v\|_{1,\Omega} \leqslant 1 \right\},$$

and prove that for every $u \in \mathrm{H}_0^1(\Omega)$ there exists a unique $g \in S$ such that

$$\int_\Omega \kappa(x)\, |\nabla u(x) - \nabla g(x)|^2\, dx = \min_{v \in S} \int_\Omega \kappa(x)\, |\nabla u(x) - \nabla v(x)|^2\, dx\,.$$

Recall here that $|\cdot|$ stands for the euclidean norm of $\mathbb{R}^n$.

▸ **4.14** ([39]) [*Aubin-Nitsche Lemma*]. *Let $(H, \langle\cdot,\cdot\rangle_H)$ and $(V, \langle\cdot,\cdot\rangle_V)$ be Hilbert spaces such that $V \subseteq H$ and the identity operator $\mathbf{i} : V \to H$ is continuous. Let $A : V \times V \to \mathbb{R}$ be a bounded and V-elliptic bilinear form, and consider the operator $\mathbf{P} : H \to V$, where for all $g \in H$, $\mathbf{P}(g)$ is the unique element in V satisfying $A(v, \mathbf{P}(g)) = \langle g, v\rangle_H$ for all $v \in V$. Given $F \in V'$ and V_h a finite-dimensional subspace of V, denote by $u \in V$ and $u_h \in V_h$ the unique solutions of the continuous and Galerkin schemes, respectively, that is,*

$$A(u, v) = F(v) \quad \forall\, v \in V \quad \text{and} \quad A(u_h, v_h) = F(v_h) \quad \forall\, v_h \in V_h\,.$$

Prove that there exists $C > 0$ such that

$$\|u - u_h\|_H \leqslant C\, \|u - u_h\|_V \sup_{g \in H} \left\{ \frac{1}{\|g\|_H} \inf_{v_h \in V_h} \|\mathbf{P}(g) - v_h\|_V \right\}.$$

▸ **4.15** ([24]) [*Fortin's Lemma*]. *Let H, Q be Hilbert spaces, and let $b : H \times Q \to \mathbb{R}$ be a bounded bilinear form that satisfies the inf-sup condition, that is, there exists $\beta > 0$ such that*

$$\sup_{v \in H \setminus \{0\}} \frac{b(v,q)}{\|v\|_H} \geqslant \beta \, \|q\|_Q \qquad \forall\, q \in Q\,.$$

Let $\{H_n\}_{n\in\mathbb{N}}$ and $\{Q_n\}_{n\in\mathbb{N}}$ be sequences of finite-dimensional subspaces of H and Q, respectively, and assume that for each $n \in \mathbb{N}$ there exists $\mathcal{P}_n \in \mathcal{L}(H, H_n)$ such that

$$b(v - \mathcal{P}_n(v), q_n) \;=\; 0 \qquad \forall\, q_n \in Q_n\,.$$

Suppose that the family $\{\mathcal{P}_n\}_{n\in\mathbb{N}}$ is uniformly bounded, that is, there exists $C > 0$ such that $\|\mathcal{P}_n\|_{\mathcal{L}(H,H_n)} \leqslant C$ for all $n \in \mathbb{N}$, and prove that there exists $\beta^* > 0$, independent of n, such that

$$\sup_{v_n \in H_n \setminus \{0\}} \frac{b(v_n,q_n)}{\|v_n\|_H} \geqslant \beta^* \, \|q_n\|_Q \qquad \forall\, q_n \in Q_n\,, \quad \forall\, n \in \mathbb{N}\,.$$

▸ **4.16** *Let $\Omega = (a,b)$, $f \in \mathrm{L}^2(\Omega)$, and consider the boundary value problem*

$$u^{(4)} \;=\; f \quad \text{in} \quad \Omega\,, \quad u(a) = u'(a) = u(b) = u'(b) \;=\; 0\,. \tag{4.257}$$

i) *Define the auxiliary unknown $\sigma := u''$ in Ω and show that a* MIXED *variational formulation of (4.257) reduces to: Find $(\sigma, u) \in \mathrm{H}^1(\Omega) \times \mathrm{H}^1_0(\Omega)$ such that*

$$\begin{aligned} \int_\Omega \sigma\,\tau + \int_\Omega u'\,\tau' &= 0 && \forall\, \tau \in \mathrm{H}^1(\Omega)\,, \\ \int_\Omega v'\,\sigma' \qquad\qquad &= -\int_\Omega f\,v && \forall\, v \in \mathrm{H}^1_0(\Omega)\,. \end{aligned} \tag{4.258}$$

ii) *Apply the Babuška-Brezzi Theory to show that (4.258) has a unique solution that depends continuously on the data f.*

▸ **4.17** *Given $\Omega := (0,1)$ and $f \in \mathrm{L}^2(\Omega)$, we are interested in solving the following problem:*

$$-\,u'' \;=\; f \quad \text{in} \quad \Omega\,, \quad u(0) \;=\; 0\,, \quad u'(1) \;=\; 1\,. \tag{4.259}$$

a) *Define $\sigma := u'$ in Ω and show that a* MIXED *variational formulation of (4.259) reduces to: Find $(\sigma, (u,\varphi)) \in H \times Q$ such that*

$$\begin{aligned} a(\sigma,\tau) + b(\tau,(u,\varphi)) &= F(\tau) && \forall\, \tau \in H\,, \\ b(\sigma,(v,\psi)) \qquad\qquad &= G((v,\psi)) && \forall\, (v,\psi) \in Q\,, \end{aligned} \tag{4.260}$$

where $H := \mathrm{H}^1(\Omega)$, $Q := \mathrm{L}^2(\Omega) \times \mathbb{R}$, $F \in H'$, $G \in Q'$, and $a : H \times H \to \mathbb{R}$, $b : H \times Q \to \mathbb{R}$ are the bilinear forms defined by

$$\begin{aligned} a(\sigma,\tau) &:= \int_\Omega \sigma\,\tau && \forall\, \sigma, \tau \in H\,, \\ b(\tau,(v,\psi)) &:= \int_\Omega v\,\tau' + \psi\,\tau(1) && \forall\, (\tau,(v,\psi)) \in H \times Q\,. \end{aligned}$$

b) *Define F and G, and apply the Babuška-Brezzi Theory to prove that (4.260) has a unique solution.*

▸ **4.18** ([39]) [THE FOLLOWING STATEMENT IS A VARIANT OF EXERCISE 4.12] *Let H be a Hilbert space, let $A : H \times H \to \mathbb{R}$ be a bounded and H-elliptic bilinear form, and let $F \in H'$. Furthermore, let $\{H_n\}_{n\in\mathbb{N}}$ be a sequence of finite-dimensional subspaces of H, and for each $n \in \mathbb{N}$ consider a functional $F_n \in H_n'$ and a bounded bilinear form $A_n : H_n \times H_n \to \mathbb{R}$. Also suppose that there exists $\tilde{\alpha} > 0$, independent of n, such that $A_n(v_n, v_n) \geqslant \tilde{\alpha}\,\|v_n\|_H^2$ for all $v_n \in H_n$, for all $n \in \mathbb{N}$.*

a) *Prove that there exist unique $u \in H$ and $u_n \in H_n$ such that*

$$A(u,v) = F(v) \quad \forall\, v \in H \qquad \text{and} \qquad A_n(u_n,v_n) = F_n(v_n) \qquad \forall\, v_n \in H_n\,.$$

b) *Show that there exists $C > 0$, independent of $n \in \mathbb{N}$, such that*

$$\|u - u_n\|_H \leqslant C\Big\{ \inf_{v_n\in H_n} \Big\{ \|u - v_n\|_H + \|\mathbf{A}(v_n)\circ \mathbf{i}_n - \mathbf{A}_n(v_n)\|_{H_n'} \Big\} + \|F\circ \mathbf{i}_n - F_n\|_{H_n'} \Big\}, \tag{4.261}$$

where $\mathbf{A} \in \mathcal{L}(H,H')$ and $\mathbf{A}_n \in \mathcal{L}(H_n,H_n')$ are the operators induced by A and A_n, respectively, and $\mathbf{i}_n : H_n \to H$ is the canonical injection, that is, $\mathbf{i}_n(v_n) = v_n$ for all $v_n \in H_n$.
[NOTE THAT (4.261) IS ESSENTIALLY THE STRANG ESTIMATE STATED IN (4.44)]

▸ **4.19** ([57]) *Let Ω be a bounded open subset of $\mathbb{R}^n$, $f \in L^2(\Omega)$, and consider the boundary value problem: $-\Delta u = f$ in Ω, $u = 0$ on $\partial\Omega$. It can be shown that a mixed variational formulation of this problem reduces to: Find $(\boldsymbol{\sigma},u) \in \mathbf{H}(\mathrm{div};\Omega)\times L^2(\Omega)$ such that*

$$\int_\Omega \boldsymbol{\sigma}\cdot\boldsymbol{\tau} + \int_\Omega u\,\mathrm{div}(\boldsymbol{\tau}) - \int_\Omega v\,\mathrm{div}(\boldsymbol{\sigma}) = \int_\Omega f\,v \tag{4.262}$$

for all $(\boldsymbol{\tau},v) \in \mathbf{H}(\mathrm{div};\Omega)\times L^2(\Omega)$, where $\boldsymbol{\sigma} := \nabla u$ in Ω represents an additional unknown. Furthermore, from this relation, and given $\delta \in \mathbb{R}$, it follows that

$$\delta \int_\Omega (\nabla u - \boldsymbol{\sigma})\cdot(\nabla v + \boldsymbol{\tau}) = 0 \quad \forall\,(\boldsymbol{\tau},v) \in \mathbf{H} := \mathbf{H}(\mathrm{div};\Omega)\times H_0^1(\Omega), \tag{4.263}$$

and also

$$\int_\Omega \mathrm{div}(\boldsymbol{\sigma})\,\mathrm{div}(\boldsymbol{\tau}) = -\int_\Omega f\,\mathrm{div}(\boldsymbol{\tau}) \quad \forall\,\boldsymbol{\tau} \in \mathbf{H}(\mathrm{div};\Omega)\,. \tag{4.264}$$

Thus, by adding (4.262), (4.263) and (4.264), one obtains a modified mixed variational formulation, which takes the form: Find $(\boldsymbol{\sigma},u) \in \mathbf{H}$ such that

$$A((\boldsymbol{\sigma},u),(\boldsymbol{\tau},v)) = F(\boldsymbol{\tau},v) \qquad \forall\,(\boldsymbol{\tau},v) \in \mathbf{H}, \tag{4.265}$$

where $A : \mathbf{H}\times\mathbf{H} \to \mathbb{R}$ is a bilinear form and $F : \mathbf{H} \to \mathbb{R}$ is a linear functional. Show that, by choosing δ appropriately, (4.265) has a unique solution $(\boldsymbol{\sigma},u) \in \mathbf{H} := \mathbf{H}(\mathrm{div};\Omega)\times H_0^1(\Omega)$.

▸ **4.20** *Let V be a Hilbert space, let $a : V\times V \to \mathbf{R}$ be a bounded, symmetric, and weakly coercive bilinear form, and let $A : V \to V'$ be the linear and bounded operator induced by a. Given $f \in V'$ and U a nonempty convex and closed subset of V, it is known, by the Lions-Stampacchia Theorem (cf. Theorem 4.17), that there exists a unique $u \in U$ such that*

$$a(u, v-u) \geqslant f(v-u) \quad \forall\, v \in U.$$

Now, let V_h be a finite-dimensional subspace of V, let U_h be a nonempty, closed, and convex subset of V_h, and let $u_h \in U_h$ denote the unique solution of the discrete problem:

$$a(u_h, v_h - u_h) \geqslant f(v_h - u_h) \qquad \forall\, v_h \in U_h.$$

Moreover, let H be a Hilbert space such that $V \subseteq H$, and the injection $i : V \to H$ is continuous and dense. Prove that if $Au - f \in H$, then there exists a constant C, independent of V_h and U_h, such that

$$\|u - u_h\|_V \leqslant C\Big\{ \inf_{v_h\in U_h} \Big(\|u - v_h\|_V^2 + \|Au - f\|_H\,\|u - v_h\|_H \Big) + \|Au - f\|_H \inf_{v\in U} \|u_h - v\|_H \Big\}^{1/2}.$$

▸ **4.21** *Let Ω be a bounded open subset of $\mathbb{R}^n$, and let $\kappa : \Omega \to \mathbb{R}$ be a continuous function for which there exist constants $M, \beta > 0$, such that $\beta \leqslant \kappa(x) \leqslant M$ for all $x \in \Omega$. Prove, using the Lax-Milgram Lemma (cf. Lemma 4.4), that for every $f \in \mathrm{L}^2(\Omega)$ and for every constant $\delta \in \left(0, \min\left\{2\beta, \frac{2}{nM}\right\}\right)$, there exists a unique $u \in \mathrm{H}^1(\Omega)$ such that*

$$\int_\Omega \left\{ \kappa \nabla u \cdot \nabla v + \frac{1}{\kappa} u v - \delta \sum_{i=1}^n u \frac{\partial v}{\partial x_i} \right\} = \int_\Omega f v \qquad \forall v \in \mathrm{H}^1(\Omega), \quad \text{and}$$

$$\|u\|_{1,\Omega} \leqslant \frac{1}{\min\left\{\beta - \frac{\delta}{2}, \frac{1}{M} - \frac{n\delta}{2}\right\}} \|f\|_{0,\Omega}.$$

▸ **4.22** ([17]) *Let H_1, H_2, Q_1, and Q_2 be Hilbert spaces over $\mathbb{R}$, and let $a : H_1 \times H_2 \to \mathbb{R}$ and $b_j : H_j \times Q_j \to \mathbb{R}$, $j \in \{1,2\}$, be bounded bilinear forms with induced operators $\mathbf{A} \in \mathcal{L}(H_1, H_2)$ and $\mathbf{B}_j \in \mathcal{L}(H_j, Q_j)$, $j \in \{1,2\}$, respectively. Also, let K_j be the null space of $\mathbf{B}_j$, $j \in \{1,2\}$, and let Π_2 be the orthogonal projector from H_2 onto K_2. Assume that*

i) *$\Pi_2 \mathbf{A} : K_1 \to K_2$ is an isomorphism.*
ii) *there exist $\beta_1, \beta_2 > 0$ such that*

$$\|\mathbf{B}_j^*(q)\|_{H_j} := \sup_{\substack{v \in H_j \\ v \neq 0}} \frac{b_j(v,q)}{\|v\|_{H_j}} \geqslant \beta_j \|q\|_{Q_j} \quad \forall q \in Q_j, \quad \forall j \in \{1,2\}.$$

Prove that, given $F \in H_2'$ and $G \in Q_1'$, there exists a unique $(u,p) \in H_1 \times Q_2$ such that

$$\begin{aligned} a(u,v) + b_2(v,p) &= F(v) \quad \forall v \in H_2, \\ b_1(u,q) &= G(q) \quad \forall q \in Q_1. \end{aligned}$$

▸ **4.23** ([24]) *Let X_1, M_1, X_2, M_2 and Q be real Hilbert spaces, and define the product space $H := X_1 \times M_1 \times X_2 \times M_2$. In turn, consider operators $A_1 \in \mathcal{L}(X_1, X_1)$, $B_1 \in \mathcal{L}(X_1, M_1)$, $A_2 \in \mathcal{L}(X_2, X_2)$, $B_2 \in \mathcal{L}(X_2, M_2)$, and $B \in \mathcal{L}(H, Q)$, and define the matrix operators $A : H \to H$ and $T : H \times Q \to H \times Q$ as*

$$A := \left(\begin{array}{cc|cc} A_1 & B_1^* & & \\ B_1 & 0 & & \\ \hline & & A_2 & B_2^* \\ & & B_2 & 0 \end{array} \right) \quad \text{and} \quad T := \begin{pmatrix} A & B^* \\ B & 0 \end{pmatrix}.$$

a) *Apply the* BABUŠKA-BREZZI THEORY *as many times as required to establish necessary and sufficient conditions that guarantee the bijectivity of T.*
b) *Establish a* GALERKIN SCHEME *associated with T and assume additional hypotheses allowing to demonstrate the corresponding* CÉA ESTIMATE.

▸ **4.24** *Given Ω a bounded open subset of $\mathbb{R}^n$ with boundary Γ of class C^1, $f \in \mathrm{L}^2(\Omega)$, and constants $\kappa_1, \kappa_2 \in \mathbb{R}$ with $\kappa_2 \neq 0$, consider the problem:*

$$-\Delta u + \kappa_1 \sum_{j=1}^n \frac{\partial u}{\partial x_j} + \kappa_2 u = f \quad \text{in} \quad \Omega, \qquad u = 0 \quad \text{on} \quad \Gamma. \tag{4.266}$$

Define the auxiliary unknown $\boldsymbol{\sigma} = \nabla u$ in Ω and derive the following mixed variational formulation of (4.266): Find $\boldsymbol{\sigma} := (\sigma_1, \sigma_2, ..., \sigma_n) \in \mathbf{H}(\mathrm{div}; \Omega)$ such that

$$\kappa_2 \int_\Omega \boldsymbol{\sigma} \cdot \boldsymbol{\tau} + \int_\Omega \mathrm{div}(\boldsymbol{\sigma})\, \mathrm{div}(\boldsymbol{\tau}) - \kappa_1 \sum_{j=1}^n \int_\Omega \sigma_j\, \mathrm{div}(\boldsymbol{\tau}) = -\int_\Omega f\, \mathrm{div}(\boldsymbol{\tau}) \quad \forall \boldsymbol{\tau} \in \mathbf{H}(\mathrm{div}; \Omega).$$

Draw the region $S := \left\{ (\kappa_1, \kappa_2) \in \mathbb{R}^2 : \ \kappa_2 > 0 \text{ and } |\kappa_1| < 2 \min\left\{\kappa_2, \frac{1}{n}\right\} \right\}$, and prove that for every $(\kappa_1, \kappa_2) \in S$ the previous problem has a unique solution $\boldsymbol{\sigma} \in \mathbf{H}(\mathrm{div}; \Omega)$ that depends continuously on the data f.

▸ **4.25** ([8]) *Let* $(H, \langle\cdot,\cdot\rangle_H, \|\cdot\|_H)$ *be a real Hilbert space, and let* $A : H \times H \to \mathbb{R}$ *be a bounded bilinear form with induced operator* $\mathbf{A} \in \mathcal{L}(H,H)$. *Assume that there exist* $\mathbf{S}_1$, $\mathbf{S}_2 \in \mathcal{L}(H,H)$ *and constants* α_1, $\alpha_2 > 0$ *such that*

$$\langle \mathbf{S}_1^* \mathbf{A}(\tau), \tau\rangle_H \geqslant \alpha_1 \|\tau\|_H^2 \quad \text{and} \quad \langle \mathbf{A}\mathbf{S}_2(\tau), \tau\rangle_H \geqslant \alpha_2 \|\tau\|_H^2 \quad \forall\, \tau \in H\,. \tag{4.267}$$

a) *Prove that for every* $F \in H'$ *there exists a unique* $\boldsymbol{\sigma} \in H$ *such that*

$$A(\boldsymbol{\sigma}, \tau) = F(\tau) \qquad \forall\, \tau \in H\,,$$

and deduce the existence of $C > 0$, *independent of* F, *such that*

$$\|\boldsymbol{\sigma}\|_H \leqslant C\, \|F\|_{H'}\,.$$

b) *Let* $\{H_h\}_{h>0}$ *be a countable family of finite-dimensional subspaces of* H *such that*

$$\lim_{h\to 0} \operatorname{dist}(\boldsymbol{\tau}, H_h) = 0 \qquad \forall\, \boldsymbol{\tau} \in H\,,$$

and given $F \in H'$, *consider the Galerkin scheme: Find* $\boldsymbol{\sigma}_h \in H_h$ *such that*

$$A(\boldsymbol{\sigma}_h, \tau_h) = F(\tau_h) \qquad \forall\, \tau_h \in H_h\,. \tag{4.268}$$

Assume that for $i = 1$ *or for* $i = 2$ *(but not for both), there exist injective operators* $\mathbf{S}_{i,h} \in \mathcal{L}(H_h, H_h)$ *for all* $h > 0$, *and constants* C_i, $\delta > 0$, *independent of* h, *such that*

$$\|\mathbf{S}_i(\tau_h) - \mathbf{S}_{i,h}(\tau_h)\|_H \leqslant C_i\, h^\delta\, \|\mathbf{S}_i(\tau_h)\|_H \quad \forall\, \tau_h \in H_h\,.$$

Show that there exists $h_0 > 0$ *such that for every* $h \leqslant h_0$ *the problem (4.268) has a unique solution, is stable, and satisfies Céa's estimate.*

c) *What can you say about the hypotheses required for a) and b) if* A *is symmetric ? In particular, is it possible to simplify (4.267) ?*

▸ **4.26** ([96]) *Given real Hilbert spaces* H, Q_1, *and* Q_2, *define* $Q := Q_1 \times Q_2$ *and consider bounded bilinear forms* $b : H \times Q \to \mathbb{R}$, $b_1 : H \times Q_1 \to \mathbb{R}$, *and* $b_2 : H \times Q_2 \to \mathbb{R}$, *with induced operators denoted by* $\mathbf{B}$, $\mathbf{B}_1$, *and* $\mathbf{B}_2$, *respectively, such that*

$$b(\boldsymbol{\tau}, (v, \psi)) = b_1(\boldsymbol{\tau}, v) + b_2(\boldsymbol{\tau}, \psi) \qquad \forall\, \boldsymbol{\tau} \in H\,, \quad \forall\, (v, \psi) \in Q\,.$$

Prove that $\mathbf{B}(\boldsymbol{\tau}) = (\mathbf{B}_1(\boldsymbol{\tau}), \mathbf{B}_2(\boldsymbol{\tau})) \in Q$ *for all* $\boldsymbol{\tau} \in H$, *introduce the null spaces* $V_1 := N(\mathbf{B}_1)$ *and* $V_2 := N(\mathbf{B}_2)$, *and then show that the following statements are equivalent*

a) *there exists* $\beta > 0$ *such that*

$$\sup_{\substack{\boldsymbol{\tau}\in H\\ \boldsymbol{\tau}\neq 0}} \frac{b(\boldsymbol{\tau}, (v, \psi))}{\|\boldsymbol{\tau}\|_H} \geqslant \beta\, \|(v, \psi)\|_Q \qquad \forall\, (v, \psi) \in Q\,.$$

b) *there exists* $\beta > 0$ *such that*

$$\sup_{\substack{\boldsymbol{\tau}\in V_1\\ \boldsymbol{\tau}\neq 0}} \frac{b_2(\boldsymbol{\tau}, \psi)}{\|\boldsymbol{\tau}\|_H} \geqslant \beta\, \|\psi\|_{Q_2} \quad \forall\, \psi \in Q_2\,, \quad \text{and} \quad \sup_{\substack{\boldsymbol{\tau}\in V_2\\ \boldsymbol{\tau}\neq 0}} \frac{b_1(\boldsymbol{\tau}, v)}{\|\boldsymbol{\tau}\|_H} \geqslant \beta\, \|v\|_{Q_1} \quad \forall\, v \in Q_1\,.$$

▸ **4.27** ([60], [64], [65], [66]) *Let* $(X_1, \langle\cdot,\cdot\rangle_{X_1})$, $(X_2, \langle\cdot,\cdot\rangle_{X_2})$, *and* $(Y, \langle\cdot,\cdot\rangle_Y)$ *be Hilbert spaces, define the product* $X := X_1 \times X_2$, *and consider bounded linear operators* $\mathbf{P} : X \to X$, $\mathbf{Q} : X \to Y$, $A : X_1 \to X_1$, $B : X_1 \to X_2$, *and* $C : X_2 \to X_2$, *such that*

$$\mathbf{P} := \begin{pmatrix} A & B^* \\ B & -C \end{pmatrix}.$$

Let $V := V_1 \times V_2$ *be the kernel of* $\mathbf{Q}$, *where* $V_1 \subseteq X_1$ *and* $V_2 \subseteq X_2$, *and suppose that*

i) *there exists* $\alpha > 0$ *such that* $\langle A(x_1), x_1\rangle_{X_1} \geqslant \alpha \|x_1\|_{X_1}^2$ *for all* $x_1 \in V_1$,

ii) *there exists* $\beta > 0$ *such that* $\sup\limits_{x_1 \in V_1 \backslash 0} \dfrac{\langle B(x_1), x_2\rangle_{X_2}}{\|x_1\|_{X_1}} \geqslant \beta \|x_2\|_{X_2} \quad \forall x_2 \in V_2$,

iii) $\langle C(x_2), x_2\rangle_{X_2} \geqslant 0$ *for all* $x_2 \in V_2$,

iv) *and there exists* $\widetilde{\beta} > 0$ *such that* $\|\mathbf{Q}^*(y)\|_X \geqslant \widetilde{\beta} \|y\|_Y$ *for all* $y \in Y$.

a) *Prove that for every* $(f, g) \in X \times Y$ *there exists a unique* $(x, y) \in X \times Y$ *such that*

$$\begin{aligned} \mathbf{P}(x) + \mathbf{Q}^*(y) &= f\,, \\ \mathbf{Q}(x) \qquad\quad &= g\,, \end{aligned} \tag{4.269}$$

and find an explicit constant $C > 0$ *such that*

$$\|(x, y)\|_{X \times Y} \leqslant C\Big\{ \|f\|_X + \|g\|_Y \Big\}.$$

b) *Define a Galerkin scheme for (4.269) and establish sufficient conditions ensuring its unique solvability and stability.*

c) *Prove the Céa estimate for the scheme defined in* b).

▸ **4.28** *Prove the equivalence result given by Lemma 4.39.*

▸ **4.29** ([39], [111]) *Given* Ω *a bounded open subset of* $\mathbb{R}^n$ *with boundary* Γ *of class* C^1, $f \in \mathrm{L}^2(\Omega)$, $g \in \mathrm{H}^{-1/2}(\Gamma)$, *and constants* $\kappa_1, \kappa_2 \in \mathbb{R}$, *consider the problem*

$$-\Delta u + \kappa_1 \sum_{j=1}^{n} \frac{\partial u}{\partial x_j} + \kappa_2 u = f \quad \text{in} \quad \Omega\,, \qquad \nabla u \cdot \mathbf{n} = g \quad \text{on} \quad \Gamma\,, \tag{4.270}$$

where $\mathbf{n}$ *is the normal vector on* Γ. *Deduce the primal formulation of* (4.270) *and find the largest feasible region for* $(\kappa_1, \kappa_2) \in \mathbb{R}^2$ *that ensures the ellipticity of the resulting bilinear form. In turn, define the associated Galerkin scheme and establish the corresponding Céa estimate in terms of* κ_1 *and* κ_2. *Then, replace the Neumann datum with a homogeneous Dirichlet one and prove in that case that the indicated ellipticity depends only on* κ_2.

▸ **4.30** *Let* Ω *be a bounded open subset of* $\mathbb{R}^2$ *with boundary* Γ *of class* $C^{0,1}$, *and let* $\Gamma_D, \Gamma_N \subseteq \Gamma$ *be such that* $\Gamma_D \cap \Gamma_N = \varnothing$, $|\Gamma_D| > 0$ *and* $\Gamma = \bar{\Gamma}_D \cup \bar{\Gamma}_N$. *Given* $f \in \mathrm{L}^2(\Omega)$ *and* $g \in \mathrm{H}^{1/2}(\Gamma_D)$, *consider the boundary value problem*

$$-\Delta u = f \ \text{ in } \Omega\,, \quad \gamma_0(u)|_{\Gamma_D} = g \ \text{ on } \Gamma_D\,, \quad \gamma_1(u) = 0 \ \text{ on } \Gamma_N\,. \tag{4.271}$$

i) *Use the definitions and results of Section 4.5.5 to prove that a primal-mixed formulation of* (4.271) *reduces to: Find* $(u, \lambda) \in H \times Q$ *such that*

$$\begin{aligned} a(u, v) + b(v, \lambda) &= F(v) \quad \forall\, v \in H\,, \\ b(u, \xi) \qquad\qquad &= G(\xi) \quad \forall\, \xi \in Q\,, \end{aligned} \tag{4.272}$$

where $H := \mathrm{H}^1(\Omega)$, $Q := \mathrm{H}^{-1/2}(\Gamma_D)$, $a : H \times H \to \mathbb{R}$ *and* $b : H \times Q \to \mathbb{R}$ *are the bilinear forms given by*

$$a(u, v) := \int_\Omega \nabla u \cdot \nabla v \quad \text{and} \quad b(v, \xi) := \langle \xi, \gamma_0(v)\rangle_{\Gamma_D} \quad \forall\, u, v \in H, \ \forall\, \xi \in Q\,,$$

and the functionals $F \in H'$ *and* $G \in Q'$ *depend on* f *and* g, *respectively.*

ii) *Apply the Babuška-Brezzi Theory to prove that there exists a unique solution of (4.272), which depends continuously on the data f and g.*

iii) *Let H_h be a finite-dimensional subspace of H and, given a partition $\{e_1,\dots,e_n\}$ of Γ_D, define*

$$Q_h := \Big\{\xi_h \in \mathrm{L}^2(\Gamma_D): \quad \xi_h|_{e_j} \in P_0(e_j) \quad \forall\, j \in \{1,2,...,n\}\Big\}.$$

Consider the Galerkin system associated with (4.272), assume that the bilinear form b satisfies the discrete inf-sup condition with a constant $\beta > 0$ independent of the dimensions of H_h and Q_h, and prove that this discrete scheme has a unique solution $(u_h,\lambda_h) \in H_h \times Q_h$. Also, establish the Céa estimate and comment on whether the error $\|u-u_h\|_{1,\Omega}$ depends or not on $\mathrm{dist}(\lambda, Q_h)$.

▸ **4.31** ([62], [63]) *Let $(H,\langle\cdot,\cdot\rangle_H)$ and $(Q,\langle\cdot,\cdot\rangle_Q)$ be Hilbert spaces such that $H \subseteq Q$, and suppose that there exists $A \in \mathcal{L}(H,Q)$ such that*

$$\langle\boldsymbol{\zeta},\boldsymbol{\tau}\rangle_H = \langle\boldsymbol{\zeta},\boldsymbol{\tau}\rangle_Q + \langle A(\boldsymbol{\zeta}),A(\boldsymbol{\tau})\rangle_Q \qquad \forall\,\boldsymbol{\zeta},\ \boldsymbol{\tau} \in H\,.$$

Moreover, given $\boldsymbol{\sigma} \in H$ and $\{H_n\}_{n\in\mathbb{N}}$ a family of finite-dimensional subspaces of H, consider for each $n \in \mathbb{N}$ an approximation $\boldsymbol{\sigma}_n \in H_n$ of $\boldsymbol{\sigma}$ such that $\sup\limits_{n\in\mathbb{N}} \|\boldsymbol{\sigma}-\boldsymbol{\sigma}_n\|_Q < +\infty$. Then, assume that $A(\boldsymbol{\sigma})$ is explicitly known, and define a **postprocessed approximation** *of $\boldsymbol{\sigma}$ as the unique element $\boldsymbol{\sigma}_n^* \in H_n$ (if it exists) such that*

$$\langle\boldsymbol{\sigma}_n^*,\boldsymbol{\tau}_n\rangle_H = \langle\boldsymbol{\sigma}_n,\boldsymbol{\tau}_n\rangle_Q + \langle A(\boldsymbol{\sigma}),A(\boldsymbol{\tau}_n)\rangle_Q \qquad \forall\,\boldsymbol{\tau}_n \in H_n\,.$$

Prove that $\boldsymbol{\sigma}_n^*$ is indeed well defined and conclude that

$$\|\boldsymbol{\sigma}-\boldsymbol{\sigma}_n^*\|_H \leqslant \|\boldsymbol{\sigma}-\Pi_n(\boldsymbol{\sigma})\|_H + \|\boldsymbol{\sigma}-\boldsymbol{\sigma}_h\|_Q\,,$$

where $\Pi_n : H \to H_n$ is the orthogonal projector.

▸ **4.32** *Let Ω be a bounded open subset of $\mathbb{R}^n$ with boundary Γ of class C^1, let $f \in \mathrm{L}^2(\Omega)$, let $\mathbf{K} \in \mathbb{C}(\bar\Omega) := [C(\bar\Omega)]^{n\times n}$ be a symmetric, uniformly positive definite matrix-valued function, and let $\kappa_1,\,\kappa_2 \in \mathbb{R}$. Consider the problem*

$$-\operatorname{div}\big(\mathbf{K}\,\nabla u\big) + \kappa_1 \sum_{j=1}^{n} \frac{\partial u}{\partial x_j} + \kappa_2\, u = f \quad in \quad \Omega, \qquad \mathbf{K}\,\nabla u\cdot\mathbf{n} = 0 \quad on \quad \Gamma, \tag{4.273}$$

where $\mathbf{n}$ is the unit normal vector on Γ. Derive the primal formulation of (4.273) and find the largest feasible region for $(\kappa_1,\kappa_2) \in \mathbb{R}^2$ that guarantees, by means of the classical Lax-Milgram Lemma, that such formulation has a unique solution. Moreover, define the associated Galerkin scheme and establish the Céa estimate in terms of $\mathbf{K}$, κ_1, and κ_2.

▸ **4.33** *Let X, M, and Q be real Hilbert spaces, define $H := X \times M$, and consider operators $A_1 \in \mathcal{L}(X,X)$, $B_1 \in \mathcal{L}(X,M)$, and $B \in \mathcal{L}(H,Q)$. Moreover, let $A : H \to H$ and $T : H\times Q \to H\times Q$ be the operators defined in matrix form by*

$$A := \begin{pmatrix} A_1 & B_1^* \\ B_1 & 0 \end{pmatrix} \qquad \text{and} \qquad T := \begin{pmatrix} A & B^* \\ B & 0 \end{pmatrix}.$$

a) *Apply the Babuška-Brezzi Theory to establish necessary and sufficient conditions guaranteeing the bijectivity of T.*

b) *Establish a Galerkin scheme associated with the operator T, and assume additional hypotheses that allow you to prove the corresponding Céa estimate.*

▸ **4.34** *Let Ω be a bounded open subset of $\mathbb{R}^n$ with smooth boundary Γ, and let*

$$H = \mathbb{H}(\mathbf{div};\Omega) := \Big\{\boldsymbol{\tau} \in \mathbb{L}^2(\Omega) : \ \mathbf{div}(\boldsymbol{\tau}) \in \mathbf{L}^2(\Omega)\Big\} \quad \text{and} \quad Q := \mathbf{L}^2(\Omega)\,,$$

be the Hilbert spaces with inner products and induced norms denoted, respectively, by $\langle\cdot,\cdot\rangle_{\mathbf{div};\Omega}$, $\langle\cdot,\cdot\rangle_{0,\Omega}$, $\|\cdot\|_{\mathbf{div};\Omega}$ and $\|\cdot\|_{0,\Omega}$. Consider the operator $P : H \to H$ that assigns to each $\boldsymbol{\sigma} \in H$ the element $P(\boldsymbol{\sigma}) = \bar{\boldsymbol{\sigma}}$, where $(\bar{\boldsymbol{\sigma}}, \bar{\mathbf{u}}) \in H \times Q$ is the solution of the problem

$$\begin{aligned} \int_\Omega \bar{\boldsymbol{\sigma}} : \boldsymbol{\tau} + \int_\Omega \bar{\mathbf{u}}\cdot\mathbf{div}(\boldsymbol{\tau}) &= 0 && \forall\,\boldsymbol{\tau} \in H\,, \\ \int_\Omega \mathbf{v}\cdot\mathbf{div}(\bar{\boldsymbol{\sigma}}) &= \int_\Omega \mathbf{v}\cdot\mathbf{div}(\boldsymbol{\sigma}) && \forall\,\mathbf{v} \in Q\,. \end{aligned} \tag{4.274}$$

a) *Apply the Babuška–Brezzi Theory to prove that $P \in \mathcal{L}(X)$.*
b) *Define the closed subspace of H given by*

$$V := \Big\{\boldsymbol{\tau} \in H : \quad \mathbf{div}(\boldsymbol{\tau}) = 0 \quad in \quad \Omega\Big\}\,,$$

and prove that $V = N(P)$, $P^2 = P$, and $H = V \oplus R(P)$.
c) *Deduce from b) that $V^\perp = R(P)$ (orthogonality in H), and show that there exists a constant $C \geqslant 1$ such that*

$$\|\boldsymbol{\tau}\|_{\mathbf{div};\Omega} \leqslant C\,\|\mathbf{div}(\boldsymbol{\tau})\|_{0,\Omega} \qquad \forall\,\boldsymbol{\tau} \in V^\perp\,,$$

thus concluding that $\|\cdot\|_{\mathbf{div};\Omega}$ and $\|\mathbf{div}(\cdot)\|_{0,\Omega}$ are equivalent on $V^\perp$.

▸ **4.35** ([85]) *Let Ω be a bounded open subset of $\mathbb{R}^n$ with boundary $\Gamma = \bar{\Gamma}_D \cup \bar{\Gamma}_N$ such that $\Gamma_D \cap \Gamma_N = \emptyset$ and $|\Gamma_D| > 0$, and let $\boldsymbol{\nu}$ be the normal vector to Γ. Given $\mathbf{f} \in \mathbf{C}(\bar{\Omega}) := [C(\bar{\Omega})]^n$ and $g \in \mathrm{H}^{1/2}(\Gamma_D)$, the* DARCY PROBLEM WITH PRESSURE DEPENDING ON POROSITY *consists of finding the velocity $\mathbf{u}$ and the pressure p of a fluid such that*

$$\begin{aligned} \mathbf{u} = p\,\mathbf{f} + \nabla p \quad in \quad \Omega\,, \quad \mathrm{div}(\mathbf{u}) = 0 \quad in \quad \Omega\,, \\ p = g \quad on \quad \Gamma_D\,, \quad \mathbf{u}\cdot\boldsymbol{\nu} = 0 \quad on \quad \Gamma_N\,. \end{aligned} \tag{4.275}$$

a) *Show that the mixed variational formulation of (4.275) reduces to: Find $(\mathbf{u}, p) \in H \times Q$ such that*

$$\begin{aligned} \mathbf{a}(\mathbf{u},\mathbf{v}) + \mathbf{b}(\mathbf{v},p) &= \langle \mathbf{v}\cdot\boldsymbol{\nu}, g\rangle_D + \int_\Omega p\,\mathbf{f}\cdot\mathbf{v} && \forall\,\mathbf{v} \in H\,, \\ \mathbf{b}(\mathbf{u},q) &= 0 && \forall\,q \in Q\,, \end{aligned} \tag{4.276}$$

where $H := \Big\{\mathbf{v} \in \mathbf{H}(\mathrm{div};\Omega) : \quad \mathbf{v}\cdot\boldsymbol{\nu} = 0 \quad on \quad \Gamma_N\Big\}$, $Q := \mathrm{L}^2(\Omega)$, $\mathbf{a} : H \times H \to \mathbb{R}$ and $\mathbf{b} : H \times Q \to \mathbb{R}$ are the bilinear forms given by

$$\mathbf{a}(\mathbf{u},\mathbf{v}) := \int_\Omega \mathbf{u}\cdot\mathbf{v} \quad and \quad \mathbf{b}(\mathbf{v},q) := \int_\Omega q\,\mathrm{div}(\mathbf{v}) \qquad \forall\,\mathbf{u},\mathbf{v} \in H\,, \quad \forall\,q \in Q\,,$$

and $\langle\cdot,\cdot\rangle_D$ denotes the dual pairing between $\mathrm{H}^{-1/2}(\Gamma_D)$ and $\mathrm{H}^{1/2}(\Gamma_D)$.
b) *Show that (4.276) can be equivalently rewritten as the fixed-point equation $T(\mathbf{u},p) = (\mathbf{u},p)$, where, thanks to the Babuška–Brezzi Theory, $T : H \times Q \to H \times Q$ is a well-defined nonlinear operator. In fact, employ the superposition principle to realize that T is actually an affine mapping. Then, apply the Banach Fixed Point Theorem (cf. Theorem 1.8) to conclude that if $\|\mathbf{f}\|_{\infty,\Omega} := \sup\limits_{x\in\Omega} \|\mathbf{f}(x)\|$ is sufficiently small, then the problem (4.276) has a unique solution.*
Prove in that case that there exists $C > 0$, depending on $\|\mathbf{f}\|_{\infty,\Omega}$, such that

$$\|\mathbf{u}\|_{\mathrm{div};\Omega} + \|p\|_{0,\Omega} \leqslant C\,\|g\|_{1/2,\Gamma_D}\,.$$

▸ **4.36** *Let Ω be a bounded open subset of $\mathbb{R}^n$ with boundary Γ of class $C^{0,1}$, and let $\boldsymbol{\nu}$ be the outward normal vector to Γ. Given $\mathbf{f} \in \mathbf{L}^2(\Omega)$ and $\mathbf{g} \in \mathbf{H}^{1/2}(\Gamma)$, consider the elasticity problem:*

$$\boldsymbol{\sigma} = \mathcal{C}\,\mathbf{e}(\mathbf{u}) \quad in \quad \Omega, \quad \mathbf{div}(\boldsymbol{\sigma}) = -\mathbf{f} \quad in \quad \Omega, \quad \gamma_0(\mathbf{u}) = \mathbf{g} \quad on \quad \Gamma\,. \tag{4.277}$$

a) *Define a suitable auxiliary unknown and prove that the primal-mixed formulation of* (4.277) *reduces to: Find* $(\mathbf{u}, \boldsymbol{\xi}) \in \mathbf{H}^1(\Omega) \times \mathbf{H}^{-1/2}(\Gamma)$ *such that*

$$\begin{aligned} \int_\Omega \mathcal{C}\,\mathbf{e}(\mathbf{u}) : \mathbf{e}(\mathbf{v}) + \langle \boldsymbol{\xi}, \gamma_0(\mathbf{v})\rangle_\Gamma &= \int_\Omega \mathbf{f}\cdot\mathbf{v} \quad \forall\, \mathbf{v} \in \mathbf{H}^1(\Omega)\,, \\ \langle \boldsymbol{\lambda}, \gamma_0(\mathbf{u})\rangle_\Gamma &= \langle \boldsymbol{\lambda}, \mathbf{g}\rangle_\Gamma \quad \forall\, \boldsymbol{\lambda} \in \mathbf{H}^{-1/2}(\Gamma)\,, \end{aligned} \tag{4.278}$$

where $\langle\cdot,\cdot\rangle_\Gamma$ *denotes the dual pairing between* $\mathbf{H}^{-1/2}(\Gamma)$ *and* $\mathbf{H}^{1/2}(\Gamma)$.

b) *Apply the Babuška–Brezzi Theory to prove that* (4.278) *has a unique solution, which depends continuously on the data* $\mathbf{f}$ *and* $\mathbf{g}$.

▸ **4.37** ([4], [43]) *Let* Ω *be a smooth open subset of* $\mathbb{R}^n$ *with boundary* Γ, *and denote by* $\langle\cdot,\cdot\rangle_{\mathrm{div};\Omega}$ *and* $\langle\cdot,\cdot\rangle_{1,\Omega}$ *the usual inner products of* $\mathbf{H}(\mathrm{div};\Omega)$ *and* $\mathrm{H}^1(\Omega)$, *respectively. Then, given* $f \in \mathrm{L}^2(\Omega)$, $g \in \mathrm{L}^2(\Gamma)$, *and constants* $a,\, b \in \mathbb{R}$, *consider the variational problem: Find* $(\boldsymbol{\sigma}, u, \phi) \in \mathbf{H}(\mathrm{div};\Omega) \times \mathrm{H}^1(\Omega) \times \mathrm{H}^1_0(\Omega)$ *such that*

$$\begin{aligned} \langle \boldsymbol{\sigma}, \boldsymbol{\tau}\rangle_{\mathrm{div};\Omega} + \langle u, v\rangle_{1,\Omega} - a\int_\Omega \phi\, v &= \int_\Omega f\,\mathrm{div}(\boldsymbol{\tau})\,, \\ \int_\Omega \nabla\phi\cdot\nabla\psi - b\int_\Omega \phi\,\psi - b\int_\Omega \nabla u\cdot\nabla\psi &= \int_\Gamma g\,\psi\,, \end{aligned} \tag{4.279}$$

for all $(\boldsymbol{\tau}, v, \psi) \in \mathbf{H}(\mathrm{div};\Omega) \times \mathrm{H}^1(\Omega) \times \mathrm{H}^1_0(\Omega)$.

i) *Introduce suitable well-defined linear operators* $S := (S_1, S_2) : \mathrm{H}^1_0(\Omega) \to \mathbf{H}(\mathrm{div};\Omega) \times \mathrm{H}^1(\Omega)$ *and* $\widetilde{S} : \mathrm{H}^1_0(\Omega) \times \mathrm{H}^1(\Omega) \to \mathrm{H}^1_0(\Omega)$, *such that* (4.279) *is equivalently reduced to the fixed-point equation: Find* $\phi \in \mathrm{H}^1_0(\Omega)$ *such that* $T(\phi) = \phi$, *where*

$$T(\phi) := \widetilde{S}(\phi, S_2(\phi)) \quad \forall\, \phi \in \mathrm{H}^1_0(\Omega)\,.$$

ii) *Prove that there exist constants* $C(a)$, $C(b) \geqslant 0$ *such that*

$$\|S_1(\phi) - S_1(\varphi)\|_{\mathrm{div};\Omega} + \|S_2(\phi) - S_2(\varphi)\|_{1,\Omega} \leqslant C(a)\,\|\phi - \varphi\|_{0,\Omega}\,,$$

$$\|\widetilde{S}(\phi, u) - \widetilde{S}(\varphi, w)\|_{1,\Omega} \leqslant C(b)\Big\{\|\phi - \varphi\|_{0,\Omega} + |u - w|_{1,\Omega}\Big\}\,,$$

for all $\phi,\ \varphi \in \mathrm{H}^1_0(\Omega)$, $u,\ w \in \mathrm{H}^1(\Omega)$.

iii) *Deduce from i) and ii) that there exists a constant* $C(a,b) \geqslant 0$ *such that*

$$\|T(\phi) - T(\varphi)\|_{1,\Omega} \leqslant C(a,b)\,\|\phi - \varphi\|_{0,\Omega} \quad \forall\, \phi,\ \varphi \in \mathrm{H}^1_0(\Omega)\,,$$

then use Theorem 10.18 to prove that T *is compact, and finally conclude that for sufficiently small* a *and* b, *the original problem* (4.279) *has a unique solution.*

▸ **4.38** *Let* Ω *be an* **annular domain** *of* $\mathbb{R}^2$ *with inner and outer boundaries, both Lipschitz continuous, given by* Σ *and* Γ, *respectively, and consider data* $\mathbf{f} \in \mathbf{L}^2(\Omega)$ *and* $\mathbf{g} \in \mathbf{H}^{-1/2}(\Gamma)$. *The* Stokes Problem *with mixed boundary conditions consists in finding the velocity* $\mathbf{u}$ *and the pressure* p *of a fluid occupying the region* Ω, *such that*

$$\begin{aligned} -\mu\,\Delta\mathbf{u} + \nabla p &= \mathbf{f} \quad \text{in } \Omega\,, \\ \mathrm{div}(\mathbf{u}) &= 0 \quad \text{in } \Omega\,, \\ \big(\mu\,\nabla\mathbf{u} - p\,\mathbb{I}\big)\,\mathbf{n} &= \mathbf{g} \quad \text{on } \Gamma\,, \\ \mathbf{u} &= 0 \quad \text{on } \Sigma\,, \end{aligned} \tag{4.280}$$

where $\mu > 0$ *is the viscosity,* $\mathbb{I}$ *is the identity matrix in* $\mathbb{R}^{2\times2}$, *and* $\mathbf{n}$ *is the normal vector to* Γ. *Introduce the auxiliary unknowns*

$$\varphi := -\mathbf{u} \quad on \quad \Gamma \quad and \quad \boldsymbol{\sigma} := \mu \nabla \mathbf{u} - p\mathbb{I} \quad in \quad \Omega,$$

eliminate p, and then apply the Babuška–Brezzi Theory to analyze the solvability of the resulting mixed variational formulation.

▸ **4.39** ([30]) *Let Ω be a bounded open subset of $\mathbb{R}^n$ with smooth boundary Γ. Then, given $\mathbf{f} \in \mathbf{L}^2(\Omega)$ and parameters κ_1 and κ_2 to be chosen appropriately, an augmented mixed formulation of the Navier–Stokes problem reduces to: Find $(\boldsymbol{\sigma}, \mathbf{u}) \in \mathrm{H} := \mathbb{H}(\mathbf{div}; \Omega) \times \mathbf{H}_0^1(\Omega)$ such that*

$$A((\boldsymbol{\sigma}, \mathbf{u}), (\boldsymbol{\tau}, \mathbf{v})) + B(\mathbf{u}; (\boldsymbol{\sigma}, \mathbf{u}), (\boldsymbol{\tau}, \mathbf{v})) = F(\boldsymbol{\tau}, \mathbf{v}) \qquad \forall\, (\boldsymbol{\tau}, \mathbf{v}) \in \mathrm{H}, \tag{4.281}$$

where A, B, and F are defined by

$$\begin{aligned} A((\boldsymbol{\sigma}, \mathbf{u}), (\boldsymbol{\tau}, \mathbf{v})) &:= \int_\Omega \boldsymbol{\sigma} : \boldsymbol{\tau} + \kappa_1 \int_\Omega \mathbf{div}(\boldsymbol{\sigma}) \cdot \mathbf{div}(\boldsymbol{\tau}) + \int_\Omega \mathbf{u} \cdot \mathbf{div}(\boldsymbol{\tau}) \\ &\quad - \int_\Omega \mathbf{v} \cdot \mathbf{div}(\boldsymbol{\sigma}) + \kappa_2 \int_\Omega \{\nabla \mathbf{u} - \boldsymbol{\sigma}\} : \nabla \mathbf{v}, \\ B(\mathbf{w}; (\boldsymbol{\sigma}, \mathbf{u}), (\boldsymbol{\tau}, \mathbf{v})) &:= \int_\Omega (\mathbf{w} \otimes \mathbf{u}) : \{\boldsymbol{\tau} - \kappa_2 \nabla \mathbf{v}\}, \\ F(\boldsymbol{\tau}, \mathbf{v}) &:= -\kappa_1 \int_\Omega \mathbf{f} \cdot \mathbf{div}(\boldsymbol{\tau}) + \int_\Omega \mathbf{f} \cdot \mathbf{v}, \end{aligned}$$

for all $\mathbf{w} \in \mathbf{H}_0^1(\Omega)$, $(\boldsymbol{\sigma}, \mathbf{u})$, $(\boldsymbol{\tau}, \mathbf{v}) \in \mathrm{H}$. Recall here that, given vectors $\mathbf{w}$, $\mathbf{u}$ and tensors $\boldsymbol{\zeta}$, $\boldsymbol{\tau}$, we define $\mathbf{w} \otimes \mathbf{u} := (w_i\, u_j)_{i,j=1}^n$ and $\boldsymbol{\zeta} : \boldsymbol{\tau} := \sum_{i,j=1}^n \zeta_{ij}\, \tau_{ij}$.

a) *Prove that for $\kappa_1 > 0$ and $\kappa_2 \in (0, 2)$, the bilinear form A is elliptic in H with a constant α depending on both parameters.*
b) *Apply the continuous embedding of $\mathbf{H}^1(\Omega)$ into $\mathbf{L}^4(\Omega)$ (cf. [111, Theorem 1.3.4]) to deduce the existence of a constant $c(\Omega, \kappa_2) > 0$, which depends on Ω and κ_2, such that*

$$\big|B(\mathbf{w}; (\boldsymbol{\sigma}, \mathbf{u}), (\boldsymbol{\tau}, \mathbf{v}))\big| \leqslant c(\Omega, \kappa_2)\, \|\mathbf{w}\|_{1,\Omega}\, \|\mathbf{u}\|_{1,\Omega}\, \|(\boldsymbol{\tau}, \mathbf{v})\|_{\mathrm{H}}.$$

c) *Prove that there exists $r > 0$ such that for each $\mathbf{w} \in \mathbf{V}(r) := \Big\{\mathbf{w} \in \mathbf{H}_0^1(\Omega) : \|\mathbf{w}\|_{1,\Omega} \leqslant r\Big\}$, there exists a unique $(\boldsymbol{\sigma}, \mathbf{u}) \in \mathrm{H}$ such that*

$$A((\boldsymbol{\sigma}, \mathbf{u}), (\boldsymbol{\tau}, \mathbf{v})) + B(\mathbf{w}; (\boldsymbol{\sigma}, \mathbf{u}), (\boldsymbol{\tau}, \mathbf{v})) = F(\boldsymbol{\tau}, \mathbf{v}) \qquad \forall\, (\boldsymbol{\tau}, \mathbf{v}) \in \mathrm{H}. \tag{4.282}$$

In such case define the operator $\mathbf{S} : \mathbf{V}(r) \to \mathbf{H}_0^1(\Omega)$ by $\mathbf{S}(\mathbf{w}) := \mathbf{u}$, and deduce the continuous dependence of (4.282) (according to the Lax–Milgram Lemma).
d) *Observe that (4.281) can be equivalently reformulated as the fixed-point equation: Find $\mathbf{u} \in \mathbf{V}(r)$ such that $\mathbf{S}(\mathbf{u}) = \mathbf{u}$, and then apply the ellipticity of the bilinear form $A(\cdot,\cdot) + B(\mathbf{w};\cdot,\cdot)$ and the Banach Fixed-Point Theorem to conclude that, under suitable assumptions on $\mathbf{f}$, the problem (4.281) has a unique solution $(\boldsymbol{\sigma}, \mathbf{u}) \in \mathrm{H}$ with $\mathbf{u} \in \mathbf{V}(r)$.*

▸ **4.40** ([74]) *Let Ω be a bounded domain of $\mathbb{R}^2$ with Lipschitz continuous boundary Γ. Given $\mathbf{f} \in \mathbf{L}^2(\Omega)$ and $\mathbf{g} \in \mathbf{H}^{-1/2}(\Gamma)$, the* ELASTICITY PROBLEM *with Neumann boundary conditions consists in finding a symmetric tensor $\boldsymbol{\sigma}$ (stresses) and a vector $\mathbf{u}$ (displacements), such that*

$$\boldsymbol{\sigma} = \mathcal{C}\, \mathbf{e}(\mathbf{u}), \quad \mathbf{div}(\boldsymbol{\sigma}) = -\mathbf{f} \quad in \quad \Omega, \quad \boldsymbol{\sigma\nu} = g \quad on \quad \Gamma. \tag{4.283}$$

Here, $\mathbf{e}(\mathbf{u}) := \frac{1}{2}\left(\nabla \mathbf{u} + (\nabla \mathbf{u})^{\mathrm{t}}\right)$ is the small strain tensor, $\mathcal{C}$ is the elasticity operator given by Hooke's law, with Lamé constants λ, $\mu > 0$, $\boldsymbol{\nu}$ is the unit normal vector to Γ, and the data satisfy the compatibility condition:

$$\int_\Omega \mathbf{f} \cdot \boldsymbol{\chi} + \langle g, \boldsymbol{\chi} \rangle = 0 \qquad \forall\, \boldsymbol{\chi} \in \mathbb{RM}(\Omega), \tag{4.284}$$

where $\langle \cdot, \cdot \rangle$ *denotes the duality pairing between* $\mathbf{H}^{-1/2}(\Gamma)$ *and* $\mathbf{H}^{1/2}(\Gamma)$*, and*

$$\mathbb{RM}(\Omega) := [\mathbb{P}_0(\Omega)]^2 \oplus \mathbb{P}_0(\Omega) \begin{pmatrix} x_2 \\ -x_1 \end{pmatrix}$$

is the space of rigid body motions in Ω*. It can be shown that (4.284) is a necessary and sufficient condition for the existence of a solution to (4.283).*

a) *Show that, in the first instance, the mixed variational formulation of (4.283) reduces to: Find* $(\boldsymbol{\sigma}, \vec{\mathbf{u}}) := (\boldsymbol{\sigma}, (\mathbf{u}, \varphi, \gamma)) \in \mathbf{H}(\mathrm{div}; \Omega) \times \mathbf{Q}$ *such that*

$$\begin{aligned} \int_\Omega \mathcal{C}^{-1}\boldsymbol{\sigma} : \boldsymbol{\tau} + \mathbf{b}(\boldsymbol{\tau}, \vec{\mathbf{u}}) &= 0 & \forall\, \boldsymbol{\tau} \in \mathbf{H}(\mathrm{div}; \Omega)\,, \\ \mathbf{b}(\boldsymbol{\sigma}, \vec{\mathbf{v}}) &= -\int_\Omega \mathbf{f} \cdot \mathbf{v} + \langle g, \psi \rangle & \forall\, \vec{\mathbf{v}} := (\mathbf{v}, \psi, \boldsymbol{\eta}) \in \mathbf{Q}\,, \end{aligned} \tag{4.285}$$

where $\mathbf{Q} := \mathbf{L}^2(\Omega) \times \mathbf{H}^{1/2}(\Gamma) \times \mathbb{L}^2_{\mathrm{skew}}(\Omega)$*, and*

$$\mathbf{b}(\boldsymbol{\tau}, \vec{\mathbf{v}}) := \int_\Omega \mathbf{v} \cdot \mathbf{div}(\boldsymbol{\tau}) + \langle \boldsymbol{\tau}\boldsymbol{\nu}, \psi \rangle + \int_\Omega \boldsymbol{\tau} : \boldsymbol{\eta} \quad \forall\, (\boldsymbol{\tau}, \vec{\mathbf{v}}) \in \mathbf{H}(\mathrm{div}; \Omega) \times \mathbf{Q}\,.$$

b) *Prove that the solution space of the homogeneous problem associated to (4.285) is given by*

$$\Big\{(\boldsymbol{\sigma}, \vec{\mathbf{u}}) : \quad \boldsymbol{\sigma} = 0\,, \quad \vec{\mathbf{u}} = (\boldsymbol{\chi}, -\boldsymbol{\chi}|_\Gamma, \nabla\boldsymbol{\chi})\,, \quad \boldsymbol{\chi} \in \mathbb{RM}(\Omega)\Big\}\,.$$

c) *Add to (4.285) the uniqueness condition* $\int_\Omega \boldsymbol{\chi} \cdot \mathbf{u} = 0$ *for all* $\boldsymbol{\chi} \in \mathbb{RM}(\Omega)$*, and prove that the resulting variational formulation is equivalent to: Find* $((\boldsymbol{\sigma}, \boldsymbol{\rho}), \vec{\mathbf{u}}) := ((\boldsymbol{\sigma}, \boldsymbol{\rho}), (\mathbf{u}, \varphi, \gamma)) \in \mathbf{H} \times \mathbf{Q}$ *such that*

$$\begin{aligned} \int_\Omega \mathcal{C}^{-1}\boldsymbol{\sigma} : \boldsymbol{\tau} + \int_\Omega \boldsymbol{\rho} \cdot \boldsymbol{\chi} + \mathbf{b}(\boldsymbol{\tau}, \vec{\mathbf{u}}) + \int_\Omega \boldsymbol{\chi} \cdot \mathbf{u} &= 0\,, \\ \mathbf{b}(\boldsymbol{\sigma}, \vec{\mathbf{v}}) + \int_\Omega \boldsymbol{\rho} \cdot \mathbf{v} &= -\int_\Omega \mathbf{f} \cdot \mathbf{v} + \langle g, \psi \rangle\,, \end{aligned} \tag{4.286}$$

for all $(\boldsymbol{\tau}, \boldsymbol{\chi}) \in \mathbf{H}$*, and for all* $\vec{\mathbf{v}} := (\mathbf{v}, \psi, \boldsymbol{\eta}) \in \mathbf{Q}$*, where* $\mathbf{H} := \mathbf{H}(\mathrm{div}; \Omega) \times \mathbb{RM}(\Omega)$*.*

d) *Use the Babuška-Brezzi Theory to prove that (4.286) is well-posed.*

▸ **4.41** [THE PRESENT IS A VARIANT OF EXERCISE 4.37]. ([4], [43]) *Let* Ω *be a smooth open subset of* $\mathbb{R}^n$ *with boundary* Γ*, and denote by* $\langle \cdot, \cdot \rangle_{\mathrm{div};\Omega}$ *and* $\langle \cdot, \cdot \rangle_{1,\Omega}$ *the usual inner products of* $\mathbf{H}(\mathrm{div}; \Omega)$ *and* $\mathrm{H}^1(\Omega)$*, respectively. Also, let* $\langle \cdot, \cdot \rangle_\Gamma$ *be the duality pairing between* $\mathrm{H}^{-1/2}(\Gamma)$ *and* $\mathrm{H}^{1/2}(\Gamma)$*. Then, given* $f \in \mathrm{L}^2(\Omega)$*,* $g \in \mathrm{H}^{-1/2}(\Gamma)$*, and constants* $a, b \in \mathbb{R}$*, consider the variational problem: Find* $(\boldsymbol{\sigma}, u, \phi) \in \mathbf{H}(\mathrm{div}; \Omega) \times \mathrm{H}^1(\Omega) \times \mathrm{H}^1_0(\Omega)$ *such that*

$$\begin{aligned} \langle \boldsymbol{\sigma}, \boldsymbol{\tau} \rangle_{\mathrm{div};\Omega} + \langle u, v \rangle_{1,\Omega} - a \int_\Omega \phi v &= \int_\Omega f\, \mathrm{div}(\boldsymbol{\tau})\,, \\ \int_\Omega \nabla\phi \cdot \nabla\psi - b \sum_{i=1}^n \int_\Omega \left\{ \phi - u + \frac{\partial u}{\partial x_i} \right\} \frac{\partial \psi}{\partial x_i} &= \langle g, \psi \rangle_\Gamma\,, \end{aligned} \tag{4.287}$$

for all $(\boldsymbol{\tau}, v, \psi) \in \mathbf{H}(\mathrm{div}; \Omega) \times \mathrm{H}^1(\Omega) \times \mathrm{H}^1_0(\Omega)$*.*

i) *Introduce well-defined linear operators* $S := (S_1, S_2) : \mathrm{H}^1_0(\Omega) \to \mathbf{H}(\mathrm{div}; \Omega) \times \mathrm{H}^1(\Omega)$ *and* $\widetilde{S} : \mathrm{H}^1_0(\Omega) \times \mathrm{H}^1(\Omega) \to \mathrm{H}^1_0(\Omega)$*, such that (4.287) is equivalently reduced to the fixed-point equation: Find* $\phi \in \mathrm{H}^1_0(\Omega)$ *such that* $T(\phi) = \phi$*, where*

$$T(\phi) := \widetilde{S}(\phi, S_2(\phi)) \quad \forall\, \phi \in \mathrm{H}^1_0(\Omega)\,.$$

ii) *Show that there exist constants* $C(a)$, $C(b) \geqslant 0$ *such that*

$$\|S_1(\phi) - S_1(\varphi)\|_{\mathrm{div};\Omega} + \|S_2(\phi) - S_2(\varphi)\|_{1,\Omega} \leqslant C(a)\,\|\phi - \varphi\|_{0,\Omega}\,,$$

$$\|\widetilde{S}(\phi,u) - \widetilde{S}(\varphi,w)\|_{1,\Omega} \leqslant C(b)\left\{\|\phi - \varphi\|_{0,\Omega} + \|u - w\|_{1,\Omega}\right\},$$

for all ϕ, $\varphi \in \mathrm{H}_0^1(\Omega)$, u, $w \in \mathrm{H}^1(\Omega)$.

iii) *Deduce from* i) *and* ii) *that there exists a constant* $C(a,b) \geqslant 0$ *such that*

$$\|T(\phi) - T(\varphi)\|_{1,\Omega} \leqslant C(a,b)\,\|\phi - \varphi\|_{0,\Omega} \quad \forall\,\phi,\ \varphi \in \mathrm{H}_0^1(\Omega)\,,$$

then employ again Theorem 10.18 to show that T *is compact, and finally conclude that for* a *and* b *sufficiently small, the original problem (4.287) has a unique solution.*

▸ **4.42** *Let* H *and* Q *be real Hilbert spaces, and let* $a : H \times H \to \mathbb{R}$ *and* $b : H \times Q \to \mathbb{R}$ *be bilinear forms satisfying the hypotheses of the Continuous Babuška–Brezzi Theorem (cf. Theorem 4.19).*

a) *Consider the operator* $T : H \times Q \to (H \times Q)' \equiv H' \times Q'$ *that assigns to each* $(\zeta, w) \in H \times Q$ *the element* $T(\zeta, w) := (F_{\zeta,w}, G_\zeta)$, *where* $F_{\zeta,w} : H \to \mathbb{R}$ *and* $G_\zeta : Q \to \mathbb{R}$ *are the functionals defined by*

$$F_{\zeta,w}(\tau) := a(\zeta,\tau) + b(\tau,w) \quad \forall\,\tau \in H \quad \text{and} \quad G_\zeta(v) := b(\zeta,v) \quad \forall\,v \in Q\,.$$

Show that T *is well-defined, and then apply the Bounded Inverse Theorem (cf. Theorem 3.34) to deduce the existence of a constant* $C > 0$ *such that, for all* $(\zeta, w) \in H \times Q$,

$$C\,\|(\zeta,w)\|_{H\times Q} \leqslant \sup_{\substack{(\tau,v)\in H\times Q\\ (\tau,v)\neq(0,0)}} \frac{|a(\zeta,\tau) + b(\tau,w) + b(\zeta,v)|}{\|(\tau,v)\|_{H\times Q}}\,. \tag{4.288}$$

b) *Given* $F \in H'$ *and* $G \in Q'$, *denote by* $(\sigma, u) \in H \times Q$ *the unique solution of*

$$\begin{aligned} a(\sigma,\tau) + b(\tau,u) &= F(\tau) \quad \forall\,\tau \in H\,,\\ b(\sigma,v) &= G(v) \quad \forall\,v \in Q\,, \end{aligned} \tag{4.289}$$

and let $(\sigma_h, u_h) \in H_h \times Q_h$ *be the associated Galerkin solution, where* H_h *and* Q_h *are finite-dimensional subspaces of* H *and* Q, *respectively, satisfying the hypotheses of the discrete Babuška–Brezzi Theory (cf. Theorem 4.22). Then apply the result obtained in* a), *and deduce the a posteriori error estimate*

$$\|(\sigma,u) - (\sigma_h,u_h)\|_{H\times Q} \leqslant c\left\{\|\mathcal{R}_h\|_{H'} + \|\mathcal{S}_h\|_{Q'}\right\},$$

where $c > 0$ *is independent of* h, *and* $\mathcal{R}_h \in H'$ *and* $\mathcal{S}_h \in Q'$ *are given by*

$$\begin{aligned} \mathcal{R}_h(\tau) &:= F(\tau) - a(\sigma_h,\tau) - b(\tau,u_h) \quad \forall\,\tau \in H\,,\\ \mathcal{S}_h(v) &:= G(v) - b(\sigma_h,v) \quad \forall\,v \in Q\,. \end{aligned}$$

▸ **4.43** ([61]) *Let* Ω *be a bounded domain in* $\mathbb{R}^2$ *with a Lipschitz continuous boundary* Γ, *and consider data* $\mathbf{f} \in \mathbf{L}^2(\Omega)$ *and* $\mathbf{g} \in \mathbf{H}^{-1/2}(\Gamma)$. *The* BRINKMAN MODEL *for flows in porous media with Neumann boundary conditions consists in finding the velocity* $\mathbf{u}$ *and the pressure* p *of a fluid that occupies the region* Ω, *such that*

$$\begin{aligned} \alpha\,\mathbf{u} - \mu\,\Delta\mathbf{u} + \nabla p &= \mathbf{f} \quad \text{in} \ \ \Omega\,,\\ \mathrm{div}(\mathbf{u}) &= 0 \quad \text{in} \ \ \Omega\,,\\ (\mu\,\nabla\mathbf{u} - p\,\mathbb{I})\mathbf{n} &= \mathbf{g} \quad \text{on} \ \ \Gamma\,, \end{aligned} \tag{4.290}$$

where $\mu > 0$ *is the viscosity of the fluid,* α *is a positive parameter given by the ratio between viscosity and permeability,* $\mathbb{I}$ *is the identity matrix in* $\mathbb{R}^{2\times 2}$, *and* $\mathbf{n}$ *is the unit normal vector to* Γ.

a) *Introduce the auxiliary unknowns*

$$\boldsymbol{\varphi} := -\mathbf{u} \quad \text{on} \quad \Gamma \quad \text{and} \quad \boldsymbol{\sigma} := \mu \nabla \mathbf{u} - p\mathbb{I} \quad \text{in} \quad \Omega\,,$$

and show that, by eliminating p and $\mathbf{u}$, one obtains a mixed variational formulation of the following form: Find $(\boldsymbol{\sigma}, \boldsymbol{\varphi}) \in H \times Q$ such that

$$\begin{aligned} a(\boldsymbol{\sigma},\boldsymbol{\tau}) \,+\, b(\boldsymbol{\tau},\boldsymbol{\varphi}) &= F(\boldsymbol{\tau}) \quad \forall\, \boldsymbol{\tau} \in H\,, \\ b(\boldsymbol{\sigma},\boldsymbol{\psi}) &= G(\boldsymbol{\psi}) \quad \forall\, \boldsymbol{\psi} \in Q\,, \end{aligned} \tag{4.291}$$

where $F \in H'$, $G \in Q'$, and $a : H \times H \to \mathbb{R}$ and $b : H \times Q \to \mathbb{R}$ are bounded bilinear forms.

b) *Apply the* Babuška–Brezzi Theory *to show that (4.291) has a unique solution, which depends continuously on the data $\mathbf{f}$ and $\mathbf{g}$.*

c) *Extend the previous analysis to the case of mixed boundary conditions, that is, given Γ_N and Γ_D disjoint subsets of nonzero measure of Γ, such that $\Gamma = \bar{\Gamma}_N \cup \bar{\Gamma}_D$, consider, instead of the full Neumann boundary condition on Γ,*

$$(\mu \nabla \mathbf{u} - p\mathbb{I})\mathbf{n} = \mathbf{g} \quad \text{on} \quad \Gamma_N \quad \text{and} \quad \mathbf{u} = 0 \quad \text{on} \quad \Gamma_D\,,$$

where $\mathbf{g}$ now belongs to $\mathbf{H}_{00}^{-1/2}(\Gamma_N)$.

▸ **4.44** *Let Ω be a bounded domain in $\mathbb{R}^n$, $n \in \{2,3\}$, with a Lipschitz-continuous boundary Γ, and let Γ_D and Γ_N be disjoint parts of Γ such that $\Gamma = \bar{\Gamma}_D \cup \bar{\Gamma}_N$. Given $f \in \mathrm{L}^2(\Omega)$, $g_D \in \mathrm{H}^{1/2}(\Gamma_D)$ and $g_N \in \mathrm{H}_{00}^{-1/2}(\Gamma_N)$, consider the boundary value problem*

$$\Delta u \,=\, f \quad \text{in} \quad \Omega\,, \quad \gamma_0(u) \,=\, g_D \quad \text{on} \quad \Gamma_D\,, \quad \gamma_1(u) \,=\, g_N \quad \text{on} \quad \Gamma_N\,. \tag{4.292}$$

a) *Use the results given in Section 4.5.5.3, in particular the identity (4.158), to show that a primal–mixed formulation of (4.292) reduces to: Find $(u,\xi) \in \mathrm{H}^1(\Omega) \times \mathrm{H}^{-1/2}(\Gamma_D)$ such that*

$$\begin{aligned} \int_\Omega \nabla u \cdot \nabla v \,+\, \langle \xi, \gamma_0(v)|_{\Gamma_D} \rangle_{\Gamma_D} &= F(v) \qquad \forall\, v \in \mathrm{H}^1(\Omega)\,, \\ \langle \lambda, \gamma_0(u)|_{\Gamma_D} \rangle_{\Gamma_D} &= G(\lambda) \qquad \forall\, \lambda \in \mathrm{H}^{-1/2}(\Gamma_D)\,, \end{aligned} \tag{4.293}$$

where $\langle \cdot, \cdot \rangle_{\Gamma_D}$ denotes the duality pairing between $\mathrm{H}^{-1/2}(\Gamma_D)$ and $\mathrm{H}^{1/2}(\Gamma_D)$, and F and G are bounded linear functionals depending on the pair (f, g_N) and on g_D, respectively.

b) *Similarly, show that a dual–mixed formulation of (4.292) is given by: Find $(\sigma,(u,\varphi)) \in \mathbf{H}(\mathrm{div};\Omega) \times \big(\mathrm{L}^2(\Omega) \times \mathrm{H}_{00}^{1/2}(\Gamma_N)\big)$ such that*

$$\begin{aligned} \int_\Omega \sigma \cdot \tau \,+\, \int_\Omega u \,\mathrm{div}(\tau) \,+\, \langle \gamma_{\boldsymbol{\nu}}(\tau)|_{\Gamma_N}, \varphi \rangle_{\Gamma_N} &= F(\tau)\,, \\ \int_\Omega v \,\mathrm{div}(\sigma) \,+\, \langle \gamma_{\boldsymbol{\nu}}(\sigma)|_{\Gamma_N}, \psi \rangle_{\Gamma_N} &= G(v,\psi)\,, \end{aligned} \tag{4.294}$$

for all $(\tau,(v,\psi)) \in \mathbf{H}(\mathrm{div};\Omega) \times \big(\mathrm{L}^2(\Omega) \times \mathrm{H}_{00}^{1/2}(\Gamma_N)\big)$, where $\langle \cdot, \cdot \rangle_{\Gamma_N}$ denotes the duality pairing between $\mathrm{H}_{00}^{-1/2}(\Gamma_N)$ and $\mathrm{H}_{00}^{1/2}(\Gamma_N)$, and F and G are bounded linear functionals depending on g_D and on the pair (f, g_N), respectively.

c) *Apply the* Babuška–Brezzi Theory *to prove that problems (4.293) and (4.294) are well-posed.*

▸ **4.45** *Let H be a real Hilbert space, and let $a : H \times H \to \mathbb{R}$ be a bilinear form satisfying the hypotheses of the generalized Lax–Milgram Lemma (cf. Lemma 4.8). Then, given $F \in H'$, denote by $\sigma \in H$ the unique solution of*

$$a(\sigma,\tau) \,=\, F(\tau) \quad \forall\, \tau \in H\,, \tag{4.295}$$

and let $\sigma_h \in H_h$ *be the associated Galerkin solution, where* H_h *is a finite-dimensional subspace of* H *satisfying the hypotheses of the discrete version of the generalized Lax–Milgram Lemma (cf. Lemma 4.12). Show that there exists a constant* $c > 0$*, independent of* h*, such that*

$$\|\sigma - \sigma_h\|_H \leqslant c \, \|\mathcal{R}_h\|_{H'} ,$$

where $\mathcal{R}_h \in H'$ *is given by*

$$\mathcal{R}_h(\tau) := F(\tau) - a(\sigma_h, \tau) \quad \forall \, \tau \in H .$$

▸ **4.46** *Let* Ω_1 *be a connected bounded open subset of* $\mathbb{R}^2$ *with boundary* Γ*, and let* Ω_2 *be the annular region bounded by* Γ *and a closed curve* Σ *whose interior contains* Γ*. Given* $\mathbf{f}_1 \in \mathbf{L}^2(\Omega_1)$*,* $\mathbf{f}_2 \in \mathbf{L}^2(\Omega_2)$*,* $\boldsymbol{g}_\Gamma \in \mathbf{H}^{1/2}(\Gamma)$*, and* $\boldsymbol{g}_\Sigma \in \mathbf{H}^{1/2}(\Sigma)$*, consider the* STOKES TRANSMISSION PROBLEM *for two fluids occupying the regions* Ω_1 *and* Ω_2*, governed by the following equations:*

$$\begin{aligned} \boldsymbol{\sigma}_1 \,&=\, \mu_1 \nabla \mathbf{u}_1 - p_1 \mathbb{I} \quad \text{in} \quad \Omega_1 , \quad \mathbf{div}(\boldsymbol{\sigma}_1) \,=\, \mathbf{f}_1 \quad \text{in} \quad \Omega_1 , \\ & \qquad \mathrm{div}(\mathbf{u}_1) \,=\, 0 \quad \text{in} \quad \Omega_1 , \\ \boldsymbol{\sigma}_2 \,&=\, \mu_2 \nabla \mathbf{u}_2 - p_2 \mathbb{I} \quad \text{in} \quad \Omega_2 , \quad \mathbf{div}(\boldsymbol{\sigma}_2) \,=\, \mathbf{f}_2 \quad \text{in} \quad \Omega_2 , \\ & \qquad \mathrm{div}(\mathbf{u}_2) \,=\, 0 \quad \text{in} \quad \Omega_2 , \\ \mathbf{u}_1 - \mathbf{u}_2 \,&=\, \boldsymbol{g}_\Gamma \quad \text{on} \quad \Gamma , \quad \boldsymbol{\sigma}_1 \boldsymbol{\nu} \,=\, \boldsymbol{\sigma}_2 \boldsymbol{\nu} \quad \text{on} \quad \Gamma , \quad \mathbf{u}_2 \,=\, \boldsymbol{g}_\Sigma \quad \text{on} \quad \Sigma , \end{aligned} \tag{4.296}$$

where μ_1 *and* μ_2 *are positive constants denoting the viscosities,* $\mathbf{u}_1$ *and* $\mathbf{u}_2$ *are the velocity vectors,* p_1 *and* p_2 *are the pressures, and* $\boldsymbol{\nu}$ *is the unit normal vector to* $\partial\Omega_1$ *and* $\partial\Omega_2$*, which points towards* Ω_2 *on* Γ *and outward from* Ω_2 *on* Σ*.*

a) *Show that a* MIXED VARIATIONAL FORMULATION *of (4.296) reduces to: Find* $\boldsymbol{\sigma} := (\boldsymbol{\sigma}_1, \boldsymbol{\sigma}_2) \in \mathrm{H} := \mathbb{H}_0(\mathbf{div}; \Omega_1) \times \mathbb{H}_0(\mathbf{div}; \Omega_2)$*,* $(\mathbf{u}, \boldsymbol{\xi}) := \big((\mathbf{u}_1, \mathbf{u}_2), \boldsymbol{\xi}\big) \in \mathrm{Q} := \big(\mathbf{L}^2(\Omega_1) \times \mathbf{L}^2(\Omega_2)\big) \times \mathbf{H}^{1/2}(\Gamma)$*, such that*

$$\begin{aligned} \sum_{i=1}^{2} \left\{ \frac{1}{\mu_i} \int_{\Omega_i} \boldsymbol{\sigma}_i^{\mathrm{d}} : \boldsymbol{\tau}_i^{\mathrm{d}} + \int_{\Omega_i} \mathbf{u}_i \cdot \mathbf{div}(\boldsymbol{\tau}_i) \right\} + \langle \boldsymbol{\tau}_1 \boldsymbol{\nu} - \boldsymbol{\tau}_2 \boldsymbol{\nu}, \boldsymbol{\xi} \rangle_\Gamma &= \mathbf{F}(\boldsymbol{\tau}) , \\ \sum_{i=1}^{2} \int_{\Omega_i} \mathbf{v}_i \cdot \mathbf{div}(\boldsymbol{\sigma}_i) + \langle \boldsymbol{\sigma}_1 \boldsymbol{\nu} - \boldsymbol{\sigma}_2 \boldsymbol{\nu}, \boldsymbol{\lambda} \rangle_\Gamma &= \mathbf{G}(\mathbf{v}, \boldsymbol{\lambda}), \end{aligned} \tag{4.297}$$

for all $\boldsymbol{\tau} := (\boldsymbol{\tau}_1, \boldsymbol{\tau}_2) \in \mathrm{H}$*, and for all* $(\mathbf{v}, \boldsymbol{\lambda}) := \big((\mathbf{v}_1, \mathbf{v}_2), \boldsymbol{\lambda}\big) \in \mathrm{Q}$*, where* $\langle \cdot, \cdot \rangle_\Gamma$ *denotes the duality pairing between* $\mathbf{H}^{-1/2}(\Gamma)$ *and* $\mathbf{H}^{1/2}(\Gamma)$*, and* $\mathbf{F} \in \mathrm{H}'$ *and* $\mathbf{G} \in \mathrm{Q}'$ *are functionals depending on* $\mathbf{f}_1$*,* $\mathbf{f}_2$*,* $\boldsymbol{g}_\Gamma$ *and* $\boldsymbol{g}_\Sigma$*.*

b) *Apply the* BABUŠKA–BREZZI THEORY *to prove that problem (4.297) has a unique solution* $(\boldsymbol{\sigma}, (\mathbf{u}, \boldsymbol{\xi})) \in \mathrm{H} \times \mathrm{Q}$*, which depends continuously on the data.*

▸ **4.47** *Given* Ω *a smooth domain in* $\mathbb{R}^n$*, and* $\big(X, \|\cdot\|\big)$ *a normed space, consider operators* $A \in \mathcal{L}\big(\mathrm{H}^1(\Omega), X\big)$ *and* $B \in \mathcal{L}\big(\mathbf{H}(\mathrm{div}; \Omega), X'\big)$*, such that*

i) *A is surjective,*
ii) $C_0^\infty(\Omega) \subseteq N(A)$*, and*
iii) *the following identity holds*

$$B(\boldsymbol{\tau})\big(A(v)\big) \;=\; \int_\Omega \Big\{ v \, \mathrm{div}(\boldsymbol{\tau}) \;+\; \boldsymbol{\tau} \cdot \nabla v \Big\} \quad \forall \, \boldsymbol{\tau} \in \mathbf{H}(\mathrm{div}; \Omega) , \quad \forall \, v \in \mathrm{H}^1(\Omega) .$$

The purpose of this exercise is to prove that B *is also surjective. To this end, given an arbitrary functional* $F \in X'$*, it is suggested to construct* $\boldsymbol{\zeta} \in \mathbf{H}(\mathrm{div}; \Omega)$ *such that* $B(\boldsymbol{\zeta}) = F$*, following the procedure indicated below.*

a) *Recall once again that* $\mathbf{L}^2(\Omega) := [\mathrm{L}^2(\Omega)]^n$, *consider the space* $\mathrm{L}^2(\Omega) \times \mathbf{L}^2(\Omega)$ *endowed with the inner product*

$$\langle (u, \vec{\mathbf{u}}), (v, \vec{\mathbf{v}}) \rangle_{0,\Omega} := \int_\Omega \Big\{ u\, v + \vec{\mathbf{u}} \cdot \vec{\mathbf{v}} \Big\} \qquad \forall\, (u, \vec{\mathbf{u}}),\, (v, \vec{\mathbf{v}}) \in \mathrm{L}^2(\Omega) \times \mathbf{L}^2(\Omega)\,, \tag{4.298}$$

define $T : \mathrm{H}^1(\Omega) \to \mathrm{L}^2(\Omega) \times \mathbf{L}^2(\Omega)$ *by* $T(v) := (v, \nabla v)$ *for all* $v \in \mathrm{H}^1(\Omega)$, *show that* $R(T)$ *is a closed subspace of* $\mathrm{L}^2(\Omega) \times \mathbf{L}^2(\Omega)$, *and conclude that* $T : \mathrm{H}^1(\Omega) \to R(T)$ *is a bijective isometry.*

b) *Consider the diagram* $R(T) \overset{T^{-1}}{\to} \mathrm{H}^1(\Omega) \overset{A}{\longrightarrow} X \overset{F}{\to} \mathbb{R}$, *define the functional* $\widetilde{F} := F \circ A \circ T^{-1}$, *show that there exists a unique* $(z, \nabla z) \in R(T)$ *such that*

$$\widetilde{F}\big((v, \nabla v)\big) = \langle (z, \nabla z), (v, \nabla v) \rangle_{0,\Omega} \qquad \forall\, (v, \nabla v) \in R(T)\,,$$

and deduce, denoting $\boldsymbol{\zeta} := \nabla z$, *that*

$$F\big(A(v)\big) = \int_\Omega \Big\{ z\, v + \boldsymbol{\zeta} \cdot \nabla v \Big\} \qquad \forall\, v \in \mathrm{H}^1(\Omega)\,. \tag{4.299}$$

c) *Use* ii) *and* (4.299) *to prove that* $\mathrm{div}(\boldsymbol{\zeta}) = z$ *in* $\mathcal{D}'(\Omega)$, *and then employ* iii) *to obtain* $F\big(A(v)\big) = B(\boldsymbol{\zeta})\big(A(v)\big)$ *for all* $v \in \mathrm{H}^1(\Omega)$.

d) *Utilize* i) *to arrive at* $F = B(\boldsymbol{\zeta})$, *thus proving the surjectivity of* B.

e) *Consider the particular case* $X' := \mathrm{H}^{1/2}(\Gamma)$, $A := \gamma_0$ *and* $B := \gamma_{\boldsymbol{\nu}}$, *to conclude that the operator* $\gamma_{\boldsymbol{\nu}} : \mathbf{H}(\mathrm{div}; \Omega) \to \mathrm{H}^{-1/2}(\Gamma)$ *(cf. Theorem 4.31) is surjective.*

▸ **4.48** [This is a "rot" version of Exercise 4.47]. *Given* Ω *a smooth domain in* $\mathbb{R}^2$, *and* $(X, \|\cdot\|)$ *a normed space, consider operators* $B \in \mathcal{L}\big(\mathrm{H}^1(\Omega), X\big)$ *and* $A \in \mathcal{L}\big(\mathbf{H}(\mathrm{rot}; \Omega), X'\big)$, *where*

$$\mathbf{H}(\mathrm{rot}; \Omega) := \Big\{ \boldsymbol{\tau} := (\tau_1, \tau_2) \in \mathbf{L}^2(\Omega) : \quad \mathrm{rot}(\boldsymbol{\tau}) := \frac{\partial \tau_2}{\partial x_1} - \frac{\partial \tau_1}{\partial x_2} \in \mathrm{L}^2(\Omega) \Big\}\,,$$

such that

i) B *is surjective,*
ii) $C_0^\infty(\Omega) \subseteq N(B)$, *and*
iii) *the following identity holds*

$$A(\boldsymbol{\tau})\big(B(v)\big) = \int_\Omega \Big\{ v\, \mathrm{rot}(\boldsymbol{\tau}) - \boldsymbol{\tau} \cdot \mathbf{curl}(v) \Big\} \quad \forall\, \boldsymbol{\tau} \in \mathbf{H}(\mathrm{rot}; \Omega)\,, \quad \forall\, v \in \mathrm{H}^1(\Omega)\,,$$

where $\mathbf{curl}(v) := \big(\frac{\partial v}{\partial x_2}, -\frac{\partial v}{\partial x_1}\big)$.

The purpose of this exercise is to prove that A *is also surjective. To this end, given an arbitrary functional* $F \in X'$, *one is suggested to construct* $\boldsymbol{\zeta} \in \mathbf{H}(\mathrm{rot}; \Omega)$ *such that* $A(\boldsymbol{\zeta}) = F$, *following the reasoning outlined below.*

a) *As in* (4.298), *bear in mind the space* $\mathrm{L}^2(\Omega) \times \mathbf{L}^2(\Omega)$ *endowed with the inner product*

$$\langle (u, \vec{\mathbf{u}}), (v, \vec{\mathbf{v}}) \rangle_{0,\Omega} := \int_\Omega \Big\{ u\, v + \vec{\mathbf{u}} \cdot \vec{\mathbf{v}} \Big\} \qquad \forall\, (u, \vec{\mathbf{u}}),\, (v, \vec{\mathbf{v}}) \in \mathrm{L}^2(\Omega) \times \mathbf{L}^2(\Omega)\,,$$

define $T : \mathrm{H}^1(\Omega) \to \mathrm{L}^2(\Omega) \times \mathbf{L}^2(\Omega)$ *by* $T(v) := (v, \mathbf{curl}(v))$ *for all* $v \in \mathrm{H}^1(\Omega)$, *show that* $R(T)$ *is a closed subspace of* $\mathrm{L}^2(\Omega) \times \mathbf{L}^2(\Omega)$, *and conclude that* $T : \mathrm{H}^1(\Omega) \to R(T)$ *is a bijective isometry.*

b) *Consider the diagram* $R(T) \overset{T^{-1}}{\to} \mathrm{H}^1(\Omega) \overset{B}{\to} X \overset{F}{\to} \mathbb{R}$, *define the functional* $\widetilde{F} := F \circ B \circ T^{-1}$, *show that there exists a unique* $(z, \mathbf{curl}(z)) \in R(T)$ *such that*

$$\widetilde{F}\big((v, \mathbf{curl}(v))\big) = \langle (z, \mathbf{curl}(z)), (v, \mathbf{curl}(v)) \rangle_{0,\Omega} \qquad \forall\, (v, \mathbf{curl}(v)) \in R(T)\,,$$

and deduce, denoting $\boldsymbol{\zeta} := -\,\mathbf{curl}(z)$, *that*

$$F\big(B(v)\big) = \int_\Omega \Big\{ z\, v - \boldsymbol{\zeta} \cdot \mathbf{curl}(v) \Big\} \qquad \forall\, v \in \mathrm{H}^1(\Omega)\,. \tag{4.300}$$

c) *Use* ii) *and* (4.299) *to prove that* $\mathrm{rot}(\boldsymbol{\zeta}) = z$ *in* $\mathcal{D}'(\Omega)$, *and then employ* iii) *to deduce that* $F\big(B(v)\big) = A(\boldsymbol{\zeta})\big(B(v)\big)$ *for all* $v \in \mathrm{H}^1(\Omega)$.

d) *Finally, make use of* i) *to conclude that* $F = A(\boldsymbol{\zeta})$, *thus proving the surjectivity of* A.

5

Variational Problems in Banach Spaces

In this chapter, we extend most of the results presented in Chapter 4 to the case of variational formulations stated in terms of Banach spaces. More precisely, we address the continuous and discrete versions of the generalized Lax–Milgram Lemma, as well as the classical and generalized Babuška–Brezzi theories, and we develop corresponding analyses for schemes exhibiting a perturbed saddle-point structure.

5.1 The generalized Lax-Milgram Lemma

In this section we generalize the Hilbert version of the Lax-Milgram Lemma (cf. Lemmas 4.4 and 4.8, Section 4.1) to the Banach case.

5.1.1 Continuous analysis

Let $\big(X, \|\cdot\|_X\big)$ and $\big(Y, \|\cdot\|_Y\big)$ be real Banach spaces, and let $A : X \times Y \to \mathbb{R}$ be a bounded bilinear form, that is, $A(x,\cdot) : Y \to \mathbb{R}$ and $A(\cdot, y) : X \to \mathbb{R}$ are linear functionals for every $x \in X$ and $y \in Y$, respectively, and there exists a constant $C_A > 0$ such that

$$|A(x,y)| \leqslant C_A \|x\|_X \|y\|_Y \qquad \forall\, (x,y) \in X \times Y\,. \tag{5.1}$$

Note that the latter actually means that $A(x,\cdot) \in Y'$ and $A(\cdot,y) \in X'$, with

$$\|A(x,\cdot)\| \leqslant C_A \|x\| \quad \forall\, x \in X\,, \quad \text{and} \quad \|A(\cdot,y)\| \leqslant C_A \|y\| \quad \forall\, y \in Y\,.$$

Whenever there is no risk of confusion, the subscripts of the norms used will be omitted throughout the analysis. On the other hand, let $\mathbf{A} : X \to Y'$ be the linear and bounded operator induced by A, that is, for each $x \in X$ we define $\mathbf{A}(x) := A(x,\cdot) \in Y'$, or more precisely

$$\mathbf{A}(x)(y) := A(x,y) \qquad \forall\, y \in Y\,,$$

which, according to (5.1), suggests using from now on $\|\mathbf{A}\|$ instead of C_A. Now, given a functional $G \in Y'$, we consider the problem: Find $x \in X$ such that

$$A(x,y) = G(y) \qquad \forall\, y \in Y\,, \tag{5.2}$$

and we aim to deduce necessary and sufficient conditions on A under which (5.2) is well-posed. For this purpose, we note that (5.2) can be equivalently reformulated as: Find $x \in X$ such that

G. N. Gatica, *Introduction to Functional Analysis*,
https://doi.org/10.1007/978-3-032-37415-8_5

$$\mathbf{A}(x) = G, \tag{5.3}$$

and therefore (5.3) is well-posed for each $G \in Y'$ if and only if $\mathbf{A}$ is bijective. In what follows, we proceed to characterize the latter in two different ways. Indeed, we begin by observing that $\mathbf{A}$ is injective if and only if for each $x \in X$, $x \neq 0$, we have $\mathbf{A}(x) \neq 0$, which means that there must exist $y \in Y$ such that $\mathbf{A}(x)(y) > 0$, whence

$$\sup_{y \in Y} \mathbf{A}(x)(y) = \sup_{y \in Y} A(x, y) > 0 \qquad \forall\, x \in X,\ x \neq 0. \tag{5.4}$$

In turn, $\mathbf{A}$ is surjective if and only if its adjoint operator $\mathbf{A}' : Y'' \to X'$ is injective and has closed range (cf. Theorem 3.86), which is equivalent to the existence of a positive constant α such that

$$\|\mathbf{A}'(\mathcal{G})\|_{X'} \geqslant \alpha \, \|\mathcal{G}\|_{Y''} \qquad \forall\, \mathcal{G} \in Y'', \tag{5.5}$$

that is,

$$\sup_{\substack{x \in X \\ x \neq 0}} \frac{\mathbf{A}'(\mathcal{G})(x)}{\|x\|} = \sup_{\substack{x \in X \\ x \neq 0}} \frac{\mathcal{G}\big(\mathbf{A}(x)\big)}{\|x\|} \geqslant \alpha \, \|\mathcal{G}\|_{Y''} \qquad \forall\, \mathcal{G} \in Y''. \tag{5.6}$$

In order to transform (5.6) into an assumption that explicitly involves the bilinear form A, we now assume that Y is reflexive, that is, the operator $\mathcal{J}_Y : Y \to Y''$ (cf. (3.24)) is bijective, and therefore (5.6) can be equivalently rewritten as

$$\sup_{\substack{x \in X \\ x \neq 0}} \frac{\mathcal{J}_Y(y)\big(\mathbf{A}(x)\big)}{\|x\|} = \sup_{\substack{x \in X \\ x \neq 0}} \frac{\mathbf{A}(x)(y)}{\|x\|} \geqslant \alpha \, \|\mathcal{J}_Y(y)\|_{Y''} \qquad \forall\, y \in Y,$$

that is,

$$\sup_{\substack{x \in X \\ x \neq 0}} \frac{A(x, y)}{\|x\|} \geqslant \alpha \, \|y\| \qquad \forall\, y \in Y. \tag{5.7}$$

Alternatively, we can characterize the bijectivity of $\mathbf{A}$ through that of $\mathbf{A}'$. In this case, $\mathbf{A}'$ is injective if and only if for each $\mathcal{G} \in Y''$, $\mathcal{G} \neq 0$, we have $\mathbf{A}'(\mathcal{G}) \neq 0$, which means that there must exist $x \in X$ such that $\mathbf{A}'(\mathcal{G})(x) = \mathcal{G}\big(\mathbf{A}(x)\big) > 0$, whence

$$\sup_{x \in X} \mathcal{G}\big(\mathbf{A}(x)\big) > 0 \qquad \forall\, \mathcal{G} \in Y'',\ \mathcal{G} \neq 0.$$

Then, applying again the reflexivity of Y, we see that the above inequality reduces to

$$\sup_{x \in X} \mathcal{J}_Y(y)\big(\mathbf{A}(x)\big) = \sup_{x \in X} A(x, y) > 0 \qquad \forall\, y \in Y,\ y \neq 0. \tag{5.8}$$

On the other hand, $\mathbf{A}'$ is surjective if and only if $\mathbf{A}$ is injective and has closed range (cf. Theorem 3.87), which reduces to the existence of a positive constant $\tilde{\alpha}$ such that

$$\|\mathbf{A}(x)\|_{Y'} \geqslant \tilde{\alpha} \, \|x\| \qquad \forall\, x \in X, \tag{5.9}$$

that is,

$$\sup_{\substack{y \in Y \\ y \neq 0}} \frac{\mathbf{A}(x)(y)}{\|y\|} = \sup_{\substack{y \in Y \\ y \neq 0}} \frac{A(x, y)}{\|y\|} \geqslant \tilde{\alpha} \, \|x\| \qquad \forall\, x \in X. \tag{5.10}$$

Now, in relation to the inf-sup conditions (5.7) and (5.10), we observe from (5.5) and (5.9), respectively, that, under the assumption that $\mathbf{A}$ (equivalently, $\mathbf{A}'$) is bijective, they reduce to

$$\|(\mathbf{A}')^{-1}\| \leqslant \frac{1}{\alpha} \tag{5.11}$$

and

$$\|\mathbf{A}^{-1}\| \leqslant \frac{1}{\widetilde{\alpha}}, \tag{5.12}$$

so that, recalling that

$$\|\mathbf{A}^{-1}\| = \|(\mathbf{A}^{-1})'\| = \|(\mathbf{A}')^{-1}\|, \tag{5.13}$$

we conclude that they are equivalent, and therefore the constants α and $\widetilde{\alpha}$ may be assumed to coincide.

Summarizing, and keeping in mind the joint hypotheses (5.4) + (5.7) and (5.8) + (5.10), we have proved the following main result, known as the generalized Lax-Milgram Lemma or the Banach-Nečas-Babuška Theorem.

Lemma 5.1. *Let $(X, \|\cdot\|_X)$ and $(Y, \|\cdot\|_Y)$ be real Banach spaces such that Y is reflexive, and let $A : X \times Y \to \mathbb{R}$ be a bounded bilinear form. Suppose that one of the following pairs of conditions is satisfied*

i) $\sup\limits_{y\in Y} A(x,y) > 0 \quad \forall\, x \in X,\ x \neq 0,$	i)' $\sup\limits_{x\in X} A(x,y) > 0 \quad \forall\, y \in Y,\ y \neq 0,$
ii) *there exists a constant* $\alpha > 0$ *such that* $$\sup_{\substack{x\in X\\ x\neq 0}} \frac{A(x,y)}{\|x\|} \geqslant \alpha\,\|y\| \quad \forall\, y \in Y.$$	ii)' *there exists a constant* $\alpha > 0$ *such that* $$\sup_{\substack{y\in Y\\ y\neq 0}} \frac{A(x,y)}{\|y\|} \geqslant \alpha\,\|x\| \quad \forall\, x \in X.$$

Then, for each $G \in Y'$ there exists a unique $x \in X$ solution of problem (5.2), that is, such that $A(x,y) = G(y)$ for all $y \in Y$. Moreover, we have

$$\|x\| \leqslant \frac{1}{\alpha}\,\|G\|_{Y'}. \tag{5.14}$$

Furthermore, each of the above pairs is a necessary condition for (5.2) to be well-posed.

Proof. The existence of a unique $x \in X$ solution of (5.2) is guaranteed by each of the pairs of assumptions i)-ii) and i)'-ii)', and by the previous analysis. In turn, as already observed, and since $\mathbf{A}$ is bijective, we have that ii) and ii)' are equivalent, so it follows that

$$\|x\| \leqslant \frac{1}{\alpha} \sup_{\substack{y\in Y\\ y\neq 0}} \frac{A(x,y)}{\|y\|} = \frac{1}{\alpha} \sup_{\substack{y\in Y\\ y\neq 0}} \frac{G(y)}{\|y\|} = \frac{1}{\alpha}\,\|G\|_{Y'},$$

which proves (5.14). Finally, the necessity of each pair of assumptions is also a consequence of the previous discussion.

□

We emphasize here that, regardless of the fact that the pairs i) - ii) and i)' - ii)' are equivalent, the hypotheses ii) and ii)' by themselves, even with different constants, are sufficient for the occurrence of i)' and i), respectively. Accordingly, we can state the following result, which provides sufficient conditions for (5.2) to be well-posed.

Lemma 5.2. *Let $(X, \|\cdot\|_X)$ and $(Y, \|\cdot\|_Y)$ be real Banach spaces such that Y is reflexive, and let $A : X \times Y \to \mathbb{R}$ be a bounded bilinear form. Suppose that*

a) *there exists a constant* $\alpha > 0$ *such that*

$$\sup_{\substack{x \in X \\ x \neq 0}} \frac{A(x,y)}{\|x\|} \geqslant \alpha \, \|y\| \quad \forall\, y \in Y \,,$$

b) *there exists a constant* $\widetilde{\alpha} > 0$ *such that*

$$\sup_{\substack{y \in Y \\ y \neq 0}} \frac{A(x,y)}{\|y\|} \geqslant \widetilde{\alpha} \, \|x\| \quad \forall\, x \in X \,.$$

Then, for each $G \in Y'$ *there exists a unique solution* $x \in X$ *to* (5.2), *and the a priori bound* (5.14) *holds.*

We now consider the particular case in which Y is the product space of N Banach spaces, say $Y = Y_1 \times Y_2 \times \cdots \times Y_N$, with $N \geqslant 2$, and for each $x \in X$ we introduce the functionals $G_{x,i} \in Y_i'$, $i \in \{1, 2, \ldots, N\}$, defined by

$$G_{x,i}(y_i) := A(x, y^{(i)}) \qquad \forall\, y_i \in Y_i \,, \tag{5.15}$$

where $y^{(i)} \in Y$ is such that its i-th component is given by y_i, while all the remaining ones are the corresponding zero vectors. Then, we have that

$$A(x,y) = \sum_{i=1}^{N} G_{x,i}(y_i) \qquad \forall\, x \in X \,, \quad \forall\, y := (y_1, y_2, \ldots, y_N) \in Y \,.$$

It follows that

$$\sup_{\substack{y \in Y \\ y \neq 0}} \frac{A(x,y)}{\|y\|} \geqslant \sup_{\substack{y_i \in Y_i \\ y_i \neq 0}} \frac{A(x,y^{(i)})}{\|y_i\|} = \sup_{\substack{y_i \in Y_i \\ y_i \neq 0}} \frac{G_{x,i}(y_i)}{\|y_i\|} = \|G_{x,i}\|_{Y_i'} \qquad \forall\, i \in \{1, 2, \ldots, N\} \,,$$

whence, summing over $i \in \{1, 2, \ldots, N\}$, we obtain

$$\sup_{\substack{y \in Y \\ y \neq 0}} \frac{A(x,y)}{\|y\|} \geqslant \frac{1}{N} \sum_{i=1}^{N} \|G_{x,i}\|_{Y_i'} \qquad \forall\, x \in X \,. \tag{5.16}$$

In turn, using that certainly $\|y\| \geqslant \|y_i\|$, we find that

$$\frac{A(x,y)}{\|y\|} = \sum_{i=1}^{N} \frac{G_{x,i}(y_i)}{\|y\|} \leqslant \sum_{i=1}^{N} \sup_{\substack{y_i \in Y_i \\ y_i \neq 0}} \frac{G_{x,i}(y_i)}{\|y_i\|} = \sum_{i=1}^{N} \|G_{x,i}\|_{Y_i'} \qquad \forall\, x \in X \,,$$

and therefore

$$\sup_{\substack{y \in Y \\ y \neq 0}} \frac{A(x,y)}{\|y\|} \leqslant \sum_{i=1}^{N} \|G_{x,i}\|_{Y_i'} \qquad \forall\, x \in X \,. \tag{5.17}$$

Thus, having from (5.16) and (5.17) that

$$\frac{1}{N} \sum_{i=1}^{N} \|G_{x,i}\|_{Y_i'} \leqslant \sup_{\substack{y \in Y \\ y \neq 0}} \frac{A(x,y)}{\|y\|} \leqslant \sum_{i=1}^{N} \|G_{x,i}\|_{Y_i'} \qquad \forall\, x \in X \,, \tag{5.18}$$

we conclude that establishing hypothesis ii)' of Lemma 5.1 (or b) of Lemma 5.2) is equivalent to proving the existence of a constant $\widehat{\alpha} > 0$ such that

$$\sum_{i=1}^{N} \|G_{x,i}\|_{Y_i'} \geqslant \widehat{\alpha}\,\|x\| \qquad \forall\, x \in X\,. \tag{5.19}$$

Similarly, suppose now that there exist Banach spaces X_i, $i \in \{1, ..., N\}$, with $N \geqslant 2$, such that $X = X_1 \times X_2 \times \cdots \times X_N$, and for each $y \in Y$ we introduce the functionals $F_{y,i} \in X_i'$, $i \in \{1, 2, \ldots, N\}$, given by

$$F_{y,i}(x_i) := A(x^{(i)}, y) \qquad \forall\, x_i \in X_i\,, \tag{5.20}$$

where $x^{(i)} \in X$ is such that its i-th component is x_i, and all the others are the corresponding zero vectors. Then, proceeding analogously to the previous case, we deduce that

$$\frac{1}{N}\sum_{i=1}^{N} \|F_{y,i}\|_{X_i'} \leqslant \sup_{\substack{x\in X\\ x\neq 0}} \frac{A(x,y)}{\|x\|} \leqslant \sum_{i=1}^{N} \|F_{y,i}\|_{X_i'} \qquad \forall\, y \in Y\,, \tag{5.21}$$

so that establishing hypothesis ii) of Lemma 5.1 (or a) of Lemma 5.2) is equivalent to showing the existence of a constant $\widehat{\alpha} > 0$ such that

$$\sum_{i=1}^{N} \|F_{y,i}\|_{X_i'} \geqslant \widehat{\alpha}\,\|y\| \qquad \forall\, y \in Y\,. \tag{5.22}$$

5.1.2 Discrete analysis

The purpose of this section is to introduce a Galerkin scheme for (5.2), prove that it is well-posed, and provide the associated a priori error estimates. To this end, let $\{X_h\}_{h>0}$ and $\{Y_h\}_{h>0}$ be families of finite-dimensional subspaces of X and Y, respectively, and for each $h > 0$ let $A_h : X_h \times Y_h \to \mathbb{R}$ be a bounded bilinear form that approximates $A|_{X_h\times Y_h}$, with induced operator $\mathbf{A}_h \in \mathcal{L}(X_h, Y_h')$. Then, given a functional $G_h \in Y_h'$ that approximates $G|_{Y_h}$, we consider the discrete system: Find $x_h \in X_h$ such that

$$A_h(x_h, y_h) = G_h(y_h) \qquad \forall\, y_h \in Y_h\,. \tag{5.23}$$

Then, applying Lemma 5.1 to (5.23), we obtain the following preliminary result.

Lemma 5.3. *In addition to the previous notations and definitions, suppose that one of the following equivalent pairs of conditions is satisfied*

i) $\sup\limits_{y_h\in Y_h} A_h(x_h, y_h) > 0$ $\quad\forall\, x_h \in X_h\backslash\{0\}$,	i)' $\sup\limits_{x_h\in X_h} A_h(x_h, y_h) > 0$ $\quad\forall\, y_h \in Y_h\backslash\{0\}$,
ii) *there exists a constant* $\alpha_{\mathrm{d}} > 0$ *such that* $\sup\limits_{\substack{x_h\in X_h\\ x_h\neq 0}} \dfrac{A_h(x_h, y_h)}{\|x_h\|} \geqslant \alpha_{\mathrm{d}}\,\|y_h\|$ $\quad\forall\, y_h \in Y_h\,.$	ii)' *there exists a constant* $\alpha_{\mathrm{d}} > 0$ *such that* $\sup\limits_{\substack{y_h\in Y_h\\ y_h\neq 0}} \dfrac{A_h(x_h, y_h)}{\|y_h\|} \geqslant \alpha_{\mathrm{d}}\,\|x_h\|$ $\quad\forall\, x_h \in X_h\,.$

Then, for each $G_h \in Y_h'$ *there exists a unique* $x_h \in X_h$ *solution of* (5.23), *and we have*

$$\|x_h\| \leqslant \frac{1}{\alpha_{\mathrm{d}}} \|G_h\|_{Y'_h}.$$

Moreover, each of the above pairs is a necessary condition for (5.23) *to be well-posed.*

We stress here that, by employing the equivalence result given by (5.11), (5.12), and (5.13) in the present discrete context, we can deduce that the constants in the inf-sup conditions ii) and ii)' in Lemma 5.3 may in fact be considered equal, thus establishing a complete analogy with Lemma 5.1. In turn, as it was mentioned in the paragraph preceding Lemma 4.12 for the Hilbert case of the Galerkin scheme (5.23), we observe here that, in fact, a necessary condition for each one of the pairs of hypotheses of Lemma 5.3 is that

$$\dim X_h = \dim Y_h,$$

which is assumed from now on. Thus, the Fredholm alternative in finite dimension, which states that a linear operator between spaces of equal dimension is surjective if and only if it is injective, allows us to conclude that the assumptions i) and i)' of the previous lemma are redundant with respect to ii) and ii)', respectively, and therefore we retain only the latter. More precisely, we can establish the following definitive result.

Lemma 5.4. *Suppose that* $\dim X_h = \dim Y_h$ *for all* $h > 0$, *and that there exists a constant* $\alpha_{\mathrm{d}} > 0$ *such that one of the following equivalent conditions is satisfied*

$$\sup_{\substack{x_h \in X_h \\ x_h \neq 0}} \frac{A_h(x_h, y_h)}{\|x_h\|} \geqslant \alpha_{\mathrm{d}} \|y_h\| \quad \forall\, y_h \in Y_h. \qquad \Big| \qquad \sup_{\substack{y_h \in Y_h \\ y_h \neq 0}} \frac{A_h(x_h, y_h)}{\|y_h\|} \geqslant \alpha_{\mathrm{d}} \|x_h\| \quad \forall\, x_h \in X_h.$$

Then, for each $G_h \in Y'_h$ *there exists a unique* $x_h \in X_h$ *solution of* (5.23). *Moreover, we have*

$$\|x_h\| \leqslant \frac{1}{\alpha_{\mathrm{d}}} \|G_h\|_{Y'_h}.$$

Furthermore, each of the above constitutes a necessary condition for (5.23) *to be well-posed.*

We now provide the a priori error estimates, also known as Strang estimates, for the unique solutions of (5.2) and (5.23).

Lemma 5.5. *Assume the hypotheses of Lemmas* 5.1 (*or* 5.2) *and* 5.4, *and let* $x \in X$ *and* $x_h \in X_h$ *be the unique solutions of* (5.2) *and* (5.23), *respectively. Then, we have*

$$\begin{aligned} \|x - x_h\| \leqslant \inf_{z_h \in X_h} \left\{ \left(1 + \frac{\|\mathbf{A}\|}{\alpha_{\mathrm{d}}}\right) \|x - z_h\| + \frac{1}{\alpha_{\mathrm{d}}} \|(A - A_h)(z_h, \cdot)\|_{Y'_h} \right\} \\ + \frac{1}{\alpha_{\mathrm{d}}} \|G - G_h\|_{Y'_h}, \end{aligned} \tag{5.24}$$

where the consistency terms are given by

$$\|(A - A_h)(z_h, \cdot)\|_{Y'_h} := \sup_{\substack{y_h \in Y_h \\ y_h \neq 0}} \frac{(A - A_h)(z_h, y_h)}{\|y_h\|} \quad \forall\, z_h \in X_h,$$

and

$$\|G - G_h\|_{Y'_h} := \sup_{\substack{y_h \in Y_h \\ y_h \neq 0}} \frac{(G - G_h)(y_h)}{\|y_h\|}.$$

Proof. We first observe, thanks to the triangle inequality and the second discrete inf-sup condition in Lemma 5.4, that for each $z_h \in X_h$ we have

$$\|x - x_h\| \leqslant \|x - z_h\| + \|x_h - z_h\|, \tag{5.25}$$

and

$$\alpha_{\mathtt{d}} \|x_h - z_h\| \leqslant \sup_{\substack{y_h \in Y_h \\ y_h \neq 0}} \frac{A_h(x_h - z_h, y_h)}{\|y_h\|}, \tag{5.26}$$

respectively. Then, according to (5.2) and (5.23), we obtain

$$A_h(x_h - z_h, y_h) = G_h(y_h) - A_h(z_h, y_h) - G(y_h) + A(x, y_h), \tag{5.27}$$

so that, subtracting and adding z_h in the first component of $A(x, y_h)$, we get

$$A_h(x_h - z_h, y_h) = \big(G_h - G\big)(y_h) + A(x - z_h, y_h) + \big(A - A_h\big)(z_h, y_h). \tag{5.28}$$

Then, replacing (5.28) into (5.26), and recalling from Section 5.1.1 that $\mathbf{A} \in \mathcal{L}(X, Y')$ is the operator induced by A, we easily deduce that

$$\|x_h - z_h\| \leqslant \frac{1}{\alpha_{\mathtt{d}}} \|G - G_h\|_{Y_h'} + \frac{\|\mathbf{A}\|}{\alpha_{\mathtt{d}}} \|x - z_h\| + \frac{1}{\alpha_{\mathtt{d}}} \|\big(A - A_h\big)(z_h, \cdot)\|_{Y_h'}, \tag{5.29}$$

which, together with (5.25), and after taking the infimum with respect to $z_h \in X_h$, leads to (5.24) and completes the proof.

□

On the other hand, when $A_h(x, y_h)$ is computable for all $y_h \in Y_h$, the estimate (5.24) in Lemma 5.5 can be simplified so that both consistency terms appear separately from the infima. Indeed, subtracting and adding x in the first component of $A_h(z_h, y_h)$ (cf. (5.27)), it follows that (5.28) can be rewritten as

$$A_h(x_h - z_h, y_h) = \big(G_h - G\big)(y_h) - A_h(z_h - x, y_h) + \big(A - A_h\big)(x, y_h),$$

which, together with (5.26), implies instead of (5.29)

$$\begin{aligned} \|x_h - z_h\| \leqslant\; & \frac{1}{\alpha_{\mathtt{d}}} \|G - G_h\|_{Y_h'} + \frac{\|\mathbf{A}_h\|}{\alpha_{\mathtt{d}}} \|x - z_h\| \\ & + \frac{1}{\alpha_{\mathtt{d}}} \|\big(A - A_h\big)(x, \cdot)\|_{Y_h'}. \end{aligned} \tag{5.30}$$

Consequently, we are in a position to state the following result.

Lemma 5.6. *Assume the hypotheses of Lemmas* 5.1 (*or* 5.2) *and* 5.4, *and let* $x \in X$ *and* $x_h \in X_h$ *be the unique solutions of* (5.2) *and* (5.23), *respectively. Moreover, suppose that* $A_h(x, y_h)$ *is computable for all* $y_h \in Y_h$. *Then, we have*

$$\begin{aligned} \|x - x_h\| \leqslant\; & \left(1 + \frac{\|\mathbf{A}_h\|}{\alpha_{\mathtt{d}}}\right) \operatorname{dist}\big(x, X_h\big) + \frac{1}{\alpha_{\mathtt{d}}} \|G - G_h\|_{Y_h'} \\ & + \frac{1}{\alpha_{\mathtt{d}}} \|\big(A - A_h\big)(x, \cdot)\|_{Y_h'}. \end{aligned} \tag{5.31}$$

Proof. Similarly to the proof of Lemma 5.5, it follows from (5.25) and (5.30), and then taking the infimum with respect to $z_h \in X_h$.

□

Finally, if the bilinear form A_h and the functional G_h coincide with A and G restricted to $X_h \times Y_h$ and Y_h, respectively, then Lemmas 5.5 and 5.6 become the classical result providing the usual Céa estimate. More precisely, we obtain the following lemma.

Lemma 5.7. *Assume the hypotheses of Lemmas* 5.1 (*or* 5.2) *and* 5.4, *and let* $x \in X$ *and* $x_h \in X_h$ *be the unique solutions of* (5.2) *and* (5.23), *respectively, with* $A_h = A$ *and* $G_h = G$. *Then, we have*

$$\|x - x_h\| \leqslant \left(1 + \frac{\|\mathbf{A}\|}{\alpha_{\mathrm{d}}}\right) \operatorname{dist}(x, X_h)\,.$$

Proof. It follows directly from both (5.24) and (5.31), noting in this latter inequality that certainly $\|\mathbf{A}_h\| = \|\mathbf{A}\|$.

□

5.2 Babuška-Brezzi Theory

In what follows, we extend the results of Section 4.4 to the case of Banach spaces, which will allow us to introduce and analyze mixed variational formulations and their respective Galerkin approximations, for a greater variety of boundary value problems, including, in particular, nonlinear and coupled models.

5.2.1 The operator equation

Let $(H, \|\cdot\|_H)$ and $(Q, \|\cdot\|_Q)$ be Banach spaces over $\mathbb{R}$, and let $a : H \times H \to \mathbb{R}$ and $b : H \times Q \to \mathbb{R}$ be bounded bilinear forms. Then, given $F \in H'$ and $G \in Q'$, we are again interested in the problem (4.69) (cf. Section 4.4.1), that is: Find $(\sigma, u) \in H \times Q$ such that

$$\begin{aligned} a(\sigma,\tau) + b(\tau,u) &= F(\tau) \qquad \forall\, \tau \in H\,, \\ b(\sigma,v) \qquad\qquad &= G(v) \qquad \forall\, v \in Q\,. \end{aligned} \tag{5.32}$$

Now, let $\mathcal{A} \in \mathcal{L}(H, H')$ and $\mathcal{B} \in \mathcal{L}(H, Q')$ be the operators induced by a and b, respectively, that is,

$$\mathcal{A}(\zeta)(\tau) := a(\zeta,\tau) \qquad \forall\, \zeta,\, \tau \in H\,, \tag{5.33}$$

and

$$\mathcal{B}(\tau)(v) := b(\tau, v) \qquad \forall\, \tau \in H\,, \quad \forall\, v \in Q\,. \tag{5.34}$$

It follows that $\mathcal{B}' \in \mathcal{L}(Q'', H')$ is defined by

$$\mathcal{B}'(\mathcal{G})(\tau) := \mathcal{G}(\mathcal{B}(\tau)) \qquad \forall\, \mathcal{G} \in Q''\,, \quad \forall\, \tau \in H\,.$$

In particular, if $\mathcal{J}_Q : Q \to Q''$ is the injective operator defined by (cf. Section 7.1) $\mathcal{J}_Q(v)(G) := G(v)$ for all $G \in Q'$, for all $v \in Q$, we have that

$$\mathcal{B}'(\mathcal{J}_Q(u))(\tau) := \mathcal{J}_Q(u)(\mathcal{B}(\tau)) = \mathcal{B}(\tau)(u) = b(\tau, u)\,,$$

and therefore (5.32) reduces to: Find $(\sigma, u) \in H \times Q$ such that

$$\begin{aligned} \mathcal{A}(\sigma) + \mathcal{B}'(\mathcal{J}_Q(u)) &= F\,, \\ \mathcal{B}(\sigma) &= G\,. \end{aligned} \tag{5.35}$$

Moreover, if we now assume that Q is reflexive, that is, the operator $\mathcal{J}_Q : Q \to Q''$ (cf. (3.24)) is bijective, then (5.35) can be reformulated as: Find $(\sigma, \mathcal{G}) \in H \times Q''$ such that

$$\begin{aligned} \mathcal{A}(\sigma) + \mathcal{B}'(\mathcal{G}) &= F\,, \\ \mathcal{B}(\sigma) &= G\,, \end{aligned} \tag{5.36}$$

and then define $u := \mathcal{J}_Q^{-1}(\mathcal{G})$. According to the above analysis, in the next section we analyze the solvability of systems of operator equations of the type given by (5.36).

5.2.2 A preliminary result

Let $(X, \|\cdot\|_X)$ and $(Y, \|\cdot\|_Y)$ be Banach spaces over $\mathbb{R}$, and let $\mathbf{A} \in \mathcal{L}(X, X')$ and $\mathbf{B} \in \mathcal{L}(X, Y)$, so that $\mathbf{B}' \in \mathcal{L}(Y', X')$. Then, given $F \in X'$ and $y \in Y$, we are interested in: Find $(\sigma, U) \in X \times Y'$ such that

$$\begin{aligned} \mathbf{A}(\sigma) + \mathbf{B}'(U) &= F\,, \\ \mathbf{B}(\sigma) &= y\,. \end{aligned} \tag{5.37}$$

For the purpose of the solvability analysis of (5.37), and proceeding in an analogous way to that developed in Section 4.4, we define the kernel of $\mathbf{B}$

$$Z := \big\{ \tau \in X : \quad \mathbf{B}(\tau) = 0 \big\}\,,$$

and the operator $\Pi\mathbf{A} : Z \to Z'$ given by

$$\big(\Pi\mathbf{A}\big)(\tau)(\zeta) := \mathbf{A}(\tau)(\zeta) \qquad \forall\, \tau, \zeta \in Z\,. \tag{5.38}$$

Then, the following theorem, whose proof essentially follows the same guidelines as Theorems 4.19 and 4.20, but which, instead of Lemma 4.18, is based on Lemmas 3.64 and 3.65, provides necessary and sufficient conditions for problem (5.37) to be well-posed.

Theorem 5.8. *For each pair $(F, y) \in X' \times Y$, problem (5.37) has a unique solution that depends continuously on (F, y) if and only if*

i) $\Pi\mathbf{A} : Z \to Z'$ *is an isomorphism, and*
ii) $\mathbf{B} : X \to Y$ *is surjective.*

Proof. Suppose that i) and ii) hold, and let $(F, y) \in X' \times Y$. Since $\mathbf{B}$ is surjective, we know from Lemma 3.64 (or from the first implication of Lemma 3.65) that there exist a constant $\beta > 0$, independent of the given y, and $\sigma_y \in X$ such that $\mathbf{B}(\sigma_y) = y$, and

$$\|\sigma_y\| \leqslant \frac{1}{\beta}\,\|y\|\,. \tag{5.39}$$

It follows that $F - \mathbf{A}(\sigma_y) \in X'$, and then $\big(F - \mathbf{A}(\sigma_y)\big)\big|_Z \in Z'$. Therefore, applying i) we deduce that there exists a unique $\sigma_0 \in Z$ such that $\big(\Pi\mathbf{A}\big)(\sigma_0) = \big(F - \mathbf{A}(\sigma_y)\big)\big|_Z$, that is,

$$\mathbf{A}(\sigma_0)(\tau) = F(\tau) - \mathbf{A}(\sigma_y)(\tau) \qquad \forall\, \tau \in Z\,,$$

or equivalently $\big(F-\mathbf{A}(\sigma_0+\sigma_y)\big)(\tau) = 0$ for all $\tau \in Z$, which means that $F-\mathbf{A}(\sigma_0+\sigma_y) \in Z^\circ$. In turn, a direct application of the Bounded Inverse Theorem (cf. Theorem 3.34) implies that

$$\begin{aligned} \|\sigma_0\| &\leqslant \|(\Pi\mathbf{A})^{-1}\| \, \|(F-\mathbf{A}(\sigma_y))|_Z\| \leqslant \|(\Pi\mathbf{A})^{-1}\| \, \|F-\mathbf{A}(\sigma_y)\| \\ &\leqslant \|(\Pi\mathbf{A})^{-1}\| \, \{\|F\| + \|\mathbf{A}\| \, \|\sigma_y\|\} . \end{aligned} \tag{5.40}$$

Now, since $\mathbf{B}$ is surjective, we have that $R(\mathbf{B}')$ is closed (because $R(\mathbf{B})$ obviously is), and therefore $R(\mathbf{B}') = N(\mathbf{B})^\circ = Z^\circ$, from which it follows that there exists $U \in Y'$ such that $\mathbf{B}'(U) = F - \mathbf{A}(\sigma_0+\sigma_y)$. Moreover, using part ii) of Lemma 3.65, we have that

$$\|U\| \leqslant \frac{1}{\beta}\|F-\mathbf{A}(\sigma_0+\sigma_y)\| \leqslant \frac{1}{\beta}\Big\{\|F\| + \|\mathbf{A}\|\,\big(\|\sigma_0\| + \|\sigma_y\|\big)\Big\} . \tag{5.41}$$

Thus, defining $\sigma = \sigma_0+\sigma_y \in X$, and observing that $\mathbf{B}(\sigma) = y$, we have that $(\sigma, U) \in X \times Y'$ is a solution of (5.37) for the given pair (F, y). Furthermore, in view of (5.39), (5.40), and (5.41), it is easily shown that

$$\|\sigma\| \leqslant \|(\Pi\mathbf{A})^{-1}\| \, \|F\| + \frac{1}{\beta}\Big(1 + \|(\Pi\mathbf{A})^{-1}\| \, \|\mathbf{A}\|\Big)\|y\| , \tag{5.42}$$

and

$$\|U\| \leqslant \frac{1}{\beta}\Big(1 + \|(\Pi\mathbf{A})^{-1}\| \, \|\mathbf{A}\|\Big)\|F\| + \frac{\|\mathbf{A}\|}{\beta^2}\Big(1 + \|(\Pi\mathbf{A})^{-1}\| \, \|\mathbf{A}\|\Big)\|y\| . \tag{5.43}$$

Now, to see that $(\sigma, U) \in X \times Y'$ is the unique solution, consider $(\widetilde{\sigma}, \widetilde{U}) \in X \times Y'$ such that

$$\begin{aligned} \mathbf{A}(\widetilde{\sigma}) + \mathbf{B}'(\widetilde{U}) &= 0 \in X' , \\ \mathbf{B}(\widetilde{\sigma}) &= 0 \in Y , \end{aligned}$$

and we next show that $(\widetilde{\sigma}, \widetilde{U}) = (0, 0)$. Indeed, from the second row of the above equation we clearly have that $\widetilde{\sigma} \in Z$, and evaluating the first in each $\tau \in Z$ yields

$$0 = \mathbf{A}(\widetilde{\sigma})(\tau) + \mathbf{B}'(\widetilde{U})(\tau) = (\Pi\mathbf{A})(\widetilde{\sigma})(\tau) + \widetilde{U}(\mathbf{B}(\tau)) = (\Pi\mathbf{A})(\widetilde{\sigma})(\tau) ,$$

which indicates that $(\Pi\mathbf{A})(\widetilde{\sigma}) = 0 \in Z'$. Thus, using that $\Pi\mathbf{A} : Z \to Z'$ is a bijection, we conclude that $\widetilde{\sigma} = 0$, and therefore we have $\mathbf{B}'(\widetilde{U}) = 0$. Finally, since $N(\mathbf{B}') = R(\mathbf{B})^\circ = \{0\}$, it follows that $\widetilde{U} = 0 \in Y'$. Conversely, suppose that (5.37) is well-posed. Then, given $y \in Y$, there exists a unique pair $(\widehat{\sigma}, \widehat{U}) \in X \times Y'$ such that

$$\begin{aligned} \mathbf{A}(\widehat{\sigma}) + \mathbf{B}'(\widehat{U}) &= 0 \in X' , \\ \mathbf{B}(\widehat{\sigma}) &= y \in Y , \end{aligned}$$

from whose second row it clearly follows that $\mathbf{B}$ is surjective, that is, ii). On the other hand, for i), let us first show the corresponding surjectivity. Indeed, given $\widetilde{F}_0 \in Z'$, we know from the Hahn-Banach Theorem in normed spaces (cf. Theorem 2.21) that there exists $\widetilde{F} \in X'$ such that $\widetilde{F}\big|_Z = \widetilde{F}_0$ and $\|\widetilde{F}\|_{X'} = \|\widetilde{F}_0\|_{Z'}$. Then, by hypothesis, there exists a unique pair $(\widetilde{\sigma}, \widetilde{U}) \in X \times Y'$ such that

$$\begin{aligned} \mathbf{A}(\widetilde{\sigma}) + \mathbf{B}'(\widetilde{U}) &= \widetilde{F} \in X' , \\ \mathbf{B}(\widetilde{\sigma}) &= 0 \in Y , \end{aligned}$$

from which it follows that $\tilde{\sigma} \in Z$. Then, for each $\tau \in Z$ we obtain

$$\big(\Pi \mathbf{A}\big)(\tilde{\sigma})(\tau) = \mathbf{A}(\tilde{\sigma})(\tau) = \big(\tilde{F} - \mathbf{B}'(\tilde{U})\big)(\tau) = \tilde{F}_0(\tau) - \tilde{U}(\mathbf{B}(\tau)) = \tilde{F}_0(\tau),$$

that is, $\big(\Pi \mathbf{A}\big)(\tilde{\sigma}) = \tilde{F}_0$, thus confirming that $\Pi \mathbf{A} : Z \to Z'$ is surjective. On the other hand, let $\bar{\sigma} \in Z$ be such that $\big(\Pi \mathbf{A}\big)(\bar{\sigma}) = 0$, or, equivalently, $\mathbf{A}(\bar{\sigma})(\tau) = 0$ for all $\tau \in Z$, which means that $\mathbf{A}(\bar{\sigma}) \in Z^\circ$. Then, since $\mathbf{B}$ is surjective, we have that $R(\mathbf{B}')$ is also closed, so that $R(\mathbf{B}') = N(\mathbf{B})^\circ = Z^\circ$, and therefore there exists $\bar{U} \in Y'$ such that $\mathbf{B}'(\bar{U}) = -\mathbf{A}(\bar{\sigma})$. In this way we obtain the system

$$\begin{aligned} \mathbf{A}(\bar{\sigma}) \,+\, \mathbf{B}'(\bar{U}) &= 0 \in X', \\ \mathbf{B}(\bar{\sigma}) \qquad\qquad &= 0 \in Y, \end{aligned}$$

and consequently, applying the hypothesis of the present implication, it follows that $(\bar{\sigma}, \bar{U}) = (0,0)$, thus proving the injectivity of $\Pi \mathbf{A} : Z \to Z'$, which completes the proof. □

5.2.3 The main result

Since our problem of interest (5.32), equivalently (5.36), is a particular case of (5.37) with $X = H$ and $Y = Q'$, a direct application of Theorem 5.8 allows us to conclude that (5.32) is well-posed if and only if

i) $\Pi \mathcal{A} : V \to V'$ is an isomorphism, and
ii) $\mathcal{B} : H \to Q'$ is surjective,

where

$$V := \big\{\tau \in H : \quad \mathcal{B}(\tau) = 0\big\} = \big\{\tau \in H : \quad b(\tau, v) = 0 \quad \forall\, v \in Q\big\}, \tag{5.44}$$

and, following (5.38), $\Pi \mathcal{A} : V \to V'$ is the operator defined by

$$\big(\Pi \mathcal{A}\big)(\tau)(\zeta) := \mathcal{A}(\tau)(\zeta) \qquad \forall\, \tau,\, \zeta \in V. \tag{5.45}$$

In what follows, and analogously to what was done in Section 4.4.3 for the Hilbert case, we proceed to deduce assumptions equivalent to i) and ii), which are expressed in terms of inf-sup conditions for the bilinear forms a and b. Indeed, first applying Theorem 3.86 to the operator $\mathcal{B}$, we have that ii) is equivalent to the existence of a constant $\beta > 0$ such that

$$\|\mathcal{B}'(\mathcal{G})\| \geqslant \beta\, \|\mathcal{G}\| \qquad \forall\, \mathcal{G} \in Q'', \tag{5.46}$$

which means that $\mathcal{B}' : Q'' \to H'$ is injective and has closed range (see also Theorem 3.58). Then, using that Q is reflexive, and recalling that $\mathcal{J}_Q \in \mathcal{L}(Q, Q'')$ denotes the respective isometric bijection, inequality (5.46) can be rewritten as

$$\|\mathcal{B}'\big(\mathcal{J}_Q(v)\big)\| \geqslant \beta\, \|\mathcal{J}_Q(v)\| \qquad \forall\, v \in Q,$$

that is,

$$\sup_{\substack{\tau \in H \\ \tau \neq 0}} \frac{\mathcal{B}'\big(\mathcal{J}_Q(v)\big)(\tau)}{\|\tau\|} \geqslant \beta\, \|v\| \qquad \forall\, v \in Q,$$

or, using that $\mathcal{B}'\big(\mathcal{J}_Q(v)\big)(\tau) = \mathcal{J}_Q(v)\big(\mathcal{B}(\tau)\big) = \mathcal{B}(\tau)(v) = b(\tau, v)$, we obtain

$$\sup_{\substack{\tau \in H \\ \tau \neq 0}} \frac{b(\tau, v)}{\|\tau\|} \geqslant \beta\, \|v\| \qquad \forall\, v \in Q. \tag{5.47}$$

On the other hand, it is clear that i) decomposes as

i-1) $\Pi\mathcal{A} : V \to V'$ is injective, and
i-2) $\Pi\mathcal{A} : V \to V'$ is surjective.

In this regard, i-1) means that for all $\zeta \in V$, $\zeta \neq 0$, we have that $(\Pi\mathcal{A})(\zeta) \neq 0 \in V'$, which is equivalent to establishing that

$$\sup_{\tau \in V} (\Pi\mathcal{A})(\zeta)(\tau) > 0 \qquad \forall\, \zeta \in V,\ \zeta \neq 0\,,$$

or, using (5.45),

$$\sup_{\tau \in V} a(\zeta, \tau) > 0 \qquad \forall\, \zeta \in V,\ \zeta \neq 0\,. \tag{5.48}$$

In turn, analogously to the analysis of ii), and again applying Theorem 3.86, we have that i-2) is equivalent to the existence of a constant $\alpha > 0$ such that

$$\|(\Pi\mathcal{A})'(\mathcal{F})\| \geqslant \alpha\,\|\mathcal{F}\| \qquad \forall\, \mathcal{F} \in V''\,, \tag{5.49}$$

which, by virtue of the bijectivity of $\Pi\mathcal{A}$, and therefore of $(\Pi\mathcal{A})'$, implies that

$$\|(\Pi\mathcal{A})^{-1}\| = \|((\Pi\mathcal{A})^{-1})'\| = \|((\Pi\mathcal{A})')^{-1}\| \leqslant \frac{1}{\alpha}\,. \tag{5.50}$$

Then, now assuming that H is reflexive, and noting that V is a closed subspace of H, it follows from Lemma 7.5 (to be proved later in Chapter 7) that V is also reflexive. In this way, denoting by $\mathcal{J}_V \in \mathcal{L}(V, V'')$ the corresponding isometry, inequality (5.49) reduces to

$$\sup_{\substack{\zeta \in V \\ \zeta \neq 0}} \frac{(\Pi\mathcal{A})'(\mathcal{J}_V(\tau))(\zeta)}{\|\zeta\|} \geqslant \alpha\,\|\mathcal{J}_V(\tau)\| \qquad \forall\, \tau \in V\,,$$

or, since $(\Pi\mathcal{A})'(\mathcal{J}_V(\tau))(\zeta) = \mathcal{J}_V(\tau)(\Pi\mathcal{A})(\zeta) = (\Pi\mathcal{A})(\zeta)(\tau) = a(\zeta, \tau)$, we obtain

$$\sup_{\substack{\zeta \in V \\ \zeta \neq 0}} \frac{a(\zeta, \tau)}{\|\zeta\|} \geqslant \alpha\,\|\tau\| \qquad \forall\, \tau \in V\,. \tag{5.51}$$

Alternatively, using that $\Pi\mathcal{A} : V \to V'$ is an isomorphism if and only if $(\Pi\mathcal{A})' : V'' \to V'$ also is, which follows from Theorem 3.90, we can consider, instead of i), the assumptions

i-1)' $(\Pi\mathcal{A})' : V'' \to V'$ is injective, and
i-2)' $(\Pi\mathcal{A})' : V'' \to V'$ is surjective,

or, using Theorem 3.87 for i-2)', it follows

i-1)' $(\Pi\mathcal{A})' : V'' \to V'$ is injective, and
i-2)' $\Pi\mathcal{A} : V \to V'$ is injective and has closed range.

In this case, proceeding analogously to the derivation of (5.48) and (5.51), we obtain that i-1)' and i-2)' are equivalent to

$$\sup_{\zeta \in V} a(\zeta, \tau) > 0 \qquad \forall\, \tau \in V,\ \tau \neq 0\,, \tag{5.52}$$

and to the existence of $\alpha > 0$ such that

$$\|(\Pi \mathcal{A})(\zeta)\| \,=\, \sup_{\substack{\tau\in V\\ \tau\neq 0}} \frac{a(\zeta,\tau)}{\|\tau\|} \,\geqslant\, \alpha\,\|\zeta\| \qquad \forall\,\zeta\in V\,, \tag{5.53}$$

respectively. Note, in particular, that from (5.53) it directly follows the estimate (5.50) for $\|(\Pi \mathcal{A})^{-1}\|$.

According to the previous analysis, we have proved the following theorem.

Theorem 5.9. *Let $(H, \|\cdot\|_H)$ and $(Q, \|\cdot\|_Q)$ be reflexive Banach spaces over $\mathbb{R}$, and let $a : H \times H \to \mathbb{R}$ and $b : H \times Q \to \mathbb{R}$ be bounded bilinear forms with induced operators given by $\mathcal{A} \in \mathcal{L}(H, H')$ (cf. (5.33)) and $\mathcal{B} \in \mathcal{L}(H, Q')$ (cf. (5.34)), respectively, such that $V := N(\mathcal{B})$ (cf. (5.44)). Assume that*

i) *one of the following pairs of conditions on a holds*

i-1) $\sup\limits_{\tau\in V} a(\zeta,\tau) \,>\, 0$ $\quad\forall\,\zeta\in V,\ \zeta\neq 0\,,$	i-1)' $\sup\limits_{\zeta\in V} a(\zeta,\tau) \,>\, 0$ $\quad\forall\,\tau\in V,\ \tau\neq 0\,,$
i-2) *there exists $\alpha > 0$ such that* $\displaystyle\sup_{\substack{\zeta\in V\\ \zeta\neq 0}} \frac{a(\zeta,\tau)}{\|\zeta\|} \,\geqslant\, \alpha\,\|\tau\| \quad \forall\,\tau\in V\,,$	i-2)' *there exists $\alpha > 0$ such that* $\displaystyle\sup_{\substack{\tau\in V\\ \tau\neq 0}} \frac{a(\zeta,\tau)}{\|\tau\|} \,\geqslant\, \alpha\,\|\zeta\| \quad \forall\,\zeta\in V\,,$

ii) *there exists $\beta > 0$ such that*

$$\sup_{\substack{\tau\in H\\ \tau\neq 0}} \frac{b(\tau,v)}{\|\tau\|} \,\geqslant\, \beta\,\|v\| \qquad \forall\,v\in Q\,.$$

Then, for each pair $(F, G) \in H' \times Q'$ there exists a unique solution $(\sigma, u) \in H \times Q$ of problem (5.32), and the following a priori estimates hold:

$$\|\sigma\| \,\leqslant\, \frac{1}{\alpha}\,\|F\| \,+\, \frac{1}{\beta}\Big(1 + \frac{\|\mathcal{A}\|}{\alpha}\Big)\|G\|\,, \tag{5.54}$$

$$\|u\| \,\leqslant\, \frac{1}{\beta}\Big(1 + \frac{\|\mathcal{A}\|}{\alpha}\Big)\|F\| \,+\, \frac{\|\mathcal{A}\|}{\beta^2}\,\Big(1 + \frac{\|\mathcal{A}\|}{\alpha}\Big)\|G\|\,. \tag{5.55}$$

Moreover, assumptions i) *and* ii) *are also necessary.*

Proof. As stated at the beginning of this section, the proof follows from a direct application of Theorem 5.8, keeping in mind the equivalences between i) and the pairs of conditions (5.48) - (5.51) and (5.52) - (5.53), and between ii) and (5.47). In turn, (5.54) and (5.55) are obtained from (5.42) and (5.43), respectively, replacing in both $\|(\Pi\mathcal{A})^{-1}\|$ by $\frac{1}{\alpha}$, according to what is stipulated in (5.50).

□

5.2.4 Discrete analysis

In this section we introduce a Galerkin scheme for (5.32), prove that it is well-posed, and provide the associated a priori error estimates. In this regard, we note in advance that most of the corresponding theorems are obtained simply as particular cases of the general analysis to be developed later in Section 5.5.4. Accordingly, we recommend that the reader remain in the present section when mainly interested in the results, and move to Section 5.5.4 when the details of the corresponding proofs are required.

To begin, let $\{H_h\}_{h>0}$ and $\{Q_h\}_{h>0}$ be families of finite-dimensional subspaces of H and Q, respectively, and let $a_h : H_h \times H_h \to \mathbb{R}$ and $b_h : H_h \times Q_h \to \mathbb{R}$ be bounded bilinear forms approximating a and b, with induced operators $\mathcal{A}_h \in \mathcal{L}(H_h, H_h')$ and $\mathcal{B}_h \in \mathcal{L}(H_h, Q_h')$, respectively. Then, given $F_h \in H_h'$ and $G_h \in Q_h'$ approximating F and G, respectively, we are interested in: Find $(\sigma_h, u_h) \in H_h \times Q_h$ such that

$$\begin{aligned} a_h(\sigma_h, \tau_h) + b_h(\tau_h, u_h) &= F_h(\tau_h) \qquad \forall\, \tau_h \in H_h\,, \\ b_h(\sigma_h, v_h) &= G_h(v_h) \qquad \forall\, v_h \in Q_h\,. \end{aligned} \tag{5.56}$$

Then, by directly applying Theorem 5.9 to the present discrete context, we arrive at the following result.

Theorem 5.10. *In addition to the above notations and definitions, let*

$$V_h := N(\mathcal{B}_h) = \big\{\tau_h \in H_h : \quad b_h(\tau_h, v_h) = 0 \quad \forall\, v_h \in Q_h\big\}\,, \tag{5.57}$$

and assume that

i) *there exists $\alpha_{\mathrm{d}} > 0$ such that one of the following hypotheses on a_h holds*

$$\sup_{\substack{\zeta_h \in V_h \\ \zeta_h \neq 0}} \frac{a_h(\zeta_h, \tau_h)}{\|\zeta_h\|} \geqslant \alpha_{\mathrm{d}}\, \|\tau_h\| \quad \forall\, \tau_h \in V_h\,, \qquad \Bigg| \qquad \sup_{\substack{\tau_h \in V_h \\ \tau_h \neq 0}} \frac{a_h(\zeta_h, \tau_h)}{\|\tau_h\|} \geqslant \alpha_{\mathrm{d}}\, \|\zeta_h\| \quad \forall\, \zeta_h \in V_h\,,$$

ii) *there exists a constant $\beta_{\mathrm{d}} > 0$ such that*

$$\sup_{\substack{\tau_h \in H_h \\ \tau_h \neq 0}} \frac{b_h(\tau_h, v_h)}{\|\tau_h\|} \geqslant \beta_{\mathrm{d}}\, \|v_h\| \qquad \forall\, v_h \in Q_h\,.$$

Then, for each pair $(F_h, G_h) \in H_h' \times Q_h'$ there exists a unique $(\sigma_h, u_h) \in H_h \times Q_h$ solution of problem (5.56)*, and the following a priori estimates hold*

$$\|\sigma_h\| \leqslant \frac{1}{\alpha_{\mathrm{d}}}\, \|F_h\| + \frac{1}{\beta_{\mathrm{d}}}\Big(1 + \frac{\|\mathcal{A}_h\|}{\alpha_{\mathrm{d}}}\Big)\|G_h\|\,,$$

$$\|u_h\| \leqslant \frac{1}{\beta_{\mathrm{d}}}\Big(1 + \frac{\|\mathcal{A}_h\|}{\alpha_{\mathrm{d}}}\Big)\|F_h\| + \frac{\|\mathcal{A}_h\|}{\beta_{\mathrm{d}}^2}\Big(1 + \frac{\|\mathcal{A}_h\|}{\alpha_{\mathrm{d}}}\Big)\|G_h\|\,.$$

Moreover, assumptions i) *and* ii) *are also necessary.*

Proof. It suffices to note that, since V_h is a finite-dimensional subspace, the discrete version of i-1) (resp. i-1)') in Theorem 5.9 is equivalent to the discrete version of i-2) (resp. i-2)'), and therefore we retain only the latter.

□

Alternatively, Theorem 5.10 corresponds to a particular case of the more general result to be presented later in Theorem 5.27 (cf. Section 5.5.4). Analogously, as a direct consequence of Theorem 5.30 (to be detailed later in Section 5.5.4), we obtain the following Strang-type estimates for the error between the solutions of (5.32) and (5.56).

Theorem 5.11. *Assume the hypotheses of Theorems* 5.9 *and* 5.10*, let* $(\sigma, u) \in H \times Q$ *and* $(\sigma_h, u_h) \in H_h \times Q_h$ *be the unique solutions of* (5.32) *and* (5.56)*, respectively, and define the constant* (*cf.* (5.202))

$$C_{\mathtt{d}} := \left(1 + \frac{\|\mathcal{A}_h\|}{\alpha_{\mathtt{d}}}\right) \frac{1}{\beta_{\mathtt{d}}}.$$

Then, one has

$$\begin{aligned} \|\sigma - \sigma_h\| \leqslant & \inf_{\tau_h \in H_h} \left\{ \left(1 + \frac{\|\mathcal{A}\|}{\alpha_{\mathtt{d}}} + \left(1 + \frac{\|\mathcal{A}_h\|}{\alpha_{\mathtt{d}}}\right) \frac{\|\mathcal{B}\|}{\beta_{\mathtt{d}}}\right) \|\sigma - \tau_h\| \right. \\ + & \left. \left(1 + \frac{\|\mathcal{A}_h\|}{\alpha_{\mathtt{d}}}\right) \frac{1}{\beta_{\mathtt{d}}} \|(b - b_h)(\tau_h, \cdot)\|_{Q_h'} + \frac{1}{\alpha_{\mathtt{d}}} \|(a - a_h)(\tau_h, \cdot)\|_{V_h'} \right\} \\ + & \inf_{v_h \in Q_h} \left\{ \frac{\|\mathcal{B}\|}{\alpha_{\mathtt{d}}} \|u - v_h\| + \frac{1}{\alpha_{\mathtt{d}}} \|(b - b_h)(\cdot, v_h)\|_{V_h'} \right\} \\ + & \frac{1}{\alpha_{\mathtt{d}}} \|F - F_h\|_{V_h'} + \left(1 + \frac{\|\mathcal{A}_h\|}{\alpha_{\mathtt{d}}}\right) \frac{1}{\beta_{\mathtt{d}}} \|G - G_h\|_{Q_h'}, \end{aligned} \tag{5.58}$$

and

$$\begin{aligned} \|u - u_h\| \leqslant C_{\mathtt{d}} & \inf_{\tau_h \in H_h} \left\{ \left(\|\mathcal{A}\| + \frac{\|\mathcal{A}_h\| \, \|\mathcal{B}\|}{\beta_{\mathtt{d}}}\right) \|\sigma - \tau_h\| \right. \\ & + \left. \frac{\|\mathcal{A}_h\|}{\beta_{\mathtt{d}}} \|(b - b_h)(\tau_h, \cdot)\|_{Q_h'} + \|(a - a_h)(\tau_h, \cdot)\|_{H_h'} \right\} \\ & + \inf_{v_h \in Q_h} \left\{ \left(1 + C_{\mathtt{d}} \|\mathcal{B}\|\right) \|u - v_h\| + C_{\mathtt{d}} \|(b - b_h)(\cdot, v_h)\|_{H_h'} \right\} \\ & + C_{\mathtt{d}} \left(\|F - F_h\|_{H_h'} + \frac{\|\mathcal{A}_h\|}{\beta_{\mathtt{d}}} \|G - G_h\|_{Q_h'} \right), \end{aligned} \tag{5.59}$$

Now, according to the estimate to be given later in (5.209), we observe here that when $V_h \subseteq V$ (cf. (5.44), (5.57)), the Strang bound (5.58) reduces to

$$\begin{aligned} \|\sigma - \sigma_h\| \leqslant & \inf_{\tau_h \in H_h} \left\{ \left(1 + \frac{\|\mathcal{A}\|}{\alpha_{\mathtt{d}}} + \left(1 + \frac{\|\mathcal{A}_h\|}{\alpha_{\mathtt{d}}}\right) \frac{\|\mathcal{B}\|}{\beta_{\mathtt{d}}}\right) \|\sigma - \tau_h\| \right. \\ + & \left. \left(1 + \frac{\|\mathcal{A}_h\|}{\alpha_{\mathtt{d}}}\right) \frac{1}{\beta_{\mathtt{d}}} \|(b - b_h)(\tau_h, \cdot)\|_{Q_h'} + \frac{1}{\alpha_{\mathtt{d}}} \|(a - a_h)(\tau_h, \cdot)\|_{V_h'} \right\} \\ + & \frac{1}{\alpha_{\mathtt{d}}} \|F - F_h\|_{V_h'} + \left(1 + \frac{\|\mathcal{A}_h\|}{\alpha_{\mathtt{d}}}\right) \frac{1}{\beta_{\mathtt{d}}} \|G - G_h\|_{Q_h'}, \end{aligned}$$

whereas (5.59) remains unchanged.

On the other hand, when $a_h(\sigma, \zeta_h)$, $b_h(\zeta_h, u)$, and $b_h(\sigma, w_h)$ are computable for all $(\zeta_h, w_h) \in H_h \times Q_h$, the result specified later in Theorem 5.31 (cf. Section 5.5.4) becomes, in the present case, the following a priori error estimates.

Theorem 5.12. *Assume the hypotheses of Theorems* 5.9 *and* 5.10*, and let* $(\sigma, u) \in H \times Q$ *and* $(\sigma_h, u_h) \in H_h \times Q_h$ *be the unique solutions of* (5.32) *and* (5.56)*, respectively. Assume further that* $a_h(\sigma, \zeta_h)$*,* $b_h(\zeta_h, u)$*, and* $b_h(\sigma, w_h)$ *are computable for all* $(\zeta_h, w_h) \in H_h \times Q_h$*. Then, one has*

$$\begin{aligned}\|\sigma - \sigma_h\| \leqslant & \left(1 + \frac{\|\mathcal{A}_h\|}{\alpha_{\mathrm{d}}}\right)\left(1 + \frac{\|\mathcal{B}_h\|}{\beta_{\mathrm{d}}}\right) \operatorname{dist}(\sigma, H_h) \\ + & \frac{\|\mathcal{B}_h\|}{\alpha_{\mathrm{d}}} \operatorname{dist}(u, Q_h) + \frac{1}{\alpha_{\mathrm{d}}} \|(b - b_h)(\cdot, u)\|_{V_h'} \\ + & \left(1 + \frac{\|\mathcal{A}_h\|}{\alpha_{\mathrm{d}}}\right) \frac{1}{\beta_{\mathrm{d}}} \|(b - b_h)(\sigma, \cdot)\|_{Q_h'} + \frac{1}{\alpha_{\mathrm{d}}} \|(a - a_h)(\sigma, \cdot)\|_{V_h'} \\ + & \frac{1}{\alpha_{\mathrm{d}}} \|F - F_h\|_{V_h'} + \left(1 + \frac{\|\mathcal{A}_h\|}{\alpha_{\mathrm{d}}}\right) \frac{1}{\beta_{\mathrm{d}}} \|G - G_h\|_{Q_h'},\end{aligned}$$

and

$$\begin{aligned}\|u - u_h\| \leqslant & \ C_{\mathrm{d}} \|\mathcal{A}_h\| \left(1 + \frac{\|\mathcal{B}_h\|}{\beta_{\mathrm{d}}}\right) \operatorname{dist}(\sigma, H_h) \\ & + \Big(1 + C_{\mathrm{d}} \|\mathcal{B}_h\|\Big) \operatorname{dist}(u, Q_h) + C_{\mathrm{d}} \|(b - b_h)(\cdot, u)\|_{H_h'} \\ & + C_{\mathrm{d}} \left(\frac{\|\mathcal{A}_h\|}{\beta_{\mathrm{d}}} \|(b - b_h)(\sigma, \cdot)\|_{Q_h'} + \|(a - a_h)(\sigma, \cdot)\|_{H_h'}\right) \\ & + C_{\mathrm{d}} \left(\|F - F_h\|_{H_h'} + \frac{\|\mathcal{A}_h\|}{\beta_{\mathrm{d}}} \|G - G_h\|_{Q_h'}\right).\end{aligned}$$

Finally, when the discrete bilinear forms and functionals coincide with their respective continuous counterparts, Theorems 5.11 and 5.12 become the following result with the usual Céa estimates.

Theorem 5.13. *Assume the hypotheses of Theorems* 5.9 *and* 5.10, *and let* $(\sigma, u) \in H \times Q$ *and* $(\sigma_h, u_h) \in H_h \times Q_h$ *be the unique solutions of* (5.32) *and* (5.56), *respectively, with* $a = a_h$, $b = b_h$, $F = F_h$, *and* $G = G_h$. *Then, one has*

$$\|\sigma - \sigma_h\| \leqslant \left(1 + \frac{\|\mathcal{A}\|}{\alpha_{\mathrm{d}}}\right)\left(1 + \frac{\|\mathcal{B}\|}{\beta_{\mathrm{d}}}\right) \operatorname{dist}(\sigma, H_h) + \frac{\|\mathcal{B}\|}{\alpha_{\mathrm{d}}} \operatorname{dist}(u, Q_h)$$

and

$$\begin{aligned}\|u - u_h\| \leqslant & \left(1 + \frac{\|\mathcal{A}\|}{\alpha_{\mathrm{d}}}\right)\left(1 + \frac{\|\mathcal{B}\|}{\beta_{\mathrm{d}}}\right) \frac{\|\mathcal{A}\|}{\beta_{\mathrm{d}}} \operatorname{dist}(\sigma, H_h) \\ + & \left(1 + \frac{\|\mathcal{B}\|}{\beta_{\mathrm{d}}} + \frac{\|\mathcal{A}\|\,\|\mathcal{B}\|}{\alpha_{\mathrm{d}}\,\beta_{\mathrm{d}}}\right) \operatorname{dist}(u, Q_h).\end{aligned}$$

Note that the previous result coincides with the Hilbertian case given by Theorem 4.24 (see also [58, Theorem 2.6]).

5.3 Perturbed saddle-point problems

In this section, we essentially follow the approach of [44] and [80, Theorems 3.2 and 4.1] to analyze the solvability of the continuous and discrete schemes that arise from perturbed saddle-point problems in Banach spaces. More precisely, given reflexive Banach spaces H and Q, bounded bilinear forms $a : H \times H \to \mathbb{R}$, $b : H \times Q \to \mathbb{R}$, and $c : Q \times Q \to \mathbb{R}$, and functionals $F \in H'$ and $G \in Q'$, we are interested in finding $(\sigma, u) \in H \times Q$ such that

$$\begin{aligned} a(\sigma,\tau) + b(\tau,u) &= F(\tau) \quad \forall\, \tau \in H\,, \\ b(\sigma,v) - c(u,v) &= G(v) \quad \forall\, v \in Q\,. \end{aligned} \tag{5.60}$$

In the particular case when H and Q are Hilbert spaces, the solvability of (5.60) and its associated Galerkin scheme, as well as the corresponding error estimates, are addressed in [24, Theorem 1.2, Section II.1.2] and [24, Proposition 2.11, Section II.2.4]. Moreover, when in such a Hilbertian context a is symmetric and c vanishes, the solution of (5.60) coincides with the unique saddle-point of a functional $\mathcal{J}$ that depends on a and b (cf. Exercise 4.7). This fact explains why, upon incorporating now a nonzero bilinear form c, the resulting variational formulation acquires the additional adjective "perturbed" (precisely due to c).

We begin the analysis of (5.60) with some preliminary results.

5.3.1 Preliminaries

Let $\mathbf{B} : H \to Q'$ and $\mathbf{B}^{\mathrm{t}} : Q \to H'$ be the linear and bounded operators induced by b, that is,

$$\mathbf{B}(\tau)(v) \,:=\, b(\tau,v) \quad \forall\, \tau \in H\,,\ \forall\, v \in Q \tag{5.61}$$

and

$$\mathbf{B}^{\mathrm{t}}(v)(\tau) \,:=\, b(\tau,v) \quad \forall\, v \in Q\,,\ \forall\, \tau \in H\,, \tag{5.62}$$

and introduce the respective null spaces

$$V \,:=\, N(\mathbf{B}) \,:=\, \big\{ \tau \in H : \quad b(\tau,v) = 0 \quad \forall\, v \in Q \big\} \tag{5.63}$$

and

$$W \,:=\, N(\mathbf{B}^{\mathrm{t}}) \,:=\, \big\{ v \in Q : \quad b(\tau,v) = 0 \quad \forall\, \tau \in H \big\}\,. \tag{5.64}$$

Moreover, suppose that V and W admit topological complements, that is, there exist closed subspaces $V^\perp$ and $W^\perp$ of H and Q, respectively, such that (cf. Definition 3.54)

$$H \,=\, V \oplus V^\perp \qquad \text{and} \qquad Q \,=\, W \oplus W^\perp\,, \tag{5.65}$$

and let $i : V^\perp \to H$ and $j : W^\perp \to Q$ be the respective injections. Clearly, in the present discussion we do not have inner products, and therefore not the concept of orthogonality either. The symbol $^\perp$ is used here only to preserve the analogy with the Orthogonal Decomposition Theorem for Hilbert spaces (cf. Theorem 2.10). However, we emphasize that, unlike that case, the norms of the oblique projectors that arise from the decompositions given by (5.65) are not necessarily equal to 1. In fact, a direct application of the Open Mapping Theorem (cf. Theorem 3.39) implies the existence of positive constants C_H and C_Q, depending only on H and Q, respectively, such that

$$\|\tau_0\|_H + \|\bar{\tau}\|_H \,\leqslant\, C_H\, \|\tau\|_H \quad \text{and} \quad \|v_0\|_Q + \|\bar{v}\|_Q \,\leqslant\, C_Q\, \|v\|_Q\,, \tag{5.66}$$

for all $\tau = \tau_0 + \bar{\tau} \in V \oplus V^\perp$, and for all $v = v_0 + \bar{v} \in W \oplus W^\perp$.

The above boundedness properties imply the following result.

Lemma 5.14. *There hold*

$$\frac{1}{C_H}\, \|\tau\|_H \,\leqslant\, \operatorname{dist}(\tau, V) \,\leqslant\, \|\tau\|_H \qquad \forall\, \tau \in V^\perp\,, \tag{5.67}$$

and

$$\frac{1}{C_Q}\, \|v\|_Q \,\leqslant\, \operatorname{dist}(v, W) \,\leqslant\, \|v\|_Q \qquad \forall\, v \in W^\perp\,. \tag{5.68}$$

Proof. We first observe that the upper bounds of (5.67) and (5.68) are straightforward and indeed valid for all $(\tau, v) \in H \times Q$. We now provide the proof of the lower bound in (5.67). The proof for (5.68) proceeds analogously. To this end, we recall that if T is a closed subspace of the dual of a reflexive Banach space X, then one has (cf. Lemma 3.78 and (3.44))

$$\operatorname{dist}(x, {}^{\circ}T) = \sup_{\substack{F \in T \\ F \neq 0}} \frac{|F(x)|}{\|F\|_{X'}} \qquad \forall\, x \in X\,.$$

Applying this identity to $X = H$ and $T = V^{\circ}$, and using the fact that ${}^{\circ}(V^{\circ}) = V$ (cf. Lemma 3.24), we obtain

$$\operatorname{dist}(\tau, V) = \sup_{\substack{F \in V^{\circ} \\ F \neq 0}} \frac{|F(\tau)|}{\|F\|_{H'}} \qquad \forall\, \tau \in H\,. \tag{5.69}$$

Now we restrict to $\tau \in V^{\perp}$. Then, given $G \in H'$, we introduce the functional $g : H \to \mathbb{R}$ defined by $g(\zeta) := G(\bar{\zeta})$ for all $\zeta = \zeta_0 + \bar{\zeta} \in H = V \oplus V^{\perp}$. It is clear that g is linear, and, using (5.66), that

$$|g(\zeta)| = |G(\bar{\zeta})| \leqslant \|G\|_{H'} \|\bar{\zeta}\|_H \leqslant C_H \|G\|_{H'} \|\zeta\|_H \quad \forall\, \zeta \in H\,,$$

which shows that g is bounded, with $\|g\|_{H'} \leqslant C_H \|G\|_{H'}$, and since $g|_V \equiv 0$, we deduce that $g \in V^{\circ}$. Thus, using (5.69), and observing that $g(\tau) = G(\tau)$, we find that

$$\operatorname{dist}(\tau, V) \geqslant \frac{|g(\tau)|}{\|g\|_{H'}} \geqslant \frac{|G(\tau)|}{C_H \|G\|_{H'}}\,,$$

from which, taking the supremum with respect to $G \in H'$, we obtain

$$\operatorname{dist}(\tau, V) \geqslant \frac{1}{C_H} \|\tau\|_H \qquad \forall\, \tau \in V^{\perp}\,,$$

thus completing the proof of (5.67).

□

On the other hand, $\mathbf{B}$ and $\mathbf{B}^{\mathrm{t}}$ are related through the equivalence properties established below.

Lemma 5.15. *The following statements are equivalent*

i) $\mathbf{B}^{\mathrm{t}} \circ j : W^{\perp} \to H'$ *is injective and has closed range, that is, there exists a constant* $\widetilde{\beta} > 0$ *such that*

$$\|\mathbf{B}^{\mathrm{t}}(v)\|_{H'} := \sup_{\substack{\tau \in H \\ \tau \neq 0}} \frac{b(\tau, v)}{\|\tau\|_H} \geqslant \widetilde{\beta}\, \|v\|_Q \qquad \forall\, v \in W^{\perp}\,. \tag{5.70}$$

ii) $j' \circ \mathbf{B} : H \to \left(W^{\perp}\right)'$ *is surjective.*

iii) $\mathbf{B} \circ i : V^{\perp} \to Q'$ *is injective and has closed range, that is, there exists a constant* $\widehat{\beta} > 0$ *such that*

$$\|\mathbf{B}(\tau)\|_{Q'} := \sup_{\substack{v \in Q \\ v \neq 0}} \frac{b(\tau, v)}{\|v\|_Q} \geqslant \widehat{\beta}\, \|\tau\|_H \qquad \forall\, \tau \in V^{\perp}\,. \tag{5.71}$$

iv) $i' \circ \mathbf{B}^{\mathrm{t}} : Q \to \left(V^{\perp}\right)'$ *is surjective.*

Proof. Let $\mathcal{J}_H : H \to H''$ and $\mathcal{J}_Q : Q \to Q''$ be the linear, bijective, and isometric operators defined by

$$\mathcal{J}_H(\tau)(F) := F(\tau) \quad \forall\, \tau \in H\,, \quad \forall\, F \in H'$$

and

$$\mathcal{J}_Q(v)(G) := G(v) \quad \forall\, v \in Q\,, \quad \forall\, G \in Q'\,.$$

Then, as suggested by the diagrams

$$H \overset{\mathcal{J}_H}{\rightarrow} H'' \overset{(\mathbf{B}^{\mathrm{t}})'}{\rightarrow} Q' \quad \text{and} \quad Q \overset{\mathcal{J}_Q}{\rightarrow} Q'' \overset{\mathbf{B}'}{\rightarrow} H'\,,$$

we conjecture that

$$\mathbf{B} = (\mathbf{B}^{\mathrm{t}})' \circ \mathcal{J}_H \quad \text{and} \quad \mathbf{B}^{\mathrm{t}} = \mathbf{B}' \circ \mathcal{J}_Q\,. \tag{5.72}$$

Indeed, given $\tau \in H$ and $v \in Q$, we easily obtain

$$\big((\mathbf{B}^{\mathrm{t}})' \circ \mathcal{J}_H\big)(\tau)(v) = (\mathbf{B}^{\mathrm{t}})'\big(\mathcal{J}_H(\tau)\big)(v) = \mathcal{J}_H(\tau)\big(\mathbf{B}^{\mathrm{t}}(v)\big) = \mathbf{B}^{\mathrm{t}}(v)(\tau) = \mathbf{B}(\tau)(v)\,,$$

which shows the first identity in (5.72). The second follows similarly or as a consequence of the first after interchanging $\mathbf{B}$ with $\mathbf{B}^{\mathrm{t}}$ and the roles of the spaces H and Q. It follows from (5.72) that

$$j' \circ \mathbf{B} = \big(j' \circ (\mathbf{B}^{\mathrm{t}})'\big) \circ \mathcal{J}_H = \big(\mathbf{B}^{\mathrm{t}} \circ j\big)' \circ \mathcal{J}_H\,, \tag{5.73}$$

and therefore, invoking the bijectivity of $\mathcal{J}_H$, we deduce that $j' \circ \mathbf{B} : H \to \left(W^{\perp}\right)'$ is surjective if and only if $\left(\mathbf{B}^{\mathrm{t}} \circ j\right)' : H'' \to \left(W^{\perp}\right)'$ is also surjective, which, in turn, is equivalent to stating that (cf. Theorem 3.87) $\mathbf{B}^{\mathrm{t}} \circ j : W^{\perp} \to H'$ is injective and has closed range. The above analysis leads to the equivalence between i) and ii). Analogously, using the second identity in (5.72), namely

$$i' \circ \mathbf{B}^{\mathrm{t}} = \left(i' \circ \mathbf{B}'\right) \circ \mathcal{J}_Q = \left(\mathbf{B} \circ i\right)' \circ \mathcal{J}_Q\,, \tag{5.74}$$

we can prove that iii) and iv) are equivalent. To conclude the proof, it suffices to show, for instance, that i) and iii) share the same property, which we address next. Indeed, suppose that i) holds. Then, since $\mathbf{B}^{\mathrm{t}} \circ j$ has closed range, we have $R(\mathbf{B}^{\mathrm{t}} \circ j) = {}^{\circ}N\big((\mathbf{B}^{\mathrm{t}} \circ j)'\big)$, where, according to (5.73), $N\big((\mathbf{B}^{\mathrm{t}} \circ j)'\big) = \mathcal{J}_H\big(N(j' \circ \mathbf{B})\big)$. A simple computation gives

$$N(j' \circ \mathbf{B}) = \big\{\tau \in H : \quad j'\big(\mathbf{B}(\tau)\big)(v) = \mathbf{B}(\tau)(v) = 0 \quad \forall\, v \in W^{\perp}\big\} = V\,,$$

and then $R(\mathbf{B}^{\mathrm{t}} \circ j) = {}^{\circ}\mathcal{J}_H(V) = V^{\circ}$. Thus, we conclude that $\mathbf{B}^{\mathrm{t}} \circ j : W^{\perp} \to V^{\circ}$ is bijective, and (5.70) implies that $\|(\mathbf{B}^{\mathrm{t}} \circ j)^{-1}\| \leqslant \frac{1}{\widetilde{\beta}}$. It follows that $(\mathbf{B}^{\mathrm{t}} \circ j)' : (V^{\circ})' \to (W^{\perp})'$ is also bijective, and

$$\big\|\big((\mathbf{B}^{\mathrm{t}} \circ j)'\big)^{-1}\big\| = \big\|\big((\mathbf{B}^{\mathrm{t}} \circ j)^{-1}\big)'\big\| = \|(\mathbf{B}^{\mathrm{t}} \circ j)^{-1}\| \leqslant \frac{1}{\widetilde{\beta}}\,,$$

which means, equivalently, that

$$\|(\mathbf{B}^{\mathrm{t}} \circ j)'(\mathcal{G})\|_{(W^{\perp})'} \geqslant \widetilde{\beta}\, \|\mathcal{G}\|_{(V^{\circ})'} \qquad \forall\, \mathcal{G} \in (V^{\circ})'\,. \tag{5.75}$$

Now, taking in particular $\mathcal{G} = \mathcal{J}_H(\tau)|_{V^{\circ}}$, with $\tau \in H$, we obtain

$$\begin{aligned}\|(\mathbf{B}^{\mathrm{t}} \circ j)'(\mathcal{G})\|_{(W^\perp)'} &= \sup_{\substack{v \in W^\perp \\ v \neq 0}} \frac{(\mathbf{B}^{\mathrm{t}} \circ j)'(\mathcal{J}_H(\tau))(v)}{\|v\|_Q} \\ &= \sup_{\substack{v \in W^\perp \\ v \neq 0}} \frac{\mathcal{J}_H(\tau)(\mathbf{B}^{\mathrm{t}}(v))}{\|v\|_Q} = \sup_{\substack{v \in W^\perp \\ v \neq 0}} \frac{b(\tau, v)}{\|v\|_Q},\end{aligned} \tag{5.76}$$

whereas

$$\|\mathcal{J}_H(\tau)|_{V^\circ}\|_{(V^\circ)'} := \sup_{\substack{F \in V^\circ \\ F \neq 0}} \frac{\mathcal{J}_H(\tau)(F)}{\|F\|_{H'}} = \sup_{\substack{F \in V^\circ \\ F \neq 0}} \frac{F(\tau)}{\|F\|_{H'}} = \operatorname{dist}(\tau, V), \tag{5.77}$$

where (5.69) has been used in the last equality. Therefore, replacing (5.76) and (5.77) into (5.75), we conclude that

$$\|\mathbf{B}(\tau)\|_{Q'} \geqslant \sup_{\substack{v \in W^\perp \\ v \neq 0}} \frac{b(\tau, v)}{\|v\|_Q} \geqslant \widetilde{\beta} \operatorname{dist}(\tau, V) \qquad \forall\, \tau \in H, \tag{5.78}$$

which, together with the lower bound in (5.67), leads to iii) (cf. (5.71)) with $\widehat{\beta} := \frac{\widetilde{\beta}}{C_H}$. Conversely, suppose that iii) holds. Then, to prove i), we proceed analogously to the reverse implication. In particular, now using (5.74), we deduce that $R(\mathbf{B} \circ i) = W^\circ$, so that $\mathbf{B} \circ i : V^\perp \to W^\circ$ and $(\mathbf{B} \circ i)' : (W^\circ)' \to (V^\perp)'$ are bijective with $\|((\mathbf{B} \circ i)')^{-1}\| = \|(\mathbf{B} \circ i)^{-1}\| \leqslant \frac{1}{\widehat{\beta}}$. Thus, we obtain the analogue of (5.78), namely

$$\|\mathbf{B}^{\mathrm{t}}(v)\|_{H'} \geqslant \sup_{\substack{\tau \in V^\perp \\ \tau \neq 0}} \frac{b(\tau, v)}{\|\tau\|_H} \geqslant \widehat{\beta} \operatorname{dist}(v, W) \qquad \forall\, v \in Q, \tag{5.79}$$

from which, together with (5.68), we arrive at (5.70) with $\widetilde{\beta} := \frac{\widehat{\beta}}{C_Q}$. Further details are omitted.

□

We emphasize that the equivalence between the inf-sup conditions (5.70) (cf. i)) and (5.71) (cf. iii)) holds with different, although related, constants $\widetilde{\beta}$ and $\widehat{\beta}$. Indeed, it is clear from the corresponding discussion in the previous proof that

$$\begin{aligned}&\sup_{\substack{\tau \in H \\ \tau \neq 0}} \frac{b(\tau, v)}{\|\tau\|_H} \geqslant \widetilde{\beta}\, \|v\|_Q \quad \forall\, v \in W^\perp \quad \Rightarrow \quad \sup_{\substack{v \in Q \\ v \neq 0}} \frac{b(\tau, v)}{\|v\|_Q} \geqslant \widetilde{\beta} \operatorname{dist}(\tau, V) \quad \forall\, \tau \in H \\ &\quad \Rightarrow \quad \sup_{\substack{v \in Q \\ v \neq 0}} \frac{b(\tau, v)}{\|v\|_Q} \geqslant \widehat{\beta}\, \|\tau\|_H \quad \forall\, \tau \in V^\perp \quad \text{with} \quad \widehat{\beta} := \frac{\widetilde{\beta}}{C_H},\end{aligned} \tag{5.80}$$

and conversely

$$\begin{aligned}&\sup_{\substack{v \in Q \\ v \neq 0}} \frac{b(\tau, v)}{\|v\|_Q} \geqslant \widehat{\beta}\, \|\tau\|_H \quad \forall\, \tau \in V^\perp \quad \Rightarrow \quad \sup_{\substack{\tau \in H \\ \tau \neq 0}} \frac{b(\tau, v)}{\|\tau\|_H} \geqslant \widehat{\beta} \operatorname{dist}(v, W) \quad \forall\, v \in Q \\ &\quad \Rightarrow \quad \sup_{\substack{\tau \in H \\ \tau \neq 0}} \frac{b(\tau, v)}{\|\tau\|_H} \geqslant \widetilde{\beta}\, \|v\|_Q \quad \forall\, v \in W^\perp \quad \text{with} \quad \widetilde{\beta} := \frac{\widehat{\beta}}{C_Q}.\end{aligned} \tag{5.81}$$

On the contrary, when H and Q are Hilbert spaces, it is easy to see that

$$\operatorname{dist}(\tau, V) = \|\tau\|_H \qquad \forall\, \tau \in V^\perp,$$

and

$$\operatorname{dist}(v, W) = \|v\|_Q \qquad \forall\, v \in W^\perp,$$

so that replacing these identities in the right hand sides of (5.78) and (5.79), respectively, we find that the equivalence between i) and iii) does hold with the same constant $\widetilde{\beta} = \widehat{\beta}$. For further details on the Hilbertian case, we refer to [19, eq. (4.3.18), Theorem 4.3.1], [58, Lemma 2.1], and [24, Proposition 1.2 and eqs. (1.15) and (1.16), Chapter II].

5.3.2 Continuous analysis

In this section we establish sufficient conditions for the unique solvability of the perturbed saddle-point formulation (5.60). To this end, and in order to employ Lemma 5.2 together with the equivalent assumptions provided by (5.19) and (5.22), we first note that (5.60) can be equivalently rewritten as: Find $(\sigma, u) \in H \times Q$ such that

$$\mathbf{A}\big((\sigma, u), (\tau, v)\big) \;=\; \mathbf{F}(\tau, v) \qquad \forall\, (\tau, v) \in H \times Q\,, \tag{5.82}$$

where $\mathbf{A} : (H \times Q) \times (H \times Q) \to \mathbb{R}$ and $\mathbf{F} : H \times Q \to \mathbb{R}$ are the bilinear form and the linear functional, respectively, defined by

$$\mathbf{A}\big((\zeta, w), (\tau, v)\big) \;:=\; a(\zeta, \tau) + b(\tau, w) + b(\zeta, v) - c(w, v) \quad \forall\, (\zeta, w),\, (\tau, v) \in H \times Q\,, \tag{5.83}$$

and

$$\mathbf{F}(\tau, v) \;:=\; f(\tau) + g(v) \qquad \forall\, (\tau, v) \in H \times Q\,.$$

Throughout the rest of this section we consider the product norm

$$\|(\tau, v)\|_{H\times Q} \;:=\; \|\tau\|_H + \|v\|_Q \qquad \forall\, (\tau, v) \in H \times Q\,.$$

Moreover, from now on we denote by $\|a\|$, $\|b\|$, and $\|c\|$ the smallest positive constants such that

$$\begin{aligned}
|a(\zeta, \tau)| \;&\leqslant\; \|a\|\, \|\zeta\|_H\, \|\tau\|_H \qquad \forall\, (\zeta, \tau) \in H \times H\,,\\
|b(\tau, v)| \;&\leqslant\; \|b\|\, \|\tau\|_H\, \|v\|_Q \qquad \forall\, (\tau, v) \in H \times Q\,,\\
|c(w, v)| \;&\leqslant\; \|c\|\, \|w\|_Q\, \|v\|_Q \qquad \forall\, (w, v) \in Q \times Q\,.
\end{aligned} \tag{5.84}$$

Then, we have the following abstract result.

Theorem 5.16. *Let H and Q be reflexive Banach spaces, and let $a : H \times H \to \mathbb{R}$, $b : H \times Q \to \mathbb{R}$, and $c : Q \times Q \to \mathbb{R}$ be bounded bilinear forms (cf. (5.84)). Moreover, let $\mathbf{B} \in \mathcal{L}(H, Q')$ and $\mathbf{B}^{\mathrm{t}} \in \mathcal{L}(Q, H')$ be the operators induced by b (cf. (5.61), (5.62)), and let $V := N(\mathbf{B})$ and $W := N(\mathbf{B}^{\mathrm{t}})$ be the respective null spaces (cf. (5.63), (5.64)). Assume that*

i) *there exist closed subspaces $V^\perp$ and $W^\perp$ of H and Q, respectively, such that $H = V \oplus V^\perp$ and $Q = W \oplus W^\perp$,*
ii) *a and c are positive semi-definite, that is*

$$a(\tau, \tau) \geqslant 0 \quad \forall\, \tau \in H \quad \textit{and} \quad c(v, v) \geqslant 0 \quad \forall\, v \in Q\,,$$

and c is symmetric,

iii) *there exists a constant* $\widetilde{\alpha} > 0$ *such that*

$$\sup_{\substack{\tau \in V \\ \tau \neq 0}} \frac{a(\vartheta,\tau)}{\|\tau\|_H} \geqslant \widetilde{\alpha}\,\|\vartheta\|_H \qquad \forall\,\vartheta \in V\,, \tag{5.85}$$

and

$$\sup_{\substack{\vartheta \in V \\ \vartheta \neq 0}} \frac{a(\vartheta,\tau)}{\|\vartheta\|_H} \geqslant \widetilde{\alpha}\,\|\tau\|_H \qquad \forall\,\tau \in V\,, \tag{5.86}$$

iv) *there exists a constant* $\widetilde{\beta} > 0$ *such that (cf.* (5.70))

$$\sup_{\substack{\tau \in H \\ \tau \neq 0}} \frac{b(\tau,v)}{\|\tau\|_H} \geqslant \widetilde{\beta}\,\|v\|_Q \qquad \forall\,v \in W^{\perp}\,, \tag{5.87}$$

v) *and there exists a constant* $\widetilde{\gamma} > 0$ *such that*

$$\sup_{\substack{v \in W \\ v \neq 0}} \frac{c(z,v)}{\|v\|_Q} \geqslant \widetilde{\gamma}\,\|z\|_Q \qquad \forall\,z \in W\,. \tag{5.88}$$

Then, for each pair $(f,g) \in H' \times Q'$ *there exists a unique solution* $(\sigma,u) \in H \times Q$ *of* (5.60) *(equivalently,* (5.82)*). Moreover, there exists a constant* $\widetilde{C} > 0$*, depending only on* $\|a\|$, $\|c\|$, $\widetilde{\alpha}$, $\widetilde{\beta}$, C_H *(cf.* (5.66)*), and* $\widetilde{\gamma}$*, such that*

$$\|(\sigma,u)\|_{H\times Q} \leqslant \widetilde{C}\left\{\|f\|_{H'} + \|g\|_{Q'}\right\}.$$

Proof. Given $\vec{\zeta} := (\zeta,w) \in H \times Q$, we define the functionals $G_{\vec{\zeta},1} \in H'$ and $G_{\vec{\zeta},2} \in Q'$ as indicated in (5.15), that is

$$G_{\vec{\zeta},1}(\tau) := \mathbf{A}\big(\vec{\zeta},(\tau,0)\big) = a(\zeta,\tau) + b(\tau,w) \qquad \forall\,\tau \in H\,, \tag{5.89}$$

and

$$G_{\vec{\zeta},2}(v) := \mathbf{A}\big(\vec{\zeta},(0,v)\big) = b(\zeta,v) - c(w,v) \qquad \forall\,v \in Q\,, \tag{5.90}$$

and we aim to prove the analogue of (5.19), that is, the existence of a positive constant C, independent of $\vec{\zeta}$, such that

$$\|\vec{\zeta}\| = \|(\zeta,w)\| \leqslant C \sum_{i=1}^{2} \|G_{\vec{\zeta},i}\|\,. \tag{5.91}$$

According to i), we first consider the decompositions of ζ and w as

$$\zeta = \zeta_0 + \bar{\zeta} \quad \text{and} \quad w = w_0 + \bar{w}\,, \tag{5.92}$$

with $\zeta_0 \in V$, $\bar{\zeta} \in V^{\perp}$, $w_0 \in W$, and $\bar{w} \in W^{\perp}$, so that the rest of the proof consists in bounding each of the four components specified in (5.92). For this, we first observe from (5.89) that $G_{\vec{\zeta},1}(\tau) = a(\zeta,\tau)$ for all $\tau \in V$, so that applying (5.85) (cf. iii)) to $\vartheta = \zeta_0$, we obtain

$$\widetilde{\alpha}\,\|\zeta_0\|_H \leqslant \sup_{\substack{\tau \in V \\ \tau \neq 0}} \frac{a(\zeta_0,\tau)}{\|\tau\|_H} = \sup_{\substack{\tau \in V \\ \tau \neq 0}} \frac{G_{\vec{\zeta},1}(\tau) - a(\bar{\zeta},\tau)}{\|\tau\|_H}\,,$$

from which it easily follows that

$$\|\zeta_0\|_H \leqslant \frac{1}{\widetilde{\alpha}} \|G_{\vec{\zeta},1}\|_{H'} + \frac{\|a\|}{\widetilde{\alpha}} \|\bar{\zeta}\|_H . \tag{5.93}$$

In turn, to bound $\bar{\zeta}$, we use the equivalence between i) and iii) of Lemma 5.15, thanks to which and to (5.87) (cf. iv)), we have (cf. (5.71))

$$\widehat{\beta}\, \|\tau\|_H \leqslant \sup_{\substack{v \in Q \\ v \neq 0}} \frac{b(\tau, v)}{\|v\|_Q} \qquad \forall\, \tau \in V^\perp ,$$

with $\widehat{\beta} := \frac{\widetilde{\beta}}{C_H}$. Then, noting from (5.90) that $G_{\vec{\zeta},2}(v) = b(\bar{\zeta}, v) - c(w, v)$ for all $v \in Q$, and applying the previous inequality to $\tau = \bar{\zeta} \in V^\perp$, we find that

$$\widehat{\beta}\, \|\bar{\zeta}\|_H \leqslant \sup_{\substack{v \in Q \\ v \neq 0}} \frac{b(\bar{\zeta}, v)}{\|v\|_Q} = \sup_{\substack{v \in Q \\ v \neq 0}} \frac{G_{\vec{\zeta},2}(v) + c(w, v)}{\|v\|_Q} . \tag{5.94}$$

Now, it is easy to see that the assumptions on c allow us to show that

$$c(w, v) \leqslant c(w, w)^{1/2}\, c(v, v)^{1/2} \qquad \forall\, w,\, v \in Q , \tag{5.95}$$

which constitutes a kind of Cauchy–Schwarz inequality for c, so that $|\cdot|_c := c(\cdot, \cdot)^{1/2}$ defines a seminorm in Q. Moreover, using the boundedness of c, it follows from (5.95) that $c(w, v) \leqslant \|c\|^{1/2} |w|_c \|v\|_Q$, and thus (5.94) leads to

$$\|\bar{\zeta}\|_H \leqslant \frac{1}{\widehat{\beta}} \|G_{\vec{\zeta},2}\|_{Q'} + \frac{\|c\|^{1/2}}{\widehat{\beta}} |w|_c . \tag{5.96}$$

Therefore, as a direct consequence of (5.93) and (5.96), we have the following preliminary bound

$$\|\zeta\|_H \leqslant \frac{1}{\widetilde{\alpha}} \|G_{\vec{\zeta},1}\|_{H'} + \left(1 + \frac{\|a\|}{\widetilde{\alpha}}\right) \frac{1}{\widehat{\beta}} \|G_{\vec{\zeta},2}\|_{Q'} + \left(1 + \frac{\|a\|}{\widetilde{\alpha}}\right) \frac{\|c\|^{1/2}}{\widehat{\beta}} |w|_c . \tag{5.97}$$

It remains to bound $|w|_c$ in terms of $\|G_{\vec{\zeta},1}\|_{H'}$ and $\|G_{\vec{\zeta},2}\|_{Q'}$, which will be done later. In the meantime, we address the estimation of $\|w\|_Q$. Indeed, from the definition of $G_{\vec{\zeta},2}$ (cf. (5.90)) we have $G_{\vec{\zeta},2}(v) = -c(w, v) = -c(w_0, v) - c(\bar{w}, v)$ for all $v \in W$, and therefore, applying (5.88) (cf. v)) to $z = w_0 \in W$, we obtain

$$\widetilde{\gamma}\, \|w_0\|_Q \leqslant \sup_{\substack{v \in W \\ v \neq 0}} \frac{c(w_0, v)}{\|v\|_Q} = \sup_{\substack{v \in W \\ v \neq 0}} \frac{-G_{\vec{\zeta},2}(v) - c(\bar{w}, v)}{\|v\|_Q} ,$$

which implies

$$\|w_0\|_Q \leqslant \frac{1}{\widetilde{\gamma}} \|G_{\vec{\zeta},2}\|_{Q'} + \frac{\|c\|}{\widetilde{\gamma}} \|\bar{w}\|_Q . \tag{5.98}$$

On the other hand, it is clear from (5.89) that $G_{\vec{\zeta},1}(\tau) = a(\zeta, \tau) + b(\tau, \bar{w})$ for all $\tau \in H$, so that making use of (5.87) (cf. iv)) with $v = \bar{w} \in W^\perp$, we obtain

$$\widetilde{\beta}\, \|\bar{w}\|_Q \leqslant \sup_{\substack{\tau \in H \\ \tau \neq 0}} \frac{b(\tau, \bar{w})}{\|\tau\|_H} = \sup_{\substack{\tau \in H \\ \tau \neq 0}} \frac{G_{\vec{\zeta},1}(\tau) - a(\zeta, \tau)}{\|\tau\|_H} ,$$

which yields

$$\|\bar{w}\|_Q \leqslant \frac{1}{\widetilde{\beta}}\,\|G_{\vec{\zeta},1}\|_{H'} + \frac{\|a\|}{\widetilde{\beta}}\,\|\zeta\|_H\,. \tag{5.99}$$

In this way, (5.98) and (5.99) lead to

$$\|w\|_Q \leqslant \Big(1+\frac{\|c\|}{\widetilde{\gamma}}\Big)\frac{1}{\widetilde{\beta}}\,\|G_{\vec{\zeta},1}\|_{H'} + \frac{1}{\widetilde{\gamma}}\,\|G_{\vec{\zeta},2}\|_{Q'} + \Big(1+\frac{\|c\|}{\widetilde{\gamma}}\Big)\frac{\|a\|}{\widetilde{\beta}}\,\|\zeta\|_H\,. \tag{5.100}$$

The goal now is to bound $|w|_c^2 := c(w,w)$. Indeed, evaluating $G_{\vec{\zeta},1}$ (cf. (5.89)) and $G_{\vec{\zeta},2}$ (cf. (5.90)) at ζ and w, respectively, and subtracting the resulting expressions, we arrive at

$$a(\zeta,\zeta) + c(w,w) = G_{\vec{\zeta},1}(\zeta) - G_{\vec{\zeta},2}(w)\,,$$

from which, using that a is positive semi-definite (cf. ii)), we deduce that

$$|w|_c^2 \leqslant \|G_{\vec{\zeta},1}\|_{H'}\,\|\zeta\|_H + \|G_{\vec{\zeta},2}\|_{Q'}\,\|w\|_Q\,. \tag{5.101}$$

Moreover, using the bounds for $\|\zeta\|_H$ and $\|w\|_Q$ provided by (5.97) and (5.100), applying Young's inequality appropriately, and performing several algebraic manipulations, we conclude from (5.101) that

$$|w|_c^2 \leqslant C_1\,\|G_{\vec{\zeta},1}\|_{H'}^2 + C_2\,\|G_{\vec{\zeta},2}\|_{Q'}^2 + \frac{1}{2}\,|w|_c^2\,, \tag{5.102}$$

where C_1 and C_2, positive constants depending on $\|a\|$, $\|c\|$, $\widetilde{\alpha}$, $\widetilde{\beta}$, $\widehat{\beta}$, and $\widetilde{\gamma}$, are explicitly given by

$$C_1 := \Big(1+\frac{\|a\|}{\widetilde{\alpha}}\Big)\Big\{\Big(1+\frac{\|a\|}{\widetilde{\alpha}}\Big)\frac{\|c\|}{\widehat{\beta}^2} + \Big(1+\frac{\|c\|}{\widetilde{\gamma}}\Big)\frac{1}{2\widehat{\beta}} + \frac{1}{2\widetilde{\beta}}\Big\} + \frac{1}{\widetilde{\alpha}}\,,$$

and

$$C_2 := \Big(1+\frac{\|a\|}{\widetilde{\alpha}}\Big)\Big(1+\frac{\|c\|}{\widetilde{\gamma}}\Big)\Big\{\Big(1+\frac{\|a\|}{\widetilde{\alpha}}\Big)\Big(1+\frac{\|c\|}{\widetilde{\gamma}}\Big)\frac{\|a\|^2\|c\|}{\widetilde{\beta}^2\widehat{\beta}^2} + \frac{\|a\|}{\widetilde{\beta}\widehat{\beta}} + \frac{1}{2\widetilde{\beta}}\Big\} + \Big(1+\frac{\|a\|}{\widetilde{\alpha}}\Big)\frac{1}{2\widehat{\beta}} + \frac{1}{\widetilde{\gamma}}\,.$$

It is straightforward to see from (5.102) that

$$|w|_c \leqslant \Big(2\max\{C_1,C_2\}\Big)^{1/2}\Big\{\|G_{\vec{\zeta},1}\|_{H'} + \|G_{\vec{\zeta},2}\|_{Q'}\Big\}\,,$$

which, when substituted into (5.97), completes the upper bound for $\|\zeta\|_H$. In turn, using this latter estimate in (5.100) yields the corresponding one for $\|w\|_Q$, thus achieving the required inequality (5.91), with which the present bilinear form $\mathbf{A}$ (cf. (5.83)) satisfies hypothesis b) of Lemma 5.2. Finally, for the fulfillment of assumption a) of Lemma 5.2, we consider $\vec{\tau} := (\tau, v) \in H \times Q$ and, following (5.20), define the functionals $F_{\vec{\tau},1} \in H'$ and $F_{\vec{\tau},2} \in Q'$ as

$$F_{\vec{\tau},1}(\zeta) := \mathbf{A}\big((\zeta,0),\vec{\tau}\big) = a(\zeta,\tau) + b(\zeta,v) \qquad \forall\,\zeta \in H\,, \tag{5.103}$$

and

$$F_{\vec{\tau},2}(w) := \mathbf{A}\big((0,w),\vec{\tau}\big) = b(\tau,w) - c(w,v) \qquad \forall\,w \in Q\,. \tag{5.104}$$

Then, the analogue of (5.22), that is, the existence of a positive constant C, independent of $\vec{\tau}$, such that

$$\|\vec{\tau}\| = \|(\tau, v)\| \leqslant C \sum_{i=1}^{2} \|F_{\vec{\tau},i}\|, \tag{5.105}$$

is proved in a similar manner to (5.91), with the only difference being the use of (5.86) instead of (5.85) to estimate the component $\tau_0 \in V$ of τ. Additional details are omitted, and the proof concludes by a direct application of Lemma 5.2.

□

It is important to emphasize here, thanks to (5.18) applied to the functionals defined by (5.89) and (5.90), with which $N = 2$, that the first part of the previous proof (derivation of (5.91)) establishes, equivalently, that the global inf-sup condition for $\mathbf{A}$ with respect to its second component holds, that is

$$\sup_{\substack{(\tau,v)\in H\times Q\\(\tau,v)\neq 0}} \frac{\mathbf{A}\big((\zeta, w), (\tau, v)\big)}{\|(\tau, v)\|_{H\times Q}} \geqslant \frac{1}{2C} \|(\zeta, w)\|_{H\times Q} \qquad \forall\, (\zeta, w) \in H \times Q\,. \tag{5.106}$$

Analogously, by virtue now of an application of (5.21) to the functionals defined by (5.103) and (5.104), the second part of the previous proof (derivation of (5.105)) establishes, equivalently, that the global inf-sup condition for $\mathbf{A}$ with respect to its first component holds, that is

$$\sup_{\substack{(\zeta,w)\in H\times Q\\(\zeta,w)\neq 0}} \frac{\mathbf{A}\big((\zeta, w), (\tau, v)\big)}{\|(\zeta, w)\|_{H\times Q}} \geqslant \frac{1}{2C} \|(\tau, v)\|_{H\times Q} \qquad \forall\, (\tau, v) \in H \times Q\,.$$

On the other hand, and regarding the observation made just after Lemma 5.15 about the constants $\widetilde{\beta}$ and $\widehat{\beta}$ that appear in the inf-sup conditions (5.70) and (5.71), respectively, we emphasize here that the fact that they do not coincide does not imply any difficulty in the solvability analysis provided by Theorem 5.16. This is certainly due to the fact that the difference between these constants is determined by the reciprocals of C_H and C_Q, which depend on the continuous spaces H and Q. However, this issue is more delicate for the associated Galerkin scheme, to be addressed later, since, when varying the finite element subspaces, the respective constants could also change, in particular according to their dimensions. Accordingly, in this case we cannot employ the equivalence as such between i) and iii) of Lemma 5.15, but rather assume (which actually means to prove, when dealing with specific subspaces) that both discrete inf-sup conditions are satisfied with constants independent of the subspaces and their dimensions.

Now we assume the particular case $W = \{0\}$, equivalently $W^{\perp} = Q$, which means that hypothesis iv) of Theorem 5.16 reduces to the existence of a constant $\widetilde{\beta} > 0$ such that

$$\|\mathbf{B}^{\mathtt{t}}(v)\|_{H'} := \sup_{\substack{\tau\in H\\ \tau\neq 0}} \frac{b(\tau, v)}{\|\tau\|_H} \geqslant \widetilde{\beta}\, \|v\|_Q \qquad \forall\, v \in Q\,. \tag{5.107}$$

Moreover, recalling from (5.72) that $\mathbf{B}^{\mathtt{t}} = \mathbf{B}' \circ \mathcal{J}_Q$, and using the reflexivity of Q and the fact that $\mathcal{J}_Q$ is an isometry, we observe that (5.107) can be rewritten, equivalently, as

$$\|\mathbf{B}'(\mathcal{G})\|_{H'} \geqslant \widetilde{\beta}\, \|\mathcal{G}\|_{Q''} \qquad \forall\, \mathcal{G} \in Q''\,. \tag{5.108}$$

Note that the above establishes that $\mathbf{B}' : Q'' \to H'$ is injective and has closed range, which is equivalent to saying that $\mathbf{B} : H \to Q'$ is surjective (cf. Theorem 3.86). Then, by applying the reciprocal implication of Lemma 3.65, we deduce from (5.108) that for each $G \in Q'$ there exists $\vartheta \in H$ such that

$$\mathbf{B}(\vartheta) = G \qquad \text{and} \qquad \|\vartheta\|_H \leqslant \frac{1}{\widetilde{\beta}} \|G\|_{Q'} . \tag{5.109}$$

As a consequence of (5.109), we can improve the statement of Theorem 5.16 as indicated below, emphasizing beforehand that now no topological complement of V nor a continuous inf-sup condition for c is required.

Theorem 5.17. *Let H and Q be reflexive Banach spaces, and let $a : H \times H \to \mathbb{R}$, $b : H \times Q \to \mathbb{R}$, and $c : Q \times Q \to \mathbb{R}$ be bounded bilinear forms (cf. (5.84)). Moreover, let $\mathbf{B} : H \to Q'$ be one of the bounded linear operators induced by b (cf. (5.61)), and let $V := N(\mathbf{B})$ be the respective null space (cf. (5.63)). Assume that*

i) *a and c are positive semi-definite, that is*

$$a(\tau,\tau) \geqslant 0 \quad \forall\, \tau \in H \quad \text{and} \quad c(v,v) \geqslant 0 \quad \forall\, v \in Q ,$$

and c is symmetric,

ii) *there exists a constant $\widetilde{\alpha} > 0$ such that*

$$\sup_{\substack{\tau \in V\\ \tau \neq 0}} \frac{a(\vartheta,\tau)}{\|\tau\|_H} \geqslant \widetilde{\alpha}\, \|\vartheta\|_H \qquad \forall\, \vartheta \in V , \tag{5.110}$$

and

$$\sup_{\substack{\vartheta \in V\\ \vartheta \neq 0}} \frac{a(\vartheta,\tau)}{\|\vartheta\|_H} \geqslant \widetilde{\alpha}\, \|\tau\|_H \qquad \forall\, \tau \in V , \tag{5.111}$$

iii) *and there exists a constant $\widetilde{\beta} > 0$ such that*

$$\sup_{\substack{\tau \in H\\ \tau \neq 0}} \frac{b(\tau,v)}{\|\tau\|_H} \geqslant \widetilde{\beta}\, \|v\|_Q \qquad \forall\, v \in Q . \tag{5.112}$$

Then, for each pair $(f,g) \in H' \times Q'$ there exists a unique solution $(\sigma,u) \in H \times Q$ of (5.60) (equivalently (5.82)). Moreover, there exists a constant $\widetilde{C} > 0$, which depends only on $\|a\|$, $\|c\|$, $\widetilde{\alpha}$, and $\widetilde{\beta}$, such that

$$\|(\sigma,u)\|_{H\times Q} \leqslant \widetilde{C} \left\{ \|f\|_{H'} + \|g\|_{Q'} \right\} .$$

Proof. We proceed analogously to the proof of Theorem 5.16, although with a substantial difference in the decomposition to be introduced below. Indeed, given $(\zeta, w) \in H \times Q$, we start by defining the functionals $G_{\vec{\zeta},1} \in H'$ and $G_{\vec{\zeta},2} \in Q'$ as done in (5.89) and (5.90), and we aim to establish inequality (5.91). Then, keeping iii) in mind, we apply (5.109) to $G := \mathbf{B}(\zeta) \in Q'$, thus obtaining the existence of $\bar{\zeta} \in H$ such that

$$\mathbf{B}(\bar{\zeta}) = \mathbf{B}(\zeta) \qquad \text{and} \qquad \|\bar{\zeta}\|_H \leqslant \frac{1}{\widetilde{\beta}} \|\mathbf{B}(\zeta)\|_{Q'} = \frac{1}{\widetilde{\beta}} \|\mathbf{B}(\bar{\zeta})\|_{Q'} . \tag{5.113}$$

In this way, ζ can be decomposed as

$$\zeta = \zeta_0 + \bar{\zeta},$$

with $\zeta_0 := \zeta - \bar{\zeta} \in V$. As announced earlier, we emphasize here that there is no need to identify a topological complement to which $\bar{\zeta}$ belongs, but rather to be able to bound $\|\bar{\zeta}\|_H$, which is guaranteed by the inequality in (5.113). Then, observing from (5.89) that $G_{\vec{\zeta},1}(\tau) = a(\zeta,\tau)$ for all $\tau \in V$, and applying (5.110) (cf. i)) to $\vartheta = \zeta_0$, we deduce, exactly as for the derivation of (5.93), that

$$\|\zeta_0\|_H \leqslant \frac{1}{\tilde{\alpha}} \|G_{\vec{\zeta},1}\|_{H'} + \frac{\|a\|}{\tilde{\alpha}} \|\bar{\zeta}\|_H . \tag{5.114}$$

Next, noting from (5.90) that

$$b(\bar{\zeta}, v) = G_{\vec{\zeta},2}(v) + c(w,v) \qquad \forall\, v \in Q,$$

it follows from the inequality in (5.113) that

$$\|\bar{\zeta}\|_H \leqslant \frac{1}{\tilde{\beta}} \|\mathbf{B}(\bar{\zeta})\|_{Q'} = \frac{1}{\tilde{\beta}} \sup_{\substack{v\in Q\\ v\neq 0}} \frac{b(\bar{\zeta},v)}{\|v\|_Q} = \frac{1}{\tilde{\beta}} \sup_{\substack{v\in Q\\ v\neq 0}} \frac{G_{\vec{\zeta},2}(v) + c(w,v)}{\|v\|_Q},$$

from which, in a manner similar to the deduction of (5.96), we obtain

$$\|\bar{\zeta}\|_H \leqslant \frac{1}{\tilde{\beta}} \|G_{\vec{\zeta},2}\|_{Q'} + \frac{\|c\|^{1/2}}{\tilde{\beta}} |w|_c , \tag{5.115}$$

and, therefore, thanks to (5.114) and (5.115), the analogue of (5.97) becomes

$$\|\zeta\|_H \leqslant \frac{1}{\tilde{\alpha}} \|G_{\vec{\zeta},1}\|_{H'} + \left(1 + \frac{\|a\|}{\tilde{\alpha}}\right) \frac{1}{\tilde{\beta}} \|G_{\vec{\zeta},2}\|_{Q'} + \left(1 + \frac{\|a\|}{\tilde{\alpha}}\right) \frac{\|c\|^{1/2}}{\tilde{\beta}} |w|_c . \tag{5.116}$$

Then, we know from (5.89) that

$$b(\tau, w) = G_{\vec{\zeta},1}(\tau) - a(\zeta,\tau) \qquad \forall\, \tau \in H,$$

so that by applying (5.112) (cf. iii)) with $v = w \in Q$, we easily deduce that

$$\|w\|_Q \leqslant \frac{1}{\tilde{\beta}} \|G_{\vec{\zeta},1}\|_{H'} + \frac{\|a\|}{\tilde{\beta}} \|\zeta\|_H . \tag{5.117}$$

The rest of the proof of (5.91) proceeds exactly as was done for Theorem 5.16. In particular, one has that (cf. (5.101))

$$|w|_c^2 \leqslant \|G_{\vec{\zeta},1}\|_{H'} \|\zeta\|_H + \|G_{\vec{\zeta},2}\|_{Q'} \|w\|_Q ,$$

and then, using the bounds for $\|\zeta\|_H$ and $\|w\|_Q$ provided by (5.116) and (5.117), and conveniently applying Young's inequality, we arrive at

$$|w|_c \leqslant \left(2 \max\left\{\tilde{C}_1, \tilde{C}_2\right\}\right)^{1/2} \left\{\|G_{\vec{\zeta},1}\|_{H'} + \|G_{\vec{\zeta},2}\|_{Q'}\right\}, \tag{5.118}$$

where

$$\tilde{C}_1 := \frac{1}{\tilde{\alpha}} + \frac{1}{\tilde{\beta}}\left(1 + \frac{\|a\|}{\tilde{\alpha}}\right) + \frac{\|c\|}{\tilde{\beta}^2}\left(1 + \frac{\|a\|}{\tilde{\alpha}}\right)^2, \quad \text{and}$$

$$\widetilde{C}_2 := \frac{1}{\widetilde{\beta}}\Big(1 + \frac{\|a\|}{\widetilde{\alpha}}\Big)\Big\{1 + \frac{\|a\|}{\widetilde{\beta}} + \frac{\|a\|^2\,\|c\|}{\widetilde{\beta}^3}\Big(1 + \frac{\|a\|}{\widetilde{\alpha}}\Big)\Big\}\,.$$

Thus, (5.118), (5.116), and (5.117) complete the deduction of (5.91), which constitutes hypothesis a) of Lemma 5.2. On the other hand, for the corresponding hypothesis b), we consider $\vec{\tau} := (\tau, v) \in H \times Q$, and define $F_{\vec{\tau},1} \in H'$ and $F_{\vec{\tau},2} \in Q'$ as in (5.103) and (5.104), and proceed similarly to (5.91) to obtain (5.105), but now using (5.111) instead of (5.110) to estimate the component $\tau_0 \in V$ of τ. Finally, as in Theorem 5.16, the present proof concludes thanks to another application of Lemma 5.2.

□

5.3.3 Discrete analysis

We now consider arbitrary families of finite-dimensional subspaces of H and Q denoted by $\{H_h\}_{h>0}$ and $\{Q_h\}_{h>0}$, respectively, and we introduce the Galerkin scheme associated with (5.60): Find $(\sigma_h, u_h) \in H_h \times Q_h$ such that

$$\begin{aligned} a(\sigma_h, \tau_h) + b(\tau_h, u_h) &= f(\tau_h) \quad \forall\, \tau_h \in H_h\,, \\ b(\sigma_h, v_h) - c(u_h, v_h) &= g(v_h) \quad \forall\, v_h \in Q_h\,, \end{aligned} \tag{5.119}$$

or, equivalently: Find $(\sigma_h, u_h) \in H_h \times Q_h$ such that

$$\mathbf{A}\big((\sigma_h, u_h), (\tau_h, v_h)\big) = \mathbf{F}(\tau_h, v_h) \qquad \forall\, (\tau_h, v_h) \in H_h \times Q_h\,. \tag{5.120}$$

In turn, let $\mathbf{B}_h : H_h \to Q_h'$ and $\mathbf{B}_h^{\mathtt{t}} : Q_h \to H_h'$ be the discrete versions of the linear and bounded operators induced by b (cf. (5.61) - (5.62)), and define the respective null spaces

$$V_h := N(\mathbf{B}_h) := \big\{\tau_h \in H_h : \quad b(\tau_h, v_h) = 0 \quad \forall\, v_h \in Q_h\big\}\,,$$

and

$$W_h := N(\mathbf{B}_h^{\mathtt{t}}) := \big\{v_h \in Q_h : \quad b(\tau_h, v_h) = 0 \quad \forall\, \tau_h \in H_h\big\}\,.$$

We emphasize here that the existence of closed subspaces $V_h^{\perp}$ and $W_h^{\perp}$ of H_h and Q_h, respectively, satisfying the decompositions

$$H_h = V_h \oplus V_h^{\perp} \quad \text{and} \quad Q_h = W_h \oplus W_h^{\perp}\,,$$

is guaranteed by the fact that H_h and Q_h are finite-dimensional. Then, the solvability result for (5.119), which is established below, does not need to incorporate the mentioned existence as an assumption (see hypothesis i) in Theorem 5.16) but rather as a fact. In this way, the discrete version of Theorem 5.16 is given as follows. From now on, the expression "independent of h" means that it does not depend on the subspaces H_h or Q_h.

Theorem 5.18. *In addition to the previous notations and definitions, assume that*

i) *a and c are positive semi-definite, and c is symmetric,*

ii) *there exists a constant $\widetilde{\alpha}_{\mathtt{d}} > 0$, independent of h, such that*

$$\sup_{\substack{\tau_h \in V_h \\ \tau_h \neq 0}} \frac{a(\vartheta_h, \tau_h)}{\|\tau_h\|_H} \geq \widetilde{\alpha}_{\mathtt{d}}\,\|\vartheta_h\|_H \qquad \forall\, \vartheta_h \in V_h\,, \tag{5.121}$$

or else

$$\sup_{\substack{\vartheta_h \in V_h \\ \vartheta_h \neq 0}} \frac{a(\vartheta_h, \tau_h)}{\|\vartheta_h\|_H} \geq \widetilde{\alpha}_{\mathtt{d}}\,\|\tau_h\|_H \qquad \forall\, \tau_h \in V_h\,, \tag{5.122}$$

iii) *there exists a constant* $\widetilde{\beta}_{\mathsf{d}} > 0$, *independent of* h, *such that*

$$\sup_{\substack{\tau_h \in H_h \\ \tau_h \neq 0}} \frac{b(\tau_h, v_h)}{\|\tau_h\|_H} \geqslant \widetilde{\beta}_{\mathsf{d}} \, \|v_h\|_Q \qquad \forall\, v_h \in W_h^{\perp} \,,$$

iv) *there exists a constant* $\widehat{\beta}_{\mathsf{d}} > 0$, *independent of* h, *such that*

$$\sup_{\substack{v_h \in Q_h \\ v_h \neq 0}} \frac{b(\tau_h, v_h)}{\|v_h\|_Q} \geqslant \widehat{\beta}_{\mathsf{d}} \, \|\tau_h\|_H \qquad \forall\, \tau_h \in V_h^{\perp} \,,$$

v) *and there exists a constant* $\widetilde{\gamma}_{\mathsf{d}} > 0$, *independent of* h, *such that*

$$\sup_{\substack{v_h \in W_h \\ v_h \neq 0}} \frac{c(z_h, v_h)}{\|v_h\|_Q} \geqslant \widetilde{\gamma}_{\mathsf{d}} \, \|z_h\|_Q \qquad \forall\, z_h \in W_h \,.$$

Then, for each pair $(f, g) \in H' \times Q'$ *there exists a unique solution* $(\sigma_h, u_h) \in H_h \times Q_h$ *of* (5.119). *Moreover, there exists a constant* $\widetilde{C}_{\mathsf{d}} > 0$, *which depends only on* $\|a\|$, $\|c\|$, $\widetilde{\alpha}_{\mathsf{d}}$, $\widetilde{\beta}_{\mathsf{d}}$, $\widehat{\beta}_{\mathsf{d}}$, *and* $\widetilde{\gamma}_{\mathsf{d}}$, *such that*

$$\|\sigma_h\|_H + \|u_h\|_Q \leqslant \widetilde{C}_{\mathsf{d}} \left\{ \|f\|_{H'} + \|g\|_{Q'} \right\} . \tag{5.123}$$

Proof. It follows analogously to the proof of Theorem 5.16, except for the fact that, instead of considering iv) as a consequence of iii), both are assumed here independently.

□

It is important to note here that only the discrete analogue of (5.85) (as detailed in ii), (5.121)) or that of (5.86) (as indicated in ii), (5.122)) is required, but not both. The reason for this is the fact that $H_h \times Q_h$ is the space to which both the unknowns and the test functions of (5.119) belong, in which case, as stipulated in the discrete version of the Generalized Lax-Milgram Lemma (cf. Lemma 5.4), one only needs to establish the global discrete inf-sup condition for $\mathbf{A}$ with respect to its second or its first component, but not to both. In particular, according to (5.16) and (5.123), we find that the first option above reduces to

$$\sup_{\substack{(\tau_h, v_h) \in H_h \times Q_h \\ (\tau_h, v_h) \neq 0}} \frac{\mathbf{A}\big((\zeta_h, w_h), (\tau_h, v_h)\big)}{\|(\tau_h, v_h)\|_{H \times Q}} \geqslant \frac{1}{2\,\widetilde{C}_{\mathsf{d}}} \|(\zeta_h, w_h)\|_{H \times Q} \tag{5.124}$$

for all $(\zeta_h, w_h) \in H_h \times Q_h$, which is used in what follows.

In fact, having established that (5.60) and (5.119) (equivalently, (5.82) and (5.120)) are well posed, we now prove the corresponding Céa estimate. For this purpose, given a subspace X_h of a generic Banach space $(X, \|\cdot\|_X)$, we define

$$\operatorname{dist}(x, X_h) := \inf_{x_h \in X_h} \|x - x_h\|_X \qquad \forall\, x \in X \,.$$

Theorem 5.19. *Assume the hypotheses of Theorems* 5.16 *and* 5.18, *and let* $(\sigma, u) \in H \times Q$ *and* $(\sigma_h, u_h) \in H_h \times Q_h$ *be the unique solutions of* (5.60) *and* (5.119), *respectively. Then, there exists a constant* $\widehat{C}_{\mathsf{d}} > 0$, *depending only on* $\|a\|$, $\|b\|$, $\|c\|$, $\widetilde{\alpha}_{\mathsf{d}}$, $\widetilde{\beta}_{\mathsf{d}}$, $\widehat{\beta}_{\mathsf{d}}$, *and* $\widetilde{\gamma}_{\mathsf{d}}$, *such that*

$$\|\sigma - \sigma_h\|_H + \|u - u_h\|_Q \leqslant \widehat{C}_{\mathsf{d}} \left\{ \operatorname{dist}(\sigma, H_h) + \operatorname{dist}(u, Q_h) \right\} . \tag{5.125}$$

Proof. The derivation of (5.125) proceeds in the usual way for formulations of this type. Indeed, applying the triangle inequality we first obtain that

$$\|(\sigma,u)-(\sigma_h,u_h)\|_{H\times Q} \leqslant \|(\sigma,u)-(\zeta_h,w_h)\|_{H\times Q} + \|(\sigma_h,u_h)-(\zeta_h,w_h)\|_{H\times Q},$$

for each $(\zeta_h,w_h)\in H_h\times Q_h$, then we apply the global discrete inf-sup condition (5.124), which implies

$$\|(\sigma_h,u_h)-(\zeta_h,w_h)\|_{H\times Q} \leqslant 2\,\widetilde{C}_{\mathtt{d}} \sup_{\substack{(\tau_h,v_h)\in H_h\times Q_h\\(\tau_h,v_h)\neq 0}} \frac{\mathbf{A}\big((\sigma_h,u_h)-(\zeta_h,w_h),(\tau_h,v_h)\big)}{\|(\tau_h,v_h)\|_{H\times Q}},$$

and finally we use the identity $\mathbf{A}\big((\sigma_h,u_h),(\tau_h,v_h)\big) = \mathbf{A}\big((\sigma,u),(\tau_h,v_h)\big)$ together with the boundedness of $\mathbf{A}$. In this way we easily arrive at the estimate (5.125) with the constant $\widehat{C}_{\mathtt{d}} := 1+2\,\widetilde{C}_{\mathtt{d}}\,\|\mathbf{A}\|$.

□

On the other hand, similarly to the analysis in the second part of Section 5.3.2, we now assume that $W_h = \{0\}$, which means that hypothesis iii) of Theorem 5.18 reduces to the existence of a constant $\widetilde{\beta}_{\mathtt{d}}>0$, independent of h, such that

$$\|\mathbf{B}_h^{\mathtt{t}}(v_h)\|_{H_h'} := \sup_{\substack{\tau_h\in H_h\\ \tau_h\neq 0}} \frac{b(\tau_h,v_h)}{\|\tau_h\|_H} \geqslant \widetilde{\beta}_{\mathtt{d}}\,\|v_h\|_Q \qquad \forall\, v_h\in Q_h. \tag{5.126}$$

Therefore, noting that the discrete version of the corresponding identity in (5.72) becomes $\mathbf{B}_h^{\mathtt{t}} = \mathbf{B}_h'\circ\mathcal{J}_{Q_h}$, we realize that (5.126) is equivalent to establishing

$$\|\mathbf{B}_h'(\mathcal{G}_h)\|_{H_h'} \geqslant \widetilde{\beta}_{\mathtt{d}}\,\|\mathcal{G}_h\|_{Q_h''} \qquad \forall\,\mathcal{G}_h\in Q_h'',$$

so that, by applying again the reciprocal implication of Lemma 3.65, we conclude that for each $G_h\in Q_h'$ there exists $\vartheta_h\in H_h$ such that

$$\mathbf{B}_h(\vartheta_h) = G_h \qquad \text{and} \qquad \|\vartheta_h\|_H \leqslant \frac{1}{\widetilde{\beta}_{\mathtt{d}}}\,\|G_h\|_{Q_h'}. \tag{5.127}$$

Consequently, we can now present the discrete version of Theorem 5.17.

Theorem 5.20. *Assume that*

i) *a and c are positive semi-definite, and c is symmetric,*
ii) *there exists a constant $\widetilde{\alpha}_{\mathtt{d}}>0$, independent of h, such that*

$$\sup_{\substack{\tau_h\in V_h\\ \tau_h\neq 0}} \frac{a(\vartheta_h,\tau_h)}{\|\tau_h\|_H} \geqslant \widetilde{\alpha}_{\mathtt{d}}\,\|\vartheta_h\|_H \qquad \forall\,\vartheta_h\in V_h,$$

or else

$$\sup_{\substack{\vartheta_h\in V_h\\ \vartheta_h\neq 0}} \frac{a(\vartheta_h,\tau_h)}{\|\vartheta_h\|_H} \geqslant \widetilde{\alpha}_{\mathtt{d}}\,\|\tau_h\|_H \qquad \forall\,\tau_h\in V_h,$$

iii) *and there exists a constant $\widetilde{\beta}_{\mathtt{d}}>0$, independent of h, such that*

$$\sup_{\substack{\tau_h\in H_h\\ \tau_h\neq 0}} \frac{b(\tau_h,v_h)}{\|\tau_h\|_H} \geqslant \widetilde{\beta}_{\mathtt{d}}\,\|v_h\|_Q \qquad \forall\, v_h\in Q_h.$$

Then, for each pair $(f, g) \in H' \times Q'$ there exists a unique solution $(\sigma_h, u_h) \in H_h \times Q_h$ of (5.119). Moreover, there exists a constant $\widetilde{C}_{\mathtt{d}} > 0$, depending only on $\|a\|$, $\|c\|$, $\widetilde{\alpha}_{\mathtt{d}}$, and $\widetilde{\beta}_{\mathtt{d}}$, such that

$$\|\sigma_h\|_H + \|u_h\|_Q \leqslant \widetilde{C}_{\mathtt{d}} \left\{ \|f\|_{H'} + \|g\|_{Q'} \right\}.$$

Proof. It follows analogously to the proof of Theorem 5.17, bearing in mind that, instead of (5.109), we apply (5.127). Thus, given $(\zeta_h, w_h) \in H_h \times Q_h$, we deduce the existence of $\bar{\zeta}_h \in H_h$ such that

$$\mathbf{B}_h(\bar{\zeta}_h) = \mathbf{B}_h(\zeta_h) \qquad \text{and} \qquad \|\bar{\zeta}_h\|_H \leqslant \frac{1}{\widetilde{\beta}_{\mathtt{d}}} \|\mathbf{B}_h(\zeta_h)\|_{Q'} = \frac{1}{\widetilde{\beta}_{\mathtt{d}}} \|\mathbf{B}_h(\bar{\zeta}_h)\|_{Q'},$$

so that ζ_h can be decomposed as

$$\zeta_h = \zeta_{0,h} + \bar{\zeta}_h,$$

with $\zeta_{0,h} := \zeta_h - \bar{\zeta}_h \in V_h$. The rest of the proof is as in Theorem 5.17. Additional details are omitted.

□

We conclude this section with the corresponding Céa estimate, whose proof is exactly the same as that of Theorem 5.19.

Theorem 5.21. *Assume the hypotheses of Theorems* 5.17 *and* 5.20, *and let $(\sigma, u) \in H \times Q$ and $(\sigma_h, u_h) \in H_h \times Q_h$ be the unique solutions of* (5.60) *and* (5.119), *respectively. Then, there exists a constant $\widehat{C}_{\mathtt{d}} > 0$, depending only on $\|a\|$, $\|b\|$, $\|c\|$, $\widetilde{\alpha}_{\mathtt{d}}$, and $\widetilde{\beta}_{\mathtt{d}}$, such that*

$$\|\sigma - \sigma_h\|_H + \|u - u_h\|_Q \leqslant \widehat{C}_{\mathtt{d}} \left\{ \operatorname{dist}(\sigma, H_h) + \operatorname{dist}(u, Q_h) \right\}.$$

5.3.4 An application example

In this section, we consider a problem arising from fluid mechanics to illustrate the applicability of the abstract theory developed in Section 5.3.2.

5.3.4.1 The model

The coupling of the Stokes and Poisson–Nernst–Planck equations is an electrodynamic model that describes the steady flow of an incompressible Newtonian fluid occupying a domain $\Omega \subseteq \mathbb{R}^n$, $n \in \{2, 3\}$, with polygonal (resp. polyhedral) boundary Γ in $\mathbb{R}^2$ (resp. $\mathbb{R}^3$) (see, e.g. [93], [94]). The dynamics of the model are determined by the concentration of ionized particles ξ_1 and ξ_2, the electric current field $\boldsymbol{\varphi}$, and the fluid velocity $\mathbf{u}$ and pressure p. In particular, knowing the vector fields $\boldsymbol{\varphi}$ and $\mathbf{u}$, a simplified version of the decoupled Nernst–Planck equation for a specific ionic species, whose concentration is denoted by ξ, and for which the dielectric and diffusion coefficients are assumed equal to 1, is expressed in mixed form as

$$\begin{gathered} \boldsymbol{\sigma} = \nabla \xi + \xi (\boldsymbol{\varphi} - \mathbf{u}) \quad \text{in} \quad \Omega, \\ \xi - \operatorname{div}(\boldsymbol{\sigma}) = f \quad \text{in} \quad \Omega, \quad \xi = g \quad \text{on} \quad \Gamma, \end{gathered} \tag{5.128}$$

where f and g are given data belonging to suitable function spaces. Regarding the latter, we adopt the standard notation for Lebesgue spaces $\mathrm{L}^t(\Omega)$, with $t \in (1, +\infty)$, and for Sobolev spaces $\mathrm{H}^m(\Omega)$ (cf. Exercise 1.10, eq. (1.37)) and $\mathrm{H}_0^m(\Omega)$, with integer $m \geqslant 0$, whose corresponding norms and seminorms (in the case of the latter), for both scalar and vector versions, are denoted by $\|\cdot\|_{0,t;\Omega}$, $\|\cdot\|_{m,\Omega}$, and $|\cdot|_{m,\Omega}$, respectively. Indeed, $\mathrm{H}_0^m(\Omega)$ is the closure of $C_0^\infty(\Omega)$ in $\mathrm{H}^m(\Omega)$ with respect to $\|\cdot\|_{m,\Omega}$. On the other hand, we recall from Sections 4.5.2 and 4.5.4 that $\mathrm{H}^{1/2}(\Gamma)$ and $\mathrm{H}^{-1/2}(\Gamma)$ are the trace spaces of $\mathrm{H}^1(\Omega)$ and its dual, with norms $\|\cdot\|_{1/2,\Gamma}$ and $\|\cdot\|_{-1/2,\Gamma}$, respectively, and that $\langle\cdot,\cdot\rangle$ represents the corresponding duality pairing. In turn, given any scalar functional space S, we denote by **S** its vector-valued counterpart.

5.3.4.2 The variational formulation

To derive the variational formulation of (5.128), we note that the appropriate spaces where the unknowns will be sought are mainly determined by the term depending on φ and $\mathbf{u}$. Indeed, using Hölder's and Cauchy–Schwarz inequalities, we observe that

$$\left|\int_\Omega \xi\,(\varphi - \mathbf{u})\cdot\boldsymbol{\tau}\right| \leqslant \|\xi\|_{0,\rho;\Omega}\left(\|\varphi\|_{0,r;\Omega} + \|\mathbf{u}\|_{0,r;\Omega}\right)\|\boldsymbol{\tau}\|_{0,\Omega}\,, \tag{5.129}$$

for all $\xi \in \mathrm{L}^\rho(\Omega)$, φ, $\mathbf{u} \in \mathbf{L}^r(\Omega)$, and $\boldsymbol{\tau} \in \mathbf{L}^2(\Omega)$, where $\rho = 2\ell$ and $r = 2j$, with ℓ, $j \in (1,+\infty)$ being conjugate to each other, i.e., such that $\frac{1}{\ell} + \frac{1}{j} = 1$. Then, denoting by $\varrho \in (1,+\infty)$ the conjugate of ρ, we introduce the Banach space

$$\mathbf{H}(\mathrm{div}_\varrho;\Omega) := \left\{\boldsymbol{\tau} \in \mathbf{L}^2(\Omega) : \quad \mathrm{div}(\boldsymbol{\tau}) \in \mathrm{L}^\varrho(\Omega)\right\}, \tag{5.130}$$

which is endowed with the norm

$$\|\boldsymbol{\tau}\|_{\mathrm{div}_\varrho;\Omega} := \|\boldsymbol{\tau}\|_{0,\Omega} + \|\mathrm{div}(\boldsymbol{\tau})\|_{0,\varrho;\Omega} \qquad \forall\,\boldsymbol{\tau} \in \mathbf{H}(\mathrm{div}_\varrho;\Omega)\,, \tag{5.131}$$

and we observe from Exercise 5.2 (see also [42, Section 3.1], [29, Section 4.1], or [79, eq. (2.11)]) that for

$$\varrho \in \begin{cases} (1,+\infty] & \text{if } n = 2 \\ [6/5,+\infty] & \text{if } n = 3 \end{cases}, \tag{5.132}$$

one has

$$\langle \boldsymbol{\tau}\cdot\boldsymbol{\nu}, v\rangle = \int_\Omega \left\{\boldsymbol{\tau}\cdot\nabla v + v\,\mathrm{div}(\boldsymbol{\tau})\right\} \qquad \forall\,(\boldsymbol{\tau}, v) \in \mathbf{H}(\mathrm{div}_\varrho;\Omega)\times \mathrm{H}^1(\Omega)\,, \tag{5.133}$$

where $\boldsymbol{\nu}$ denotes the unit normal vector to Γ. Note that the integration-by-parts formula (5.133) implicitly establishes that $\boldsymbol{\tau}\cdot\boldsymbol{\nu} \in \mathrm{H}^{-1/2}(\Gamma)$ for each $\boldsymbol{\tau} \in \mathbf{H}(\mathrm{div}_\varrho;\Omega)$.

Moreover, since $\rho = 2\ell > 2$, it follows that $\varrho \in (1,2)$, and therefore the feasible range for ϱ is reduced to $(1,2)$ and $[6/5,2)$ in 2D and 3D, respectively. Then, testing the first equation of (5.128) against $\boldsymbol{\tau} \in \mathbf{H}(\mathrm{div}_\varrho;\Omega)$, and then applying (5.133) with $v = \xi$, which requires assuming that originally $\xi \in \mathrm{H}^1(\Omega)$ and $g \in \mathrm{H}^{1/2}(\Gamma)$, we obtain

$$\int_\Omega \boldsymbol{\sigma}\cdot\boldsymbol{\tau} + \int_\Omega \xi\,\mathrm{div}(\boldsymbol{\tau}) - \int_\Omega \xi\,(\varphi - \mathbf{u})\cdot\boldsymbol{\tau} = \langle\boldsymbol{\tau}\cdot\boldsymbol{\nu}, g\rangle\,. \tag{5.134}$$

In turn, assuming that $f \in \mathrm{L}^\varrho(\Omega)$, and testing the second equation of (5.128) against $\eta \in \mathrm{L}^\rho(\Omega)$, we have

$$\int_\Omega \eta\,\mathrm{div}(\boldsymbol{\sigma}) - \int_\Omega \xi\,\eta = -\int_\Omega f\,\eta\,. \tag{5.135}$$

Thus, combining (5.134) and (5.135), we arrive at the following mixed variational formulation of (5.128): Find $(\boldsymbol{\sigma}, \xi) \in H \times Q$ such that

$$\begin{aligned} a(\boldsymbol{\sigma}, \boldsymbol{\tau}) + b(\boldsymbol{\tau}, \xi) - \int_\Omega \xi\,(\boldsymbol{\varphi} - \mathbf{u}) \cdot \boldsymbol{\tau} &= F(\boldsymbol{\tau}) \quad \forall\, \boldsymbol{\tau} \in H\,, \\ b(\boldsymbol{\sigma}, \eta) - c(\xi, \eta) &= G(\eta) \quad \forall\, \eta \in Q\,, \end{aligned} \tag{5.136}$$

where

$$H := \mathbf{H}(\mathrm{div}_\varrho; \Omega)\,, \qquad Q := \mathrm{L}^\rho(\Omega)\,,$$

and the bilinear forms $a : H \times H \to \mathbb{R}$, $b : H \times Q \to \mathbb{R}$, and $c : Q \times Q \to \mathbb{R}$, and the functionals $F : H \to \mathbb{R}$ and $G : Q \to \mathbb{R}$, are defined, respectively, by

$$a(\boldsymbol{\zeta}, \boldsymbol{\tau}) := \int_\Omega \boldsymbol{\zeta} \cdot \boldsymbol{\tau} \qquad \forall\, (\boldsymbol{\zeta}, \boldsymbol{\tau}) \in H \times H\,, \tag{5.137}$$

$$b(\boldsymbol{\tau}, \eta) := \int_\Omega \eta\, \mathrm{div}(\boldsymbol{\tau}) \qquad \forall\, (\boldsymbol{\tau}, \eta) \in H \times Q\,, \tag{5.138}$$

$$c(\lambda, \eta) := \int_\Omega \lambda\, \eta \qquad \forall\, (\lambda, \eta) \in Q \times Q\,, \tag{5.139}$$

$$F(\boldsymbol{\tau}) := \langle \boldsymbol{\tau} \cdot \boldsymbol{\nu}, g \rangle \qquad \forall\, \boldsymbol{\tau} \in H\,, \tag{5.140}$$

and

$$G(\eta) := -\int_\Omega f\, \eta \qquad \forall\, \eta \in Q\,. \tag{5.141}$$

Equivalently, introducing the bilinear forms $\mathbf{A}$, $\mathbf{A}_{\boldsymbol{\varphi},\mathbf{u}} : (H \times Q) \times (H \times Q) \to \mathbb{R}$ given by

$$\mathbf{A}\big((\boldsymbol{\zeta}, \lambda), (\boldsymbol{\tau}, \eta)\big) := a(\boldsymbol{\zeta}, \boldsymbol{\tau}) + b(\boldsymbol{\tau}, \lambda) + b(\boldsymbol{\zeta}, \eta) - c(\lambda, \eta)\,, \tag{5.142}$$

and

$$\mathbf{A}_{\boldsymbol{\varphi},\mathbf{u}}\big((\boldsymbol{\zeta}, \lambda), (\boldsymbol{\tau}, \eta)\big) := \mathbf{A}\big((\boldsymbol{\zeta}, \lambda), (\boldsymbol{\tau}, \eta)\big) - \int_\Omega \lambda\,(\boldsymbol{\varphi} - \mathbf{u}) \cdot \boldsymbol{\tau}\,, \tag{5.143}$$

for all $(\boldsymbol{\zeta}, \lambda)$, $(\boldsymbol{\tau}, \eta) \in H \times Q$, we deduce that (5.136) can be rewritten as: Find $(\boldsymbol{\sigma}, \xi) \in H \times Q$ such that

$$\mathbf{A}_{\boldsymbol{\varphi},\mathbf{u}}\big((\boldsymbol{\sigma}, \xi), (\boldsymbol{\tau}, \eta)\big) = F(\boldsymbol{\tau}) + G(\eta) \qquad \forall\, (\boldsymbol{\tau}, \eta) \in H \times Q\,. \tag{5.144}$$

According to the above, in what follows we show that the bilinear forms that are part of $\mathbf{A}$ satisfy the hypotheses of Theorem 5.17. Subsequently, we combine this fact with the effect produced by the extra term that completes the definition of $\mathbf{A}_{\boldsymbol{\varphi},\mathbf{u}}$, so that, by applying Lemma 5.1, we conclude the solvability of (5.136) (or (5.144)).

5.3.4.3 Solvability analysis

We begin by observing that the reflexivity of $\mathbf{L}^2(\Omega)$, $\mathrm{L}^\varrho(\Omega)$, and $\mathrm{L}^\rho(\Omega)$ implies that both H and Q are reflexive Banach spaces. Furthermore, direct applications of the Hölder and Cauchy-Schwarz inequalities show that a, b, and c are all bounded with respective constants $\|a\| \leqslant 1$, $\|b\| \leqslant 1$, and $\|c\| \leqslant |\Omega|^{(\rho-2)/\rho}$. It is also clear from (5.137) and (5.139) that a and c are symmetric and positive semi-definite (assumption i) of Theorem 5.17). Moreover, recalling the definitions of b (cf. (5.138)) and of the null space V of the operator $\mathbf{B}$ induced by b (cf. (5.63)), we find that

$$V \;=\; \big\{ \boldsymbol{\tau} \in \mathbf{H}(\mathrm{div}_\varrho;\Omega) : \quad \mathrm{div}(\boldsymbol{\tau}) = 0 \big\}\,, \tag{5.145}$$

and therefore

$$a(\boldsymbol{\tau},\boldsymbol{\tau}) \;=\; \|\boldsymbol{\tau}\|_{0,\Omega}^2 \;=\; \|\boldsymbol{\tau}\|_{\mathrm{div}_\varrho;\Omega}^2 \qquad \forall\, \boldsymbol{\tau} \in V\,, \tag{5.146}$$

from which it easily follows that a satisfies the continuous inf-sup condition (5.110) (and hence (5.111) as well because of its symmetry) with constant $\widetilde{\alpha} = 1$ (assumption ii) of Theorem 5.17). It remains to show that b satisfies the continuous inf-sup condition (5.112) (assumption iii) of Theorem 5.17), for which we appeal to the corresponding result available in [79, Lemma 2.9] and some of its references. More precisely, we have the following lemma.

Lemma 5.22. *There exists a constant $\widetilde{\beta} > 0$, depending only on Ω and ρ, such that*

$$\sup_{\substack{\boldsymbol{\tau}\in H\\ \boldsymbol{\tau}\neq 0}} \frac{b(\boldsymbol{\tau},\eta)}{\|\boldsymbol{\tau}\|_H} \;\geqslant\; \widetilde{\beta}\,\|\eta\|_Q \qquad \forall\, \eta \in Q\,. \tag{5.147}$$

Proof. Given $\eta \in Q := \mathrm{L}^\rho(\Omega)$, we define $\eta_\varrho := |\eta|^{\rho-2}\,\eta$ and observe, by simple algebraic calculations, that $\eta_\varrho \in \mathrm{L}^\varrho(\Omega)$ and

$$\int_\Omega \eta\,\eta_\varrho \;=\; \|\eta\|_{0,\rho;\Omega}\,\|\eta_\varrho\|_{0,\varrho;\Omega}\,. \tag{5.148}$$

Next, let $\widetilde{\boldsymbol{\tau}} := \nabla z \in \mathbf{L}^2(\Omega)$, where $z \in \mathrm{H}_0^1(\Omega)$ is the unique solution of the variational formulation

$$\int_\Omega \nabla z\cdot\nabla w \;=\; -\int_\Omega \eta_\varrho\, w \qquad \forall\, w \in \mathrm{H}_0^1(\Omega)\,. \tag{5.149}$$

Note that the Hölder inequality and the continuous embedding $i_\rho : \mathrm{H}^1(\Omega) \to \mathrm{L}^\rho(\Omega)$ (cf. [111, Theorem 1.3.4]) guarantee that the right-hand side of (5.149) defines a functional in $\mathrm{H}_0^1(\Omega)'$, so that the Lax-Milgram Lemma (cf. Lemma 4.4) ensures the unique solvability of this problem. It follows from (5.149) that

$$\mathrm{div}(\widetilde{\boldsymbol{\tau}}) \;=\; \eta_\varrho \quad \text{in} \quad \Omega\,, \tag{5.150}$$

which shows that $\widetilde{\boldsymbol{\tau}} \in \mathbf{H}(\mathrm{div}_\varrho;\Omega)$. Moreover, according to the continuous dependence result for (5.149) and the resulting estimate for the aforementioned functional, we deduce the existence of a constant $c_\rho > 0$, depending on $\|i_\rho\|$, such that $\|z\|_{1,\Omega} \leqslant c_\rho\,\|\eta_\varrho\|_{0,\varrho;\Omega}$, and therefore

$$\|\widetilde{\boldsymbol{\tau}}\|_{\mathrm{div}_\varrho;\Omega} \;=\; |z|_{1,\Omega} + \|\eta_\varrho\|_{0,\varrho;\Omega} \;\leqslant\; \big(1 + c_\rho\big)\,\|\eta_\varrho\|_{0,\varrho;\Omega}\,. \tag{5.151}$$

Finally, invoking the definition of b (cf. (5.138)), and using (5.150), (5.148), and (5.151), we obtain

$$\sup_{\substack{\boldsymbol{\tau}\in H\\ \boldsymbol{\tau}\neq 0}} \frac{b(\boldsymbol{\tau},\eta)}{\|\boldsymbol{\tau}\|_H} \;\geqslant\; \frac{b(\widetilde{\boldsymbol{\tau}},\eta)}{\|\widetilde{\boldsymbol{\tau}}\|_H} \;=\; \frac{\|\eta\|_{0,\rho;\Omega}\,\|\eta_\varrho\|_{0,\varrho;\Omega}}{\|\widetilde{\boldsymbol{\tau}}\|_H} \;\geqslant\; \widetilde{\beta}\,\|\eta\|_{0,\rho;\Omega}\,,$$

with $\widetilde{\beta} := \big(1 + c_\rho\big)^{-1}$, thereby proving the continuous inf-sup condition (5.147).

□

Now, knowing that a, b, and c satisfy the hypotheses of Theorem 5.17, it follows that the bilinear form $\mathbf{A}$ (cf. (5.142)) satisfies the global inf-sup condition (5.106), which in this case means that there exists a constant $\widehat{c} > 0$, depending only on $\|a\|$ ($\leqslant 1$), $\|c\|$ ($\leqslant |\Omega|^{(\rho-2)/\rho}$), $\widetilde{\alpha} = 1$, and $\widetilde{\beta} = \big(1 + c_\rho\big)^{-1}$, such that

$$\sup_{\substack{(\boldsymbol{\tau},\eta)\neq 0\\(\boldsymbol{\tau},\eta)\in H\times Q}} \frac{\mathbf{A}\big((\boldsymbol{\zeta},\lambda),(\boldsymbol{\tau},\eta)\big)}{\|(\boldsymbol{\tau},\eta)\|_{H\times Q}} \geqslant \widehat{c}\,\|(\boldsymbol{\zeta},\lambda)\|_{H\times Q} \qquad \forall\,(\boldsymbol{\zeta},\lambda)\in H\times Q\,. \tag{5.152}$$

Therefore, it follows easily from (5.143), (5.129), and (5.152) that

$$\sup_{\substack{(\boldsymbol{\tau},\eta)\neq 0\\(\boldsymbol{\tau},\eta)\in H\times Q}} \frac{\mathbf{A}_{\boldsymbol{\varphi},\mathbf{u}}\big((\boldsymbol{\zeta},\lambda),(\boldsymbol{\tau},\eta)\big)}{\|(\boldsymbol{\tau},\eta)\|_{H\times Q}} \geqslant \Big\{\widehat{c} - \big(\|\boldsymbol{\varphi}\|_{0,r;\Omega} + \|\mathbf{u}\|_{0,r;\Omega}\big)\Big\}\,\|(\boldsymbol{\zeta},\lambda)\|_{H\times Q}$$

for all $(\boldsymbol{\zeta},\lambda)\in H\times Q$, from which, assuming that

$$\|\boldsymbol{\varphi}\|_{0,r;\Omega} + \|\mathbf{u}\|_{0,r;\Omega} \leqslant \frac{\widehat{c}}{2}\,, \tag{5.153}$$

we conclude that

$$\sup_{\substack{(\boldsymbol{\tau},\eta)\neq 0\\(\boldsymbol{\tau},\eta)\in H\times Q}} \frac{\mathbf{A}_{\boldsymbol{\varphi},\mathbf{u}}\big((\boldsymbol{\zeta},\lambda),(\boldsymbol{\tau},\eta)\big)}{\|(\boldsymbol{\tau},\eta)\|_{H\times Q}} \geqslant \frac{\widehat{c}}{2}\,\|(\boldsymbol{\zeta},\lambda)\|_{H\times Q} \quad \forall\,(\boldsymbol{\zeta},\lambda)\in H\times Q\,. \tag{5.154}$$

Similarly, using the symmetry of $\mathbf{A}$ and (5.152), and under the same hypothesis on $\boldsymbol{\varphi}$ and $\mathbf{u}$, we obtain

$$\sup_{\substack{(\boldsymbol{\zeta},\lambda)\neq 0\\(\boldsymbol{\zeta},\lambda)\in H\times Q}} \frac{\mathbf{A}_{\boldsymbol{\varphi},\mathbf{u}}\big((\boldsymbol{\zeta},\lambda),(\boldsymbol{\tau},\eta)\big)}{\|(\boldsymbol{\zeta},\lambda)\|_{H\times Q}} \geqslant \frac{\widehat{c}}{2}\,\|(\boldsymbol{\tau},\eta)\|_{H\times Q} \qquad \forall\,(\boldsymbol{\tau},\eta)\in H\times Q\,. \tag{5.155}$$

On the other hand, recalling from the proof of Lemma 5.22 that i_ρ is the continuous embedding of $\mathrm{H}^1(\Omega)$ into $\mathrm{L}^\rho(\Omega)$, it is easy to see from (5.133) that there exists a constant $C_\rho > 0$, depending on $\|i_\rho\|$, such that $\|\boldsymbol{\tau}\cdot\boldsymbol{\nu}\|_{-1/2,\Gamma} \leqslant C_\rho\,\|\boldsymbol{\tau}\|_{\mathrm{div}_\varrho;\Omega}$ for all $\boldsymbol{\tau}\in\mathbf{H}(\mathrm{div}_\varrho;\Omega)$. Hence, from (5.140) we deduce that $F\in H'$ with $\|F\|_{H'} \leqslant C_\rho\,\|g\|_{-1/2,\Gamma}$. In turn, (5.141) and the Hölder inequality imply that $G\in Q'$ with $\|G\|_{Q'} \leqslant \|f\|_{0,\varrho;\Omega}$.

Consequently, we are now in position to establish that (5.136) (equivalently (5.144)) is well-posed.

Theorem 5.23. *Let* $\boldsymbol{\varphi}$, $\mathbf{u}\in\mathbf{L}^r(\Omega)$ *be such that* (5.153) *is satisfied. Then, there exists a unique solution* $(\boldsymbol{\sigma},\xi)\in H\times Q$ *of* (5.136), *and*

$$\|\boldsymbol{\sigma}\|_{\mathrm{div}_\varrho;\Omega} + \|\xi\|_{0,\rho;\Omega} \leqslant \frac{2}{\widehat{c}}\,\max\big\{1, C_\rho\big\}\Big\{\|g\|_{-1/2,\Gamma} + \|f\|_{0,\varrho;\Omega}\Big\}\,.$$

Proof. Thanks to (5.154), (5.155), and the boundedness of F and G, the proof follows directly from an application of Lemma 5.2.

□

We conclude this section by emphasizing that the continuous and discrete analyses of the complete model given by the coupling of the Stokes and Poisson–Nernst–Planck equations, including of course that of (5.136), can be found in [46]. In turn, for the extension to the case of coupling with the Navier–Stokes equations, we refer to [45].

5.4 Perturbed saddle-point problems: a variant

In this section, we present a variant of Theorem 5.17, in which the symmetry of the perturbation c is not required, but the bilinear form a must be elliptic in the whole space H. More precisely, we have the following result.

Theorem 5.24. *Let H and Q be reflexive Banach spaces, and let $a : H \times H \to \mathbb{R}$, $b : H \times Q \to \mathbb{R}$, and $c : Q \times Q \to \mathbb{R}$ be bounded bilinear forms with respective constants denoted by $\|a\|$, $\|b\|$, and $\|c\|$, respectively. Suppose that*

i) *c is positive semi-definite, that is, $c(v,v) \geqslant 0$ for all $v \in Q$,*
ii) *a is H-elliptic, that is, there exists a positive constant α such that*

$$a(\tau,\tau) \geqslant \alpha \|\tau\|_H^2 \quad \forall \tau \in H, \quad \textit{and}$$

iii) *b satisfies the inf-sup condition, that is, there exists a positive constant β such that*

$$\sup_{\substack{\tau\in H\\ \tau\neq 0}} \frac{b(\tau,v)}{\|\tau\|_H} \geqslant \beta \|v\|_Q \quad \forall v \in Q .$$

Then, for each pair $(F,G) \in H' \times Q'$ there exists a unique $(\sigma,u) \in H \times Q$ such that

$$\begin{aligned} a(\sigma,\tau) + b(\tau,u) &= F(\tau) \quad \forall \tau \in H, \\ b(\sigma,v) - c(u,v) &= G(v) \quad \forall v \in Q. \end{aligned} \tag{5.156}$$

Moreover, there exists a positive constant C, depending only on $\|a\|$, $\|b\|$, α, and β, such that

$$\|\sigma\|_H + \|u\|_Q \leqslant C\big\{\|F\|_{H'} + \|G\|_{Q'}\big\}. \tag{5.157}$$

Proof. This result reduces to a natural simplification of the analysis developed in [2, Section 3] for a nonlinear version of (5.156). Indeed, for the existence of a solution we first observe, thanks to ii) and the Banach–Nečas–Babuška Theorem (cf. Lemma 5.1), that there exists a unique $\sigma_0 \in H$ such that

$$a(\sigma_0,\tau) = F(\tau) \qquad \forall \tau \in H, \tag{5.158}$$

and that for each $w \in Q$ there exists a unique $\sigma_w \in H$ such that

$$a(\sigma_w,\tau) = -\,b(\tau,w) \qquad \forall \tau \in H. \tag{5.159}$$

In turn, the corresponding a priori estimates are given by

$$\|\sigma_0\|_H \leqslant \frac{1}{\alpha}\|F\|_{H'} \quad \text{and} \quad \|\sigma_w\|_H \leqslant \frac{\|b\|}{\alpha}\|w\|_Q \qquad \forall w \in Q. \tag{5.160}$$

Next, using iii) and (5.159), we obtain for each $w \in Q$ that

$$\beta\|w\|_Q \leqslant \sup_{\substack{\tau\in H\\ \tau\neq 0}} \frac{b(\tau,w)}{\|\tau\|_H} = \sup_{\substack{\tau\in H\\ \tau\neq 0}} \frac{a(\sigma_w,\tau)}{\|\tau\|_H},$$

from which it follows that

$$\frac{\beta}{\|a\|}\|w\|_Q \leqslant \|\sigma_w\|_H \qquad \forall w \in Q. \tag{5.161}$$

Now, noting from (5.159) that σ_w depends linearly on w, we can introduce the bilinear form

$$\Theta(w,v) := c(w,v) - b(\sigma_w, v) \qquad \forall\, w,\, v \in Q\,,$$

which is clearly bounded due to the same property of c and b, and to the second estimate in (5.160). Moreover, according to (5.159), i), ii), and (5.161), we deduce that for each $v \in Q$ we have

$$\Theta(v,v) \;=\; c(v,v) - b(\sigma_v, v) \;=\; c(v,v) + a(\sigma_v,\sigma_v) \;\geqslant\; \alpha\, \|\sigma_v\|_H^2 \;\geqslant\; \frac{\alpha\,\beta^2}{\|a\|^2}\, \|v\|_Q^2\,,$$

which proves that Θ is Q-elliptic. Then, applying again Lemma 5.1, we conclude that there exists a unique $u \in Q$ such that

$$\Theta(u,v) \;=\; b(\sigma_0, v) - G(v) \qquad \forall\, v \in Q\,,$$

that is,

$$c(u,v) - b(\sigma_u, v) \;=\; b(\sigma_0, v) - G(v) \qquad \forall\, v \in Q\,,$$

or equivalently,

$$b(\sigma_0 + \sigma_u, v) - c(u,v) \;=\; G(v) \qquad \forall\, v \in Q\,. \tag{5.162}$$

Then, defining $\sigma := \sigma_0 + \sigma_u \in H$, it follows from (5.158) and (5.159) that

$$a(\sigma,\tau) \;=\; a(\sigma_0,\tau) + a(\sigma_u,\tau) \;=\; F(\tau) - b(\tau,u)\,,$$

that is,

$$a(\sigma,\tau) + b(\tau,u) \;=\; F(\tau) \qquad \forall\, \tau \in H\,,$$

which, together with (5.162), shows that $(\sigma,u) \in H \times Q$ is a solution of (5.156). In turn, the a priori estimate for u becomes

$$\|u\|_Q \;\leqslant\; \frac{\|a\|^2}{\alpha\,\beta^2}\,\big\{\|b\|\,\|\sigma_0\|_H + \|G\|_{Q'}\big\}\,,$$

and using the first inequality in (5.160), it becomes

$$\|u\|_Q \;\leqslant\; \frac{\|a\|^2\,\|b\|}{\alpha^2\,\beta^2}\,\|F\|_{H'} + \frac{\|a\|^2}{\alpha\,\beta^2}\,\|G\|_{Q'}\,, \tag{5.163}$$

while, using the estimates from (5.160) and (5.163), we obtain

$$\|\sigma\|_H \;\leqslant\; \frac{1}{\alpha}\left(1 + \frac{\|a\|^2\,\|b\|^2}{\alpha^2\,\beta^2}\right)\|F\|_{H'} + \frac{\|a\|^2\,\|b\|}{\alpha^2\,\beta^2}\,\|G\|_{Q'}\,. \tag{5.164}$$

Having proved the existence of a solution (σ,u) to (5.156) that satisfies (5.163) and (5.164), it only remains to show uniqueness. To this end, let $(\tilde{\sigma},\tilde{u}) \in H \times Q$ be such that

$$\begin{aligned} a(\tilde{\sigma},\tau) + b(\tau,\tilde{u}) &= 0 \quad \forall\, \tau \in H\,,\\ b(\tilde{\sigma},v) - c(\tilde{u},v) &= 0 \quad \forall\, v \in Q\,. \end{aligned} \tag{5.165}$$

Taking $\tau = \tilde{\sigma}$ and $v = \tilde{u}$ in (5.165), and then subtracting the resulting equations and using ii), we obtain

$$0 = a(\tilde{\sigma},\tilde{\sigma}) + c(\tilde{u},\tilde{u}) \;\geqslant\; \alpha\,\|\tilde{\sigma}\|_H^2\,,$$

whence $\tilde{\sigma} = 0$. On the other hand, it is clear from the first row of (5.165) and from (5.159) that $\tilde{\sigma}_{\tilde{u}} = \tilde{\sigma} = 0$, so that, by virtue of (5.161), we conclude that $\tilde{u} = 0$, thus confirming the uniqueness of the solution to (5.156). Finally, (5.163) and (5.164) imply (5.157), which completes the proof.

□

5.5 Babuška–Brezzi Theory: general case

Here we extend the results of Section 5.2 to the general case of variational formulations of the Babuška–Brezzi type, in which three bilinear forms are involved and the spaces of the unknowns and test functions do not necessarily coincide. Although the logical sequence of the analysis is basically the same as in the mentioned section, all the corresponding details are provided below for the sake of clarity.

5.5.1 The operator equation

Let $(X_2, \|\cdot\|_{X_2})$, $(M_1, \|\cdot\|_{M_1})$, $(X_1, \|\cdot\|_{X_1})$, and $(M_2, \|\cdot\|_{M_2})$ be Banach spaces, and let $a : X_2 \times X_1 \to \mathbb{R}$ and $b_i : X_i \times M_i \to \mathbb{R}$, $i \in \{1, 2\}$, be bounded bilinear forms. Then, given $F \in X_1'$ and $G \in M_2'$, we consider the problem: Find $(\sigma, u) \in X_2 \times M_1$ such that

$$\begin{aligned} a(\sigma,\tau) \,+\, b_1(\tau,u) &= F(\tau) \qquad \forall\, \tau \,\in\, X_1\,, \\ b_2(\sigma,v) &= G(v) \qquad \forall\, v \,\in\, M_2\,. \end{aligned} \tag{5.166}$$

Similarly to (5.33) and (5.34), we denote by $\mathcal{A} \in \mathcal{L}(X_2, X_1')$ and $\mathcal{B}_i \in \mathcal{L}(X_i, M_i')$ the linear and bounded operators induced by a and b_i, $i \in \{1, 2\}$, respectively, so that, in particular, $\mathcal{B}_1' \in \mathcal{L}(M_1'', X_1')$ is defined as

$$\mathcal{B}_1'(\mathcal{G})(\tau) \,:=\, \mathcal{G}(\mathcal{B}_1(\tau)) \qquad \forall\, \mathcal{G} \in M_1'', \quad \forall\, \tau \in X_1\,.$$

Now, if $\mathcal{J}_1 : M_1 \to M_1''$ is the injective operator defined as $\mathcal{J}_1(v)(G) := G(v)$ for all $G \in M_1'$, for all $v \in M_1$, we have that

$$\mathcal{B}_1'(\mathcal{J}_1(u))(\tau) \,:=\, \mathcal{J}_1(u)(\mathcal{B}_1(\tau)) \,=\, \mathcal{B}_1(\tau)(u) \,=\, b_1(\tau, u)\,,$$

and, therefore, (5.166) can be rewritten equivalently as: Find $(\sigma, u) \in X_2 \times M_1$ such that

$$\begin{aligned} \mathcal{A}(\sigma) \,+\, \mathcal{B}_1'(\mathcal{J}_1(u)) &= F\,, \\ \mathcal{B}_2(\sigma) &= G\,. \end{aligned} \tag{5.167}$$

Moreover, assuming from now on that M_1 is reflexive, which means that the operator $\mathcal{J}_1 : M_1 \to M_1''$ is bijective, problem (5.167) can be reformulated as: Find $(\sigma, \mathcal{G}) \in X_2 \times M_1''$ such that

$$\begin{aligned} \mathcal{A}(\sigma) \,+\, \mathcal{B}_1'(\mathcal{G}) &= F\,, \\ \mathcal{B}_2(\sigma) &= G\,, \end{aligned} \tag{5.168}$$

and then define $u := \mathcal{J}_1^{-1}(\mathcal{G})$. Similarly to what was done in Section 5.2.2, the solvability of operator equations of the type given by (5.168) is analyzed below.

5.5.2 A preliminary result

Let now $(Z_2, \|\cdot\|_{Z_2})$, $(Z_1, \|\cdot\|_{Z_1})$, $(Y_2, \|\cdot\|_{Y_2})$, and $(Y_1, \|\cdot\|_{Y_1})$ be real Banach spaces, and consider operators $\mathbf{A} \in \mathcal{L}(Z_2, Z_1')$ and $\mathbf{B}_i \in \mathcal{L}(Z_i, Y_i)$, $i \in \{1, 2\}$, so that $\mathbf{B}_i' \in \mathcal{L}(Y_i', Z_i')$. Then, given $F \in Z_1'$ and $y \in Y_2$, we are interested in the problem: Find $(\sigma, U) \in Z_2 \times Y_1'$ such that

$$\begin{aligned} \mathbf{A}(\sigma) \,+\, \mathbf{B}_1'(U) &= F\,, \\ \mathbf{B}_2(\sigma) &= y\,. \end{aligned} \tag{5.169}$$

It is easy to see that (5.168) is a particular case of (5.169) with the spaces

$$Z_2 = X_2\,, \quad Z_1 = X_1\,, \quad Y_2 = M_2'\,, \quad \text{and} \quad Y_1 = M_1'\,.$$

In turn, for each $i \in \{1,2\}$ we introduce the null space of $\mathbf{B}_i$, that is

$$K_i := N(\mathbf{B}_i) := \big\{\tau \in Z_i : \quad \mathbf{B}_i(\tau) = 0\big\}\,.$$

Moreover, we denote by $\Pi\mathbf{A} : K_2 \to K_1'$ the operator defined by

$$\big(\Pi\mathbf{A}\big)(\tau)(\zeta) := \mathbf{A}(\tau)(\zeta) \qquad \forall\,\tau \in K_2\,, \quad \forall\,\zeta \in K_1\,. \tag{5.170}$$

Then, necessary and sufficient conditions for problem (5.169) to be well-posed are detailed in the following theorem.

Theorem 5.25. *For each pair $(F, y) \in Z_1' \times Y_2$, problem (5.169) has a unique solution that depends continuously on (F, y) if and only if*

i) $\Pi\mathbf{A} : K_2 \to K_1'$ *is an isomorphism, and*
ii) *for each $i \in \{1,2\}$, $\mathbf{B}_i : Z_i \to Y_i$ is surjective.*

Proof. Assume first i) and ii). Then, given $(F, y) \in Z_1' \times Y_2$, the surjectivity of $\mathbf{B}_2$ (cf. ii)) and Lemma 3.65, part i), guarantee the existence of a constant β_2, so that for such y there exists $\sigma_y \in Z_2$ such that $\mathbf{B}_2(\sigma_y) = y$ and

$$\|\sigma_y\| \leqslant \frac{1}{\beta_2}\,\|y\|\,. \tag{5.171}$$

Then, since $\big(F - \mathbf{A}(\sigma_y)\big)|_{K_1} \in K_1'$, it follows from i) that there exists a unique $\sigma_0 \in K_2$ such that $\big(\Pi\mathbf{A}\big)(\sigma_0) = \big(F - \mathbf{A}(\sigma_y)\big)|_{K_1}$, that is, $\mathbf{A}(\sigma_0)(\tau) = F(\tau) - \mathbf{A}(\sigma_y)(\tau)$ for all $\tau \in K_1$, or equivalently $F - \mathbf{A}(\sigma) \in K_1^\circ$, where $\sigma := \sigma_0 + \sigma_y$. Moreover, thanks to the Bounded Inverse Theorem (cf. Theorem 3.34), we obtain

$$\|\sigma_0\| = \|\big(\Pi\mathbf{A}\big)^{-1}\big(F - \mathbf{A}(\sigma_y)\big)|_{K_1}\| \leqslant \|\big(\Pi\mathbf{A}\big)^{-1}\|\,\Big\{\|F\| + \|\mathbf{A}\|\,\|\sigma_y\|\Big\}\,. \tag{5.172}$$

Since $R(\mathbf{B}_1)$ is closed, $R(\mathbf{B}_1')$ is also closed, and therefore $R(\mathbf{B}_1') = K_1^\circ$, from which it follows that there exists $U \in Y_1'$ such that $\mathbf{B}_1'(U) = F - \mathbf{A}(\sigma)$. Thus, noting that $\mathbf{B}_2(\sigma) = \mathbf{B}_2(\sigma_y) = y$, it follows easily that $(\sigma, U) \in Z_2 \times Y_1'$ is a solution of (5.169). On the other hand, thanks to the surjectivity of $\mathbf{B}_1$ (cf. ii)) and Lemma 3.65, part ii), there exists a positive constant β_1 such that

$$\|U\| \leqslant \frac{1}{\beta_1}\,\|\mathbf{B}_1'(U)\| = \frac{1}{\beta_1}\,\|F - \mathbf{A}(\sigma)\| \leqslant \frac{1}{\beta_1}\,\Big\{\|F\| + \|\mathbf{A}\|\,\|\sigma\|\Big\}\,. \tag{5.173}$$

Thus, according to (5.171), (5.172), and (5.173), we arrive at

$$\|\sigma\| \leqslant \|(\Pi\mathbf{A})^{-1}\|\,\|F\| + \frac{1}{\beta_2}\Big(1 + \|(\Pi\mathbf{A})^{-1}\|\,\|A\|\Big)\|y\|\,, \tag{5.174}$$

and

$$\|U\| \leqslant \frac{1}{\beta_1}\Big(1 + \|(\Pi\mathbf{A})^{-1}\|\,\|A\|\Big)\|F\| + \frac{\|A\|}{\beta_1\,\beta_2}\Big(1 + \|(\Pi\mathbf{A})^{-1}\|\,\|A\|\Big)\|y\|\,, \tag{5.175}$$

which shows that the solution (σ, U) depends continuously on the data (F, y). To establish uniqueness, let $(\widetilde{\sigma}, \widetilde{U}) \in Z_2 \times Y_1'$ be a solution of the associated homogeneous system, that is

$$\begin{aligned} \mathbf{A}(\widetilde{\sigma}) + \mathbf{B}_1'(\widetilde{U}) &= 0 \in Z_1', \\ \mathbf{B}_2(\widetilde{\sigma}) \qquad\qquad &= 0 \in Y_2. \end{aligned} \tag{5.176}$$

It follows from the second equation that $\widetilde{\sigma} \in K_2$, and evaluating the first with $\tau \in K_1$, we obtain

$$0 = \mathbf{A}(\widetilde{\sigma})(\tau) + \mathbf{B}_1'(\widetilde{U})(\tau) = (\Pi\mathbf{A})(\widetilde{\sigma})(\tau) + \widetilde{U}(\mathbf{B}_1(\tau)) = \big(\Pi\mathbf{A}\big)(\widetilde{\sigma})(\tau),$$

which says that $\big(\Pi\mathbf{A}\big)(\widetilde{\sigma}) = 0 \in K_1'$. Then, thanks to hypothesis i), we deduce that $\widetilde{\sigma} = 0$, and, therefore, the first equation of (5.176) reduces to $\mathbf{B}_1'(\widetilde{U}) = 0 \in Z_1'$. Now noting that $N(\mathbf{B}_1') = R(\mathbf{B}_1)^\circ = Y_1^\circ = \{0\} \subseteq Y_1'$, we conclude that $\widetilde{U} = 0 \in Y_1'$, which completes the proof of sufficiency of i) and ii). Conversely, suppose that there exists a constant of continuous dependence $C_{\mathsf{cd}} > 0$ such that for each $(F, y) \in Z_1' \times Y_2$ there exists a unique $(\sigma, U) \in Z_2 \times Y_1'$ solution of (5.169) satisfying

$$\|\sigma\| + \|U\| \leqslant C_{\mathsf{cd}} \Big\{ \|F\| + \|y\| \Big\}.$$

Equivalently, we could say that the matrix operator induced by the left-hand side of (5.169), that is $\begin{pmatrix} \mathbf{A} & \mathbf{B}_1' \\ \mathbf{B}_2 & 0 \end{pmatrix}$, belongs to $\mathcal{L}\big(Z_2 \times Y_1', Z_1' \times Y_2\big)$ and is bijective, so that, according to the Bounded Inverse Theorem (cf. Theorem 3.34), its inverse is bounded as well with norm $\leqslant C_{\mathsf{d}}$. In particular, if F is the zero functional, it follows that for each $y \in Y_2$ there exists a unique $(\widehat{\sigma}, \widehat{U}) \in Z_2 \times Y_1'$ such that

$$\begin{aligned} \mathbf{A}(\widehat{\sigma}) + \mathbf{B}_1'(\widehat{U}) &= 0, \\ \mathbf{B}_2(\widehat{\sigma}) \qquad\qquad &= y, \end{aligned}$$

so that from the second equation above we deduce that $\mathbf{B}_2$ is surjective. In turn, given $U \in Y_1'$, we define $F := \mathbf{B}_1'(U) \in Z_1'$ and deduce that there exists a unique $(\bar{\sigma}, \bar{U}) \in Z_2 \times Y_1'$ such that

$$\begin{aligned} \mathbf{A}(\bar{\sigma}) + \mathbf{B}_1'(\bar{U}) &= \mathbf{B}_1'(U), \\ \mathbf{B}_2(\bar{\sigma}) \qquad\qquad &= 0, \end{aligned}$$

and

$$\|\bar{\sigma}\| + \|\bar{U}\| \leqslant C_{\mathsf{cd}} \|\mathbf{B}_1'(U)\|.$$

However, it is clear from the uniqueness of the solution to (5.169) that $\bar{\sigma} = 0$ and $\bar{U} = U$, so that the previous inequality reduces to

$$\|U\| \leqslant C_{\mathsf{cd}} \|\mathbf{B}_1'(U)\| \qquad \forall\, U \in Y_1',$$

which says that $\mathbf{B}_1'$ is injective with closed range, or equivalently, that $\mathbf{B}_1$ is surjective, thus completing the proof of ii). It remains to prove i), that is, that the operator $\Pi\mathbf{A}$: $K_2 \to K_1'$ is bijective. Indeed, for surjectivity, take $\widetilde{F}_0 \in K_1'$ and first observe, thanks to the Hahn–Banach Theorem (cf. Theorem 2.21), that there exists $\widetilde{F} \in Z_1'$ such that $\widetilde{F}|_{K_1} = \widetilde{F}_0$ and $\|\widetilde{F}\|_{Z_1'} = \|\widetilde{F}_0\|_{K_1'}$. Then, there exists a unique $(\widetilde{\sigma}, \widetilde{U}) \in Z_2 \times Y_1'$ such that

$$\begin{aligned} \mathbf{A}(\widetilde{\sigma}) + \mathbf{B}_1'(\widetilde{U}) &= \widetilde{F}\,, \\ \mathbf{B}_2(\widetilde{\sigma}) &= 0\,, \end{aligned}$$

from which it is clear that $\widetilde{\sigma} \in K_2$. Then, it follows easily from the first equation of the above system that $\widetilde{F}_0(\tau) = \widetilde{F}(\tau) = \mathbf{A}(\widetilde{\sigma})(\tau)$ for all $\tau \in K_1$, that is, $\big(\Pi\mathbf{A}\big)(\widetilde{\sigma}) = \widetilde{F}_0$, which proves that $\Pi\mathbf{A}$ is surjective. For the injectivity of $\Pi\mathbf{A}$, consider $\bar{\sigma} \in K_2$ such that $\big(\Pi\mathbf{A}\big)(\bar{\sigma}) = 0$, that is, $\mathbf{A}(\bar{\sigma})(\tau) = 0$ for all $\tau \in K_1$, which means that $\mathbf{A}(\bar{\sigma}) \in K_1^\circ$. Knowing that $\mathbf{B}_1$ is surjective, we have that $R(\mathbf{B}_1')$ is also closed and $R(\mathbf{B}_1') = K_1^\circ$, from which there exists $\bar{U} \in Y_1'$ such that $\mathbf{B}_1'(\bar{U}) = -\mathbf{A}(\bar{\sigma})$. Thus, we can write

$$\begin{aligned} \mathbf{A}(\bar{\sigma}) + \mathbf{B}_1'(\bar{U}) &= 0\,, \\ \mathbf{B}_2(\bar{\sigma}) &= 0\,, \end{aligned}$$

from which, in particular, $\bar{\sigma} = 0$, completing the proof.

□

5.5.3 The main result

We begin by observing that the problem of interest (5.166), equivalently (5.168), is a particular case of (5.169) with $Z_i = X_i$ and $Y_i = M_i'$ for each $i \in \{1,2\}$. Then, applying Theorem 5.25, it follows that (5.166) is well-posed if and only if

i) $\Pi\mathcal{A} : \mathcal{K}_2 \to \mathcal{K}_1'$ is an isomorphism, and
ii) for each $i \in \{1,2\}$, $\mathcal{B}_i : X_i \to M_i'$ is surjective,

where

$$\begin{aligned} \mathcal{K}_i := N(\mathcal{B}_i) &= \big\{\tau \in X_i : \quad \mathcal{B}_i(\tau) = 0\big\} \\ &= \big\{\tau \in X_i : \quad b_i(\tau, v) = 0 \quad \forall\, v \in M_i\big\}\,, \end{aligned} \tag{5.177}$$

and, according to (5.170), $\Pi\mathcal{A} : \mathcal{K}_2 \to \mathcal{K}_1'$ is the operator defined by

$$\big(\Pi\mathcal{A}\big)(\tau)(\zeta) := \mathcal{A}(\tau)(\zeta) \qquad \forall\, \tau \in \mathcal{K}_2\,, \quad \forall\, \zeta \in \mathcal{K}_1\,. \tag{5.178}$$

The next goal is to rewrite i) and ii) in terms of inf-sup conditions for the bilinear forms a, b_1, and b_2. In this regard, we first note that the surjectivity of $\mathcal{B}_i : X_i \to M_i'$ is equivalent to requiring that $\mathcal{B}_i' : M_i'' \to X_i'$ be injective with closed range (cf. Theorem 3.86), that is, that there exists a constant $\beta_i > 0$ such that

$$\|\mathcal{B}_i'(\mathcal{G})\| \geqslant \beta_i\, \|\mathcal{G}\| \qquad \forall\, \mathcal{G} \in M_i''\,. \tag{5.179}$$

Then, recalling from Section 5.5.1 that, for the purpose of establishing (5.168), it was assumed that M_1 was reflexive, and assuming from now on that M_2 is also reflexive, we can proceed in a similar way to what was done in Section 5.2.3 to conclude that (5.179) finally reduces to

$$\sup_{\substack{\tau \in X_i \\ \tau \neq 0}} \frac{b_i(\tau, v)}{\|\tau\|} \geqslant \beta_i\, \|v\| \qquad \forall\, v \in M_i\,. \tag{5.180}$$

In turn, i) decomposes as

i-1) $\Pi\mathcal{A} : \mathcal{K}_2 \to \mathcal{K}_1'$ is injective, and
i-2) $\Pi\mathcal{A} : \mathcal{K}_2 \to \mathcal{K}_1'$ is surjective.

Thus, i-1) means that for each $\tau \in \mathcal{K}_2$, $\tau \neq 0$, we have $(\Pi\mathcal{A})(\tau) \neq 0 \in \mathcal{K}_1'$, which, using (5.178), reduces to

$$\sup_{\zeta \in \mathcal{K}_1} a(\tau, \zeta) > 0 \qquad \forall\, \tau \in \mathcal{K}_2,\ \tau \neq 0\,. \tag{5.181}$$

On the other hand, similarly to the analysis of ii), i-2) is equivalent to the existence of a positive constant α such that

$$\|(\Pi\mathcal{A})'(\mathcal{F})\| \geq \alpha\,\|\mathcal{F}\| \qquad \forall\, \mathcal{F} \in \mathcal{K}_1''\,, \tag{5.182}$$

so that, using the bijectivity of $(\Pi\mathcal{A})'$, which follows from that of $\Pi\mathcal{A}$, (5.182) implies that

$$\|(\Pi\mathcal{A})^{-1}\| \leq \frac{1}{\alpha}\,. \tag{5.183}$$

Thus, assuming now that X_1, and therefore its closed subspace $\mathcal{K}_1$ as well, is reflexive (cf. Lemma 7.5), we easily find that (5.182) can be rewritten as

$$\sup_{\substack{\tau \in \mathcal{K}_2 \\ \tau \neq 0}} \frac{a(\tau, \zeta)}{\|\tau\|} \geq \alpha\,\|\zeta\| \qquad \forall\, \zeta \in \mathcal{K}_1\,. \tag{5.184}$$

Alternatively, we can characterize the bijectivity of $\Pi\mathcal{A} : \mathcal{K}_2 \to \mathcal{K}_1'$ through that of $(\Pi\mathcal{A})' : \mathcal{K}_1'' \to \mathcal{K}_2'$ (cf. Theorem 3.90), and therefore consider, instead of i), the assumptions

i-1)' $(\Pi\mathcal{A})' : \mathcal{K}_1'' \to \mathcal{K}_2'$ is injective, and
i-2)' $(\Pi\mathcal{A})' : \mathcal{K}_1'' \to \mathcal{K}_2'$ is surjective.

Consequently, again assuming that X_1, and therefore $\mathcal{K}_1$, is reflexive, and noting, analogously to i-2), that i-2)' is equivalent to $\Pi\mathcal{A} : \mathcal{K}_2 \to \mathcal{K}_1'$ being injective with closed range, we proceed similarly to the derivation of (5.181) and (5.184) to conclude that i-1)' and i-2)' become, respectively,

$$\sup_{\tau \in \mathcal{K}_2} a(\tau, \zeta) > 0 \qquad \forall\, \zeta \in \mathcal{K}_1,\ \zeta \neq 0\,, \tag{5.185}$$

and the existence of a constant $\alpha > 0$ such that

$$\|(\Pi\mathcal{A})(\tau)\| = \sup_{\substack{\zeta \in \mathcal{K}_1 \\ \zeta \neq 0}} \frac{a(\tau, \zeta)}{\|\zeta\|} \geq \alpha\,\|\tau\| \qquad \forall\, \tau \in \mathcal{K}_2\,. \tag{5.186}$$

We emphasize here that, in the present case, the estimate (5.183) for $\|(\Pi\mathcal{A})^{-1}\|$ follows directly from (5.186).

By virtue of the above analysis, we are now in a position to establish the following main result.

Theorem 5.26. *Let $(X_2, \|\cdot\|_{X_2})$, $(M_1, \|\cdot\|_{M_1})$, $(X_1, \|\cdot\|_{X_1})$, and $(M_2, \|\cdot\|_{M_2})$ be real Banach spaces such that M_1, X_1, and M_2 are reflexive, and let $a : X_2 \times X_1 \to \mathbb{R}$ and $b_i : X_i \times M_i \to \mathbb{R}$, $i \in \{1,2\}$, be bounded bilinear forms with induced operators $\mathcal{A} \in \mathcal{L}(X_2, X_1')$ and $\mathcal{B}_i \in \mathcal{L}(X_i, M_i')$, respectively. Moreover, for each $i \in \{1,2\}$ let $\mathcal{K}_i := N(\mathcal{B}_i)$ (cf. (5.177)), and suppose that*

i) *one of the following pairs of conditions on* a *is satisfied*

i-1) $\sup_{\zeta\in\mathcal{K}_1} a(\tau,\zeta) > 0 \quad \forall\,\tau\in\mathcal{K}_2,\ \tau\neq 0,$	i-1)' $\sup_{\tau\in\mathcal{K}_2} a(\tau,\zeta) > 0 \quad \forall\,\zeta\in\mathcal{K}_1,\ \zeta\neq 0,$
i-2) *there exists* $\alpha>0$ *such that* $\sup_{\substack{\tau\in\mathcal{K}_2\\ \tau\neq 0}} \frac{a(\tau,\zeta)}{\|\tau\|} \geqslant \alpha\,\|\zeta\| \quad \forall\,\zeta\in\mathcal{K}_1,$	i-2)' *there exists* $\alpha>0$ *such that* $\sup_{\substack{\zeta\in\mathcal{K}_1\\ \zeta\neq 0}} \frac{a(\tau,\zeta)}{\|\zeta\|} \geqslant \alpha\,\|\tau\| \quad \forall\,\tau\in\mathcal{K}_2,$

ii) *for each* $i\in\{1,2\}$ *there exists a constant* $\beta_i>0$ *such that*

$$\sup_{\substack{\tau\in X_i\\ \tau\neq 0}} \frac{b_i(\tau,v)}{\|\tau\|} \geqslant \beta_i\,\|v\| \qquad \forall\,v\in M_i\,.$$

Then, for each pair $(F,G)\in X_1'\times M_2'$ *there exists a unique solution* $(\sigma,u)\in X_2\times M_1$ *of problem* (5.166), *and the following a priori bounds hold*

$$\|\sigma\| \leqslant \frac{1}{\alpha}\,\|F\| + \frac{1}{\beta_2}\Big(1+\frac{\|\mathcal{A}\|}{\alpha}\Big)\|G\|\,, \tag{5.187}$$

$$\|u\| \leqslant \frac{1}{\beta_1}\Big(1+\frac{\|\mathcal{A}\|}{\alpha}\Big)\|F\| + \frac{\|\mathcal{A}\|}{\beta_1\,\beta_2}\Big(1+\frac{\|\mathcal{A}\|}{\alpha}\Big)\|G\|\,. \tag{5.188}$$

Moreover, assumptions i) *and* ii) *are also necessary.*

Proof. According to what was indicated at the beginning of this section, the proof follows directly by applying Theorem 5.25, bearing in mind the equivalences between i) and the pairs of assumptions (5.181)–(5.184) and (5.185)–(5.186), and between ii) and (5.180). In turn, the a priori bounds (5.187) and (5.188) follow from (5.174) and (5.175), respectively, replacing in both cases $\|(\varPi\mathcal{A})^{-1}\|$ by $\frac{1}{\alpha}$, as established in (5.183).

5.5.4 Discrete analysis: general case

In what follows, we introduce a general Galerkin scheme associated with (5.166), prove that it is well-posed, and derive the corresponding a priori error estimates. For this purpose, let $\{X_{2,h}\}_{h>0}$, $\{M_{1,h}\}_{h>0}$, $\{X_{1,h}\}_{h>0}$, and $\{M_{2,h}\}_{h>0}$ be families of finite-dimensional subspaces of X_2, M_1, X_1, and M_2, respectively, and let $a_h : X_{2,h}\times X_{1,h}\to\mathbb{R}$ and $b_{i,h} : X_{i,h}\times M_{i,h}\to\mathbb{R}$, $i\in\{1,2\}$, be bounded bilinear forms that approximate a and b_i, with induced operators $\mathcal{A}_h\in\mathcal{L}(X_{2,h},X_{1,h}')$ and $\mathcal{B}_{i,h}\in\mathcal{L}(X_{i,h},M_{i,h}')$, respectively. Then, given approximations $F_h\in X_{1,h}'$ and $G_h\in M_{2,h}'$ of the functionals F and G, respectively, we consider the discrete system: Find $(\sigma_h,u_h)\in X_{2,h}\times M_{1,h}$ such that

$$\begin{aligned} a_h(\sigma_h,\tau_h) + b_{1,h}(\tau_h,u_h) &= F_h(\tau_h) \qquad \forall\,\tau_h\in X_{1,h}\,,\\ b_{2,h}(\sigma_h,v_h) &= G_h(v_h) \qquad \forall\,v_h\in M_{2,h}\,. \end{aligned} \tag{5.189}$$

Then, applying Theorem 5.26 directly to the present discrete context, we obtain the following main result.

Theorem 5.27. *In addition to the previous notations and definitions, for each* $i \in \{1,2\}$ *define*

$$\mathcal{K}_{i,h} := N(\mathcal{B}_{i,h}) = \left\{\tau_h \in X_{i,h} : \quad b_{i,h}(\tau_h, v_h) = 0 \quad \forall\, v_h \in M_{i,h}\right\},$$

and suppose that

i) *one of the following equivalent pairs of hypotheses on* a_h *is satisfied*

i-1) $\sup\limits_{\zeta_h \in \mathcal{K}_{1,h}} a_h(\tau_h, \zeta_h) > 0 \quad \forall\, \tau_h \in \mathcal{K}_{2,h} \backslash \{0\}$,

i-2) *there exists* $\alpha_{\mathtt{d}} > 0$ *such that*

$$\sup_{\substack{\tau_h \in \mathcal{K}_{2,h} \\ \tau_h \neq 0}} \frac{a_h(\tau_h, \zeta_h)}{\|\tau_h\|} \geqslant \alpha_{\mathtt{d}} \|\zeta_h\| \quad \forall\, \zeta_h \in \mathcal{K}_{1,h},$$

i-1)' $\sup\limits_{\tau_h \in \mathcal{K}_{2,h}} a_h(\tau_h, \zeta_h) > 0 \quad \forall\, \zeta_h \in \mathcal{K}_{1,h} \backslash \{0\}$,

i-2)' *there exists* $\alpha_{\mathtt{d}} > 0$ *such that*

$$\sup_{\substack{\zeta_h \in \mathcal{K}_{1,h} \\ \zeta_h \neq 0}} \frac{a_h(\tau_h, \zeta_h)}{\|\zeta_h\|} \geqslant \alpha_{\mathtt{d}} \|\tau_h\| \quad \forall\, \tau_h \in \mathcal{K}_{2,h},$$

ii) *for each* $i \in \{1,2\}$ *there exists a constant* $\beta_{i,\mathtt{d}} > 0$ *such that*

$$\sup_{\substack{\tau_h \in X_{i,h} \\ \tau_h \neq 0}} \frac{b_{i,h}(\tau_h, v_h)}{\|\tau_h\|} \geqslant \beta_{i,\mathtt{d}} \|v_h\| \qquad \forall\, v_h \in M_{i,h}.$$

Then, for each pair $(F_h, G_h) \in X'_{1,h} \times M'_{2,h}$ *there exists a unique* $(\sigma_h, u_h) \in X_{2,h} \times M_{1,h}$ *solution of problem* (5.189), *and the following a priori bounds hold*

$$\|\sigma_h\| \leqslant \frac{1}{\alpha_{\mathtt{d}}} \|F_h\| + \frac{1}{\beta_{2,\mathtt{d}}} \left(1 + \frac{\|\mathcal{A}_h\|}{\alpha_{\mathtt{d}}}\right) \|G_h\|,$$

$$\|u_h\| \leqslant \frac{1}{\beta_{1,\mathtt{d}}} \left(1 + \frac{\|\mathcal{A}_h\|}{\alpha_{\mathtt{d}}}\right) \|F_h\| + \frac{\|\mathcal{A}_h\|}{\beta_{1,\mathtt{d}}\, \beta_{2,\mathtt{d}}} \left(1 + \frac{\|\mathcal{A}_h\|}{\alpha_{\mathtt{d}}}\right) \|G_h\|.$$

Moreover, assumptions i) *and* ii) *are also necessary.*

We observe here that in the case where $\dim \mathcal{K}_{1,h} = \dim \mathcal{K}_{2,h}$, the finite-dimensional Fredholm alternative, which ensures that a linear operator is surjective if and only if it is injective, allows us to conclude that in Theorem 5.27 the assumptions i-1) and i-1)' are redundant with respect to i-2) and i-2)', respectively, and therefore it is sufficient to assume only the latter.

After establishing necessary and sufficient conditions for (5.166) (cf. Theorem 5.26) and (5.189) (cf. Theorem 5.27) to be well-posed, our next objective is to derive the corresponding a priori error estimates. In this regard, although they were originally obtained in [17, Section 2.3], we proceed here in a slightly different way, providing a more direct proof that does not make use of any shifted discrete kernel, but only of suitable upper bounds for $\|\sigma_h - \tau_h\|$ and $\|u_h - v_h\|$ for each $(\tau_h, v_h) \in X_{2,h} \times M_{1,h}$. In a certain sense, our approach, extracted from [59], can be seen as a generalization of the technique employed in the proof of [52, Lemma 2.44], where the particular case, in which $X_2 = X_1$, $M_2 = M_1$, $b_1 = b_2$, $X_{2,h} = X_{1,h}$, $M_{2,h} = M_{1,h}$, and the discrete bilinear forms a_h and $b_{1,h} = b_{2,h}$, as well as the discrete functionals F_h and G_h, coincide with the restriction of the continuous ones to the respective discrete spaces, was considered.

The upper bounds mentioned above are precisely given by the following lemmas.

Lemma 5.28. *Assume the hypotheses of Theorems* 5.26 *and* 5.27, *and let* $(\sigma, u) \in X_2 \times M_1$ *and* $(\sigma_h, u_h) \in X_{2,h} \times M_{1,h}$ *be the unique solutions of* (5.166) *and* (5.189), *respectively. Then, for each* $(\tau_h, v_h) \in X_{2,h} \times M_{1,h}$ *one has*

$$
\begin{aligned}
\|\sigma_h - \tau_h\| \leqslant & \left(\frac{\|\mathcal{A}\|}{\alpha_{\mathrm{d}}} + \left(1 + \frac{\|\mathcal{A}_h\|}{\alpha_{\mathrm{d}}}\right) \frac{\|\mathcal{B}_2\|}{\beta_{2,\mathrm{d}}} \right) \|\sigma - \tau_h\| \\
+ & \left(1 + \frac{\|\mathcal{A}_h\|}{\alpha_{\mathrm{d}}}\right) \frac{1}{\beta_{2,\mathrm{d}}} \|(b_2 - b_{2,h})(\tau_h, \cdot)\|_{M'_{2,h}} + \frac{1}{\alpha_{\mathrm{d}}} \|(a - a_h)(\tau_h, \cdot)\|_{\mathcal{K}'_{1,h}} \\
+ & \frac{\|\mathcal{B}_1\|}{\alpha_{\mathrm{d}}} \|u - v_h\| + \frac{1}{\alpha_{\mathrm{d}}} \|(b_1 - b_{1,h})(\cdot, v_h)\|_{\mathcal{K}'_{1,h}} \\
+ & \frac{1}{\alpha_{\mathrm{d}}} \|F - F_h\|_{\mathcal{K}'_{1,h}} + \left(1 + \frac{\|\mathcal{A}_h\|}{\alpha_{\mathrm{d}}}\right) \frac{1}{\beta_{2,\mathrm{d}}} \|G - G_h\|_{M'_{2,h}}
\end{aligned}
\tag{5.190}
$$

where the consistency terms in (5.190) *are defined as*

$$\|(b_2 - b_{2,h})(\tau_h, \cdot)\|_{M'_{2,h}} := \sup_{\substack{w_h \in M_{2,h} \\ w_h \neq 0}} \frac{(b_2 - b_{2,h})(\tau_h, w_h)}{\|w_h\|}, \tag{5.191}$$

$$\|(a - a_h)(\tau_h, \cdot)\|_{\mathcal{K}'_{1,h}} := \sup_{\substack{\zeta_h \in \mathcal{K}_{1,h} \\ \zeta_h \neq 0}} \frac{(a - a_h)(\tau_h, \zeta_h)}{\|\zeta_h\|}, \tag{5.192}$$

$$\|(b_1 - b_{1,h})(\cdot, v_h)\|_{\mathcal{K}'_{1,h}} := \sup_{\substack{\zeta_h \in \mathcal{K}_{1,h} \\ \zeta_h \neq 0}} \frac{(b_1 - b_{1,h})(\zeta_h, v_h)}{\|\zeta_h\|},$$

$$\|F - F_h\|_{\mathcal{K}'_{1,h}} := \sup_{\substack{\zeta_h \in \mathcal{K}_{1,h} \\ \zeta_h \neq 0}} \frac{(F - F_h)(\zeta_h)}{\|\zeta_h\|}, \quad \textit{and} \tag{5.193}$$

$$\|G - G_h\|_{M'_{2,h}} := \sup_{\substack{w_h \in M_{2,h} \\ w_h \neq 0}} \frac{(G - G_h)(w_h)}{\|w_h\|}. \tag{5.194}$$

Proof. We begin by noting, thanks to the obvious reflexivity of $M_{2,h}$, that the inf-sup condition for $b_{2,h}$ (cf. ii) in Theorem 5.27) is equivalent to stating that

$$\|\mathcal{B}'_{2,h}(\mathcal{G}_h)\| \geqslant \beta_{2,\mathrm{d}} \|\mathcal{G}_h\| \qquad \forall\, \mathcal{G}_h \in M''_{2,h}.$$

Then, applying the converse of Lemma 3.65, it follows that for each $\widetilde{G}_h \in M'_{2,h}$ there exists $\widetilde{\tau}_h \in X_{2,h}$ such that

$$\mathcal{B}_{2,h}(\widetilde{\tau}_h) = \widetilde{G}_h \quad \text{and} \quad \|\widetilde{\tau}_h\| \leqslant \frac{1}{\beta_{2,\mathrm{d}}} \|\widetilde{G}_h\|_{M'_{2,h}}.$$

In particular, given $\tau_h \in X_{2,h}$, there exists $\widetilde{\tau}_h \in X_{2,h}$ such that

$$\mathcal{B}_{2,h}(\widetilde{\tau}_h) = b_{2,h}(\tau_h, \cdot) - G_h \in M'_{2,h},$$

that is,

$$b_{2,h}(\widetilde{\tau}_h, w_h) = b_{2,h}(\tau_h, w_h) - G_h(w_h) \qquad \forall\, w_h \in M_{2,h}, \tag{5.195}$$

and

$$\|\widetilde{\tau}_h\| \leqslant \frac{1}{\beta_{2,\mathtt{d}}} \|b_{2,h}(\tau_h,\cdot) - G_h\|_{M'_{2,h}} := \frac{1}{\beta_{2,\mathtt{d}}} \sup_{\substack{w_h \in M_{2,h} \\ w_h \neq 0}} \frac{b_{2,h}(\tau_h, w_h) - G_h(w_h)}{\|w_h\|} . \tag{5.196}$$

Now, using from the second equation of (5.166) that $b_2(\sigma, w_h) = G(w_h)$, and then subtracting and adding τ_h in the first component of $b_2(\sigma, w_h)$, we obtain

$$b_{2,h}(\tau_h, w_h) - G_h(w_h) = -b_2(\sigma - \tau_h, w_h) + (b_{2,h} - b_2)(\tau_h, w_h) + (G - G_h)(w_h), \tag{5.197}$$

which, substituted back into (5.196), implies

$$\|\widetilde{\tau}_h\| \leqslant \frac{\|\mathcal{B}_2\|}{\beta_{2,\mathtt{d}}} \|\sigma - \tau_h\| + \frac{1}{\beta_{2,\mathtt{d}}} \|(b_2 - b_{2,h})(\tau_h,\cdot)\|_{M'_{2,h}} + \frac{1}{\beta_{2,\mathtt{d}}} \|G - G_h\|_{M'_{2,h}} . \tag{5.198}$$

In turn, it is clear from (5.195) and the second equation of (5.189) that $\sigma_h - (\tau_h - \widetilde{\tau}_h) \in \mathcal{K}_{2,h}$, and therefore, according to i-2)' of Theorem 5.27, we get that

$$\alpha_{\mathtt{d}} \|\sigma_h - (\tau_h - \widetilde{\tau}_h)\| \leqslant \sup_{\substack{\zeta_h \in \mathcal{K}_{1,h} \\ \zeta_h \neq 0}} \frac{a_h\big(\sigma_h - (\tau_h - \widetilde{\tau}_h), \zeta_h\big)}{\|\zeta_h\|} . \tag{5.199}$$

Then, using the first rows of (5.189) and (5.166), we have for each $\zeta_h \in \mathcal{K}_{1,h}$

$$\begin{aligned} a_h\big(\sigma_h - (\tau_h - \widetilde{\tau}_h), \zeta_h\big) &= F_h(\zeta_h) - a_h(\tau_h - \widetilde{\tau}_h, \zeta_h) + a(\sigma, \zeta_h) + b_1(\zeta_h, u) - F(\zeta_h) \\ &= a(\sigma - \tau_h, \zeta_h) + (a - a_h)(\tau_h, \zeta_h) + a_h(\widetilde{\tau}_h, \zeta_h) \\ &\quad + b_1(\zeta_h, u - v_h) + (b_1 - b_{1,h})(\zeta_h, v_h) - (F - F_h)(\zeta_h), \end{aligned} \tag{5.200}$$

where we have subtracted and added an arbitrary $v_h \in M_{1,h}$ in the second component of $b_1(\zeta_h, u)$. We note here that in general one cannot assume that $\mathcal{K}_{1,h}$ is contained in $\mathcal{K}_1$, but if this were the case, then $b_1(\zeta_h, u)$ would vanish, there would be no need to consider the mentioned $v_h \in M_{1,h}$, and the final estimate (5.190) would not include its third line. We shall return to this observation later. On the other hand, note that, although $b_{1,h}(\zeta_h, v_h)$ is certainly zero, we keep it in the expression above to emphasize that $(b_1 - b_{1,h})(\zeta_h, v_h)$ is a consistency term. Thus, substituting (5.200) back into (5.199), and performing the corresponding estimates, we arrive at

$$\begin{aligned} \|\sigma_h - (\tau_h - \widetilde{\tau}_h)\| &\leqslant \frac{\|\mathcal{A}\|}{\alpha_{\mathtt{d}}} \|\sigma - \tau_h\| + \frac{1}{\alpha_{\mathtt{d}}} \|(a - a_h)(\tau_h,\cdot)\|_{\mathcal{K}'_{1,h}} \\ &\quad + \frac{\|\mathcal{A}_h\|}{\alpha_{\mathtt{d}}} \|\widetilde{\tau}_h\| + \frac{\|\mathcal{B}_1\|}{\alpha_{\mathtt{d}}} \|u - v_h\| \\ &\quad + \frac{1}{\alpha_{\mathtt{d}}} \|(b_1 - b_{1,h})(\cdot, v_h)\|_{\mathcal{K}'_{1,h}} + \frac{1}{\alpha_{\mathtt{d}}} \|F - F_h\|_{\mathcal{K}'_{1,h}} . \end{aligned} \tag{5.201}$$

Therefore, applying the triangle inequality and (5.201), we obtain

$$\begin{aligned} \|\sigma_h - \tau_h\| &\leqslant \frac{\|\mathcal{A}\|}{\alpha_{\mathtt{d}}} \|\sigma - \tau_h\| + \frac{1}{\alpha_{\mathtt{d}}} \|(a - a_h)(\tau_h,\cdot)\|_{\mathcal{K}'_{1,h}} \\ &\quad + \left(1 + \frac{\|\mathcal{A}_h\|}{\alpha_{\mathtt{d}}}\right) \|\widetilde{\tau}_h\| + \frac{\|\mathcal{B}_1\|}{\alpha_{\mathtt{d}}} \|u - v_h\| \\ &\quad + \frac{1}{\alpha_{\mathtt{d}}} \|(b_1 - b_{1,h})(\cdot, v_h)\|_{\mathcal{K}'_{1,h}} + \frac{1}{\alpha_{\mathtt{d}}} \|F - F_h\|_{\mathcal{K}'_{1,h}} , \end{aligned}$$

which, together with estimate (5.198), leads to (5.190) and concludes the proof.

□

Lemma 5.29. *Assume the hypotheses of Theorems* 5.26 *and* 5.27, *and let* $(\sigma, u) \in X_2 \times M_1$ *and* $(\sigma_h, u_h) \in X_{2,h} \times M_{1,h}$ *be the unique solutions of* (5.166) *and* (5.189), *respectively. Then, denoting*

$$C_{\mathtt{d}} := \left(1 + \frac{\|\mathcal{A}_h\|}{\alpha_{\mathtt{d}}}\right) \frac{1}{\beta_{1,\mathtt{d}}}, \tag{5.202}$$

one has for each $(\tau_h, v_h) \in X_{2,h} \times M_{1,h}$

$$\begin{aligned} \|u_h - v_h\| &\leqslant C_{\mathtt{d}} \left(\|\mathcal{A}\| + \frac{\|\mathcal{A}_h\| \, \|\mathcal{B}_2\|}{\beta_{2,\mathtt{d}}}\right) \|\sigma - \tau_h\| \\ &+ C_{\mathtt{d}} \left(\frac{\|\mathcal{A}_h\|}{\beta_{2,\mathtt{d}}} \|(b_2 - b_{2,h})(\tau_h, \cdot)\|_{M'_{2,h}} + \|(a - a_h)(\tau_h, \cdot)\|_{X'_{1,h}}\right) \\ &+ C_{\mathtt{d}} \left(\|\mathcal{B}_1\| \, \|u - v_h\| + \|(b_1 - b_{1,h})(\cdot, v_h)\|_{X'_{1,h}}\right) \\ &+ C_{\mathtt{d}} \left(\|F - F_h\|_{X'_{1,h}} + \frac{\|\mathcal{A}_h\|}{\beta_{2,\mathtt{d}}} \|G - G_h\|_{M'_{2,h}}\right), \end{aligned} \tag{5.203}$$

where the consistency terms in (5.203) *are defined as in Lemma* 5.28 (*cf.* (5.191)–(5.194)), *except that in* (5.192)–(5.193) $\mathcal{K}_{1,h}$ *is replaced by* $X_{1,h}$.

Proof. From the inf-sup condition for $b_{1,h}$ (cf. ii) in Theorem 5.27) we first note that

$$\beta_{1,\mathtt{d}} \|u_h - v_h\| \leqslant \sup_{\substack{\zeta_h \in X_{1,h} \\ \zeta_h \neq 0}} \frac{b_{1,h}(\zeta_h, u_h - v_h)}{\|\zeta_h\|}. \tag{5.204}$$

Then, according to the first equations of (5.189) and (5.166), one has

$$\begin{aligned} b_{1,h}(\zeta_h, u_h - v_h) &= F_h(\zeta_h) - a_h(\sigma_h, \zeta_h) - b_{1,h}(\zeta_h, v_h) \\ &= \big(F_h - F\big)(\zeta_h) + a(\sigma, \zeta_h) + b_1(\zeta_h, u) - a_h(\sigma_h, \zeta_h) - b_{1,h}(\zeta_h, v_h), \end{aligned}$$

from which, subtracting and adding $\tau_h \in X_{2,h}$ (resp. $v_h \in M_{1,h}$) in the first component of $a(\sigma, \zeta_h)$ (resp. second component of $b_1(\zeta_h, u)$), we obtain

$$\begin{aligned} b_{1,h}(\zeta_h, u_h - v_h) &= \big(F_h - F\big)(\zeta_h) + a(\sigma - \tau_h, \zeta_h) + (a - a_h)(\tau_h, \zeta_h) \\ &- a_h(\sigma_h - \tau_h, \zeta_h) + b_1(\zeta_h, u - v_h) + (b_1 - b_{1,h})(\zeta_h, v_h). \end{aligned} \tag{5.205}$$

Next, substituting (5.205) back into (5.204), and proceeding similarly to the proof of Lemma 5.28, one easily deduces that

$$\begin{aligned} \|u_h - v_h\| &\leqslant \frac{\|\mathcal{A}\|}{\beta_{1,\mathtt{d}}} \|\sigma - \tau_h\| + \frac{1}{\beta_{1,\mathtt{d}}} \|(a - a_h)(\tau_h, \cdot)\|_{X'_{1,h}} + \frac{\|\mathcal{A}_h\|}{\beta_{1,\mathtt{d}}} \|\sigma_h - \tau_h\| \\ &+ \frac{\|\mathcal{B}_1\|}{\beta_{1,\mathtt{d}}} \|u - v_h\| + \frac{1}{\beta_{1,\mathtt{d}}} \|(b_1 - b_{1,h})(\cdot, v_h)\|_{X'_{1,h}} + \frac{1}{\beta_{1,\mathtt{d}}} \|F - F_h\|_{X'_{1,h}}. \end{aligned} \tag{5.206}$$

Finally, using (5.190) (cf. Lemma 5.28) in the above inequality, performing minor algebraic manipulations, and bounding $\|(a - a_h)(\tau_h, \cdot)\|_{\mathcal{K}'_{1,h}}$, $\|(b_1 - b_{1,h})(\cdot, v_h)\|_{\mathcal{K}'_{1,h}}$, and $\|F - F_h\|_{\mathcal{K}'_{1,h}}$ by $\|(a - a_h)(\tau_h, \cdot)\|_{X'_{1,h}}$, $\|(b_1 - b_{1,h})(\cdot, v_h)\|_{X'_{1,h}}$, and $\|F - F_h\|_{X'_{1,h}}$, respectively, yields (5.203) and completes the proof.

□

We are now in a position to state the main result of this section.

Theorem 5.30. *Assume the hypotheses of Theorems* 5.26 *and* 5.27*, and let* $(\sigma, u) \in X_2 \times M_1$ *and* $(\sigma_h, u_h) \in X_{2,h} \times M_{1,h}$ *be the unique solutions of* (5.166) *and* (5.189)*, respectively. Then one has*

$$
\begin{aligned}
\|\sigma - \sigma_h\| \leqslant & \inf_{\tau_h \in X_{2,h}} \Bigg\{ \left(1 + \frac{\|\mathcal{A}\|}{\alpha_{\mathtt{d}}} + \left(1 + \frac{\|\mathcal{A}_h\|}{\alpha_{\mathtt{d}}}\right) \frac{\|\mathcal{B}_2\|}{\beta_{2,\mathtt{d}}}\right) \|\sigma - \tau_h\| \\
+ & \left(1 + \frac{\|\mathcal{A}_h\|}{\alpha_{\mathtt{d}}}\right) \frac{1}{\beta_{2,\mathtt{d}}} \|(b_2 - b_{2,h})(\tau_h, \cdot)\|_{M'_{2,h}} + \frac{1}{\alpha_{\mathtt{d}}} \|(a - a_h)(\tau_h, \cdot)\|_{\mathcal{K}'_{1,h}} \Bigg\} \\
+ & \inf_{v_h \in M_{1,h}} \left\{ \frac{\|\mathcal{B}_1\|}{\alpha_{\mathtt{d}}} \|u - v_h\| + \frac{1}{\alpha_{\mathtt{d}}} \|(b_1 - b_{1,h})(\cdot, v_h)\|_{\mathcal{K}'_{1,h}} \right\} \\
+ & \frac{1}{\alpha_{\mathtt{d}}} \|F - F_h\|_{\mathcal{K}'_{1,h}} + \left(1 + \frac{\|\mathcal{A}_h\|}{\alpha_{\mathtt{d}}}\right) \frac{1}{\beta_{2,\mathtt{d}}} \|G - G_h\|_{M'_{2,h}},
\end{aligned} \tag{5.207}
$$

and, with the constant $C_{\mathtt{d}}$ *of* (5.202),

$$
\begin{aligned}
\|u - u_h\| \leqslant C_{\mathtt{d}} & \inf_{\tau_h \in X_{2,h}} \Bigg\{ \left(\|\mathcal{A}\| + \frac{\|\mathcal{A}_h\| \, \|\mathcal{B}_2\|}{\beta_{2,\mathtt{d}}}\right) \|\sigma - \tau_h\| \\
+ & \frac{\|\mathcal{A}_h\|}{\beta_{2,\mathtt{d}}} \|(b_2 - b_{2,h})(\tau_h, \cdot)\|_{M'_{2,h}} + \|(a - a_h)(\tau_h, \cdot)\|_{X'_{1,h}} \Bigg\} \\
+ & \inf_{v_h \in M_{1,h}} \left\{ \left(1 + C_{\mathtt{d}} \|\mathcal{B}_1\|\right) \|u - v_h\| + C_{\mathtt{d}} \|(b_1 - b_{1,h})(\cdot, v_h)\|_{X'_{1,h}} \right\} \\
+ & C_{\mathtt{d}} \left(\|F - F_h\|_{X'_{1,h}} + \frac{\|\mathcal{A}_h\|}{\beta_{2,\mathtt{d}}} \|G - G_h\|_{M'_{2,h}} \right),
\end{aligned} \tag{5.208}
$$

Proof. Applying the triangle inequality we have

$$
\|\sigma - \sigma_h\| \leqslant \|\sigma - \tau_h\| + \|\sigma_h - \tau_h\| \qquad \forall\, \tau_h \in X_{2,h},
$$

and

$$
\|u - u_h\| \leqslant \|u - v_h\| + \|u_h - v_h\| \qquad \forall\, v_h \in M_{1,h},
$$

whence, together with the estimates for $\|\sigma_h - \tau_h\|$ and $\|u_h - v_h\|$ provided by (5.190) and (5.203), respectively, and after taking infima with respect to $\tau_h \in X_{2,h}$ and $v_h \in M_{1,h}$, one easily deduces (5.207) and (5.208).

□

As announced in the proof of Lemma 5.28, we now emphasize that when $\mathcal{K}_{1,h} \subseteq \mathcal{K}_1$, Strang's estimate (5.207) reduces to

$$
\begin{aligned}
\|\sigma - \sigma_h\| \leqslant & \inf_{\tau_h \in X_{2,h}} \Bigg\{ \left(1 + \frac{\|\mathcal{A}\|}{\alpha_{\mathtt{d}}} + \left(1 + \frac{\|\mathcal{A}_h\|}{\alpha_{\mathtt{d}}}\right) \frac{\|\mathcal{B}_2\|}{\beta_{2,\mathtt{d}}}\right) \|\sigma - \tau_h\| \\
+ & \left(1 + \frac{\|\mathcal{A}_h\|}{\alpha_{\mathtt{d}}}\right) \frac{1}{\beta_{2,\mathtt{d}}} \|(b_2 - b_{2,h})(\tau_h, \cdot)\|_{M'_{2,h}} + \frac{1}{\alpha_{\mathtt{d}}} \|(a - a_h)(\tau_h, \cdot)\|_{\mathcal{K}'_{1,h}} \Bigg\} \\
+ & \frac{1}{\alpha_{\mathtt{d}}} \|F - F_h\|_{\mathcal{K}'_{1,h}} + \left(1 + \frac{\|\mathcal{A}_h\|}{\alpha_{\mathtt{d}}}\right) \frac{1}{\beta_{2,\mathtt{d}}} \|G - G_h\|_{M'_{2,h}},
\end{aligned} \tag{5.209}
$$

while (5.208) remains unchanged.

On the other hand, a significant simplification arises when the expressions $a_h(\sigma,\zeta_h)$, $b_{1,h}(\zeta_h,u)$, and $b_{2,h}(\sigma,w_h)$ are computable for all $(\zeta_h,w_h)\in X_{1,h}\times M_{2,h}$. Indeed, in this case we first observe that (5.197) can be rewritten as

$$b_{2,h}(\tau_h,w_h)-G_h(w_h) \;=\; -b_{2,h}(\sigma-\tau_h,w_h) \;-\; (b_2-b_{2,h})(\sigma,w_h) \;+\; (G-G_h)(w_h)\,,$$

which, when replaced in (5.196), leads, instead of (5.198), to

$$\|\widetilde{\tau}_h\| \,\leqslant\, \frac{\|\mathcal{B}_{2,h}\|}{\beta_{2,\mathrm{d}}}\,\|\sigma-\tau_h\| \,+\, \frac{1}{\beta_{2,\mathrm{d}}}\,\|(b_2-b_{2,h})(\sigma,\cdot)\|_{M'_{2,h}} \,+\, \frac{1}{\beta_{2,\mathrm{d}}}\,\|G-G_h\|_{M'_{2,h}}\,. \tag{5.210}$$

In turn, we also note that (5.200) becomes

$$\begin{aligned} a_h\big(\sigma_h-(\tau_h-\widetilde{\tau}_h),\zeta_h\big) \;&=\; a_h(\sigma-\tau_h,\zeta_h) \,+\, (a-a_h)(\sigma,\zeta_h) \,+\, a_h(\widetilde{\tau}_h,\zeta_h)\\ &\quad+\, b_{1,h}(\zeta_h,u-v_h) \,+\, (b_1-b_{1,h})(\zeta_h,u) \,-\, (F-F_h)(\zeta_h)\,, \end{aligned}$$

so that, together with (5.199), it implies, instead of (5.201), that

$$\begin{aligned} \|\sigma_h-(\tau_h-\widetilde{\tau}_h)\| \,&\leqslant\, \frac{\|\mathcal{A}_h\|}{\alpha_{\mathrm{d}}}\,\|\sigma-\tau_h\| + \frac{1}{\alpha_{\mathrm{d}}}\,\|(a-a_h)(\sigma,\cdot)\|_{\mathcal{K}'_{1,h}} + \frac{\|\mathcal{A}_h\|}{\alpha_{\mathrm{d}}}\,\|\widetilde{\tau}_h\|\\ &+\, \frac{\|\mathcal{B}_{1,h}\|}{\alpha_{\mathrm{d}}}\,\|u-v_h\| + \frac{1}{\alpha_{\mathrm{d}}}\,\|(b_1-b_{1,h})(\cdot,u)\|_{\mathcal{K}'_{1,h}} + \frac{1}{\alpha_{\mathrm{d}}}\,\|F-F_h\|_{\mathcal{K}'_{1,h}}\,. \end{aligned} \tag{5.211}$$

Furthermore, (5.205) can now be expressed as

$$\begin{aligned} b_{1,h}(\zeta_h,u_h-v_h) \;&=\; \big(F_h-F\big)(\zeta_h) + a_h(\sigma-\tau_h,\zeta_h) + (a-a_h)(\sigma,\zeta_h)\\ &\quad+ (b_1-b_{1,h})(\zeta_h,u) + b_{1,h}(\zeta_h,u-v_h) + a_h(\sigma_h-\tau_h,\zeta_h)\,, \end{aligned}$$

whence, in combination with (5.204), one deduces, instead of (5.206), that

$$\begin{aligned} \|u_h-v_h\| \,&\leqslant\, \frac{\|\mathcal{A}_h\|}{\beta_{1,\mathrm{d}}}\,\|\sigma-\tau_h\| \,+\, \frac{1}{\beta_{1,\mathrm{d}}}\,\|(a-a_h)(\sigma,\cdot)\|_{X'_{1,h}} \,+\, \frac{\|\mathcal{A}_h\|}{\beta_{1,\mathrm{d}}}\,\|\sigma_h-\tau_h\|\\ &+\, \frac{\|\mathcal{B}_{1,h}\|}{\beta_{1,\mathrm{d}}}\,\|u-v_h\| \,+\, \frac{1}{\beta_{1,\mathrm{d}}}\,\|(b_1-b_{1,h})(\cdot,u)\|_{X'_{1,h}} \,+\, \frac{1}{\beta_{1,\mathrm{d}}}\,\|F-F_h\|_{X'_{1,h}}\,. \end{aligned} \tag{5.212}$$

In this way, using (5.210) and (5.211), we find that for each $(\tau_h,v_h)\in X_{2,h}\times M_{1,h}$ one has

$$\begin{aligned} \|\sigma_h-\tau_h\| \,&\leqslant\, \left(\frac{\|\mathcal{A}_h\|}{\alpha_{\mathrm{d}}} + \left(1+\frac{\|\mathcal{A}_h\|}{\alpha_{\mathrm{d}}}\right)\frac{\|\mathcal{B}_{2,h}\|}{\beta_{2,\mathrm{d}}}\right)\|\sigma-\tau_h\|\\ &+\, \left(1+\frac{\|\mathcal{A}_h\|}{\alpha_{\mathrm{d}}}\right)\frac{1}{\beta_{2,\mathrm{d}}}\|(b_2-b_{2,h})(\sigma,\cdot)\|_{M'_{2,h}} + \frac{1}{\alpha_{\mathrm{d}}}\,\|(a-a_h)(\sigma,\cdot)\|_{\mathcal{K}'_{1,h}}\\ &+\, \frac{\|\mathcal{B}_{1,h}\|}{\alpha_{\mathrm{d}}}\,\|u-v_h\| + \frac{1}{\alpha_{\mathrm{d}}}\,\|(b_1-b_{1,h})(\cdot,u)\|_{\mathcal{K}'_{1,h}}\\ &+\, \frac{1}{\alpha_{\mathrm{d}}}\,\|F-F_h\|_{\mathcal{K}'_{1,h}} + \left(1+\frac{\|\mathcal{A}_h\|}{\alpha_{\mathrm{d}}}\right)\frac{1}{\beta_{2,\mathrm{d}}}\,\|G-G_h\|_{M'_{2,h}}\,. \end{aligned} \tag{5.213}$$

Similarly, thanks to (5.212) and (5.213), and by bounding the terms $\|(a-a_h)(\sigma,\cdot)\|_{\mathcal{K}'_{1,h}}$, $\|(b_1-b_{1,h})(\cdot,u)\|_{\mathcal{K}'_{1,h}}$, and $\|F-F_h\|_{\mathcal{K}'_{1,h}}$ by $\|(a-a_h)(\sigma,\cdot)\|_{X'_{1,h}}$, $\|(b_1-b_{1,h})(\cdot,u)\|_{X'_{1,h}}$, and $\|F-F_h\|_{X'_{1,h}}$, respectively, we also obtain that for each $(\tau_h,v_h)\in X_{2,h}\times M_{1,h}$ one has

$$
\begin{aligned}
\|u_h - v_h\| \leqslant\ & C_{\mathtt{d}} \, \|\mathcal{A}_h\| \left(1 + \frac{\|\mathcal{B}_{2,h}\|}{\beta_{2,\mathtt{d}}}\right) \|\sigma - \tau_h\| \\
& + C_{\mathtt{d}} \left(\frac{\|\mathcal{A}_h\|}{\beta_{2,\mathtt{d}}} \, \|(b_2 - b_{2,h})(\sigma, \cdot)\|_{M'_{2,h}} + \|(a - a_h)(\sigma, \cdot)\|_{X'_{1,h}} \right) \\
& + C_{\mathtt{d}} \left(\|\mathcal{B}_{1,h}\| \, \|u - v_h\| + \|(b_1 - b_{1,h})(\cdot, u)\|_{X'_{1,h}} \right) \\
& + C_{\mathtt{d}} \left(\|F - F_h\|_{X'_{1,h}} + \frac{\|\mathcal{A}_h\|}{\beta_{2,\mathtt{d}}} \|G - G_h\|_{M'_{2,h}} \right),
\end{aligned}
\tag{5.214}
$$

where $C_{\mathtt{d}}$ is the constant defined in (5.202). Note here that (5.213) and (5.214) constitute the analogues of (5.190) and (5.203), respectively.

As a consequence of the previous estimates, we obtain the following result, in which the new consistency terms appear separated from the corresponding infima, and therefore can be handled independently of them.

Theorem 5.31. *Assume the hypotheses of Theorems 5.26 and 5.27, and let $(\sigma, u) \in X_2 \times M_1$ and $(\sigma_h, u_h) \in X_{2,h} \times M_{1,h}$ be the unique solutions of (5.166) and (5.189), respectively. Assume, moreover, that $a_h(\sigma, \zeta_h)$, $b_{1,h}(\zeta_h, u)$, and $b_{2,h}(\sigma, w_h)$ are computable for all $(\zeta_h, w_h) \in X_{1,h} \times M_{2,h}$. Then one has*

$$
\begin{aligned}
\|\sigma - \sigma_h\| \leqslant\ & \left(1 + \frac{\|\mathcal{A}_h\|}{\alpha_{\mathtt{d}}}\right) \left(1 + \frac{\|\mathcal{B}_{2,h}\|}{\beta_{2,\mathtt{d}}}\right) \operatorname{dist}\big(\sigma, X_{2,h}\big) \\
+ \ & \frac{\|\mathcal{B}_{1,h}\|}{\alpha_{\mathtt{d}}} \operatorname{dist}(u, M_{1,h}) + \frac{1}{\alpha_{\mathtt{d}}} \|(b_1 - b_{1,h})(\cdot, u)\|_{\mathcal{K}'_{1,h}} \\
+ \ & \left(1 + \frac{\|\mathcal{A}_h\|}{\alpha_{\mathtt{d}}}\right) \frac{1}{\beta_{2,\mathtt{d}}} \|(b_2 - b_{2,h})(\sigma, \cdot)\|_{M'_{2,h}} + \frac{1}{\alpha_{\mathtt{d}}} \|(a - a_h)(\sigma, \cdot)\|_{\mathcal{K}'_{1,h}} \\
+ \ & \frac{1}{\alpha_{\mathtt{d}}} \|F - F_h\|_{\mathcal{K}'_{1,h}} + \left(1 + \frac{\|\mathcal{A}_h\|}{\alpha_{\mathtt{d}}}\right) \frac{1}{\beta_{2,\mathtt{d}}} \|G - G_h\|_{M'_{2,h}},
\end{aligned}
$$

and

$$
\begin{aligned}
\|u - u_h\| \leqslant\ & C_{\mathtt{d}} \, \|\mathcal{A}_h\| \left(1 + \frac{\|\mathcal{B}_{2,h}\|}{\beta_{2,\mathtt{d}}}\right) \operatorname{dist}\big(\sigma, X_{2,h}\big) \\
& + \Big(1 + C_{\mathtt{d}} \, \|\mathcal{B}_{1,h}\|\Big) \operatorname{dist}\big(u, M_{1,h}\big) + C_{\mathtt{d}} \, \|(b_1 - b_{1,h})(\cdot, u)\|_{X'_{1,h}} \\
& + C_{\mathtt{d}} \left(\frac{\|\mathcal{A}_h\|}{\beta_{2,\mathtt{d}}} \|(b_2 - b_{2,h})(\sigma, \cdot)\|_{M'_{2,h}} + \|(a - a_h)(\sigma, \cdot)\|_{X'_{1,h}} \right) \\
& + C_{\mathtt{d}} \left(\|F - F_h\|_{X'_{1,h}} + \frac{\|\mathcal{A}_h\|}{\beta_{2,\mathtt{d}}} \|G - G_h\|_{M'_{2,h}} \right).
\end{aligned}
$$

Proof. It follows in the same way as the proof of Theorem 5.30, now using the estimates (5.213) and (5.214).

□

On the other hand, when the discrete bilinear forms and functionals coincide with the continuous ones, Theorems 5.30 and 5.31 become the following result with the usual Céa-type estimates.

Theorem 5.32. *Assume the hypotheses of Theorems* 5.26 *and* 5.27*, and let* $(\sigma, u) \in X_2 \times M_1$ *and* $(\sigma_h, u_h) \in X_{2,h} \times M_{1,h}$ *be the unique solutions of* (5.166) *and* (5.189)*, respectively, with* $a = a_h$*,* $b_i = b_{i,h}$*,* $i \in \{1, 2\}$*,* $F = F_h$*, and* $G = G_h$*. Then one has*

$$\|\sigma - \sigma_h\| \leqslant \left(1 + \frac{\|\mathcal{A}\|}{\alpha_{\mathrm{d}}}\right)\left(1 + \frac{\|\mathcal{B}_2\|}{\beta_{2,\mathrm{d}}}\right) \operatorname{dist}(\sigma, X_{2,h}) + \frac{\|\mathcal{B}_1\|}{\alpha_{\mathrm{d}}} \operatorname{dist}(u, M_{1,h})$$

and

$$\begin{aligned} \|u - u_h\| \leqslant & \left(1 + \frac{\|\mathcal{A}\|}{\alpha_{\mathrm{d}}}\right)\left(1 + \frac{\|\mathcal{B}_2\|}{\beta_{2,\mathrm{d}}}\right) \frac{\|\mathcal{A}\|}{\beta_{1,\mathrm{d}}} \operatorname{dist}(\sigma, X_{2,h}) \\ & + \left(1 + \frac{\|\mathcal{B}_1\|}{\beta_{1,\mathrm{d}}} + \frac{\|\mathcal{A}\|\,\|\mathcal{B}_1\|}{\alpha_{\mathrm{d}}\,\beta_{1,\mathrm{d}}}\right) \operatorname{dist}(u, M_{1,h}) . \end{aligned}$$

Proof. It follows from the corresponding estimates of Theorems 5.30 and 5.31, replacing $\|\mathcal{A}_h\|$, $\|\mathcal{B}_{1,h}\|$, and $\|\mathcal{B}_{2,h}\|$ by $\|\mathcal{A}\|$, $\|\mathcal{B}_1\|$, and $\|\mathcal{B}_2\|$, respectively, and performing minor algebraic manipulations.

□

5.5.5 An application example

We now apply the classical and general Babuška–Brezzi theories developed in Section 5.2 and in the present Section 5.5, respectively, to analyze the continuous solvability of a heat transfer model in a porous medium.

5.5.5.1 The model

The temperature distribution φ of a fluid in a porous medium occupying a simply connected and Lipschitz-continuous domain $\Omega \subset \mathbb{R}^n$, $n \in \{2, 3\}$, is modeled by coupling Darcy's law with a convection–diffusion equation depending on the fluid velocity $\mathbf{u}$ (see, e.g., [18], [79]). More precisely, denoting $\Gamma := \partial\Omega$ with associated unit normal vector $\boldsymbol{\nu}$, the corresponding system of equations is given by

$$\begin{gathered} \mu(\varphi)\,\mathbf{u} + \nabla p = \mathbf{f} \quad \text{in} \quad \Omega, \quad \operatorname{div}(\mathbf{u}) = 0 \quad \text{in} \quad \Omega, \quad \mathbf{u}\cdot\boldsymbol{\nu} = 0 \quad \text{on} \quad \Gamma, \\ -\kappa\,\Delta\varphi + \mathbf{u}\cdot\nabla\varphi = f \quad \text{in} \quad \Omega, \quad \varphi = 0 \quad \text{on} \quad \Gamma, \end{gathered} \tag{5.215}$$

where μ is the coefficient (temperature-dependent) representing the porosity times the dynamic viscosity, divided by the permeability, and will hereafter simply be referred to as the scaled viscosity; p is the pressure, $\mathbf{f}$ is an external body force, κ is the thermal conductivity coefficient, which is certainly positive, and f denotes an external scalar heat source (per unit volume of the porous medium). Throughout the following analysis, appropriate assumptions on the data $\mathbf{f}$ and f will be given. We also assume that the scaled viscosity $\mu : \mathbb{R} \to \mathbb{R}^+$ is uniformly bounded and Lipschitz continuous, which means that there exist positive constants μ_0, μ_1, and L_μ such that

$$\mu_0 \leqslant \mu(t) \leqslant \mu_1 \quad \forall\, t \in \mathbb{R} \quad \text{and} \quad |\mu(t) - \mu(\tilde{t})| \leqslant L_\mu\,|t - \tilde{t}| \quad \forall\, t, \tilde{t} \in \mathbb{R}. \tag{5.216}$$

Since our goal is to employ mixed formulations for both the Darcy and heat equations, we now introduce as auxiliary unknown the pseudo-heat flux given by the difference between the conductive and convective heat fluxes, that is

$$\boldsymbol{\sigma} := \kappa\,\nabla\varphi - \varphi\,\mathbf{u} \quad \text{in} \quad \Omega,$$

and observe that, thanks to the incompressibility condition given by the second equation in the first row of (5.215), we obtain

$$\operatorname{div}(\boldsymbol{\sigma}) \,=\, \kappa\,\Delta\varphi \,-\, \mathbf{u}\cdot\nabla\varphi \quad \text{in} \quad \Omega\,.$$

Note that the identity $\operatorname{div}(\varphi\,\mathbf{u}) = \mathbf{u}\cdot\nabla\varphi$, which was first utilized in Example 4.34 (cf. Chapter 4), is actually justified later on in Example 9.51 (cf. Chapter 9), where it is established in the distributional sense (cf. (9.33)). Therefore, (5.215) can be equivalently rewritten as the first-order nonlinear system

$$\begin{aligned} \mu(\varphi)\,\mathbf{u} + \nabla p = \mathbf{f} \quad \text{in} \quad \Omega\,, \quad \operatorname{div}(\mathbf{u}) = 0 \quad \text{in} \quad \Omega\,, \quad \mathbf{u}\cdot\boldsymbol{\nu} = 0 \quad \text{on} \quad \Gamma\,,\\ \kappa\,\nabla\varphi - \varphi\,\mathbf{u} = \boldsymbol{\sigma} \quad \text{in} \quad \Omega\,, \quad \operatorname{div}(\boldsymbol{\sigma}) = -f \quad \text{in} \quad \Omega\,, \quad \varphi = 0 \quad \text{on} \quad \Gamma\,. \end{aligned} \tag{5.217}$$

We stress that one of the advantages of also using a mixed formulation for the heat equation is the possibility of directly computing another physically meaningful variable, namely the temperature gradient, through the formula

$$\nabla\varphi \,=\, \kappa^{-1}\left(\boldsymbol{\sigma} \,+\, \varphi\,\mathbf{u}\right).$$

5.5.5.2 The variational formulation

We begin the derivation of the weak formulation of (5.217) by testing the first equation in its second row against a vector function $\boldsymbol{\tau}$, which formally gives

$$\int_\Omega \boldsymbol{\sigma}\cdot\boldsymbol{\tau} \,-\, \kappa\int_\Omega \nabla\varphi\cdot\boldsymbol{\tau} \,+\, \int_\Omega \varphi\,\mathbf{u}\cdot\boldsymbol{\tau} \,=\, 0\,. \tag{5.218}$$

Then, using the Cauchy–Schwarz and Hölder inequalities, in a manner similar to the derivation of (5.129), we find that for all $\ell, \jmath \in (1,+\infty)$ conjugate to each other, that is, such that $\frac{1}{\ell} + \frac{1}{\jmath} = 1$, we have

$$\left|\int_\Omega \varphi\,\mathbf{u}\cdot\boldsymbol{\tau}\right| \,\leqslant\, \|\varphi\|_{0,2\ell;\Omega}\,\|\mathbf{u}\|_{0,2\jmath;\Omega}\,\|\boldsymbol{\tau}\|_{0,\Omega}\,, \tag{5.219}$$

which shows that the third term on the left-hand side of (5.218) is well defined if $\varphi \in \mathrm{L}^\rho(\Omega)$, $\mathbf{u} \in \mathbf{L}^r(\Omega)$, and $\boldsymbol{\tau} \in \mathbf{L}^2(\Omega)$, where

$$\rho \,:=\, 2\ell \quad \text{and} \quad r \,:=\, 2\jmath\,. \tag{5.220}$$

Then, knowing the space to which $\boldsymbol{\tau}$ must belong, the first and second terms on the left-hand side of (5.218) are, in turn, well defined if $\boldsymbol{\sigma} \in \mathbf{L}^2(\Omega)$ and $\nabla\varphi \in \mathbf{L}^2(\Omega)$, respectively. Note that this latter fact, together with $\rho > 2$, justifies assuming that φ is sought initially in $\mathrm{H}^1(\Omega)$. Moreover, for consistency with its previous membership in $\mathrm{L}^\rho(\Omega)$, we need $\mathrm{H}^1(\Omega)$ to embed continuously into $\mathrm{L}^\rho(\Omega)$, which, denoting by $\varrho \in (1,+\infty)$ the conjugate of ρ, is ensured by the range established for ϱ in (5.132). Then, recalling from (5.130) the Banach space

$$\mathbf{H}(\operatorname{div}_\varrho;\Omega) \,:=\, \Big\{\boldsymbol{\tau} \in \mathbf{L}^2(\Omega) : \quad \operatorname{div}(\boldsymbol{\tau}) \in \mathrm{L}^\varrho(\Omega)\Big\}\,,$$

endowed with the norm defined in (5.131), we can apply the integration by parts formula given in (5.133) to obtain

$$\int_\Omega \nabla\varphi\cdot\boldsymbol{\tau} \,=\, -\int_\Omega \varphi\operatorname{div}(\boldsymbol{\tau}) \,+\, \langle\boldsymbol{\tau}\cdot\boldsymbol{\nu},\varphi\rangle \,=\, -\int_\Omega \varphi\operatorname{div}(\boldsymbol{\tau}) \quad \forall\,\boldsymbol{\tau} \in \mathbf{H}(\operatorname{div}_\varrho;\Omega)\,. \tag{5.221}$$

Although the corresponding boundary term vanishes due to the homogeneous Dirichlet condition on φ, we recall once again, as stated in (5.133), that $\boldsymbol{\tau}\cdot\boldsymbol{\nu}$ belongs to $\mathrm{H}^{-1/2}(\Gamma)$ and that $\langle\cdot,\cdot\rangle$ denotes the duality pairing between $\mathrm{H}^{-1/2}(\Gamma)$ and $\mathrm{H}^{1/2}(\Gamma)$. Now, substituting the resulting expression from (5.221) into (5.218), we obtain

$$\int_\Omega \boldsymbol{\sigma}\cdot\boldsymbol{\tau} + \kappa\int_\Omega \varphi\,\mathrm{div}(\boldsymbol{\tau}) + \int_\Omega \varphi\,\mathbf{u}\cdot\boldsymbol{\tau} = 0 \qquad \forall\,\boldsymbol{\tau}\in\mathbf{H}(\mathrm{div}_\varrho;\Omega)\,. \tag{5.222}$$

In turn, testing the second equation of the second row of (5.217) with $\psi\in\mathrm{L}^\rho(\Omega)$, which implicitly requires the unknown $\boldsymbol{\sigma}$ to belong to $\mathbf{H}(\mathrm{div}_\varrho;\Omega)$ and the data f to belong to $\mathrm{L}^\varrho(\Omega)$, and multiplying by the constant κ, we obtain

$$\kappa\int_\Omega \psi\,\mathrm{div}(\boldsymbol{\sigma}) = -\kappa\int_\Omega f\,\psi \qquad \forall\,\psi\in\mathrm{L}^\rho(\Omega)\,. \tag{5.223}$$

Thus, given $\mathbf{u}\in\mathbf{L}^r(\Omega)$, and defining

$$H := \mathbf{H}(\mathrm{div}_\varrho;\Omega) \quad\text{and}\quad Q := \mathrm{L}^\rho(\Omega)\,, \tag{5.224}$$

the mixed variational formulation of the convection–diffusion model reduces to (5.222) and (5.223), that is: Find $(\boldsymbol{\sigma},\varphi)\in H\times Q$ such that

$$\begin{aligned} a(\boldsymbol{\sigma},\boldsymbol{\tau}) + b(\boldsymbol{\tau},\varphi) + \int_\Omega \varphi\,\mathbf{u}\cdot\boldsymbol{\tau} = 0 \qquad & \forall\,\boldsymbol{\tau}\in H\,,\\ b(\boldsymbol{\sigma},\psi) = -\kappa\int_\Omega f\,\psi \qquad & \forall\,\psi\in Q\,, \end{aligned} \tag{5.225}$$

where $a: H\times H\to\mathbb{R}$ and $b: H\times Q\to\mathbb{R}$ are the bilinear forms defined as

$$a(\boldsymbol{\zeta},\boldsymbol{\tau}) := \int_\Omega \boldsymbol{\zeta}\cdot\boldsymbol{\tau} \qquad \forall\,(\boldsymbol{\zeta},\boldsymbol{\tau})\in H\times H\,, \tag{5.226a}$$

$$b(\boldsymbol{\tau},\psi) := \kappa\int_\Omega \psi\,\mathrm{div}(\boldsymbol{\tau}) \qquad \forall\,(\boldsymbol{\tau},\psi)\in H\times Q\,. \tag{5.226b}$$

It is easy to see that a and b are bounded with respect to the usual norms of the spaces $H := \mathbf{H}(\mathrm{div}_\varrho;\Omega)$ (cf. (5.131)) and $Q := \mathrm{L}^\rho(\Omega)$, and the corresponding continuity constants are given by

$$\|a\| = 1 \qquad\text{and}\qquad \|b\| = \kappa\,.$$

On the other hand, since we already know that $\mathbf{u}$ must be sought in $\mathbf{L}^r(\Omega)$, and keeping in mind the boundary and incompressibility conditions of the Darcy equation (cf. the first row of (5.217)), whose mixed variational formulation we now derive, we introduce for each $t\in(1,+\infty)$ the Banach space

$$\mathbf{H}^t(\mathrm{div}_t;\Omega) := \Big\{\mathbf{w}\in\mathbf{L}^t(\Omega):\quad \mathrm{div}(\mathbf{w})\in\mathrm{L}^t(\Omega)\Big\}\,,$$

endowed with the norm

$$\|\boldsymbol{\tau}\|_{t,\mathrm{div}_t;\Omega} := \|\boldsymbol{\tau}\|_{0,t;\Omega} + \|\mathrm{div}(\boldsymbol{\tau})\|_{0,t;\Omega} \qquad \forall\,\boldsymbol{\tau}\in\mathbf{H}^t(\mathrm{div}_t;\Omega)\,.$$

Then, denoting by $s\in(1,+\infty)$ the conjugate of r, we define the subspaces

$$X_2 = \mathbf{H}^r_0(\mathrm{div}_r;\Omega) := \Big\{\mathbf{w}\in\mathbf{H}^r(\mathrm{div}_r;\Omega):\quad \mathbf{w}\cdot\boldsymbol{\nu} = 0 \quad\text{on}\quad \Gamma\Big\}\,,\quad\text{and} \tag{5.227}$$

$$X_1 = \mathbf{H}_0^s(\mathrm{div}_s;\Omega) := \Big\{ \mathbf{w} \in \mathbf{H}^s(\mathrm{div}_s;\Omega) : \quad \mathbf{w}\cdot\boldsymbol{\nu} = 0 \quad \text{on} \quad \Gamma \Big\}. \tag{5.228}$$

Moreover, we recall from [52, Corollary B.57] the integration by parts formula stating that

$$\langle \boldsymbol{\tau}\cdot\boldsymbol{\nu}, v\rangle_\Gamma = \int_\Omega \Big\{ \boldsymbol{\tau}\cdot\nabla v + v\,\mathrm{div}(\boldsymbol{\tau}) \Big\} \qquad \forall\,(\boldsymbol{\tau},v) \in \mathbf{H}^s(\mathrm{div}_s;\Omega)\times \mathrm{W}^{1,r}(\Omega)\,, \tag{5.229}$$

where

$$\mathrm{W}^{1,r}(\Omega) := \Big\{ v \in \mathrm{L}^r(\Omega) : \quad \nabla v \in \mathbf{L}^r(\Omega) \Big\}.$$

Note that $\langle\cdot,\cdot\rangle_\Gamma$ denotes the duality pairing between $\mathrm{W}^{-1/s,s}(\Gamma)$ and $\mathrm{W}^{1/s,r}(\Gamma)$, where the latter is the trace space of $\mathrm{W}^{1,r}(\Omega)$, and the former is its dual.

Then, given $\varphi \in \mathrm{L}^\rho(\Omega)$, and assuming that the datum $\mathbf{f}$ belongs to $\mathbf{L}^r(\Omega)$, we test the first equation of the first row of (5.217) with $\mathbf{v} \in X_1$. Applying (5.229) to $\boldsymbol{\tau} = \mathbf{v} \in \mathbf{H}^s(\mathrm{div}_s;\Omega)$ and $v = p \in \mathrm{W}^{1,r}(\Omega)$, we obtain

$$\int_\Omega \mu(\varphi)\,\mathbf{u}\cdot\mathbf{v} - \int_\Omega p\,\mathrm{div}(\mathbf{v}) = \int_\Omega \mathbf{f}\cdot\mathbf{v} \qquad \forall\,\mathbf{v}\in X_1\,. \tag{5.230}$$

Next, we observe that the second term on the left-hand side of (5.230) vanishes when p is constant, and therefore, for uniqueness of the solution, the pressure unknown will henceforth be sought in the space

$$M_1 := \mathrm{L}_0^r(\Omega) := \Big\{ q \in \mathrm{L}^r(\Omega) : \quad \int_\Omega q = 0 \Big\}.$$

Related to this, and thanks to the decomposition $\mathrm{L}^s(\Omega) = \mathrm{L}_0^s(\Omega) \oplus \mathbb{R}$ and the boundary condition satisfied by $\mathbf{u}$, we realize that testing the incompressibility condition (second equation of the first row of (5.217)) with $q \in \mathrm{L}^s(\Omega)$ is equivalent to doing it with $q \in \mathrm{L}_0^s(\Omega)$, so that the associated test function space reduces to $M_2 := \mathrm{L}_0^s(\Omega)$. Consequently, the mixed variational formulation of the Darcy problem can be summarized as: Find $(\mathbf{u},p) \in X_2 \times M_1$ such that

$$\begin{aligned} a_\varphi(\mathbf{u},\mathbf{v}) + b_1(\mathbf{v},p) &= \int_\Omega \mathbf{f}\cdot\mathbf{v} \qquad \forall\,\mathbf{v}\in X_1\,,\\ b_2(\mathbf{u},q) &= 0 \qquad\qquad\quad \forall\,q \in M_2\,, \end{aligned} \tag{5.231}$$

where, given $\psi \in \mathrm{L}^\rho(\Omega)$, $a_\psi : X_2 \times X_1 \to \mathbb{R}$, $b_1 : X_1 \times M_1 \to \mathbb{R}$, and $b_2 : X_2 \times M_2 \to \mathbb{R}$ are the bilinear forms defined by

$$\begin{aligned} a_\psi(\mathbf{w},\mathbf{v}) &:= \int_\Omega \mu(\psi)\,\mathbf{w}\cdot\mathbf{v} \qquad \forall\,(\mathbf{w},\mathbf{v}) \in X_2\times X_1\,,\\ b_i(\mathbf{v},q) &:= -\int_\Omega q\,\mathrm{div}(\mathbf{v}) \qquad \forall\,(\mathbf{v},q)\in X_i\times M_i\,, \quad \forall\, i\in\{1,2\}\,. \end{aligned} \tag{5.232}$$

As with a and b, we note that under the assumptions on μ (cf. (5.216)), a_ψ is bounded with corresponding constant $\|a_\psi\| = \mu_1$ for all $\psi \in \mathrm{L}^\rho(\Omega)$, while b_1 and b_2 are also bounded with $\|b_1\| = \|b_2\| = 1$. Moreover, it is interesting to observe that, although b_1 and b_2 coincide algebraically, they differ topologically (as they are defined on spaces with different norms), and therefore they are essentially distinct.

We summarize the previous discussion with the resulting variational formulation for the full coupled problem (5.217), obtained by combining (5.225) and (5.231), namely: Find $(\boldsymbol{\sigma},\varphi) \in H \times Q$ and $(\mathbf{u},p) \in X_2 \times M_1$ such that

$$
\begin{aligned}
a(\boldsymbol{\sigma},\boldsymbol{\tau}) + b(\boldsymbol{\tau},\varphi) + \int_\Omega \varphi\,\mathbf{u}\cdot\boldsymbol{\tau} &= 0 && \forall\,\boldsymbol{\tau}\in H\,,\\
b(\boldsymbol{\sigma},\psi) &= -\kappa\int_\Omega f\,\psi && \forall\,\psi\in Q\,,\\
a_\varphi(\mathbf{u},\mathbf{v}) + b_1(\mathbf{v},p) &= \int_\Omega \mathbf{f}\cdot\mathbf{v} && \forall\,\mathbf{v}\in X_1\,,\\
b_2(\mathbf{u},q) &= 0 && \forall\,q\in M_2\,.
\end{aligned}
\tag{5.233}
$$

5.5.5.3 Solvability analysis

To study the existence of a solution to (5.233), we employ a fixed-point strategy similar to the one used in several related works (see, e.g., [14], [15], [42], [81], [85], and the references therein). For this purpose, let $\widehat{T}:\mathbf{L}^r(\Omega)\to H\times Q$ be the operator defined for each $\mathbf{w}\in\mathbf{L}^r(\Omega)$ as $\widehat{T}(\mathbf{w}) = (\widehat{T}_1(\mathbf{w}),\widehat{T}_2(\mathbf{w})) := (\widehat{\boldsymbol{\sigma}},\widehat{\varphi})$, where $(\widehat{\boldsymbol{\sigma}},\widehat{\varphi})\in H\times Q$ is the unique solution (to be proven later) of (5.225) with $\mathbf{w}$ instead of $\mathbf{u}$, that is,

$$
\begin{aligned}
a(\widehat{\boldsymbol{\sigma}},\boldsymbol{\tau}) + b(\boldsymbol{\tau},\widehat{\varphi}) + \int_\Omega \widehat{\varphi}\,\mathbf{w}\cdot\boldsymbol{\tau} &= 0 && \forall\,\boldsymbol{\tau}\in H\,,\\
b(\widehat{\boldsymbol{\sigma}},\psi) &= -\kappa\int_\Omega f\,\psi && \forall\,\psi\in Q\,.
\end{aligned}
\tag{5.234}
$$

Similarly, let $\widetilde{T}:\mathrm{L}^\rho(\Omega)\to X_2\times M_1$ be the operator defined for each $\psi\in\mathrm{L}^\rho(\Omega)$ as $\widetilde{T}(\psi) = (\widetilde{T}_1(\psi),\widetilde{T}_2(\psi)) := (\widetilde{\mathbf{u}},\widetilde{p})$, where $(\widetilde{\mathbf{u}},\widetilde{p})\in X_2\times M_1$ is the unique solution (to be confirmed later as well) of (5.231) with ψ instead of φ, that is,

$$
\begin{aligned}
a_\psi(\widetilde{\mathbf{u}},\mathbf{v}) + b_1(\mathbf{v},\widetilde{p}) &= \int_\Omega \mathbf{f}\cdot\mathbf{v} && \forall\,\mathbf{v}\in X_1\,,\\
b_2(\widetilde{\mathbf{u}},q) &= 0 && \forall\,q\in M_2\,.
\end{aligned}
\tag{5.235}
$$

Hence, defining the composite operator $T: X_2\to X_2$ as

$$
T(\mathbf{w}) := \widetilde{T}_1\big(\widehat{T}_2(\mathbf{w})\big) \qquad \forall\,\mathbf{w}\in X_2\,,
$$

it follows that solving (5.233) is equivalent to finding a fixed point of T, that is, $\mathbf{u}\in X_2$ such that

$$
T(\mathbf{u}) = \mathbf{u}\,. \tag{5.236}
$$

For all details regarding the solvability of (5.236), which include, among other aspects, proving continuity properties of $\widehat{T}$ and $\widetilde{T}$, and hence of T, we refer to [79, Section 2.5].

In what follows, we focus only on proving that $\widehat{T}$ and $\widetilde{T}$ are well defined, which is equivalent to showing that the decoupled problems given by (5.234) and (5.235) are well posed. In this regard, it is important to note first that, given $\mathbf{w}\in\mathbf{L}^r(\Omega)$, and after omitting the third term in the first row of (5.234), the resulting structure corresponds to that of the model analyzed by the classical Babuška–Brezzi Theory in Section 5.2. Likewise, given $\varphi\in\mathrm{L}^\rho(\Omega)$, the structure of (5.235) coincides with that of the problem analyzed by the general theory in Section 5.5. Consequently, our next goal is to verify the hypotheses of the main results of those sections, namely Theorem 5.9, for H, Q, a, and b, and Theorem 5.26, for X_2, M_1, X_1, M_2, a_ψ, b_1, and b_2.

As noted in Section 5.3.4, we first observe that, thanks to the reflexivity of $\mathbf{L}^2(\Omega)$ and $\mathrm{L}^t(\Omega)$, with $t\in(1,+\infty)$, it can be shown that H, Q, X_2, X_1, M_1, and M_2 are all reflexive

Banach spaces. In particular, Example 7.7 (cf. Chapter 7) provides the corresponding proof for $\mathbf{H}^t(\mathrm{div}_t;\Omega)$. Nevertheless, it is worth recalling that Theorem 5.26 does not require this property for X_2, but it does for M_1, X_1, and M_2. Now, to verify the hypotheses of Theorem 5.9, we proceed as in (5.145) and first define the null space V of the operator induced by the bilinear form b, that is,

$$V := \left\{\boldsymbol{\tau} \in H: \quad b(\boldsymbol{\tau},\psi) = 0 \quad \forall\, \psi \in Q\right\},$$

which, according to the definitions of b (cf. (5.226b)) and the spaces H and Q (cf. (5.224)), reduces to

$$V := \left\{\boldsymbol{\tau} \in \mathbf{H}(\mathrm{div}_\varrho;\Omega): \quad \mathrm{div}(\boldsymbol{\tau}) = 0\right\}.$$

Then, as for (5.146), it follows directly from the definitions of a (cf. (5.226a)) and the norm of $\mathbf{H}(\mathrm{div}_\varrho;\Omega)$ (cf. (5.131)) that

$$a(\boldsymbol{\tau},\boldsymbol{\tau}) = \|\boldsymbol{\tau}\|^2_{\mathrm{div}_\varrho;\Omega} \qquad \forall\, \boldsymbol{\tau} \in V,$$

from which it easily follows that a satisfies the assumptions required by i) in Theorem 5.9, in particular with constant $\widehat{\alpha} = 1$ in i-2) (or in i-2)'). On the other hand, we already know from Lemma 5.22 (cf. Section 5.3.4) that the bilinear form b satisfies hypothesis ii) of Theorem 5.9 with constant $\widehat{\beta} = \widetilde{\beta}$.

Consequently, a direct application of Theorem 5.9, and in particular of the a priori bounds given by (5.54) and (5.55), allows us to conclude, by virtue of the general equivalence established in (5.18), that the bounded bilinear form $A : \big(H \times Q\big) \times \big(H \times Q\big) \to \mathbb{R}$, which arises from (5.234) after summing the left-hand sides of the equations, but without including the term that depends on the given $\mathbf{w}$, satisfies a global inf–sup condition in its second component. More precisely, defining

$$A\big((\boldsymbol{\zeta},\phi),(\boldsymbol{\tau},\psi)\big) := a(\boldsymbol{\zeta},\boldsymbol{\tau}) + b(\boldsymbol{\tau},\phi) + b(\boldsymbol{\zeta},\psi)$$

for all $(\boldsymbol{\zeta},\phi)$, $(\boldsymbol{\tau},\psi) \in H \times Q$, there exists a positive constant $\alpha_{\widehat{T}}$, depending only on $\widehat{\alpha}$, $\widehat{\beta}$, and $\|a\|$, such that

$$\sup_{\substack{(\boldsymbol{\tau},\psi)\in H\times Q\\ (\boldsymbol{\tau},\psi)\neq 0}} \frac{A\big((\boldsymbol{\zeta},\phi),(\boldsymbol{\tau},\psi)\big)}{\|(\boldsymbol{\tau},\psi)\|_{H\times Q}} \geqslant \alpha_{\widehat{T}}\,\|(\boldsymbol{\zeta},\phi)\|_{H\times Q} \qquad \forall\,(\boldsymbol{\zeta},\phi) \in H \times Q. \tag{5.237}$$

Now, let $A_{\mathbf{w}} : \big(H \times Q\big) \times \big(H \times Q\big) \to \mathbb{R}$ be the bilinear form obtained by summing the complete left-hand sides of (5.234), which is clearly bounded thanks to the boundedness of A and to estimate (5.219). Then, formulation (5.234) can be equivalently rewritten as: Find $(\widehat{\boldsymbol{\sigma}},\widehat{\varphi}) \in H \times Q$ such that

$$A_{\mathbf{w}}\big((\widehat{\boldsymbol{\sigma}},\widehat{\varphi}),(\boldsymbol{\tau},\psi)\big) = G\big((\boldsymbol{\tau},\psi)\big) \qquad \forall\,(\boldsymbol{\tau},\psi) \in H \times Q, \tag{5.238}$$

where $G \in (H \times Q)'$ is defined as

$$G\big((\boldsymbol{\tau},\psi)\big) := -\kappa \int_\Omega f\,\psi \qquad \forall\,(\boldsymbol{\tau},\psi) \in H \times Q. \tag{5.239}$$

Hence, according to the definition of $A_{\mathbf{w}}$, and using (5.237) and (5.219), we deduce that

$$\sup_{\substack{(\boldsymbol{\tau},\psi)\in H\times Q\\(\boldsymbol{\tau},\psi)\neq 0}} \frac{A_{\mathbf{w}}\big((\boldsymbol{\zeta},\phi),(\boldsymbol{\tau},\psi)\big)}{\|(\boldsymbol{\tau},\psi)\|_{H\times Q}} \geqslant \alpha_{\widehat{T}}\,\|(\boldsymbol{\zeta},\phi)\|_{H\times Q} - \|\phi\|_{0,\rho;\Omega}\,\|\mathbf{w}\|_{0,r;\Omega}$$
$$\geqslant \Big\{\alpha_{\widehat{T}} - \|\mathbf{w}\|_{0,r;\Omega}\Big\}\,\|(\boldsymbol{\zeta},\phi)\|_{H\times Q} \quad \forall\,(\boldsymbol{\zeta},\phi)\in H\times Q\,,$$

so that, assuming $\|\mathbf{w}\|_{0,r;\Omega} \leqslant \dfrac{\alpha_{\widehat{T}}}{2}$, we obtain

$$\sup_{\substack{(\boldsymbol{\tau},\psi)\in H\times Q\\(\boldsymbol{\tau},\psi)\neq 0}} \frac{A_{\mathbf{w}}\big((\boldsymbol{\zeta},\phi),(\boldsymbol{\tau},\psi)\big)}{\|(\boldsymbol{\tau},\psi)\|_{H\times Q}} \geqslant \frac{\alpha_{\widehat{T}}}{2}\,\|(\boldsymbol{\zeta},\phi)\|_{H\times Q} \qquad \forall\,(\boldsymbol{\zeta},\phi)\in H\times Q\,. \tag{5.240}$$

Similarly, noting that A is symmetric, using again (5.237) and (5.219), and under the same assumption $\|\mathbf{w}\|_{0,r;\Omega} \leqslant \dfrac{\alpha_{\widehat{T}}}{2}$, it can be shown that

$$\sup_{\substack{(\boldsymbol{\zeta},\phi)\in H\times Q\\(\boldsymbol{\zeta},\phi)\neq 0}} \frac{A_{\mathbf{w}}\big((\boldsymbol{\zeta},\phi),(\boldsymbol{\tau},\psi)\big)}{\|(\boldsymbol{\zeta},\phi)\|_{H\times Q}} \geqslant \frac{\alpha_{\widehat{T}}}{2}\,\|(\boldsymbol{\tau},\psi)\|_{H\times Q} \qquad \forall\,(\boldsymbol{\tau},\psi)\in H\times Q\,. \tag{5.241}$$

We are now in a position to establish that $\widehat{T}$ (cf. (5.234)) is well defined.

Theorem 5.33. *For every $\mathbf{w}\in\mathbf{L}^r(\Omega)$ such that $\|\mathbf{w}\|_{0,r;\Omega} \leqslant \dfrac{\alpha_{\widehat{T}}}{2}$, there exists a unique solution $(\widehat{\boldsymbol{\sigma}},\widehat{\varphi})\in H\times Q$ to problem (5.238) (equivalently (5.234)), and therefore we can define $\widehat{T}(\mathbf{w}) := (\widehat{\boldsymbol{\sigma}},\widehat{\varphi})$. Moreover, the following a priori bound holds*

$$\|\widehat{T}(\mathbf{w})\|_{H\times Q} = \|\widehat{\boldsymbol{\sigma}}\|_{\mathrm{div}_\varrho;\Omega} + \|\widehat{\varphi}\|_{0,\rho;\Omega} \leqslant \frac{2}{\alpha_{\widehat{T}}}\,|\kappa|\,\|f\|_{0,\varrho;\Omega}\,. \tag{5.242}$$

Proof. It suffices to observe, according to (5.240) and (5.241), that $A_{\mathbf{w}}$ satisfies assumptions a) and b) of Lemma 5.2 (or equivalently i) and ii) of Lemma 5.1), with $\alpha = \widetilde{\alpha} = \dfrac{\alpha_{\widehat{T}}}{2}$, and therefore the existence of a unique solution follows as a consequence of either result. Accordingly, estimate (5.242) follows from (5.14), together with the fact that (5.239) and the Hölder inequality imply $\|G\| \leqslant |\kappa|\,\|f\|_{0,\varrho;\Omega}$.

□

The next goal is to verify the hypotheses of Theorem 5.26 for X_2, M_1, X_1, M_2, a_ψ, b_1, and b_2, in order to prove that, given $\psi\in \mathrm{L}^\rho(\Omega)$, problem (5.235) is well posed; equivalently, that the operator $\widetilde{T}$ is well defined. To this end, we need some preliminary results.

We begin with a theorem that establishes the solvability in $\mathrm{W}^{1,t}(\Omega)$, with $t\in(1,+\infty)$ in a range to be specified, of a second-order elliptic equation in divergence form with Neumann boundary conditions. In this regard, recall from Section 5.5.5.2 that, given t, $t'\in(1,+\infty)$ conjugate to each other, $\langle\cdot,\cdot\rangle_\Gamma$ denotes the duality pairing between $\mathrm{W}^{-1/t,t}(\Gamma)$ and $\mathrm{W}^{1/t,t'}(\Gamma)$.

Theorem 5.34. *Let Ω be a bounded, Lipschitz-continuous, simply connected domain of $\mathbb{R}^n$, $n\in\{2,3\}$, and let $g\in\mathrm{L}^t(\Omega)$, $\mathbf{g}\in\mathbf{L}^t(\Omega)$, and $g_N\in\mathrm{W}^{-1/t,t}(\Gamma)$, with $t\in(1,+\infty)$, such that g and g_N satisfy the compatibility condition*

$$\int_\Omega g = \langle g_N, 1\rangle_\Gamma\,. \tag{5.243}$$

Let $\xi : \Omega \to \mathbb{R}$ be a measurable function for which there exist constants ξ_0, ξ_1 such that $0 < \xi_0 \leqslant \xi(x) \leqslant \xi_1$ for all $x \in \Omega$. Then, for each $t \in [4/3, 4]$ when $n = 2$, and for each $t \in [3/2, 3]$ when $n = 3$, there exists a unique $u \in \widetilde{\mathrm{W}}^{1,t}(\Omega) := \Big\{ v \in \mathrm{W}^{1,t}(\Omega) : \quad \int_\Omega v = 0 \Big\}$, such that

$$\operatorname{div}\Big(\frac{1}{\xi}\,\nabla u - \mathbf{g}\Big) = g \quad \text{in} \quad \Omega \quad \text{and} \quad \Big(\frac{1}{\xi}\,\nabla u - \mathbf{g}\Big) \cdot \boldsymbol{\nu} = g_N \quad \text{on} \quad \Gamma\,. \tag{5.244}$$

Moreover, there exists a constant $C > 0$, depending only on n, t, ξ_0, ξ_1, and Ω, such that

$$\|u\|_{1,t;\Omega} \leqslant C \Big\{ \|g\|_{0,t;\Omega} + \|\mathbf{g}\|_{0,t;\Omega} + \|g_N\|_{-1/t,t;\Gamma} \Big\}\,. \tag{5.245}$$

Proof. It follows as a particular case of the more general result given in [89, Theorem 1.2]. □

As a consequence of the previous theorem, we can now establish the existence of an operator that will later be used to prove the continuous inf–sup conditions for a_ψ required by Theorem 5.26. More precisely, we have the following lemma.

Lemma 5.35. *Let $\xi : \Omega \to \mathbb{R}$ be as in Theorem 5.34, and let r, $s \in (1, +\infty)$ be conjugate to each other, such that r (and therefore s as well) satisfy the ranges stipulated by that theorem. Then, there exists a linear and bounded operator $D_s : \mathbf{L}^s(\Omega) \to \mathbf{L}^s(\Omega)$, with $\|D_s\|$ depending only on n, s, ξ_0, ξ_1, and Ω, such that*

$$\operatorname{div}\big(D_s(\mathbf{w})\big) = 0 \quad \text{in} \quad \Omega \quad \text{and} \quad D_s(\mathbf{w}) \cdot \boldsymbol{\nu} = 0 \quad \text{on} \quad \Gamma \qquad \forall\, \mathbf{w} \in \mathbf{L}^s(\Omega)\,. \tag{5.246}$$

Furthermore, for each $\mathbf{z} \in \mathbf{L}^r(\Omega)$ such that $\operatorname{div}(\mathbf{z}) = 0$ in Ω and $\mathbf{z} \cdot \boldsymbol{\nu} = 0$ on Γ, one has

$$\int_\Omega \xi\,\mathbf{z} \cdot D_s(\mathbf{w}) = \int_\Omega \xi\,\mathbf{z} \cdot \mathbf{w} \qquad \forall\, \mathbf{w} \in \mathbf{L}^s(\Omega)\,. \tag{5.247}$$

Proof. Given $\mathbf{w} \in \mathbf{L}^s(\Omega)$, we apply Theorem 5.34 with $t = s$, $g = 0$, $\mathbf{g} = \mathbf{w}$, and $g_N = 0$, to deduce that there exists a unique $u \in \widetilde{\mathrm{W}}^{1,s}(\Omega)$ such that

$$\operatorname{div}\Big(\frac{1}{\xi}\,\nabla u - \mathbf{w}\Big) = 0 \quad \text{in} \quad \Omega \quad \text{and} \quad \Big(\frac{1}{\xi}\,\nabla u - \mathbf{w}\Big) \cdot \boldsymbol{\nu} = 0 \quad \text{on} \quad \Gamma\,, \tag{5.248}$$

and a positive constant C, depending only on n, s, ξ_0, ξ_1, and Ω, such that

$$\|u\|_{1,s;\Omega} \leqslant C\, \|\mathbf{w}\|_{0,s;\Omega}\,. \tag{5.249}$$

Then, defining $D_s(\mathbf{w}) := \mathbf{w} - \dfrac{1}{\xi}\,\nabla u$, it is clear that D_s is linear, and that, according to (5.248), it satisfies the identities stated in (5.246). In turn, by using the triangle inequality, the lower bound of ξ, the fact that $\|\nabla u\|_{0,s;\Omega} = |u|_{1,s;\Omega} \leqslant \|u\|_{1,s;\Omega}$, and estimate (5.249), we easily obtain

$$\|D_s(\mathbf{w})\|_{0,s;\Omega} \leqslant \|\mathbf{w}\|_{0,s;\Omega} + \xi_0^{-1}\,\|u\|_{1,s;\Omega} \leqslant \big(1 + C\,\xi_0^{-1}\big)\,\|\mathbf{w}\|_{0,s;\Omega}\,,$$

which proves the required boundedness of D_s. Finally, for $\mathbf{z}$ as stated in the lemma, we have

$$\int_\Omega \xi\,\mathbf{z} \cdot D_s(\mathbf{w}) = \int_\Omega \xi\,\mathbf{z} \cdot \mathbf{w} - \int_\Omega \mathbf{z} \cdot \nabla u\,,$$

and applying the integration by parts formula (5.229) to $\boldsymbol{\tau} = \mathbf{z} \in \mathbf{H}^r(\mathrm{div}_r;\Omega)$ and $v = u \in \mathrm{W}^{1,s}(\Omega)$, we deduce that

$$\int_\Omega \mathbf{z}\cdot\nabla u = -\int_\Omega u\,\mathrm{div}(\mathbf{z}) + \langle \mathbf{z}\cdot\boldsymbol{\nu}, u\rangle_\Gamma = 0\,,$$

so that from the two previous equations we conclude (5.247), thus completing the proof. □

On the other hand, given r, $s \in (1,+\infty)$ conjugate to each other, we define for each $\mathbf{z} \in \mathbf{L}^r(\Omega)$ the function

$$\mathbf{z}_s := \begin{cases} |\mathbf{z}|^{r-2}\,\mathbf{z} & \text{if } \mathbf{z} \neq 0\,, \\ 0 & \text{if } \mathbf{z} = 0\,, \end{cases} \tag{5.250}$$

and observe, by simple algebraic computations, that $\mathbf{z}_s \in \mathbf{L}^s(\Omega)$ and

$$\int_\Omega \mathbf{z}\cdot\mathbf{z}_s = \|\mathbf{z}\|_{0,r;\Omega}^r = \|\mathbf{z}_s\|_{0,s;\Omega}^s = \|\mathbf{z}\|_{0,r;\Omega}\,\|\mathbf{z}_s\|_{0,s;\Omega}\,. \tag{5.251}$$

We are now in a position to prove the continuous inf–sup conditions required by Theorem 5.26. Indeed, let $\mathcal{K}_i$, $i \in \{1,2\}$, be the kernel of the bilinear form b_i (cf. (5.232)), that is,

$$\mathcal{K}_i := \Big\{ \mathbf{v} \in X_i : \quad b_i(\mathbf{v},q) = 0 \qquad \forall\, q \in M_i \Big\}\,,$$

which, according to the definitions of X_1 (cf. (5.228)), X_2 (cf. (5.227)), and b_i (cf. (5.232)), leads to

$$\mathcal{K}_1 := \Big\{ \mathbf{v} \in \mathbf{H}_0^s(\mathrm{div}_s;\Omega) : \quad \mathrm{div}(\mathbf{v}) = 0 \quad \text{in} \quad \Omega \Big\}\,, \quad \text{and}$$
$$\mathcal{K}_2 := \Big\{ \mathbf{w} \in \mathbf{H}_0^r(\mathrm{div}_r;\Omega) : \quad \mathrm{div}(\mathbf{w}) = 0 \quad \text{in} \quad \Omega \Big\}\,.$$

It is important to note here, by virtue of Lemma 5.35 (cf. (5.246)), that

$$D_s(\mathbf{w}) \in \mathcal{K}_1 \quad \forall\, \mathbf{w} \in \mathbf{L}^s(\Omega) \quad \text{and} \quad D_r(\mathbf{w}) \in \mathcal{K}_2 \quad \forall\, \mathbf{w} \in \mathbf{L}^r(\Omega)\,. \tag{5.252}$$

Then, the first announced result is given as follows.

Lemma 5.36. *There exists a constant $\widetilde{\alpha} > 0$ such that for each $\psi \in \mathrm{L}^\rho(\Omega)$ one has*

$$\sup_{\substack{\mathbf{v}\in\mathcal{K}_1\\ \mathbf{v}\neq 0}} \frac{a_\psi(\mathbf{w},\mathbf{v})}{\|\mathbf{v}\|_{X_1}} \geqslant \widetilde{\alpha}\,\|\mathbf{w}\|_{X_2} \qquad \forall\, \mathbf{w} \in \mathcal{K}_2\,, \tag{5.253}$$

and

$$\sup_{\mathbf{w}\in\mathcal{K}_2} a_\psi(\mathbf{w},\mathbf{v}) > 0 \qquad \forall\, \mathbf{v} \in \mathcal{K}_1\,,\ \mathbf{v} \neq 0\,. \tag{5.254}$$

Proof. Let $\psi \in \mathrm{L}^\rho(\Omega)$ and introduce $\xi := \mu(\psi)$, which satisfies, by (5.216), the hypotheses of Lemma 5.35 with $\xi_i = \mu_i$, $i \in \{0,1\}$. Then, given $\mathbf{w} \in \mathcal{K}_2$, $\mathbf{w} \neq 0$, we let $\mathbf{z} := \xi\,\mathbf{w}$ and observe that $\mathbf{z} \in \mathbf{L}^r(\Omega)$, so that, recalling that s is the conjugate of r, we define, according to (5.250), $\mathbf{z}_s \in \mathbf{L}^s(\Omega)$. Then, bounding below the supremum in (5.253) with $\mathbf{v} := D_s(\mathbf{z}_s) \in \mathcal{K}_1$ (cf. (5.252)), and using (5.247), (5.251), the boundedness of D_s, and the fact that $\xi \geqslant \mu_0$, we obtain

$$\sup_{\substack{\mathbf{v}\in\mathcal{K}_1\\ \mathbf{v}\neq 0}} \frac{a_\psi(\mathbf{w},\mathbf{v})}{\|\mathbf{v}\|_{X_1}} = \sup_{\substack{\mathbf{v}\in\mathcal{K}_1\\ \mathbf{v}\neq 0}} \frac{\int_\Omega \xi\,\mathbf{w}\cdot\mathbf{v}}{\|\mathbf{v}\|_{X_1}} \geqslant \frac{\int_\Omega \xi\,\mathbf{w}\cdot D_s(\mathbf{z}_s)}{\|D_s(\mathbf{z}_s)\|_{X_1}} = \frac{\int_\Omega \mathbf{z}\cdot\mathbf{z}_s}{\|D_s(\mathbf{z}_s)\|_{X_1}}$$

$$= \frac{\|\mathbf{z}\|_{0,r;\Omega}\,\|\mathbf{z}_s\|_{0,s;\Omega}}{\|D_s(\mathbf{z}_s)\|_{X_1}} \geqslant \frac{\|\mathbf{z}\|_{0,r;\Omega}}{\|D_s\|} = \frac{\|\xi\,\mathbf{w}\|_{0,r;\Omega}}{\|D_s\|} \geqslant \frac{\mu_0}{\|D_s\|}\,\|\mathbf{w}\|_{0,r;\Omega}\,,$$

which, noting that $\|\mathbf{w}\|_{0,r;\Omega} = \|\mathbf{w}\|_{X_2}$, proves (5.253) with $\widetilde{\alpha} := \frac{\mu_0}{\|D_s\|}$. For (5.254), the argument proceeds analogously by interchanging the roles of $\mathcal{K}_1$ and $\mathcal{K}_2$ (equivalently, those of $\mathbf{w}$ and $\mathbf{v}$), and using the first (instead of the last) equality in (5.251). Thus, given $\mathbf{v}\in\mathcal{K}_1$, $\mathbf{v}\neq 0$, we define $\mathbf{z} := \xi\,\mathbf{v}\in\mathbf{L}^s(\Omega)$, $\mathbf{z}_r\in\mathbf{L}^r(\Omega)$ (according to (5.250)), and bound below the supremum with $D_r(\mathbf{z}_r)$ to finally obtain

$$\sup_{\mathbf{w}\in\mathcal{K}_2} a_\psi(\mathbf{w},\mathbf{v}) \geqslant \|\xi\,\mathbf{v}\|^s_{0,s;\Omega} \geqslant \mu_0^s\,\|\mathbf{v}\|^s_{0,s;\Omega} > 0\,,$$

thus completing the proof. □

We now prove the continuous inf–sup conditions for b_1 and b_2.

Lemma 5.37. *There exist constants $\widetilde{\beta}_1,\,\widetilde{\beta}_2 > 0$ such that for each $i\in\{1,2\}$ one has*

$$\sup_{\substack{\mathbf{v}\in X_i\\ \mathbf{v}\neq 0}} \frac{b_i(\mathbf{v},q)}{\|\mathbf{v}\|_{X_i}} \geqslant \widetilde{\beta}_i\,\|q\|_{M_i} \qquad \forall\, q\in M_i\,. \tag{5.255}$$

Proof. We carry out the proof only for $i = 1$, since the case $i = 2$ follows verbatim by interchanging the roles of r and s. We begin by noting that (5.250) and (5.251) are also valid for the corresponding scalar version, which we employ below. Indeed, given $q\in M_1 = \mathrm{L}^r_0(\Omega)$, we introduce

$$q_s := \begin{cases} |q|^{r-2}\,q & \text{if } q\neq 0\,,\\ 0 & \text{if } q = 0\,,\end{cases}, \tag{5.256}$$

and notice that $q_s\in\mathrm{L}^s(\Omega)$ and

$$\int_\Omega q\,q_s = \|q\|^r_{0,r;\Omega} = \|q_s\|^s_{0,s;\Omega} = \|q\|_{0,r;\Omega}\,\|q_s\|_{0,s;\Omega}\,. \tag{5.257}$$

Next, defining $q^0_s := q_s - \frac{1}{|\Omega|}\int_\Omega q_s$, which belongs to $\mathrm{L}^s_0(\Omega)$, we apply Theorem 5.34 to deduce that there exists a unique $u\in\widetilde{\mathrm{W}}^{1,s}(\Omega)$ solution to problem (5.244) with $g = q^0_s$, $\mathbf{g} = 0$, and $g_N = 0$, that is,

$$\Delta u = q^0_s \quad\text{in}\quad \Omega\,,\quad \nabla u\cdot\boldsymbol{\nu} = 0 \quad\text{on}\quad \Gamma\,,\quad \int_\Omega u = 0\,. \tag{5.258}$$

Note, in particular, that g and g_N satisfy the compatibility condition (5.243). The continuous dependence result for (5.258) (cf. (5.245)) implies the existence of a constant $C_s > 0$, depending only on n, s, ξ_0, ξ_1, and Ω, such that $\|u\|_{1,s;\Omega} \leqslant C_s\,\|q^0_s\|_{0,s;\Omega}$. In turn, it is easy to see that there exists a constant $\widetilde{C}_s > 0$ such that $\|q^0_s\|_{0,s;\Omega} \leqslant \widetilde{C}_s\,\|q_s\|_{0,s;\Omega}$. Then, defining $\bar{\mathbf{v}} := -\nabla u \in \mathbf{L}^s(\Omega)$, we obtain $\operatorname{div}(\bar{\mathbf{v}}) = -q_s^0$ in Ω and $\bar{\mathbf{v}} \cdot \boldsymbol{\nu} = 0$ on Γ, so that $\bar{\mathbf{v}} \in X_1$ (cf. (5.228)) and

$$\begin{aligned}\|\bar{\mathbf{v}}\|_{X_1} &= \|\bar{\mathbf{v}}\|_{0,s;\Omega} + \|\operatorname{div}(\bar{\mathbf{v}})\|_{0,s;\Omega} = |u|_{1,s;\Omega} + \|q_s^0\|_{0,s;\Omega} \\ &\leqslant (1 + C_s)\,\|q_s^0\|_{0,s;\Omega} \leqslant (1 + C_s)\,\widetilde{C}_s\,\|q_s\|_{0,s;\Omega}\,.\end{aligned}$$

Thus, using that $\displaystyle\int_\Omega q\,q_s^0 = \int_\Omega q\,q_s$, we obtain

$$\sup_{\substack{\mathbf{v}\in X_1\\ \mathbf{v}\neq 0}} \frac{b_1(\mathbf{v},q)}{\|\mathbf{v}\|_{X_1}} \geqslant \frac{b_1(\bar{\mathbf{v}},q)}{\|\bar{\mathbf{v}}\|_{X_1}} = \frac{\displaystyle\int_\Omega q\,q_s}{\|\bar{\mathbf{v}}\|_{X_1}} = \frac{\|q\|_{0,r;\Omega}\,\|q_s\|_{0,s;\Omega}}{\|\bar{\mathbf{v}}\|_{X_1}} \geqslant \big((1 + C_s)\,\widetilde{C}_s\big)^{-1}\,\|q\|_{0,r;\Omega}\,,$$

which proves (5.255) for $i = 1$ with $\widetilde{\beta}_1 = \big((1 + C_s)\widetilde{C}_s\big)^{-1}$. As noted at the beginning of this proof, the continuous inf–sup condition for b_2 is obtained, given $q \in M_2 = \mathrm{L}_0^s(\Omega)$, by introducing $q_r \in \mathrm{L}^r(\Omega)$ and $q_r^0 \in \mathrm{L}_0^r(\Omega)$, analogously to (5.256) and the definition of q_s^0, respectively, and then taking $u \in \widetilde{\mathrm{W}}^{1,r}(\Omega)$ as the unique solution to (5.244) with $g = q_r^0$, $\mathbf{g} = 0$, and $g_N = 0$.

□

We remark here that for the value of r in the present example (cf. (5.220)), which is clearly greater than 2, the definitions given by (5.250) and (5.256) can certainly be simplified to $\mathbf{z}_s := |\mathbf{z}|^{r-2}\,\mathbf{z}$ and $q_s := |q|^{r-2}\,q$, respectively. Those more general definitions are actually required when applied to s instead of r, since, being $s - 2 < 0$, and in order to define $\mathbf{z}_r$ and q_r, one must distinguish whether the given functions $\mathbf{z}$ and q vanish or not.

Now, let $F \in X_1'$ be given by the right-hand side of the first row of (5.235), that is,

$$F(\mathbf{v}) := \int_\Omega \mathbf{f} \cdot \mathbf{v} \qquad \forall\, \mathbf{v} \in X_1\,,$$

which, thanks to Hölder's inequality, clearly satisfies

$$\|F\|_{X_1'} \leqslant \|\mathbf{f}\|_{0,r;\Omega}\,. \tag{5.259}$$

The following result establishes that the operator $\widetilde{T}$ (cf. (5.235)) is well defined.

Theorem 5.38. *For each $\psi \in \mathrm{L}^\rho(\Omega)$ there exists a unique solution $(\tilde{\mathbf{u}}, \tilde{p}) \in X_2 \times M_1$ to (5.235), and therefore one can define $\widetilde{T}(\psi) := (\tilde{\mathbf{u}}, \tilde{p})$. Moreover, the following a priori estimates hold:*

$$\begin{aligned}\|\widetilde{T}_1(\psi)\|_{X_2} &= \|\tilde{\mathbf{u}}\|_{X_2} \leqslant \frac{1}{\widetilde{\alpha}}\,\|\mathbf{f}\|_{0,r;\Omega} \quad \text{and} \\ \|\widetilde{T}_2(\psi)\|_{M_1} &= \|\tilde{p}\|_{M_1} \leqslant \frac{1}{\widetilde{\beta}_1}\Big(1 + \frac{\mu_1}{\widetilde{\alpha}}\Big)\,\|\mathbf{f}\|_{0,r;\Omega}\,.\end{aligned} \tag{5.260}$$

Proof. Let $\psi \in \mathrm{L}^\rho(\Omega)$. Then, thanks to Lemmas 5.36 and 5.37, and knowing that a_ψ, b_1, and b_2 are bounded, that X_2 is Banach, and that X_1, M_2, and M_1 are reflexive, the proof follows by a direct application of Theorem 5.26. In particular, the estimates established in (5.260) follow from (5.187) and (5.188), together with the upper bound for $\|F\|_{X_1'}$ given by (5.259), and the fact that the corresponding functional from the second row of (5.235) is zero.

□

We conclude this section by mentioning that the rest of the continuous analysis of the mixed variational formulation (5.233) of the coupled problem (5.215), as well as the entire analysis of an associated Galerkin scheme, can be found in [79].

5.6 An equivalence result on inf-sup conditions

In this section we establish an equivalence result that has shown to be very useful to prove continuous and discrete inf-sup conditions of bilinear forms acting on Banach spaces. In addition, a sufficiency hypothesis for one of the statements involved is also provided.

In what follows, given $\mathbf{X}$ and $\mathbf{Y}$ reflexive Banach spaces and a bounded bilinear form $\mathbf{b} : \mathbf{X} \times \mathbf{Y} \longrightarrow \mathrm{R}$, we proceed as in Section 5.3.1 and let $\mathbf{B} : \mathbf{X} \longrightarrow \mathbf{Y}'$ and $\mathbf{B}^{\mathrm{t}} : \mathbf{Y} \longrightarrow \mathbf{X}'$ be the bounded linear operators induced by $\mathbf{b}$, which are defined by

$$\mathbf{B}(x)(y) := \mathbf{b}(x,y) \quad \text{and} \quad \mathbf{B}^{\mathrm{t}}(y)(x) := \mathbf{b}(x,y) \qquad \forall\,(x,y) \in \mathbf{X} \times \mathbf{Y}\,.$$

In this regard, we recall from (5.72) that

$$\mathbf{B} \,=\, \left(\mathbf{B}^{\mathrm{t}}\right)' \circ \mathcal{J}_{\mathbf{X}} \quad \text{and} \quad \mathbf{B}^{\mathrm{t}} \,=\, \mathbf{B}' \circ \mathcal{J}_{\mathbf{Y}}\,, \tag{5.261}$$

where $\mathbf{B}' \in \mathcal{L}(\mathbf{Y}'', \mathbf{X}')$ and $\left(\mathbf{B}^{\mathrm{t}}\right)' \in \mathcal{L}(\mathbf{X}'', \mathbf{Y}')$ are the adjoint operators of $\mathbf{B}$ and $\mathbf{B}^{\mathrm{t}}$, and $\mathcal{J}_{\mathbf{X}} : \mathbf{X} \to \mathbf{X}''$ and $\mathcal{J}_{\mathbf{Y}} : \mathbf{Y} \to \mathbf{Y}''$ are the bijective mappings characterizing the reflexivity of $\mathbf{X}$ and $\mathbf{Y}$, respectively.

Then, we have the following result.

Lemma 5.39. *Let X, Y, Y_1, Y_2, and Z be reflexive Banach spaces with Y_1 and Y_2 being closed subspaces of Y such that $Y = Y_1 \oplus Y_2$, and assume that the norm of Y can be redefined, equivalently, but with constants independent of Y_1 and Y_2, as $\|y\| := \|y_1\| + \|y_2\|$ $\forall\, y = y_1 + y_2 \in Y$, with $y_j \in Y_j$ for $j \in \{1,2\}$. In addition, let $b : \big(X \times Y\big) \times Z \longrightarrow \mathrm{R}$ be a bounded bilinear form with boundedness constant denoted by $\|b\|$, and define the following subspaces:*

$$V := \Big\{(x,y) \in X \times Y : \quad b\big((x,y),z\big) = 0 \quad \forall\, z \in Z\Big\}, \tag{5.262}$$

and

$$Z_0 := \Big\{z \in Z : \quad b\big((x,y_2),z\big) = 0 \quad \forall\,(x,y_2) \in X \times Y_2\Big\}. \tag{5.263}$$

Then, the following statements are equivalent:

1) *there exist positive constants β_1 and β_2 such that*

$$\sup_{\substack{(x,y)\in X\times Y\\ (x,y)\neq 0}} \frac{b\big((x,y),z\big)}{\|(x,y)\|} \,\geqslant\, \beta_1\,\|z\| \qquad \forall\, z \in Z\,, \tag{5.264}$$

and

$$\|y_1\| \,\geqslant\, \beta_2\,\|(x,y_2)\| \quad \forall\,(x,y) \in V, \quad \textit{with} \quad y = y_1 + y_2 \in Y_1 \oplus Y_2 = Y\,. \tag{5.265}$$

2) *there exist positive constants β_3 and β_4 such that*

$$\sup_{\substack{z\in Z\\ z\neq 0}} \frac{b\big((x,y_2),z\big)}{\|z\|} \,\geqslant\, \beta_3\,\|(x,y_2)\| \qquad \forall\,(x,y_2) \in X \times Y_2\,, \tag{5.266}$$

and

$$\sup_{\substack{y_1 \in Y_1 \\ y_1 \neq 0}} \frac{b\big((0, y_1), z\big)}{\|y_1\|} \geqslant \beta_4 \|z\| \qquad \forall\, z \in Z_0 \,. \tag{5.267}$$

Moreover, in the implication 1) $\Rightarrow$ 2), the constants β_3 and β_4 depend only on β_1 and β_2, whereas in the converse implication 2) $\Rightarrow$ 1), the constants β_1 and β_2 depend only on β_3, β_4, and $\|b\|$.

Proof. Let us assume 1) and show 2). Indeed, taking $z \in Z_0$ in (5.264), which means that $b\big((x, y_2), z\big) = 0 \;\; \forall\,(x, y_2) \in X \times Y_2$, and using that certainly there holds $\|(x, y)\| \geqslant \|y_1\|$, we first obtain

$$\beta_1 \|z\| \leqslant \sup_{\substack{(x,y) \in X \times Y \\ (x,y) \neq 0}} \frac{b\big((x, y), z\big)}{\|(x, y)\|} = \sup_{\substack{(x,y) \in X \times Y \\ (x,y) \neq 0}} \frac{b\big((0, y_1), z\big)}{\|(x, y)\|} \leqslant \sup_{\substack{y_1 \in Y_1 \\ y_1 \neq 0}} \frac{b\big((0, y_1), z\big)}{\|y_1\|} \,,$$

which shows (5.267) with $\beta_4 = \beta_1$. Now, letting $B : X \times Y \to Z'$ and $B^{\mathtt{t}} : Z \to \big(X \times Y\big)'$ be the bounded linear operators induced by b, and observing from (5.262) that the null space $N(B) = V$, we realize that (5.264) establishes equivalently that $B^{\mathtt{t}}$, and hence B' as well (cf. (5.261)), is injective and of closed range, that is that B is surjective. The later means, in particular, that $R(B)$ is closed, which is equivalent (cf. Theorem 3.61) to the inf-sup condition

$$\|B\big((x, y)\big)\|_{Z'} := \sup_{\substack{z \in Z \\ z \neq 0}} \frac{b\big((x, y), z\big)}{\|z\|} \geqslant \beta_1 \operatorname{dist}\big((x, y), V\big) \quad \forall\,(x, y) \in X \times Y \,. \tag{5.268}$$

Note that the constant β_1 in (5.268) can in fact be chosen as the same from (5.264) thanks to the first implication in (5.80). Thus, taking $(x, y) = (x, y_2) \in X \times Y_2$, applying (5.265) to an arbitrary $(w, s) \in V$, with $s = s_1 + s_2 \in Y_1 \oplus Y_2 = Y$, and employing the triangle inequality, we get

$$\begin{aligned}
\operatorname{dist}\big((x, y_2), V\big) &:= \inf_{(w,s) \in V} \|(x, y_2) - (w, s)\| = \inf_{(w,s) \in V} \|(x - w, y_2 - s_2 - s_1)\| \\
&= \inf_{(w,s) \in V} \Big\{ \|x - w\| + \|y_2 - s_2\| + \|s_1\| \Big\} = \inf_{(w,s) \in V} \Big\{ \|(x, y_2) - (w, s_2)\| + \|s_1\| \Big\} \\
&\geqslant \min\{1, \beta_2\} \inf_{(w,s) \in V} \Big\{ \|(x, y_2) - (w, s_2)\| + \|(w, s_2)\| \Big\} \geqslant \min\{1, \beta_2\} \, \|(x, y_2)\| \,,
\end{aligned}$$

which, along with (5.268), yields (5.266) with $\beta_3 = \beta_1 \min\{1, \beta_2\}$. Conversely, in what follows we suppose 2) and prove 1). In fact, denoting by $B_1 : Y_1 \to Z_0'$ and $B_1^{\mathtt{t}} : Z_0 \to Y_1'$ the bounded linear operators induced by $b\big((0, \cdot), \cdot\big) : Y_1 \times Z_0 \longrightarrow \mathrm{R}$, we first realize from (5.267), similarly as done for B before, that $B_1^{\mathtt{t}}$, and hence B_1' as well, is injective and of closed range, that is that B_1 is surjective. Actually, using from the second identity in (5.261) that $B_1^{\mathtt{t}} = B_1' \circ \mathcal{J}_{Z_0}$, it is easy to see that (5.267) can be rewritten as $\|B_1'(\mathcal{G})\| \geqslant \beta_4 \|\mathcal{G}\|$ $\forall\, \mathcal{G} \in Z_0''$. In this way, given $G \in Z'$, we obviously have that $G|_{Z_0} \in Z_0'$, and hence, applying the converse implication of Lemma 3.65, we deduce that there exists $y_1 \in Y_1$ such that $B_1(y_1) = G|_{Z_0}$, that is

$$b\big((0, y_1), z\big) = G(z) \qquad \forall\, z \in Z_0 \,, \tag{5.269}$$

and $\beta_4 \|y_1\| \leqslant \|G|_{Z_0}\| \leqslant \|G\|$. On the other hand, denoting by $B_2 : X \times Y_2 \to Z'$ and $B_2^{\mathtt{t}} :$ $Z \to (X \times Y_2)'$ the bounded linear operators induced by $b|_{(X \times Y_2) \times Z}$, we see from (5.266) that B_2 is injective and of closed range. It follows that $R(B_2) = {}^{\circ}N(B_2')$, and according to the second identity from (5.261) applied to B_2, there holds $N(B_2') := \mathcal{J}_Z\big(N(B_2^{\mathtt{t}})\big)$, where

$$N(B_2^{\mathtt{t}}) := \Big\{z \in Z: \quad b\big((x, y_2), z\big) = 0 \quad \forall\, (x, y_2) \in X \times Y_2\Big\} = Z_0 \,,$$

whence

$$R(B_2) = {}^{\circ}\mathcal{J}_Z\big(Z_0\big) = \Big\{F \in Z': \quad \mathcal{J}_Z(z)(F) := F(z) = 0 \quad \forall\, z \in Z_0\Big\} = Z_0^{\circ}\,.$$

Thanks to the above, and since it is clear from (5.269) that $G - b\big((0, y_1), \cdot\big) \in Z_0^{\circ} = R(B_2)$, we deduce the existence of a unique $(x_G, y_2) \in X \times Y_2$ such that $B_2(x_G, y_2) = G - b\big((0, y_1), \cdot\big)$, that is $b\big((x_G, y_2), z\big) = G(z) - b\big((0, y_1), z\big) \;\; \forall\, z \in Z$, and

$$\beta_3 \, \|(x_G, y_2)\| \leqslant \|G - b\big((0, y_1), \cdot\big)\| \leqslant \|G\| + \|B_1\| \, \|y_1\| \leqslant \big(1 + \beta_4^{-1} \|b\|\big) \, \|G\| \,.$$

Thus, defining $y_G := y_1 + y_2 \in Y$, we readily see that $b\big((x_G, y_G), z\big) = G(z) \;\; \forall\, z \in Z$, and

$$\|(x_G, y_G)\| \leqslant \Big\{\beta_3^{-1}\big(1 + \beta_4^{-1}\|b\|\big) + \beta_4^{-1}\Big\} \, \|G\| \,.$$

Next, given arbitrary $z \in Z$ and $G \in Z'$, and defining $\beta_1 := \Big\{\beta_3^{-1}\big(1 + \beta_4^{-1}\|b\|\big) + \beta_4^{-1}\Big\}^{-1}$, we find that

$$\sup_{\substack{(x,y) \in X \times Y \\ (x,y) \neq 0}} \frac{b\big((x,y), z\big)}{\|(x,y)\|} \geqslant \frac{\big|b\big((x_G, y_G), z\big)\big|}{\|(x_G, y_G)\|} \geqslant \beta_1 \, \frac{|G(z)|}{\|G\|} \,,$$

from which, taking supremum on all $G \in Z'$, we conclude (5.264). It remains to prove (5.265). To this end, given $(x, y) \in V$, with $y = y_1 + y_2 \in Y_1 \oplus Y_2 = Y$, we first recall that for each $z \in Z$ there holds $0 = b\big((x, y), z\big) = b\big((x, y_2), z\big) + b\big((0, y_1), z\big)$, that is $b\big((x, y_2), z\big) = -b\big((0, y_1), z\big)$. Hence, employing (5.266) we deduce that

$$\beta_3 \, \|(x, y_2)\| \leqslant \sup_{\substack{z \in Z \\ z \neq 0}} \frac{\big|b\big((x, y_2), z\big)\big|}{\|z\|} = \sup_{\substack{z \in Z \\ z \neq 0}} \frac{\big|b\big((0, y_1), z\big)\big|}{\|z\|} \leqslant \|b\| \, \|y_1\| \,,$$

which yields (5.265) with $\beta_2 = \beta_3 \, \|b\|^{-1}$.

∎

Having established the equivalence given by Lemma 5.39, we now provide sufficient conditions for the inf-sup condition (5.266).

Lemma 5.40. *In addition to the notations and assumptions from Lemma* 5.39, *we now introduce the subspace*

$$Z_1 := \Big\{z \in Z: \quad b\big((x, 0), z\big) = 0 \qquad \forall\, x \in X\Big\}, \tag{5.270}$$

and assume that there exist positive constants β_5 *and* β_6 *such that*

$$\sup_{\substack{z \in Z \\ z \neq 0}} \frac{b\big((x, 0), z\big)}{\|z\|} \geqslant \beta_5 \, \|x\| \qquad \forall\, x \in X \,, \tag{5.271}$$

and

$$\sup_{\substack{z \in Z_1 \\ z \neq 0}} \frac{b\big((0, y_2), z\big)}{\|z\|} \geqslant \beta_6 \, \|y_2\| \qquad \forall\, y_2 \in Y_2 \,. \tag{5.272}$$

Then, the inf-sup condition (5.266) *is satisfied.*

Proof. Given $(x, y_2) \in X \times Y_2$, we begin by noticing that (5.271) and (5.272) guarantee the existence of $\tilde{z} \in Z$ and $\hat{z} \in Z_1$, respectively, satisfying $\|\tilde{z}\| = \|\hat{z}\| = 1$, and the inequalities

$$b\big((x,0),\tilde{z}\big) \geqslant \frac{\beta_5}{2}\,\|x\| \quad \text{and} \quad b\big((0,y_2),\hat{z}\big) \geqslant \frac{\beta_6}{2}\,\|y_2\|\,.$$

Then, defining $\bar{z} := C_1\,\tilde{z} + C_2\,\hat{z}$, with positive constants C_1 and C_2 to be chosen later on, which yields $\|\bar{z}\| \leqslant C_1 + C_2$, and using that $b\big((x,0),\hat{z}\big) = 0$, we obtain

$$\begin{aligned}
\sup_{\substack{z\in Z\\ z\neq 0}} \frac{b\big((x,y_2),z\big)}{\|z\|} &\geqslant \frac{\big|b\big((x,y_2),\bar{z}\big)\big|}{\|\bar{z}\|} = \frac{\big|b\big((x,0),\bar{z}\big) + b\big((0,y_2),\bar{z}\big)\big|}{\|\bar{z}\|}\\
&= \frac{\big|C_1\,b\big((x,0),\tilde{z}\big) + C_2\,b\big((0,y_2),\hat{z}\big) + C_1\,b\big((0,y_2),\tilde{z}\big)\big|}{\|\bar{z}\|}\\
&\geqslant \frac{1}{C_1 + C_2}\left\{\frac{C_1\,\beta_5}{2}\,\|x\| + \Big(\frac{C_2\,\beta_6}{2} - C_1\,\|b\|\Big)\,\|y_2\|\right\},
\end{aligned}$$

from which, choosing C_1 and C_2 such that $C_1 > 0$ and $\dfrac{C_2\,\beta_6}{2} > C_1\,\|b\|$, we arrive at (5.266) with β_3 depending on C_1, C_2, β_5, β_6, and $\|b\|$. For instance, taking $C_1 = 1$ and $C_2 = 4\|b\|\beta_6^{-1}$, we get $\beta_3 = \frac{1}{\left(1+4\|b\|\beta_6^{-1}\right)}\min\left\{\frac{\beta_5}{2}, \|b\|\right\}$.

∎

At this point we remark that a particular case of the equivalence between the statements (1) and (3) in [96, Theorem 3.1] would imply that actually (5.266) and the pair (5.271) - (5.272) are equivalent. However, we believe that the necessity of (5.271) - (5.272), and particularly that of (5.272), requires additionally that the kernel of the bilinear form $b\big((0,\cdot),\cdot\big) : Y_2 \times Z_1 \longrightarrow \mathrm{R}$ be the null space, that is that

$$\Big\{y_2 \in Y_2: \quad b\big((0,y_2),z\big) = 0 \quad \forall\, z \in Z_1\Big\} = \Big\{0\Big\}, \tag{5.273}$$

which is not included in the statement of [96, Theorem 3.1]. In any case, and though (5.273) clearly follows from (5.272), in all the applications of Lemma 5.39 that are described throughout the rest of this book we do not need neither such equivalence nor (5.273) as such, but only the sufficiency provided by Lemma 5.40.

We end this section by stating the particular version of Lemma 5.39 that arises when $Y = Y_1$, which, as shown throughout the rest of this book, is explicitly applied in several problems.

Lemma 5.41. *Let X, Y, and Z be reflexive Banach spaces, and let $b : \big(X \times Y\big) \times Z \longrightarrow \mathrm{R}$ be a bounded bilinear form with boundedness constant denoted by $\|b\|$. In addition, define the following subspaces:*

$$V := \Big\{(x,y) \in X \times Y: \quad b\big((x,y),z\big) = 0 \quad \forall\, z \in Z\Big\}, \tag{5.274}$$

and

$$Z_0 := \Big\{z \in Z: \quad b\big((x,0),z\big) = 0 \quad \forall\, x \in X\Big\}. \tag{5.275}$$

Then, the following statements are equivalent:

1) *there exist positive constants* β_1 *and* β_2 *such that*

$$\sup_{\substack{(x,y)\in X\times Y\\ (x,y)\neq 0}} \frac{b\big((x,y),z\big)}{\|(x,y)\|} \geqslant \beta_1 \|z\| \qquad \forall\, z\in Z\,, \tag{5.276}$$

and

$$\|y\| \geqslant \beta_2 \|x\| \quad \forall\,(x,y)\in V\,. \tag{5.277}$$

2) *there exist positive constants* β_3 *and* β_4 *such that*

$$\sup_{\substack{z\in Z\\ z\neq 0}} \frac{b\big((x,0),z\big)}{\|z\|} \geqslant \beta_3 \|x\| \qquad \forall\, x\in X\,, \tag{5.278}$$

and

$$\sup_{\substack{y\in Y\\ y\neq 0}} \frac{b\big((0,y),z\big)}{\|y\|} \geqslant \beta_4 \|z\| \qquad \forall\, z\in Z_0\,. \tag{5.279}$$

Moreover, in the implication 1) $\Rightarrow$ 2)*, the constants* β_3 *and* β_4 *depend only on* β_1 *and* β_2*, whereas in the converse implication* 2) $\Rightarrow$ 1)*, the constants* β_1 *and* β_2 *depend only on* β_3*,* β_4*, and* $\|b\|$*.*

5.7 Exercises.

▸ **5.1** *Given* $r\in(1,+\infty)$*,* Ω *a bounded open subset of* $\mathbb{R}^n$*, and* $\mathbb{K}=\mathbb{R}$*, we recall from Exercise 1.10 that* $\mathrm{L}^r(\Omega)$ *denotes the real space of all (equivalence classes of) measurable functions* $v:\Omega\to\mathbb{R}$ *such that*

$$\int_\Omega |v|^r < \infty\,,$$

which, endowed with the norm

$$\|v\|_{0,r;\Omega} := \left(\int_\Omega |v|^r\right)^{1/r} \qquad \forall\, v\in \mathrm{L}^r(\Omega)\,,$$

is known to be a reflexive Banach space. Next, given $p,\,q\in(1,+\infty)$ *such that* $\frac{1}{p}+\frac{1}{q}=1$*, the Hölder inequality establishes that*

$$\left|\int_\Omega wv\right| \leqslant \|w\|_{0,p;\Omega}\,\|v\|_{0,q;\Omega} \qquad \forall\, w\in \mathrm{L}^p(\Omega)\,,\quad \forall\, v\in \mathrm{L}^q(\Omega)\,.$$

a) *Here we aim to justify the identities (5.148) and (5.257), which certainly extends easily to (5.251). Indeed, given* $w\in \mathrm{L}^p(\Omega)$*, define*

$$w_q := \begin{cases} |w|^{p-2}\, w & \text{if } w\neq 0\\ 0 & \text{if } w=0\end{cases},$$

deduce that $w_q\in \mathrm{L}^q(\Omega)$*, and prove that*

$$\int_\Omega w\, w_q = \|w\|^p_{0,p;\Omega} = \|w_q\|^q_{0,q;\Omega} = \|w\|_{0,p;\Omega}\,\|w_q\|_{0,q;\Omega}\,.$$

b) *Apply the Generalized Lax–Milgram Theorem (cf. Lemma 5.1) to show that for all* $F\in \mathrm{L}^q(\Omega)'$ *there exists a unique* $u\in \mathrm{L}^p(\Omega)$ *such that* $F(v)=\int_\Omega uv$ *for all* $v\in \mathrm{L}^q(\Omega)$*, and conclude, furthermore, that* $\|F\|=\|u\|_{0,p;\Omega}$*. Deduce, from the above, that* $\mathrm{L}^q(\Omega)'$ *(resp.* $\mathrm{L}^p(\Omega)'$*) can be identified, through an isomorphism, with* $\mathrm{L}^p(\Omega)$ *(resp.* $\mathrm{L}^q(\Omega)$*).*

▸ **5.2** ([42]) *Let Ω be a bounded open subset of $\mathbb{R}^n$, $n \in \{2,3\}$, with boundary Γ of class $C^{0,1}$, and let $p,\, q \in (1,+\infty)$ be such that $\frac{1}{p} + \frac{1}{q} = 1$. One of the classical Sobolev embedding theorems (cf. [111, Theorem 1.3.4]) guarantees that $\mathrm{H}^1(\Omega) \subseteq \mathrm{L}^q(\Omega)$, and that the injection $i_q : \mathrm{H}^1(\Omega) \to \mathrm{L}^q(\Omega)$ is bounded, for all $q \in \begin{cases} [1,+\infty) & \text{if } n = 2 \\ [1,6] & \text{if } n = 3 \end{cases}$. On the other hand, define the Banach space*

$$\mathbf{H}(\mathrm{div}_p;\Omega) := \left\{\tau \in \mathbf{L}^2(\Omega) : \quad \mathrm{div}(\tau) \in \mathrm{L}^p(\Omega)\right\},$$

endowed with the induced norm

$$\|\tau\|_{\mathrm{div}_p;\Omega} := \|\tau\|_{0,\Omega} + \|\mathrm{div}(\tau)\|_{0,p;\Omega} \qquad \forall\, \tau \in \mathbf{H}(\mathrm{div}_p;\Omega),$$

where, as in Exercise 5.1, $\|\cdot\|_{0,t;\Omega}$ denotes the $\mathrm{L}^t(\Omega)$ norm for each $t \in (1,+\infty)$.

a) *Proceed analogously to what was done in Section 4.5.4 (cf. Lemma 4.32) for the space $\mathbf{H}(\mathrm{div};\Omega)$ ($p = 2$), and use the previous remarks to prove that for each $p \in \begin{cases} (1,+\infty] & \text{if } n = 2 \\ [6/5,+\infty] & \text{if } n = 3 \end{cases}$, one can also define $\gamma_{\boldsymbol{\nu}} \in \mathcal{L}\big(\mathbf{H}(\mathrm{div}_p;\Omega), \mathrm{H}^{-1/2}(\Gamma)\big)$.*

b) *Conclude from a) that in such case there holds*

$$\langle \gamma_{\boldsymbol{\nu}}(\tau), \gamma_0(v)\rangle = \int_\Omega \tau \cdot \nabla v + \int_\Omega v\,\mathrm{div}(\tau) \qquad \forall\, \tau \in \mathbf{H}(\mathrm{div}_p;\Omega), \quad \forall\, v \in \mathrm{H}^1(\Omega).$$

▸ **5.3** *Let Ω be a bounded open subset of $\mathbb{R}^2$, and let ε, δ be given positive constants. Let $r,\, s \in (1,+\infty)$ be such that $\frac{1}{r} + \frac{1}{s} = 1$, and define the spaces*

$$\mathrm{V}_r := \mathrm{H}^1(\Omega) \times \mathrm{L}^r(\Omega) \quad \text{and} \quad \mathrm{V}_s := \mathrm{H}^1(\Omega) \times \mathrm{L}^s(\Omega).$$

Then, for each $\mathrm{F} \in \mathrm{V}_s'$, consider the problem: Find $(\sigma,u) \in \mathrm{V}_r$ such that

$$\mathrm{B}\big((\sigma,u),(\tau,v)\big) = \mathrm{F}\big((\tau,v)\big) \qquad \forall\, (\tau,v) \in \mathrm{V}_s, \tag{5.280}$$

where

$$\mathrm{B}\big((\zeta,w),(\tau,v)\big) := \mathrm{A}\big((\zeta,w),(\tau,v)\big) - \varepsilon \int_\Omega \zeta\, v, \quad \text{and}$$

$$\mathrm{A}\big((\zeta,w),(\tau,v)\big) := \int_\Omega \big\{\zeta\,\tau + \delta\,\nabla\zeta\cdot\nabla\tau\big\} + \int_\Omega w\, v,$$

for all $\big((\zeta,w),(\tau,v)\big) \in \mathrm{V}_r \times \mathrm{V}_s$.

a) *Use the fact that $\mathrm{H}^1(\Omega)$ is contained in $\mathrm{L}^r(\Omega)$ and that the injection $i_r : \mathrm{H}^1(\Omega) \to \mathrm{L}^r(\Omega)$ is bounded, to prove that there exists a positive constant C_r, which depends on $\|i_r\|$, such that*

$$\sup_{\substack{(\tau,v)\in \mathrm{V}_s \\ (\tau,v)\neq 0}} \frac{\mathrm{B}\big((\zeta,w),(\tau,v)\big)}{\|(\tau,v)\|_{\mathrm{V}_s}} \geqslant \sup_{\substack{(\tau,v)\in \mathrm{V}_s \\ (\tau,v)\neq 0}} \frac{\mathrm{A}\big((\zeta,w),(\tau,v)\big)}{\|(\tau,v)\|_{\mathrm{V}_s}} - C_r\,\varepsilon\,\|(\zeta,w)\|_{\mathrm{V}_r}$$

for all $(\zeta,w) \in \mathrm{V}_r$.

b) *From a) and additional computations, prove that for ε sufficiently small, and for each $\mathrm{F} \in \mathrm{V}_s'$, problem (5.280) has a unique solution $(\sigma,u) \in \mathrm{V}_r$. Also prove that there exists a constant $c > 0$, depending on δ, such that*

$$\|(\sigma,u)\|_{\mathrm{V}_r} \leqslant c\,\|\mathrm{F}\|_{\mathrm{V}_s'}.$$

▸ **5.4** *Let Ω be a bounded open subset of $\mathbb{R}^n$, $\kappa \in C(\Omega)$, and suppose there exist constants κ_0, $\kappa_1 > 0$ such that $\kappa_0 \leqslant \kappa(\mathbf{x}) \leqslant \kappa_1$ for all $\mathbf{x} \in \Omega$. Let r, $s \in (1, +\infty)$ be such that $\frac{1}{r} + \frac{1}{s} = 1$, and define the spaces*

$$\mathbf{X}_r := \mathbf{H}(\mathrm{div}_0; \Omega) \times \mathrm{L}^r(\Omega) \quad \textit{and} \quad \mathbf{X}_s := \mathbf{H}(\mathrm{div}_0; \Omega) \times \mathrm{L}^s(\Omega)\,,$$

where

$$\mathbf{H}(\mathrm{div}_0; \Omega) := \Big\{ \boldsymbol{\tau} \in \mathbf{H}(\mathrm{div}; \Omega) : \quad \mathrm{div}(\boldsymbol{\tau}) = 0 \quad in \quad \Omega \Big\}.$$

Prove that for each $\mathbf{F} \in \mathbf{X}_s'$ there exists a unique $(\boldsymbol{\sigma}, u) \in \mathbf{X}_r$ such that

$$\int_\Omega \kappa\, \boldsymbol{\sigma} \cdot \boldsymbol{\tau} - \int_\Omega u\, v = \mathbf{F}\big((\boldsymbol{\tau}, v)\big) \qquad \forall\, (\boldsymbol{\tau}, v) \in \mathbf{X}_s\,.$$

Also show that there exists a constant $c > 0$, depending on κ_0, such that

$$\|(\boldsymbol{\sigma}, u)\|_{\mathbf{X}_r} \leqslant c\, \|\mathbf{F}\|_{\mathbf{X}_s'}\,.$$

▸ **5.5** [THIS EXERCISE IS A VARIANT OF EXERCISE 5.4]. *Let Ω be a bounded open subset of $\mathbb{R}^2$, $\kappa \in C(\Omega)$, and suppose that there exist constants κ_0, $\kappa_1 > 0$ such that $\kappa_0 \leqslant \kappa(\mathbf{x}) \leqslant \kappa_1$ for all $\mathbf{x} \in \Omega$. Let r, $s \in (1, +\infty)$ be such that $\frac{1}{r} + \frac{1}{s} = 1$, and define the spaces*

$$\mathbf{U}_r := \mathrm{L}^r(\Omega) \times \mathbf{H}(\mathrm{rot}; \Omega) \quad \textit{and} \quad \mathbf{U}_s := \mathrm{L}^s(\Omega) \times \mathbf{H}(\mathrm{rot}; \Omega)\,,$$

where

$$\mathbf{H}(\mathrm{rot}; \Omega) := \Big\{ \boldsymbol{\tau} \in \mathbf{L}^2(\Omega) := [\mathrm{L}^2(\Omega)]^2 : \quad \mathrm{rot}(\boldsymbol{\tau}) := \frac{\partial \tau_2}{\partial x_1} - \frac{\partial \tau_1}{\partial x_2} \in \mathrm{L}^2(\Omega) \Big\}.$$

Prove that for each $\mathbf{F} \in \mathbf{U}_s'$ there exists a unique $(u, \boldsymbol{\sigma}) \in \mathbf{U}_r$ such that

$$\int_\Omega u\, v - \int_\Omega \Big\{ \kappa\, \mathrm{rot}(\boldsymbol{\sigma})\, \mathrm{rot}(\boldsymbol{\tau}) + \boldsymbol{\sigma} \cdot \boldsymbol{\tau} \Big\} = \mathbf{F}\big((v, \boldsymbol{\tau})\big) \qquad \forall\, (v, \boldsymbol{\tau}) \in \mathbf{U}_s\,.$$

Also prove that there exists a constant $c > 0$, depending on κ_0, such that

$$\|(\boldsymbol{\sigma}, u)\|_{\mathbf{U}_r} \leqslant c\, \|\mathbf{F}\|_{\mathbf{U}_s'}\,.$$

▸ **5.6** *Let Ω be a bounded open subset of $\mathbb{R}^n$, $\kappa \in C(\Omega)$, and suppose there exist constants κ_0, $\kappa_1 > 0$ such that $\kappa_0 \leqslant \kappa(\mathbf{x}) \leqslant \kappa_1$ for all $\mathbf{x} \in \Omega$. Let r, $s \in (1, +\infty)$ be such that $\frac{1}{r} + \frac{1}{s} = 1$, and define the spaces*

$$\mathbf{X}_r := \mathbf{H}(\mathrm{div}; \Omega) \times \mathrm{L}^r(\Omega)\,, \quad \mathbf{X}_s := \mathbf{H}(\mathrm{div}; \Omega) \times \mathrm{L}^s(\Omega)\,, \quad \textit{and} \quad \mathrm{Q} := \mathrm{L}^2(\Omega)\,.$$

Given the pair $(\mathbf{F}, \mathrm{G}) \in \mathbf{X}_s' \times \mathrm{Q}'$, consider the problem: Find $\big((\boldsymbol{\sigma}, u), \varphi\big) \in \mathbf{X}_r \times \mathrm{Q}$ such that

$$\begin{aligned} \int_\Omega \kappa\, \boldsymbol{\sigma} \cdot \boldsymbol{\tau} - \int_\Omega u\, v + \int_\Omega \varphi\, \mathrm{div}(\boldsymbol{\tau}) &= \mathbf{F}\big((\boldsymbol{\tau}, v)\big) \ \forall\, (\boldsymbol{\tau}, v) \in \mathbf{X}_s\,, \\ \int_\Omega \psi\, \mathrm{div}(\boldsymbol{\sigma}) \qquad\qquad\qquad &= \quad \mathrm{G}(\psi) \quad \forall\, \psi \in \mathrm{Q}\,. \end{aligned} \tag{5.281}$$

Identify the respective spaces X_2, M_1, X_1, and M_2, and the associated bilinear forms a, b_1, and b_2, to prove, using the Generalized Babuška–Brezzi Theory, that (5.281) has a unique solution that depends continuously on the data $\mathbf{F}$ and G.

▸ **5.7** ([27], [42]) *Let Ω be a bounded open subset of $\mathbb{R}^n$ with smooth boundary Γ, and let p, $q \in (1, +\infty)$ be such that $1/p + 1/q = 1$. The vector version of the Sobolev embedding mentioned in Exercise 5.2 states that $\mathbf{H}^1(\Omega) \subseteq \mathbf{L}^q(\Omega)$, and that the injection $\mathbf{i}_q : \mathbf{H}^1(\Omega) \to \mathbf{L}^q(\Omega)$ is bounded, for all $q \in \begin{cases} [1, +\infty) & if \ \ n = 2 \\ [1, 6] & if \ \ n = 3 \end{cases}$. Define the Banach space*

$$\mathbb{H}(\mathbf{div}_p;\Omega) := \left\{\boldsymbol{\tau} \in \mathbb{L}^2(\Omega) : \quad \mathbf{div}(\boldsymbol{\tau}) \in \mathbf{L}^p(\Omega)\right\},$$

and its subspace

$$\mathbb{H}_0(\mathbf{div}_p;\Omega) := \left\{\boldsymbol{\tau} \in \mathbb{H}(\mathbf{div}_p;\Omega) : \quad \int_\Omega \operatorname{tr}(\boldsymbol{\tau}) = 0\right\},$$

and adapt the proof of Lemma 4.38 to show that there exists a constant $\tilde{c}_1 > 0$ such that

$$\|\boldsymbol{\tau}^{\mathrm{d}}\|_{0,\Omega}^2 + \|\mathbf{div}(\boldsymbol{\tau})\|_{0,p;\Omega}^2 \geqslant \tilde{c}_1 \|\boldsymbol{\tau}\|_{0,\Omega}^2 \qquad \forall\, \boldsymbol{\tau} \in \mathbb{H}_0(\mathbf{div}_p;\Omega).$$

▸ **5.8** ([27]) *The motion of an incompressible fluid in a region Ω of $\mathbb{R}^n$, $n \in \{2,3\}$, with boundary Γ, is governed by the following* NAVIER–STOKES EQUATIONS:

$$\begin{aligned} -\mu\,\Delta\mathbf{u} + (\nabla\mathbf{u})\,\mathbf{u} + \nabla p &= \mathbf{f} \quad \text{in} \quad \Omega, \\ \operatorname{div}(\mathbf{u}) = 0 \quad \text{in} \quad \Omega, &\quad \mathbf{u} = 0 \quad \text{on} \quad \Gamma, \end{aligned} \tag{5.282}$$

where μ, $\mathbf{u}$, and p denote the viscosity, velocity, and pressure, respectively, div *is the usual scalar divergence operator, and $\mathbf{f}$ is a given external force. Note that a more general case of* (5.282)*, in which μ depends on $|\nabla\mathbf{u}|$, is studied in [16]. In what follows, and as before,* **div** *denotes the operator* div *acting along the rows of a matrix.*

a) *Given a vector function $\mathbf{v} = (v_i)_{i=\overline{1,n}}$, define $(\mathbf{v}\otimes\mathbf{v}) := (v_i\,v_j)_{i,j=\overline{1,n}}$, and show that there holds $\mathbf{div}(\mathbf{v}\otimes\mathbf{v}) = (\nabla\mathbf{v})\,\mathbf{v}$ when $\operatorname{div}(\mathbf{v}) = 0$.*
b) *Introduce $\boldsymbol{\sigma} := \mu\,\nabla\mathbf{u} - (\mathbf{u}\otimes\mathbf{u}) - p\mathbb{I}$ in Ω, where $\mathbb{I}$ is the identity matrix of $\mathbb{R}^n$, apply the matrix trace* tr *and the deviatoric operator ^{d} to show that* (5.282) *is equivalently transformed into:*

$$\begin{aligned} \boldsymbol{\sigma}^{\mathrm{d}} := \mu\,\nabla\mathbf{u} - (\mathbf{u}\otimes\mathbf{u})^{\mathrm{d}} \quad \text{in} \quad \Omega, &\quad \mathbf{div}(\boldsymbol{\sigma}) = -\mathbf{f} \quad \text{in} \quad \Omega, \\ \mathbf{u} = 0 \quad \text{on} \quad \Gamma, &\quad p = -\frac{1}{n}\operatorname{tr}\big(\boldsymbol{\sigma} + (\mathbf{u}\otimes\mathbf{u})\big). \end{aligned} \tag{5.283}$$

c) *Define the space $\mathbb{H}_0(\mathbf{div}_{4/3};\Omega) := \left\{\boldsymbol{\tau} \in \mathbb{H}(\mathbf{div}_{4/3};\Omega) : \int_\Omega \operatorname{tr}(\boldsymbol{\tau}) = 0\right\}$, and utilize that $\mathbb{H}(\mathbf{div}_{4/3};\Omega) = \mathbb{H}_0(\mathbf{div}_{4/3};\Omega)\oplus\mathbb{R}\,\mathbb{I}$ to deduce that the mixed variational formulation of* (5.283) *reduces to: Find $(\boldsymbol{\sigma},\mathbf{u}) \in \mathbb{H}_0(\mathbf{div}_{4/3};\Omega) \times \mathbf{L}^4(\Omega)$ such that*

$$\begin{aligned} \int_\Omega \boldsymbol{\sigma}^{\mathrm{d}} : \boldsymbol{\tau}^{\mathrm{d}} + \mu \int_\Omega \mathbf{u}\cdot\mathbf{div}(\boldsymbol{\tau}) + \int_\Omega (\mathbf{u}\otimes\mathbf{u})^{\mathrm{d}} : \boldsymbol{\tau} &= \quad 0, \\ \mu \int_\Omega \mathbf{v}\cdot\mathbf{div}(\boldsymbol{\sigma}) &= -\mu \int_\Omega \mathbf{f}\cdot\mathbf{v}, \end{aligned} \tag{5.284}$$

for all $(\boldsymbol{\tau},\mathbf{v}) \in \mathbb{H}_0(\mathbf{div}_{4/3};\Omega) \times \mathbf{L}^4(\Omega)$.
d) *Replace $(\mathbf{u}\otimes\mathbf{u})$ in* (5.284) *by $(\mathbf{z}\otimes\mathbf{u})$, with $\mathbf{z} \in \mathbf{L}^4(\Omega)$ given, and show that for $\|\mathbf{z}\|_{0,4;\Omega}$ sufficiently small, the resulting linear problem has a unique solution.*

▸ **5.9** [ALTERNATIVE PROOF OF THE HILBERT CASE OF THEOREM 5.9] *Let $(H,\langle\cdot,\cdot\rangle_H)$ and $(Q,\langle\cdot,\cdot\rangle_Q)$ be Hilbert spaces with norms $\|\cdot\|_H$ and $\|\cdot\|_Q$, respectively, let $a : H\times H \to \mathbb{R}$ and $b : H\times Q \to \mathbb{R}$ be bounded bilinear forms with corresponding constants $\|a\|$ and $\|b\|$, and let V be the null space of the operator induced by b. Suppose that*

i) *there exists a constant $\alpha > 0$ such that* $\displaystyle\sup_{\substack{\tau\in V\\ \tau\neq 0}} \frac{a(\zeta,\tau)}{\|\tau\|_H} \geqslant \alpha\,\|\zeta\|_H \qquad \forall\,\zeta \in V,$

ii) *for each $\tau \in V$, $\tau \neq 0$, one has* $\displaystyle\sup_{\zeta\in V} a(\zeta,\tau) > 0$, *and*

iii) *there exists a constant $\beta > 0$ such that* $\displaystyle\sup_{\substack{\tau\in H\\ \tau\neq 0}} \frac{b(\tau,v)}{\|\tau\|_H} \geqslant \beta\,\|v\|_Q \qquad \forall\,v \in Q.$

Furthermore, let $A : (H \times Q) \times (H \times Q) \to \mathbb{R}$ *be the bilinear form defined by*

$$A\big((\zeta, w), (\tau, v)\big) := a(\zeta, \tau) + b(\tau, w) + b(\zeta, v) \qquad \forall\, (\zeta, w),\, (\tau, v) \in H \times Q\,,$$

and for each $(\zeta, w) \in H \times Q$ *introduce* $S(\zeta, w) := \sup\limits_{\substack{(\tau,v)\in H\times Q\\ (\tau,v)\neq 0}} \dfrac{A\big((\zeta, w), (\tau, v)\big)}{\|(\tau, v)\|_{H\times Q}}$ *and the functionals* $F_{\zeta,w} \in H'$ *and* $G_{\zeta,w} \in Q'$ *given by*

$$F_{\zeta,w}(\tau) := A\big((\zeta, w), (\tau, 0)\big) \quad \forall\, \tau \in H \text{ and } G_{\zeta,w}(v) := A\big((\zeta, w), (0, v)\big) \quad \forall\, v \in Q\,.$$

a) *Prove that for each* $(\zeta, w) \in H \times Q$ *one has [see Section 5.1.1 for the general case]*

$$\frac{1}{2}\Big\{\|F_{\zeta,w}\|_{H'} + \|G_{\zeta,w}\|_{Q'}\Big\} \leqslant S(\zeta, w) \leqslant \|F_{\zeta,w}\|_{H'} + \|G_{\zeta,w}\|_{Q'}\,. \tag{5.285}$$

b) *Proceeding similarly to the proofs of Theorems 5.16 and 5.17, use i), iii), and the fact that each* $\zeta \in H$ *decomposes as* $\zeta = \zeta_0 + \bar{\zeta}$, *with* $\zeta_0 \in V$ *and* $\bar{\zeta} \in V^{\perp}$ *uniquely, to prove that for each* $(\zeta, w) \in H \times Q$ *one has*

$$\begin{aligned} \|\zeta\|_H &\leqslant \frac{1}{\alpha}\,\|F_{\zeta,w}\|_{H'} + \frac{1}{\beta}\Big(1 + \frac{\|a\|}{\alpha}\Big)\|G_{\zeta,w}\|_{Q'} \quad \text{and}\\ \|w\|_Q &\leqslant \frac{1}{\beta}\Big(1 + \frac{\|a\|}{\alpha}\Big)\|F_{\zeta,w}\|_{H'} + \frac{\|a\|}{\beta^2}\Big(1 + \frac{\|a\|}{\alpha}\Big)\|G_{\zeta,w}\|_{Q'}\,, \end{aligned}$$

and deduce, using (5.285), that there exists a constant $\tilde{\alpha} > 0$, *depending on* α, β, *and* $\|a\|$, *such that* $S(\zeta, w) \geqslant \tilde{\alpha}\,\|(\zeta, w)\|_{H\times Q}$ *for all* $(\zeta, w) \in H \times Q$.

c) *Show that* $\sup\limits_{(\zeta,w)\in H\times Q} A\big((\zeta, w), (\tau, v)\big) > 0$ *for each* $(\tau, v) \in H \times Q$, $(\tau, v) \neq 0$.

d) *Conclude, using b), c), and the Generalized Lax–Milgram Lemma, that for each pair* $(f, g) \in H' \times Q'$ *there exists a unique* $(\sigma, u) \in H \times Q$ *solution of the problem*

$$A\big((\sigma, u), (\tau, v)\big) = f(\tau) + g(v) \qquad \forall\, (\tau, v) \in H \times Q\,,$$

which satisfies $\|(\sigma, u)\|_{H\times Q} \leqslant \tilde{\alpha}^{-1}\Big\{\|f\|_{H'} + \|g\|_{Q'}\Big\}$.

HINT: For b) also keep in mind the inf-sup condition equivalent to iii) (cf. Lemma 4.18), and for c) recall the pair of inf-sup conditions equivalent to the pair i) - ii).

▸ **5.10** ([71]) *Given* Ω *a bounded open subset of* $\mathbb{R}^n$, *star-shaped with respect to a ball, it is known that for each* $t \in (1, +\infty)$ *the operator* $\mathrm{div} : \mathbf{W}^{1,t}(\Omega) \to \mathrm{L}^t_0(\Omega)$ *is surjective (cf. [52, Lemma B.69], [20], [49]), where*

$$\mathbf{W}^{1,t}(\Omega) := \Big\{z \in \mathrm{L}^t(\Omega) : \quad \nabla z \in \mathbf{L}^t(\Omega)\Big\} \quad \text{and} \quad \mathrm{L}^t_0(\Omega) := \Big\{v \in \mathrm{L}^t(\Omega) : \quad \int_\Omega v = 0\Big\}.$$

a) *Apply the Open Mapping Theorem (cf. Theorem 3.39), or directly Lemma 3.64)or Lemma 3.65), to prove that for each* $t \in (1, +\infty)$ *there exists a constant* $C_t > 0$, *such that for each* $v \in \mathrm{L}^t_0(\Omega)$ *there exists* $z_v \in \mathbf{W}^{1,t}(\Omega)$ *satisfying*

$$\mathrm{div}(z_v) = v \quad \text{and} \quad \|z_v\|_{1,t;\Omega} \leqslant C_t\,\|v\|_{0,t;\Omega}\,.$$

b) *Consider the spaces*

$$\begin{aligned} \mathbb{H}^t(\mathbf{div}_t; \Omega) &:= \Big\{\boldsymbol{\tau} \in \mathbb{L}^t(\Omega) : \quad \mathbf{div}(\boldsymbol{\tau}) \in \mathbf{L}^t(\Omega)\Big\} \quad \text{and}\\ \mathbb{H}^t_0(\mathbf{div}_t; \Omega) &= \Big\{\boldsymbol{\tau} \in \mathbb{H}^t(\mathbf{div}_t; \Omega) : \quad \mathrm{tr}(\boldsymbol{\tau}) \in \mathrm{L}^t_0(\Omega)\Big\}, \end{aligned}$$

and use a) *and the duality result obtained in part* b) *of Exercise 5.1, to prove that for each* $t \in (1, +\infty)$ *there exist constants* $\widetilde{C}_t$, $\bar{C}_t > 0$, *such that*

$$\begin{aligned} \|\mathrm{tr}(\boldsymbol{\tau})\|_{0,t;\Omega} &\leqslant \widetilde{C}_t \Big\{ \|\boldsymbol{\tau}^{\mathrm{d}}\|_{0,t;\Omega} + \|\mathbf{div}(\boldsymbol{\tau})\|_{0,t;\Omega} \Big\} \quad \forall\, \boldsymbol{\tau} \in \mathbb{H}_0^t(\mathbf{div}_t;\Omega)\,, \text{ and} \\ \|\boldsymbol{\tau}\|_{0,t;\Omega} &\leqslant \bar{C}_t \Big\{ \|\boldsymbol{\tau}^{\mathrm{d}}\|_{0,t;\Omega} + \|\mathbf{div}(\boldsymbol{\tau})\|_{0,t;\Omega} \Big\} \quad \forall\, \boldsymbol{\tau} \in \mathbb{H}_0^t(\mathbf{div}_t;\Omega)\,. \end{aligned} \tag{5.286}$$

Note that the inequalities corresponding to the case $t = 2$ were established in Lemma 4.38, so that (5.286) constitutes the generalization of those results to each $t \in (1,+\infty)$.

▸ **5.11** ([80]) *Using the notation of Exercise 5.8, a viscous fluid inside a porous medium Ω is governed by the* NAVIER–STOKES–BRINKMAN MODEL, *which, in a simplified version, reduces to finding the velocity $\mathbf{u} : \Omega \to \mathbb{R}^n$ and the pressure $p : \Omega \to \mathbb{R}$ such that (cf. [82, 98]):*

$$\begin{aligned} \eta\,\mathbf{u} \,-\, \lambda\,\mathbf{div}(\boldsymbol{\epsilon}(\mathbf{u})) \,+\, (\nabla \mathbf{u})\mathbf{u} \,+\, \nabla p = \mathbf{f} \quad & \text{in} \quad \Omega\,, \\ \mathrm{div}(\mathbf{u}) = 0 \quad & \text{in} \quad \Omega\,, \\ \int_\Omega p = 0\,, \quad \mathbf{u} = \mathbf{u}_D \quad & \text{on} \quad \Gamma\,, \end{aligned} \tag{5.287}$$

where η and λ are positive constants, $\boldsymbol{\epsilon}(\mathbf{u}) := \frac{1}{2}\big(\nabla\mathbf{u} + (\nabla\mathbf{u})^{\mathtt{t}}\big)$, $\mathbf{f}$ is an external force, and $\mathbf{u}_D$ is a Dirichlet datum.

a) *Denoting once again by $\mathbb{I}$ the identity matrix of $\mathbb{R}^n$, introduce the auxiliary unknowns*

$$\mathbf{t} := \boldsymbol{\epsilon}(\mathbf{u}) = \nabla\mathbf{u} - \boldsymbol{\gamma}\,, \quad \boldsymbol{\gamma} := \frac{1}{2}\big(\nabla\mathbf{u} - (\nabla\mathbf{u})^{\mathtt{t}}\big)\,, \quad \boldsymbol{\sigma} := \lambda\,\mathbf{t} - (\mathbf{u}\otimes\mathbf{u}) - p\,\mathbb{I}\,,$$

and deduce that (5.287) can be reformulated as: Find $\mathbf{t}$, $\boldsymbol{\sigma}$, $\mathbf{u}$, and $\boldsymbol{\gamma}$ in spaces to be defined later, such that

$$\begin{aligned} \mathbf{t} \,+\, \boldsymbol{\gamma} = \nabla\mathbf{u} \quad & \text{in} \quad \Omega\,, \\ \lambda\,\mathbf{t} \,-\, (\mathbf{u}\otimes\mathbf{u})^{\mathrm{d}} = \boldsymbol{\sigma}^{\mathrm{d}} \quad & \text{in} \quad \Omega\,, \\ \eta\,\mathbf{u} \,-\, \mathbf{div}(\boldsymbol{\sigma}) = \mathbf{f} \quad & \text{in} \quad \Omega\,, \\ \int_\Omega \mathrm{tr}\big(\boldsymbol{\sigma} + (\mathbf{u}\otimes\mathbf{u})\big) = 0\,, \quad \mathbf{u} = \mathbf{u}_D \quad & \text{on} \quad \Gamma\,. \end{aligned} \tag{5.288}$$

b) *Define the spaces*

$$\mathbb{L}^2_{\mathrm{tr}}(\Omega) := \Big\{ \mathbf{s} \in \mathbb{L}^2(\Omega) : \quad \mathrm{tr}(\mathbf{s}) = 0 \Big\}\,, \quad \text{and} \quad \mathbb{L}^2_{\mathtt{skew}}(\Omega) := \Big\{ \boldsymbol{\delta} \in \mathbb{L}^2(\Omega) : \quad \boldsymbol{\delta}^{\mathtt{t}} = -\boldsymbol{\delta} \Big\}\,,$$

and prove that the mixed variational formulation of (5.288) reduces to: Find $\vec{\mathbf{t}} := (\mathbf{t},\boldsymbol{\sigma}) \in \mathbf{H} := \mathbb{L}^2_{\mathrm{tr}}(\Omega) \times \mathbb{H}_0(\mathbf{div}_{4/3};\Omega)$ and $\vec{\mathbf{u}} := (\mathbf{u},\boldsymbol{\gamma}) \in \mathbf{Q} := \mathbf{L}^4(\Omega) \times \mathbb{L}^2_{\mathtt{skew}}(\Omega)$, such that

$$\begin{aligned} a(\mathbf{t},\mathbf{s}) + b_1(\mathbf{s},\boldsymbol{\sigma}) \qquad\qquad + b(\mathbf{u};\mathbf{u},\mathbf{s}) &= 0\,, \\ b_2(\mathbf{t},\boldsymbol{\tau}) \qquad\qquad + \mathbf{b}(\vec{\mathbf{s}},\vec{\mathbf{u}}) &= \langle \boldsymbol{\tau}\,\boldsymbol{\nu}, \mathbf{u}_D \rangle\,, \\ \mathbf{b}(\vec{\mathbf{t}},\vec{\mathbf{v}}) \qquad - \mathbf{c}(\vec{\mathbf{u}},\vec{\mathbf{v}}) &= -\int_\Omega \mathbf{f}\cdot\mathbf{v}\,, \end{aligned} \tag{5.289}$$

for all $(\vec{\mathbf{s}},\vec{\mathbf{v}}) := \big((\mathbf{s},\boldsymbol{\tau}),(\mathbf{v},\boldsymbol{\delta})\big) \in \mathbf{H}\times\mathbf{Q}$, with $\mathbf{f} \in \mathbf{L}^{4/3}(\Omega)$ and $\mathbf{u}_D \in \mathbf{H}^{1/2}(\Gamma)$, where $a : \mathbb{L}^2_{\mathrm{tr}}(\Omega) \times \mathbb{L}^2_{\mathrm{tr}}(\Omega) \to \mathbb{R}$, $b_i : \mathbb{L}^2_{\mathrm{tr}}(\Omega) \times \mathbb{H}_0(\mathbf{div}_{4/3};\Omega) \to \mathbb{R}$, $i \in \{1,2\}$, $\mathbf{b} : \mathbf{H}\times\mathbf{Q} \to \mathbb{R}$, and $\mathbf{c} : \mathbf{Q}\times\mathbf{Q} \to \mathbb{R}$, are the bilinear forms defined as

$$\begin{aligned} a(\mathbf{r},\mathbf{s}) &:= \lambda \int_\Omega \mathbf{r} : \mathbf{s} \qquad \forall\, \mathbf{r},\, \mathbf{s} \in \mathbb{L}^2_{\mathrm{tr}}(\Omega)\,, \\ b_1(\mathbf{s},\boldsymbol{\tau}) &:= -\int_\Omega \mathbf{s} : \boldsymbol{\tau}\,, \quad b_2(\mathbf{s},\boldsymbol{\tau}) := \int_\Omega \mathbf{s} : \boldsymbol{\tau}\,, \quad \forall\,(\mathbf{s},\boldsymbol{\tau}) \in \mathbb{L}^2_{\mathrm{tr}}(\Omega) \times \mathbb{H}_0(\mathbf{div}_{4/3};\Omega)\,, \\ \mathbf{b}(\vec{\mathbf{s}},\vec{\mathbf{v}}) &:= \int_\Omega \boldsymbol{\delta} : \boldsymbol{\tau} + \int_\Omega \mathbf{v}\cdot\mathbf{div}(\boldsymbol{\tau}) \qquad \forall\,(\vec{\mathbf{s}},\vec{\mathbf{v}}) \in \mathbf{H}\times\mathbf{Q}\,, \\ \mathbf{c}(\vec{\mathbf{w}},\vec{\mathbf{v}}) &:= \eta \int_\Omega \mathbf{w}\cdot\mathbf{v} \qquad \forall\,\vec{\mathbf{w}},\,\vec{\mathbf{v}} \in \mathbf{Q}\,, \end{aligned}$$

and for each $\mathbf{w} \in \mathbf{L}^4(\Omega)$, $b(\mathbf{w};\cdot,\cdot) : \mathbf{L}^4(\Omega) \times \mathbb{L}^2_{\mathrm{tr}}(\Omega) \to \mathrm{R}$ *is the bilinear form given by*

$$b(\mathbf{w};\mathbf{v},\mathbf{s}) := -\int_\Omega (\mathbf{w}\otimes\mathbf{v}) : \mathbf{s} \qquad \forall\,(\mathbf{v},\mathbf{s}) \in \mathbf{L}^4(\Omega) \times \mathbb{L}^2_{\mathrm{tr}}(\Omega)\,.$$

c) *Define the bilinear form* $\mathbf{a} : \mathbf{H} \times \mathbf{H} \to \mathrm{R}$ *by*

$$\mathbf{a}(\vec{\mathbf{r}},\vec{\mathbf{s}}) := a(\mathbf{r},\mathbf{s}) + b_1(\mathbf{s},\boldsymbol{\zeta}) + b_2(\mathbf{r},\boldsymbol{\tau}) \qquad \forall\,\vec{\mathbf{r}},\,\vec{\mathbf{s}} \in \mathbf{H}\,,$$

and observe that (5.289) can be rewritten as: Find $(\vec{\mathbf{t}},\vec{\mathbf{u}}) \in \mathbf{H} \times \mathbf{Q}$ *such that*

$$\begin{aligned} \mathbf{a}(\vec{\mathbf{t}},\vec{\mathbf{s}}) + \mathbf{b}(\vec{\mathbf{s}},\vec{\mathbf{u}}) + b(\mathbf{u};\mathbf{u},\mathbf{s}) &= \langle \boldsymbol{\tau}\,\boldsymbol{\nu}, \mathbf{u}_D\rangle \quad \forall\,\vec{\mathbf{s}} \in \mathbf{H}\,,\\ \mathbf{b}(\vec{\mathbf{t}},\vec{\mathbf{v}}) - \mathbf{c}(\vec{\mathbf{u}},\vec{\mathbf{v}}) &= -\int_\Omega \mathbf{f}\cdot\mathbf{v} \quad \forall\,\vec{\mathbf{v}} \in \mathbf{Q}\,. \end{aligned} \tag{5.290}$$

Then, denote by $\mathbf{V}$ *the kernel of the operator induced by the bilinear form* $\mathbf{b}$, *define it explicitly, and apply Theorem 5.26 to prove that* $\mathbf{a}$ *satisfies the global inf-sup conditions in* $\mathbf{V}$.

d) *Replace* $b(\mathbf{u};\mathbf{u},\mathbf{s})$ *in (5.290) by* $b(\mathbf{w};\mathbf{u},\mathbf{s})$, *with* $\mathbf{w} \in \mathbf{L}^4(\Omega)$ *given, and apply the theory of perturbed saddle point problems (cf. Theorem 5.17) together with the Banach–Nečas–Babuška Theorem (cf. Lemma 5.1), to prove that, when* $\|\mathbf{w}\|_{0,4;\Omega}$ *is sufficiently small, the resulting linear problem has a unique solution depending continuously on the data* $\mathbf{f}$ *and* $\mathbf{u}_D$.

▸ **5.12** ([71], [72]) *With the same notation as in Exercise 5.10, recall from [72] that the analysis of a* STRESS-ASSISTED DIFFUSION MODEL *requires studying an elasticity problem (with Lamé constants given by* λ *and* μ*) formulated in Banach spaces. More precisely, let* r, $s \in (1,+\infty)$ *be conjugate exponents, and first introduce the spaces*

$$X_2 := \mathbb{H}^r_0(\mathbf{div}_r;\Omega)\,, \quad M_1 := \mathbf{L}^r(\Omega)\,, \quad X_1 := \mathbb{H}^s_0(\mathbf{div}_s;\Omega)\,, \quad \text{and} \quad M_2 := \mathbf{L}^s(\Omega)\,. \tag{5.291}$$

Then, given $F \in X_1'$ *and* $G \in M_2'$, *we are interested in: Find* $(\boldsymbol{\sigma},\mathbf{u}) \in X_2 \times M_1$ *such that*

$$\begin{aligned} a(\boldsymbol{\sigma},\boldsymbol{\tau}) + b_1(\boldsymbol{\tau},\mathbf{u}) &= F(\boldsymbol{\tau}) \quad \forall\,\boldsymbol{\tau} \in X_1,\\ b_2(\boldsymbol{\sigma},\mathbf{v}) &= G(\mathbf{v}) \quad \forall\,\mathbf{v} \in M_2, \end{aligned} \tag{5.292}$$

where $a : X_2 \times X_1 \to \mathrm{R}$ *and* $b_i : X_i \times M_i \to \mathrm{R}$, $i \in \{1,2\}$, *are the bilinear forms defined by*

$$\begin{aligned} a(\boldsymbol{\zeta},\boldsymbol{\tau}) &:= \frac{1}{\mu}\int_\Omega \boldsymbol{\zeta}^{\mathrm{d}} : \boldsymbol{\tau}^{\mathrm{d}} + \frac{1}{n(n\lambda + (n+1)\mu)}\int_\Omega \mathrm{tr}(\boldsymbol{\zeta})\,\mathrm{tr}(\boldsymbol{\tau}) \qquad \forall\,(\boldsymbol{\zeta},\boldsymbol{\tau}) \in X_2 \times X_1\,,\\ b_i(\boldsymbol{\tau},\mathbf{v}) &:= \int_\Omega \mathbf{v}\cdot\mathbf{div}(\boldsymbol{\tau}) \qquad \forall\,(\boldsymbol{\tau},\mathbf{v}) \in X_i \times M_i\,. \end{aligned}$$

Assume there exists $D_r \in \mathcal{L}\big(\mathbb{L}^r(\Omega),\mathbb{L}^r(\Omega)\big)$ *such that for all* $\boldsymbol{\tau} \in \mathbb{L}^r(\Omega)$ *it holds that*

$$\mathbf{div}\big(D_r(\boldsymbol{\tau})\big) = 0\,, \quad \int_\Omega \mathrm{tr}\big(D_r(\boldsymbol{\tau})\big) = \int_\Omega \mathrm{tr}(\boldsymbol{\tau})\,, \quad \text{and} \quad \int_\Omega \boldsymbol{\zeta}^{\mathrm{d}} : \big(D_r(\boldsymbol{\tau})\big)^{\mathrm{d}} = \int_\Omega \boldsymbol{\zeta}^{\mathrm{d}} : \boldsymbol{\tau}^{\mathrm{d}}\,,$$

for each $\boldsymbol{\zeta} \in \mathbb{L}^s(\Omega)$ *such that* $\mathbf{div}(\boldsymbol{\zeta}) = 0$. *In turn, assume that for each* $\mathbf{v} \in \mathbf{L}^r(\Omega)$ *there exists a unique solution* $\mathbf{z} \in \mathbf{W}^{1,r}_0(\Omega) := \big\{\mathbf{w} \in \mathbf{W}^{1,r}(\Omega) : \ \mathbf{w} = 0 \ \text{ on } \ \partial\Omega\big\}$ *of the boundary value problem:* $\Delta\mathbf{z} = \mathbf{v}$ *in* Ω, $\mathbf{z} = 0$ *on* $\partial\Omega$, *such that* $\|\mathbf{z}\|_{1,r;\Omega} \leqslant C_r\,\|\mathbf{v}\|_{0,r;\Omega}$, *with a positive constant* C_r *independent of* $\mathbf{z}$. *Then, keeping in mind that the previous hypotheses are also valid when interchanging* r *and* s, *apply the Generalized Babuška–Brezzi Theory (cf. Theorem 5.26), and use, in particular, the inequalities of Exercise 5.10 (cf. (5.286)), to conclude that, when* λ *is sufficiently large, (5.292) has a unique solution that depends continuously on the data* F *and* G.

▸ **5.13** *Prove that the lower bounds of the inequalities (5.67) and (5.68) (cf. Lemma 5.14) also follow from (5.66).*

▸ **5.14** *In this final proposal for practicing the theories developed in this chapter, the reader is invited to "carry out a reverse process", that is, to take the author's place this time. In other words, the invitation is to dare to formulate new problems, similar to those posed by Exercises 5.8, 5.12, and 5.11, whose solution also makes use of the theorems studied here. For this purpose, and merely as an initial suggestion, without prejudice to other equally valid options, it is recommended to begin this experience by consulting the models and variational formulations proposed, for instance, in the following references: [5], [26], [28], [31], [32], [33], [34], [35], [36], [37], [38], [91], and [103].*

6

Compact Operators

The goal of this chapter is to study properties of compact operators, which constitute a generalization of operators whose range has finite dimension. In particular, the main results to be considered here include the Fredholm alternative, the passage from weak to strong convergence, and the passage from pointwise convergence to convergence in norm. Much of what follows is based on the approaches in [101] and [115].

6.1 Preliminaries

We begin with some classical results in finite dimension.

Theorem 6.1. *Let $(X, \|\cdot\|)$ be a normed vector space of finite dimension. Then, all norms on X are equivalent.*

Proof. Let $\{x_1, x_2, \ldots, x_n\}$ be a basis of X. Then we consider the following norm

$$\|\cdot\|_1 : X \to \mathbb{R}$$

$$x = \sum_{j=1}^{n} \alpha_j \, x_j \longrightarrow \|x\|_1 := \sum_{j=1}^{n} |\alpha_j|,$$

and next we prove that any other norm $\|\cdot\|$ defined on X is equivalent to $\|\cdot\|_1$. Indeed, note first that

$$\|x\| = \Big\| \sum_{j=1}^{n} \alpha_j \, x_j \Big\| \leqslant \sum_{j=1}^{n} |\alpha_j| \, \|x_j\|$$

$$\leqslant \Big\{ \max_{j \in \{1, \cdots, n\}} \|x_j\| \Big\} \sum_{j=1}^{n} |\alpha_j| = \Big\{ \max_{j \in \{1, \cdots, n\}} \|x_j\| \Big\} \|x\|_1 \qquad \forall \, x \in X.$$

Certainly, for the other inequality between $\|\cdot\|$ and $\|\cdot\|_1$ one could apply directly the result, consequence of the Bounded Inverse Theorem (cf. Theorem 3.34) given by Example 3.42. However, the idea here is not to use such "*heavy machinery*" but rather to follow the classical proof from real analysis texts. Indeed, suppose by contradiction that there does not exist $C > 0$ such that

$$\|x\|_1 \leqslant C \, \|x\| \qquad \forall \, x \in X.$$

It follows, in particular, that for each $k \in \mathbb{N}$ there exists $x_k := \sum_{j=1}^{n} \alpha_{k,j} \, x_j \in X$ such that

G. N. Gatica, *Introduction to Functional Analysis*,
https://doi.org/10.1007/978-3-032-37415-8_6

$$\|x_k\|_1 > k\,\|x_k\|,$$

that is,

$$\left\|\sum_{j=1}^{n} \alpha_{k,j}\, x_j\right\| < \frac{1}{k}\sum_{j=1}^{n} |\alpha_{k,j}|. \tag{6.1}$$

Then, defining

$$\beta_{k,i} := \frac{\alpha_{k,i}}{\sum_{j=1}^{n} |\alpha_{k,j}|} \qquad \forall\, i \in \{1,2,\ldots,n\} \quad \text{and} \quad y_k := \sum_{i=1}^{n} \beta_{k,i}\, x_i,$$

we obtain $\|y_k\|_1 = \sum_{i=1}^{n} |\beta_{k,i}| = 1$ and

$$\|y_k\| < \frac{1}{k} \qquad \forall\, k \in \mathbb{N}. \tag{6.2}$$

Notice here that the strict inequality (6.1) guarantees that $\sum_{j=1}^{n} |\alpha_{k,j}| \neq 0$. Now, it is clear from (6.2) that $y_k \overset{k\to\infty}{\to} 0$ in the norm $\|\cdot\|$. On the other hand, applying the Bolzano–Weierstrass Theorem, which states that every bounded sequence in $\mathbb{R}$ has a convergent subsequence, it follows that there exist $\left\{\beta_{k,1}^{(1)}\right\}_{k\in\mathbb{N}} \subseteq \left\{\beta_{k,1}\right\}_{k\in\mathbb{N}}$ and $\beta_1 \in \mathbb{R}$ such that $\beta_{k,1}^{(1)} \overset{k\to\infty}{\to} \beta_1$. In the same way, there exist $\left\{\beta_{k,2}^{(2)}\right\}_{k\in\mathbb{N}} \subseteq \left\{\beta_{k,2}^{(1)}\right\}_{k\in\mathbb{N}}$ and $\beta_2 \in \mathbb{R}$ such that $\beta_{k,2}^{(2)} \overset{k\to\infty}{\to} \beta_2$. Continuing with this inductive reasoning we obtain the existence of a subsequence $\left\{\beta_{k,n}^{(n)}\right\}_{k\in\mathbb{N}} \subseteq \left\{\beta_{k,n}^{(n-1)}\right\}_{k\in\mathbb{N}}$ and scalars $\{\beta_1, \beta_2, \ldots, \beta_n\} \subset \mathbb{R}$ such that $\beta_{k,i}^{(n)} \overset{k\to\infty}{\to} \beta_i$ for all $i \in \{1,2,\ldots,n\}$. Then, defining $y_k^{(n)} := \sum_{i=1}^{n} \beta_{k,i}^{(n)}\, x_i$ and $y := \sum_{i=1}^{n} \beta_i\, x_i$, we obtain

$$\|y\|_1 := \sum_{i=1}^{n} |\beta_i| = \sum_{i=1}^{n} \lim_{k\to\infty} |\beta_{k,i}^{(n)}| = \lim_{k\to\infty} \sum_{i=1}^{n} |\beta_{k,i}^{(n)}| = 1$$

and

$$\|y_k^{(n)} - y\| = \left\|\sum_{i=1}^{n} \left(\beta_{k,i}^{(n)} - \beta_i\right) x_i\right\| \leq \sum_{i=1}^{n} \left|\beta_{k,i}^{(n)} - \beta_i\right| \|x_i\| \overset{k\to\infty}{\to} 0,$$

which indicates that $y_k^{(n)} \overset{k\to\infty}{\to} y$ in the norm $\|\cdot\|$. By uniqueness of the limit, it follows that $y = 0$, and since $\{x_1, x_2, \ldots, x_n\}$ is a basis of X, necessarily $\beta_i = 0$ for all $i \in \{1,2,\ldots,n\}$, which contradicts that $\|y\|_1 = 1$.

□

Corollary 6.2. *Every normed vector space of finite dimension is complete, and every finite-dimensional subspace of a normed vector space is closed.*

Definition 6.3. *A subset S of a normed vector space X is said to be* COMPACT *if every sequence in S has a subsequence converging to an element of S.*

Definition 6.4. *A subset S of a normed vector space X is said to be* RELATIVELY COMPACT *if every sequence in S has a subsequence converging to an element of X. Equivalently, S is relatively compact if and only if $\bar{S}$ is compact.*

Definition 6.5. *A subset S of a normed vector space X is said to be* TOTALLY BOUNDED *if for every $\varepsilon > 0$ there exist $N \in \mathbb{N}$ and $\{x_1, \ldots, x_N\} \subseteq S$ such that $S \subseteq \bigcup_{j=1}^{N} B(x_j, \varepsilon)$.*

Corollary 6.6. *Let X be a finite-dimensional normed vector space, and let S be a closed and bounded subset of X. Then S is compact.*

Lemma 6.7 (Riesz Lemma). *Let S be a closed subspace of a normed vector space X such that $S \neq X$. Then, for each $\varepsilon \in (0,1)$ there exists $x_\varepsilon \in X$ such that $\|x_\varepsilon\| = 1$ and* $\operatorname{dist}(x_\varepsilon, S) \geqslant \varepsilon$.

Proof. Since $S \neq X$, there exists $x_1 \in X$ such that $x_1 \notin S$ and then, since S is closed, $d := \operatorname{dist}(x_1, S) > 0$. It follows that for each $\varepsilon \in (0,1)$ there exists $x_0 \in S$ such that

$$\|x_1 - x_0\| < d + \frac{d(1-\varepsilon)}{\varepsilon} = \frac{d}{\varepsilon}.$$

Then, defining

$$x_\varepsilon := \frac{x_1 - x_0}{\|x_1 - x_0\|},$$

we obtain $\|x_\varepsilon\| = 1$, and for all $x \in S$

$$\|x_\varepsilon - x\| = \frac{1}{\|x_1 - x_0\|} \left\| x_1 - x_0 - \|x_1 - x_0\| \, x \right\|$$

$$= \frac{1}{\|x_1 - x_0\|} \left\| x_1 - (x_0 + \|x_1 - x_0\| \, x) \right\| \geqslant \frac{d}{\|x_1 - x_0\|} > \varepsilon,$$

whence $\operatorname{dist}(x_\varepsilon, S) \geqslant \varepsilon$.

□

Theorem 6.8. *Let $(X, \|\cdot\|)$ be a normed vector space. Then the unit sphere*

$$\partial B(0,1) := \{x \in X : \quad \|x\| = 1\}$$

is compact if and only if X is finite-dimensional.

Proof. The converse implication follows from Corollary 6.6 and from the trivial fact that $\partial B(0,1)$ is clearly a closed and bounded set. Now assume that $\partial B(0,1)$ is compact and let us prove that X is finite-dimensional. Let $x_1 \in X$ such that $\|x_1\| = 1$ and define $M_1 := \langle \{x_1\} \rangle$. If $X = M_1$ then the proof is complete. If $X \neq M_1$ then we apply Riesz's Lemma (see Lemma 6.7) with $\varepsilon = 1/2$ and deduce that there exists $x_2 \in X$ such that $\|x_2\| = 1$ and $\operatorname{dist}(x_2, M_1) \geqslant 1/2$. If $X = M_2 := \langle \{x_1, x_2\} \rangle$ then the proof is complete. If $X \neq M_2$ then we apply again Lemma 6.7 and conclude the existence of $x_3 \in X$ such that $\|x_3\| = 1$ and $\operatorname{dist}(x_3, M_2) \geqslant 1/2$. Then we define $M_3 := \langle \{x_1, x_2, x_3\} \rangle$ and continue this reasoning. If this inductive process never stops, we construct in this way a sequence

$$\{x_n\}_{n \in \mathbb{N}} \subseteq \partial B(0,1) \subseteq X$$

such that

$$\|x_n - x_m\| \geqslant 1/2 \quad \forall\, n > m\,.$$

This last inequality indicates that $\{x_n\}_{n\in\mathbb{N}}$ does not have convergent subsequences, which contradicts the fact that $\partial B(0,1)$ is compact. Therefore, the above process must necessarily stop at some stage, which means that there exists $k \in \mathbb{N}$ such that $X = M_k$, and thus X is finite-dimensional.

□

6.2 Finite-rank operators

6.2.1 Preliminaries

Let $(X, \|\cdot\|)$ and $(Y, \|\cdot\|)$ be normed vector spaces over $\mathbb{K}$, and consider the sets $\{F_1, F_2, \ldots, F_n\} \subseteq X'$ and $\{y_1, y_2, \ldots, y_n\} \subseteq Y$. Then, we define $K : X \to Y$ by

$$K(x) := \sum_{j=1}^{n} F_j(x)\, y_j \qquad \forall\, x \in X\,. \tag{6.3}$$

Notice that the operator K is linear (since each F_j is) and bounded because

$$\|K(x)\| \leqslant \sum_{j=1}^{n} |F_j(x)|\, \|y_j\| \leqslant \left\{ \sum_{j=1}^{n} \|F_j\|\, \|y_j\| \right\} \|x\| \qquad \forall\, x \in X.$$

Moreover, $R(K)$ is finite-dimensional since

$$R(K) \subseteq \langle\{y_1, y_2, \cdots, y_n\}\rangle.$$

The following lemma establishes the converse of the above result.

Lemma 6.9. *Let $(X, \|\cdot\|)$ and $(Y, \|\cdot\|)$ be normed vector spaces, and let $K \in \mathcal{L}(X,Y)$ be such that $R(K)$ is finite-dimensional. Then K is of the form* (6.3).

Proof. Let $\{y_1, y_2, \ldots, y_n\} \subseteq Y$ be a basis of $R(K)$. Then, for each $x \in X$ there exist scalars $\alpha_1(x), \alpha_2(x), \ldots, \alpha_n(x)$ such that

$$K(x) = \sum_{j=1}^{n} \alpha_j(x)\, y_j\,.$$

This induces the definition of functionals $F_j : X \to \mathbb{K}$ given by $F_j(x) := \alpha_j(x)$ for all $x \in X$. It is clear that each F_j is linear and bounded. Indeed, for the second property we deduce, thanks to the equivalence of norms in finite dimension (see Theorem 6.1), applied in this case to the space $R(K)$, that there exists $C > 0$ such that

$$\|K(x)\|_1 := \sum_{j=1}^{n} |F_j(x)| \leqslant C\, \|K(x)\| \quad \forall\, x \in X.$$

Then, by the boundedness of K, we obtain

$$|F_i(x)| \leqslant \sum_{j=1}^{n} |F_j(x)| \leqslant C\, \|K(x)\| \leqslant C\, \|K\|\, \|x\| \quad \forall\, x \in X,$$

which shows that $F_i \in X'$ for all $i \in \{1, 2, \ldots, n\}$, and thus K is of the form (6.3).

□

Definition 6.10. *An operator $K \in \mathcal{L}(X,Y)$ having the form (6.3) is called a* FINITE-RANK OPERATOR.

6.2.2 The Fredholm alternative

The following theorem establishes the classical FREDHOLM ALTERNATIVE for finite-rank operators. Later we will extend this result to the case of compact operators.

Theorem 6.11 (Fredholm's alternative). *Let $(X, \|\cdot\|)$ be a normed vector space, let $K \in \mathcal{L}(X)$ be a finite-rank operator, and define $A := I - K$, where $I : X \to X$ is the identity operator. Then $R(A)$ is closed in X, and*

$$\dim N(A) = \dim N(A') < +\infty. \tag{6.4}$$

Proof. Let $n \in \mathbb{N}$, let $\{x_1, \ldots, x_n\} \subseteq X$, and let $\{F_1, F_2, \ldots, F_n\} \subseteq X'$ be such that

$$K(x) = \sum_{j=1}^{n} F_j(x)\, x_j \qquad \forall\, x \in X.$$

In order to characterize $R(A)$, we first consider, for a given $y \in X$, the fundamental equation: Find $x \in X$ such that

$$A(x) = y. \tag{6.5}$$

According to the definitions of A and K, $x \in X$ is a solution of (6.5) if and only if

$$x = y + \sum_{j=1}^{n} \beta_j\, x_j, \tag{6.6}$$

with $\beta_j := F_j(x)$ for all $j \in \{1, 2, \ldots, n\}$. Then, evaluating F_i in (6.6) we obtain

$$\beta_i - \sum_{j=1}^{n} F_i(x_j)\, \beta_j = F_i(y) \qquad \forall\, i \in \{1, 2, \ldots, n\},$$

or equivalently,

$$\sum_{j=1}^{n} \Big(\delta_{ij} - F_i(x_j)\Big)\beta_j = F_i(y) \qquad \forall\, i \in \{1, 2, \ldots, n\}. \tag{6.7}$$

Conversely, it is easy to see that if the vector $(\beta_1, \beta_2, \ldots, \beta_n)^{\mathrm{t}} \in \mathbb{R}^n$ is a solution of the linear system (6.7), then x defined by (6.6) satisfies $F_i(x) = \beta_i$ for all $i \in \{1, 2, \ldots, n\}$, and therefore x is a solution of (6.5). In this way we have proved that $y \in R(A)$ (equivalently, (6.5) has a solution) if and only if (6.7) has a solution. But from the theory of linear systems, we know that the latter occurs if and only if $(F_1(y), F_2(y), \ldots, F_n(y))^{\mathrm{t}}$ is orthogonal to all the solutions of the transposed homogeneous system, that is,

$$\sum_{j=1}^{n} \alpha_j\, F_j(y) = 0 \tag{6.8}$$

for every vector $(\alpha_1, \alpha_2, \ldots, \alpha_n)^{\mathrm{t}} \in \mathbb{R}^n$ such that

$$\sum_{j=1}^{n} \Big(\delta_{ji} - F_j(x_i)\Big)\alpha_j = 0 \qquad \forall\, i \in \{1, 2, \ldots, n\}. \tag{6.9}$$

On the other hand, a simple calculation shows that the adjoint operator $A' \in \mathcal{L}(X')$ is given by

$$A'(G) = G - \sum_{j=1}^{n} G(x_j)\, F_j \qquad \forall\, G \in X'.$$

Moreover, we have

$$N(A') = \Big\{ G = \sum_{j=1}^{n} \alpha_j\, F_j : \quad (\alpha_1, \alpha_2, \ldots, \alpha_n)^{\mathrm{t}} \quad \text{solution of (6.9)} \Big\}.$$

Indeed, if $G \in N(A')$, then $G = \sum_{j=1}^{n} \alpha_j\, F_j$ with $\alpha_j := G(x_j)$ for all $j \in \{1, 2, \ldots, n\}$. Evaluating at x_i yields

$$\alpha_i = \sum_{j=1}^{n} \alpha_j\, F_j(x_i) \qquad \forall\, i \in \{1, 2, \ldots, n\},$$

or equivalently,

$$\sum_{j=1}^{n} (\delta_{j\,i} - F_j(x_i))\alpha_j = 0 \qquad \forall\, i \in \{1, 2, \ldots, n\},$$

which shows that $(\alpha_1, \alpha_2, \ldots, \alpha_n)^{\mathrm{t}} \in \mathbb{R}^n$ is a solution of (6.9). Conversely, if this vector is a solution of (6.9), then $G := \sum_{j=1}^{n} \alpha_j\, F_j$ satisfies

$$G(x_i) = \sum_{j=1}^{n} \alpha_j\, F_j(x_i) = \alpha_i \qquad \forall\, i \in \{1, 2, \ldots, n\},$$

and hence $G \in N(A')$. According to the above, condition (6.8) can be rewritten as

$$G(y) = 0 \qquad \forall\, G \in N(A'),$$

that is, $y \in {}^{\circ}N(A')$, from which we conclude that $R(A) = {}^{\circ}N(A')$, and therefore $R(A)$ is closed. Finally, from the analysis of the first part of this proof, taking $y = 0$, we easily deduce that

$$N(A) = \Big\{ x = \sum_{j=1}^{n} \beta_j\, x_j : \quad (\beta_1, \beta_2 \ldots, \beta_n)^{\mathrm{t}} \quad \text{solution of the homogeneous (6.7)} \Big\}.$$

Thus, since the homogeneous systems (6.7) and (6.9) have the same number of linearly independent solutions (as one is the transpose of the other), (6.4) is satisfied.

□

We find it important to emphasize now, as suggested at the beginning of the proof of Theorem 6.11, that the concept of the FREDHOLM ALTERNATIVE concerns the resolution of the fundamental equation (6.5), that is, given $y \in X$, find $x \in X$ such that $A(x) = y$. To understand this, note first that the fact that $R(A)$ is closed and that the dimensions of $N(A)$ and $N(A')$ are finite and coincide, implies that the eventual surjectivity of A is equivalent to its injectivity. Indeed, if $R(A) = X$, then $N(A') = R(A)^{\circ} = \{0\} \subseteq X'$, whence $N(A) = \{0\} \subseteq X$. Conversely, if $N(A) = \{0\}$, it is clear that $N(A') = \{0\}$, and since $R(A)$ is closed, we have $R(A) = {}^\circ N(A') = X$. This means that only one of the following alternatives can occur:

$$
\begin{gathered}
R(A) = X \quad \text{and} \quad N(A) = \{0\} \\
\textbf{or else} \\
R(A) \neq X \quad \text{and} \quad N(A) \neq \{0\},
\end{gathered}
\tag{6.10}
$$

and the crossed options are not possible. In other words, the equation $A(x) = y$ has a unique solution for every $y \in X$ **or else** the homogeneous equation $A(x) = 0$ admits a finite number m of linearly independent solutions. In this case, $A(x) = y$ has a solution if and only if $y \in {}^\circ N(A')$. Equivalently, if and only if $F_j(y) = 0$ for all $j \in \{1, 2, \ldots, m\}$, where $\{F_1, F_2, \ldots, F_m\}$ is a basis of $N(A')$.

Next, we extend the Fredholm alternative to linear and bounded operators that can be written as the norm limit of a sequence of finite-rank operators. More precisely, we have the following result.

Theorem 6.12. *Let X be a Banach space, let $K \in \mathcal{L}(X)$, and let $\{K_n\}_{n\in\mathbb{N}}$ be a sequence of finite-rank operators converging to K in $\mathcal{L}(X)$, that is,*

$$\|K - K_n\|_{\mathcal{L}(X)} \overset{n\to\infty}{\longrightarrow} 0.$$

If $A := I - K$, then $R(A)$ is closed in X and $\dim N(A) = \dim N(A') < +\infty$.

Proof. Let $N \in \mathbb{N}$ be large enough such that $\|K - K_N\| < 1$, and set

$$A = I - K = I - (K - K_N) - K_N = B_N - K_N,$$

with $B_N := I - (K - K_N)$. It follows, according to a classical result in Banach spaces (see Exercise 3.12 in Chapter 3), that B_N^{-1} is defined on all of X. Moreover, since B_N is bounded, the Bounded Inverse Theorem (cf. Theorem 3.34) implies that $B_N^{-1} \in \mathcal{L}(X)$. Then, we can write $A = B_N\left(I - B_N^{-1} K_N\right)$, where $B_N^{-1} K_N$, being the product of a bounded operator and a finite-rank operator, is also a finite-rank operator. Thus, the equation $A(x) = y$ is equivalent to

$$\left(I - B_N^{-1} K_N\right)(x) = B_N^{-1}(y),$$

from which, applying Theorem 6.11, we obtain

$$\dim N(A) = \dim N(I - B_N^{-1} K_N) = \dim N\left((I - B_N^{-1} K_N)'\right) < +\infty.$$

In turn, since $A' = \left(I - B_N^{-1} K_N\right)' B_N'$ and B_N' is invertible with $(B_N')^{-1} = (B_N^{-1})'$, it follows that $\dim N(A') = \dim N\left((I - B_N^{-1} K_N)'\right) = \dim N(A)$. Let us show that $R(A)$ is closed. Indeed, let $\{x_k\}_{k\in\mathbb{N}} \subseteq X$ and $y \in X$ such that $A(x_k) \overset{k\to\infty}{\longrightarrow} y$. It follows that $\left(I - B_N^{-1} K_N\right)(x_k) \to B_N^{-1}(y)$, and since $R\left(I - B_N^{-1} K_N\right)$ is closed, we deduce that there exists $x \in X$ such that $\left(I - B_N^{-1} K_N\right)(x) = B_N^{-1}(y)$, from which, applying B_N, we obtain $A(x) = y$, and thus $y \in R(A)$. This shows that $R(A)$ is in fact closed. Alternatively, one can certainly apply the characterization result given by Theorem 3.61. Indeed, since $R\left(I - B_N^{-1} K_N\right)$ is closed, there exists $C > 0$ such that

$$\|(I - B_N^{-1} K_N)(x)\| \geqslant C \operatorname{dist}\left(x, N(I - B_N^{-1} K_N)\right) \qquad \forall\, x \in X,$$

whence, the boundedness of B_N^{-1} and the fact that $N(A) = N(I - B_N^{-1} K_N)$, yield

$$\begin{aligned} \|A(x)\| &= \|B_N (I - B_N^{-1} K_N)(x)\| \geqslant \frac{1}{\|B_N^{-1}\|} \|(I - B_N^{-1} K_N)(x)\| \\ &\geqslant \frac{C}{\|B_N^{-1}\|} \operatorname{dist}\big(x, N(A)\big) \qquad \forall\, x \in X\,, \end{aligned}$$

thus proving that $R(A)$ is closed.

□

It is important to note here, as it is clear from the respective proof, that the previous theorem does not actually require a sequence of finite-rank operators converging to K, but only one operator K_N with that property such that $\|K - K_N\| < 1$. Naturally, this fact is ensured by the existence of the convergent sequence, but it is not essential that the latter occurs.

6.3 Compact operators in normed spaces

6.3.1 Preliminaries

Definition 6.13. *Let X and Y be normed vector spaces. A linear operator $K : X \to Y$ is said to be* COMPACT *(or* COMPLETELY CONTINUOUS*) if $\mathcal{D}(K) = X$ and for every bounded sequence $\{x_n\}_{n\in\mathbb{N}} \subseteq X$, the sequence $\{K(x_n)\}_{n\in\mathbb{N}}$ has a convergent subsequence in Y. Equivalently, such an operator K is said to be compact if for every bounded subset S of X, $K(S)$ is relatively compact in Y, that is, $\overline{K(S)}$ is compact in Y. The set of compact operators from X to Y is denoted by $\mathcal{K}(X,Y)$.*

The first thing to observe is that every linear compact operator is bounded, that is, $\mathcal{K}(X,Y) \subseteq \mathcal{L}(X,Y)$. Indeed, let $K \in \mathcal{K}(X,Y)$ and suppose, by contradiction, that K is not bounded. Then, for each $n \in \mathbb{N}$ there exists $x_n \in X$ such that

$$\|x_n\| < \frac{1}{n} \|K(x_n)\|\,.$$

Thus, defining $z_n := \frac{n\, x_n}{\|K(x_n)\|}$ we observe that

$$\|z_n\| < 1 \quad \text{and} \quad \|K(z_n)\| = n \quad \forall\, n \in \mathbb{N}\,,$$

which contradicts the compactness of K since $\{K(z_n)\}_{n\in\mathbb{N}}$ has no convergent subsequences. Similarly, using only the definition of compactness, one can easily prove that the sum of compact operators and the product of a scalar by a compact operator are both compact. Moreover, every finite-rank operator is compact, and the composition of a compact operator with a bounded operator, both on the left and on the right, is also compact. More precisely, we have the following lemma.

Lemma 6.14. *Let X, Y, and Z be normed vector spaces. Then*

i) *$\mathcal{K}(X,Y)$ is a subspace of $\mathcal{L}(X,Y)$,*
ii) *$K \in \mathcal{L}(X,Y)$ of finite-rank $\Rightarrow$ $K \in \mathcal{K}(X,Y)$,*
iii) *$A \in \mathcal{L}(X,Y)$, $K \in \mathcal{K}(Y,Z)$ $\Rightarrow$ $K\,A \in \mathcal{K}(X,Z)$, and*

iv) $K \in \mathcal{K}(X,Y)$, $A \in \mathcal{L}(Y,Z) \Rightarrow AK \in \mathcal{K}(X,Z)$.

Another interesting result, as established in the following theorem, is the fact that $\mathcal{K}(X,Y)$ is a closed subspace of $\mathcal{L}(X,Y)$.

Theorem 6.15. *Let X and Y be normed vector spaces, with Y Banach, and let $\{K_n\}_{n\in\mathbb{N}} \subseteq \mathcal{K}(X,Y)$ and $K \in \mathcal{L}(X,Y)$ be such that*

$$\|K_n - K\|_{\mathcal{L}(X,Y)} \overset{n\to\infty}{\to} 0 .$$

Then $K \in \mathcal{K}(X,Y)$.

Proof. Let $\{x_n\}_{n\in\mathbb{N}}$ be a bounded sequence in X. Since K_1 is compact, there exists a subsequence $\{x_n^{(1)}\}_{n\in\mathbb{N}} \subseteq \{x_n\}_{n\in\mathbb{N}}$ such that $\{K_1(x_n^{(1)})\}_{n\in\mathbb{N}}$ is convergent in Y. In turn, since K_2 is also compact, there exists a subsequence $\{x_n^{(2)}\}_{n\in\mathbb{N}} \subseteq \{x_n^{(1)}\}_{n\in\mathbb{N}}$ such that $\{K_2(x_n^{(2)})\}_{n\in\mathbb{N}}$ is convergent in Y, and then $\{K_i(x_n^{(2)})\}_{n\in\mathbb{N}}$ is convergent in Y for all $i \in \{1,2\}$. Analogously, due to the compactness of K_3, there exists a subsequence $\{x_n^{(3)}\}_{n\in\mathbb{N}} \subseteq \{x_n^{(2)}\}_{n\in\mathbb{N}}$ such that $\{K_i(x_n^{(3)})\}_{n\in\mathbb{N}}$ is convergent in Y for all $i \in \{1,2,3\}$. Following this inductive reasoning, we deduce that for each $k \in \mathbb{N}$ there exists a subsequence $\{x_n^{(k)}\}_{n\in\mathbb{N}} \subseteq \{x_n^{(k-1)}\}_{n\in\mathbb{N}}$ such that

$$\{K_i(x_n^{(k)})\}_{n\in\mathbb{N}} \quad \text{is convergent in } Y \quad \forall\, i \in \{1,2,\ldots,k\} .$$

The above suggests the "*diagonal*" argument, in which, given the sequence of subsequences shown in Figure 6.1, we define the subsequence that is generated precisely on the diagonal of this arrangement, that is formed by the terms enclosed by a circle:

$$z_n := x_n^{(n)} \qquad \forall\, n \in \mathbb{N} .$$

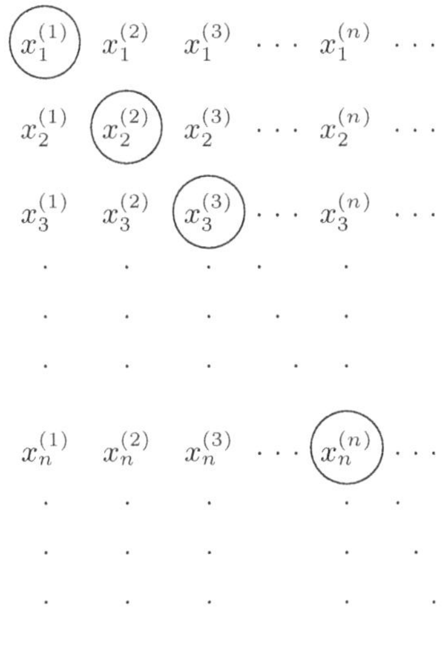

Fig. 6.1. *Illustration of the "diagonal" argument*

It is clear that $\{z_n\}_{n\in\mathbb{N}}$ is a subsequence of the original sequence $\{x_n\}_{n\in\mathbb{N}}$, and moreover $\{K_k(z_n)\}_{n\in\mathbb{N}}$ is convergent in Y for all $k\in\mathbb{N}$. Indeed, given $k\in\mathbb{N}$, $\{z_n\}_{n\geqslant k} = \{x_n^{(n)}\}_{n\geqslant k}$ is a subsequence of $\{x_n^{(k)}\}_{n\in\mathbb{N}}$ and therefore $\{K_k(z_n)\}_{n\in\mathbb{N}}$ is convergent. Now, for $m, n\in\mathbb{N}, m\neq n$, and for every $k\in\mathbb{N}$, we have

$$\begin{aligned}\|K(z_n)-K(z_m)\| &\leqslant \|K(z_n)-K_k(z_n)\|\\ &+\|K_k(z_n)-K_k(z_m)\|+\|K_k(z_m)-K(z_m)\|\\ &\leqslant \|K-K_k\|\,\|z_n\|+\|K_k(z_n)-K_k(z_m)\|+\|K-K_k\|\,\|z_m\|\\ &\leqslant 2C\,\|K-K_k\|+\|K_k(z_n)-K_k(z_m)\|,\end{aligned}$$

where $C>0$ is such that $\|x_n\|\leqslant C$ for all $n\in\mathbb{N}$. In particular, given $\varepsilon>0$, we have

$$\|K(z_n)-K(z_m)\|\leqslant 2C\varepsilon+\|K_N(z_n)-K_N(z_m)\|,$$

where $N\in\mathbb{N}$ is such that $\|K-K_N\|\leqslant\varepsilon$. Then, since $\{K_N(z_n)\}_{n\in\mathbb{N}}$ is convergent in Y, there exists $N_1\in\mathbb{N}$ such that

$$\|K_N(z_n)-K_N(z_m)\|\leqslant\varepsilon\quad\forall m, n\geqslant N_1,$$

and therefore

$$\|K(z_n)-K(z_m)\|\leqslant(2C+1)\,\varepsilon\quad\forall m, n\geqslant N_1.$$

This shows that $\{K(z_n)\}_{n\in\mathbb{N}}$ is a Cauchy sequence in the Banach space Y, and consequently $\{K(z_n)\}_{n\in\mathbb{N}}$ is convergent in Y.

□

Note that the previous theorem establishes, in particular, that if $K\in\mathcal{L}(X,Y)$ and $\{K_n\}_{n\in\mathbb{N}}\subseteq\mathcal{L}(X,Y)$ is a sequence of operators of finite-rank that converges to K, that is such that $\|K_n-K\|_{\mathcal{L}(X,Y)}\overset{n\to\infty}{\to}0$, then $K\in\mathcal{K}(X,Y)$. This simple fact provides, in some cases, a very simple way to prove that a given operator is compact. On the other hand, one might ask whether the converse is also true. In general, this is not the case unless, as shown below, Y is a Hilbert space.

Theorem 6.16. *Let X be a normed vector space and Y a Hilbert space, and let $K\in\mathcal{K}(X,Y)$. Then there exists a sequence $\{K_n\}_{n\in\mathbb{N}}\subseteq\mathcal{L}(X,Y)$ of finite-rank operators such that $\|K_n-K\|_{\mathcal{L}(X,Y)}\overset{n\to\infty}{\to}0$.*

Proof. We use the equivalent definition of a compact operator which states that the image of a bounded set has compact closure. Thus, given the unit ball $\bar{B}(0,1)$ of X, the set $S:=\overline{K(\bar{B}(0,1))}$ is compact in Y, and therefore, given $n\in\mathbb{N}$, there exists a finite index set I_n and vectors $\{y_i:\ i\in I_n\}\subseteq Y$, such that

$$S\subseteq\bigcup_{i\in I_n}B(y_i,1/n).\tag{6.11}$$

Then, we define $Y_n:=\langle\{y_i\}_{i\in I_n}\rangle$ and consider the orthogonal projector $\varPi_n:Y\to Y_n$ (cf. Section 2.2). It follows that the operator $K_n:=\varPi_n K: X\to Y$ belongs to $\mathcal{L}(X,Y)$ and is of finite-rank. Moreover, given $x\in\bar{B}(0,1)$, we deduce from the inclusion (6.11) that there exists $j \in I_n$ such that $K(x) \in B(y_j, 1/n)$, that is, $\|K(x) - y_j\| < 1/n$. Consequently, we obtain

$$\begin{aligned}\|K_n(x) - K(x)\| &\leqslant \|K_n(x) - y_j\| + \|K(x) - y_j\| \\ &= \|\Pi_n(K(x)) - \Pi_n(y_j)\| + \|K(x) - y_j\| \\ &= \|\Pi_n(K(x) - y_j)\| + \|K(x) - y_j\| \\ &\leqslant 2\|K(x) - y_j\| \leqslant 2/n \quad \forall\, x \in \bar{B}(0,1)\,,\end{aligned}$$

which proves that $\|K_n - K\|_{\mathcal{L}(X,Y)} \overset{n\to\infty}{\to} 0$.

□

Regarding the above result, we would like to remark that the question of whether every compact operator between arbitrary Banach spaces X and Y is a norm limit of a sequence of finite-rank operators was known for a long time as THE APPROXIMATION PROBLEM. Closely related to it is the APPROXIMATION PROPERTY of the space X, which says that for every compact set $S \subset X$ and every $\varepsilon > 0$, there exists a finite-rank operator $T \in \mathcal{K}(X, X)$ such that

$$\|T(x) - x\| < \varepsilon \qquad \forall\, x \in S\,.$$

In fact, denoting by $\mathcal{F}(X, Y)$ the subspace of $\mathcal{K}(X, Y)$ formed by the finite-rank operators, it can be shown that if X has the approximation property, then $\mathcal{F}(X, Y)$ is dense in $\mathcal{K}(X, Y)$ for every Banach space Y. Conversely, if this latter property holds for all Banach spaces Y, then X has the approximation property. It was later shown by the Swedish mathematician Per Enflo in [50] that there exists a separable and reflexive Banach space X without the approximation property, thus establishing the existence of a compact operator in $\mathcal{K}(X, X)$ that is not the norm limit of a sequence of finite-rank operators.

6.3.2 The Fredholm alternative

Let us first note, according to the analysis in the previous section, that the Fredholm alternative for an operator $K \in \mathcal{K}(X, X)$ with X Hilbert is a simple corollary of Theorems 6.12 and 6.16. More precisely, we have the following theorem.

Theorem 6.17. *Let X be a Hilbert space, and let $K \in \mathcal{K}(X, X)$. If $A := I - K$, then $R(A)$ is closed in X and* $\dim N(A) = \dim N(A') < +\infty$.

Our next goal is to extend the Fredholm alternative to the more general case of a Banach space. To this end, we proceed through several intermediate results that will ultimately constitute the main theorem.

Lemma 6.18. *Let X be a Banach space, and let $K \in \mathcal{K}(X, X)$. If $A := I - K$, then $R(A)$ is closed in X.*

Proof. According to the characterization result given by Theorem 3.61, it suffices to prove that there exists $C > 0$ such that

$$\operatorname{dist}(x, N(A)) \leqslant C\,\|A(x)\| \quad \forall\, x \in X. \tag{6.12}$$

To this end, note first that given $x \in X$ and $\varepsilon > 0$, there exists $x_1 \in N(A)$ such that

$$\operatorname{dist}(x, N(A)) \leqslant \|x - x_1\| < \operatorname{dist}(x, N(A)) + \varepsilon,$$

and then, defining $x_0 := x - x_1$, we have

$$A(x_0) = A(x) \quad \text{and} \quad \text{dist}(x, N(A)) \leqslant \|x_0\| < \text{dist}\,(x, N(A)) + \varepsilon\,. \tag{6.13}$$

Now, suppose by contradiction that (6.12) does not hold. It follows, in particular, that for each $n \in \mathbb{N}$ there exists $y_n \in X$ such that

$$\text{dist}(y_n, N(A)) > n\,\|A(y_n)\|\,,$$

and then, denoting

$$x_n := \frac{y_n}{\text{dist}(y_n, N(A))} \qquad \forall\, n \in \mathbb{N},$$

we obtain

$$\text{dist}(x_n, N(A)) = 1 \quad \text{and} \quad \|A(x_n)\| < \frac{1}{n} \qquad \forall\, n \in \mathbb{N}.$$

In turn, applying (6.13) to $x = x_n$ and $\varepsilon = 1$ for each $n \in \mathbb{N}$, we deduce the existence of a sequence $\{z_n\}_{n\in\mathbb{N}} \subseteq X$ such that

$$A(z_n) = A(x_n) \quad \text{and} \quad 1 \leqslant \|z_n\| < 2 \qquad \forall\, n \in \mathbb{N}.$$

In addition, since $x_n - z_n \in N(A)$, there also hold

$$\text{dist}(z_n, N(A)) = \text{dist}(x_n, N(A)) = 1$$

and $\|A(z_n)\| = \|A(x_n)\| \overset{n\to\infty}{\longrightarrow} 0$, which shows that $A(z_n) \overset{n\to\infty}{\longrightarrow} 0$. Thus, since K is compact, there exist a subsequence $\{z_n^{(1)}\}_{n\in\mathbb{N}} \subseteq \{z_n\}_{n\in\mathbb{N}}$ and $w \in X$ such that

$$K(z_n^{(1)}) \overset{n\to\infty}{\longrightarrow} w\,,$$

which yields

$$z_n^{(1)} = K(z_n^{(1)}) + A(z_n^{(1)}) \overset{n\to\infty}{\longrightarrow} w. \tag{6.14}$$

From the above and by the continuity of A, we find that

$$A(w) = \lim_{n\to\infty} A(z_n^{(1)}) = 0,$$

that is $w \in N(A)$, and hence

$$1 = \text{dist}(z_n^{(1)}, N(A)) = \inf_{z\in N(A)} \|z_n^{(1)} - z\| \leqslant \|z_n^{(1)} - w\| \qquad \forall\, n \in \mathbb{N}\,,$$

which contradicts (6.14). Therefore, there must necessarily exist $C > 0$ such that (6.12) holds, which proves that $R(A)$ is closed.

□

Lemma 6.19. *Let X be a Banach space, and let $K \in \mathcal{K}(X, X)$. If $A := I - K$, then both $N(A)$ and $N(A')$ are finite-dimensional.*

Proof. In order to conclude that $\dim N(A) < +\infty$, it suffices to prove, invoking Theorem 6.8, that the unit sphere in $N(A)$, namely $S := \{x \in N(A) : \|x\| = 1\}$, is compact. Indeed, given $\{x_n\}_{n\in\mathbb{N}} \subseteq S$, the compactness of K guarantees the existence of a subsequence $\{x_n^{(1)}\}_{n\in\mathbb{N}} \subseteq \{x_n\}_{n\in\mathbb{N}}$ such that $\{K(x_n^{(1)})\}_{n\in\mathbb{N}}$ is convergent in X. Thus, since $0 = A(x_n^{(1)}) = x_n^{(1)} - K(x_n^{(1)})$, we readily deduce that $\{x_n^{(1)}\}_{n\in\mathbb{N}}$ is also convergent in X, and therefore S is compact. On the other hand, regarding $N(A')$, we will later show that, being K compact, its adjoint $K' \in \mathcal{L}(X', X')$ is compact as well, and consequently the same proof as above applies to $A' := I' - K'$.

□

Lemma 6.20. *Let X be a Banach space, and let $K \in \mathcal{K}(X, X)$. If $A := I - K$, then $N(A) = \{0\} \subseteq X$ if and only if $N(A') = \{0\} \subseteq X'$.*

Proof.
($\Leftarrow$) Let us first assume that $N(A') = \{0\}$. Since $R(A)$ is closed (cf. Lemma 6.18), there holds $R(A) = {}^{\circ}N(A') = X$. Suppose now by contradiction that $N(A)$ is not the null space, which means that there exists $x_1 \in X$, $x_1 \neq 0$, such that $A(x_1) = 0$. The fact that A is surjective implies that there exists $x_2 \in X$ such that $A(x_2) = x_1$, so that proceeding recursively in this way, we construct a sequence $\{x_n\}_{n\in\mathbb{N}} \subseteq X$ such that $A(x_n) = x_{n-1}$ for all $n \geqslant 2$. Since A^n is clearly a bounded operator, $N(A^n)$ is a closed subspace of X. Moreover, $N(A^{n-1}) \subsetneqq N(A^n)$ for all $n \geqslant 2$ because

$$A^n(x_n) = A^{n-1}(x_{n-1}) = A^{n-2}(x_{n-2}) = \cdots = A(x_1) = 0$$

and

$$A^{n-1}(x_n) = A^{n-2}(x_{n-1}) = A^{n-3}(x_{n-2}) = \cdots = A(x_2) = x_1 \neq 0,$$

which shows that $x_n \in N(A^n)$ and $x_n \notin N(A^{n-1})$. Then, applying Riesz's Lemma (Lemma 6.7) with $X = N(A^n)$, $S = N(A^{n-1})$ and $\varepsilon = 1/2$, we deduce the existence of $z_n \in N(A^n)$ such that $\|z_n\| = 1$ and $\operatorname{dist}(z_n, N(A^{n-1})) \geqslant 1/2$. Next, given n, $m \in \mathbb{N}$ such that $n > m$, we notice that $N(A^m) \subseteq N(A^{n-1}) \subseteq N(A^n)$, and thus

$$A^{n-1}\big(z_m - A(z_m) + A(z_n)\big) = A^{n-1}(z_m) - A^n(z_m) + A^n(z_n) = 0,$$

which says that $\big(z_m - A(z_m) + A(z_n)\big) \in N(A^{n-1})$. Therefore,

$$\begin{aligned}\|K(z_n) - K(z_m)\| &= \|z_n - A(z_n) - z_m + A(z_m)\| = \big\|z_n - \big(z_m - A(z_m) + A(z_n)\big)\big\| \\ &\geqslant \operatorname{dist}\big(z_n, N(A^{n-1})\big) \geqslant 1/2\,,\end{aligned}$$

and consequently $\{K(z_n)\}_{n\in\mathbb{N}}$ has no convergent subsequence, which contradicts the fact that K is compact. Hence, necessarily $N(A) = \{0\}$.

($\Rightarrow$) Conversely, assume now that $N(A) = \{0\}$. Since $R(A)$ is closed (cf. Lemma 6.18), it follows from Theorem 3.62 that $R(A')$ is closed as well, and hence $R(A') = N(A)^{\circ} = X'$. Then, since $A' = I' - K'$ and K' is compact, the same reasoning as in the previous implication shows that $N(A') = \{0\}$, thus completing the proof.

□

Now, knowing from Lemma 6.19 that the null spaces have finite dimensions, it only remains to show that they are equal. For this we need the following preliminary result.

Lemma 6.21. *Let X be a Banach space, and let $\{x_1, \ldots, x_n\}$ and $\{F_1, \ldots, F_m\}$ be linearly independent sets in X and X', respectively. Then there exist $F \in X'$ and $x_0 \in X$ such that*

i) $F(x_j) = 0$ *for all* $j \in \{1, \ldots, n-1\}$, *and* $F(x_n) \neq 0$.
ii) $F_j(x_0) = 0$ *for all* $j \in \{1, \ldots, m-1\}$, *and* $F_m(x_0) \neq 0$.

Proof.

i) Let $M = \langle\{x_1, x_2, \cdots, x_{n-1}\}\rangle$. It is clear that M is a closed subspace of X and that $x_n \notin M$. It follows that $\operatorname{dist}(x_n, M) > 0$, and then, applying a consequence of the Hahn-Banach Theorem (cf. Theorem 2.26), there exists $F \in X'$ such that $\|F\| = 1$, $F(x_n) = \operatorname{dist}(x_n, M)$, and $F(x) = 0$ for all $x \in M$. In particular, $F(x_j) = 0$ for all $j \in \{1, \cdots, n-1\}$ and clearly $F(x_n) \neq 0$.

ii) Here we apply finite induction on m, the number of functionals. For $m = 1$ the result is immediate since $F_1 \neq 0$. Suppose the statement holds for $m = k-1$, and consider a set of k linearly independent functionals $\{F_1, \cdots, F_k\}$. Applying the induction hypothesis to each of the sets of $k-1$ functionals given by $S_i := \Big\{F_j : j \in \{1, \cdots, k-1\}, j \neq i\Big\} \cup \{F_i\}$ for all $i \in \{1, \cdots, k-1\}$, we deduce that there exist vectors $x_0^{(1)}, x_0^{(2)}, \cdots, x_0^{(k-1)}$ such that $F_j(x_0^{(i)}) = 0$ for all $i, j \in \{1, \cdots, k-1\}$ with $i \neq j$, and $F_j(x_0^{(j)}) \neq 0$ for all $j \in \{1, \cdots, k-1\}$. Normalizing if necessary, we may assume that $F_j(x_0^{(j)}) = 1$ for all $j \in \{1, 2, \cdots, k-1\}$, so that for each $x \in X$ we get

$$F_j\left(x - \sum_{i=1}^{k-1} F_i(x) x_0^{(i)}\right) = 0 \qquad \forall\, j \in \{1, 2, \cdots, k-1\}.$$

Moreover, there must exist $\widetilde{x} \in X$ such that

$$F_k\left(\widetilde{x} - \sum_{i=1}^{k-1} F_i(\widetilde{x}) x_0^{(i)}\right) \neq 0.$$

Otherwise, we would have

$$F_k(x) = \sum_{i=1}^{k-1} F_k(x_0^{(i)})\, F_i(x) \qquad \forall\, x \in X,$$

from which $F_k \in \langle\{F_1, \cdots, F_{k-1}\}\rangle$, contradicting the linear independence of the set $\{F_1, F_2, \cdots, F_k\}$. Hence, choosing $x_0 = \widetilde{x} - \sum_{i=1}^{k-1} F_i(\widetilde{x}) x_0^{(i)}$, the induction is completed.

□

Before continuing with the sequence of announced intermediate results, we note that it is possible to provide an alternative proof for part ii) of the previous lemma when X is a REFLEXIVE Banach space. Indeed, in this case (cf. Lemma 3.74) we have $({}^{\circ}T)^{\circ} = T$ for every closed subspace T of X'. Hence, assuming by contradiction that there does not exist $x_0 \in X$ satisfying ii) of Lemma 6.21 is equivalent to saying that

$${}^{\circ}\langle\{F_1, \ldots, F_{m-1}\}\rangle \subseteq {}^{\circ}\langle\{F_m\}\rangle,$$

and then, applying right annihilators, we obtain

$$\langle\{F_m\}\rangle \subseteq \langle\{F_1, \ldots, F_{m-1}\}\rangle,$$

which contradicts the fact that $\{F_1, \ldots, F_m\}$ is linearly independent, and therefore such an $x_0 \in X$ does indeed exist.

Lemma 6.22. *Let X be a Banach space, let $K \in \mathcal{K}(X, X)$, and let $A := I - K$. Then* $\dim N(A) = \dim N(A')$.

Proof. Let $\{x_1, \ldots, x_n\} \subseteq X$ and $\{F_1, \ldots, F_m\} \subseteq X'$ be bases of $N(A)$ and $N(A')$, respectively, and let $F \in X'$ and $x_0 \in X$ be the elements whose existence is guaranteed by Lemma 6.21. We then define the operator $A_1 := A - K_0 = I - (K + K_0)$, with K_0 the finite-rank operator given by $K_0(x) = F(x)\, x_0$ for all $x \in X$, and show next that

$$N(A_1) = \langle\{x_1, \ldots, x_{n-1}\}\rangle \tag{6.15}$$

and

$$N(A_1') = \langle\{F_1, \ldots, F_{m-1}\}\rangle. \tag{6.16}$$

Indeed, given $x \in N(A_1)$ we have $A(x) = K_0(x) = F(x)\, x_0$, and since $x_0 \notin {}^\circ N(A') = R(A)$ because $F_m(x_0) \neq 0$, it follows that necessarily $F(x) = 0$, whence $A(x) = 0$. Thus, there exist scalars $\alpha_1, \ldots, \alpha_n$ such that $x = \sum_{j=1}^{n} \alpha_j x_j$, and then we get

$$0 = F(x) = \sum_{j=1}^{n} \alpha_j\, F(x_j) = \alpha_n\, F(x_n),$$

which implies $\alpha_n = 0$ and therefore $x \in \langle\{x_1, \ldots, x_{n-1}\}\rangle$. Conversely, every element x of the form $\sum_{j=1}^{n-1} \alpha_j x_j$ belongs to $N(A_1)$ since it clearly belongs to $N(A)$ and also $F\Big(\sum_{j=1}^{n-1} \alpha_j x_j\Big) = 0$. Thus we conclude identity (6.15), and therefore $\dim N(A_1) = n - 1$. On the other hand, it is easy to see that $A_1' = A' - K_0'$, where $K_0'(G) = G(x_0)\, F$ for all $G \in X'$. Then, given $G \in N(A_1')$ we have $A'(G) = K_0'(G) = G(x_0)\, F$, and since $F \notin N(A)^\circ = R(A')$ because $F(x_n) \neq 0$, it follows that $G(x_0) = 0$, whence $A'(G) = 0$. Therefore, there exist scalars $\beta_1, \ldots, \beta_m$ such that $G = \sum_{j=1}^{m} \beta_j\, F_j$, so that we obtain

$$0 = G(x_0) = \sum_{j=1}^{m} \beta_j\, F_j(x_0) = \beta_m\, F_m(x_0),$$

which implies $\beta_m = 0$ and consequently $G \in \langle\{F_1, \ldots, F_{m-1}\}\rangle$. Conversely, if there holds $G = \sum_{j=1}^{m-1} \beta_j\, F_j$, then $G \in N(A_1')$ since certainly $G \in N(A')$ and also

$$G(x_0) = \sum_{j=1}^{m-1} \beta_j\, F_j(x_0) = 0.$$

This proves (6.16), and therefore $\dim N(A_1') = m - 1$. Thus, since $A_1 = I - (K + K_0)$, with $K + K_0$ compact, we can repeat this process and successively construct operators $A_k := A_{k-1} - K_{k-1}$ for all $k \in \{2, \ldots, n\}$, with each K_{k-1} compact, such that

$$\dim N(A_k) = n - k \quad \text{and} \quad \dim N(A_k') = m - k.$$

Here we have assumed, without loss of generality, that $n \leqslant m$. In particular, for $k = n$ we obtain $N(A_n) = \{0\}$ and therefore, according to Lemma 6.20, $N(A_n') = \{0\}$, from which we conclude finally that $n = m$.

□

We note here that the identities $R(A) = {}^\circ N(A')$ and $R(A') = N(A)^\circ$ used in the previous proof follow from the fact that $R(A)$ is closed (cf. Lemma 6.18) and from Theorems 3.28 and 3.62.

We are now in a position to establish the Fredholm alternative for a Banach space.

Theorem 6.23. *Let X be a Banach space, let $K \in \mathcal{K}(X,X)$, and let $A := I - K$. Then $R(A)$ is closed in X and* $\dim N(A) = \dim N(A') < +\infty$.

Proof. This follows directly from Lemmas 6.18, 6.19, 6.20, and 6.22.

□

Moreover, the previous result can be extended to the case of an operator A given by a compact perturbation, not only of the identity I, but of any invertible operator. The corresponding theorem, which is very easy to prove, is stated as follows.

Theorem 6.24. *Let X and Y be Banach spaces, let $K \in \mathcal{K}(X,Y)$, let $B \in \mathcal{L}(X,Y)$ be bijective, and let $A := B-K$. Then $R(A)$ is closed in Y, and* $\dim N(A) = \dim N(A') < +\infty$.

Proof. First note, according to the Bounded Inverse Theorem (cf. Theorem 3.34), that $B^{-1} \in \mathcal{L}(Y,X)$. It follows, by part iv) of Lemma 6.14, that $B^{-1}K \in \mathcal{K}(X,X)$, whence, thanks to Theorem 6.23, $R(I-B^{-1}K)$ is closed in X. Now, given a sequence $\{x_n\}_{n\in\mathbb{N}} \subseteq X$ such that $\{A(x_n)\}_{n\in\mathbb{N}}$ converges to a vector $y \in Y$, we deduce, by the continuity of B^{-1}, that

$$\big(I - B^{-1}K\big)(x_n) \,=\, B^{-1}A(x_n) \overset{n\to\infty}{\to} B^{-1}(y) \in X\,,$$

from which, invoking the closedness of $R(I - B^{-1}K)$, there exists $x \in X$ such that

$$\big(I - B^{-1}K\big)(x) \,=\, B^{-1}(y)\,.$$

Applying B to this equation we obtain $A(x) = y$, which proves that $R(A)$ is closed in Y. Certainly, as in the last part of the proof of Theorem 6.12, one could proceed here by applying the characterization result given by Theorem 3.61. On the other hand, from the identities $A = B\big(I - B^{-1}K\big)$ and $A' = \big(I - B^{-1}K\big)'B'$, and from the fact that B and B' are bijective, we obtain

$$N(A) \,=\, N(I - B^{-1}K) \quad \text{and} \quad N(A') \,=\, (B')^{-1}\,N\big((I - B^{-1}K)'\big)\,,$$

which, using Theorem 6.23, gives

$$\dim N(I - B^{-1}K) \,=\, \dim N\big((I - B^{-1}K)'\big) < +\infty,$$

and hence

$$\dim N(A) = \dim N(I - B^{-1}K) = \dim N\big((I - B^{-1}K)'\big) = \dim N(A') < +\infty,$$

thus completing the proof.

□

6.4 The adjoint of a compact operator

As previously announced, we now prove that the adjoint of a compact operator is itself compact. To this end, and bearing in mind Definitions 6.4 and 6.5, we require the following lemma.

Lemma 6.25. *If S is a relatively compact subset of a normed vector space X, then S is totally bounded. Conversely, if S is a totally bounded subset of a Banach space X, then S is relatively compact.*

Proof. Let S be a relatively compact subset of a normed vector space X, and let $\varepsilon > 0$. Given $x_1 \in S$, the possible inclusion $S \subseteq B(x_1, \varepsilon)$ ends the proof. If this does not occur, then there exists $x_2 \in S$ such that $\|x_2 - x_1\| \geqslant \varepsilon$. Now, the possible inclusion $S \subseteq B(x_1, \varepsilon) \cup B(x_2, \varepsilon)$ also ends the proof. If this does not happen, there must exist $x_3 \in S$ such that $\|x_3 - x_1\| \geqslant \varepsilon$ and $\|x_3 - x_2\| \geqslant \varepsilon$. If this process continues indefinitely, then we construct a sequence $\{x_n\}_{n\in\mathbb{N}} \subseteq S$ such that $\|x_j - x_k\| \geqslant \varepsilon$ for all $j \neq k$, which implies that $\{x_n\}_{n\in\mathbb{N}}$ has no convergent subsequences. Since this contradicts the hypothesis on S (cf. Definition 6.4), there must exist $n \in \mathbb{N}$ such that $S \subseteq \bigcup_{j=1}^{N} B(x_j, \varepsilon)$, thus proving that S is totally bounded. Conversely, let S be a totally bounded subset of a Banach space X, and let $\{x_n\}_{n\in\mathbb{N}} \subseteq S$. For $\varepsilon = 1$, there exist $N_1 \in \mathbb{N}$ and $\{z_1^{(1)}, \ldots, z_{N_1}^{(1)}\} \subseteq S$ such that $S \subseteq \bigcup_{j=1}^{N_1} B(z_j^{(1)}, 1)$. Hence there exists a subsequence $\{x_n^{(1)}\}_{n\in\mathbb{N}} \subseteq \{x_n\}_{n\in\mathbb{N}}$ such that $\{x_n^{(1)}\}_{n\in\mathbb{N}} \subseteq B(z_i^{(1)}, 1)$ for some $i \in \{1, \ldots, N_1\}$. Likewise, there exist $N_2 \in \mathbb{N}$ and $\{z_1^{(2)}, \ldots, z_{N_2}^{(2)}\} \subseteq S$ such that $S \subseteq \bigcup_{j=1}^{N_2} B(z_j^{(2)}, 1/2)$, and then there exists a subsequence $\{x_n^{(2)}\}_{n\in\mathbb{N}} \subseteq \{x_n^{(1)}\}_{n\in\mathbb{N}}$ such that $\{x_n^{(2)}\}_{n\in\mathbb{N}} \subseteq B(z_i^{(2)}, 1/2)$ for some $i \in \{1, \ldots, N_2\}$. Following this reasoning, one deduces that for each $k \in \mathbb{N}$ there exist $N_k \in \mathbb{N}$, $\{z_1^{(k)}, \ldots, z_{N_k}^{(k)}\} \subseteq S$, and a subsequence $\{x_n^{(k)}\}_{n\in\mathbb{N}} \subseteq \{x_n^{(k-1)}\}_{n\in\mathbb{N}}$ such that $S \subseteq \bigcup_{j=1}^{N_k} B(z_j^{(k)}, 1/k)$ and $\{x_n^{(k)}\}_{n\in\mathbb{N}} \subseteq B(z_i^{(k)}, 1/k)$ for some $i \in \{1, \ldots, N_k\}$. Then, defining $y_n := x_n^{(n)}$ for all $n \in \mathbb{N}$, we obtain that $\{y_n\}_{n\in\mathbb{N}}$ is a subsequence of $\{x_n\}_{n\in\mathbb{N}}$, and for $\varepsilon > 0$ there exists N such that $1/N < \varepsilon/2$, so that

$$\|y_n - y_m\| \leqslant \|y_n - z_i^{(N)}\| + \|y_m - z_i^{(N)}\| < \varepsilon \qquad \forall\, n, m \geqslant N\,.$$

This proves that $\{y_n\}_{n\in\mathbb{N}}$ is a Cauchy sequence and therefore convergent in X, from which we conclude that S is relatively compact.

□

We now prove the result mentioned at the beginning of this section.

Theorem 6.26. *Let X and Y be normed vector spaces, and let $K \in \mathcal{K}(X, Y)$. Then $K' \in \mathcal{K}(Y', X')$.*

Proof. Since X' is Banach, it suffices to show, by the previous lemma, that for every bounded set W of Y', the set $K'(W)$ is totally bounded in X'. Indeed, note first that, by the compactness of K, the set $K(\bar{B}(0,1))$ is totally bounded in Y, and therefore, given $\varepsilon > 0$, there exist $N \in \mathbb{N}$ and $\{x_1, \ldots, x_N\} \subseteq \bar{B}(0,1)$ such that $K(\bar{B}(0,1)) \subseteq \bigcup_{j=1}^{N} B(K(x_j), \varepsilon)$. Then, define the operator $T : Y' \longrightarrow \mathbb{R}^N$ by

$$T(G) := (G(K(x_1)), \ldots, G(K(x_N)))^{\mathrm{t}} \quad \forall\, G \in Y'\,,$$

which, being of finite-rank, is clearly compact. Then, if W is a bounded set in Y', we have that $T(W)$ is totally bounded, which implies that there exist $m \in \mathbb{N}$ and $\{G_1, \ldots, G_m\} \subseteq W$ such that

$$T(W) \subseteq \bigcup_{j=1}^{m} B(T(G_j), \varepsilon).$$

Thus, given $G \in W$, there exists $k \in \{1, \ldots, m\}$ such that $\|T(G) - T(G_k)\| < \varepsilon$, that is,

$$|G(K(x_i)) - G_k(K(x_i))| < \varepsilon \qquad \forall\, i \in \{1, 2, \cdots, N\}.$$

Moreover, given $x \in \bar{B}(0, 1)$, there exists $i \in \{1, \ldots, N\}$ such that

$$\|K(x) - K(x_i) \;| < \varepsilon,$$

and therefore

$$\begin{aligned}
&|K'(G)(x) - K'(G_k)(x)| = |G(K(x)) - G_k(K(x))| \\
&\leqslant |G(K(x)) - G(K(x_i))| + |G(K(x_i)) - G_k(K(x_i))| + |G_k(K(x_i)) - G_k(K(x))| \\
&\leqslant \|G\|\, \|K(x) - K(x_i)\| + \varepsilon + \|G_k\|\, \|K(x_i) - K(x)\| \\
&\leqslant 2 C_0 \varepsilon + \varepsilon =: c\varepsilon,
\end{aligned}$$

where $C_0 > 0$ is such that $\|\widetilde{G}\| \leqslant C_0$ for all $\widetilde{G} \in W$. It follows that $\|K'(G) - K'(G_k)\| < c\varepsilon$, from which we deduce that

$$K'(W) \subseteq \bigcup_{j=1}^{m} B(K'(G_j), c\varepsilon),$$

and consequently, $K'(W)$ is totally bounded.

□

The converse of the previous theorem is also true, that is, if X and Y are normed vector spaces and $K' \in \mathcal{K}(Y', X')$, then $K \in \mathcal{L}(X, Y)$. The details of the corresponding proof can be found in [23, Théorème VI.4].

6.5 Convergences, compactness, and the preadjoint

In this section we prove some results on the effect of compactness on certain types of convergence. First, we establish that a compact operator transforms weak convergence into strong convergence, and pointwise convergence into convergence in norm. Then, we extend the first consequence above to the case of weak* convergence, and through this we obtain, as a by-product, the characterization of the existence of what we later define as the preadjoint of an operator.

We begin the discussion by introducing precisely the meanings of weak and weak* convergence. Although we will use some topological notations that will be explained later in Chapter 7, the essential ideas for their understanding are given by the following definitions.

Definition 6.27. *A sequence $\{x_n\}_{n\in\mathbb{N}}$ in a normed vector space X is said to converge to $x \in X$ with respect to the weak topology $\sigma(X, X')$ of X, which we write as $x_n \overset{w}{\to} x$, if*

$$\lim_{n\to\infty} F(x_n) = F(x) \qquad \forall\, F \in X'.$$

Definition 6.28. *A sequence* $\{G_n\}_{n\in\mathbb{N}}$ *in the dual* Y' *of a normed vector space* Y *is said to converge to* $G \in Y'$ *with respect to the weak* topology* $\sigma(Y',Y)$ *of* Y'*, which we write as* $G_n \overset{w^*}{\to} G$*, if*

$$\lim_{n\to\infty} G_n(y) = G(y) \qquad \forall\, y \in Y\,.$$

Note that the above concepts are more general than strong convergence (in norm), since the latter, both from $\{x_n\}_{n\in\mathbb{N}}$ to $x \in X$, and from $\{G_n\}_{n\in\mathbb{N}}$ to $G \in Y'$, implies weak and weak* convergence, respectively. On the other hand, it is easy to see that weak and weak* limits are unique. Indeed, the uniqueness of the former follows from the corollary of the Hahn-Banach Theorem given by Lemma 2.24, while that of the latter follows from the equality of functionals. Moreover, by applying the Banach–Steinhaus Theorem (cf. Theorem 3.46) one can conclude that every sequence that converges weak or weak* is bounded. In particular, given $\{x_n\}_{n\in\mathbb{N}} \subseteq X$ and $x \in X$ such that $x_n \overset{w}{\to} x$, we consider the map $\mathcal{J}_X : X \to X''$, which is defined as (cf. (3.24)) $\mathcal{J}_X(x)(F) := F(x)$ for all $x \in X$, for all $F \in X'$, and observe that

$$\sup_{n\in\mathbb{N}} |\mathcal{J}_X(x_n)(F)| = \sup_{n\in\mathbb{N}} |F(x_n)| < +\infty \quad \forall\, F \in X'.$$

Then, applying the aforementioned theorem (cf. Theorem 3.46) in the context

$$X \leftarrow X', \qquad Y \leftarrow \mathbb{R}, \qquad I \leftarrow \mathbb{N}, \qquad A_i \leftarrow \mathcal{J}_X(x_n),$$

we deduce that there exists $M > 0$ such that $\|\mathcal{J}_X(x_n)\| \leqslant M$ for all $n \in \mathbb{N}$, that is, $\|x_n\| \leqslant M$ for all $n \in \mathbb{N}$. Analogously, it can be proved that a sequence $\{G_n\}_{n\in\mathbb{N}} \subseteq Y'$ that converges weak* is bounded.

In Chapter 7 we will show versions of the converse of these results, namely, every bounded sequence in a reflexive space possesses a subsequence that converges weakly (cf. Theorem 7.13), and every bounded sequence in the dual of a separable space possesses a subsequence that converges weak* (cf. Theorem 7.11).

The first result announced at the beginning of this section is now established.

Lemma 6.29. *Let* X *and* Y *be normed vector spaces, and let* $\{x_n\}_{n\in\mathbb{N}}$ *be a sequence in* X *such that* $x_n \overset{w}{\to} x \in X$*. Then, for all* $K \in \mathcal{K}(X,Y)$ *we have* $K(x_n) \overset{n\to\infty}{\longrightarrow} K(x)$*.*

Proof. Given $G \in Y'$, we clearly have that $G \circ K \in X'$, and therefore

$$G(K(x_n)) \overset{n\to\infty}{\longrightarrow} G(K(x))\,,$$

which proves that $K(x_n) \overset{w}{\longrightarrow} K(x)$. Now, suppose by contradiction that the sequence $\{K(x_n)\}_{n\in\mathbb{N}}$ does not converge in norm to $K(x)$. It follows that there exist $\varepsilon > 0$ and a subsequence $\{x_n^{(1)}\}_{n\in\mathbb{N}} \subseteq \{x_n\}_{n\in\mathbb{N}}$ such that

$$\|K(x_n^{(1)}) - K(x)\| \geqslant \varepsilon \quad \forall\, n \in \mathbb{N}\,. \tag{6.17}$$

Since $\{x_n^{(1)}\}_{n\in\mathbb{N}}$ is bounded and K is compact, there exists a subsequence $\{x_n^{(2)}\}_{n\in\mathbb{N}} \subseteq \{x_n^{(1)}\}_{n\in\mathbb{N}}$ such that $K(x_n^{(2)}) \to y \in Y$. Then $K(x_n^{(2)}) \overset{w}{\longrightarrow} y$, and by the uniqueness of the weak limit we obtain $y = K(x)$. Hence we can write $K(x_n^{(2)}) \to K(x)$, and, by (6.17), $\|K(x_n^{(2)}) - K(x)\| \geqslant \varepsilon$ for all $n \in \mathbb{N}$, which is clearly a contradiction. Consequently, it must be that $K(x_n) \overset{n\to\infty}{\longrightarrow} K(x)$.

□

Furthermore, we have the following important result, whose schematic representation is shown in Figure 6.2.

Lemma 6.30. *Let X, Y, and Z be normed vector spaces, and let $\{A_n\}_{n\in\mathbb{N}} \subseteq \mathcal{L}(Y,Z)$ and $A \in \mathcal{L}(Y,Z)$ be such that $A_n(y) \overset{n\to\infty}{\longrightarrow} A(y)$ for all $y \in Y$. Assume moreover that there exists $M > 0$ such that $\|A_n\| \leqslant M$ for all $n \in \mathbb{N}$. Then, for every $K \in \mathcal{K}(X,Y)$ we have that $\|A_nK - AK\| \overset{n\to\infty}{\longrightarrow} 0$.*

Proof. Suppose by contradiction that, for a given $K \in \mathcal{K}(X,Y)$, $\{\|A_nK - AK\|\}_{n\in\mathbb{N}}$ does not converge to 0. In that case, there must exist $\varepsilon > 0$ and a subsequence $\{A_n^{(1)}\}_{n\in\mathbb{N}} \subseteq \{A_n\}_{n\in\mathbb{N}}$ such that

$$\|A_n^{(1)}K - AK\| > \varepsilon \quad \forall n \in \mathbb{N}.$$

It follows that for each $n \in \mathbb{N}$ there exists $x_n^{(1)} \in X$ with $\|x_n^{(1)}\| = 1$ such that

$$\|A_n^{(1)}K(x_n^{(1)}) - AK(x_n^{(1)})\| > \varepsilon. \tag{6.18}$$

Since K is compact, there exist a subsequence $\{x_n^{(2)}\}_{n\in\mathbb{N}} \subseteq \{x_n^{(1)}\}_{n\in\mathbb{N}}$ and $y \in Y$ such that $K(x_n^{(2)}) \to y$. Now, by the triangle inequality we clearly have

$$\begin{aligned}
&\|A_n^{(2)}K(x_n^{(2)}) - AK(x_n^{(2)})\| \\
&\quad \leqslant \|A_n^{(2)}K(x_n^{(2)}) - A_n^{(2)}(y)\| + \|A_n^{(2)}(y) - A(y)\| + \|A(y) - AK(x_n^{(2)})\| \\
&\quad \leqslant M\,\|K(x_n^{(2)}) - y\| + \|A_n^{(2)}(y) - A(y)\| + \|A\|\,\|y - K(x_n^{(2)})\|,
\end{aligned}$$

from which it follows, by the previous analysis, that

$$\lim_{n\to\infty} \|A_n^{(2)}K(x_n^{(2)}) - AK(x_n^{(2)})\| = 0.$$

This result contradicts (6.18) and completes the proof.

□

It is important to note here that the uniform boundedness of the operators $\{A_n\}_{n\in\mathbb{N}}$ can be omitted from the statement of the lemma when Y is Banach, since in that case such a hypothesis is a direct consequence of the Banach–Steinhaus Theorem (cf. Theorem 3.46). Accordingly, we can establish the following result.

Lemma 6.31. *Let X, Y, and Z be normed vector spaces, with Y Banach, and let $\{A_n\}_{n\in\mathbb{N}} \subseteq \mathcal{L}(Y,Z)$ and $A \in \mathcal{L}(Y,Z)$ be such that $A_n(y) \overset{n\to\infty}{\longrightarrow} A(y)$ for all $y \in Y$. Then, for every $K \in \mathcal{K}(X,Y)$, $\|A_nK - AK\| \overset{n\to\infty}{\longrightarrow} 0$.*

Corollary 6.32. *Let X, Y, and Z be normed vector spaces, and let $\{A_n\}_{n\in\mathbb{N}} \subseteq \mathcal{L}(X,Y)$ and $A \in \mathcal{L}(X,Y)$ be such that $A_n'(G) \overset{n\to\infty}{\longrightarrow} A'(G)$ for all $G \in Y'$. Then, for every $K \in \mathcal{K}(Y,Z)$, $\|KA_n - KA\| \overset{n\to\infty}{\longrightarrow} 0$.*

Proof. It suffices to observe that $\|KA_n - KA\| = \|(KA_n - KA)'\| = \|A_n'K' - A'K'\|$, recall that $K' \in \mathcal{K}(Z',Y')$, and then apply Lemma 6.31 directly.

□

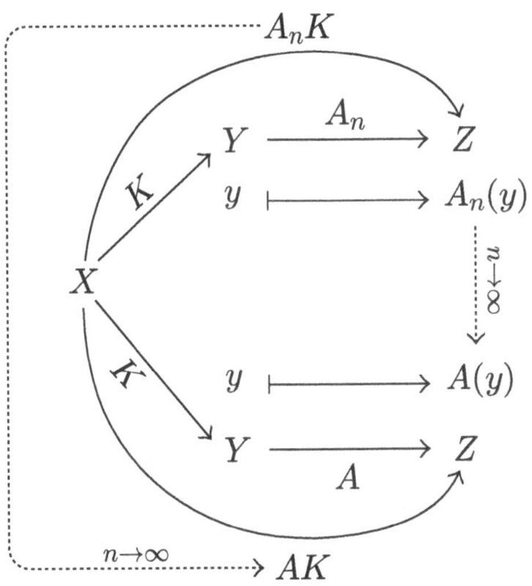

Fig. 6.2. *Schematic representation of Lemma* 6.30

The next goal is to extend the result given by Lemma 6.29 to the case of sequences that converge weak*. More precisely, if X, Y, and Z are normed vector spaces over the same field $\mathbb{K}$ ($\mathbb{R}$ or $\mathbb{C}$), we are interested in deriving sufficient conditions under which, given an arbitrary sequence $\{G_n\}_{n\in\mathbb{N}} \subseteq Y'$ that converges weak* to $G \in Y'$, there holds

$$K(G_n) \stackrel{n\to\infty}{\longrightarrow} K(G) \qquad \forall\, K \in \mathcal{K}(Y', Z)\,. \tag{6.19}$$

To this end, and in order to employ, initially, arguments similar to those used in the proof of Lemma 6.29, we introduce below the concept of preadjoint.

Definition 6.33. *An operator $B \in \mathcal{L}(Y', X')$ is said to possess a preadjoint if there exists $A \in \mathcal{L}(X, Y)$ such that $A' = B$.*

We are now in a position to present the following result.

Lemma 6.34. *Let X and Y be Banach spaces, and let $\{G_n\}_{n\in\mathbb{N}}$ be a sequence in Y' that converges weak* to $G \in Y'$. Then, for every $K \in \mathcal{K}(Y', X')$ that possesses a preadjoint, we have*

$$K(G_n) \stackrel{n\to\infty}{\longrightarrow} K(G)\,.$$

Proof. Given $A \in \mathcal{L}(X, Y)$ such that $A' = K$, we first observe, thanks to the weak* convergence of $\{G_n\}_{n\in\mathbb{N}}$, that for each $x \in X$ we have

$$K(G_n)(x) = A'(G_n)(x) = G_n\big(A(x)\big) \stackrel{n\to\infty}{\longrightarrow} G\big(A(x)\big) = A'(G)(x) = K(G)(x)\,,$$

which shows that $K(G_n) \stackrel{w^*}{\longrightarrow} K(G) \in X'$. Now, suppose by contradiction that the sequence $\{K(G_n)\}_{n\in\mathbb{N}}$ does not converge strongly to $K(G)$ in X'. It follows that there exist $\delta > 0$ and a subsequence $\{G_n^{(1)}\}_{n\in\mathbb{N}} \subseteq \{G_n\}_{n\in\mathbb{N}}$ such that

$$\|K(G_n^{(1)}) - K(G)\|_{X'} \geqslant \delta \qquad \forall\, n \in \mathbb{N}\,. \tag{6.20}$$

Moreover, the boundedness of $\{G_n^{(1)}\}_{n\in\mathbb{N}}$ and the compactness of K guarantee the existence of a subsequence $\{G_n^{(2)}\}_{n\in\mathbb{N}} \subseteq \{G_n^{(1)}\}_{n\in\mathbb{N}}$ and $F \in X'$ such that

$$\|K(G_n^{(2)}) - F\|_{X'} \overset{n\to\infty}{\longrightarrow} 0. \tag{6.21}$$

This last statement clearly implies that $K(G_n^{(2)}) \overset{w^*}{\longrightarrow} F \in X'$, so that by the uniqueness of the weak* limit we must have $F = K(G)$. Therefore, (6.21) reduces to

$$\|K(G_n^{(2)}) - K(G)\|_{X'} \overset{n\to\infty}{\longrightarrow} 0,$$

which contradicts (6.20) and finishes the proof.

□

The preceding result now suggests investigating under what conditions an operator $B \in \mathcal{L}(Y', X')$ possesses a preadjoint $A \in \mathcal{L}(X, Y)$. In other words, denoting by

$$T : \mathcal{L}(X, Y) \longrightarrow \mathcal{L}(Y', X')$$

the linear operator that assigns to each $A \in \mathcal{L}(X, Y)$ its adjoint $A' \in \mathcal{L}(Y', X')$, which is clearly injective, we now focus on determining the hypotheses under which T is also surjective. The answer, given by the following lemma, is far from unexpected.

Lemma 6.35. *T is bijective if and only if Y is reflexive. Moreover, in this case one has $T^{-1}(K) \in \mathcal{K}(X, Y)$ for every $K \in \mathcal{K}(Y', X')$.*

Proof. Suppose first that T is bijective, and let $\mathcal{G} \in Y''$ be arbitrary. Then, given any $F \in X'$, $F \neq 0$, we define the operator $B \in \mathcal{L}(Y', X')$ by

$$B(G) := \mathcal{G}(G)\, F \qquad \forall\, G \in Y'. \tag{6.22}$$

Note that the linearity and boundedness of B follow from the same properties of $\mathcal{G}$. Then, denoting $A := T^{-1}(B) \in \mathcal{L}(X, Y)$, we obviously have $A' = B$, so that (6.22) implies

$$\mathcal{G}(G)\, F(x) = A'(G)(x) = G(A(x)) \qquad \forall\, G \in Y', \quad \forall\, x \in X. \tag{6.23}$$

Thus, taking any $\widetilde{x} \in X$ such that $F(\widetilde{x}) \neq 0$, which is possible by our choice of F, and defining $\widetilde{y} := \dfrac{A(\widetilde{x})}{F(\widetilde{x})}$, we deduce from (6.23) that $\mathcal{G}(G) = G(\widetilde{y})$ for all $G \in Y'$. This identity shows that $\mathcal{G} = \mathcal{J}_Y(\widetilde{y})$, and therefore Y is reflexive. Conversely, assume now that Y is reflexive and show that, given any $B \in \mathcal{L}(Y', X')$, there exists $A \in \mathcal{L}(X, Y)$ such that $A' = B$. Indeed, since $B' \in \mathcal{L}(X'', Y'')$ and $\mathcal{J}_Y$ is bijective, we consider the following diagram:

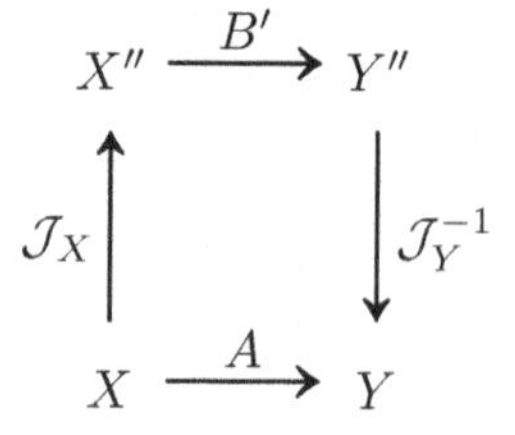

in which we define $A := \mathcal{J}_Y^{-1} \circ B' \circ \mathcal{J}_X \in \mathcal{L}(X, Y)$. To prove that $A' = B$, note first that $G(\mathcal{J}_Y^{-1}(\mathcal{G})) = \mathcal{G}(G)$ for all $G \in Y'$, for all $\mathcal{G} \in Y''$. Thus, we find that

$$
\begin{aligned}
A'(G)(x) &= G(A(x)) = G\big(\mathcal{J}_Y^{-1}(B'(\mathcal{J}_X(x)))\big) \\
&= B'(\mathcal{J}_X(x))(G) = \mathcal{J}_X(x)\big(B(G)\big) = B(G)(x) \qquad \forall\, G \in Y', \quad \forall x \in X,
\end{aligned}
$$

from which it follows that $A'(G) = B(G)$ for all $G \in Y'$, i.e., $A' = B$. Finally, in the case where $B \in \mathcal{K}(Y', X')$, by Theorem 6.26 we have that $B' \in \mathcal{K}(X'', Y'')$, and therefore $T^{-1}(B) := \mathcal{J}_Y^{-1} \circ B' \circ \mathcal{J}_X$ is also compact.

□

As a direct consequence of Lemmas 6.34 and 6.35, we obtain the following particular version of (6.19).

Lemma 6.36. *Let X and Y be Banach spaces such that Y is reflexive, and let $\{G_n\}_{n\in\mathbb{N}}$ be a sequence in Y' that converges weak* to $G \in Y'$. Then we have*

$$
K(G_n) \overset{n\to\infty}{\longrightarrow} K(G) \qquad \forall\, K \in \mathcal{K}(Y', X').
$$

Moreover, we are now in a position to extend the previous lemma to the more general version of (6.19).

Lemma 6.37. *Let Y and Z be Banach spaces such that Y is reflexive, and let $\{G_n\}_{n\in\mathbb{N}}$ be a sequence in Y' that converges weak* to $G \in Y'$. Then we have*

$$
K(G_n) \overset{n\to\infty}{\longrightarrow} K(G) \qquad \forall\, K \in \mathcal{K}(Y', Z). \tag{6.24}
$$

Proof. Given $K \in \mathcal{K}(Y', Z)$, we consider the following diagram:

$$
\begin{array}{ccc}
Y' & \overset{K}{\longrightarrow} & Z \\
 & \widetilde{K} \searrow & \downarrow \mathcal{J}_Z \\
 & & Z''
\end{array} \tag{6.25}
$$

in which we define $\widetilde{K} := \mathcal{J}_Z \circ K \in \mathcal{K}(Y', Z'') = \mathcal{K}\big(Y', (Z')'\big)$. Then, recalling that $\mathcal{J}_Z$ is an isometry, and applying Lemma 6.36 in the context $X \longleftarrow Z'$, $Y \longleftarrow Y$, and $K \longleftarrow \widetilde{K}$, we deduce that for each $\{G_n\}_{n\in\mathbb{N}}$ and G as indicated, we have

$$
\|K(G_n) - K(G)\|_Z = \|\mathcal{J}_Z\big(K(G_n) - K(G)\big)\|_{Z''} = \|\widetilde{K}(G_n) - \widetilde{K}(G)\|_{(Z')'} \overset{n\to\infty}{\longrightarrow} 0, \tag{6.26}
$$

which proves (6.24) and completes the proof.

□

At this point, we note that, to the best of the author's knowledge, the possible necessity of the reflexivity of Y for (6.24) remains an open problem.

On the other hand, it is important to observe that the same diagram (6.25) suggests the following extension of Lemma 6.34 to the case $K \in \mathcal{K}(Y', Z)$.

Lemma 6.38. *Let Y and Z be Banach spaces, and let $\{G_n\}_{n\in\mathbb{N}}$ be a sequence in Y' that converges weak* to $G \in Y'$. Moreover, let $K \in \mathcal{K}(Y', Z)$ for which there exists $A \in \mathcal{L}(Z', Y)$ such that $A' = \widetilde{K} := \mathcal{J}_Z \circ K \in \mathcal{K}(Y', Z'') = \mathcal{K}\big(Y', (Z')'\big)$. Then we have*

$$
K(G_n) \overset{n\to\infty}{\longrightarrow} K(G).
$$

Proof. Similarly to the proof of Lemma 6.37, it suffices to use (6.26) and then apply Lemma 6.34 in the context $X \longleftarrow Z'$, $Y \longleftarrow Y$, and $K \longleftarrow \widetilde{K}$.

□

Finally, we emphasize that the main result of this section, namely Lemma 6.37, is proven in a simpler way in Chapter 7 using the topological concepts introduced there (see Lemma 7.49 at the end of Section 7.4.3).

6.6 Galerkin method and Fredholm alternative

In this section we address a very useful result for the analysis of the stability of Galerkin methods in Hilbert spaces. Essentially, it establishes the equivalence of this property for Galerkin schemes associated with bijective operators whose difference is compact. The extension to a Banach spaces framework is left as an exercise (cf. Exercise 6.22).

6.6.1 Main result

Let $(X, \langle\cdot,\cdot\rangle_X)$ and $(Y, \langle\cdot,\cdot\rangle_Y)$ be Hilbert spaces over $\mathbb{C}$ (or $\mathbb{R}$), and let $A \in \mathcal{L}(X, Y)$ be the operator induced by a bounded sesquilinear (or bilinear) form $a : X \times Y \longrightarrow \mathbb{C}$ (or $\mathbb{R}$), that is:

$$\langle A(x), y\rangle_Y = a(x, y) \quad \forall\, x \in X, \quad \forall\, y \in Y.$$

In what follows we assume that A is bijective. Then, for each $F \in Y'$ there exists a unique $x \in X$ such that $A(x) = R_Y(F)$, where $R_Y : Y' \longrightarrow Y$ is the Riesz mapping. Equivalently, for each $F \in Y'$ there exists a unique $x \in X$ such that

$$a(x, y) = \overline{F(y)} \quad \forall\, y \in Y.$$

Now, let $\{X_n\}_{n\in\mathbb{N}}$ and $\{Y_n\}_{n\in\mathbb{N}}$ be sequences of finite-dimensional subspaces of X and Y, respectively, such that $\dim X_n = \dim Y_n$ for all $n \in \mathbb{N}$, and consider the associated Galerkin scheme: Find $x_n \in X_n$ such that

$$a(x_n, y_n) = \overline{F(y_n)} \quad \forall\, y_n \in Y_n. \tag{6.27}$$

Note that (6.27) can be restated as

$$a(x_n, y_n) = a(x, y_n) \quad \forall\, y_n \in Y_n,$$

or equivalently,

$$\langle A(x_n), y_n\rangle_Y = \langle A(x), y_n\rangle_Y \quad \forall\, y_n \in Y_n,$$

which implies that the Galerkin scheme is equivalently reduced to: Find $x_n \in X_n$ such that

$$\mathcal{P}_n\, A(x_n) = \mathcal{P}_n\, A(x), \tag{6.28}$$

where $\mathcal{P}_n : Y \longrightarrow Y_n$ is the orthogonal projector. At the same time, if we introduce the operator $A_n := \mathcal{P}_n\, A|_{X_n}$, then (6.28) can be rewritten as: Find $x_n \in X_n$ such that

$$A_n(x_n) = \mathcal{P}_n\, A(x). \tag{6.29}$$

The collection of Galerkin schemes (6.29) for all $n \in \mathbb{N}$ and $x \in X$ (equivalently, for all $F \in Y'$), is called the Galerkin system and will be denoted from now on by $\mathcal{G}\big(A; X_n, Y_n\big)_{n\in\mathbb{N}}$.

Definition 6.39. *A Galerkin system $\mathcal{G}\big(A; X_n, Y_n\big)_{n\in\mathbb{N}}$ is said to be* CONVERGENT *if there exists $N \in \mathbb{N}$ such that for each $y \in A(X)$, the equation $A_n(x_n) = \mathcal{P}_n(y)$ has a unique solution $x_n \in X_n$ for all $n \geqslant N$, and moreover $x_n \stackrel{n\to\infty}{\longrightarrow} x$, where $x \in X$ is the unique element satisfying $A(x) = y$. In other words, $\mathcal{G}\big(A; X_n, Y_n\big)_{n\in\mathbb{N}}$ is said to be* CONVERGENT *if there exists $N \in \mathbb{N}$ such that for every $n \geqslant N$ the operator $A_n : X_n \longrightarrow Y_n$ is* BIJECTIVE *and $A_n^{-1}\mathcal{P}_n A(x) \stackrel{n\to\infty}{\longrightarrow} x$ for all $x \in X$.*

In general, a necessary condition for the convergence of the Galerkin system is that

$$\lim_{n\to\infty} \operatorname{dist}(z, X_n) = 0 \quad \forall\, z \in X, \tag{6.30}$$

which we will assume from now on.

Lemma 6.40. *The Galerkin system $\mathcal{G}\big(A; X_n, Y_n\big)_{n\in\mathbb{N}}$ is convergent if and only if there exist $N \in \mathbb{N}$ and $M > 0$ such that for every $n \geqslant N$ the operator $A_n : X_n \longrightarrow Y_n$ is bijective and $\|A_n^{-1}\mathcal{P}_n A\|_{\mathcal{L}(X,X)} \leqslant M$.*

Proof. Suppose that $\mathcal{G}\big(A; X_n, Y_n\big)_{n\in\mathbb{N}}$ is convergent. This means that there exists $N \in \mathbb{N}$ such that for each $n \geqslant N$, the operator $A_n : X_n \longrightarrow Y_n$ is bijective and

$$A_n^{-1}\mathcal{P}_n A(x) \stackrel{n\to\infty}{\longrightarrow} x \qquad \forall\, x \in X\,.$$

It follows that the sequence $\big\{A_n^{-1}\mathcal{P}_n A(x)\big\}_{n\in\mathbb{N}}$ is bounded for each $x \in X$, and therefore, by applying the Banach-Steinhaus Theorem (cf. Theorem 3.46), we deduce that there exists $M > 0$ such that $\|A_n^{-1}\mathcal{P}_n A\| \leqslant M$ for all $n \geqslant N$. Conversely, suppose that there exist $N \in \mathbb{N}$ and $M > 0$ such that the operator $A_n : X_n \longrightarrow Y_n$ is bijective and $\|A_n^{-1}\mathcal{P}_n A\| \leqslant M$ for every $n \geqslant N$. Since $A_n^{-1}\mathcal{P}_n A(u_n) = u_n$ for all $u_n \in X_n$, it follows that for each $x \in X$ we have

$$\begin{aligned}
\|x - A_n^{-1}\mathcal{P}_n A(x)\| &= \|(I - A_n^{-1}\mathcal{P}_n A)(x)\| \\
&= \|(I - A_n^{-1}\mathcal{P}_n A)(x - u_n)\| && \forall\, u_n \in X_n \\
&\leqslant \|I - A_n^{-1}\mathcal{P}_n A\|\, \|x - u_n\| && \forall\, u_n \in X_n \\
&\leqslant (1 + M)\, \|x - u_n\| && \forall\, u_n \in X_n\,,
\end{aligned}$$

and hence

$$\|x - A_n^{-1}\mathcal{P}_n A(x)\| \leqslant \big(1 + M\big) \operatorname{dist}(x, X_n)\,. \tag{6.31}$$

Hereafter, I stands for a generic identity operator. Alternatively, proceeding as in the analysis given at the end of Section 4.2.2 (after the proof of Lemma 4.13), noting that $A_n^{-1}\mathcal{P}_n A$ coincides with the Galerkin projector G_n defined there, and using again the result given by Exercise 3.26, we have $\|I - A_n^{-1}\mathcal{P}_n A\| = \|A_n^{-1}\mathcal{P}_n A\|$, and therefore

$$\|x - A_n^{-1}\mathcal{P}_n A(x)\| \leqslant M \operatorname{dist}(x, X_n). \tag{6.32}$$

In any case, by applying assumption (6.30), we conclude that $A_n^{-1}\mathcal{P}_n A(x) \stackrel{n\to\infty}{\longrightarrow} x$.

□

We are now in a position to prove the main result of this section.

Theorem 6.41. *Let X and Y be Hilbert spaces over $\mathbb{C}$ (or $\mathbb{R}$), and let $A \in \mathcal{L}(X, Y)$ and $K \in \mathcal{K}(X, Y)$ be such that*

i) *A is bijective*
ii) $T := A + K$ *is injective.*

Let $\{X_n\}_{n\in\mathbb{N}}$ *and* $\{Y_n\}_{n\in\mathbb{N}}$ *be sequences of finite-dimensional subspaces of* X *and* Y*, respectively, such that* (6.30) *holds, and suppose that the Galerkin system* $\mathcal{G}(A; X_n, Y_n)_{n\in\mathbb{N}}$ *is convergent. Then, the Galerkin system* $\mathcal{G}(T; X_n, Y_n)_{n\in\mathbb{N}}$ *is also convergent.*

Proof. First note, according to the Bounded Inverse Theorem (cf. Theorem 3.34), that $A^{-1} \in \mathcal{L}(Y, X)$, and therefore, thanks to Lemma 6.14, part iv), we deduce that $A^{-1}K \in \mathcal{K}(X)$. Moreover, the convergence of the Galerkin scheme $\mathcal{G}(A; X_n, Y_n)_{n\in\mathbb{N}}$ ensures that there exists $N_1 \in \mathbb{N}$ such that A_n is bijective for all $n \geqslant N_1$, and $A_n^{-1}\mathcal{P}_n A(x) \overset{n\to\infty}{\longrightarrow} x$ for all $x \in X$. It follows, by composing on the right with the compact operator $A^{-1}K$ and applying Lemma 6.31, that

$$\|A_n^{-1}\mathcal{P}_n K - A^{-1}K\| \overset{n\to\infty}{\longrightarrow} 0. \tag{6.33}$$

Also, by virtue of Lemma 6.40, there exists $M > 0$ such that

$$\|A_n^{-1}\mathcal{P}_n A\| \leqslant M \quad \forall\, n \geqslant N_1. \tag{6.34}$$

On the other hand, the Fredholm alternative implies that $T := A + K$ is in fact bijective, and therefore

$$I + A^{-1}K = A^{-1}(A + K) = A^{-1}T : X \to X$$

is also bijective. Accordingly, by adding and subtracting $A^{-1}K$, we obtain

$$I + A_n^{-1}\mathcal{P}_n K = (I + A^{-1}K)\Big\{I + (I + A^{-1}K)^{-1}(A_n^{-1}\mathcal{P}_n K - A^{-1}K)\Big\},$$

and, according to the convergence given by (6.33), there exists $N_2 \in \mathbb{N}$ such that

$$\|(I + A^{-1}K)^{-1}\|\,\|A_n^{-1}\mathcal{P}_n K - A^{-1}K\| < \frac{1}{2} \quad \forall\, n \geqslant N_2.$$

Thus, applying the result of Exercise 3.12, we deduce that the operator

$$I + (I + A^{-1}K)^{-1}\,(A_n^{-1}\mathcal{P}_n K - A^{-1}K) : X \to X$$

is bijective for each $n \geqslant N_2$, and moreover

$$\left\|\Big(I + (I + A^{-1}K)^{-1}\,(A_n^{-1}\mathcal{P}_n K - A^{-1}K)\Big)^{-1}\right\| \leqslant 2 \quad \forall\, n \geqslant N_2,$$

which implies that $I + A_n^{-1}\mathcal{P}_n K : X \to X$ is also bijective and

$$\left\|(I + A_n^{-1}\mathcal{P}_n K)^{-1}\right\| \leqslant 2\left\|(I + A^{-1}K)^{-1}\right\| \quad \forall\, n \geqslant N_2. \tag{6.35}$$

The proof is completed by verifying that $\mathcal{G}(T; X_n, Y_n)_{n\in\mathbb{N}}$ satisfies the equivalent statement of Lemma 6.40. Indeed, for $n \geqslant N := \max\{N_1, N_2\}$, we have

$$\begin{aligned} T_n = \mathcal{P}_n T|_{X_n} &= (\mathcal{P}_n A + \mathcal{P}_n K)|_{X_n} = A_n + \mathcal{P}_n K|_{X_n} \\ &= A_n\,(I + A_n^{-1}\mathcal{P}_n K)|_{X_n} : X_n \to Y_n\,, \end{aligned}$$

which is bijective because both $A_n : X_n \to Y_n$ and $(I + A_n^{-1}\mathcal{P}_n K)|_{X_n} : X_n \to X_n$ are bijective. Moreover,

$$\begin{aligned} T_n^{-1}\mathcal{P}_n T &= (I + A_n^{-1}\mathcal{P}_n K)^{-1}\, A_n^{-1}\mathcal{P}_n\,(A+K) \\ &= (I + A_n^{-1}\mathcal{P}_n K)^{-1}\, A_n^{-1}\mathcal{P}_n\, A\,(I + A^{-1}K)\,, \end{aligned}$$

from which, using estimates (6.34) and (6.35), we deduce that

$$\|T_n^{-1}\mathcal{P}_n T\| \leqslant 2\,M\,\|(I + A^{-1}K)^{-1}\|\,\|I + A^{-1}K\|\,,$$

and thus a straightforward application of Lemma 6.40 completes the proof.

□

We end this section by stressing that the extension of Theorem 6.41 to the corresponding Banach setting is provided later on in Exercise 6.22.

6.6.2 Application example

In this section we illustrate the applicability of Theorem 6.41 through the finite element method for solving a one-dimensional boundary value problem. In fact, given $\Omega = (0,1)$, $f \in \mathrm{L}^2(\Omega)$ and a constant $k \in \mathbb{R}^+$, we consider the equation $u'' + k\,u = f$ in Ω with Dirichlet boundary conditions $u(0) = u(1) = 0$. It is easy to see that the corresponding variational formulation is: Find $u \in \mathrm{H}_0^1(\Omega)$ such that

$$\int_\Omega \left\{ u'\,v' - k\,u\,v \right\} = -\int_\Omega f\,v \qquad \forall\, v \in \mathrm{H}_0^1(\Omega)\,. \tag{6.36}$$

Then, introducing the space $H := \mathrm{H}_0^1(\Omega)$ and the bilinear forms $a : H \times H \to \mathbb{R}$ and $b : H \times H \to \mathbb{R}$ defined by

$$a(w,v) := \int_\Omega w'\,v' \quad \text{and} \quad b(w,v) := -k\int_\Omega w\,v \qquad \forall\, w, v \in H,$$

problem (6.36) can be equivalently rewritten as: Find $u \in H$ such that

$$(A + B)(u) = g,$$

where $A : H \to H$ and $B : H \to H$ are the bounded linear operators induced by a and b, respectively, that is,

$$\langle A(w), v\rangle_{1,\Omega} = a(w,v) \quad \text{and} \quad \langle B(w), v\rangle_{1,\Omega} = b(w,v) \qquad \forall\, w, v \in H,$$

with $\langle\cdot,\cdot\rangle_{1,\Omega}$ denoting the usual inner product of $\mathrm{H}^1(\Omega)$, and $g \in H$ the unique element, guaranteed by the Riesz Representation Theorem (cf. Theorem 2.14), such that

$$-\int_\Omega f v = \langle v, g\rangle_{1,\Omega} \quad \forall\, v \in H\,.$$

Moreover, according to the Friedrichs–Poincaré inequality (cf. Lemma 4.5), it follows that, unlike $a + b$, which defines the variational formulation (6.36) and is not H-elliptic, the bilinear form a is indeed H-elliptic (cf. Definition 4.3), which implies, thanks to the Lax–Milgram Lemma (cf. Lemma 4.4), that the operator A is bijective. On the other hand, from the observation made after the proof of the Rellich Compact Embedding Theorem (cf. Theorem 10.18) and from the compactness result given by Theorem 6.26, we deduce that the embeddings $i : \mathrm{H}_0^1(\Omega) \to \mathrm{L}^2(\Omega)$ and $i' : \mathrm{L}^2(\Omega)' \to \mathrm{H}_0^1(\Omega)'$ are compact. Moreover, the operator $\mathcal{B} : \mathrm{H}_0^1(\Omega) \to \mathrm{L}^2(\Omega)'$ defined by

$$\mathcal{B}(w)(v) := b(w,v) \qquad \forall\, w \in \mathrm{H}_0^1(\Omega), \quad \forall\, v \in \mathrm{L}^2(\Omega)$$

is clearly bounded, and therefore, denoting by $\mathcal{R}$ the Riesz map of $\mathrm{H}_0^1(\Omega)$, we conclude from the diagram

$$H := \mathrm{H}_0^1(\Omega) \xrightarrow{\mathcal{B}} \mathrm{L}^2(\Omega)' \xrightarrow{i'} \mathrm{H}_0^1(\Omega)' \xrightarrow{\mathcal{R}} \mathrm{H}_0^1(\Omega)$$

that the operator $B := \mathcal{R} \circ i' \circ \mathcal{B} : H \to H$ is compact. Let us now show that $T := A + B$ is injective. Indeed, let $u \in H := \mathrm{H}_0^1(\Omega)$ be such that $(A+B)(u) = 0$, that is,

$$\int_\Omega \Big\{ u'\, v' \,-\, k\, u\, v \Big\} = 0 \qquad \forall\, v \in \mathrm{H}_0^1(\Omega).$$

Taking $v = u$ we obtain $|u|_{1,\Omega}^2 = k\,\|u\|_{0,\Omega}^2$, and thanks to (4.9) (cf. Lemma 4.5), we have $\|u\|_{0,\Omega}^2 \leqslant \frac{1}{2}|u|_{1,\Omega}^2$, which gives $\big(1 - \frac{k}{2}\big)|u|_{1,\Omega}^2 \leqslant 0$. Hence, since that lemma finally states that $\|\cdot\|_{1,\Omega}$ and $|\cdot|_{1,\Omega}$ are equivalent in $\mathrm{H}_0^1(\Omega)$, it follows that necessarily $u = 0$ if we additionally assume that $k \in (0,2)$. We have thus proved that for this range of k, the hypotheses of Theorem 6.41 hold. Consequently, any Galerkin scheme convergent for A, whose respective finite element space satisfies (6.30), also provides a convergent Galerkin scheme for our operator $A + B$. In particular, from finite element theory it is known that this occurs with the subspace of H given for each $n \in \mathbb{N}$ by

$$H_n := \Big\{ v \in C(\bar{\Omega}) : \quad v(0) = v(1) = 0, \quad v|_{]x_{j-1},x_j[} \in P_1(]x_{j-1},x_j[) \ \ \forall\, j \in \{1,\cdots,n\} \Big\},$$

where $0 = x_0 < x_1 < \cdots < x_{n-1} < x_n = 1$ is a partition of Ω and $P_1(]x_{j-1},x_j[)$ is the space of polynomials of degree $\leqslant 1$ over the interval $]x_{j-1},x_j[$ (cf. Example 4.10). Thus, by virtue of the conclusion of Theorem 6.41, we deduce that there exists $N \in \mathbb{N}$ such that for each $n \geqslant N$ there exists a unique $u_n \in H_n$ satisfying

$$\int_\Omega \Big\{ u_n'\, v_n' \,-\, k\, u_n\, v_n \Big\} = -\int_\Omega f\, v_n \qquad \forall\, v_n \in H_n.$$

Moreover, in virtue of (6.31) (or (6.32)), there exists $C > 0$, independent of n, such that

$$\|u - u_n\|_{1,\Omega} \leqslant C\,\mathrm{dist}(u, H_n) \qquad \forall\, n \geqslant N, \tag{6.37}$$

where $u \in \mathrm{H}_0^1(\Omega)$ is the solution of the original problem: $u'' + k\,u = f$ in Ω, with Dirichlet boundary conditions $u(0) = u(1) = 0$. Moreover, according to the approximation properties of the subspace H_n, there exists $C > 0$ such that $\mathrm{dist}(v, H_n) \leqslant \frac{C}{n}\,\|v''\|_{0,\Omega}$ for all $v \in \mathrm{H}^2(\Omega)$. Then, since in our case $u \in \mathrm{H}_0^1(\Omega)$ and $u'' = f - k\,u \in \mathrm{L}^2(\Omega)$, it is clear that $u \in \mathrm{H}^2(\Omega)$, and therefore, using that $\|u\|_{0,\Omega} \leqslant \|u\|_{1,\Omega} \leqslant C\,\|f\|_{0,\Omega}$, which follows from the continuous dependence result for (6.36), we arrive at $\mathrm{dist}(u, H_n) \leqslant \frac{C}{n}\,\|f\|_{0,\Omega}$. Combining this estimate with (6.37) we finally obtain

$$\|u - u_n\|_{1,\Omega} \leqslant \frac{C}{n}\,\|f\|_{0,\Omega} \qquad \forall\, n \geqslant N\,.$$

For other interesting applications of Theorem 6.41 related to the numerical analysis of various boundary value and transmission problems, we refer the reader to [25], [48], [69], [73], [75], [76], [77], [78], and [83]. In particular, a situation very similar to some aspects of the analysis in [73] and [75] is proposed below as Exercise 6.10.

6.7 EXERCISES.

▸ **6.1** *Let X be the vector space of continuous functions on $\bar{\Omega} := [0,1]$ endowed with the uniform norm. Given u_1, u_2, $y \in X$, consider the integral equation: Find $u \in X$ such that*

$$u(t) - \int_0^1 u_1(t)\, s^2\, u(s)\, ds - \int_0^1 u_2(t)\, \big(1 - \tfrac{3}{2}s\big)\, u(s)\, ds = y(t) \quad \forall\, t \in \bar{\Omega}. \tag{6.38}$$

i) *If $u_1(t) = 4t$ and $u_2(t) = 1$ for all $t \in \Omega$, deduce a necessary and sufficient condition for equation (6.38) to have at least one solution.*

ii) *Show that if $u_1(t) = t$ and $u_2(t) = 1$, then for each $y \in X$ the integral equation (6.38) has a unique solution.*

▸ **6.2** *Let X, Y, and Z be normed vector spaces.*

i) *Show that if $K \in \mathcal{L}(X,Y)$ is of finite-rank, then K is compact.*

ii) *Prove that if $A \in \mathcal{L}(X,Y)$ and $K \in \mathcal{K}(Y,Z)$, then KA is also compact.*

▸ **6.3** *Let X be an* INFINITE*-dimensional normed vector space and let $K \in \mathcal{K}(X,X)$ be an injective operator. Prove that $K^{-1} \notin \mathcal{L}(X,X)$.*

▸ **6.4** ([40]) *This exercise aims to establish the converse of Theorem 6.1, thus concluding that a normed vector space is of finite dimension if and only if all its norms are equivalent. To this end, it is suggested to employ the concept of Hamel basis to prove that any infinite-dimensional vector space possesses norms that are not equivalent. Indeed, we first recall that, given a normed vector space X, a Hamel basis of X is any family $\{e_i\}_{i\in I} \subseteq X$, where I is an arbitrary set of indexes, such that*

i) *for any finite set $J \subseteq I$, $\{e_j\}_{j\in J}$ is linearly independent, and*

ii) *for each $x \in X$ there exists a finite set $J(x) \subseteq I$ such that $x \in \langle \{e_j\}_{j\in J(x)} \rangle$, that is there exist scalars α_j, $j \in J(x)$, such that $x = \sum_{j\in J(x)} \alpha_j\, e_j$.*

Then, using that any normed vector space X has a Hamel basis $\{e_i\}_{i\in I}$ (cf. [40, Theorem 2.1-1]), and bearing in mind the decomposition from ii), we consider the mappings $\|\cdot\|_1 : X \to \mathbb{R}$ and $\|\cdot\|_\infty : X \to \mathbb{R}$ defined by

$$\|x\|_1 := \sum_{j\in J(x)} |\alpha_j| \quad \text{and} \quad \|x\|_\infty := \max_{j\in J(x)} |\alpha_j| \qquad \forall\, x \in X\,.$$

In turn, assuming now that X is infinite-dimensional, it is readily seen that there exists a sequence $\{f_k\}_{k\in\mathbb{N}} \subseteq \{e_i\}_{i\in I}$.

a) *Prove that $\|\cdot\|_1$ and $\|\cdot\|_\infty$ are in fact norms on X.*

b) *Prove that for each $n \in \mathbb{N}$ there exists a suitable vector $x_n \in \langle \{f_1, \ldots, f_n\} \rangle$ such that $\|x_n\|_1 = 1$ and $\|x_n\|_\infty = \frac{1}{n}$, and conclude from this that $\|\cdot\|_1$ and $\|\cdot\|_\infty$ are not equivalent in X.*

▸ **6.5** *Let X and Y be Banach spaces, and let $A \in \mathcal{L}(X,Y)$. Suppose that there exist sequences $\{F_n\}_{n\in\mathbb{N}} \subseteq X'$, $\{y_n\}_{n\in\mathbb{N}} \subseteq Y$, $\{\lambda_n\}_{n\in\mathbb{N}} \subseteq \mathbb{R}$ such that*

$$\sup_{n\in\mathbb{N}} \|F_n\|_{X'} < +\infty, \quad \sup_{n\in\mathbb{N}} \|y_n\|_Y < +\infty, \quad \sum_{n=1}^{\infty} |\lambda_n| < +\infty,$$

and

$$A(x) = \lim_{m\to\infty} \sum_{n=1}^{m} \lambda_n\, F_n(x)\, y_n \qquad \forall\, x \in X.$$

Prove that A is compact.

▸ **6.6** *Determine,* **with justification**, *whether the following statements are true or false:*

i) *Every linear, bounded, and bijective operator from a Banach space X to a Banach space Y is compact.*
ii) *There exist compact operators that are not closed.*

▸ **6.7** *Let $p > 1$ and consider the operator $K : l_p(\mathbb{R}) \to l_p(\mathbb{R})$ defined by*

$$K(x_1, x_2, \ldots, x_k, \ldots) = (x_1, \tfrac{x_2}{2}, \ldots, \tfrac{x_k}{k}, \ldots) \qquad \forall\, (x_1, x_2, \ldots, x_k, \ldots) \in l_p(\mathbb{R}).$$

Prove that K is compact.

▸ **6.8** ([68]) *Let H and V be two Hilbert spaces such that $H \subseteq V$ and the embedding $i : H \to V$ is compact.*

a) *A bounded bilinear form A defined on $H \times H$ is said to satisfy the* GÅRDING INEQUALITY *if there exist $\alpha > 0$ and a bounded bilinear form $K : H \times V \to \mathbb{R}$ such that*

$$A(v, v) \geqslant \alpha \|v\|_H^2 - K(v, v) \qquad \forall\, v \in H.$$

Given $F \in H'$, consider the following variational problem: Find $u \in H$ such that

$$A(u, v) = F(v) \qquad \forall\, v \in H. \tag{6.39}$$

Prove that if A satisfies the Gårding inequality, then (6.39) has a solution for each $F \in H'$ IF AND ONLY IF *$u = 0$ is the unique solution of the corresponding homogeneous problem.*

b) *A is said to satisfy the* GENERALIZED GÅRDING INEQUALITY *if there exist $\alpha > 0$, a bounded bilinear form $K : H \times V \to \mathbb{R}$, and an isomorphism $S : H \to H$ such that*

$$A(v, Sv) \geqslant \alpha \|v\|_H^2 - K(v, Sv) \qquad \forall\, v \in H.$$

Prove that if A satisfies the generalized Gårding inequality, then (6.39) has a solution for each $F \in H'$ IF AND ONLY IF *$u = 0$ is the unique solution of the corresponding homogeneous problem.*

▸ **6.9** *Let $K : [0, 1] \times [0, 1] \to \mathbb{R}$ be a continuous function, and let $\mathbf{K} : C[0, 1] \to C[0, 1]$ be the operator defined by*

$$\mathbf{K}(u)(t) := \int_0^1 K(t, s)\, u(s)\, ds \qquad \forall\, t \in [0, 1], \quad \forall\, u \in C[0, 1].$$

Apply the Arzelá–Ascoli Theorem (cf. [3, Theorem 4.12]) to prove that $\mathbf{K}$ is compact.

▸ **6.10** ([73], [75]) *Let Ω be a convex polygonal domain in $\mathbb{R}^2$ with boundary Γ, and given the datum $f \in \mathrm{L}^2(\Omega)$, consider the Helmholtz equation with homogeneous Dirichlet boundary condition*

$$\Delta u + u = f \quad \text{in} \quad \Omega, \quad u = 0 \quad \text{on} \quad \Gamma. \tag{6.40}$$

i) *Introduce the auxiliary unknown $\boldsymbol{\sigma} := \nabla u$ in Ω and show that a mixed variational formulation of (6.40) reduces to: Find $\boldsymbol{\sigma} \in \mathbf{H}(\mathrm{div}; \Omega)$ such that*

$$\int_\Omega \boldsymbol{\sigma} \cdot \boldsymbol{\tau} - \int_\Omega \mathrm{div}(\boldsymbol{\sigma})\, \mathrm{div}(\boldsymbol{\tau}) = - \int_\Omega f\, \mathrm{div}(\boldsymbol{\tau}) \quad \forall\, \boldsymbol{\tau} \in \mathbf{H}(\mathrm{div}; \Omega). \tag{6.41}$$

ii) *Define the operator $P : \mathbf{H}(\mathrm{div}; \Omega) \to \mathbf{L}^2(\Omega)$ by $P(\boldsymbol{\tau}) := \nabla z$ for all $\boldsymbol{\tau} \in \mathbf{H}(\mathrm{div}; \Omega)$, where $z \in \mathrm{H}_0^1(\Omega) \cap \mathrm{H}^2(\Omega)$ is the unique solution of the problem:*

$$\Delta z = \mathrm{div}(\boldsymbol{\tau}) \quad \text{in} \quad \Omega, \quad z = 0 \quad \text{on} \quad \Gamma.$$

Prove that P is a compact projector, and conclude that (cf. Exercise 3.23)

$$\mathbf{H}(\mathrm{div}; \Omega) = P(\mathbf{H}(\mathrm{div}; \Omega)) \oplus (I - P)(\mathbf{H}(\mathrm{div}; \Omega)).$$

iii) *Use the above decomposition of* $\mathbf{H}(\mathrm{div};\Omega)$ *to show that (6.41) can be equivalently written as*

$$A(\boldsymbol{\sigma},\boldsymbol{\tau}) + K(\boldsymbol{\sigma},\boldsymbol{\tau}) = F(\boldsymbol{\tau}) \qquad \forall\,\boldsymbol{\tau}\in\mathbf{H}(\mathrm{div};\Omega),$$

where

$$A(\boldsymbol{\sigma},\boldsymbol{\tau}) := -\int_\Omega P(\boldsymbol{\sigma})\cdot P(\boldsymbol{\tau}) - \int_\Omega \mathrm{div}(P(\boldsymbol{\sigma}))\,\mathrm{div}(P(\boldsymbol{\tau})) + \int_\Omega (I-P)(\boldsymbol{\sigma})\cdot(I-P)(\boldsymbol{\tau}),$$

$$K(\boldsymbol{\sigma},\boldsymbol{\tau}) := 2\int_\Omega P(\boldsymbol{\sigma})\cdot P(\boldsymbol{\tau}) + \int_\Omega P(\boldsymbol{\sigma})\cdot(I-P)(\boldsymbol{\tau}) + \int_\Omega (I-P)(\boldsymbol{\sigma})\cdot P(\boldsymbol{\tau}),$$

and F *is the functional defined by the right-hand side of (6.41).*

iv) *Let* $\mathbf{A}\in\mathcal{L}\big(\mathbf{H}(\mathrm{div};\Omega),\mathbf{H}(\mathrm{div};\Omega)\big)$ *and* $\mathbf{K}\in\mathcal{L}\big(\mathbf{H}(\mathrm{div};\Omega),\mathbf{H}(\mathrm{div};\Omega)\big)$ *be the operators induced by the bilinear forms* A *and* K*, respectively. Prove that* $\mathbf{A}$ *is bijective and that* $\mathbf{K}$ *is compact.* HINT. *Define the operator* $S(\boldsymbol{\tau}) := (I-2P)(\boldsymbol{\tau})$ *and consider the expression* $A(\boldsymbol{\tau},S(\boldsymbol{\tau}))$ *to show that* A *satisfies the continuous inf–sup condition.*

▸ **6.11**

a) *Let* X *and* Y *be Hilbert spaces, and let* $K\in\mathcal{L}(X,Y)$ *be such that* K *maps weakly convergent sequences in* X *to convergent sequences in* Y*. Prove that* K *is compact.*

b) *Let* X *be a Hilbert space, and let* $\{x_n\}_{n\in\mathbb{N}}\subseteq X$ *and* $x\in X$ *be such that* $x_n\overset{w}{\to}x$ *and* $\|x_n\|\to\|x\|$*. Prove that* $x_n\to x$*.*

c) *Let* X *and* Y *be Hilbert spaces, and let* $K\in\mathcal{L}(X,Y)$ *be a compact operator. Apply a) and b) to prove that* $K^*\in\mathcal{L}(Y,X)$ *is also compact.*

▸ **6.12** ([67]) *Let* X *be a Banach space, and let* $T: X\to X'$ *be a* NONLINEAR *operator. We say that* T *satisfies property* (S) *if for every sequence* $\{x_n\}_{n\in\mathbb{N}}\subseteq X$ *such that*

$$x_n\overset{w}{\to}x \quad\text{and}\quad (T(x_n))\,(x_n-x)\;-\;(T(x))\,(x_n-x)\to 0\,,$$

we have $x_n\to x$*. Now let* H*,* Q*,* V *be Banach spaces such that* $H\subseteq V$ *and the injection* $\mathbf{i}: H\to V$ *is compact. Set* $X := H\times Q$*, and consider a bounded bilinear form* $B: X\times X\to\mathbb{R}$ *and a nonlinear operator* $A: H\to H'$ *satisfying*

i) *there exists* $C_0>0$ *such that*

$$B((u,\lambda),(u,\lambda))\;\geqslant\;C_0\,\|\lambda\|_Q^2 \qquad \forall\,(u,\lambda)\in X\,,$$

ii) *and there exist* $C_1, C_2>0$ *such that for all* $u,v\in H$*,*

$$(A(u))\,(u-v)\;-\;(A(v))\,(u-v)\;\geqslant\;C_1\,\|u-v\|_H^2\;+\;R(u,v)\,,$$

where

$$|R(u,v)|\;\leqslant\;C_2\,\{\,1+\|u\|_H+\|v\|_H\,\}\,\|u-v\|_V\,.$$

Prove that the nonlinear operator $T: X\to X'$ *defined by*

$$(T(u,\lambda))\,(v,\mu)\;:=\;(A(u))(v)\;+\;B((u,\lambda),(v,\mu))\,,$$

satisfies property (S)*.*

▸ **6.13** *Let* Ω *be a bounded open interval of* $\mathbb{R}$*. Assume that the embedding* $i: \mathrm{H}^1(\Omega)\to\mathrm{L}^2(\Omega)$ *is compact, and prove that, for every integer* $m\geqslant 2$*, the embedding* $i: \mathrm{H}^m(\Omega)\to\mathrm{H}^{m-1}(\Omega)$ *is also compact. The assumption and the result required here are established later in Chapter 10 (cf. Theorems 10.18 and 10.20) for the more general case of a domain in* $\mathbb{R}^n$*.*

▸ **6.14** *Let $(H,\langle\cdot,\cdot\rangle)$ be a real Hilbert space, and let $a : H \times H \to \mathbb{R}$ be a bounded bilinear form whose induced operator $A \in \mathcal{L}(H)$ is bijective. Given $\{u_1,\dots,u_N\} \subseteq H$ and $F \in H'$, consider the variational formulation: Find $u \in H$ such that*

$$a(u,v) - \sum_{j=1}^{N} \langle u,u_j\rangle\langle v,u_j\rangle = F(v) \qquad \forall\, v \in H\,. \tag{6.42}$$

Derive a necessary and sufficient condition on $\{u_1,\dots,u_N\}$ that guarantees that for each $F \in H'$ problem (6.42) admits a unique solution.

▸ **6.15** *Let $(H,\langle\cdot,\cdot\rangle)$ be a real Hilbert space, and let $a : H \times H \to \mathbb{R}$ and $b : H \times H \to \mathbb{R}$ be bounded bilinear forms such that a is H-elliptic and b is symmetric. Given $\{u_1,\dots,u_N\} \subseteq H$ and $F \in H'$, consider the variational formulation: Find $u \in H$ such that*

$$a(u,v) + \sum_{j=1}^{N} a(u,u_j)\, b(v,u_j) = F(v) \qquad \forall\, v \in H\,. \tag{6.43}$$

Derive a necessary and sufficient condition on $\{u_1,\dots,u_N\}$ that guarantees that for each $F \in H'$ problem (6.43) admits a unique solution. In particular, what happens if there exists $c \in \mathbb{R}$ such that

$$b(u_i,u_j)\, c\,\delta_{ij} \qquad \forall\, i,\, j \in \{1,2,\dots,N\}\,?$$

On the other hand, what could you conclude if $a = b$?

▸ **6.16** ([102], [111]) *Let Ω be a bounded open subset of $\mathbb{R}^n$ with a $C^{0,1}$ boundary, and consider the Sobolev space $\mathrm{H}^2(\Omega)$, with inner product $\langle\cdot,\cdot\rangle_{2,\Omega}$, induced norm $\|\cdot\|_{2,\Omega}$, and seminorm $|\cdot|_{2,\Omega}$. Moreover, let $P_1(\Omega)$ be the space of polynomials on Ω of degree $\leqslant 1$ with basis $\{p_0,\dots,p_n\}$, where $p_0(x) = 1$ and $p_i(x) = x_i$ for all $i \in \{1,\dots,n\}$, and $x := (x_1,\dots,x_n)^{\mathrm{t}} \in \Omega$.*

a) [Generalized Poincaré Inequality]. *(cf. Theorem 10.21 and Example 10.22). Define the mapping*

$$|\!|\!| v |\!|\!| := \left\{ |v|_{2,\Omega}^2 + \sum_{i=0}^{n} |\langle v,p_i\rangle_{2,\Omega}|^2 \right\}^{1/2} \qquad \forall\, v \in \mathrm{H}^2(\Omega)\,,$$

and show that there exist C_1, $C_2 > 0$ such that

$$C_1\, |\!|\!| v |\!|\!| \leqslant \|v\|_{2,\Omega} \leqslant C_2\, |\!|\!| v |\!|\!| \quad \forall\, v \in \mathrm{H}^2(\Omega)\,.$$

Hint: *For the second inequality, proceed by contradiction, i.e., suppose that for each $n \in \mathbb{N}$ there exists $v_n \in \mathrm{H}^2(\Omega)$ such that $\|v_n\|_{2,\Omega} > n\, |\!|\!| v_n |\!|\!|$. Then define*

$$w_n := \frac{v_n}{\|v_n\|_{2,\Omega}}\,,$$

observe that $\|w_n\|_{2,\Omega} = 1$, note that $|\!|\!| w_n |\!|\!| < \frac{1}{n}$, and apply the fact that the embedding of $\mathrm{H}^2(\Omega)$ into $\mathrm{H}^1(\Omega)$ is compact (cf. Theorem 10.20).

b) *[Deny–Lions Lemma]. Consider the quotient space $\mathrm{H}^2(\Omega)/P_1(\Omega)$ with norm*

$$\|[v]\|_{\mathrm{H}^2(\Omega)/P_1(\Omega)} := \inf_{p\in P_1(\Omega)} \|v-p\|_{2,\Omega}\,,$$

and define

$$|[v]|_{\mathrm{H}^2(\Omega)/P_1(\Omega)} := |v|_{2,\Omega} \qquad \forall\, [v] \in \mathrm{H}^2(\Omega)/P_1(\Omega)\,.$$

Prove that $|\cdot|_{\mathrm{H}^2(\Omega)/P_1(\Omega)}$ is well-defined and that there exists $C > 0$ such that

$$|[v]|_{\mathrm{H}^2(\Omega)/P_1(\Omega)} \leqslant \|[v]\|_{\mathrm{H}^2(\Omega)/P_1(\Omega)}$$

$$\leqslant C\, |[v]|_{\mathrm{H}^2(\Omega)/P_1(\Omega)} \qquad \forall\, [v] \in \mathrm{H}^2(\Omega)/P_1(\Omega)\,.$$

Hint: *Note that for every $p \in P_1(\Omega)$ and every multi-index α with $|\alpha| = 2$, we have $\partial^\alpha p = 0$. Apply the generalized Poincaré inequality.*

c) *[Bramble–Hilbert Lemma]. Let $\Pi \in \mathcal{L}(\mathrm{H}^2(\Omega), \mathrm{H}^1(\Omega))$ be such that $\Pi(p) = p$ for all $p \in P_1(\Omega)$. Prove that there exists $C > 0$ such that*

$$\|v - \Pi(v)\|_{1,\Omega} \leqslant C\,|v|_{2,\Omega} \qquad \forall\, v \in \mathrm{H}^2(\Omega)\,.$$

HINT: *Note that $v - \Pi(v) = v - p - \Pi(v - p)$ for all $p \in P_1(\Omega)$, and then apply the Deny–Lions Lemma.*

▸ **6.17** *Extend the analysis of Section 6.6.2 to the case of the boundary value problem:*

$$u'' + k\,u = f \quad \text{in} \quad \Omega := (0,1)\,, \quad u'(0) = a\,, \quad u'(1) = b\,,$$

with $f \in \mathrm{L}^2(\Omega)$, $k \in \mathbb{R}^+$, and $a, b \in \mathbb{R}$.

▸ **6.18** *Let X and Y be Hilbert spaces, and let $B \in \mathcal{L}(Y', X')$.*

i) *Show that there exists a unique $A \in \mathcal{L}(X, Y)$ such that $A' = B$.*
ii) *Use the fact that the Hilbert adjoint of a compact operator is also compact, to prove that $A \in \mathcal{K}(X, Y)$ whenever $B \in \mathcal{K}(Y', X')$.*

▸ **6.19** *In the context of the proof of Lemma 6.35, prove that, in fact, when Y is reflexive, we have $A = \mathcal{J}_Y^{-1} \circ B' \circ \mathcal{J}_X$* **if and only if** *$A' = B$.*

▸ **6.20** ([84]) *This exercise aims to show that if a symmetric positive definite bilinear form on a Hilbert space becomes elliptic after adding another bilinear form induced by a compact operator, then it is itself elliptic. More precisely, the following result and its application hold.*

a) *Let X be a real Hilbert space, and let $a : X \times X \to \mathbb{R}$ and $k : X \times X \to \mathbb{R}$ be bounded bilinear forms induced by operators $A \in \mathcal{L}(X, X)$ and $K \in \mathcal{K}(X, X)$, respectively. Assume that a is symmetric and positive definite, and that there exists a constant $\alpha > 0$ such that*

$$a(x, x) + k(x, x) \geqslant \alpha\,\|x\|^2 \qquad \forall\, x \in X.$$

Using the Fredholm alternative (cf. (6.10), Theorem 6.17) and Lemma 8.8 (cf. Chapter 8), show that there exists a constant $\tilde{\alpha} > 0$ such that

$$a(x, x) \geqslant \tilde{\alpha}\,\|x\|^2 \qquad \forall\, x \in X.$$

b) *Recall from (4.116) and (4.128) that, given a bounded open set $\Omega \subset \mathbb{R}^n$ with sufficiently smooth boundary Γ, we can define the Hilbert space*

$$\mathbf{H}(\mathrm{div}; \Omega) := \Big\{ \boldsymbol{\tau} := (\tau_1, \ldots, \tau_n) \in \mathbf{L}^2(\Omega) : \quad \mathrm{div}(\boldsymbol{\tau}) := \sum_{i=1}^{n} \frac{\partial \tau_i}{\partial x_i} \in \mathrm{L}^2(\Omega) \Big\},$$

whose norm $\|\cdot\|_{\mathrm{div};\Omega}$ is given by (4.117), and its subspace

$$\mathbf{H}(\mathrm{div}_{\mathbb{R}}; \Omega) := \Big\{ \boldsymbol{\tau} \in \mathbf{H}(\mathrm{div}; \Omega) : \quad \mathrm{div}(\boldsymbol{\tau}) = c \in \mathbb{R} \text{ in } \mathcal{D}'(\Omega) \Big\}.$$

Observe that for all $\boldsymbol{\tau} \in \mathbf{H}(\mathrm{div}_{\mathbb{R}}; \Omega)$ we have

$$\|\boldsymbol{\tau}\|^2_{\mathrm{div};\Omega} = \|\boldsymbol{\tau}\|^2_{0,\Omega} + k(\boldsymbol{\tau}, \boldsymbol{\tau}),$$

where $k : \mathbf{H}(\mathrm{div}; \Omega) \times \mathbf{H}(\mathrm{div}; \Omega) \to \mathbb{R}$ is the bilinear form given by

$$k(\boldsymbol{\tau}, \boldsymbol{\zeta}) := \frac{1}{|\Omega|} \Big\{ \int_\Omega \mathrm{div}(\boldsymbol{\tau}) \Big\} \Big\{ \int_\Omega \mathrm{div}(\boldsymbol{\zeta}) \Big\} \qquad \forall\, \boldsymbol{\tau}, \boldsymbol{\zeta} \in \mathbf{H}(\mathrm{div}; \Omega),$$

and then apply a) to conclude that there exists a constant $\beta > 0$ such that

$$\|\boldsymbol{\tau}\|_{0,\Omega} \geqslant \beta\,\|\boldsymbol{\tau}\|_{\mathrm{div};\Omega} \qquad \forall\, \boldsymbol{\tau} \in \mathbf{H}(\mathrm{div}_{\mathbb{R}}; \Omega)\,.$$

▸ **6.21** *Use a suitable integration by parts formula, and then apply the analysis on the Fredholm alternative and the Galerkin method to study the discrete solvability of the boundary value problem:*

$$\Delta u + k\,u = f \quad \text{in} \quad \Omega := (0,1)^2\,, \qquad \frac{\partial u}{\partial \nu} = g \quad \text{on} \quad \Gamma := \partial\Omega\,,$$

where $k \in [-1, +\infty[$, $f \in \mathrm{L}^2(\Omega)$, ν *is the outward normal vector on* Γ, *and* $g \in \mathrm{H}^{-1/2}(\Gamma)$.

▸ **6.22** *Let* X *and* Y *be Banach spaces, and let* $A : X \times Y \to \mathbb{R}$ *be a bounded bilinear form with induced operator* $\mathbf{A} \in \mathcal{L}(X, Y')$, *that is, for each* $z \in X$, $\mathbf{A}(z)$ *is the functional in* Y' *defined by* $\mathbf{A}(z)(y) := A(z,y)$ *for all* $y \in Y$. *Suppose that* A *satisfies the hypotheses of the Banach–Nečas–Babuška Theorem (or generalized Lax–Milgram Lemma) (cf. Lemma 5.1), equivalently, that* $\mathbf{A}$ *is bijective, so that for each* $F \in Y'$ *there exists a unique* $x \in X$ *such that* $\mathbf{A}(x) = F$, *that is,*

$$A(x,y) = F(y) \qquad \forall\, y \in Y\,.$$

Moreover, given sequences $\{X_n\}_{n\in\mathbb{N}}$ *and* $\{Y_n\}_{n\in\mathbb{N}}$ *of finite-dimensional subspaces of* X *and* Y, *respectively, such that* $\dim X_n = \dim Y_n$ *for all* $n \in \mathbb{N}$, *consider for each* $n \in \mathbb{N}$ *the Galerkin scheme: Find* $x_n \in X_n$ *such that*

$$A(x_n, y_n) = F(y_n) \qquad \forall\, y_n \in Y_n\,, \tag{6.44}$$

and denote the whole family by $\mathcal{G}(\mathbf{A}; X_n, Y_n)_{n\in\mathbb{N}}$. *The idea of this exercise is to extend what was developed in Section 6.6, on the Galerkin method and the Fredholm alternative in Hilbert spaces, to the present Banach setting, for which it is suggested to proceed as follows.*

a) *Given the continuous injection* $i_n : Y_n \to Y$ *with respective adjoint operator* $i_n' : Y' \to Y_n'$, *show that, for each* $x \in X$, (6.44) *can be rewritten as: Find* $x_n \in X_n$ *such that*

$$\mathbf{A}_n(x_n) = \big(i_n'\,\mathbf{A}\big)(x)\,,$$

where $\mathbf{A}_n := \big(i_n'\,\mathbf{A}\big)|_{X_n} : X_n \to Y_n'$. *In this case, the family* $\mathcal{G}(\mathbf{A}; X_n, Y_n)_{n\in\mathbb{N}}$ *is said to be convergent if there exists* $N \in \mathbb{N}$ *such that* $\mathbf{A}_n : X_n \to Y_n'$ *is bijective for all* $n \geqslant N$, *and* $\big(\mathbf{A}_n^{-1}\,i_n'\,\mathbf{A}\big)(x) \overset{n\to\infty}{\longrightarrow} x$ *for all* $x \in X$.

b) *Assume that* $\lim\limits_{n\to\infty} \operatorname{dist}(z, X_n) = 0$ *for all* $z \in X$, *and prove that* $\mathcal{G}(\mathbf{A}; X_n, Y_n)_{n\in\mathbb{N}}$ *is convergent if and only if there exist* $N \in \mathbb{N}$ *and* $M > 0$ *such that* $\mathbf{A}_n : X_n \to Y_n'$ *is bijective for all* $n \geqslant N$, *and* $\|\mathbf{A}_n^{-1}\,i_n'\,\mathbf{A}\| \leqslant M$ *for all* $n \in \mathbb{N}$.

c) *Given* $\mathbf{K} \in \mathcal{K}(X, Y')$, *define* $\mathbf{T} := \mathbf{A} + \mathbf{K}$, *assume that* $\mathbf{T}$ *is injective, and adapt the arguments of the proof of Theorem 6.41 to show that* $\mathcal{G}(\mathbf{A}; X_n, Y_n)_{n\in\mathbb{N}}$ *is convergent if and only if* $\mathcal{G}(\mathbf{T}; X_n, Y_n)_{n\in\mathbb{N}}$ *is convergent.*

7

Reflexivity and Separability

In this chapter we introduce the concepts of reflexive and separable spaces, we prove some of the most important results associated with them, and we establish their various connections with weak topologies. Our main references on these topics are [23], [104], and [115].

7.1 Preliminaries

In what follows, given a normed vector space X over $\mathbb{K}$, we denote by X'' the dual space of X', which is usually called the bidual, that is

$$X'' := \Big\{\mathcal{F} : X' \to \mathbb{K} : \quad \mathcal{F} \quad \text{is linear and bounded}\Big\}.$$

Then, we let $\mathcal{J} : X \to X''$ be the linear operator given by

$$\mathcal{J}(x)(F) := \ F(x) \quad \forall F \in X', \quad \forall x \in X\,, \tag{7.1}$$

which is easily seen to be well defined since

$$|\mathcal{J}(x)(F)| \ = \ |F(x)| \ \leqslant \ \|x\|\,\|F\| \quad \forall F \in X'\,.$$

This proves that $\|\mathcal{J}(x)\|_{X''} \leqslant \|x\|$ for all $x \in X$, and thus $\mathcal{J}(x) \in X''$ for all $x \in X$. In other words, the above shows that every $x \in X$ induces a linear and bounded functional on X''. On the other hand, thanks to a result from Chapter 2 (cf. Lemma 2.25), which is a consequence of the Hahn-Banach Theorem, it follows that

$$\|x\| \ = \ \sup_{\substack{F \in X' \\ F \neq 0}} \frac{|F(x)|}{\|F\|} \ = \ \sup_{\substack{F \in X' \\ F \neq 0}} \frac{|\mathcal{J}(x)(F)|}{\|F\|} \ = \ \|\mathcal{J}(x)\|_{X''}\,,$$

that is,

$$\|\mathcal{J}(x)\| \ = \ \|x\| \qquad \forall x \in X\,, \tag{7.2}$$

which, in addition to establishing that $\mathcal{J}$ is an isometry and that $\|\mathcal{J}\|_{\mathcal{L}(X,X'')} = 1$, says that $\mathcal{J}$ is injective. We remark here that the fact that X is not necessarily Banach stop us of applying Theorem 3.58 (or Theorem 3.61) to conclude from (7.2) that $R(\mathcal{J})$ is closed.

Definition 7.1. *A Banach space X is said to be* REFLEXIVE *if $\mathcal{J}$ is surjective, that is, if for each $\mathcal{F} \in X''$ there exists $x \in X$ such that $\mathcal{F}(F) = F(x)$ for all $F \in X'$.*

G. N. Gatica, *Introduction to Functional Analysis*,
https://doi.org/10.1007/978-3-032-37415-8_7

Note that the concept of reflexivity only makes sense for Banach spaces since, as X'' is always complete (because it is the dual of X') and since $\mathcal{J} : X \to X''$ constitutes an isometric bijection, necessarily X must also be Banach.

Example 7.2 Every Hilbert space $(X, \langle\cdot,\cdot\rangle_X)$ is reflexive. Indeed, assume for simplicity that X is real, and denote by $\mathcal{R} : X' \to X$ the corresponding Riesz map. Then, defining $\langle\cdot,\cdot\rangle_{X'} : X' \times X' \to \mathbb{R}$ by

$$\langle F, G\rangle_{X'} := \langle \mathcal{R}(F), \mathcal{R}(G)\rangle_X \qquad \forall F, G \in X',$$

we have that $\langle\cdot,\cdot\rangle_{X'}$ is an inner product on X' and, thanks to the Riesz Representation Theorem (cf. Theorem 2.14), its induced norm coincides with the usual norm of X', from which it follows that $(X', \langle\cdot,\cdot\rangle_{X'})$ is a Hilbert space as well. Let then $\widetilde{\mathcal{R}} : X'' \to X'$ be the associated Riesz map. It follows that for each $\mathcal{F} \in X''$ and $F \in X'$ there holds

$$\mathcal{F}(F) = \langle F, \widetilde{\mathcal{R}}(\mathcal{F})\rangle_{X'} = \langle \mathcal{R}(F), \mathcal{R}(\widetilde{\mathcal{R}}(\mathcal{F}))\rangle_X = F(\mathcal{R}(\widetilde{\mathcal{R}}(\mathcal{F}))) = F(x),$$

with $x := \mathcal{R}(\widetilde{\mathcal{R}}(\mathcal{F})) \in X$, which shows that $\mathcal{F} = \mathcal{J}(x)$ and therefore $\mathcal{J}$ is surjective.

Definition 7.3. *A normed vector space X is said to be* SEPARABLE *if it contains a dense countable subset.*

7.2 Connections with duals and subspaces

In this section we present several results that connect the possible properties of reflexivity and separability of a Banach space with those of its duals and subspaces. It is also shown that these properties are preserved through isomorphic spaces.

Lemma 7.4. *X is a reflexive Banach space if and only if its dual X' is also reflexive.*

Proof. Suppose that X is a reflexive Banach space and let $\mathcal{J}_1 : X' \to X'''$ be the map given by

$$\mathcal{J}_1(G)(\mathcal{F}) = \mathcal{F}(G) \qquad \forall G \in X', \quad \forall \mathcal{F} \in X'',$$

where X''' is the dual of X''. Then, given $\mathbb{F} \in X'''$, we define $F : X \to \mathbb{R}$ by

$$F(x) := \mathbb{F}(\mathcal{J}(x)) \qquad \forall x \in X, \tag{7.3}$$

where $\mathcal{J} : X \to X''$ is, by hypothesis, bijective. It is clear that F is linear (since both $\mathbb{F}$ and $\mathcal{J}$ are), and moreover

$$|F(x)| \leqslant \|\mathbb{F}\| \, \|\mathcal{J}(x)\| = \|\mathbb{F}\| \, \|x\|,$$

which proves that $F \in X'$. It follows from (7.3) that

$$\mathbb{F}(\mathcal{J}(x)) = \mathcal{J}(x)(F) \qquad \forall x \in X,$$

which, by virtue of the reflexivity of X, is equivalent to saying that

$$\mathbb{F}(\mathcal{F}) = \mathcal{F}(F) \qquad \forall \mathcal{F} \in X'',$$

or equivalently,

$$\mathbb{F}(\mathcal{F}) = \mathcal{J}_1(F)(\mathcal{F}) \qquad \forall \mathcal{F} \in X'',$$

and hence $\mathbb{F} := \mathcal{J}_1(F)$. This shows that $\mathcal{J}_1$ is surjective, and therefore X' is reflexive. Conversely, let X be a Banach space such that X' is reflexive, that is, $\mathcal{J}_1 : X' \to X'''$ is bijective, and let us prove that $\mathcal{J} : X \to X''$ is surjective. Note first that the facts that $\mathcal{J}$ is an isometry and that X is complete imply that $\mathcal{J}(X)$ is a closed subspace of X''. Suppose now, by contradiction, that $\mathcal{J}(X) \neq X''$. This means that there exists $\mathcal{F} \in X''$ such that $\mathcal{F} \notin \mathcal{J}(X)$, and then, by applying a consequence of the Hahn-Banach Theorem (cf. Theorem 2.26), there exists $\mathbb{F} \in X'''$ such that $\|\mathbb{F}\| = 1$, $\mathbb{F}(\mathcal{F}) = \mathrm{dist}\,(\mathcal{F}, \mathcal{J}(X)) > 0$ and $\mathbb{F}(\mathcal{G}) = 0$ for all $\mathcal{G} \in \mathcal{J}(X)$. By the reflexivity of X', there exists $F \in X'$ such that $\mathcal{J}_1(F) = \mathbb{F}$, and therefore, for each $x \in X$ we have

$$0 = \mathbb{F}(\mathcal{J}(x)) = \mathcal{J}_1(F)(\mathcal{J}(x)) = \mathcal{J}(x)(F) = F(x),$$

whence F is the zero functional. It follows that $\mathbb{F}$ also vanishes, which contradicts the fact that $\mathbb{F}(\mathcal{F}) > 0$, and hence necessarily $\mathcal{J}(X) = X''$, thus proving that X is reflexive.

□

Lemma 7.5. *Let Z be a closed subspace of a reflexive Banach space X. Then Z is also reflexive.*

Proof. We know that $\mathcal{J} : X \to X''$ is surjective and we would like to prove that the corresponding map $\tilde{\mathcal{J}} : Z \to Z''$ is also surjective. Given $\tilde{\mathcal{F}} \in Z''$, we define $\mathcal{F} : X' \to \mathbb{R}$ by

$$\mathcal{F}(F) = \tilde{\mathcal{F}}(F_r) \qquad \forall F \in X',$$

where $F_r := F|_Z \in Z'$. It is clear that

$$|\mathcal{F}(F)| \leqslant \|\tilde{\mathcal{F}}\|_{Z''} \, \|F_r\|_{Z'} \leqslant \|\tilde{\mathcal{F}}\|_{Z''} \, \|F\|_{X'} \quad \forall F \in X',$$

and therefore $\mathcal{F} \in X''$. It follows by hypothesis that there exists $x \in X$ such that $\mathcal{J}(x) = \mathcal{F}$, and then

$$\tilde{\mathcal{F}}(F_r) = F(x) \qquad \forall F \in X'. \tag{7.4}$$

Let us now show that $x \in Z$. Suppose, by contradiction, that this is not the case. Then, applying again the consequence of the Hahn-Banach Theorem given by Theorem 2.26, we deduce that there exists $G \in X'$ such that $\|G\| = 1$, $G(x) = \mathrm{dist}\,(x, Z) > 0$ and $G(z) = 0$ for all $z \in Z$. It follows from (7.4) that

$$G(x) = \tilde{\mathcal{F}}(G_r) = \tilde{\mathcal{F}}(0) = 0,$$

which is a contradiction. Consequently, we must have $x \in Z$. Finally, given $\tilde{G} \in Z'$, the Hahn-Banach Theorem ensures that there exists $F \in X'$ such that $F_r = \tilde{G}$, whence

$$\tilde{\mathcal{F}}(\tilde{G}) = \tilde{\mathcal{F}}(F_r) = F(x) = \tilde{G}(x) = \tilde{\mathcal{J}}(x)(\tilde{G}),$$

and therefore $\tilde{\mathcal{F}} = \tilde{\mathcal{J}}(x)$, thus proving that Z is reflexive.

□

The following lemma proves that reflexivity is preserved through isomorphisms. The corresponding analogue for separability is stated in Exercise 7.4, part a).

Lemma 7.6. *Let X and Y be Banach spaces for which there exists a bijective $T \in \mathcal{L}(X, Y)$. Then X is reflexive if and only if Y is reflexive.*

Proof. We begin by recalling from Theorem 3.90 that the bijectivity of T is successively transferred to $T' \in \mathcal{L}(Y', X')$, and from the latter to $T'' \in \mathcal{L}(X'', Y'')$ as well. Suppose that Y is reflexive and notice, as we did in the proof of Theorem 3.66 (cf. (3.25)), that the following diagram commutes

$$\begin{array}{ccc} X & \xrightarrow{\;T\;} & Y \\ \mathcal{J}_X \big\downarrow & & \big\downarrow \mathcal{J}_Y \\ X'' & \xrightarrow{\;T''\;} & Y'' \end{array}$$

Indeed, given arbitrary $x \in X$ and $G \in Y'$, we obtain

$$\begin{aligned} \Big((T'' \circ \mathcal{J}_X)(x) \Big)(G) \,&=\, T''(\mathcal{J}_X(x))(G) \,=\, (T')'(\mathcal{J}_X(x))(G) \,=\, \mathcal{J}_X(x)(T'(G)) \\ &=\, T'(G)(x) \,=\, G(T(x)) \,=\, \mathcal{J}_Y(T(x))(G) \,=\, \Big((\mathcal{J}_Y \circ T)(x) \Big)(G)\,, \end{aligned}$$

which confirms what was claimed, that is, $T'' \circ \mathcal{J}_X = \mathcal{J}_Y \circ T$. Hence it follows that $\mathcal{J}_X = (T'')^{-1} \circ \mathcal{J}_Y \circ T$ is bijective, thus proving the reflexivity of X. For the converse, it suffices to reverse the roles of X and Y, using T^{-1} instead of T, in the preceding argument. □

Example 7.7 Given Ω a bounded open subset of R^n, $n \in \{2,3\}$, and $t \in (1, +\infty)$, we consider the Lebesgue space $\mathrm{L}^t(\Omega)$ and its vector version $\mathbf{L}^t(\Omega) := [\mathrm{L}^t(\Omega)]^n$, whose norms in both cases we denote by $\|\cdot\|_{0,t;\Omega}$, to illustrate the applicability of some of the previous results. More precisely, we introduce the space

$$\mathbf{H}^t(\mathrm{div}_t;\Omega) \,:=\, \Big\{ \boldsymbol{\tau} \in \mathbf{L}^t(\Omega) : \quad \mathrm{div}(\boldsymbol{\tau}) \in \mathrm{L}^t(\Omega) \Big\}\,,$$

which is endowed with the norm

$$\|\boldsymbol{\tau}\|_{t,\mathrm{div}_t;\Omega} \,:=\, \|\boldsymbol{\tau}\|_{0,t;\Omega} \,+\, \|\mathrm{div}(\boldsymbol{\tau})\|_{0,t;\Omega} \qquad \forall\, \boldsymbol{\tau} \in \mathbf{H}^t(\mathrm{div}_t;\Omega)\,,$$

and we use the fact that $\mathrm{L}^t(\Omega)$ (cf. [40, Theorem 5.14-2, part (e)]), and therefore $\mathbf{L}^t(\Omega)$ (cf. Exercise 7.4, part b)), are reflexive, to prove that $\mathbf{H}^t(\mathrm{div}_t;\Omega)$ is as well. Indeed, by virtue again of what is stated in Exercise 7.4, part b), we now note that the product space $\mathbf{L}^t(\Omega) \times \mathrm{L}^t(\Omega)$, endowed with the norm

$$\|(\mathbf{v}, z)\|_{t,t;\Omega} \,:=\, \|\mathbf{v}\|_{0,t;\Omega} \,+\, \|z\|_{0,t;\Omega} \qquad \forall\, (\mathbf{v}, z) \in \mathbf{L}^t(\Omega) \times \mathrm{L}^t(\Omega)\,,$$

is itself reflexive. Then we define the operator $T : \mathbf{H}^t(\mathrm{div}_t;\Omega) \longrightarrow \mathbf{L}^t(\Omega) \times \mathrm{L}^t(\Omega)$ by

$$T(\boldsymbol{\tau}) \,:=\, \big(\boldsymbol{\tau}, \mathrm{div}(\boldsymbol{\tau})\big) \qquad \forall\, \boldsymbol{\tau} \in \mathbf{H}^t(\mathrm{div}_t;\Omega)\,,$$

which is clearly linear and bounded, and in particular we observe that

$$\|T(\boldsymbol{\tau})\|_{t,t;\Omega} \,=\, \|\boldsymbol{\tau}\|_{t,\mathrm{div}_t;\Omega} \qquad \forall\, \boldsymbol{\tau} \in \mathbf{H}^t(\mathrm{div}_t;\Omega)\,,$$

which shows, by virtue of Theorem 3.58, that T is injective and has closed range $R(T)$. It follows from Lemma 7.5 that $R(T)$ is reflexive, and since $T : \mathbf{H}^t(\mathrm{div}_t;\Omega) \longrightarrow R(T)$ is obviously bijective, we conclude, by Lemma 7.6, that $\mathbf{H}^t(\mathrm{div}_t;\Omega)$ is reflexive.

Lemma 7.8. *Let X be a normed vector space such that X' is separable. Then X is also separable.*

Proof. Let $\{F_n\}_{n\in\mathbb{N}}$ be a dense sequence in X'. We claim that for each $n \in \mathbb{N}$ there exists $x_n \in X$ such that $\|x_n\| = 1$ and $|F_n(x_n)| \geqslant \frac{\|F_n\|}{2}$. Otherwise, if we assume that $|F_n(x)| < \frac{\|F_n\|}{2}$ for all $x \in X$, $\|x\| = 1$, it follows that $\|F_n\| \leqslant \frac{\|F_n\|}{2}$, whence $\|F_n\| = 0$, which contradicts the strict inequality above. Now, let $M := \langle \{x_n\}_{n\in\mathbb{N}} \rangle$, and let us prove that $\overline{M} = X$. Indeed, suppose by contradiction that there exists $x_0 \in X$ such that $x_0 \notin \overline{M}$. Then, by Theorem 2.26 we know that there exists $F \in X'$ such that $\|F\| = 1$, $F(x_0) = \operatorname{dist}(x_0, \overline{M}) > 0$ and $F(x) = 0$ for all $x \in \overline{M}$, and hence, in particular, $F(x_n) = 0$ for all $n \in \mathbb{N}$. Thus we have

$$\frac{\|F_n\|}{2} \leqslant |F_n(x_n)| = |F_n(x_n) - F(x_n)| \leqslant \|F_n - F\|,$$

and then

$$1 = \|F\| \leqslant \|F - F_n\| + \|F_n\| \leqslant 3\|F - F_n\| \qquad \forall n \in \mathbb{N},$$

which contradicts the fact that $\{F_n\}_{n\in\mathbb{N}}$ is dense in X'. Consequently, it must be that $\overline{M} = X$. Now, define

$$M_{\mathbb{Q}} := \left\{ x = \sum_{j=1}^{N} \alpha_j z_j : \quad N \in \mathbb{N}, \quad \alpha_j \in \mathbb{Q} \quad \forall j \in \{1, \cdots, N\}, \right.$$
$$\left. \{z_1, z_2, \cdots, z_n\} \subseteq \{x_n\}_{n\in\mathbb{N}} \right\},$$

and observe, by virtue of the density of $\mathbb{Q}$ in $\mathbb{R}$, that $M \subseteq \overline{M_{\mathbb{Q}}}$. It follows that there holds $X = \overline{M} = \overline{M_{\mathbb{Q}}}$, and since $M_{\mathbb{Q}}$ is countable, X is therefore separable.

□

The converse of the previous lemma requires the reflexivity of the space X. Later we will give an example of a separable normed space X such that X' is not separable.

Lemma 7.9. *Let X be a reflexive and separable Banach space. Then X' is also separable.*

Proof. Let $\{x_n\}_{n\in\mathbb{N}}$ be a dense sequence in X, and let $\mathcal{F} \in X''$. Since X is reflexive, there exists $x \in X$ such that $\mathcal{J}(x) = \mathcal{F}$. Moreover, for each $\varepsilon > 0$ there exists $n \in \mathbb{N}$ such that $\|x - x_n\| < \varepsilon$. It follows that

$$\|\mathcal{F} - \mathcal{J}(x_n)\| = \|\mathcal{J}(x) - \mathcal{J}(x_n)\| = \|x - x_n\| < \varepsilon,$$

and therefore $\{\mathcal{J}(x_n)\}_{n\in\mathbb{N}}$ is dense in X''. This means that X'' is separable, and thus by Lemma 7.8, X' is also separable.

□

Note that the above proof can be simplified by using the result given in Exercise 7.4, part a). Indeed, since $\mathcal{J} : X \to X''$ is bijective and X is separable, the separability of X'' follows immediately.

Definition 7.10. *Let X be a normed vector space. A sequence $\{F_n\}_{n\in\mathbb{N}}$ in X' is said to be weak* convergent to $F \in X'$, and we write $F_n \overset{\omega^*}{\longrightarrow} F$, if*

$$F_n(x) \overset{n\to\infty}{\longrightarrow} F(x) \qquad \forall\, x \in X\,.$$

Note that strong (norm) convergence in X' implies weak* convergence. Also observe, according to the definition of equality of functionals in X', that the weak*-limit is unique. Moreover, assuming now that X is Banach, and applying the Banach–Steinhaus Theorem (cf. Theorem 3.46) in the same way as at the beginning of Section 6.5, one proves that every weak*-convergent sequence in X' is bounded. A sort of converse of this result is given by the following classical theorem.

Theorem 7.11. *Let X be a separable normed vector space. Then every bounded sequence in X' has a weak*-convergent subsequence.*

Proof. Let $\{x_n\}_{n\in\mathbb{N}}$ be a dense sequence in X, and let $\{F_n\}_{n\in\mathbb{N}}$ be a bounded sequence in X'. Since $\{F_n(x_1)\}_{n\in\mathbb{N}}$ is a bounded sequence in $\mathbb{R}$, there exists $\{F_n^{(1)}\}_{n\in\mathbb{N}} \subseteq \{F_n\}_{n\in\mathbb{N}}$ such that $\{F_n^{(1)}(x_1)\}_{n\in\mathbb{N}}$ converges in $\mathbb{R}$. Similarly, there exists $\{F_n^{(2)}\}_{n\in\mathbb{N}} \subseteq \{F_n^{(1)}\}_{n\in\mathbb{N}}$ such that $\{F_n^{(2)}(x_2)\}_{n\in\mathbb{N}}$ converges in $\mathbb{R}$. Continuing in this way, we deduce that for each $k \in \mathbb{N}$ there exists a subsequence $\{F_n^{(k)}\}_{n\in\mathbb{N}} \subseteq \{F_n^{(k-1)}\}_{n\in\mathbb{N}}$ such that $\{F_n^{(k)}(x_j)\}_{n\in\mathbb{N}}$ converges in $\mathbb{R}$ for all $j \in \{1, 2, \ldots, k\}$. Applying a diagonal argument used previously (see, for example, the proof of Theorem 6.15), we then define the sequence $G_n := F_n^{(n)}$ for all $n \in \mathbb{N}$. It follows that $\{G_n\}_{n\in\mathbb{N}}$ is a subsequence of $\{F_n\}_{n\in\mathbb{N}}$ for which $\{G_n(x_k)\}_{n\in\mathbb{N}}$ converges in $\mathbb{R}$ for all $k \in \mathbb{N}$ since $\{G_n(x_k)\}_{n\geqslant k} \subseteq \{F_n^{(k)}(x_k)\}_{n\in\mathbb{N}}$. We now show that $\{G_n\}_{n\in\mathbb{N}}$ is weak*-convergent. Indeed, given $x \in X$ and $\varepsilon > 0$, there exists $k \in \mathbb{N}$ such that $\|x - x_k\| < \varepsilon$. Then, with $M > 0$ such that $\|F_n\| \leqslant M$ for all $n \in \mathbb{N}$, we have

$$\begin{aligned}|G_n(x) - G_m(x)| &\leqslant |G_n(x) - G_n(x_k)| + |G_n(x_k) - G_m(x_k)| + |G_m(x_k) - G_m(x)| \\ &\leqslant 2\,M\,\varepsilon + |G_n(x_k) - G_m(x_k)|\,.\end{aligned}$$

Since $\{G_n(x_k)\}_{n\in\mathbb{N}}$ converges in $\mathbb{R}$, there exists $N \in \mathbb{N}$ such that

$$|G_n(x_k) - G_m(x_k)| < \varepsilon \quad \forall\, n, m \geqslant N,$$

and hence

$$|G_n(x) - G_m(x)| \leqslant (2M + 1)\,\varepsilon \qquad \forall\, n, m \geqslant N.$$

This shows that $\{G_n(x)\}_{n\in\mathbb{N}}$ converges in $\mathbb{R}$ for all $x \in X$, and therefore, defining

$$G(x) := \lim_{n\to\infty} G_n(x) \quad \forall\, x \in X\,,$$

we have that $G \in X'$ and $G_n \xrightarrow{\omega^*} G$.

□

The following result is the analogue (in the present analysis of separability) of Lemma 7.5 (for the analysis of reflexivity).

Lemma 7.12. *Let Z be a subspace of a separable normed space X. Then Z is also separable.*

Proof. Let $\{x_n\}_{n\in\mathbb{N}}$ be a dense sequence in X. Then, for each pair $(j, k) \in \mathbb{N}\times\mathbb{N}$ we choose, if it exists, an element $x_{jk} \in Z$ such that $\|x_{jk} - x_k\| < 1/j$. If such x_{jk} does not exist, the corresponding pair is not considered. In other words, for each pair $(j, k) \in \mathbb{N}\times\mathbb{N}$ we check whether the ball $B(x_k, 1/j)$ intersects Z or not. If it does, we choose any element of this intersection, denoted by x_{jk}, and we declare the pair (j, k) admissible (see Figure 7.1 for an illustration of both cases). We then define the subset of Z given by

$$S := \left\{x_{jk} : \quad (j,k) \text{ admissible}\right\},$$

which is clearly countable, and we now prove that $Z = \bar{S}$. Indeed, given $z \in Z$ and $\varepsilon > 0$, we take $j \in \mathbb{N}$ such that $\frac{1}{j} < \frac{\varepsilon}{2}$. Since $\left\{x_n\right\}_{n\in\mathbb{N}}$ is dense in X, there exists $k \in \mathbb{N}$ such that $\|z - x_k\| < \frac{1}{j}$. It follows that the pair (j,k) is admissible and moreover

$$\|z - x_{jk}\| \leqslant \|z - x_k\| + \|x_k - x_{jk}\| < \frac{2}{j} < \varepsilon,$$

which completes the proof. □

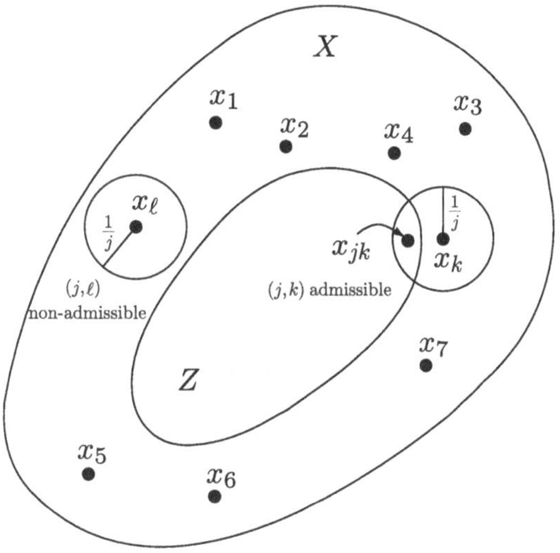

Fig. 7.1. *Definition of S in the proof of Lemma 7.12*

The next goal is to prove the result also announced in Section 6.5, which refers to the existence of a weakly convergent subsequence in every bounded sequence of a reflexive Banach space. Recall that the concept of weak convergence was given in Definition 6.27 (cf. Section 6.5).

Theorem 7.13. *Let X be a reflexive Banach space. Then every bounded sequence in X has a weakly convergent subsequence.*

Proof. Let $\left\{x_n\right\}_{n\in\mathbb{N}}$ be a bounded sequence in X, and let $Z := \overline{\langle\left\{x_n\right\}_{n\in\mathbb{N}}\rangle}$, that is, Z is the closed subspace generated by the sequence in question. It is easy to prove that Z is separable. Indeed, it suffices to note that $Z = \overline{\langle\left\{x_n\right\}_{n\in\mathbb{N}}\rangle_{\mathbb{Q}}}$, where $\langle\left\{x_n\right\}_{n\in\mathbb{N}}\rangle_{\mathbb{Q}}$ is the countable set given by the finite linear combinations (with coefficients in $\mathbb{Q}$) of the elements of $\left\{x_n\right\}_{n\in\mathbb{N}}$. Moreover, since Z is a closed subspace of a reflexive space, Lemma 7.5 implies that Z is also reflexive. It follows from Lemma 7.9 that Z' is separable as well. Now, given $\mathcal{J} : Z \to Z''$ the corresponding bijective mapping, and noting that $\left\{\mathcal{J}(x_n)\right\}_{n\in\mathbb{N}}$ is bounded in Z'', Theorem 7.11 ensures the existence of a weak*-convergent subsequence $\left\{\mathcal{J}(x_n^{(1)})\right\}_{n\in\mathbb{N}}$.

This means that there exists $x \in Z$ such that $\mathcal{J}(x_n^{(1)})(F) \overset{n\to\infty}{\to} \mathcal{J}(x)(F)$ for all $F \in Z'$, that is, $F(x_n^{(1)}) \overset{n\to\infty}{\to} F(x)$ for all $F \in Z'$. Finally, given $G \in X'$, we denote $\widetilde{G} := G|_Z \in Z'$. It follows that $G(x_n^{(1)}) = \widetilde{G}(x_n^{(1)}) \overset{n\to\infty}{\to} \widetilde{G}(x) = G(x)$, which proves that $x_n^{(1)} \overset{\omega}{\to} x$.

□

An interesting consequence of the previous theorem is the fact that weak convergence is not equivalent to strong convergence in an infinite-dimensional reflexive Banach space X. Indeed, suppose by contradiction that they were equivalent, and let $\{x_n\}_{n\in\mathbb{N}} \subseteq X$ be such that $\|x_n\| = 1$. It follows from Theorem 7.13 that there exist $\{x_n^{(1)}\}_{n\in\mathbb{N}} \subseteq \{x_n\}_{n\in\mathbb{N}}$ and $x \in X$ such that $x_n^{(1)} \overset{\omega}{\longrightarrow} x$. Thus we have $x_n^{(1)} \overset{n\to\infty}{\longrightarrow} x$, and therefore the unit sphere $S := \{x \in X : \quad \|x\| = 1\}$ is compact, which contradicts the dimension of X (cf. Theorem 6.8).

Example 7.14 Here we give an example of a sequence that converges weakly but does not converge in norm. Let $\Omega = (0,1)$ and $X := L^2(\Omega)$ (cf. Example 1.20, part 4), which, endowed with the inner product

$$\langle f, g\rangle := \int_\Omega fg \qquad \forall\, f, g \in X\,,$$

is a Hilbert space. Later, in the chapter on distributions (cf. Theorem 9.10, Chapter 9), we will see that $C_0^\infty(\Omega)$ is dense in X. Then we consider the sequence $\{x_n\}_{n\in\mathbb{N}} \subseteq X$ given by

$$x_n(t) := \sqrt{2}\, \sin(n\pi\, t) \qquad \forall\, t \in \Omega.$$

It is easy to see that

$$\|x_n\|^2 = \int_0^1 2\, \sin^2(n\pi\, t)\, dt = 1 \quad \forall n \in \mathbb{N}.$$

Moreover, for every $\varphi \in C_0^\infty(\Omega)$ we have

$$\begin{aligned}\langle \varphi, x_n\rangle &= \sqrt{2} \int_0^1 \varphi(t)\, \sin(n\pi\, t)\, dt \\ &= \sqrt{2} \left\{ \frac{-1}{n\pi} \cos(n\pi\, t)\, \varphi(t)\Big|_0^1 + \int_0^1 \frac{1}{n\pi} \cos(n\pi\, t)\, \varphi'(t)\, dt \right\} \\ &= \frac{\sqrt{2}}{n\pi} \int_0^1 \varphi'(t)\, \cos(n\pi\, t) dt\,,\end{aligned}$$

from which it follows that

$$|\langle \varphi, x_n\rangle| \leqslant \frac{\sqrt{2}}{n\pi} \int_0^1 |\varphi'(t)|\, dt \overset{n\to\infty}{\longrightarrow} 0,$$

that is,

$$\langle \varphi, x_n\rangle \overset{n\to\infty}{\longrightarrow} 0 \quad \forall \varphi \in C_0^\infty(\Omega).$$

Now, given $f \in X$ and $\varepsilon > 0$, there exists $\varphi_\varepsilon \in C_0^\infty(\Omega)$ such that $\|f - \varphi_\varepsilon\|_X < \varepsilon/2$. Moreover, for this φ_ε and the same $\varepsilon > 0$, there exists $N \in \mathbb{N}$ such that $|\langle \varphi_\varepsilon, x_n\rangle| < \varepsilon/2$ for all $n \geqslant N$. Hence, for each $n \geqslant N$ we have

$$|\langle f, x_n\rangle| = |\langle f - \varphi_\varepsilon, x_n\rangle + \langle \varphi_\varepsilon, x_n\rangle| \leqslant \|f - \varphi_\varepsilon\|\, \|x_n\| + |\langle \varphi_\varepsilon, x_n\rangle| < \frac{\varepsilon}{2} + \frac{\varepsilon}{2} = \varepsilon\,,$$

which shows that $\langle f, x_n\rangle \overset{n\to\infty}{\longrightarrow} 0$ for all $f \in X$. This is equivalent to saying that $x_n \overset{\omega}{\longrightarrow} 0$, and yet $\{x_n\}_{n\in\mathbb{N}}$ does not converge strongly to 0 since $\|x_n\| = 1$.

Lemma 7.15. *Let X be a finite-dimensional Banach space. Then the concept of weak convergence is equivalent to strong convergence.*

Proof. Let $\{x_n\}_{n\in\mathbb{N}}$ be a sequence in X that converges weakly to $x \in X$, and let $N \in \mathbb{N}$ be the dimension of X, which certainly coincides with that of X'. Thus, letting $\{F_1, \ldots, F_N\}$ be a basis of X', for each $F \in X'$ there exist $\alpha_1, \ldots, \alpha_N \in \mathbb{R}$ such that $F = \sum_{j=1}^{N} \alpha_j F_j$. Moreover, since all norms are equivalent in X' (cf. Theorem 6.1), we may take $\|F\| := \sum_{j=1}^{N} |\alpha_j|$. In turn, since $\{x_n\}_{n\in\mathbb{N}}$ converges weakly to x, it follows that for each $\varepsilon > 0$ and for each $j \in \{1, \ldots, N\}$, there exists $N_j \in \mathbb{N}$ such that

$$|F_j(x - x_k)| < \varepsilon \qquad \forall\, k \geqslant N_j.$$

Then, defining $m := \max\{N_j : \quad j \in \{1, \ldots, N\}\}$, we have

$$|F_j(x - x_k)| < \varepsilon \quad \forall\, k \geqslant m, \quad \forall j \in \{1, 2, \cdots, N\},$$

and hence, for every $F \in X'$ there holds

$$|F(x - x_k)| \leqslant \sum_{j=1}^{N} |\alpha_j|\, |F_j(x - x_k)| < \|F\|\, \varepsilon \qquad \forall\, k \geqslant m,$$

from which we obtain

$$\|x - x_k\| = \sup_{\substack{F \in X' \\ F \neq 0}} \frac{|F(x - x_k)|}{\|F\|} \leqslant \varepsilon \qquad \forall\, k \geqslant m.$$

This shows that $x_n \xrightarrow{n\to\infty} x$.

□

Alternatively, one can use results on compact operators to prove Lemma 7.15. Indeed, it suffices to recall that when X is finite-dimensional, the identity mapping $i : X \longrightarrow X$ is compact, and then apply Lemma 6.29, which states that a compact operator transforms weak convergence into strong convergence.

We now illustrate part of the results obtained so far on reflexivity and separability.

Example 7.16 Let X be a reflexive Banach space, and let $\{x_n\}_{n\in\mathbb{N}}$ be a bounded sequence in X with a unique weak limit point $\bar{x}$. We prove that $x_n \xrightarrow{\omega} \bar{x}$. Indeed, if we assume by contradiction that this does not occur, then there exists $F \in X'$ such that $F(x_n) \overset{n\to\infty}{\nrightarrow} F(\bar{x})$. This implies that there exist $\varepsilon > 0$ and $\{x_n^{(1)}\}_{n\in\mathbb{N}} \subseteq \{x_n\}_{n\in\mathbb{N}}$ such that $|F(x_n^{(1)}) - F(\bar{x})| \geqslant \varepsilon$ for all $n \in \mathbb{N}$. However, thanks to the boundedness of $\{x_n^{(1)}\}_{n\in\mathbb{N}}$, Theorem 7.13, and the uniqueness of the weak limit point, it follows that there exists $\{x_n^{(2)}\}_{n\in\mathbb{N}} \subseteq \{x_n^{(1)}\}_{n\in\mathbb{N}}$ such that $x_n^{(2)} \xrightarrow{\omega} \bar{x}$. In particular, it follows that $F(x_n^{(2)}) \xrightarrow{n\to\infty} F(\bar{x})$, which contradicts the previously obtained inequality.

Analogously to the previous example, one can prove that if X is a separable normed vector space and $\{F_n\}_{n\in\mathbb{N}}$ is a bounded sequence in X' with a unique weak* limit point $\bar{F} \in X'$, then $F_n \xrightarrow{\omega *} \bar{F}$.

Example 7.17 Recall first that, given $p \in [1, +\infty)$, one defines

$$\ell_p(\mathbb{R}) := \left\{ (x_n)_{n\in\mathbb{N}} : \quad x_n \in \mathbb{R} \quad \forall\, n \in \mathbb{N} \quad \text{and} \quad \sum_{n\in\mathbb{N}} |x_n|^p < +\infty \right\},$$

which is endowed with the norm

$$\|(x_n)_{n\in\mathbb{N}}\|_{\ell_p(\mathbb{R})} := \left\{ \sum_{n\in\mathbb{N}} |x_n|^p \right\}^{1/p} \qquad \forall\, (x_n)_{n\in\mathbb{N}} \in \ell_p(\mathbb{R}),$$

and we prove next that $\ell_p(\mathbb{R})$ is separable. Indeed, we define the set

$$W := \left\{ (x_n)_{n\in\mathbb{N}} \in \ell_p(\mathbb{R}) : \quad x_n \in \mathbb{Q} \quad \forall\, n \in \mathbb{N} \quad \text{and} \quad \exists\, N \in \mathbb{N} \text{ such that } x_n = 0 \quad \forall\, n > N \right\}.$$

It is easy to prove, using the countability of $\mathbb{Q}$, and the fact that a countable union of countable sets is also countable, that W is countable. Now, given $x := (x_n)_{n\in\mathbb{N}} \in \ell_p(\mathbb{R})$ and $\varepsilon > 0$, there exists $N \in \mathbb{N}$ such that

$$\sum_{n=N+1}^{\infty} |x_n|^p < \frac{\varepsilon}{2}.$$

Moreover, for each $n \leqslant N$, there exists $\widetilde{x}_n \in \mathbb{Q}$ such that

$$|x_n - \widetilde{x}_n|^p < \frac{\varepsilon}{2N}.$$

Then, defining $\widetilde{x} := (\widetilde{x}_n)_{n\in\mathbb{N}}$ with $\widetilde{x}_n = 0$ for all $n \geqslant N+1$, we obtain $\widetilde{x} \in W$ and

$$\|x - \widetilde{x}\|^p_{\ell_p(\mathbb{R})} = \sum_{n=1}^{N} |x_n - \widetilde{x}_n|^p + \sum_{n=N+1}^{\infty} |x_n|^p < \frac{\varepsilon}{2} + \frac{\varepsilon}{2} = \varepsilon,$$

which shows that W is dense in $\ell_p(\mathbb{R})$, and therefore $\ell_p(\mathbb{R})$ is separable.

In connection with the previous example, it can also be proved that $\ell_1'(\mathbb{R}) = \ell_\infty(\mathbb{R})$ and that $\ell_\infty(\mathbb{R})$ is not separable, where

$$\ell_\infty(\mathbb{R}) := \left\{ (x_n)_{n\in\mathbb{N}} : \quad x_n \in \mathbb{R} \quad \forall\, n \in \mathbb{N} \quad \text{and} \quad \sup_{n\in\mathbb{N}} |x_n| < +\infty \right\},$$

provided with the norm

$$\|(x_n)_{n\in\mathbb{N}}\|_{\ell_\infty(\mathbb{R})} := \sup_{n\in\mathbb{N}} |x_n| \qquad \forall\, (x_n)_{n\in\mathbb{N}} \in \ell_\infty(\mathbb{R}).$$

In particular, to see the non-separability of $\ell_\infty(\mathbb{R})$, it suffices to consider an arbitrary sequence $\{a^{(n)}\}_{n\in\mathbb{N}} \subseteq \ell_\infty(\mathbb{R})$ with $a^{(n)} := (a_k^n)_{k\in\mathbb{N}}$, and then, given $r \in \mathbb{R}$, $r \neq 0$, define $b := (b_k)_{k\in\mathbb{N}} \in \ell_\infty(\mathbb{R})$ by

$$b_k := \begin{cases} a_k^{(k)} + r & \text{if } |a_k^{(k)}| < |r| \\ 0 & \text{if } |a_k^{(k)}| \geqslant |r| \end{cases} \qquad \forall\, k \in \mathbb{N}.$$

It follows that for each $n \in \mathbb{N}$

$$\|b - a^{(n)}\|_{\ell_\infty(\mathbb{R})} := \sup_{k\in\mathbb{N}} |b_k - a_k^n| \geqslant |b_n - a_n^n| \geqslant |r|,$$

which shows that $\{a^{(n)}\}_{n\in\mathbb{N}}$ cannot be dense in $\ell_\infty(\mathbb{R})$.

According to the above, and in view of Lemma 7.9 and Example 7.17, it follows that $\ell_1(\mathbb{R})$ is not reflexive. Another way to see the same result is through the following example and the fact that weak convergence cannot be equivalent to strong convergence in an infinite-dimensional reflexive Banach space.

Example 7.18 Let us again consider the space $\ell_1(\mathbb{R})$ (see definition in Example 7.17), and prove that weak and strong convergence are in fact equivalent here. To this end, we use that $f \in \ell_1'(\mathbb{R})$ if and only if there exists $z \in \ell_\infty(\mathbb{R})$ such that $f(x) = \sum_{k=1}^{\infty} z_k\, x_k$ for all $x := (x_k)_{k\in\mathbb{N}} \in \ell_1(\mathbb{R})$. Then, if we assume that the aforementioned convergences are not equivalent, there exists a sequence $\{x^{(n)}\}_{n\in\mathbb{N}} \subseteq \ell_1(\mathbb{R})$, with $x^{(n)} := (x_k^{(n)})_{k\in\mathbb{N}}$, that converges weakly to $x = 0$ but does not converge strongly to $x = 0$. This means that

$$\sum_{k=1}^{\infty} z_k\, x_k^{(n)} \overset{n\to\infty}{\longrightarrow} 0 \quad \forall\, z := (z_k)_{k\in\mathbb{N}} \in \ell_\infty(\mathbb{R}), \tag{7.5}$$

and that there exists $\varepsilon > 0$ such that

$$\|x^{(n)}\| := \sum_{k=1}^{\infty} |x_k^{(n)}| \geqslant \varepsilon \quad \forall n \in \mathbb{N}. \tag{7.6}$$

It follows from (7.5), taking for each $m \in \mathbb{N}$ the element $z = (\delta_{mk})_{k\in\mathbb{N}} \in \ell_\infty(\mathbb{R})$, that

$$x_m^{(n)} \overset{n\to\infty}{\longrightarrow} 0 \quad \forall\, m \in \mathbb{N}.$$

Now let $n_0 = 0$, $m_0 = 1$, and define the sequences $\{n_k\}_{k\in\mathbb{N}}$ and $\{m_k\}_{k\in\mathbb{N}}$ as follows:

$$n_k := \text{the smallest integer} > n_{k-1} \text{ such that } \sum_{m=1}^{m_{k-1}} |x_m^{(n_k)}| < \frac{\varepsilon}{5},$$

and

$$m_k := \text{the smallest integer} > m_{k-1} \text{ such that } \sum_{m=m_k}^{\infty} |x_m^{(n_k)}| < \frac{\varepsilon}{5}.$$

Note that n_k and m_k exist because $\sum_{m=1}^{m_{k-1}} |x_m^{(n)}| \overset{n\to\infty}{\longrightarrow} 0$ and $\|x^{(n_k)}\| := \sum_{m=1}^{\infty} |x_m^{(n_k)}| < \infty$, respectively. Now let $z := (z_m)_{m\in\mathbb{N}} \in \ell_\infty(\mathbb{R})$ be defined by

$$z_m := \operatorname{sign}\big(x_m^{(n_k)}\big) \quad \text{if} \quad m_{k-1} < m \leqslant m_k,$$

where $\operatorname{sign}(\alpha) := \begin{cases} 0 & \text{if} \quad \alpha = 0 \\ \dfrac{\alpha}{|\alpha|} & \text{if} \quad \alpha \neq 0. \end{cases}$ We thus have

$$\left|\sum_{m=1}^{\infty} \Big\{z_m\, x_m^{(n_k)} - |x_m^{(n_k)}|\Big\}\right| = \left|\sum_{m=1}^{m_{k-1}} \Big\{z_m\, x_m^{(n_k)} - |x_m^{(n_k)}|\Big\} + \sum_{m=m_k}^{\infty} \Big\{z_m\, x_m^{(n_k)} - |x_m^{(n_k)}|\Big\}\right|$$

since $z_m\, x_m^{(n_k)} - |x_m^{(n_k)}| = 0$ for $m \in]m_{k-1}, m_k]$. It follows that

$$\left|\sum_{m=1}^{\infty} \Big\{z_m\, x_m^{(n_k)} - |x_m^{(n_k)}|\Big\}\right| \leqslant 2 \sum_{m=1}^{m_{k-1}} |x_m^{(n_k)}| + 2 \sum_{m=m_k}^{\infty} |x_m^{(n_k)}| \leqslant \frac{4\varepsilon}{5}.$$

In turn, using from (7.6) that $\|x^{(n)}\| \geqslant \varepsilon$ for all $n \in \mathbb{N}$, we obtain that

$$\left|\sum_{m=1}^{\infty} z_m\, x_m^{(n_k)}\right| = \left|\sum_{m=1}^{\infty} \Big\{z_m\, x_m^{(n_k)} - |x_m^{(n_k)}|\Big\} + \|x^{(n_k)}\|\right| \geqslant \|x^{(n_k)}\| - \frac{4\varepsilon}{5} \geqslant \frac{\varepsilon}{5} \quad \forall k \in \mathbb{N},$$

which contradicts (7.5).

Example 7.19 Let us prove here that $C([a,b])$, endowed with the uniform norm (cf. (2.27))

$$\|u\| := \max_{t\in[a,b]} |u(t)| \qquad \forall\, u \in C([a,b])\,, \tag{7.7}$$

is separable. Indeed, let us define the set

$$W := \Big\{x \in C([a,b]) : \quad \exists\, a = t_0 < t_1 < \cdots < t_n = b\,, \text{ with } t_j \in \mathbb{Q} \quad \forall\, j \in \{1,\cdots,n-1\}$$

$$\text{and } x(t) := s_k + \Big(\frac{s_{k+1}-s_k}{t_{k+1}-t_k}\Big)(t-t_k) \quad \forall\, t \in [t_k,t_{k+1}]\,, \text{ with } s_k \in \mathbb{Q} \quad \forall\, k \in \{0,1,\cdots,n\}\Big\}.$$

It is easy to see, thanks to the countability of $\mathbb{Q}$, that W is countable. Now, given $u \in C([a,b])$ and $\varepsilon > 0$, there exists $\delta > 0$ such that

$$|u(t) - u(s)| < \frac{\varepsilon}{4} \quad \forall\, |t-s| < \delta\,.$$

Moreover, let $a = t_0 < t_1 < \cdots < t_n = b$ with $t_j \in \mathbb{Q}$ for all $j \in \{1,\ldots,n-1\}$, such that

$$\max_{j\in\{1,\cdots,n\}} |t_j - t_{j-1}| < \delta,$$

and let $s_0, s_1, \cdots, s_n \in \mathbb{Q}$ be such that

$$|u(t_k) - s_k| < \frac{\varepsilon}{4} \quad \forall\, k \in \{0,1,\cdots,n\}\,.$$

Then, letting x as in the definition of W, we have for each $t \in [t_k, t_{k+1}]$

$$\begin{aligned} x(t) - u(t) &= s_k + \Big(\frac{s_{k+1}-s_k}{t_{k+1}-t_k}\Big)(t-t_k) - \Big(\frac{t_{k+1}-t_k}{t_{k+1}-t_k}\Big)\,u(t) \\ &= \Big(\frac{t_{k+1}-t}{t_{k+1}-t_k}\Big)(s_k - u(t)) + \Big(\frac{t-t_k}{t_{k+1}-t_k}\Big)(s_{k+1} - u(t))\,, \end{aligned}$$

from which it follows that

$$\begin{aligned} |x(t) - u(t)| &\leqslant |s_k - u(t)| + |s_{k+1} - u(t)| \\ &\leqslant |s_k - u(t_k)| + |u(t_k) - u(t)| + |s_{k+1} - u(t_{k+1})| + |u(t_{k+1}) - u(t)| \\ &\leqslant \frac{\varepsilon}{4} + \frac{\varepsilon}{4} + \frac{\varepsilon}{4} + \frac{\varepsilon}{4} = \varepsilon\,. \end{aligned}$$

This proves that $\|x-u\| \leqslant \varepsilon$ and hence W is dense in $C([a,b])$.

7.3 Saturated subspaces and total subsets

Let us first recall that if M is a closed subspace of a normed vector space X, then (cf. Theorem 2.26) for each $x \in X - M$ there exists $F \in M^\circ$ such that $F(x) = \operatorname{dist}(X, M) > 0$, which is equivalent to saying that ${}^\circ(M^\circ) \subseteq M$. Certainly, since the inclusion $M \subseteq {}^\circ(M^\circ)$ always occurs, we conclude that $M = {}^\circ(M^\circ)$ (as was proved in Lemma 3.24). In turn, given a subspace W of X', it is also clear that $W \subseteq ({}^\circ W)^\circ$, yet the reverse inclusion is not necessarily true, unless X is reflexive and W is closed (as it follows from Lemma 3.74). The above leads to the following definition.

Definition 7.20. *A subspace W of X' is said to be* SATURATED *if $({}^\circ W)^\circ \subseteq W$. Equivalently, if for every $F \in X' - W$ there exists $x \in {}^\circ W$ such that $F(x) \neq 0$, or simply if the identity $W = ({}^\circ W)^\circ$ holds.*

In particular, if X is reflexive, Lemma 3.74 guarantees that every closed subspace of X' is saturated. In fact, we will later prove (cf. Theorem 7.24) that this property characterizes reflexive spaces.

The following result connects saturated subspaces with the concept of weak* topology (see more details in Section 7.4).

Lemma 7.21. *Every saturated subspace of X' is weak*-closed.*

Proof. Let W be a saturated subspace of X', that is such that $W = ({}^\circ W)^\circ$, and let $\{F_n\}_{n\in\mathbb{N}} \subseteq W$ be such that $F_n \overset{\omega^*}{\to} F \in X'$, that is, $F_n(x) \overset{n\to\infty}{\to} F(x)$ for all $x \in X$. If we assume by contradiction that $F \notin W$, it follows that there exists $x_0 \in {}^\circ W$ such that $F(x_0) \neq 0$ (cf. Definition 7.20), and yet

$$F(x_0) = \lim_{n\to\infty} F_n(x_0) = \lim_{n\to\infty} 0 = 0\,.$$

Thus, necessarily $F \in W$, and therefore W is weak*-closed.

□

On the other hand, the announced characterization of reflexive Banach spaces (cf. Theorem 7.24) and some of its consequences require a couple of preliminary lemmas.

Lemma 7.22. *Every finite-dimensional subspace of X' is saturated.*

Proof. Let $\{F_1,\ldots,F_n\}$ be a basis of a subspace W of X', and consider $F \notin W$. Since the set $\{F_1,\ldots,F_n,F\}$ is clearly linearly independent, a direct application of part ii) of Lemma 6.21 implies the existence of $x_0 \in X$ such that $F_j(x_0) = 0$ for all $j \in \{1,\ldots,n\}$ and $F(x_0) \neq 0$. It follows that $x_0 \in {}^\circ W$, and therefore W is saturated.

□

Lemma 7.23. *Given a normed vector space X, $F \in X'$ and $x_0 \in X$ such that $F(x_0) \neq 0$, define the subspaces $U := {}^\circ\langle\{F\}\rangle$ and $V := \langle\{x_0\}\rangle$. Then $X = U \oplus V$.*

Proof. It is clear that $U \cap V = \{0\}$ since if $x = \alpha\, x_0$, with $\alpha \in \mathbb{R}$, and $0 = F(x) = \alpha\, F(x_0)$, necessarily $\alpha = 0$ and hence $x = 0$. Moreover, for any $x \in X$, it is easy to see that, defining $\alpha := \frac{F(x)}{F(x_0)}$, one has

$$x = (x - \alpha\, x_0) + \alpha\, x_0 \in U + V,$$

which completes the proof.

□

We are now in a position to establish the following result.

Theorem 7.24. *A Banach space X is reflexive if and only if every closed subspace of X' is saturated.*

Proof. It remains only to verify the converse implication, that is, suppose that all closed subspaces of X' are saturated, and prove that X is reflexive. Indeed, let $\mathcal{F} \in X''$, $\mathcal{F} \neq 0$, and let us show that there exists $\bar{x} \in X$ such that $\mathcal{F} = \mathcal{J}(\bar{x})$, where $\mathcal{J} : X \to X''$ is the injective and isometric mapping defined in Section 7.1. For this purpose, first define

$$W := \Big\{F \in X' : \quad \mathcal{F}(F) = 0\Big\} = \mathcal{F}^{-1}(\{0\}),$$

and note that W is closed and therefore saturated. Moreover, since $\mathcal{F} \neq 0$, W is a proper subspace of X', so we can consider a functional $G \in X' - W$. Thus, since W is saturated, there exists $x_1 \in {}^\circ W$ such that $G(x_1) \neq 0$. In turn, applying Lemma 7.23 to the context

$$X \longleftarrow X', \quad F \longleftarrow \mathcal{F}, \quad x_0 \longleftarrow G,$$

it follows that $X' = U \oplus V$, where

$$U := W \quad \text{and} \quad V := \langle \{G\} \rangle.$$

Then, defining $\mathcal{H} \in X''$ by

$$\mathcal{H}(F) := G(x_1)\,\mathcal{F}(F) - \mathcal{F}(G)\,F(x_1) \quad \forall\, F \in X', \tag{7.8}$$

we obtain that

$$\mathcal{H}(F) = 0 \quad \forall\, F \in W \quad \text{and} \quad \mathcal{H}(G) = 0,$$

whence, by virtue of the above decomposition of X', $\mathcal{H}(F) = 0$ for all $F \in X'$. It follows from (7.8) that

$$\mathcal{F}(F) = \frac{\mathcal{F}(G)\,F(x_1)}{G(x_1)} \qquad \forall\, F \in X',$$

or equivalently, letting $\beta := \frac{\mathcal{F}(G)}{G(x_1)} \in \mathbb{R}$, we have

$$\mathcal{F}(F) = F(\beta\, x_1) = \mathcal{J}(\beta\, x_1)(F) \qquad \forall\, F \in X',$$

thus proving that $\mathcal{F} = \mathcal{J}(\beta\, x_1)$, and therefore the reflexivity of X.

□

Corollary 7.25. *Every finite-dimensional Banach space X is reflexive.*

Proof. Since $\dim X' = \dim X$, all subspaces of X' are finite-dimensional, and therefore, by Lemma 7.22, they are saturated. Hence, Theorem 7.24 confirms the reflexivity of X.

□

The following concept is needed in order to later prove another characterization of reflexive spaces.

Definition 7.26. *A subset W of X' is said to be* TOTAL *if ${}^\circ W = \{0\}$.*

In fact, the next goal is to show that a Banach space X is reflexive if and only if every total subspace of X' is dense in X'. The first implication of this equivalence follows from the two lemmas we prove below.

Lemma 7.27. *Let X be a Banach space, and let W be a subspace of X'. Then W is dense in X' if and only if*

$$W^\circ := \Big\{\mathcal{F} \in X'' : \quad \mathcal{F}(F) = 0 \quad \forall\, F \in W\Big\} = \{0\}.$$

Proof. Suppose that W is a dense subspace of X', that is $\overline{W} = X'$ and let $\mathcal{F} \in W^\circ$. Then, given $G \in X'$, there exists $\{G_n\}_{n\in\mathbb{N}} \subseteq W$ such that $\|G_n - G\|_{X'} \overset{n\to\infty}{\longrightarrow} 0$, and therefore

$$\mathcal{F}(G) = \lim_{n\to\infty} \mathcal{F}(G_n) = \lim_{n\to\infty} 0 = 0,$$

which proves that $\mathcal{F}$ is the zero functional in X''. Conversely, suppose that $W^\circ = \{0\}$ and let us prove that $\overline{W} = X'$. Assume by contradiction that there exists $G \in X'$ such that $G \notin \overline{W}$. Then, applying the corollary of the Hahn-Banach Theorem given by Theorem 2.26, it follows that there exists $\mathcal{F} \in X''$ such that $\|\mathcal{F}\| = 1$, $\mathcal{F}(G) = \text{dist}(G, \overline{W}) > 0$ and $\mathcal{F} \in \overline{W}^\circ = W^\circ$, which contradicts the fact that $W^\circ = \{0\}$.

□

Note that the previous lemma is valid not only for every subspace W of the dual of a Banach space, but also for any subspace W of an arbitrary Banach space, in which case the identity $\overline{W}^\circ = W^\circ$ employed at the end of the previous proof is always valid, as proved by Lemma 3.26.

Lemma 7.28. *Let X be a reflexive Banach space, and let $W \subseteq X'$. Then W is total if and only if $W^\circ = \{0\}$.*

Proof. Suppose that W is total and let $\widetilde{\mathcal{F}} \in W^\circ$, that is $\widetilde{\mathcal{F}} \in X''$ and $\widetilde{\mathcal{F}}(F) = 0$ for all $F \in W$. The reflexivity of X implies that there exists $\widetilde{x} \in X$ such that $\widetilde{\mathcal{F}} = \mathcal{J}(\widetilde{x})$. Then, for each $F \in W$ one has $F(\widetilde{x}) = \mathcal{J}(\widetilde{x})(F) = \widetilde{\mathcal{F}}(F) = 0$, which indicates that $\widetilde{x} \in {}^\circ W = \{0\}$. Thus $\widetilde{x} = 0$ and therefore $\widetilde{\mathcal{F}}$ is the zero functional of X''. Conversely, suppose that $W^\circ = \{0\}$ and let us prove that W is total, that is ${}^\circ W = \{0\}$. Indeed, given $x_0 \in {}^\circ W$, one has for each $F \in W$ that $\mathcal{J}(x_0)(F) = F(x_0) = 0$, which shows that $\mathcal{J}(x_0) \in W^\circ$, and hence the injectivity of $\mathcal{J}$ implies that $x_0 = 0$.

□

The announced equivalence can now be proved.

Theorem 7.29. *A Banach space X is reflexive if and only if every total subspace of X' is dense in X'.*

Proof. As already mentioned, the first implication follows from Lemmas 7.27 and 7.28. Conversely, assume that every total subspace of X' is dense in X', and suppose by contradiction that X is not reflexive. It follows that there exists $\widetilde{\mathcal{F}} \in X''$ that does not belong to $\mathcal{J}(X)$, with which we define

$$W := {}^\circ\langle\{\widetilde{\mathcal{F}}\}\rangle = \Big\{F \in X' : \quad \widetilde{\mathcal{F}}(F) = 0\Big\}.$$

Since $\widetilde{\mathcal{F}} \neq 0$, it is clear that W is not dense in X'. Nevertheless, we now show that W is total, which will yield the desired contradiction. To this end, note first that this property of W is equivalent to saying that for each $x \in X$, $x \neq 0$, there exists $F \in W$ such that $F(x) \neq 0$. Therefore, in order to prove this fact, we will consider $\widetilde{G} \in X'$ such that $\widetilde{\mathcal{F}}(\widetilde{G}) \neq 0$. Thus, according to Lemma 7.23, we have the decomposition

$$X' = W \oplus \langle\{\widetilde{G}\}\rangle. \tag{7.9}$$

Now let $x \in X$, $x \neq 0$. If $\widetilde{G}(x) = 0$, take $F \in X'$ such that $F(x) \neq 0$. Then, applying (7.9), there exist $\widetilde{F} \in W$ and $\alpha \in \mathbb{R}$ such that $F = \widetilde{F} + \alpha\,\widetilde{G}$, whence $\widetilde{F}(x) = F(x) \neq 0$. On the other hand, if $\widetilde{G}(x) \neq 0$, define $\beta := \frac{\widetilde{\mathcal{F}}(\widetilde{G})}{\widetilde{G}(x)} \neq 0$ and observe that $\widetilde{G}(\beta x) = \widetilde{\mathcal{F}}(\widetilde{G}) \neq 0$. Moreover, since $\widetilde{\mathcal{F}} \notin \mathcal{J}(X)$, there must exist $F \in X'$ such that $F(\beta\, x) \neq \widetilde{\mathcal{F}}(F)$, because otherwise we would have $\widetilde{\mathcal{F}} = \mathcal{J}(\beta x)$. Thus, using again (7.9), we have $F = \widetilde{F} + \alpha\widetilde{G}$ with $\widetilde{F} \in W$ and $\alpha := \frac{\widetilde{\mathcal{F}}(F)}{\widetilde{\mathcal{F}}(\widetilde{G})}$, from which it follows that

$$\widetilde{F}(\beta x) = F(\beta x) - \alpha\widetilde{G}(\beta x) = F(\beta x) - \alpha\widetilde{\mathcal{F}}(\widetilde{G}) = F(\beta x) - \widetilde{\mathcal{F}}(F) \neq 0,$$

and therefore, since $\beta \neq 0$, we have $\widetilde{F}(x) \neq 0$, thus confirming, in both cases regarding $\widetilde{G}(x)$, that W becomes total.

□

7.4 Weak topologies

In this section we discuss the main theoretical aspects that give rise to the concepts of weak and weak* convergence used in this and in some previous chapters (cf. Definitions 6.27 and 7.10).

7.4.1 Preliminaries

Let X be a set, let $\{(Y_i, \mathcal{T}_i)\}_{i\in I}$ be a family of topological spaces, and for each $i \in I$ let $\varphi_i : X \to Y_i$ be a mapping. Our goal is to endow X with a topology $\mathcal{T}$ that makes each φ_i continuous. Moreover, we seek such a topology $\mathcal{T}$ with the smallest possible collection of open sets, that is, one that is the "*most economical*" among all topologies having this property.In this sense, the "*least economical*" topology is the so-called discrete one, in which every subset of X is open. Furthermore, if all sets of the form $\varphi_i^{-1}(A_i)$, with $i \in I$ and A_i open in Y_i, are collected into the family $\{B_\lambda\}_{\lambda\in\Lambda}$, the topology $\mathcal{T}$ we are looking for is the most economical one such that each B_λ is open for all $\lambda \in \Lambda$. In other words, we want to construct a family of open sets $\mathcal{F}$ "*with the smallest possible number*" of elements, such that $B_\lambda \in \mathcal{F}$ for all $\lambda \in \Lambda$, and in such a way that finite intersections and arbitrary unions of members of $\mathcal{F}$ remain in $\mathcal{F}$.

If one constructs $\mathcal{F}$ by first taking arbitrary unions of the elements B_λ, $\lambda \in \Lambda$, and then finite intersections of the resulting sets, then $\mathcal{F}$ is closed only under finite intersections. However, if this procedure is carried out in the reverse order, it can be proved that $\mathcal{F}$ is stable under both operations. This fact suggests to endow X with the topology $\mathcal{T}$ defined as follows.

Definition 7.30. *The open sets of $\mathcal{T}$ are obtained by first taking finite intersections of sets of the form $\varphi_i^{-1}(A_i)$, with $i \in I$ and A_i open in Y_i, and then taking arbitrary unions of the resulting sets.*

According to the above definition, a neighborhood base of $x \in X$ for the topology $\mathcal{T}$ is given by the sets of the form $\bigcap_{i\in\widetilde{I}} \varphi_i^{-1}(A_i)$, where $\widetilde{I}$ is a finite subset of I and each A_i is a neighborhood of $\varphi_i(x)$ in Y_i.

Our first result establishes the notion of convergence of sequences in $(X, \mathcal{T})$.

Lemma 7.31. *Let $\{x_n\}_{n\in\mathbb{N}}$ be a sequence in $(X, \mathcal{T})$, and let $x \in X$. Then $x_n \overset{n\to\infty}{\to} x$ if and only if $\varphi_i(x_n) \to \varphi_i(x)$ for all $i \in I$.*

Proof. Suppose that $\{x_n\}_{n\in\mathbb{N}}$ converges to $x \in X$ in the topology $\mathcal{T}$. Since each mapping φ_i is continuous from $(X,\mathcal{T})$ into $(Y_i,\mathcal{T}_i)$, one obviously has $\varphi_i(x_n) \overset{n\to\infty}{\to} \varphi_i(x)$ for all $i \in I$. Conversely, suppose that $\varphi_i(x_n) \longrightarrow \varphi_i(x)$ for all $i \in I$ and let V be a neighborhood of $\{x\}$. It follows that there exists a finite subset $\widetilde{I}$ of I and neighborhoods $A_i \in \mathcal{T}_i$ of $\varphi_i(x)$ for each $i \in \widetilde{I}$ such that $x \in \bigcap_{i\in\widetilde{I}} \varphi_i^{-1}(A_i) \subseteq V$. According to the convergence of $\{\varphi_i(x_n)\}_{n\in\mathbb{N}}$ to $\varphi_i(x)$, it follows that for each $i \in \widetilde{I}$ there exists $N_i \in \mathbb{N}$ such that $\varphi_i(x_n) \in A_i$ for all $n \geqslant N_i$. Then, denoting $N := \max\{N_i : \ i \in \widetilde{I}\}$, we have that $\varphi_i(x_n) \in \bigcap_{i\in\widetilde{I}} A_i$ for all $n \geqslant N$, whence

$$x_n \in \bigcap_{i\in\widetilde{I}} \varphi_i^{-1}(A_i) \subseteq V \quad \forall\, n \geqslant N\,.$$

This proves that $x_n \overset{n\to\infty}{\to} x$, which completes the proof.

□

The following result characterizes the continuity of functions whose range lies in $(X,\mathcal{T})$.

Lemma 7.32. *Let $(Z,\mathcal{T}_Z)$ be a topological space, and let $\psi : (Z,\mathcal{T}_Z) \to (X,\mathcal{T})$. Then ψ is continuous if and only if $\varphi_i \circ \psi : (Z,\mathcal{T}_Z) \to (Y_i,\mathcal{T}_i)$ is continuous for all $i \in I$.*

Proof. Since the composition of continuous functions is also continuous, it is clear that the continuity of ψ implies the continuity of $\varphi_i \circ \psi$ for all $i \in I$. Conversely, suppose that $\varphi_i \circ \psi$ is continuous for each $i \in I$, and let V be an open subset of X. We want to prove that $\psi^{-1}(V)$ is open in Z. Indeed, we know from Definition 7.30 that there exist an index set S and finite sets $\widetilde{I}_s \subseteq I$, for all $s \in S$, such that $V = \bigcup_{s\in S} \bigcap_{i\in\widetilde{I}_s} \varphi_i^{-1}(A_i)$ with each A_i open in Y_i. Then, since the continuity of $\varphi_i \circ \psi$ guarantees that $(\varphi_i \circ \psi)^{-1}(A_i) \equiv \psi^{-1} \circ \varphi_i^{-1}(A_i)$ is open in Z, it follows that

$$\psi^{-1}(V) = \bigcup_{s\in S} \bigcap_{i\in\widetilde{I}_s} \psi^{-1}(\varphi_i^{-1}(A_i)) = \bigcup_{s\in S} \bigcap_{i\in\widetilde{I}_s} (\varphi_i \circ \psi)^{-1}(A_i)$$

also belongs to $\mathcal{T}_Z$, thus proving that ψ is continuous.

□

7.4.2 The weak topology $\sigma(X, X')$

Let X be a Banach space, and consider the index set $I := X'$ together with the family $\{\varphi_f\}_{f\in I} \equiv \{f\}_{f\in X'}$. Then, the topology $\mathcal{T}$ resulting from the construction in Section 7.4.1, applied to the setting

$$X \longleftarrow X, \quad I \longleftarrow X', \quad Y_i \longleftarrow \mathbb{R}\,,$$

is called the weak topology of X and is denoted by $\sigma(X,X')$. Equivalently, we have the following definition.

Definition 7.33. *Given a Banach space X, the weak topology $\sigma(X,X')$ on X is the "most economical" topology that makes all elements of the family $\{\varphi_f\}_{f\in X'}$ continuous.*

Next, we prove a series of important results about the weak topology. We begin with its separability (in the sense of Hausdorff).

Lemma 7.34. *$\sigma(X,X')$ is separable.*

Proof. Let x_1, $x_2 \in X$, $x_1 \neq x_2$, and let us prove that there exist disjoint open sets V_1, $V_2 \in \sigma(X,X')$ such that $x_1 \in V_1$ and $x_2 \in V_2$. Indeed, by applying the second geometric form of the Hahn-Banach Theorem (cf. Theorem 2.33) to $A := \{x_1\}$ and $B := \{x_2\}$, it follows that there exist $f \in X'$ and $\alpha \in \mathbb{R}$ such that $f(x_1) < \alpha < f(x_2)$. Then it is clear that the sets

$$V_1 := \Big\{x \in X : \quad f(x) < \alpha\Big\} = \varphi_f^{-1}(]-\infty,\alpha[)$$

and

$$V_2 := \Big\{x \in X : \quad f(x) > \alpha\Big\} = \varphi_f^{-1}(]\alpha,+\infty[)$$

are open in $\sigma(X,X')$ and satisfy the required condition.

□

The following lemma provides local neighborhood bases for $\sigma(X,X')$.

Lemma 7.35. *Given $x_0 \in X$, $\varepsilon > 0$ and a finite set $\mathcal{M} \subseteq X'$, define*

$$V(x_0,\varepsilon;\mathcal{M}) := \Big\{x \in X : \quad |f(x-x_0)| < \varepsilon \quad \forall\, f \in \mathcal{M}\Big\}. \tag{7.10}$$

Then, all sets of the form (7.10), *with ε and $\mathcal{M}$ variable, constitute a neighborhood base of $\{x_0\}$ for the topology $\sigma(X,X')$.*

Proof. First note that every set $V(x_0,\varepsilon;\mathcal{M})$ is open in $\sigma(X,X')$, since it can be rewritten as $\bigcap_{f\in\mathcal{M}} \varphi_f^{-1}(]f(x_0)-\varepsilon, f(x_0)+\varepsilon[)$, and clearly the interval $]f(x_0)-\varepsilon, f(x_0)+\varepsilon[$ is open in $Y_i = \mathbb{R}$. Now, let U be a neighborhood of x_0 in $\sigma(X,X')$. We know, according to the discussion in Section 7.4.1 (just after Definition 7.30), that there exist a finite set $\mathcal{M} \subseteq X'$ and neighborhoods A_f of $f(x_0)$ in $Y_i = \mathbb{R}$, for each $f \in \mathcal{M}$, such that

$$x_0 \in \bigcap_{f\in\mathcal{M}} \varphi_f^{-1}(A_f) \subseteq U.$$

Thus, for each $f \in \mathcal{M}$ there exists $\varepsilon_f > 0$ such that $]f(x_0)-\varepsilon_f, f(x_0)+\varepsilon_f[\subseteq A_f$, and then, defining $\varepsilon := \min\{\varepsilon_f : f \in \mathcal{M}\}$, we obtain

$$x_0 \in V(x_0,\varepsilon;\mathcal{M}) := \bigcap_{f\in\mathcal{M}} \varphi_f^{-1}(]f(x_0)-\varepsilon, f(x_0)+\varepsilon[) \subseteq \bigcap_{f\in\mathcal{M}} \varphi_f^{-1}(A_f) \subseteq U,$$

which completes the proof.

□

In what follows, the convergence of $\{x_n\}_{n\in\mathbb{N}} \subseteq X$ to $x \in X$ with respect to the topology $\sigma(X,X')$ is called WEAK CONVERGENCE (cf. Definition 6.27) and is denoted by $x_n \overset{\omega}{\to} x$ (or equivalently, $x_n \overset{n\to\infty}{\rightharpoonup} x$). In turn, the convergence with respect to the norm topology of the space X is called STRONG CONVERGENCE and is denoted in the usual way, that is $x_n \overset{n\to\infty}{\to} x$. The following theorem establishes several properties and consequences of both types of convergence.

Theorem 7.36. *Let $\{x_n\}_{n\in\mathbb{N}}$ be a sequence in X. Then the following implications hold:*

i) $x_n \overset{\omega}{\to} x \Leftrightarrow f(x_n) \overset{n\to\infty}{\to} f(x)$ *for all* $f \in X'$.
ii) $x_n \overset{n\to\infty}{\to} x \Rightarrow x_n \overset{\omega}{\to} x$.
iii) $x_n \overset{\omega}{\to} x \Rightarrow \{x_n\}_{n\in\mathbb{N}}$ *is bounded and* $\|x\| \leqslant \liminf_{n\to\infty} \|x_n\|$.
iv) $\{x_n \overset{\omega}{\to} x$ *and* $f_n \to f\} \Rightarrow f_n(x_n) \overset{n\to\infty}{\to} f(x)$.

Proof.

i) This follows directly from the definition of $\sigma(X, X')$ and from Lemma 7.31.
ii) If $\{x_n\}_{n\in\mathbb{N}}$ converges strongly to $x \in X$, the continuity of each $f \in X'$ implies that $f(x_n) \overset{n\to\infty}{\to} f(x)$, which means, by virtue of i), that $x_n \overset{\omega}{\to} x$.
iii) For the first statement we rely, although in a more concise way, on the corresponding proof given at the beginning of Section 6.5 (just after Definition 6.27). Indeed, suppose that $x_n \overset{\omega}{\to} x$, that is, $f(x_n) \overset{n\to\infty}{\to} f(x)$ for all $f \in X'$. It follows that for each $f \in X'$ the set $\{f(x_n)\}_{n\in\mathbb{N}}$ is bounded in $\mathbb{R}$. Then, applying the corollary of the Banach–Steinhaus Theorem given by Lemma 3.48 to the setting

$$X \longleftarrow X, \quad S \longleftarrow \{x_n\}_{n\in\mathbb{N}},$$

we deduce that $\{x_n\}_{n\in\mathbb{N}}$ is bounded. Now, given $f \in X'$, one clearly has $|f(x_n)| \leqslant \|f\| \, \|x_n\|$ for all $n \in \mathbb{N}$, whence, taking the lim inf as $n \to \infty$, it follows that

$$|f(x)| \leqslant \|f\| \liminf_{n\to\infty} \|x_n\|.$$

Finally, applying the consequence of the analytic Hahn-Banach Theorem given by Lemma 2.25, we conclude that

$$\|x\| = \sup_{\substack{f\in X' \\ f\neq 0}} \frac{|f(x)|}{\|f\|} \leqslant \liminf_{n\to\infty} \|x_n\|.$$

iv) Suppose that $x_n \overset{\omega}{\to} x$ and $f_n \longrightarrow f$. Then, using iii) and i), we deduce that

$$\begin{aligned} |f_n(x_n) - f(x)| &\leqslant |(f_n - f)(x_n)| + |f(x_n) - f(x)| \\ &\leqslant \|f_n - f\| \, \|x_n\| + |f(x_n) - f(x)| \\ &\leqslant C \, \|f_n - f\| + |f(x_n) - f(x)| \overset{n\to\infty}{\to} 0, \end{aligned}$$

which completes the proof.

□

The following result complements what was proved in Lemma 7.15 in that it establishes the full coincidence (not only for sequence convergence) between the weak topology and the usual topology when the space X is finite-dimensional.

Theorem 7.37. *Let X be a finite-dimensional normed vector space. Then the weak topology $\sigma(X, X')$ and the topology of X coincide. In particular $x_n \overset{\omega}{\to} x$ if and only if $x_n \overset{n\to\infty}{\to} x$.*

Proof. Note first that in any space X (regardless of its dimension) every open set of $\sigma(X, X')$ is also a strong open set of X. Indeed, since every $f \in X'$ is continuous with respect to the usual topology of X, it is clear that $\varphi_f^{-1}(A_i) \equiv f^{-1}(A_i)$ is open in the strong topology of X for every open set A_i of $\mathbb{R}$. Conversely, let U be a strong open subset of X, and let us prove that $U \in \sigma(X, X')$. To this end, it suffices to consider $x_0 \in U$ and construct a weak neighborhood of x_0 contained in U. Equivalently, according to Lemma 7.35, we must find a finite set $\mathcal{M} \subseteq X'$ and $\varepsilon > 0$ such that

$$V(x_0, \varepsilon; \mathcal{M}) := \Big\{ x \in X : \quad |f(x - x_0)| < \varepsilon \quad \forall f \in \mathcal{M} \Big\} \subseteq U.$$

Indeed, let $r > 0$ be such that $B(x_0, r) \subseteq U$, and let $\{x_1, \ldots, x_N\}$ be a basis of X with $\|x_j\| = 1$ for all $j \in \{1, \ldots, N\}$. Thus, given $x \in X$, there exist unique scalars $\alpha_1, \ldots, \alpha_N$ such that $x = \sum_{j=1}^{N} \alpha_j x_j$, which induces the definition of $f_i : X \to \mathbb{R}$ by

$$f_i(x) := \alpha_i \qquad \forall x \in X, \quad \forall i \in \{1, \ldots, N\}.$$

It is easy to see that $f_i \in X'$ for all $i \in \{1, \ldots, N\}$, and therefore we consider the finite set $\mathcal{M} := \{f_1, \ldots, f_N\}$ and introduce, with $\varepsilon > 0$ to be determined, the neighborhood $V(x_0, \varepsilon; \mathcal{M})$. In this way, for each $x \in V(x_0, \varepsilon; \mathcal{M})$ we have

$$\|x - x_0\| = \left\| \sum_{i=1}^{N} f_i(x - x_0)\, x_j \right\| \leqslant \sum_{i=1}^{N} \big| f_i(x - x_0) \big| \, \|x_j\| = \sum_{i=1}^{N} \big| f_i(x - x_0) \big| < N \varepsilon,$$

from which, choosing $\varepsilon = \frac{r}{N}$, it follows that $x \in B(x_0, r)$, and hence $V(x_0, \varepsilon; \mathcal{M}) \subseteq B(x_0, r)$, thus proving that $V(x_0, \varepsilon; \mathcal{M}) \subseteq U$.

□

It is important to observe here that when X is infinite-dimensional, the weak topology $\sigma(X, X')$ contains "*fewer open sets*" than the strong topology. In other words, there exist strong open (resp. closed) sets that are not weakly open (resp. weakly closed). The following two examples illustrate these facts.

Example 7.38 In what follows we illustrate the case of a strong closed set that is not weakly closed. Indeed, let $S := \{x \in X : \quad \|x\| = 1\}$, with X infinite-dimensional, and let us show that $\bar{S}^{\sigma(X,X')}$ (the closure of S with respect to $\sigma(X, X')$) is given by $\{x \in X : \quad \|x\| \leqslant 1\}$. To this end, let $x_0 \in X$ be such that $\|x_0\| < 1$, and let us prove that $x_0 \in \bar{S}^{\sigma(X,X')}$, which is equivalent to showing that for every neighborhood U of x_0 in $\sigma(X, X')$ one has $U \cap S \neq \emptyset$. Without loss of generality, by virtue of Lemma 7.35, we can consider $U = V(x_0, \varepsilon; \mathcal{M})$ with $\mathcal{M}$ a finite subset of X'. We claim that there exists $y_0 \in X$, $y_0 \neq 0$, such that $f(y_0) = 0$ for all $f \in \mathcal{M}$, since otherwise the mapping $\psi : X \to \mathbb{R}^N$, with $N := \mathrm{card}(\mathcal{M})$, defined by $\psi(x) := \{f(x)\}_{f \in \mathcal{M}}$ for all $x \in X$, would be injective, whence $\dim X = \dim \psi(X) \leqslant N$, which contradicts the hypothesis on X. Then we define

$$g : [0, +\infty) \to [0, +\infty) \quad \text{by} \quad g(t) := \|x_0 + t\, y_0\| \quad \forall t \geqslant 0.$$

It is clear that g is continuous and $g(0) = \|x_0\| < 1$. Moreover, since $\|x_0 + t\, y_0\| \geqslant t\, \|y_0\| - \|x_0\|$, it follows easily that $\lim_{t \to +\infty} g(t) = +\infty$. Thus, there exists $t_0 > 0$ such that $g(t_0) := \|x_0 + t_0\, y_0\| = 1$, that is, $x_0 + t_0\, y_0 \in S$. Furthermore, for each $f \in \mathcal{M}$ we have

$$|f(x_0 + t_0\, y_0 - x_0)| = t_0\, |f(y_0)| = 0,$$

which shows that $x_0 + t_0\, y_0 \in V(x_0, \varepsilon; \mathcal{M})$, and consequently $V(x_0, \varepsilon; \mathcal{M}) \cap S \neq \emptyset$. We have thus proved that

$$\{x \in X : \quad \|x\| \leqslant 1\} \subseteq \bar{S}^{\sigma(X,X')},$$

and the announced equality follows from Lemma 7.40 below, which states that weakly closed and strongly closed concepts coincide for a convex set.

It is important to note from the development of the previous example that if X is an infinite-dimensional normed vector space, then every weak neighborhood V of $x_0 \in X$ contains a line passing through x_0. Indeed, following the same notation as in Example 7.38, we obtain that

$$x_0 + t\, y_0 \in V(x_0, \varepsilon; \mathcal{M}) \quad \forall\, t \in \mathbb{R}\,.$$

In fact, this observation motivates the following example.

Example 7.39 Here we show the case of a strong open set that is not weakly open. Indeed, given a Banach space X of infinite dimension, consider $U := \big\{x \in X : \quad \|x\| < 1\big\}$ and let us prove that the weak interior of U is $\varnothing$. Suppose by contradiction that there exist $x_0 \in U$ and a weak neighborhood $V(x_0, \varepsilon; \mathcal{M})$ of x_0, with $\varepsilon > 0$ and $\mathcal{M}$ a finite subset of X', such that $V(x_0, \varepsilon; \mathcal{M}) \subseteq U$. It follows from the previous comment that there exists $y_0 \in X$, $y_0 \neq 0$, such that the line $\big\{x_0 + t\, y_0 : \quad t \in \mathbb{R}\big\} \subseteq V(x_0, \varepsilon; \mathcal{M}) \subseteq U$, which is a contradiction since $\|x_0 + t\, y_0\| \overset{t \to +\infty}{\longrightarrow} +\infty$. Thus, necessarily $\mathrm{int}(U)^{\boldsymbol{\sigma}(X,X')} = \varnothing$ and therefore U is not weakly open.

Recall here, as illustrated through Example 7.14, that if X is infinite-dimensional, there generally exist sequences that converge weakly but do not converge in the strong topology. Moreover, according to what was observed just before that example, we know that in every infinite-dimensional reflexive Banach space there are situations of this kind. In other words, if weak and strong convergence are equivalent in a certain infinite-dimensional normed vector space X (see a particular case in Example 7.18), then necessarily X is not reflexive. On the other hand, since every weakly open set is also strongly open, it follows that every weakly closed set is obviously strongly closed. In turn, Example 7.38 shows that the converse is false when the space is infinite-dimensional. However, the converse is true, as shown in the following lemma, when the set in question is convex.

Lemma 7.40. *Let S be a convex subset of a Banach space X. Then S is weakly closed ($S^c \in \boldsymbol{\sigma}(X, X')$) if and only if S is strongly closed.*

Proof. It only remains to prove the reverse implication. Indeed, let S be a convex and strongly closed subset of X and let us prove that $S^c \in \boldsymbol{\sigma}(X, X')$. To that end, given $x_0 \in S^c$, we apply the second geometric form of the Hahn-Banach Theorem (cf. Theorem 2.33) to $A := S$ and $B := \big\{x_0\big\}$, and we deduce that there exist $f \in X'$ and $\alpha \in \mathbb{R}$, such that

$$f(x_0) < \alpha < f(x) \qquad \forall\, x \in S.$$

Then, defining the weak open set

$$V := f^{-1}\big(]-\infty, \alpha[\big)\,,$$

it follows that $x_0 \in V \subseteq S^c$, which shows that S^c is weakly open.

□

Next, we present an important consequence of the previous lemma.

Theorem 7.41 (Mazur's Theorem). *Let $\big\{x_n\big\}_{n\in\mathbb{N}}$ be a sequence in a Banach space X such that it converges weakly to $x \in X$. Then, there exists a sequence of convex combinations of elements of $\big\{x_n\big\}_{n\in\mathbb{N}}$ that converges strongly to x.*

Proof. It suffices to define the convex hull S of the elements of the sequence $\big\{x_n\big\}_{n\in\mathbb{N}}$ and then observe that $x \in \bar{S}^{\boldsymbol{\sigma}(X,X')} = \bar{S}$.

□

On the other hand, the following theorem establishes the equivalence of the continuity of operators with respect to the strong and weak topologies.

Theorem 7.42. *Let X and Y be Banach spaces, and let $A : X \to Y$ be a linear operator. Then $A \in \mathcal{L}(X,Y)$ if and only if $A : \big(X, \sigma(X,X')\big) \to \big(Y, \sigma(Y,Y')\big)$ is continuous.*

Proof. Suppose that $A \in \mathcal{L}(X,Y)$. According to Lemma 7.32, it suffices to prove that for each $g \in Y'$ the mapping $g \circ A : \big(X, \sigma(X,X')\big) \to \mathbb{R}$ is continuous. Indeed, since $g \circ A \in X'$, this follows from the fact that $\sigma(X,X')$ is the coarsest topology making $f : \big(X, \sigma(X,X')\big) \to \mathbb{R}$ continuous for every $f \in X'$. Conversely, suppose that $A : \big(X, \sigma(X,X')\big) \to \big(Y, \sigma(Y,Y')\big)$ is continuous. It follows that the operator $A : \big(X, \sigma(X,X')\big) \to \big(Y, \sigma(Y,Y')\big)$ is obviously closed with respect to these weak topologies, which means (see the discussion following Definition 3.32) that $\mathcal{G}(A)$ (the graph of A) is closed in $X \times Y$ with respect to the topology $\sigma(X \times Y, X' \times Y')$. Naturally, here we identify $(X \times Y)'$ with $X' \times Y'$. Then, since every weakly closed set is also strongly closed, it follows that $\mathcal{G}(A)$ is also closed in the strong topology of $X \times Y$, and therefore, the Closed Graph Theorem (cf. Theorem 3.35) and the fact that $X \times Y$ is Banach imply that $A \in \mathcal{L}(X,Y)$.

□

A similar argument to that of the second implication of the previous theorem allows us to prove that if $A : X \to \big(Y, \sigma(Y,Y')\big)$ is linear and continuous, then $A \in \mathcal{L}(X,Y)$ (cf. Exercise 7.11).

7.4.3 The weak* topology $\sigma(X', X)$

Let X be a Banach space, let X' be its dual, and let $X'' := (X')'$ be its bidual. We know from the analysis in Section 7.1 that the mapping $\mathcal{J} : X \to X''$ defined by

$$\mathcal{J}(x)(F) := F(x) \quad \forall\, F \in X', \quad \forall\, x \in X,$$

is a linear injective and isometric mapping, and therefore X can be identified with the subspace $\mathcal{J}(X)$ of X''. Recall also (cf. Definition 7.1) that X is said to be reflexive if $\mathcal{J}(X) = X''$. The objective of this section is to introduce and analyze a new topology on X', distinct from the strong topology of X' (induced by its norm $\|\cdot\|_{X'}$) and also different from the usual weak topology $\sigma(X', X'')$ on X' (as discussed in Section 7.4.2). More precisely, we define the weak* topology on X', denoted $\sigma(X', X)$, as the *coarsest* topology that makes all the mappings $\big\{\mathcal{J}(x)\big\}_{x \in X} \subseteq X''$ continuous. It follows, in particular, that

$$\mathcal{J}(x) : \big(X', \sigma(X', X)\big) \to \mathbb{R} \quad \text{is continuous} \quad \forall\, x \in X\,.$$

Moreover, since in general X (which is identified with $\mathcal{J}(X)$) is strictly contained in X'', it follows that $\sigma(X', X)$ is *coarser* than $\sigma(X', X'')$, which, in turn, has fewer open sets than the strong topology of X'. In this regard, it is important to mention here that if a topology has fewer open sets, then it will have more compact sets. Later we will prove, for instance, that the unit ball of X' is compact with respect to $\sigma(X', X)$. Meanwhile, we first prove some preliminary results about $\sigma(X', X)$. Analogously to Lemma 7.34, we begin with the separability of $\sigma(X', X)$ (in the sense of Hausdorff as well).

Lemma 7.43. *$\sigma(X', X)$ is separable.*

Proof. Let $f_1,\ f_2 \in X'$, $f_1 \neq f_2$, and let us prove that there exist disjoint open sets U_1, $U_2 \in \boldsymbol{\sigma}(X',X)$ such that $f_1 \in U_1$ and $f_2 \in U_2$. Indeed, let $\widetilde{x} \in X$ be such that $f_1(\widetilde{x}) \neq f_2(\widetilde{x})$, and assume, without loss of generality, that $f_1(\widetilde{x}) < f_2(\widetilde{x})$. Then, given $\alpha \in \mathbb{R}$ such that $f_1(\widetilde{x}) < \alpha < f_2(\widetilde{x})$, define

$$U_1 := \mathcal{J}(\widetilde{x})^{-1}(]-\infty,\alpha[) = \Big\{ f \in X' : \quad f(\widetilde{x}) < \alpha \Big\}$$

and

$$U_2 := \mathcal{J}(\widetilde{x})^{-1}(]\alpha,+\infty[) = \Big\{ f \in X' : \quad f(\widetilde{x}) > \alpha \Big\}.$$

It is clear that $U_1,\ U_2 \in \boldsymbol{\sigma}(X',X)$, $f_1 \in U_1$, $f_2 \in U_2$, and $U_1 \cap U_2 = \emptyset$, which completes the proof.

□

Similarly to Lemma 7.35, the following result—whose proof is left as an exercise (cf. Exercise 7.12)—provides local neighborhood bases for $\boldsymbol{\sigma}(X',X)$.

Lemma 7.44. *Given $f_0 \in X'$, $\varepsilon > 0$ and a finite set $\mathcal{M} \subseteq X$, define*

$$V(f_0,\varepsilon;\mathcal{M}) := \Big\{ f \in X' : \quad |f(x) - f_0(x)| < \varepsilon \quad \forall\, x \in \mathcal{M} \Big\},$$

or, equivalently,

$$V(f_0,\varepsilon;\mathcal{M}) := \Big\{ f \in X' : \quad |\mathcal{J}(x)(f - f_0)| < \varepsilon \quad \forall\, x \in \mathcal{M} \Big\}.$$

Then all sets $V(f_0,\varepsilon;\mathcal{M})$ with variable ε and $\mathcal{M}$ constitute a neighborhood base of $\{f_0\}$ for the topology $\boldsymbol{\sigma}(X',X)$.

In turn, the analogue of Theorem 7.36, whose proof is also left as an exercise (cf. Exercise 7.13), is stated below. In what follows, the convergence of sequences with respect to the topology $\boldsymbol{\sigma}(X',X)$ is denoted by $\overset{\omega^*}{\to}$ (cf. Definition 7.10).

Theorem 7.45. *Let $\{f_n\}_{n\in\mathbb{N}}$ be a sequence in X'. Then the following implications hold:*

i) $f_n \overset{\omega^*}{\to} f \Leftrightarrow \mathcal{J}(x)(f_n) = f_n(x) \to \mathcal{J}(x)(f) = f(x)$ *for all* $x \in X$.

ii) $f_n \overset{n\to\infty}{\to} f \Rightarrow f_n \overset{\omega^*}{\to} f$.

iii) $f_n \overset{\omega}{\to} f \Rightarrow f_n \overset{\omega^*}{\to} f$.

iv) $\Big\{ f_n \overset{\omega^*}{\to} f \;\; \text{and} \;\; x_n \overset{n\to\infty}{\to} x \Big\} \Rightarrow f_n(x_n) \overset{n\to\infty}{\longrightarrow} f(x)$.

Note here that statement i) refers precisely to the concept of weak* convergence given in Definition 7.10. In turn, observe that the first convergence in iii) is with respect to the topology $\boldsymbol{\sigma}(X',X'')$. It is also important to mention that the proof of iv) makes use of the fact that every weak*-convergent sequence in X' is bounded (see the remark just before Theorem 7.11). Finally, it is interesting to note that replacing $x_n \overset{n\to\infty}{\longrightarrow} x$ by $x_n \overset{\omega}{\longrightarrow} x$ in iv) makes the respective implication no longer necessarily true. In fact, this also fails if we simultaneously replace $f_n \overset{\omega^*}{\longrightarrow} f$ by $f_n \overset{\omega}{\longrightarrow} f$.

On the other hand, it is not difficult to prove that if X is finite-dimensional, then the three topologies on X'—namely, the strong topology, $\boldsymbol{\sigma}(X',X)$, and $\boldsymbol{\sigma}(X',X'')$—coincide. In particular, since in this case $\mathcal{J}$ is surjective, it follows easily that $\boldsymbol{\sigma}(X',X) = \boldsymbol{\sigma}(X',X'')$.

The next goal is to characterize the linear and continuous functionals on the topological vector space $(X', \boldsymbol{\sigma}(X',X))$ and the closed hyperplanes that it contains. For the first of these results, we require the lemma established below, which is proved using the second geometric form of the Hahn-Banach Theorem (cf. Theorem 2.33).

Lemma 7.46. *Let $\{\varphi, \varphi_1, \ldots, \varphi_n\}$ be a set of linear forms on the normed vector space X such that*

$$\Big\{x \in X : \quad \varphi_i(x) = 0 \quad \forall i \in \{1,\ldots,n\}\Big\} \subseteq \Big\{x \in X : \quad \varphi(x) = 0\Big\}.$$

Then, there exist scalars $\lambda_1, \ldots, \lambda_n \in \mathbb{R}$ such that $\varphi = \sum\limits_{i=1}^{n} \lambda_i \varphi_i$.

Proof. First note that if we adopt the notation introduced in Definition 3.21, temporarily ignoring the fact that the linear forms are not necessarily in X', then the hypothesis of the present lemma can be rewritten as

$$^{\circ}\Big\{\varphi_1, \varphi_2, \cdots, \varphi_n\Big\} \subseteq {}^{\circ}\{\varphi\}.$$

Now, let $F : X \to \mathbb{R}^{n+1}$ be the mapping defined by

$$F(u) := (\varphi_1(u), \ldots, \varphi_n(u), \varphi(u))^{\mathrm{t}} \quad \forall u \in X.$$

It follows from the hypothesis that the vector $x_0 := (0,\ldots,0,1)^{\mathrm{t}} \in \mathbb{R}^{n+1}$ does not belong to $R(F)$. Then, applying the second geometric form of the Hahn-Banach Theorem (cf. Theorem 2.33) to the context $E = \mathbb{R}^{n+1}$, $A = R(F)$, and $B = \{x_0\}$, where A and B are clearly closed and compact, respectively, we deduce that there exist a linear functional $f : \mathbb{R}^{n+1} \to \mathbb{R}$ and $\alpha \in \mathbb{R}$ such that the affine hyperplane $[f = \alpha]$ strictly separates A and B. More precisely, there exist scalars $\widetilde{\lambda}_1, \ldots, \widetilde{\lambda}_n, \widetilde{\lambda}_{n+1}$ such that

$$f(x) = \sum_{j=1}^{n+1} \widetilde{\lambda}_j \, x_j \quad \forall x := (x_1, \ldots, x_n, x_{n+1})^{\mathrm{t}} \in \mathbb{R}^{n+1}$$

and

$$f(x_0) < \alpha < f(F(u)) \quad \forall u \in X,$$

that is

$$\widetilde{\lambda}_{n+1} < \alpha < f(F(u)) \quad \forall u \in X. \tag{7.11}$$

From this inequality it follows necessarily that

$$f(F(u)) := \sum_{j=1}^{n} \widetilde{\lambda}_j \, \varphi_j(u) + \widetilde{\lambda}_{n+1} \, \varphi(u) = 0 \quad \forall u \in X. \tag{7.12}$$

Otherwise, if there exists $\bar{u} \in X$ such that $f(F(\bar{u})) \neq 0$, then defining $\widetilde{u} := \dfrac{\alpha \, \bar{u}}{f(F(\bar{u}))} \in X$, we obtain $f(F(\widetilde{u})) = \alpha$, which contradicts (7.11). Next, we deduce from (7.11) that there holds $\widetilde{\lambda}_{n+1} < \alpha < 0$, whence $\widetilde{\lambda}_{n+1} \neq 0$, and therefore (7.12) yields

$$\varphi(u) = -\sum_{j=1}^{n} \frac{\widetilde{\lambda}_j}{\widetilde{\lambda}_{n+1}}\, \varphi_j(u) \quad \forall\, u \in X,$$

from which, defining $\lambda_j := -\dfrac{\widetilde{\lambda}_j}{\widetilde{\lambda}_{n+1}}$ for all $j \in \{1, \ldots, n\}$, the proof is completed.

□

Now we are in a position to characterize the continuous linear forms on $(X', \boldsymbol{\sigma}(X', X))$. More precisely, the following result establishes that they are of the form $\mathcal{J}(x)$ (cf. Section 7.1) with $x \in X$, which constitutes a kind of representation theorem for such forms.

Theorem 7.47. *Let $\varphi : (X', \boldsymbol{\sigma}(X', X)) \to \mathbb{R}$ be a linear and continuous form. Then there exists $x_0 \in X$ such that $\varphi = \mathcal{J}(x_0)$, that is, $\varphi(F) = F(x_0)$ for all $F \in X'$.*

Proof. The aforementioned continuity of φ guarantees that there exists a neighborhood V in $\boldsymbol{\sigma}(X', X)$ of the zero functional in X' such that

$$|\varphi(F)| < 1 \quad \forall\, F \in V. \tag{7.13}$$

More precisely, by virtue of Lemma 7.44 we may assume that there exist $\varepsilon > 0$ and a finite set $\mathcal{M} \subseteq X$ such that

$$V := \Big\{ F \in X' : \quad |F(x)| < \varepsilon \quad \forall\, x \in \mathcal{M} \Big\}.$$

Now consider the finite family of linear forms on X' given by $\{\mathcal{J}(x)\}_{x \in \mathcal{M}}$, and let $F \in X'$ such that $\mathcal{J}(x)(F) := F(x) = 0$ for all $x \in \mathcal{M}$. It follows that $\mathcal{J}(x)(mF) = 0$ for all $x \in \mathcal{M}$, for all $m \in \mathbb{N}$, which proves that $mF \in V$ for all $m \in \mathbb{N}$, and therefore, according to (7.13), $|\varphi(mF)| < 1$, or equivalently $|\varphi(F)| < \frac{1}{m}$ for all $m \in \mathbb{N}$, whence necessarily $\varphi(F) = 0$. Thus, applying Lemma 7.46 to the forms $\{\mathcal{J}(x)\}_{x \in \mathcal{M}} \cup \{\varphi\}$ on X', it follows that there exist scalars $\{\lambda_x\}_{x \in \mathcal{M}} \subseteq \mathbb{R}$ such that $\varphi = \sum\limits_{x \in \mathcal{M}} \lambda_x\, \mathcal{J}(x)$. Hence, for each $F \in X'$ we have

$$\varphi(F) = \sum_{x \in \mathcal{M}} \lambda_x\, \mathcal{J}(x)(F) = \sum_{x \in \mathcal{M}} \lambda_x\, F(x) = F\Big(\sum_{x \in \mathcal{M}} \lambda_x\, x \Big),$$

that is, $\varphi(F) = \mathcal{J}(x_0)(F)$ with $x_0 := \sum\limits_{x \in \mathcal{M}} \lambda_x\, x$, which completes the proof.

□

Continuing with what was announced earlier, we now characterize the closed hyperplanes of $(X', \boldsymbol{\sigma}(X', X))$.

Lemma 7.48. *Let H be a closed hyperplane of X' with respect to the topology $\boldsymbol{\sigma}(X', X)$. Then there exist $x_0 \in X$, $x_0 \neq 0$ and $\alpha \in \mathbb{R}$, such that*

$$H = \Big\{ F \in X' : \quad F(x_0) = \alpha \Big\}.$$

Proof. From the original definition of hyperplane (cf. Definition 2.8) we know that there exist a nonzero linear form $\varphi : X' \to \mathbb{R}$ and $\alpha \in \mathbb{R}$ such that

$$H = \Big\{ F \in X' : \quad \varphi(F) = \alpha \Big\}.$$

Then, by virtue of Theorem 7.47, we only need to show that φ is continuous with respect to $\boldsymbol{\sigma}(X',X)$. For this purpose, and in order to prove that φ is continuous at the zero functional of X' (which will be sufficient for its continuity on all of X'), consider an element $F_0 \in X'$ such that $F_0 \notin H$. Thus, since H is closed with respect to $\boldsymbol{\sigma}(X',X)$, there exists a neighborhood V of F_0 such that $V \subseteq H^c$. We may assume, by virtue of Lemma 7.44, that there exist $\varepsilon > 0$ and a finite set $\mathcal{M} \subseteq X$ such that

$$V := \Big\{F \in X' : \quad |F(x) - F_0(x)| < \varepsilon \quad \forall\, x \in \mathcal{M}\Big\}.$$

Then, since V is clearly convex, we have that either

$$\varphi(F) < \alpha \qquad \forall\, F \in V\,, \tag{7.14}$$

or

$$\varphi(F) > \alpha \qquad \forall\, F \in V\,. \tag{7.15}$$

Indeed, if we suppose on the contrary that there exist F, $\widetilde{F} \in V$ such that

$$\varphi(F) < \alpha < \varphi(\widetilde{F})\,,$$

then we can choose $\lambda \in (0,1)$ such that $\lambda\varphi(F) + (1-\lambda)\,\varphi(\widetilde{F}) = \alpha$, whence there holds $\varphi\big(\lambda F + (1-\lambda)\widetilde{F}\big) = \alpha$. It follows that $\lambda F + (1-\lambda)\widetilde{F} \in H$, and moreover, by the convexity of V, $\lambda F + (1-\lambda)\widetilde{F} \in V$, which is a contradiction. Then, assuming the validity of (7.14) (the argument for (7.15) is analogous), we deduce that

$$\varphi(G) < \alpha - \varphi(F_0) \quad \forall\, G \in W := V - \{F_0\},$$

and since $-W = W$, we deduce that

$$|\varphi(G)| < |\alpha - \varphi(F_0)| \quad \forall\, G \in W. \tag{7.16}$$

Since W is a neighborhood of the zero functional in X', inequality (7.16) proves that φ is continuous there, and therefore $\varphi : (X', \boldsymbol{\sigma}(X',X)) \to \mathbb{R}$ is continuous, thus completing the proof.

□

In what follows we provide some additional explanations about the last part of the previous proof. First note that the identity $-W = W$ follows from the fact that, given $G = F - F_0 \in W$ with $F \in V$, we have

$$-G = F_0 - F = (-F + 2F_0) - F_0,$$

with $-F + 2F_0 \in V$ since

$$|(-F + 2F_0 - F_0)(x)| = |(F_0 - F)(x)| < \varepsilon \qquad \forall\, x \in \mathcal{M}.$$

Next, let us see that the continuity of φ at the zero functional follows precisely from (7.16). Indeed, given $\delta > 0$ and $\widetilde{F} \in \varphi^{-1}(]-\delta,\delta[)$, there exists $\rho \in [0,1[$ such that $|\varphi(\widetilde{F})| = \rho\,\delta$. Then, consider the neighborhood of $\widetilde{F}$ given by $V(\widetilde{F}, \widetilde{\varepsilon}, \mathcal{M})$ (cf. Lemma 7.44) with

$$\widetilde{\varepsilon} := \frac{\varepsilon\,\delta}{|\alpha - \varphi(F_0)|\, m}\,,$$

where $m \in \mathbb{N}$ will be specified later, and observe that for each $F \in V(\widetilde{F}, \widetilde{\varepsilon}, \mathcal{M})$ we have that $\frac{\varepsilon}{\widetilde{\varepsilon}}(F - \widetilde{F}) \in W$. It follows from (7.16) that

$$|\varphi(F - \widetilde{F})| < \frac{\widetilde{\varepsilon}}{\varepsilon} |\alpha - \varphi(F_0)| = \frac{\delta}{m},$$

and therefore

$$|\varphi(F)| \leqslant |\varphi(F - \widetilde{F})| + |\varphi(\widetilde{F})| \leqslant \left\{\frac{1}{m} + \rho\right\} \delta < \delta,$$

for m sufficiently small. This proves that

$$V(\widetilde{F}, \widetilde{\varepsilon}, \mathcal{M}) \subseteq \varphi^{-1}(]-\delta, \delta[) \qquad \forall\, \widetilde{F} \in \varphi^{-1}(]-\delta, \delta[),$$

and therefore $\varphi^{-1}(]-\delta, \delta[)$ is open in $\boldsymbol{\sigma}(X', X)$.

We conclude this section with a more simplified proof of a result initially provided in Chapter 6, namely Lemma 6.37, which makes use of the meaning of the weak and weak* topologies.

Lemma 7.49. *Let Y and Z be Banach spaces such that Y is reflexive, and let $\{G_n\}_{n\in\mathbb{N}}$ be a sequence in Y' that converges weak* to $G \in Y'$. Then we have*

$$K(G_n) \overset{n\to\infty}{\to} K(G) \qquad \forall\, K \in \mathcal{K}(Y', Z). \tag{7.17}$$

Proof. It suffices to observe that when Y is reflexive, the weak* topology $\sigma(Y', Y)$ of Y' coincides with its weak topology $\sigma(Y', Y'')$, so that $G_n \overset{w*}{\to} G$ is equivalent to saying $G_n \overset{w}{\to} G$, and therefore (7.17) follows from a direct application of Lemma 6.29 to the context $X \longleftarrow Y'$ and $Y \longleftarrow Z$.

□

7.4.4 Connections with compactness and density

In this section we address some interesting results that connect the weak topologies $\boldsymbol{\sigma}(X, X')$ and $\boldsymbol{\sigma}(X', X)$ with the eventual compactness or density of the unit balls and spheres of the spaces involved. We begin with the classical Banach-Alaoglu-Bourbaki Theorem.

Theorem 7.50 (Banach-Alaoglu-Bourbaki Theorem). *Let X be a normed vector space. Then, the closed unit ball*

$$\bar{B}_{X'}(0,1) := \left\{F \in X' : \quad \|F\| \leqslant 1\right\}$$

is compact with respect to the weak topology $\boldsymbol{\sigma}(X', X)$.*

Proof. We begin by defining the product space

$$Y := \mathbb{R}^X = \left\{w := \{w_x\}_{x\in X} : \quad w_x \in \mathbb{R} \quad \forall\, x \in X\right\},$$

which is endowed with the *coarsest* topology that makes continuous the linear mappings $\varphi_x : Y \to \mathbb{R}$ defined by

$$\varphi_x(w) := w_x \qquad \forall\, w \in Y, \quad \forall\, x \in X.$$

In turn, let $\Phi : (X', \sigma(X', X)) \to Y$ be the mapping given by

$$\Phi(F) := \{F(x)\}_{x \in X} \qquad \forall\, F \in X',$$

and for each $x \in X$ consider the composition $\varphi_x \circ \Phi : X' \to \mathbb{R}$, that is

$$(\varphi_x \circ \Phi)(F) = \varphi_x(\Phi(F)) = F(x) \qquad \forall\, F \in X.$$

According to the definition of $\sigma(X', X)$ (cf. Section 7.4.3) we know that $\varphi_x \circ \Phi = \mathcal{J}(x)$ is continuous for all $x \in X$, and hence Lemma 7.32 guarantees that Φ is also continuous. Let us now show that Φ is a homeomorphism from X' onto $\Phi(X')$. Indeed, from the definition of equality of functionals in X' it is clear that Φ is injective. Furthermore, for the continuity of $\Phi^{-1} : \Phi(X') \to X'$ we again apply Lemma 7.32, which requires proving first that for each $x \in X$, we have that $\mathcal{J}(x) \circ \Phi^{-1} : \Phi(X') \to \mathbb{R}$ is continuous. This follows from the fact that $\mathcal{J}(x) \circ \Phi^{-1} = (\varphi_x \circ \Phi) \circ \Phi^{-1} = \varphi_x|_{\Phi(X')}$ and from the definition of the topology of Y. Having established this, we now proceed to prove that $K := \Phi(\bar{B}_{X'}(0,1))$ is compact in Y, which will then imply the compactness of $\bar{B}_{X'}(0,1)$ in $\sigma(X', X)$. To this end, we first observe that $K = K_1 \cap K_2$, where

$$K_1 := \Big\{ w := \{w_x\}_{x \in X} \in Y : \quad |w_x| \leqslant \|x\| \quad \forall\, x \in X \Big\}$$

and

$$K_2 := \Big\{ w := \{w_x\}_{x \in X} \in Y : \quad w_{x+y} = w_x + w_y$$

$$\text{and} \quad w_{\lambda x} = \lambda w_x \quad \forall\, \lambda \in \mathbb{R}, \quad \forall\, x,\, y \in X \Big\}.$$

Since the arbitrary intersection of compact sets is again compact, it follows easily that $K_1 = \bigcap_{x \in X} \varphi_x^{-1}([-\|x\|, \|x\|])$ is compact. Equivalently, the set $K_1 = \Pi_{x \in X}[-\|x\|, \|x\|]$ is compact because it is the Cartesian product of compact sets of $\mathbb{R}$. On the other hand, since

$$K_2 = \Big\{ \bigcap_{x,y \in X} A_{x,y} \Big\} \bigcap \Big\{ \bigcap_{\substack{x \in X \\ \lambda \in \mathbb{R}}} B_{\lambda, x} \Big\},$$

where the sets

$$A_{x,y} := (\varphi_{x+y} - \varphi_x - \varphi_y)^{-1}(\{0\}) \quad \text{and} \quad B_{\lambda,x} := (\varphi_{\lambda x} - \lambda \varphi_x)^{-1}(\{0\})$$

are closed (by definition of the topology of Y), we deduce that K_2 is closed as well and consequently K is compact.

□

As a historical remark, we point out that the Banach–Alaoglu–Bourbaki Theorem was proved by the Canadian-American mathematician Leonidas Alaoglu in [1]. An earlier proof had been given by Banach in [11], but under the additional assumption that the normed vector space was separable.

It should be noted here that the importance of the previous result is reinforced, among other reasons, by Theorem 6.8, which establishes that the closed unit ball is compact in the strong topology only when the normed space is finite-dimensional. In other words, the relevance in this case of the topology $\sigma(X', X)$ lies in the fact that it guarantees the compactness of this ball regardless of the dimension of the space. Moreover, it will be shown later (cf. Section 7.4.5) that if X is a separable Banach space, then $\bar{B}_{X'}(0,1)$ is metrizable with respect to the weak* topology, which means that there exists a metric on $\bar{B}_{X'}(0,1)$ whose topology coincides with $\boldsymbol{\sigma}(X',X)$. This result, together with the Banach-Alaoglu-Bourbaki Theorem and the fact that every sequence in a compact set of a metric space has a convergent subsequence, will then provide an alternative proof of Theorem 7.11 (every bounded sequence in the dual of a separable space has a weak*-convergent subsequence).

The next objective is to prove Kakutani's Theorem (cf. Theorem 7.55), which establishes the equivalence between the reflexivity of a Banach space and the weak compactness of the closed unit ball. We begin with the first respective implication.

Theorem 7.51. *Let X be a reflexive Banach space. Then, the closed unit ball*

$$\bar{B}_X(0,1) := \Big\{x \in X : \quad \|x\| \leqslant 1\Big\}$$

is compact with respect to the weak topology $\boldsymbol{\sigma}(X,X')$.

Proof. Since X is reflexive, we have that $\mathcal{J} : X \rightarrow X''$ is a bijective isometry (cf. Definition 7.1) and therefore

$$\mathcal{J}(\bar{B}_x(0,1)) = \bar{B}_{X''}(0,1) := \Big\{\mathcal{F} \in X'' : \quad \|\mathcal{F}\| \leqslant 1\Big\}.$$

In turn, applying Theorem 7.50 to the dual space X', it follows that $\bar{B}_{X''}(0,1)$ is compact with respect to $\boldsymbol{\sigma}(X'',X')$, and therefore, to conclude the desired result it suffices to prove that $\mathcal{J}^{-1} : (X'',\boldsymbol{\sigma}(X'',X')) \rightarrow (X,\boldsymbol{\sigma}(X,X'))$ is continuous. To this end, we must verify, according again to Lemma 7.32, that for each $F \in X'$ the mapping $F \circ \mathcal{J}^{-1} : (X'',\boldsymbol{\sigma}(X'',X')) \rightarrow \mathbb{R}$ is continuous. Indeed, given $F \in X'$ we have for each $\mathcal{F} \in X''$ that

$$(F \circ \mathcal{J}^{-1})(\mathcal{F}) = F(\mathcal{J}^{-1}(\mathcal{F})) = F(x) = \mathcal{J}(x)(F) = \mathcal{F}(F),$$

where $x \in X$ is such that $\mathcal{J}(x) = \mathcal{F}$. The above shows that $F \circ \mathcal{J}^{-1} = \mathcal{J}'(F)$, where $\mathcal{J}' : X' \rightarrow (X')'' \equiv (X'')'$ is the mapping that to each $F \in X'$ assigns $\mathcal{J}'(F) \in (X'')'$ defined by $\mathcal{J}'(F)(\mathcal{F}) := \mathcal{F}(F) \quad \forall\, \mathcal{F} \in X''$. Then, since $\boldsymbol{\sigma}(X'',X')$ is precisely the coarsest topology that makes the mappings $\big\{\mathcal{J}'(F)\big\}_{F\in X'}$ continuous, we conclude that $F \circ \mathcal{J}^{-1} = \mathcal{J}'(F)$ is continuous for all $F \in X'$, which completes the proof.

□

Analogously to what was discussed after the Banach-Alaoglu-Bourbaki Theorem (cf. Theorem 7.50), we anticipate here that in Section 7.4.5 it will also be shown that if X is a Banach space whose dual X' is separable, then $\bar{B}_X(0,1)$ is metrizable with respect to the weak topology. In that case, this result and Theorem 7.51 will provide a first attempt at an alternative proof of Theorem 7.13 (every bounded sequence in a reflexive Banach space has a weakly convergent subsequence). Nevertheless, as we already know from the standard proof of this theorem, the separability of X' is not required, and therefore we shall finally give a second proof, also based on arguments of metrizability, which requires only the reflexivity of X.

Now, in order to prove the second implication that completes Kakutani's Theorem, we need a pair of preliminary lemmas. The first one is rather technical in nature, while the second establishes a density property of $\mathcal{J}(\bar{B}_X(0,1))$ with respect to $\boldsymbol{\sigma}(X'',X')$.

Lemma 7.52 (Helly's Lemma). *Let X be a normed vector space, let $N \in \mathbb{N}$, and let $\{F_1, \dots, F_N\} \subseteq X'$ and $\{\alpha_1, \dots, \alpha_N\} \subseteq \mathbb{R}$. Then, the following statements are equivalent:*

a) *for each $\varepsilon > 0$ there exists $x_\varepsilon \in X$ such that $\|x_\varepsilon\| \leqslant 1$ and*

$$|F_j(x_\varepsilon) - \alpha_j| < \varepsilon \qquad \forall\, j \in \{1, \dots, N\}.$$

b)

$$\left| \sum_{j=1}^{N} \beta_j\, \alpha_j \right| \leqslant \left\| \sum_{j=1}^{N} \beta_j\, F_j \right\| \qquad \forall\, (\beta_1, \dots, \beta_N)^{\mathrm{t}} \in \mathbb{R}^N .$$

Proof.
a) $\Rightarrow$ b) Assume a) holds and let $(\beta_1, \dots, \beta_N)^{\mathrm{t}} \in \mathbb{R}^N$. Then, given $\varepsilon > 0$ we apply the consequence of a) and obtain

$$\left| \sum_{j=1}^{N} \beta_j\, F_j(x_\varepsilon) - \sum_{j=1}^{N} \beta_j\, \alpha_j \right| = \left| \sum_{j=1}^{N} \beta_j\, \big(F_j(x_\varepsilon) - \alpha_j\big) \right| < \sum_{j=1}^{N} |\beta_j|\, \varepsilon.$$

In turn, using that $\|x_\varepsilon\| \leqslant 1$, we get

$$\begin{aligned} \left| \sum_{j=1}^{N} \beta_j\, \alpha_j \right| &\leqslant \left| \sum_{j=1}^{N} \beta_j\, \alpha_j - \sum_{j=1}^{N} \beta_j\, F_j(x_\varepsilon) \right| + \left| \sum_{j=1}^{N} \beta_j\, F_j(x_\varepsilon) \right| \\ &\leqslant \sum_{j=1}^{N} |\beta_j|\, \varepsilon + \left\| \sum_{j=1}^{N} \beta_j\, F_j \right\|, \end{aligned}$$

whence, taking the limit as $\varepsilon \to 0^+$, b) follows.

b) $\Rightarrow$ a) Assume statement b) holds and let $A : X \to \mathbb{R}^N$ be the bounded linear operator defined by

$$A(x) := \big(F_1(x), \dots, F_N(x)\big)^{\mathrm{t}} \quad \forall\, x \in X.$$

Thus, property a) is equivalent to saying that the vector $\vec{\alpha} := (\alpha_1, \dots, \alpha_N)^{\mathrm{t}}$ belongs to $\overline{A(\bar{B}_X(0,1))}$. Suppose then, by contradiction, that $\vec{\alpha} := (\alpha_1, \dots, \alpha_N)^{\mathrm{t}} \notin \overline{A(\bar{B}_X(0,1))}$. Then, applying the second geometric version of the Hahn-Banach Theorem (cf. Theorem 2.33) to the compact set $\{\vec{\alpha}\}$ and the closed set $\overline{A(\bar{B}_X(0,1))}$, we deduce that there exists an affine hyperplane $[f = \gamma]$, with $f \neq 0$, such that

$$f(z) < \gamma < f(\vec{\alpha}) \qquad \forall\, z \in \overline{A(\bar{B}_X(0,1))}.$$

More precisely, representing f by $\vec{\beta} \in \mathbb{R}^N$, and taking in particular $z \in A(\bar{B}_X(0,1))$, we obtain

$$\vec{\beta} \cdot A(x) < \gamma < \vec{\beta} \cdot \vec{\alpha} \qquad \forall\, x \in \bar{B}_X(0,1)\,.$$

It follows that

$$\left| \left(\sum_{j=1}^{N} \beta_j\, F_j \right)(x) \right| = \left| \vec{\beta} \cdot A(x) \right| < \gamma < \vec{\beta} \cdot \vec{\alpha} \qquad \forall\, x \in \bar{B}_X(0,1),$$

and therefore

$$\left\| \sum_{j=1}^{N} \beta_j\, F_j \right\| \leqslant \gamma < \vec{\alpha} \cdot \vec{\beta}\,,$$

which contradicts b). □

Lemma 7.53 (Goldstine's Lemma). *Let X be a Banach space. Then, $\mathcal{J}(\bar{B}_X(0,1))$ is dense in $\bar{B}_{X''}(0,1)$ with respect to the topology $\boldsymbol{\sigma}(X'', X')$.*

Proof. Let $\mathcal{F} \in \bar{B}_{X''}(0,1)$, and let V be a neighborhood of $\mathcal{F}$ for the topology $\boldsymbol{\sigma}(X'', X')$. It can be assumed, by virtue of Lemma 7.44 applied to X', that there exist $\varepsilon > 0$ and a finite set $\mathcal{M} \subseteq X'$ such that

$$V := \Big\{\mathcal{G} \in X'' : \quad |\mathcal{G}(F) - \mathcal{F}(F)| < \varepsilon \quad \forall F \in \mathcal{M}\Big\}.$$

We will show that $\mathcal{J}(\bar{B}_{X'}(0,1)) \cap V \neq \varnothing$, which is equivalent to finding $\widetilde{x} \in \bar{B}_{X'}(0,1)$ such that

$$\Big|\mathcal{J}(\widetilde{x})(F) - \mathcal{F}(F)\Big| < \varepsilon \qquad \forall F \in \mathcal{M},$$

that is,

$$\Big|F(\widetilde{x}) - \mathcal{F}(F)\Big| < \varepsilon \qquad \forall F \in \mathcal{M}.$$

To this end, let $N = \operatorname{card}(\mathcal{M})$ and $\vec{\alpha} := (\mathcal{F}(F))_{F\in\mathcal{M}} \in \mathbb{R}^N$. Then, for each $\vec{\beta} := (\beta_F)_{F\in\mathcal{M}} \in \mathbb{R}^N$, and noting that $\|\mathcal{F}\| \leqslant 1$, we have

$$\Big|\vec{\beta}\cdot\vec{\alpha}\Big| = \Big|\mathcal{F}\Big(\sum_{F\in\mathcal{M}} \beta_F\, F\Big)\Big| \leqslant \Big\|\sum_{F\in\mathcal{M}} \beta_F\, F\Big\|,$$

which yields b) of Helly's Lemma (cf. Lemma 7.52). Thus we obtain part a) of this lemma, that is, there exists $\widetilde{x} \in \bar{B}_X(0,1)$ such that

$$\Big|F(\widetilde{x}) - \alpha_F\Big| = \Big|F(\widetilde{x}) - \mathcal{F}(F)\Big| < \varepsilon \qquad \forall F \in \mathcal{M},$$

which proves what was required and completes the proof. □

It is important to recall here that $\mathcal{J}(\bar{B}_X(0,1))$ is closed in $\bar{B}_{X''}(0,1)$ for the strong topology of X'', which follows from the fact that $\bar{B}_X(0,1)$ is closed and that the operator $\mathcal{J}$ is an isometry. However, in general, $\mathcal{J}(\bar{B}_X(0,1))$ is not dense in $\bar{B}_{X''}(0,1)$ for the strong topology, unless X is reflexive, in which case obviously $\mathcal{J}(\bar{B}_X(0,1)) = \bar{B}_{X''}(0,1)$.

Having proved the previous lemmas, we are now in a position to prove the following result.

Theorem 7.54. *Let X be a Banach space such that $\bar{B}_X(0,1)$ is compact with respect to the topology $\boldsymbol{\sigma}(X, X')$. Then X is reflexive.*

Proof. First recall that $\mathcal{J} : X \to X''$ is continuous with respect to the strong topologies of X and X'', and therefore, by virtue of Theorem 7.42, it follows that

$$\mathcal{J} : (X, \boldsymbol{\sigma}(X, X')) \to \big(X'', \boldsymbol{\sigma}(X'', X''')\big), \quad \text{with} \quad X''' \equiv (X'')' \equiv (X')'',$$

is also continuous. Now, since $\boldsymbol{\sigma}(X'', X''')$ and $\boldsymbol{\sigma}(X'', X')$ are the coarsest topologies that make continuous the families of mappings $\{\varphi_{\mathbb{F}}\}_{\mathbb{F}\in X'''}$ and $\{\mathcal{J}'(F)\}_{F\in X'}$, respectively, where

$$\varphi_{\mathbb{F}}(\mathcal{F}) := \mathbb{F}(\mathcal{F}) \quad \forall\, \mathcal{F} \in X'' \quad \text{and} \quad \mathcal{J}' : X' \to X'''$$

is such that $\mathcal{J}'(F)(\mathcal{F}) = \mathcal{F}(F)$ for all $\mathcal{F} \in X''$, it follows that the open sets of $\boldsymbol{\sigma}(X'', X')$ are open in $\boldsymbol{\sigma}(X'', X''')$ as well. Consequently, $\mathcal{J} : \big(X, \boldsymbol{\sigma}(X, X')\big) \to \big(X'', \boldsymbol{\sigma}(X'', X')\big)$

is also continuous, and therefore $\mathcal{J}(\bar{B}_X(0,1))$ is compact with respect to the topology $\boldsymbol{\sigma}(X'', X')$. Then, since $\mathcal{J}(\bar{B}_X(0,1))$ is dense in $\bar{B}_{X''}(0,1)$ for $\boldsymbol{\sigma}(X'', X')$ (cf. Lemma 7.53), we conclude that $\mathcal{J}(\bar{B}_X(0,1)) = \bar{B}_{X''}(0,1)$, from which it follows that $\mathcal{J}(X) = X''$, and thus X is reflexive.

□

Thanks to Theorems 7.51 and 7.54, we have proved the following result.

Theorem 7.55 (Kakutani's Theorem). *Let X be a Banach space. Then X is reflexive if and only if $\bar{B}_X(0,1)$ is compact with respect to the topology $\boldsymbol{\sigma}(X, X')$.*

In what follows we review some of the consequences of Kakutani's Theorem, even though several of them have already been proved by other arguments in previous sections. The first refers to the reflexivity of any closed subspace Z of a reflexive Banach space X. In this regard, it is first necessary to observe (cf. Exercise 7.14) that

$$\boldsymbol{\sigma}(Z, Z') = \boldsymbol{\sigma}(X, X')|_Z := \Big\{ A \cap Z : \quad A \in \boldsymbol{\sigma}(X, X') \Big\}.$$

Then we have the following result, whose statement and proof were already given in Section 7.2 (cf. Lemma 7.5). Here, we present an alternative proof.

Lemma 7.56. *Let Z be a closed subspace of a reflexive Banach space X. Then Z is also reflexive.*

Proof. According to Theorem 7.55, it suffices to show that $\bar{B}_Z(0,1) := \big\{ z \in Z : \quad \|z\| \leqslant 1 \big\}$ is compact with respect to $\boldsymbol{\sigma}(Z, Z')$. Indeed, note first, by virtue of the same theorem, that the reflexivity of X guarantees that $\bar{B}_X(0,1)$ is compact with respect to $\boldsymbol{\sigma}(X, X')$. Furthermore, according to Lemma 7.40, we know that Z is closed with respect to $\boldsymbol{\sigma}(X, X')$, and therefore $\bar{B}_Z(0,1) = \bar{B}_X(0,1) \cap Z$ is compact with respect to $\boldsymbol{\sigma}(X, X')$, from which it follows that it is also compact for $\boldsymbol{\sigma}(Z, Z')$.

□

Next, we recall and prove once again Lemma 7.4, this time using Kakutani's Theorem instead of the analytic form of the Hahn-Banach Theorem.

Lemma 7.57. *X is a reflexive Banach space if and only if its dual X' is also reflexive.*

Proof. Suppose that X is a reflexive Banach space. It follows that $\boldsymbol{\sigma}(X', X) = \boldsymbol{\sigma}(X', X'')$, and therefore, since the Banach-Alaoglu-Bourbaki Theorem (cf. Theorem 7.50) ensures that $\bar{B}_{X'}(0,1)$ is compact in $\boldsymbol{\sigma}(X', X)$, we deduce that this closed ball is also compact in $\boldsymbol{\sigma}(X', X'')$. Hence, applying Kakutani's Theorem to X', we conclude that X' is reflexive. Conversely, suppose that X is Banach and that X' is reflexive. From the same implication in the previous part of this proof it follows that X'' is reflexive, and since $\mathcal{J}(X)$ is a closed subspace of X'' (because X is Banach), it follows, by virtue of Lemma 7.5 (or equivalently Lemma 7.56), that $\mathcal{J}(X)$ is reflexive as well. Finally, since reflexivity is a property that is preserved under isomorphisms (cf. part a) of Exercise 7.4), we conclude that X is reflexive.

□

A third application of Kakutani's Theorem is given as follows.

Lemma 7.58. *Let X be a reflexive Banach space, and let $K \subseteq X$ be convex, closed, and bounded. Then K is compact with respect to $\boldsymbol{\sigma}(X, X')$.*

Proof. Since K is convex and strongly closed, Lemma 7.40 implies that K is closed with respect to $\boldsymbol{\sigma}(X, X')$. Moreover, since K is bounded, there exists $M > 0$ such that $\|x\| \leqslant M$ for all $x \in K$, which means that $K \subseteq M\bar{B}_X(0,1)$. Since Kakutani's Theorem (cf. Theorem 7.55) guarantees that $\bar{B}_X(0,1)$, and thus $M\,\bar{B}_X(0,1)$, are compact for $\boldsymbol{\sigma}(X, X')$, we conclude that K, being closed and contained in a compact set, is also compact.

□

We conclude this section with a theorem that complements the analysis of the adjoint of unbounded operators (cf. Section 3.12.2). In fact, the corresponding result was announced in that section, just before Theorem 3.81. Although it does not coincide exactly with the topics of the theorems and lemmas discussed in the present section, it nevertheless concerns a density property in the context of reflexive spaces. More precisely, we have the following theorem.

Theorem 7.59. *Let X and Y be reflexive Banach spaces, and let $A : \mathcal{D}(A) \subseteq X \to Y$ be a closed linear operator such that $\overline{\mathcal{D}(A)} = X$. Then $\mathcal{D}(A')$ is dense in Y' and therefore the adjoint operator $A'' : \mathcal{D}(A'') \subseteq X'' \to Y''$ exists, and it can be identified with A.*

Proof. Recall first from Definition 3.80 that

$$\mathcal{D}(A') := \Big\{ G \in Y' : \quad \exists\, c \geqslant 0 \quad \text{such that} \quad |G(A(x))| \leqslant c\,\|x\| \qquad \forall\, x \in \mathcal{D}(A) \Big\}.$$

To prove that $\mathcal{D}(A')$ is dense in Y', it is enough to show that $\mathcal{D}(A')^{\circ}$ is the zero functional of Y''. Indeed, if we assume that $\overline{\mathcal{D}(A')} \neq Y'$, then Theorem 2.26 guarantees the existence of $\mathcal{F} \in Y''$, $\mathcal{F} \neq 0$, such that $\mathcal{F}(G) = 0$ for all $G \in \mathcal{D}(A')$, which would yield a contradiction. Let therefore $\mathcal{F} \in Y''$ be such that $\mathcal{F}(G) = 0$ for all $G \in \mathcal{D}(A')$, and let us prove that $\mathcal{F}$ is the zero functional. Since Y is reflexive, there exists $\varphi \in Y$ such that $\mathcal{F}(G) = G(\varphi)$ for all $G \in Y'$, and in particular $G(\varphi) = 0$ for all $G \in \mathcal{D}(A')$. If we assume by contradiction that $\varphi \neq 0$, then clearly $(0, \varphi) \notin \mathcal{G}(A) := \Big\{ (x, A(x)) \in X \times Y : \quad x \in \mathcal{D}(A) \Big\}$. Since $\mathcal{G}(A)$ is a closed subspace of $X \times Y$ (because A is closed), and $\big\{(0, \varphi)\big\}$ is a compact subset of $X \times Y$, the second geometric form of the Hahn-Banach Theorem (cf. Theorem 2.33) implies that there exist $(\widetilde{F}, \widetilde{G}) \in X' \times Y' \equiv (X \times Y)'$, $(\widetilde{F}, \widetilde{G}) \neq 0$, and $\alpha \in \mathbb{R}$ such that

$$\widetilde{F}(x) + \widetilde{G}(A(x)) < \alpha < \widetilde{G}(\varphi) \qquad \forall\, x \in \mathcal{D}(A). \tag{7.18}$$

It follows that $\widetilde{F}(x) + \widetilde{G}(A(x)) = 0$ for all $x \in \mathcal{D}(A)$ since otherwise, if we assume the existence of $\widetilde{x} \in \mathcal{D}(A)$ for which $\widetilde{F}(\widetilde{x}) + \widetilde{G}(A(\widetilde{x})) \neq 0$, then by taking $x = \frac{\alpha\,\widetilde{x}}{\widetilde{F}(\widetilde{x}) + \widetilde{G}(A(\widetilde{x}))}$ in inequality (7.18), we would obtain $\alpha < \alpha$. Thus, since the left-hand side of (7.18) vanishes, we conclude that $\widetilde{G}(\varphi) > 0$. However, we also have

$$|\widetilde{G}(A(x))| = |-\widetilde{F}(x)| \leqslant \|\widetilde{F}\|\,\|x\| \qquad \forall\, x \in \mathcal{D}(A),$$

which shows that $\widetilde{G} \in \mathcal{D}(A')$, and therefore $\widetilde{G}(\varphi) = 0$. From this contradiction it follows that necessarily $\varphi = 0$, and hence $\mathcal{F}$ is the zero functional of Y''. On the other hand, let us now see that A'' and A can be identified via an isomorphism. Indeed, recall from Lemma 3.82 that if $\boldsymbol{\mathcal{S}} : Y' \times X' \to X' \times Y'$ and $\bar{\boldsymbol{\mathcal{S}}} : Y'' \times X'' \to X'' \times Y''$ are defined as

$$\boldsymbol{\mathcal{S}}(G, F) := (-F, G) \quad \forall\, (G, F) \in Y' \times X',$$

and

$$\bar{\boldsymbol{S}}(\mathcal{G},\mathcal{F}) := (-\mathcal{F},\mathcal{G}) \quad \forall\,(\mathcal{G},\mathcal{F}) \in Y'' \times X'',$$

then

$$\boldsymbol{S}(\mathcal{G}(A')) = \mathcal{G}(A)^\circ \quad \text{and} \quad \bar{\boldsymbol{S}}(\mathcal{G}(A'')) = \mathcal{G}(A')^\circ .$$

Since $\boldsymbol{S}$ and $\bar{\boldsymbol{S}}$ are bijective and $\mathcal{G}(A)$ is closed, we arrive at

$$\mathcal{G}(A'') = \left(\bar{\boldsymbol{S}}\right)^{-1}\left(\mathcal{G}(A')^\circ\right) = \left(\bar{\boldsymbol{S}}\right)^{-1}\left(\left(\boldsymbol{S}^{-1}(\mathcal{G}(A)^\circ)\right)^\circ\right) \equiv {}^\circ\left(\mathcal{G}(A)^\circ\right) = \mathcal{G}(A),$$

which completes the proof.

□

7.4.5 Metrizability of weak topologies

In this section we characterize the possible metrizability, with respect to the weak and weak* topologies, of the closed unit balls of a Banach space X and its dual X', in terms of the separability of X' and X, respectively. Subsequently, we use these results to provide alternative proofs of the classical Theorems 7.11 and 7.13. We begin the analysis with the following result.

Theorem 7.60. *Let X be a separable Banach space. Then $\bar{B}_{X'}(0,1)$ is metrizable with respect to $\boldsymbol{\sigma}(X',X)$, which means that there exists a metric on $\bar{B}_{X'}(0,1)$ whose associated topology coincides with $\boldsymbol{\sigma}(X',X)$ on $\bar{B}_{X'}(0,1)$. Conversely, if $\bar{B}_{X'}(0,1)$ is metrizable for $\boldsymbol{\sigma}(X',X)$, then X is separable.*

Proof. Let $\{y_n\}_{n\in\mathbb{N}}$ be dense in X, and let us show that $\{x_n\}_{n\in\mathbb{N}} := \{y_n\}_{n\in\mathbb{N}} \cap \bar{B}_X(0,1)$ is dense in $\bar{B}_X(0,1)$. Indeed, given $x \in \bar{B}_X(0,1)$ and $\varepsilon > 0$, consider $x_\varepsilon \in B_X(0,1)$ such that $\|x - x_\varepsilon\| < \varepsilon/2$. It follows that there exists $k \in \mathbb{N}$ such that $\|y_k\| \leqslant 1$ and $\|x_\varepsilon - y_k\| < \varepsilon/2$ (see Figure 7.2). Thus, $y_k \in \{x_n\}_{n\in\mathbb{N}}$ and clearly

$$\|x - y_k\| \leqslant \|x - x_\varepsilon\| + \|x_\varepsilon - y_k\| < \varepsilon .$$

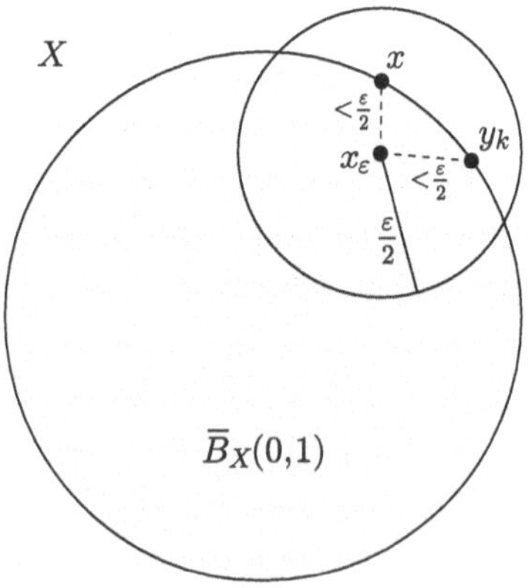

Fig. 7.2. *Illustration of the density of $\{x_n\}_{n\in\mathbb{N}}$*

Next, define the mapping $d : \bar{B}_{X'}(0,1) \times \bar{B}_{X'}(0,1) \to \mathbb{R}$ by

$$d(F,G) := \sum_{n=1}^{\infty} \frac{1}{2^n} |(F-G)(x_n)| \qquad \forall\, F,\, G \in \bar{B}_{X'}(0,1), \tag{7.19}$$

and let us show that d is a metric on $\bar{B}_{X'}(0,1)$. Indeed, for each $N \in \mathbb{N}$ we have

$$\begin{aligned}\sum_{n=1}^{N} \frac{1}{2^n} |(F-G)(x_n)| &\leqslant \sum_{n=1}^{N} \frac{1}{2^n} \|F-G\|\,\|x_n\| \\ &\leqslant \sum_{n=1}^{N} \frac{1}{2^n} \Big\{ \|F\| + \|G\| \Big\} \leqslant 2 \sum_{n=1}^{N} \frac{1}{2^n} < 2,\end{aligned}$$

which shows that the series in (7.19) is convergent, and thus $d(F,G)$ is well defined for all $F,\, G \in \bar{B}_{X'}(0,1)$. Furthermore, if $d(F,G) = 0$, it follows that $F(x_n) = G(x_n)$ for all $n \in \mathbb{N}$, whence, by the density of $\{x_n\}_{n\in\mathbb{N}}$, we have $F(x) = G(x)$ for all $x \in \bar{B}_X(0,1)$, and therefore $F = G$. Finally, it is easy to verify that d satisfies the triangle inequality. Let us now show that the topology induced by d coincides with $\boldsymbol{\sigma}(X',X)$ on $\bar{B}_{X'}(0,1)$. Let $F_0 \in \bar{B}_{X'}(0,1)$ and let V be a neighborhood of F_0 in $\boldsymbol{\sigma}(X',X)|_{\bar{B}_{X'}(0,1)}$. Without loss of generality, it can be assumed, by virtue of Lemma 7.44, that there exist $\varepsilon > 0$ and a finite set $\mathcal{M} \subseteq X$, such that

$$V := \Big\{ F \in \bar{B}_{X'}(0,1) : \quad |(F-F_0)(x)| < \varepsilon \quad \forall\, x \in \mathcal{M} \Big\}. \tag{7.20}$$

Moreover, redefining ε, if necessary, as $\dfrac{\varepsilon}{\max\big\{\|x\| : x \in \mathcal{M}, x \neq 0\big\}}$, we may also assume that $\|x\| \leqslant 1$ for all $x \in \mathcal{M}$. We want to prove that there exists $r > 0$ such that

$$U := \Big\{ F \in \bar{B}_{X'}(0,1) : \quad d(F,F_0) < r \Big\} \subseteq V.$$

Indeed, the density of $\{x_n\}_{n\in\mathbb{N}}$ in $\bar{B}_X(0,1)$ guarantees that for each $x \in \mathcal{M}$ there exists $n(x) \in \mathbb{N}$ such that $\|x - x_{n(x)}\| < \varepsilon/4$. Then, take $r > 0$ such that $2^{n(x)} r < \varepsilon/2$ for all $x \in \mathcal{M}$, that is, $r < \frac{\varepsilon}{2}\Big\{ \max\{2^{n(x)} : x \in \mathcal{M}\}\Big\}^{-1}$. Thus, given $F \in \bar{B}_{X'}(0,1)$ such that $d(F,F_0) < r$, we have

$$\frac{1}{2^{n(x)}} \Big| (F-F_0)(x_{n(x)}) \Big| < r \qquad \forall\, x \in \mathcal{M},$$

whence

$$\Big| (F-F_0)(x_{n(x)}) \Big| < \frac{\varepsilon}{2} \qquad \forall\, x \in \mathcal{M}.$$

It follows that for each $x \in \mathcal{M}$

$$\begin{aligned}\Big| (F-F_0)(x) \Big| &\leqslant \Big| (F-F_0)(x - x_{n(x)}) \Big| + \Big| (F-F_0)(x_{n(x)}) \Big| \\ &< \Big\{ \|F\| + \|F_0\| \Big\} \frac{\varepsilon}{4} + \frac{\varepsilon}{2} \leqslant \varepsilon,\end{aligned}$$

which shows that $F \in V$ and therefore $U \subseteq V$. Conversely, let $F_0 \in \bar{B}_{X'}(0,1)$, and let $U := \big\{ F \in \bar{B}_{X'}(0,1) : \quad d(F,F_0) < r \big\}$ be a neighborhood of F in the topology induced by d. We want to show that there exists a neighborhood V of F_0 in $\boldsymbol{\sigma}(X',X)|_{\bar{B}_{X'}(0,1)}$ such that $V \subseteq U$. To this end, we look for V in the form (7.20) with $\mathcal{M} = \{x_1, \ldots, x_N\}$ and $N \in \mathbb{N}$. In this case, for $F \in V$ we have

$$d(F, F_0) = \sum_{n=1}^{N} \frac{1}{2^n} \left|(F - F_0)(x_n)\right| + \sum_{n=N+1}^{\infty} \frac{1}{2^n} \left|(F - F_0)(x_n)\right|$$

$$< \varepsilon + 2 \sum_{n=N+1}^{\infty} \frac{1}{2^n} = \varepsilon + \frac{1}{2^{N-1}},$$

so that by choosing $\varepsilon > 0$ and $N \in \mathbb{N}$ such that $\varepsilon < r/2$ and $\frac{1}{2^{N-1}} < r/2$, it follows that $d(F, F_0) < r$, whence $F \in U$ and thus $V \subseteq U$. We have therefore shown that $\bar{B}_{X'}(0,1)$ is metrizable with respect to $\boldsymbol{\sigma}(X', X)$. Conversely, suppose that this last statement holds and let us prove that X is separable. To this end, for each $n \in \mathbb{N}$ define

$$U_n := \left\{F \in \bar{B}_{X'}(0,1) : \quad d(F, 0) < \frac{1}{n}\right\}$$

and denote by V_n a neighborhood of 0 in $\boldsymbol{\sigma}(X', X)$ such that $V_n \subseteq U_n$. More precisely, let $\varepsilon_n > 0$ and $\mathcal{M}_n$ be a finite subset of X such that

$$V_n := \left\{F \in \bar{B}_{X'}(0,1) : \quad |F(x)| < \varepsilon_n \quad \forall\, x \in \mathcal{M}_n\right\}.$$

It follows that $S := \bigcup_{n \in \mathbb{N}} \mathcal{M}_n$ is countable and that $\bigcap_{n \in \mathbb{N}} V_n = \{0\}$. Hence, if $F \in \bar{B}_{X'}(0,1)$ is such that $F(x) = 0$ for all $x \in S$, we obtain that $F \in \bigcap_{n \in \mathbb{N}} V_n$, whence $F = 0$. This proves that $\langle S \rangle_{\mathbb{R}}$, the vector space generated by S, is dense in X. Finally, if we denote by $\langle S \rangle_{\mathbb{Q}}$ the set of all finite linear combinations of elements of S with coefficients in $\mathbb{Q}$, we deduce that $\langle S \rangle_{\mathbb{Q}}$ is countable (since S and $\mathbb{Q}$ are countable) and that $\overline{\langle S \rangle_{\mathbb{Q}}} = \overline{\langle S \rangle_{\mathbb{R}}} = X$, thus concluding that X is separable.

□

The analogue of Theorem 7.60 for the case of the closed unit ball of X is stated below. Note, however, that we omit the arguments of the reciprocal implication since, being of a much more technical nature, they go beyond the main objectives of this text.

Theorem 7.61. *Let X be a Banach space such that X' is separable. Then $\bar{B}_X(0,1)$ is metrizable with respect to $\boldsymbol{\sigma}(X, X')$. Conversely, if $\bar{B}_X(0,1)$ is metrizable for $\boldsymbol{\sigma}(X, X')$, then X' is separable.*

Proof. For the first implication, we proceed as in the proof of Theorem 7.60, interchanging the roles of X and X'. More precisely, given a dense sequence $\{F_n\}_{n \in \mathbb{N}}$ in X', define $\{G_n\}_{n \in \mathbb{N}} := \{F_n\}_{n \in \mathbb{N}} \cap \bar{B}_{X'}(0,1)$, which is clearly dense in $\bar{B}_{X'}(0,1)$, and then introduce $d : \bar{B}_X(0,1) \times \bar{B}_X(0,1) \to \mathbb{R}$ as

$$d(x, y) := \sum_{n=1}^{\infty} \frac{1}{2^n} \left|G_n(x - y)\right| \quad \forall\, x, y \in \bar{B}_X(0,1).$$

It is easily shown that d is a metric, so that the rest of the proof is analogous to that of the theorem mentioned. In particular, in this case one must use the characterization given by Lemma 7.35 of the neighborhoods in $\boldsymbol{\sigma}(X, X')$. Additional details are omitted, and we refer to Exercise 7.15.

□

Having proved the previous theorems, we now proceed to provide alternative proofs of the classical results announced at the beginning of this section. We begin with the first of them (cf. Theorem 7.11).

Theorem 7.62. *Let X be a separable normed vector space. Then every bounded sequence in X' has a weak*-convergent subsequence.*

Proof. Recall first that a subset of a metric space is compact if and only if every sequence in it has convergent subsequence. Then, we know first from the Banach-Alaoglu-Bourbaki Theorem (cf. Theorem 7.50) that $\bar{B}_{X'}(0,1)$ is compact with respect to $\boldsymbol{\sigma}(X', X)$. In turn, according to Theorem 7.60, $\bar{B}_{X'}(0,1)$ is metrizable for that topology, and therefore every sequence in $\bar{B}_{X'}(0,1)$ has a weak*-convergent subsequence.

□

On the other hand, we now present the alternative proof of Theorem 7.13, which, however, begins in the same way as the one given previously.

Theorem 7.63. *Let X be a reflexive Banach space. Then every bounded sequence in X has a weakly convergent subsequence.*

Proof. Let $\{x_n\}_{n\in\mathbb{N}}$ be a bounded sequence in X, and let Z be the closed subspace generated by $\{x_n\}_{n\in\mathbb{N}}$. According to the first part of the proof of Theorem 7.13, in which Lemmas 7.5 and 7.9 are used, we know that Z is separable and reflexive, and that Z' is separable. Then, by virtue of Theorems 7.60 and 7.61 applied to Z' and Z, respectively, it follows that $\bar{B}_{Z''}(0,1)$ and $\bar{B}_Z(0,1)$ are metrizable for $\boldsymbol{\sigma}(Z'', Z')$ and $\boldsymbol{\sigma}(Z, Z')$. Moreover, thanks to Theorem 7.50, $\bar{B}_{Z''}(0,1)$ is compact for $\boldsymbol{\sigma}(Z'', Z')$, and since Z is reflexive, we can identify $\bar{B}_Z(0,1)$ with $\bar{B}_{Z''}(0,1)$ and $\boldsymbol{\sigma}(Z'', Z')$ with $\boldsymbol{\sigma}(Z, Z')$. It then follows that $\bar{B}_Z(0,1)$ is compact and metrizable for $\boldsymbol{\sigma}(Z, Z')$, from which the bounded sequence $\{x_n\}_{n\in\mathbb{N}}$ has a subsequence that converges weakly for $\boldsymbol{\sigma}(Z, Z')$, and therefore also converges weakly for $\boldsymbol{\sigma}(X, X')$.

□

We conclude this section by noting that the reciprocal of the previous theorem is also valid. More precisely, if every bounded sequence has a weakly convergent subsequence, then necessarily X is reflexive. This result is known as the Eberlein-Šmulian Theorem.

7.5 EXERCISES

▸ **7.1** *Given a normed vector space X, prove that there exist another normed vector space $\widetilde{X}$ and a Banach space Y such that X is isomorphic to $\widetilde{X}$, $\widetilde{X} \subseteq Y$, $\|x\|_Y = \|x\|_{\widetilde{X}}$ for all $x \in \widetilde{X}$, and $\widetilde{X}$ is dense in Y.*

▸ **7.2** *Let X, Y be Banach spaces, and let $T \in \mathcal{L}(X, Y)$. We say that T is* WEAKLY COMPACT *if for every bounded sequence $\{x_n\}_{n\in\mathbb{N}}$ in X, there exists a subsequence $\{x_n^{(1)}\}_{n\in\mathbb{N}}$ such that $\{Tx_n^{(1)}\}_{n\in\mathbb{N}}$ converges weakly in Y. Prove that if X or Y is reflexive, then every operator $T \in \mathcal{L}(X, Y)$ is weakly compact.*

▸ **7.3** *Let $a, b \in \mathbb{R}$ be such that $a < b$. Assume that $C[a, b]$ is separable, and prove that, for every nonnegative integer k, the space $C^k[a, b]$ is also separable.*

▸ **7.4**

a) *Let X, Y be Banach spaces, and suppose that there exists a bijective operator $T \in \mathcal{L}(X, Y)$. Prove that X is separable if and only if Y is separable (note that the analogous equivalence for reflexivity was established in Lemma 7.6).*

b) *Let X, Y be separable (reflexive) Banach spaces. Prove that the product space $X \times Y$ is also separable (reflexive).*

c) *Given an open set $\Omega \subset \mathbb{R}^N$, $p \in (1, +\infty)$, and $\mathbb{K} = \mathbb{R}$, we recall once again from Exercise 1.10 (see also Exercise 5.1) that $\mathrm{L}^p(\Omega)$ denotes the real space of all (equivalence classes of) measurable functions $v : \Omega \to \mathbb{R}$ such that*

$$\int_\Omega |v|^p < \infty,$$

which, endowed with the norm

$$\|v\|_{0,p;\Omega} := \left\{ \int_\Omega |v|^p \right\}^{1/p} \qquad \forall\, v \in \mathrm{L}^p(\Omega),$$

is known, for the purposes of the present exercise, to be a separable Banach space. On the other hand, the Sobolev space $\mathrm{W}^{1,p}(\Omega)$ is given by (cf. Exercise 1.10)

$$\mathrm{W}^{1,p}(\Omega) := \Big\{ u \in \mathrm{L}^p(\Omega) : \quad \text{there exist } g_i \in \mathrm{L}^p(\Omega),\, i = \overline{1, N} \text{ such that}$$
$$\int_\Omega u \frac{\partial \phi}{\partial x_i} = - \int_\Omega g_i \phi \qquad \forall\, \phi \in C_0^\infty(\Omega) \Big\}.$$

In this case we write $\frac{\partial u}{\partial x_i} = g_i$, and define the norm

$$\|u\|_{1,p;\Omega} := \|u\|_{0,p;\Omega} + \sum_{i=1}^N \|g_i\|_{0,p;\Omega}.$$

Knowing that the space $(\mathrm{W}^{1,p}(\Omega), \|\cdot\|_{1,p;\Omega})$ is Banach (cf. Exercise 1.10), prove that it is separable.

▸ **7.5** *Let X, Y be normed vector spaces, and let $T : X \to Y$ be a compact operator. Prove that $\mathcal{R}(T)$ is separable.*

HINT: *Write $X = \bigcup_{n=1}^{\infty} \{x \in X : \|x\| \leqslant n\}$, and then use that every relatively compact subset of a metric space is separable.*

▸ **7.6** *An important result states that every uniformly convex Banach space (cf. Exercise 1.5) is reflexive. Assume now that Ω is an open subset of $\mathbb{R}^n$ and that $p \geqslant 2$.*

a) *From the fact that $\alpha^p + \beta^p \leqslant (\alpha^2 + \beta^2)^{p/2}$ for all $\alpha, \beta \geqslant 0$, prove Clarkson's inequality:*

$$\left\| \frac{f+g}{2} \right\|_{0,p;\Omega}^p + \left\| \frac{f-g}{2} \right\|_{0,p;\Omega}^p \leqslant \frac{1}{2} \left\{ \|f\|_{0,p;\Omega}^p + \|g\|_{0,p;\Omega}^p \right\} \quad \forall\, f, g \in \mathrm{L}^p(\Omega).$$

b) *Use the previous inequality to prove that $\mathrm{L}^p(\Omega)$, $2 \leqslant p < \infty$, is uniformly convex and therefore reflexive. Moreover, deduce that $\mathrm{W}^{1,p}(\Omega)$ is also reflexive.*

▸ **7.7** *Given $p \geqslant 1$, consider the vector space*

$$\ell_p(\mathbb{C}) := \Big\{ \mathbf{x} := \{x_n\}_{n \in N} : \quad x_n \in \mathbb{C} \quad \forall\, n \in \mathbb{N} \quad \text{and} \quad \sum_{n \in \mathbb{N}} |x_n|^p < +\infty \Big\},$$

which is endowed with the norm $\|\mathbf{x}\| := \left\{ \sum_{n \in \mathbb{N}} |x_n|^p \right\}^{1/p}$, and prove that $\ell_p(\mathbb{C})$ is separable.

▸ **7.8** *Given an open set* Ω *of* $\mathbb{R}^n$, *consider once again the Hilbert space* $(\mathbf{H}(\mathrm{div};\Omega), \langle\cdot,\cdot\rangle_{\mathrm{div};\Omega})$ *(cf. Exercise 2.9), where*

$$\mathbf{H}(\mathrm{div};\Omega) := \left\{\boldsymbol{\tau} \in \mathbf{L}^2(\Omega) : \quad \mathrm{div}(\boldsymbol{\tau}) \in \mathrm{L}^2(\Omega)\right\},$$

and

$$\langle\boldsymbol{\sigma},\boldsymbol{\tau}\rangle_{\mathrm{div};\Omega} := \int_\Omega \left\{\boldsymbol{\sigma}\cdot\boldsymbol{\tau} + \mathrm{div}(\boldsymbol{\sigma})\,\mathrm{div}(\boldsymbol{\tau})\right\} \qquad \forall\,\boldsymbol{\sigma},\,\boldsymbol{\tau} \in \mathbf{H}(\mathrm{div};\Omega).$$

Assume that $\mathrm{L}^2(\Omega)$ *is separable and prove that* $\mathbf{H}(\mathrm{div};\Omega)$ *is also separable.*

▸ **7.9** *Let* X *and* Y *be real Banach spaces, and let* $\mathcal{A} : X \times Y \to \mathbb{R}$ *be a bounded bilinear form. Define the operators* $\mathbf{A} \in \mathcal{L}(X, Y')$ *and* $\mathbf{B} \in \mathcal{L}(Y, X')$ *by*

$$\mathbf{A}(x)(y) := \mathcal{A}(x,y) \quad \text{and} \quad \mathbf{B}(y)(x) := \mathcal{A}(x,y) \qquad \forall\, x \in X, \quad \forall\, y \in Y.$$

a) *Define explicitly the adjoint operators* $\mathbf{A}'$ *and* $\mathbf{B}'$, *and prove that*

$$\mathbf{B} = \mathbf{A}' \circ J_Y \quad \text{and} \quad \mathbf{A} = \mathbf{B}' \circ J_X\,,$$

where $J_Y : Y \to Y''$ *and* $J_X : X \to X''$ *are the respective injections.*

b) *Prove that under any of the pairs of hypotheses* i)-ii) *or* iii)-iv) *given below, and assuming that* Y *is reflexive, it follows that* $\mathbf{A}$, $\mathbf{A}'$, *and* $\mathbf{B}$ *are bijective.*

i) *there exists* $\alpha > 0$ *such that* $\displaystyle\sup_{\substack{x\in X\\ x\neq 0}} \frac{\mathcal{A}(x,y)}{\|x\|} \geqslant \alpha\,\|y\| \quad \forall\, y \in Y,$

ii) $\displaystyle\sup_{y\in Y} \mathcal{A}(x,y) > 0 \quad \forall\, x \in X,\ x \neq 0,$

iii) *there exists* $\alpha > 0$ *such that* $\displaystyle\sup_{\substack{y\in Y\\ y\neq 0}} \frac{\mathcal{A}(x,y)}{\|y\|} \geqslant \alpha\,\|x\| \quad \forall\, x \in X,$

iv) $\displaystyle\sup_{x\in X} \mathcal{A}(x,y) > 0 \quad \forall\, y \in Y,\ y \neq 0.$

c) *Prove that under any of the pairs of hypotheses* i)-ii) *or* iii)-iv) *given in* b), *and assuming that* X *is reflexive, it follows that* $\mathbf{B}$, $\mathbf{B}'$, *and* $\mathbf{A}$ *are bijective.*

▸ **7.10** *Let* M *be a proper closed subspace of a reflexive Banach space* X, *and let* $x_0 \in X - M$. *Apply the second geometric form of the Hahn-Banach Theorem to prove that there exists* $F \in M^o$ *such that* $\|F\| = 1$ *and* $|F(x_0)| \leqslant \mathrm{dist}(x_0, M)$.

▸ **7.11** *Prove that if* X *and* Y *are Banach spaces and* $A : X \to (Y, \boldsymbol{\sigma}(Y, Y'))$ *is linear and continuous, then* $A \in \mathcal{L}(X, Y)$.

▸ **7.12** *Prove Lemma 7.44.*

▸ **7.13** *Prove Theorem 7.45.*

▸ **7.14** *Prove that if* Z *is a closed subspace of a normed vector space* X, *then*

$$\boldsymbol{\sigma}(Z, Z') = \boldsymbol{\sigma}(X, X')|_Z := \left\{A \cap Z : \quad A \in \boldsymbol{\sigma}(X, X')\right\}.$$

▸ **7.15** *Complete the proof of Theorem 7.61 following the model used in the proof of Theorem 7.60, and using the characterization given by Lemma 7.35.*

▸ **7.16** *Let* X, Y, *and* Z *be Banach spaces such that* Z *is reflexive.*

a) *Define the operator* $\mathcal{S} : X \to \mathcal{L}\big(\mathcal{L}(X,Y), Y\big)$ *by*

$$\mathcal{S}(x)(A) := A(x) \qquad \forall\, A \in \mathcal{L}(X,Y), \quad \forall\, x \in X,$$

and prove, using in particular Lemma 3.10, that $\mathcal{S}$ *is linear, bounded, injective, and has closed range.*

b) *Let* $\{x_n\}_{n\in\mathrm{N}} \subseteq X$, *and let* $\{B_n\}_{n\in\mathrm{N}} \subseteq \mathcal{L}(X,Z)$ *be such that*
   i) $\{A(x_n)\}_{n\in\mathrm{N}}$ *converges in* Y *for all* $A \in \mathcal{L}(X,Y)$, *and*
   ii) $\{B_n(x)\}_{n\in\mathrm{N}}$ *converges in* Z *for all* $x \in X$.

   Prove that there exist $\{x_n^{(1)}\}_{n\in\mathrm{N}} \subseteq \{x_n\}_{n\in\mathrm{N}}$, $\{B_n^{(1)}\}_{n\in\mathrm{N}} \subseteq \{B_n\}_{n\in\mathrm{N}}$, *and* $z \in Z$, *such that* $G(B_n^{(1)}(x_n^{(1)})) \overset{n\to+\infty}{\longrightarrow} G(z)$ *for all* $G \in Z'$.

c) *Explain what a) and b) reduce to in the particular case* $Y = \mathbb{R}$.

▸ **7.17** *The goal of this problem is to prove the Brezis–Lieb Lemma, whose respective hypotheses and conclusion are given below at the beginning of ii) and within iii). For this purpose, it is suggested to proceed as indicated below.*

i) *Given* $p \in (1,+\infty)$, *prove that*

$$C_p := \sup_{|t|\leqslant 1}\left\{\frac{\big||t+1|^p - |t|^p - 1\big|}{|t|^{p-1} + |t|}\right\} < \infty,$$

and from this, taking $t = a/b$ *or* $t = b/a$ *when* $a\,b \neq 0$, *show that*

$$\big||a+b|^p - |a|^p - |b|^p\big| \leqslant C_p\left\{|a|^{p-1}|b| + |a||b|^{p-1}\right\} \qquad \forall\, a,b \in \mathrm{R}. \tag{7.21}$$

ii) *Let* $\{f_n\}_{n\in\mathrm{N}} \subseteq \mathrm{L}^p(\Omega)$, *and let* $f \in \mathrm{L}^p(\Omega)$ *be such that*

$$f_n(x) \to f(x) \quad \text{a.e. in } \Omega \quad \text{and} \quad \|f_n\|_{0,p;\Omega} \to \|f\|_{0,p;\Omega}.$$

Recall, for example from Exercise 5.1, that if $q \in (1,+\infty)$ *is such that* $\frac{1}{p} + \frac{1}{q} = 1$, *the dual of* $\mathrm{L}^p(\Omega)$ *identifies with* $\mathrm{L}^q(\Omega)$. *Prove then that*

$$|f_n - f| \overset{w}{\to} 0 \quad \text{in} \quad \mathrm{L}^p(\Omega) \quad \text{and} \quad |f_n - f|^{p-1} \overset{w}{\to} 0 \quad \text{in} \quad \mathrm{L}^q(\Omega).$$

Then apply (7.21) with $a = (f_n - f)(x)$ *and* $b = f(x)$, *integrate over* Ω, *and deduce that*

$$\lim_{n\to\infty}\int_\Omega \left\{|f_n|^p - |f_n - f|^p\right\} = \int_\Omega |f|^p. \tag{7.22}$$

iii) *Finally conclude from (7.22) that*

$$\|f_n - f\|_{0,p;\Omega} \longrightarrow 0.$$

Hint: *For the first weak convergence requested in ii) you may use a variant of Mazur's Theorem, which states that for every sequence* $\{u_n\}_{n\in\mathrm{N}}$ *converging weakly in a Banach space* X, *there exists a sequence* $\{v_n\}_{n\in\mathrm{N}}$ *that converges strongly to the same limit, where each* v_n *is a convex combination of elements in* $\{u_k\}_{k\geqslant n}$. *Also, keep in mind that every convergent sequence in* $\mathrm{L}^p(\Omega)$ *has a subsequence that converges pointwise a.e. in* Ω. *Finally, recall that since* $\mathrm{L}^q(\Omega)$ *is separable, the unit ball of* $\mathrm{L}^p(\Omega)$ *is metrizable with respect to the weak topology.*

▸ **7.18** *Let* H *be a real Hilbert space, and let* $\mathcal{R} : H' \to H$ *be its associated Riesz operator. Prove that* A *is weak* open in* H' *if and only if* $\mathcal{R}(A)$ *is weakly open in* H, *that is*

$$A \in \sigma(H',H) \Leftrightarrow \mathcal{R}(A) \in \sigma(H,H').$$

▸ **7.19** *Let* Ω *be a domain of* $\mathbb{R}^2$, *and let* $\{\boldsymbol{\sigma}_n\}_{n\in\mathrm{N}} \subseteq \mathbf{H}(\mathrm{rot};\Omega)$ *and* $\boldsymbol{\sigma} \in \mathbf{H}(\mathrm{rot};\Omega)$ *be such that* $\lim_{n\to+\infty}\langle\boldsymbol{\sigma}_n,\boldsymbol{\tau}\rangle_{\mathrm{rot};\Omega} = \langle\boldsymbol{\sigma},\boldsymbol{\tau}\rangle_{\mathrm{rot};\Omega}$ *for all* $\boldsymbol{\tau} \in \mathbf{H}(\mathrm{rot};\Omega)$. *Prove that for each* $\boldsymbol{\tau} \in \mathbf{H}(\mathrm{rot};\Omega)$ *there holds*

$$\lim_{n\to+\infty}\int_\Omega \boldsymbol{\sigma}_n\cdot\boldsymbol{\tau} = \int_\Omega \boldsymbol{\sigma}\cdot\boldsymbol{\tau} \quad \text{and} \quad \lim_{n\to+\infty}\int_\Omega \mathrm{rot}(\boldsymbol{\sigma}_n)\,\mathrm{rot}(\boldsymbol{\tau}) = \int_\Omega \mathrm{rot}(\boldsymbol{\sigma})\,\mathrm{rot}(\boldsymbol{\tau}).$$

Hint: *Recall that the space* $\mathbf{H}(\text{rot};\Omega)$, *given by*

$$\mathbf{H}(\text{rot};\Omega) := \left\{ \boldsymbol{\zeta} := (\zeta_1,\zeta_2) \in \mathbf{L}^2(\Omega) : \quad \text{rot}(\boldsymbol{\zeta}) := \frac{\partial \zeta_2}{\partial x_1} - \frac{\partial \zeta_1}{\partial x_2} \in \mathrm{L}^2(\Omega) \right\},$$

and endowed with the inner product

$$\langle \boldsymbol{\zeta}, \boldsymbol{\tau} \rangle_{\text{rot};\Omega} := \int_\Omega \boldsymbol{\zeta} \cdot \boldsymbol{\tau} \, dx + \int_\Omega \text{rot}(\boldsymbol{\zeta}) \, \text{rot}(\boldsymbol{\tau}) \, dx \quad \forall \, \boldsymbol{\zeta}, \, \boldsymbol{\tau} \in \mathbf{H}(\text{rot};\Omega),$$

is a Hilbert space.

8

Notions of Spectral Theory

The purpose of this chapter, mainly based on the approach and results given in [47] and [115], is to study the main properties of the spectrum of bounded and unbounded operators, including, in particular, the compact and self-adjoint cases.

8.1 Motivation

Given $k \in \mathbb{R}$ and $f \in \mathrm{L}^2(\Omega)$, where Ω is a bounded open subset of $\mathbb{R}^n, n \in \{2,3\}$, with boundary Γ, we consider the Helmholtz equation:

$$\triangle u + k^2 u = f \quad \text{in} \quad \Omega, \qquad u = 0 \quad \text{on} \quad \Gamma, \tag{8.1}$$

which arises when assuming a harmonic behavior of the solution of the wave equation. The corresponding weak formulation reduces to: Find $u \in \mathrm{H}_0^1(\Omega)$ such that

$$\int_\Omega \nabla u \cdot \nabla v \,-\, k^2 \int_\Omega u\, v = -\int_\Omega f\, v \quad \forall\, v \in \mathrm{H}_0^1(\Omega). \tag{8.2}$$

Then, we define the operators $A : \mathrm{H}_0^1(\Omega) \to \mathrm{H}_0^1(\Omega)$ and $K : \mathrm{H}_0^1(\Omega) \to \mathrm{H}_0^1(\Omega)$ according to the diagrams

$$\mathrm{H}_0^1(\Omega) \xrightarrow{\mathcal{A}} \mathrm{H}_0^1(\Omega)' \xrightarrow{\mathcal{R}} \mathrm{H}_0^1(\Omega)$$

and

$$\mathrm{H}_0^1(\Omega) \xrightarrow{\mathcal{K}} \mathrm{L}^2(\Omega)' \xrightarrow{i'} \mathrm{H}_0^1(\Omega)' \xrightarrow{\mathcal{R}} \mathrm{H}_0^1(\Omega),$$

that is, $A := \mathcal{R} \circ \mathcal{A}$ and $K := \mathcal{R} \circ i' \circ \mathcal{K}$, where

$$\mathcal{A}(u)(v) := \int_\Omega \nabla u \cdot \nabla v \qquad \forall\, u, v \in \mathrm{H}_0^1(\Omega),$$
$$\mathcal{K}(u)(v) := k^2 \int_\Omega u\, v \qquad \forall\, u \in \mathrm{H}_0^1(\Omega), \quad \forall\, v \in \mathrm{L}^2(\Omega),$$

$\mathcal{R} : \mathrm{H}_0^1(\Omega)' \to \mathrm{H}_0^1(\Omega)$ is the usual Riesz map, and $i : \mathrm{H}_0^1(\Omega) \to \mathrm{L}^2(\Omega)$ is the compact injection (cf. Lemma 10.19) with adjoint operator $i' : \mathrm{L}^2(\Omega)' \to \mathrm{H}_0^1(\Omega)'$. It follows, thanks to the Lax–Milgram Lemma (cf. Lemma 4.4), that A is bijective, and it is also easy to see, according to the compactness of i' (cf. Theorem 6.26), that K is compact. With these operators, the formulation (8.2) can be rewritten as: Find $u \in \mathrm{H}_0^1(\Omega)$ such that

G. N. Gatica, *Introduction to Functional Analysis*,
https://doi.org/10.1007/978-3-032-37415-8_8

$$(A - K)(u) = \mathcal{R}(F), \tag{8.3}$$

where $F \in H_0^1(\Omega)'$ is defined by $F(v) := -\int_\Omega f v$ for all $v \in H_0^1(\Omega)$. Then, a direct application of Theorem 6.24 allows one to conclude that the Fredholm alternative holds for problem (8.3) (equivalently (8.2) or (8.1)). This means that, in order to study the solvability of (8.2), it suffices to analyze the associated homogeneous problem:

$$\Delta u + k^2 u = 0 \quad \text{in} \quad \Omega, \qquad u = 0 \quad \text{on} \quad \Gamma. \tag{8.4}$$

In particular, if $u = 0$ is the only solution of (8.4), then the original Helmholtz equation (8.1) has a unique solution for every $f \in L^2(\Omega)$. To better understand (8.4), let us now define the operator $T : L^2(\Omega) \to H_0^1(\Omega)$ that assigns to each $g \in L^2(\Omega)$ the unique solution $z := T(g) \in H_0^1(\Omega)$, which is guaranteed by the Lax–Milgram Lemma, of the Poisson problem:

$$-\Delta z = g \quad \text{in} \quad \Omega, \qquad z = 0 \quad \text{on} \quad \Gamma. \tag{8.5}$$

It follows that T is linear and bounded. In particular, the continuous dependence result states that $\|T(g)\|_{1,\Omega} := \|z\|_{1,\Omega} \leqslant C\,\|g\|_{0,\Omega}$, which gives precisely the boundedness of T. Moreover, the elliptic regularity of (8.5) establishes the existence of $\delta \in (0, 1/2)$ such that, in fact, $z \in H^{1+\delta}(\Omega)$. Then, since the injection $i : H^{1+\delta}(\Omega) \to H^1(\Omega)$ is compact, we have that $i : H^{1+\delta}(\Omega) \cap H_0^1(\Omega) \to H_0^1(\Omega)$, and hence $T : L^2(\Omega) \to H_0^1(\Omega)$, are compact as well. In addition, while this property of T certainly implies the compactness of $T : L^2(\Omega) \to L^2(\Omega)$, we stress that this latter result does not require the additional regularity of (8.5), but only the boundedness of $T : L^2(\Omega) \to H_0^1(\Omega)$ and the compactness of $i : H_0^1(\Omega) \to L^2(\Omega)$ (cf. Lemma 10.19). On the other hand, from the weak formulation of (8.5) we have that for each $g \in L^2(\Omega)$ there holds

$$\int_\Omega \nabla T(g) \cdot \nabla v = \int_\Omega g\, v \quad \forall\, v \in H_0^1(\Omega),$$

and thus, taking $v = T(f)$, it follows that

$$\int_\Omega \nabla T(g) \cdot \nabla T(f) = \int_\Omega g\, T(f) \quad \forall\, f \in L^2(\Omega),$$

so that, exchanging the roles of f and g, we get

$$\int_\Omega f\, T(g) = \int_\Omega \nabla T(f) \cdot \nabla T(g) = \int_\Omega \nabla T(g) \cdot \nabla T(f) = \int_\Omega g\, T(f),$$

which proves that $T : L^2(\Omega) \longrightarrow L^2(\Omega)$ is self-adjoint. Thus, using the operator T, the homogeneous equation (8.4) can be rewritten as $u = T(k^2 u)$, that is, $T(u) = \frac{1}{k^2}\, u$, which means that (8.4) has nontrivial solutions if $\frac{1}{k^2}$ is an eigenvalue of T. Precisely, one of the topics to be studied in the following sections concerns the spectrum of compact and self-adjoint operators.

8.2 Preliminaries

Throughout the rest of this chapter, I stands for a generic identity operator.

Definition 8.1. *Let X be a Banach space, and let $A \in \mathcal{L}(X)$. The* RESOLVENT *of A is defined as the set*

$$\rho(A) := \Big\{\lambda \in \mathbb{C}: \quad (A - \lambda I) : X \to X \text{ is bijective}\Big\},$$

and the SPECTRUM *of A as the complement of the resolvent, that is,*

$$\boldsymbol{\sigma}(A) := \mathbb{C} - \rho(A).$$

Definition 8.2. *The* POINT SPECTRUM *of A is defined as the set*

$$\boldsymbol{\sigma}_p(A) := \Big\{\lambda \in \mathbb{C}: \quad N(A - \lambda I) \neq \{0\}\Big\}.$$

In this case, for $\lambda \in \boldsymbol{\sigma}_p(A)$, the space $N(A - \lambda I)$ is called the EIGENSPACE *(or simply* PROPER SPACE*) associated with the eigenvalue λ.*

Note, according to the Bounded Inverse Theorem (cf. Theorem 3.34), that the operator $(A - \lambda I)^{-1} \in \mathcal{L}(X)$ for all $\lambda \in \rho(A)$. Moreover, the inclusion $\boldsymbol{\sigma}_p(A) \subseteq \boldsymbol{\sigma}(A)$ is, in general, strict, unless X is finite-dimensional, in which case both sets coincide. In other words, there exist $\lambda \in \boldsymbol{\sigma}(A)$ such that $N(A - \lambda I) = \{0\}$ and $R(A - \lambda I) \neq X$. As an example, consider the operator $A : l^2(\mathbb{R}) \to l^2(\mathbb{R})$ defined by

$$A(u) := (0, u_1, u_2, \ldots) \qquad \forall\, u := (u_n)_{n\in\mathbb{N}} \in l^2(\mathbb{R}).$$

It is clear that $N(A - 0 \cdot I) = N(A) = \{0\}$, which shows that $0 \notin \boldsymbol{\sigma}_p(A)$, and yet $0 \in \boldsymbol{\sigma}(A)$ since $A - 0 \cdot I = A$ is not surjective.

Lemma 8.3. *Let X be a complex Banach space, and let $A \in \mathcal{L}(X)$. Then the spectrum $\boldsymbol{\sigma}(A)$ is a nonempty closed set contained in*

$$\bar{B}(0, \|A\|) := \Big\{\lambda \in \mathbb{C}: \quad |\lambda| \leqslant \|A\|\Big\}. \tag{8.6}$$

Proof. Let $\lambda \in \mathbb{C}$, with $|\lambda| > \|A\|$, and let us show that $\lambda \in \rho(A)$, that is, that $A - \lambda I$ is bijective. Indeed, given $f \in X$, consider the equation: Find $u \in X$ such that $(A - \lambda I)(u) = f$, or equivalently, such that

$$u = T(u) := \frac{1}{\lambda}\big(A(u) - f\big).$$

It is clear that

$$\|T(v) - T(w)\| = \left\|\frac{1}{\lambda}\big(A(v) - A(w)\big)\right\| \leqslant \frac{\|A\|}{|\lambda|}\,\|v - w\| \quad \forall\, v, w \in X,$$

which shows that T is a contraction, since $\dfrac{\|A\|}{|\lambda|} < 1$. Therefore, by the Banach Fixed Point Theorem (cf. Theorem 1.8), there exists a unique $u \in X$ such that $u = T(u)$, or equivalently, $(A - \lambda I)(u) = f$. This proves that $A - \lambda I$ is bijective. Next, we prove that $\rho(A)$ is open, which is equivalent to saying that $\boldsymbol{\sigma}(A)$ is closed. Indeed, let $\lambda_0 \in \rho(A)$, that is, $\lambda_0 \in \mathbb{C}$ and $(A - \lambda_0 I) : X \to X$ is a bijection. Then, given $\lambda \in \mathbb{C}$ and $f \in X$, consider again the equation $(A - \lambda I)(u) = f$, which can be written equivalently as

$$(A - \lambda_0 I)(u) = f + (\lambda - \lambda_0)\, u,$$

or

$$u = T(u) := (A - \lambda_0 I)^{-1}\Big\{f + (\lambda - \lambda_0)u\Big\}.$$

It follows that

$$\|T(v) - T(w)\| = \Big\|(A - \lambda_0 I)^{-1}\Big\{(\lambda - \lambda_0)(v - w)\Big\}\Big\|$$
$$\leqslant \|(A - \lambda_0 I)^{-1}\| \, |\lambda - \lambda_0| \, \|v - w\| \quad \forall\, v, w \in X\,,$$

which proves that if

$$|\lambda - \lambda_0| < \|(A - \lambda_0 I)^{-1}\|^{-1},$$

then T is a contraction, and thus, invoking again Theorem 1.8, $(A - \lambda I)$ is bijective. This shows that $B(\lambda_0, \|(A-\lambda_0 I)^{-1}\|^{-1}) \subseteq \rho(A)$, and therefore $\rho(A)$ is open. It remains to prove that $\boldsymbol{\sigma}(A)$ is nonempty. To this end, assume by contradiction that $\rho(A) = \mathbb{C}$. Then we can define the resolvent mapping $\mathfrak{R}_A : \mathbb{C} \to \mathcal{L}(X)$ by

$$\mathfrak{R}_A(\lambda) := (A - \lambda I)^{-1} \qquad \forall\, \lambda \in \mathbb{C}\,,$$

which is holomorphic throughout $\mathbb{C}$, meaning that it is complex differentiable. Indeed, given λ, $\mu \in \rho(A) = \mathbb{C}$, we find, after simple algebraic manipulations, that

$$\mathfrak{R}_A(\lambda) - \mathfrak{R}_A(\mu) = (A - \lambda I)^{-1}\Big((A - \mu I) - (A - \lambda I)\Big)(A - \mu I)^{-1}$$
$$= \mathfrak{R}_A(\lambda)\Big((\lambda - \mu)\, I\Big)\mathfrak{R}_A(\mu) = (\lambda - \mu)\,\mathfrak{R}_A(\lambda)\,\mathfrak{R}_A(\mu)\,,$$

that is

$$\mathfrak{R}_A(\lambda) - \mathfrak{R}_A(\mu) = (\lambda - \mu)\,\mathfrak{R}_A(\lambda)\,\mathfrak{R}_A(\mu) \qquad \forall\, \lambda,\ \mu \in \mathbb{C}\,,$$

which is known as the resolvent identity. In this way, dividing by $\lambda - \mu$, with $\lambda \neq \mu$, and letting μ tend to λ, we deduce that

$$\mathfrak{R}'_A(\lambda) = \big(\mathfrak{R}_A(\lambda)\big)^2 \qquad \forall\, \lambda \in \mathbb{C}\,,$$

thus confirming the aforementioned differentiability. Next, we have

$$(A - \lambda I)^{-1} = -\frac{1}{\lambda}\left(I - \frac{1}{\lambda}A\right)^{-1},$$

from which, taking $|\lambda| > \|A\|$ and employing the estimate provided by Exercise 3.12, we find that

$$\|(A - \lambda I)^{-1}\| \leqslant \frac{1}{|\lambda| - \|A\|}\,,$$

and hence

$$\|(A - \lambda I)^{-1}\| \to 0 \qquad \text{as } |\lambda| \to +\infty\,.$$

Thus, the resolvent mapping is bounded on $\mathbb{C}$. By Liouville's Theorem (cf. [113, Theorem 10.23]), it must be constant, and since it tends to 0 at infinity, it is identically zero, which contradicts the invertibility of $(A - \lambda I)^{-1}$. Therefore, necessarily, $\boldsymbol{\sigma}(A) \neq \emptyset$.

□

As a consequence of Lemma 8.3, and assuming again that X is a complex Banach space, we define the SPECTRAL RADIUS of an operator $A \in \mathcal{L}(X)$ by

$$r_{\boldsymbol{\sigma}}(A) := \sup\big\{|\lambda| : \quad \lambda \in \boldsymbol{\sigma}(A)\big\}\,, \tag{8.7}$$

from which it is clear that $\boldsymbol{\sigma}(A) \subseteq \bar{B}(0, r_{\boldsymbol{\sigma}}(A))$. In addition, it follows immediately from (8.6) that $r_{\boldsymbol{\sigma}}(A) \leqslant \|A\|$.

Lemma 8.4. *Let X be a Banach space, let $A \in \mathcal{L}(X)$, and let*

$$\{\lambda_1, \ldots, \lambda_N\} \subseteq \boldsymbol{\sigma}_p(A) \quad \text{and} \quad \{e_1, \ldots, e_N\} \subseteq X$$

be sets of N eigenvalues and N corresponding eigenvectors, respectively, such that $\lambda_i \neq \lambda_j$ for all $i \neq j$. Then $\{e_1, \ldots, e_N\}$ is linearly independent.

Proof. By finite induction on N. For $N = 1$ it is immediate since $e_1 \neq 0$. Now assume that the statement holds for $N = k$, that is, $\{e_1, \ldots, e_k\}$ is linearly independent, and let us prove that $\{e_1, \ldots, e_k, e_{k+1}\}$ is as well. Let $\alpha_1, \ldots, \alpha_{k+1} \in \mathbb{C}$ be such that $\sum_{j=1}^{k+1} \alpha_j\, e_j = 0$. If $\alpha_{k+1} = 0$, then the induction hypothesis implies $\alpha_1 = \cdots = \alpha_k = 0$, which completes the proof. Otherwise, we obtain $e_{k+1} = \sum_{j=1}^{k} \widetilde{\alpha}_j\, e_j$ with $\widetilde{\alpha}_j := -\alpha_j/\alpha_{k+1}$ for all $j \in \{1, \ldots, k\}$, and by applying A we get

$$\sum_{j=1}^{k} \widetilde{\alpha}_j\, \lambda_j\, e_j = A(e_{k+1}) = \lambda_{k+1}\, e_{k+1} = \sum_{j=1}^{k} \widetilde{\alpha}_j\, \lambda_{k+1}\, e_j,$$

that is,

$$\sum_{j=1}^{k} \widetilde{\alpha}_j\, (\lambda_j - \lambda_{k+1})\, e_j = 0,$$

which, thanks to the induction hypothesis, implies that $\widetilde{\alpha}_j\, (\lambda_j - \lambda_{k+1}) = 0$ for all $j \in \{1, \ldots, k\}$. Since the elements of $\{\lambda_1, \ldots, \lambda_{k+1}\}$ are all distinct, it follows that $\widetilde{\alpha}_j = 0$ for all $j \in \{1, \ldots, k\}$, and therefore $e_{k+1} = 0$, which is a contradiction. Thus, necessarily $\alpha_{k+1} = 0$, and we return to the previous argument, which completes the induction.

□

8.3 Spectrum of a compact operator

The following three lemmas describe the main characteristics of the spectrum of a compact operator.

Lemma 8.5. *Let X be a Banach space with* $\dim X = +\infty$, *and let $K \in \mathcal{K}(X)$. Then*

i) $0 \in \boldsymbol{\sigma}(K)$.
ii) $\boldsymbol{\sigma}(K) \backslash \{0\} = \boldsymbol{\sigma}_p(K) \backslash \{0\}$.

Proof. Suppose, by contradiction, that $0 \in \rho(K)$. It follows that K is bijective and hence, by the Bounded Inverse Theorem (cf. Theorem 3.34), $K^{-1} \in \mathcal{L}(X)$, which yields $I = K\,K^{-1} \in \mathcal{K}(X)$. This implies that the unit sphere $\partial B(0,1) = \overline{I(\partial B(0,1))}$ is compact, which contradicts the infinite dimension of X. Thus, necessarily, $0 \in \boldsymbol{\sigma}(K)$. On the other hand, let $\lambda \in \boldsymbol{\sigma}(K) \backslash \{0\}$ and suppose, again by contradiction, that $\lambda \notin \boldsymbol{\sigma}_p(K)$, that is, $N(K - \lambda I) = \{0\}$. It follows, by the Fredholm Alternative (cf. Theorem 6.24), that $K - \lambda I$ is bijective, which means that $\lambda \in \rho(K)$, contradicting the assumption that $\lambda \in \boldsymbol{\sigma}(K)$.

□

Note that part ii) of the previous lemma establishes, equivalently, that

$$\boldsymbol{\sigma}(K) = \boldsymbol{\sigma}_p(K) \cup \{0\},$$

from which it follows that $\boldsymbol{\sigma}(K)$ and $\boldsymbol{\sigma}_p(K)$ coincide only if $0 \in \boldsymbol{\sigma}_p(K)$.

Lemma 8.6. *Let X be a Banach space with $\dim X = +\infty$, and let $K \in \mathcal{K}(X)$. Moreover, let $\{\lambda_n\}_{n\in\mathbb{N}}$ be a sequence of distinct complex numbers such that*

$$\lambda_n \in \boldsymbol{\sigma}(K)\setminus\{0\} \quad \forall n \in \mathbb{N} \quad \textit{and} \quad \lambda_n \overset{n\to\infty}{\to} \lambda \in \mathbb{C}.$$

Then $\lambda = 0$.

Proof. Since $\boldsymbol{\sigma}(K)\setminus\{0\} = \boldsymbol{\sigma}_p(K)\setminus\{0\}$ (by ii) of Lemma 8.5), we have that $\lambda_n \in \boldsymbol{\sigma}_p(K)$ for all $n \in \mathbb{N}$, and therefore we can choose eigenvectors $e_n \in N(K - \lambda_n I)$, $e_n \neq 0$, for all $n \in \mathbb{N}$. Then, we define the subspace $X_n := \langle\{e_1, e_2, \ldots, e_n\}\rangle$ for all $n \in \mathbb{N}$, and observe, thanks to Lemma 8.4, that $\dim X_n = n$. It follows that X_n is a proper subspace of X_{n+1}, and it is clear that $(K - \lambda_n I)(X_n) \subseteq X_{n-1}$ for all $n \in \mathbb{N}$, since $e_n \in N(K - \lambda_n I)$. Then, applying Riesz's Lemma (cf. Lemma 6.7) to the pair of subspaces (X_{n-1}, X_n) with $\varepsilon = 1/2$, one constructs a sequence $\{u_n\}_{n\in\mathbb{N}} \subseteq X$ such that $\|u_n\| = 1$ and $\operatorname{dist}(u_n, X_{n-1}) \geqslant 1/2$ for all $n \in \mathbb{N}$. Then, for $n > m \geqslant 2$, we have $X_{m-1} \subsetneq X_m \subseteq X_{n-1} \subsetneq X_n$, and hence

$$\begin{aligned}
&\left\|\frac{K(u_n)}{\lambda_n} - \frac{K(u_m)}{\lambda_m}\right\| = \left\|\frac{(K(u_n) - \lambda_n u_n)}{\lambda_n} - \frac{(K(u_m) - \lambda_m u_m)}{\lambda_m} + u_n - u_m\right\| \\
&= \left\|\left\{\frac{(K - \lambda_n I)(u_n)}{\lambda_n} - \frac{(K - \lambda_m I)(u_m)}{\lambda_m} - u_m\right\} + u_n\right\| \\
&\geqslant \operatorname{dist}(u_n, X_{n-1}) \geqslant 1/2.
\end{aligned}$$

Finally, if we assume by contradiction that $\lambda \neq 0$, it follows that there exist $N \in \mathbb{N}$ and $\delta > 0$ such that $|\lambda_n| \geqslant \delta$ for all $n \geqslant N$. Thus, the sequence $\left\{\frac{u_n}{\lambda_n}\right\}_{n\geqslant N}$ is bounded, and yet $\left\|K\left(\frac{u_n}{\lambda_n}\right) - K\left(\frac{u_m}{\lambda_m}\right)\right\| \geqslant 1/2$ for all $n > m \geqslant N$, which contradicts the compactness of K, and therefore, necessarily, $\lambda = 0$.

□

Having established the two previous lemmas, we can now prove the following result.

Lemma 8.7. *Let X be a Banach space with $\dim X = +\infty$, and let $K \in \mathcal{K}(X)$. Then one of the following alternatives holds*

a) $\boldsymbol{\sigma}(K) = \{0\}$, *or*
b) $\boldsymbol{\sigma}(K)\setminus\{0\}$ *is finite, or*
c) $\boldsymbol{\sigma}(K)\setminus\{0\}$ *is a sequence converging to* 0.

Proof. For each $n \in \mathbb{N}$ we define the set

$$\boldsymbol{\sigma}_n(K) := \boldsymbol{\sigma}(K) \cap \left\{\lambda \in \mathbb{C} : \quad |\lambda| \geqslant 1/n\right\},$$

which, by virtue of Lemma 8.6, is either empty or contains a finite number of elements. Indeed, suppose by contradiction that $\boldsymbol{\sigma}_n(K)$ has infinitely many elements. Then, since $\boldsymbol{\sigma}_n(K)$ is compact (being the intersection of a compact set with a closed one), it has an accumulation point, contradicting Lemma 8.6. Hence

$$\boldsymbol{\sigma}(K)\setminus\{0\} = \cup\left\{\boldsymbol{\sigma}_n(K): \quad n \in \mathbb{N}\right\},$$

and therefore $\boldsymbol{\sigma}(K)\setminus\{0\}$ is finite or countable. If $\boldsymbol{\sigma}(K)\setminus\{0\} = \{\lambda_k\}_{k\in\mathbb{N}}$, then the fact that $\boldsymbol{\sigma}_n(K)$ is finite implies that for each $n \in \mathbb{N}$ there exists $N \in \mathbb{N}$ such that $\lambda_k \notin \boldsymbol{\sigma}_n(K)$ for all $k \geqslant N$, that is, $|\lambda_k| < 1/n$ for all $k \geqslant N$, which proves that $\lambda_k \overset{n\to\infty}{\to} 0$.

□

A kind of reciprocal to part c) of Lemma 8.7 consists in asking whether, given a sequence $\{\lambda_n\}_{n\in\mathbb{N}} \subseteq \mathbb{C}\setminus\{0\}$ converging to 0, is it possible to construct a compact operator K such that $\boldsymbol{\sigma}(K) = \{\lambda_n\}_{n\in\mathbb{N}} \cup \{0\}$? The answer is indeed affirmative, for which we may consider, for example, the space

$$X = l^2(\mathbb{C}) := \left\{u := (u_n)_{n\in\mathbb{N}},\ u_n \in \mathbb{C} \quad \forall\, n \in \mathbb{N} \quad \text{and} \quad \sum_{n\in\mathbb{N}} |u_n|^2 < +\infty\right\}$$

and the operator $K : X \to X$ defined by

$$K(u) := (\lambda_n\, u_n)_{n\in\mathbb{N}} \quad \forall\, u := (u_n)_{n\in\mathbb{N}} \in X.$$

Indeed, note first that, being convergent, $\{\lambda_n\}_{n\in\mathbb{N}}$ is bounded, and therefore $K(u) \in X$ for all $u \in X$. Moreover, K is clearly compact since $\|K - K_n\| \overset{n\to\infty}{\to} 0$, where $K_n : X \to X$ is the finite-rank operator defined by

$$K_n(u) := (\lambda_1\, u_1, \lambda_2\, u_2, \ldots, \lambda_n\, u_n, 0, 0, \ldots, 0) \quad \forall\, u \in X, \quad \forall_n \in \mathbb{N}.$$

Note that here we have applied Theorem 6.15, which establishes that $\mathcal{K}(X)$ is a closed subspace of $\mathcal{L}(X)$. Thus, since $\dim X = +\infty$, Lemma 8.5 implies that $0 \in \boldsymbol{\sigma}(K)$. Now, let $\lambda \in \mathbb{C}$ such that $\lambda \neq 0$ and $\lambda \neq \lambda_n$ for all $n \in \mathbb{N}$, and let us prove that $\lambda \in \rho(K)$. Indeed, given $u \in N(K - \lambda\, I)$, we have $K(u) = \lambda\, u$, hence $(\lambda_n - \lambda)\, u_n = 0$ for all $n \in \mathbb{N}$, which implies necessarily that $u_n = 0$ for all $n \in \mathbb{N}$, that is, $u = 0$, and thus $N(K - \lambda\, I) = \{0\}$. On the other hand, given $v \in X$, define $u := (u_n)_{n\in\mathbb{N}}$ with $u_n := \dfrac{1}{(\lambda_n - \lambda)}\, v_n$ for all $n \in \mathbb{N}$. Since the sequence $\left\{\dfrac{1}{\lambda_n - \lambda}\right\}_{n\in\mathbb{N}}$ converges to $-\dfrac{1}{\lambda}$, it is bounded, and therefore $u \in X$. Moreover, it is clear that $(K - \lambda\, I)(u) = v$, which shows that $R(K - \lambda\, I) = X$, and hence $\lambda \in \rho(K)$. We have thus proved that $\boldsymbol{\sigma}(K) \subseteq \{\lambda_n\}_{n\in\mathbb{N}} \cup \{0\}$. Finally, let us show that $\lambda_n \in \boldsymbol{\sigma}_p(K)$ for all $n \in \mathbb{N}$. Indeed, given $n \in \mathbb{N}$ we define the sequence $u^{(n)} := (\delta_{nk})_{k\in\mathbb{N}} = (0, 0, \ldots 0, 1, 0, \ldots)$, with 1 in the n-th position, which certainly belongs to X and satisfies $K(u^{(n)}) = \lambda_n\, u^{(n)}$, that is, $u^{(n)} \in N(K - \lambda_n\, I)$, from which it follows that $\lambda_n \in \boldsymbol{\sigma}_p(K)$.

We now consider the case of a Hilbert space X and prove a result that connects the spectrum of a compact and self-adjoint operator $A \in \mathcal{L}(X)$ with the existence of a Hilbert basis of X. For this, we first need some preliminary results. We begin by observing (to be proved later) that if X is a Hilbert space over $(\mathbb{C})$ and $A \in \mathcal{L}(X)$ is self-adjoint, then $\boldsymbol{\sigma}(A) \subseteq \mathbb{R}$. In particular, the inclusion $\boldsymbol{\sigma}_p(A) \subseteq \mathbb{R}$ is quite simple to verify. Indeed, let $\lambda \in \mathbb{C}$ be such that $\lambda \in \boldsymbol{\sigma}_p(A)$. It follows that there exists $u \neq 0$ such that $A(u) = \lambda\, u$, and then

$$\lambda\, \|u\|^2 = \langle A(u), u\rangle = \langle u, A(u)\rangle = \bar{\lambda}\, \|u\|^2,$$

from which $\lambda = \bar{\lambda}$, that is, $\lambda \in \mathbb{R}$. The full proof that $\boldsymbol{\sigma}(A) \subseteq \mathbb{R}$ will be given in Section 8.5 for the more general case of an unbounded self-adjoint operator.

Lemma 8.8. *Let $(X, \langle\cdot,\cdot\rangle)$ be a real Hilbert space, let $A \in \mathcal{L}(X)$ be self-adjoint, and define*

$$m := \inf_{\|u\|=1} \langle A(u), u\rangle \quad \text{and} \quad M := \sup_{\|u\|=1} \langle A(u), u\rangle .$$

Then $\boldsymbol{\sigma}(A) \subseteq [m, M]$, and $m,\ M \in \boldsymbol{\sigma}(A)$.

Proof. Let us first show that if $\lambda > M$, then $\lambda \in \rho(A)$. Indeed, from the inequality

$$\langle A(u), u\rangle \leqslant M \|u\|^2 \qquad \forall\, u \in X,$$

we obtain

$$\langle(\lambda I - A)(u), u\rangle = \lambda \|u\|^2 - \langle A(u), u\rangle \geqslant (\lambda - M)\|u\|^2 \quad \forall\, u \in X,$$

which shows that $(\lambda I - A)$ induces an X-elliptic bilinear form, and therefore, by the Lax–Milgram Lemma (cf. Lemma 4.4), $(\lambda I - A)$ is bijective. Similarly, one shows that if $\lambda < m$, then $\lambda \in \rho(A)$, and thus $\boldsymbol{\sigma}(A) \subseteq [m, M]$. We now prove that $M \in \boldsymbol{\sigma}(A)$. To this end, define the bilinear form $a : X \times X \to \mathbb{R}$ by

$$a(u, v) := \langle Mu - A(u), v\rangle \quad \forall\, u,\ v \in X,$$

which, thanks to the hypotheses on A and the definition of M, is bilinear, bounded, symmetric, and positive semidefinite. It follows that a satisfies the Cauchy–Schwarz inequality, which means that

$$\begin{aligned}|\langle Mu - A(u), v\rangle| &=: |a(u, v)| \leqslant a(u, u)^{1/2} a(v, v)^{1/2} \\ &= \langle Mu - A(u), u\rangle^{1/2} \langle Mv - A(v), v\rangle^{1/2} \quad \forall\, u,\ v \in X.\end{aligned}$$

In particular, for $v = Mu - A(u)$ we obtain

$$\begin{aligned}\|Mu - A(u)\|^2 &\leqslant \langle Mu - A(u), u\rangle^{1/2} \langle (MI - A)(Mu - A(u)), Mu - A(u)\rangle^{1/2} \\ &\leqslant \|MI - A\|^{1/2} \langle Mu - A(u), u\rangle^{1/2} \|Mu - A(u)\| ,\end{aligned}$$

from which we deduce

$$\|Mu - A(u)\| \leqslant C \langle Mu - A(u), u\rangle^{1/2} \quad \forall\, u \in X, \tag{8.8}$$

with $C := \|MI - A\|^{1/2}$. Moreover, by the definition of M, there exists a sequence $\{u_n\}_{n\in\mathbb{N}} \subseteq X$ such that $\|u_n\| = 1$ for all $n \in \mathbb{N}$ and $M = \lim\limits_{n\to\infty} \langle A(u_n), u_n\rangle$. Then, applying inequality (8.8) to u_n, we obtain

$$\begin{aligned}\|Mu_n - A(u_n)\| &\leqslant C \Big\{M \|u_n\|^2 - \langle A(u_n), u_n\rangle\Big\}^{1/2} \\ &= C \Big\{M - \langle A(u_n), u_n\rangle\Big\}^{1/2} \overset{n\to\infty}{\longrightarrow} 0,\end{aligned}$$

which indicates that $(MI - A)(u_n) \overset{n\to\infty}{\longrightarrow} 0$. If we now suppose by contradiction that $M \in \rho(A)$, then $(MI - A)$ is bijective, and hence invertible, so the above convergence implies that

$$u_n = (MI - A)^{-1} (MI - A)(u_n) \overset{n\to\infty}{\longrightarrow} 0,$$

contradicting the fact that $\|u_n\| = 1$ for all $n \in \mathbb{N}$. This proves that, necessarily, $M \in \boldsymbol{\sigma}(A)$, and by an analogous argument one concludes that m also belongs to $\boldsymbol{\sigma}(A)$.

□

Corollary 8.9. *Let $(X, \langle \cdot, \cdot \rangle)$ be a real Hilbert space, and let $A \in \mathcal{L}(X)$ be self-adjoint such that $\boldsymbol{\sigma}(A) = \{0\}$. Then $A \equiv 0$.*

Proof. From the previous lemma, it is clear that $m = M = 0$, which implies that $\langle A(u), u \rangle = 0$ for all $u \in X$. Thus, for each $u, v \in X$ we deduce that

$$2 \langle A(u), v \rangle = \langle A(u+v), u+v \rangle - \langle A(u), u \rangle - \langle A(v), v \rangle = 0,$$

from which it follows that $A(u) = 0$ for all $u \in X$.

□

Definition 8.10. *Let X be a normed vector space. A sequence $\{x_n\}_{n \in \mathbb{N}}$ in X is said to form a* COMPLETE SYSTEM *if $X = \overline{\langle \{x_n\}_{n \in \mathbb{N}} \rangle}$.*

Definition 8.11. *Let $(X, \langle \cdot, \cdot \rangle)$ be a Hilbert space, and let $\{x_n\}_{n \in \mathbb{N}} \subseteq X$. This sequence is said to form a* HILBERT BASIS *(or* COMPLETE ORTHONORMAL SYSTEM*) if it is a complete system in X such that $\langle x_n, x_m \rangle = \delta_{nm}$ for all $n, m \in \mathbb{N}$.*

Regarding this last concept, it is easy to prove the following results.

Theorem 8.12. *Let $\{x_n\}_{n \in \mathbb{N}}$ be a Hilbert basis of a Hilbert space $(X, \langle \cdot, \cdot \rangle)$. Then, for each $x \in X$ there exists a unique series of the form $\sum_{n \in \mathbb{N}} a_n x_n$ that converges to x.*

Theorem 8.13. *A Hilbert space $(X, \langle \cdot, \cdot \rangle)$ is separable if and only if it possesses a Hilbert basis.*

Definition 8.14. *Let $\{X_n\}_{n \in \mathbb{N}}$ be a sequence of subspaces of a Hilbert space $(X, \langle \cdot, \cdot \rangle)$. We say that X is the* HILBERT SUM *of $\{X_n\}_{n \in \mathbb{N}}$ if $X_n \perp X_m$ for all $n \neq m$ and*

$$X = \overline{\langle \cup \{X_n : \quad n \in \mathbb{N}\} \rangle}.$$

We can now state and prove the announced result.

Theorem 8.15. *Let $(X, \langle \cdot, \cdot \rangle)$ be a separable real Hilbert space, and let $A \in \mathcal{L}(X)$ be compact and self-adjoint. Then X admits a Hilbert basis formed by eigenvectors of A.*

Proof. If $\boldsymbol{\sigma}(A) = \{0\}$, then being A self-adjoint, it follows from Corollary 8.9 that $A \equiv 0$, and therefore all elements of X are eigenvectors of A. Now, if $\boldsymbol{\sigma}(A) \neq \{0\}$, we let $\{\lambda_n\}_{n \in \mathbb{N}}$ be the (finite or countable) sequence of distinct nonzero eigenvalues of A, and define the spaces $X_0 := N(A)$ and $X_n := N(A - \lambda_n I)$ for all $n \in \mathbb{N}$. Recall that, by the compactness of A, $\dim X_n < +\infty$ for all $n \in \mathbb{N}$, and let us now prove that X is the Hilbert sum of $\{X_n\}_{n \geqslant 0}$. Indeed, given $u \in X_n$, $u \neq 0$, $v \in X_m$, $v \neq 0$, with $n \neq m$, we have $A(u) = \lambda_n u$ and $A(v) = \lambda_m v$, where $\lambda_0 = 0$. Then, since A is self-adjoint, it follows that

$$0 = \langle A(u), v \rangle - \langle u, A(v) \rangle = \lambda_n \langle u, v \rangle - \lambda_m \langle u, v \rangle = (\lambda_n - \lambda_m) \langle u, v \rangle,$$

from which, necessarily, $\langle u, v \rangle = 0$, and hence $X_n \perp X_m$. On the other hand, let

$$Y := \langle \cup \{X_n : \quad n \in \mathbb{N} \cup \{0\}\} \rangle$$

and let us prove that Y is dense in X. To this end, note first, using that $A(Y) \subseteq Y$, that $A(Y^{\perp}) \subseteq Y^{\perp}$. Indeed, for $u \in Y^{\perp}$ and $v \in Y$, we have $A(v) \in Y$, and therefore

$$\langle A(u), v\rangle = \langle u, A(v)\rangle = 0,$$

which shows that $A(u) \in Y^{\perp}$. It follows that $A_0 := A|_{Y^{\perp}} \in \mathcal{L}(Y^{\perp})$ is also compact and self-adjoint. Now let $\lambda \in \boldsymbol{\sigma}(A_0)\backslash\{0\}$. By Lemma 8.5 we know that $\lambda \in \boldsymbol{\sigma}_p(A_0)$, and thus there exists $u \in Y^{\perp}$, $u \neq 0$, such that $A_0(u) = \lambda u$, that is, $A(u) = \lambda u$. This implies that $\lambda \in \boldsymbol{\sigma}_p(A)$, that is, there exists $m \in \mathbb{N}$ such that $\lambda = \lambda_m$, and therefore $u \in X_m \subseteq Y$. Then, $u \in Y^{\perp} \cap Y = \{0\}$, which is a contradiction, and consequently, necessarily $\boldsymbol{\sigma}(A_0) = \{0\}$. It follows from Corollary 8.9 that $A_0 \equiv 0$, that is, $A|_{Y^{\perp}} \equiv 0$, which implies that $Y^{\perp} \subseteq N(A) =: X_0 \subseteq Y$, and therefore $Y^{\perp} = \{0\}$, proving thus the density of Y. Finally, since each X_n is a closed subspace of the separable Hilbert space X, all of them are also separable, and therefore each possesses a Hilbert basis. The union of all these bases forms a Hilbert basis of X.

□

On the other hand, a result related to Lemma 8.8 in the case of a complex Hilbert space is given by the following theorem.

Theorem 8.16 (Courant's Theorem). *Let $(X, \langle\cdot,\cdot\rangle)$ be a complex Hilbert space, let $A \in \mathcal{L}(X)$ be self-adjoint, and define*

$$m := \inf_{\|u\|=1} \langle A(u), u\rangle$$

and

$$M := \sup_{\|u\|=1} \langle A(u), u\rangle .$$

Then

$$\|A\| = \max\Big\{|m|, |M|\Big\} = \sup_{\|u\|=1} |\langle A(u), u\rangle|.$$

Proof. Let $S := \sup_{\|u\|=1} |\langle A(u), u\rangle|$, which clearly coincides with $\max\{|m|, |M|\}$ since, being A self-adjoint, there holds $\langle A(u), u\rangle \in \mathbb{R}$ for all $u \in X$. Then, applying the Cauchy-Schwarz inequality, we have for each $u \in X$, with $\|u\| = 1$, that $|\langle A(u), u\rangle| \leqslant \|A(u)\|$, whence

$$S \leqslant \sup_{\|u\|=1} \|A(u)\|,$$

that is, $S \leqslant \|A\|$. Now, if $A(u) = 0$ for all u such that $\|u\| = 1$, it follows clearly that A is the zero operator, and in that case the required identity is trivial. Let us now consider $u \in X$ such that $\|u\| = 1$ and $A(u) \neq 0$, and define $v := \|A(u)\|^{1/2} u$ and $\omega := \|A(u)\|^{-1/2} A(u)$. It follows that $\|v\| = \|A(u)\|^{1/2} = \|\omega\|$, and for $z_1 := v + \omega$ and $z_2 := v - \omega$ we obtain

$$\begin{aligned}\langle A(z_1), z_1\rangle - \langle A(z_2), z_2\rangle &= 2\Big\{\langle A(v), \omega\rangle + \langle A(\omega), v\rangle\Big\} \\ &= 2\Big\{\langle A(u), A(u)\rangle + \langle A^2(u), u\rangle\Big\} = 4\|A(u)\|^2 .\end{aligned} \tag{8.9}$$

Moreover, for each $z \neq 0$ we have

$$|\langle A(z), z\rangle| = \|z\|^2 \left|\left\langle A\left(\frac{z}{\|z\|}\right), \frac{z}{\|z\|}\right\rangle\right| \leqslant \|z\|^2 S,$$

from which it follows, using (8.9), that

$$4\,\|A(u)\|^2 \;=\; \langle A(z_1), z_1\rangle \;-\; \langle A(z_2), z_2\rangle \;\leqslant\; \Big\{\|z_1\|^2 \;+\; \|z_2\|^2\Big\}\,S$$
$$=\; 2\Big\{\|v\|^2 \;+\; \|\omega\|^2\Big\}\,S \;=\; 4\,\|A(u)\|\,S\,,$$

and hence $\|A(u)\| \leqslant S$ for all $u \in X$, with $\|u\| = 1$, that is, $\|A\| \leqslant S$.

□

An important consequence of Theorem 8.16, in the case of a compact and self-adjoint operator A, is that the corresponding point spectrum contains either $\|A\|$ or $-\|A\|$. More precisely, we have the following result.

Lemma 8.17. *Let X be a complex Hilbert space, and let $A \in \mathcal{L}(X)$ be a compact and self-adjoint operator. Then, there exists $\lambda \in \boldsymbol{\sigma}_p(A)$ such that $|\lambda| = \|A\|$.*

Proof. According to the previous lemma, it is clear that

$$\|A\| \;=\; \sup_{\|u\|=1}\langle A(u), u\rangle \quad \text{or else} \quad -\,\|A\| \;=\; \inf_{\|u\|=1}\langle A(u), u\rangle\,.$$

Suppose the first case holds (the other one is analogous). Then there exists a sequence $\{u_n\}_{n\in\mathbb{N}} \subseteq X$ such that $\|u_n\| = 1$ for all $n \in \mathbb{N}$ and

$$\lim_{n\to\infty}\langle A(u_n), u_n\rangle \;=\; \|A\|\,.$$

Since A is compact, there exist a subsequence $\{u_n^{(1)}\}_{n\in\mathbb{N}} \subseteq \{u_n\}_{n\in\mathbb{N}}$ and an element $v \in X$ such that $A(u_n^{(1)}) \overset{n\to\infty}{\longrightarrow} v$. Moreover, since obviously $\lim\limits_{n\to\infty}\langle A(u_n^{(1)}), u_n^{(1)}\rangle = \|A\|$, and

$$\langle A(u_n^{(1)}), u_n^{(1)}\rangle \;\leqslant\; \|A(u_n^{(1)})\| \;\leqslant\; \|A\| \quad \forall\, n \in \mathbb{N}\,,$$

it follows that $\|A\| = \lim\limits_{n\to\infty}\|A(u_n^{(1)})\| = \|v\|$. Let us now show that $u_n^{(1)} \overset{n\to\infty}{\longrightarrow} \|A\|^{-1}\,v$. Indeed, observe first that

$$|\langle v, u_n^{(1)}\rangle \;-\; \langle A(u_n^{(1)}), u_n^{(1)}\rangle| \;\leqslant\; \|v - A(u_n^{(1)})\| \overset{n\to\infty}{\longrightarrow} 0,$$

which implies that

$$\lim_{n\to\infty}\langle v, u_n^{(1)}\rangle \;=\; \lim_{n\to\infty}\langle A(u_n^{(1)}), u_n^{(1)}\rangle \;=\; \|A\|\,,$$

and therefore

$$\big\|\|A\|^{-1}\,v - u_n^{(1)}\big\| \;=\; \frac{\|v\|^2}{\|A\|^2} + \|u_n^{(1)}\|^2 \;-\; 2\,\|A\|^{-1}\,\mathrm{Re}\langle v, u_n^{(1)}\rangle$$
$$=\; 2 \;-\; 2\,\|A\|^{-1}\,\mathrm{Re}\langle v, u_n^{(1)}\rangle \overset{n\to\infty}{\longrightarrow} 0.$$

Finally,

$$v \;=\; \lim_{n\to\infty} A(u_n^{(1)}) \;=\; A(\|A\|^{-1}\,v) \;=\; \|A\|^{-1}\,A(v),$$

and hence $A(v) = \|A\|\,v$, which shows that $\|A\| \in \boldsymbol{\sigma}_p(A)$.

□

The main results derived in the present section can be summarized as follows.

Theorem 8.18 (Spectral Theorem for Compact Self-Adjoint Operators). *Let X be a separable real Hilbert space, and let $A \in \mathcal{L}(X)$ be compact and self-adjoint. Then*

i) $\sigma(A)\backslash\{0\}$ *is either finite or a sequence converging to* 0;
ii) *every nonzero element of* $\sigma(A)$ *is an eigenvalue of finite multiplicity;*
iii) X *admits a Hilbert basis formed by eigenvectors of* A;
iv) *there exists* $\lambda \in \sigma_p(A)$ *such that*

$$|\lambda| = \|A\|.$$

Proof. It suffices to combine Lemmas 8.5–8.7, which describe the spectrum of compact operators, with Theorem 8.15 and Lemma 8.17. In particular, Theorem 8.15 establishes that if $\{\lambda_n\}_{n\in I}$ denotes the family of distinct nonzero eigenvalues of A, then

$$X = N(A) \oplus \bigoplus_{n\in I} N(A - \lambda_n I),$$

as a Hilbert sum. Hence, by choosing a Hilbert basis in each of these mutually orthogonal eigenspaces, one obtains a Hilbert basis of X consisting entirely of eigenvectors of A. □

8.4 Other general results

In this section we complement what was presented in Section 8.2 with other general results.

Lemma 8.19. *Let X be a complex normed vector space, and let $A \in \mathcal{L}(X)$. Then, for each $n \in \mathbb{N}$ we have*

$$\Big\{\lambda^n : \quad \lambda \in \boldsymbol{\sigma}(A)\Big\} \subseteq \boldsymbol{\sigma}(A^n).$$

Proof. Given $\lambda \in \boldsymbol{\sigma}(A)$ and $n \in \mathbb{N}$, suppose by contradiction that $\lambda^n \in \rho(A^n)$. It is easy to prove by induction that

$$A^n - \lambda I = (A - \lambda I)B = B(A - \lambda I),$$

where

$$B := A^{n-1} + \lambda\, A^{n-2} + \lambda^2\, A^{n-3} + \cdots + \lambda^{n-2}\, A + \lambda^{n-1}\, I.$$

Then, since $(A^n - \lambda^n I)$ is bijective, it follows from the identity $A^n - \lambda^n I = B(A - \lambda I)$ that $N(A - \lambda I) = \{0\}$, while from the relation $A^n - \lambda^n I = (A - \lambda I)B$ we deduce that $R(A - \lambda I) = X$. This shows that $\lambda \in \rho(A)$, which contradicts the initial hypothesis, and therefore, necessarily, $\lambda^n \in \boldsymbol{\sigma}(A)$. □

Lemma 8.20. *Let X be a complex Banach space, let $A \in \mathcal{L}(X)$, and define*

$$r(A) := \inf\Big\{\|A^n\|^{1/n} : \quad n \in \mathbb{N}\Big\}. \tag{8.10}$$

Then

$$\Big\{\lambda \in \mathbb{C}: \quad |\lambda| > r(A)\Big\} \subseteq \rho(A),$$

or equivalently, $\boldsymbol{\sigma}(A) \subseteq \bar{B}(0, r(A))$.

Proof. Given $\lambda \in \boldsymbol{\sigma}(A)$, we have from the previous lemma that $\lambda^n \in \boldsymbol{\sigma}(A^n)$ for all $n \in \mathbb{N}$, which, by virtue of Lemma 8.1, implies that $|\lambda^n| = |\lambda|^n \leqslant \|A^n\|$. It follows immediately that $|\lambda| \leqslant \|A^n\|^{1/n}$ for all $n \in \mathbb{N}$, and hence $|\lambda| \leqslant r(A)$.

□

We emphasize that one could equivalently define $r(A)$ as

$$r(A) := \lim_{n\to\infty} \|A^n\|^{1/n} . \tag{8.11}$$

Furthermore, it follows from the previous lemma and (8.7) that $r_{\boldsymbol{\sigma}}(A) \leqslant r(A)$. Moreover, the Gelfand spectral radius formula states (cf. [105, Theorem 5.15], [114, Theorem 10.13]) that the spectral radius $r_{\boldsymbol{\sigma}}(A)$ (cf. (8.7)) and the quantity $r(A)$, defined by either (8.10) or (8.11), coincide. Needless to say, the proofs given in [105, Theorem 5.15] and [114, Theorem 10.13] are carried out in the more general setting of complex Banach algebras.

Now, given a polynomial p of the form $p(t) = \sum_{k=0}^{n} a_k t^k$, with $a_k \in \mathbb{C}$ for all $k \in \{0, 1, \ldots, n\}$, we are interested in identifying the spectrum of the polynomial operator (denoting $A^0 = I$)

$$p(A) := \sum_{k=0}^{n} a_k A^k \qquad \forall\, A \in \mathcal{L}(X) .$$

In this regard, we first have the following result.

Lemma 8.21. *Let X be a normed vector space over $\mathbb{C}$ or $\mathbb{R}$, and let $A \in \mathcal{L}(X)$. Then, for each polynomial p, we have*

$$p(\boldsymbol{\sigma}(A)) \subseteq \boldsymbol{\sigma}(p(A)).$$

Proof. The proof reduces to showing that if $\lambda \in \mathbb{C}$ is such that $p(\lambda) \in \rho(p(A))$, then $\lambda \in \rho(A)$. Indeed, since λ is clearly a root of the polynomial g defined by

$$g(t) := p(t) - p(\lambda) \qquad \forall\, t \in \mathbb{C} ,$$

there exists a polynomial q of degree at most $n - 1$ (assuming that p has degree at most n) such that

$$p(t) - p(\lambda) = (t - \lambda)\, q(t) = q(t)(t - \lambda) \qquad \forall\, t \in \mathbb{C} .$$

It follows that

$$p(A) - p(\lambda)I = (A - \lambda I)\, q(A) = q(A)(A - \lambda I) ,$$

and thus the bijectivity of $p(A) - p(\lambda)I$ yields that of $A - \lambda I$, whence $\lambda \in \rho(A)$.

□

The inclusion rather than an equality in the previous lemma suggests that there may exist scalars in $\boldsymbol{\sigma}(p(A))$ that are not of the form $p(\lambda)$ for some $\lambda \in \boldsymbol{\sigma}(A)$. Indeed, as an example consider the real vector space $X := \mathbb{R}^2$ and define the operator $A \in \mathcal{L}(X)$ by $A(x, y) := (y, -x)$ for all $(x, y) \in X$. Then, given $\lambda \in \mathbb{R}$, the equation $(A - \lambda I)(x, y) = (0, 0)$ reduces to the linear system

$$\begin{pmatrix} -\lambda & 1 \\ -1 & -\lambda \end{pmatrix} \begin{pmatrix} x \\ y \end{pmatrix} = \begin{pmatrix} 0 \\ 0 \end{pmatrix} ,$$

whose determinant is $1+\lambda^2 > 0$, and hence its only solution is the zero vector. This proves that $N(A-\lambda I) = \{0\}$, and therefore $\lambda \in \rho(A)$ for all $\lambda \in \mathbb{R}$, or equivalently, $\boldsymbol{\sigma}(A) = \varnothing$. However, it is easy to see that the operator A^2 has $\lambda = -1$ as an eigenvalue. Indeed, for each $(x,y) \in \mathbb{R}^2$ we have

$$A^2(x,y) \;=\; A(A(x,y)) \;=\; A(y,-x) \;=\; (-x,-y) \;=\; -1 \cdot (x,y)\,,$$

which implies that $-1 \in \boldsymbol{\sigma}(A^2)$ and $N(A^2+I) = X$. The key to this example is the fact that the underlying field is $\mathbb{R}$, so that complex numbers cannot be considered as eigenvalues.

Let us now suppose that we wish to solve an equation of the form

$$p(A)\,x \;=\; y \qquad \text{with} \qquad x, y \in X\,, \tag{8.12}$$

where $A \in \mathcal{L}(X)$ and p is a polynomial of degree $\leqslant n$. Clearly, the unique solvability of (8.12) is guaranteed for each $y \in X$ if $0 \notin \boldsymbol{\sigma}(p(A))$ (equivalently, $0 \in \rho(p(A))$), since in that case $p(A)$ is bijective. The following theorem complements what was given in Lemma 8.21 and allows us to characterize the spectrum of $p(A)$ in terms of that of A.

Theorem 8.22 (Spectral Mapping Theorem). *Let X be a complex normed vector space, let $A \in \mathcal{L}(X)$, and let p be a polynomial over $\mathbb{C}$ of degree $\leqslant n$. Then*

$$\boldsymbol{\sigma}(p(A)) = p(\boldsymbol{\sigma}(A)),$$

that is, $\mu \in \boldsymbol{\sigma}(p(A))$ if and only if $\mu = p(\lambda)$ for some $\lambda \in \boldsymbol{\sigma}(A)$.

Proof. It remains only to prove the inclusion $\boldsymbol{\sigma}(p(A)) \subseteq p(\boldsymbol{\sigma}(A))$. To this end, we let $\mu \in \boldsymbol{\sigma}(p(A))$ and aim to show next that there exists $\lambda \in \boldsymbol{\sigma}(A)$ such that $\mu = p(\lambda)$. Indeed, let $\gamma_1, \gamma_2, \ldots, \gamma_n$ be the complex roots of the polynomial $p(t) - \mu$. This means that there exists $c \in \mathbb{C}\backslash\{0\}$ such that

$$p(t) - \mu \;=\; c(t-\gamma_1)\,(t-\gamma_2)\cdots(t-\gamma_n) \quad \forall\, t \in \mathbb{C}\,,$$

and hence

$$p(A) - \mu I \;=\; c(A-\gamma_1 I)\,(A-\gamma_2 I)\cdots(A-\gamma_n I)\,.$$

Then, if we suppose that $\gamma_j \in \rho(A)$ for all $j \in \{1,\ldots,n\}$, it follows that each operator $(A-\gamma_j I)$ is bijective for all $j \in \{1,\ldots,n\}$, and therefore $(p(A)-\mu I)$ would also be bijective, which contradicts the fact that $\mu \in \boldsymbol{\sigma}(p(A))$. Thus, necessarily, there exists $i \in \{1,\ldots n\}$ such that $\gamma_i \in \boldsymbol{\sigma}(A)$, and then, since $p(\gamma_i) - \mu = 0$, we obviously have $\mu = p(\gamma_i)$, which completes the proof.

□

The following result is a direct corollary of the previous theorem.

Lemma 8.23. *Let X be a complex normed vector space, let $A \in \mathcal{L}(X)$, and let p be a polynomial over $\mathbb{C}$ of degree $\leqslant n$. Then the equation $p(A)\,x = y$ has a unique solution for each $y \in X$ if and only if $p(\lambda) \neq 0$ for all $\lambda \in \boldsymbol{\sigma}(A)$.*

Proof. It suffices to note that $0 \notin \boldsymbol{\sigma}(p(A))$ if and only if $p(\lambda) \neq 0$ for all $\lambda \in \boldsymbol{\sigma}(A)$.

□

On the other hand, we now prove that the spectra of an operator and its adjoint coincide.

Theorem 8.24. *Let X be a complex Banach space, and let $A \in \mathcal{L}(X)$. Then $\boldsymbol{\sigma}(A) = \boldsymbol{\sigma}(A')$ (equivalently, $\rho(A) = \rho(A')$).*

Proof. Let $\lambda \in \rho(A)$, that is, $\lambda \in \mathbb{C}$ and $(A - \lambda I)$ is bijective. It follows, by the Bounded Inverse Theorem (cf. Theorem 3.34), that there exists $B := (A - \lambda I)^{-1} \in \mathcal{L}(X)$ such that $(A - \lambda I)\, B = B(A - \lambda I) = I$ in X. Then, taking adjoints we obtain

$$B'(A' - \lambda I') = (A' - \lambda I')\, B' = I' \quad \text{in} \quad X',$$

which shows that $(A' - \lambda I')$ is invertible, with $(A' - \lambda I')^{-1} = B'$, and therefore $\lambda \in \rho(A')$. Conversely, let $\lambda \in \rho(A')$, that is, $\lambda \in \mathbb{C}$ and $(A' - \lambda I')$ is bijective. Then, applying again Theorem 3.34, we deduce that there exists $C := (A' - \lambda I')^{-1} \in \mathcal{L}(X')$ such that $(A' - \lambda I')C = C(A' - \lambda I') = I'$ in X'. Then, taking adjoints we obtain

$$C'(A'' - \lambda I'') = (A'' - \lambda I'')C' = I'' \quad \text{in} \quad X'',$$

where $I'' : X'' \to X''$ is the identity operator. Now, let $J : X \to X''$ be the usual isometry (cf. (3.24), (7.1)) defined by $J(x)(F) := F(x)$ for all $x \in X$, for all $F \in X'$. It follows that

$$A'' J(x)(F) = (A')' J(x)(F) = J(x)(A'(F)) = A'(F)(x) = F(A(x)) = J(A(x))(F),$$

and hence

$$A''\, J(x) = J(A(x)) \qquad \forall\, x \in X.$$

Moreover, from the identity $I'' = C'(A'' - \lambda I'')$ in X'', we have that for each $x \in X$

$$J(x) = C'(A''\, J(x) - \lambda J(x)) = C'(J(A(x)) - \lambda J(x)) = C' J((A - \lambda I)(x)),$$

and consequently

$$\|x\| = \|J(x)\| \leqslant \|C' J\|\, \|(A - \lambda I)(x)\| \qquad \forall\, x \in X.$$

It follows that $N(A - \lambda I) = \{0\}$ and $R(A - \lambda I)$ is closed in X, from which we deduce that

$$R(A - \lambda I) = {}^{\circ}(R(A - \lambda I)^{\circ}) = {}^{\circ}N(A' - \lambda I') = {}^{\circ}\{0\} = X.$$

Thus, $(A - \lambda I)$ is bijective, that is, $\lambda \in \rho(A)$, which completes the proof. □

The last result in this part concerns the process of including a real Banach space within a complex Banach space, which is called the COMPLEXIFICATION. Indeed, given $(X, \|\cdot\|)$ a Banach space over $\mathbb{R}$, we define $Z := X \times X$ endowed with the operations

$$(x_1, y_1) + (x_2, y_2) := (x_1 + x_2, y_1 + y_2) \quad \forall\, (x_1, y_1), (x_2, y_2) \in Z$$

and

$$(\alpha + i\,\beta) \cdot (x, y) := (\alpha\, x - \beta\, y, \beta\, x + \alpha\, y) \quad \forall\, \alpha + i\,\beta \in \mathbb{C}, \quad \forall\, (x, y) \in Z.$$

In addition, we define

$$\|(x, y)\|_Z := \|x\| + \|y\| \qquad \forall\, (x, y) \in Z.$$

Then, it is easy to see that $(Z, +, \cdot, \|\cdot\|_Z)$ is a Banach space over $\mathbb{C}$, and that X can be identified with the subspace of Z given by

$$Z_0 := \Big\{(x,0): \quad x \in X\Big\}.$$

Moreover, given $A \in \mathcal{L}(X)$, we define the operator $\widehat{A}: Z \to Z$ by

$$\widehat{A}(x,y) := (A(x), A(y)) \quad \forall\,(x,y) \in Z,$$

which turns out to be linear and bounded with $\|\widehat{A}\| = \|A\|$. Now, given $\lambda \in \mathbb{R}$, we have, with $\widehat{I}: Z \to Z$ the identity operator on Z, that

$$(\widehat{A} - \lambda\widehat{I})(x,y) = ((A-\lambda I)(x), (A-\lambda I)(y)),$$

which shows that $\lambda \in \rho(\widehat{A})$ if and only if $\lambda \in \rho(A)$. Furthermore, if p is a polynomial over $\mathbb{R}$, we obtain that

$$p(\widehat{A})(x,y) = (p(A)(x), p(A)(y)),$$

from which we deduce that $p(\widehat{A})$ is bijective in $\mathcal{L}(Z)$ if and only if $p(A)$ is bijective in $\mathcal{L}(X)$. According to this and to the Spectral Mapping Theorem (cf. Theorem 8.22), we conclude that the equation

$$p(A)(x) = y, \quad x,\, y \in X,$$

has a unique solution for each $y \in X$ if and only if $p(\lambda) \neq 0$ for all $\lambda \in \boldsymbol{\sigma}(\widehat{A})$.

Example 8.25 Let us consider again $X = \mathbb{R}^2$ and the operator $A: X \to X$ given by

$$A(\mathbf{x}) := (y, -x) \qquad \forall\,\mathbf{x} := (x,y) \in X.$$

It follows that $Z := X \times X \equiv \mathbb{R}^4$ and $\widehat{A}: Z \to Z$ is defined by

$$\widehat{A}((x,y,u,v)) := (y,-x,v,-u) \quad \forall\,(x,y,u,v) \in Z.$$

Then, it is easy to see that, for $\lambda := \lambda_1 + i\,\lambda_2 \in \mathbb{C}$, the equation

$$(\widehat{A} - \lambda\widehat{I})((x,y,u,v)) = (0,0,0,0)$$

reduces to the homogeneous system

$$\begin{pmatrix} -\lambda_1 & 1 & \lambda_2 & 0 \\ -1 & -\lambda_1 & 0 & \lambda_2 \\ -\lambda_2 & 0 & -\lambda_1 & 1 \\ 0 & -\lambda_2 & -1 & \lambda_1 \end{pmatrix} \begin{pmatrix} x \\ y \\ u \\ v \end{pmatrix} = \begin{pmatrix} 0 \\ 0 \\ 0 \\ 0 \end{pmatrix},$$

whose determinant is given by $\lambda_1^2(\lambda_1^2 + 2\lambda_2^2 + 2) + (\lambda_2^2 - 1)^2$, and hence, the only possibility for it to vanish is that $\lambda_1 = 0$ and $\lambda_2^2 = 1$. This implies that $\boldsymbol{\sigma}(\widehat{A}) = \{-i, i\}$, and therefore, taking in particular $p(t) = t^2 + 1$, we have that the equation $(A^2 + I)(\mathbf{x}) = \mathbf{y}$ has a unique solution if and only if $p(\lambda) = \lambda^2 + 1 \neq 0$ for all $\lambda \in \boldsymbol{\sigma}(\widehat{A})$, which is not true since, given $i \in \boldsymbol{\sigma}(\widehat{A})$, we have $p(i) = 0$, and thus $N(A^2 + I) \neq \{0\}$.

8.5 Spectrum of unbounded operators

In this section we extend some of the concepts and results given previously to the case of a linear operator not necessarily bounded.

Definition 8.26. *Let X be a complex Banach space, and let $A : \mathcal{D}(A) \subseteq X \to X$ be a linear operator. We say that $\lambda \in \mathbb{C}$ belongs to the resolvent set of A, also denoted by $\rho(A)$, if*

i) $A_\lambda := (A - \lambda I) : \mathcal{D}(A) \subseteq X \to X$ *is injective, that is* $N(A_\lambda) = \{0\}$,

ii) $A_\lambda(\mathcal{D}(A))$ *is dense in* X, *and*

iii) $A_\lambda^{-1} \in \mathcal{L}(A_\lambda(\mathcal{D}(A)), X)$.

In turn, we denote by $\boldsymbol{\sigma}(A)$, and call it the SPECTRUM of A, the complement of $\rho(A)$, that is, $\boldsymbol{\sigma}(A) := \mathbb{C}\backslash\rho(A)$. Note that $\boldsymbol{\sigma}(A)$ is given by the union of the following disjoint sets:

POINT SPECTRUM of A

$$\boldsymbol{\sigma}_p(A) := \Big\{\lambda \in \mathbb{C} : \quad N(A_\lambda) \neq \{0\}\Big\},$$

CONTINUOUS SPECTRUM of A

$$\boldsymbol{\sigma}_c(A) := \Big\{\lambda \in \mathbb{C} : \quad N(A_\lambda) = \{0\}, \quad A_\lambda(\mathcal{D}(A)) \text{ dense in } X, \quad A_\lambda^{-1} \notin \mathcal{L}\big(A_\lambda(\mathcal{D}(A)), X\big)\Big\},$$

RESIDUAL SPECTRUM of A

$$\boldsymbol{\sigma}_r(A) := \Big\{\lambda \in \mathbb{C} : \quad N(A_\lambda) = \{0\}, \quad A_\lambda(\mathcal{D}(A)) \text{ is not dense in } X\Big\}.$$

Now, in the particular case where A is closed, the characterization of its resolvent $\rho(A)$ simplifies as indicated in the following lemma.

Lemma 8.27. *Let X be a complex Banach space, and let $A : \mathcal{D}(A) \subseteq X \to X$ be a closed linear operator. Then*

$$\rho(A) = \Big\{\lambda \in \mathbb{C} : \quad N(A_\lambda) = \{0\}, \quad A_\lambda(\mathcal{D}(A)) = X \quad \textit{and} \quad A_\lambda^{-1} \in \mathcal{L}(X)\Big\}. \tag{8.13}$$

Proof. Let us denote by $\widetilde{\rho}(A)$ the set appearing on the right-hand side of (8.13). Then, it is clear that $\widetilde{\rho}(A) \subseteq \rho(A)$. Conversely, let $\lambda \in \rho(A)$. Since $A_\lambda^{-1} \in \mathcal{L}\big(A_\lambda(\mathcal{D}(A)), X\big)$, there exists $C > 0$ such that

$$\|A_\lambda^{-1}(y)\| \leqslant C\,\|y\| \quad \forall\, y \in A_\lambda(\mathcal{D}(A)),$$

or equivalently,

$$\|A_\lambda(x)\| \geqslant c\,\|x\| \quad \forall\, x \in \mathcal{D}(A).$$

Then, since A_λ is closed (because A is), it follows from the foregoing inequality that (cf. Theorem 3.58) $R(A_\lambda)$ is closed in X, and hence

$$X = \overline{A_\lambda(\mathcal{D}(A))} = A_\lambda(\mathcal{D}(A)),$$

which shows that $\lambda \in \widetilde{\rho}(A)$.

□

Next, we present two examples illustrating the characteristics of the spectrum of some unbounded operators. We begin with the case of an operator whose spectrum is purely continuous.

Example 8.28 Let $X := \mathrm{L}^2(\mathbb{R})$ with $\mathbb{K} = \mathbb{C}$, endowed with the usual inner product

$$\langle f, g \rangle := \int_{\mathbb{R}} f \bar{g} \qquad \forall f, g \in X,$$

with which X is a Hilbert space over $\mathbb{C}$, and let $A : \mathcal{D}(A) \subseteq X \to X$ be the operator given by

$$A(f)(x) := x\, f(x) \qquad \forall x \in \mathbb{R}, \quad \forall f \in X.$$

It is clear that $\mathcal{D}(A) := \Big\{ f \in X : \quad x\, f \in X \Big\}$, and let us show that

$$\boldsymbol{\sigma}_p(A) = \boldsymbol{\sigma}_r(A) = \phi, \quad \text{and} \quad \boldsymbol{\sigma}_c(A) = \mathbb{R},$$

so that $\rho(A) = \mathbb{C} - \mathbb{R}$. Indeed, given $\lambda \in \mathbb{C}$, the homogeneous equation $A_\lambda(f) = 0$ is equivalent to $(x - \lambda) f(x) = 0$ for all $x \in \mathbb{R}$ a.e., which implies that $f(x) = 0$ for all $x \in \mathbb{R}$ a.e., thus $f \equiv 0$. This shows that $N(A_\lambda) = \{0\}$ for all $\lambda \in \mathbb{C}$, and consequently $\boldsymbol{\sigma}_p(A) = \varnothing$. It follows that A_λ is injective for all $\lambda \in \mathbb{C}$ and therefore the operator $A_\lambda^{-1} : A_\lambda\big(\mathcal{D}(A)\big) \to X$ exists. In particular, let us see that for each $\lambda \in \mathbb{R}$, $A_\lambda(\mathcal{D}(A))$ contains the set

$$U := \Big\{ \varphi \in X : \quad \varphi \equiv 0 \quad \text{on} \quad [\lambda - \delta, \lambda + \delta] \quad \text{for some} \quad \delta > 0 \Big\}.$$

To see this, given $\varphi \in U$, define the function

$$\psi(x) := \begin{cases} 0 & \text{on } [\lambda - \delta, \lambda + \delta], \\ \dfrac{\varphi(x)}{(x - \lambda)} & \text{on } \mathbb{R} \backslash [\lambda - \delta, \lambda + \delta], \end{cases}$$

and note that $\psi \in \mathcal{D}(A)$ and $A_\lambda(\psi) = \varphi$. The second is immediate, and for the first we note that

$$x\, \psi(x) = \varphi(x) + \widetilde{\varphi}(x) \qquad \forall x \in \mathbb{R},$$

with

$$\widetilde{\varphi}(x) := \begin{cases} 0 & \text{on } [\lambda - \delta, \lambda + \delta], \\ \dfrac{\lambda\, \varphi(x)}{(x - \lambda)} & \text{on } \mathbb{R} \backslash [\lambda - \delta, \lambda + \delta]. \end{cases}$$

Then, since $\left| \dfrac{\lambda}{x - \lambda} \right| \leqslant \dfrac{|\lambda|}{\delta}$ for all $x \in \mathbb{R} \backslash [\lambda - \delta, \lambda + \delta]$, it follows that $\widetilde{\varphi} \in X$ and therefore $x\psi \in X$, that is, $\psi \in \mathcal{D}(A)$. Now, given a function $g \in X := \mathrm{L}^2(\mathbb{R})$, define for each $n \in \mathbb{N}$

$$g_n(x) := \begin{cases} 0 \ \text{ on } [\lambda - 1/n, \lambda + 1/n], \\ g(x) \ \text{ on } \mathbb{R} \backslash [\lambda - 1/n, \lambda + 1/n]. \end{cases}$$

It is clear that $g_n \in U \subseteq A_\lambda\big(\mathcal{D}(A)\big) = \mathcal{D}(A_\lambda^{-1})$ and, by applying Lebesgue's Dominated Convergence Theorem (cf. [113, Theorem 1.34], [120, Theorem 5.36]), we obtain

$$\|g - g_n\|_X^2 = \int_{\lambda - 1/n}^{\lambda + 1/n} |g(x)|^2 \, dx = \int_{\mathbb{R}} \mathbf{1}_{[\lambda - 1/n, \lambda + 1/n]} \, |g(x)|^2 \, dx \overset{n \to \infty}{\longrightarrow} 0,$$

which proves that $A_\lambda(\mathcal{D}(A))$ is dense in X. Moreover, it is easy to see that for $\lambda \in \mathbb{R}$, A_λ^{-1} is not bounded. Indeed, given $\lambda \in \mathbb{R}$ and $n \geqslant 2$, define (see Figure 8.1)

$$\varphi_n(x) := \begin{cases} 1 \ \text{on } [\lambda - 1, \lambda - 1/n] \cup [\lambda + 1/n, \lambda + 1], \\ 0 \ \text{on } \mathbb{R} \backslash ([\lambda - 1, \lambda - 1/n] \cup [\lambda + 1/n, \lambda + 1]), \end{cases}$$

and

$$\psi_n(x) := \begin{cases} 0 & \text{on } [\lambda - 1/n, \lambda + 1/n]\,, \\ \dfrac{\varphi_n(x)}{(x-\lambda)} & \text{on } \mathbb{R}\backslash[\lambda - 1/n, \lambda + 1/n]\,. \end{cases}$$

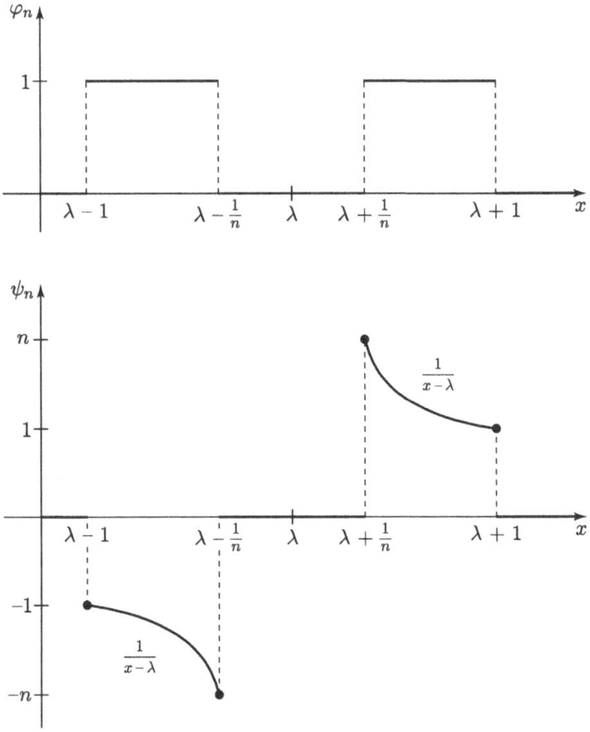

Fig. 8.1. *Graphs of φ_n and ψ_n*

It is clear, according to previous calculations, that $A_\lambda^{-1}(\varphi_n) = \psi_n$. It follows that

$$\|A_\lambda^{-1}(\varphi_n)\|_X^2 = \|\psi_n\|_X^2 = \int_{\lambda-1}^{\lambda-1/n} \frac{dx}{(x-\lambda)^2} + \int_{\lambda+1/n}^{\lambda+1} \frac{dx}{(x-\lambda)^2} = 2(n-1)\,,$$

and

$$\|\varphi_n\|_X^2 = \int_{\lambda-1}^{\lambda-1/n} dx + \int_{\lambda+1/n}^{\lambda+1} dx = 2\,\frac{(n-1)}{n}\,,$$

from which

$$\frac{\|A_\lambda^{-1}(\varphi_n)\|_X^2}{\|\varphi_n\|_X^2} = n \overset{n\to\infty}{\longrightarrow} \infty\,,$$

and thus A_λ^{-1} is not bounded. This proves that $\lambda \in \boldsymbol{\sigma}_c(A)$ for all $\lambda \in \mathbb{R}$. Finally, let us prove that $\mathbb{C} - \mathbb{R} \subseteq \rho(A)$. Indeed, given $\lambda \in \mathbb{C} - \mathbb{R}$ and $\psi \in X$, we have that the function φ defined by $\varphi(x) = \dfrac{\psi(x)}{x-\lambda}$ for all $x \in \mathbb{R}$, belongs to $\mathcal{D}(A)$ and $A_\lambda(\varphi) = \psi$. The second statement is direct, and for the first one it suffices to note that

$$x\varphi(x) = \psi(x) + \widetilde{\psi}(x) \quad \forall\, x \in \mathbb{R}\,,$$

with $\widetilde{\psi}(x) = \dfrac{\lambda}{x-\lambda}\,\psi(x)$ for all $x \in \mathbb{R}$. Then, since $\left|\dfrac{\lambda}{x-\lambda}\right| \leqslant \dfrac{|\lambda|}{|\operatorname{Im}\lambda|}$ for all $x \in \mathbb{R}$, we have that $\widetilde{\psi} \in X$ and therefore $\varphi \in \mathcal{D}(A)$. In this way, we prove that $A_\lambda\big(\mathcal{D}(A)\big) = X$. Furthermore, it is easy to see that A_λ^{-1} is bounded with $\|A_\lambda^{-1}\| \leqslant \dfrac{1}{|\operatorname{Im}\lambda|}$ since

$$\|A_\lambda^{-1}(\psi)\|_X^2 = \int_{\mathbb{R}} \big(A_\lambda^{-1}(\psi)(x)\big)^2\,dx = \int_{\mathbb{R}} \frac{|\psi^2(x)|}{|x-\lambda|^2}\,dx \leqslant \frac{1}{|\operatorname{Im}\lambda|^2}\,\|\psi\|_X^2 < +\infty \qquad \forall\,\psi \in X\,.$$

Thus $\mathbb{C} - \mathbb{R} \subseteq \rho(A)$, and since $\mathbb{R} \subseteq \boldsymbol{\sigma}_c(A)$, we conclude that $\rho(A) = \mathbb{C} - \mathbb{R}$, $\boldsymbol{\sigma}_c(A) = \mathbb{R}$ and $\boldsymbol{\sigma}_r(A) = \varnothing$.

The following example considers the case of an operator with an empty spectrum.

Example 8.29 Let $\Omega := (0,1)$ and $X := \mathrm{L}^2(\Omega)$. Then we define $A : \mathcal{D}(A) \subseteq X \to X$ as $A(u) := u'$ (distributional derivative) for all $u \in \mathcal{D}(A)$. In fact, without loss of generality, we may consider

$$\mathcal{D}(A) := \Big\{u \in \mathrm{L}^2(\Omega) : \quad u' \in \mathrm{L}^2(\Omega) \quad \text{and} \quad u(0) = 0\Big\}\,,$$

that is

$$\mathcal{D}(A) := \Big\{u \in \mathrm{H}^1(\Omega) : \quad u(0) = 0\Big\}\,.$$

Then, given $\lambda \in \mathbb{C}$ and $f \in X$, the equation $(A - \lambda I)u = f$, with $u \in X$, can be rewritten as

$$u' - \lambda u = f \quad \text{in} \quad \Omega, \qquad u(0) = 0\,,$$

whose unique solution is given by

$$u(t) = \int_0^t e^{\lambda(t-s)}\,f(s)\,ds \qquad \forall\,t \in \Omega\,.$$

It follows that $A_\lambda : \mathcal{D}(A) \subseteq X \to X$ is invertible and

$$A_\lambda\big(\mathcal{D}(A)\big) = A_\lambda(X) = X = \mathrm{L}^2(\Omega) \qquad \forall\,\lambda \in \mathbb{C}\,.$$

Moreover, for each $f \in X := \mathrm{L}^2(\Omega)$ we have

$$\begin{aligned} |A_\lambda^{-1}f(t)| &= \left|\int_0^t e^{\lambda(t-s)} f(s)\,ds\right| \leqslant \int_0^t |e^{\lambda(t-s)}|\,|f(s)|\,ds \\ &\leqslant \left\{\int_0^1 |e^{\lambda(t-s)}|^2\,dt\right\}^{1/2} \|f\|_X \leqslant C_\lambda\,\|f\|_X \quad \forall\,t \in \Omega\,, \end{aligned}$$

where $C_\lambda := \max\limits_{t,s\in[0,1]} |e^{\lambda(t-s)}|$. Thus we obtain that $\|A_\lambda^{-1}(f)\|_X \leqslant C_\lambda\,\|f\|_X$ for all $f \in X$, which shows that A_λ^{-1} is bounded and consequently $\rho(A) = \mathbb{C}$.

Finally, although this section is devoted to unbounded operators, we make an exception in order to present an example of a bounded operator with non-empty residual spectrum.

Example 8.30 Let

$$X := \ell^2(\mathbb{C}) := \Big\{u := (u_n)_{n\in\mathbb{N}} : \quad u_n \in \mathbb{C} \quad \forall\,n \in \mathbb{N} \quad \text{and} \quad \sum_{n\in\mathbb{N}} |u_n|^2 < +\infty\Big\}\,,$$

which, endowed with the inner product

$$\langle u, v\rangle := \sum_{n\in\mathbb{N}} u_n\,\bar{v}_n \quad \forall\,u := (u_n)_{n\in\mathbb{N}}\,, \quad v := (v_n)_{n\in\mathbb{N}} \in X\,,$$

is a Hilbert space. Then, we define the operator $A : \mathcal{D}(A) \subseteq X \to X$ by

$$A(u) := v := (v_n)_{n\in\mathbb{N}}, \quad \text{with} \quad v_1 = 0 \quad \text{and} \quad v_n = u_{n-1} \quad \forall\, n \geqslant 2.$$

It is easy to see that $\mathcal{D}(A) = X$. Moreover, given $\lambda \in \mathbb{C}$, the equation $(A - \lambda I)(u) = 0$ is equivalent to $\lambda u_1 = 0$, $\lambda u_n = u_{n-1}$ for all $n \geqslant 2$. Then, if $\lambda = 0$, from $\lambda u_2 = u_1$ onward we deduce that $u = 0$. Likewise, if $\lambda \neq 0$, from $\lambda u_1 = 0$ we also obtain $u = 0$. This shows that $N(A_\lambda) = \{0\}$ for all $\lambda \in \mathbb{C}$ and therefore $\boldsymbol{\sigma}_p(A) = \varnothing$. On the other hand, given $\lambda \in \mathbb{C}$, consider the space $A_\lambda(\mathcal{D}(A)) = A_\lambda(X)$, which reduces to

$$A_\lambda(X) := \Big\{(u_{n-1} - \lambda u_n)_{n\in\mathbb{N}} : \quad u := (u_n)_{n\in\mathbb{N}} \in X, \quad u_0 = 0\Big\}.$$

It follows that $v \in A_\lambda(X)^\perp$ if and only if $v \in X$ and

$$0 = \langle A_\lambda(u), v\rangle = \sum_{n\in\mathbb{N}} (u_{n-1} - \lambda u_n)\bar{v}_n \quad \forall\, u \in X.$$

Then, given $v \in A_\lambda(X)^\perp$, we have in particular

$$\begin{array}{llll} u := (u_1, 0, 0, \cdots, 0) & \Rightarrow \quad 0 = -\lambda\, u_1\, \bar{v}_1 + u_1\, \bar{v}_2 & \forall\, u_1 \in \mathbb{C} & \Rightarrow \quad \bar{v}_2 = \lambda\, \bar{v}_1 \\ u := (0, u_2, 0, \cdots, 0) & \Rightarrow \quad 0 = -\lambda\, u_2\, \bar{v}_2 + u_2\, \bar{v}_3 & \forall\, u_2 \in \mathbb{C} & \Rightarrow \quad \bar{v}_3 = \lambda\, \bar{v}_2 \\ u := (0, 0, u_3, \cdots, 0) & \Rightarrow \quad 0 = -\lambda\, u_3\, \bar{v}_3 + u_3\, \bar{v}_4 & \forall\, u_3 \in \mathbb{C} & \Rightarrow \quad \bar{v}_4 = \lambda\, \bar{v}_3 \end{array}$$

and so on, yielding $\bar{v}_n = \lambda \bar{v}_{n-1}$ for all $n \geqslant 2$, or equivalently $\bar{v}_n = \lambda^{n-1}\bar{v}_1$ for all $n \geqslant 2$. Since v must belong to X, we also require that $\sum_{n\in\mathbb{N}} |v_n|^2 = |v_1|^2 \sum_{n=0}^{\infty} |\lambda|^{2n} < +\infty$, which is only possible if $|\lambda| < 1$ or $v_1 = 0$. Conversely, suppose that $|\lambda| < 1$ and consider an element $v := (v_n)_{n\in\mathbb{N}} \in X$ of the form $\bar{v}_n = \lambda^{n-1}\,\bar{v}_1$ for all $n \geqslant 2$, with $v_1 \in \mathbb{C}$ arbitrary. Then, given $u := (u_n)_{n\in\mathbb{N}} \in X$ and $N \in \mathbb{N}$, we have

$$S_N := \sum_{n=1}^{N} (u_{n-1} - \lambda u_n)\, \bar{v}_n = \sum_{n=1}^{N} \bar{v}_1\, (\lambda^{n-1} u_{n-1} - \lambda^n u_n) = -\bar{v}_1\, \lambda^N u_N \xrightarrow{N\to\infty} 0,$$

which shows that

$$\langle A_\lambda(u), v\rangle = \sum_{n\in\mathbb{N}} (u_{n-1} - \lambda u_n)\, \bar{v}_n = \lim_{N\to\infty} S_N = 0,$$

and therefore $v \in A_\lambda(X)^\perp$. We have thus proved that if $|\lambda| < 1$, then $A_\lambda(X)^\perp$ is given by

$$A_\lambda(X)^\perp = \Big\{v := (\lambda^{n-1} v_1)_{n\in\mathbb{N}} : \quad v_1 \in \mathbb{C}\Big\},$$

which is clearly nontrivial, and consequently $A_\lambda(\mathcal{D}(A)) = A_\lambda(X)$ is not dense in X for all $\lambda \in \mathbb{C}$ such that $|\lambda| < 1$. This proves that $B(0,1) := \Big\{\lambda \in \mathbb{C} : \quad |\lambda| < 1\Big\} \subseteq \boldsymbol{\sigma}_r(A)$. Moreover, from the same previous analysis it follows that if $|\lambda| \geqslant 1$ then necessarily $A_\lambda(\mathcal{D}(A))^\perp = \{0\}$, that is, $A_\lambda(\mathcal{D}(A))$ is dense in X, and therefore $\boldsymbol{\sigma}_r(A) = \Big\{\lambda \in \mathbb{C} : \quad |\lambda| < 1\Big\}$.

The next goal is to provide some results on the spectrum of unbounded and self-adjoint operators, for which we first need to define what we mean here by the adjoint operator. In fact, although this concept was already introduced in Chapter 3 for the general case of Banach spaces (cf. Definition 3.80), we now particularize it to the present context of Hilbert spaces.

Definition 8.31. *Let $(X, \langle\cdot,\cdot\rangle)$ be a complex Hilbert space, and let $A : \mathcal{D}(A) \subseteq X \to X$ be a linear operator such that $\mathcal{D}(A)$ is dense in X. Then we define the* HILBERT ADJOINT OPERATOR *of A as $A^* : \mathcal{D}(A^*) \subseteq X \to X$, where*

$$\mathcal{D}(A^*) := \Big\{y \in X : \quad \exists\, y^* \in X \quad \text{such that} \quad \langle A(x), y\rangle = \langle x, y^*\rangle \quad \forall\, x \in \mathcal{D}(A)\Big\},$$

and in such a case we set $A^*(y) = y^*$ for all $y \in \mathcal{D}(A^*)$. In other words, $y \in X$ belongs to $\mathcal{D}(A^*)$ if and only if the mapping $\mathcal{D}(A) \ni x \to \langle A(x), y\rangle \in \mathbb{C}$ can be represented in the form $\mathcal{D}(A) \ni x \longrightarrow \langle x, y^*\rangle \in \mathbb{C}$ for some $y^* \in X$. Note that the density of $\mathcal{D}(A)$ in X is key to obtaining the uniqueness of y^*. Indeed, let $y^*, y^{**} \in X$ be such that

$$\langle A(x), y\rangle = \langle x, y^*\rangle = \langle x, y^{**}\rangle \quad \forall\, x \in \mathcal{D}(A).$$

It follows that $\langle x, y^* - y^{**}\rangle = 0$ for all $x \in \mathcal{D}(A)$, and then $y^* - y^{**} \in \mathcal{D}(A)^\perp = \overline{\mathcal{D}(A)}^\perp = X^\perp = \{0\}$, and thus $y^* = y^{**}$.

It is easy to prove from the previous definition that A^* is linear. Moreover, as indicated below, the concept of self-adjointness is analogous to that given for bounded operators.

Definition 8.32. *Let $(X, \langle\cdot,\cdot\rangle)$ be a Hilbert space over $\mathbb{C}$, and let $A : \mathcal{D}(A) \subseteq X \to X$ be a linear operator with $\overline{\mathcal{D}(A)} = X$. We say that A is* SELF-ADJOINT *if $A = A^*$, that is,*

$$\mathcal{D}(A) = \mathcal{D}(A^*) \quad \text{and} \quad \langle A(x), y\rangle = \langle x, A(y)\rangle \quad \forall\, x,\, y \in \mathcal{D}(A).$$

The following lemma establishes that the adjoint is always closed and therefore every self-adjoint operator is necessarily closed. The corresponding Banach space version of this result was proved in Theorem 3.81.

Lemma 8.33. *Let $(X, \langle\cdot,\cdot\rangle)$ be a complex Hilbert space, and let $A : \mathcal{D}(A) \subseteq X \to X$ be a linear operator with $\overline{\mathcal{D}(A)} = X$. Then, A^* is closed.*

Proof. Let $\{x_n\}_{n\in\mathbb{N}} \subseteq \mathcal{D}(A^*)$, and let $x,\, u \in X$ be such that $x_n \overset{n\to\infty}{\longrightarrow} x$ and $A^*(x_n) \overset{n\to\infty}{\longrightarrow} u$. Then, for each $y \in \mathcal{D}(A)$ we have

$$\langle A(y), x\rangle = \lim_{n\to\infty} \langle A(y), x_n\rangle = \lim_{n\to\infty} \langle y, A^*(x_n)\rangle \overset{n\to\infty}{\longrightarrow} \langle y, u\rangle,$$

which proves that $x \in \mathcal{D}(A^*)$ and $A^*(x) = u$, thus confirming that A^* is closed. □

We are now in a position to establish the first main result.

Theorem 8.34. *Let $(X, \langle\cdot,\cdot\rangle)$ be a complex Hilbert space, and let $A : \mathcal{D}(A) \subseteq X \to X$ be a linear operator with $\overline{\mathcal{D}(A)} = X$ such that A is self-adjoint. Then,*

i) $\mathrm{Im}\,\langle A_\lambda(x), x\rangle = -(\mathrm{Im}\,\lambda)\, \|x\|^2$ *for all* $x \in \mathcal{D}(A)$, *for all* $\lambda \in \mathbb{C}$.
ii) $\|A_\lambda^{-1}\| \leqslant \dfrac{1}{|\mathrm{Im}\,\lambda|}$ *for all* $\lambda \in \mathbb{C} - \mathbb{R}$.
iii) $\mathbb{C} - \mathbb{R} \subseteq \rho(A)$ (equivalently $\boldsymbol{\sigma}(A) \subseteq \mathbb{R}$).

Proof. Given $x \in \mathcal{D}(A)$, the fact that A is self-adjoint implies that

$$\langle A(x), x\rangle = \langle x, A(x)\rangle = \overline{\langle A(x), x\rangle},$$

which means that $\langle A(x), x\rangle \in \mathbb{R}$. Then, given $\lambda \in \mathbb{C}$ we have for each $x \in \mathcal{D}(A)$

$$\operatorname{Im}\langle A_\lambda(x), x\rangle = \operatorname{Im}\{\langle A(x), x\rangle - \langle \lambda x, x\rangle\} = \operatorname{Im}\{ -\lambda \|x\|^2\} = -(\operatorname{Im}\lambda)\|x\|^2,$$

which proves (i). Moreover, note that the same assumption on A implies, thanks to Lemma 8.33, that A is closed, and therefore A_λ is also closed for any $\lambda \in \mathbb{C}$. Now, from (i) we deduce that

$$|\operatorname{Im}\lambda|\,\|x\|^2 \leqslant |\operatorname{Im}\langle A_\lambda(x), x\rangle| \leqslant |\langle A_\lambda(x), x\rangle| \leqslant \|A_\lambda(x)\|\,\|x\| \quad \forall\, x \in \mathcal{D}(A),$$

and hence

$$\|A_\lambda(x)\| \geqslant |\operatorname{Im}\lambda|\,\|x\| \quad \forall\, x \in \mathcal{D}(A).$$

The foregoing inequality and the characterization result given by Theorem 3.58 show that if $\operatorname{Im}\lambda \neq 0$, that is, if $\lambda \in \mathbb{C}-\mathbb{R}$, then A_λ is injective and $R(A_\lambda) := A_\lambda(\mathcal{D}(A))$ is closed in X. It follows that A_λ is invertible for all $\lambda \in \mathbb{C}-\mathbb{R}$, and according to (8.13), $\|A_\lambda^{-1}\| \leqslant \dfrac{1}{|\operatorname{Im}\lambda|}$, which proves (ii). Finally, in order to conclude (iii) it only remains to show, by virtue of Lemma 8.33 and the fact that A is closed (since it is self-adjoint), that $A_\lambda(\mathcal{D}(A)) = X$ for all $\lambda \in \mathbb{C}-\mathbb{R}$. Indeed, let $y \in X$ such that $y \perp A_\lambda(\mathcal{D}(A))$. This means that

$$0 = \langle A(x), y\rangle = \langle A(x) - \lambda x, y\rangle = \langle A(x), y\rangle - \lambda\langle x, y\rangle,$$

and hence

$$\langle A(x), y\rangle = \lambda\langle x, y\rangle = \langle x, \bar{\lambda}\, y\rangle \quad \forall\, x \in \mathcal{D}(A),$$

which proves that $y \in \mathcal{D}(A^*)$ and $A^*(y) = \bar{\lambda}\, y$, or equivalently, since A is self-adjoint, $A(y) = \bar{\lambda}\, y$. It follows that $\bar{\lambda}\,\|y\|^2 = \bar{\lambda}\langle y, y\rangle = \langle A(y), y\rangle \in \mathbb{R}$, so that necessarily $y = 0$ since otherwise $\bar{\lambda} = 0$, which would contradict the fact that $\operatorname{Im}\lambda \neq 0$. This proves that $A_\lambda(\mathcal{D}(A)^\perp) = \{0\}$ and consequently $X = \overline{A_\lambda(\mathcal{D}(A))} = A_\lambda(\mathcal{D}(A))$. Thus, we have proved that $\mathbb{C}-\mathbb{R} \subseteq \rho(A)$.

□

The following lemma provides a more precise characterization of the resolvent (and therefore the spectrum) of a self-adjoint operator.

Lemma 8.35. *Let $(X, \langle\cdot,\cdot\rangle)$ be a complex Hilbert space, and let $A : \mathcal{D}(A) \subseteq X \to X$ be a linear operator with $\overline{\mathcal{D}(A)} = X$ such that A is self-adjoint. Then, $\lambda \in \rho(A)$ if and only if there exists $C > 0$ such that*

$$\|A_\lambda(x)\| \geqslant C\,\|x\| \quad \forall\, x \in \mathcal{D}(A).$$

Proof. Let $\lambda \in \rho(A)$, that is, $\lambda \in \mathbb{C}$, $N(A_\lambda) = \{0\}$, $A_\lambda(\mathcal{D}(A))$ is dense in X, and $A_\lambda^{-1} \in \mathcal{L}(A_\lambda(\mathcal{D}(A)), X)$. From this boundedness property, it follows that there exists $C > 0$ such that $\|A_\lambda^{-1}(y)\| \leqslant C\,\|y\|$ for all $y \in A_\lambda(\mathcal{D}(A))$, which is equivalent, taking $y = A_\lambda(x)$, $x \in \mathcal{D}(A)$, to

$$\|x\| \leqslant C\,\|A_\lambda(x)\| \quad \forall\, x \in \mathcal{D}(A).$$

Conversely, let $\lambda \in \mathbb{C}$ be such that there exists $C > 0$ satisfying

$$\|A_\lambda(x)\| \geqslant C\,\|x\| \qquad \forall\, x \in \mathcal{D}(A)\,. \tag{8.14}$$

Since A_λ is closed (because A is self-adjoint) (cf. Lemma 8.33), the above implies that $N(A_\lambda) = \{0\}$ and $A_\lambda(\mathcal{D}(A))$ is closed in X. It follows that A_λ is invertible and therefore (8.14) can be rewritten, with $y = A_\lambda(x)$, as

$$\|A_\lambda^{-1}(y)\| \leqslant \frac{1}{C}\,\|y\| \quad \forall\, y \in A_\lambda(\mathcal{D}(A))\,,$$

which shows that $A_\lambda^{-1} \in \mathcal{L}\big(A_\lambda(\mathcal{D}(A)), X\big)$. To conclude that $\lambda \in \rho(A)$, it only remains to show, by virtue of Lemma 8.27, that $\overline{A_\lambda(\mathcal{D}(A))} = X$. Indeed, given $y \in X$ such that $y \perp A_\lambda(\mathcal{D}(A))$, it follows, reasoning in the same way as in the proof of Theorem 8.34, that $y \in \mathcal{D}(A^*) = \mathcal{D}(A)$ and $A(y) = \overline{\lambda}\,y$. It follows that $\langle A(y), y\rangle = \overline{\lambda}\,\|y\|^2$, from which $y = 0$ or else $\lambda \in \mathbb{R}$. In this second case we have $A(y) = \lambda\,y$, that is $y \in N(A_\lambda) = \{0\}$, and therefore $y = 0$. Thus $A_\lambda(\mathcal{D}(A))^\perp = \{0\}$, which implies that $X = \overline{A_\lambda(\mathcal{D}(A))} = A_\lambda(\mathcal{D}(A))$.

□

An equivalent way to state the previous result is through the following lemma (cf. Exercise 8.14).

Lemma 8.36. *Let $(X, \langle\cdot,\cdot\rangle)$ be a complex Hilbert space, and let $A : \mathcal{D}(A) \subseteq X \to X$ be a linear operator with $\overline{\mathcal{D}(A)} = X$ such that A is self-adjoint. Then, $\lambda \in \boldsymbol{\sigma}(A)$ if and only if there exists $\{x_n\}_{n\in\mathbb{N}} \subseteq \mathcal{D}(A)$ such that $\|x_n\| = 1$ and $\lim\limits_{n\to\infty} \|A_\lambda(x_n)\| = 0$.*

The following theorems provide additional properties of the spectrum of unbounded and self-adjoint operators . We begin with a more general version (in the self-adjoint case) of Lemma 8.1.

Theorem 8.37. *Let $(X, \langle\cdot,\cdot\rangle)$ be a complex Hilbert space, and let $A : \mathcal{D}(A) \subseteq X \to X$ be a linear operator with $\overline{\mathcal{D}(A)} = X$ such that A is self-adjoint. Then, $\boldsymbol{\sigma}(A)$ is a closed subset of $\mathbb{R}$.*

Proof. Since $\boldsymbol{\sigma}(A) \subseteq \mathbb{R}$ (cf. Theorem 8.34), it suffices to prove that $\rho(A) \cap \mathbb{R}$ is open in $\mathbb{R}$. Indeed, let $\lambda_0 \in \rho(A) \cap \mathbb{R}$ and consider $\lambda \in \mathbb{R}$. Then, for each $x \in \mathcal{D}(A)$ we have

$$\|A_{\lambda_0}(x)\| \leqslant \|A_\lambda(x)\| + |\lambda - \lambda_0|\,\|x\|,$$

or equivalently

$$\|A_\lambda(x)\| \geqslant \|A_{\lambda_0}(x)\| - |\lambda - \lambda_0|\,\|x\| \quad \forall\, x \in \mathcal{D}(A).$$

Applying Lemma 8.35, there exists $C > 0$ such that $\|A_{\lambda_0}(x)\| \geqslant C\,\|x\|$ for all $x \in \mathcal{D}(A)$, and then

$$\|A_\lambda(x)\| \geqslant (C - |\lambda - \lambda_0|)\,\|x\| \quad \forall\, x \in \mathcal{D}(A).$$

In particular, for each $\lambda \in \mathbb{R}$ such that $|\lambda - \lambda_0| < C/2$ we have

$$\|A_\lambda(x)\| \geqslant \frac{C}{2}\,\|x\| \quad \forall\, x \in \mathcal{D}(A),$$

which shows that $\lambda \in \rho(A)$, and thus $\rho(A) \cap \mathbb{R}$ is open in $\mathbb{R}$.

□

Theorem 8.38. *Let $(X, \langle\cdot,\cdot\rangle)$ be a complex Hilbert space, and let $A : \mathcal{D}(A) \subseteq X \to X$ be a linear operator with $\overline{\mathcal{D}(A)} = X$ such that A is self-adjoint. Then $\boldsymbol{\sigma}_r(A) = \varnothing$.*

Proof. Suppose, by contradiction, that there exists $\lambda \in \boldsymbol{\sigma}_r(A)$. It follows that $A_\lambda(\mathcal{D}(A))$ is not dense in X, equivalently $A_\lambda(\mathcal{D}(A))^\perp \neq \{0\}$, and thus there exists $y \neq 0$ such that $\langle A_\lambda(x), y\rangle = 0$ for all $x \in \mathcal{D}(A)$. Then, reasoning again in the same way as in the proof of Theorem 8.34, and noting that $\boldsymbol{\sigma}(A) \subseteq \mathbb{R}$, we deduce that $A(y) = \lambda y$, which shows that $\lambda \in \boldsymbol{\sigma}_p(A)$, thereby contradicting the fact that $\boldsymbol{\sigma}_p(A) \cap \boldsymbol{\sigma}_r(A) = \varnothing$.

□

Note that the same proof as above, with minor modifications, serves to show that if $A : \mathcal{D}(A) \subseteq X \to X$ is linear with $\overline{\mathcal{D}(A)} = X$, then $\bar{\lambda} \in \boldsymbol{\sigma}_p(A^*)$ for all $\lambda \in \boldsymbol{\sigma}_r(A)$, that is

$$\left\{\bar{\lambda} : \quad \lambda \in \boldsymbol{\sigma}_r(A)\right\} \subseteq \boldsymbol{\sigma}_p(A^*).$$

8.6 EXERCISES.

▸ **8.1** *Let X be a complex Banach space, and let $A,\, B \in \mathcal{L}(X,X)$ be such that $0 \in \rho(A)$ and*

$$\|A - B\| < \frac{1}{\|A^{-1}\|}.$$

Show that $0 \in \rho(B)$ and

$$\|B^{-1}\| \leqslant \frac{\|A^{-1}\|}{(1 - \|A^{-1}\|\,\|A - B\|)}$$

▸ **8.2** *Let X be a complex Hilbert space, and let $A : \mathcal{D}(A) \subseteq X \to X$ be a linear operator such that $\mathcal{D}(A)$ is dense in X. Prove that $\lambda \in \rho(A^*)$ if and only if $(A^* - \lambda I)$ is injective and there exists $C > 0$ such that $\|(A^* - \lambda I)^{-1}(x)\| \leqslant C\,\|x\|$ for all $x \in X$.*

▸ **8.3** *Let X be a complex Hilbert space, and let $A \in \mathcal{L}(X,X)$ be a compact self-adjoint operator.*

i) *Prove that there exists a sequence $\{A_n\}_{n\in\mathbb{N}} \subseteq \mathcal{L}(X,X)$ of finite-rank operators such that*

$$\lim_{n\to\infty} \|A - A_n\|_{\mathcal{L}(X,X)} = 0.$$

ii) *Show that the equation $Au = f$ admits a solution u if and only if*

$$\sum_{\lambda\in\sigma_p(A)} \frac{|f_\lambda|^2}{|\lambda|^2} < \infty \quad \text{where} \quad f = \sum_{\lambda\in\sigma_p(A)} f_\lambda\,, \quad \text{with} \quad f_\lambda \in N(A_\lambda)\,.$$

iii) *Given $\mu \notin \sigma(A)$ and $f \in X$, solve the equation: $\mu u - Au = f$, in terms of a Hilbertian basis of X.*

iv) *Apply the previous result to the following integral equation*

$$u \in X\,, \qquad 3\,u(x) - \int_\Omega K(x,t)\,u(t)\,dt = f(x) \quad \forall\, x \in \Omega\,,$$

where $\Omega := (0,1)$, $X := \mathrm{L}^2(\Omega)$, $f \in X$, and K is the function on $\Omega \times \Omega$ defined by

$$K(x,t) := \begin{cases} x(1-t) & \text{if} \quad t \leqslant x\,, \\[2ex] t(1-x) & \text{if} \quad t > x\,. \end{cases}$$

▸ **8.4** *Let X be the complex Hilbert space $\mathrm{L}^2(\Omega)$, with $\Omega := (0,1)$, and consider the linear operator $K : X \to X$ defined by*

$$(K\,u)(t) := \int_0^1 k(t,s)\,u(s)\,ds \qquad \forall\, t \in \Omega\,, \quad \forall\, u \in X\,,$$

where $k \in \mathrm{L}^2(\Omega \times \Omega)$ is given by $k(t,s) := \min\{t,s\}$ for all $(t,s) \in \Omega \times \Omega$.

i) *Show that K is self-adjoint and then use an orthonormal basis of* $\mathrm{L}^2(\Omega)$ *to conclude that K is compact.*
ii) *Find the spectrum of K. To this end, given $\lambda \in \sigma_p(K)$ and $v \in N(K - \lambda I)\backslash 0$, differentiate twice the equation $K(v) - \lambda v = 0$ and obtain the boundary value problem*

$$\lambda v'' + v = 0 \quad in \quad \Omega, \quad v'(1) = v(0) = 0.$$

Observe that $\lambda \neq 0$. Then, multiply the differential equation by $\bar{v}$, integrate over Ω, and deduce from the resulting identity that $\lambda > 0$. Next, prove that the solution of the differential equation is given by $v(t) := C_1 \cos\left(\frac{t}{\sqrt{\lambda}}\right) + C_2 \sin\left(\frac{t}{\sqrt{\lambda}}\right)$ for all $t \in \Omega$, and use the boundary conditions to show that the eigenvalues of K are given by $\lambda_j := \frac{4}{\pi^2 (2j-1)^2}$ for all $j \in \mathbb{N}$, with eigenfunctions

$$v_j(t) := \sin\left(\frac{\pi(2j-1)\,t}{2}\right) \qquad \forall\, t \in \Omega.$$

iii) *Conclude that* $\|K\| = \frac{4}{\pi^2}$.

▸ **8.5** *Let H be a complex Hilbert space. We say that $A \in \mathcal{L}(H,H)$ is a normal operator if A and A^* commute, that is, $A^* A = A A^*$. Prove in this case that*

a) *$\lambda \in \sigma_p(A)$ if and only if $\bar{\lambda} \in \sigma_p(A^*)$. Conclude furthermore that*

$$N(A_\lambda) := N(A - \lambda I) = N(A^*_{\bar{\lambda}}) := N(A^* - \bar{\lambda} I).$$

b) *If $\lambda, \mu \in \sigma_p(A)$, $\lambda \neq \mu$, then $N(A_\lambda) \perp N(A^*_{\bar{\mu}})$.*

▸ **8.6** *Let $A : \mathcal{D}(A) \subseteq \mathrm{L}^2(-1,1) \to \mathrm{L}^2(-1,1)$ be the operator defined by*

$$(A\,u)(x) = x\,u(x) + \delta \int_{-1}^{1} u(t)\,dt,$$

where $\delta \in \mathbb{R}$. Obtain as much information as possible about the spectrum of A.

▸ **8.7** *Let $\Omega := (0,1)$, let $p \in C^1(\bar{\Omega})$ and $q \in C(\bar{\Omega})$, with $p(x) \geqslant \alpha > 0$ for all $x \in \Omega$. Prove that there exist a Hilbert basis $\{e_n\}_{n\in\mathbb{N}}$ of $\mathrm{L}^2(\Omega)$ and a sequence $\{\lambda_n\}_{n\in\mathbb{N}}$ of positive real numbers, with $\lambda_n \to \infty$ as $n \to \infty$, such that $e_n \in H_0^1(\Omega) \cap \mathrm{H}^2(\Omega)$ and*

$$-(p e_n')' + q e_n = \lambda_n e_n \quad in \quad \Omega.$$

▸ **8.8** *Let H be a separable Hilbert space. We say that $T : H \to H$ is a* HILBERT-SCHMIDT OPERATOR *if there exists a basis $\{e_n\}_{n\in\mathbb{N}}$ of H such that*

$$\|T\|^2 := \sum_{n\in\mathbb{N}} \|T(e_n)\|^2 < \infty.$$

Let p and q be as in the previous problem. Prove that the linear operator $T : \mathrm{L}^2(\Omega) \to \mathrm{L}^2(\Omega)$ that assigns to each $f \in \mathrm{L}^2(\Omega)$ the unique solution $u := T(f)$ of

$$-(p u')' + q u = f \quad in \quad \Omega := (0,1),$$
$$u(0) = u(1) = 0,$$

is a Hilbert-Schmidt operator.

▸ **8.9** *Let $(H, \langle\cdot,\cdot\rangle_H)$ be a complex Hilbert space, and let $A : \mathcal{D}(A) \subseteq H \to H$ be a linear operator such that $\mathcal{D}(A)$ is dense in H and A is self-adjoint. Prove that $\mathcal{D}(A)$, endowed with the inner product $\langle u, v\rangle_{\mathcal{D}(A)} := \langle u, v\rangle_H + \langle A(u), A(v)\rangle_H$, is a Hilbert space. Now suppose that $\mathbf{i} : \mathcal{D}(A) \to H$ is a compact injection, and that $A^{-1} \in \mathcal{L}(H, \mathcal{D}(A))$. Prove that*

i) $\sigma(A) = \sigma_p(A) \subseteq \mathbb{R}$.

ii) *The elements of* $\sigma_p(A)$ *can be ordered in a sequence* $\{\lambda_n\}_{n\in\mathbb{N}}$ *such that* $|\lambda_n| \to \infty$ *as* $n \to \infty$.
iii) *The eigenspaces* $N(A_{\lambda_n}) := N(A - \lambda_n I)$ *are of finite dimension, mutually orthogonal, and* $H = \bigoplus\{N(A_{\lambda_n}): \quad n \in \mathbb{N}\}$.

▸ **8.10** *Let* H *be a nontrivial complex Hilbert space, and let* $A \in \mathcal{L}(H, H)$ *be a bijective operator such that* $A^* = A^{-1}$ (UNITARY OPERATOR). *Prove that* $\|A\| = 1$, *and conclude that*

$$\sigma(A) \subseteq \{\lambda \in \mathbb{C}: \quad |\lambda| = 1\}.$$

▸ **8.11** *Let* X *be a complex Hilbert space, and let* $A \in \mathcal{L}(X, X)$ *be a self-adjoint operator. Prove that for every* $\lambda \in \mathbb{C}$ *there holds* $X = N(A - \lambda I) \oplus \overline{R(A - \lambda I)}$.

▸ **8.12** [EIGENVALUES OF Δ]. *Let* Ω *be a bounded open subset of* $\mathbb{R}^n$ *with Lipschitz-continuous boundary. Prove that there exist a Hilbertian basis* $\{e_n\}_{n\in\mathbb{N}}$ *of* $\mathrm{L}^2(\Omega)$, *and a sequence* $\{\lambda_n\}_{n\in\mathbb{N}}$ *of real numbers with* $\lambda_n > 0$, *and* $\lambda_n \to +\infty$ *as* $n \to +\infty$, *such that* $e_n \in \mathrm{H}_0^1(\Omega) \cap \mathrm{H}^2(\Omega)$, *and* $-\Delta e_n = \lambda_n e_n$ *in* Ω. *We say here that the* λ_n *are the eigenvalues of the Laplacian (with Dirichlet condition), and that the* e_n *are the associated eigenfunctions.*

▸ **8.13** [SELF-ADJOINT SPECTRUM]. *Let* H *be a complex Hilbert space, and let* $A \in \mathcal{L}(H, H)$ *be a self-adjoint operator. Given a closed subspace* S *of* H *invariant with respect to* A, *denote by* $\sigma(A, S)$ *and* $\sigma(A, S^\perp)$ *the spectra of the restrictions* $A|_S$ *and* $A|_{S^\perp}$, *respectively. Prove that*

$$\sigma(A) = \sigma(A, S) \cup \sigma(A, S^\perp).$$

▸ **8.14** *Prove Lemma 8.36.*

▸ **8.15** *Let* X *be a complex Hilbert space, and let* $A : \mathcal{D}(A) \subseteq X \to X$ *be a linear operator such that* $\mathcal{D}(A)$ *is dense in* X. *Prove that*

$$\left\{\bar{\lambda}: \quad \lambda \in \boldsymbol{\sigma}_r(A)\right\} \subseteq \boldsymbol{\sigma}_p(A^*).$$

In particular, consider the space $X := \mathrm{L}^2(\Omega)$, *with* $\Omega := (0, 1)$, *and prove that* $\boldsymbol{\sigma}_r(A) = \emptyset$, *where* $A : X \to X$ *is defined by*

$$(A u)(t) := \int_0^1 (t^2 s + s^2 t)\, u(s)\, ds \qquad \forall\, t \in \Omega, \quad \forall\, u \in X.$$

9

Theory of Distributions

In this chapter we continue the discussion initiated in Section 1.5, where some basic notions of distributions were introduced. Most of the results presented below follow the approaches developed in [53] and [95]. In addition, similarly to other sections of the book (see, e.g., Exercise 1.10, Section 5.3.4.1), given a domain $\mathcal{O} \subset \mathbb{R}^n$, $m \in \mathbb{N}$, and $t \in [1, \infty]$, we denote by $\mathrm{L}^t(\mathcal{O})$ and $\mathrm{W}^{m,t}(\mathcal{O})$ the standard Lebesgue and Sobolev spaces with norms $\|\cdot\|_{0,t;\mathcal{O}}$ and $\|\cdot\|_{m,t;\mathcal{O}}$, respectively. When $t = 2$, these norms simplify to $\|\cdot\|_{0,\mathcal{O}}$ and $\|\cdot\|_{m,\mathcal{O}}$, respectively. In turn, for each scalar functional space S, we denote by **S** its vector version. In particular, $\mathbf{L}^t(\mathcal{O}) := [\mathrm{L}^t(\mathcal{O})]^n$ and we also use $\|\cdot\|_{0,t;\mathcal{O}}$, or $\|\cdot\|_{0,\mathcal{O}}$ when $t = 2$, to denote its norm. Analogously, $\|\cdot\|_{m,t;\mathcal{O}}$ stands also for the norm of $\mathbf{W}^{m,t}(\mathcal{O}) := [\mathrm{W}^{m,t}(\mathcal{O})]^n$. Furthermore, we recall from Definition 1.43 and (1.15) that the space of locally integrable functions on $\mathcal{O}$, denoted $\mathrm{L}^1_{loc}(\mathcal{O})$, is defined as the set of all (equivalence classes of) measurable functions $f : \mathcal{O} \to \mathbb{C}$ such that

$$\int_K |f| < \infty \quad \forall \, \text{compact } K \subseteq \mathcal{O}, \tag{9.1}$$

and that the support of a function $\varphi \in C(\Omega)$ is defined by

$$\operatorname{supp}(\varphi) := \overline{\{x \in \Omega : \quad \varphi(x) \neq 0\}}. \tag{9.2}$$

9.1 Preliminaries

We begin by recalling that a linear form $u : \mathcal{D}(\Omega) \to \mathbb{C}$ is called a DISTRIBUTION ON Ω (cf. Definitions 1.39 and 1.41) if it is continuous (equivalently, sequentially continuous), and that the concept of convergence of a sequence in $\mathcal{D}(\Omega)$ is provided in Definition 1.40. In turn, we also recall (cf. Definition 1.45) that a regular distribution is one generated by a locally integrable function. In this regard, our first goal in this section is to prove the uniqueness of such a function. In particular, we now show that this occurs for distributions generated by continuous functions. More precisely, we have the following result.

Lemma 9.1. *Let Ω be an open subset of $\mathbb{R}^n$, and let $h \in C(\Omega)$ be such that*

$$\int_\Omega h\varphi = 0 \quad \forall \, \varphi \in C_0^\infty(\Omega).$$

Then $h(x) = 0$ for all $x \in \Omega$.

G. N. Gatica, *Introduction to Functional Analysis*,
https://doi.org/10.1007/978-3-032-37415-8_9

Proof. By separating the real and imaginary parts of h, we can assume, without loss of generality, that h is real. Then, suppose by contradiction that there exists $x_0 \in \Omega$ such that $h(x_0) > 0$ (the case $h(x_0) < 0$ is analogous). It follows, by continuity of h, that there exists $\delta > 0$ such that $\bar{B}(x_0, \delta) \subseteq \Omega$ and $h(x) > 0$ for all $x \in B(x_0, \delta)$. Next, we consider the function $\varphi_{\varepsilon,x_0} \in C_0^\infty(\Omega)$ (cf. Example 1.35) with $\varepsilon = \delta/2$, that is

$$\varphi_{\varepsilon,x_0}(x) := \begin{cases} e^{-\varepsilon^2/(\varepsilon^2 - \|x-x_0\|^2)} & \text{if } \|x - x_0\| < \varepsilon, \\ 0 & \text{if } \|x - x_0\| \geqslant \varepsilon, \end{cases}$$

and we observe that

$$\int_\Omega h\,\varphi_{\varepsilon,x_0} = \int_{B(x_0,\varepsilon)} h\,\varphi_{\varepsilon,x_0} > 0,$$

which contradicts the hypothesis.

□

In order to extend the above result to the case $h \in L^1_{loc}(\Omega)$, we now resort to a consequence of the Lebesgue Differentiation Theorem (cf. [120, Section 7.2, Theorem 7.16]), which states that for each $h \in L^1_{loc}(\Omega)$ there holds

$$\lim_{\varepsilon \to 0^+} \varepsilon^{-n} \int_{B(x,\varepsilon)} |h(x) - h(y)|\, dy = 0 \quad \forall\, x \in \Omega \quad \text{a.e.}$$

Then, we have the following result.

Lemma 9.2. *Let Ω be an open subset of $\mathbb{R}^n$, and let $h \in L^1_{loc}(\Omega)$ be such that*

$$\int_\Omega h\,\varphi = 0 \quad \forall\, \varphi \in C_0^\infty(\Omega)\,.$$

Then $h(x) = 0$ for all $x \in \Omega$ a.e.

Proof. Let $\varphi \in C_0^\infty(\mathbb{R}^n)$ be such that $\displaystyle\int_{\mathbb{R}^n} \varphi = 1$, and let $\text{supp}(\varphi) = \bar{B}(0,1)$, for instance

$$\varphi = \frac{1}{c}\,\varphi_{1,0} \quad \text{with} \quad c := \int_{B(0,1)} \varphi_{1,0}\,.$$

Then, for $x \in \Omega$ and $\varepsilon < \text{dist}(x, \partial\Omega)$, we have

$$\varepsilon^{-n} \int_\Omega \varphi\left(\frac{x-y}{\varepsilon}\right) dy = \int_{B(0,1)} \varphi(z)\, dz = 1\,,$$

and hence we can write

$$h(x) = \varepsilon^{-n} \int_\Omega h(x)\,\varphi\left(\frac{x-y}{\varepsilon}\right) dy,$$

or equivalently,

$$\begin{aligned} h(x) = {} & \varepsilon^{-n} \int_\Omega \Big\{h(x) - h(y)\Big\}\,\varphi\left(\frac{x-y}{\varepsilon}\right) dy \\ & + \varepsilon^{-n} \int_\Omega h(y)\varphi\left(\frac{x-y}{\varepsilon}\right) dy\,. \end{aligned}$$

It is clear that the second term above vanishes by hypothesis. Moreover, we have

$$\begin{aligned}
&\left|\varepsilon^{-n}\int_{\Omega}\Big\{h(x)-h(y)\Big\}\,\varphi\left(\frac{x-y}{\varepsilon}\right)dy\right| \\
&\quad\leqslant \varepsilon^{-n}\int_{B(x,\varepsilon)}|h(x)-h(y)|\left|\varphi\left(\frac{x-y}{\varepsilon}\right)\right|dy \\
&\quad\leqslant M\,\varepsilon^{-n}\int_{B(x,\varepsilon)}|h(x)-h(y)|\,dy\,,
\end{aligned}$$

with $M := \sup\limits_{z\in\bar{B}(0,1)} |\varphi(z)|$, which, thanks again to the aforementioned consequence of the Lebesgue Differentiation Theorem (cf. [120, Section 7.2, Theorem 7.16]), tends to 0 for all $x \in \Omega$ a.e. when $\varepsilon \to 0^+$. In this way, we conclude that $h(x) = 0$ for all $x \in \Omega$ a.e.

□

We conclude this section with some nontrivial examples of distributions on $\mathbb{R}$. For this purpose, we need the following preliminary lemma.

Lemma 9.3. *Let $\varphi \in C_0^\infty(\mathbb{R})$ and let $M > 0$ be such that* $\operatorname{supp}(\varphi) \subseteq [-M, M]$. *For each $n \in \mathbb{N} \cup \{0\}$, define*

$$\widetilde{\varphi}_n(x) := \begin{cases} \dfrac{1}{x^{n+1}}\left\{\varphi(x) - \displaystyle\sum_{j=0}^{n}\frac{x^j}{j!}\,\varphi^{(j)}(0)\right\} & \text{if} \quad x \neq 0, \\[2ex] \dfrac{1}{(n+1)!}\,\varphi^{(n+1)}(0) & \text{if} \quad x = 0. \end{cases}$$

Then $\widetilde{\varphi}_n \in C(\mathbb{R})$ and there exists $c > 0$ such that

$$\sup_{x\in\mathbb{R}} |\widetilde{\varphi}_n(x)| \leqslant c \sup_{x\in[-M,M]} |\varphi^{(n+1)}(x)|\,.$$

Proof. The Taylor formula with integral remainder applied to φ around $x = 0$, that is the Maclaurin formula, is given for each $n \in \mathbb{N} \cup \{0\}$ by

$$\varphi(x) = \sum_{j=0}^{n}\frac{x^j}{j!}\,\varphi^{(j)}(0) + \frac{x^{n+1}}{n!}\int_0^1 (1-t)^n\,\varphi^{(n+1)}(xt)\,dt \quad \forall\, x \in \mathbb{R},$$

from which it follows that

$$\widetilde{\varphi}_n(x) = \frac{1}{n!}\int_0^1 (1-t)^n\,\varphi^{(n+1)}(xt)\,dt \quad \forall\, x \neq 0.$$

Taking the limit as $x \to 0$ yields

$$\lim_{x\to 0}\widetilde{\varphi}_n(x) = \frac{\varphi^{(n+1)}(0)}{n!}\int_0^1 (1-t)^n\,dt = \frac{1}{(n+1)!}\,\varphi^{(n+1)}(0) = \widetilde{\varphi}_n(0)\,,$$

which proves that $\widetilde{\varphi}_n$ is continuous at $x = 0$, and thus $\widetilde{\varphi}_n \in C(\mathbb{R})$. Moreover, it is clear that

$$\begin{aligned}
|\widetilde{\varphi}_n(x)| &\leqslant \frac{1}{n!}\sup_{y\in[-M,M]}|\varphi^{(n+1)}(y)|\int_0^1 (1-t)^n\,dt \\
&= \frac{1}{(n+1)!}\sup_{y\in[-M,M]}|\varphi^{(n+1)}(y)| \quad \forall\, x \neq 0\,,
\end{aligned}$$

and since $\widetilde{\varphi}_n(0) = \dfrac{1}{(n+1)!}\,\varphi^{(n+1)}(0)$, we conclude that

$$\sup_{x\in\mathbb{R}} |\widetilde{\varphi}_n(x)| \leqslant \frac{1}{(n+1)!} \sup_{x\in[-M,M]} |\varphi^{(n+1)}(x)|,$$

which completes the proof.

□

Example 9.4 Let $u : \mathcal{D}(\mathbb{R}) \to \mathbb{R}$ be defined by

$$\langle u, \varphi\rangle := \left\langle FP\Big(\frac{H(x)}{x}\Big), \varphi\right\rangle := \lim_{\varepsilon\to 0^+} \left\{ \int_\varepsilon^{+\infty} \frac{\varphi(x)}{x}\, dx + \varphi(0)\ln(\varepsilon)\right\} \quad \forall\, \varphi \in \mathcal{D}(\mathbb{R}).$$

The goal of this example is to show that $u \in \mathcal{D}'(\mathbb{R})$. The notation $FP\big(\frac{H(x)}{x}\big)$ refers to the so-called finite part (FP) of the Heaviside function

$$H(x) := \begin{cases} 1 & \text{if } x > 0 \\ 0 & \text{if } x \leqslant 0 \end{cases},$$

divided by x. Let K be a compact subset of $\mathbb{R}$, and let $M > 0$ be such that $K \subseteq [-M, M]$. Then, given $\varphi \in C_0^\infty(\mathbb{R})$ with $\operatorname{supp}(\varphi) \subseteq K$, the previous lemma for $n = 0$ gives

$$\varphi(x) = \varphi(0) + x\,\widetilde{\varphi}_0(x) \quad \forall\, x \in \mathbb{R},$$

with $\widetilde{\varphi}_0 \in C(\mathbb{R})$ and $\sup\limits_{x\in\mathbb{R}} |\widetilde{\varphi}_0(x)| \leqslant c \sup\limits_{x\in[-M,M]} |\varphi'(x)|$. It follows that for each $\varepsilon > 0$

$$\begin{aligned}
\int_\varepsilon^M \frac{\varphi(x)}{x}\, dx + \varphi(0)\ln(\varepsilon) &= \int_\varepsilon^M \left\{\frac{\varphi(0)}{x} + \widetilde{\varphi}_0(x)\right\} dx + \varphi(0)\ln(\varepsilon) \\
&= \varphi(0)\big(\ln(M) - \ln(\varepsilon)\big) + \int_\varepsilon^M \widetilde{\varphi}_0(x)\, dx + \varphi(0)\ln(\varepsilon) \\
&= \varphi(0)\ln(M) + \int_\varepsilon^M \widetilde{\varphi}_0(x)\, dx,
\end{aligned}$$

from which

$$\lim_{\varepsilon\to 0^+} \left\{ \int_\varepsilon^{+\infty} \frac{\varphi(x)}{x}\, dx + \varphi(0)\ln(\varepsilon)\right\} = \varphi(0)\ln(M) + \int_0^M \widetilde{\varphi}_0(x)\, dx,$$

which shows that u is well defined and is a linear functional. Moreover, we have

$$\begin{aligned}
|\langle u, \varphi\rangle| &\leqslant |\ln(M)|\,|\varphi(0)| + M \sup_{x\in\mathbb{R}} |\widetilde{\varphi}_0(x)| \\
&\leqslant |\ln(M)| \sup_{x\in[-M,M]} |\varphi(x)| + Mc \sup_{x\in[-M,M]} |\varphi'(x)| \\
&\leqslant \max\Big\{|\ln(M)|, Mc\Big\} \sum_{j=0}^{1} \sup_{x\in K} |\varphi^{(j)}(x)|,
\end{aligned}$$

which, thanks to Theorem 1.42, implies that indeed $u \in \mathcal{D}'(\mathbb{R})$.

Example 9.5 Let $u : \mathcal{D}(\mathbb{R}) \to \mathbb{R}$ be defined by

$$\langle u, \varphi\rangle := \left\langle FP\Big(\frac{H(x)}{x^2}\Big), \varphi\right\rangle := \lim_{\varepsilon\to 0^+} \left\{ \int_\varepsilon^{+\infty} \frac{\varphi(x)}{x^2} - \frac{\varphi(0)}{\varepsilon} + \varphi'(0)\ln(\varepsilon)\right\} \quad \forall\, \varphi \in \mathcal{D}(\mathbb{R}),$$

and let us also show, using the previous lemma, that $u \in \mathcal{D}'(\mathbb{R})$. Indeed, let K be a compact subset of $\mathbb{R}$, and let $M > 0$ be such that $K \subseteq [-M, M]$. Then, given $\varphi \in C_0^\infty(\mathbb{R})$ with $\operatorname{supp}(\varphi) \subseteq K$, applying Lemma 9.3 to the case $n = 1$ gives

$$\varphi(x) = \varphi(0) + x\varphi'(0) + x^2\, \widetilde{\varphi}_1(x) \quad \forall\, x \in \mathbb{R},$$

with $\widetilde{\varphi}_1 \in C(\mathbb{R})$ and $\sup\limits_{x\in\mathbb{R}} |\widetilde{\varphi}_1(x)| \leqslant c \sup\limits_{x\in[-M,M]} |\varphi''(x)|$. Then, for each $\varepsilon > 0$ we have

$$\begin{aligned}
&\int_\varepsilon^M \frac{\varphi(x)}{x^2}\, dx - \frac{\varphi(0)}{\varepsilon} + \varphi'(0)\, \ln(\varepsilon) \\
&\quad = \int_\varepsilon^M \left\{ \frac{\varphi(0)}{x^2} + \frac{\varphi'(0)}{x} \right\} dx + \int_\varepsilon^M \widetilde{\varphi}_1(x)\, dx - \frac{\varphi(0)}{\varepsilon} + \varphi'(0)\, \ln(\varepsilon) \\
&\quad = -\frac{\varphi(0)}{M} + \varphi'(0)\, \ln(M) + \int_\varepsilon^M \widetilde{\varphi}_1(x)\, dx,
\end{aligned}$$

from which

$$\lim_{\varepsilon\to 0^+} \left\{ \int_\varepsilon^{+\infty} \frac{\varphi(x)}{x^2} - \frac{\varphi(0)}{\varepsilon} + \varphi'(0)\, \ln(\varepsilon) \right\} = -\frac{\varphi(0)}{M} + \varphi'(0)\, \ln(M) + \int_0^M \widetilde{\varphi}_1(x)\, dx,$$

thus showing that u is well defined and clearly a linear functional. Furthermore, it is clear that

$$\begin{aligned}
|\langle u, \varphi\rangle| &\leqslant \frac{1}{M} |\varphi(0)| + |\ln(M)|\, |\varphi'(0)| + M \sup_{x\in\mathbb{R}} |\widetilde{\varphi}_1(x)| \\
&\leqslant \frac{1}{M} \sup_{x\in[-M,M]} |\varphi(x)| + |\ln(M)| \sup_{x\in[-M,M]} |\varphi'(x)| + Mc \sup_{x\in[-M,M]} |\varphi''(x)| \\
&\leqslant \max\left\{ \frac{1}{M}, |\ln(M)|, Mc \right\} \sum_{j=0}^{2} \sup_{x\in K} |\varphi^{(j)}(x)|,
\end{aligned}$$

which, once again by virtue of Theorem 1.42, shows that $u \in \mathcal{D}'(\mathbb{R})$.

9.2 Convolution and regularization

Given $u, v \in C(\mathbb{R}^n)$ such that at least one of them has compact support, the convolution $u * v$ is defined as the continuous function given by

$$(u * v)(x) := \int_{\mathbb{R}^n} u(x-y)\, v(y)\, dy \quad \forall\, x \in \mathbb{R}^n.$$

Note that the compactness of one of the supports, say that of v, guarantees that $(u * v)(x)$ is well defined since

$$\begin{aligned}
\left| \int_{\mathbb{R}^n} u(x-y)\, v(y)\, dy \right| &= \left| \int_{\operatorname{supp}(v)} u(x-y)\, v(y)\, dy \right| \leqslant \int_{\operatorname{supp}(v)} |u(x-y)|\, |v(y)|\, dy \\
&\leqslant \max_{y\in\operatorname{supp}(v)} |u(x-y)| \max_{y\in\operatorname{supp}(v)} |v(y)|\, |\operatorname{supp}(v)| < +\infty.
\end{aligned}$$

Moreover, it is easy to see that the Lebesgue Dominated Convergence Theorem (cf. [113, Theorem 1.34], [120, Theorem 5.36]) allows us to prove the continuity of $u * v$. Indeed, given a sequence $\{x_n\}_{n\in\mathbb{N}} \subseteq \mathbb{R}^n$ that converges to $\bar{x} \in \mathbb{R}^n$, we have

$$(u * v)(x_n) := \int_{\mathbb{R}^n} u(x_n - y)\, v(y)\, dy = \int_K f_n(y)\, dy,$$

where $K := \mathrm{supp}(v)$ and $\{f_n\}_{n\in\mathbb{N}}$ is defined by $f_n(y) := u(x_n - y)\, v(y)$ for all $y \in K$. The continuity of u implies that $f_n(y) \overset{n\to\infty}{\to} f(y) := u(\bar{x} - y)\, v(y)$ for all $y \in K$ a.e. Moreover, since $\{x_n\}_{n\in\mathbb{N}}$ is clearly bounded, there exists $\delta > 0$ such that

$$\{x_n\}_{n\in\mathbb{N}} \cup \{\bar{x}\} \subseteq B(0,\delta).$$

It follows that

$$|f_n(y)| \leqslant g(y) := \max_{z\in\bar{B}(0,\delta)} |u(z-y)|\,|v(y)| \quad \forall\, y \in K,$$

whence

$$\int_K g(y)\, dy \leqslant \max_{z\in\bar{B}(0,\delta)-K} |u(z)| \max_{y\in K} |v(y)|\,|K| < +\infty,$$

which proves that $g \in \mathrm{L}^1(K)$. Thus, by applying the mentioned result, it follows that

$$(u * v)(x_n) = \int_K f_n(y)\, dy \overset{n\to\infty}{\to} \int_K f(y)\, dy = \int_K u(\bar{x} - y)\, v(y)\, dy = (u * v)(\bar{x}),$$

which confirms the continuity of $u * v$.

Notice that $u * v$ is also well defined and continuous if u is continuous and has compact support in $\mathbb{R}^n$, that is, $u \in C_0(\mathbb{R}^n)$, and $v \in \mathrm{L}^1_{loc}(\mathbb{R}^n)$, or vice versa. Indeed, in such case we have

$$\begin{aligned} \left|\int_{\mathbb{R}^n} u(x-y)\, v(y)\, dy\right| &= \left|\int_{\{x\}-\mathrm{supp}(u)} u(x-y)\, v(y)\, dy\right| \\ &\leqslant \int_{\{x\}-\mathrm{supp}(u)} |u(x-y)|\,|v(y)|\, dy \\ &\leqslant \max_{z\in\{x\}-\mathrm{supp}(u)} |u(z)| \int_{\{x\}-\mathrm{supp}(u)} |v(y)|\, dy < +\infty, \end{aligned}$$

since $\{x\} - \mathrm{supp}(u)$ is certainly compact and $v \in \mathrm{L}^1_{loc}(\mathbb{R}^n)$. At the same time, proceeding in an analogous way to the previous case, one can prove, again using the Lebesgue Dominated Convergence Theorem (cf. [113, Theorem 1.34], [120, Theorem 5.36]), that $u * v$ is continuous. Indeed, given $\{x_n\}_{n\in\mathbb{N}} \subseteq \mathbb{R}^n$ that converges to $\bar{x} \in \mathbb{R}^n$, there exists $\delta > 0$ such that $\{x_n\}_{n\in\mathbb{N}} \cup \{\bar{x}\} \subseteq B(0,\delta)$, and then

$$(u * v)(x_n) := \int_{\mathbb{R}^n} u(x_n - y)\, v(y)\, dy = \int_{\mathbb{R}^n} f_n(y)\, dy = \int_{B(0,\delta)-K} f_n(y)\, dy,$$

where $K := \mathrm{supp}(u)$ and $\{f_n\}_{n\in\mathbb{N}}$ is given by

$$f_n(y) = u(x_n - y)\, v(y) \quad \forall\, y \in B(0,\delta) - K.$$

It follows that

$$|f_n(y)| \leqslant g(y) = \max_{z\in K} |u(z)|\,|v(y)| \quad \forall\, y \in B(0,\delta) - K,$$

whence

$$\int_{B(0,\delta)-K} g(y)\, dy \leqslant \max_{z\in K} |u(z)| \int_{B(0,\delta)-K} |v(y)|\, dy < +\infty,$$

which proves that $g \in \mathrm{L}^1(B(0,\delta) - K)$. Then, applying the Lebesgue Dominated Convergence Theorem (cf. [113, Theorem 1.34], [120, Theorem 5.36]), it follows that

$$(u * v)(x_n) := \int_{B(0,\delta)-K} f_n(y)\, dy \overset{n\to\infty}{\longrightarrow} \int_{B(0,\delta)-K} \lim_{n\to\infty} f_n(y)\, dy$$
$$= \int_{B(0,\delta)-K} u(\bar{x} - y)\, v(y)\, dy = \int_{\mathbb{R}^n} u(\bar{x} - y)\, v(y)\, dy = (u * v)(\bar{x}),$$

thus proving the continuity of $u * v$.

On the other hand, introducing $z := x - y$ as a new integration variable, we obtain

$$(u * v)(x) := \int_{\mathbb{R}^n} u(x - y)\, v(y)\, dy = \int_{\mathbb{R}^n} u(z)\, v(x - z)\, dz = (v * u)(x) \quad \forall\, x \in \mathbb{R},$$

which proves the COMMUTATIVITY of the convolution, that is $u * v = v * u$. Another consequence of the definition of $(u * v)$ is that

$$\operatorname{supp}(u * v) \subseteq \operatorname{supp}(u) + \operatorname{supp}(v) := \Big\{ x + y : \quad x \in \operatorname{supp}(u), \quad y \in \operatorname{supp}(v) \Big\}. \tag{9.3}$$

Moreover, given u, v, $w \in C(\mathbb{R}^n)$ such that at least two of them have compact support, one can prove the ASSOCIATIVITY of the convolution, that is

$$u * (v * w) = (u * v) * w.$$

Now, if $u \in C^1(\mathbb{R}^n)$ and $v \in C(\mathbb{R}^n)$, with one of them having compact support, we may differentiate under the integral sign in the definition of $(u * v)$ and deduce that

$$\frac{\partial}{\partial x_i}(u * v) = \frac{\partial u}{\partial x_i} * v \in C(\mathbb{R}^n) \quad \forall\, i \in \{1, 2, \ldots, n\},$$

which shows that $u * v \in C^1(\mathbb{R}^n)$. Similarly, if $u \in C_0^1(\mathbb{R}^n)$ and $v \in \mathrm{L}^1_{loc}(\mathbb{R}^n)$, the previous argument implies the same conclusion. Furthermore, if $u \in C^j(\mathbb{R}^n)$ and $v \in C^k(\mathbb{R}^n)$, one of them with compact support, the commutativity allows us to show that $(u * v) \in C^{j+k}(\mathbb{R}^n)$. Indeed, given α and β multi-indices with $|\alpha| = k$ and $|\beta| = j$, we have

$$\partial^{\alpha+\beta}(u * v) = \partial^\alpha(\partial^\beta(u * v)) = \partial^\alpha(\partial^\beta\, u * v)$$
$$= \partial^\alpha(v * \partial^\beta\, u) = \partial^\alpha v * \partial^\beta\, u = \partial^\beta\, u * \partial^\alpha v \in C^{j+k}(\mathbb{R}^n).$$

The above can be summarized in the following lemma.

Lemma 9.6.
i) $u \in C_0^j(\mathbb{R}^n)$, $v \in \mathrm{L}^1_{loc}(\mathbb{R}^n) \implies u * v \in C^j(\mathbb{R}^n)$.

ii) $u \in C_0^j(\mathbb{R}^n)$, $v \in C^k(\mathbb{R}^n) \implies u * v \in C^{j+k}(\mathbb{R}^n)$.

We are now in a position to introduce the concept of REGULARIZATION OF FUNCTIONS. To this end, let

$$\rho_\varepsilon := \frac{1}{C}\,\varphi_{\varepsilon,0} \quad \text{where} \quad \varphi_{\varepsilon,0}(x) := \begin{cases} e^{-\varepsilon^2/(\varepsilon^2 - \|x\|^2)} & \text{if } \|x\| < \varepsilon \\ 0 & \text{if } \|x\| \geqslant \varepsilon \end{cases} \tag{9.4}$$

and

$$C := \int_{\mathbb{R}^n} \varphi_{\varepsilon,0}(x)\,dx = \int_{B(0,\varepsilon)} \varphi_{\varepsilon,0}(x)\,dx\,. \tag{9.5}$$

It is clear that

$$\operatorname{supp}(\rho_\varepsilon) = \bar{B}(0,\varepsilon)\,, \quad \rho_\varepsilon \in C_0^\infty(\mathbb{R}^n)\,, \quad \text{and} \quad \int_{\mathbb{R}^n} \rho_\varepsilon(x)\,dx = 1\,.$$

Definition 9.7. *Let Ω be an open subset of $\mathbb{R}^n$, and let $f \in \mathrm{L}^1_{loc}(\Omega)$. The regularization of f is defined as*

$$f_\varepsilon(x) := \int_\Omega f(y)\,\rho_\varepsilon\,(x-y)\,dy \qquad \forall\, x \in \mathbb{R}^n\,, \quad \forall\, \varepsilon > 0.$$

Note that if f is extended by zero outside Ω, then

$$f_\varepsilon(x) = \int_{\mathbb{R}^n} f(y)\,\rho_\varepsilon(x-y)\,dy = (\rho_\varepsilon * f)(x) = (f * \rho_\varepsilon)(x)$$
$$= \int_{\mathbb{R}^n} f(x-y)\,\rho_\varepsilon(y)\,dy = \int_{B(0,\varepsilon)} f(x-y)\,\rho_\varepsilon(y)\,dy \quad \forall\, x \in \mathbb{R}^n\,,$$

which, thanks to Lemma 9.6, implies that $f_\varepsilon \in C^\infty(\mathbb{R}^n)$.

Other important properties of the regularization of functions are summarized in the following theorem.

Theorem 9.8. *Let Ω be an open subset of $\mathbb{R}^n$, and let $f \in \mathrm{L}^1(\Omega)$.*

a) *If* $\operatorname{supp}(f)$ *is a compact subset K of Ω, then*

$$\operatorname{supp}(f_\varepsilon) \subseteq K_\varepsilon := \Big\{ x \in \mathbb{R}^n : \quad \operatorname{dist}(x,K) \leqslant \varepsilon \Big\}\,.$$

In particular, if $\varepsilon < \operatorname{dist}(\mathrm{K}, \partial\Omega)$, then $f_\varepsilon \in C_0^\infty(\Omega)$.

b) *If $f \in \mathrm{L}^p(\Omega)$, with $1 < p < \infty$, then $\lim\limits_{\varepsilon\to 0} \|f - f_\varepsilon\|_{0,p;\Omega} = 0$.*

c) *If $f \in C_0^k(\Omega)$, then $\partial^\alpha f_\varepsilon$ converges uniformly to $\partial^\alpha f$ as $\varepsilon \to 0$, for all α such that $|\alpha| \leqslant k$.*

Proof.

a) According to property (9.3) we clearly have

$$\operatorname{supp}(f_\varepsilon) = \operatorname{supp}(f * \rho_\varepsilon) \subseteq \operatorname{supp}(f) + \operatorname{supp}(\rho_\varepsilon) = K + \bar{B}(0,\varepsilon) \subseteq K_\varepsilon.$$

Then, if $\varepsilon < \operatorname{dist}(K, \partial\Omega)$, it follows that $K_\varepsilon \subseteq \Omega$, and since $f_\varepsilon \in C^\infty(\mathbb{R}^n)$, we deduce that $f_\varepsilon \in C_0^\infty(\Omega)$.

b) Let $f \in \mathrm{L}^p(\Omega)$, with $1 < p < +\infty$, and extend f by zero outside Ω. Also, let $q \in (1, +\infty)$ be such that $\dfrac{1}{p} + \dfrac{1}{q} = 1$. Then, applying Hölder's inequality in the third line below, we obtain

$$\begin{aligned}\int_\Omega |f_\varepsilon(x)|^p\,dx &= \int_\Omega \left|\int_{\mathbb{R}^n} f(y)\,\rho_\varepsilon(x-y)\,dy\right|^p dx\\ &\leqslant \int_\Omega \left\{\int_{\mathbb{R}^n} |f(y)|\,\rho_\varepsilon^{1/p}(x-y)\,\rho_\varepsilon^{1/q}(x-y)\,dy\right\}^p dx\\ &\leqslant \int_\Omega \left\{\int_{\mathbb{R}^n} |f(y)|^p\,\rho_\varepsilon(x-y)\,dy\right\}\left\{\int_{\mathbb{R}^n} \rho_\varepsilon(x-y)\,dy\right\}^{p/q} dx.\\ &= \int_\Omega \left\{\int_{\mathbb{R}^n} |f(y)|^p\,\rho_\varepsilon(x-y)\,dy\right\} dx\,,\end{aligned}$$

where we have also used that

$$\begin{aligned}\int_{\mathbb{R}^n} \rho_\varepsilon(x-y)\,dy &= (\rho_\varepsilon * 1)(x) = (1 * \rho_\varepsilon)(x)\\ &= \int_{\mathbb{R}^n} \rho_\varepsilon(y)\,dy = 1 \qquad \forall\, x \in \mathbb{R}^n\,.\end{aligned} \tag{9.6}$$

Then, using Fubini's and Tonelli's Theorems (cf. [120, Theorems 6.1 and 6.10]), which allow us to interchange the order of integration, and noting that

$$\int_\Omega \rho_\varepsilon(x-y)\,dx \leqslant \int_{\mathbb{R}^n} \rho_\varepsilon(x-y)\,dx = 1\,,$$

we arrive at

$$\int_\Omega |f_\varepsilon(x)|^p\,dx \leqslant \int_{\mathbb{R}^n} |f(y)|^p \left\{\int_\Omega \rho_\varepsilon(x-y)\,dx\right\} dy \leqslant \int_{\mathbb{R}^n} |f(y)|^p\,dy = \|f\|_{0,p;\Omega}^p\,.$$

Thus, $\|f_\varepsilon\|_{0,p;\Omega} \leqslant \|f\|_{0,p;\Omega}$ for all $\varepsilon > 0$, and therefore $f_\varepsilon \in \mathrm{L}^p(\Omega)$ for all $\varepsilon > 0$. On the other hand, employing (9.6) once again, we can write

$$f_\varepsilon(x) - f(x) = \int_{\mathbb{R}^n} \Big\{f(y) - f(x)\Big\}\,\rho_\varepsilon(x-y)\,dy,$$

whence, applying Hölder's inequality, making the change of variable $z = x - y$, and applying once again Fubini's and Tonelli's Theorems (cf. [120, Theorems 6.1 and 6.10]), we deduce that

$$\begin{aligned}\|f_\varepsilon - f\|_{0,p;\Omega}^p &\leqslant \int_\Omega \left\{\int_{\mathbb{R}^n} |f(y) - f(x)|^p\,\rho_\varepsilon(x-y)\,dy\right\} dx\\ &= \int_\Omega \left\{\int_{\mathbb{R}^n} |f(x-z) - f(x)|^p\,\rho_\varepsilon(z)\,dz\right\} dx\\ &= \int_{\mathbb{R}^n} \rho_\varepsilon(z) \left\{\int_\Omega |f(x-z) - f(x)|^p\,dx\right\} dz\\ &= \int_{B(0,\varepsilon)} \rho_\varepsilon(z) \left\{\int_\Omega |f(x-z) - f(x)|^p\,dx\right\} dz\\ &\leqslant \sup_{|z| \leqslant \varepsilon} \int_\Omega |f(x-z) - f(x)|^p\,dx\,.\end{aligned}$$

Finally, Kolmogorov's compactness criterion (see details after the present proof) allows us to conclude that $\|f_\varepsilon - f\|_{0,p;\Omega} \overset{\varepsilon \to 0}{\to} 0$.

c) Given $f \in C_0^k(\Omega)$, its extension by zero in $\mathbb{R}^n \backslash \Omega$ constitutes an element of $C_0^k(\mathbb{R}^n)$. In addition, since $\mathrm{supp}(f)$ is a compact set contained in Ω, we have that

$$\mathrm{supp}(f_\varepsilon) \subseteq \Omega \qquad \forall\, \varepsilon < \mathrm{dist}\big(\mathrm{supp}(f), \partial\Omega\big)\,.$$

It follows that for each $x \in \Omega$ and for each $\varepsilon < \mathrm{dist}\big(\mathrm{supp}(f), \partial\Omega\big)$ there holds

$$\begin{aligned}\partial^\alpha f_\varepsilon(x) &= \partial^\alpha (f * \rho_\varepsilon)(x) = (\partial^\alpha f * \rho_\varepsilon)(x)\\ &= \int_{\mathbb{R}^n} \partial_x^\alpha f(x-y)\, \rho_\varepsilon(y)\, dy = \int_{B(0,\varepsilon)} \partial_x^\alpha f(x-y)\, \rho_\varepsilon(y)\, dy,\end{aligned}$$

and since $\displaystyle\int_{B(0,\varepsilon)} \rho_\varepsilon(y)\, dy = 1$, we get

$$\partial^\alpha f_\varepsilon(x) - \partial^\alpha f(x) = \int_{B(0,\varepsilon)} \Big\{\partial_x^\alpha f(x-y) - \partial^\alpha f(x)\Big\}\, \rho_\varepsilon(y)\, dy,$$

whence

$$|\partial^\alpha f_\varepsilon(x) - \partial^\alpha f(x)| \leqslant \sup_{|y| \leqslant \varepsilon} |\partial_x^\alpha f(x-y) - \partial^\alpha f(x)|,$$

which, by the uniform continuity of $\partial^\alpha f$ in Ω, for all α, $|\alpha| \leqslant k$, yields

$$\sup_{x \in \Omega} |\partial^\alpha f_\varepsilon(x) - \partial^\alpha f(x)| \overset{\varepsilon\to 0}{\longrightarrow} 0\,.$$

□

To complement part b) of the previous proof, we now recall that the **Kolmogorov compactness criterion** (cf. [121, Chapter I, p. 4]) states that a subset M of $\mathrm{L}^p(\Omega)$, with $p \in [1, +\infty)$, is relatively compact (cf. [115, p. 100]), equivalently $\overline{M}$ compact, if and only if

i) M is bounded

ii) $\displaystyle\lim_{z\to 0} \int_\Omega |g(x-z) - g(x)|^p\, dx = 0$ uniformly for $g \in M$.

iii) $\displaystyle\lim_{\alpha\to+\infty} \int_{\Omega \cap \{x:\, \|x\| > \alpha\}} |g(x)|^p = 0$ uniformly for $g \in M$.

Thus, applying part ii) to the set given by $M = \{f\}$, which is clearly relatively compact, we conclude that

$$\lim_{\varepsilon\to 0} \sup_{|z| \leqslant \varepsilon} \int_\Omega |f(x-z) - f(x)|^p\, dx = 0, \tag{9.7}$$

which completes the proof of b) in the previous theorem. Indeed, given $\delta > 0$, the limit in ii) implies that there exists $\varepsilon_0 > 0$ such that

$$\int_\Omega |f(x-z) - f(x)|^p\, dx < \delta \quad \forall\, z,\ \|z\| \leqslant \varepsilon_0,$$

whence

$$\sup_{|z| \leqslant \varepsilon} \int_\Omega |f(x-z) - f(x)|^p\, dx \leqslant \sup_{|z| \leqslant \varepsilon_0} \int_\Omega |f(x-z) - f(x)|^p\, dx < \delta \quad \forall\, \varepsilon \leqslant \varepsilon_0,$$

thus proving (9.7).

On the other hand, let us observe that the hypothesis required on supp(f), namely that it is a compact subset of Ω, is strictly necessary for the uniform convergence established in c). Indeed, we have the following example.

Example 9.9 Let $\Omega = (0,1)^2 \subseteq \mathbb{R}^2$, and define $f : \Omega \to \mathbb{R}$ by $f(x) = 1$ for all $x \in \Omega$ (which clearly does not satisfy this hypothesis). Then, extending f by 0 outside Ω, we have

$$f_\varepsilon(x) = \int_{\mathbb{R}^2} f(y)\, \rho_\varepsilon(x-y)\, dy = \int_\Omega \rho_\varepsilon(x-y)\, dy \quad \forall\, x \in \mathbb{R}^2 .$$

In particular, for $\bar{x} = (1/2, 1) \in \partial\Omega$ and $\varepsilon \in (0, 1/2)$, we have

$$f_\varepsilon(\bar{x}) = \int_\Omega \rho_\varepsilon(\bar{x}-y)\, dy = \int_{\Omega \cap B(\bar{x},\varepsilon)} \rho_\varepsilon(\bar{x}-y)\, dy = \int_{B_i(\bar{x},\varepsilon)} \rho_\varepsilon(\bar{x}-y)\, dy = \frac{1}{2},$$

where $B_i(\bar{x},\varepsilon) := \Big\{ y := (y_1, y_2) \in B(\bar{x},\varepsilon) : \quad y_2 < 1 \Big\}$ (see Figure 9.1). Now, let $\{x_n\}_{n\in\mathbb{N}} \subseteq \Omega$ be such that $\|x_n - \bar{x}\| \overset{n\to\infty}{\longrightarrow} 0$. Since f_ε is continuous in $\mathbb{R}^2$, it is clear that

$$\lim_{n\to\infty} f_\varepsilon(x_n) = f_\varepsilon(\bar{x}) = 1/2,$$

and then, given $\eta > 0$, there exists $N \in \mathbb{N}$ such that

$$|f_\varepsilon(x_N) - 1/2| < \eta.$$

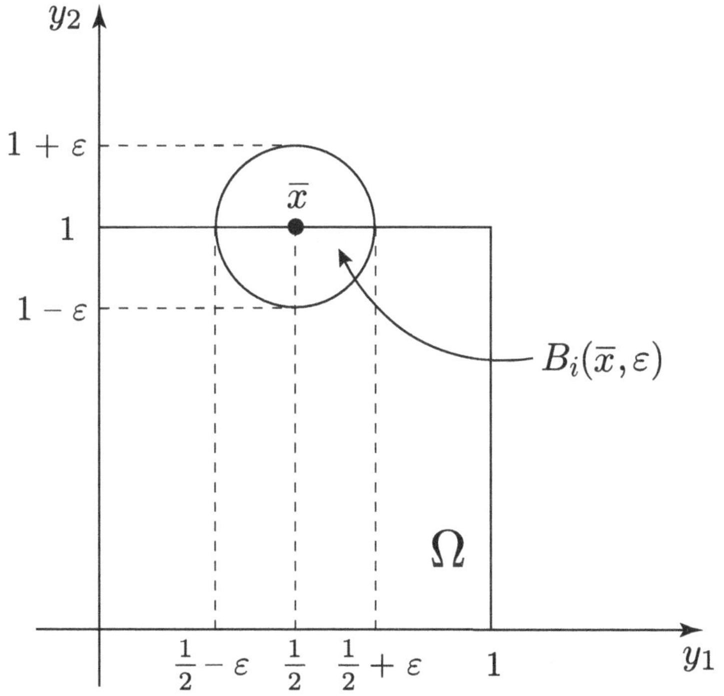

Fig. 9.1. *Geometry of Example 9.9*

It follows that

$$\begin{aligned}\sup_{x\in\bar{\Omega}} |f(x) - f_\varepsilon(x)| &\geqslant |f(x_N) - f_\varepsilon(x_N)| \\ &= |1 - f_\varepsilon(x_N)| = \left|\frac{1}{2} + \left(\frac{1}{2} - f_\varepsilon(x_N)\right)\right| \\ &\geqslant \frac{1}{2} - \eta \qquad \forall \eta > 0,\end{aligned}$$

and therefore

$$\sup_{x\in\bar{\Omega}} |f(x) - f_\varepsilon(x)| \geqslant 1/2 \qquad \forall \varepsilon \in (0, 1/2).$$

We now prove one of the main density properties of the space $C_0^\infty(\Omega)$.

Theorem 9.10. *Let Ω be an open subset of $\mathbb{R}^n$, and let $p \in (1, +\infty)$. Then $C_0^\infty(\Omega)$ is dense in $\mathrm{L}^p(\Omega)$.*

Proof. Let $f \in \mathrm{L}^p(\Omega)$, with $1 < p < +\infty$, and suppose first that Ω is bounded. Then, for each $n \in \mathbb{N}$, we define the set

$$\Omega_n := \left\{x \in \Omega : \quad \mathrm{dist}(x, \partial\Omega) < 1/n\right\}$$

and the function

$$f_n(x) := \begin{cases} 0 & \text{if } x \in \bar{\Omega}_n \\ f(x) & \text{if } x \in \Omega\backslash\Omega_n \end{cases}$$

It follows that

$$\|f - f_n\|_{0,p;\Omega}^p = \int_{\Omega_n} |f(x)|^p \, dx = \int_\Omega \mathbf{1}_{\Omega_n} |f(x)|^p \, dx,$$

and applying the Lebesgue Dominated Convergence Theorem (cf. [113, Theorem 1.34], [120, Theorem 5.36]), noting that $|f(\cdot)|^p$ belongs to $\mathrm{L}^1(\Omega)$ since $f \in \mathrm{L}^p(\Omega)$, we deduce that

$$\|f - f_n\|_{0,p;\Omega} \overset{n\to\infty}{\longrightarrow} 0.$$

Thus, given $\eta > 0$, there exists $N \in \mathbb{N}$ such that $\|f - f_N\|_{0,p;\Omega} < \eta/2$. Then, since $f_N \in \mathrm{L}^p(\Omega)$, it follows from part b) of Theorem 9.8 that $\|f_N - f_{N,\varepsilon}\|_{0,p;\Omega} \overset{\varepsilon\to 0}{\longrightarrow} 0$, and therefore, for the same η, there exists $\varepsilon_0 > 0$ such that $\|f_N - f_{N,\varepsilon}\|_{0,p;\Omega} < \eta/2$ for all $\varepsilon \in (0, \varepsilon_0)$. In particular, for $\varepsilon = \frac{1}{2}\min\left\{\varepsilon_0, \frac{1}{N}\right\}$ it follows from part a) of Theorem 9.8 that $f_{N,\varepsilon} \in C_0^\infty(\Omega)$, and moreover

$$\|f - f_{N,\varepsilon}\|_{0,p;\Omega} \leqslant \|f - f_N\|_{0,p;\Omega} + \|f_N - f_{N,\varepsilon}\|_{0,p;\Omega} < \eta.$$

In summary, given $f \in \mathrm{L}^p(\Omega)$ and $\eta > 0$, there exists a function $g := f_{N,\varepsilon} \in C_0^\infty(\Omega)$ such that $\|f - g\|_{0,p;\Omega} < \eta$, which proves that $\mathrm{L}^p(\Omega) = \overline{C_0^\infty(\Omega)}^{\|\cdot\|_{0,p;\Omega}}$. Now suppose that Ω is unbounded. Then, for each $n \in \mathbb{N}$, we define $\Omega_n := \Omega \cap B(0, n)$ (see Figure 9.2 for an example of these domains). It is clear that

i) Ω_n is open, bounded, and contained in Ω for all $n \in \mathbb{N}$,
ii) $\Omega_n \subseteq \Omega_{n+1}$ for all $n \in \mathbb{N}$, and
iii) $\Omega = \bigcup_{n\in\mathbb{N}} \Omega_n$.

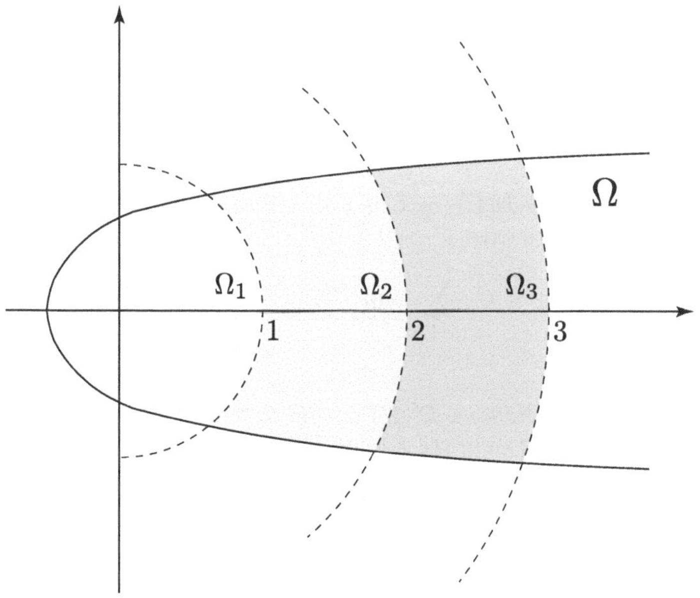

Fig. 9.2. *Example of domains Ω_n*

Then, defining for each $n \in \mathbb{N}$ the function $f_n : \Omega \to \mathbb{R}$ given by $f_n := \mathbf{1}_{\Omega_n}|f|^p$, which is clearly measurable, we observe, thanks to ii) and iii), that

$$0 \leqslant f_n \leqslant f_{n+1} \quad \forall\, n \in \mathbb{N} \quad \text{and} \quad \lim_{n\to\infty} f_n(x) = |f(x)|^p \quad \forall\, x \in \Omega\,,$$

from which, using the Lebesgue Monotone Convergence Theorem (cf. [113, Theorem 1.26], [120, Theorem 5.32]), it follows that

$$\lim_{n\to\infty} \int_\Omega f_n(x)\,dx = \int_\Omega |f(x)|^p\,dx,$$

that is,

$$\lim_{n\to\infty} \int_{\Omega_n} |f(x)|^p\,dx = \int_\Omega |f(x)|^p\,dx < +\infty\,.$$

Thus, given $\varepsilon > 0$, there exists $N \in \mathbb{N}$ such that

$$\int_{\Omega\backslash\Omega_N} |f(x)|^p\,dx < \varepsilon/2\,.$$

On the other hand, since $\mathbf{1}_{\Omega_N} f \in \mathrm{L}^p(\Omega_N)$ and Ω_N is bounded, it follows from the previous case that there exists $\varphi \in C_0^\infty(\Omega_N)$ such that

$$\|\mathbf{1}_{\Omega_N} f - \varphi\|_{0,p;\Omega_N}^p = \int_{\Omega_N} |f(x) - \varphi(x)|^p\,dx < \varepsilon/2\,.$$

Finally, extending φ by 0 on $\Omega\backslash\Omega_N$, we have that $\varphi \in C_0^\infty(\Omega)$ and

$$\|f - \varphi\|_{0,p;\Omega}^p = \int_{\Omega_N} |f(x) - \varphi(x)|^p\,dx + \int_{\Omega\backslash\Omega_N} |f(x)|^p < \varepsilon\,,$$

which completes the proof. □

9.3 Distributions of finite order

Recall that a linear form $u : \mathcal{D}(\Omega) \to \mathbb{C}$ is a distribution on an open set Ω of $\mathbb{R}^n$ if for every compact $K \subseteq \Omega$, there exist $C > 0$ and $N \in \mathbb{N} \cup \{0\}$ such that

$$|\langle u, \varphi\rangle| \leqslant C \sum_{|\alpha|\leqslant N} \sup_{x\in K} |\partial^\alpha \varphi(x)| \quad \forall \varphi \in \mathcal{D}_K(\Omega)\,. \tag{9.8}$$

Definition 9.11. *A distribution $u \in \mathcal{D}'(\Omega)$ is said to be of finite order if the same integer N can be used in (9.8) for all compact subsets $K \subseteq \Omega$. The smallest such N is called the* ORDER *of u. In what follows we denote by $\mathcal{D}'_k(\Omega)$ the space of distributions on Ω of order $\leqslant k$.*

Example 9.12 The distribution induced by $f \in L^1_{loc}(\Omega)$, that is,

$$\langle f, \varphi\rangle := \int_\Omega f\varphi \quad \forall \varphi \in \mathcal{D}(\Omega),$$

is clearly of order 0. Moreover, given $x_0 \in \Omega$ and $\alpha \in (\mathbb{N} \cup \{0\})^n$, the distribution

$$\langle u, \varphi\rangle := \partial^\alpha \varphi(x_0) \quad \forall \varphi \in \mathcal{D}(\Omega)\,,$$

is of order $|\alpha|$.

Definition 9.13. *Given $k \in \mathbb{N}\cup\{0\}$, we say that a sequence $\{\varphi_m\}_{m\in\mathbb{N}} \subseteq C_0^k(\Omega)$ converges to $\varphi \in C_0^k(\Omega)$ if there exists a compact $K \subseteq \Omega$ such that $\operatorname{supp}(\varphi_m) \subseteq K$ for all $m \in \mathbb{N}$, $\operatorname{supp}(\varphi) \subseteq K$ and*

$$\lim_{m\to\infty} \sum_{|\alpha|\leqslant k} \sup_{x\in K} |\partial^\alpha(\varphi_m - \varphi)(x)| = 0.$$

The following result identifies the distributions in $\mathcal{D}'_k(\Omega)$ with sequentially continuous linear forms on $C_0^k(\Omega)$.

Theorem 9.14. *If $u \in \mathcal{D}'_k(\Omega)$, then u can be extended, in a unique way, to a sequentially continuous linear form on $C_0^k(\Omega)$. Conversely, if u is a sequentially continuous linear form on $C_0^k(\Omega)$, then $u|_{C_0^\infty(\Omega)} \in \mathcal{D}'_k(\Omega)$.*

Proof. Given $u \in \mathcal{D}'_k(\Omega)$, for every compact $K \subseteq \Omega$ there exists $C(K) > 0$ such that

$$|\langle u, \varphi\rangle| \leqslant C(K) \sum_{|\alpha|\leqslant k} \sup_{x\in K} |\partial^\alpha \varphi(x)| \quad \forall \varphi \in \mathcal{D}_K(\Omega). \tag{9.9}$$

Now, given $\varphi \in C_0^k(\Omega)$ and $m \in \mathbb{N}$, we define (cf. (9.4), (9.5))

$$\varepsilon_m := \frac{1}{2m} \operatorname{dist}\big(\operatorname{supp}(\varphi), \partial\Omega\big) \quad \text{and} \quad \widetilde{\varphi}_m := \varphi_{\varepsilon_m} := \varphi * \rho_{\varepsilon_m}\,.$$

It follows that

$$\operatorname{supp}(\widetilde{\varphi}_m) \subseteq \operatorname{supp}(\varphi) + B(0,\varepsilon_m) \subseteq \operatorname{supp}(\varphi) + B(0,\varepsilon_1) =: K_0 \subseteq \Omega,$$

so that $\widetilde{\varphi}_m \in C_0^\infty(\Omega)$ for all $m \in \mathbb{N}$. Moreover, applying part c) of Theorem 9.8, we deduce that

$$\lim_{m\to\infty} \sum_{|\alpha|\leqslant k} \sup_{x\in K_0} |\partial^\alpha(\varphi - \widetilde{\varphi}_m)(x)| = 0, \tag{9.10}$$

which indicates that $\{\widetilde{\varphi}_m\}_{m\in\mathbb{N}}$ converges to φ in $C_0^k(\Omega)$. Furthermore, it is clear from (9.9) that

$$|\langle u, \widetilde{\varphi}_m\rangle - \langle u, \widetilde{\varphi}_\ell\rangle| = |\langle u, \widetilde{\varphi}_m - \widetilde{\varphi}_\ell\rangle| \leqslant C(K_0) \sum_{|\alpha|\leqslant k} \sup_{x\in K_0} |\partial^\alpha(\widetilde{\varphi}_m - \widetilde{\varphi}_\ell)(x)|,$$

which, by virtue of identity (9.10), tends to 0 as $m, \ell \to +\infty$. It thus follows that $\{\langle u, \widetilde{\varphi}_m\rangle\}_{m\in\mathbb{N}}$ is convergent in $\mathbb{C}$, and it is easy to show that this limit is independent of the sequence $\{\widetilde{\varphi}_m\}_{m\in\mathbb{N}} \subseteq C_0^\infty(\Omega)$ that converges to φ in $C_0^k(\Omega)$. This suggests defining the linear form $\widetilde{u} : C_0^k(\Omega) \to \mathbb{C}$ by

$$\langle \widetilde{u}, \varphi\rangle := \lim_{m\to\infty} \langle u, \widetilde{\varphi}_m\rangle \quad \forall\, \varphi \in C_0^k(\Omega),$$

which is sequentially continuous. Indeed, applying (9.9) to $\widetilde{\varphi}_m \in \mathcal{D}_{K_0}(\Omega)$, we obtain

$$|\langle u, \widetilde{\varphi}_m\rangle| \leqslant C(K_0) \sum_{|\alpha|\leqslant k} \sup_{x\in K_0} |\partial^\alpha \widetilde{\varphi}_m(x)|,$$

from which, taking the limit as $m \to \infty$, and using (9.10), we get

$$|\langle \widetilde{u}, \varphi\rangle| \leqslant C(K_0) \sum_{|\alpha|\leqslant k} \sup_{x\in K_0} |\partial^\alpha \varphi(x)| \quad \forall\, \varphi \in C_0^k(\Omega) \quad \text{such that} \quad \operatorname{supp}(\varphi) \subseteq K_0.$$

Finally, let us show that this extension $\widetilde{u}$ is unique. Let $\bar{u} : C_0^k(\Omega) \to \mathbb{C}$ be another sequentially continuous linear form such that $\langle \bar{u}, \varphi\rangle = \langle u, \varphi\rangle$ for all $\varphi \in \mathcal{D}(\Omega)$. Then, for each $\varphi \in C_0^k(\Omega)$, using the sequential continuity of $\bar{u}$, we have

$$\langle \widetilde{u}, \varphi\rangle = \lim_{m\to\infty} \langle u, \widetilde{\varphi}_m\rangle = \lim_{m\to\infty} \langle \bar{u}, \widetilde{\varphi}_m\rangle = \langle \bar{u}, \varphi\rangle,$$

from which it is clear that $\widetilde{u} = \bar{u}$. Now, let u be a sequentially continuous linear form on $C_0^k(\Omega)$, and suppose by contradiction that $u|_{C_0^\infty(\Omega)} \notin \mathcal{D}'_k(\Omega)$. This means that there exists a compact $K \subseteq \Omega$ such that for all $C > 0$, there exists $\varphi_C \in \mathcal{D}_K(\Omega)$ such that

$$|\langle u, \varphi_C\rangle| > C \sum_{|\alpha|\leqslant k} \sup_{x\in K} |\partial^\alpha \varphi_C(x)|.$$

In particular, for each $n \in \mathbb{N}$ there exists $\varphi_n \in \mathcal{D}_K(\Omega)$ such that

$$|\langle u, \varphi_n\rangle| > n \sum_{|\alpha|\leqslant k} \sup_{x\in K} |\partial^\alpha \varphi_n(x)|,$$

and defining $\psi_n := \dfrac{\varphi_n}{|\langle u, \varphi_n\rangle|}$, we deduce that

$$|\langle u, \psi_n\rangle| = 1 \quad \text{and} \quad \sum_{|\alpha|\leqslant k} \sup_{x\in K} |\partial^\alpha \psi_n(x)| < \frac{1}{n} \quad \forall\, n \in \mathbb{N}.$$

It follows that $\{\psi_n\}_{n\in\mathbb{N}}$ converges to the zero function in $C_0^k(\Omega)$, and yet $\langle u, \psi_n\rangle$ does not converge to 0, which contradicts the sequential continuity of u in $C_0^k(\Omega)$.

□

The following example illustrates the case of a distribution that is not of finite order.

Example 9.15 Define $u : \mathcal{D}(\mathbb{R}) \to \mathbb{R}$ by

$$\langle u, \varphi \rangle := \sum_{m=0}^{\infty} \varphi^{(m)}(m) \quad \forall \varphi \in \mathcal{D}(\mathbb{R}).$$

It is clear that u is linear. Moreover, let K be a compact subset of $\mathbb{R}$, and let $M > 0$ be such that $K \subseteq [-M, M]$. Then, given $\varphi \in \mathcal{D}_K(\Omega)$, we have $\varphi^{(m)}(m) = 0$ for all $m > M$, and hence

$$|\langle u, \varphi \rangle| = \left| \sum_{j=0}^{M} \varphi^{(j)}(j) \right| \leqslant \sum_{j=0}^{M} |\varphi^{(j)}(j)| \leqslant \sum_{j=0}^{M} \sup_{x \in K} |\varphi^{(j)}(x)|,$$

which proves that $u \in \mathcal{D}'(\mathbb{R})$. Now suppose, by contradiction, that u is of order $k \in \mathbb{N}$ and choose $\varphi \in \mathcal{D}(-1, 1)$ such that $\varphi^{(k+1)}(0) = 1$. Then, for $\varepsilon \in (0, 1)$ define

$$\psi_\varepsilon(x) := \varphi\left(\frac{x - (k+1)}{\varepsilon}\right) \quad \forall x \in \mathbb{R}.$$

It follows that $\operatorname{supp}(\psi_\varepsilon) \subseteq (k + 1 - \varepsilon, k + 1 + \varepsilon) \subseteq (k, k + 2) =: K$. Moreover, since

$$\psi_\varepsilon^{(j)}(x) = \frac{1}{\varepsilon^j} \varphi^{(j)}\left(\frac{x - (k+1)}{\varepsilon}\right) \quad \forall x \in \mathbb{R},$$

we have that $\psi_\varepsilon^{(k+1)}(k + 1) = \frac{\varphi^{(k+1)}(0)}{\varepsilon^{k+1}} = \frac{1}{\varepsilon^{k+1}}$, and then

$$\langle u, \psi_\varepsilon \rangle = \sum_{m=0}^{\infty} \psi_\varepsilon^{(m)}(m) = \psi_\varepsilon^{(k+1)}(k + 1) = \frac{1}{\varepsilon^{k+1}}.$$

On the other hand, since u is assumed to be of order k, for the compact K there exists $C(K) > 0$ such that

$$|\langle u, \varphi \rangle| \leqslant C(K) \sum_{j=0}^{k} \sup_{x \in K} |\varphi^{(j)}(x)| \quad \forall \varphi \in \mathcal{D}_K(\mathbb{R}),$$

and in particular, for $\varphi = \psi_\varepsilon$ we obtain

$$\begin{aligned} \frac{1}{\varepsilon^{k+1}} &\leqslant C(K) \sum_{j=0}^{k} \sup_{x \in K} \frac{1}{\varepsilon^j} \left| \varphi^{(j)}\left(\frac{x - (k+1)}{\varepsilon}\right) \right| \\ &\leqslant \frac{C(K)}{\varepsilon^k} \sum_{j=0}^{k} \sup_{y \in [-1,1]} |\varphi^{(j)}(y)| = \frac{C_1}{\varepsilon^k}, \end{aligned}$$

from which it follows that $\varepsilon > \dfrac{1}{C_1}$ for all $\varepsilon \in (0, 1)$, which is obviously a contradiction.

9.4 Localization and partition of unity

Let Ω be an open subset of $\mathbb{R}^n$, and let Ω_1 be an open subset of Ω. Given $\varphi \in C_0^\infty(\Omega_1)$, its zero extension on $\Omega \backslash \Omega_1$ obviously belongs to $C_0^\infty(\Omega)$, and thus we can formally write

$$C_0^\infty(\Omega_1) \subseteq C_0^\infty(\Omega).$$

Hence, if $u \in \mathcal{D}'(\Omega)$, it is easy to see that $u_{\Omega_1} : C_0^\infty(\Omega_1) \to \mathbb{C}$ defined by

$$\langle u_{\Omega_1}, \varphi \rangle := \langle u, \varphi \rangle \quad \forall \varphi \in C_0^\infty(\Omega_1),$$

is an element of $\mathcal{D}'(\Omega_1)$. In such case, we say that u_{Ω_1} is the RESTRICTION of u to Ω_1.

Definition 9.16. *Two distributions* u, $v \in \mathcal{D}'(\Omega)$ *are said to be equal on* $\Omega_1 \subseteq \Omega$ *if and only if* $u_{\Omega_1} = v_{\Omega_1}$, *that is,*

$$\langle u, \varphi \rangle = \langle v, \varphi \rangle \quad \forall\, \varphi \in C_0^\infty(\Omega_1)\,.$$

The restriction of $u \in \mathcal{D}'(\Omega)$ to an open subset of Ω is also called the LOCALIZATION of u. The main goal of this section is to prove that every $u \in \mathcal{D}'(\Omega)$ is uniquely determined by its localizations. For this purpose, several preliminary results are needed.

Lemma 9.17. *Let* Ω *be an open subset of* $\mathbb{R}^n$, *and let* K *be a compact subset of* Ω. *Then there exists* $\varphi \in C_0^\infty(\Omega)$ *such that* $0 \leqslant \varphi \leqslant 1$ *in* Ω *and* $\varphi \equiv 1$ *in a neighborhood of* K.

Proof. Let $\varepsilon = \dfrac{1}{4}\,\mathrm{dist}(K, \partial\Omega)$, or $\varepsilon > 0$ arbitrary if $\Omega = \mathbb{R}^n$. For each $\delta > 0$, define the δ-neighborhood of K as

$$K_\delta := \Big\{ x \in \mathbb{R}^n : \quad \mathrm{dist}(x, K) \leqslant \delta \Big\}\,.$$

Then, letting v be the characteristic function of $K_{2\varepsilon}$ and its regularization $v_\varepsilon := v * \rho_\varepsilon$ (cf. (9.4), (9.5)), we know, according to part (a) of Theorem 9.8, that $\mathrm{supp}(v_\varepsilon) \subseteq K_{3\varepsilon} \subseteq \Omega$, which implies $v_\varepsilon \in C_0^\infty(\Omega)$. Moreover, for each $x \in \Omega$ we have

$$\begin{aligned} 0 \leqslant v_\varepsilon(x) &= \int_{K_{2\varepsilon}} \rho_\varepsilon(x-y)\,dy = \int_{B(x,\varepsilon)\cap K_{2\varepsilon}} \rho_\varepsilon(x-y)\,dy \\ &\leqslant \int_{B(x,\varepsilon)} \rho_\varepsilon(x-y)\,dy = 1\,. \end{aligned}$$

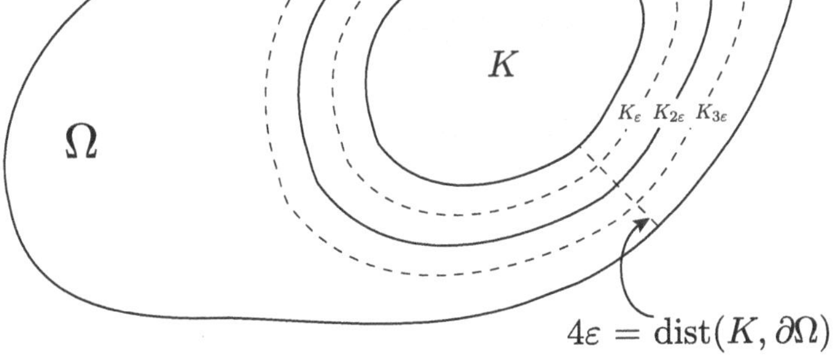

Fig. 9.3. *Geometry of the proof of Lemma 9.17*

In particular, taking $x \in K_\varepsilon$, we have $B(x,\varepsilon) \cap K_{2\varepsilon} = B(x,\varepsilon)$, and therefore (see Figure 9.3)

$$v_\varepsilon(x) = \int_{B(x,\varepsilon)} \rho_\varepsilon(x-y)\,dy = 1,$$

which completes the proof with $\varphi := v_\varepsilon$.

□

The function v_ε constructed in the proof of the previous lemma is called a CUT-OFF FUNCTION. In addition to the properties given by this lemma, we now also deduce the behavior of its partial derivatives. Indeed, recall first that

$$\rho_\varepsilon(x) = \frac{1}{C_\varepsilon}\varphi_{\varepsilon,0}(x) = \frac{1}{C_\varepsilon}\varphi_{1,0}\left(\frac{x}{\varepsilon}\right) \quad \forall\, x \in \mathbb{R}^n,$$

with

$$C_\varepsilon := \int_{B(0,\varepsilon)} \varphi_{\varepsilon,0}(x) = \int_{B(0,\varepsilon)} \varphi_{1,0}\left(\frac{x}{\varepsilon}\right)dx = \varepsilon^n \int_{B(0,1)} \varphi_{1,0}(z)\,dz,$$

and hence $\rho_\varepsilon(x) = \frac{\varepsilon^{-n}}{c}\varphi_{1,0}\left(\frac{x}{\varepsilon}\right)$ for all $x \in \mathbb{R}^n$, with $c := \int_{B(0,1)} \varphi_{1,0}(z)\,dz$. It follows that

$$\partial_x^\alpha \rho_\varepsilon(x-y) = \frac{\varepsilon^{-n}}{c}\,\partial_x^\alpha \varphi_{1,0}\left(\frac{x-y}{\varepsilon}\right)\varepsilon^{-|\alpha|} \quad \forall\, x \in \mathbb{R}^n,$$

and then, making the change of variable $z = \dfrac{x-y}{\varepsilon}$, we get

$$\begin{aligned}
|\partial^\alpha v_\varepsilon(x)| &= \left|\int_{\mathbb{R}^n} \partial_x^\alpha \rho_\varepsilon(x-y)\,dy\right| \leqslant \int_{\mathbb{R}^n} |\partial_x^\alpha \rho_\varepsilon(x-y)|\,dy \\
&= \frac{\varepsilon^{-n}}{c}\int_{\mathbb{R}^n} \varepsilon^{-|\alpha|}\left|\partial_x^\alpha \varphi_{1,0}\left(\frac{x-y}{\varepsilon}\right)\right|dy \\
&= \frac{\varepsilon^{-n}\,\varepsilon^{-|\alpha|}}{c}\int_{B(0,1)} \varepsilon^n\,|\partial^\alpha \varphi_{1,0}(z)|\,dz \\
&= \left\{\frac{1}{c}\int_{B(0,1)} |\partial^\alpha \varphi_{1,0}(z)|\,dz\right\}\varepsilon^{-|\alpha|},
\end{aligned}$$

that is, denoting $C_\alpha := \frac{1}{c}\int_{B(0,1)} |\partial^\alpha \varphi_{1,0}(z)|\,dz$, we obtain

$$|\partial^\alpha v_\varepsilon(x)| \leqslant C_\alpha\,\varepsilon^{-|\alpha|} \quad \forall\, x \in \mathbb{R}^n.$$

Lemma 9.18. *Let $X_1, X_2, \ldots, X_m$ be open subsets of $\mathbb{R}^n$, and let $\varphi \in C_0^\infty\left(\bigcup_{j=1}^m X_j\right)$. Then, there exist $\varphi_j \in C_0^\infty(X_j)$ for all $j \in \{1,2,\ldots,m\}$, such that $\varphi = \sum_{j=1}^m \varphi_j$. Moreover, if $\varphi \geqslant 0$, then also $\varphi_j \geqslant 0$ for all $j \in \{1,2,\ldots,m\}$.*

Proof. Since $\mathrm{supp}(\varphi) \subseteq \bigcup_{j=1}^{m} X_j$, for each $x \in \mathrm{supp}(\varphi)$ there exist a constant $\varepsilon(x) > 0$ and $k(x) \in \{1, 2, \ldots, m\}$ such that $\bar{B}(x, \varepsilon(x)) \subseteq X_{k(x)}$, and hence

$$\mathrm{supp}(\varphi) \subseteq \bigcup_{x \in \mathrm{supp}(\varphi)} B(x, \varepsilon(x)).$$

Since $\mathrm{supp}(\varphi)$ is compact, there exists a finite set $\{x_1, x_2, \ldots, x_N\} \subseteq \mathrm{supp}(\varphi)$ such that

$$\mathrm{supp}(\varphi) \subseteq \bigcup_{i=1}^{N} B(x_i, \varepsilon(x_i)).$$

Then, for each $j \in \{1, 2, \ldots, m\}$, we define the compact set

$$K_j := \bigcup_{k(x_i)=j} \bar{B}(x_i, \varepsilon(x_i)).$$

Should K_j happen to be empty, we instead define it as any closed ball contained in X_j. In this way we obtain $K_j \subseteq X_j$ for all $j \in \{1, 2, \ldots, m\}$ and $\mathrm{supp}(\varphi) \subseteq \bigcup_{j=1}^{m} K_j$. Then, applying Lemma 9.17 to X_j and K_j for all $j \in \{1, 2, \ldots, m\}$, we deduce that there exist $\psi_j \in C_0^\infty(X_j)$ such that $0 \leqslant \psi_j \leqslant 1$ and $\psi_j \equiv 1$ in a neighborhood of K_j. It follows that the functions defined by

$$\begin{aligned} \varphi_1 &:= \varphi\,\psi_1 \\ \varphi_2 &:= \varphi\,\psi_2(1-\psi_1) \\ \varphi_3 &:= \varphi\,\psi_3(1-\psi_1)(1-\psi_2) \\ &\vdots \\ \varphi_j &:= \varphi\,\psi_j(1-\psi_1)(1-\psi_2)\cdots(1-\psi_{j-1}) \end{aligned} \tag{9.11}$$

for each $j \in \{1, 2, \ldots, m\}$, satisfy the required properties. In particular, the fact that $\psi_j \in C_0^\infty(X_j)$ clearly implies that $\varphi_j \in C_0^\infty(X_j)$. Moreover, elementary calculations and induction on m show that

$$\begin{aligned} \varphi - \sum_{j=1}^{m} \varphi_j &= \varphi\Big\{1 - \psi_1 - \psi_2(1-\psi_1) - \cdots - \psi_m(1-\psi_1)(1-\psi_2)\cdots(1-\psi_{m-1})\Big\} \\ &= \varphi(1-\psi_1)(1-\psi_2)\cdots(1-\psi_m), \end{aligned}$$

from which we deduce that

$$\varphi = \sum_{j=1}^{m} \varphi_j.$$

Indeed, if $x \in \mathrm{supp}(\varphi)$ then $x \in K_i$ for some $i \in \{1, 2, \ldots, m\}$, and therefore $(1-\psi_i)(x) = 0$. On the other hand, if $x \notin \mathrm{supp}(\varphi)$ then $\varphi(x) = 0$. Thus, in both cases we obtain

$$\varphi\,(1-\psi_1)\,(1-\psi_2)\,\cdots\,(1-\psi_m) \equiv 0 \quad \text{in} \quad \bigcup_{j=1}^{m} X_j.$$

Finally, it follows from (9.11) and the stipulated ranges of the ψ_j that, whenever $\varphi \geqslant 0$, one also has $\varphi_j \geqslant 0$ for all $j \in \{1, \ldots, m\}$, which completes the proof.

□

An interesting consequence of the two previous lemmas is given by the following result, which introduces the concept of a PARTITION OF UNITY.

Lemma 9.19. *Let $X_1, X_2, \ldots, X_m$ be open subsets of $\mathbb{R}^n$, and let K be a compact subset of $\bigcup_{j=1}^{m} X_j$. Then there exist functions $\varphi_j \in C_0^\infty(X_j)$, $j \in \{1, 2, \ldots, m\}$, such that*

$$\varphi_j \geqslant 0, \quad 0 \leqslant \sum_{j=1}^{m} \varphi_j \leqslant 1 \quad \textit{and} \quad \sum_{j=1}^{m} \varphi_j \equiv 1 \quad \textit{in a neighborhood of } K\,.$$

Proof. Applying Lemma 9.17 to $\Omega = \bigcup_{j=1}^{m} X_j$ and to the compact set K, we deduce that there exists $\varphi \in C_0^\infty\left(\bigcup_{j=1}^{m} X_j\right)$ such that $0 \leqslant \varphi \leqslant 1$ and $\varphi \equiv 1$ in a neighborhood of K. Then, Lemma 9.18 guarantees the existence of

$$\varphi_j \in C_0^\infty(X_j) \quad \text{such that} \quad \varphi_j \geqslant 0 \quad \forall\, j \in \{1, 2, \ldots, m\} \quad \text{and} \quad \varphi = \sum_{j=1}^{m} \varphi_j\,.$$

It follows that $0 \leqslant \sum_{j=1}^{m} \varphi_j \leqslant 1$ and $\sum_{j=1}^{m} \varphi_j \equiv 1$ in a neighborhood of K.

□

Definition 9.20. *The set $\{\varphi_j\}_{j=1}^m$ (cf. Lemma 9.19) is said to form a* PARTITION OF UNITY *on K, subordinated to the covering $\{X_j\}_{j=1}^m$ of K.*

Definition 9.21. *Let Ω be an open subset of $\mathbb{R}^n$, and let $u \in \mathcal{D}'(\Omega)$. The support of u, denoted by* supp(u), *is defined as the complement of the following open set*

$$\Big\{x \in \Omega : \quad \exists \textit{ a neighborhood } V(x) \textit{ of } x \textit{ such that } u_{V(x)} = 0\Big\}.$$

In other words, supp(u) *is the set of points in Ω that do not admit a neighborhood on which the restriction of u vanishes.*

Note, for example, that $\operatorname{supp}(\delta) = \{0\}$ since $\delta \equiv 0$ in $\mathbb{R}^n \backslash \{0\}$.

Lemma 9.22. *Let $u \in \mathcal{D}'(\Omega)$, and let $\varphi \in C_0^\infty(\Omega)$ be such that* $\operatorname{supp}(u) \cap \operatorname{supp}(\varphi) = \varnothing$. *Then $\langle u, \varphi \rangle = 0$.*

Proof. Let $K = \operatorname{supp}(\varphi)$. Since $\operatorname{supp}(u) \cap K = \varnothing$, we have that for each $x \in K$ there exists $\varepsilon(x) > 0$ such that $u_{B(x,\varepsilon(x))} = 0$, that is, $\langle u, \psi \rangle = 0$ for all $\psi \in C_0^\infty(B(x, \varepsilon(x)))$. The compactness of K and the obvious fact that $K \subseteq \bigcup_{x \in K} B(x, \varepsilon(x))$ imply the existence of $\{x_1, x_2, \ldots, x_N\} \subseteq K$ such that

$$K \subseteq \sum_{j=1}^{N} B(x_j, \varepsilon(x_j)) .$$

According to Lemma 9.19, there exist functions $\psi_j \in C_0^\infty(B(x_j, \varepsilon(x_j)))$ such that

$$\psi_j \geqslant 0 , \quad 0 \leqslant \sum_{j=1}^{N} \psi_j \leqslant 1 \quad \text{and} \quad \sum_{j=1}^{N} \psi_j \equiv 1 \text{ in a neighborhood of} K.$$

Consequently,

$$\langle u, \varphi \rangle = \langle u, \varphi \sum_{j=1}^{N} \psi_j \rangle = \sum_{j=1}^{N} \langle u, \varphi \psi_j \rangle,$$

and since $\varphi \psi_j \in C_0^\infty(B(x_j, \varepsilon(x_j)))$, we have $\langle u, \varphi \psi_j \rangle = 0$, and hence $\langle u, \varphi \rangle = 0$.

□

Corollary 9.23. *Let $u \in \mathcal{D}'(\Omega)$ be such that, for each $x \in \Omega$, there exists a neighborhood $V(x)$ of x in Ω for which $u_{V(x)} \equiv 0$. Then $u \equiv 0$ in Ω.*

Proof. It suffices to observe that $\operatorname{supp}(u) = \varnothing$ and then apply the previous lemma.

□

We are now ready to state the theorem that characterizes a distribution in terms of its localizations.

Theorem 9.24. *Let $\{\Omega_i\}_{i \in I}$ be a family of open subsets of $\mathbb{R}^n$, and let $\Omega := \bigcup_{i \in I} \Omega_i$. In turn, let $\{u_i\}_{i \in I}$ be an associated family of distributions such that $u_i \in \mathcal{D}'(\Omega_i)$ for all $i \in I$ and $u_i = u_j$ in $\Omega_i \cap \Omega_j$ for all $i, j \in I$. Then, there exists a unique $u \in \mathcal{D}'(\Omega)$ such that $u_{\Omega_i} = u_i$ for all $i \in I$.*

Proof. First note that the uniqueness of such u, if it exists, follows from the previous corollary. Indeed, if $u, v \in \mathcal{D}'(\Omega)$ are such that $u_{\Omega_i} = v_{\Omega_i}$ for all $i \in I$, then, since for each $x \in \Omega$ there exists $j \in I$ such that $x \in \Omega_j$, it follows that Ω_j is a neighborhood of x where $u - v$ vanishes. Thus, necessarily, $u \equiv v$ in Ω. Let us now prove that u actually exists. Given $\varphi \in \mathcal{D}(\Omega)$, the compactness of $\operatorname{supp}(\varphi)$ implies the existence of a finite set $\{\widetilde{\Omega}_1, \widetilde{\Omega}_2, \ldots, \widetilde{\Omega}_N\} \subseteq \{\Omega_i\}_{i \in I}$ such that $\operatorname{supp}(\varphi) \subseteq \bigcup_{j=1}^{N} \widetilde{\Omega}_j$. The corresponding distributions are denoted by $\{\widetilde{u}_1, \widetilde{u}_2, \ldots, \widetilde{u}_N\} \subseteq \{u_i\}_{i \in I}$. According to Lemma 9.18, there exist functions $\varphi_j \in C_0^\infty(\widetilde{\Omega}_j)$ for all $j \in \{1, 2, \ldots, N\}$ such that

$$\varphi_j \geqslant 0 , \quad 0 \leqslant \sum_{j=1}^{N} \varphi_j \leqslant 1 \quad \text{and} \quad \sum_{j=1}^{N} \varphi_j \equiv 1 \quad \text{in a neighborhood of} \quad \operatorname{supp}(\varphi) .$$

Thus, noting that $\varphi = \sum_{j=1}^{N} \varphi \varphi_j$ on $\operatorname{supp}(\varphi)$ and that $\varphi \varphi_j \in C_0^\infty(\widetilde{\Omega}_j)$, we can define

$$\langle u, \varphi \rangle := \sum_{j=1}^{N} \langle \widetilde{u}_j, \varphi \varphi_j \rangle . \tag{9.12}$$

To see that u is well defined, consider $\{\widehat{\Omega}_1, \widehat{\Omega}_2, \ldots, \widehat{\Omega}_M\} \subseteq \{\Omega_i\}_{i\in I}$ another open covering of $\mathrm{supp}(\varphi)$ with $\{\widehat{u}_1, \widehat{u}_2, \ldots, \widehat{u}_M\} \subseteq \{u_i\}_{i\in I}$ as the corresponding distributions, and denote by $\{\psi_k\}_{k=1}^M$ the respective functions obtained from Lemma 9.18. Then, according to the hypothesis and the fact that $\mathrm{supp}(\varphi\,\varphi_j\,\psi_k) \subseteq \widetilde{\Omega}_j \cap \widehat{\Omega}_k$, we have

$$\sum_{j=1}^{N}\langle\widetilde{u}_j, \varphi\,\varphi_j\rangle = \sum_{j=1}^{N}\sum_{k=1}^{M}\langle\widetilde{u}_j, \varphi\,\varphi_j\,\psi_k\rangle = \sum_{j=1}^{N}\sum_{k=1}^{M}\langle\widehat{u}_k, \varphi\,\varphi_j\,\psi_k\rangle = \sum_{k=1}^{M}\langle\widehat{u}_k, \varphi\,\psi_k\rangle,$$

which proves that the computation of $\langle u, \varphi\rangle$ given by (9.12) is independent of the finite covering of $\mathrm{supp}(\varphi)$. Using arguments similar to those above and applying the Leibniz formula, it can be easily shown that $u \in \mathcal{D}'(\Omega)$. Finally, given $i \in I$ and $\varphi \in C_0^\infty(\Omega_i)$, it is clear that $\{\Omega_i\}$ is a covering of $\mathrm{supp}(\varphi)$. Then, if $\psi_i \in C_0^\infty(\Omega_i)$ satisfies $0 \leqslant \psi_i \leqslant 1$ and $\psi_i \equiv 1$ in a neighborhood of $\mathrm{supp}(\varphi)$, we have, according to (9.12),

$$\langle u, \varphi\rangle := \langle u_i, \varphi\,\psi_i\rangle = \langle u_i, \varphi\rangle,$$

thus concluding that $u_{\Omega_i} = u_i$.

□

9.5 More on partition of unity

In this section, we extend the analysis of Section 9.4 (cf. Lemma 9.19, Definition 9.20) by introducing a more general idea of partition of unity, which is called a LOCALLY FINITE PARTITION OF UNITY. For this purpose, we need some preliminary concepts and results.

Definition 9.25. *Let Ω be an open subset of $\mathbb{R}^n$, and let $\{\Omega_i\}_{i\in I}$ be an open covering of Ω. We say that the family of open sets $\{U_j\}_{j\in J}$ forms a refinement of $\{\Omega_i\}_{i\in I}$ if $\Omega = \bigcup_{j\in J} U_j$ and for each $j \in J$ there exists $i(j) \in I$ such that $U_j \subseteq \Omega_{i(j)}$.*

Definition 9.26. *Let Ω be an open subset of $\mathbb{R}^n$, and let $\{\Omega_i\}_{i\in I}$ be an open covering of Ω. We say that this covering is* LOCALLY FINITE *if for each $x \in \Omega$ there exists a ball $B(x, \delta)$ that intersects only a finite number of open sets Ω_i.*

The following theorem establishes an interesting connection between the previous concepts in the case $\Omega = \mathbb{R}^n$.

Theorem 9.27. *Let $\{\Omega_i\}_{i\in I}$ be an open covering of $\mathbb{R}^n$. Then there exists a locally finite refinement of open sets $\{U_j\}_{j\in\mathbb{N}}$ with $\bar{U}_j$ compact for all $j \in \mathbb{N}$.*

Proof. Let $B_N := B(0, N)$ for all $N \in \mathbb{N}$. Then it is clear that the compact annuli

$$\bar{B}_1,\ \bar{B}_2\backslash B_1,\ \bar{B}_3\backslash B_2,\ \cdots,\ \bar{B}_{k+1}\backslash B_k,\ \cdots$$

cover the whole $\mathbb{R}^n$. Now, for $x \in \bar{B}_1$, there exists $i(x) \in I$ such that $x \in \Omega_{i(x)}$, and then we can choose $\delta(x) < 1/2$ small enough so that $\bar{U}_x := \bar{B}(x, \delta(x)) \subseteq \Omega_{i(x)}$. Since $\bar{B}_1$ is compact, there exists a finite set $\{x_1, x_2, \ldots, x_m\} \subseteq \bar{B}_1$ such that $\bar{B}_1 \subseteq \bigcup_{j=1}^m U_{x_j}$, so that we define the sets $U_i := U_{x_i}$ for all $i \in \{1, 2, \ldots, m\}$, and observe that $\{U_i\}_{i=1}^m$ forms an open covering of $\bar{B}_1$. Similarly, for each $x \in \bar{B}_2\backslash B_1$, choose $V_x := B(x, \delta(x))$

with $\delta(x) < \frac{1}{2}$ and $\bar{V}_x := \bar{B}(x, \delta(x)) \subseteq \Omega_{i(x)}$ for some $i(x) \in I$. Thus, using now the compactness of $\bar{B}_2 \backslash B_1$, we deduce the existence of a finite set $\{\tilde{x}_1, \tilde{x}_2, \ldots, \tilde{x}_l\} \subseteq \bar{B}_2 \backslash B_1$ such that $\bar{B}_2 \backslash B_1 \subseteq \bigcup_{j=1}^{l} V_{\tilde{x}_j}$. If we then eliminate the open sets $V_{\tilde{x}_j}$ that coincide with some U_i, $i \in \{1, \ldots, m\}$, and denote the remaining open sets $V_{\tilde{x}_j}$ by $U_{m+1}, U_{m+2}, \ldots, U_M$, it follows that $\left\{U_1, U_2, \ldots, U_M\right\}$ constitutes an open covering of $\bar{B}_2$. Continuing recursively in this manner, we obtain an open covering $\{U_j\}_{j\in\mathbb{N}}$ of $\mathbb{R}^n$, which, by construction, is a refinement of $\{\Omega_i\}_{i\in I}$. Moreover, it is clear that the closed balls $\bar{U}_j$ are compact sets for all $j \in \mathbb{N}$. Finally, let us show that $\{U_j\}_{j\in\mathbb{N}}$ is locally finite. Indeed, let $x \in \bar{B}_{k+1} \backslash B_k$ with $k \in \mathbb{N}$, and consider the ball $B(x, 1/4)$. Since the radii of the balls U_j are all smaller than 1/2, we observe that only the sets U_j covering $\bar{B}_{k+1} \backslash B_k$ and the neighboring rings $\bar{B}_{k+2} \backslash B_{k+1}$ and $\bar{B}_k \backslash B_{k-1}$ may possibly intersect $B(x, 1/4)$, and these certainly form a finite number of open sets.

□

The extension of Theorem 9.27 to the case of an arbitrary open set Ω of $\mathbb{R}^n$ is proved below.

Theorem 9.28. *Let Ω be an open subset of $\mathbb{R}^n$, and let $\{\Omega_i\}_{i\in I}$ be an open covering of Ω. Then, there exists a locally finite refinement of open sets $\{U_j\}_{j\in\mathbb{N}}$ with $\bar{U}_j$ compact for all $j \in \mathbb{N}$.*

Proof. The result follows using arguments similar to those of Theorem 9.27. Indeed, consider a sequence of compact sets $\{K_j\}_{j\in\mathbb{N}}$ such that $K_j \subseteq K_{j+1}$ for all $j \in \mathbb{N}$ and satisfying $\bigcup_{j\in\mathbb{N}} K_j = \Omega$, and then cover Ω with the compact sets

$$K_1, K_2 \backslash \overset{\circ}{K}_1, \ldots, K_k \backslash \overset{\circ}{K}_{k-1} \quad \forall\, k \geqslant 2\,.$$

The remainder of the proof is analogous to that of Theorem 9.27.

□

For the main result of this section, which is stated and proved below, we need the following definition.

Definition 9.29. *Let Ω be an open subset of $\mathbb{R}^n$, and let $\{\Omega_i\}_{i\in I}$ be an open covering of Ω. We say that the family of functions $\{\varphi_j\}_{j\in J} \subseteq \mathcal{D}(\Omega)$ is a* LOCALLY FINITE PARTITION OF UNITY *over Ω subordinated to the covering $\{\Omega_i\}_{i\in I}$ if*

a) $\{\overset{\circ}{\operatorname{supp}}(\varphi_j)\}_{j\in J}$ *constitutes a locally finite refinement of* $\{\Omega_i\}_{i\in I}$,
b) $0 \leqslant \varphi_j \leqslant 1$ *for all* $j \in J$ *and* $\sum_{j\in J} \varphi_j(x) = 1$ *for all* $x \in \Omega$.

Theorem 9.30 (Locally Finite Partition of Unity). *Let $\{\Omega_i\}_{i\in I}$ be an open covering of the open set Ω in $\mathbb{R}^n$. Then there exists a locally finite partition of unity $\{\varphi_j\}_{j\in J}$ over Ω subordinated to $\{\Omega_i\}_{i\in I}$.*

Proof. According to Theorem 9.28, there exists a locally finite refinement of $\{\Omega_i\}_{i\in I}$ consisting of open sets

$$\{U_j\}_{j\in\mathbb{N}} \quad \text{with} \quad U_j := B(x_j, \delta_j), \quad x_j \in \Omega, \quad \delta_j \in (0, 1/2) \quad \text{and} \quad \bar{U}_j \subseteq \Omega_{i(j)}$$

for some $i(j) \in I$. Then, we define the functions

$$\psi_j(x) := \begin{cases} e^{-\delta_j^2/(\delta_j^2 - \|x-x_j\|^2)} & \text{if } \|x - x_j\| < \delta_j \\ 0 & \text{if } \|x - x_j\| \geqslant \delta_j, \end{cases}$$

which satisfy $\psi_j \in C_0^\infty(\Omega)$, $\mathrm{supp}(\psi_j) = \bar{U}_j$, and $\psi_j(x) > 0$ for all $x \in U_j$, for all $j \in \mathbb{N}$. Moreover, since the refinement $\{U_j\}_{j\in\mathbb{N}}$ is locally finite, for each $x \in \Omega$ there exists a ball $B(x,\delta)$ that intersects only a finite number of open sets U_j, and therefore the restriction of the series $\psi(x) := \sum_{j\in\mathbb{N}} \psi_j(x)$ to the ball $B(x,\delta)$ has only a finite number of nonzero terms. It follows that ψ is well defined on Ω and that $\psi \in C^\infty(\Omega)$. Furthermore, since $\Omega = \bigcup_{j\in\mathbb{N}} U_j$, for each $x \in \Omega$ there exists $j(x) \in \mathbb{N}$ such that $x \in U_{j(x)}$, and hence

$$\psi(x) = \sum_{j\in\mathbb{N}} \psi_j(x) \geqslant \psi_{j(x)}(x) > 0.$$

Finally, defining

$$\varphi_j := \frac{\psi_j}{\psi} \qquad \forall\, j \in \mathbb{N},$$

the proof is completed.

□

9.6 Convergence of distributions

Let Ω be an open subset of $\mathbb{R}^n$, and let $\{u_j\}_{j\in\mathbb{N}}$ be a sequence in $\mathcal{D}'(\Omega)$. We say that $\{u_j\}_{j\in\mathbb{N}}$ converges to $u \in \mathcal{D}'(\Omega)$ if

$$\lim_{j\to\infty} \langle u_j, \varphi\rangle = \langle u, \varphi\rangle \quad \forall\, \varphi \in \mathcal{D}(\Omega).$$

It is important to note that if the $u_j \in \mathcal{D}'(\Omega)$ are determined by continuous, locally integrable, or other functions, one must clearly distinguish between pointwise convergence and convergence in $\mathcal{D}'(\Omega)$. In this regard, we have the following example.

Example 9.31 In $\Omega = \mathbb{R}$, for each $j \in \mathbb{N}$ consider the distribution $u_j \in \mathcal{D}'(\Omega)$ induced by the function

$$u_j(x) := e^{\imath jx} = \cos(jx) + \imath \sin(jx) \quad \forall\, x \in \mathbb{R}.$$

It is easy to see that $\{u_j(x)\}_{j\in\mathbb{N}}$ is a convergent sequence in $\mathbb{C}$ only when $x = 2m\pi, m \in \mathbb{Z}$. However, $\{u_j\}_{j\in\mathbb{N}}$ converges to the zero distribution in $\mathcal{D}'(\Omega)$. Indeed, for each $\varphi \in \mathcal{D}(\Omega)$, integrating by parts, we have

$$\langle u_j, \varphi\rangle = \int_{\mathbb{R}} e^{\imath jx}\, \varphi(x)\, dx = -\frac{1}{\imath j} \int_{\mathbb{R}} e^{\imath jx}\, \varphi'(x)\, dx\,,$$

and hence

$$|\langle u_j, \varphi\rangle| \leqslant \frac{1}{j} \int_{\mathbb{R}} |\varphi'(x)|\, dx = \frac{1}{j} \int_{\mathrm{supp}(\varphi)} |\varphi'(x)|\, dx \overset{j\to\infty}{\longrightarrow} 0.$$

Alternatively, the above also follows from the Riemann-Lebesgue Lemma (cf. [120, Theorem 12.21]), which states that for integrable g, in particular for $g \in C[a,b]$, we have

$$\lim_{j\to\infty} \int_a^b g(x)\, \cos(jx)\, dx = \lim_{j\to\infty} \int_a^b g(x)\, \sin(jx)\, dx = 0\,.$$

The previous example confirms that convergence in $\mathcal{D}'(\Omega)$ does not necessarily imply pointwise convergence. Regarding the converse, we have the following result, which resembles (and also makes use of) the Lebesgue Dominated Convergence Theorem (cf. [113, Theorem 1.34], [120, Theorem 5.36]).

Lemma 9.32. *Let Ω be an open subset of $\mathbb{R}^n$, and let $\{u_j\}_{j\in\mathbb{N}} \subseteq \mathrm{L}^1_{loc}(\Omega)$ be such that*

$$u_j(x) \overset{j\to\infty}{\to} u(x) \quad \forall\, x \in \Omega \text{ a.e.}$$

Assume that there exists $g \in \mathrm{L}^1_{loc}(\Omega)$ such that $|u_j| \leqslant g$ for all $j \in \mathbb{N}$. Then $u \in \mathrm{L}^1_{loc}(\Omega)$ and $u_j \overset{j\to\infty}{\to} u$ in $\mathcal{D}'(\Omega)$.

Proof. First, let us show that $u \in \mathrm{L}^1_{loc}(\Omega)$. Indeed, given K_0 a compact subset of Ω, the Lebesgue Dominated Convergence Theorem (cf. [113, Theorem 1.34], [120, Theorem 5.36]) implies that

$$\int_{K_0} |u| = \lim_{j\to\infty} \int_{K_0} |u_j|,$$

and since $\displaystyle\int_{K_0} |u_j| \leqslant \int_{K_0} g < +\infty$, it follows that $\displaystyle\int_{K_0} |u|$ is also finite. On the other hand, for $\varphi \in \mathcal{D}(\Omega)$, we apply again the same theorem and deduce that

$$\lim_{j\to\infty} \langle u_j, \varphi\rangle = \lim_{j\to\infty} \int_{\mathrm{supp}(\varphi)} u_j\,\varphi = \int_{\mathrm{supp}(\varphi)} u\,\varphi = \int_\Omega u\,\varphi = \langle u\,,\varphi\rangle,$$

which proves that u_j indeed converges to u in $\mathcal{D}'(\Omega)$.

□

Example 9.33 For each $j \in \mathbb{N}$, consider $u_j \in \mathcal{D}'(\mathbb{R})$ the distribution induced by the locally integrable function

$$u_j(x) := \frac{\sin(jx)}{x} \quad \forall\, x \in \mathbb{R},$$

and let us now prove that $u_j \overset{j\to\infty}{\to} \pi\delta$ in $\mathcal{D}'(\mathbb{R})$. Indeed, let $\varphi \in \mathcal{D}(\mathbb{R})$ and let $M > 0$ be such that $\mathrm{supp}(\varphi) \subseteq [-M, M]$. Then, according to Lemma 9.3 we have

$$\varphi(x) = \varphi(0) + x\,\widetilde{\varphi}_0(x) \quad \forall\, x \in \mathbb{R},$$

with $\widetilde{\varphi}_0 \in C(\mathbb{R})$ and $\sup\limits_{x\in\mathbb{R}} |\widetilde{\varphi}_0(x)| \leqslant C \sup\limits_{x\in[-M,M]} |\varphi'(x)|$ for some $C > 0$. It follows that

$$\begin{aligned} \langle u_j, \varphi\rangle &= \int_{\mathbb{R}} \frac{\sin(jx)}{x}\,\varphi(x)\,dx = \int_{-M}^{M} \frac{\sin(jx)}{x}\,\varphi(x)\,dx \\ &= \int_{-M}^{M} \frac{\sin(jx)}{x}\left\{\varphi(0) + x\,\widetilde{\varphi}_0(x)\right\} dx \\ &= \varphi(0)\int_{-M}^{M} \frac{\sin(jx)}{x}\,dx + \int_{-M}^{M} \widetilde{\varphi}_0(x)\,\sin(jx)\,dx\,. \end{aligned} \tag{9.13}$$

Making the change of variable $y = \dfrac{jx}{M}$, we obtain

$$\int_{-M}^{M} \frac{\sin(jx)}{x}\,dx = \int_{-j}^{j} \frac{\sin(My)}{y}\,dy = 2\int_0^j \frac{\sin(My)}{y}\,dy,$$

and then, setting $z = M\,y$, we get

$$\lim_{j\to\infty}\int_{-M}^{M}\frac{\sin(jx)}{x}\,dx = 2\int_0^\infty\frac{\sin(My)}{y}\,dy = 2\int_0^\infty\frac{\sin(z)}{z}\,dz = \pi\,. \tag{9.14}$$

Moreover, the Riemann-Lebesgue Lemma (cf. [120, Theorem 12.21]) guarantees that

$$\lim_{j\to\infty}\int_{-M}^{M}\widetilde{\varphi}_0(x)\,\sin(jx)\,dx = 0\,. \tag{9.15}$$

Thus, from (9.13), (9.14) and (9.15), we conclude that

$$\lim_{j\to\infty}\langle u_j,\varphi\rangle = \varphi(0)\,\pi = \langle\pi\delta,\varphi\rangle,$$

which confirms the announced convergence.

Example 9.34 For each $j\in\mathbb{N}$ define $u_j\in\mathcal{D}'(\mathbb{R})$ by

$$\langle u_j,\varphi\rangle := \frac{j}{\sqrt{2\pi}}\int_{\mathbb{R}} e^{-j^2x^2/2}\,\varphi(x)\,dx \quad \forall\,\varphi\in\mathcal{D}(\mathbb{R}),$$

and let us prove that $u_j \overset{j\to\infty}{\to} \delta$ in $\mathcal{D}'(\mathbb{R})$. Indeed, let $\varphi\in\mathcal{D}(\mathbb{R})$, and let $M>0$ be such that $\operatorname{supp}(\varphi)\subseteq[-M,M]$, and let $\widetilde{\varphi}_1\in C(\mathbb{R})$, given by Lemma 9.3, such that

$$\varphi(x) = \varphi(0) + x\varphi'(0) + x^2\widetilde{\varphi}_1(x) \quad \forall\,x\in\mathbb{R},$$

and

$$\sup_{x\in\mathbb{R}}|\widetilde{\varphi}_1(x)| \leqslant C\sup_{x\in[-M,M]}|\varphi''(x)|.$$

It follows that

$$\begin{aligned}\langle u_j,\,\varphi\rangle &= \frac{\varphi(0)j}{\sqrt{2\pi}}\int_{-M}^{M}e^{-j^2x^2/2}\,dx + \frac{\varphi'(0)j}{\sqrt{2\pi}}\int_{-M}^{M}x\,e^{-j^2x^2/2}\,dx\\ &\quad + \frac{j}{\sqrt{2\pi}}\int_{-M}^{M}x^2\,\widetilde{\varphi}_1(x)\,e^{-j^2x^2/2}\,dx\,.\end{aligned} \tag{9.16}$$

Since the function in the second integral is odd, that integral vanishes. On the other hand, making the change of variable $u = jx/\sqrt{2}$ in the first one, and using that e^{-u^2} is even, we obtain

$$\frac{j}{\sqrt{2\pi}}\int_{-M}^{M}e^{-j^2x^2/2}\,dx = \frac{1}{\sqrt{\pi}}\int_{-\frac{jM}{\sqrt{2}}}^{\frac{jM}{\sqrt{2}}}e^{-u^2}\,du = \frac{2}{\sqrt{\pi}}\int_0^{\frac{jM}{\sqrt{2}}}e^{-u^2}\,du\,,$$

which implies that

$$\lim_{j\to\infty}\frac{j}{\sqrt{2\pi}}\int_{-M}^{M}e^{-j^2x^2/2}\,dx = \frac{2}{\sqrt{\pi}}\int_0^\infty e^{-u^2}\,du = 1. \tag{9.17}$$

Finally, noting that $e^{-u}\leqslant\dfrac{1}{u}$ for all $u>0$, it follows that

$$\left|\frac{j}{\sqrt{2\pi}}\int_{-M}^{M}x^2\,\widetilde{\varphi}_1(x)\,e^{-j^2x^2/2}\,dx\right| \leqslant \frac{1}{j}\frac{\sqrt{2}}{\sqrt{\pi}}\int_{-M}^{M}|\widetilde{\varphi}_1(x)|\,dx \leqslant \frac{c}{j}\overset{j\to\infty}{\to}0\,. \tag{9.18}$$

Thus, (9.16), (9.17), and (9.18) imply that $\lim_{j\to\infty}\langle u_j,\varphi\rangle = \varphi(0) = \langle\delta,\varphi\rangle$, that is, $u_j\overset{j\to\infty}{\to}\delta$ in $\mathcal{D}'(\mathbb{R})$.

9.7 Distributions with compact support

Let Ω be an open subset of $\mathbb{R}^n$. Then, similarly as for Definition 1.36, we now introduce the following concept.

Definition 9.35. *We let $\mathcal{E}(\Omega)$ denote the topological vector space of functions $\varphi \in C^\infty(\Omega)$, endowed with the topology induced by the uncountable family of seminorms*

$$\Big\{p_{K,m}: \quad K \text{ compact subset of } \Omega\,, \quad m \in \mathbb{N} \cup \{0\}\Big\}\,,$$

where (cf. (1.31))

$$p_{K,m}(\varphi) := \sum_{|\alpha| \leqslant m} \sup_{x \in K} |\partial^\alpha \varphi(x)| \qquad \forall\, \varphi \in C^\infty(\Omega)\,. \tag{9.19}$$

Moreover, given a sequence $\{K_i\}_{i \in \mathbb{N}}$ of compact subsets of Ω such that $K_i \subseteq K_{i+1}$ for all $i \in \mathbb{N}$, and $\Omega = \bigcup_{i \in \mathbb{N}} K_i$, the topology of $\mathcal{E}(\Omega)$ coincides with the one induced by the countable family of seminorms $\Big\{p_{K_i,m}: \quad i \in \mathbb{N}\,, \quad m \in \mathbb{N} \cup \{0\}\Big\}$.

The countability of the family introduced at the end of the previous definition guarantees that the topology of $\mathcal{E}(\Omega)$ is metrizable (cf. [114, Chapter 1, Section 1.46, and Chapter 6]). In addition, letting $\{|\cdot|_{j,\mathcal{E}}\}_{j \in \mathbb{N}}$ be an enumeration of these seminorms, a compatible metric is given by

$$d_{\mathcal{E}}(\varphi, \psi) := \sum_{j=1}^{\infty} \frac{1}{2^j} \left\{ \frac{|\varphi - \psi|_{j,\mathcal{E}}}{1 + |\varphi - \psi|_{j,\mathcal{E}}} \right\} \qquad \forall\, \varphi,\, \psi \in \mathcal{E}(\Omega)\,. \tag{9.20}$$

It then readily follows that the continuity of a mapping from $\mathcal{E}(\Omega)$ into any other topological space is equivalent to its sequential continuity.

Hence, although convergence in $\mathcal{E}(\Omega)$ could be defined in terms of the metric $d_{\mathcal{E}}$, we next introduce a simpler characterization, which follows from (9.20) and the summability of the series $\sum_{j=1}^{\infty} \frac{1}{2^j}$.

Definition 9.36. *A sequence $\{\varphi_k\}_{k \in \mathbb{N}} \subseteq \mathcal{E}(\Omega)$ is said to converge to the zero function in $\mathcal{E}(\Omega)$ if, for each $\alpha \in (\mathbb{N} \cup \{0\})^n$, the sequence $\{\partial^\alpha \varphi_k\}_{k \in \mathbb{N}}$ converges uniformly to the zero function on every compact subset of Ω. Equivalently, it suffices to require this convergence on each of the compact sets K_i, $i \in \mathbb{N}$.*

The space of continuous (i.e., sequentially continuous) linear forms on $\mathcal{E}(\Omega)$ is denoted by $\mathcal{E}'(\Omega)$. The following theorem characterizes the elements of $\mathcal{E}'(\Omega)$ in a manner analogous to the characterization of $\mathcal{D}'(\Omega)$ given by Theorem 1.42.

Theorem 9.37. *A linear form u on $\mathcal{E}(\Omega)$ belongs to $\mathcal{E}'(\Omega)$ if and only if there exist a compact set $K_0 \subseteq \Omega$, $C > 0$ and $N \in \mathbb{N} \cup \{0\}$ such that*

$$|\langle u, \varphi \rangle| \leqslant C \sum_{|\alpha| \leqslant N} \sup_{x \in K_0} |\partial^\alpha \varphi(x)| \qquad \forall\, \varphi \in \mathcal{E}(\Omega)\,. \tag{9.21}$$

Proof. Let $u : \mathcal{E}(\Omega) \to \mathbb{C}$ be a linear form, and suppose that there exist a compact set $K_0 \subseteq \Omega$, $C > 0$ and $N \in \mathbb{N} \cup \{0\}$ such that (9.21) holds. Let $\{\varphi_k\}_{k\in\mathbb{N}}$ be a sequence in $\mathcal{E}(\Omega)$ converging to 0 in $\mathcal{E}(\Omega)$. Then, for each $\alpha \in (\mathbb{N} \cup \{0\})^n$ and, in particular, for the compact set K_0, we have

$$\lim_{k\to\infty} \sup_{x\in\mathbb{K}_0} |\partial^\alpha \varphi_k(x)| = 0\,,$$

which, together with inequality (9.21), implies that $\langle u, \varphi_k\rangle \overset{k\to\infty}{\to} 0$, thereby showing that $u \in \mathcal{E}'(\Omega)$. Conversely, let $u \in \mathcal{E}'(\Omega)$, and let $\{K_j\}_{j\in\mathbb{N}}$ be an increasing sequence of compact subsets of Ω whose union is Ω. For example, for each $j \in \mathbb{N}$ we can take (see Figure 9.4 for examples of such compact sets)

$$K_j := \left\{x \in \Omega : \quad \operatorname{dist}(x, \partial\Omega) \geqslant \frac{1}{j}\right\} \cap \bar{B}(0, j)\,.$$

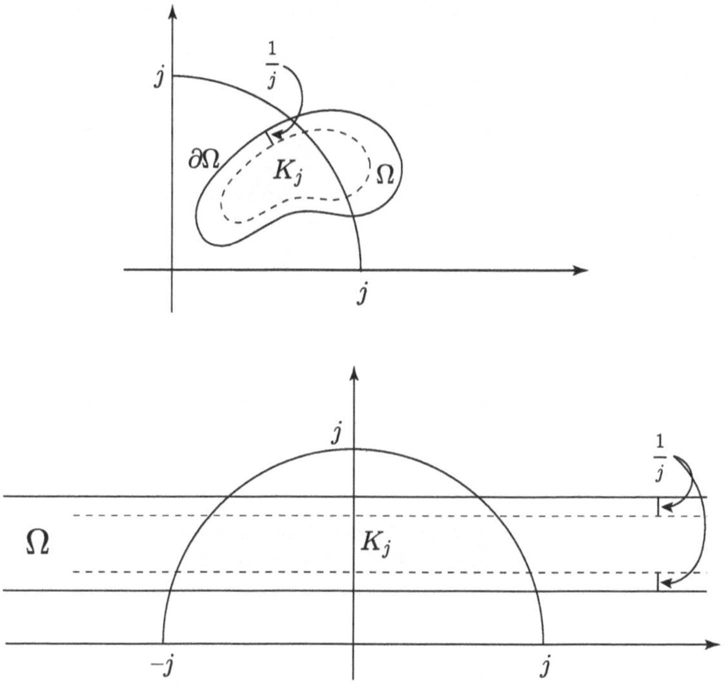

Fig. 9.4. *Compact sets K_j for bounded and unbounded Ω*

We proceed by contradiction and assume that for every compact $K \subseteq \Omega$, $C > 0$ and $N \in \mathbb{N} \cup \{0\}$, there exists $\varphi_{K,C,N} \in C^\infty(\Omega)$ such that

$$|\langle u, \varphi_{K,C,N}\rangle| > C \sum_{|\alpha|\leqslant N} \sup_{x\in K} |\partial^\alpha \varphi_{K,C,N}(x)|\,.$$

In particular, for all $N \in \mathbb{N}$ (taking $K = K_N$ and $C = N$) there exists $\varphi_N \in C^\infty(\Omega)$ such that

$$|\langle u, \varphi_N\rangle| > N \sum_{|\alpha|\leqslant N} \sup_{x\in K_N} |\partial^\alpha \varphi_N(x)|\,.$$

Thus, defining

$$\psi_N := \frac{\varphi_N}{|\langle u, \varphi_N\rangle|} \quad \forall\, N \in \mathbb{N},$$

we have $|\langle u, \psi_N\rangle| = 1$ and

$$\sum_{|\alpha|\leqslant N} \sup_{x\in K_N} |\partial^\alpha \psi_N(x)| < \frac{1}{N} \qquad \forall\, N \in \mathbb{N}.$$

Now, for $\beta \in (\mathbb{N} \cup \{0\})^n$ and K a compact subset of Ω, it is clear that there exists $M \in \mathbb{N}$ such that $K \subseteq K_N$ for all $N > M$, and therefore for each $N > \max\big\{M, |\beta|\big\}$ we have

$$\sup_{x\in K} |\partial^\beta \psi_N(x)| \leqslant \sup_{x\in K_N} |\partial^\beta \psi_N(x)| \leqslant \sum_{|\alpha|\leqslant N} \sup_{x\in K_N} |\partial^\alpha \psi_N(x)| < \frac{1}{N} \overset{N\to\infty}{\to} 0,$$

showing that $\psi_N \overset{N\to\infty}{\to} 0$ in $\mathcal{E}(\Omega)$, while $|\langle u, \psi_N\rangle| = 1$ for all $N \in \mathbb{N}$. This contradicts the sequential continuity of u and completes the proof.

□

It is easy to see, from the theorem above, that if $u \in \mathcal{E}'(\Omega)$ then its restriction to $C_0^\infty(\Omega)$ is an element of $\mathcal{D}'(\Omega)$. In fact, given a compact $K \subseteq \Omega$ and $\varphi \in \mathcal{D}_K(\Omega)$, Theorem 9.37 implies that

$$\begin{aligned} |\langle u, \varphi\rangle| &\leqslant C \sum_{|\alpha|\leqslant N} \sup_{x\in K_0} |\partial^\alpha \varphi(x)| \\ &= C \sum_{|\alpha|\leqslant N} \sup_{x\in K_0\cap K} |\partial^\alpha \varphi(x)| \leqslant C \sum_{|\alpha|\leqslant N} \sup_{x\in K} |\partial^\alpha \varphi(x)|, \end{aligned}$$

which shows that $u \in \mathcal{D}'(\Omega)$ and is of finite order $\leqslant N$. Moreover, it is clear that if $\mathrm{supp}(\varphi) \cap K_0 = \varnothing$ then $\langle u, \varphi\rangle = 0$, implying that $\mathrm{supp}(u)$, with u regarded as an element of $\mathcal{D}'(\Omega)$, is contained in K_0, and thus u has compact support. Furthermore, the following theorem provides the converse result.

Theorem 9.38. *Let $u \in \mathcal{D}'(\Omega)$ with compact support $\subseteq \Omega$. Then there exists a unique $v \in \mathcal{E}'(\Omega)$ such that $v|_{C_0^\infty(\Omega)} = u$.*

Proof. Let $\rho \in C_0^\infty(\Omega)$ be such that $\rho \equiv 1$ in a neighborhood V_ρ of $\mathrm{supp}(u)$. Define a linear form v on $C^\infty(\Omega)$ by

$$\langle v, \varphi\rangle := \langle u, \rho\,\varphi\rangle \quad \forall\, \varphi \in C^\infty(\Omega).$$

In particular, for each $\varphi \in C_0^\infty(\Omega)$ we have

$$\langle v, \varphi\rangle = \langle u, \rho\,\varphi\rangle = \langle u, \varphi\rangle - \langle u, (1-\rho)\,\varphi\rangle,$$

and since $\mathrm{supp}(u) \cap \mathrm{supp}\big((1-\rho)\varphi\big) = \varnothing$ because $(1-\rho)\varphi \equiv 0$ in a neighborhood of $\mathrm{supp}(u)$, we deduce that $\langle u, (1-\rho)\varphi\rangle = 0$, and thus

$$\langle v, \varphi\rangle = \langle u, \varphi\rangle,$$

proving that $v|_{C_0^\infty(\Omega)} = u$. Let us now show that this definition of v is independent of the chosen cutoff function ρ. Indeed, let $\psi \in C_0^\infty(\Omega)$ such that $\psi \equiv 1$ in a neighborhood V_ψ of $\mathrm{supp}(u)$. Then, for each $\varphi \in C^\infty(\Omega)$ we have

$$\langle u, \rho\varphi\rangle - \langle u, \psi\varphi\rangle = \langle u, (\rho - \psi)\varphi\rangle = 0$$

because $(\rho - \psi)\varphi$ vanishes in the neighborhood $V_\rho \cap V_\psi$ of $\operatorname{supp}(u)$, implying that

$$\operatorname{supp}(u) \cap \operatorname{supp}\big((\rho - \psi)\varphi\big) = \varnothing .$$

Let us now prove that $v \in \mathcal{E}'(\Omega)$. Indeed, since $\operatorname{supp}(\rho\varphi) \subseteq \operatorname{supp}(\rho) =: K_0$ for all $\varphi \in C^\infty(\Omega)$, the fact that $u \in \mathcal{D}'(\Omega)$ guarantees that for this compact K_0 there exist $C_0 > 0$ and $N_0 \in \mathbb{N} \cup \{0\}$ such that

$$|\langle v, \varphi\rangle| = |\langle u, \rho\varphi\rangle| \leqslant C_0 \sum_{|\alpha| \leqslant N_0} \sup_{x \in K_0} |\partial^\alpha(\rho\varphi)(x)| \quad \forall\, \varphi \in C^\infty(\Omega) .$$

Applying the Leibniz formula we have

$$\partial^\alpha(\rho\varphi)(x) = \sum_{\beta \leqslant \alpha} \binom{\alpha}{\beta} \partial^{\alpha-\beta} \rho(x)\, \partial^\beta \varphi(x),$$

and then, defining $C_\alpha := \max\limits_{\beta \leqslant \alpha} \sup\limits_{x \in K_0} \binom{\alpha}{\beta} |\partial^{\alpha-\beta} \rho(x)|$, $C_1 := \max\Big\{C_\alpha : |\alpha| \leqslant N_0\Big\}$, and $C_2 := \operatorname{card}\Big\{\alpha : |\alpha| \leqslant N_0\Big\}$, we deduce that

$$\begin{aligned}
\sum_{|\alpha| \leqslant N_0} \sup_{x \in K_0} |\partial^\alpha(\rho\varphi)(x)| &\leqslant \sum_{|\alpha| \leqslant N_0} C_\alpha \sum_{\beta \leqslant \alpha} \sup_{x \in K_0} |\partial^\beta \varphi(x)| \\
&\leqslant C_1 \sum_{|\alpha| \leqslant N_0} \sum_{\beta \leqslant \alpha} \sup_{x \in K_0} |\partial^\beta \varphi(x)| \\
&\leqslant C_1 \sum_{|\alpha| \leqslant N_0} \sum_{|\beta| \leqslant N_0} \sup_{x \in K_0} |\partial^\beta \varphi(x)| \\
&= C_1\, C_2 \sum_{|\beta| \leqslant N_0} \sup_{x \in K_0} |\partial^\beta \varphi(x)| ,
\end{aligned}$$

and therefore

$$|\langle v, \varphi\rangle| \leqslant C_0\, C_1\, C_2 \sum_{|\beta| \leqslant N_0} \sup_{x \in K_0} |\partial^\beta \varphi(x)| \quad \forall\, \varphi \in C^\infty(\Omega) ,$$

proving that $v \in \mathcal{E}'(\Omega)$. Finally, the uniqueness of v follows from the fact that $C_0^\infty(\Omega)$ is dense in $C^\infty(\Omega)$, which, as shown next in Lemma 9.39, means that for each $\varphi \in C^\infty(\Omega)$ there exists a sequence $\{\varphi_k\}_{k \in \mathbb{N}} \subseteq C_0^\infty(\Omega)$ such that $\varphi_k \overset{k\to\infty}{\to} \varphi$ in $\mathcal{E}(\Omega)$.

□

We have thus shown that $\mathcal{E}'(\Omega)$ can be identified with the subspace of $\mathcal{D}'(\Omega)$ consisting of distributions with compact support (which are of finite order) through the linear and injective mapping $\mathcal{R} : \mathcal{E}'(\Omega) \to \mathcal{D}'(\Omega)$ defined by

$$\mathcal{R}(v) := v|_{C_0^\infty(\Omega)} \quad \forall\, v \in \mathcal{E}'(\Omega) .$$

In fact, given $k \in \mathbb{N} \cup \{0\}$, we denote by $\mathcal{E}'_k(\Omega)$ the subspace of distributions in $\mathcal{E}'(\Omega)$ of order $\leqslant k$. Equivalently, recalling the definition of $\mathcal{D}'_k(\Omega)$ given at the beginning of Section 9.3, we may also write

$$\mathcal{E}'_k(\Omega) := \Big\{u \in \mathcal{E}'(\Omega) : \quad u|_{C_0^\infty(\Omega)} \in \mathcal{D}'_k(\Omega)\Big\} .$$

Lemma 9.39. *$C_0^\infty(\Omega)$ is dense in $C^\infty(\Omega)$.*

Proof. Although this proof can be carried out using the concepts and results concerning locally finite partitions of unity introduced in Section 9.5, here we adopt an alternative analysis that only uses Lemma 9.17 (cf. Section 9.4). Indeed, let $\{K_j\}_{j\in\mathbb{N}}$ be an increasing sequence of compact sets contained in Ω such that $\Omega = \bigcup_{j\in\mathbb{N}} K_j$ (see an example in the proof of Theorem 9.37). Then, for each $j \in \mathbb{N}$, consider $\chi_j \in C_0^\infty(\Omega)$ such that $\chi_j \equiv 1$ in a neighborhood of K_j. Then, given $\varphi \in C^\infty(\Omega)$, define the sequence $\{\varphi_j\}_{j\in\mathbb{N}} \subseteq C_0^\infty(\Omega)$ by $\varphi_j := \chi_j\,\varphi$ for all $j \in \mathbb{N}$. Thus, given a compact $K \subseteq \Omega$, there exists $N \in \mathbb{N}$ such that $K \subseteq K_j$ for all $j \geqslant N$, and therefore, for each α we have

$$\sup_{x\in K} |\partial^\alpha(\varphi - \varphi_j)(x)| \leqslant \sup_{x\in K_j} |\partial^\alpha(\varphi - \varphi_j)(x)| = 0 \quad \forall\, j \geqslant N\,,$$

from which it is clear that

$$\lim_{j\to\infty} \sup_{x\in K} |\partial^\alpha(\varphi - \varphi_j)(x)| = 0\,,$$

which proves that $\varphi_j \overset{j\to\infty}{\to} \varphi$ in $\mathcal{E}(\Omega)$.

□

On the other hand, the simplest case of distributions in $\mathcal{E}'(\Omega)$ is when the support reduces to a point. In order to characterize such distributions as linear combinations of derivatives of Dirac delta distributions, we need the following preliminary result.

Theorem 9.40. *Let $u \in \mathcal{E}'(\Omega)$ be of order N, and let $\varphi \in \mathcal{E}(\Omega)$ be such that*

$$\partial^\alpha\varphi(x) = 0 \quad \forall\, x \in \mathrm{supp}(u)\,, \quad \forall\, \alpha, |\alpha| \leqslant N\,.$$

Then $\langle u, \varphi\rangle = 0$.

Proof. Since $u \in \mathcal{E}'(\Omega)$ is of order N, there exist a compact set $K_0 \subseteq \Omega$ and $C > 0$ such that

$$|\langle u, \psi\rangle| \leqslant C \sum_{|\alpha|\leqslant N} \sup_{x\in K_0} |\partial^\alpha\psi(x)| \quad \forall\, \psi \in \mathcal{E}(\Omega)\,. \tag{9.22}$$

Moreover, from Lemma 9.17 we know that for $\varepsilon > 0$, $\varepsilon < \frac{1}{4}\,\mathrm{dist}(\mathrm{supp}(u), \partial\Omega)$, there exists $\chi_\varepsilon \in C_0^\infty(\Omega)$ such that $\chi_\varepsilon \equiv 1$ in the neighborhood K_ε of $K := \mathrm{supp}(u)$ and $\chi_\varepsilon \equiv 0$ outside $K_{3\varepsilon}$, where $K_\delta := \Big\{y \in \Omega : \quad \mathrm{dist}(y, K) \leqslant \delta\Big\}$ for each $\delta > 0$. Furthermore, the following estimate holds

$$|\partial^\alpha\chi_\varepsilon(x)| \leqslant a_\alpha\,\varepsilon^{-|\alpha|} \quad \forall\, x \in \Omega, \quad \forall\, \alpha\,, \quad |\alpha| \leqslant N\,. \tag{9.23}$$

Now, it is clear that

$$\langle u, \varphi\rangle = \langle u, \chi_\varepsilon\,\varphi\rangle + \langle u, (1 - \chi_\varepsilon)\,\varphi\rangle\,.$$

Let $\{\varphi_j\}_{j\in\mathbb{N}} \subseteq C_0^\infty(\Omega)$, whose existence is guaranteed by Lemma 9.39, such that $\varphi_j \overset{j\to\infty}{\to} \varphi$ in $\mathcal{E}(\Omega)$. Then, using the Leibniz formula, one easily shows that $(1-\chi_\varepsilon)\,\varphi_j \overset{j\to\infty}{\to} (1-\chi_\varepsilon)\,\varphi$ in $\mathcal{E}(\Omega)$, and since $\mathrm{supp}(u) \cap \mathrm{supp}\big((1 - \chi_\varepsilon)\,\varphi_j\big) = \varnothing$, it follows that

$$\langle u, (1-\chi_\varepsilon)\,\varphi\rangle = \lim_{j\to\infty}\langle u, (1-\chi_\varepsilon)\,\varphi_j\rangle = 0\,,$$

which implies that

$$\langle u, \varphi\rangle = \langle u, \chi_\varepsilon\,\varphi\rangle.$$

Applying (9.22) and the fact that $\operatorname{supp}(\chi_\varepsilon\,\varphi) \subseteq K_{3\varepsilon}$, we obtain

$$|\langle u, \varphi\rangle| = |\langle u, \chi_\varepsilon\,\varphi\rangle| \leqslant C \sum_{|\alpha|\leqslant N} \sup_{x\in K_0} |\partial^\alpha(\chi_\varepsilon\,\varphi)(x)| \leqslant C \sum_{|\alpha|\leqslant N} \sup_{x\in K_{3\varepsilon}} |\partial^\alpha(\chi_\varepsilon\,\varphi)(x)|\,.$$

Then, using once again the Leibniz formula, together with the bound (9.23), denoting the constants

$$C_\alpha := \max_{\beta\leqslant\alpha} \binom{\alpha}{\beta}, \quad \widetilde{C}_\alpha := \max_{\beta\leqslant\alpha} a_{\alpha-\beta}\,,$$

$$C_1 := \max_{|\alpha|\leqslant N} C_\alpha\,\widetilde{C}_\alpha\,, \quad C_2 := \operatorname{card}\Big\{\alpha : |\alpha| \leqslant \mathrm{N}\Big\},$$

and observing that $\varepsilon^{-|\alpha|} \leqslant \varepsilon^{-N}$ for $\varepsilon < 1$ and $|\alpha| \leqslant N$, it follows that

$$\begin{aligned}
|\langle u, \varphi\rangle| &\leqslant C \sum_{|\alpha|\leqslant N} \sup_{x\in K_{3\varepsilon}} \left|\sum_{\beta\leqslant\alpha}\binom{\alpha}{\beta}\,\partial^{\alpha-\beta}\,\chi_\varepsilon(x)\,\partial^\beta\,\varphi(x)\right| \\
&\leqslant C \sum_{|\alpha|\leqslant N} C_\alpha \sum_{\beta\leqslant\alpha} \sup_{x\in K_{3\varepsilon}} |\partial^{\alpha-\beta}\,\chi_\varepsilon(x)|\,|\partial^\beta\,\varphi(x)| \\
&\leqslant C \sum_{|\alpha|\leqslant N} C_\alpha \sum_{\beta\leqslant\alpha} a_{\alpha-\beta}\,\varepsilon^{|\beta|-|\alpha|} \sup_{x\in K_{3\varepsilon}} |\partial^\beta\,\varphi(x)| \\
&\leqslant C\,C_1 \sum_{|\alpha|\leqslant N}\sum_{\beta\leqslant\alpha} \varepsilon^{|\beta|-N} \sup_{x\in K_{3\varepsilon}} |\partial^\beta\,\varphi(x)| \\
&\leqslant C\,C_1 \sum_{|\alpha|\leqslant N}\sum_{|\beta|\leqslant N} \varepsilon^{|\beta|-N} \sup_{x\in K_{3\varepsilon}} |\partial^\beta\,\varphi(x)|,
\end{aligned}$$

from which, setting $\widetilde{C} := C\,C_1\,C_2$ and replacing β by α, we deduce that

$$|\langle u, \varphi\rangle| \leqslant \widetilde{C} \sum_{|\alpha|\leqslant N} \varepsilon^{|\alpha|-N} \sup_{x\in K_{3\varepsilon}} |\partial^\alpha\,\varphi(x)|\,. \tag{9.24}$$

If $|\alpha| = N$ then $\varepsilon^{|\alpha|-N} \sup_{x\in K_{3\varepsilon}} |\partial^\alpha\varphi(x)| = \sup_{x\in K_{3\varepsilon}} |\partial^\alpha\varphi(x)|$, which tends to 0 as $\varepsilon \to 0$ because $\partial^\alpha\varphi$ is uniformly continuous on compact sets and vanishes on $K = \operatorname{supp}(u)$. On the other hand, if $|\alpha| < N$, we apply the Taylor formula with integral remainder to $\partial^\alpha\varphi$ and obtain, for $x \in K$, $y \in \Omega$

$$\begin{aligned}
\partial^\alpha\varphi(y) &= \sum_{|\beta|\leqslant N-|\alpha|-1} \partial^{\beta+\alpha}\,\varphi(x)\,\frac{(y-x)^\beta}{\beta!} \\
&+ (N-|\alpha|)\int_0^1 \left\{(1-t)^{N-|\alpha|-1} \sum_{|\beta|=N-|\alpha|} \partial^{\beta+\alpha}\,\varphi(x+t(y-x))\,\frac{(y-x)^\beta}{\beta!}\right\} dt\,.
\end{aligned} \tag{9.25}$$

Recall that, in general, for $\psi \in C^k(\Omega)$ and $k \in \mathbb{N}$ we have the Taylor formula with integral remainder

$$\psi(y) = \sum_{|\beta|\leqslant k-1} \partial^\beta \psi(x) \frac{(y-x)^\beta}{\beta!}$$

$$+ k \int_0^1 \left\{ (1-t)^{k-1} \sum_{|\beta|=k} \partial^\beta \psi(x+t(y-x)) \frac{(y-x)^\beta}{\beta!} \right\} dt\,.$$

Now, for $|\beta| \leqslant N-|\alpha|-1$ we have $|\beta+\alpha| \leqslant N-1$, and therefore $\partial^{\beta+\alpha}\varphi(x) = 0$ for all $x \in K := \operatorname{supp}(u)$, so that (9.25) reduces to:

$$\partial^\alpha\varphi(y) = (N-|\alpha|) \int_0^1 \left\{ (1-t)^{N-|\alpha|-1} \sum_{|\beta|=N-|\alpha|} \partial^{\beta+\alpha}\varphi(x+t\,(y-x)) \frac{(y-x)^\beta}{\beta!} \right\} dt\,,$$

and thus, setting $\widetilde{C}_\alpha := \max\limits_{|\beta|=N-|\alpha|} \dfrac{1}{\beta!}$, we obtain

$$\frac{|\partial^\alpha\varphi(y)|}{(N-|\alpha|)\,\widetilde{C}_\alpha} \leqslant \sup_{t\in[0,1]} \left\{ \sum_{|\beta|=N-|\alpha|} |\partial^{\beta+\alpha}\varphi(x+t(y-x))|\, \|y-x\|^\beta \right\} \int_0^1 (1-t)^{N-|\alpha|-1} dt\,.$$

Then, for $y \in K_{3\varepsilon}$, we can choose $x \in K$ such that $\|y-x\| \leqslant 3\varepsilon$, and thus the inequality above yields

$$|\partial^\alpha\varphi(y)| \leqslant \bar{C}_\alpha\, \varepsilon^{N-|\alpha|} \sup_{t\in[0,1]} \sum_{|\beta|=N} |\partial^\beta\varphi(x+t(y-x))|\,,$$

where

$$\bar{C}_\alpha := \widetilde{C}_\alpha\,(N-|\alpha|)\,3^{N-|\alpha|} \int_0^1 (1-t)^{N-|\alpha|-1}\,dt\,.$$

But since $x + t\,(y-x) \in K_{3\varepsilon}$, it follows that

$$|\partial^\alpha\varphi(y)| \leqslant \bar{C}_\alpha\, \varepsilon^{N-|\alpha|} \sum_{|\beta|=N} \sup_{x\in K_{3\varepsilon}} |\partial^\beta\varphi(x)| \quad \forall\, y \in K_{3\varepsilon}\,,$$

and therefore

$$\sup_{x\in K_{3\varepsilon}} |\partial^\alpha\varphi(x)| \leqslant \bar{C}_\alpha\, \varepsilon^{N-|\alpha|} \sum_{|\beta|=N} \sup_{x\in K_{3\varepsilon}} |\partial^\beta\varphi(x)|\,,$$

or equivalently, for $|\alpha| < N$,

$$\varepsilon^{|\alpha|-N} \sup_{x\in K_{3\varepsilon}} |\partial^\alpha\varphi(x)| \leqslant \bar{C}_\alpha \sum_{|\beta|=N} \sup_{x\in K_{3\varepsilon}} |\partial^\beta\varphi(x)|\,.$$

Since this last term also tends to 0 as $\varepsilon \to 0$, we conclude that

$$\lim_{\varepsilon\to 0} \sum_{|\alpha|\leqslant N} \varepsilon^{|\alpha|-N} \sup_{x\in K_{3\varepsilon}} |\partial^\alpha\varphi(x)| = 0\,,$$

and consequently, (9.24) implies that $\langle u, \varphi\rangle = 0$.

□

Theorem 9.41. *Let $\Omega \subseteq \mathbb{R}^n$ be open, and let $u \in \mathcal{E}'(\Omega)$ be such that* $\operatorname{supp}(u) = \{y\}$*, $y \in \Omega$. Then there exist constants $N \in \mathbb{N} \cup \{0\}$ and $c_\alpha \in \mathbb{C}$(or $\mathbb{R}$) for all $\alpha, |\alpha| \leqslant N$, such that $u = \sum_{|\alpha|\leqslant N} c_\alpha\, \partial^\alpha \delta_y$, that is,*

$$\langle u, \varphi\rangle = \sum_{|\alpha|\leqslant N} c_\alpha\,(-1)^{|\alpha|}\,\langle \delta_y, \partial^\alpha\varphi\rangle = \sum_{|\alpha|\leqslant N} c_\alpha\,(-1)^{|\alpha|}\,\partial^\alpha\varphi(y) \quad \forall\, \varphi \in \mathcal{E}(\Omega)\,.$$

Proof. Let $\varphi \in \mathcal{E}(\Omega)$. Then the Taylor series expansion of φ gives, for each $n \in \mathbb{N}$ and for $x \in \Omega$,

$$\varphi(x) = \sum_{|\alpha| \leqslant n} \partial^\alpha \varphi(y) \frac{(x-y)^\alpha}{\alpha!} - \psi_n(x),$$

where ψ_n is the integral remainder. In particular, given N = the order of u and $|\beta| \leqslant N$, we have

$$\partial^\beta \psi_N(x) = \sum_{\substack{\alpha \geqslant \beta \\ |\alpha| \leqslant N}} \frac{\partial^\alpha \varphi(y)(x-y)^{\alpha-\beta}}{(\alpha-\beta)!} - \partial^\beta \varphi(x) \quad \forall\, x \in \Omega,$$

and hence, for $x = y$, it follows that $\partial^\beta \psi_N(y) = 0$, that is, $\partial^\beta \psi_N$ vanishes on $\operatorname{supp}(u)$ for all β, $|\beta| \leqslant N$. This shows, according to Theorem 9.40, that $\langle u, \psi_N \rangle = 0$, and therefore

$$\langle u, \varphi \rangle = \sum_{|\alpha| \leqslant N} \left\langle u, \frac{(\cdot - y)^\alpha}{\alpha!} \right\rangle \partial^\alpha \varphi(y) = \sum_{|\alpha| \leqslant N} c_\alpha (-1)^{|\alpha|} \partial^\alpha \varphi(y),$$

with $c_\alpha := \left\langle u, \frac{(\cdot - y)^\alpha}{\alpha!} \right\rangle (-1)^{|\alpha|}$.

□

9.8 Derivation and multiplication by functions

In this section, we study in greater detail the properties of distributional differentiation and introduce and analyze multiplication by smooth functions. To this end, we begin by recalling from Section 1.5 (cf. Definition 1.47) the former concept.

Definition 9.42. *Let $u \in \mathcal{D}'(\Omega)$, and let α be a multi-index. The α-th derivative of u, denoted $\partial^\alpha u$, is defined as the distribution on $\mathcal{D}(\Omega)$ given by*

$$\langle \partial^\alpha u, \varphi \rangle = (-1)^{|\alpha|} \langle u, \partial^\alpha \varphi \rangle \qquad \forall\, \varphi \in \mathcal{D}(\Omega).$$

The fact that $\partial^\alpha u$ is indeed an element of $\mathcal{D}'(\Omega)$ was proved at the end of Chapter 1, right after Definition 1.47.

9.8.1 Additional results on distributional derivatives

A first elementary result concerning distributional derivatives is given by their sequential continuity. More precisely, the following lemma holds.

Lemma 9.43. *Let $\{u_j\}_{j \in \mathbb{N}}$ be a sequence in $\mathcal{D}'(\Omega)$ converging to $u \in \mathcal{D}'(\Omega)$. Then $\{\partial^\alpha u_j\}_{j \in \mathbb{N}}$ converges to $\partial^\alpha u$ in $\mathcal{D}'(\Omega)$ for all $\alpha \in (\mathbb{N} \cup \{0\})^n$.*

Proof. For each $\varphi \in \mathcal{D}(\Omega)$, it follows, according to the definition of convergence of distributions (cf. Section 9.6), that

$$\langle \partial^\alpha u_j, \varphi \rangle = (-1)^{|\alpha|} \langle u_j, \partial^\alpha \varphi \rangle \overset{j \to +\infty}{\rightarrow} (-1)^{|\alpha|} \langle u, \partial^\alpha \varphi \rangle = \langle \partial^\alpha u, \varphi \rangle.$$

□

We now give some classical examples of distributional derivatives.

Example 9.44

i) Let $H : \mathbb{R} \to \mathbb{R}$ be the Heaviside function, that is,

$$H(x) = \begin{cases} 1 & \text{if } x > 0\,, \\ 0 & \text{if } x \leqslant 0\,. \end{cases}$$

Then, we have

$$\langle H', \varphi \rangle = -\langle H, \varphi' \rangle = -\int_0^{\infty} \varphi'(x)\,dx = \varphi(0) = \langle \delta, \varphi \rangle \quad \forall\, \varphi \in \mathcal{D}(\mathbb{R})\,,$$

which shows that $H' = \delta$ in $\mathcal{D}'(\mathbb{R})$.

ii) Now let $u = FP\left(\dfrac{H(x)}{x}\right)$ be the distribution defined in Example 9.4, that is,

$$\langle u, \varphi \rangle = \lim_{\varepsilon \to 0^+} \left\{ \int_{\varepsilon}^{\infty} \frac{\varphi(x)}{x}\,dx + \varphi(0)\,\ln(\varepsilon) \right\} \qquad \forall\, \varphi \in \mathcal{D}(\mathbb{R})\,.$$

Then, the computation of u' gives

$$\begin{aligned}
\langle u', \varphi \rangle &:= -\langle u, \varphi' \rangle = -\lim_{\varepsilon \to 0^+} \left\{ \int_{\varepsilon}^{\infty} \frac{\varphi'(x)}{x}\,dx + \varphi'(0)\,\ln(\varepsilon) \right\} \\
&= -\lim_{\varepsilon \to 0^+} \left\{ \frac{\varphi(x)}{x}\bigg|_{\varepsilon}^{\infty} + \int_{\varepsilon}^{\infty} \frac{\varphi(x)}{x^2}\,dx + \varphi'(0)\,\ln(\varepsilon) \right\} \\
&= -\lim_{\varepsilon \to 0^+} \left\{ \int_{\varepsilon}^{\infty} \frac{\varphi(x)}{x^2}\,dx - \frac{\varphi(\varepsilon)}{\varepsilon} + \varphi'(0)\ln(\varepsilon) \right\} \\
&= -\lim_{\varepsilon \to 0^+} \left\{ \int_{\varepsilon}^{\infty} \frac{\varphi(x)}{x^2}\,dx - \frac{(\varphi(\varepsilon) - \varphi(0))}{\varepsilon} - \frac{\varphi(0)}{\varepsilon} + \varphi'(0)\ln(\varepsilon) \right\} \\
&= -\lim_{\varepsilon \to 0^+} \left\{ \int_{\varepsilon}^{\infty} \frac{\varphi(x)}{x^2}\,dx - \frac{\varphi(0)}{\varepsilon} + \varphi'(0)\ln(\varepsilon) \right\} + \varphi'(0)\,,
\end{aligned}$$

from which, using Example 9.5, we obtain

$$\langle u', \varphi \rangle = \left\langle -FP\left(\frac{H(x)}{x^2}\right), \varphi \right\rangle - \langle \delta', \varphi \rangle \qquad \forall\, \varphi \in \mathcal{D}(\mathbb{R})\,,$$

or equivalently,

$$u' = -FP\left(\frac{H(x)}{x^2}\right) - \delta'\,.$$

iii) Let $u = FP\left(\dfrac{H(-x)}{x}\right)$, that is,

$$\langle u, \varphi \rangle = \lim_{\varepsilon \to 0^+} \left\{ \int_{-\infty}^{-\varepsilon} \frac{\varphi(x)}{x} - \varphi(0)\,\ln(\varepsilon) \right\} \qquad \forall\, \varphi \in \mathcal{D}(\mathbb{R})\,.$$

Then, similarly to the previous calculation, one obtains

$$\langle u', \varphi \rangle = -\lim_{\varepsilon \to 0^+} \left\{ \int_{-\infty}^{-\varepsilon} \frac{\varphi(x)}{x^2}\,dx - \frac{\varphi(0)}{\varepsilon} - \varphi'(0)\,\ln(\varepsilon) \right\} - \varphi'(0)\,,$$

which implies that

$$u' = -FP\left(\frac{H(-x)}{x^2}\right) + \delta'\,.$$

iv) Consider $u = PV\left(\frac{1}{x}\right)$, the Cauchy principal value of $\frac{1}{x}$, that is,

$$\langle u, \varphi \rangle = \lim_{\varepsilon \to 0^+} \left\{ \int_{|x| \geqslant \varepsilon} \frac{\varphi(x)}{x} \, dx \right\} \qquad \forall \, \varphi \in \mathcal{D}(\mathbb{R}) .$$

It is easy to see that

$$u = FP\left(\frac{H(-x)}{x}\right) + FP\left(\frac{H(x)}{x}\right),$$

from which, using ii) and iii), we have

$$u' = -FP\left(\frac{H(-x)}{x^2}\right) - FP\left(\frac{H(x)}{x^2}\right),$$

that is,

$$\langle u', \varphi \rangle = \lim_{\varepsilon \to 0^+} \left\{ \int_{|x| \geqslant \varepsilon} \frac{\varphi(x)}{x^2} \, dx - \frac{2\varphi(0)}{\varepsilon} \right\} \qquad \forall \, \varphi \in \mathcal{D}(\mathbb{R}) ,$$

or equivalently,

$$\left(PV\left(\frac{1}{x}\right)\right)' = -FP\left(\frac{1}{x^2}\right).$$

v) Finally, one can show (left as an exercise) that $\left(\ln(|x|)\right)' = PV\left(\frac{1}{x}\right)$.

The following lemmas establish some basic results on distributional derivatives, which include jump properties and the implicit idea of "integration" of distributions.

Lemma 9.45. *Let Ω be an open subset of $\mathbb{R}$, and let $u \in C^1(\Omega \backslash \{x_0\})$, for some $x_0 \in \Omega$. If the function v defined by $v(x) = u'(x)$ for $x \neq x_0$, is integrable in a neighborhood of x_0, then the limits*

$$u(x_0^{\pm}) = \lim_{x \to x_0^{\pm}} u(x)$$

exist and

$$u' = v + \left\{u(x_0^+) - u(x_0^-)\right\} \delta_{x_0}$$

in the sense of distributions.

Proof. Let $y \in \Omega$ be such that $x_0 < y$ and $[x_0, y] \subseteq \Omega$. Then, from the relation

$$u(x) = u(y) - \int_x^y v(t) \, dt \quad \forall \, x \in (x_0, y) ,$$

it follows that the limit $u(x_0^+)$ exists. Similarly, one proves the existence of $u(x_0^-)$. This shows that u is piecewise continuous and therefore defines a distribution $u \in \mathcal{D}'(\Omega)$. Then, for each $\varphi \in \mathcal{D}(\Omega)$ we have

$$\langle u', \varphi \rangle = -\langle u, \varphi' \rangle = -\lim_{\varepsilon \to 0^+} \int_{|x - x_0| \geqslant \varepsilon} u(x) \, \varphi'(x) \, dx$$

$$= \lim_{\varepsilon \to 0^+} \left\{ u(x_0 + \varepsilon)\, \varphi(x_0 + \varepsilon) - u(x_0 - \varepsilon)\, \varphi(x_0 - \varepsilon) + \int_{|x - x_0| \geqslant \varepsilon} u'(x)\, \varphi(x)\, dx \right\}$$

$$= \lim_{\varepsilon \to 0^+} \left\{ u(x_0 + \varepsilon)\, \varphi(x_0 + \varepsilon) - u(x_0 - \varepsilon)\, \varphi(x_0 - \varepsilon) + \int_{|x - x_0| \geqslant \varepsilon} v(x)\, \varphi(x)\, dx \right\}$$

$$= \left\{ u(x_0^+) - u(x_0^-) \right\} \varphi(x_0) + \langle v, \varphi \rangle,$$

which completes the proof.

□

It is important to note here, according to the hypotheses on u and v in the previous lemma, that $u' - v \equiv 0$ in $\Omega \backslash \{x_0\}$, which indicates that $u' - v$ induces a distribution with support in $\{x_0\}$. It follows, by virtue of Theorem 9.41, that $u' - v$ is a finite linear combination of δ_{x_0} and some of its derivatives. However, what Lemma 9.45 actually proves is much more precise, since it establishes that $u' - v$ is only a multiple of δ_{x_0}.

Theorem 9.46. *Let Ω be an open subset of $\mathbb{R}$, and let $u \in \mathcal{D}'(\Omega)$ be such that $u' = 0$. Then u is a constant.*

Proof. According to the hypothesis we have

$$\langle u, \psi' \rangle = -\langle u', \psi \rangle = 0 \qquad \forall\, \psi \in \mathcal{D}(\Omega)\,.$$

Now, given $\varphi \in \mathcal{D}(\Omega)$, the equation $\psi' = \varphi$ has the unique solution $\psi \in C^\infty(\Omega)$ defined by

$$\psi(x) = \int_{-\infty}^{x} \varphi(t)\, dt \quad \forall\, x \in \Omega\,,$$

which certainly vanishes to the left of $\mathrm{supp}(\varphi)$. For ψ to have compact support, that is, $\psi \in C_0^\infty(\Omega)$, it is further required that $\mathbf{I}(\varphi) := \int_{-\infty}^{\infty} \varphi(x)\, dx$ be zero. In this way, if $\varphi \in C_0^\infty(\Omega)$ is such that $\mathbf{I}(\varphi) = 0$, then $\langle u, \varphi \rangle = 0$. Now, if $\varphi_0 \in C_0^\infty(\Omega)$ is chosen such that $\mathbf{I}(\varphi_0) := \int_{-\infty}^{\infty} \varphi_0(x)\, dx = 1$, then for each $\varphi \in C_0^\infty(\Omega)$, the function $\varphi - \mathbf{I}(\varphi)\, \varphi_0$ satisfies $\mathbf{I}\big(\varphi - \mathbf{I}(\varphi)\, \varphi_0\big) = 0$, and therefore

$$\langle u, \varphi - \mathbf{I}(\varphi)\, \varphi_0 \rangle = 0\,,$$

that is,

$$\langle u, \varphi \rangle = \mathbf{I}(\varphi) \langle u, \varphi_0 \rangle = \langle u, \varphi_0 \rangle \int_{-\infty}^{\infty} \varphi(x)\, dx\,,$$

or equivalently,

$$\langle u, \varphi \rangle = \langle \langle u, \varphi_0 \rangle, \varphi \rangle\,,$$

which proves that u coincides with the distribution induced by the constant value $\langle u, \varphi_0 \rangle$.

□

A result analogous to the previous one for the case of $\Omega \subseteq \mathbb{R}^n$, is given below in Lemma 9.48, for which the following preliminary result is required.

Lemma 9.47. *Let $\eta \in C^{\infty}(\widetilde{\Omega} \times I)$, and suppose that there exists a compact set $K \subseteq \widetilde{\Omega}$ such that $\eta(\widetilde{x}, y) = 0$ for all $\widetilde{x} \notin K$ and for all $y \in I$. Then, for each $\widetilde{u} \in \mathcal{D}'(\widetilde{\Omega})$, the function $\widetilde{\omega}$ defined by $\widetilde{\omega}(y) := \langle \widetilde{u}, \eta(\cdot, y)\rangle$ for all $y \in I$, belongs to $C^{\infty}(I)$, and for every multi-index α one has*

$$\partial^{\alpha}\widetilde{\omega}(y) = \langle \widetilde{u}, \partial_y^{\alpha}\eta(\cdot, y)\rangle .$$

Lemma 9.48. *Let $\widetilde{\Omega}$ be an open subset of $\mathbb{R}^{n-1}$, and let I be an open interval in $\mathbb{R}$. Suppose that $u \in \mathcal{D}'(\widetilde{\Omega} \times I)$ satisfies $\dfrac{\partial u}{\partial x_n} = 0$. Then there exists $\widetilde{u} \in \mathcal{D}'(\widetilde{\Omega})$ such that*

$$\langle u, \varphi\rangle = \int_I \langle \widetilde{u}, \varphi(\cdot, x_n)\rangle\, dx_n \qquad \forall\, \varphi \in \mathcal{D}(\widetilde{\Omega} \times I),$$

whence u is a distribution independent of x_n.

Proof. Let $\varphi_0 \in C_0^{\infty}(I)$ be such that $\int_I \varphi_0(t)\, dt = 1$, and define $\widetilde{u} : \mathcal{D}(\widetilde{\Omega}) \to \mathbb{C}$ by

$$\langle \widetilde{u}, \widetilde{\phi}\rangle := \langle u, \phi\rangle \quad \forall\, \widetilde{\phi} \in \mathcal{D}(\widetilde{\Omega}),$$

where $\phi(x) = \widetilde{\phi}(\widetilde{x})\,\varphi_0(x_n)$ for all $x := (\widetilde{x}, x_n) \in \widetilde{\Omega} \times I$. It is easy to verify that $\widetilde{u} \in \mathcal{D}'(\widetilde{\Omega})$. Now, given $\varphi \in \mathcal{D}(\widetilde{\Omega} \times I)$ we define

$$\mathbf{J}(\varphi)(\widetilde{x}) = \int_I \varphi(\widetilde{x}, x_n)\, dx_n \quad \forall\, \widetilde{x} \in \widetilde{\Omega},$$

$$\psi(x) = \varphi(x) - \mathbf{J}(\varphi)(\widetilde{x})\,\varphi_0(x_n) \quad \forall\, x := (\widetilde{x}, x_n) \in \widetilde{\Omega} \times I,$$

and observe that $\mathbf{J}(\psi)(\widetilde{x}) = 0$ for all $\widetilde{x} \in \widetilde{\Omega}$. Then, by arguments similar to those in the previous lemma, it follows that there exists $\Psi \in \mathcal{D}(\widetilde{\Omega} \times I)$ such that $\dfrac{\partial \Psi}{\partial x_n} = \psi$, and thus, since $\dfrac{\partial u}{\partial x_n} = 0$, it results that $\langle u, \psi\rangle = 0$. Hence, letting $\mathbf{J}(\varphi)\,\varphi_0 : \widetilde{\Omega} \times I \to \mathbb{R}$ be the function defined as $\big(\mathbf{J}(\varphi)\,\varphi_0\big)(\widetilde{x}, x_n) := \mathbf{J}(\varphi)(\widetilde{x})\,\varphi_0(x_n)$ for all $(\widetilde{x}, x_n) \in \widetilde{\Omega} \times I$, we find that

$$\langle u, \varphi\rangle = \langle u, \mathbf{J}(\varphi)\,\varphi_0\rangle = \langle \widetilde{u}, \mathbf{J}(\varphi)\rangle$$

$$= \langle \widetilde{u}, \int_I \varphi(\cdot, x_n)\, dx_n\rangle = \int_I \langle \widetilde{u}, \varphi(\cdot, x_n)\rangle\, dx_n ,$$

where the last equality follows from applying Lemma 9.47 to the function

$$\eta(\widetilde{x}, x_n) := \int_{-\infty}^{x_n} \varphi(\widetilde{x}, t)\, dt .$$

Indeed, with this function η it follows that

$$\frac{\partial}{\partial x_n}\langle \widetilde{u}, \eta(\cdot, x_n)\rangle = \langle \widetilde{u}, \frac{\partial}{\partial x_n}\eta(\cdot, x_n)\rangle,$$

that is,

$$\frac{\partial}{\partial x_n}\langle \widetilde{u}, \eta(\cdot, x_n)\rangle = \langle \widetilde{u}, \varphi(\cdot, x_n)\rangle,$$

and, integrating with respect to x_n in I, we conclude that

$$\langle \widetilde{u}, \int_I \varphi(\cdot, x_n)\, dx_n\rangle = \int_I \langle \widetilde{u}, \varphi(\cdot, x_n)\rangle\, dx_n .$$

□

An important corollary of the previous lemma is given by the following result.

Lemma 9.49. *Let Ω be an open subset of $\mathbb{R}^n$, and let $u,\ f \in C(\Omega)$ be such that $\frac{\partial u}{\partial x_j} = f$ in $\mathcal{D}'(\Omega)$ for a given $j \in \{1, \ldots, n\}$. Then $\frac{\partial u}{\partial x_j}$ exists in the classical sense in Ω and is equal to f.*

Proof. Without loss of generality, we may assume that $j = n$ and that $\Omega = \widetilde{\Omega} \times I$, where $\widetilde{\Omega}$ and I are as in Lemma 9.48. Let $t_0 \in I$ be fixed, and define

$$v(x) := \int_{t_0}^{x_n} f(\widetilde{x}, t)\, dt \quad \forall\, x := (\widetilde{x}, x_n) := (x_1, \ldots, x_{n-1}, x_n) \in \widetilde{\Omega} \times I\,.$$

Then, it is clear that $\dfrac{\partial v}{\partial x_n} = f$ and that $\dfrac{\partial}{\partial x_n}(u - v) = 0$ in the distributional sense. Therefore, according to Lemma 9.48, it follows that there exists $\widetilde{u} \in \mathcal{D}'(\widetilde{\Omega})$ such that

$$\langle u - v, \varphi \rangle = \int_I \langle \widetilde{u}, \varphi(\cdot, x_n) \rangle\, dx_n \qquad \forall\, \varphi \in \mathcal{D}(\widetilde{\Omega} \times I)\,,$$

where

$$\langle \widetilde{u}, \widetilde{\phi} \rangle = \langle u - v, \phi \rangle \qquad \forall\, \widetilde{\phi} \in \mathcal{D}(\widetilde{\Omega})\,,$$

with $\phi(\widetilde{x}, x_n) = \widetilde{\phi}(\widetilde{x})\, \varphi_0(x_n)$. Recall here, according to the proof of Lemma 9.48, that φ_0 is a function in $C_0^\infty(I)$ such that $\displaystyle\int_I \varphi_0(t)\, dt = 1$. It follows that

$$\langle u - v, \varphi \rangle = \int_I \langle u - v, \widetilde{\varphi} \rangle\, dx_n\,,$$

where $\widetilde{\varphi}(\widetilde{x}, t) = \varphi(\widetilde{x}, x_n)\, \varphi_0(t)$. Then, by applying Fubini's and Tonelli's Theorems (cf. [120, Theorems 6.1 and 6.10]), we obtain

$$\begin{aligned}
\langle u - v, \varphi \rangle &= \int_I \left\{ \int_{\widetilde{\Omega}} \int_I \Big\{ u(\widetilde{x}, t) - v(\widetilde{x}, t) \Big\}\, \varphi(\widetilde{x}, x_n)\, \varphi_0(t)\, d\widetilde{x}\, dt \right\} dx_n \\
&= \int_{\widetilde{\Omega}} \int_I \left\{ \int_I \Big\{ u(\widetilde{x}, t) - v(\widetilde{x}, t) \Big\}\, \varphi_0(t)\, dt \right\} \varphi(\widetilde{x}, x_n)\, d\widetilde{x}\, dx_n,
\end{aligned}$$

which implies that $\langle u - v, \varphi \rangle = \langle u_0, \varphi \rangle$ for all $\varphi \in \mathcal{D}(\widetilde{\Omega} \times I)$, where $u_0 \in C(\widetilde{\Omega})$ is given by

$$u_0(x) = u_0(\widetilde{x}) := \int_I \Big\{ u(\widetilde{x}, t) - v(\widetilde{x}, t) \Big\}\, \varphi_0\,(t)\, dt \quad \forall\, x \in \widetilde{\Omega}\,.$$

Thus, $u(x) - v(x) = u_0(\widetilde{x})$, and therefore, pointwise there holds

$$\frac{\partial u}{\partial x_n}(x) = \frac{\partial v}{\partial x_n}(x) + \frac{\partial}{\partial x_n}\, u_0(\widetilde{x}) = \frac{\partial v}{\partial x_n}(x) = f(x) \quad \forall\, x \in \widetilde{\Omega}\,.$$

□

9.8.2 Product of a distribution and a smooth function

Let Ω be an open subset of $\mathbb{R}^n$, and let u, $f \in C(\Omega)$. Then the product $fu \in C(\Omega)$ and clearly fu induces the distribution

$$\langle fu, \varphi\rangle = \int_\Omega fu\,\varphi\,dx = \int_\Omega u\,(f\varphi)\,dx \quad \forall\,\varphi \in \mathcal{D}(\Omega).$$

Now, if $f \in C^\infty(\Omega)$, then $f\varphi \in C_0^\infty(\Omega)$ and $\mathrm{supp}(f\varphi) \subseteq \mathrm{supp}(\varphi)$ for all $\varphi \in \mathcal{D}(\Omega)$. In this case, the previous equation can be rewritten as

$$\langle fu, \varphi\rangle = \langle u, f\varphi\rangle \quad \forall\,\varphi \in \mathcal{D}(\Omega). \tag{9.26}$$

Moreover, using the Leibniz formula, it is easy to see that, given a compact set $K \subseteq \Omega$, for each $N \in \mathbb{N}$ there holds

$$\sum_{|\alpha|\leqslant N} \sup_{x\in K} |\partial^\alpha(f\varphi)(x)| \leqslant C_N \sum_{|\alpha|\leqslant N} \sup_{x\in K} |\partial^\alpha\varphi(x)| \quad \forall\,\varphi \in \mathcal{D}_K(\Omega),$$

where C_N is a positive constant depending on K and f. This fact and (9.26) prove that if $u \in \mathcal{D}'(\Omega)$, then the mapping $\mathcal{D}(\Omega) \ni \varphi \to \langle u, f\varphi\rangle \in \mathbb{C}$ is a distribution, which leads to the following definition.

Definition 9.50. *Let $u \in \mathcal{D}'(\Omega)$, and let $f \in C^\infty(\Omega)$. Then the distribution defined by*

$$\langle fu, \varphi\rangle := \langle u, f\varphi\rangle \qquad \forall\,\varphi \in \mathcal{D}(\Omega),$$

is called the product of f and u, or the product of u and f when defining likewise

$$\langle uf, \varphi\rangle := \langle u, f\varphi\rangle \qquad \forall\,\varphi \in \mathcal{D}(\Omega).$$

The foregoing definition is illustrated in the following example by combining it with the distributional derivative concept. To this end, we now introduce the space

$$C^\infty(\bar\Omega) := \Big\{ v|_\Omega : \quad v \in C_0^\infty(\mathbb{R}^n)\Big\}. \tag{9.27}$$

Example 9.51 Let Ω be an open set of $\mathbb{R}^n$, and let initially $\mathbf{v} := (v_1, \ldots, v_n) \in \mathbf{L}^1_{loc}(\Omega)$. Then, bearing in mind Definition 1.47, we observe that for each $\varphi \in C_0^\infty(\Omega)$ there holds

$$\langle \mathrm{div}(\mathbf{v}), \varphi\rangle = \sum_{i=1}^n \Big\langle \frac{\partial v_i}{\partial x_i}, \varphi\Big\rangle = -\sum_{i=1}^n \Big\langle v_i, \frac{\partial \varphi}{\partial x_i}\Big\rangle = -\int_\Omega \mathbf{v}\cdot\nabla\varphi. \tag{9.28}$$

In particular, taking $\mathbf{v} := \phi\,\mathbf{u}$, with $\phi \in C^\infty(\bar\Omega) \subseteq C^\infty(\Omega)$ and $\mathbf{u} := (u_1, \ldots, u_n) \in \mathbf{L}^1_{loc}(\Omega)$, it is clear that $\mathbf{v}$ belongs to $\mathbf{L}^1_{loc}(\Omega)$ as well, whence, employing now Definition 9.50, and assuming additionally that $\mathrm{div}(\mathbf{u}) = 0$ in $\mathcal{D}'(\Omega)$, we deduce that

$$\begin{aligned} -\int_\Omega \phi\,\mathbf{u}\cdot\nabla\varphi &= -\sum_{i=1}^n \Big\langle u_i, \phi\,\frac{\partial\varphi}{\partial x_i}\Big\rangle = -\sum_{i=1}^n \Big\langle u_i, \frac{\partial}{\partial x_i}(\phi\,\varphi) - \varphi\,\frac{\partial\phi}{\partial x_i}\Big\rangle \\ &= \langle \mathrm{div}(\mathbf{u}), \phi\,\varphi\rangle + \sum_{i=1}^n \Big\langle u_i, \varphi\,\frac{\partial\phi}{\partial x_i}\Big\rangle = \Big\langle \sum_{i=1}^n u_i\,\frac{\partial\phi}{\partial x_i}, \varphi\Big\rangle, \end{aligned}$$

that is, according to (9.28),

$$\langle \mathrm{div}(\phi\,\mathbf{u}), \varphi\rangle = \langle \mathbf{u}\cdot\nabla\phi, \varphi\rangle \qquad \forall\,\varphi \in C_0^\infty(\Omega). \tag{9.29}$$

Now, given $\phi \in \mathrm{H}^1(\Omega)$, we know from the density of $C^\infty(\bar{\Omega})$ in $\mathrm{H}^1(\Omega)$ (cf. Theorem 10.14) that there exists a sequence $\{\phi_k\}_{k\in\mathbb{N}} \subseteq C^\infty(\bar{\Omega})$ such that $\|\phi - \phi_k\|_{1,\Omega} \overset{n\to\infty}{\to} 0$. In this way, employing (9.28), (9.29), and the fact that $\mathbf{u} \cdot \nabla\phi_k \in \mathrm{L}^1_{loc}(\Omega)$, we deduce that

$$- \int_\Omega \phi_k \, \mathbf{u} \cdot \nabla\varphi = \int_\Omega \mathbf{u} \cdot \nabla\phi_k \, \varphi \quad \forall\, k \in \mathbb{N}, \quad \forall\, \varphi \in C_0^\infty(\Omega). \tag{9.30}$$

Moreover, assuming in what follows that $\mathbf{u} \in \mathbf{L}^2(\Omega)$, and applying Cauchy-Schwarz's inequality, we easily find that

$$\left| \int_\Omega (\phi - \phi_k)\, \mathbf{u} \cdot \nabla\varphi \right| \leqslant \|\phi - \phi_k\|_{0,\Omega} \, \|\mathbf{u}\|_{0,\Omega} \, \|\nabla\varphi\|_{\infty,\Omega} \overset{k\to+\infty}{\to} 0,$$

and similarly

$$\left| \int_\Omega \mathbf{u} \cdot (\nabla\phi - \nabla\phi_k)\, \varphi \right| \leqslant |\phi - \phi_k|_{1,\Omega} \, \|\mathbf{u}\|_{0,\Omega} \, \|\varphi\|_{\infty,\Omega} \overset{k\to+\infty}{\to} 0,$$

so that taking the limit as $k \to \infty$ in (9.30), we conclude that

$$- \int_\Omega \phi\, \mathbf{u} \cdot \nabla\varphi = \int_\Omega \mathbf{u} \cdot \nabla\phi\, \varphi \qquad \forall\, \varphi \in C_0^\infty(\Omega). \tag{9.31}$$

Consequently, noting by Cauchy-Schwarz's inequality that $\phi\,\mathbf{u} \in \mathbf{L}^1(\Omega)$ and $\mathbf{u} \cdot \nabla\phi \in \mathrm{L}^1(\Omega)$ when $\phi \in \mathrm{H}^1(\Omega)$, which certainly yields $\phi\,\mathbf{u} \in \mathbf{L}^1_{loc}(\Omega)$ and $\mathbf{u} \cdot \nabla\phi \in \mathrm{L}^1_{loc}(\Omega)$, respectively, and using again (9.28), the foregoing identity becomes

$$\langle \operatorname{div}(\phi\,\mathbf{u}), \varphi \rangle = \langle \mathbf{u} \cdot \nabla\phi, \varphi \rangle \quad \forall\, \varphi \in C_0^\infty(\Omega), \tag{9.32}$$

thus showing that

$$\operatorname{div}(\phi\,\mathbf{u}) = \mathbf{u} \cdot \nabla\phi \quad \text{in} \quad \mathcal{D}'(\Omega) \quad \forall\, \mathbf{u} \in \mathbf{L}^2(\Omega), \quad \forall\, \phi \in \mathrm{H}^1(\Omega). \tag{9.33}$$

Furthermore, assume now that $\phi \in \mathrm{H}^1_0(\Omega)$, and let us keep $\mathbf{u}$ in $\mathbf{L}^2(\Omega)$. Then, thanks to the density of $C_0^\infty(\Omega)$ in $\mathrm{H}^1_0(\Omega)$, which actually defines this latter space, there exists a sequence $\{\phi_k\}_{k\in\mathbb{N}} \subseteq C_0^\infty(\Omega)$ such that $\|\phi - \phi_k\|_{1,\Omega} \overset{n\to\infty}{\to} 0$. Thus, using again Cauchy-Schwarz's inequality along with (9.28) and (9.33), we find that for each $\varphi \in C_0^\infty(\Omega)$ there holds

$$\begin{aligned} \int_\Omega \varphi\, \mathbf{u} \cdot \nabla\phi &= \lim_{k\to\infty} \int_\Omega \varphi\, \mathbf{u} \cdot \nabla\phi_k = - \lim_{k\to\infty} \langle \operatorname{div}(\varphi\,\mathbf{u}), \phi_k \rangle \\ &= - \lim_{k\to\infty} \int_\Omega \mathbf{u} \cdot \nabla\varphi\, \phi_k = - \int_\Omega \mathbf{u} \cdot \nabla\varphi\, \phi, \end{aligned}$$

that is

$$\int_\Omega \varphi\, \mathbf{u} \cdot \nabla\phi = - \int_\Omega \phi\, \mathbf{u} \cdot \nabla\varphi \qquad \forall\, \phi \in \mathrm{H}^1_0(\Omega), \quad \forall\, \varphi \in C_0^\infty(\Omega).$$

Moreover, extending the above by density again to $\varphi \in \mathrm{H}^1_0(\Omega)$, we arrive at

$$\int_\Omega \varphi\, \mathbf{u} \cdot \nabla\phi = - \int_\Omega \phi\, \mathbf{u} \cdot \nabla\varphi \qquad \forall\, \mathbf{u} \in \mathbf{L}^2(\Omega), \quad \forall\, \phi,\, \varphi \in \mathrm{H}^1_0(\Omega), \tag{9.34}$$

which constitutes the skew-symmetric property already employed in Example 4.34.

We now aim to define the product of two distributions $u,\, v \in \mathcal{D}'(\Omega)$, for which we previously introduce the following definition.

Definition 9.52. *Given $u \in \mathcal{D}'(\Omega)$, the* SINGULAR SUPPORT *of u, denoted by* $\operatorname{sing\,supp}(u)$, *is defined as the set of points $x \in \Omega$ that do not have a neighborhood in which the restriction of u is a C^∞ function.*

The above definition implies, by virtue of Theorem 9.24 (cf. Section 9.4), that the restriction of u to $\Omega\backslash\operatorname{sing\,supp}(u)$ is a C^∞ function. Hence, if u, $v \in \mathcal{D}'(\Omega)$ are such that

$$\operatorname{sing\,supp}(u) \cap \operatorname{sing\,supp}(v) = \varnothing,$$

we can define the product $uv \in \mathcal{D}'(\Omega)$ in the following way. Let $\{\Omega_i\}_{i\in I}$ be a family of open sets contained in Ω such that $\Omega = \bigcup_{i\in I} \Omega_i$, and such that for each $i \in I$, either $u|_{\Omega_i} \in C^\infty(\Omega_i)$ or $v|_{\Omega_i} \in C^\infty(\Omega_i)$. Then, for each $\varphi \in C_0^\infty(\Omega_i)$ we define

$$\langle (uv)_i, \varphi\rangle := \begin{cases} \langle u, v\varphi\rangle & \text{if } \; v \in C^\infty(\Omega_i), \\ \langle v, u\varphi\rangle & \text{if } \; u \in C^\infty(\Omega_i). \end{cases}$$

In the case that both $u|_{\Omega_i}$ and $v|_{\Omega_i}$ belong to $C^\infty(\Omega_i)$, the two definitions coincide. Thus, one has a well-defined product $(uv)_i \in \mathcal{D}'(\Omega_i)$. Finally, by the Localization Theorem (cf. Theorem 9.24), it follows that there exists a unique $uv \in \mathcal{D}'(\Omega)$ whose restriction to Ω_i coincides with $(uv)_i$ for all $i \in I$.

Let us now prove that the Leibniz formula is also valid for the product of a function $f \in C^\infty(\Omega)$ and a distribution $u \in \mathcal{D}'(\Omega)$. To this end, we first recall from Definition 1.34 that, given two n-tuples α, β, we say that $\alpha \leqslant \beta$ if $\alpha_i \leqslant \beta_i$ for all $i \in \{1, \ldots, n\}$.

Theorem 9.53. *Let Ω be an open subset of $\mathbb{R}^n$, and let $f \in C^\infty(\Omega)$ and $u \in \mathcal{D}'(\Omega)$. Then, for each multi-index α one has*

$$\partial^\alpha(fu) = \sum_{\beta\leqslant\alpha} \binom{\alpha}{\beta} \partial^\beta f\, \partial^{\alpha-\beta} u\,.$$

Proof. For $|\alpha| = 0$ the result is obvious. For $|\alpha| = 1$, one obtains

$$\begin{aligned}
\left\langle \frac{\partial}{\partial x_i}(fu), \varphi \right\rangle &= -\left\langle fu, \frac{\partial \varphi}{\partial x_i}\right\rangle = -\left\langle u, f\frac{\partial \varphi}{\partial x_i}\right\rangle \\
&= -\left\langle u, \frac{\partial}{\partial x_i}(f\varphi) - \frac{\partial f}{\partial x_i}\varphi \right\rangle \\
&= -\left\langle u, \frac{\partial}{\partial x_i}(f\varphi) \right\rangle + \left\langle u, \frac{\partial f}{\partial x_i}\varphi \right\rangle \\
&= \left\langle \frac{\partial u}{\partial x_i}, f\varphi \right\rangle + \left\langle \frac{\partial f}{\partial x_i} u, \varphi \right\rangle \\
&= \left\langle f\frac{\partial u}{\partial x_i} \varphi \right\rangle + \left\langle \frac{\partial f}{\partial x_i} u, \varphi \right\rangle \quad \forall \varphi \in \mathcal{D}(\Omega),
\end{aligned}$$

that is,

$$\frac{\partial}{\partial x_i}(fu) = f\frac{\partial u}{\partial x_i} + \frac{\partial f}{\partial x_i} u \quad \text{in} \quad \mathcal{D}'(\Omega) \quad \forall\, i \in \{1, \cdots, n\}.$$

It follows easily by induction that

$$\partial^\alpha(fu) = \sum_{\beta\leqslant\alpha} C_\beta^\alpha\, \partial^\beta f\, \partial^{\alpha-\beta} u \quad \text{with} \quad C_\beta^\alpha > 0 \quad \forall \alpha, \forall \beta.$$

Let us now show that $C^\alpha_\beta = \binom{\alpha}{\beta}$. To this end, consider the functions $f(x) = e^{x\cdot\xi}$ and $u(x) = e^{x\cdot\eta}$, with $\xi, \eta \in \mathbb{R}^n$, and observe that

$$\partial^\alpha_x \left(e^{x\cdot(\xi+\eta)}\right) = \sum_{\beta\leqslant\alpha} C^\alpha_\beta \, \partial^\beta_x (e^{x\cdot\xi}) \, \partial^{\alpha-\beta}_x (e^{x\cdot\eta}),$$

that is,

$$(\xi+\eta)^\alpha \, e^{x\cdot(\xi+\eta)} = \sum_{\beta\leqslant\alpha} C^\alpha_\beta \, \xi^\beta \, e^{x\cdot\xi} \, \eta^{\alpha-\beta} \, e^{x\cdot\eta} = e^{x\cdot(\xi+\eta)} \sum_{\beta\leqslant\alpha} C^\alpha_\beta \, \xi^\beta \, \eta^{\alpha-\beta},$$

hence

$$(\xi+\eta)^\alpha = \sum_{\beta\leqslant\alpha} C^\alpha_\beta \, \xi^\beta \, \eta^{\alpha-\beta} \quad \forall\, \xi, \eta \in \mathbb{R}^n . \tag{9.35}$$

On the other hand, recall that the Taylor formula with integral remainder states that

$$g(x+y) = \sum_{|\beta|\leqslant k-1} \partial^\beta g(x) \frac{y^\beta}{\beta!} + k \int_0^1 (1-t)^{k-1} \sum_{|\beta|=k} \partial^\beta g(x+t\,y) \frac{y^\beta}{\beta!} \, dt \quad \forall\, g \in C^k(\mathbb{R}^n).$$

Then, applying this formula to $g(x) = x^\alpha$ and $k = |\alpha|+1$, we obtain

$$(x+y)^\alpha = \sum_{|\beta|\leqslant\alpha} \frac{\alpha!\, x^{\alpha-\beta}}{(\alpha-\beta)!\beta!} y^\beta = \sum_{\beta\leqslant\alpha} \binom{\alpha}{\beta} x^{\alpha-\beta} y^\beta \quad \forall\, x, y \in \mathbb{R}^n ,$$

from which it follows, by comparison with (9.35), that $C^\alpha_\beta = \binom{\alpha}{\beta}$.

□

Definition 9.54. *Let $\{a_\alpha\}_{|\alpha|\leqslant m}$ be a finite set of functions in $C^\infty(\Omega)$ such that a_α is nonzero for at least one multi-index α with $|\alpha| = m$. Then*

$$P(x,\partial) := \sum_{|\alpha|\leqslant m} a_\alpha(x)\, \partial^\alpha$$

is called a LINEAR DIFFERENTIAL OPERATOR *of order m on Ω, with C^∞ coefficients. In fact, P must be understood as the mapping*

$$\begin{aligned} P : \mathcal{D}'(\Omega) &\to \mathcal{D}'(\Omega) \\ v \to P(v) &= \sum_{|\alpha|\leqslant m} a_\alpha \, \partial^\alpha v , \end{aligned}$$

which is sequentially continuous Moreover, for each $\varphi \in C^\infty_0(\Omega)$ one obtains

$$\begin{aligned} \langle P(v), \varphi\rangle &= \sum_{|\alpha|\leqslant m} \langle a_\alpha \, \partial^\alpha v, \varphi\rangle = \sum_{|\alpha|\leqslant m} \langle \partial^\alpha v, a_\alpha \, \varphi\rangle \\ &= \sum_{|\alpha|\leqslant m} (-1)^{|\alpha|} \langle v, \partial^\alpha (a_\alpha\varphi)\rangle = \langle v, P^{\mathrm{t}}(\varphi)\rangle \end{aligned}$$

where $P^{\mathrm{t}}(\varphi) := \sum_{|\alpha|\leqslant m} (-1)^{|\alpha|} \partial^\alpha (a_\alpha\varphi)$ is called the ADJOINT OPERATOR *of P.*

Theorem 9.55. *Let Ω be an open subset of $\mathbb{R}$, and consider the differential operator defined by $P(v) := v' + av$ for all $v \in \mathcal{D}'(\Omega)$, with $a \in C^\infty(\Omega)$. If $u \in \mathcal{D}'(\Omega)$ and $P(u) = f \in C(\Omega)$, then $u \in C^1(\Omega)$, which implies that $u' + au = f$ in the classical sense.*

Proof. Assume first that $a \equiv 0$, that is, $P(v) = v'$ for all $v \in \mathcal{D}'(\Omega)$. Then, since f has an antiderivative $v \in C^1(\Omega)$ and $(u - v)' = u' - v' = 0$ in the distributional sense, it follows from Theorem 9.46 that $u - v = C$ (a constant), and therefore $u = v + C \in C^1(\Omega)$. Now, for $a \neq 0$, let E be the function defined by $E(x) = \exp\left\{ \int_{x_0}^{x} a(t)\, dt \right\}$ for all $x \in \Omega$, with fixed $x_0 \in \Omega$. It is clear that $E \in C^\infty(\Omega)$ and that $E' = E\, a$, so that, using the Leibniz formula, one deduces that

$$(Eu)' = E'u + Eu' = Eau + Eu' = E(u' + au) = E\big(P(u)\big) = Ef \in C(\Omega)$$

in the distributional sense. Thus, in any case one obtains $Eu \in C^1(\Omega)$, and therefore $u \in C^1(\Omega)$.

□

Note that the previous theorem is also valid if $u = (u_1, \cdots, u_k)$, $f = (f_1, \cdots, f_k)$, and a is a $k \times k$ matrix with C^∞ coefficients. Moreover, since ordinary differential equations of order > 1 can be reduced to first-order systems, one obtains the following more general result.

Theorem 9.56. *Let Ω be an open subset of $\mathbb{R}$, and let $u \in \mathcal{D}'(\Omega)$ be such that*

$$u^{(m)} + a_{m-1}\, u^{(m-1)} + \cdots + a_1\, u' + a_0\, u = f \in C(\Omega),$$

where $a_j \in C^\infty(\Omega)$ for all $j \in \{0, 1, \cdots, m-1\}$. Then $u \in C^m(\Omega)$, which means that the ordinary differential equation is satisfied in the classical sense.

Next, we introduce one of the most important concepts related to distributional derivatives.

Definition 9.57. *Let $P = \sum_{|\alpha| \leqslant m} a_\alpha\, \partial^\alpha$ be a differential operator with constant coefficients a_α. We say that $E \in \mathcal{D}'(\mathbb{R}^n)$ is a* FUNDAMENTAL SOLUTION *of P if it satisfies $P(E) = \delta$.*

Fundamental solutions play a key role in the study of existence and regularity properties of solutions to differential equations. These aspects will be addressed in the next two sections when introducing the concept of convolution of distributions and discussing the role of such solutions. In particular, it will be proved in Section 9.8.4 that, given $f \in \mathcal{E}'(\mathbb{R}^n)$, the function $u := E * f$ constitutes a solution of the differential equation $P(u) = f$.

Now, we give some examples of fundamental solutions, through the following theorems.

Theorem 9.58. *Let $P := \sum_{j=1}^{n} \frac{\partial^2}{\partial x_j^2}$ be the Laplace operator Δ, and define*

$$E(x) = \begin{cases} (2\,\pi)^{-1} \log|x| & \text{if} \quad x \in \mathbb{R}^2 \backslash \{0\}, \\ -\dfrac{|x|^{2-n}}{(n-2)\, c_n} & \text{if} \quad x \in \mathbb{R}^n \backslash \{0\},\ n \geqslant 3, \end{cases}$$

where c_n is the area of the unit sphere. Then E is a fundamental solution of P, that is, $P(E) = \delta$.

Proof. Let $E(x) = \log|x|$ for all $x \in \mathbb{R}^2$. We must prove that $\Delta E = 2\pi\,\delta$, for which we first need to verify that $E \in \mathcal{D}'(\mathbb{R}^2)$. Indeed, using polar coordinates (r, ϑ) we obtain

$$\int_{|x|\leqslant 1} |E(x)|\,dx = \int_0^{2\pi}\int_0^1 |\ln(r)|\,r\,dr\,d\vartheta = -2\pi\int_0^1 r\ln(r)\,dr = \frac{\pi}{2},$$

which proves that E is locally integrable in $\mathbb{R}^2$. It follows that

$$\langle \Delta E, \varphi\rangle = \langle E, \Delta\varphi\rangle \quad \forall\,\varphi \in \mathcal{D}(\mathbb{R}^2),$$

that is,

$$\langle \Delta E, \varphi\rangle = \int_{\mathbb{R}^2} E(x)\,\Delta\varphi(x)\,dx \quad \forall\,\varphi \in \mathcal{D}(\mathbb{R}^2).$$

Now, for $\varepsilon > 0$, define

$$I_\varepsilon(\varphi) = \int_{|x|\geqslant\varepsilon} E(x)\,\Delta\varphi(x)\,dx = \int_{\mathbb{R}^2}\Big\{1 - \mathbf{1}_{B(0,\varepsilon)}\Big\}\,E(x)\,\Delta\varphi(x)\,dx,$$

where $\mathbf{1}_{B(0,\varepsilon)}$ denotes the characteristic function of $B(0,\varepsilon)$. Since

$$|\{1 - \mathbf{1}_{B(0,\varepsilon)}\}\,E(x)\,\Delta\varphi(x)| \leqslant |E(x)\,\Delta\varphi(x)|$$

and $E\,\Delta\varphi \in \mathrm{L}^1_{\mathrm{loc}}(\mathbb{R}^2)$, it follows by the Lebesgue Dominated Convergence Theorem (cf. [113, Theorem 1.34], [120, Theorem 5.36]) that

$$\lim_{\varepsilon\to 0} I_\varepsilon(\varphi) = \int_{\mathbb{R}^2} E(x)\,\Delta\varphi(x)\,dx \qquad \forall\,\varphi \in \mathcal{D}(\mathbb{R}^2),$$

that is

$$\langle \Delta E, \varphi\rangle = \lim_{\varepsilon\to 0} I_\varepsilon(\varphi) \quad \forall\,\varphi \in \mathcal{D}(\mathbb{R}^2). \tag{9.36}$$

Next, we compute the value of the limit in (9.36). To that end, given $\varphi \in \mathcal{D}(\mathbb{R}^2)$, consider $R > 0$ such that $\operatorname{supp}(\varphi) \subseteq B(0,R)$. Then, applying Green's formula in the annular domain $\Omega_\varepsilon := \Big\{x \in \mathbb{R}^2 : \quad \varepsilon < |x| < R\Big\}$ (see Figure 9.5), and using that φ vanishes in a neighborhood of $\partial B(0,R)$, we obtain

$$I_\varepsilon(\varphi) := \int_{\Omega_\varepsilon} E(x)\,\Delta\varphi(x)\,dx = \int_{\Omega_\varepsilon} \Delta E(x)\,\varphi(x)\,dx + \int_{|x|=\varepsilon}\left\{E(x)\,\frac{\partial\varphi}{\partial\boldsymbol{\nu}} - \varphi\,\frac{\partial E}{\partial\boldsymbol{\nu}}\right\}ds,$$

where $\boldsymbol{\nu}$ is the inward normal vector to $\partial B(0,\varepsilon)$. Elementary calculations show that

$$\frac{\partial E}{\partial x_j} = -\frac{x_j}{|x|^2}, \quad \frac{\partial^2 E}{\partial x_1^2} = \frac{x_1^2 - x_2^2}{|x|^4}, \quad \text{and} \quad \frac{\partial^2 E}{\partial x_2^2} = \frac{x_2^2 - x_1^2}{|x|^4} \quad \forall\,x \neq 0,$$

from which it follows that $\Delta E = 0$ for all $x \neq 0$, and therefore

$$I_\varepsilon(\varphi) = \int_{|x|=\varepsilon}\left\{E(x)\,\frac{\partial\varphi}{\partial\boldsymbol{\nu}} - \varphi\,\frac{\partial E}{\partial\boldsymbol{\nu}}\right\}ds.$$

Now, on the circle $|x| = \varepsilon$ we clearly have $\boldsymbol{\nu}(x) = -\dfrac{x}{\varepsilon}$, so that the previous expression reduces to

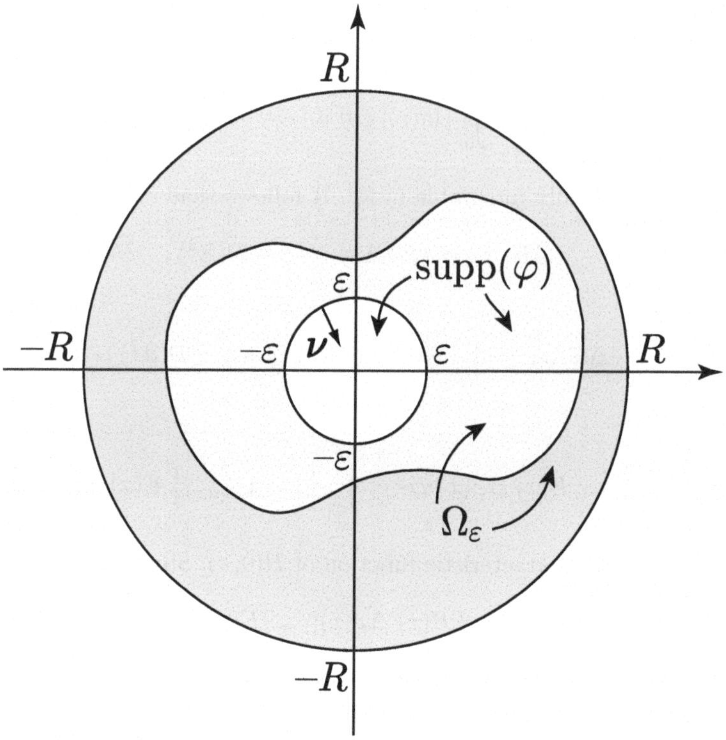

Fig. 9.5. *Geometry for Green's formula*

$$I_\varepsilon(\varphi) \;=\; \frac{1}{\varepsilon}\int_{|x|=\varepsilon} \varphi(x)\,\nabla E(x)\cdot x\,ds \;-\; \frac{1}{\varepsilon}\int_{|x|=\varepsilon} E(x)\,\nabla\varphi\cdot x\,ds\,.$$

Moreover, since $\nabla E(x) \;=\; -\dfrac{x}{|x|^2}$ we have $\nabla E(x)\cdot x \;=\; -1$, and thus

$$\frac{1}{\varepsilon}\int_{|x|=\varepsilon} \varphi(x)\,\nabla\,E(x)\cdot x\,ds \;=\; -\frac{1}{\varepsilon}\int_{|x|=\varepsilon} \varphi(x)\,ds$$

$$=\; \frac{1}{\varepsilon}\int_0^{2\pi} \varphi\big(\varepsilon\cos(\vartheta),\varepsilon\sin(\vartheta)\big)\,\varepsilon\,d\vartheta \;=\; \int_0^{2\pi} \varphi\big(\varepsilon\cos(\vartheta),\varepsilon\sin(\vartheta)\big)\,d\vartheta \;\overset{\varepsilon\to 0}{\rightarrow}\; 2\pi\,\varphi(0)\,.$$

In turn,

$$\left|\frac{1}{\varepsilon}\int_{|x|=\varepsilon} E(x)\,\nabla\varphi\cdot x\,ds\right| \;=\; \left|\frac{1}{2\pi\varepsilon}\,\log(\varepsilon)\int_{|x|=\varepsilon} \nabla\varphi\cdot x\,ds\right|$$

$$\leqslant\; \frac{1}{2\pi\varepsilon}\,|\log(\varepsilon)|\int_{|x|=\varepsilon} \varepsilon\,ds\;\sup_{x\in\mathbb{R}^2}\|\nabla\varphi(x)\| \;=\; c\,\varepsilon\,|\log(\varepsilon)| \;\overset{\varepsilon\to 0}{\rightarrow}\; 0\,.$$

From the previous computations it follows that

$$\lim_{\varepsilon\to 0} I_\varepsilon(\varphi) \;=\; 2\pi\varphi(0)\,,$$

and consequently $\Delta E \;=\; 2\pi\delta$. The case $n \geqslant 3$ is similar and is left as an exercise.

□

Theorem 9.59. *Let* $P := \frac{\partial}{\partial t} - \Delta = \frac{\partial}{\partial t} - \sum_{j=1}^{n} \frac{\partial^2}{\partial x_j^2}$ *be the heat operator in* $\mathbb{R}^n \times \mathbb{R}$*, and define*

$$E(x,t) = \begin{cases} (4\pi t)^{-n/2} \exp(-|x|^2/4t) & \textit{if} \quad t > 0\,, \\ 0 & \textit{if} \quad t \leqslant 0\,. \end{cases}$$

Then $P(E) = \delta$ *in* $\mathcal{D}'\big(\mathbb{R}^n \times \mathbb{R}\big)$.

Proof. For simplicity, and since the procedure for $n \geqslant 2$ is similar, let us consider only the case $n = 1$. Then, we have

$$E(x,t) = \frac{H(t)\, e^{-x^2/4t}}{2\sqrt{\pi t}} \leqslant \frac{H(t)}{2\sqrt{\pi t}} \quad \forall\, x \in \mathbb{R}, \quad \forall\, t > 0\,,$$

where H is the Heaviside function

$$H(t) = \begin{cases} 1 & \text{if} \quad t > 0\,, \\ 0 & \text{if} \quad t \leqslant 0\,. \end{cases}$$

Since $\frac{H(t)}{2\sqrt{\pi t}}$ is locally integrable in $\mathbb{R}^2$, it follows that $E \in \mathrm{L}^1_{loc}(\mathbb{R}^2)$, and thus $E \in \mathcal{D}'(\mathbb{R}^2)$. Now, let $\varphi \in \mathcal{D}(\mathbb{R}^2)$, and let $\mathtt{r} > 0$ be such that $\mathrm{supp}(\varphi) \subseteq \Big\{(x,t) : \quad |x|,\, |t| < \mathtt{r}\Big\}$. We must prove that

$$\langle P(E), \varphi\rangle := -\langle E, \frac{\partial \varphi}{\partial t} + \frac{\partial^2 \varphi}{\partial x^2}\rangle = \langle \delta, \varphi\rangle = \varphi(0)\,.$$

Indeed, since

$$|G_\varepsilon(x,t)| := \left| \mathbf{1}_{\mathbb{R}\times[\varepsilon,+\infty)} \frac{e^{-x^2/4t}}{2\sqrt{\pi t}} \frac{\partial \varphi}{\partial t}(x,t) \right| \leqslant \frac{C\left|\frac{\partial \varphi}{\partial t}(x,t)\right|}{\sqrt{t}} := G(x,t)$$

and

$$|H_\varepsilon(x,t)| := \left| \mathbf{1}_{\mathbb{R}\times[\varepsilon,+\infty)} \frac{e^{-x^2/4t}}{2\sqrt{\pi t}} \frac{\partial^2 \varphi}{\partial x^2}(x,t) \right| \leqslant \frac{C\left|\frac{\partial^2 \varphi}{\partial x^2}(x,t)\right|}{\sqrt{t}} := H(x,t)\,,$$

where $G,\, H \in \mathrm{L}^1_{loc}(\mathbb{R}^2)$, we may apply Fubini's and Tonelli's Theorems (cf. [120, Theorems 6.1 and 6.10]) and the Lebesgue Dominated Convergence Theorem (cf. [113, Theorem 1.34], [120, Theorem 5.36]) to deduce that

$$\langle E, \frac{\partial \varphi}{\partial t} + \frac{\partial^2 \varphi}{\partial x^2}\rangle = \lim_{\varepsilon \to 0} \Big\{ I_\varepsilon(\varphi) + J_\varepsilon(\varphi) \Big\}\,, \tag{9.37}$$

where

$$I_\varepsilon(\varphi) := \int_{-\infty}^{\infty} \left\{ \int_{\varepsilon}^{+\infty} \frac{e^{-x^2/4t}}{2\sqrt{\pi t}} \frac{\partial \varphi}{\partial t}\, dt \right\} dx$$

and

$$J_\varepsilon(\varphi) := \int_{\varepsilon}^{+\infty} \left\{ \int_{-\infty}^{\infty} \frac{e^{-x^2/4t}}{2\sqrt{\pi t}} \frac{\partial^2 \varphi}{\partial x^2}\, dx \right\} dt\,.$$

Then, integrating by parts, and recalling that $\mathrm{supp}(\varphi) \subseteq \Big\{(x,t): \quad |x|, |t| < \mathtt{r}\Big\}$, we have

$$\begin{aligned}\int_{\varepsilon}^{\infty} \frac{e^{-x^2/4t}}{2\sqrt{\pi t}} \frac{\partial \varphi}{\partial t}\, dt &= \int_{\varepsilon}^{\mathtt{r}} \frac{e^{-x^2/4t}}{2\sqrt{\pi t}} \frac{\partial \varphi}{\partial t}\, dt \\ &= -\frac{e^{-x^2/4\varepsilon}}{2\sqrt{\pi \varepsilon}}\, \varphi(x,\varepsilon) - \int_{\varepsilon}^{\mathtt{r}} \frac{\partial}{\partial t} \left\{ \frac{e^{-x^2/4t}}{2\sqrt{\pi t}} \right\} \varphi(x,t)\, dt,\end{aligned}$$

and noting that

$$\frac{\partial}{\partial t} \left\{ \frac{e^{-x^2/4t}}{2\sqrt{\pi t}} \right\} = \frac{e^{-x^2/4t}}{4\sqrt{\pi}} \left(\frac{x^2}{2\, t^{5/2}} - \frac{1}{t^{3/2}} \right),$$

we get

$$\begin{aligned}I_\varepsilon(\varphi) &= -\frac{1}{4\sqrt{\pi}} \int_{-\mathtt{r}}^{\mathtt{r}} \int_{\varepsilon}^{\mathtt{r}} \left(\frac{x^2}{2\, t^{5/2}} - \frac{1}{t^{3/2}} \right) e^{-x^2/4t}\, \varphi(x,t)\, dt\, dx \\ &\qquad - \int_{-\mathtt{r}}^{\mathtt{r}} \frac{e^{-x^2/4\varepsilon}}{2\sqrt{\pi\varepsilon}}\, \varphi(x,\varepsilon)\, dx\,.\end{aligned} \tag{9.38}$$

On the other hand, integrating by parts again and using that φ vanishes in a neighborhood of $\big\{(x,t): \quad |x| = |t| = \mathtt{r}\big\}$, we obtain

$$\begin{aligned}\int_{-\mathtt{r}}^{\mathtt{r}} \frac{e^{-x^2/4t}}{2\sqrt{\pi t}} \frac{\partial^2 \varphi}{\partial x^2}\, dx &= -\int_{-\mathtt{r}}^{\mathtt{r}} \frac{\partial}{\partial x} \left(\frac{e^{-x^2/4t}}{2\sqrt{\pi t}} \right) \frac{\partial \varphi}{\partial x}\, dx \\ &= \int_{-\mathtt{r}}^{\mathtt{r}} \varphi(x,t)\, \frac{\partial^2}{\partial x^2} \left(\frac{e^{-x^2/4t}}{2\sqrt{\pi t}} \right) dx \\ &= \frac{1}{4\sqrt{\pi}} \int_{-\mathtt{r}}^{\mathtt{r}} \left(\frac{x^2}{2t^{5/2}} - \frac{1}{t^{3/2}} \right) e^{-x^2/4t}\, \varphi(x,t)\, dx,\end{aligned}$$

which, when substituted into the definition of J_ε, implies that

$$J_\varepsilon(\varphi) = \frac{1}{4\sqrt{\pi}} \int_{\varepsilon}^{\mathtt{r}} \left\{ \int_{-\mathtt{r}}^{\mathtt{r}} \left(\frac{x^2}{2t^{5/2}} - \frac{1}{t^{3/2}} \right) e^{-x^2/4t}\, \varphi(x,t)\, dx \right\} dt\,,$$

or equivalently, using Fubini's and Tonelli's Theorems (cf. [120, Theorems 6.1 and 6.10]),

$$J_\varepsilon(\varphi) = \frac{1}{4\sqrt{\pi}} \int_{-\mathtt{r}}^{\mathtt{r}} \int_{\varepsilon}^{+\infty} \left(\frac{x^2}{2t^{5/2}} - \frac{1}{t^{3/2}} \right) e^{-x^2/4t}\, \varphi(x,t)\, dx\, dt\,. \tag{9.39}$$

It then follows from (9.38) and (9.39) that

$$I_\varepsilon(\varphi) + J_\varepsilon(\varphi) = -\int_{-\mathtt{r}}^{\mathtt{r}} \frac{e^{-x^2/4\varepsilon}}{2\sqrt{\pi\varepsilon}}\, \varphi(x,\varepsilon)\, dx,$$

and therefore, replacing it back into (9.37), we obtain

$$-\langle E, \frac{\partial \varphi}{\partial t} + \frac{\partial^2 \varphi}{\partial x^2} \rangle = \lim_{\varepsilon \to 0} \int_{-\mathtt{r}}^{\mathtt{r}} \frac{e^{-x^2/4\varepsilon}}{2\sqrt{\pi\varepsilon}}\, \varphi(x,\varepsilon)\, dx = \lim_{\varepsilon \to 0} K_\varepsilon(\varphi)\,,$$

where

$$K_\varepsilon(\varphi) := \int_{-\mathrm{r}}^{\mathrm{r}} \frac{e^{-x^2/4\varepsilon}}{2\sqrt{\pi\varepsilon}}\, \varphi(x,\varepsilon)\, dx\,.$$

Now, performing the change of variables $y = \dfrac{x}{2\sqrt{\varepsilon}}$ gives

$$K_\varepsilon(\varphi) = \frac{1}{\sqrt{\pi}} \int_{-\infty}^{\infty} e^{-y^2}\, \varphi(2\sqrt{\varepsilon}\, y, \varepsilon)\, dy\,,$$

and since

$$|e^{-y^2}\, \varphi(2\sqrt{\varepsilon} y, \varepsilon)| \leqslant \sup_{(x,t)\in\mathbb{R}^2} |\varphi(x,t)|\, e^{-y^2} =: g(y)\,,$$

where $g \in \mathrm{L}^1(\mathbb{R})$, we apply the Lebesgue Dominated Convergence Theorem (cf. [113, Theorem 1.34], [120, Theorem 5.36]) to conclude that

$$\begin{aligned}\lim_{\varepsilon\to 0} K_\varepsilon(\varphi) &= \frac{1}{\sqrt{\pi}} \int_{-\infty}^{\infty} \left\{\lim_{\varepsilon\to 0} e^{-y^2}\, \varphi(2\sqrt{\varepsilon}\, y, \varepsilon)\right\} dy \\ &= \frac{1}{\sqrt{\pi}}\, \varphi(0) \int_{-\infty}^{\infty} e^{-y^2}\, dy = \varphi(0) = \langle \delta, \varphi\rangle.\end{aligned}$$

This proves that $P(E) = \delta$ and completes the proof.

□

Theorem 9.60. *Let* $P = \dfrac{\partial^2}{\partial t^2} - \dfrac{\partial^2}{\partial x^2}$ *be the wave operator in* $\mathbb{R}^2$, *and define*

$$E(x,t) = \begin{cases} \frac{1}{2} & \text{if } |x| < t\,, \\ 0 & \text{if } |x| > t\,, \end{cases} \qquad \forall\, (x,t) \in \mathbb{R}^2\,.$$

Then $P(E) = \delta$ *in* $\mathcal{D}'(\mathbb{R}^2)$.

The proof of Theorem 9.60 is left as an exercise (cf. Exercise 9.18).

9.8.3 Convolution of distributions

In order to define the convolution of distributions it is first necessary to introduce the concept of the convolution of a distribution with a smooth function. Indeed, given $u \in \mathcal{D}'(\mathbb{R}^n)$ and $\varphi \in C_0^\infty(\mathbb{R}^n)$, we define the function

$$(u * \varphi)\,(x) = \langle u, \varphi(x - \cdot)\rangle \qquad \forall\, x \in \mathbb{R}^n\,.$$

Observe that when u is a continuous function, this definition coincides with the usual convolution of functions (cf. Section 9.2). Moreover, we have the following theorem.

Theorem 9.61. *Let* $u \in \mathcal{D}'(\mathbb{R}^n)$ *and* $\varphi \in C_0^\infty(\mathbb{R}^n)$. *Then* $u * \varphi \in C^\infty(\mathbb{R}^n)$, *and for each multi-index* α *we have*

$$\partial^\alpha(u * \varphi) = (\partial^\alpha u) * \varphi = u * \partial^\alpha \varphi\,. \tag{9.40}$$

Moreover, $\mathrm{supp}(u * \varphi) \subseteq \mathrm{supp}(u) + \mathrm{supp}(\varphi)$.

Proof. It follows easily from Lemma 9.47 that $u * \varphi \in C^\infty(\mathbb{R}^n)$ and that

$$\partial^\alpha(u * \varphi) = u * \partial^\alpha \varphi .$$

Furthermore, using that $\partial_y^\alpha \varphi(x-y) = (-1)^{|\alpha|} \partial^\alpha \varphi(x-y)$, we obtain

$$\begin{aligned}(u * \partial^\alpha \varphi)(x) &= \langle u, \partial^\alpha \varphi(x-\cdot)\rangle = (-1)^{|\alpha|} \langle u, \partial_y^\alpha \varphi(x-\cdot)\rangle \\ &= \langle \partial^\alpha u, \varphi(x-\cdot)\rangle = (\partial^\alpha u * \varphi)(x),\end{aligned}$$

which completes the identity (9.40). On the other hand, if $x \in \overbrace{\operatorname{supp}(u * \varphi)}^{\circ}$, then it is clear that $(u * \varphi)(x) \neq 0$. Moreover, since $\operatorname{supp}(u) \cap \operatorname{supp}(\psi) = \varnothing$ necessarily implies that $\langle u, \psi \rangle = 0$, it follows that there must exist $y \in \operatorname{supp}(u)$ such that $x - y \in \operatorname{supp}(\varphi)$, which means that $x \in \operatorname{supp}(u) + \operatorname{supp}(\varphi)$, and therefore

$$\overbrace{\operatorname{supp}(u * \varphi)}^{\circ} \subseteq \operatorname{supp}(u) + \operatorname{supp}(\varphi) .$$

Finally, since $\operatorname{supp}(u) + \operatorname{supp}(\varphi)$ is clearly closed, the required inclusion follows, which completes the proof.

□

It is important to note here that commutativity does not make sense in the present concept of convolution. However, the following associative property does hold.

Theorem 9.62. *Let* $u \in \mathcal{D}'(\mathbb{R}^n)$, *and let* $\varphi, \psi \in C_0^\infty(\mathbb{R}^n)$. *Then*

$$u * (\varphi * \psi) = (u * \varphi) * \psi .$$

Proof. We start by defining, for each $h > 0$, the function $S_h \in C^\infty(\mathbb{R}^n)$ given by

$$S_h(x) = \sum_{k \in \mathbb{Z}^n} \varphi(x - hk)\, h^n\, \psi(hk) \quad \forall\, x \in \mathbb{R}^n ,$$

which constitutes the Riemann sum for the integral

$$(\varphi * \psi)(x) := \int_{\mathbb{R}^n} \varphi(x-y)\, \psi(y)\, dy \quad \forall\, x \in \mathbb{R}^n .$$

Since the support of S_h is contained in the compact set $\operatorname{supp}(\varphi) + \operatorname{supp}(\psi)$, it is clear that $S_h \in C_0^\infty(\mathbb{R}^n)$. Moreover, since the function $\chi(x,y) := \varphi(x-y)\,\psi(y)$ is uniformly continuous, it follows that the sum S_h converges uniformly to $(\varphi * \psi)(x)$ as $h \to 0$. Likewise, since

$$\partial^\alpha S_h(x) = \sum_{k \in \mathbb{Z}^n} \partial^\alpha \varphi(x - hk)\, h^n\, \psi(hk) ,$$

we obtain that $\partial^\alpha S_h$ converges uniformly to $(\partial^\alpha \varphi) * \psi = \partial^\alpha(\varphi * \psi)$ as $h \to 0$, for each multi-index α, which proves that $S_h \to \varphi * \psi$ in $\mathcal{D}(\mathbb{R}^n)$ as $h \to 0$. According to the above, it follows that

$$
\begin{aligned}
\big(u * (\varphi * \psi)\big)(x) &= \langle u, (\varphi * \psi)(x - \cdot)\rangle = \langle u, \lim_{h\to 0} S_h(x-\cdot)\rangle = \lim_{h\to 0}\langle u, S_h(x-\cdot)\rangle \\
&= \lim_{h\to 0}\langle u, \sum_{k\in\mathbb{Z}^n} \varphi(x - \cdot - hk)\, h^n\, \psi(hk)\rangle \\
&= \lim_{h\to 0}\sum_{k\in\mathbb{Z}^n} \langle u, \varphi(x - \cdot - hk)\, h^n\, \psi(hk) \\
&= \lim_{h\to 0}\sum_{k\in\mathbb{Z}^n} (u*\varphi)(x - hk)\, h^n\, \psi(hk) \\
&= \int_{\mathbb{R}^n} (u*\varphi)\,(x-y)\,\psi(y)\,dy \\
&= \big((u*\varphi)*\psi\big)(x) \quad \forall\, x \in \mathbb{R}^n,
\end{aligned}
$$

which proves the required identity.

□

Note that the previous theorem also holds if ψ is only a function in $C(\mathbb{R}^n)$ with compact support, that is, $\psi \in C_0(\mathbb{R}^n)$. Next, we establish a regularization-type result. To this end, for each $\varphi \in C_0^\infty(\mathbb{R}^n)$, we define

$$
\widetilde{\varphi}(x) := \varphi(-x) \quad \forall\, x \in \mathbb{R}^n. \tag{9.41}
$$

Theorem 9.63. *Let $\{\varphi_j\}_{j\in\mathbb{N}} \subseteq C_0^\infty(\mathbb{R}^n)$ be such that*

$$
\begin{gathered}
0 \leqslant \varphi_j, \quad \int_{\mathbb{R}^n} \varphi_j = 1, \quad \operatorname{supp}(\varphi_j) \subseteq B\big(0, \tfrac{1}{j}\big), \\
\textit{and} \qquad \operatorname{supp}(\varphi_{j+1}) \subseteq \operatorname{supp}(\varphi_j),
\end{gathered}
\tag{9.42}
$$

and, given $u \in \mathcal{D}'(\mathbb{R}^n)$, define the regularizations (cf. Theorem 9.61)

$$
u_j := u * \varphi_j \in C^\infty(\mathbb{R}^n) \qquad \forall\, j \in \mathbb{N}.
$$

Then, $u_j \to u$ in $\mathcal{D}'(\mathbb{R}^n)$ as $j \to +\infty$.

Proof. Given $\psi \in C_0^\infty(\mathbb{R}^n)$, using (9.41) we obtain

$$
(u * \widetilde{\psi})(0) = \langle u, \widetilde{\psi}(0 - \cdot)\rangle = \langle u, \psi(\cdot - 0)\rangle = \langle u, \psi\rangle. \tag{9.43}
$$

Moreover, it is easy to see that $\varphi_j * \widetilde{\psi} = \widetilde{(\widetilde{\varphi}_j * \psi)}$ for all $j \in \mathbb{N}$, for all $\psi \in C_0^\infty(\mathbb{R}^n)$. It then follows, using the associative property given by Theorem 9.62, that

$$
\begin{aligned}
\langle u_j, \psi\rangle &= (u_j * \widetilde{\psi})(0) = ((u*\varphi_j) * \widetilde{\psi})(0) \\
&= (u * (\varphi_j * \widetilde{\psi}))(0) \\
&= (u * \widetilde{(\widetilde{\varphi}_j * \psi)}\,)(0) \\
&= \langle u, \widetilde{\varphi}_j * \psi\rangle.
\end{aligned}
$$

In turn, it is not difficult to prove that $\widetilde{\varphi}_j * \psi \to \psi$ in $\mathcal{D}(\mathbb{R}^n)$ as $j \to +\infty$. Therefore, for each $\psi \in C_0^\infty(\mathbb{R}^n)$ we have

$$
\lim_{j\to+\infty} \langle u_j, \psi\rangle = \lim_{j\to+\infty} \langle u, \widetilde{\varphi}_j * \psi\rangle = \langle u, \psi\rangle,
$$

which shows that $u_j \to u$ in $\mathcal{D}'(\mathbb{R}^n)$ as $j \to +\infty$.

□

The previous theorem clearly shows that $C^\infty(\mathbb{R}^n)$, viewed as a subspace of $\mathcal{D}'(\mathbb{R}^n)$, is dense in it. In fact, we have the following more general result.

Theorem 9.64. *Let Ω be an open subset of $\mathbb{R}^n$ and let $u \in \mathcal{D}'(\Omega)$. Then, there exists a sequence $\{u_j\}_{j\in\mathbb{N}} \subseteq C_0^\infty(\Omega)$ such that $u_j \to u$ in $\mathbb{D}'(\Omega)$.*

Proof. Let $\{K_j\}_{j\in\mathbb{N}}$ be a sequence of compact sets such that $K_j \subseteq K_{j+1}$ and $\Omega = \bigcup_{j\in\mathbb{N}} K_j$ (see Figure 9.4 for an example of such compacts). In addition, let $\{\chi_j\}_{j\in\mathbb{N}}$ be a sequence in $C_0^\infty(\Omega)$ such that $\chi_j \equiv 1$ in a neighborhood of K_j (cf. Lemma 9.17). Also, consider a sequence $\{\varphi_j\}_{j\in\mathbb{N}} \subseteq C_0^\infty(\mathbb{R}^n)$ as in (9.42), and such that

$$\operatorname{supp}(\varphi_j) + \operatorname{supp}(\chi_j) \subseteq \Omega \quad \forall\, j \in \mathbb{N}\,.$$

Then, since $\chi_j\, u \in \mathcal{E}'(\Omega) \subseteq \mathcal{E}'(\mathbb{R}^n)$, the function $u_j := (\chi_j\, u) * \varphi_j$ is an element of $C^\infty(\mathbb{R}^n)$, and moreover satisfies

$$\operatorname{supp}(u_j) \subseteq \operatorname{supp}(\chi_j\, u) + \operatorname{supp}(\varphi_j) \subseteq \operatorname{supp}(\chi_j) + \operatorname{supp}(\varphi_j) \subseteq \Omega\,,$$

which indicates that $u_j \in C_0^\infty(\Omega)$. We now prove that $u_j \to u$ in $\mathcal{D}'(\Omega)$. Indeed, applying again the identity given by (9.43) and the associative property (cf. Theorem 9.62), we obtain for each $\psi \in C_0^\infty(\Omega)$

$$\begin{aligned}\langle u_j, \psi\rangle &= \langle (\chi_j\, u) * \varphi_j, \psi\rangle = ((\chi_j\, u * \varphi_j) * \widetilde{\psi})(0) = (\chi_j\, u * (\varphi_j * \widetilde{\psi}))(0)\\ &= (\chi_j\, u * \widetilde{(\widetilde{\varphi}_j * \psi)}\,)(0) = \langle \chi_j\, u, \widetilde{\varphi}_j * \psi\rangle = \langle u, \chi_j\,(\widetilde{\varphi}_j * \psi)\rangle\,.\end{aligned}$$

But it is clear that $\operatorname{supp}(\widetilde{\varphi}_j * \psi) \subseteq \operatorname{supp}(\widetilde{\varphi}_j) + \operatorname{supp}(\psi)$, which is contained in a compact set K_0 of Ω, for j sufficiently large ($j > N_0 := \frac{1}{\operatorname{dist}(\operatorname{supp}(\psi), \partial\Omega)}$). Moreover, there exists $N_1 \in \mathbb{N}$ such that $K_0 \subseteq K_j$ for all $j > N_1$, and therefore $\chi_j \equiv 1$ in K_0 for all $j > N_1$. Taking $N = \max\{N_0, N_1\}$ we obtain

$$\chi_j(\widetilde{\varphi}_j * \psi) = \widetilde{\varphi}_j * \psi \quad \forall\, j > N\,,$$

and therefore, from the previous calculation we have that

$$\langle u_j, \psi\rangle = \langle u, \chi_j(\widetilde{\varphi}_j * \psi)\rangle = \langle u, \widetilde{\varphi}_j * \psi\rangle\,.$$

Finally, it follows that $\widetilde{\varphi}_j * \psi \overset{j\to+\infty}{\longrightarrow} \psi$ in $\mathcal{D}(\Omega)$, which implies that $\langle u_j, \psi\rangle \overset{j\to+\infty}{\longrightarrow} \langle u, \psi\rangle$, thus proving that $u_j \to u$ in $\mathcal{D}'(\Omega)$.

□

We now introduce the TRANSLATION OPERATOR and then establish its connection with the concept of convolution.

Definition 9.65. *Given $h \in \mathbb{R}^n$ and $\varphi \in C(\mathbb{R}^n)$, we define the* TRANSLATION OPERATOR *τ_h by*

$$(\tau_h\, \varphi)(x) = \varphi(x - h) \qquad \forall\, x \in \mathbb{R}^n\,.$$

Observe that, denoting by $\delta_h \in \mathcal{D}'(\mathbb{R}^n)$ the delta distribution at h, we have

$$\tau_h \varphi = \delta_h * \varphi \quad \text{and} \quad u * (\tau_h\, \varphi) = \tau_h(u * \varphi) \quad \forall\, \varphi \in C_0^\infty(\mathbb{R}^n)\,, \quad \forall\, u \in \mathcal{D}'(\mathbb{R}^n)\,,$$

which means that the operation $u*$ commutes with the translation operator. Conversely, we have the following result.

Theorem 9.66. *Let v be a linear and sequentially continuous mapping from $C_0^\infty(\mathbb{R}^n)$ into $C^\infty(\mathbb{R}^n)$, and suppose that v commutes with all translation operators, that is,*

$$v(\boldsymbol{\tau}_h \varphi) = \boldsymbol{\tau}_h(v(\varphi)) \quad \forall\, h \in \mathbb{R}^n, \quad \forall\, \varphi \in C_0^\infty(\mathbb{R}^n).$$

Then, there exists a unique $u \in \mathcal{D}'(\mathbb{R}^n)$ such that $v(\varphi) = u * \varphi$ for all $\varphi \in C_0^\infty(\mathbb{R}^n)$.

Proof. First, observe that if such a u exists, then (cf. (9.41))

$$v(\varphi)(0) = (u * \varphi)(0) = \langle u, \varphi(0 - \cdot)\rangle = \langle u, \widetilde{\varphi}\rangle \quad \forall\, \varphi \in C_0^\infty(\mathbb{R}^n).$$

This fact suggests defining the linear functional

$$\mathcal{D}(\mathbb{R}^n) \ni \varphi \to \langle u, \varphi\rangle = v(\widetilde{\varphi})(0) \in \mathbb{C}.$$

It is not difficult to see that u defined in this way is an element of $\mathcal{D}'(\mathbb{R}^n)$. Moreover, it is clear that

$$(u * \varphi)(0) = \langle u, \varphi(0 - \cdot)\rangle = \langle u, \widetilde{\varphi}\rangle = v(\varphi)(0).$$

Then, using the commutativity of v with $\boldsymbol{\tau}_h$ for each $h \in \mathbb{R}^n$, we have

$$\begin{aligned}(v(\varphi))(-h) &= \boldsymbol{\tau}_h(v(\varphi))(0) = (v(\boldsymbol{\tau}_h \varphi))(0) \\ &= (u * (\boldsymbol{\tau}_h \varphi))(0) = (u * \varphi)(-h) \quad \forall\, \varphi \in C_0^\infty(\mathbb{R}^n),\end{aligned}$$

which proves that $v(\varphi) = u * \varphi$. Finally, uniqueness follows from the fact that if

$$u * \varphi = 0 \quad \forall\, \varphi \in C_0^\infty(\mathbb{R}^n),$$

then

$$\langle u, \widetilde{\varphi}\rangle = (u * \varphi)(0) = 0 \quad \forall\, \widetilde{\varphi} \in C_0^\infty(\mathbb{R}^n),$$

and hence $u \equiv 0$ in $\mathcal{D}'(\mathbb{R}^n)$.

□

Let us note here that if $u \in \mathcal{E}'(\mathbb{R}^n)$, then $\varphi \to u * \varphi$ is a continuous mapping from $C_0^\infty(\mathbb{R}^n)$ into $C_0^\infty(\mathbb{R}^n)$. We also observe that, in fact, the convolution $u * \varphi$ is well defined for every $\varphi \in C^\infty(\mathbb{R}^n)$, and it clearly constitutes a continuous mapping from $C^\infty(\mathbb{R}^n)$ into $C^\infty(\mathbb{R}^n)$. This observation suggests a unique way to define the convolution of two distributions u_1 and u_2 – when one of them has compact support – so that the following associative property holds

$$(u_1 * u_2) * \varphi = u_1 * (u_2 * \varphi)$$

for every $\varphi \in \mathcal{D}(\mathbb{R}^n)$. Indeed, let u_1, $u_2 \in \mathcal{D}'(\mathbb{R}^n)$ be such that u_1 or u_2 belongs to $\mathcal{E}'(\mathbb{R}^n)$. Then, the mapping

$$C_0^\infty(\mathbb{R}^n) \ni \varphi \to v(\varphi) := u_1 * (u_2 * \varphi) \in C^\infty(\mathbb{R}^n)$$

is clearly linear and continuous. Moreover, for each $h \in \mathbb{R}^n$ we have

$$\begin{aligned}v(\boldsymbol{\tau}_h \varphi) &:= u_1 * (u_2 * \boldsymbol{\tau}_h \varphi) = u_1 * (\boldsymbol{\tau}_h(u_2 * \varphi)) \\ &= \boldsymbol{\tau}_h(u_1 * (u_2 * \varphi)) = \boldsymbol{\tau}_h(v(\varphi)) \qquad \forall\, \varphi \in C_0^\infty(\mathbb{R}^n),\end{aligned}$$

which proves that v commutes with translations. Therefore, according to Theorem 9.66, there exists a unique $u \in \mathcal{D}'(\mathbb{R}^n)$ such that

$$u_1 * (u_2 * \varphi) = u * \varphi \quad \forall\, \varphi \in \mathcal{D}(\mathbb{R}^n).$$

Motivated by the above, we now introduce the notion of convolution of distributions.

Definition 9.67. *Let u_1, $u_2 \in \mathcal{D}'(\mathbb{R}^n)$ be such that one of them belongs to $\mathcal{E}'(\mathbb{R}^n)$. Then, we define the convolution $u_1 * u_2$ as the unique distribution $u \in \mathcal{D}'(\mathbb{R}^n)$ such that*

$$u_1 * (u_2 * \varphi) = u * \varphi \qquad \forall\, \varphi \in \mathcal{D}(\mathbb{R}^n).$$

We note, based on the associative property proved above (cf. Theorem 9.62), that Definition 9.67 is consistent with the original one for $u_2 \in C_0^\infty(\mathbb{R}^n)$ and $u_1 \in \mathcal{D}'(\mathbb{R}^n)$. Moreover, according to the proof of Theorem 9.66, it follows that the convolution $u_1 * u_2$ can be computed explicitly by the following formula

$$\langle u_1 * u_2, \varphi\rangle = (u_1 * (u_2 * \widetilde{\varphi}))(0) \qquad \forall\, \varphi \in C_0^\infty(\mathbb{R}^n),$$

that is,

$$\langle u_1 * u_2, \varphi\rangle = \langle u_1, (u_2 * \widetilde{\varphi})(0 - \cdot)\rangle = \langle u_1, \widetilde{(u_2 * \widetilde{\varphi})}\rangle,$$

where, according to the identity

$$(u_2 * \widetilde{\varphi})(x) = \langle u_2, \widetilde{\varphi}(x - \cdot)\rangle = \langle u_2, \varphi(\cdot - x)\rangle,$$

we have that $\widetilde{(u_2 * \widetilde{\varphi})}(x) = \langle u_2, \varphi(\cdot + x)\rangle$, and therefore

$$\langle u_1 * u_2, \varphi\rangle = \langle u_1, \psi\rangle \quad \text{with} \quad \psi(x) := \langle u_2, \varphi(\cdot + x)\rangle \quad \forall\, x \in \mathbb{R}^n. \tag{9.44}$$

On the other hand, it is not difficult to see that the convolution of distributions is associative, that is,

$$u_1 * (u_2 * u_3) = (u_1 * u_2) * u_3,$$

if all the distributions u_j, except possibly one of them, have compact support. Indeed, by virtue of Definition 9.67, it readily follows that

$$\begin{aligned}
\big(u_1 * (u_2 * u_3)\big) * \varphi &= u_1 * \big((u_2 * u_3) * \varphi\big) \\
&= u_1 * \big(u_2 * (u_3 * \varphi)\big) \\
&= (u_1 * u_2) * (u_3 * \varphi) \\
&= \big((u_1 * u_2) * u_3\big) * \varphi \qquad \forall\, \varphi \in C_0^\infty(\mathbb{R}^n).
\end{aligned}$$

Furthermore, the following theorem establishes the commutativity of the convolution of distributions.

Theorem 9.68. *Let u_1, $u_2 \in \mathcal{D}'(\mathbb{R}^n)$ be such that u_1 or u_2 has compact support. Then*

$$u_1 * u_2 = u_2 * u_1 \quad \textit{and} \quad \operatorname{supp}(u_1 * u_2) \subseteq \operatorname{supp}(u_1) + \operatorname{supp}(u_2).$$

Proof. First, observe that in order to prove that two distributions v_1, $v_2 \in \mathcal{D}'(\mathbb{R}^n)$ are equal, it suffices to show that

$$v_1 * (\varphi * \psi) = v_2 * (\varphi * \psi) \qquad \forall\, \varphi,\, \psi \in C_0^\infty(\mathbb{R}^n).$$

Indeed, if the previous equality holds, then by associativity we have

$$(v_1 * \varphi) * \psi = (v_2 * \varphi) * \psi \qquad \forall\, \varphi,\, \psi \in C_0^\infty(\mathbb{R}^n),$$

from which

$$(v_1 * \varphi) = (v_2 * \varphi) \qquad \forall\, \varphi \in C_0^\infty(\mathbb{R}^n),$$

which implies that $v_1 = v_2$ in $\mathcal{D}'(\mathbb{R}^n)$. Now, in the particular case of the present proof, we apply Definition 9.67 and the corresponding associative and commutative properties to deduce that

$$\begin{aligned}(u_1 * u_2) * (\varphi * \psi) &= u_1 * \big(u_2 * (\varphi * \psi)\big) \\ &= u_1 * \big((u_2 * \varphi) * \psi\big) \\ &= u_1 * \big(\psi * (u_2 * \varphi)\big) \\ &= (u_1 * \psi) * (u_2 * \varphi) .\end{aligned}$$

In a similar way, and using the previous identity, we obtain

$$\begin{aligned}(u_2 * u_1) * (\varphi * \psi) &= (u_2 * u_1) * (\psi * \varphi) \\ &= (u_2 * \varphi) * (u_1 * \psi) \\ &= (u_1 * \psi) * (u_2 * \varphi) \\ &= (u_1 * u_2) * (\varphi * \psi) ,\end{aligned}$$

which allows us to conclude, by virtue of the initial observation, that $u_1 * u_2 = u_2 * u_1$. On the other hand, let $\varphi \in C_0^\infty(\mathbb{R}^n)$ be such that $0 \leqslant \varphi$ and $\displaystyle\int_{\mathbb{R}^n} \varphi = 1$. Then, since $(u_1 * u_2) * \varphi = u_1 * (u_2 * \varphi)$, it follows from Theorem 9.61 that

$$\operatorname{supp}\big((u_1 * u_2) * \varphi\big) \subseteq \operatorname{supp}(u_1) + \operatorname{supp}(u_2 * \varphi) \subseteq \operatorname{supp}(u_1) + \operatorname{supp}(u_2) + \operatorname{supp}(\varphi) .$$

Letting $\operatorname{supp}(\varphi) \to \{0\}$, we obtain, according to Theorem 9.63, that $(u_1 * u_2) * \varphi \to (u_1 * u_2)$ in $\mathcal{D}'(\mathbb{R}^n)$, and therefore we conclude that

$$\operatorname{supp}(u_1 * u_2) \subseteq \operatorname{supp}(u_1) + \operatorname{supp}(u_2) .$$

□

Alternatively, we now present another proof for the second part of the previous theorem, namely that $\operatorname{supp}(u_1 * u_2) \subseteq \operatorname{supp}(u_1) + \operatorname{supp}(u_2)$. Recall first from (9.44) that

$$\langle u_1 * u_2, \varphi \rangle = \langle u_1, \psi \rangle , \quad \text{where } \psi(x) = \langle u_2, \varphi(x + \cdot) \rangle \quad \forall\, x \in \mathbb{R}^n, \quad \forall\, \varphi \in C_0^\infty(\mathbb{R}^n) .$$

In turn, it is not difficult to prove that if $A \subseteq \mathbb{R}^n$ is compact and $B \subseteq \mathbb{R}^n$ is closed, then $A + B$ is closed. Thus, $\operatorname{supp}(u_1) + \operatorname{supp}(u_2)$ is a closed subset of $\mathbb{R}^n$. Then, to obtain the desired result, it suffices to show that the restriction of $u_1 * u_2$ to $\mathbb{R}^n \backslash (\operatorname{supp}(u_1) + \operatorname{supp}(u_2))$ is zero. Let $\varphi \in C_0^\infty(\mathbb{R}^n)$ be such that

$$\operatorname{supp}(\varphi) \subseteq \mathbb{R}^n \backslash (\operatorname{supp}(u_1) + \operatorname{supp}(u_2)) .$$

In what follows, given $\delta > 0$ and $U \subseteq \mathbb{R}^n$, we denote by U_δ the δ-neighborhood of U, that is, $U_\delta := \big\{x \in \mathbb{R}^n : \ \operatorname{dist}(x, U) < \delta\big\}$. We must prove that $\langle u_1 * u_2, \varphi \rangle = 0$. First, note that there exists $\varepsilon > 0$ such that

$$\varphi(z) = 0 \qquad \forall\, z \in \big(\operatorname{supp}(u_1) + \operatorname{supp}(u_2)\big)_\varepsilon . \tag{9.45}$$

Then, choose $\delta > 0$ small enough so that

$$\big(\operatorname{supp}(u_1)\big)_\delta + \big(\operatorname{supp}(u_2)\big)_\delta \subseteq \big(\operatorname{supp}(u_1) + \operatorname{supp}(u_2)\big)_\varepsilon .$$

Now, suppose by contradiction that $\langle u_1 * u_2, \varphi\rangle \neq 0$, that is, $\langle u_1, \psi\rangle \neq 0$. Then, there exists $x \in (\text{supp}(u_1))_\delta$ such that $\psi(x) \neq 0$. Otherwise, if $\psi(x) = 0$ for all $x \in (\text{supp}(u_1))_\delta$, we would have $\text{supp}(\psi) \subseteq (\text{supp}(u_1))^c$, which implies $\langle u_1, \psi\rangle = 0$. Hence,

$$\psi(x) = \langle u_2, \varphi(x + \cdot)\rangle \neq 0 \qquad \forall\, x \in \mathbb{R}^n .$$

Therefore, there exists $y \in (\text{supp}(u_2))_\delta$ such that $\varphi(x+y) \neq 0$. Otherwise, if $\varphi(x+y) = 0$ for all $y \in (\text{supp}(u_2))_\delta$, we would have $\text{supp}(\widetilde{\varphi}) \subseteq (\text{supp}(u_2))^c$, which implies $\langle u_2, \widetilde{\varphi}\rangle = 0$, where $\widetilde{\varphi}(y) = \varphi(x+y)$. Finally, we have

$$z := x + y \in (\text{supp}(u_1))_\delta + (\text{supp}(u_2))_\delta \subseteq (\text{supp}(u_1) + \text{supp}(u_2))_\varepsilon \quad \text{and} \quad \varphi(z) \neq 0 ,$$

which contradicts (9.45).

Let us note, on the other hand, that the differentiation of distributions can also be interpreted as a convolution, since one obtains that

$$\partial^\alpha u = (\partial^\alpha \delta) * u \quad \forall\, u \in \mathcal{D}'(\mathbb{R}^n) . \tag{9.46}$$

Indeed, in view of Theorem 9.61 and the identity $\delta * \partial^\alpha \varphi = \partial^\alpha \varphi$, and using Definition 9.67 together with the commutativity property, we obtain for each $\varphi \in C_0^\infty(\mathbb{R}^n)$

$$\begin{aligned}\partial^\alpha u * \varphi &= u * \partial^\alpha \varphi = u * (\delta * \partial^\alpha \varphi)\\ &= u * (\partial^\alpha \delta * \varphi) = (u * \partial^\alpha \delta) * \varphi = (\partial^\alpha \delta * u) * \varphi ,\end{aligned}$$

which implies the identity (9.46). We also observe that convolution with δ is the identity operator. Indeed,

$$(u * \delta) * \varphi = u * (\delta * \varphi) = u * \varphi,$$

from which $u * \delta = u$ for all $u \in \mathcal{D}'(\mathbb{R}^n)$. Moreover, if $u_1, u_2 \in \mathcal{D}'(\mathbb{R}^n)$, one of them in $\mathcal{E}'(\mathbb{R}^n)$, then

$$\partial^\alpha (u_1 * u_2) = \partial^\alpha u_1 * u_2 = u_1 * \partial^\alpha u_2 .$$

Furthermore, we have the following general result.

Theorem 9.69. *Let* $P := \sum_{|\alpha|\leqslant m} a_\alpha \partial^\alpha$ *be a partial differential operator with constant coefficients* $a_\alpha \in \mathbb{C}$. *Then*

i) $P(u) = P(\delta) * u$ *for all* $u \in \mathcal{D}'(\mathbb{R}^n)$, *and*
ii) $P(u_1 * u_2) = P(u_1) * u_2 = u_1 * P(u_2)$ *for all* $u_1 \in \mathcal{D}'(\mathbb{R}^n)$, *for all* $u_2 \in \mathcal{E}'(\mathbb{R}^n)$.

Proof. We have, according to the definition of P and to identity (9.46), that

$$P(u) = \sum_{|\alpha|\leqslant m} a_\alpha \partial^\alpha u = \sum_{|\alpha|\leqslant m} a_\alpha(\partial^\alpha \delta * u) = \left\{ \sum_{|\alpha|\leqslant m} a_\alpha \partial^\alpha \delta \right\} * u = P(\delta) * u,$$

which proves i). In turn, since $P(u_1 * u_2) = \sum_{|\alpha|\leqslant m} a_\alpha \partial^\alpha (u_1 * u_2)$, it remains only to show that $\partial^\alpha (u_1 * u_2) = \partial^\alpha u_1 * u_2 = u_1 * \partial^\alpha u_2$ to conclude ii). Indeed,

$$\begin{aligned}\partial^\alpha (u_1 * u_2) &= \partial^\alpha \delta * (u_1 * u_2) = (\partial^\alpha \delta * u_1) * u_2 = \partial^\alpha u_1 * u_2\\ &= (u_1 * \partial^\alpha \delta) * u_2 = u_1 * (\partial^\alpha \delta * u_2) = u_1 * \partial^\alpha u_2 ,\end{aligned}$$

which completes the proof.

□

Another important result related to the singular support (cf. Definition 9.52) of the convolution of two distributions is the following.

Theorem 9.70. *Let u_1, $u_2 \in \mathcal{D}'(\mathbb{R}^n)$, at least one of them with compact support. Then*

$$\operatorname{sing\,supp}(u_1 * u_2) \subseteq \operatorname{sing\,supp}(u_1) + \operatorname{sing\,supp}(u_2).$$

Proof. Suppose, without loss of generality, that $u_2 \in \mathcal{E}'(\mathbb{R}^n)$, that is, $\operatorname{supp}(u_2)$ is a compact subset of $\mathbb{R}^n$. Since $u_2|_{(\operatorname{supp}(u_2))^c} = 0 \in C^\infty$, it follows that

$$(\operatorname{supp}(u_2))^c \subseteq (\operatorname{sing\,supp}(u_2))^c,$$

and hence $\operatorname{sing\,supp}(u_2) \subseteq \operatorname{supp}(u_2)$. This implies that $\operatorname{sing\,supp}(u_2)$ is a compact set in $\mathbb{R}^n$. Thus, for each $\varepsilon > 0$, one can choose $\varphi_\varepsilon \in C_0^\infty(\mathbb{R}^n)$ such that $\varphi_\varepsilon \equiv 1$ in the ε-neighborhood of $\operatorname{sing\,supp}(u_2)$ and $\varphi_\varepsilon \equiv 0$ in the complement of the corresponding 2ε-neighborhood. Then, it is clear that $u_2 = v_\varepsilon + \omega_\varepsilon$, with $v_\varepsilon := \varphi_\varepsilon u_2$ and $\omega_\varepsilon := (1 - \varphi_\varepsilon) u_2$. Since $\omega_\varepsilon \equiv 0$ in the ε-neighborhood of $\operatorname{sing\,supp}(u_2)$, and $u_2 \in C^\infty\big((\operatorname{sing\,supp}(u_2))^c\big)$, it clearly follows that $\omega_\varepsilon \in C_0^\infty(\mathbb{R}^n)$. In this way, we deduce by Theorem 9.61 that there holds $u_1 * \omega_\varepsilon \in C^\infty(\mathbb{R}^n)$, which implies that

$$\operatorname{sing\,supp}(u_1 * u_2) = \operatorname{sing\,supp}(u_1 * v_\varepsilon + u_1 * \omega_\varepsilon) = \operatorname{sing\,supp}(u_1 * v_\varepsilon).$$

On the other hand, define the set

$$A := \Big\{ z \in \mathbb{R}^n : \quad \{z\} - \operatorname{supp}(v_\varepsilon) \subseteq (\operatorname{sing\,supp}(u_1))^c \Big\}.$$

Our goal is to prove that $A \subseteq (\operatorname{sing\,supp}(u_1 * v_\varepsilon))^c$. To conclude this, and since

$$\langle u_1 * v_\varepsilon, \varphi \rangle = \langle u_1, \widehat{\varphi} \rangle \quad \text{with} \quad \widehat{\varphi}(x) = \langle v_\varepsilon, \varphi(x + \cdot) \rangle \quad \forall\, \varphi \in C_0^\infty(\mathbb{R}^n),$$

it suffices to show that $\operatorname{supp}(\widehat{\varphi}) \subseteq (\operatorname{sing\,supp}(u_1))^c$ for all $\varphi \in C_0^\infty(A)$. Indeed, for each $\varphi \in C_0^\infty(A)$ we have

$$\begin{aligned}
x \in \overbrace{\operatorname{supp}(\widehat{\varphi})}^{\circ} &\Rightarrow \langle v_\varepsilon, \varphi(x + \cdot) \rangle \neq 0 \\
&\Rightarrow \exists\, y \in \operatorname{supp}(v_\varepsilon) \quad \text{such that} \quad z := x + y \in \operatorname{supp}(\varphi) \\
&\Rightarrow x = (z - y) \in \{z\} - \operatorname{supp}(v_\varepsilon) \subseteq (\operatorname{sing\,supp}(u_1))^c \\
&\Rightarrow x \in (\operatorname{sing\,supp}(u_1))^c,
\end{aligned}$$

which shows that $\operatorname{supp}(\widehat{\varphi}) \subseteq (\operatorname{sing\,supp}(u_1))^c$. Thus, $u_1 \in C^\infty(B)$, with

$$B = \cup \big\{ \operatorname{supp}(\widehat{\varphi}) : \quad \varphi \in C_0^\infty(A) \big\},$$

and therefore $u_1 * v_\varepsilon = v_\varepsilon * u_1 \in C^\infty(A)$. Consequently, we have

$$\begin{aligned}
x \in \operatorname{sing\,supp}(u_1 * v_\varepsilon) &\Rightarrow x \notin A \\
&\Rightarrow \exists\, y \in \operatorname{supp}(v_\varepsilon) \quad \text{such that} \quad x - y \in \operatorname{sing\,supp}(u_1) \\
&\Rightarrow x \in \operatorname{sing\,supp}(u_1) + \operatorname{supp}(v_\varepsilon),
\end{aligned}$$

which indicates that

$$\operatorname{sing\,supp}(u_1 * v_\varepsilon) \subseteq \operatorname{sing\,supp}(u_1) + \operatorname{supp}(v_\varepsilon) \subseteq \operatorname{sing\,supp}(u_1) + \operatorname{supp}(\varphi_\varepsilon).$$

Finally, since $\operatorname{supp}(\varphi_\varepsilon) \overset{\varepsilon\to 0}{\to} \operatorname{sing\,supp}(u_2)$, we have that

$$\operatorname{sing\,supp}(u_1 * v_\varepsilon) \subseteq \operatorname{sing\,supp}(u_1) + \operatorname{sing\,supp}(u_2),$$

which completes the proof.

□

9.8.4 The role of fundamental solutions

Recall that a distribution $E \in \mathcal{D}'(\mathbb{R}^n)$ is said to be a fundamental solution of the differential operator $P := \sum_{|\alpha|\leqslant m} a_\alpha \partial^\alpha$, $a_\alpha \in \mathbb{C}$, if $P(E) = \delta$. The importance of fundamental solutions follows from the following identities, which are basically a consequence of the fact that $u * \delta = \delta * u = u$ for all $u \in \mathcal{D}'(\mathbb{R}^n)$, and from property ii) of Theorem 9.69. More precisely, one has

$$E * P(u) = u \quad \forall\, u \in \mathcal{E}'(\mathbb{R}^n) \quad \text{and} \quad P(E * f) = f \qquad \forall\, f \in \mathcal{E}'(\mathbb{R}^n).$$

The second identity implies that the equation $P(u) = f$ has a solution for each $f \in \mathcal{E}'(\mathbb{R}^n)$, while the first one shows that it is possible to obtain information about the singularities of u from those of $P(u)$. More precisely, one has the following result.

Theorem 9.71. *Assume that the differential operator P admits a fundamental solution E such that* $\operatorname{sing\,supp}(E) = \{0\}$. *Then, for each open set $\Omega \subseteq \mathbb{R}^n$, we have*

$$\operatorname{sing\,supp}(u) = \operatorname{sing\,supp}\big(P(u)\big) \quad \forall\, u \in \mathcal{D}'(\Omega).$$

Proof. We first observe that one always has $\operatorname{sing\,supp}\big(P(u)\big) \subseteq \operatorname{sing\,supp}(u)$. Indeed, if $x \in \big(\operatorname{sing\,supp}(u)\big)^c$, then there exists an open set V containing x such that

$$V \subseteq \big(\operatorname{sing\,supp}(u)\big)^c \quad \text{and} \quad u \in C^\infty(V).$$

It follows that $P(u) \in C^\infty(V)$ and therefore $x \in \big(\operatorname{sing\,supp}\big(P(u)\big)\big)^c$. Now, suppose that $u \in \mathcal{E}'(\Omega)$ and let $\psi \in C_0^\infty(\Omega)$ be such that $\psi \equiv 1$ in a neighborhood of $\operatorname{supp}(u)$. Then u can be easily extended to a distribution $u \in \mathcal{E}'(\mathbb{R}^n)$ by defining

$$\langle u, \varphi\rangle = \langle u, \psi\varphi\rangle \qquad \forall\, \varphi \in C_0^\infty(\mathbb{R}^n).$$

Then, since $u = E * P(u)$, we have, by Theorem 9.70, that

$$\begin{aligned}\operatorname{sing\,supp}(u) &= \operatorname{sing\,supp}\big(E * P(u)\big) \subseteq \operatorname{sing\,supp}(E) + \operatorname{sing\,supp}\big(P(u)\big)\\ &= \operatorname{sing\,supp}\big(P(u)\big),\end{aligned}$$

and therefore

$$\operatorname{sing\,supp}(u) = \operatorname{sing\,supp}\big(P(u)\big) \qquad \forall\, u \in \mathcal{E}'(\Omega). \tag{9.47}$$

Now let $u \in \mathcal{D}'(\Omega)$, and consider an increasing sequence of compact sets $\{K_j\}_{j\in\mathbb{N}}$ such that $\Omega = \bigcup_{j\in\mathbb{N}} K_j$ (see Figure 9.4 for an example of them). Moreover, let $\{\varphi_j\}_{j\in\mathbb{N}} \subseteq C_0^\infty(\Omega)$ be such that $\varphi_j \equiv 1$ in a neighborhood of K_j (cf. Lemma 9.17). Then, using the fact that $\varphi_j\, u \in \mathcal{E}'(\Omega)$ and the identity (9.47), we have for each $j \in \mathbb{N}$ that

$$\begin{aligned}\mathring{K}_j \cap \operatorname{sing\,supp}\big(P(u)\big) &= \mathring{K}_j \cap \operatorname{sing\,supp}\big(P(\varphi_j u)\big)\\ &= \mathring{K}_j \cap \operatorname{sing\,supp}(\varphi_j u) = \mathring{K}_j \cap \operatorname{sing\,supp}(u)\,,\end{aligned}$$

and then, taking $\bigcup\limits_{j\in\mathbb{N}}$ in the previous identity, we deduce that

$$\operatorname{sing\,supp}\big(P(u)\big) = \operatorname{sing\,supp}(u) \qquad \forall\, u \in \mathcal{D}'(\Omega)\,.$$

□

It is important to observe here that the Laplace and heat operators (cf. Theorems 9.58 and 9.59) constitute examples for which the previous theorem can be applied. Conversely, if $\Omega = \mathbb{R}^n$, and if $\operatorname{sing\,supp}(u) = \operatorname{sing\,supp}\big(P(u)\big)$ for all $u \in \mathcal{D}'(\mathbb{R}^n)$, then it can be proved that every fundamental solution E of P satisfies

$$\operatorname{sing\,supp}(E) = \operatorname{sing\,supp}\big(P(E)\big) = \operatorname{sing\,supp}(\delta) = \{0\}\,.$$

Corollary 9.72. *Let P be a differential operator with a fundamental solution E such that $\operatorname{sing\,supp}(E) = \{0\}$. If $\{u_j\}_{j\in\mathbb{N}} \subseteq C^\infty(\Omega)$ is such that*

$$P(u_j) = 0 \quad \forall\, j \in \mathbb{N} \quad \text{and} \quad u_j \overset{j\to+\infty}{\longrightarrow} u \quad \text{in} \quad \mathcal{D}'(\Omega)\,,$$

then $u \in C^\infty(\Omega)$ and $u_j \overset{j\to+\infty}{\longrightarrow} u$ in $\mathcal{E}(\Omega)$.

Proof. Since $P(u) = P(\lim u_j) = \lim\limits_{j\to+\infty} P(u_j) = 0$, we have that

$$\operatorname{sing\,supp}(u) = \operatorname{sing\,supp}\big(P(u)\big) = \varnothing\,,$$

and therefore $u \in C^\infty(\Omega)$. Thus, there is no loss of generality in assuming that $u \equiv 0$. Then, if $\{u_j\}_{j\in\mathbb{N}} \subseteq C^\infty(\Omega)$ is such that $P(u_j) = 0$ for all $j \in \mathbb{N}$ and $u_j \overset{j\to+\infty}{\longrightarrow} 0$ in $\mathcal{D}'(\Omega)$, we must show that $u_j \to 0$ in $\mathcal{E}(\Omega)$. Indeed, let Ω_0 be a bounded open subset of Ω, and let $\psi \in C_0^\infty(\Omega)$ be such that $\psi \equiv 1$ in a neighborhood of $\bar{\Omega}_0$. Define $f_j = P(\psi u_j)$. It is clear that $f_j \in C_0^\infty(\Omega)$ and that $\operatorname{supp}(f_j) \subseteq \Omega \backslash \Omega_0$, since in a neighborhood of $\bar{\Omega}_0$ we have $f_j = P(\psi u_j) = P(u_j) = 0$. Thus, $f_j \in \mathcal{E}'(\Omega)$ and it extends easily to a distribution $f_j \in \mathcal{E}'(\mathbb{R}^n)$. Then, for each $\varphi \in C^\infty(\mathbb{R}^n)$ we have

$$\begin{aligned}\langle f_j, \varphi\rangle &= (f_j * \tilde{\varphi})(0) = \big(P(\psi u_j) * \tilde{\varphi}\big)(0)\\ &= \big(\psi u_j * P(\tilde{\varphi})\big)(0) = \langle \psi u_j,\ \widetilde{P(\tilde{\varphi})}\ \rangle = \langle u_j, \psi\ \widetilde{P(\tilde{\varphi})}\ \rangle.\end{aligned}$$

Since $\psi\ \widetilde{P(\tilde{\varphi})} \in C_0^\infty(\Omega)$, and $u_j \overset{j\to+\infty}{\longrightarrow} 0$ in $\mathcal{D}'(\Omega)$, we conclude that $\langle f_j, \varphi\rangle \overset{j\to+\infty}{\longrightarrow} 0$, that is, $f_j \to 0$ in $\mathcal{E}'(\mathbb{R}^n)$. On the other hand, given a compact set $K \subseteq \Omega_0$, we have for each $x \in K$, and using that $\psi \equiv 1$ on K, that

$$\begin{aligned}(\partial^\alpha u_j)(x) &= \partial^\alpha(\psi\, u_j)(x) \\ &= \partial^\alpha(\delta * \psi\, u_j)(x) = \partial^\alpha\big(P(E) * \psi\, u_j\big)(x) \\ &= \partial^\alpha\big(E * P(\psi\, u_j)\big)(x) = \partial^\alpha(E * f_j)(x) \\ &= (\partial^\alpha E * f_j)(x) = (f_j * \partial^\alpha E)(x) = \langle f_j, \partial^\alpha E(x - \cdot)\rangle,\end{aligned}$$

which, using that $f_j \overset{j\to\infty}{\to} 0$ in $\mathcal{E}'(\Omega)$, allows us to conclude that $\partial^\alpha u_j$ converges uniformly to 0 on K.

□

On the other hand, in order to prove next an existence result for our differential operator P, we need the following theorem.

Theorem 9.73 (Runge Approximation Theorem). *Suppose that P has a fundamental solution $E \in \mathcal{D}'(\mathbb{R}^n)$ which is real analytic in $\mathbb{R}^n\backslash\{0\}$, and let $\widehat{\Omega} \subseteq \Omega \subseteq \mathbb{R}^n$ be open sets such that $\Omega\backslash\widehat{\Omega}$ is not a disjoint union $F \cup K$, with K compact $\neq \varnothing$ and F closed in Ω. Then, every solution $u \in C^\infty(\widehat{\Omega})$ of the equation $P(u) = 0$ is the limit in $\mathcal{E}(\widehat{\Omega})$ of the restrictions to $\widehat{\Omega}$ of solutions of the same equation in Ω.*

Proof. See [95, Theorem 4.45, p. 112].

□

Theorem 9.74. *Suppose that the operator P admits a fundamental solution $E \in \mathcal{D}'(\mathbb{R}^n)$ which is real analytic in $\mathbb{R}^n\backslash 0$, and let $\Omega \subset \mathbb{R}^n$ be an open set. Then, for every $f \in \mathcal{D}'(\Omega)$, there exists $u \in \mathcal{D}'(\Omega)$ such that $P(u) = f$.*

Proof. For each $j \in \mathbb{N}$ define

$$\Omega_j = \big\{x \in \Omega : \quad |x| < j \quad \text{and} \quad \operatorname{dist}(x, \Omega^c) > \tfrac{1}{j}\big\},$$

which coincides with the set $\overset{\circ}{K}_j$ of Figure 9.4. Then, the hypotheses of Theorem 9.73 are satisfied for $\widehat{\Omega} = \Omega_j$. Indeed, suppose by contradiction that $\Omega\backslash\Omega_j = F \cup K$, with $F \cap K = \varnothing$, K compact $\neq \varnothing$, and F closed in Ω. Then, given $x \in K$ and $y \in \Omega^c$, the segment $S(x,y)$ joining these two points must necessarily contain some element $z \in \Omega_j$. If this were not the case, $S(x,y)\backslash\big(K \cup \Omega^c\big)$ would be contained in F, which would imply that $\partial K \cap \partial F \neq \varnothing$, thus contradicting that K and F are disjoint. Moreover, it is clear that x belongs to a segment with endpoints $z_1, z_2 \in \Omega_j$, and hence $x = (1-t)z_1 + tz_2$ for some $t \in (0,1)$. It then follows (see Figure 9.6 for a geometric interpretation of these facts) that

$$\operatorname{dist}(x,y) \geqslant \operatorname{dist}(z,y) \geqslant \operatorname{dist}(z,\Omega^c) > \frac{1}{j} \quad \text{and} \quad |x| \leqslant (1-t)|z_1| + t|z_2| < j\,,$$

which shows that $x \in \Omega_j$, thus yielding the desired contradiction. Now, since $\bar{\Omega}_j$ is clearly compact, we can choose $\varphi_j \in C_0^\infty(\Omega)$ such that $\varphi_j \equiv 1$ in a neighborhood of Ω_j, and then define the distribution

$$v_j = E * (\varphi_j\, f)\,.$$

Then, using Theorem 9.69, we clearly have

$$P(v_j) = P(E) * \varphi_j f = \delta * \varphi_j f = \varphi_j f\,,$$

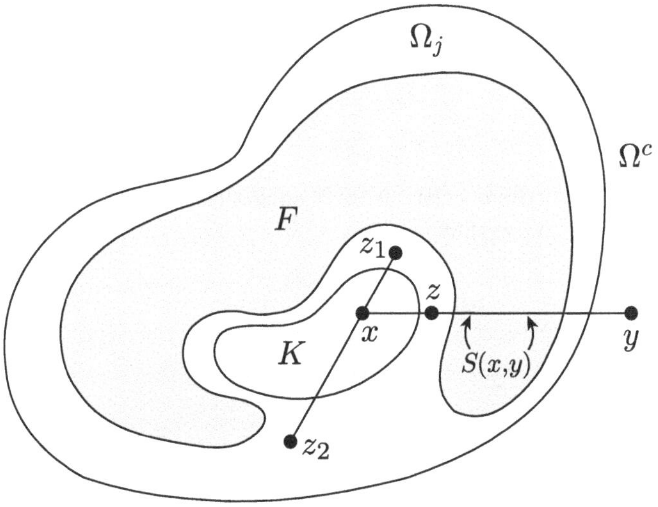

Fig. 9.6. *Geometry for the verification of the hypotheses of Theorem 9.74*

from which

$$P(v_j) = f \quad \text{in} \quad \Omega_j\,,$$

that is, $\langle P(v_j), \varphi\rangle = \langle f, \varphi\rangle$ for all $\varphi \in C_0^\infty(\Omega_j)$. This equality, together with the fact that $\Omega_j \subseteq \Omega_{j+1}$, implies that $(v_{j+1} - v_j)$ satisfies the equation

$$P(v_{j+1} - v_j) = 0 \quad \text{in} \quad \Omega_j\,.$$

Moreover, according to Theorem 9.71, we have that

$$\operatorname{sing\,supp}(v_{j+1} - v_j) = \operatorname{sing\,supp}\big(P(v_{j+1} - v_j)\big) = \varnothing\,,$$

which means that $(v_{j+1} - v_j) \in C^\infty(\Omega_j)$. Thus, applying the Runge Approximation Theorem (cf. Theorem 9.73), it follows that there exists a sequence $\{\omega^{(m)}\}_{m\in\mathbb{N}} \subseteq C^\infty(\Omega)$, with $P(\omega^{(m)}) = 0$ in $\mathcal{D}'(\Omega)$, such that

$$\omega^{(m)}|_{\Omega_j} \overset{m\to+\infty}{\to} v_{j+1} - v_j \quad \text{in} \quad \mathcal{E}(\Omega_j)\,.$$

This means (cf. Definition 9.36) that for every compact $K \subseteq \Omega_j$ and for every multi-index α,

$$\sup_{x\in K} |\partial^\alpha(v_{j+1} - v_j - \omega^{(m)})(x)| \overset{m\to+\infty}{\to} 0\,,$$

and in particular, for $K = \bar{\Omega}_{j-1}$ and $\alpha = (0, \cdots, 0)$, we have

$$\lim_{m\to+\infty} \sup_{x\in\bar{\Omega}_{j-1}} |(v_{j+1} - v_j - \omega^{(m)})(x))| = 0\,.$$

Consequently, there exists $\omega_j \in \{\omega^{(m)}\}_{m\in\mathbb{N}}$, that is, $\omega_j \in C^\infty(\Omega)$ with $P(\omega_j) = 0$ in $\mathcal{D}'(\Omega)$, such that

$$\sup_{x \in \bar{\Omega}_{j-1}} |(v_{j+1} - v_j - \omega_j)(x)| < 2^{-j} \qquad \forall j \in \mathbb{N}.$$

Now consider the sequence $\{u_j\}_{j \in \mathbb{N}} \subseteq \mathcal{D}'(\Omega)$ defined as

$$u_j = v_j - \sum_{i<j} \omega_i \qquad \forall j \in \mathbb{N}.$$

Then, given $\varphi \in C_0^\infty(\Omega)$, there exists $N \in \mathbb{N}$ such that $\operatorname{supp}(\varphi) \subseteq \Omega_j$ for all $j \geqslant N$. Hence, for $j > k+1 > N$, we have

$$\langle u_j, \varphi \rangle - \langle u_{k+1}, \varphi \rangle = \sum_{i=k+1}^{j-1} \langle v_{i+1} - v_i - \omega_i, \varphi \rangle,$$

from which

$$\begin{aligned} |\langle u_j, \varphi \rangle - \langle u_{k+1}, \varphi \rangle| &\leqslant \sum_{i=k+1}^{j-1} \int_{\Omega_N} |(v_{i+1} - v_i - \omega_i)(x)| \, |\varphi(x)| \, dx \\ &\leqslant \sup_{x \in \Omega_N} |\varphi(x)| \, |\Omega_N| \sum_{i=k+1}^{j-1} 2^{-i}, \end{aligned}$$

which shows that the sequence $\{\langle u_j, \varphi \rangle\}_{j \in \mathbb{N}}$ is convergent in $\mathbb{C}$. Therefore, defining

$$\langle u, \varphi \rangle := \lim_{j \to +\infty} \langle u_j, \varphi \rangle \quad \forall \varphi \in C_0^\infty(\Omega),$$

it follows that u is an element of $\mathcal{D}'(\Omega)$. Moreover, since obviously $u = \lim_{k \to +\infty} u_k$ in $\mathcal{D}'(\Omega)$, we have $P(u) = \lim_{k \to +\infty} P(u_k)$, from which it follows easily that $P(u) = f$ in $\mathcal{D}'(\Omega)$. Indeed, given $\varphi \in C_0^\infty(\Omega)$, there exists $N \in \mathbb{N}$ such that $\operatorname{supp}(\varphi) \subseteq \Omega_{k-1}$ for all $k > N$. Then, the fact that $P(v_k) = f$ in Ω_{k-1} implies that

$$\langle P(u_k), \varphi \rangle = \langle P(v_k), \varphi \rangle - \sum_{i<k} \langle P(\omega_i), \varphi \rangle = \langle f, \varphi \rangle \quad \forall k > N,$$

and thus

$$\langle P(u), \varphi \rangle = \lim_{k \to +\infty} \langle P(u_k), \varphi \rangle = \langle f, \varphi \rangle,$$

which completes the proof.

□

We conclude this section with the following result on the representation of distributions, in which we use the concepts of locally finite partition of unity developed in Section 9.5.

Theorem 9.75. *Let $\Omega \subset \mathbb{R}^n$ be an open set, and let $f \in \mathcal{D}'(\Omega)$. Then there exist functions $f_\alpha \in C(\Omega)$, for each multi-index α, such that*

$$\langle f, \varphi \rangle = \sum_\alpha (-1)^{|\alpha|} \int_\Omega f_\alpha \, \partial^\alpha \varphi \, dx \qquad \forall \varphi \in C_0^\infty(\Omega),$$

that is,

$$f = \sum_\alpha \partial^\alpha f_\alpha.$$

Moreover, $\{\operatorname{supp}(f_\alpha)\}_\alpha$ is locally finite, and if f is of finite order, the sum $\sum_\alpha$ has a finite number of terms.

Proof. For each nonnegative integer m, we define the function on $\mathbb{R}$

$$x_+^m = \begin{cases} x^m & \text{if} \quad x > 0 \\ 0 & \text{if} \quad x \leqslant 0 . \end{cases}$$

Then, it is not difficult to see that $\dfrac{x_+^m}{m!}$ is a fundamental solution of the operator $\left(\dfrac{d}{dx}\right)^{m+1}$ in $\mathbb{R}$. It then follows that the function E_m defined by

$$E_m(x) = \frac{x_{1+}^m \, x_{2+}^m \cdots x_{n+}^m}{(m!)^n} \qquad \forall\, x := (x_1, \dots x_n) \in \mathbb{R}^n ,$$

is a fundamental solution of the partial differential operator

$$P_m := \left(\frac{\partial}{\partial x_1} \frac{\partial}{\partial x_2} \cdots \frac{\partial}{\partial x_n} \right)^{m+1} \quad \text{in} \quad \mathbb{R}^n .$$

Moreover, it is easy to see that $E_m \in C^k(\mathbb{R}^n)$ if $m > k$. Now, if $g \in \mathcal{E}_k'(\mathbb{R}^n)$ and $m > k$, then, using Riemann sums as in Theorem 9.62, one proves that $u := g * E_m \in C(\mathbb{R}^n)$, and clearly (cf. Theorem 9.69)

$$P_m(u) = g * P_m(E_m) = g * \delta = g .$$

On the other hand, let $\{\psi_j\}_{j \in \mathbb{N}}$ be a locally finite partition of unity on Ω (cf. Theorem 9.30), and for each $j \in \mathbb{N}$, let $\chi_j \in C_0^\infty(\Omega)$ be such that $\chi_j \equiv 1$ in a neighborhood of $\operatorname{supp}(\psi_j)$. Since every compactly supported distribution is a distribution of finite order (cf. Section 9.7), it follows that there exists a nonnegative integer k_j such that $\chi_j f \in \mathcal{E}_{k_j}'(\mathbb{R}^n)$, for all $j \in \mathbb{N}$. Thus, if $m_j > k_j$, then according to the previous analysis, there exists $u_j \in C(\mathbb{R}^n)$ such that

$$P_{m_j}(u_j) = \chi_j f .$$

Consequently, for each $\varphi \in C_0^\infty(\Omega)$ we have

$$\begin{aligned} \langle f, \varphi \rangle &= \langle f, \sum_{j \in \mathbb{N}} \psi_j \varphi \rangle = \sum_{j \in \mathbb{N}} \langle f, \psi_j \varphi \rangle = \sum_{j \in \mathbb{N}} \langle f, \chi_j \psi_j \varphi \rangle = \sum_{j \in \mathbb{N}} \langle \chi_j f, \psi_j \varphi \rangle \\ &= \sum_{j \in \mathbb{N}} \langle P_{m_j}(u_j), \psi_j \varphi \rangle = \sum_{j \in \mathbb{N}} \langle u_j, P_{m_j}^*(\psi_j \varphi) \rangle = \sum_{j \in \mathbb{N}} \langle u_j, (-1)^{|\alpha_j|} \, \partial^{\alpha_j}(\psi_j \varphi) \rangle \end{aligned}$$

where $P_{m_j}^*$ is the adjoint of P_{m_j} and $\alpha_j = (m_j + 1, \dots, m_j + 1)$. But, applying the Leibniz formula we obtain

$$\partial^{\alpha_j}(\psi_j \varphi) = \sum_{\beta \leqslant \alpha_j} \binom{\alpha_j}{\beta} \partial^{\alpha_j - \beta} \psi_j \, \partial^\beta \varphi ,$$

and then

$$\langle f, \varphi \rangle = \sum_{j \in \mathbb{N}} (-1)^{|\alpha_j|} \sum_{\beta \leqslant \alpha_j} \binom{\alpha_j}{\beta} \int_\Omega u_j \, \partial^{\alpha_j - \beta} \psi_j \, \partial^\beta \varphi \, dx .$$

Now set

$$N(\beta) = \left\{ j \in \mathbb{N} : \quad \beta \leqslant \alpha_j \right\} \qquad \forall\, \beta .$$

Then, it is not difficult to see that

$$\langle f, \varphi \rangle = \sum_{\beta} (-1)^{|\beta|} \int_{\Omega} f_\beta \, \partial^\beta \varphi \, dx \,,$$

with

$$f_\beta = (-1)^{|\beta|} \sum_{j \in N(\beta)} (-1)^{|\alpha_j|} \binom{\alpha_j}{\beta} u_j \, \partial^{\alpha_j - \beta} \psi_j \in C(\Omega).$$

Moreover, since $\operatorname{supp}(f_\beta) \subseteq \bigcup_{j \in N(\beta)} \operatorname{supp}(\psi_j)$ and $\{\operatorname{supp}(\psi_j)\}_{j \in \mathbb{N}}$ is locally finite (cf. Definition 9.26), it follows that $\{\operatorname{supp}(f_\beta)\}_\beta$ is also locally finite. Finally, if f is of finite order, then there exists a non-negative integer k such that $\chi_j f \in \mathcal{E}'_k(\mathbb{R}^n)$ for all $j \in \mathbb{N}$. Thus, we can choose $m_j = m > k$ for all $j \in \mathbb{N}$, and therefore

$$\alpha_j = \alpha = (m+1, \ldots, m+1) \quad \forall\, j \in \mathbb{N} \,.$$

In this case, we obtain

$$\langle f, \varphi \rangle = \sum_{j \in \mathbb{N}} (-1)^{|\alpha|} \sum_{\beta \leqslant \alpha} \binom{\alpha}{\beta} \int_{\Omega} u_j \partial^{\alpha - \beta} \psi_j \, \partial^\beta \varphi \, dx \,,$$

or equivalently,

$$\langle f, \varphi \rangle = \sum_{\beta \leqslant \alpha} (-1)^{|\beta|} \int_{\Omega} f_\beta \, \partial^\beta \varphi \, dx \,,$$

with

$$f_\beta = (-1)^{|\beta|} \sum_{j \in \mathbb{N}} (-1)^{|\alpha|} \binom{\alpha}{\beta} u_j \, \partial^{\alpha - \beta} \psi_j \in C(\Omega) \,,$$

which completes the proof. □

9.9 Fourier transform and tempered distributions

Given a function $f : \mathbb{R}^n \to \mathbb{R}$, its Fourier transform $\widehat{f}$ (if it exists) is defined by

$$\widehat{f}(\xi) := \int_{\mathbb{R}^n} f(x) \, e^{-\imath x \cdot \xi} \, dx \quad \forall\, \xi \in \mathbb{R}^n \,.$$

Clearly, to ensure the existence of the above integral, one needs some assumptions on f, which are specified in the following result.

Theorem 9.76.

i) *Let $f \in \mathrm{L}^1(\mathbb{R}^n)$. Then its Fourier transform $\widehat{f}$ exists, is continuous on $\mathbb{R}^n$, and*

$$|\widehat{f}(\xi)| \leqslant \|f\|_{0,1;\mathbb{R}^n} \qquad \forall\, \xi \in \mathbb{R}^n \,.$$

ii) *Let $f, g \in \mathrm{L}^1(\mathbb{R}^n)$. Then*

$$\int_{\mathbb{R}^n} f \widehat{g} = \int_{\mathbb{R}^n} \widehat{f} g \,.$$

iii) *Let* $f,\, g \in \mathrm{L}^1(\mathbb{R}^n)$. *Then the convolution* $f * g$ *exists* a.e. *in* $\mathbb{R}^n$, $f * g \in \mathrm{L}^1(\mathbb{R}^n)$ *and*

$$\widehat{f * g} \,=\, \widehat{f}\,\widehat{g}\,.$$

Proof. Since $|f(x)\, e^{-\imath x\cdot\xi}| \,=\, |f(x)|$ for all $x,\, \xi \in \mathbb{R}^n$ and for all $f \in \mathrm{L}^1(\mathbb{R}^n)$, it is straightforward to see that $\widehat{f}(\xi) \,<\, +\infty$ for all $\xi \in \mathbb{R}^n$ since $|\widehat{f}(\xi)| \,\leqslant\, \|f\|_{0,1;\mathbb{R}^n}$. Moreover, if $\{\xi_j\}_{j\in\mathbb{N}} \subseteq \mathbb{R}^n$ is such that $\xi_j \overset{j\to\infty}{\longrightarrow} \xi \in \mathbb{R}^n$, then

$$\widehat{f}(\xi_j) \,=\, \int_{\mathbb{R}^n} f(x)\, e^{-\imath x\cdot\xi_j}\, dx \,=\, \int_{\mathbb{R}^n} g_j(x)\, dx\,,$$

where

$$g_j(x) \,:=\, f(x)\, e^{-\imath x\cdot\xi_j} \quad \forall\, x \in \mathbb{R}^n\,, \quad \forall\, j \in \mathbb{N}\,.$$

Then, since each g_j is measurable, $g_j(x) \overset{j\to\infty}{\longrightarrow} g(x) := f(x)\, e^{-\imath x\cdot\,\xi}$ and $|g_j(x)| \,\leqslant\, |f(x)|$ for all $x \in \mathbb{R}^n$, with $f \in \mathrm{L}^1(\mathbb{R}^n)$, a direct application of the Lebesgue Dominated Convergence Theorem (cf. [113, Theorem 1.34], [120, Theorem 5.36]) implies that $\widehat{f}(\xi_j) \overset{j\to\infty}{\longrightarrow} \widehat{f}(\xi)$, thus proving that $\widehat{f}$ is continuous, which completes the proof of i). On the other hand, given $f, g \in \mathrm{L}^1(\mathbb{R}^n)$, we have

$$\widehat{f},\, \widehat{g} \in C(\mathbb{R}^n)\,, \quad |\widehat{f}(\xi)| \,\leqslant\, \|f\|_{0,1;\mathbb{R}^n} \quad \text{and} \quad |\widehat{g}(\xi)| \,\leqslant\, \|g\|_{0,1;\mathbb{R}^n} \quad \forall\, \xi \in \mathbb{R}^n\,.$$

It follows that $f\,\widehat{g}$ and $\widehat{f}\,g$ belong to $\mathrm{L}^1(\mathbb{R}^n)$, and therefore, by applying the classical Fubini's and Tonelli's Theorems (cf. [120, Theorems 6.1 and 6.10]), we obtain

$$\begin{aligned}\int_{\mathbb{R}^n} f(x)\,\widehat{g}(x)\, dx \,&=\, \int_{\mathbb{R}^n} f(x) \left\{ \int_{\mathbb{R}^n} g(y)\, e^{-\imath x\cdot y}\, dy \right\} dx \\ &=\, \int_{\mathbb{R}^n} g(y) \left\{ \int_{\mathbb{R}^n} f(x)\, e^{-\imath x\cdot y}\, dx \right\} dy \,=\, \int_{\mathbb{R}^n} g(y)\, \widehat{f}(y)\, dy\,,\end{aligned}$$

which proves ii). Finally, given $f,\, g \in \mathrm{L}^1(\mathbb{R}^n)$, consider $F : \mathbb{R}^n \times \mathbb{R}^n \to \mathbb{R}$ defined by $F(x, y) \,:=\, f(y)\, g(x - y)$ for all $x,\, y \in \mathbb{R}^n$, which is also measurable. Moreover, it is clear that

$$\begin{aligned}\int_{\mathbb{R}^n} \left\{ \int_{\mathbb{R}^n} |F(x, y)|\, dx \right\} dy \,&=\, \int_{\mathbb{R}^n} |f(y)| \left\{ \int_{\mathbb{R}^n} |g(x - y)|\, dx \right\} dy \\ &=\, \|f\|_{0,1;\mathbb{R}^n}\, \|g\|_{0,1;\mathbb{R}^n} \,<\, +\infty\,,\end{aligned}$$

which shows, by Tonelli's Theorem (cf. [120, Theorem 6.10]), that $F \in \mathrm{L}^1(\mathbb{R}^n \times \mathbb{R}^n)$. Then, applying Fubini's Theorem (cf. [120, Theorem 6.1, part ii)]), we deduce that the function

$$\mathbb{R}^n \ni x \to \int_{\mathbb{R}^n} F(x, y)\, dy \,=\, \int_{\mathbb{R}^n} f(y)\, g(x - y)\, dy \,=:\, (f * g)(x)$$

exists a.e. in $\mathbb{R}^n$ and belongs to $\mathrm{L}^1(\mathbb{R}^n)$. Then, again by the Fubini's and Tonelli's Theorems (cf. [120, Theorems 6.1 and 6.10]), we get

$$\begin{aligned}\widehat{(f * g)}(\xi) \,&=\, \int_{\mathbb{R}^n} (f * g)(x)\, e^{-\imath x\cdot\xi}\, dx \,=\, \int_{\mathbb{R}^n} \left\{ \int_{\mathbb{R}^n} f(y)\, g(x - y)\, dy \right\} e^{-\imath x\cdot\xi}\, dx \\ &=\, \int_{\mathbb{R}^n} f(y) \left\{ \int_{\mathbb{R}^n} g(x - y)\, e^{-\imath x\cdot\xi}\, dx \right\} dy\,.\end{aligned}$$

Performing the change of variable $z = x - y$, it follows that

$$\widehat{(f * g)}(\xi) = \int_{\mathbb{R}^n} f(y)\, e^{-\imath y\cdot\xi} \left\{ \int_{\mathbb{R}^n} g(z)\, e^{-\imath z\cdot\xi}\, dz \right\} dy$$
$$= \int_{\mathbb{R}^n} f(y)\, e^{-\imath y\cdot\xi}\, \widehat{g}(\xi)\, dy = \widehat{f}(\xi)\, \widehat{g}(\xi) \quad \forall\, \xi \in \mathbb{R}^n\,,$$

which proves iii) and completes the proof.

□

It is important to note here that if $f \in \mathrm{L}^1(\mathbb{R}^n)$ and $g \in C_0^\infty(\mathbb{R}^n)$, the identity ii) of the previous theorem states that the distribution induced by $\widehat{f} \in C(\mathbb{R}^n)$ is given in terms of f, that is,

$$\langle \widehat{f}, g\rangle := \int_{\mathbb{R}^n} \widehat{f}\, g = \int_{\mathbb{R}^n} f\, \widehat{g}\,.$$

However, $\widehat{g}$ is not necessarily an element of $C_0^\infty(\mathbb{R}^n)$, and therefore the right-hand side of the previous equation cannot be written as $\langle f, \widehat{g}\rangle$ (distribution induced by f evaluated at $\widehat{g}$). This observation motivates the introduction of a new space of test functions, namely $\mathcal{S}(\mathbb{R}^n) \subset C^\infty(\mathbb{R}^n)$, such that $\widehat{g} \in \mathcal{S}(\mathbb{R}^n)$ whenever $g \in \mathcal{S}(\mathbb{R}^n)$. In this way, the continuous linear functionals on $\mathcal{S}(\mathbb{R}^n)$ constitute a subspace $\mathcal{S}'(\mathbb{R}^n)$ of $\mathcal{D}(\mathbb{R}^n)$.

Definition 9.77. *Given multi-indices α, $\beta \in (\mathbb{N} \cup \{0\})^n$, define*

$$|\varphi|_{\alpha,\beta} := \sup_{x\in\mathbb{R}^n} |x^\alpha\, \partial^\beta \varphi(x)| \qquad \forall\, \varphi \in C^\infty(\mathbb{R}^n)\,,$$

and introduce the space

$$\mathcal{S}(\mathbb{R}^n) := \Big\{ \varphi \in C^\infty(\mathbb{R}^n) : \quad |\varphi|_{\alpha,\beta} < +\infty \quad \forall\, \alpha,\, \beta \in (\mathbb{N} \cup \{0\})^n \Big\}\,.$$

The space $\mathcal{S}(\mathbb{R}^n)$ is known as the Schwartz space, and its elements are referred to as rapidly decreasing test functions. The reason for this terminology is that if $\varphi \in \mathcal{S}(\mathbb{R}^n)$, then, for every multi-index α, the derivative $\partial^\alpha\varphi$ tends to 0 faster than $\|x\|^{-N}$ for every $N \in \mathbb{N}$ as $\|x\| \to +\infty$. Indeed, given $\varphi \in \mathcal{S}(\mathbb{R}^n)$ and $N \in \mathbb{N}$, we consider $\|x\| := \|x\|_\infty = |x_j|$ and $\beta = N\, e_j$, where e_j denotes the jth vector of the canonical basis of $\mathbb{R}^n$, and observe that

$$|\partial^\alpha \varphi(x)|\, \|x\|^N = |\partial^\alpha \varphi(x)\, x_j^N| = |\partial^\alpha \varphi(x)\, x^\beta| \leqslant |\varphi|_{\beta,\alpha} < +\infty\,,$$

hence

$$|\partial^\alpha \varphi(x)| \leqslant |\varphi|_{\beta,\alpha}\, \|x\|^{-N} \qquad \forall\, x \in \mathbb{R}^n \setminus \{0\}\,. \tag{9.48}$$

A nontrivial example of an element of $\mathcal{S}(\mathbb{R}^n)$ is given by the function $\varphi(x) = e^{-a\|x\|^2}$ for all $x \in \mathbb{R}^n$, with $a > 0$.

Furthermore, similarly to the corresponding analysis for $\mathcal{E}(\Omega)$ (cf. Section 9.7), the space $\mathcal{S}(\mathbb{R}^n)$ is endowed with the topology induced by the family of seminorms given by $\Big\{ |\cdot|_{\alpha,\beta} : \alpha,\, \beta \in (\mathbb{N} \cup \{0\})^n \Big\}$, which, being countable, also induces a metrizable topology. Moreover, letting $\big\{ |\cdot|_{j,\mathcal{S}} \big\}_{j\in\mathbb{N}}$ be an enumeration of these seminorms, and analogously to (9.20), a compatible metric is given by

$$d_{\mathcal{S}}(\varphi, \psi) := \sum_{j=1}^{\infty} \frac{1}{2^j} \left\{ \frac{|\varphi - \psi|_{j,\mathcal{S}}}{1 + |\varphi - \psi|_{j,\mathcal{S}}} \right\} \qquad \forall\, \varphi,\, \psi \in \mathcal{S}(\mathbb{R}^n)\,. \tag{9.49}$$

Consequently, the continuity of a mapping from $\mathcal{S}(\mathbb{R}^n)$ into any other topological space is equivalent to its sequential continuity.

Accordingly, although convergence in $\mathcal{S}(\mathbb{R}^n)$ could simply be defined in terms of the metric $d_{\mathcal{S}}$, we next provide a more explicit characterization, which, as in the case of $\mathcal{E}(\Omega)$ (cf. Definition 9.36), follows from (9.49) and the summability of the series $\sum_{j=1}^{\infty} \frac{1}{2^j}$.

Definition 9.78. *A sequence $\{\varphi_j\}_{j \in \mathbb{N}} \subseteq \mathcal{S}(\mathbb{R}^n)$ is said to converge to the zero function in $\mathcal{S}(\mathbb{R}^n)$ if $|\varphi_j|_{\alpha,\beta} \overset{j\to\infty}{\longrightarrow} 0$ for all $\alpha,\, \beta \in (\mathbb{N} \cup \{0\})^n$.*

The following lemma will be used frequently throughout the remainder of this chapter and in the next.

Lemma 9.79. *For each $N > n/2$, the function $\left(1 + \|\cdot\|^2\right)^{-N}$ belongs to $\mathrm{L}^1(\mathbb{R}^n)$.*

Proof. Using spherical coordinates,

$$\begin{aligned}
x_1 &= r\, \cos(\vartheta_1)\\
x_2 &= r\, \sin(\vartheta_1)\, \cos(\vartheta_2)\\
x_3 &= r\, \sin(\vartheta_1)\, \sin(\vartheta_2)\, \cos(\vartheta_3)\\
&\vdots\\
x_{n-1} &= r\, \sin(\vartheta_1)\, \sin(\vartheta_2)\, \cdots\, \sin(\vartheta_{n-2})\, \cos(\vartheta_{n-1})\\
x_n &= r\, \sin(\vartheta_1)\, \sin(\vartheta_2)\, \cdots\, \sin(\vartheta_{n-2})\, \sin(\vartheta_{n-1})\,,
\end{aligned}$$

we obtain

$$\begin{aligned}
\int_{\mathbb{R}^n} (1 + \|x^2\|)^{-N}\, dx &= \int_0^{\infty} \int_0^{\pi} \int_0^{\pi} \cdots \int_0^{\pi} \int_0^{2\pi} (1 + r^2)^{-N}\, r^{n-1}\, d\vartheta_1 \cdots d\vartheta_{n-1}\, dr\\
&= 2\pi^{n-1} \int_0^{\infty} (1 + r^2)^{-N}\, r^{n-1}\, dr\\
&= 2\pi^{n-1} \left\{ \int_0^1 (1 + r^2)^{-N}\, r^{n-1}\, dr + \int_1^{+\infty} (1 + r^2)^{-N}\, r^{n-1}\, dr \right\}.
\end{aligned}$$

Note that the first integral on the right-hand side of the last line of the previous equation is finite. Moreover, for the second integral we have

$$\int_1^{+\infty} \left(1 + r^2\right)^{-N} r^{n-1}\, dr \leqslant \int_1^{+\infty} r^{\alpha}\, dr\,,$$

with $\alpha = n - 1 - 2N$ and $1 + \alpha = n - 2N < 0$, from which it follows that

$$\int_1^{+\infty} r^{\alpha}\, dr = \frac{-1}{1 + \alpha} < +\infty\,,$$

and consequently $\int_{\mathbb{R}^n} \left(1 + \|x\|^2\right)^{-N} dx < +\infty$.

□

From now on, we denote

$$D_j := -\imath \frac{\partial}{\partial x_j} \quad \forall\, j \in \{1, 2, \ldots, n\} \quad \text{and} \quad D^\alpha := (-i)^{|\alpha|}\, \partial^\alpha \quad \forall\, \alpha \in (\mathbb{N} \cup \{0\})^n . \tag{9.50}$$

Theorem 9.80. *The following results hold*

i) *$\mathcal{S}(\mathbb{R}^n)$ is stable under differentiation and multiplication by polynomials, that is, for every pair of polynomials p and q, the mapping*

$$\begin{array}{rl} \mathcal{S}(\mathbb{R}^n) \to & \mathcal{S}(\mathbb{R}^n) \\ \varphi & \to p\, q(D)(\varphi)\,, \end{array}$$

where $q(x) = \sum\limits_{|\gamma| \leqslant m} a_\gamma\, x^\gamma$, $a_\gamma \in \mathbb{K}$, and $q(D)(\varphi) := \sum\limits_{|\gamma| \leqslant m} a_\gamma\, D^\gamma \varphi$, is continuous.

ii) *$C_0^\infty(\mathbb{R}^n) \subseteq \mathcal{S}(\mathbb{R}^n)$ and the inclusion $i : C_0^\infty(\mathbb{R}^n) \to \mathcal{S}(\mathbb{R}^n)$ is continuous.*

iii) *$C_0^\infty(\mathbb{R}^n)$ is dense in $\mathcal{S}(\mathbb{R}^n)$.*

iv) *$\mathcal{S}(\mathbb{R}^n) \subseteq \mathrm{L}^1(\mathbb{R}^n)$ and the inclusion $i : \mathcal{S}(\mathbb{R}^n) \to \mathrm{L}^1(\mathbb{R}^n)$ is continuous.*

Proof. Let q be the polynomial given by $q(x) = \sum\limits_{|\gamma| \leqslant m} a_\gamma\, x^\gamma$. Then, for each $\varphi \in \mathcal{S}(\mathbb{R}^n)$, and for each pair of multi-indices α and β, we have

$$\begin{aligned} |q(D)(\varphi)|_{\alpha,\beta} &:= \sup_{x \in \mathbb{R}^n} \left| x^\alpha\, \partial^\beta \left\{ \sum_{|\gamma| \leqslant m} a_\gamma\, D^\gamma \varphi(x) \right\} \right| \\ &= \sup_{x \in \mathbb{R}^n} \left| x^\alpha \sum_{|\gamma| \leqslant m} (-\imath)^{|\gamma|}\, a_\gamma\, \partial^{\beta+\gamma} \varphi(x) \right| \\ &\leqslant \sum_{|\gamma| \leqslant m} |a_\gamma| \sup_{x \in \mathbb{R}^n} \left| x^\alpha\, \partial^{\beta+\gamma} \varphi(x) \right| \\ &= \sum_{|\gamma| \leqslant m} |a_\gamma|\, |\varphi|_{\alpha,\beta+\gamma} < +\infty\,, \end{aligned} \tag{9.51}$$

which proves that $q(D)\,\varphi \in \mathcal{S}(\mathbb{R}^n)$. Moreover, given another polynomial p of the form $p(x) = \sum\limits_{|\gamma| \leqslant N} b_\gamma\, x^\gamma$, we have, for each $\varphi \in \mathcal{S}(\mathbb{R}^n)$, and for each pair of multi-indices α and β, that

$$\begin{aligned} |p\,\varphi|_{\alpha,\beta} &:= \sup_{x \in \mathbb{R}^n} |\, x^\alpha\, \partial^\beta (p(x)\varphi(x))| \\ &= \sup_{x \in \mathbb{R}^n} \left| x^\alpha \sum_{\delta \leqslant \beta} \binom{\beta}{\delta} \partial^\delta p(x)\, \partial^{\beta-\delta} \varphi(x) \right| \\ &= \sup_{x \in \mathbb{R}^n} \left| x^\alpha \sum_{\delta \leqslant \beta} \binom{\beta}{\delta} \left\{ \sum_{\gamma \geqslant \delta} \frac{b_\gamma\, \gamma!}{(\gamma-\delta)!}\, x^{\gamma-\delta} \right\} \partial^{\beta-\delta} \varphi(x) \right| \\ &\leqslant \sum_{\delta \leqslant \beta} \binom{\beta}{\delta} \sum_{\gamma \geqslant \delta} \frac{|b_\gamma|\, \gamma!}{(\gamma-\delta)!} \sup_{x \in \mathbb{R}^n} \left| x^{\alpha+\gamma-\delta}\, \partial^{\beta-\delta} \varphi(x) \right| \\ &= \sum_{\delta \leqslant \beta} \binom{\beta}{\delta} \sum_{\gamma \geqslant \delta} \frac{|b_\gamma|\, \gamma!}{(\gamma-\delta)!}\, |\varphi|_{\alpha+\gamma-\delta,\beta-\delta} < +\infty\,, \end{aligned} \tag{9.52}$$

which shows that $p\varphi \in \mathcal{S}(\mathbb{R}^n)$. Thus, from (9.51) and (9.52) we deduce that for each pair of polynomials p and q, and for each $\varphi \in \mathcal{S}(\mathbb{R}^n)$, one has $p\,q(D)(\varphi) \in \mathcal{S}(\mathbb{R}^n)$. Furthermore, from these same inequalities it can be easily shown that, if $\{\varphi_j\}_{j\in\mathbb{N}} \subseteq \mathcal{S}(\mathbb{R}^n)$ converges to φ in $\mathcal{S}(\mathbb{R}^n)$, then $p\,q(D)(\varphi_j) \overset{j\to\infty}{\to} p\,q(D)(\varphi)$ in $\mathcal{S}(\mathbb{R}^n)$, which completes the proof of i). On the other hand, given $\varphi \in C_0^\infty(\mathbb{R}^n)$ and α, β multi-indices, it is clear that

$$|\varphi|_{\alpha,\beta} := \sup_{x\in\mathbb{R}^n} |x^\alpha\, \partial^\beta \varphi(x)| = \sup_{x\in\operatorname{supp}(\varphi)} |x^\alpha\, \partial^\beta \varphi(x)| < +\infty,$$

which means that $\varphi \in \mathcal{S}(\mathbb{R}^n)$. Moreover, if $\{\varphi_j\}_{j\in\mathbb{N}} \subseteq \mathcal{D}(\mathbb{R}^n)$ converges to the zero function, then there exists a compact set $K \subseteq \mathbb{R}^n$ such that $\operatorname{supp}(\varphi_j) \subseteq K$ for all $j \in \mathbb{N}$ and

$$\sup_{x\in K} |\partial^\alpha \varphi_j(x)| \overset{j\to\infty}{\to} 0 \quad \forall\, \alpha \in (\mathbb{N}\cup\{0\})^n.$$

It follows that for each pair of multi-indices α, β, there holds

$$|\varphi_j|_{\alpha,\beta} = \sup_{x\in\mathbb{R}^n} |x^\alpha\, \partial^\beta \varphi_j(x)| = \sup_{x\in K} |x^\alpha\, \partial^\beta \varphi_j(x)| \leqslant C_\alpha \sup_{x\in K} |\partial^\beta \varphi_j(x)| \overset{j\to\infty}{\to} 0,$$

which says that $\varphi_j \overset{j\to\infty}{\to} 0$ in $\mathcal{S}(\mathbb{R}^n)$, and therefore the injection $i : C_0^\infty(\mathbb{R}^n) \to \mathcal{S}(\mathbb{R}^n)$ is continuous, which proves ii). Let us now show that $C_0^\infty(\mathbb{R}^n)$ is dense in $\mathcal{S}(\mathbb{R}^n)$. Indeed, let $\varphi \in \mathcal{S}(\mathbb{R}^n)$ and, given $\psi \in C_0^\infty(\mathbb{R}^n)$ such that $\psi \equiv 1$ on $\bar{B}(0,1)$, define

$$\varphi_j(x) := \psi\left(\frac{x}{j}\right) \varphi(x) \quad \forall\, x \in \mathbb{R}^n, \quad \forall\, j \in \mathbb{N}.$$

It is clear that $\varphi_j \in C_0^\infty(\mathbb{R}^n)$ and that $\varphi_j \equiv \varphi$ on $\bar{B}(0,j)$. We now see that $\varphi_j \overset{j\to\infty}{\to} \varphi$ in $\mathcal{S}(\mathbb{R}^n)$. In fact, for each pair α, $\beta \in (\mathbb{N}\cup\{0\})^n$, we have

$$\begin{aligned}
|\varphi - \varphi_j|_{\alpha,\beta} &= \sup_{x\in\mathbb{R}^n} |x^\alpha\, \partial^\beta (\varphi - \varphi_j)(x)| \\
&= \sup_{\|x\|\geqslant j} \left| x^\alpha\, \partial^\beta \left\{ \varphi(x) \left(1 - \psi\left(\frac{x}{j}\right)\right)\right\}\right| \\
&= \sup_{\|x\|\geqslant j} \left| x^\alpha \sum_{\gamma\leqslant\beta} \left\{ \binom{\beta}{\gamma} \partial^\gamma \varphi(x)\, \partial_x^{\beta-\gamma} \left(1 - \psi\left(\frac{x}{j}\right)\right)\right|\right. \\
&\leqslant \sum_{\gamma\leqslant\beta} \binom{\beta}{\gamma} \sup_{\|x\|\geqslant j} \left| x^\alpha\, \partial^\gamma \varphi(x)\, \partial^{\beta-\gamma} \left(1 - \psi\left(\frac{x}{j}\right)\right) j^{-|\beta-\gamma|}\right|,
\end{aligned}$$

from which, noting that $j^{-|\beta-\gamma|} \leqslant 1$, and defining

$$C_\beta := \max_{\gamma\leqslant\beta} \left\{ \binom{\beta}{\gamma} \sup_{y\in\mathbb{R}^n} \left|\partial^{\beta-\gamma}(1 - \psi(y))\right|\right\},$$

it follows that

$$|\varphi - \varphi_j|_{\alpha,\beta} \leqslant C_\beta \sum_{\gamma\leqslant\beta} \sup_{\|x\|\geqslant j} |x^\alpha\, \partial^\gamma \varphi(x)|.$$

Now, given $x \in \mathbb{R}^n$, let $i \in \{1, 2, \ldots, n\}$ be such that $\|x\| = |x_i|$. Then we have

$$|x^\alpha\, \partial^\gamma \varphi(x)| = \|x\|^{-1}\, |x^\alpha\, x_i\, \partial^\gamma \varphi(x)| \leqslant \|x\|^{-1} \sum_{|\delta|=|\alpha|+1} |\varphi|_{\delta,\gamma},$$

which implies that

$$\sup_{\|x\|\geqslant j} |x^\alpha \, \partial^\gamma \varphi(x)| \leqslant \frac{1}{j} \sum_{|\delta|=|\alpha|+1} |\varphi|_{\delta,\gamma}\,,$$

and consequently

$$|\varphi - \varphi_j|_{\alpha,\beta} \leqslant \left\{ C_\beta \sum_{\gamma\leqslant\beta} \sum_{|\delta|=|\alpha|+1} |\varphi|_{\delta,\gamma} \right\} \frac{1}{j} \overset{j\to\infty}{\longrightarrow} 0\,,$$

thus completing the proof of iii). Finally, let $\varphi \in \mathcal{S}(\mathbb{R}^n)$. We first show that $\varphi \in \mathrm{L}^1(\mathbb{R}^n)$, and then that the injection $i : \mathcal{S}(\mathbb{R}^n) \to \mathrm{L}^1(\mathbb{R}^n)$ is continuous. Indeed, it is easy to observe first that for each $N \in \mathbb{N}$

$$\left(1 + \|x\|^2\right)^N |\varphi(x)| \leqslant C \sum_{|\alpha|\leqslant 2N} |\varphi|_{\alpha,0} < +\infty \quad \forall\, x \in \mathbb{R}^n\,.$$

Moreover, if we consider in particular $N > n/2$, it follows from Lemma 9.79 that the function $\left(1 + \|\cdot\|^2\right)^{-N}$ belongs to $\mathrm{L}^1(\mathbb{R}^n)$. Consequently, since clearly

$$\varphi(x) = \left(\left(1 + \|x\|^2\right)^N \varphi(x)\right) \left(1 + \|x\|^2\right)^{-N} =: F(x)\, G(x) \quad \forall\, x \in \mathbb{R}^n\,,$$

with F bounded in $\mathbb{R}^n$ and $G \in \mathrm{L}^1(\mathbb{R}^n)$, we deduce that $\varphi \in \mathrm{L}^1(\mathbb{R}^n)$. Furthermore, from the above bounds we have

$$\|\varphi\|_{0,1;\mathbb{R}^n} \leqslant C \left\| \left(1 + \|\cdot\|^2\right)^{-N} \right\|_{0,1;\mathbb{R}^n} \sum_{|\alpha|\leqslant 2N} |\varphi|_{\alpha,0}\,,$$

from which it follows that $i : \mathcal{S}(\mathbb{R}^n) \to \mathrm{L}^1(\mathbb{R}^n)$ is continuous, and thus iv) is proved. □

The next goal is to prove that the Fourier transform is a continuous isomorphism from $\mathcal{S}(\mathbb{R}^n)$ onto $\mathcal{S}(\mathbb{R}^n)$. For this purpose, we need some preliminary lemmas.

Lemma 9.81. *Let $\varphi \in \mathcal{S}(\mathbb{R}^n)$. Then, for each multi-index α, we have*

a) $(\widehat{D^\alpha \varphi})(\xi) = \xi^\alpha \, \widehat{\varphi}\,(\xi)$ *for all* $\xi \in \mathbb{R}^n$.
b) $(\widehat{x^\alpha \varphi})(\xi) = (-1)^{|\alpha|} D^\alpha \, \widehat{\varphi}\,(\xi)$ *for all* $\xi \in \mathbb{R}^n$.

Proof. Note first that the Fourier transforms of $D^\alpha \varphi$ and $x^\alpha\, \varphi$ exist by i) and iv) of Theorem 9.80. Then, for each $\xi \in \mathbb{R}^n$ we have

$$\begin{aligned} \xi^\alpha\, \widehat{\varphi}(\xi) &= \xi^\alpha \int_{\mathbb{R}^n} \varphi(x)\, e^{-\imath x\cdot\xi}\, dx = \int_{\mathbb{R}^n} \varphi(x)\, \xi^\alpha\, e^{-\imath x\cdot\xi}\, dx \\ &= \int_{\mathbb{R}^n} \varphi(x)\, \imath^{|\alpha|}\, \partial_x^\alpha (e^{-\imath x\cdot\xi})\, dx = \lim_{R\to+\infty} \int_{B(0,R)} \varphi(x)\, \imath^{|\alpha|}\, \partial_x^\alpha (e^{-\imath x\cdot\xi})\, dx\,. \end{aligned}$$

For $|\alpha| = 1$, we integrate by parts and obtain

$$\begin{aligned} \imath \int_{B(0,R)} \varphi(x)\, \frac{\partial}{\partial x_j} (e^{-\imath x\cdot\xi})\, dx &= (-\imath) \int_{B(0,R)} \frac{\partial \varphi(x)}{\partial x_j}\, e^{-\imath x\cdot\xi}\, dx \\ &\quad + \imath \int_{\partial B(0,R)} \varphi(x)\, e^{-\imath x\cdot\xi}\, \nu_j\, ds\,, \end{aligned} \tag{9.53}$$

where $\boldsymbol{\nu} := (\nu_1, \nu_2, \ldots, \nu_n)^{\mathrm{t}}$ is the unit normal vector on $\partial B(0,R)$. Then, since for each $N \in \mathbb{N}$ there exists $C_N > 0$ such that

$$|\varphi(x)| \leqslant C_N \|x\|^{-N} \quad \forall\, x \in \mathbb{R}^n$$

(cf. (9.48)), it follows that

$$\begin{aligned} \left| \int_{\partial B(0,R)} \varphi(x)\, e^{-\imath x\cdot\xi}\, \nu_j\, ds \right| &\leqslant C_N\, R^{-N}\, |\partial B(0,R)| \\ &= C_N\, C\, R^{-N}\, R^{n-1} = \widetilde{C}\, R^{n-1-N} \overset{R\to+\infty}{\longrightarrow} 0 \end{aligned}$$

provided that $N > n-1$, from which we conclude, according to (9.53), that

$$\imath \int_{\mathbb{R}^n} \varphi(x)\, \frac{\partial}{\partial x_j}\,(e^{-\imath x\cdot\xi})\, dx = (-\imath) \int_{\mathbb{R}^n} \frac{\partial \varphi(x)}{\partial x_j}\, e^{-\imath x\cdot\xi}\, dx\,.$$

Moreover, using induction it is easy to see that

$$\imath^{|\alpha|} \int_{\mathbb{R}^n} \varphi(x)\, \partial_x^\alpha\,(e^{-\imath x\cdot\xi})\, dx = (-\imath)^{|\alpha|} \int_{\mathbb{R}^n} \partial^\alpha \varphi(x)\, e^{-\imath x\cdot\xi}\, dx$$

for each $\alpha \in \big(\mathbb{N} \cup \{0\}\big)^n$, and consequently

$$\xi^\alpha\, \widehat{\varphi}(\xi) = \int_{\mathbb{R}^n} (-\imath)^{|\alpha|}\, \partial^\alpha \varphi(x)\, e^{-\imath x\cdot\xi}\, dx = \int_{\mathbb{R}^n} D^\alpha \varphi(x)\, e^{-\imath x\cdot\xi}\, dx = (\widehat{D^\alpha \varphi})(\xi)\,,$$

which proves a). On the other hand, differentiating under the integral sign in the expression defining $\widehat{\varphi}(\xi)$, we obtain

$$\begin{aligned} \partial^\alpha \widehat{\varphi}(\xi) &= \partial^\alpha \left\{ \int_{\mathbb{R}^n} \varphi(x)\, e^{-\imath x\cdot\xi}\, dx \right\} = \imath^{|\alpha|}\, (-1)^{|\alpha|} \int_{\mathbb{R}^n} x^\alpha\, \varphi(x)\, e^{-\imath x\cdot\xi}\, dx \\ &= \imath^{|\alpha|}\, (-1)^{|\alpha|}\, (\widehat{x^\alpha\, \varphi})(\xi)\,, \end{aligned}$$

that is,

$$(-1)^{|\alpha|}\, D^\alpha \widehat{\varphi}(\xi) = (\widehat{x^\alpha\, \varphi})(\xi) \quad \forall\, \xi \in \mathbb{R}^n\,,$$

thus proving b).

□

Lemma 9.82. *The Fourier transform $\mathcal{F} : \mathcal{S}(\mathbb{R}^n) \to \mathcal{S}(\mathbb{R}^n)$ defined by*

$$\widehat{f} := \mathcal{F}(f) \qquad \forall\, f \in \mathcal{S}(\mathbb{R}^n)\,, \tag{9.54}$$

is continuous.

Proof. Given $\varphi \in \mathcal{S}(\mathbb{R}^n)$, we let (cf. (9.54)) $\widehat{\varphi} := \mathcal{F}(\varphi)$. Then, according to part b) of Lemma 9.81, we have

$$D^\alpha \widehat{\varphi}(\xi) = (-1)^{|\alpha|} (\widehat{x^\alpha\, \varphi})(\xi) \quad \forall\, \xi \in \mathbb{R}^n\,,$$

which indicates, since $(\widehat{x^\alpha\, \varphi}) \in C(\mathbb{R}^n)$, that $D^\alpha \widehat{\varphi} \in C(\mathbb{R}^n)$ for all α, and thus $\widehat{\varphi} \in C^\infty(\mathbb{R}^n)$. Moreover, using parts a) and b) of Lemma 9.81, we deduce that for each $\alpha, \beta \in \big(\mathbb{N} \cup \{0\}\big)^n$ there holds

$$(\widehat{D^\alpha(x^\beta\varphi)})(\xi) \;=\; \xi^\alpha\,(\widehat{x^\beta\varphi})(\xi) \;=\; \xi^\alpha\,(-1)^{|\beta|}\,D^\beta\widehat{\varphi}(\xi)\,,$$

and then, noting that $D^\alpha(x^\beta\varphi) \in \mathcal{S}(\mathbb{R}^n) \subseteq \mathrm{L}^1(\mathbb{R}^n)$, and using part i) of Theorem 9.76, it follows that

$$|\widehat{\varphi}|_{\alpha,\beta} \;=\; \sup_{\xi\in\mathbb{R}^n}|\xi^\alpha\,\partial^\beta\widehat{\varphi}(\xi)| \;=\; \sup_{\xi\in\mathbb{R}^n}|(\widehat{D^\alpha(x^\beta\,\varphi)})(\xi)| \;\leqslant\; \|D^\alpha(x^\beta\,\varphi)\|_{0,1;\mathbb{R}^n} \;<+\infty\,, \qquad (9.55)$$

which allows us to conclude that $\widehat{\varphi} \in \mathcal{S}(\mathbb{R}^n)$. On the other hand, for the continuity of $\mathcal{F}$, we observe from (9.55) that for all $\alpha,\,\beta \in \big(\mathbb{N}\cup\{0\}\big)^n$, and for $N > \frac{n}{2}$ (cf. Lemma 9.79), we have

$$\begin{aligned} |\widehat{\varphi}|_{\alpha,\beta} \;&\leqslant\; \|D^\alpha(x^\beta\,\varphi)\|_{0,1;\mathbb{R}^n} \;=\; \int_{\mathbb{R}^n}|D^\alpha(x^\beta\,\varphi)(x)|\,dx \\ &=\; \int_{\mathbb{R}^n}\big(1+\|x\|^2\big)^{-N}\,\big(1+\|x\|^2\big)^{N}\,|D^\alpha(x^\beta\,\varphi)(x)|\,dx \\ &\leqslant\; \sup_{x\in\mathbb{R}^n}\Big(\big(1+\|x\|^2\big)^N\,|D^\alpha(x^\beta\,\varphi)(x)|\Big)\,\big\|\big(1+\|\cdot\|^2\big)^{-N}\big\|_{0,1;\mathbb{R}^n} \\ &\leqslant\; C\,C_{\alpha,\beta}\sum_{\substack{|\widetilde{\alpha}|\leqslant 2N+|\beta|\\ |\widetilde{\beta}|\leqslant|\alpha|}}|\varphi|_{\widetilde{\alpha},\widetilde{\beta}}\,, \end{aligned} \qquad (9.56)$$

where $C := \big\|\big(1+\|\cdot\|^2\big)^{-N}\big\|_{0,1;\mathbb{R}^n}$ and $C_{\alpha,\beta} > 0$ is computed similarly to the derivations of (9.51) and (9.52). Thus, if $\{\varphi_j\}_{j\in\mathbb{N}} \subseteq \mathcal{S}(\mathbb{R}^n)$ converges to the zero function in $\mathcal{S}(\mathbb{R}^n)$, the inequality (9.56) ensures that the same occurs for $\{\widehat{\varphi}_j\}_{j\in\mathbb{N}} \subseteq \mathcal{S}(\mathbb{R}^n)$.

□

The following lemma provides a key technical result for the subsequent analysis.

Lemma 9.83. *Let* $\rho(x) = e^{-\frac{1}{2}\|x\|^2}$ *for all* $x \in \mathbb{R}^n$. *Then*

$$\widehat{\rho}(\xi) \;=\; (2\pi)^{n/2}\,e^{-\frac{1}{2}\|\xi\|^2} \qquad \forall\,\xi\in\mathbb{R}^n\,.$$

Proof. Let us begin by observing that for $n = 1$, ρ satisfies the ordinary differential equation $\imath\,D\rho + x\,\rho = 0$, from which, by applying Lemma 9.81, we deduce that $\imath\,D\widehat{\rho}(\xi) + \xi\,\widehat{\rho}(\xi) = 0$. Note that here we use the first notation from (9.50), but for simplicity we write D instead of D_1. Solving this first-order equation yields $\widehat{\rho}(\xi) = c\,e^{-\frac{1}{2}\xi^2}$ for all $\xi \in \mathbb{R}$, with $c \in \mathbb{R}$, and since $c = \widehat{\rho}(0) = \displaystyle\int_{\mathbb{R}} e^{-\frac{1}{2}x^2}\,dx = (2\pi)^{1/2}$, we obtain $\widehat{\rho}(\xi) = (2\pi)^{1/2}\,e^{-\frac{1}{2}\xi^2}$ for all $\xi \in \mathbb{R}$. Then, for $n > 1$ we have

$$\widehat{\rho}(\xi) \;=\; \int_{\mathbb{R}^n} e^{-\frac{1}{2}\|x\|^2}\,e^{-\imath x\cdot\xi}\,dx \;=\; \int_{\mathbb{R}^n}\left\{\prod_{j=1}^n e^{-\frac{1}{2}x_j^2-\imath x_j\xi_j}\right\}dx\,,$$

which, by applying Fubini's and Tonelli's Theorems (cf. [120, Theorems 6.1 and 6.10]) and the previous computation for $n = 1$, reduces to

$$\begin{aligned} \widehat{\rho}(\xi) \;&=\; \prod_{j=1}^n\left\{\int_{\mathbb{R}} e^{-\frac{1}{2}x_j^2-\imath\,x_j\,\xi_j}\,dx_j\right\} \;=\; \prod_{j=1}^n\left\{(2\pi)^{1/2}\,e^{-\frac{1}{2}\xi_j^2}\right\} \\ &=\; (2\pi)^{n/2}\,e^{-\frac{1}{2}\|\xi\|^2} \qquad \forall\,\xi\in\mathbb{R}^n\,. \end{aligned}$$

□

We are now in a position to prove the Fourier Inversion Theorem, which recovers the function $\varphi \in \mathcal{S}(\mathbb{R}^n)$ from its transform $\widehat{\varphi} \in \mathcal{S}(\mathbb{R}^n)$.

Theorem 9.84. *For each $\varphi \in \mathcal{S}(\mathbb{R}^n)$ we have*

$$\varphi(x) = (2\pi)^{-n} \int_{\mathbb{R}^n} \widehat{\varphi}(\xi)\, e^{\imath\, x\cdot\xi}\, d\xi \quad \forall\, x \in \mathbb{R}^n .$$

Proof. Let $\varphi \in \mathcal{S}(\mathbb{R}^n)$, and for each $\varepsilon > 0$ define

$$\psi_\varepsilon(x) = e^{-\frac{1}{2}\varepsilon^2 \|x\|^2} \quad \forall\, x \in \mathbb{R}^n ,$$

that is, $\psi_\varepsilon(x) = \rho(\varepsilon x)$ for all $x \in \mathbb{R}^n$, where ρ is the function from the previous lemma. It follows, by making the change of variable $y = \varepsilon x$, that

$$\begin{aligned}\widehat{\psi}_\varepsilon(\xi) &= \int_{\mathbb{R}^n} \rho(\varepsilon x)\, e^{-\imath x\cdot\xi}\, dx = \varepsilon^{-n} \int_{\mathbb{R}^n} \rho(y)\, e^{-\imath y\cdot\frac{\xi}{\varepsilon}}\, dy \\ &= \varepsilon^{-n}\, \widehat{\rho}\left(\frac{\xi}{\varepsilon}\right) = \varepsilon^{-n}\, (2\pi)^{n/2}\, e^{-\frac{1}{2\varepsilon^2}\|\xi\|^2} \quad \forall\, \xi \in \mathbb{R} .\end{aligned}$$

Then, applying identity ii) of Theorem 9.76 and making the change of variable $\varepsilon z = x$, we obtain

$$\begin{aligned}\int_{\mathbb{R}^n} \widehat{\varphi}(\xi)\, \psi_\varepsilon(\xi)\, d\xi &= \int_{\mathbb{R}^n} \varphi(x)\, \widehat{\psi}_\varepsilon(x)\, dx \\ &= \int_{\mathbb{R}^n} \varphi(x)\, \varepsilon^{-n}\, (2\pi)^{n/2}\, e^{-\frac{1}{2\varepsilon^2}\|x\|^2}\, dx \\ &= (2\pi)^{n/2} \int_{\mathbb{R}^n} \varphi(\varepsilon z)\, e^{-\frac{1}{2}\|z\|^2}\, dz ,\end{aligned}$$

that is,

$$\int_{\mathbb{R}^n} \widehat{\varphi}(\xi)\, e^{-\frac{1}{2}\varepsilon^2\|\xi\|^2}\, d\xi = (2\pi)^{n/2} \int_{\mathbb{R}^n} \varphi(\varepsilon z)\, e^{-\frac{1}{2}\|z\|^2}\, dz .$$

For the integral on the left-hand side, we note that $\widehat{\varphi} \in \mathcal{S}(\mathbb{R}^n) \subseteq \mathrm{L}^1(\mathbb{R}^n)$ and that there holds $0 < e^{-\frac{1}{2}\varepsilon^2\|\xi\|^2} < 1$. Similarly, for the one on the right-hand side we observe that $\varphi(\varepsilon z)$ is bounded and that $e^{-\frac{1}{2}\|z\|^2} \in \mathrm{L}^1(\mathbb{R}^n)$. The Lebesgue Dominated Convergence Theorem (cf. [113, Theorem 1.34], [120, Theorem 5.36]) can then be applied to both, so that letting $\varepsilon \to 0$, we obtain

$$\int_{\mathbb{R}^n} \widehat{\varphi}(\xi)\, d\xi = (2\pi)^{n/2}\, \varphi(0) \int_{\mathbb{R}^n} e^{-\frac{1}{2}\|z\|^2}\, dz = (2\pi)^n\, \varphi(0) ,$$

or equivalently,

$$\varphi(0) = (2\pi)^{-n} \int_{\mathbb{R}^n} \widehat{\varphi}(\xi)\, d\xi . \tag{9.57}$$

On the other hand, defining $\boldsymbol{\tau}_{-h}\, \varphi(x) := \varphi(x+h)$ for all $x \in \mathbb{R}^n$, for all $h \in \mathbb{R}^n$, we easily deduce that

$$\widehat{(\boldsymbol{\tau}_{-h}\, \varphi)}(\xi) = \widehat{\varphi}(\xi)\, e^{\imath\, \xi\cdot h} \quad \forall\, \xi \in \mathbb{R}^n ,$$

and then, applying (9.57) to $\boldsymbol{\tau}_{-h}\, \varphi \in \mathcal{S}(\mathbb{R}^n)$, we conclude that

$$(\boldsymbol{\tau}_{-h}\, \varphi)(0) = (2\pi)^{-n} \int_{\mathbb{R}^n} \widehat{\varphi}(\xi)\, e^{\imath\, \xi\cdot h}\, d_\xi ,$$

that is,

$$\varphi(h) = (2\pi)^{-n} \int_{\mathbb{R}^n} \widehat{\varphi}(\xi)\, e^{\imath \xi \cdot h}\, d_\xi \qquad \forall\, h \in \mathbb{R}^n,$$

thus completing the proof.

□

By combining the previous results we obtain the following theorem.

Theorem 9.85. *The Fourier transform $\mathcal{F} : \mathcal{S}(\mathbb{R}^n) \to \mathcal{S}(\mathbb{R}^n)$ is bijective, and both $\mathcal{F}$ and its inverse $\mathcal{F}^{-1}$ are continuous isomorphisms.*

Proof. Note that the continuity of $\mathcal{F}$ was established in Lemma 9.82. Now, let $\varphi \in \mathcal{S}(\mathbb{R}^n)$ be such that $\widehat{\varphi} := \mathcal{F}(\varphi)$ is the zero function. Then, the inversion formula (cf. Theorem 9.84) implies that $\varphi \equiv 0$, and therefore $\mathcal{F}$ is injective. Furthermore, the same formula can also be rewritten, by performing the change of variable $\eta = -\xi$, as

$$\varphi(x) = (2\pi)^{-n} \int_{\mathbb{R}^n} \widehat{\varphi}(-\eta)\, e^{-\imath x \cdot \eta}\, d\eta = \mathcal{F}(\psi)(x),$$

with $\psi(\eta) = (2\pi)^{-n}\, \widehat{\varphi}(-\eta)$ for all $\eta \in \mathbb{R}^n$. Since it is clear that $\psi \in \mathcal{S}(\mathbb{R}^n)$, given that $\widehat{\varphi}$ belongs to this space, we deduce from the previous identity that $\mathcal{F}$ is a continuous isomorphism. Finally, the continuity of $\mathcal{F}^{-1}$ follows from the identity

$$\begin{aligned}(\mathcal{F}^{-1}\, \psi)(x) := (2\pi)^{-n} \int_{\mathbb{R}^n} \psi(\xi)\, e^{\imath x \cdot \xi}\, d\xi &= \int_{\mathbb{R}^n} (2\pi)^{-n}\, \psi(-\eta)\, e^{-\imath x \cdot \eta}\, d\eta \\ &= (\mathcal{F} \circ r)(\psi)(x) \qquad \forall\, x \in \mathbb{R}^n, \quad \forall\, \psi \in \mathcal{S}(\mathbb{R}^n),\end{aligned}$$

where

$$r(\psi)(\eta) := (2\pi)^{-n}\, \psi(-\eta) \qquad \forall\, \eta \in \mathbb{R}^n,$$

and from the fact that the mapping $r : \mathcal{S}(\mathbb{R}^n) \to \mathcal{S}(\mathbb{R}^n)$ is continuous.

□

Definition 9.86. *The set of linear and continuous forms $u : \mathcal{S}(\mathbb{R}^n) \to \mathbb{C}$ (or $\mathbb{R}$) is denoted by $\mathcal{S}'(\mathbb{R}^n)$ and is called the space of* TEMPERED DISTRIBUTIONS.

The following result, analogous to Theorems 1.42 and 9.37, characterizes $\mathcal{S}'(\mathbb{R}^n)$. Its proof is left as an exercise (see Exercise 9.25).

Theorem 9.87. *A linear form $u : \mathcal{S}(\mathbb{R}^n) \to \mathbb{C}$ is continuous, that is, belongs to $\mathcal{S}'(\mathbb{R}^n)$, if and only if there exist $C > 0$ and $N \in \mathbb{N} \cup \{0\}$ such that (cf. Definition 9.77)*

$$|\langle u, \varphi\rangle| \leqslant C \sum_{|\alpha|, |\beta| \leqslant N} |\varphi|_{\alpha,\beta} \qquad \forall\, \varphi \in \mathcal{S}(\mathbb{R}^n).$$

Definition 9.88. *A sequence $\{u_j\}_{j\in\mathbb{N}} \subseteq \mathcal{S}'(\mathbb{R}^n)$ is said to converge to $u \in \mathcal{S}'(\mathbb{R}^n)$ if $\langle u_j, \varphi\rangle \overset{j\to\infty}{\to} \langle u, \varphi\rangle$ for all $\varphi \in \mathcal{S}(\mathbb{R}^n)$.*

Definition 9.89. *Given a polynomial $p(x) := \sum_{|\alpha| \leqslant m} a_\alpha\, x^\alpha$ and $u \in \mathcal{S}'(\mathbb{R}^n)$, we define the linear forms $pu : \mathcal{S}(\mathbb{R}^n) \to \mathbb{C}$ and $p(D)(u) : \mathcal{S}(\mathbb{R}^n) \to \mathbb{C}$, respectively, by*

$$\langle pu, \varphi \rangle := \langle u, p\varphi \rangle \qquad \forall\, \varphi \in \mathcal{S}(\mathbb{R}^n)\,,$$

and

$$\langle p(D)(u), \varphi \rangle := \sum_{|\alpha| \leqslant m} a_\alpha \langle D^\alpha u, \varphi \rangle = \sum_{|\alpha| \leqslant m} (-1)^{|\alpha|} a_\alpha \langle u, D^\alpha \varphi \rangle \quad \forall\, \varphi \in \mathcal{S}(\mathbb{R}^n)\,.$$

Note that the stability of $\mathcal{S}(\mathbb{R}^n)$ under differentiation and multiplication by polynomials ensures that pu and $p(D)(u)$ are well defined. Moreover, using the characterization result given by Theorem 9.87, it is easily shown that both pu and $p(D)(u)$ belong to $\mathcal{S}'(\mathbb{R}^n)$, which means that this space of distributions is also stable under differentiation and multiplication by polynomials. On the other hand, it is clear that $\mathrm{L}^1(\mathbb{R}^n)$ and $\mathcal{E}'(\mathbb{R}^n)$ are contained in $\mathcal{S}'(\mathbb{R}^n)$. Indeed, for example, for $u \in \mathrm{L}^1(\mathbb{R}^n)$ and $\varphi \in \mathcal{S}(\mathbb{R}^n)$ we have

$$|\langle u, \varphi \rangle| = \left| \int_{\mathbb{R}^n} u\varphi \right| \leqslant C\, |\varphi|_{0,0} = C \sum_{|\alpha|, |\beta| \leqslant N} |\varphi|_{\alpha,\beta},$$

with $C = \displaystyle\int_{\mathbb{R}^n} |u|$, $|\varphi|_{0,0} := \displaystyle\sup_{x \in \mathbb{R}^n} |\varphi(x)|$ and $N = 0$, which proves, by virtue of Theorem 9.87, that $u \in \mathcal{S}'(\mathbb{R}^n)$.

Another typical example of tempered distributions is provided by those induced by continuous functions with polynomial growth. More precisely, we have the following.

Example 9.90 Let $f \in C(\mathbb{R}^n)$ for which there exist constants $C > 0$ and $M \in \mathbb{N} \cup \{0\}$ such that

$$|f(x)| \leqslant C\big(1 + \|x\|\big)^M \quad \forall\, x \in \mathbb{R}^n\,,$$

and define $f : \mathcal{S}(\mathbb{R}^n) \to \mathbb{C}$ by

$$\langle f, \varphi \rangle := \int_{\mathbb{R}^n} f\varphi \quad \forall\, \varphi \in \mathcal{S}(\mathbb{R}^n)\,.$$

It is clear that f is linear. Moreover, using that the function $g(x) := \big(1 + \|x\|^2\big)^{-N}$ belongs to $\mathrm{L}^1(\mathbb{R}^n)$ for every $N > n/2$ (cf. Lemma 9.79), we have

$$\begin{aligned}
|\langle f, \varphi \rangle| &\leqslant C \int_{\mathbb{R}^n} \big(1 + \|x\|\big)^M |\varphi(x)|\, dx \\
&= C \int_{\mathbb{R}^n} \big(1 + \|x\|^2\big)^{-N} \big(1 + \|x\|^2\big)^N \big(1 + \|x\|\big)^M |\varphi(x)|\, dx \\
&\leqslant C \sup_{x \in \mathbb{R}^n} \Big\{ \big(1 + \|x\|^2\big)^N \big(1 + \|x\|\big)^M |\varphi(x)| \Big\} \int_{\mathbb{R}^n} \big(1 + \|x\|^2\big)^{-N} dx \\
&= C\, \widetilde{C} \sup_{x \in \mathbb{R}^n} \Big\{ \big(1 + \|x\|^2\big)^N \big(1 + \|x\|\big)^M |\varphi(x)| \Big\},
\end{aligned}$$

with $\widetilde{C} := \|(1 + \|\cdot\|^2)^{-N}\|_{0,1;\mathbb{R}^n}$. Now, considering $\|x\| := \sum_{i=1}^n |x_i|$, we obtain

$$\big(1 + \|x\|^2\big)^N \big(1 + \|x\|\big)^M = \sum_{|\alpha| \leqslant 2N+M} a_\alpha\, x^\alpha\,,$$

and consequently

$$\sup_{x\in\mathbb{R}^n}\left\{\left(1+\|x\|^2\right)^N\left(1+\|x\|\right)^M|\varphi(x)|\right\}\leqslant \widehat{C}_{N,M}\sum_{|\alpha|\leqslant 2N+M}|\varphi|_{\alpha,0}\,,$$

with $\widehat{C}_{N,M} := \max_{|\alpha|\leqslant 2N+M}|a_\alpha|$. In this way we conclude that

$$|\langle f,\varphi\rangle|\leqslant \bar{C}\sum_{|\alpha|\leqslant 2N+M}|\varphi|_{\alpha,0}\qquad \forall\,\varphi\in\mathcal{S}(\mathbb{R}^n)\,,$$

with $\bar{C} := C\,\widetilde{C}\,\widehat{C}_{N,M}$, thus proving, by Theorem 9.87, that $f\in\mathcal{S}'(\mathbb{R}^n)$.

In particular, as a consequence of the previous example, we have that if $u\in\mathrm{L}^1(\mathbb{R}^n)$, its Fourier transform $\widehat{u}$ induces a distribution $\widehat{u}\in\mathcal{S}'(\mathbb{R}^n)$ since $\widehat{u}\in C(\mathbb{R}^n)$ and is bounded. Then, since $\widehat{\varphi}\in\mathcal{S}(\mathbb{R}^n)\subseteq\mathrm{L}^1(\mathbb{R}^n)$ for all $\varphi\in\mathcal{S}(\mathbb{R}^n)$, identity ii) of Theorem 9.76 yields

$$\int_{\mathbb{R}^n}\widehat{u}\,\varphi=\int_{\mathbb{R}^n}u\,\widehat{\varphi}\,,$$

that is,

$$\langle\widehat{u},\varphi\rangle=\langle u,\widehat{\varphi}\rangle\quad\forall\,\varphi\in\mathcal{S}(\mathbb{R}^n)\,.\tag{9.58}$$

Moreover, if $u\in\mathcal{S}'(\mathbb{R}^n)$, not necessarily induced by an element of $\mathrm{L}^1(\mathbb{R}^n)$, the linear form $v:\mathcal{S}(\mathbb{R}^n)\to\mathbb{C}$ defined by $\langle v,\varphi\rangle := \langle u,\widehat{\varphi}\rangle$ for all $\varphi\in\mathcal{S}(\mathbb{R}^n)$, also belongs to $\mathcal{S}'(\mathbb{R}^n)$, since $v=u\circ\mathcal{F}$ and both u and $\mathcal{F}:\mathcal{S}(\mathbb{R}^n)\to\mathcal{S}(\mathbb{R}^n)$ are continuous. This observation leads to the following definition.

Definition 9.91. *Given $u\in\mathcal{S}'(\mathbb{R}^n)$, its Fourier transform $\mathcal{F}$ is defined as the distribution $\widehat{u} := \mathcal{F}(u)\in\mathcal{S}'(\mathbb{R}^n)$ given by*

$$\langle\widehat{u},\varphi\rangle := \langle u,\widehat{\varphi}\rangle\qquad\forall\,\varphi\in\mathcal{S}(\mathbb{R}^n)\,.$$

Note from (9.58) that if $u\in\mathrm{L}^1(\mathbb{R}^n)$, then its usual Fourier transform coincides with its Fourier transform as an element of $\mathcal{S}'(\mathbb{R}^n)$. In what follows, we use $\mathcal{F}$ to denote both $\mathcal{F}:\mathcal{S}(\mathbb{R}^n)\to\mathcal{S}(\mathbb{R}^n)$ and $\mathcal{F}:\mathcal{S}'(\mathbb{R}^n)\to\mathcal{S}'(\mathbb{R}^n)$.

Theorem 9.92. *The Fourier transform $\mathcal{F}:\mathcal{S}'(\mathbb{R}^n)\to\mathcal{S}'(\mathbb{R}^n)$ is bijective, and both $\mathcal{F}$ and $\mathcal{F}^{-1}$ are sequentially continuous.*

Proof. Let $u\in\mathcal{S}'(\mathbb{R}^n)$ be such that $\mathcal{F}(u)=0$. This means that $\langle u,\widehat{\varphi}\rangle=0$ for all $\varphi\in\mathcal{S}(\mathbb{R}^n)$, and since $\mathcal{F}:\mathcal{S}(\mathbb{R}^n)\to\mathcal{S}(\mathbb{R}^n)$ is bijective, it follows that $\langle u,\varphi\rangle=0$ for all $\varphi\in\mathcal{S}(\mathbb{R}^n)$, which shows that $\mathcal{F}:\mathcal{S}'(\mathbb{R}^n)\to\mathcal{S}'(\mathbb{R}^n)$ is injective. Moreover, given $u\in\mathcal{S}'(\mathbb{R}^n)$, the linear form $v:\mathcal{S}(\mathbb{R}^n)\to\mathbb{C}$ defined by

$$\langle v,\psi\rangle := \langle u,\mathcal{F}^{-1}(\psi)\rangle=\langle u,(\mathcal{F}\circ r)(\psi)\rangle\quad\forall\,\psi\in\mathcal{S}(\mathbb{R}^n)\,,$$

belongs to $\mathcal{S}'(\mathbb{R}^n)$, since $u,\mathcal{F}$, and r are continuous, and satisfies $\mathcal{F}(v)=u$, which proves that $\mathcal{F}:\mathcal{S}'(\mathbb{R}^n)\to\mathcal{S}'(\mathbb{R}^n)$ is also surjective. Finally, the sequential continuity of $\mathcal{F}$ and $\mathcal{F}^{-1}:\mathcal{S}'(\mathbb{R}^n)\to\mathcal{S}'(\mathbb{R}^n)$ follows from the continuity of r, $\mathcal{F}$, and $\mathcal{F}^{-1}:\mathcal{S}(\mathbb{R}^n)\to\mathcal{S}(\mathbb{R}^n)$. □

The next goal is to extend the calculation of the Fourier transform to distributions in $\mathrm{L}^2(\mathbb{R}^n)$. For this, we need the following preliminary result.

Lemma 9.93. *There hold:*

a) $\mathcal{S}(\mathbb{R}^n) \subseteq L^2(\mathbb{R}^n)$
b) $L^2(\mathbb{R}^n) \subseteq \mathcal{S}'(\mathbb{R}^n)$.

Proof. Recall first from Lemma 9.79 that the function $g(x) := \left(1 + \|x\|^2\right)^{-N}$ for all $x \in \mathbb{R}^n$ belongs to $L^1(\mathbb{R}^n)$ for all $N > n/2$, and hence $g \in L^2(\mathbb{R}^n)$ for all $N > n/4$. Then, given $\varphi \in \mathcal{S}(\mathbb{R}^n)$ and $N > n/4$, we have

$$\begin{aligned}
\int_{\mathbb{R}^n} |\varphi(x)|^2 \, dx &= \int_{\mathbb{R}^n} \left(1 + \|x\|^2\right)^{2N} |\varphi(x)| \left(1 + \|x\|^2\right)^{-2N} |\varphi(x)| \, dx \\
&\leqslant \sup_{x \in \mathbb{R}^n} \left\{ \left(1 + \|x\|^2\right)^{2N} |\varphi(x)| \right\} \sup_{x \in \mathbb{R}^n} |\varphi(x)| \, \|\left(1 + \|\cdot\|^2\right)^{-N}\|_{0,\mathbb{R}^n}^2 \\
&\leqslant C_N \left\{ \sum_{|\alpha| \leqslant 4N} |\varphi|_{\alpha,0} \right\} |\varphi|_{0,0} \, \|\left(1 + \|\cdot\|^2\right)^{-N}\|_{0,\mathbb{R}^n}^2 < +\infty,
\end{aligned}$$

which proves that $\varphi \in L^2(\mathbb{R}^n)$, thus confirming a). On the other hand, let $u \in L^2(\mathbb{R}^n)$ and define the linear form $u : \mathcal{S}(\mathbb{R}^n) \to \mathbb{C}$ by

$$\langle u, \varphi \rangle := \int_{\mathbb{R}^n} u \, \varphi \qquad \forall \, \varphi \in \mathcal{S}(\mathbb{R}^n).$$

It follows, taking $N > n/4$ and applying the Cauchy–Schwarz inequality, that

$$\begin{aligned}
|\langle u, \varphi \rangle| &\leqslant \int_{\mathbb{R}^n} |u(x)| \, |\varphi(x)| \, dx \\
&= \int_{\mathbb{R}^n} \left(1 + \|x\|^2\right)^{-N} |u(x)| \left(1 + \|x\|^2\right)^{N} |\varphi(x)| \, dx \\
&\leqslant \sup_{x \in \mathbb{R}^n} \left\{ \left(1 + \|x\|^2\right)^{N} |\varphi(x)| \right\} \int_{\mathbb{R}^n} \left(1 + \|x\|^2\right)^{-N} |u(x)| \, dx \\
&\leqslant C_N \sum_{|\alpha| \leqslant 2N} |\varphi|_{\alpha,0} \, \|\left(1 + \|\cdot\|^2\right)^{-N}\|_{0,\mathbb{R}^n} \|u\|_{0,\mathbb{R}^n} \\
&\leqslant C \sum_{|\alpha|, |\beta| \leqslant 2N} |\varphi|_{\alpha,\beta} \qquad \forall \, \varphi \in \mathcal{S}(\mathbb{R}^n),
\end{aligned}$$

with $C := C_N \|\left(1 + \|\cdot\|^2\right)^{-N}\|_{0,\mathbb{R}^n} \|u\|_{0,\mathbb{R}^n}$, which shows, by Theorem 9.87, that $u \in \mathcal{S}'(\mathbb{R}^n)$, thus proving b).

□

Having established that $L^2(\mathbb{R}^n) \subseteq \mathcal{S}'(\mathbb{R}^n)$, we can compute the Fourier transform of $u \in L^2(\mathbb{R}^n)$ as the distribution $\widehat{u} \in \mathcal{S}'(\mathbb{R}^n)$ defined by

$$\langle \widehat{u}, \varphi \rangle := \langle u, \widehat{\varphi} \rangle \qquad \forall \, \varphi \in \mathcal{S}(\mathbb{R}^n). \tag{9.59}$$

Moreover, we have the following theorem.

Theorem 9.94 (Plancherel-Parseval Theorem). *For every $u \in L^2(\mathbb{R}^n)$, its Fourier transform $\widehat{u}$ (cf. (9.59)) also belongs to $L^2(\mathbb{R}^n)$, and satisfies the Parseval identity*

$$\|\widehat{u}\|_{0,\mathbb{R}^n}^2 = (2\pi)^n \, \|u\|_{0,\mathbb{R}^n}^2. \tag{9.60}$$

Proof. We first consider $u \in \mathcal{S}(\mathbb{R}^n)$. In this case, we already know from Theorem 9.85 and part a) of Lemma 9.93 that $\widehat{u} \in \mathcal{S}(\mathbb{R}^n) \subseteq \mathrm{L}^2(\mathbb{R}^n)$. Then, defining

$$v(\xi) := \overline{\widehat{u}(\xi)} := \int_{\mathbb{R}^n} \overline{u(x)}\, e^{\imath x\cdot\xi}\, dx \qquad \forall\, \xi \in \mathbb{R}^n\,,$$

and applying the inversion formula from Theorem 9.84 to $\varphi := v \in \mathcal{S}(\mathbb{R}^n)$, we deduce that

$$\widehat{v}(\xi) = (2\pi)^n\, \overline{u(\xi)} \qquad \forall\, \xi \in \mathbb{R}^n\,.$$

It follows, using identity ii) of Theorem 9.76, that

$$\begin{aligned} \|\widehat{u}\|_{0,\mathbb{R}^n}^2 &= \int_{\mathbb{R}^n} \widehat{u}(\xi)\, \overline{\widehat{u}(\xi)}\, dx = \int_{\mathbb{R}^n} \widehat{u}(\xi)\, v(\xi)\, d\xi \\ &= \int_{\mathbb{R}^n} u(\xi)\, \widehat{v}(\xi)\, d\xi = \int_{\mathbb{R}^n} u(\xi)\, (2\pi)^n\, \overline{u(\xi)}\, d\xi = (2\pi)^n\, \|u\|_{0,\mathbb{R}^n}^2\,. \end{aligned}$$

Let us now take $u \in \mathrm{L}^2(\mathbb{R}^n)$. Then, the Cauchy–Schwarz inequality and the previous identity imply that

$$\begin{aligned} |\langle \widehat{u}, \varphi\rangle| &= |\langle u, \widehat{\varphi}\rangle| = \left| \int_{\mathbb{R}^n} u\, \widehat{\varphi} \right| \leqslant \|u\|_{0,\mathbb{R}^n}\, \|\widehat{\varphi}\|_{0,\mathbb{R}^n} \\ &= \|u\|_{0,\mathbb{R}^n}\, (2\pi)^{n/2}\, \|\varphi\|_{0,\mathbb{R}^n} \qquad \forall\, \varphi \in \mathcal{S}(\mathbb{R}^n)\,, \end{aligned}$$

that is

$$|\langle \widehat{u}, \varphi\rangle| \leqslant (2\pi)^{n/2}\, \|u\|_{0,\mathbb{R}^n}\, \|\varphi\|_{0,\mathbb{R}^n} \qquad \forall\, \varphi \in \mathcal{S}(\mathbb{R}^n)\,. \tag{9.61}$$

Since $C_0^\infty(\mathbb{R}^n)$ is dense in $\mathrm{L}^2(\mathbb{R}^n)$ and $C_0^\infty(\mathbb{R}^n) \subseteq \mathcal{S}(\mathbb{R}^n)$, we prove next that the transform $\widehat{u} : \mathcal{S}(\mathbb{R}^n) \to \mathbb{C}$ admits an extension to a linear and continuous functional (still denoted by $\widehat{u}$) on $\mathrm{L}^2(\mathbb{R}^n)$. In fact, given $\varphi \in \mathrm{L}^2(\mathbb{R}^n)$, we consider a sequence $\{\varphi_j\}_{j\in\mathrm{N}} \subseteq C_0^\infty(\mathbb{R}^n)$ such that $\|\varphi_j - \varphi\|_{0,\mathbb{R}^n} \overset{j\to\infty}{\to} 0$, and define

$$\langle \widehat{u}, \varphi\rangle := \lim_{j\to\infty} \langle \widehat{u}, \varphi_j\rangle = \lim_{j\to\infty} \langle u, \widehat{\varphi}_j\rangle\,. \tag{9.62}$$

It is readily verified, using (9.61), that the above limit exists and is independent of the chosen sequence. Moreover, invoking (9.62) and (9.61), we find that

$$\big|\langle \widehat{u}, \varphi\rangle\big| = \lim_{j\to\infty} \big|\langle \widehat{u}, \varphi_j\rangle\big| \leqslant \lim_{j\to\infty} (2\pi)^{n/2}\, \|u\|_{0,\mathbb{R}^n}\, \|\varphi_j\|_{0,\mathbb{R}^n} = (2\pi)^{n/2}\, \|u\|_{0,\mathbb{R}^n}\, \|\varphi\|_{0,\mathbb{R}^n}\,,$$

which proves that $\widehat{u}$, defined in this way, belongs to $\mathrm{L}^2(\mathbb{R}^n)'$ and satisfies

$$\|\widehat{u}\|_{\mathrm{L}^2(\mathbb{R}^n)'} \leqslant (2\pi)^{n/2}\, \|u\|_{0,\mathbb{R}^n}\,.$$

Thus, applying the Riesz Representation Theorem (cf. Theorem 2.14), we deduce that there exists $v \in \mathrm{L}^2(\mathbb{R}^n)$ such that

$$\langle \widehat{u}, \varphi\rangle = \int_{\mathbb{R}^n} v\, \varphi \qquad \forall\, \varphi \in \mathrm{L}^2(\mathbb{R}^n)$$

and

$$\|v\|_{0,\mathbb{R}^n} = \|\widehat{u}\|_{\mathrm{L}^2(\mathbb{R}^n)'} \leqslant (2\pi)^{n/2}\, \|u\|_{0,\mathbb{R}^n}\,,$$

which yields the identification of $\widehat{u} \in \mathrm{L}^2(\mathbb{R}^n)'$ with $v \in \mathrm{L}^2(\mathbb{R}^n)$, and therefore

$$\|\widehat{u}\|_{0,\mathbb{R}^n} \leqslant (2\pi)^{n/2}\,\|u\|_{0,\mathbb{R}^n} \quad \forall\, u \in \mathrm{L}^2(\mathbb{R}^n). \tag{9.63}$$

On the other hand, applying the inversion formula again, we obtain

$$\widehat{\widehat{\varphi}}(\xi) \;=\; (2\pi)^{n/2}\,\varphi(-\xi) \quad \forall\,\xi \in \mathbb{R}^n, \quad \forall\,\varphi \in \mathcal{S}(\mathbb{R}^n),$$

and then, given $u \in \mathcal{S}'(\mathbb{R}^n)$, we have

$$\langle \widehat{\widehat{u}}, \varphi\rangle \;=\; \langle \widehat{u}, \widehat{\varphi}\rangle \;=\; \langle u, \widehat{\widehat{\varphi}}\rangle \;=\; \langle u, (2\pi)^n\,\varphi(-\cdot)\rangle \;=\; \langle (2\pi)^n\,u, \varphi(-\cdot)\rangle.$$

In particular, for $u \in \mathrm{L}^2(\mathbb{R}^n)$ we get

$$\langle \widehat{\widehat{u}}, \varphi\rangle \;=\; \int_{\mathbb{R}^n} (2\pi)^n\,u(x)\,\varphi(-x)\,dx \;=\; \int_{\mathbb{R}^n} (2\pi)^n\,u(-x)\,\varphi(x)\,dx,$$

from which it follows that $\widehat{\widehat{u}} \equiv (2\pi)^n\,u(-\cdot)$ and $\|\widehat{\widehat{u}}\|_{0,\mathbb{R}^n} \;=\; (2\pi)^n\,\|u\|_{0,\mathbb{R}^n}$. Finally, applying (9.63) to $\widehat{u} \in \mathrm{L}^2(\mathbb{R}^n)$, with $u \in \mathrm{L}^2(\mathbb{R}^n)$, we deduce that

$$(2\pi)^n\,\|u\|_{0,\mathbb{R}^n} \;\leqslant\; (2\pi)^{n/2}\,\|\widehat{u}\|_{0,\mathbb{R}^n},$$

that is,

$$(2\pi)^{n/2}\,\|u\|_{0,\mathbb{R}^n} \;\leqslant\; \|\widehat{u}\|_{0,\mathbb{R}^n} \quad \forall\, u \in \mathrm{L}^2(\mathbb{R}^n),$$

so that, from this inequality and (9.63), we conclude that

$$\|\widehat{u}\|^2_{0,\mathbb{R}^n} \;=\; (2\pi)^n\,\|u\|^2_{0,\mathbb{R}^n} \qquad \forall\, u \in \mathrm{L}^2(\mathbb{R}^n).$$

□

Corollary 9.95 (Parseval identity). *For every* $u,\, v \in \mathrm{L}^2(\mathbb{R}^n)$ *there holds*

$$\langle \widehat{u}, \widehat{v}\rangle_{0,\mathbb{R}^n} \;=\; (2\pi)^n\,\langle u, v\rangle_{0,\mathbb{R}^n}.$$

Proof. It follows easily from (9.60) and from the identity

$$\langle u, v\rangle_{0,\mathbb{R}^n} \;=\; \frac{1}{4}\Big\{\|u+v\|^2_{0,\mathbb{R}^n} - \|u-v\|^2_{0,\mathbb{R}^n} + \imath\|u+\imath v\|^2_{0,\mathbb{R}^n} - \imath\|u-\imath v\|^2_{0,\mathbb{R}^n}\Big\}.$$

□

9.10 Exercises.

▸ **9.1** *Prove that the following are distributions on* $\mathbb{R}$

i) $\langle u, \varphi\rangle \;=\; \lim\limits_{\varepsilon\to 0}\left\{\displaystyle\int_{|x|\geqslant\varepsilon} \frac{\varphi(x)}{x^2}\,dx - 2\,\frac{\varphi(0)}{\varepsilon}\right\}$ *for all* $\varphi \in \mathcal{D}(\mathbb{R})$.

ii) $\langle u, \varphi\rangle \;=\; \lim\limits_{\varepsilon\to 0}\displaystyle\int_{|x|\geqslant\varepsilon} \frac{\varphi(x)}{x}\,dx$ *for all* $\varphi \in \mathcal{D}(\mathbb{R})$.

iii) $\langle u, \varphi\rangle := \lim\limits_{\varepsilon\to 0}\left\{\displaystyle\int_{\varepsilon}^{\infty} \frac{\varphi(x)}{x^2}\,dx - \frac{\varphi(0)}{\varepsilon} + \varphi'(0)\,ln\,(\varepsilon)\right\}$ *for all* $\varphi \in \mathcal{D}(\mathbb{R})$.

▸ **9.2**

i) *Prove that* $PV(\frac{1}{x}) : \mathcal{D}(\mathbb{R}) \to \mathbb{R}$ *defined by*

$$\langle PV(\frac{1}{x}), \varphi\rangle = \lim_{\varepsilon\to 0} \int_{|x|\geqslant\varepsilon} \frac{\varphi(x)}{x}\, dx \qquad \forall\, \varphi \in \mathcal{D}(\mathbb{R})$$

is a distribution on $\mathbb{R}$.

ii) *Show that* $(\log|x|)' = PV(\frac{1}{x})$ *in* $\mathcal{D}'(\mathbb{R})$.

▸ **9.3** *Let* Ω *be an open subset of* $\mathbb{R}^n$, *and let* $f \in C_0^k(\Omega)$. *Bearing in mind Definition 9.7, prove that for each* α, $|\alpha| \leqslant k$,

$$\lim_{\varepsilon\to 0} \sup_{x\in\Omega} |\,\partial^\alpha (f_\varepsilon - f)(x)\,| = 0\,.$$

▸ **9.4** *Let* $\Omega = (0,1)\times(0,1)$, *and for each* $n \geqslant 3$ *define the sets*

$$\Omega_n := \left(\frac{1}{n}, \frac{n-1}{n}\right) \times \left(\frac{1}{n}, \frac{n-1}{n}\right).$$

Construct a sequence of functions $\{f_n\}_{n\geqslant 3} \subseteq C_0^\infty(\Omega)$ *verifying*

i) $f_n = 1$ *in* Ω_n,

ii) *for every multi-index* α *there exists a positive constant* C_α *such that* $|\partial^\alpha f_n(x)| \leqslant C_\alpha\, n^{|\alpha|}$ *for all* $x \in \Omega$, *and*

iii) $\displaystyle\lim_{n\to\infty} \int_\Omega f_n\,(2 - f_n)\, dx = 1.$

▸ **9.5** *For each* $\varepsilon \in (0,1)$ *consider* $u_\varepsilon \in \mathrm{L}^1_{loc}(\mathbb{R}^n)$ *such that*

$$\operatorname{supp}(u_\varepsilon) \subseteq \bar{B}(0,\varepsilon)\,, \quad \int_{\mathbb{R}^n} u_\varepsilon(x)\, dx = 1\,, \quad \text{and} \quad \int_{\mathbb{R}^n} |u_\varepsilon(x)|\, dx < \infty\,.$$

Prove that $\displaystyle\lim_{\varepsilon\to 0} \int_{\mathbb{R}^n} u_\varepsilon\, \varphi\, dx = \varphi(0)$ *for all* $\varphi \in \mathcal{D}(\mathbb{R}^n)$.

▸ **9.6** *Prove that* $\displaystyle\lim_{n\to\infty} \int_{\mathbb{R}} \frac{n}{1+n^2x^2}\, \varphi(x)\, dx = \pi\, \varphi(0)$ *for all* $\varphi \in \mathcal{D}(\mathbb{R})$.

▸ **9.7** *Let* $X_1, X_2, \ldots, X_m$ *be open subsets of* $\mathbb{R}^n$, *and let* K *be a compact set contained in* $\displaystyle\bigcup_{j=1}^m X_j$.

Prove that there exist functions $\varphi_j \in C_0^\infty(X_j)$ *such that*

$$0 \leqslant \varphi_j \leqslant 1 \quad \text{and} \quad \sum_{j=1}^m \varphi_j^2(x) = 1 \quad \text{in a neighborhood of } K\,.$$

▸ **9.8** *Let* $u \in \mathrm{L}^p_{loc}(\Omega)$, *with* $p > 1$, *and suppose that there exists* $v \in \mathrm{L}^p_{loc}(\Omega)$ *such that*

$$\int_\Omega v\,\varphi\, dx = (-1)^{|\alpha|} \int_\Omega u\, \partial^\alpha \varphi\, dx \qquad \forall\, \varphi \in \mathcal{D}(\Omega)\,.$$

Prove that $\partial^\alpha u_\varepsilon(x) = v_\varepsilon(x)$, *for all* $x \in \Omega$, *for all* $\varepsilon < \operatorname{dist}(x, \partial\Omega)$. *Moreover, prove that for every compact* $K \subseteq \Omega$ *there exists a sequence* $\{\varphi_j\}_{j\in\mathbb{N}} \subseteq C^{|\alpha|}(\Omega)$ *such that*

$$\|\varphi_j - u\|_{0,p;K} \to 0 \quad \text{and} \quad \|\partial^\alpha \varphi_j - v\|_{0,p;K} \to 0 \quad \text{as} \quad j \to \infty\,.$$

▸ **9.9**

i) *Prove that the injection* $\mathcal{E}'(\Omega) \hookrightarrow \mathcal{D}'(\Omega)$ *is dense.*

ii) *Let* $u \in \mathcal{E}'(\Omega)$ *and* $\varphi \in \mathcal{E}(\Omega)$ *such that* $\operatorname{supp}(u) \cap \operatorname{supp}(\varphi) = \phi$. *Prove that* $\langle u, \varphi\rangle = 0$.

▸ **9.10** *Let Ω be an open subset of $\mathbb{R}^n$, and let $x_0, x_1 \in \Omega$ be such that $x_0 \neq x_1$. Suppose that $u \in \mathcal{D}'(\Omega)$ and $\operatorname{supp}(u) \subseteq \{x_0, x_1\}$.*

i) *Prove that there exist nonnegative integers N_0, N_1, and constants $c_\alpha, d_\beta \in \mathbb{C}$, $|\alpha| \leqslant N_0$, $|\beta| \leqslant N_1$, such that for all $\varphi \in \mathcal{D}(\Omega)$*

$$\langle u, \varphi \rangle = \sum_{|\alpha| \leqslant N_0} c_\alpha \langle \partial^\alpha \delta_{x_0}, \varphi \rangle + \sum_{|\beta| \leqslant N_1} d_\beta \langle \partial^\beta \delta_{x_1}, \varphi \rangle .$$

ii) *Extend the previous result to the case in which $\operatorname{supp}(u) \subseteq \{x_0, \ldots, x_N\}$, with $x_i \neq x_j$ for all $i \neq j \in \{1, \ldots, N\}$.*

▸ **9.11**

i) *Let $u \in \mathcal{D}'(\Omega)$, where Ω is the open subset of $\mathbb{R}^2$ defined by*

$$\Omega = \{(x, y) \in \mathbb{R}^2 : \quad a < x < b, \quad c < y < d\}.$$

If $\frac{\partial u}{\partial x} = \frac{\partial u}{\partial y} = 0$ in $\mathcal{D}'(\Omega)$, prove that there exists a constant $\lambda \in \mathbb{C}$ such that

$$\langle u, \varphi \rangle = \lambda \int_\Omega \varphi \, dx \qquad \forall \varphi \in \mathcal{D}(\Omega) .$$

ii) *Let $u \in \mathcal{D}'(\Omega)$, where Ω is the open subset of $\mathbb{R}^3$ defined by*

$$\Omega = \{(x, y, z) \in \mathbb{R}^3 : \quad a < x < b, \quad c < y < d, \quad e < z < f\}.$$

If $\frac{\partial u}{\partial x} = \frac{\partial u}{\partial y} = 0$ in $\mathcal{D}'(\Omega)$, what can be said about the distribution u?

▸ **9.12** *Recall the definition of convergence of distributions given in Section 9.6 and prove that*

$$\frac{j}{\sqrt{2\pi}} e^{-\frac{j^2 x^2}{2}} \to \delta \quad \text{as} \quad j \to \infty .$$

HINT: *Note that $\int_0^\infty e^{-x^2} \, dx = \frac{\sqrt{\pi}}{2}$.*

▸ **9.13** *Let $u \in \mathrm{L}^p(\Omega)$, and let $\{u_k\}_{k \in \mathbb{N}} \subseteq \mathrm{W}^{m,p}(\Omega)$ be such that $\|u_k\|_{m,p;\Omega} \leqslant M$ for all $k \in \mathbb{N}$. Prove that if $u_k \overset{k \to \infty}{\rightarrow} u$ in $\mathcal{D}'(\Omega)$, then $u \in \mathrm{W}^{m,p}(\Omega)$.*

▸ **9.14** *Let Ω be an open subset of $\mathbb{R}^n$, let $m \in \mathbb{N}$, and let $\{u_k\}_{k \in \mathbb{N}}$ be a bounded sequence in $\mathrm{H}^m(\Omega)$ such that $u_k \overset{k \to \infty}{\rightarrow} u$ in $\mathcal{D}'(\Omega)$. Prove that $u \in \mathrm{H}^m(\Omega)$.*

▸ **9.15** *Let Ω be an open subset of $\mathbb{R}^n$, and let $g \in C^\infty(\Omega)$ and $u \in \mathcal{D}'(\Omega)$. The distribution $gu \in \mathcal{D}'(\Omega)$ is defined by $\langle gu, \varphi \rangle := \langle u, g\varphi \rangle$ for all $\varphi \in \mathcal{D}(\Omega)$. Prove that for every multi-index α, there holds the* LEIBNIZ FORMULA

$$\partial^\alpha (gu) = \sum_{\beta \leqslant \alpha} \binom{\alpha}{\beta} \partial^\beta g \, \partial^{\alpha - \beta} u .$$

▸ **9.16** *Let Ω be an open subset of $\mathbb{R}$, and consider the operator $L(v) := v' + v$ for all $v \in \mathcal{D}'(\Omega)$. Prove that if $u \in \mathcal{D}'(\Omega)$ and $L(u) = f \in C^\infty(\Omega)$, then $u \in C^\infty(\Omega)$. Thus, $L(u) = f$ in the classical sense, and the smoothness of f is transmitted to the solution u.*

▸ **9.17** *Let Ω be an open subset of $\mathbb{R}$, and consider the operator $L(v) := v' + gv$ for all $v \in \mathcal{D}'(\Omega)$, where $g \in C^\infty(\Omega)$. Prove that if $u \in \mathcal{D}'(\Omega)$ and $L(u) = f \in C(\Omega)$, then $u \in C^1(\Omega)$, and thus $L(u) = f$ in the classical sense.*

▸ **9.18** [THEOREM 9.60]. *Let $L(u) := \frac{\partial^2 u}{\partial t^2} - \frac{\partial^2 u}{\partial x^2}$ for all $u \in \mathcal{D}'(\mathbb{R}^2)$, the* WAVE OPERATOR *in the plane. Prove that $L(E) = \delta$ in $\mathcal{D}'(\mathbb{R}^2)$, where*

$$E(x,t) := \begin{cases} \frac{1}{2} & \text{if } |x| < t\,, \\ 0 & \text{if } |x| > t\,, \end{cases} \qquad \forall\,(x,t) \in \mathbb{R}^2\,.$$

▸ **9.19** *Let Ω be an open subset of $\mathbb{R}^n$, and let $u \in \mathcal{D}'(\Omega)$ be such that $\langle u, \varphi\rangle \geqslant 0$ for every nonnegative function $\varphi \in C_0^\infty(\Omega)$. Prove that u is a distribution of order 0.*

▸ **9.20** *Let Ω be an open subset of $\mathbb{R}^n$ that contains the zero vector, and let $u \in \mathcal{D}'(\Omega)$ be such that $x_j\, u = 0$ in $\mathcal{D}'(\Omega)$, for all $j \in \{1, \ldots, n\}$. Prove that there exists a constant C such that $u = C\,\delta$.*

▸ **9.21** *Let Ω be an open subset of $\mathbb{R}^n$, and let $\{f_j\}_{j\in\mathbb{N}} \subseteq \mathrm{L}^1_{loc}(\Omega)$ be such that*

$$\lim_{j\to\infty} \int_K |f_j(x)|\, dx = 0$$

for every compact $K \subseteq \Omega$. Prove that $\partial^\alpha f_j \to 0$ in $\mathcal{D}'(\Omega)$ for all α.

▸ **9.22**

i) *Let $u \in \mathcal{D}'(\Omega)$, and let K be a compact subset contained in Ω. Prove that there exists a function $f \in C(\Omega)$ and a multi-index α such that*

$$\langle u, \varphi\rangle = (-1)^{|\alpha|} \int_\Omega f(x)\, \partial^\alpha \varphi(x)\, dx \qquad \forall\, \varphi \in \mathcal{D}_K(\Omega)\,.$$

ii) *Let V and Ω be open subsets of $\mathbb{R}^n$, and let K be a compact set such that $K \subset V \subset \Omega$. Let $u \in \mathcal{D}'(\Omega)$ be of order N with $\mathrm{supp}(u) = K$. Apply* i) *to show that there exists a finite number of functions $f_\beta \in C(\Omega)$, whose supports are contained in V, such that*

$$u = \sum_\beta \partial^\beta f_\beta\,.$$

▸ **9.23** *Prove that if $\{\varphi_j\}_{j\in\mathbb{N}}$ converges to zero in $\mathcal{D}(\mathbb{R}^n)$, then for each $u \in \mathcal{D}'(\mathbb{R}^n)$, $u * \varphi_j$ converges to zero in $\mathcal{E}(\mathbb{R}^n)$.*

▸ **9.24** *Let Ω be a bounded open subset of $\mathbb{R}^n$, let $f \in \mathrm{L}^1_{loc}(\Omega)$, and let f_ε denote its regularization. Prove that*

i) $f_\varepsilon \in C^\infty(\Omega_\varepsilon)$, where $\Omega_\varepsilon := \{x \in \Omega : \ \mathrm{dist}(x, \partial\Omega) > \varepsilon\}$, and

ii) $f_\varepsilon(x) \overset{\varepsilon\to 0}{\to} f(x)$ for almost every $x \in \Omega$.

In addition, show that

iii) *if $f \in C(\Omega)$, then $f_\varepsilon \overset{\varepsilon\to 0}{\rightrightarrows} f$ uniformly on compact subsets of Ω, and*

iv) *if $f \in \mathrm{L}^p_{loc}(\Omega)$, $1 \leqslant p < \infty$, then $f_\varepsilon \overset{\varepsilon\to 0}{\to} f$ in $\mathrm{L}^p_{loc}(\Omega)$.*

▸ **9.25** [Theorem 9.87]. *Let u be a linear form on $\mathcal{S}(\mathbb{R}^n)$. Prove that u is continuous (tempered distribution) if and only if there exist $C > 0$ and a nonnegative integer N such that*

$$|\langle u, \varphi\rangle| \leqslant C \sum_{|\alpha|,|\beta|\leqslant N} \left\{ \sup_{x\in\mathbb{R}^n} |x^\alpha\, \partial^\beta \varphi(x)| \right\} \qquad \forall\, \varphi \in \mathcal{S}(\mathbb{R}^n)\,.$$

▸ **9.26** *Prove that for every $u \in \mathcal{S}'(\mathbb{R}^n)$, for every multi-index α, and for each $h \in \mathbb{R}^n$, one has*

i) $\widehat{D^\alpha u} = \xi^\alpha \, \widehat{u}$,

ii) $\widehat{x^\alpha u} = (-1)^{|\alpha|} \, D^\alpha \, \widehat{u}$,

iii) $\widehat{\tau_h u} = e^{-\imath \xi \cdot h} \, \widehat{u}$, and

iv) $\widehat{e^{\imath x \cdot h} u} = \tau_h \, \widehat{u}$.

Deduce, from these identities, that $\widehat{D^\alpha \delta} = \xi^\alpha$ *and* $\widehat{x^\alpha} = (-1)^{|\alpha|} \, D^\alpha \, \delta$, *first proving that* $\widehat{\delta} = 1$.

▸ **9.27** *Let* $\{u_n\}_{n \in \mathbb{N}}$ *be a sequence in* $\mathcal{D}'(\mathbb{R}^2)$ *such that* $x_1 \, u_n = \left(x_2 - \frac{1}{n}\right) u_n = 0$ *in* $\mathcal{D}'(\mathbb{R}^2)$ *and* $\langle u_n, 1 \rangle = \dfrac{n-1}{n+1}$ *for all* $n \in \mathbb{N}$. *Justify the consistency of these hypotheses and find, if possible, the distributional limit of* u_n *as* $n \to +\infty$.

10

Sobolev Spaces

In this chapter, we discuss some essential aspects of the theory of Sobolev spaces. Our presentation, based mainly on the approaches in [39], [90], [102], [107], and [112], is not intended to be exhaustive; rather, it focuses on results that have had—and continue to have—lasting importance and usefulness in applications.

10.1 Preliminary concepts and results

We first notice that some preliminary results, mainly oriented toward applications to the analysis of variational problems, were presented in Section 4.5. We now refine and extend some of them, and also introduce others of both practical and theoretical interest. We begin by recalling the classical SOBOLEV SPACE OF ORDER m, with $m \in \mathbb{N} \cup \{0\}$, which, given an open set Ω of $\mathbb{R}^n$, is defined as (cf. Exercise 1.10, eq. (1.37))

$$\mathrm{H}^m(\Omega) := \Big\{ v \in \mathrm{L}^2(\Omega) : \quad \partial^\alpha v \in \mathrm{L}^2(\Omega), \quad \forall\, \alpha,\ |\alpha| \leqslant m \Big\}. \tag{10.1}$$

Note that, as explained in Section 1.5 and Chapter 9, the condition $\partial^\alpha v \in \mathrm{L}^2(\Omega)$ must be understood in the distributional sense, which means that there exists $v_\alpha \in \mathrm{L}^2(\Omega)$ such that

$$\langle v, \partial^\alpha \varphi \rangle := \int_\Omega v\, \partial^\alpha \varphi = (-1)^{|\alpha|} \int_\Omega v_\alpha\, \varphi \quad \forall\, \varphi \in \mathcal{D}(\Omega).$$

Then, letting $\langle \cdot, \cdot \rangle_{0,\Omega}$ be the usual inner product of $\mathrm{L}^2(\Omega)$, with induced norm denoted $\| \cdot \|_{0,\Omega}$, we introduce the mapping $\langle \cdot, \cdot \rangle_{m,\Omega} : \mathrm{H}^m(\Omega) \times \mathrm{H}^m(\Omega) \to \mathbb{R}$ as

$$\langle v, \omega \rangle_{m,\Omega} := \sum_{|\alpha| \leqslant m} \langle \partial^\alpha v, \partial^\alpha \omega \rangle_{0,\Omega} \quad \forall\, v,\, \omega \in \mathrm{H}^m(\Omega),$$

which can easily be shown to be an inner product on $\mathrm{H}^m(\Omega)$. The respective induced norm is given by

$$\|v\|_{m,\Omega} := \left\{ \sum_{|\alpha| \leqslant m} \|\partial^\alpha v\|_{0,\Omega}^2 \right\}^{1/2} \quad \forall v \in \mathrm{H}^m(\Omega), \tag{10.2}$$

and the mapping $|\cdot| : \mathrm{H}^m(\Omega) \to \mathbb{R}$ defined by

G. N. Gatica, *Introduction to Functional Analysis*,
https://doi.org/10.1007/978-3-032-37415-8_10

$$|v|_{m,\Omega} := \left\{ \sum_{|\alpha|=m} \|\partial^\alpha v\|_{0,\Omega}^2 \right\}^{1/2} \qquad \forall\, v \in \mathrm{H}^m(\Omega),$$

constitutes a seminorm on this space.

The following theorem establishes the completeness of $\mathrm{H}^m(\Omega)$ as a consequence of the same property of $\mathrm{L}^2(\Omega)$.

Theorem 10.1. *The Sobolev space* $\left(\mathrm{H}^m(\Omega), \langle\cdot,\cdot\rangle_{m,\Omega}\right)$ *is a Hilbert space.*

Proof. Let $\{v_j\}_{j\in\mathrm{N}}$ be a Cauchy sequence in $\mathrm{H}^m(\Omega)$. Equivalently, by (10.2), for every multi-index α with $|\alpha| \leqslant m$, the sequence $\{\partial^\alpha v_j\}_{j\in\mathrm{N}}$ is a Cauchy sequence in $\mathrm{L}^2(\Omega)$. Since $\mathrm{L}^2(\Omega)$ is a Hilbert space, it follows that there exist $v \in \mathrm{L}^2(\Omega)$ and $v_\alpha \in \mathrm{L}^2(\Omega)$, for all α with $0 < |\alpha| \leqslant m$, such that $v_j \overset{j\to\infty}{\to} v$ and $\partial^\alpha v_j \overset{j\to\infty}{\to} v_\alpha$ in $\mathrm{L}^2(\Omega)$. Next, we show that $v_\alpha = \partial^\alpha v$ in $\mathcal{D}'(\Omega)$, which implies that $v \in \mathrm{H}^m(\Omega)$ and that $v_j \to v$ in $\mathrm{H}^m(\Omega)$. Indeed, given $\varphi \in \mathcal{D}(\Omega)$ and a multi-index α as above, it follows from the definition of the distributional derivative and the above convergences that

$$\begin{aligned}
\langle \partial^\alpha v, \varphi\rangle &:= (-1)^{|\alpha|}\langle v, \partial^\alpha\varphi\rangle = (-1)^{|\alpha|}\int_\Omega v\,\partial^\alpha\varphi \\
&= (-1)^{|\alpha|}\lim_{j\to\infty}\int_\Omega v_j\,\partial^\alpha\varphi = (-1)^{|\alpha|}\lim_{j\to\infty}\langle v_j, \partial^\alpha\varphi\rangle \\
&= \lim_{j\to\infty}\langle \partial^\alpha v_j, \varphi\rangle = \lim_{j\to\infty}\int_\Omega \partial^\alpha v_j\,\varphi \\
&= \int_\Omega v_\alpha\,\varphi = \langle v_\alpha, \varphi\rangle,
\end{aligned}$$

which proves that $\partial^\alpha v = v_\alpha$ in $\mathcal{D}'(\Omega)$, thus completing the proof.

□

We now prove the separability of $\mathrm{H}^m(\Omega)$, using the corresponding property of $\mathrm{L}^2(\Omega)$.

Theorem 10.2. *The Sobolev space* $(\mathrm{H}^m(\Omega), \langle\cdot,\cdot\rangle_{m,\Omega})$ *is a separable Hilbert space.*

Proof. Let $N := \mathrm{card}\{\alpha : |\alpha| \leqslant \mathrm{m}\}$, and let $T : \mathrm{H}^m(\Omega) \to \mathbf{L}^2(\Omega)$ be the operator that assigns to each $v \in \mathrm{H}^m(\Omega)$ the N-tuple $T(v) := (\partial^\alpha v)_{|\alpha|\leqslant m}$. It is clear that

$$\|T(v)\|_{0,\Omega} = \|v\|_{m,\Omega} \qquad \forall\, v \in \mathrm{H}^m(\Omega),$$

and therefore $T : \mathrm{H}^m(\Omega) \to T(\mathrm{H}^m(\Omega))$ is an isometric bijection with closed range $T(\mathrm{H}^m(\Omega))$. Moreover, since $\mathrm{L}^2(\Omega)$ is separable, it follows (cf. part b) of Exercise 7.4) that $\mathbf{L}^2(\Omega)$ is as well. Hence, applying Lemma 7.12 to the subspace of $\mathbf{L}^2(\Omega)$ given by $T(\mathrm{H}^m(\Omega))$, and taking into account part a) of Exercise 7.4, we conclude that $\mathrm{H}^m(\Omega)$ is separable.

□

An interesting subspace of $\mathrm{H}^m(\Omega)$ is given by $\mathrm{H}_0^m(\Omega)$, the closure of $C_0^\infty(\Omega)$ with respect to the norm $\|\cdot\|_{m,\Omega}$ (as already mentioned at the beginning of Section 5.3.4.1), that is, $\mathrm{H}_0^m(\Omega) := \overline{C_0^\infty(\Omega)}^{\|\cdot\|_{m,\Omega}}$, or equivalently,

$$\mathrm{H}_0^m(\Omega) = \Big\{ v \in \mathrm{H}^m(\Omega) : \quad \text{there exists} \quad \{\varphi_j\}_{j\in\mathbb{N}} \subseteq C_0^\infty(\Omega) \\ \text{such that} \quad \|v - \varphi_j\|_{m,\Omega} \overset{j\to\infty}{\longrightarrow} 0 \Big\}. \tag{10.3}$$

It is important to note here that, if Ω is bounded, then $\mathrm{H}_0^m(\Omega)$ does not coincide with $\mathrm{H}^m(\Omega)$. Indeed, recall that the Friedrichs–Poincaré inequality, proved in Lemma 4.5 for an interval Ω in $\mathbb{R}$, is also valid in a bounded open set Ω of $\mathbb{R}^n$ (cf. (4.167)). Hence, if we assume that $\mathrm{H}_0^1(\Omega) = \mathrm{H}^1(\Omega)$, it would follow that there exists a constant $c > 0$ such that

$$\|v\|_{1,\Omega} \leqslant c\,|v|_{1,\Omega} \quad \forall\, v \in \mathrm{H}^1(\Omega),$$

from which, taking $\widetilde{v} \equiv 1$ in Ω, we deduce that $|\Omega|^{1/2} =: \|\widetilde{v}\|_{1,\Omega} \leqslant c\,|\widetilde{v}|_{1,\Omega} = 0$, which is a contradiction.

Nevertheless, the following theorem shows that in the case $\Omega = \mathbb{R}^n$, the spaces $\mathrm{H}_0^1(\Omega)$ and $\mathrm{H}^1(\Omega)$ do coincide.

Theorem 10.3. *$C_0^\infty(\mathbb{R}^n)$ is dense in $\mathrm{H}^1(\mathbb{R}^n)$ with respect to the norm $\|\cdot\|_{1,\mathbb{R}^n}$.*

Proof. Recall first (cf. Definition 9.21) that, given $v \in \mathcal{D}'(\mathbb{R}^n)$, the support of v is given by the complement of the open set

$$\Big\{ x \in \mathbb{R}^n : \quad \exists \text{ a neighborhood } V(x) \text{ of } x \text{ such that } v_{V(x)} = 0 \Big\},$$

where $v_{V(x)} = 0$ means that $\langle v, \varphi\rangle = 0$ for all $\varphi \in C_0^\infty(V(x))$. With this in mind, we introduce the space

$$\widetilde{\mathrm{H}}^1(\mathbb{R}^n) := \Big\{ v \in \mathrm{H}^1(\mathbb{R}^n) : \quad \mathrm{supp}(v) \text{ is compact} \Big\},$$

which clearly satisfies

$$C_0^\infty(\mathbb{R}^n) \subseteq \widetilde{\mathrm{H}}^1(\mathbb{R}^n) \subseteq \mathrm{H}^1(\mathbb{R}^n).$$

We now prove that $\widetilde{\mathrm{H}}^1(\mathbb{R}^n)$ is dense in $\mathrm{H}^1(\mathbb{R}^n)$, that is, $\overline{\widetilde{\mathrm{H}}^1(\mathbb{R}^n)}^{\|\cdot\|_{1,\mathbb{R}^n}} = \mathrm{H}^1(\mathbb{R}^n)$. Indeed, let $v \in \mathrm{H}^1(\mathbb{R}^n)$, and consider (cf. Lemma 9.17) a function $\varphi \in C_0^\infty(\mathbb{R}^n)$ such that $\varphi(x) = 1$ if $\|x\| \leqslant 1$, $\varphi(x) = 0$ if $\|x\| \geqslant 2$, and $0 < \varphi(x) < 1$ if $1 < \|x\| < 2$. Then, for each $m \in \mathbb{N}$ define $\varphi_m \in C_0^\infty(\mathbb{R}^n)$ by $\varphi_m(x) := \varphi\left(\frac{x}{m}\right)$ for all $x \in \mathbb{R}^n$, and set $v_m := \varphi_m v$ (cf. Definition 9.50). It is clear that $v_m \in \mathrm{H}^1(\mathbb{R}^n)$ and $\mathrm{supp}(v_m) \subseteq \bar{B}(0, 2m)$ (since $\mathrm{supp}(\varphi) = \bar{B}(0,2)$ implies that $\mathrm{supp}(\varphi_m) = \bar{B}(0,2m)$), and therefore $v_m \in \widetilde{\mathrm{H}}^1(\mathbb{R}^n)$. Moreover, a simple calculation and the Lebesgue Dominated Convergence Theorem (cf. [113, Theorem 1.26], [120, Theorem 5.36]) imply that

$$\|v_m - v\|_{0,\mathbb{R}^n}^2 = \int_{\|x\|\geqslant m} \Big\{1 - \varphi_m(x)\Big\}^2 v^2(x)\,dx \leqslant \int_{\|x\|\geqslant m} v^2(x)\,dx \overset{m\to\infty}{\longrightarrow} 0. \tag{10.4}$$

On the other hand, since

$$\frac{\partial \varphi_m}{\partial x_i}(x) = \frac{1}{m}\frac{\partial \varphi}{\partial x_i}\left(\frac{x}{m}\right) \quad \forall\, x \in \mathbb{R}^n,$$

it follows that

$$\sup_{x\in\mathbb{R}^n} \left|\frac{\partial \varphi_m}{\partial x_i}(x)\right| \leqslant \frac{1}{m} \sup_{x\in\mathbb{R}^n} \left|\frac{\partial \varphi}{\partial x_i}\left(\frac{x}{m}\right)\right| \leqslant \frac{1}{m} \sup_{\|y\|\leqslant 2} \left|\frac{\partial \varphi}{\partial x_i}(y)\right| \overset{m\to\infty}{\longrightarrow} 0,$$

and using also that $\dfrac{\partial v}{\partial x_i} \in \mathrm{L}^2(\mathbb{R}^n)$, we deduce that

$$\begin{aligned}
\left\| \frac{\partial v_m}{\partial x_i} - \frac{\partial v}{\partial x_i} \right\|_{0,\mathbb{R}^n}^2 &= \left\| \varphi_m \frac{\partial v}{\partial x_i} + v \frac{\partial \varphi_m}{\partial x_i} - \frac{\partial v}{\partial x_i} \right\|_{0,\mathbb{R}^n}^2 \\
&\leqslant 2 \left\| \varphi_m \frac{\partial v}{\partial x_i} - \frac{\partial v}{\partial x_i} \right\|_{0,\mathbb{R}}^2 + 2 \left\| v \frac{\partial \varphi_m}{\partial x_i} \right\|_{0,\mathbb{R}^n}^2 \\
&= 2 \int_{\|x\| \geqslant m} \big\{ 1 - \varphi_m(x) \big\}^2 \left\{ \frac{\partial v}{\partial x_i}(x) \right\}^2 dx + 2 \int_{\mathbb{R}^n} v^2(x) \left\{ \frac{\partial \varphi_m}{\partial x_i}(x) \right\}^2 dx \\
&\leqslant 2 \int_{\|x\| \geqslant m} \left\{ \frac{\partial v}{\partial x_i}(x) \right\}^2 dx + 2 \, \|v\|_{0,\mathbb{R}^n}^2 \sup_{x \in \mathbb{R}^n} \left| \frac{\partial \varphi_m}{\partial x_i}(x) \right|^2 \overset{m \to \infty}{\longrightarrow} 0.
\end{aligned} \tag{10.5}$$

Thus, by (10.4) and (10.5), we have shown that $\|v_m - v\|_{1,\mathbb{R}^n} \overset{m\to\infty}{\longrightarrow} 0$, and therefore $\widetilde{\mathrm{H}}^1(\mathbb{R}^n)$ is dense in $\mathrm{H}^1(\mathbb{R}^n)$. The next goal is to show that $\widetilde{\mathrm{H}}^1(\mathbb{R}^n) \subseteq \overline{C_0^\infty(\mathbb{R}^n)}^{\|\cdot\|_{1,\mathbb{R}^n}}$. To this end, given $\omega \in \widetilde{\mathrm{H}}^1(\mathbb{R}^n)$, define (cf. Definition 9.7) the regularized function $\omega_\varepsilon := \omega * \rho_\varepsilon$ for all $\varepsilon > 0$. Since ω has compact support, we know from part a) of Theorem 9.8 that $\omega_\varepsilon \in C_0^\infty(\mathbb{R}^n)$, and it is also clear that

$$\frac{\partial \omega_\varepsilon}{\partial x_i} = \frac{\partial \omega}{\partial x_i} * \rho_\varepsilon = \left(\frac{\partial \omega}{\partial x_i} \right)_\varepsilon \qquad \forall \, \varepsilon > 0.$$

Moreover, it is easy to see, using part b) of the same theorem, that

$$\begin{aligned}
\|\omega - \omega_\varepsilon\|_{1,\mathbb{R}^n}^2 &= \|\omega - \omega_\varepsilon\|_{0,\mathbb{R}^n}^2 + \sum_{i=1}^n \left\| \frac{\partial \omega}{\partial x_i} - \frac{\partial \omega_\varepsilon}{\partial x_i} \right\|_{0,\mathbb{R}^n}^2 \\
&= \|\omega - \omega_\varepsilon\|_{0,\mathbb{R}^n}^2 + \sum_{i=1}^n \left\| \frac{\partial \omega}{\partial x_i} - \left(\frac{\partial \omega}{\partial x_i} \right)_\varepsilon \right\|_{0,\mathbb{R}^n}^2 \overset{\varepsilon \to 0}{\longrightarrow} 0,
\end{aligned}$$

which shows that $\omega \in \overline{C_0^\infty(\mathbb{R}^n)}^{\|\cdot\|_{1,\mathbb{R}^n}}$, and thus $\widetilde{\mathrm{H}}^1(\mathbb{R}^n) \subseteq \overline{C_0^\infty(\mathbb{R}^n)}^{\|\cdot\|_{1,\mathbb{R}^n}}$. Taking closures in this inclusion and using that $\widetilde{\mathrm{H}}^1(\mathbb{R}^n)$ is dense in $\mathrm{H}^1(\mathbb{R}^n)$, we obtain $\mathrm{H}^1(\mathbb{R}^n) \subseteq \overline{C_0^\infty(\mathbb{R}^n)}^{\|\cdot\|_{1,\mathbb{R}^n}}$, from which it follows that $\mathrm{H}^1(\mathbb{R}^n) = \overline{C_0^\infty(\mathbb{R}^n)}^{\|\cdot\|_{1,\mathbb{R}^n}} = \mathrm{H}_0^1(\mathbb{R}^n)$, which completes the proof. □

At the risk of being redundant, we emphasize, based on the second part of the proof of Theorem 10.3, that the fact that for each $\omega \in \widetilde{\mathrm{H}}^1(\mathbb{R}^n)$ there exists a sequence in $C_0^\infty(\mathbb{R}^n)$ converging to ω with respect to the norm $\|\cdot\|_{1,\mathbb{R}^n}$ does not prove that $C_0^\infty(\mathbb{R}^n)$ is dense in $\widetilde{\mathrm{H}}^1(\mathbb{R}^n)$, but merely that $\widetilde{\mathrm{H}}^1(\mathbb{R}^n) \subseteq \overline{C_0^\infty(\mathbb{R}^n)}^{\|\cdot\|_{1,\mathbb{R}^n}}$. If one were to (incorrectly) conclude that $\overline{C_0^\infty(\mathbb{R}^n)}^{\|\cdot\|_{1,\mathbb{R}^n}} = \widetilde{\mathrm{H}}^1(\mathbb{R}^n)$, then it would follow that $\mathrm{H}^1(\mathbb{R}^n) = \widetilde{\mathrm{H}}^1(\mathbb{R}^n)$, which is certainly false.

The following theorem constitutes one of the simplest results on extensions of Sobolev spaces.

Theorem 10.4. *Let Ω be an open subset of $\mathbb{R}^n$. Then, for each $v \in \mathrm{H}_0^1(\Omega)$, the extension (or prolongation) $E(v) : \mathbb{R}^n \to \mathbb{R}$ defined by*

$$E(v) := \begin{cases} v & in & \Omega, \\ 0 & in & \mathbb{R}^n \backslash \Omega, \end{cases}$$

belongs to $\mathrm{H}^1(\mathbb{R}^n)$.

Proof. Given $v \in \mathrm{H}_0^1(\Omega)$, we know from (10.3) with $m = 1$ that there exists a sequence $\{\varphi_j\}_{j\in\mathbb{N}} \subseteq C_0^\infty(\Omega)$ such that $\|v - \varphi_j\|_{1,\Omega} \overset{j\to\infty}{\to} 0$. Define for each $j \in \mathbb{N}$ the extension $\widetilde{\varphi}_j : \mathbb{R}^n \to \mathbb{R}$ given by

$$\widetilde{\varphi}_j := \begin{cases} \varphi_j & \text{in} & \Omega, \\ 0 & \text{in} & \mathbb{R}^n \backslash \Omega, \end{cases}$$

and note that $\widetilde{\varphi}_j \in C_0^\infty(\mathbb{R}^n)$. It follows easily that

$$\|\widetilde{\varphi}_j - \widetilde{\varphi}_\ell\|_{1,\mathbb{R}^n} = \|\varphi_j - \varphi_\ell\|_{1,\Omega} \overset{j,\ell\to\infty}{\to} 0,$$

and therefore, since $\mathrm{H}^1(\mathbb{R}^n)$ is a Hilbert space (cf. Theorem 10.1), there exists $\widetilde{v} \in \mathrm{H}^1(\mathbb{R}^n)$ such that $\|\widetilde{\varphi}_j - \widetilde{v}\|_{1,\mathbb{R}^n} \overset{j\to\infty}{\to} 0$. Now, it is clear that $\widetilde{v}|_\Omega \in \mathrm{H}^1(\Omega)$ and moreover

$$\|\varphi_j - \widetilde{v}|_\Omega\|_{1,\Omega} = \|\widetilde{\varphi}_j - \widetilde{v}|_\Omega\|_{1,\Omega} \leqslant \|\widetilde{\varphi}_j - \widetilde{v}\|_{1,\mathbb{R}^n} \overset{j\to\infty}{\to} 0,$$

from which, by the uniqueness of the limit in $\mathrm{H}^1(\Omega)$, it follows that $v = \widetilde{v}|_\Omega$. In particular, we have that $\|\widetilde{\varphi}_j - \widetilde{v}\|_{0,\mathbb{R}^n} \overset{j\to\infty}{\to} 0$, and therefore there exists $\{\widetilde{\varphi}_j^{(1)}\}_{j\in\mathbb{N}} \subseteq \{\widetilde{\varphi}_j\}_{j\in\mathbb{N}}$ such that $\widetilde{\varphi}_j^{(1)}(x) \overset{j\to\infty}{\to} \widetilde{v}(x)$ for all $x \in \mathbb{R}^n$ (except on a subset of measure zero), which is denoted a.e. in $\mathbb{R}^n$. Thus, $\widetilde{v}(x) = \lim_{j\to\infty} \widetilde{\varphi}_j^{(1)}(x) = 0$ a.e. in $\mathbb{R}^n \backslash \bar{\Omega}$, and consequently $\widetilde{v} = E(v)$, thus proving that $E(v) \in \mathrm{H}^1(\mathbb{R}^n)$.

□

10.2 Trace Theorems

In this section, we deepen and extend the results on traces discussed in Section 4.5. To this end, and by way of introduction, we consider a bounded open set $\Omega \subset \mathbb{R}^n$ with boundary Γ. Given $v \in \mathrm{H}^1(\Omega)$, we are interested in defining, or at least giving meaning to, $v|_\Gamma$, the restriction of v to Γ. In this regard, the following example illustrates that, for $n \geqslant 2$, elements of $\mathrm{H}^1(\Omega)$ are not necessarily induced by continuous functions.

Example 10.5 Let $\Omega := B(0, r) \subseteq \mathbb{R}^2$ with $0 < r < 1$, and, given $s \in (0, 1/2)$, consider the function $v(x) := \big(-\ln\|x\|\big)^s$ for all $x \in \Omega\backslash\{0\}$. It is clear that v is not continuous in Ω since it blows up at $x = 0$ for $s > 0$, and yet we show below that $v \in \mathrm{H}^1(\Omega)$. Indeed, a straightforward calculation gives

$$\frac{\partial v}{\partial x_i}(x) = s\big(-\ln\|x\|\big)^{s-1}\Big(-\frac{1}{\|x\|}\Big)\frac{x_i}{\|x\|}$$
$$= -\frac{s}{\|x\|^2}\big(-\ln\|x\|\big)^{s-1} x_i \quad \forall\, x \in \Omega\backslash\{0\} \quad \forall\, i \in \{1, 2\},$$

and hence $\|\nabla v(x)\|^2 = \dfrac{s^2}{\|x\|^2}\Big|\ln\|x\|\Big|^{2s-2}$ for all $x \in \Omega\backslash\{0\}$. It follows, using polar coordinates, that

$$
\begin{aligned}
\|v\|_{1,\Omega}^2 &= \int_0^{2\pi}\int_0^r |\ln\rho|^{2s}\,\rho\,d\rho\,d\vartheta + \int_0^{2\pi}\int_0^r \frac{s^2}{\rho}|\ln\rho|^{2s-2}\,d\rho\,d\vartheta \\
&= 2\pi\left(\lim_{\varepsilon\to 0}\int_\varepsilon^r |\ln\rho|^{2s}\,\rho\,d\rho + \lim_{\varepsilon\to 0} s^2\int_\varepsilon^r \frac{|\ln\rho|^{2s-2}}{\rho}\,d\rho\right).
\end{aligned}
\tag{10.6}
$$

Now, it is easy to see that for $s < 1/2$ one has $\lim_{\varepsilon\to 0}|\ln\varepsilon|^{2s}\varepsilon = 0$, so that the first limit in (10.6) is finite. Moreover,

$$
\int_\varepsilon^r \frac{|\ln\rho|^{2s-2}}{\rho}\,d\rho = -\left.\frac{|\ln\rho|^{2s-1}}{2s-1}\right|_\varepsilon^r = -\frac{|\ln r|^{2s-1}}{2s-1} + \frac{|\ln\varepsilon|^{2s-1}}{2s-1},
$$

and then, noting that $\lim_{\varepsilon\to 0}|\ln\varepsilon|^{2s-1} = 0$ for $s < 1/2$, we obtain

$$
\lim_{\varepsilon\to 0}\int_\varepsilon^r \frac{|\ln\rho|^{2s-2}}{\rho}\,d\rho = -\frac{|\ln r|^{2s-1}}{2s-1} < +\infty.
$$

According to the above, we conclude that $v \in \mathrm{H}^1(\Omega)$.

Our next goal is to prove the Trace Theorem in the upper half-space given by

$$
\mathbb{R}^n_+ := \left\{x := (\hat{x}, x_n) \in \mathbb{R}^n : \quad \hat{x} \in \mathbb{R}^{n-1}, x_n > 0\right\}
$$

(see the upper part of Figure 10.1), whose boundary is given by

$$
\partial\mathbb{R}^n_+ := \left\{x := (\hat{x}, 0) \in \mathbb{R}^n : \quad \hat{x} \in \mathbb{R}^{n-1}\right\}.
$$

To this end, given an open set Ω in $\mathbb{R}^n$ and a nonnegative integer m, we define the space

$$
C^m(\bar{\Omega}) := \left\{v|_\Omega : \quad v \in C_0^m(\mathbb{R}^n)\right\},
$$

and recall from (9.27) that

$$
C^\infty(\bar{\Omega}) := \left\{v|_\Omega : \quad v \in C_0^\infty(\mathbb{R}^n)\right\}.
$$

We then have the following lemma, whose proof is analogous to that of Theorem 10.3.

Lemma 10.6. *The space* $C^\infty(\overline{\mathbb{R}^n_+})$ *is dense in* $\mathrm{H}^1(\mathbb{R}^n_+)$.

Next, following the standard procedure suggested by the analysis in Section 4.5.1 (cf. Theorems 4.25 and 4.26), we prove the trace inequality in $C^\infty(\overline{\mathbb{R}^n_+})$. Indeed, as already indicated in that section, every trace theorem is based on two fundamental facts: the existence of a dense subspace and the validity therein of the corresponding trace inequality.

Theorem 10.7. *For each* $v \in C^\infty(\overline{\mathbb{R}^n_+})$, *we define its trace on the boundary* $\partial\mathbb{R}^n_+$ *as the mapping* $v(\cdot,0) : \mathbb{R}^{n-1} \to \mathbb{R}$ *that assigns to each* $\hat{x} \in \mathbb{R}^{n-1}$ *the value* $v(\hat{x},0)$. *Then,*

$$
\|v(\cdot,0)\|_{0,\mathbb{R}^{n-1}} \leqslant \|v\|_{1,\mathbb{R}^n_+} \quad \forall\, v \in C^\infty(\overline{\mathbb{R}^n_+}).
$$

Proof. Given $v \in C^\infty(\overline{\mathbb{R}^n_+})$, we know by the previous definition that there exists $\omega \in C_0^\infty(\mathbb{R}^n)$ such that $\omega|_{\mathbb{R}^n_+} = v$ (see the lower image in Figure 10.1). It follows that there also exists $r > 0$ such that $\mathrm{supp}(\omega) \subseteq \bar{B}(0,r)$. Then, for each $\hat{x} \in \mathbb{R}^{n-1}$ we have

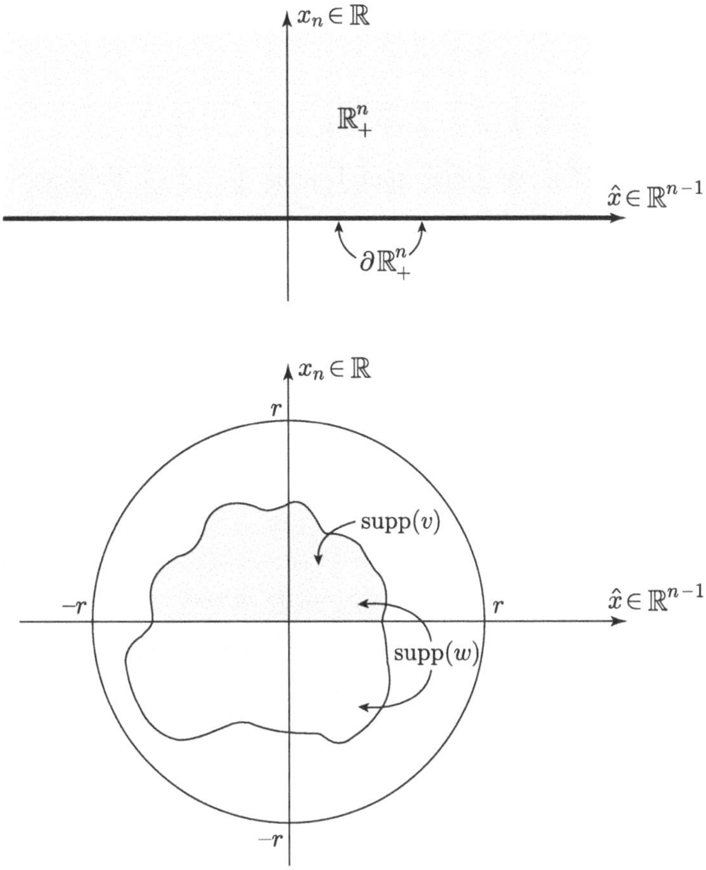

Fig. 10.1. *Geometry for the trace inequality in* $\mathbb{R}^n_+$

$$
\begin{aligned}
|v(\hat{x},0)|^2 &= |\omega(\hat{x},0)|^2 = -\int_0^r \frac{\partial}{\partial x_n}\left(|\omega(\hat{x},x_n)|^2\right) dx_n \\
&= -\int_0^r \frac{\partial}{\partial x_n}\left(|v(\hat{x},x_n)|^2\right) dx_n = -2\int_0^r v(\hat{x},x_n)\,\frac{\partial v}{\partial x_n}(\hat{x},x_n)\,dx_n, \\
&\leqslant \int_0^r \left(|v(\hat{x},x_n)|^2 + \left|\frac{\partial v}{\partial x_n}(\hat{x},x_n)\right|^2\right) dx_n\,,
\end{aligned}
$$

from which, integrating with respect to $\hat{x} \in \mathbb{R}^{n-1}$, the desired inequality follows.

□

Having proved the trace inequality in $C^\infty(\overline{\mathbb{R}^n_+})$ (cf. Theorem 10.7), and knowing that $C^\infty(\overline{\mathbb{R}^n_+})$ is dense in $\mathrm{H}^1(\mathbb{R}^n_+)$ (cf. Lemma 10.6), we now prove the Trace Theorem in $\mathrm{H}^1(\mathbb{R}^n_+)$.

Theorem 10.8. *The mapping that assigns to each* $v \in C^\infty(\overline{\mathbb{R}^n_+})$ *its trace* $v(\cdot,0)|_{\mathbb{R}^{n-1}} \in \mathcal{D}(\mathbb{R}^{n-1}) \subseteq \mathrm{L}^2(\mathbb{R}^{n-1})$ *extends by continuity to a linear operator* $\gamma_0 : \mathrm{H}^1(\mathbb{R}^n_+) \to \mathrm{L}^2(\mathbb{R}^{n-1})$, *called the* TRACE OPERATOR OF ORDER 0, *such that*

$$\gamma_0(v) = v(\cdot,0)|_{\mathbb{R}^{n-1}} \quad \forall v \in C^\infty(\overline{\mathbb{R}^n_+}),$$

and

$$\|\gamma_0(v)\|_{0,\mathbb{R}^{n-1}} \leqslant \|v\|_{1,\mathbb{R}^n_+} \quad \forall v \in \mathrm{H}^1(\mathbb{R}^n_+). \tag{10.7}$$

Proof. Given $v \in \mathrm{H}^1(\mathbb{R}^n_+)$, we know from Lemma 10.6 that there exists a sequence $\{\varphi_k\}_{k\in\mathbb{N}} \subseteq C^\infty(\overline{\mathbb{R}^n_+})$ such that $\|v - \varphi_k\|_{1,\mathbb{R}^n_+} \overset{k\to\infty}{\longrightarrow} 0$. It follows, according to the trace inequality in $C^\infty(\overline{\mathbb{R}^n_+})$ (cf. Theorem 10.7), that for all $k,\, j \in \mathbb{N}$

$$\|\varphi_k(\cdot,0) - \varphi_j(\cdot,0)\|_{0,\mathbb{R}^{n-1}} \leqslant \|\varphi_k - \varphi_j\|_{1,\mathbb{R}^n_+},$$

which proves that $\{\varphi_k(\cdot,0)\}_{k\in\mathbb{N}}$ is a Cauchy sequence in $\mathrm{L}^2(\mathbb{R}^{n-1})$. Thus, since $\mathrm{L}^2(\mathbb{R}^{n-1})$ is a Hilbert space, there exists $\varphi \in \mathrm{L}^2(\mathbb{R}^{n-1})$ such that $\|\varphi_k(\cdot,0) - \varphi\|_{0,\mathbb{R}^{n-1}} \overset{k\to\infty}{\longrightarrow} 0$. Moreover, if $\{\psi_k\}_{k\in\mathbb{N}}$ is another sequence in $C^\infty(\overline{\mathbb{R}^n_+})$ converging to v in $\mathrm{H}^1(\mathbb{R}^n_+)$, the same trace inequality implies that

$$\|\varphi_k(\cdot,0) - \psi_k(\cdot,0)\|_{0,\mathbb{R}^{n-1}} \leqslant \|\varphi_k - \psi_k\|_{1,\mathbb{R}^n_+} \overset{k\to\infty}{\longrightarrow} 0,$$

and therefore $\{\psi_k(\cdot,0)\}_{k\in\mathbb{N}}$ also converges to φ in $\mathrm{L}^2(\mathbb{R}^{n-1})$. According to the above, we define $\gamma_0(v) = \varphi$, and taking the limit as $k \to \infty$ in the inequality

$$\|\varphi_k(\cdot,0)\|_{0,\mathbb{R}^{n-1}} \leqslant \|\varphi_k\|_{1,\mathbb{R}^n_+},$$

we conclude that $\|\varphi\|_{0,\mathbb{R}^{n-1}} \leqslant \|v\|_{1,\mathbb{R}^n_+}$, which completes the proof.

□

In order to extend the Trace Theorem to the space $\mathrm{H}^1(\Omega)$, where Ω is a bounded open subset of $\mathbb{R}^n$, we need some preliminary definitions related to the regularity of the boundary Γ of Ω.

Definition 10.9. *An open set Ω in $\mathbb{R}^n$ is said to be 1-regular (or of class C^1) if Ω is bounded, lies locally on only one side of its boundary Γ, and if there exist open sets $X_0, X_1, \ldots, X_N$ in $\mathbb{R}^n$ such that*

a) $\bar{X}_0 \subseteq \Omega$,

b) $\bar{\Omega} \subseteq \bigcup_{i=0}^{N} X_i$,

c) *for each $i \in \{1,2,\ldots,N\}$ there exists an invertible function $\psi_i : X_i \to B(0,1)$ such that $\psi_i \in C^1(X_i)$ and $\psi_i^{-1} \in C^1(B(0,1))$, and*

$$\psi_i(X_i \cap \Omega) := \Big\{x := (\hat{x}, x_n) : \quad \|x\| < 1, \quad x_n > 0\Big\} = B(0,1) \cap \mathbb{R}^+_n,$$

and

$$\psi_i(X_i \cap \Gamma) := \Big\{x := (\hat{x}, x_n) : \quad \|x\| < 1, \quad x_n = 0\Big\} = B(0,1) \cap \partial\mathbb{R}^+_n.$$

In this case, $\Big\{X_i, \psi_i\Big\}_{i=1}^{N}$ is called a SYSTEM OF LOCAL CHARTS *that defines Γ (see Figure 10.2 for a geometric illustration of this definition).*

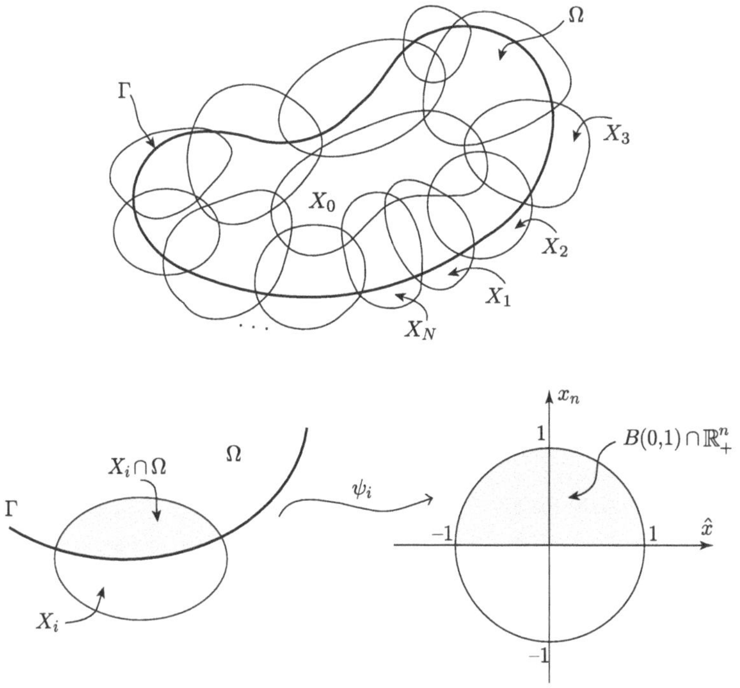

Fig. 10.2. *Geometry for Definition 10.9*

If, in the previous definition, it holds that $\psi_i,\ \psi_i^{-1} \in C^m$, then Ω is said to be a domain of class C^m (or an m-regular domain). Moreover, if $\psi_i,\ \psi_i^{-1} \in C^{0,1}$, that is, if both mappings are Lipschitz continuous, then Ω is said to be a domain of class $C^{0,1}$.

In what follows, we present a series of definitions and results for a domain Ω, which we assume to be of class C^1.

Definition 10.10. *Let Ω be a 1-regular open subset of $\mathbb{R}^n$, and let $\left\{X_i, \psi_i\right\}_{i=1}^N$ be a system of local charts for its boundary Γ. Moreover, let $\{\varphi_i\}_{i=0}^N$ be a partition of unity on $\bar{\Omega}$ subordinate to the covering of $\bar{\Omega}$ by the open sets $X_0, X_1, \ldots, X_N$, that is, $\varphi_i \in C_0^\infty(X_i)$, $\varphi_i \geqslant 0$ for all $i \in \{0, 1, \ldots, N\}$, and $\sum_{i=0}^N \varphi_i \equiv 1$ in a neighborhood of $\bar{\Omega}$* (cf. Lemma 9.19). *Then we define $\mathrm{L}^2(\Gamma)$ as the space of all (equivalence classes of) measurable functions $v : \Gamma \to \mathbb{R}$ such that*

$$\widetilde{(\varphi_i v) \circ \psi_i^{-1}}(\cdot, 0) \in \mathrm{L}^2(\mathbb{R}^{n-1}) \qquad \forall\, i \in \{1, 2, \ldots, N\},$$

where, given $\omega : \mathbb{R}^n \to \mathbb{R}$, $\tilde{\omega} : \mathbb{R}^n \to \mathbb{R}$ denotes the function

$$\tilde{\omega}(\hat{x}, x_n) := \begin{cases} \omega(\hat{x}, x_n) & \text{if } \ \|(\hat{x}, x_n)\| < 1, \\ 0 & \text{if } \ \|(\hat{x}, x_n)\| \geqslant 1. \end{cases}$$

In this case, $\mathrm{L}^2(\Gamma)$ is endowed with the norm

$$\|v\|_{0,\Gamma} := \left\{ \sum_{i=1}^{N} \left\| \widetilde{(\varphi_i v) \circ \psi_i^{-1}}(\cdot, 0) \right\|_{0,\mathbb{R}^{n-1}}^2 \right\}^{1/2}. \tag{10.8}$$

The previous definition is equivalent to the usual concept of the space $L^2(\Gamma)$ given by the set of all (equivalence classes of) measurable functions $v : \Gamma \to \mathbb{R}$ such that

$$\int_\Gamma v^2 < +\infty,$$

which is equipped with the norm

$$\|v\|_{0,\Gamma} := \left\{ \int_\Gamma v^2 \right\}^{1/2}.$$

Indeed, in order to prove the equivalence between $\|\cdot\|_{0,\Gamma}$ and $\|\cdot\|_{0,\Gamma}$, one uses the fact that $\sum_{i=1}^{N} \varphi_i \equiv 1$ in a neighborhood of Γ.

Definition 10.11. *Given $s \in \mathbb{R}$, we consider the function $p_s : \mathbb{R}^n \to \mathbb{R}$ defined by*

$$p_s(\xi) := \left(1 + |\xi|^2\right)^{s/2} \qquad \forall\, \xi \in \mathbb{R}^n, \tag{10.9}$$

and introduce the Sobolev space of order s over $\mathbb{R}^n$ as

$$H^s(\mathbb{R}^n) := \left\{ v \in L^2(\mathbb{R}^n) : \quad p_s \widehat{v} \in L^2(\mathbb{R}^n) \right\},$$

which is endowed with the norm

$$\|v\|_{s,\mathbb{R}^n} := \|p_s \widehat{v}\|_{0,\mathbb{R}^n} \qquad \forall\, v \in H^s(\mathbb{R}^n), \tag{10.10}$$

where $\widehat{v} \in L^2(\mathbb{R}^n)$ stands for the Fourier transform of v (cf. (9.59), Theorem 9.94).

Note that in the case $s = m \in \mathbb{N}$, the usual norm of $H^m(\Omega)$ given by (10.2) is equivalent to (10.10) (cf. Exercise 10.1).

Definition 10.12. *With the same assumptions and notation as in Definition 10.10, we define, for each $s \in \mathbb{R}$, the Sobolev space of order s over Γ, denoted by $H^s(\Gamma)$, as the set of all (equivalence classes of) measurable functions $v : \Gamma \to \mathbb{R}$ such that*

$$\widetilde{(\varphi_i v) \circ \psi_i^{-1}}(\cdot, 0) \in H^s(\mathbb{R}^{n-1}) \qquad \forall\, i \in \{1, 2, \ldots, N\},$$

and endowed with the norm

$$\|v\|_{s,\Gamma} := \left\{ \sum_{i=1}^{N} \left\| \widetilde{(\varphi_i v) \circ \psi_i^{-1}}(\cdot, 0) \right\|_{s,\mathbb{R}^{n-1}}^2 \right\}^{1/2}.$$

The following theorem, known as the Calderón Extension Theorem, is fundamental for the results proved below. Note that, although it is stated for Ω of class C^1, it also holds for Lipschitz domains (cf. [107, Theorem A.4, p. 313]).

Theorem 10.13. *Let Ω be a 1-regular open subset of $\mathbb{R}^n$. Then, for each $m \in \mathbb{N}$, there exists a linear and bounded operator $E : \mathrm{H}^m(\Omega) \to \mathrm{H}^m(\mathbb{R}^n)$ such that $E(v) = v$ a.e. in Ω for all $v \in \mathrm{H}^m(\Omega)$.*

The extension of Lemma 10.6 to the case of $\mathrm{H}^1(\Omega)$ is established in the following theorem.

Theorem 10.14. *Let Ω be a 1-regular open subset of $\mathbb{R}^n$. Then $C^\infty(\bar{\Omega})$ is dense in $\mathrm{H}^1(\Omega)$.*

Proof. Given $v \in \mathrm{H}^1(\Omega)$, consider the extension $E(v) \in \mathrm{H}^1(\mathbb{R}^n)$ provided by the previous theorem. By the density of $C_0^\infty(\mathbb{R}^n)$ in $\mathrm{H}^1(\mathbb{R}^n)$ (cf. Theorem 10.3), there exists a sequence $\{\varphi_k\}_{k\in\mathbb{N}} \subseteq C_0^\infty(\mathbb{R}^n)$ such that $\|E(v) - \varphi_k\|_{1,\mathbb{R}^n} \xrightarrow{k\to\infty} 0$. We then define $v_k := \varphi_k|_\Omega$ for all $k \in \mathbb{N}$, and observe that $\{v_k\}_{k\in\mathbb{N}} \subseteq C^\infty(\bar{\Omega})$. Finally, it follows that

$$\|v_k - v\|_{1,\Omega} = \|\varphi_k - E(v)\|_{1,\Omega} \leqslant \|\varphi_k - E(v)\|_{1,\mathbb{R}^n} \xrightarrow{k\to\infty} 0,$$

which shows that $v \in \overline{C^\infty(\bar{\Omega})}^{\|\cdot\|_{1,\Omega}}$, and hence $\mathrm{H}^1(\Omega) = \overline{C^\infty(\bar{\Omega})}^{\|\cdot\|_{1,\Omega}}$.

□

The following theorem extends the trace inequality in $C^\infty(\overline{\mathbb{R}^n_+})$ (cf. Theorem 10.7) to the space $C^\infty(\bar{\Omega})$. In fact, this result was already stated and used, even for domains of class $C^{0,1}$, in Chapter 4 (cf. Theorem 4.25), so we essentially adopt the same statement here.

Theorem 10.15. *Let Ω be a 1-regular open subset of $\mathbb{R}^n$, and let $\gamma_0 : C^\infty(\bar{\Omega}) \to \mathrm{L}^2(\Gamma)$ be the mapping defined by*

$$\gamma_0(\varphi) := \varphi|_\Gamma \quad \forall\, \varphi \in C^\infty(\bar{\Omega}).$$

Then there exists $C > 0$ such that

$$\|\gamma_0(\varphi)\|_{0,\Gamma} \leqslant C\,\|\varphi\|_{1,\Omega} \quad \forall\, \varphi \in C^\infty(\bar{\Omega}).$$

Proof. We make use of the system of local charts $\{X_i, \psi_i\}_{i=1}^N$ and the partition of unity $\{\varphi_i\}_{i=0}^N$ introduced in Definitions 10.9 and 10.10. Accordingly, given $v \in C^\infty(\bar{\Omega})$, for each $i \in \{1, 2, \ldots, N\}$ we consider the function

$$\omega_i := (\varphi_i\, v) \circ \psi_i^{-1} : B(0,1) \to \mathbb{R}$$

(see Figure 10.3 for the associated geometry). By Theorem 10.4, its extension $\widetilde{\omega}_i$ defined by

$$\widetilde{\omega}_i(\widehat{x}, x_n) := \begin{cases} \omega_i(\widehat{x}, x_n) & \text{if } \|(\widehat{x}, x_n)\| < 1, \\ 0 & \text{if } \|(\widehat{x}, x_n)\| \geqslant 1, \end{cases}$$

belongs to $\mathrm{H}^1(\mathbb{R}^n)$, and hence $\widetilde{\omega}_i|_{\mathbb{R}^n_+} \in \mathrm{H}^1(\mathbb{R}^n_+)$. Thus, the trace inequality in $\mathrm{H}^1(\mathbb{R}^n_+)$ (cf. (10.7) in Theorem 10.8) yields

$$\|\widetilde{\omega}_i(\cdot\,, 0)\|_{0,\mathbb{R}^{n-1}} \leqslant \|\widetilde{\omega}_i\|_{1,\mathbb{R}^n_+}\,. \tag{10.11}$$

Moreover, it is easy to see that there exists $C_i > 0$, depending on φ_i and ψ_i, such that

$$\|\widetilde{\omega}_i\|_{1,\mathbb{R}^n_+} = \|\widetilde{\omega}_i\|_{1,B^+(0,1)} \leqslant C_i\,\|v\|_{1,X_i\cap\Omega} \leqslant C_i\,\|v\|_{1,\Omega},$$

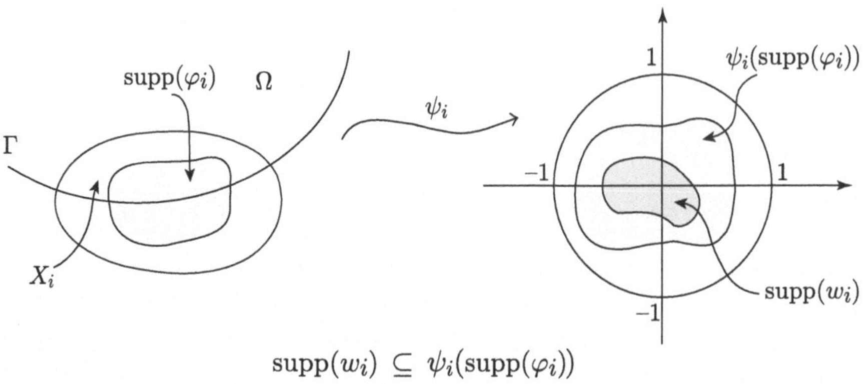

Fig. 10.3. *Illustrative geometry of the proof of Theorem 10.15*

where $B^+(0,1) := \Big\{(\widehat{x}, x_n) \in B(0,1) : \, x_n > 0\Big\}$. Combining this estimate with (10.11), we obtain

$$\|\widetilde{\omega}_i(\cdot, 0)\|_{0,\mathbb{R}^{n-1}} \leqslant C_i \, \|v\|_{1,\Omega} \qquad \forall \, i \in \{1, 2, \cdots, N\}.$$

Finally, by the definition of the norm $|\!|\!| \cdot |\!|\!|_{0,\Gamma}$ given in (10.8) and its equivalence with the usual norm $\| \cdot \|_{0,\Gamma}$ of $\mathrm{L}^2(\Gamma)$, we conclude that

$$\|\gamma_0(v)\|^2_{0,\Gamma} \leqslant C \sum_{i=1}^{N} \|\widetilde{\omega}_i(\cdot, 0)\|^2_{0,\mathbb{R}^{n-1}} \leqslant C \left\{ \sum_{i=1}^{N} C_i^2 \right\} \|v\|^2_{1,\Omega},$$

which completes the proof.

□

Having established the trace inequality in $C^\infty(\bar{\Omega})$ (cf. Theorem 10.15) and recalling from Theorem 10.14 that $C^\infty(\bar{\Omega})$ is dense in $\mathrm{H}^1(\Omega)$, it remains, as in the case of $\mathrm{H}^1(\mathbb{R}^n_+)$, to prove the Trace Theorem in $\mathrm{H}^1(\Omega)$. However, this result has already been proved in Chapter 4 (cf. Theorem 4.26), and we therefore refer the reader to that chapter. Additional concepts and results that complete the present analysis, including *the trace space* $\mathrm{H}^{1/2}(\Gamma)$, *the integration by parts formula in* $\mathrm{H}^1(\Omega)$, *Green's identities, and the normal traces of* $\mathbf{H}(\mathrm{div}; \Omega)$, can be found in detail in Sections 4.5.2, 4.5.3, and 4.5.4.

We conclude this section on traces with the following result characterizing $\mathrm{H}^1_0(\Omega)$ (cf. [112, Théorème 1.4-1]).

Theorem 10.16. *Let* Ω *be a* 1*-regular open subset of* $\mathbb{R}^n$. *Then*

$$\mathrm{H}^1_0(\Omega) = N(\gamma_0) := \Big\{ v \in \mathrm{H}^1(\Omega) : \quad \gamma_0(v) = 0 \quad on \quad \Gamma \Big\}.$$

10.3 Compact and continuous embeddings

The main goal of this section is to prove the well-known results on compact and continuous embeddings of Sobolev spaces $\mathrm{H}^m(\Omega)$. To this end, and since we will make use of the extension operator established in Theorem 10.13, we begin by recalling (cf. Definition 10.11) that $\mathrm{H}^1(\mathbb{R}^n)$ can be characterized as

$$\mathrm{H}^1(\mathbb{R}^n) := \Big\{v \in \mathrm{L}^2(\mathbb{R}^n) : \quad p_1\,\widehat{v} \in \mathrm{L}^2(\mathbb{R}^n)\Big\}, \tag{10.12}$$

where (cf. (10.9), Definition 10.11) $p_1(\xi) := \big(1 + \|\xi\|^2\big)^{1/2}$ for all $\xi \in \mathbb{R}^n$. In the following lemma, we prove the equivalence between this characterization and the original definition of $\mathrm{H}^1(\mathbb{R}^n)$. The more general case of $\mathrm{H}^m(\mathbb{R}^n)$, with $m \in \mathbb{N}$, can be established in a similar way (cf. Exercise 10.1), whereas the case $m = 0$ was already proved by the Plancherel-Parseval Theorem (cf. Theorem 9.94).

Lemma 10.17. *The definitions of* $\mathrm{H}^1(\mathbb{R}^n)$ *given by* (10.1) *(with* $m = 1$*) and* (10.12) *are equivalent.*

Proof. Let $v \in \mathrm{L}^2(\mathbb{R}^n)$ be such that $\dfrac{\partial v}{\partial x_j} \in \mathrm{L}^2(\mathbb{R}^n)$ for all $j \in \big\{1, 2, \ldots, n\big\}$. It follows (cf. Lemma 9.81 and Exercise 9.26) that $\widehat{\dfrac{\partial v}{\partial x_j}}(\xi) = \imath\,\xi_j\,\widehat{v}(\xi)$ for all $\xi \in \mathbb{R}^n$, for all $j \in \big\{1, 2, \ldots, n\big\}$. Applying the Plancherel–Parseval Theorem (cf. Theorem 9.94), we obtain

$$\begin{aligned}\|v\|_{1,\mathbb{R}^n}^2 &= \|v\|_{0,\mathbb{R}^n}^2 + \sum_{j=1}^n \left\|\frac{\partial v}{\partial x_j}\right\|_{0,\mathbb{R}^n}^2 = (2\pi)^n \left\{\|\widehat{v}\|_{0,\mathbb{R}^n}^2 + \sum_{j=1}^n \|\imath\,\xi_j\,\widehat{v}(\cdot)\|_{0,\mathbb{R}^n}^2\right\} \\ &= (2\pi)^n \int_{\mathbb{R}^n} \Big(1 + \|\xi\|^2\Big)\,|\widehat{v}(\xi)|^2\,d\xi = (2\pi)^n\,\|p_1\,\widehat{v}\|_{0,\mathbb{R}^n}^2\,,\end{aligned}$$

that is

$$\|v\|_{1,\mathbb{R}^n}^2 = (2\pi)^n\,\|p_1\,\widehat{v}\|_{0,\mathbb{R}^n}^2\,, \tag{10.13}$$

which shows that $p_1\,\widehat{v} \in \mathrm{L}^2(\mathbb{R}^n)$ for all $v \in \mathrm{H}^1(\mathbb{R}^n)$. Conversely, let $v \in \mathrm{L}^2(\mathbb{R}^n)$ be such that $p_1\,\widehat{v} \in \mathrm{L}^2(\mathbb{R}^n)$. Then, for each $j \in \big\{1, \ldots, n\big\}$,

$$\int_{\mathbb{R}^n} \xi_j^2\,|\widehat{v}(\xi)|^2\,d\xi \le \int_{\mathbb{R}^n} \big(1 + \|\xi\|^2\big)\,|\widehat{v}(\xi)|^2\,d\xi = \|p_1\,\widehat{v}\|_{0,\mathbb{R}^n}^2,$$

which implies that $\xi_j\,\widehat{v}(\cdot) \in \mathrm{L}^2(\mathbb{R}^n)$. The Fourier transform of this function is given by $\dfrac{\partial v}{\partial x_j}$ (cf. Lemma 9.81 and Exercise 9.26). Applying again the Plancherel–Parseval Theorem, we conclude that $v \in \mathrm{H}^1(\mathbb{R}^n)$ (according to definition (10.1)).

□

Next, we state and prove the classical Rellich Compact Embedding Theorem, which, in addition to the extension operator introduced above and the Fourier transform, relies on the connection between reflexivity and weak convergence (cf. Theorem 7.13). Note that the one-dimensional version of Theorem 10.18 was assumed in the statement of Exercise 6.13 (cf. Chapter 6).

Theorem 10.18 (Rellich Compact Embedding Theorem). *Let* Ω *be a* 1*-regular open subset of* $\mathbb{R}^n$*. Then the canonical embedding* $i : \mathrm{H}^1(\Omega) \to \mathrm{L}^2(\Omega)$ *is compact.*

Proof. First, note that by Theorem 10.13, there exists a linear and bounded extension operator $E : \mathrm{H}^1(\Omega) \to \mathrm{H}^1(\mathbb{R}^n)$. Next, consider a bounded open set G, sufficiently large so that $\bar{\Omega} \subseteq G$. Then, by Lemma 9.17, we can choose $\varphi \in C_0^\infty(G)$ such that $0 \leqslant \varphi \leqslant 1$ and $\varphi \equiv 1$ in a neighborhood of $\bar{\Omega}$. It follows that the operator $\widetilde{E}_1 : \mathrm{H}^1(\Omega) \to \mathrm{H}^1(\mathbb{R}^n)$ defined by $\widetilde{E}_1(v) := \varphi\, E(v)$ for all $v \in \mathrm{H}^1(\Omega)$ is also linear and bounded, satisfies $\widetilde{E}_1(v) = v$ a.e. in Ω for all $v \in \mathrm{H}^1(\Omega)$, and $\operatorname{supp}\big(\widetilde{E}_1(v)\big) \subseteq G$ for all $v \in \mathrm{H}^1(\Omega)$. Now, in order to establish the compactness of i, let $\{v_m\}_{m\in\mathbb{N}}$ be a bounded sequence in $\mathrm{H}^1(\Omega)$, and define $u_m := \widetilde{E}_1(v_m)$ for all $m \in \mathbb{N}$. The continuity of $\widetilde{E}_1$ implies that

$$\|u_m\|_{1,\mathbb{R}^n} \leqslant C\,\|v_m\|_{1,\Omega} \leqslant \overline{C} \quad \forall\, m \in \mathbb{N}\,, \tag{10.14}$$

which shows that $\{u_m\}_{m\in\mathbb{N}}$ is bounded in $\mathrm{H}^1(\mathbb{R}^n)$, and in particular,

$$\|u_m\|_{0,\mathbb{R}^n} \leqslant \bar{C} \quad \forall\, m \in \mathbb{N}\,.$$

Since $\mathrm{L}^2(\mathbb{R}^n)$ is a Hilbert space, and hence reflexive (cf. Example 7.2), it follows from Theorem 7.13 that there exists a subsequence $\{u_m^{(1)}\}_{m\in\mathbb{N}} \subseteq \{u_m\}_{m\in\mathbb{N}}$ such that $\{u_m^{(1)}\}_{m\in\mathbb{N}}$ converges weakly in $\mathrm{L}^2(\mathbb{R}^n)$. Without loss of generality, assume that $u_m^{(1)} \overset{\omega}{\to} 0$ in $\mathrm{L}^2(\mathbb{R}^n)$, and let us now show that this convergence is in fact strong, that is, $u_m^{(1)} \overset{m\to\infty}{\to} 0$ in $\mathrm{L}^2(\mathbb{R}^n)$. Indeed, applying the Plancherel–Parseval Theorem (cf. Theorem 9.94), we have, for every $M > 0$,

$$\begin{aligned}
\|u_m^{(1)}\|_{0,\mathbb{R}^n}^2 &= \int_{\mathbb{R}^n} |\widehat{u}_m^{(1)}(\xi)|^2\, d\xi \\
&= \int_{\|\xi\|\leqslant M} |\widehat{u}_m^{(1)}(\xi)|^2\, d\xi + \int_{\|\xi\|\geqslant M} \frac{(1+\|\xi\|^2)}{(1+\|\xi\|^2)}\, |\widehat{u}_m^{(1)}(\xi)|^2\, d\xi \\
&\leqslant \int_{\|\xi\|\leqslant M} |\widehat{u}_m^{(1)}(\xi)|^2\, d\xi + \frac{1}{1+M^2} \int_{\|\xi\|\geqslant M} (1+\|\xi\|^2)\, |\widehat{u}_m^{(1)}(\xi)|^2\, d\xi \\
&\leqslant \int_{\|\xi\|\leqslant M} |\widehat{u}_m^{(1)}(\xi)|^2\, d\xi + \frac{1}{1+M^2}\, \|p_1\, \widehat{u}_m^{(1)}\|_{0,\mathbb{R}^n}^2\,,
\end{aligned}$$

from which, using that $\|v\|_{1,\mathbb{R}^n}^2 = (2\pi)^n\, \|p_1\, \widehat{v}\|_{0,\mathbb{R}^n}^2$ for all $v \in \mathrm{H}^1(\mathbb{R}^n)$ (cf. (10.13), proof of Lemma 10.17), and the boundedness of $\{u_m\}_{m\in\mathbb{N}}$ (cf. (10.14)), we deduce that

$$\|u_m^{(1)}\|_{0,\mathbb{R}^n}^2 \leqslant \int_{\|\xi\|\leqslant M} |\widehat{u}_m^{(1)}(\xi)|^2\, d\xi + \frac{\widehat{C}}{1+M^2} \quad \forall\, M > 0\,,$$

with $\widehat{C} := (2\pi)^{-n}\, \bar{C}^2$. Then, given $\varepsilon > 0$, we can choose $M > 0$ such that $\dfrac{\widehat{C}}{1+M^2} \leqslant \varepsilon/2$. Moreover, since $\operatorname{supp}(u_m^{(1)}) \subseteq G$, we obtain that

$$\widehat{u}_m^{(1)}(\xi) = \int_G e^{-\imath x\cdot\xi}\, u_m^{(1)}(x)\, dx = \int_{\mathbb{R}^n} e^{-\imath x\cdot\xi}\, \mathbf{1}_G(x)\, u_m^{(1)}(x)\, dx \quad \forall\, \xi \in \mathbb{R}^n \quad \text{a.e.}\,,$$

where $\mathbf{1}_G$ denotes the characteristic function of G. Then, since for each $\xi \in \mathbb{R}^n$ the function $x \mapsto e^{-\imath x\cdot\xi}\, \mathbf{1}_G(x)$ belongs to $\mathrm{L}^2(\mathbb{R}^n)$, the weak convergence of $u_m^{(1)}$ to 0 in $\mathrm{L}^2(\mathbb{R}^n)$ implies that

$$\lim_{m\to\infty} \widehat{u}_m^{(1)}(\xi) = 0 \quad \forall\, \xi \in \mathbb{R}^n \text{ a.e.}$$

Moreover, by the Cauchy–Schwarz inequality and the boundedness of $\{u_m\}_{m\in\mathbb{N}}$ (cf. (10.14)), we have

$$|\widehat{u}_m^{(1)}(\xi)| \leqslant \int_G |u_m^{(1)}(x)|\,dx \leqslant |G|^{1/2}\,\|u_m^{(1)}\|_{0,G} \leqslant \bar{C}\,|G|^{1/2} \quad \forall\,\xi \in \mathbb{R}^n \quad \text{a.e.}, \quad \forall\, m \in \mathbb{N}.$$

Therefore, by Lebesgue's Dominated Convergence Theorem (cf. [113, Theorem 1.26], [120, Theorem 5.36]), we deduce that

$$\lim_{m\to\infty} \int_{\|\xi\|\leqslant M} |\widehat{u}_m^{(1)}(\xi)|^2\,d\xi = 0,$$

and hence we can choose $N \in \mathbb{N}$ such that

$$\int_{\|\xi\|\leqslant M} |\widehat{u}_m^{(1)}(\xi)|^2\,d\xi \leqslant \varepsilon/2 \quad \forall\, m \geqslant N.$$

Thus, we conclude that for each $\varepsilon > 0$ there exists $N \in \mathbb{N}$ such that $\|u_m^{(1)}\|_{0,\mathbb{R}^n}^2 \leqslant \varepsilon$ for all $m \geqslant N$, which proves that $u_m^{(1)} \overset{m\to\infty}{\longrightarrow} 0$ in $\mathrm{L}^2(\mathbb{R}^n)$. Consequently, $v_m^{(1)} := u_m^{(1)}|_\Omega \overset{m\to\infty}{\longrightarrow} 0$ in $\mathrm{L}^2(\Omega)$. This establishes the compactness of $i : \mathrm{H}^1(\Omega) \to \mathrm{L}^2(\Omega)$ and completes the proof. □

It is important to note that the boundedness assumption on Ω in Theorem 10.18 is essential. In fact, it can be shown, in particular, that the canonical embedding of $\mathrm{H}^1(\mathbb{R}^n)$ into $\mathrm{L}^2(\mathbb{R}^n)$ is not compact. Moreover, a simple, yet important, consequence of Theorem 10.18 is given by the following result.

Lemma 10.19. *The embedding* $i_0 : \mathrm{H}_0^1(\Omega) \to \mathrm{L}^2(\Omega)$ *is compact.*

Proof. It suffices to consider the diagram

$$\mathrm{H}_0^1(\Omega) \overset{j}{\to} \mathrm{H}^1(\Omega) \overset{i}{\to} \mathrm{L}^2(\Omega)$$

and observe that $i_0 = i \circ j$, with i compact and j continuous (cf. part (iii) of Lemma 6.14). □

On the other hand, the following theorem, whose one-dimensional version was already included in Chapter 6 as Exercise 6.13, constitutes a natural extension of Rellich Compact Embedding Theorem and is also referred to by that name.

Theorem 10.20 (Rellich Compact Embedding Theorem). *Let Ω be a 1-regular open subset of $\mathbb{R}^n$, and let $m \in \mathbb{N}$. Then the canonical embedding $i : \mathrm{H}^m(\Omega) \to \mathrm{H}^{m-1}(\Omega)$ is compact.*

Proof. We proceed by induction on m. For $m = 1$, the result follows directly from Theorem 10.18. Suppose now that $\mathrm{H}^m(\Omega)$ embeds compactly into $\mathrm{H}^{m-1}(\Omega)$, and consider a bounded sequence $\{v_k\}_{k\in\mathbb{N}}$ in $\mathrm{H}^{m+1}(\Omega)$. Since $\mathrm{H}^m(\Omega)$ is a Hilbert space (cf. Theorem 10.1), there exist a subsequence $\{v_k^{(1)}\}_{k\in\mathbb{N}} \subseteq \{v_k\}_{k\in\mathbb{N}}$ and $v \in \mathrm{H}^m(\Omega)$ such that $v_k^{(1)} \overset{\omega}{\to} v$ in $\mathrm{H}^m(\Omega)$. Then, since $i : \mathrm{H}^m(\Omega) \to \mathrm{H}^{m-1}(\Omega)$ is compact, it follows (cf. Lemma 6.29) that $v_k^{(1)} \overset{k\to\infty}{\longrightarrow} v$ in $\mathrm{H}^{m-1}(\Omega)$. Moreover, given α with $|\alpha| = m$, we have that $\{\partial^\alpha v_k^{(1)}\}_{k\in\mathbb{N}}$ is bounded in $\mathrm{H}^1(\Omega)$, and therefore there exist $\{v_k^{(2)}\}_{k\in\mathbb{N}} \subseteq \{v_k^{(1)}\}_{k\in\mathbb{N}}$ and $\omega_\alpha \in \mathrm{L}^2(\Omega)$ such that $\partial_\alpha v_k^{(2)} \overset{k\to\infty}{\longrightarrow} \omega_\alpha$ in $\mathrm{L}^2(\Omega)$. In addition, for another multi-index β with $|\beta| = m$, we reason in the same way to deduce the existence of $\{v_k^{(3)}\}_{k\in\mathbb{N}} \subseteq \{v_k^{(2)}\}_{k\in\mathbb{N}}$ and $\omega_\beta \in \mathrm{L}^2(\Omega)$ such that $\partial^\beta v_k \overset{k\to\infty}{\to} \omega_\beta$ in $\mathrm{L}^2(\Omega)$. Following this inductive reasoning we conclude, after a finite number of steps (given by the cardinality of all α such that $|\alpha| = m$), that there exist a subsequence $\{\omega_k\}_{k\in\mathbb{N}} \subseteq \{v_k^{(1)}\}_{k\in\mathbb{N}}$ and functions $\omega_\alpha \in \mathrm{L}^2(\Omega)$, for all α, $|\alpha| = m$, such that $\partial^\alpha \omega_k \overset{k\to\infty}{\to} \omega_\alpha$ in $\mathrm{L}^2(\Omega)$. Then, for each $\varphi \in C_0^\infty(\Omega)$ and for each α with $|\alpha| = m$, we have that

$$\begin{aligned}\langle \partial^\alpha v, \varphi\rangle &:= (-1)^{|\alpha|}\langle v, \partial^\alpha\varphi\rangle = (-1)^{|\alpha|}\int_\Omega v\,\partial^\alpha\varphi \\ &= (-1)^{|\alpha|}\lim_{k\to\infty}\langle \omega_k, \partial^\alpha\varphi\rangle = \lim_{k\to\infty}\langle \partial^\alpha\omega_k, \varphi\rangle \\ &= \lim_{k\to\infty}\int_\Omega \partial^\alpha\omega_k\,\varphi = \langle \omega_\alpha, \varphi\rangle,\end{aligned}$$

which proves that $\partial^\alpha v = \omega_\alpha$ in $\mathcal{D}'(\Omega)$ for all α, $|\alpha| = m$, and consequently $\omega_k \overset{k\to\infty}{\to} v$ in $\mathrm{H}^m(\Omega)$. This completes the induction and thus the proof.

□

Having proved the previous theorem, we can now establish a generalization, which has proven to be very useful in applications, of the classical Friedrichs–Poincaré inequality (cf. Lemma 4.5, (4.167), Exercise 6.16).

Theorem 10.21 (Generalized Poincaré Inequality). *Let Ω be a 1-regular open subset of $\mathbb{R}^n$, and let $m \in \mathbb{N}$. Moreover, let $\mathcal{F}$ be a finite family of functionals in $\mathrm{H}^m(\Omega)'$ such that*

$$\mathbb{P}_{m-1}(\Omega) \cap {}^\circ\mathcal{F} = \{0\}, \tag{10.15}$$

where $\mathbb{P}_{m-1}(\Omega)$ is the space of polynomials of degree $\leqslant m-1$ defined on Ω, 0 denotes the zero polynomial, and, according to Definition 3.21,

$${}^\circ\mathcal{F} := \Big\{ v \in \mathrm{H}^m(\Omega) : \quad F(v) = 0 \quad \forall\, F \in \mathcal{F} \Big\}.$$

Then, the mapping $|||\cdot||| : \mathrm{H}^m(\Omega) \to \mathbb{R}$ defined by

$$|||v||| := \Big\{ |v|_{m,\Omega}^2 + \sum_{F\in\mathcal{F}} |F(v)|^2 \Big\}^{1/2} \qquad \forall\, v \in \mathrm{H}^m(\Omega), \tag{10.16}$$

is a norm on $\mathrm{H}^m(\Omega)$ equivalent to the usual norm $\|\cdot\|_{m,\Omega}$ of this space.

Proof. Thanks to the boundedness of the elements of $\mathcal{F}$, we first observe that

$$|||v|||^2 \leqslant \|v\|_{m,\Omega}^2 + \sum_{F\in\mathcal{F}} \|F\|^2\,\|v\|_{m,\Omega}^2 = \Big\{1 + \sum_{F\in\mathcal{F}} \|F\|^2\Big\}\,\|v\|_{m,\Omega}^2 \quad \forall\, v \in \mathrm{H}^m(\Omega),$$

which establishes one side of the desired norm equivalence. Now, suppose by contradiction that there does not exist $C > 0$ such that $\|v\|_{m,\Omega} \leqslant C\,|||v|||_{m,\Omega}$ for all $v \in \mathrm{H}^m(\Omega)$. This means that for every constant $C > 0$ there exists $v_C \in \mathrm{H}^m(\Omega)$ such that $\|v_C\|_{m,\Omega} > C\,|||v_C|||$. In particular, for each $k \in \mathbb{N}$ there exists $v_k \in \mathrm{H}^m(\Omega)$ such that $\|v_k\|_{m,\Omega} > k\,|||v_k|||$ for all $k \in \mathbb{N}$, and we then define the normalized sequence

$$u_k := \frac{v_k}{\|v_k\|_{m,\Omega}} \qquad \forall\, k \in \mathbb{N}.$$

It follows that

$$\|u_k\|_{m,\Omega} = 1 \quad \text{and} \quad |||u_k||| < \frac{1}{k} \quad \forall\, k \in \mathbb{N},$$

from which, in view of (10.16), we deduce that $\partial^\alpha u_k \overset{k\to\infty}{\longrightarrow} 0$ in $\mathrm{L}^2(\Omega)$ for all α, $|\alpha| = m$, and moreover $F(u_k) \overset{k\to\infty}{\longrightarrow} 0$ for all $F \in \mathcal{F}$. Now, since $\{u_k\}_{k\in\mathbb{N}}$ is bounded in $\mathrm{H}^m(\Omega)$ and the embedding $i : \mathrm{H}^m(\Omega) \to \mathrm{H}^{m-1}(\Omega)$ is compact (cf. Theorem 10.20), there exist a subsequence $\{u_k^{(1)}\}_{k\in\mathbb{N}} \subseteq \{u_k\}_{k\in\mathbb{N}}$ and $u \in \mathrm{H}^{m-1}(\Omega)$ such that $u_k^{(1)} \overset{k\to\infty}{\longrightarrow} u$ in $\mathrm{H}^{m-1}(\Omega)$. Proceeding analogously to the proof of Theorem 10.20, one then uses this convergence together with the fact that $\partial^\alpha u_k^{(1)} \overset{k\to\infty}{\longrightarrow} 0$ in $\mathrm{L}^2(\Omega)$ for all $\alpha, |\alpha| = m$, to conclude that $\partial^\alpha u = 0$ in $\mathcal{D}'(\Omega)$ for all α, $|\alpha| = m$. This shows that $u \in \mathrm{H}^m(\Omega)$ and that $u_k^{(1)} \overset{k\to\infty}{\longrightarrow} u$ in $\mathrm{H}^m(\Omega)$, from which

$$\|u\|_{m,\Omega} = \lim_{k\to\infty} \|u_k^{(1)}\|_{m,\Omega} = 1 \quad \text{and} \quad F(u) = \lim_{k\to\infty} F(u_k^{(1)}) = 0 \quad \forall\, F \in \mathcal{F}. \tag{10.17}$$

However, since all derivatives of order m of u vanish, it follows that $u \in \mathbb{P}_{m-1}(\Omega)$, and therefore the hypothesis (10.15) implies that $u \equiv 0$. This contradicts the first identity in (10.17) and completes the proof.

□

We now present some examples of applications of the Generalized Poincaré inequality, which have already been used in previous chapters.

Example 10.22 Here we make use of Theorem 10.21 to prove the equivalence between norms and seminorms of Sobolev spaces over some of their subspaces. In what follows, Ω denotes a 1-regular open subset of $\mathbb{R}^n$ with boundary Γ.

1. Define the closed subspace of $\mathrm{H}^1(\Omega)$ given by

$$\widetilde{\mathrm{H}}^1(\Omega) := \left\{ v \in \mathrm{H}^1(\Omega) : \quad \int_\Omega v = 0 \right\},$$

and let us prove that $\|\cdot\|_{1,\Omega}$ and $|\cdot|_{1,\Omega}$ are equivalent on this space. Indeed, it suffices to consider $\mathcal{F} = \{F\} \subseteq \mathrm{H}^1(\Omega)'$ with $F(v) := \int_\Omega v$ for all $v \in \mathrm{H}^1(\Omega)$, and set

$$|||v||| := \left\{ |v|_{1,\Omega}^2 + |F(v)|^2 \right\}^{1/2} \quad \forall\, v \in \mathrm{H}^1(\Omega).$$

It is clear that $\mathbb{P}_0(\Omega) \cap {}^\circ\mathcal{F}$ reduces to the zero polynomial, and therefore a direct application of Theorem 10.21 yields the equivalence between $|||\cdot|||$ and $\|\cdot\|_{1,\Omega}$ in $\mathrm{H}^1(\Omega)$. Finally, since $|||v||| = |v|_{1,\Omega}$ for all $v \in \widetilde{\mathrm{H}}^1(\Omega)$, the result follows.

2. Let $\Gamma_D \subseteq \Gamma$ be such that $\Gamma_D \neq \Gamma$ and $|\Gamma_D| > 0$, and define

$$\mathrm{H}^1_{\Gamma_D}(\Omega) := \left\{ v \in \mathrm{H}^1(\Omega) : \quad \gamma_0(v) = 0 \quad \text{on} \quad \Gamma_D \right\},$$

where $\gamma_0 : \mathrm{H}^1(\Omega) \to \mathrm{L}^2(\Gamma)$ is the respective trace operator (cf. Theorems 10.15 and 4.26). Next, consider $\mathcal{F} = \{F\} \subseteq \mathrm{H}^1(\Omega)'$ with $F(v) := \int_{\Gamma_D} \gamma_0(v)$ for all $v \in \mathrm{H}^1(\Omega)$, and define

$$|||v||| := \left\{ |v|_{1,\Omega}^2 + |F(v)|^2 \right\}^{1/2} \quad \forall\, v \in \mathrm{H}^1(\Omega).$$

Note that the Cauchy–Schwarz and trace inequalities (cf. Theorems 4.25 and 10.15) imply that

$$|F(v)| \leqslant \int_{\Gamma_D} |\gamma_0(v)| \leqslant |\Gamma_D|^{1/2} \|\gamma_0(v)\|_{0,\Gamma_D} \leqslant |\Gamma_D|^{1/2} \|\gamma_0(v)\|_{0,\Gamma} \leqslant C\,|\Gamma_D|^{1/2} \|v\|_{1,\Omega} \quad \forall\, v \in \mathrm{H}^1(\Omega),$$

which shows that $F \in \mathrm{H}^1(\Omega)'$. Moreover, using that $|\Gamma_D| > 0$, it is easily verified that $\mathbb{P}_0(\Omega) \cap {}^\circ\mathcal{F}$ is again the zero polynomial, whereby Theorem 10.21 guarantees that $\|\!|\cdot|\!\|$ and $\|\cdot\|_{1,\Omega}$ are equivalent in $\mathrm{H}^1(\Omega)$. Finally, since $\|\!|v|\!\| = |v|_{1,\Omega}$ for all $v \in \mathrm{H}^1_{\Gamma_D}(\Omega)$, we conclude the equivalence between $\|\cdot\|_{1,\Omega}$ and $|\cdot|_{1,\Omega}$ in $\mathrm{H}^1_{\Gamma_D}(\Omega)$.

3. Now define

$$\widetilde{\mathrm{H}}^2(\Omega) := \left\{ v \in \mathrm{H}^2(\Omega) : \quad \int_\Omega p v = 0 \quad \forall\, p \in \mathbb{P}_1(\Omega) \right\},$$

and let us prove that $\|\cdot\|_{2,\Omega}$ and $|\cdot|_{2,\Omega}$ are equivalent in this subspace of $\mathrm{H}^2(\Omega)$. Indeed, consider $\mathcal{F} = \{F_0, F_1, F_2, \dots, F_n\} \subseteq \mathrm{H}^2(\Omega)'$ with

$$F_0(v) := \int_\Omega v \quad \text{and} \quad F_i(v) := \int_\Omega x_i\, v \quad \forall\, v \in \mathrm{H}^2(\Omega), \quad \forall\, i \in \{1, 2, \dots, n\},$$

where $x := (x_1, x_2, \dots, x_n)$ stands for a generic vector of $\mathbb{R}^n$, and let

$$\|\!|v|\!\| := \left\{ |v|^2_{2,\Omega} + \sum_{j=0}^{n} |F_j(v)|^2 \right\}^{1/2} \quad \forall\, v \in \mathrm{H}^2(\Omega).$$

Note that the Cauchy–Schwarz inequality ensures that the finite family of linear functionals $\mathcal{F}$ is contained in $\mathrm{H}^2(\Omega)'$. Moreover, it is easy to see that $\widetilde{\mathrm{H}}^2(\Omega) = \mathrm{H}^2(\Omega) \cap {}^\circ\mathcal{F}$ and that $\mathbb{P}_1(\Omega) \cap {}^\circ\mathcal{F}$ is the zero polynomial. It follows from Theorem 10.21 that $\|\!|\cdot|\!\|$ and $\|\cdot\|_{2,\Omega}$ are equivalent in $\mathrm{H}^2(\Omega)$, and since $\|\!|v|\!\| = |v|_{2,\Omega}$ for all $v \in \widetilde{\mathrm{H}}^2(\Omega)$, the desired equivalence follows.

We conclude this section with the classical result of Sobolev embedding.

Theorem 10.23 (Sobolev Embedding Theorem). *Let $m \in \mathbb{N}$ be such that $m > n/2$, and let Ω be a 1-regular open subset of $\mathbb{R}^n$. Then $\mathrm{H}^m(\Omega)$ is a subspace of $C(\bar{\Omega})$, and the embedding $i : \mathrm{H}^m(\Omega) \to C(\bar{\Omega})$ is bounded.*

Proof. Given $v \in \mathrm{H}^m(\Omega)$, we define $u := E(v) \in \mathrm{H}^m(\mathbb{R}^n)$, where E is the linear and bounded extension operator established by Theorem 10.13. Then, we consider the Fourier transform $\widehat{u}$ of u and write

$$\widehat{u}(\xi) = p_m^{-1}(\xi)\, p_m(\xi)\, \widehat{u}(\xi) \qquad \forall\, \xi \in \mathbb{R}^n, \tag{10.18}$$

where (cf. Definition 10.11) $p_m(\xi) := \left(1 + \|\xi\|^2\right)^{m/2}$ for all $\xi \in \mathbb{R}^n$. Recall that $p_m\, \widehat{u} \in \mathrm{L}^2(\mathbb{R}^n)$ and that $\|p_m\, \widehat{u}\|_{0,\mathbb{R}^n}$ and $\|u\|_{m,\mathbb{R}^n}$ are equivalent (cf. Lemma 10.17 for $m = 1$ and Exercise 10.1 for any $m \in \mathbb{N}$). Moreover, from Lemma 9.79 we know that for $m > n/2$ one has $p_m^{-1} \in \mathrm{L}^2(\mathbb{R}^n)$. It follows from (10.18) that $\widehat{u} \in \mathrm{L}^1(\mathbb{R}^n)$ and, applying the Cauchy–Schwarz inequality and the aforementioned equivalence of norms, we obtain

$$\|\widehat{u}\|_{0,1;\mathbb{R}^n} \leqslant \|p_m^{-1}\|_{0,\mathbb{R}^n} \|p_m\, \widehat{u}\|_{0,\mathbb{R}^n} \leqslant \widehat{C}\, \|u\|_{m,\mathbb{R}^n},$$

where $\widehat{C}$ depends on m, n, and p_m. Thus, since $\widehat{u} \in \mathrm{L}^1(\mathbb{R}^n)$, it follows that u, given by

$$u(x) := (2\pi)^{-n} \int_{\mathbb{R}^n} e^{\imath x \cdot \xi}\, \widehat{u}(\xi)\, d\xi \quad \forall\, x \in \mathbb{R}^n,$$

is continuous in $\mathbb{R}^n$ thanks to the Lebesgue Dominated Convergence Theorem (cf. [113, Theorem 1.26], [120, Theorem 5.36]), and

$$|u(x)| \leqslant (2\pi)^{-n} \|\hat{u}\|_{0,1;\mathbb{R}^n} \leqslant (2\pi)^{-n} \widehat{C} \|u\|_{m,\mathbb{R}^n} \quad \forall x \in \mathbb{R}^n .$$

Finally, it is clear that $v = u|_{\Omega} \in C(\bar{\Omega})$ and, using also the boundedness of E, we conclude that

$$\|v\|_{\infty,\Omega} := \sup_{x \in \bar{\Omega}} |v(x)| \leqslant \|u\|_{\infty,\mathbb{R}^n} \leqslant (2\pi)^{-n} \widehat{C} \|u\|_{m,\mathbb{R}^n} \leqslant \bar{C} \|v\|_{m,\Omega} ,$$

with $\bar{C} = (2\pi)^{-n} \widehat{C} \|E\|$, which confirms the boundedness of $i : \mathrm{H}^m(\Omega) \to C(\bar{\Omega})$.

□

The following lemma is an immediate corollary of the previous theorem.

Lemma 10.24. *Let Ω be a 1-regular open subset of $\mathbb{R}^n$, and let $m, j \in \mathbb{N}$ be such that $m-j > n/2$. Then $\mathrm{H}^m(\Omega)$ is a subspace of $C^j(\bar{\Omega})$ and the embedding $i : \mathrm{H}^m(\Omega) \to C^j(\bar{\Omega})$ is bounded.*

10.4 Interpolation of normed spaces

10.4.1 Preliminaries

Let $(X_0, \|\cdot\|_0)$, $(X_1, \|\cdot\|_1)$, and $(X, \|\cdot\|)$ be normed vector spaces over the same field ($\mathbb{R}$ or $\mathbb{C}$) such that X_0 and X_1 are subspaces of X. In this case, we say that X_0 and X_1 form a compatible pair. Then, we define norms on $X_0 \cap X_1$ and $X_0 + X_1$ by

$$\|u\|_{\cap} := \left\{ \|u\|_0^2 + \|u\|_1^2 \right\}^{1/2}$$

for each $u \in X_0 \cap X_1$, and

$$\|u\|_{+} := \inf \left\{ \left\{ \|u_0\|_0^2 + \|u_1\|_1^2 \right\}^{1/2} : \quad u = u_0 + u_1, \ u_0 \in X_0, \ u_1 \in X_1 \right\}$$

for each $u \in X_0 + X_1$. It follows easily that

$$\|u\|_{+} \leqslant \|u\|_j \leqslant \|u\|_{\cap} \qquad \forall u \in X_0 \cap X_1, \quad \forall j \in \{0,1\},$$

which shows that the inclusions

$$X_0 \cap X_1 \subseteq X_j \subseteq X_0 + X_1 \qquad \forall j \in \{0,1\},$$

are continuous. Note also that, in the particular case where $X_1 \subseteq X_0$, we obviously have

$$X_0 \cap X_1 = X_1 \quad \text{and} \quad X_0 + X_1 = X_0 .$$

Now, given parameters $\delta \in]0,1[$ and $q \in [1,+\infty)$, we will construct a normed vector space denoted $X_{\delta,q} := (X_0, X_1)_{\delta,q}$, which will be called the INTERPOLATED SPACE OF ORDER (δ, q) between X_0 and X_1, such that

$$X_0 \cap X_1 \subseteq X_{\delta,q} \subseteq X_0 + X_1 .$$

Moreover, if (Y_0, Y_1) is another compatible pair, and if $A_0 \in \mathcal{L}(X_0, Y_0)$ and $A_1 \in \mathcal{L}(X_1, Y_1)$ are such that $A_0(u) = A_1(u)$ for all $u \in X_0 \cap X_1$, it will also be shown that there exists $A_{\delta,q} \in \mathcal{L}(X_{\delta,q}, Y_{\delta,q})$, where $Y_{\delta,q} := (Y_0, Y_1)_{\delta,q}$ is the interpolated space of order (δ, q) between Y_0 and Y_1, such that

$$A_{\delta,q}(u) = A_0(u) = A_1(u) \qquad \forall u \in X_0 \cap X_1 .$$

10.4.2 Construction of interpolated spaces

Here we adopt the K-method for the construction of real interpolation spaces, as explained in [107, Appendix B]. Indeed, given $t > 0$ and $u \in X_0 + X_1$, we define the *interpolation K-functional*

$$K(t,u) := \inf\Big\{ \big(\|u_0\|_0^2 + t^2\,\|u_1\|_1^2\big)^{1/2} : \quad u = u_0 + u_1\,, \quad u_0 \in X_0\,, \quad u_1 \in X_1 \Big\}. \tag{10.19}$$

Equivalently,

$$K(t,u) := \inf\Big\{ \big(\|u_0\|_0^2 + t^2\,\|u - u_0\|_1^2\big)^{1/2} : \quad u_0 \in X_0, \quad u - u_0 \in X_1 \Big\}$$

or, alternatively,

$$K(t,u) := \inf\Big\{ \big(\|u - u_1\|_0^2 + t^2\,\|u_1\|_1^2\big)^{1/2} : \quad u_1 \in X_1, \quad u - u_1 \in X_0 \Big\}.$$

In case it is necessary to make the compatible pair (X_0, X_1) explicit, $K(t,u)$ is denoted by $K(t,u;X_0,X_1)$.

Other constructions include the J-method (cf. [107, Appendix B]) and the Calderón approach for real and complex interpolation, respectively. The former defines, instead of (10.19), for $t > 0$ and $u \in X_0 \cap X_1$,

$$J(t,u) := \max\big\{\|u\|_0\,,\, t\,\|u\|_1\big\}, \tag{10.20}$$

which, however, is equivalent to the K-method in the setting of Banach spaces, whereas the latter uses holomorphic functions defined on the strip $\big\{z \in \mathbb{C} : \quad 0 \leqslant \mathrm{Re}(z) \leqslant 1\big\}$.

The following lemma establishes some basic properties of K.

Lemma 10.25. *For each $s,\, t > 0$, $\lambda \in \mathbb{R}$, and $u,\, v \in X_0 + X_1$, we have*

a) $K(t,\lambda u) = |\lambda|\,K(t,u)$.
b) $K(t,u+v) \leqslant K(t,u) + K(t,v)$.
c) $\min\{1,t\}\,\|u\|_+ \leqslant K(t,u) \leqslant \max\{1,t\}\,\|u\|_+$.
d) $\min\{1,t/s\}K(s,u) \leqslant K(t,u) \leqslant \max\{1,t/s\}\,K(s,u)$.

Proof. Note first that the case $\lambda = 0$ is immediate for a). Now, given $\lambda \neq 0$ and $u \in X_0 + X_1$, let $u_0 \in X_0$ and $u_1 \in X_1$ be such that $u = u_0 + u_1$. It follows immediately that $\lambda u = \lambda u_0 + \lambda u_1$, and therefore

$$K(t,\lambda u) \leqslant \Big(\|\lambda u_0\|_0^2 + t^2\,\|\lambda u_1\|_1^2\Big)^{1/2} = |\lambda|\,\Big(\|u_0\|_0^2 + t^2\,\|u_1\|_1^2\Big)^{1/2},$$

whence, by taking the infimum with respect to u_0 and u_1, we obtain

$$K(t,\lambda u) \leqslant |\lambda|\,K(t,u)\,.$$

Conversely, let $\widetilde{u}_0 \in X_0$ and $\widetilde{u}_1 \in X_1$ be such that $\lambda u = \widetilde{u}_0 + \widetilde{u}_1$, that is, $u = \frac{1}{\lambda}\,\widetilde{u}_0 + \frac{1}{\lambda}\,\widetilde{u}_1$. It follows that

$$\Big(\|\widetilde{u}_0\|_0^2 + t^2\,\|\widetilde{u}_1\|_1^2\Big)^{1/2} = |\lambda|\,\left(\left\|\frac{\widetilde{u}_0}{\lambda}\right\|_0^2 + t^2\left\|\frac{\widetilde{u}_1}{\lambda}\right\|_1^2\right)^{1/2} \geqslant |\lambda|\,K(t,u)\,,$$

and hence, by taking the infimum with respect to $\widetilde{u}_0$ and $\widetilde{u}_1$, we obtain

$$K(t, \lambda u) \geqslant |\lambda|\, K(t, u),$$

which completes the proof of a). For b), let u_0, $v_0 \in X_0$ and u_1, $v_1 \in X_1$ be such that $u = u_0 + u_1$ and $v = v_0 + v_1$. Then, using the triangle inequality of the product norm $\|\cdot\|$ on $X_0 \times X_1$, we have

$$\begin{aligned}
K(t, u + v) &\leqslant \Big(\|u_0 + v_0\|_0^2 + t^2\, \|u_1 + v_1\|_1^2\Big)^{1/2} \\
&= \|(u_0, tu_1) + (v_0, tv_1)\| \\
&\leqslant \|(u_0, tu_1)\| + \|(v_0, tv_1)\| \\
&= \Big(\|u_0\|_0^2 + t^2\, \|u_1\|_1^2\Big)^{1/2} + \Big(\|v_0\|_0^2 + t^2\, \|v_1\|_1^2\Big)^{1/2},
\end{aligned}$$

and thus, by taking the infimum first with respect to u_0 and u_1, and then with respect to v_0 and v_1, we obtain b). In turn, by taking also the infimum with respect to u_0 and u_1 in the inequality

$$\begin{aligned}
\min\{1, t\}\Big(\|u_0\|_0^2 + \|u_1\|_1^2\Big)^{1/2} &\leqslant \Big(\|u_0\|_0^2 + t^2\, \|u_1\|_1^2\Big)^{1/2} \\
&\leqslant \max\{1, t\}\Big(\|u_0\|_0^2 + \|u_1\|_1^2\Big)^{1/2},
\end{aligned}$$

we immediately obtain c). Finally, proceeding analogously with the inequality

$$\begin{aligned}
\min\{1, t/s\}\Big(\|u_0\|_0^2 + s^2\, \|u_1\|_1^2\Big)^{1/2} &\leqslant \Big(\|u_0\|_0^2 + t^2\, \|u_1\|_1^2\Big)^{1/2} \\
&\leqslant \max\{1, t/s\}\Big(\|u_0\|_0^2 + s^2\, \|u_1\|_1^2\Big)^{1/2},
\end{aligned}$$

we deduce d), which completes the proof.

□

Note that, according to a), b), and c), for fixed $t > 0$, $K(t, \cdot)$ is a norm on $X_0 + X_1$ equivalent to $|\cdot|_+$. Likewise, it is clear from d) that, for fixed u, $K(\cdot, u)$ is a non-decreasing function of t.

Now, given $\delta \in (0, 1)$ and $q \in [1, +\infty)$, we define the weighted $\mathrm{L}^q(\mathbb{R}^+)$ norm

$$\|f\|_{\delta,q} := \left\{\int_0^\infty t^{-(1+\delta q)}\, |f(t)|^q\, dt\right\}^{1/q}. \tag{10.21}$$

It is easy to verify, by means of a simple change of variable, that the following *dilation property* holds

$$\|f(a\,\cdot)\|_{\delta,q} = a^\delta\, \|f\|_{\delta,q} \qquad \forall\, a > 0. \tag{10.22}$$

Then, the INTERPOLATED SPACE OF ORDER (δ, q) between X_0 and X_1 is defined as

$$(X_0, X_1)_{\delta,q} := \Big\{u \in X_0 + X_1 : \quad \|K(\cdot, u)\|_{\delta,q} < +\infty\Big\},$$

which is endowed with the norm

$$|||u|||_{\delta,q} := N_{\delta,q} \, \|K(\cdot,u)\|_{\delta,q} \qquad \forall\, u \in \left(X_0, X_1\right)_{\delta,q}, \tag{10.23}$$

where $N_{\delta,q}$, a constant introduced only for normalization purposes, is given by

$$N_{\delta,q} := \|\min\{1,\cdot\}\|_{\delta,q}^{-1}. \tag{10.24}$$

An explicit computation yields

$$\begin{aligned} \|\min\{1,\cdot\}\|_{\delta,q} &= \left\{ \int_0^1 t^{-(1+\delta q)+q}\, dt + \int_1^\infty t^{-(1+\delta q)}\, dt \right\}^{1/q} \\ &= [(1-\delta)\,\delta\, q]^{-1/q}, \end{aligned}$$

whence

$$N_{\delta,q} = [(1-\delta)\,\delta\, q]^{1/q}. \tag{10.25}$$

Note that the definition of K (cf. (10.19)), together with properties a) and b) of Lemma 10.25, allow one to prove, using the fact that $\|\cdot\|_{\delta,q}$ is a norm, that $|||\cdot|||_{\delta,q}$ is indeed also a norm on $\left(X_0, X_1\right)_{\delta,q}$.

The following lemma establishes the essential properties of the interpolated space $\left(X_0, X_1\right)_{\delta,q}$ (see Figure 10.4 for an illustration of the corresponding inclusions).

Lemma 10.26. *The space $X_0 \cap X_1$ is contained in $\left(X_0, X_1\right)_{\delta,q}$ and*

$$|||u|||_{\delta,q} \leqslant \|u\|_0^{1-\delta} \|u\|_1^{\delta} \leqslant \|u\|_\cap \qquad \forall\, u \in X_0 \cap X_1.$$

Moreover, it is clear that $\left(X_0, X_1\right)_{\delta,q} \subseteq X_0 + X_1$ and we have

$$K(t,u) \leqslant t^\delta\, |||u|||_{\delta,q} \quad \textit{and} \quad \|u\|_+ \leqslant |||u|||_{\delta,q} \qquad \forall\, u \in \left(X_0, X_1\right)_{\delta,q}.$$

Proof. Let $u \in X_0 \cap X_1$, $u \neq 0$, and define the constant $a := \dfrac{\|u\|_1}{\|u\|_0}$. Since it is clear that $u = u + 0 \in X_0 + X_1$ and also $u = 0 + u \in X_0 + X_1$, it follows directly from the definition of K (cf. (10.19)) that

$$K(t,u) \leqslant \|u\|_0 \quad \text{and} \quad K(t,u) \leqslant t\,\|u\|_1,$$

whence

$$K(t,u) \leqslant \min\left\{\|u\|_0, t\,\|u\|_1\right\} = \|u\|_0 \min\left\{1, at\right\}.$$

It follows, by using the dilation property (10.22), and invoking (10.24), that

$$\|K(\cdot,u)\|_{\delta,q} \leqslant \|u\|_0 \, \|\min\{1,a\cdot\}\|_{\delta,q} = \|u\|_0\, a^\delta\, \|\min\{1,\cdot\}\|_{\delta,q} = \frac{\|u\|_0^{1-\delta}\, \|u\|_1^{\delta}}{N_{\delta,q}},$$

that is,

$$|||u|||_{\delta,q} \leqslant \|u\|_0^{1-\delta}\, \|u\|_1^{\delta}. \tag{10.26}$$

In turn, applying Hölder's inequality with exponents $\bar{p} = \frac{1}{1-\delta}$ and $\bar{q} = \frac{1}{\delta}$ (such that $\frac{1}{\bar{p}} + \frac{1}{\bar{q}} = 1$), and the Cauchy–Schwarz inequality in $\mathbb{R}^2$, we obtain

$$\|u\|_0^{1-\delta}\, \|u\|_0^{\delta} \leqslant (1-\delta)\, \|u\|_0 + \delta\, \|u\|_1 \leqslant \left\{(1-\delta)^2 + \delta^2\right\}^{1/2} \|u\|_\cap \leqslant \|u\|_\cap,$$

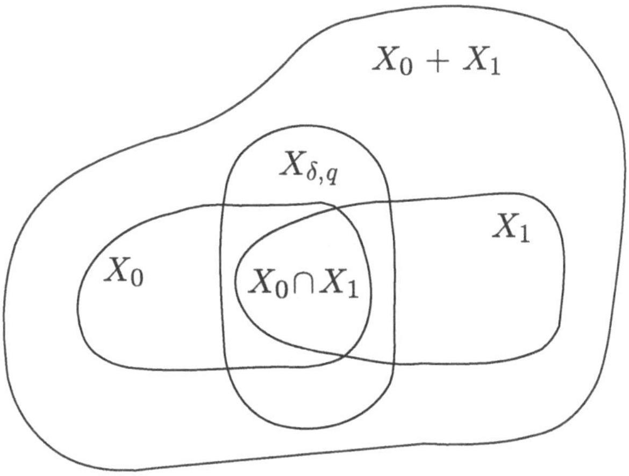

Fig. 10.4. *Illustration of the inclusions in Lemma 10.26*

which, together with (10.26), proves the first part of the lemma. On the other hand, given $u \in (X_0, X_1)_{\delta,q}$, it follows from d) in Lemma 10.25 that

$$\min\{1, s/t\}\, K(t,u) \leqslant K(s,u) \qquad \forall\, s,t > 0,$$

which, upon integrating with respect to s in the definition of $\|K(\cdot,u)\|_{\delta,q}$ and using the dilation property (10.22), implies that

$$K(t,u)\left\|\min\left\{1,\frac{\cdot}{t}\right\}\right\|_{\delta,q} = K(t,u)\, t^{-\delta}\, \|\min\{1,\cdot\}\|_{\delta,q} \leqslant \|K(\cdot,u)\|_{\delta,q},$$

that is (cf. (10.24)),

$$K(t,u)\, t^{-\delta}\, N_{\delta,q}^{-1} \leqslant \|K(\cdot,u)\|_{\delta,q},$$

or, by virtue of (10.23),

$$K(t,u) \leqslant t^{\delta}\, |\!|\!| u |\!|\!|_{\delta,q}.$$

Finally, applying the definition of the norm $|\!|\!| \cdot |\!|\!|_{\delta,q}$ in the first inequality of part c) of Lemma 10.25, we obtain

$$\|u\|_+\, N_{\delta,q}^{-1} \leqslant \|K(\cdot,u)\|_{\delta,q},$$

that is,

$$\|u\|_+ \leqslant |\!|\!| u |\!|\!|_{\delta,q},$$

thus completing the present proof. □

We now prove what is known as the *interpolation property.*

Theorem 10.27. *Let $(X_0, \|\cdot\|_0), (X_1, \|\cdot\|_1), (Y_0, \|\cdot\|_0)$ and $(Y_1, \|\cdot\|_1)$ be normed vector spaces over $\mathbb{R}$ (or $\mathbb{C}$) such that (X_0, X_1) and (Y_0, Y_1) form compatible pairs. Moreover, let $A_0 \in \mathcal{L}(X_0, Y_0)$ and $A_1 \in \mathcal{L}(X_1, Y_1)$ be such that $A_0(u) = A_1(u)$ for all $u \in X_0 \cap X_1$, and let $M_0, M_1 > 0$ satisfy $\|A_j(u)\|_j \leqslant M_j \|u\|_j$ for all $u \in X_j$, for all $j \in \{0,1\}$. Given $\delta \in (0,1)$ and $q \in [1,+\infty)$, we denote for simplicity $X_{\delta,q} := (X_0, X_1)_{\delta,q}$ and $Y_{\delta,q} := (Y_0, Y_1)_{\delta,q}$. Then, there exists an operator $A_{\delta,q} \in \mathcal{L}(X_{\delta,q}, Y_{\delta,q})$ such that*

$$A_{\delta,q}(u) = A_0(u) = A_1(u) \qquad \forall\, u \in X_0 \cap X_1 ,$$

and satisfying

$$\|A_{\delta,q}(u)\|_{\delta,q} \leqslant M_0^{1-\delta}\, M_1^{\delta}\, \|u\|_{\delta,q} \qquad \forall\, u \in X_{\delta,q} . \tag{10.27}$$

Proof. Note first that we have used the same notation $\|\cdot\|_j$ for the norms of X_j and Y_j, $j \in \{0,1\}$. Nevertheless, these norms do not necessarily coincide, and their meaning is clear from the context in which they are used. We begin the proof by defining the operator $A : X_0 + X_1 \to Y_0 + Y_1$ as $A(u) := A_0(u_0) + A_1(u_1)$ with $u_0 \in X_0$ and $u_1 \in X_1$ such that $u = u_0 + u_1$. In this regard, given $\widetilde{u}_0 \in X_0$ and $\widetilde{u}_1 \in X_1$ such that $u = \widetilde{u}_0 + \widetilde{u}_1 = u_0 + u_1$, it follows that $u_0 - \widetilde{u}_0 = \widetilde{u}_1 - u_1 \in X_0 \cap X_1$, and hence, by hypothesis, $A_0(u_0 - \widetilde{u}_0) = A_1(\widetilde{u}_1 - u_1)$. This implies that $A_0(u_0) + A_1(u_1) = A_0(\widetilde{u}_0) + A_1(\widetilde{u}_1)$, thus proving that A is well defined. Moreover, it is clear that A is linear and bounded (since A_0 and A_1 are). To see the latter (although this is not required for the proof), take $u = u_0 + u_1 \in X_0 + X_1$ with $u_0 \in X_0$ and $u_1 \in X_1$. It follows that $A(u) = A_0(u_0) + A_1(u_1)$ and, by definition of $\|\cdot\|_+$ in $Y_0 + Y_1$, we have

$$\begin{aligned}
\|A(u)\|_+ &\leqslant \Big(\|A_0(u_0)\|_0^2 + \|A_1(u_1)\|_1^2\Big)^{1/2} \\
&\leqslant \Big(\|A_0\|^2\,\|u_0\|_0^2 + \|A_1\|^2\,\|u_1\|_1^2\Big)^{1/2} \\
&\leqslant \max\Big\{\|A_0\|, \|A_1\|\Big\}\Big\{\|u_0\|_0^2 + \|u_1\|_1^2\Big\}^{1/2} ,
\end{aligned}$$

whence, taking the infimum with respect to u_0 and u_1, we obtain

$$\|A(u)\|_+ \leqslant \max\Big\{\|A_0\|, \|A_1\|\Big\}\,\|u\|_+ \qquad \forall\, u \in X_0 + X_1 .$$

Now we introduce the operator $A_{\delta,q} := A|_{X_{\delta,q}}$ and show that $A_{\delta,q} \in \mathcal{L}(X_{\delta,q}, Y_{\delta,q})$. It is clear that $A_{\delta,q}$ is linear, and by virtue of the hypothesis on A_0 and A_1 together with the definition of A, we have

$$A_{\delta,q}(u) = A_0(u) = A_1(u) \qquad \forall\, u \in X_0 \cap X_1 .$$

Then, given $u = u_0 + u_1 \in X_{\delta,q}$, with $u_0 \in X_0$ and $u_1 \in X_1$, we have

$$A_{\delta,q}(u) = A_0(u_0) + A_1(u_1) .$$

This implies, using also that $\|A_j\| \leqslant M_j$ for all $j \in \{0,1\}$, that

$$\begin{aligned}
K(t, A_{\delta,q}(u)) &\leqslant \Big(\|A_0(u_0)\|_0^2 + t^2\,\|A_1(u_1)\|_1^2\Big)^{1/2} \\
&\leqslant \Big(\|A_0\|^2\,\|u_0\|_0^2 + t^2\,\|A_1\|^2\,\|u_1\|_1^2\Big)^{1/2} \\
&\leqslant \Big(M_0^2\,\|u_0\|_0^2 + M_1^2\, t^2\,\|u_1\|_1^2\Big)^{1/2} \\
&\leqslant M_0\Big(\|u_0\|_0^2 + (a\,t)^2\,\|u_1\|_1^2\Big)^{1/2} ,
\end{aligned}$$

where $a = \dfrac{M_1}{M_0}$. It follows, by taking the infimum with respect to u_0 and u_1, that

$$K(t, A_{\delta,q}(u)) \leqslant M_0\, K(at, u) ,$$

whence, invoking the norm $\|\!|\cdot|\!\|_{\delta,q}$ (cf. (10.23)), and using the dilation property (10.22), we deduce that

$$\|\!|A_{\delta,q}(u)|\!\|_{\delta,q} := N_{\delta,q}\,\|K(\cdot, A_{\delta,q}(u))\|_{\delta,q} \leqslant N_{\delta,q}\,M_0\,\|K(a\,\cdot, u)\|_{\delta,q}$$

$$= M_0\,a^{\delta}\,N_{\delta,q}\,\|K(\cdot, u)\|_{\delta,q} = M_0^{1-\delta}\,M_1^{\delta}\,\|\!|u|\!\|_{\delta,q}\,,$$

which proves inequality (10.27), and therefore $A_{\delta,q} \in \mathcal{L}(X_{\delta,q}, Y_{\delta,q})$.

□

It is important to observe that the previous theorem can also be stated as follows: if $A \in \mathcal{L}(X_0 + X_1, Y_0 + Y_1)$ is such that $A|_{X_0} \in \mathcal{L}(X_0, Y_0)$ and $A|_{X_1} \in \mathcal{L}(X_1, Y_1)$, with $\|A|_{X_j}\| \leqslant M_j$, $j \in \{0, 1\}$, then $A_{\delta,q} := A|_{X_{\delta,q}} \in \mathcal{L}(X_{\delta,q}, Y_{\delta,q})$ with the bound given by (10.27).

10.4.3 Interpolation of Sobolev spaces

In this part, we apply the previous abstract theory to the particular case of Sobolev spaces. We begin with the following interpolation result for these spaces over the domain $\Omega = \mathbb{R}^n$ (according to Definition 10.11).

Theorem 10.28. *Given s_0, $s_1 \in \mathbb{R}$ and $\delta \in (0,1)$, define $s := (1-\delta)\,s_0 + \delta\,s_1$. Then we have*

$$\mathrm{H}^s(\mathbb{R}^n) = \big(\mathrm{H}^{s_0}(\mathbb{R}^n), \mathrm{H}^{s_1}(\mathbb{R}^n)\big)_{\delta,2}\,.$$

Moreover,

$$\|u\|_{s,\mathbb{R}^n} = \frac{S_{\delta,2}}{N_{\delta,2}}\,\|\!|u|\!\|_{\delta,2} \qquad \forall\, u \in \mathrm{H}^s(\mathbb{R}^n)\,,$$

with

$$S_{\delta,2} := \left(\frac{2\,\sin(\pi\delta)}{\pi}\right)^{1/2}. \tag{10.28}$$

Proof. By virtue of the properties of the Fourier transform $\mathcal{F} : \mathcal{S}'(\mathbb{R}^n) \to \mathcal{S}'(\mathbb{R}^n)$ provided by Theorem 9.92, Lemma 9.93, and Theorem 9.94, and bearing in mind Definition 10.11 of $\mathrm{H}^t(\mathbb{R}^n)$ for each $t \in \mathbb{R}$, we observe that $\mathcal{F}$ and $\mathcal{F}^{-1}$ define isometries from $\mathrm{H}^t(\mathbb{R}^n)$ onto $F_t(\mathbb{R}^n)$ and from $F_t(\mathbb{R}^n)$ onto $\mathrm{H}^t(\mathbb{R}^n)$, respectively, where

$$F_t(\mathbb{R}^n) := \Big\{ v \in \mathrm{L}^2(\mathbb{R}^n) : \quad p_t\,v \in \mathrm{L}^2(\mathbb{R}^n) \Big\}\,.$$

Then, given s_0, $s_1 \in \mathbb{R}$ and $\delta \in (0,1)$, the interpolation property provided by Theorem 10.27 (or by the equivalent version stated immediately after its proof) ensures that $\mathcal{F}$ and $\mathcal{F}^{-1}$ are continuous bijections from $\big(\mathrm{H}^{s_0}(\mathbb{R}^n), \mathrm{H}^{s_1}(\mathbb{R}^n)\big)_{\delta,2}$ onto $\big(F_{s_0}(\mathbb{R}^n), F_{s_1}(\mathbb{R}^n)\big)_{\delta,2}$ and from $\big(F_{s_0}(\mathbb{R}^n), F_{s_1}(\mathbb{R}^n)\big)_{\delta,2}$ onto $\big(\mathrm{H}^{s_0}(\mathbb{R}^n), \mathrm{H}^{s_1}(\mathbb{R}^n)\big)_{\delta,2}$, respectively. Thus, it follows that $\big(\mathrm{H}^{s_0}(\mathbb{R}^n), \mathrm{H}^{s_1}(\mathbb{R}^n)\big)_{\delta,2}$ coincides with the tempered distributions whose Fourier transform belongs to $\big(Fs_0(\mathbb{R}^n), F_{s_1}(\mathbb{R}^n)\big)_{\delta,2}$. Hence, to conclude the first identity of the theorem, it only remains to prove that

$$\big(F_{s_0}(\mathbb{R}^n), F_{s_1}(\mathbb{R}^n)\big)_{\delta,2} = F_s(\mathbb{R}^n)\,,$$

where $s := (1-\delta)\,s_0 + \delta\,s_1$. Indeed, given $v \in F_{s_0}(\mathbb{R}^n) + F_{s_1}(\mathbb{R}^n)$, we have, for each $t > 0$,

$$K(t,v)^2 := \inf\Big\{\|p_{s_0}\,v_0\|_{0,\mathbb{R}^n}^2 + t^2\,\|p_{s_1}\,v_1\|_{0,\mathbb{R}^n}^2 :\ v = v_0 + v_1,\ v_j \in F_{s_j}(\mathbb{R}^n),\ j \in \{0,1\}\Big\}$$
$$= \inf\Big\{\int_{\mathbb{R}^n}\Big(p_{s_0}^2(\xi)\,|v_0(\xi)|^2 + t^2\,p_{s_1}^2(\xi)\,|v_1(\xi)|^2\Big)d\xi :$$
$$v = v_0 + v_1,\quad v_j \in F_{s_j}(\mathbb{R}^n),\quad j \in \{0,1\}\Big\},$$

whence, exchanging the infimum and the integral, we obtain

$$\begin{aligned} K(t,v)^2 &= \int_{\mathbb{R}^n} \inf\Big\{\Big(p_{s_0}^2(\xi)\,|v_0(\xi)|^2 + t^2\,p_{s_1}^2(\xi)\,|v_1(\xi)|^2\Big) : \\ &\quad v(\xi) = v_0(\xi) + v_1(\xi),\quad v_j(\xi) \in \mathbb{C},\ j \in \{0,1\}\Big\}\,d\xi \\ &= \int_{\mathbb{R}^n} \frac{t^2\,p_{s_0}^2(\xi)\,p_{s_1}^2(\xi)\,|v(\xi)|^2}{p_{s_0}^2(\xi) + t^2\,p_{s_1}^2(\xi)}\,d\xi. \end{aligned} \tag{10.29}$$

Note that in this last calculation we have used that if $A_0,\ A_1 > 0$ and $z \in \mathbb{C}$, then

$$\inf\Big\{A_0\,|z_0|^2 + A_1\,|z_1|^2 :\quad z = z_0 + z_1,\ z_j \in \mathbb{C},\ j \in \{0,1\}\Big\}$$
$$= \frac{A_0\,A_1\,|z|^2}{A_0 + A_1},$$

and that this infimum is attained when

$$A_0\,z_0 = A_1\,z_1 = \frac{A_0\,A_1\,z}{A_0 + A_1}.$$

It follows from (10.29) that

$$\begin{aligned} K(t,v)^2 &= \int_{\mathbb{R}^n} \frac{t^2(1+\|\xi\|^2)^{s_0+s_1}\,|v(\xi)|^2}{(1+\|\xi\|^2)^{s_0} + t^2(1+\|\xi\|^2)^{s_1}}\,d\xi \\ &= \int_{\mathbb{R}^n} \frac{(1+\|\xi\|^2)^{s_0}\,t^2(1+\|\xi\|^2)^{s_1-s_0}\,|v(\xi)|^2}{1 + t^2(1+\|\xi\|^2)^{s_1-s_0}}\,d\xi \\ &= \int_{\mathbb{R}^n} (1+\|\xi\|^2)^{s_0}\Big(f(a(\xi)\,t)\Big)^2\,|v(\xi)|^2\,d\xi, \end{aligned}$$

where

$$\begin{aligned} a(\xi) &:= (1+\|\xi\|^2)^{(s_1-s_0)/2} \quad \forall\,\xi \in \mathbb{R}^n \quad \text{and} \\ f(s) &= \frac{s}{\sqrt{1+s^2}} \quad \forall\,s \in \mathbb{R}. \end{aligned} \tag{10.30}$$

Thus, applying the norm $\|\cdot\|_{\delta,2}$ (cf. (10.21)), using the Fubini's and Tonelli's Theorems (cf. [120, Theorems 6.1 and 6.10]) to exchange the order of integration, and again invoking the dilation property (10.22), we find that

$$\|K(\cdot,v)\|_{\delta,2}^2 = \int_0^\infty t^{-(1+2\delta)}\Big(\int_{\mathbb{R}^n}(1+\|\xi\|^2)^{s_0}\,\{f(a(\xi)\,t)\}^2\,|v(\xi)|^2\,d\xi\Big)\,dt$$

$$= \int_{\mathbb{R}^n} (1 + \|\xi\|^2)^{s_0} \left(\int_0^\infty t^{-(1+2\delta)} \left\{ f(a(\xi)\, t) \right\}^2 dt \right) |v(\xi)|^2 \, d\xi$$

$$= \int_{\mathbb{R}^n} (1 + \|\xi\|^2)^{s_0} \, \|f(a(\xi)\cdot)\|_{\delta,2}^2 \, |v(\xi)|^2 \, d\xi$$

$$= \int_{\mathbb{R}^n} (1 + \|\xi\|^2)^{s_0} \, a(\xi)^{2\delta} \, \|f\|_{\delta,2}^2 \, |v(\xi)|^2 \, d\xi$$

$$= \|f\|_{\delta,2}^2 \int_{\mathbb{R}^n} (1 + \|\xi\|^2)^{s} \, |v(\xi)|^2 \, d\xi \,,$$

that is,

$$\|K(\cdot, v)\|_{\delta,2}^2 \;=\; \|f\|_{\delta,2}^2 \, \|p_s \, v\|_{0,\mathbb{R}^n}^2 \,,$$

or equivalently,

$$|||v|||_{\delta,2} \;=\; N_{\delta,2} \, \|f\|_{\delta,2} \, \|p_s \, v\|_{0,\mathbb{R}^n} \qquad \forall\, v \in F_{s_0}(\mathbb{R}^n) + F_{s_1}(\mathbb{R}^n) \,.$$

From this identity and the definitions of $\big(F_{s_0}(\mathbb{R}^n), F_{s_1}(\mathbb{R}^n)\big)_{\delta,2}$ and $F_s(\mathbb{R}^n)$, we deduce that the two spaces coincide. Therefore, in light of the preceding discussion, we conclude that indeed

$$\mathrm{H}^s(\mathbb{R}^n) \;=\; \big(\mathrm{H}^{s_0}(\mathbb{R}^n), \mathrm{H}^{s_1}(\mathbb{R}^n)\big)_{\delta,2} \,,$$

with

$$\|u\|_{s,\mathbb{R}^n} \;=\; \|p_s \, \widehat{u}\|_{0,\mathbb{R}^n} \;=\; \frac{\|f\|_{\delta,2}^{-1}}{N_{\delta,2}} \, |||\widehat{u}|||_{\delta,2} \;=\; \frac{\|f\|_{\delta,2}^{-1}}{N_{\delta,2}} \, |||u|||_{\delta,2} \,.$$

Note that the last equality uses the fact that, by endowing $F_t(\mathbb{R}^n)$ with the norm

$$|||v|||_{t,\mathbb{R}^n} \;:=\; \|p_t \, v\|_{0,\mathbb{R}^n} \qquad \forall\, v \in F_t(\mathbb{R}^n) \,, \quad \forall\, t \in \mathbb{R} \,,$$

the Fourier transform $\mathcal{F} : \big(\mathrm{H}^{s_0}(\mathbb{R}^n), \mathrm{H}^{s_1}(\mathbb{R}^n)\big)_{\delta,2} \to \big(F_{s_0}(\mathbb{R}^n), F_{s_1}(\mathbb{R}^n)\big)_{\delta,2}$ and its inverse $\mathcal{F}^{-1}$ are isometries. Finally, the proof is completed by observing, as shown below, that $\|f\|_{\delta,2}^{-1} = S_{\delta,2}$ (cf. (10.28)).

□

Lemma 10.29. *Let $f : \mathbb{R} \to \mathbb{R}$ be the function from* (10.30)*, that is, $f(s) := \frac{s}{\sqrt{1+s^2}}$ for all $s \in \mathbb{R}$. Then $\|f\|_{\delta,2} \;=\; \left(\frac{\pi}{2\sin(\pi\delta)}\right)^{1/2}$.*

Proof. A simple exercise in complex analysis (see [106, Exercise 2.11, Chapter II, Volume II]) states that if g is a rational function without poles on the positive real axis such that, for some $r \in \mathbb{R}$,

$$\lim_{z \to 0} z^{r+1} \, g(z) \;=\; \lim_{z \to \infty} z^{r+1} \, g(z) \;=\; 0 \,,$$

then

$$\int_0^\infty t^r \, g(t) \, dt \;=\; \frac{\pi}{\sin\big((r+1)\pi\big)} \sum_{\text{poles of } g} \mathrm{Res}\big\{(-z)^r \, g(z)\big\} \,. \tag{10.31}$$

In our case we have

$$\|f\|_{\delta,2}^2 \;=\; \int_0^\infty t^{-(1+2\delta)} \, |f(t)|^2 \, dt \;=\; \int_0^\infty \frac{t^{1-2\delta}}{1+t^2} \, dt \;=\; \int_0^\infty t^r \, g(t) \, dt \,,$$

with $r = 1 - 2\delta$ and $g(z) = (1 + z^2)^{-1}$ for all $z \in \mathbb{C}$. Since $\delta \in (0, 1)$, we have

$$\lim_{z \to 0} z^{r+1} g(z) = \lim_{z \to 0} \frac{z^{2(1-\delta)}}{1 + z^2} = 0$$

and

$$\lim_{z \to \infty} z^{r+1} g(z) = \lim_{z \to \infty} \frac{z^{2(1-\delta)}}{1 + z^2} = 0 .$$

Hence, by virtue of (10.31) and the fact that $\imath$ and $-\imath$ are the poles of g, we obtain

$$\|f\|_{\delta,2}^2 = \frac{-\pi}{2 \sin(\pi\delta) \cos(\pi\delta)} \left(\lim_{z \to \imath} \frac{(-z)^{1-2\delta} (z - \imath)}{1 + z^2} + \lim_{z \to -\imath} \frac{(-z)^{1-2\delta} (z + \imath)}{1 + z^2} \right).$$

Thus, recalling that $z^r = e^{r \ln |z|} e^{\imath r \arg(z)}$ with $\arg(z) \in]-\pi, \pi[$, straightforward algebraic computations complete the proof.

□

We now extend the interpolation of Sobolev spaces to the case of an arbitrary open set $\Omega \subset \mathbb{R}^n$. To this end, we introduce the Sobolev space $\mathrm{H}^s(\Omega)$, with $s \in \mathbb{R}$, in terms of the corresponding space on $\mathbb{R}^n$ given in Definition 10.11.

Definition 10.30. *Given a nonempty open subset Ω of $\mathbb{R}^n$, we define, for each $s \in \mathbb{R}$,*

$$\mathrm{H}^s(\Omega) := \Big\{ u \in \mathcal{D}'(\Omega) : \quad u = U|_\Omega \quad \textit{for some} \quad U \in \mathrm{H}^s(\mathbb{R}^n) \Big\},$$

which is endowed with the norm

$$\|u\|_{s,\Omega} := \min \Big\{ \|U\|_{s,\mathbb{R}^n} : \quad U \in \mathrm{H}^s(\mathbb{R}^n),\ U|_\Omega = u \Big\} \quad \forall\, u \in \mathrm{H}^s(\Omega).$$

Note that the minimum in the definition of $\|\cdot\|_{s,\Omega}$ is actually attained, and uniquely so, thanks to the Second Best Approximation Theorem, (cf. Theorem 2.7) as applied to the convex and closed subset $\Big\{ U \in \mathrm{H}^s(\mathbb{R}^n) : \quad U|_\Omega = u \Big\}$ of the Hilbert space $\mathrm{H}^s(\mathbb{R}^n)$.

Theorem 10.31. *Given $s_0,\, s_1 \in \mathbb{R}$ and $\delta \in (0,1)$, define $s := (1 - \delta)\, s_0 + \delta\, s_1$. Then we have*

$$\mathrm{H}^s(\Omega) = (\mathrm{H}^{s_0}(\Omega), \mathrm{H}^{s_1}(\Omega))_{\delta,2}$$

and (*cf.* (10.25), (10.28))

$$\|u\|_{s,\Omega} = \frac{S_{\delta,2}}{N_{\delta,2}} \|u\|_{\delta,2} \qquad \forall\, u \in \mathrm{H}^s(\Omega).$$

Proof. Given $u \in \mathrm{H}^s(\Omega)$, we consider the unique $U \in \mathrm{H}^s(\mathbb{R}^n)$ such that $U|_\Omega = u$ and $\|u\|_{s,\Omega} = \|U\|_{s,\mathbb{R}^n}$. Since $\mathrm{H}^s(\mathbb{R}^n) = \big(\mathrm{H}^{s_0}(\mathbb{R}), \mathrm{H}^{s_1}(\mathbb{R}^n)\big)_{\delta,2}$ (cf. Theorem 10.28), we can take $U_j \in \mathrm{H}^{s_j}(\mathbb{R}^n)$, $j \in \{0,1\}$, such that $U = U_0 + U_1$. It then follows clearly that $u = u_0 + u_1$ with $u_j := U_j|_\Omega \in \mathrm{H}^{s_j}(\Omega)$, $j \in \{0,1\}$, and then, according to the definition of $\|\cdot\|_{s,\Omega}$, we have

$$K(t,u)^2 \leqslant \|u_0\|_{s_0,\Omega}^2 + t^2 \|u_1\|_{s_1,\Omega}^2 \leqslant \|U_0\|_{s_0,\mathbb{R}^n}^2 + t^2 \|U_1\|_{s_1,\mathbb{R}^n}^2 ,$$

whence, taking the infimum with respect to U_0 and U_1, we deduce that

$$K(t,u)^2 \leqslant K(t,U)^2 .$$

This inequality and Theorem 10.28 imply that $u \in \left(\mathrm{H}^{s_0}(\Omega), \mathrm{H}^{s_1}(\Omega)\right)_{\delta,2}$ and

$$\|u\|_{\delta,2} \leqslant \|U\|_{\delta,2} = \frac{N_{\delta,2}}{S_{\delta,2}} \|U\|_{s,\mathbb{R}^n} = \frac{N_{\delta,2}}{S_{\delta,2}} \|u\|_{s,\Omega},$$

that is,

$$\|u\|_{\delta,2} \leqslant \frac{N_{\delta,2}}{S_{\delta,2}} \|u\|_{s,\Omega}. \tag{10.32}$$

Conversely, let $u \in \left(\mathrm{H}^{s_0}(\Omega), \mathrm{H}^{s_1}(\Omega)\right)_{\delta,2}$ and consider a decomposition $u = u_0 + u_1$, with $u_j \in \mathrm{H}^{s_j}(\Omega)$, $j \in \{0,1\}$. Also, let $U_j \in \mathrm{H}^{s_j}(\mathbb{R}^n)$ be such that

$$U_j|_\Omega = u_j \quad \text{and} \quad \|U_j\|_{s_j,\mathbb{R}^n} = \|u_j\|_{s_j,\Omega}, \quad j \in \{0,1\},$$

and define $U := U_0 + U_1$. It follows that

$$K(t,U)^2 \leqslant \|U_0\|^2_{s_0,\mathbb{R}^n} + t^2 \|U_1\|^2_{s_1,\mathbb{R}^n} = \|u_0\|^2_{s_0,\Omega} + t^2 \|u_1\|^2_{s_1,\Omega},$$

whence, taking the infimum with respect to u_0 and u_1, we obtain

$$K(t,U)^2 \leqslant K(t,u)^2,$$

and therefore $\|U\|_{\delta,2} \leqslant \|u\|_{\delta,2}$, which proves that $U \in \left(\mathrm{H}^{s_0}(\mathbb{R}^n), \mathrm{H}^{s_1}(\mathbb{R}^n)\right)_{\delta,2} = \mathrm{H}^s(\mathbb{R}^n)$. Thus, since $u = U|_\Omega$, we have $u \in \mathrm{H}^s(\Omega)$ and

$$\|u\|_{s,\Omega} \leqslant \|U\|_{s,\mathbb{R}^n} = \frac{S_{\delta,2}}{N_{\delta,2}} \|U\|_{\delta,2} \leqslant \frac{S_{\delta,2}}{N_{\delta,2}} \|u\|_{\delta,2}. \tag{10.33}$$

Finally, from (10.32) and (10.33) we conclude that

$$\|u\|_{s,\Omega} = \frac{S_{\delta,2}}{N_{\delta,2}} \|u\|_{\delta,2},$$

which completes the proof.

□

Example 10.32 In this example we illustrate the applicability of the previous analysis to a very common situation in the theory of finite element methods. More precisely, given a bounded open set Ω of $\mathbb{R}^n$ with polyhedral boundary Γ and a finite-dimensional subspace X_h of $\mathrm{H}^1(\Omega)$, where h denotes the mesh size associated with a triangulation of $\bar{\Omega}$, it is usually the case that the orthogonal projector $\Pi_h : \mathrm{H}^1(\Omega) \to X_h$ satisfies

$$\|v - \Pi_h(v)\|_{1,\Omega} \leqslant C\,h\,\|v\|_{2,\Omega} \qquad \forall\, v \in \mathrm{H}^2(\Omega), \tag{10.34}$$

with a constant $C > 0$ independent of h. The objective is therefore to extend the above result to the space $\mathrm{H}^{1+\delta}(\Omega)$ with $\delta \in (0,1)$. For this purpose, we first observe, according to the properties of the orthogonal projector $(I - \Pi_h)$, that there holds

$$\|v - \Pi_h(v)\|_{1,\Omega} \leqslant \|v\|_{1,\Omega} \qquad \forall\, v \in \mathrm{H}^1(\Omega). \tag{10.35}$$

Then, defining the spaces $X_0 := \mathrm{H}^2(\Omega)$, $X_1 := \mathrm{H}^1(\Omega)$, and $Y_0 = Y_1 := \mathrm{H}^1(\Omega)$, as well as the operators $A_0 : X_0 \to Y_0$ and $A_1 : X_1 \to Y_1$ given by $A_0 = A_1 := I - \Pi_h$, it follows from (10.34) and (10.35) that $A_j \in \mathcal{L}(X_j, Y_j)$ and $\|A_j(v)\|_j \leqslant M_j \|v\|_j$, for all $v \in X_j$, $j \in \{0,1\}$, with $M_0 = C\,h$ and $M_1 = 1$. Moreover, it is clear that $(Y_0, Y_1)_{1-\delta,2} = \mathrm{H}^1(\Omega)$ and, in accordance with Theorem 10.31, we have $(X_0, X_1)_{1-\delta,2} = \big(\mathrm{H}^2(\Omega), \mathrm{H}^1(\Omega)\big)_{1-\delta,2} = \mathrm{H}^{1+\delta}(\Omega)$. Consequently, applying Theorem 10.27 directly, we conclude that $(I - \Pi_h) \in \mathcal{L}\big(\mathrm{H}^{1+\delta}(\Omega), \mathrm{H}^1(\Omega)\big)$ and

$$\|v - \Pi_h(v)\|_{1,\Omega} \leqslant M_0^\delta\, M_1^{1-\delta}\, \|v\|_{1+\delta,\Omega} \qquad \forall\, v \in \mathrm{H}^{1+\delta}(\Omega)\,,$$

that is,

$$\|v - \Pi_h(v)\|_{1,\Omega} \leqslant \widetilde{C}\, h^\delta\, \|v\|_{1+\delta,\Omega} \qquad \forall\, v \in \mathrm{H}^{1+\delta}(\Omega)\,,$$

with $\widetilde{C} > 0$ independent of h. In this way, it has been shown that the classical estimate (10.34) is also valid for fractional Sobolev spaces of orders between 1 and 2.

10.5 Exercises.

▸ **10.1** *Prove that* (10.2) *and* (10.10) *are equivalent for* $s = m \in \mathbb{N}$.

▸ **10.2** *Let* Ω^- *be a bounded and connected open subset of* $\mathbb{R}^2$ *with boundary* Γ, *and let* Ω^+ *be the annular region bounded by* Γ *and a closed curve* Σ *whose interior contains* Γ. *Moreover, let* $\gamma_0^- : \mathrm{H}^1(\Omega^-) \to \mathrm{H}^{1/2}(\Gamma)$ *and* $\gamma_0^+ : \mathrm{H}^1(\Omega^+) \to \mathrm{H}^{1/2}(\partial\Omega^+) \equiv \mathrm{H}^{1/2}(\Gamma) \times \mathrm{H}^{1/2}(\Sigma)$ *be the respective trace operators, and denote* $\Omega := \Omega^- \cup \Gamma \cup \Omega^+$.

a) *Show that* $v \in \mathrm{H}^1(\Omega)$ *if and only if:*

$$v \in \mathrm{L}^2(\Omega)\,,\; v|_{\Omega^-} \in \mathrm{H}^1(\Omega^-)\,,\; v|_{\Omega^+} \in \mathrm{H}^1(\Omega^+)\,,\; \text{and } \gamma_0^-(v|_{\Omega^-}) = \gamma_0^+(v|_{\Omega^+}) \text{ on } \Gamma\,.$$

Given $f^- \in \mathrm{L}^2(\Omega^-)$, $f^+ \in \mathrm{L}^2(\Omega^+)$, $g_\Gamma \in \mathrm{H}^{-1/2}(\Gamma)$, *and* $g_\Sigma \in \mathrm{H}^{-1/2}(\Sigma)$, *consider the transmission problem: Find* $(u^-, u^+) \in \mathrm{H}^1(\Omega^-) \times \mathrm{H}^1(\Omega^+)$ *such that*

$$\begin{gathered} -\Delta u^- = f^- \text{ in } \Omega^-\,, \quad -\Delta u^+ = f^+ \text{ in } \Omega^+\,, \\ \gamma_0^-(u^-) = \gamma_0^+(u^+) \text{ in } \Gamma\,, \quad \gamma_{\boldsymbol{\nu}}^-(\nabla u^-) - \gamma_{\boldsymbol{\nu}}^+(\nabla u^+) = g_\Gamma \text{ in } \Gamma\,, \\ \gamma_{\boldsymbol{\nu}}^+(\nabla u^+) = g_\Sigma \text{ in } \Sigma\,, \quad \text{and} \quad \int_{\Omega^-} u^- + \int_{\Omega^+} u^+ = 0\,, \end{gathered} \tag{10.36}$$

where $\gamma_{\boldsymbol{\nu}}^- : \mathbf{H}(\mathrm{div};\Omega^-) \to \mathrm{H}^{-1/2}(\Gamma)$ *and* $\gamma_{\boldsymbol{\nu}}^+ : \mathbf{H}(\mathrm{div};\Omega^+) \to \mathrm{H}^{-1/2}(\partial\Omega^+) \equiv \mathrm{H}^{-1/2}(\Gamma) \times \mathrm{H}^{-1/2}(\Sigma)$ *are the respective normal trace operators* ($\boldsymbol{\nu}$ *points towards* Ω^+ *on* Γ *and outward from* Ω^+ *on* Σ) *(cf. Section 4.5.4).*

b) *Use Green's identities in appropriate Sobolev spaces and deduce a variational formulation of* (10.36) *with the unknown in a closed subspace* V *of* $\mathrm{H}^1(\Omega)$.
c) *Identify a compatibility condition on the data, and show in that case that the formulation obtained in b) admits a unique solution, which depends continuously on* f^-, f^+, g_Γ, *and* g_Σ.
d) *Prove that the associated Galerkin scheme is convergent for any countable family* $\{V_h\}_{h>0}$ *of finite-dimensional subspaces of* V *such that* $\lim\limits_{h\to 0} \mathrm{dist}(v, V_h) = 0$ *for all* $v \in V$.
e) *Show that the formulation obtained in b) is equivalent to a mixed variational formulation with unknown in* $\mathrm{H}^1(\Omega) \times \mathbb{R}$, *and verify that it satisfies the hypotheses of the Babuška-Brezzi Theory (cf. Section 4.4).*

▸ **10.3** *Let* Ω *be a bounded open subset of* $\mathbb{R}^n$ *with* C^1 *boundary* Γ, *and define*

$$\mathcal{D}(\Gamma) := \big\{ v|_\Gamma : \quad v \in C_0^\infty(\mathbb{R}^n) \big\}\,.$$

Prove that $\mathrm{H}^{1/2}(\Gamma) \subseteq \overline{\mathcal{D}(\Gamma)}^{\|\cdot\|_{0,\Gamma}}$ *and* $\mathrm{H}^{1/2}(\Gamma) = \overline{\mathcal{D}(\Gamma)}^{\|\cdot\|_{1/2,\Gamma}}$.

▸ **10.4** *Let Ω be a bounded open subset of $\mathbb{R}^n$ with C^1 boundary Γ, and consider the mapping $\|\!| \cdot |\!\| : \mathrm{H}^1(\Omega) \to \mathbb{R}$ defined by*

$$|\!|\!| v |\!|\!| := \left\{ |v|_{1,\Omega}^2 + \|\gamma_0(v)\|_{0,\Gamma}^2 \right\}^{1/2} \qquad \forall\, v \in \mathrm{H}^1(\Omega),$$

where $\gamma_0 : \mathrm{H}^1(\Omega) \to \mathrm{H}^{1/2}(\Gamma)$ is the usual trace operator. Using an argument analogous to the proof of the generalized Poincaré inequality (cf. Theorem 10.21), show that $\|\cdot\|_{1,\Omega}$ and $|\!|\!|\cdot|\!|\!|$ are equivalent norms on $\mathrm{H}^1(\Omega)$.

▸ **10.5** *Prove that $C^\infty(\overline{\mathbb{R}^n_+})$ is dense in $\mathrm{H}^1(\mathbb{R}^n_+)$.*

▸ **10.6** *Let Ω be a bounded open subset of $\mathbb{R}^2$ with C^1 boundary Γ. The goal of this problem is to show the first Korn inequality*

$$\|\mathbf{e}(\mathbf{v})\|_{0,\Omega}^2 \geqslant \frac{1}{2}\,|\mathbf{v}|_{1,\Omega}^2 \qquad \forall\, \mathbf{v} \in \mathbf{H}_0^1(\Omega), \tag{10.37}$$

where $\mathbf{e}(\mathbf{v}) := \dfrac{1}{2}\left\{\nabla\mathbf{v} + (\nabla\mathbf{v})^{\mathtt{t}}\right\}$. To this end, proceed as follows.

i) *Given $\boldsymbol{\sigma} := (\sigma_{ij})$ and $\boldsymbol{\tau} := (\tau_{ij})$ in $\mathbb{R}^{2\times 2}$, define the tensor product $\boldsymbol{\sigma} : \boldsymbol{\tau} := \sum_{i,j=1}^{2} \sigma_{ij}\,\tau_{ij}$ and introduce the subspaces*

$$\mathbb{R}^{2\times 2}_{\mathtt{sym}} := \left\{\boldsymbol{\tau} \in \mathbb{R}^{2\times 2} : \quad \boldsymbol{\tau}^{\mathtt{t}} = \boldsymbol{\tau}\right\} \quad \text{and} \quad \mathbb{R}^{2\times 2}_{\mathtt{skew}} := \left\{\boldsymbol{\tau} \in \mathbb{R}^{2\times 2} : \quad \boldsymbol{\tau}^{\mathtt{t}} = -\boldsymbol{\tau}\right\}.$$

Show that $\boldsymbol{\sigma} : \boldsymbol{\tau} = 0$ for all $\boldsymbol{\sigma} \in \mathbb{R}^{2\times 2}_{\mathtt{sym}}$, for all $\boldsymbol{\tau} \in \mathbb{R}^{2\times 2}_{\mathtt{skew}}$.

ii) *Note that $\nabla\mathbf{v} = \mathbf{e}(\mathbf{v}) + \mathbf{w}(\mathbf{v})$, with $\mathbf{w}(\mathbf{v}) := \frac{1}{2}\left\{\nabla\mathbf{v} - (\nabla\mathbf{v})^{\mathtt{t}}\right\}$, and recall that $\|\boldsymbol{\tau}\|_{0,\Omega)}^2 = \int_\Omega \boldsymbol{\tau} : \boldsymbol{\tau}$ for all $\boldsymbol{\tau} \in \mathbb{L}^2(\Omega)$, to show that*

$$\|\mathbf{e}(\mathbf{v})\|_{0,\Omega}^2 - \|\mathbf{w}(\mathbf{v})\|_{0,\Omega}^2 = \int_\Omega \nabla\mathbf{v} : (\nabla\mathbf{v})^{\mathtt{t}}$$

and

$$\|\mathbf{e}(\mathbf{v})\|_{0,\Omega}^2 + \|\mathbf{w}(\mathbf{v})\|_{0,\Omega}^2 = \int_\Omega \nabla\mathbf{v} : \nabla\mathbf{v}\,.$$

iii) *Deduce the identity $\nabla\mathbf{v} : (\nabla\mathbf{v})^{\mathtt{t}} = \operatorname{div}\Big(\nabla\mathbf{v}\,\mathbf{v} - \operatorname{div}(\mathbf{v})\,\mathbf{v}\Big) + (\operatorname{div}(\mathbf{v}))^2$ and conclude inequality (10.37).*

▸ **10.7** ([90]) *This problem constitutes a particular version of the Peetre-Tartar Lemma.*

a) *Let $(X_j, \|\cdot\|_j)$, $j \in \{1,2,3\}$, be Banach spaces, and consider operators $A \in \mathcal{L}(X_1, X_2)$ and $B \in \mathcal{L}(X_1, X_3)$ such that B is compact. Moreover, assume that there exists $c > 0$ such that*

$$\|x\|_1 \leqslant c\left\{\|A(x)\|_2 + \|B(x)\|_3\right\} \qquad \forall\, x \in X_1\,.$$

Prove that $N(A)$ is finite-dimensional, and then argue by contradiction to show that there exists $C > 0$ such that

$$\operatorname{dist}(x, N(A)) \leqslant C\,\|A(x)\|_2 \qquad \forall\, x \in X_1\,.$$

b) *Let Ω be a bounded open subset of $\mathbb{R}^n$ with smooth boundary Γ, and consider the decomposition $\mathrm{H}^1(\Omega) = \widetilde{\mathrm{H}}^1(\Omega) \oplus \mathrm{P}_0(\Omega)$, where $\mathrm{P}_0(\Omega)$ is the space of constant functions over Ω and*

$$\widetilde{\mathrm{H}}^1(\Omega) := \left\{ v \in \mathrm{H}^1(\Omega) : \quad \int_\Omega v = 0 \right\}.$$

Prove that there exists $\alpha > 0$ such that

$$\alpha\,\|v\|_{1,\Omega} \leqslant |v|_{1,\Omega} \qquad \forall\, v \in \widetilde{\mathrm{H}}^1(\Omega)\,. \tag{10.38}$$

c) *Apply* a), (10.38), *and the Trace Theorem (cf. Theorem 4.26) to show that there exist positive constants* C_1, C_2 *such that*

$$C_1 \|v\|_{1,\Omega}^2 \leqslant |v|_{1,\Omega}^2 + \|v\|_{0,\Gamma}^2 \leqslant C_2 \|v\|_{1,\Omega}^2 \qquad \forall\, v \in \mathrm{H}^1(\Omega). \tag{10.39}$$

HINT: *for the lower inequality (10.39) define appropriate operators* A *and* B *using the spaces*

$$X_1 := \mathrm{H}^1(\Omega), \quad X_2 := \mathbf{L}^2(\Omega) \times \mathrm{L}^2(\Gamma), \quad \text{and} \quad X_3 := \mathrm{P}_0(\Omega).$$

▸ **10.8** ([90]) *Let* Ω *be a bounded and connected open subset of* $\mathbb{R}^n$ *of class* C^1, *and let* $\mathrm{H}^{-1}(\Omega)$ *(resp.* $\mathbf{H}^{-1}(\Omega)$*) be the dual of* $\mathrm{H}_0^1(\Omega)$ *(resp.* $\mathbf{H}_0^1(\Omega)$*), with the corresponding norms denoted in both cases by* $\|\cdot\|_{-1,\Omega}$. *Using that* $C_0^\infty(\Omega)$ *(resp.* $[C_0^\infty(\Omega)]^n$*) is dense in* $\mathrm{H}_0^1(\Omega)$ *(resp.* $\mathbf{H}_0^1(\Omega)$*), prove that the identity mapping* $i : \mathrm{L}^2(\Omega) \to \mathcal{D}'(\Omega)$ *and the gradient mapping* $\nabla : L^2(\Omega) \to [\mathcal{D}'(\Omega)]^n$ *extend by density to operators* $i : \mathrm{L}^2(\Omega) \to \mathrm{H}^{-1}(\Omega)$ *and* $\nabla : \mathrm{L}^2(\Omega) \to \mathbf{H}^{-1}(\Omega)$, *respectively. In this case, one can prove that there exists a constant* $c_1 > 0$, *depending only on* Ω, *such that*

$$\|p\|_{0,\Omega} \leqslant c_1 \left\{ \|p\|_{-1,\Omega} + \|\nabla p\|_{-1,\Omega} \right\} \quad \forall\, p \in \mathrm{L}^2(\Omega).$$

a) *Consider the operator* $B := \mathcal{R}\nabla$, *where* $\mathcal{R} : \mathbf{H}^{-1}(\Omega) \to \mathbf{H}_0^1(\Omega)$ *is the associated Riesz operator, and prove that the range of* B *is closed in* $\mathbf{H}_0^1(\Omega)$.

b) *Identify the adjoint operator* $B^* : \mathbf{H}_0^1(\Omega) \to \mathrm{L}^2(\Omega)$ *and use Exercise 10.7 to prove that* $\mathrm{div} : \mathrm{W}^\perp \to \mathrm{L}_0^2(\Omega)$ *is an isomorphism, where*

$$\mathrm{W} := \left\{ \mathbf{v} \in \mathbf{H}_0^1(\Omega) : \quad \mathrm{div}(\mathbf{v}) = 0 \right\}, \qquad \mathbf{H}_0^1(\Omega) = \mathrm{W} \oplus \mathrm{W}^\perp,$$

and

$$\mathrm{L}_0^2(\Omega) := \left\{ p \in \mathrm{L}^2(\Omega) : \quad \int_\Omega p = 0 \right\}.$$

▸ **10.9** *Given* $\Omega := (0,1)$ *and* $m \in \mathbb{N}$, *introduce the Sobolev space of order* m

$$\mathrm{H}^m(\Omega) := \left\{ v \in \mathrm{L}^2(\Omega) : \quad v^{(j)} \in \mathrm{L}^2(\Omega) \quad \forall\, j \in \{1, \ldots, m\} \right\},$$

which is a Hilbert space with inner product $\langle \cdot, \cdot \rangle_{m,\Omega}$ *and induced norm* $\|\cdot\|_{m,\Omega}$ *defined, respectively, as*

$$\langle u, v \rangle_{m,\Omega} := \sum_{j=0}^{m} \int_0^1 u^{(j)}\, v^{(j)}$$

and

$$\|v\|_{m,\Omega}^2 = \sum_{j=0}^{m} \|v^{(j)}\|_{0,\Omega}^2 \qquad \forall\, u,\, v \in \mathrm{H}^m(\Omega).$$

Then, define $\mathrm{H}_0^m(\Omega) := \overline{C_0^\infty(\Omega)}^{\|\cdot\|_{m,\Omega}}$, $|v|_{m,\Omega} := \|v^{(m)}\|_{0,\Omega}$ *for all* $v \in \mathrm{H}^m(\Omega)$, *and prove that*

$$\|v\|_{m,\Omega}^2 \leqslant \left\{ 2 - \frac{1}{2^m} \right\} |v|_{m,\Omega}^2 \qquad \forall\, v \in \mathrm{H}_0^m(\Omega).$$

▸ **10.10** *Let* Ω *be a bounded open subset of* $\mathbb{R}^N$ *with smooth boundary* Γ, *and let* $w \colon \Omega \mapsto \mathbb{R}$ *be a function satisfying the hypotheses:*

i) $w > 0$ *almost everywhere.*
ii) $1/w \in \mathrm{L}_{\mathrm{loc}}^1(\Omega)$; *that is, given any open set* Ω_0 *whose closure is compactly contained in* Ω, $1/w \in \mathrm{L}^1(\Omega_0)$.

a) *Prove that* $\mathrm{L}^2_w(\Omega)$, *the space of all (equivalence classes of) measurable functions* $u : \Omega \to \mathbb{R}$ *such that*

$$\int_\Omega w\,|u|^2 < \infty\,,$$

equipped with the inner product

$$\langle u, v\rangle_{0,\Omega;w} := \int_\Omega w\,u\,v$$

is a Hilbert space.

b) *Prove that*

$$\mathrm{H}^1_w(\Omega) := \left\{u \in \mathrm{L}^2_w(\Omega) : \quad \frac{\partial u}{\partial x_i} \in \mathrm{L}^2_w(\Omega) \quad \forall\, i \in \{1, \dots, N\}\right\},$$

equipped with the norm

$$\|u\|_{1,\Omega;w} := \left\{\|u\|^2_{0,\Omega;w} + \sum_{i=1}^N \left\|\frac{\partial u}{\partial x_i}\right\|^2_{0,\Omega;w}\right\}^{1/2},$$

is a Hilbert space.

▸ **10.11** *Let* Ω, w, $\mathrm{L}^2_w(\Omega)$, *and* $\mathrm{H}^1_w(\Omega)$ *be as in Exercise 10.10, and let* $\{\Omega_n\}_{n\in\mathbb{N}}$ *be a sequence of open subsets of* $\mathbb{R}^N$ *such that* $\Omega_n \subseteq \Omega_{n+1}$ *for all* $n \in \mathbb{N}$ *and* $\Omega = \bigcup_{n\in\mathbb{N}} \Omega_n$. *Assume moreover the following additional hypothesis on* w:

iii) *For every* $n \in \mathbb{N}$, *the embedding of* $\mathrm{H}^1_w(\Omega_n)$ *into* $\mathrm{L}^2_w(\Omega_n)$ *is compact.*

Prove that the following two statements are equivalent

a) *For each* $\varepsilon > 0$ *there exists* $m \in \mathbb{N}$ *such that*

$$\|u\|^2_{0,\Omega;w} \leqslant \varepsilon\,\|u\|^2_{1,\Omega;w} + \|u\|^2_{0,\Omega_m;w} \quad \forall\, u \in \mathrm{H}^1_w(\Omega)\,.$$

b) *The embedding of* $\mathrm{H}^1_w(\Omega)$ *into* $\mathrm{L}^2_w(\Omega)$ *is compact.*

References

1. Alaoglu, L., *Weak topologies of normed linear spaces.* Annals of Mathematics, vol. 41, pp. 252–267, (1940).
2. Almonacid, J.A., Díaz, H.S., Gatica, G.N. and Márquez, A., *A fully mixed finite element method for the coupling of the Stokes and Darcy-Forchheimer problems.* IMA Journal of Numerical Analysis, vol. 40, 2, pp. 1454–1502, (2019).
3. Alt, H.W., Linear Functional Analysis. An Application-Oriented Introduction. Translated from the German edition by Robert Nürnberg. Universitext Springer-Verlag London, Ltd., London, 2016.
4. Álvarez, M., Gatica, G.N. and Ruiz-Baier, R., *An augmented mixed-primal finite element method for a coupled flow-transport problem.* ESAIM: Mathematical Modelling and Numerical Analysis, vol. 49, 5, pp. 1399–1427, (2015).
5. Angelo, L., Camaño, J. and Caucao, S., *A five-field mixed formulation for stationary magnetohydrodynamic flows in porous media.* Computer Methods in Applied Mechanics and Engineering, vol. 414, Paper No. 116158, (2023).
6. Aubin, J.P., Applied Functional Analysis. Wiley-Interscience, 1979.
7. Azerad, P., Analyse des Équations de Navier-Stokes en Bassin peu Profond et de l' Équation de Transport. Thèse de l'Université de Neuchâtel, 1995.
8. Babuška, I. and Aziz, A.K., *Survey lectures on the mathematical foundations of the finite element method.* In: The Mathematical Foundations of the Finite Element Method with Applications to Partial Differential Equations (Proc. Sympos., University of Maryland, Baltimore, USA, 1972), pp. 1–359. Academic Press, New York, 1972.
9. Banach, S., *Sur les opérations dans les ensembles abstraits et leur application aux équations intégrales.* Fundamenta Mathematicae, vol. 3, pp. 133–181, (1922).
10. Banach, S., *Sur les fonctionelles linéaires.* Studia Mathematica, vol. 1, pp. 211–216, (1929).
11. Banach, S., *Sur les fonctionelles linéaires* II. Studia Mathematica, vol. 1, pp. 223–239, (1929).
12. Banach, S., Théorie des Opérations Linéaires. Monografiee Matematyczne 1, Warszawa 1932.
13. Banach, S. and Steinhaus, H., *Sur le principle de la condensation de singularités.* Fundamenta Mathematicae, vol. 9, pp. 50–61, (1927).
14. Benavides, G.A., Caucao, S., Gatica, G.N. and Hopper, A., *A Banach spaces-based analysis of a new mixed-primal finite element method for a coupled flow-transport problem.* Computer Methods in Applied Mechanics and Engineering, vol. 371, Paper No. 113285, (2020).
15. Benavides, G.A., Caucao, S., Gatica, G.N. and Hopper, A., *A new non-augmented and momentum-conserving fully-mixed finite element method for a coupled flow-transport problem.* Calcolo, vol. 59, 1, article: 6, (2022).
16. Bermúdez, I., Correa, C.I., Gatica, G.N. and Silva, J.P., *A perturbed twofold saddle point-based mixed finite element method for the Navier-Stokes equations with variable viscosity.* Applied Numerical Mathematics, vol. 201, pp. 465–487, (2024).
17. Bernardi, C., Canuto, C. and Maday, Y., *Generalized inf-sup conditions for Chebyshev spectral approximation of the Stokes problem.* SIAM Journal on Numerical Analysis, vol. 25, 6, pp. 1237–1271, (1998).

G. N. Gatica, *Introduction to Functional Analysis*,
https://doi.org/10.1007/978-3-032-37415-8

18. Bernardi, C., Dib, S., Girault, V., Hecht, F., Murat, F. and Sayah, T., *Finite element methods for Darcy's problem coupled with the heat equation.* Numerische Mathematik, vol. 139, 2, pp. 315–348, (2018).
19. Boffi, D., Brezzi, F. and Fortin, M., Mixed Finite Element Methods and Applications. Springer Series in Computational Mathematics, vol. 44, 2013.
20. Bogovskiĭ, M.E., *Solutions of some problems of vector analysis, associated with the operators div and grad.* (Russian) Theory of cubature formulas and the application of functional analysis to problems of mathematical physics, pp. 5–40, 149 Proc. Sobolev Sem., No. 1, 1980 [Proceedings of the S. L. Sobolev Seminar] Akad. Nauk SSSR Sibirsk. Otdel., Inst. Mat., Novosibirsk, 1980.
21. Bohnenblust, F. and Sobczyk, A., *Extensions of functionals on complex linear spaces.* Bulletin of the American Mathematical Society, vol. 44, pp. 91–93, (1938).
22. Brenner, S.C. and Scott, L.R., The Mathematical Theory of Finite Element Methods. Springer-Verlag, 1994.
23. Brezis, H., Analyse Fonctionnelle. Théorie at Applications. Masson, Paris, 1983.
24. Brezzi, F. and Fortin, M., Mixed and Hybrid Finite Element Methods. Springer-Verlag, 1991.
25. Buffa, A., *Remarks on the discretization of some non-coercive operator with applications to heterogeneous Maxwell equations.* SIAM Journal on Numerical Analysis, vol. 43, 1, pp. 1–18, (2005).
26. Bustos, A.J., Caucao, S., Gatica, G.N. and Venegas, B.N., *New fully mixed finite element methods for the coupled convective Brinkman-Forchheimer and nonlinear transport equations.* Journal of Scientific Computing, vol. 104, 2, article: 64, (2025).
27. Camaño, J., García, C. and Oyarzúa, R., *Analysis of a momentum conservative mixed-FEM for the stationary Navier-Stokes problem.* Numerical Methods for Partial Differential Equations, vol. 37, 5, pp. 2895–2923, (2021).
28. Camaño, J., García, C. and Oyarzúa, R., *Analysis of a new mixed FEM for stationary incompressible magneto-hydrodynamics.* Computer & Mathematics with Applications, vol. 127, pp. 65–79, (2022).
29. Camaño, J., Muñoz, C. and Oyarzúa, R., *Numerical analysis of a dual-mixed problem in non-standard Banach spaces.* Electronic Transactions on Numerical Analysis, vol. 48, pp. 114–130, (2018).
30. Camaño, J., Oyarzúa, R. and Tierra, G., *Analysis of an augmented mixed-FEM for the Navier-Stokes problem.* Mathematics of Computation, vol. 86, 304, pp. 589–615, (2017).
31. Careaga, J., Gatica, G.N., Inzunza, C. and Ruiz-Baier, R., *New Banach spaces-based mixed finite element methods for the coupled poroelasticity and heat equations.* IMA Journal of Numerical Analysis, vol. 45, 4, pp. 1936–1984, (2025).
32. Carrasco, S., Caucao, S. and Gatica, G.N., *New mixed finite element methods for the coupled convective Brinkman-Forchheimer and double-diffusion equations.* Journal of Scientific Computing, vol. 97, 3, Paper No. 61, (2023).
33. Carrasco, S., Caucao, S. and Gatica, G.N., *A twofold perturbed saddle point-based fully mixed finite element method for the coupled Brinkman–Forchheimer Darcy problem.* ESAIM: Mathematical Modelling and Numerical Analysis, vol. 60, 2, pp. 785–826, (2026).
34. Caucao, S., Colmenares, E., Gatica, G.N. and Inzunza, C., *A Banach spaces-based fully-mixed finite element method for the stationary chemotaxis-Navier-Stokes problem.* Computer & Mathematics with Applications, vol. 145, pp. 65–89, (2023).
35. Caucao, S. and Discacciati, M., *A mixed* FEM *for the coupled Brinkman-Forchheimer/Darcy problem.* Applied Numerical Mathematics, vol. 190, pp. 138–154, (2023).
36. Caucao, S., Discacciati, M., Gatica, G.N. and Oyarzúa, R., *A conforming mixed finite element method for the Navier–Stokes/Darcy–Forchheimer coupled problem.* ESAIM: Mathematical Modelling and Numerical Analysis, vol. 54, 5, pp. 1689–1723, (2020).
37. Caucao, S., Gatica, G.N. and Gatica, L.F., *A Banach spaces-based mixed finite element method for the stationary convective Brinkman-Forchheimer problem.* Calcolo, vol. 60, 4, Paper No. 51, (2023).

38. Caucao, S., Oyarzúa, R. and Villa-Fuentes, S., *A new mixed-FEM for steady-state natural convection models allowing conservation of momentum and thermal energy.* Calcolo, vol. 57, 4, Paper No. 36, (2020).
39. Ciarlet, P.G., The Finite Element Method for Elliptic Problems. North-Holland, 1978.
40. Ciarlet, P.G. Linear and Nonlinear Functional Analysis with Applications. Society for Industrial and Applied Mathematics, Philadelphia, PA, 2013.
41. Ciesielski, K. and Moslehian, M.S., *Some remarks on the history of functional analysis.* Annals of Functional Analysis, vol. 1, 1, pp. 1–12, (2010).
42. Colmenares, E., Gatica, G.N. and Moraga, S., *A Banach spaces-based analysis of a new fully-mixed finite element method for the Boussinesq problem.* ESAIM: Mathematical Modelling and Numerical Analysis, vol. 54, 5, pp. 1525–1568, (2020).
43. Colmenares, E., Gatica, G.N. and Oyarzúa, R., *Analysis of an augmented mixed-primal formulation for the stationary Boussinesq problem.* Numerical Methods for Partial Differential Equations, vol. 32, 2, pp. 445–478, (2016).
44. Correa, C.I. and Gatica, G.N., *On the continuous and discrete well-posedness of perturbed saddle-point formulations in Banach spaces.* Computer and Mathematics with Applications, vol. 117, pp. 14–23, (2022).
45. Correa, C.I., Gatica, G.N., Henríquez, E., Ruiz-Baier, R. and Solano, M., *Banach spaces-based mixed finite element methods for the coupled Navier–Stokes and Poisson–Nernst–Planck equations.* Calcolo, vol. 61, article: 31, (2024).
46. Correa, C.I., Gatica, G.N. and Ruiz-Baier, R., *New mixed finite element methods for the coupled Stokes and Poisson-Nernst-Planck equations in Banach spaces.* ESAIM: Mathematical Modelling and Numerical Analysis, vol. 57, 3, pp. 1511–1551, (2023).
47. Dautray, R. and Lions, J.L., Mathematical Analysis and Numerical Methods for Science and Technology, vols. 2–4. Springer Verlag, 1990.
48. Domínguez, C., Gatica, G.N., Meddahi, S. and Oyarzúa, R., *A priori error analysis of a fully-mixed finite element method for a two-dimensional fluid-solid interaction problem.* ESAIM: Mathematical Modelling and Numerical Analysis, vol. 47, 2, pp. 471–506, (2013).
49. Durán, R.G. and Muschietti, M.A., *An explicit right inverse of the divergence operator which is continuous in weighted norms.* Studia Mathematica, vol. 148, 3, pp. 207–219, (2001).
50. Enflo, P.H., *A counterexample to the approximation property in Banach spaces.* Acta Mathematica, vol. 130, pp. 309–317, (1973).
51. Enflo, P.H., Moslehian, M.S. and Seoane-Sepúlveda, J.B., *A history of solving some famous problems in mathematical analysis.* British Journal for the History of Mathematics, vol. 37, 1, pp. 64–80, (2022).
52. Ern, A. and Guermond, J.-L., Theory and Practice of Finite Elements. Applied Mathematical Sciences, 159. Springer-Verlag, New York, 2004.
53. Friedlander, F.G. and Joshi, M., Introduction to the Theory of Distributions. Cambridge University Press, Second Edition, 1999.
54. Friedman, A., Foundations of Modern Analysis. Dover Publications, 1982.
55. Galvis, J. and Sarkis, M., *Non-matching mortar discretization analysis for the coupling Stokes-Darcy equations.* Electronic Transactions on Numerical Analysis, vol. 26, pp. 350–384, (2007).
56. Gatica, G.N., *On the coerciveness property of the biharmonic operator.* Proyecciones, vol. 10, 17, pp. 27–34, (1991).
57. Gatica, G.N., *Analysis of a new augmented mixed finite element method for linear elasticity allowing RT_0-P_1-P_0 approximations.* ESAIM: Mathematical Modelling and Numerical Analysis, vol. 40, 1, pp. 1–28, (2006).
58. Gatica, G.N., A Simple Introduction to the Mixed Finite Element Method. Theory and Applications. SpringerBriefs in Mathematics. Springer, Cham, 2014.
59. Gatica, G.N., *A note on the generalized Babuška-Brezzi theory: revisiting the proof of the associated Strang error estimates.* Applied Mathematics Letters, vol. 157, Paper No. 109197, (2024).

60. Gatica, G.N. and Gatica, L.F., *On the a-priori and a-posteriori error analysis of a two-fold saddle point approach for nonlinear incompressible elasticity.* International Journal for Numerical Methods in Engineering, vol. 68, 8, pp. 861–892, (2006).
61. Gatica, G.N., Gatica, L.F. and Márquez, A., *Analysis of a pseudostress-based mixed finite element method for the Brinkman model of porous media flow.* Numerische Mathematik, vol. 126, 4, pp. 635–677, (2014).
62. Gatica, G.N., Gatica, L.F. and Sequeira, F., *A* RT_k - P_k *approximation for linear elasticity yielding a broken* $H(\mathrm{div})$ *convergent postprocessed stress.* Applied Mathematics Letters, vol. 49, pp. 133–140, (2015).
63. Gatica, G.N., Gatica, L.F. and Sequeira, F., *A priori and a posteriori error analyses of a pseudostress-based mixed formulation for linear elasticity.* Computers & Mathematics with Applications, vol. 71, 2, pp. 585–614, (2016).
64. Gatica, G.N., Gatica, L.F. and Stephan, E.P., *A dual-mixed finite element method for nonlinear incompressible elasticity with mixed boundary conditions.* Computer Methods in Applied Mechanics and Engineering, vol. 196, 35-36, pp. 3348–3369, (2007).
65. Gatica, G.N. and Heuer, N., *A dual-dual formulation for the coupling of mixed-FEM and BEM in hyperelasticity.* SIAM Journal on Numerical Analysis, vol. 38, 2, pp. 380–400, (2000).
66. Gatica, G.N., Heuer, N. and Meddahi, S., *On the numerical analysis of nonlinear two-fold saddle point problems.* IMA Journal of Numerical Analysis, vol. 23, 2, pp. 301–330, (2003).
67. Gatica, G.N. and Hsiao, G.C., *The coupling of boundary integral and finite element methods for nonmonotone nonlinear problems.* Numerical Functional Analysis and Optimization, vol. 13, 5&6, pp. 431–447, (1992).
68. Gatica, G.N. and Hsiao, G.C., *A Gårding's inequality for variational problems with constraints.* Applicable Analysis, vol. 54, 1–2, pp. 73–90, (1994).
69. Gatica, G.N., Hsiao, G.C. and Meddahi, S., *A coupled mixed finite element method for the interaction problem between an electromagnetic field and an elastic body.* SIAM Journal on Numerical Analysis, vol. 48, 4, pp. 1338–1368, (2010).
70. Gatica, G.N., Hsiao, G.C. and Meddahi, S., *Further developments on boundary-field equation methods for nonlinear transmission problems.* Journal of Mathematical Analysis and Applications, vol. 502, 2, Art. Num. 125262, (2021).
71. Gatica, G.N. and Inzunza, C., *On the well-posedness of Banach spaces-based mixed formulations for the nearly incompressible Navier-Lame and Stokes equations.* Computers & Mathematics with Applications, vol. 102, pp. 87–94, (2021).
72. Gatica, G.N., Inzunza, C. and Sequeira, F., *A pseudostress-based mixed-primal finite element method for stress-assisted diffusion problems in Banach spaces.* Journal of Scientific Computing, vol. 92, 3, article: 103, (2022).
73. Gatica, G.N., Márquez, A. and Meddahi, S., *Analysis of the coupling of primal and dual-mixed finite element methods for a two-dimensional fluid-solid interaction problem.* SIAM Journal on Numerical Analysis, vol. 45, 5, pp. 2072–2097, (2007).
74. Gatica, G.N., Márquez, A. and Meddahi, S., *A new dual-mixed finite element method for the plane linear elasticity problem with pure traction boundary conditions.* Computer Methods in Applied Mechanics and Engineering, vol. 197, 9-12, pp. 1115–1130, (2008).
75. Gatica, G.N., Márquez, A. and Meddahi, S., *A new coupling of mixed finite element and boundary element methods for an exterior Helmholtz problem in the plane.* Advances in Computational Mathematics, vol. 30, 3, pp, 281–301, (2009).
76. Gatica, G.N., Márquez, A. and Meddahi, S., *Analysis of the coupling of BEM, FEM, and mixed-FEM for a two-dimensional fluid-solid interaction problem.* Applied Numerical Mathematics, vol. 59, 11, pp. 2735–2750, (2009).
77. Gatica, G.N., Márquez, A. and Meddahi, S., *Analysis of the coupling of Lagrange and Arnold-Falk-Winther finite elements for a fluid-solid interaction problem in* 3D. SIAM Journal on Numerical Analysis, vol. 50, 3, pp. 1648–1674, (2012).
78. Gatica, G.N. and Meddahi, S., *Finite element analysis of a time harmonic Maxwell problem with an impedance boundary condition.* IMA Journal of Numerical Analysis, vol. 32, 2, pp. 534–552, (2012).

79. Gatica, G.N., Meddahi, S. and Ruiz-Baier, R., *An L^p spaces-based formulation yielding a new fully mixed finite element method for the coupled Darcy and heat equations.* IMA Journal of Numerical Analysis, vol. 42, 4, pp. 3154–3206, (2022).
80. Gatica, G.N., Núñez, N. and Ruiz-Baier, R., *New non-augmented mixed finite element methods for the Navier-Stokes-Brinkman equations using Banach spaces.* Journal of Numerical Mathematics, vol. 31, 4, pp. 343–373, (2023).
81. Gatica, G.N., Oyarzúa, R., Ruiz-Baier, R. and Sobral, Y.D., *Banach spaces-based analysis of a fully-mixed finite element method for the steady-state model of fluidized beds.* Computer and Mathematics with Applications, vol. 84, pp. 244–276, (2021).
82. Gatica, L.F., Oyarzúa, R. and Sánchez, N., *A priori and a posteriori error analysis of an augmented mixed-FEM for the Navier-Stokes-Brinkman problem.* Computer & Mathematics with Applications, vol. 75, 7, pp. 2420–2444, (2018).
83. Gatica, G.N., Oyarzúa, R. and Sayas, F.-J., *Convergence of a family of Galerkin discretizations for the Stokes-Darcy coupled problem.* Numerical Methods for Partial Differential Equations, vol. 27, 3, pp. 721–748, (2011).
84. Gatica, G.N., Oyarzúa, R. and Sayas, F.-J., *Analysis of fully-mixed finite element methods for the Stokes-Darcy coupled problem.* Mathematics of Computation, vol. 80, 276, pp. 1911–1948, (2011).
85. Gatica, G.N., Ruiz-Baier, R. and Tierra, G., *A mixed finite element method for the Darcy equations with pressure-dependent porosity.* Mathematics of Computation, vol. 85, 297, pp. 1–33, (2016).
86. Gatica, G.N. and Sayas, F.-J., *Characterizing the inf-sup condition on product spaces.* Numerische Mathematik, vol. 109, 2, pp. 209–231, (2008).
87. Gatica, G.N. and Wendland, W.L., *Coupling of mixed finite elements and boundary elements for linear and nonlinear elliptic problems.* Applicable Analysis, vol. 63, pp. 39–75, (1996).
88. Gatica, G.N. and Wendland, W.L., *Coupling of mixed finite elements and boundary elements for a hyperelastic interface problem.* SIAM Journal on Numerical Analysis, vol. 34, 6, pp. 2335–2356, (1997).
89. Geng, J., *$W^{1,p}$ estimates for elliptic problems with Neumann boundary conditions in Lipschitz domains.* Advances in Mathematics, vol. 229, 4, pp. 2427–2448, (2012).
90. Girault, V. and Raviart, P.-A., Finite Element Methods for Navier-Stokes Equations. Theory and Algorithms. Springer-Verlag, 1986.
91. Gómez-Vargas, B., Mardal, K.-A., Ruiz-Baier, R. and Vinje, V., *Twofold saddle-point formulation of Biot poroelasticity with stress-dependent diffusion.* SIAM Journal on Numerical Analysis, vol. 61, 3, pp. 1449–1481, (2023).
92. Hahn, H., *Uber lineare Gleichungssysteme in linearen Räumen.* Journal für die Reine und Angewandte Mathematik, vol. 157, pp. 214–229, (1927).
93. He, M. and Sun, P., *Mixed finite element analysis for the Poisson-Nernst-Planck/Stokes coupling.* Journal of Computational and Applied Mathematics, vol. 341, pp. 61–79, (2018).
94. He, M. and Sun, P., *Mixed finite element method for modified Poisson-Nernst-Planck/Navier-Stokes equations.* Journal of Scientific Computing, vol. 87, 3, Paper No. 80, (2021).
95. Hormander, L., The Analysis of Linear Partial Differential Equations I. Springer Verlag, 1983.
96. Howell, J.S. and Walkington, N.J., *Inf-sup conditions for twofold saddle point problems.* Numerische Mathematik, vol. 118, 4, pp. 663–693, (2011).
97. Hsiao, G.C., *A modified Galerkin scheme for elliptic equations with natural boundary conditions.* Numerical Mathematics and Applications (Oslo, 1985), 193–197, IMACS Trans. Sci. Comput.-85, I, North Holland, Amsterdam, 1986.
98. Ingram, R., *Finite element approximation of nonsolenoidal, viscous flows around porous and solid obstacles.* SIAM Journal on Numerical Analysis, vol. 49, 2, pp. 491–520, (2011).
99. Johnson, C., Numerical Solution of Partial Differential Equations by the Finite Element Method. Cambridge University Press, Cambridge, 1987.
100. Kreiszig, E., Introductory Functional Analysis with Applications. John Wiley, 1978.
101. Kress, R., Linear Integral Equations. Springer-Verlag New York Inc., Second Edition, 1999.

102. KUFNER, A., JOHN, O. AND FUČÍK, S., Function Spaces. Monographs and Textbooks on Mechanics of Solids and Fluids; Mechanics: Analysis. Noordhoff International Publishing, Leyden; Academia, Prague, 1977.
103. LAMICHHANE, B.P., RUIZ-BAIER, R. AND VILLA-FUENTES, S., *New twofold saddle-point formulations for Biot poroelasticity with porosity-dependent permeability.* Results in Applied Mathematics, vol. 21, Paper No. 100438, (2024).
104. LAX, P., Functional Analysis. John Wiley & Sons, Inc., 2002.
105. MACCLUER, B., Elementary Functional Analysis. Graduate Texts in Mathematics, 253. Springer, New York, 2009.
106. MARKUSHEVICH, A.I., Theory of Functions of a Complex Variable: Three volumes in one. Chelsea Publishing Company, New York, 1965, 1967.
107. MC LEAN, W., Strongly Elliptic Systems and Boundary Integral Equations. Cambridge University Press, 2000.
108. MOSLEHIAN, M.S., MUÑOZ-FERNÁNDEZ, G.A., PERALTA, A.M. AND SEOANE-SEPÚLVEDA, J.B., *Similarities and differences between real and complex Banach spaces: an overview and recent developments.* Revista de la Real Academia de Ciencias Exactas, Físicas y Naturales. Serie A. Matemáticas, vol. 116, article: 88, (2022).
109. NEČAS, J., Introduction to the Theory of Nonlinear Elliptic Equations. Reprint of the 1983 edition. A Wiley-Interscience Publication. John Wiley & Sons, Ltd., Chichester, 1986.
110. ODEN, J.T. AND DEMKOWICZ, L., Applied Functional Analysis. CRS Press, 1996.
111. QUARTERONI, A. AND VALLI, A., Numerical Approximation of Partial Differential Equations. Springer-Verlag, 1994.
112. RAVIART, P.-A. AND THOMAS, J.-M., Introduction á L'Analyse Numérique des Equations aux Dérivées Partielles. Masson, 1983.
113. RUDIN, W., Real and complex analysis. Second edition. McGraw-Hill Series in Higher Mathematics. McGraw-Hill Book Co., New York-Düsseldorf-Johannesburg, 1974.
114. RUDIN, W., Functional Analysis. Second edition. Internat. Ser. Pure Appl. Math. McGraw-Hill, Inc., New York, 1991.
115. SCHECHTER, M., Principles of Functional Analysis. Academic Press, 1971.
116. SCHEURER, B., *Existence et approximation de points selles pour certains problèmes non linéaires.* RAIRO Analyse Numérique, vol. 11, no. 4, pp. 369–400, (1977).
117. SOKAL, A.D., *A really simple elementary proof of the uniform boundedness theorem.* The American Mathematical Monthly, vol. 118, 5, pp. 450–451, (2011).
118. SOUKHOMLINOV, G.A., *Uber fortsetzung von linearen funktionalen in linearen komplexen räumen und linearen quaternionenräumen.* Matematicheskii Sbornik, vol. 3, pp. 353–358, (1938).
119. STAKGOLD, I. AND HOLST, M.J., Green's Functions and Boundary Value Problems. Pure and Applied Mathematics: A Wiley Series of Texts, Monographs and Tracts, 3rd (third) Edition, 2011.
120. WHEEDEN, R.L. AND ZYGMUND, A., Measure and Integral. An Introduction to Real Analysis. Marcel Dekker, Inc., 1977.
121. WLOKA, J., Partial Differential Equations. Cambridge University Press, 1987.
122. XU, J. AND ZIKATANOV, L., *Some observations on Babuška and Brezzi theories.* Numerische Mathematik, vol. 94, 1, pp. 195–202, (2003).
123. ZEIDLER, E., Applied Functional Analysis. Applications to Mathematical Physics. Springer Verlag New York, Inc., 1995.

Index

G. N. Gatica, *Introduction to Functional Analysis*,
https://doi.org/10.1007/978-3-032-37415-8

The manufacturer's authorised representative in the EU is Springer Nature Customer Service Centre GmbH, Europaplatz 3, 69115 Heidelberg, Germany. If you have any concerns regarding our products, please contact ProductSafety@springernature.com

Printed and bound by CPI Group (UK) Ltd, Croydon, CR0 4YY
20/10/2026
02236388-0020